[Written according to the latest syllabus Calcutta University, Burdwan University, Kalyani University, North Bengal University, West Bengal State University, Vidyasagar University, Gaur Banga University, Tripura University and other Indian Universities for the students of B.Sc. Zoology (General) and B.Sc. Zoology (Honours) and for all those who are interested in Practical Zoology.]

ADVANCED
PRACTICAL ZOOLOGY

Jayanta Sinha
M.Sc., Ph.D., PGDEE, FZS (Cal), FZAB, FICCE.
Head (Former) and Associate Professor, Department of Zoology, B.B. College, Asansol, West Bengal.
Associate Professor in Zoology, Department of Zoology, Vivekananda College, Thakurpukur, Kolkata.

Arun Kumar Chatterjee, M.Sc., Ph.D.
Reader and Head (Retired), Department of Zoology Bangabasi College, Kolkata, West Bengal.

Piyali Chattopadhyay, M.Sc., Ph.D.
Deputy Director
Zoological Garden, Alipore, Kolkata, West Bengal.

BOOKS AND ALLIED (P) LTD.
NO.1-E(1) "SHUBHAM PLAZA" (1st Floor)
83/1, BELIAGHATA MAIN ROAD, KOLKATA 700010 (INDIA)
Tel : (033)23740062, 8274085530
e-mail : booksandallied1960@gmail.com • booksmktng@gmail.com

ADVANCED
PRACTICAL ZOOLOGY

First Published : August 2010
Second Revised Edition : 2011
Third Edition (thoroughly revised & updated) **: 2014**
Reprinted : 2015, 2016,
Reprinted : 2019

Publisher : Ruba Sen
BOOKS AND ALLIED (P) LTD.
8/1 Chintamoni Das Lane, Kolkata 700009

Typesetter : BOOKS AND ALLIED (P) LTD.
8/1 Chintamoni Das Lane, Kolkata 700009

Printer : EUREKA
91/A Baithakkhana Road, Kolkata 700009

ISBN **978-81-87134-42-8**

Price : **₹ 745.00**

Dedicated to

The Pioneers of Zoology

Preface to the Third Edition

After the first edition and second edition of the book, it is a very happy moment for us to find that this book is proving to be useful and helpful for the students of Practical zoology and also being appreciated by many learned and respected teachers of Zoology. A book needs constant modification, inclusion and editing for making it more up-to-date, relevant and comprehensive. Keeping this in mind and stimulated by feedbacks from different readers we thought for a new edition.

The new edition include changes in Section 1 (General Zoology) where some of the phyla are classified and described following Ruppert and Barnes, 1994 in addition to the previous classification style (as narrated in the previous editions).Inclusions are made to describe Bursa Fabricus and Mammalian lymph node in Section 4; Histology and Histochemical Procedures. Inclusion is also made in respect to 'Effect of temperature on the activity of Ptyalin' in Section 7; Biochemistry and Physiology. A chapter entitled 'Introduction to Field based study' is appended in Section 13.This chapter will help the students to understand how to carry on a Field based study.

We are always indebted to all our readers who are providing us constant inputs and thus triggering us for continuous quality enrichment of the book.

We hope we will always be bestowed with unceasing support from the students and respected teachers as usual, so that we can further bring out more editions of this book in future.

We also request all readers of this edition to feel free and communicate to us with their suggestions, comments, views and positive criticism, that will immensely help us in the steady evolution of this book.

Dr. Jayanta Sinha

Mahalaya, 2013 **Dr. Arun Kumar Chatterjee**

Kolkata **Dr. Piyali Chattopadhyay**

Preface to the First Edition

Zoology is an enormous and intriguing subject that deals with various aspects of structure, function, evolution, ecology, classification, behaviour of different animals. Like other science subjects zoology is also making rapid progress, which is reflected through various scholastic books, research papers, reviews and other scientific literatures. Zoology being a developing subject has modern inputs from other subjects like physics, chemistry, biochemistry, statistics, and behavioural sciences thus is a dwelling ground of people from various disciplines. For in depth understanding and study of the subject practical approach to it is indispensable, as visualizing is always better retained in the brain than reading. Thus practical zoology includes hands on training of the students on various avenues of zoology that encompasses both classical and modern in addition to basic and applied aspects of the subject.

This book aims towards catering the various facets of today's practical zoology. The book is written as per the B.Sc. General and Honours undergraduate syllabus of practical zoology of different universities of West Bengal and India. During writing the book utmost care has been taken to present the subject in lucid, well illustrated and in an up-to-date fashion. The book is divided into thirteen sections. Each section is divided into many chapters. Sections 1 to 4 are devoted to classical and basic topics like General Zoology, Osteology, Developmental Biology, Histology & Histochemical techniques. Sections 5 to 7 are devoted to modern topics like Cytology, Genetics and Biochemistry & Physiology. Sections 8 to 10 are committed to topics like Ecology, Ethology & Toxicology, Parasitology, Immunology & Microbiology and Economic Zoology. Sections 11 and 12 deal with applied topics such as Biostatistics and Biological Instrumentation. Section 13 is completely dedicated to Field training one of the essential components of Zoology. The book has got a very diverse and informative Appendix and very exhaustive Glossary.

Though special care is taken to provide correct data and information and to make the book free of typological errors still there may remain some slip of pens, lacunae and mistakes in this book for which the authors extend their due apology.

Suggestion, comments and advice for the improvement of the book shall be very much solicited from the end of respected teachers, researchers and keen students.

Rathjatra, 2010
Kolkata

J. Sinha
A. K. Chatterjee
P. Chattopadhyay

Acknowledgement

The authors feel great pleasure to acknowledge all staff members and officials of Department of Forests, Government of West Bengal. Authors express their sincere acknowledgement to Director, Zoological Garden, Alipore. Authors also acknowledge Mr. S. Sau, Deputy Secretary, Department of Land and Land Reforms, Government of West Bengal.

JS acknowledges the constant encouragement and support from all his departmental colleagues of Banwarilal Bhalotia College.

The authors also acknowledge the advice from Dr. Soumen Bhattacharya, Department of Zoology, North Bengal University in preparation of the manuscript. Acknowledgement is also due to Ms. Amrita Pal, Department of Economics, Banwarilal Bhalotia College. Authors also acknowledge Ms. Tandra Sarkar for her encouragement during edition of the book.

The authors also thank their friends and family members who supported, encouraged and tolerated them during the preparation of the manuscript.

Acknowledgement is also due to my (JS) students Mr. Sourav Deoghoria, Ms. Anumita Das and Ms. Oindrilla Dutta.

Authors would also like to thank artist Mr. Narayan Sur who have drawn most of the diagrams of the book. Acknowledgements are also due to Mr. Abhijit Ghosal, Mr. Partha Protim SinghaRoy and other staff members of Books & Allied (P) Ltd.

The authors also thank Mr. Arunabha Sen, Director, Books & Allied (P) Ltd. who provided constant encouragement and help and without whose active endeavour the book would not have seen the light of the day.

About the Book

There is an enormous demand for practical books beside the theoretical books in Zoology because no science can be complete without practical studies. The current practical syllabus of Zoology has been framed in all Universities of West Bengal as well as in national level. These syllabi include classical Zoology as well as applications of various modern techniques in the subject, which is useful in studying the subject thoroughly. As there are no such books which cover different modern methods and practical studies widely used in Histochemistry, Physiology, Haematology, Biochemistry, Histology, Embryology, Cytology and Genetics, Parasitology and Microbiology, Osteology, Ecology, Ethology & toxicology, Biostatistics, Economic zoology, Biological instrumentation and Field training. Further the book also has a very informative Appendix that includes, preparation of general reagents & fixatives, SI units, simple solutions & dilutions, pH indicators and concentration of acid & bases, fixatives, stains & buffers, laboratory safety guidelines, statistical tables, etc. We thought to prepare such a book for students, scholars and teachers as well as for those people who have scientific aptitude and interest for the subject. The book is richly illustrated with figures, and photographs.

Further the book has the following features : (1) Theoretical background, (2) Bold face key terms, (3) Glossary, (4) Coloured plates, (5) References and (6) Index.

About the Authors

Dr. Jayanta Sinha is presently working as the Associate Professor and Head (Former), Post Graduate Department of Zoology, Banwarilal Bhalotia College, Asansol, West Bengal. He stood first class first in M.Sc. Zoology from Calcutta University and earned his Ph.D. from Jadavpur University working in CSIR–Indian Institute of Chemical Biology, Kolkata. He is also attached with the Department of Environment and Water Management in his college. He is the Life member of Zoological Society of India (MZSI) and Zoological Society, Kolkata. He is also the Fellow Member, International Congress of Chemistry and Environment (FICCE) and Fellow, Zoological Society of Kolkata (FZS, Cal) and also the Fellow, Zoological Association of Burdwan (FZAB). He has many research publications in leading International and National scientific journals. He is also co-author of a book on Immunology. He is an awardee of M.M. Chakraborty Memorial Award in Protozoology and Parasitology (2003) of Zoological Society, Kolkata.

Dr. Arun Kumar Chatterjee was in the Department of Zoology; Bangabasi College under the University Calcutta and retired as Reader and Head in 1999. He has over 36 years of experience in teaching and research and taught thousands of undergraduates. He obtained his Ph.D. working on the problems of differentiation in vertebrates from Kalyani University. He has several publications to his credit having published in various National and International journals. He is life member/member of scientific organizations and societies. He was co-author of Zoology textbook and practical book for General and Honours studies of Calcutta, Burdwan, Vidyasagar, Kalyani and Tripura University.

Dr. Piyali Chattopadhyay an ex-Presidencian completed her undergraduate studies in 1991 and awarded postgraduate degree in Zoology in 1992 from Calcutta University. Being a research scholar in the Zoological Survey of India during 1994 -1999 she has worked on the soil protozoa of North and North East India for which a Memoir and a State Fauna series have been published by Zoological Survey of India. She got her Ph.D. degree in 2002 from Calcutta University. An awardee of M.M. Chakraborty Memorial Award of Zoological Society, Kolkata in the field of Protozoology and Parasitology in 2004, her research works has been reflected in number of reputed journals. She is life member of Zoological Society, Kolkata. She is now the Deputy Director of Zoological Garden, Alipore.

Contents

COLOUR PLATES (Section 4)

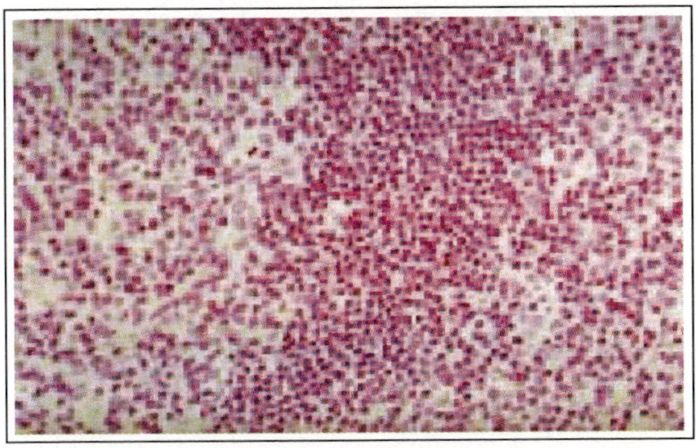

Fig. 4.3.1 : Feulgen reaction of DNA of tissue.

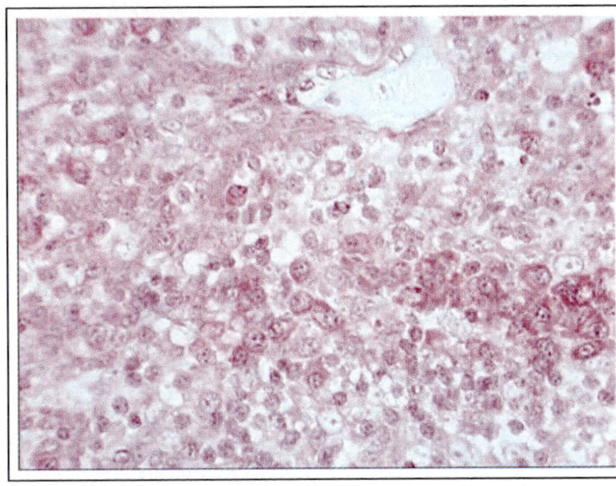

Fig. 4.3.2 : Pyronin Y for RNA in tonsil.

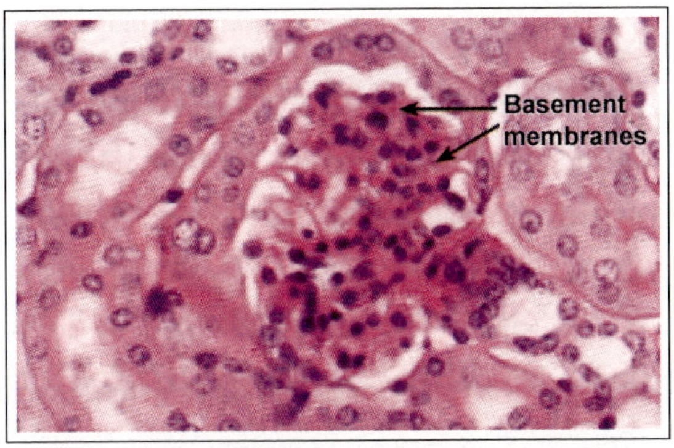

Fig. 4.3.3 : PAS stain for glycogen of kidney.

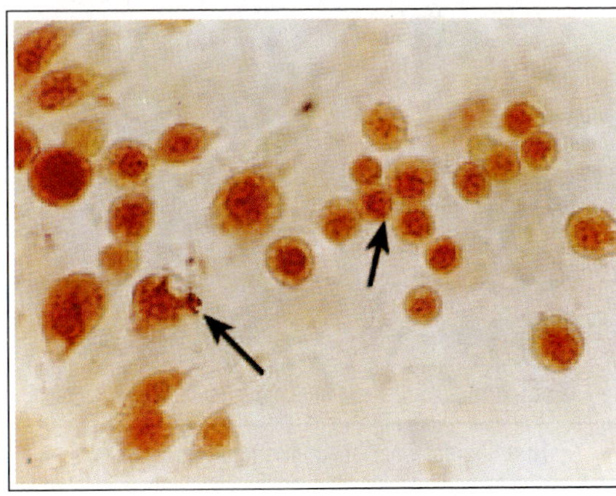

Fig. 4.3.4 : Acid phosphatase stained osteoclast.

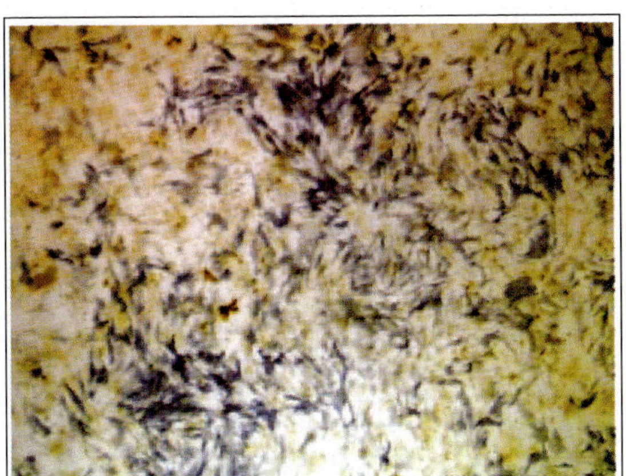

Fig. 4.3.5 : Alkaline phosphatase stain in osteogenic differention.

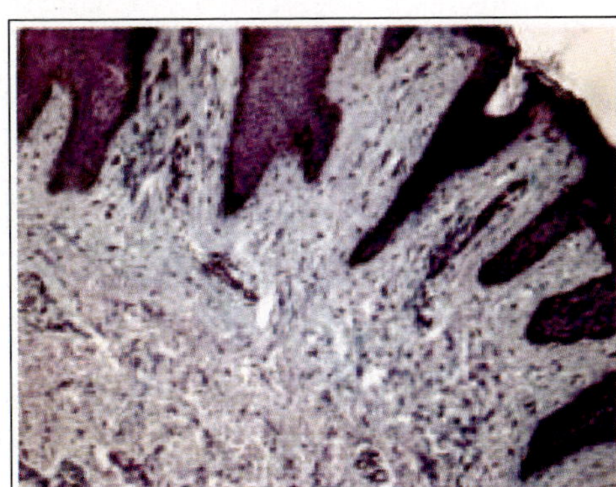

Fig. 4.3.6 : Metachromatic stain.

COLOUR PLATES (Section 4)

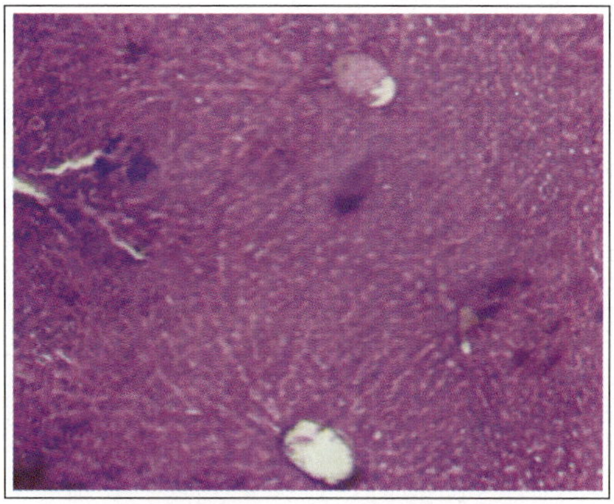

Histological section of mammalian liver

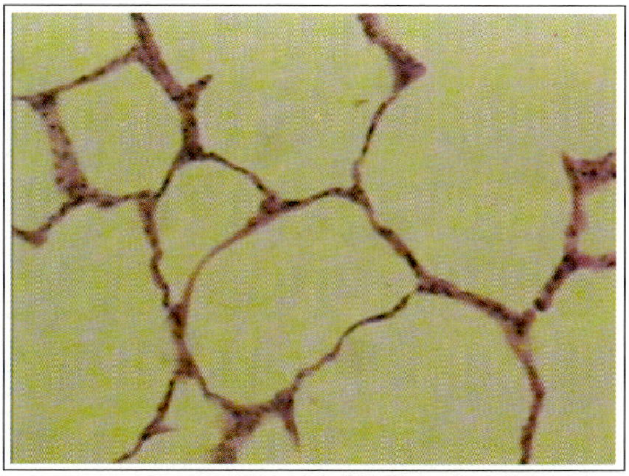

Histological section of mammalian lung

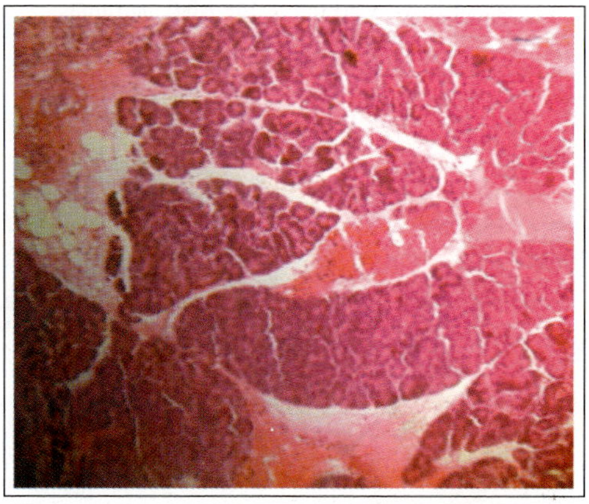

Histological section of mammalian pancreas

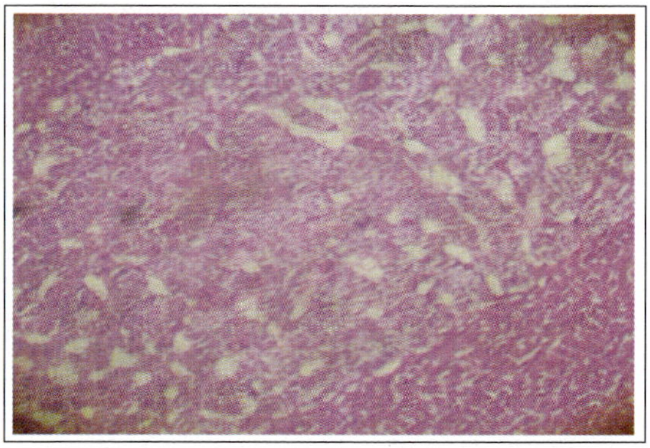

Histological section of mammalian adrenal

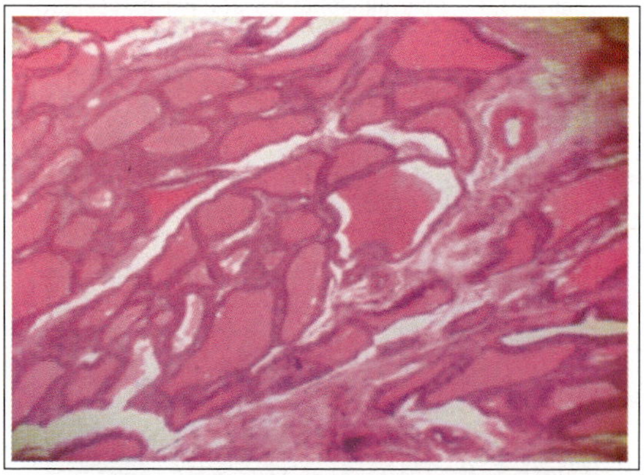

Histological section of mammalian thyroid

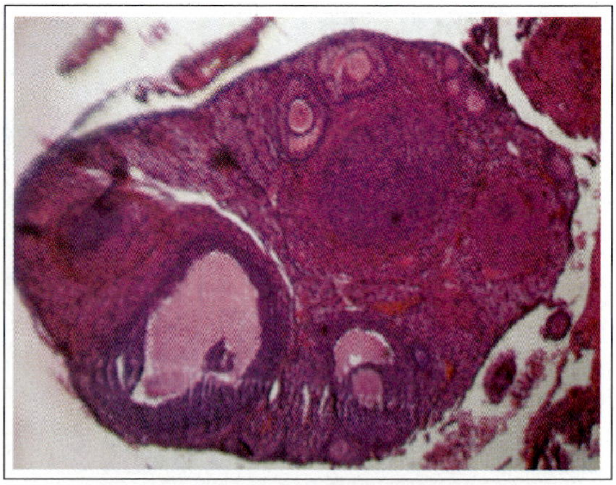

Histological section of mammalian ovary

Courtesy : Mr. Sourav Deoghoria, Ms. Oindrilla Dutta and Ms. Anumita Das

COLOUR PLATES (Section 4)

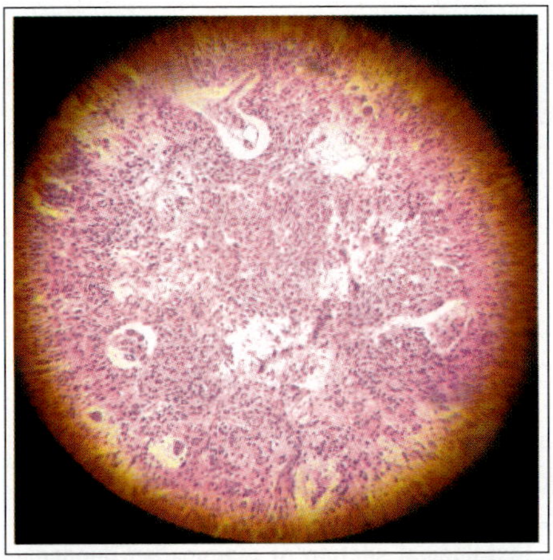

Histological section of mammalian pituitary

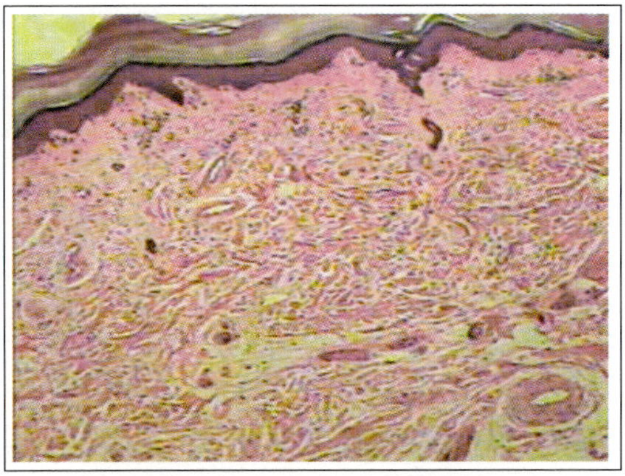

Histological section of mammalian skin

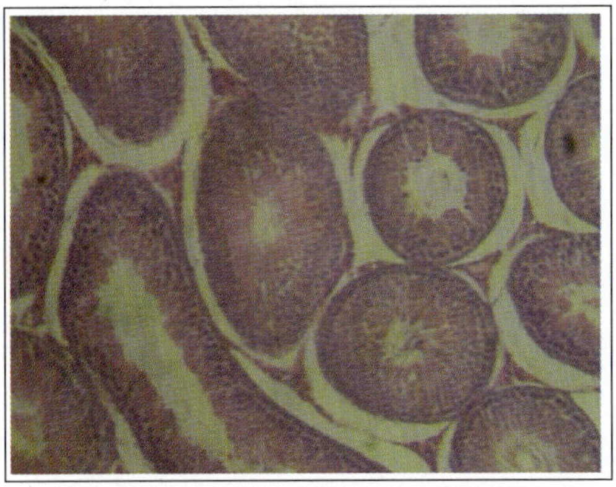

Histological section of mammalian testis

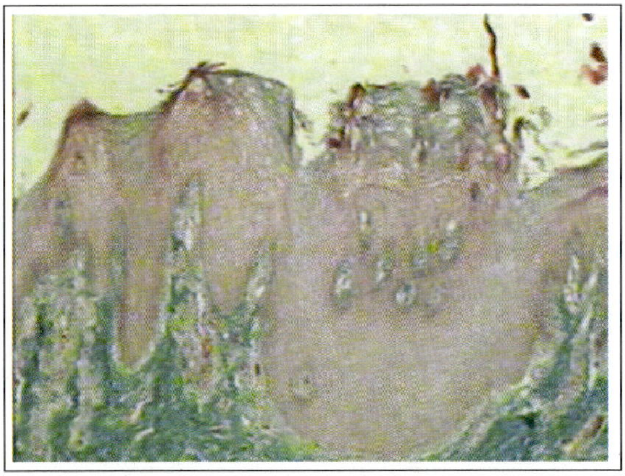

Histological section of mammalian tongue

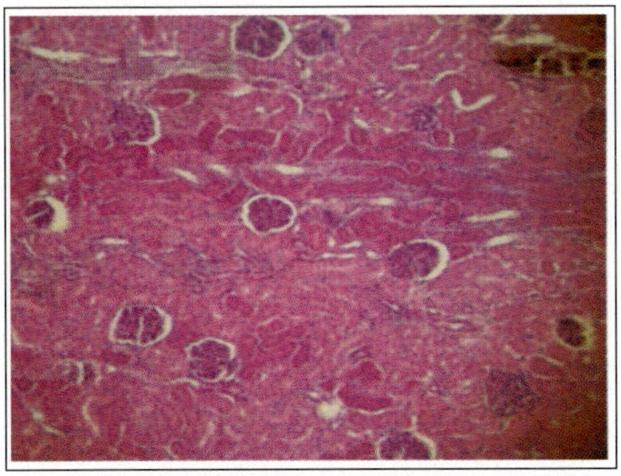

Histological section of kidney

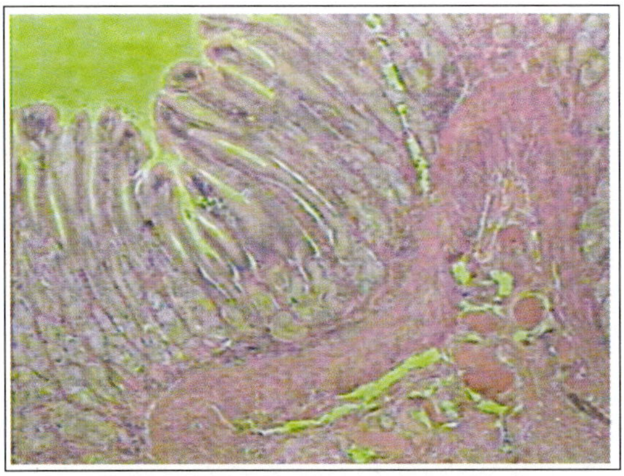

Histological section of mammalian spleen

Courtesy : Mr. Sourav Deoghoria, Ms. Oindrilla Dutta and Ms. Anumita Das

adapted from http://www.unm.edu/~vscience/microscopy.htm

COLOUR PLATES (Section 4)

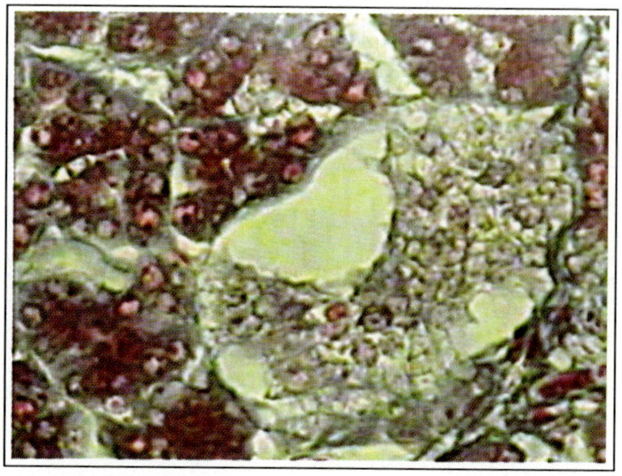

Histological section of mammalian cardiac stomach

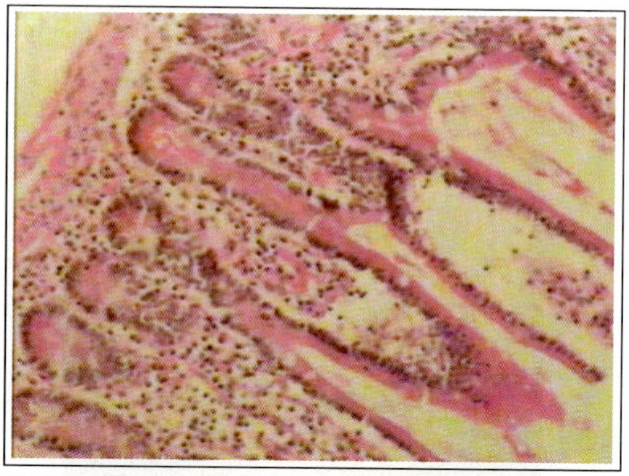

Histological section of mammalian duodenum

adapted from http://www.unm.edu/~vscience/microscopy.htm

COLOUR PLATES (Section 6)

(a)

(b)

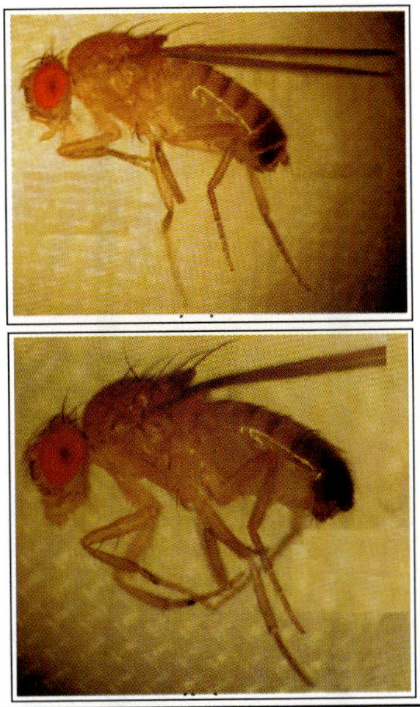

Fig. 6.1.3.1 : Wild type *Drosophila melanogaster (male and female).*

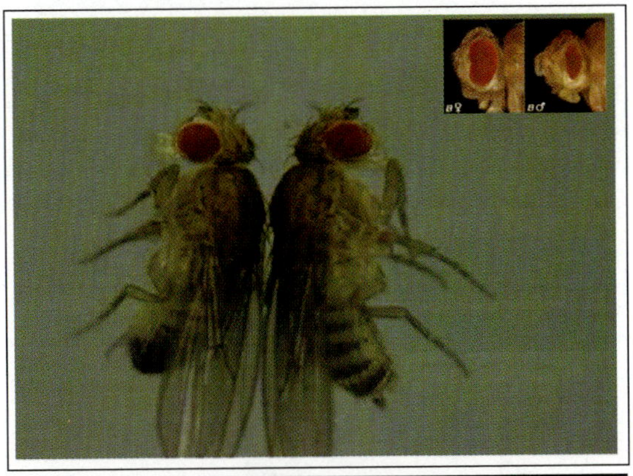

Fig. 6.1.3.2(i) : Bar eye mutant of *Drosophila melanogaster.*

Fig. 6.1.3.2(a) : Yellow bodied mutant *Drosophila melanogaster.*

Fig. 6.1.3.2(j) : White eye mutant of *Drosophila melanogaster.*

Fig. 6.1.3.2(h) : Shaven bristles mutant of *Drosophila melanogaster.*

Fig. 6.1.3.2(k) : Vermillion eye mutant of *Drosophila melanogaster.*

COLOUR PLATES (Section 7)

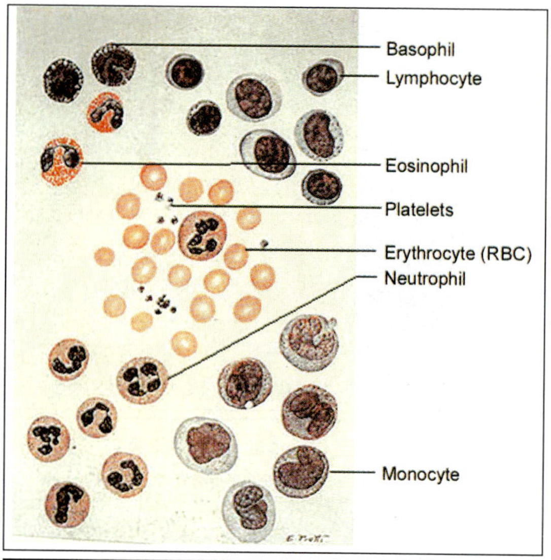

Fig. 7.3.1.2 : Various types of blood cells in a blood smear of human.

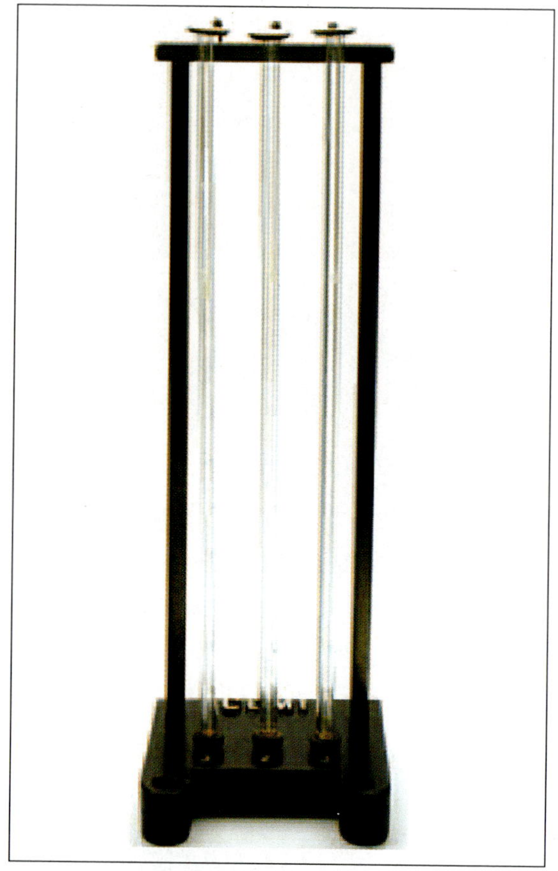

Erythrocytic Sedimentation Rate (ESR) equipment stand (Adapted from www.museumboerhaave.nl)

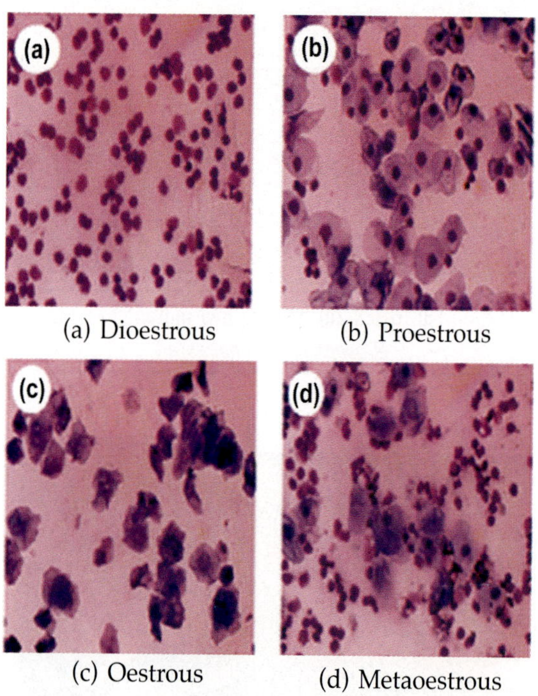

Various phases of Oestrous Cycle in mice determined through study of cells obtained from Giemsa stained vaginal smears by light microscopy. Adapted from Mettus and Rane, *Oncogene* (2003) 22, 8413–8421.

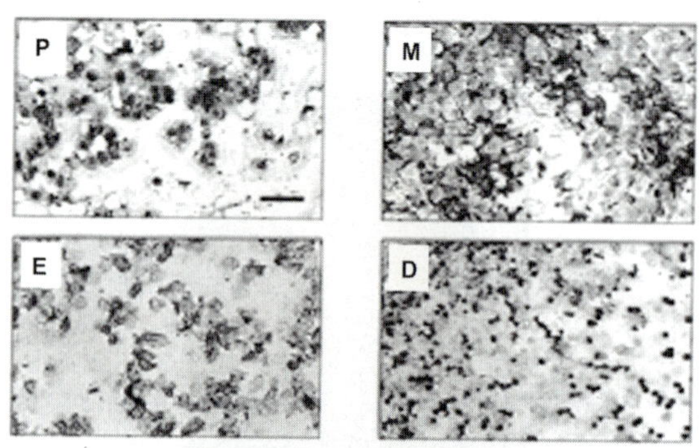

Vaginal smears of rat during Oestrous Cycle. P (Proestrous); E (Oestrous); M (Metestrous); and D (Diestrous). Adapted from Haruyama *et. al. Journal of Dental Research* (2002) 81,406-410.

COLOUR PLATES (Section 9)

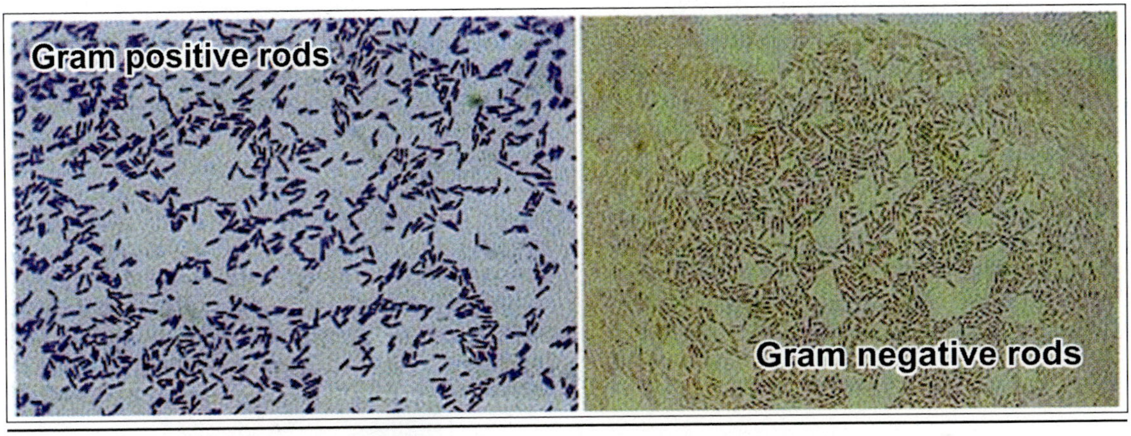

Figure 9.3.1.1 : Gram stained positive and negative rods.

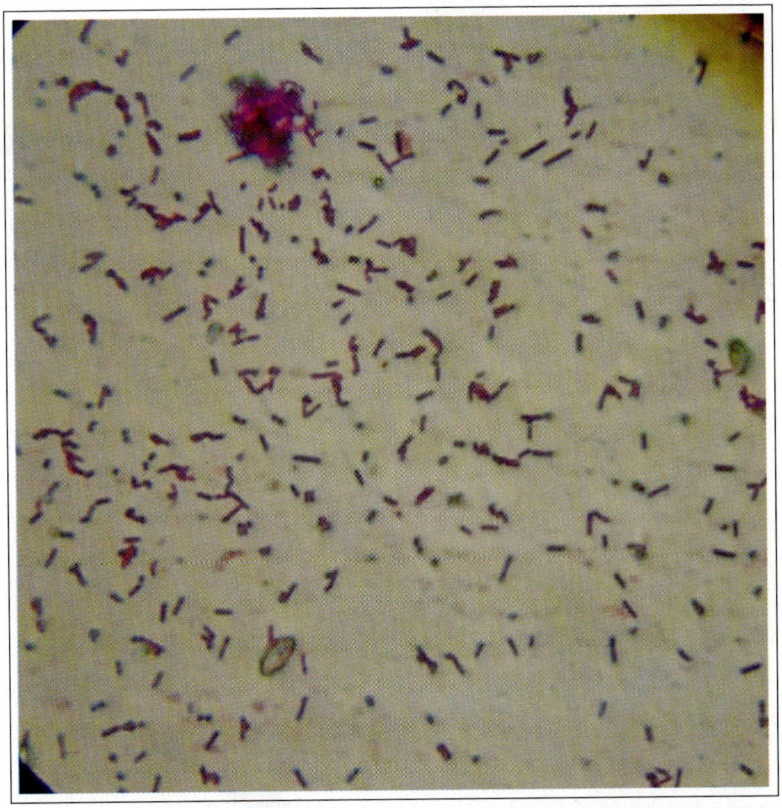

Fig. 9.3.2.1 : Carbol Fucshin simple staining of bacteria. *Courtesy : Ms. Anumita Das*

COLOUR PLATES (Section 10)

Table 10.3.1 : Different types of fish diseases and causative agents

Epistyis sp., (ciliated protozoan)

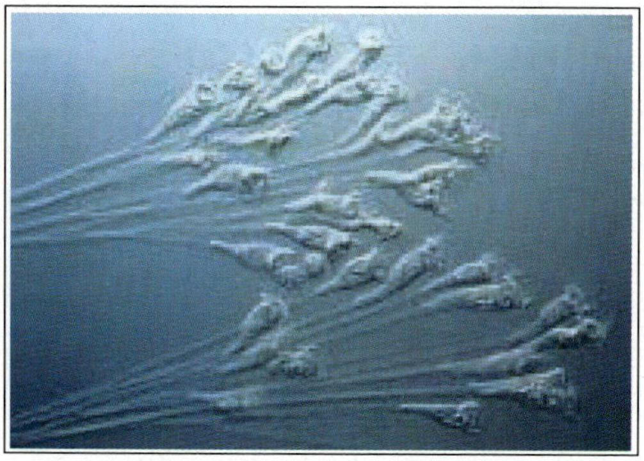

Myxobolus cerebralis

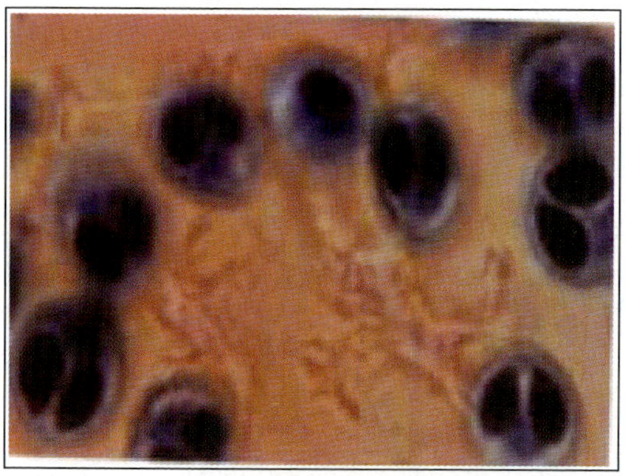

Pleistophora sp.

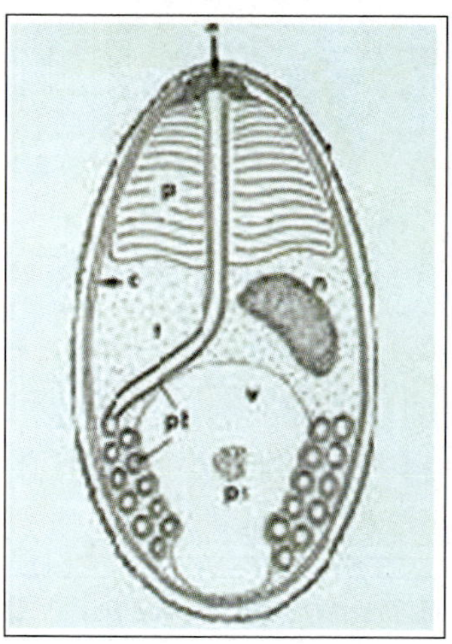

Saprolegnia sp.

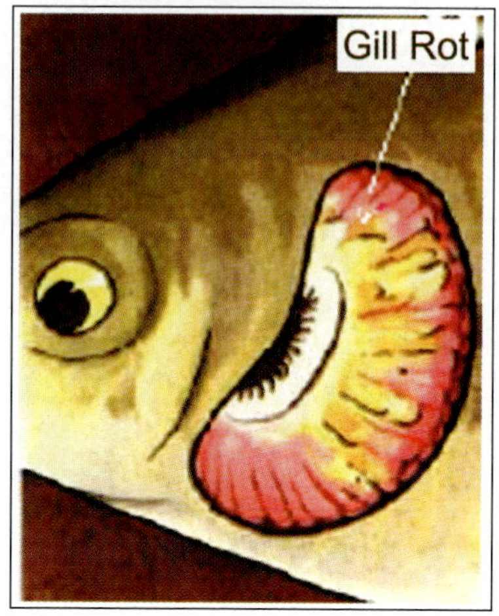

Diplostomum sp.

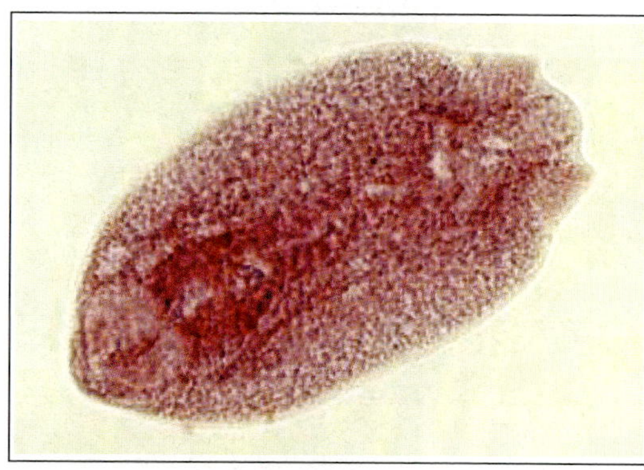

Lernea sp.

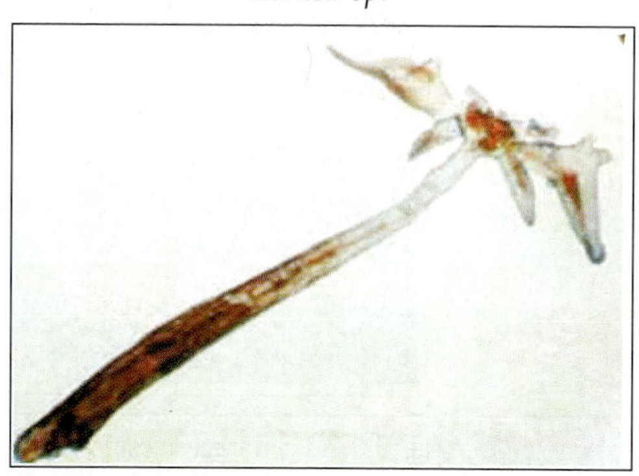

Section 1

GENERAL ZOOLOGY

SECTION AGENDA

INTRODUCTION

Living things can be grouped into five kingdoms : **Monera, Protista, Plantae, Fungi and Animalia.** Monera are prokaryotic organisms that have no organized muscles viz., blue green algae etc. **Protista** are muscular eukaryotic organisms, with an organized muscle viz., Protozoa, protophytans and protofungi. **Plantae** are multicellular with cell walls of chitin and they generally obtain nourishment by means of chemical break down of material (Osmotropic — autotropic). **Animalia** are multicellular with no cell walls and generally consume other organism or take in dissolved nutrients. Plantae, Fungi and Animalia are all eukaryotic multicellular and generally organized into tissue (specialized association of cells) to achieve some specific function. There is no longer formal taxonomic category called the Protista, although it is useful general term for referring the multicellular eukaryotes.

The aim of zoological study is to know about the life of the animals concerned. About one million animal species have been named and described taxonomically and the number is constantly increasing every year. These animals are terrestrial, fresh water, marine and parasitic forms. It is rather impossible to become familiar with this vast number of animals. Therefore, they must be sorted out into groups. This can be done only after ascertaining the degrees of similarity and dissimilarity that will define each group. Thus by studying one or a few representative from each group, a clear picture of the entire animal population may be obtained.

Invertebrates have been grouped in various ways. The basic division is based upon whether individuals are 'single celled' or 'many celled'. True animals are multicellular, diploid organisms. They develop from blastula and are collectively called **Metazoa** or **Metazoans.** Other invertebrates are considered either **'unicellular'** (single celled) or **'acellular'** (without cell) and never develop from a blastula.

Animals may be classified based on their general body form. Two types of body symmetry found amongst most metazoans. Some are **bilaterally symmetrical**, i.e., right and left sides are approximate mirror images of each other (animal possess distinct anterior and posterior ends). Others **are radially symmetrical**, in which body can divided into a number of radial planes i.e., an animal can be cut into two approximately equal halves by any plain that passes through the center of the organism. Therefore most animals belong to either the **Radiata** or the **Bilateria.** Animals like us are bilaterally symmetrical. Multicellular invertebrates have been divided into two groups viz., **diploblastic** and **triploblastic** animals. Diploblastic animals possess two germ layers. The outer most layers of cells are called the **ectoderm** and the innermost layer of cells are called **endoderm.** The other group is termed as triploblastic animals having a third germ layer, the **mesoderm.** The mesoderm layer lies between the ectoderm and endodermal tissue.

Triploblastic animals according to the nature of their body cavities can be classified into three major groups. In the first group triploblastic animals without an internal body, the region between the digestive tract and the epidermis is completely filled with **mesenchyme** and muscles fibres, with no body space present.

In the second group of animals there is a fluid filled cavity in between the outer body wall musculature and the endoderm after the gut. This type of body cavity is termed as **pseudocoele**, and the animal belonging to it is said to be **pseudocoelomate.**

The third group of triploblastic animals, the body cavity is true **coelom**: an internal fluid filled body cavity lying between the gut and the outer body wall musculature are lined by mesoderm tissue. The animals having such a body cavity are called **coelomates.**

All animals above ctenophora have this third germ layer. The triploblastic condition is added more weight to the phylogenetic relationship of the chordates with the invertebrates. The chordate is a vast group, which includes diverse animals ranging from *Ascidia* to man. The chordate occupies a greater variety of habitats and show more complicated mechanisms of self maintenance than any other group in the whole animal kingdom. The animals, which fall into this great group, possess, at some stage of their life, a few characteristics like the '**notochord, dorsal tabular nerve chord** and **pharygeal gill-slits**'. The chordates differ from the invertebrates significantly by the position of the nerve cord. The alimentary canal is dorsal and above the nerve cord in invertebrates, while the nerve cord is dorsal and above the gut in chordates.

The system of classifying animals on a scientific basis is called **Animal Taxonomy**. Taxonomy involves two aspects – naming (**nomenclature**) and orderly grouping of animals (**classification**) on the basis of their relationship.

Modern classification is the expanded form of the Linnean classification. Carolus Linnaeus (1758) who for the first time proposed five categories in classification i.e., Kingdom, Class, Order, Genus and Species. Later on Phylum and Family have been added.

Kingdom
Phylum
Class
Order
Family
Genus
Species

Many subgroups are super-imposed upon this basic framework, for example, subphylum, subclass, superorder, suborder, infraorder etc. When classifying an animal, every animal must be put into a Phylum, Class, Order, Family, Genus and Species. Any named group of organisms that is sufficiently distinct to be allotted to such a category is called a **Taxon**. In our book phylum form the highest taxonomic level and species the lowest one.

The smallest basic unit of taxonomy is the **species**, which is the group of interbreeding individuals reproductively isolated from other such groups. Species is the most important category in taxonomic hierarchy. The scientific name of a species consists of two words and thus is called binomial. The first word is the name of the **genus** (generic name) to which the species belongs and second is the species name. The generic and specific name are usually italicized in print and underlined in writing. The generic name is always a noun and written with an initial capital letter, while specific name is generally an objective is always in small letter. Once the generic name is spelled out, it may be abbreviated when used succeeding. In case several names given to an animal by different workers, the earliest name is to be considered valid and it should be retained (**Law of Priority**), duplicate names are **synonyms**. Larger taxonomic categories have uniform endings (rules of naming) thus :

Superfamilies	oidea
Families	idea
Subfamilies	inae
Tribe	ini
Subtribes	ina

Sometimes an additional name often follows the species name of an organism. This is the name of the person who first described the organism. The name is capitalized but is not italicized and may be placed with in Parenthesis. Sometimes a persons name is followed by a date, identifying the year in which the species is described.

Linnaeus in the tenth edition of book 'systema naturae' give the system of classification, is known as 'Father of Taxonomy'.

Chapter 1.1

Identification of Important Specimens

Zoological classification is the ordering of animals into groups on the basis of their relationship. But in identification, an animal is categorized up to species level on the basis of characters. Here generally the external characters are considered ordinarily, the internal characters are not mentioned in identification, as these cannot be studied in an undissected animal. Sometimes internal characters are recorded and considered as reported characters.

To identify an animal two divisions are considered.

A. Spot identification

In this case only the specific important characters of an animal are recorded followed by systematic position and scientific name.

B. Identification with reasons

It is a step wise procedure.. Here the systematic position of an animal is noted on the basis of peculiar characters the category possess.

The classification schemes in biology always seem tidy. The system of classification is the interpretation by various authors, as there are problems in taxonomic studies. The classification we have used in the following edition is the conventional and long accepted one, but we have not hesitated to leave them where new information has provided possible improvements.

Chapter 1.1.1

Nonchordates

PROTOZOA

(G.PROTO = the very first; ZOA = animals)

Distinguishing Characteristics : 1. Unicellular eukaryotes lacking collagen and chitinous cell walls. 2. Non-photosynthetic in primitive condition.

1. *AMOEBA PROTEUS*

Habit and Habitats

Amoeba proteus generally lives in fresh water, ponds, lakes, ditches, and slow running streams, often found in shallow water on the underside of aquatic vegetation and among the sediments on the bottom. They are particle feeders, feeds on bacteria, algae, protozoans, rotifer etc.

Systematic Position

Subkingdom	**Protozoa**
Phylum	**Sarcomastigophora**
Sub Phylum	**Sarcodina**
Super class	**Rhizopoda**
Class	**Lobosea**
Subclass	**Gymnamoebia**
Order	**Amoebida**
Type	***Amoeba proteus***

Distinctive Features

1. Irregular shape with branched **pseudopodia**.

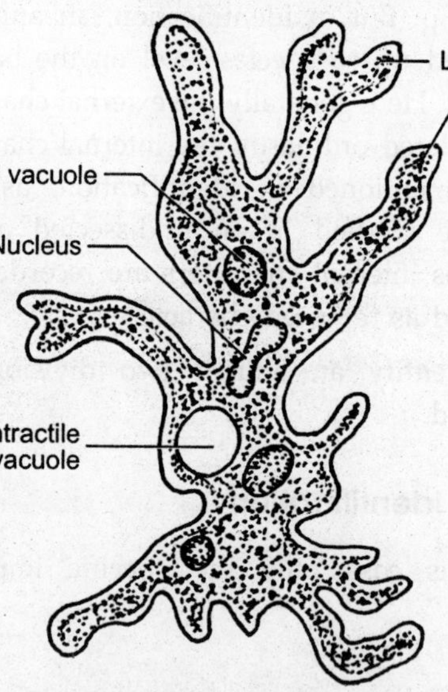

Fig. 1.1.1.1 : *Amoeba proteus*

2. Cytoplasm is distinguished into ectoplasm and endoplasm.

3. Single contractile vacuole and food vacuoles many.

4. Uninucleate.

Classification is based on Levine, N.D. et. al., 1980.

Identification with Reasons

SubKingdom—*Protozoa* :

1. Single celled, eukaryotes having distinct nuclei and other membrane-bound organelles.
2. Movement by pseudopodia or **cilia** or **flagella**.
3. Microscopic animals usually occur as single, one-celled individuals, many other species are colonial.

Phylum—*Sarcomastigophora* :

1. Locomotory organelles are pseudopodia or flagella or both.
2. Single type of nucleus (monomorphic).

Sub Phylum—*Sarcodina* :

1. Organelles of locomotion is pseudopodia, or locomotive protoplasmic flow without discrete pseudopodia.
2. Flagella, when present, usually restricted to developmental stages.
3. Body naked, or with external or internal test or skeleton.
4. Mostly free living.

Super class—*Rhizopoda* :

1. Generally creeping forms.
2. Locomotion by **lobopodia, filopodia** or **reticulopodia** or by protoplasmic flow without production of distinct pseudopodia.

Class—*Lobosea* :

1. Pseudopodia are typically lobose or more or less filiform but produced from broader hyaline lobe.
2. Usually uninucleate, multinucleate forms not flattened or anastomosing plasmodia.

Subclass—*Gymnamoebia* :

1. Test absent.

Order—*Amoebida* :

1. Typically uninucleate.
2. No flagellate stage.

Type—*Amoeba proteus* :

1. Irregular shape with many pseudopodia, pseudopodia lobose type.
2. Ectoplasm and endoplasm are clearly differentiated.
3. Single large nucleus in the endoplasm.
4. Contractile vacuole single, food vacuoles are many.

2. *ENTAMOEBA HISTOLYTICA*

Habit and Habitats

It lives as a pathogenic endoparasite in the upper part of large intestine and frequently invading the appendix and occasionally into the lower part of small intestine of man. It feeds on R.B.C and of the intestinal wall.

Economic Status

E. histolytica causes amoebic desentry and abscesses in liver, lung etc. In **amoebic desentry** it secrets tissues dissolving proteolytic enzyme (cytolysine), which destroys the intestinal lining, causing necrosis and abscess which finally breaks down, leading to the development of the ulcer that bleed. Frequently the term **amoebiasis** is used clinically to denote all those conditions which are produced in human by infection with ***E. histolytica*** at different areas it invade.

Systematic Position

Subkingdom	**Protozoa**
Phylum	**Sarcomastigophora**
Sub Phylum	**Sarcodina**
Super class	**Rhizopoda**
Class	**Lobosea**
Subclass	**Gymnamoebia**
Order	**Amoebida**
Suborder	**Tubulina**
Type	***Entamoeba histolytica***

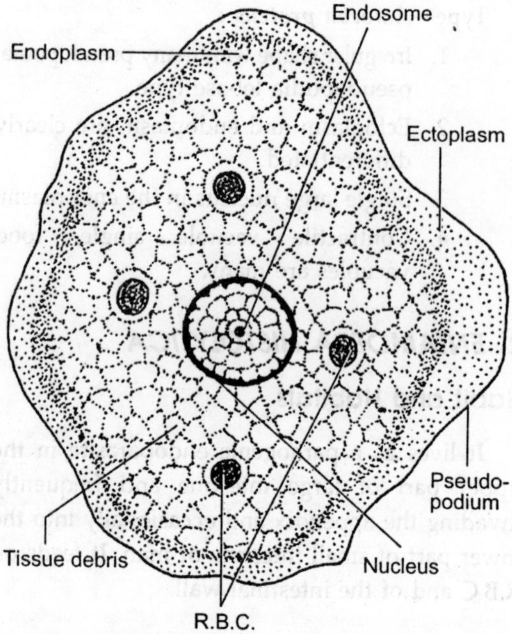

Fig. 1.1.1.2A : Entamoeba histolytica (Trophozoite)

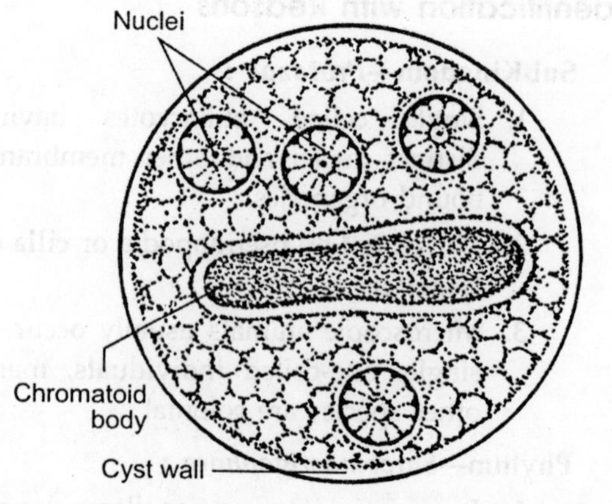

Fig. 1.1.1.2B : Entamoeba histolytica (Cystic stage)

Distinctive features

1. Cytoplasm differentiated into ectoplasm and endoplasm.
2. Nucleus is spherical.
3. Contractile vacuole is absent.
4. Red-blood cells are present.
5. Single large pseudopodium.

Identification with reasons

State classification and characters up to—

Order—Amoebida as in *Amoeba.*

Trophic form (Trophozoite stage) of *Entamoeba histolytica* :

1. Cytoplasam is clearly differentiated into ectoplasm and granular endoplasm.
2. Endoplasm contains nucleus, food particles, RBC and tissue debris.
3. The nucleus is rounded. In the centre of the nucleus is a small endosome or karyosome, fine chromatin granules line the delicate nuclear membrane.
4. Contractile vacuole is absent.
5. Single large pseudopodium is formed by clear and hyaline ectoplasm.

Cystic stage of *Entamoeba histolytica* :

1. Spherical with clear and hyaline cytoplasm, surrounded by transparent cell wall.
2. Nucleus are two to four with central karyosome.
3. Rod like **chromatoid bodies** with round ends in the cytoplasm.

3. *ELPHIDIUM*

Habit and Habitat

Primarily marine and live on bottom sediments as benthic organisms. They can be found at a depth of shallower than about 2000 meters. They can occasionally found on the shore. *Elphidium* feeds primarily upon diatoms, but flagellates, unicellular algae, bacteria and organic particles many also be ingested.

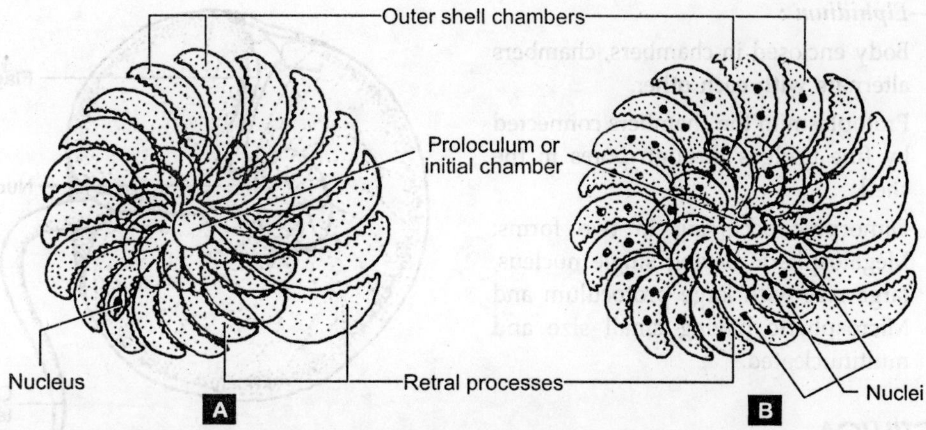

Fig. 1.1.1.3 : *Elphidium* (Polystomella). A. Megalosphric, B. Microspheric

Economic Status

From the paleontological point of view the foraminiferans are a very important group of organism. Geologist have been interested in certain foraminiferan fossils as an indicator of climates in past geological ages. Fossil foraminiferans are extremely useful in determining the approximate age of sediment as a fairly reliable indicator of likely places to drill for oil.

The extensive fossilized remains of foraminiferans have a considerable economic importance. Cement and black board chalk are foraminiferan - containing elements.

Systematic Position

SubKingdom	**Protozoa**
Phylum	**Sarcomastigophora**
Subphylum	**Sarcodina**
Super class	**Rhizopoda**
Class	**Granuloreticulosia**
Order	**Foraminiferida**
Type	***Elphidium***

Distinctive features

1. Body covered with hard and translucent shell.

2. Shell multilocular and perforated.

3. Dimorphic, occurring two forms; **Megalosphric** having large size, with single nucleus and large **proloculum; Microspheric** with multinuclei and small protoculum.

4. Chambers of shell alternate with each other. From the chambers several cytoplasmic processes protruded through the fine pores, or **foramina.**

Identification with reasons

State the characters up to—

Super class—*Rhizopoda* as in *Amoeba.*

Class—*Granuloreticulosia* : Delicate, reticular and finely granular but non-anastomosing pseudopodia.

Order—*Foraminiferida* **:**

1. Test with one to many chambers, chambers alternate with each other and with fine pores.
2. Pseudopodia protruding from aperture or wall perforations or both.
3. Nuclear dimorphism in developmental stages. Megalospheric with a single nucleus and microspheric with many nuclei.

Type—*Elphidium* :

1. Body enclosed in chambers, chambers alternate with each other.
2. Protoplasm in the chambers connected by fine bridges through pores in the shell.
3. Dimorphic occurring in two forms: Megalospheric having single nucleus, large size with large protoculum and Microspheric having small size and multinucleated.

4. *NOCTILUCA*

Habit and Habitat

Marine, pelagic, found in abundance on water surface near shore and feeds on other protozoans. It exhibits **Bioluminescence.**

Special features

Noctiluca emits greenish – white or bluish-white light when irritated.

N. scintillans is so plentiful that the sea is coloured either red or yellow by day and at night it glows by strong bioluminescent light.

The light is due to **luciferin** which owing to the influence of an enzyme **luciferase** emits light.

Systematic Position

SubKingdom	**Protozoa**
Phylum	**Sarcomastigophora**
Subphylum	**Mastigophora**
Class	**Phytomastigophorea**
Order	**Dinoflagellida**
Type	***Noctiluca***

Distinctive features

1. Body is spherical jelly-like and is covered by a stout pellicle.

2. Presence of **cytostome** and **cytopharynx.**

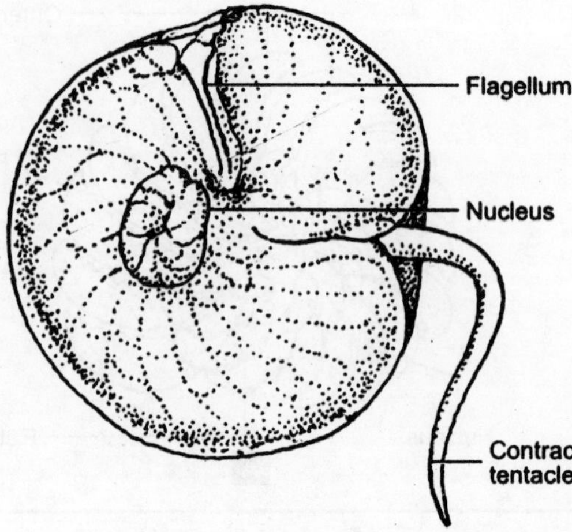

Fig. 1.1.1.4 : *Noctiluca*

3. Single nucleus and from it delicate protoplasmic strands extend up to periphery.

4. A small flagellum and a large tentacle are present.

Identification with reasons

State the characters up to—

Subphylum—*Sarcomastigophora* as in *Amoeba.*

Subphylum—Mastigophora :

1. One or more flagella in adult form.
2. Body covered by pellicle.

Class—Phytomastigophorea :

1. Mostly free living.
2. Typically with chloroplasts (**Chromatophores**).
3. Usually with one flagella.

Order—Dinoflagellida :

1. More or less spherical body with distinct groove, and is covered by a stout pellicle.
2. Cytostome opens into a cytopharynx.
3. Two **heterodynamic flagella,** one ribbon shaped with paraxial rod and other smooth.

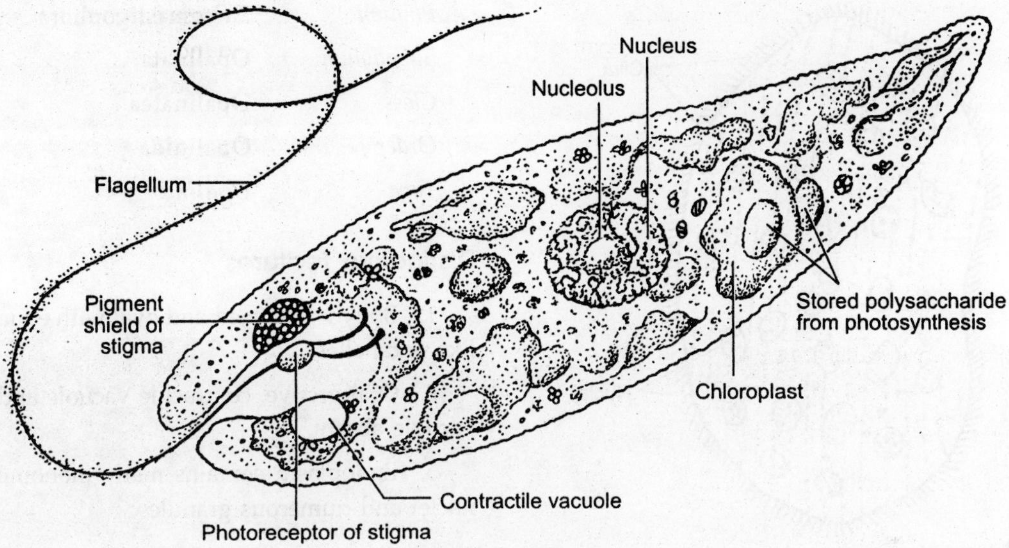

Fig. 1.1.1.5 : *Euglena*

4. Chloroplast typically green or golden yellow.

Type—*Noctiluca* :

1. Body is spherical jelly like mass.
2. A small flagellum and a large tentacle are present.
3. Nucleus is in the central protoplasmic mass and from it delicate protoplasmic strands extend upto periphery.
4. Midventrally there is a cytostome which opens into cytopharynx.

5. *EUGLENA*

Habit and Habitat

Euglena is found in fresh water ponds, but is abundant in stagnant water such as ditches, pools rich in decaying organic matter. They are autotroph as long as in the light but saprophytic in complete darkness where they utilized organic nutrients in solution.

Systematic Position

SubKingdom	**Protozoa**
Phylum	**Sarcomastigophora**
Subphylum	**Mastigophora**
Class	**Phytomastigophorea**
Order	**Euglenida**
Type	***Euglena***

Distinctive features

1. Spindle-shaped body having rounded anterior and pointed posterior end.

2. Anterior end contains funnel like cytopharynx from which flagellum arise by two roots.

3. Stigma usually anterior and nucleus posterior.

4. **Paramylum** and **pyrenoids** are stored as food—material.

Identification with reasons

State the characters up to—

Class—*Phytomastigophorea* as in *Noctiluca.*

Order—Euglenida :

1. Spindle-shaped body with rounded anterior end and pointed posterior end.

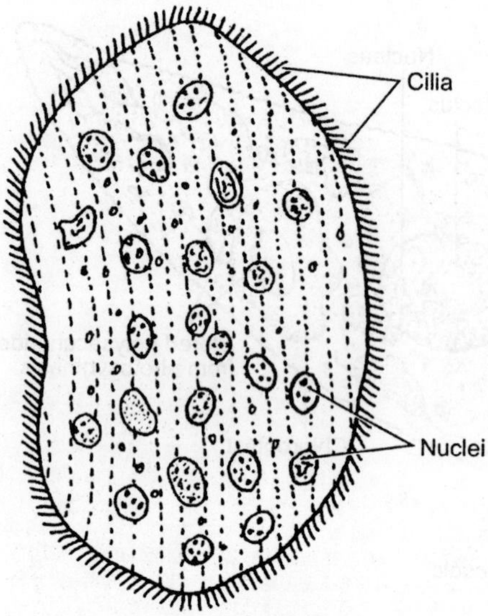

Fig. 1.1.1.6 : *Opalina*

2. Two flagella, one or both emerging from the anterior reservoir.
3. Spirally marked complex pellicle.
4. Presence of mouth and cytopharynx.

Type—*Euglena* :

1. Spindle shaped body.
2. Flagellum arises by two roots from inside the cytopharynx.
3. Stigma, **contractile vacuole** and chromatophores present.
4. Paramylum and pyrenoids present.
5. Nucleus posterior.

6. *OPALINA*

Habit and Habitat

Parasitic or commensal within the guts or rectum of frogs, toads, and few fishes. Absorb nutrients from the gut or rectum of host through general body surface.

Systematic Position

SubKingdom	**Protozoa**
Phylum	**Sarcomastigophora**
Subphylum	**Opalinata**
Class	**Opalinatea**
Order	**Opalinida**
Type	***Opalina***

Distinctive features

1. Body is flattened and oval with equal cilia like organelles.
2. Oral groove, contractile vacuole and food vacuole absent.
3. Endoplasm contains many monomorphic nuclei and numerous granules.

Identification with reasons

State the characters up to—

Phylum—*Sarcomastigophora* as in ***Amoeba.***

Subphylum—*Opalinata* :

1. The entire body surface is covered with numerous cilia in oblique rows.
2. Cytostome absent.
3. The ectoplasm is differentiated into outer thin **pellicle** and inner layer with contractile **myonemes.**

Class—*Opalinatea* :

1. Contractile vacuole and oral groove absent.
2. More than one nuclei of same type.

Order—*Opalinida* : Character similar as those of class.

Type—*Opalina* : State distinctive features as described previously.

7. *TRYPANOSOMA*

Habit and Habitat

It is polymorphic, parasite found in the blood, lymphatic system and cerebrospinal fluid of man. These parasites also found in chordates and in intestine of Leeches and blood sucking invertebrates.

Economic Status

Trypanosoma gambiense and *T.rhodesiense* a parasite of the blood plasma of man cause **Mid- and West-African sleeping sickness** or **Gambian trypanosomiasis** and **East African sleeping sickness** respectively. *Trypanosoma cruzi,* a parasite of muscular and nervous tissues of man causes **South American trypanosomiasis** or **Chagas' disease**. *T. brucei* causes "**nagana**" in domestic animals of tropical africa. *T. evensi* causes "**surra**" in horses and mules of India. Trypanosoma in Man is transmitted by the blood sucking insect *Glossina palpalis* (tsetse fly). The parasite reaches cerebrospinal fluid and causes sleeping sickness in which an infected man goes on sleeping and ultimately dies if not treated in preliminary stages *i.e.,* **stage of trypanosoma fever.**

Systematic Position

SubKingdom	**Protozoa**
Phylum	**Sarcomastigophora**
Subphylum	**Mastigophora**
Class	**Zoomastigophorea**
Order	**Kinetoplastida**
Type	***Trypanosoma***

Distinctive features

1. Microscopic, irregularly elongated flattened and tapering at both ends with a single nucleus.
2. Body is covered by thin pellicle.
3. On one side the edges of the body is thrown into irregular folds called undulating membrane.
4. Single flagellum originating from a basal granule and attached to the body by undulating membrane.
5. **Kinetoplast** relatively small and compact.
6. Parasite on vertebrate host.

Identification with reasons

State the characters up to—

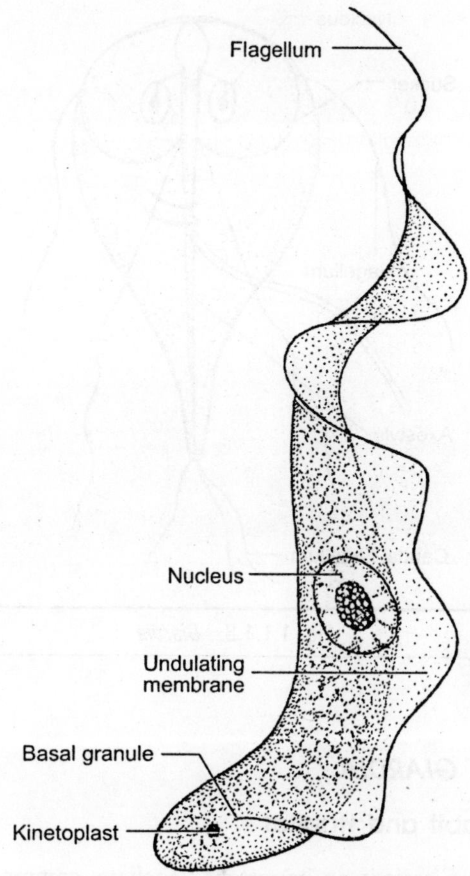

Fig. 1.1.1.7 : *Trypanosoma*

Subphylum—*Mastigophora* as in *Noctiluca.*

Class—*Zoomastigophorea* :

1. Chloroplast absent.
2. One to many flagella present.
3. Amoeboid forms with or without flagella.
4. Often with undulating membrane.

Order—*Kinetoplastida* :

1. One or two flagella arising from depression.
2. Kinetoplast located near flagellar **kinetosomes.**
3. Parasite or free living.

Type—*Trypanosoma* : State distinctive features as described previously.

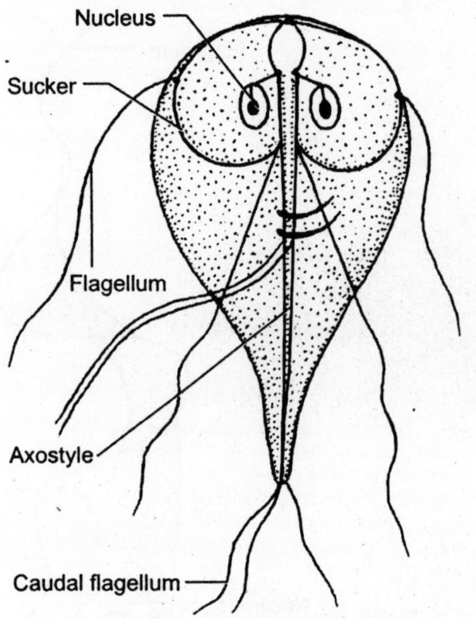

Fig. 1.1.1.8 : *Giardia*

8. *GIARDIA*

Habit and Habitat

Giardia is an important flagellate, commonly found in the intestines of rat, mice, rabbit, guineapigs, dogs and frogs. *G.intestinalis* lives as parasite in the duodenum and the upper part of the ileum of man causes chronic diarrhoea without loss of blood.

Systematic Position

SubKingdom	**Protozoa**
Phylum	**Sarcomastigophora**
Subphylum	**Mastigophora**
Class	**Zoomastigophorea**
Order	**Diplomonadida**
Type	***Giardia***

Distinctive features

1. Body oval shaped with broad anterior end and narrow posterior end. Ventral surface slightly concave and bears adhesive disc.

2. Two nuclei, two **parabasal bodies**, two **Karyo-mastigonts** and individual mastigonts with four flagella.

3. Single axostyle with four flagella forms the median longitudinal axis of the body.

4. Parasitic.

Identification with reasons

State the characters up to—

Class—*Zoomastigophorea* as those of ***Trypanosoma*.**

Order—*Diplomonadida* :

1. Free living or parasitic.
2. Cyst present.
3. One or two karyomastigonts and individual mastigonts with one to four flagella.
4. Mitochondria and Golgi apparatus absent.
5. Set of accessory organelles are present.

Type—*Giardia* : States distinctive features as described previously.

9. *LEISHMANIA*

Habit and Habitat

Leishmania is a pathogenic protomonad causing serious disease among man, cattle, dog, sheep, and horse. The disease is called **Leishmaniasis.** It is intercellular parasite found in the cells of liver, spleen, bone marrow, intestine and lymph gland in the **reticuloendothelial cells.**

Economic Status

Leishmania, parasites of man are three in number, *Leishmania donovani,* causing **Kala-azar.** The disease Kala-azar is also called **visceral leishmaniasis.** *L. tropica* causes oriental sore. The infection is limited to a local lesion of the skin

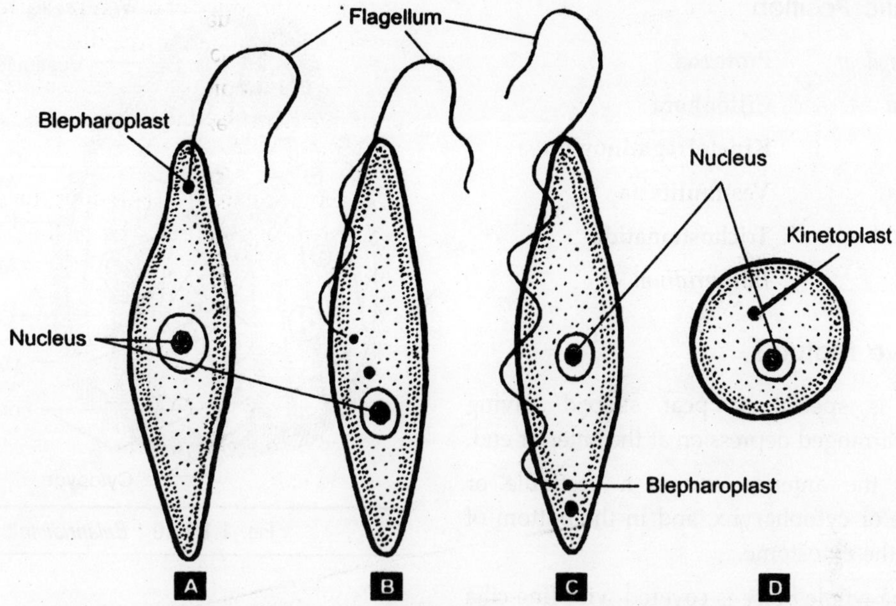

Fig. 1.1.1.9 : *Leishmania.* Stages : A. Promastigote (Leptomonas), B. Epimastigote (Crithidia), C. Opisthomastigote (Trypanosoma), D. Amastigote (Leishmania).

and subcutaneous tissues. *L.brasiliensis* causes **espundia or naso-pharyngeal leishmaniasis.** The infection is limited to a local lesion of the skin and nasopharyngeal mucous membrane.

Systematic Position

SubKingdom	**Protozoa**
Phylum	**Sarcomastigophora**
Subphylum	**Mastigophora**
Class	**Zoomastigophorea**
Order	**Kinetoplastida.**
Type	**Leishmania**

Distinctive features

1. Microscopic, rounded or oval form.
2. Body is covered by a pellicle.
3. Endoplasm contains an oval nucleus, rod-shaped or dot-like kinetoplast and a parabasal body.
4. No flagellum.
5. It is parasitic and polymorphic.

Identification with reasons

State the characters up to—

Order—*Kinetoplastida* as in *Trypanosoma.*

Type—*Leishmania* : State distinctive features as described previously.

10. *BALANTIDIUM*

Habit and Habitat

Balantidium is parasitic of the large intestine of pig, sheep, camel, monkey, cockroach and man. It is also found in the rectal content of the frog.

Economic Status

The pathogenic species *Balantidium coli* found in the large intestine of man cause ulceration of the large intestine and invade the tissues of the walls. The colon is sometimes ulcerated. Majority of infected persons suffer only from diarrhoea, only a small number develop severe or fatal dysentry.

Systematic Position

SubKingdom	**Protozoa**
Phylum	**Ciliophora**
Class	**Kinetofragminophorea**
Subclass	**Vestibuliferia**
Order	**Trichostomatida**
Type	***Balantidium***

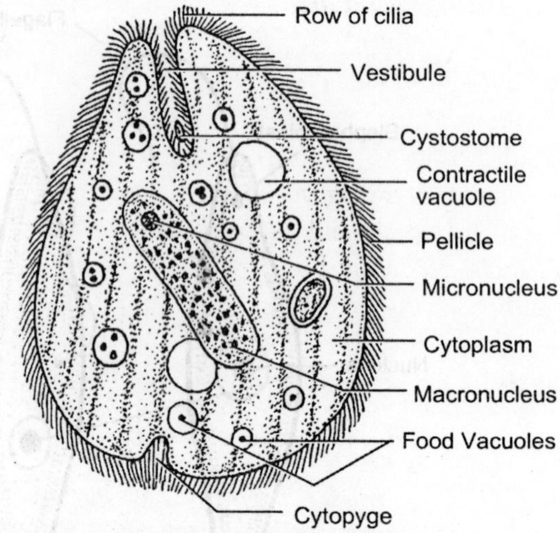

Fig. 1.1.1.10 : *Balantidium*

Distinctive features

1. It is special or pear shaped having obliquely arranged depression at the anterior end.

2. At the anterior end is a vestibule or peristome or cytopharynx, and in the bottom of which is the cytostome.

3. The whole body is covered with fine cilia arranged in rows.

4. The **macronucleus** is very slightly curved with a slight concavity on either side; **micronucleus** is very small and inconspicuous.

5. Two contractile vacuoles and food vacuoles are present in the endoplasm.

Identification with reasons

State the characters up to—

Subkingdom—Protozoa as in *Amoeba.*

Phylum—*Ciliophora* :

1. Simple cilia or compound ciliary organelles at least in one stage of life cycle.
2. Individual cilia are connected below the body surface through a complex cord of fibres (infraciliature).
3. Two types of nuclei i.e., micro and macronucleus present (heterokaryotic).
4. Presence of contractile vacuole.
5. Binary fission transverse, but budding and multiple fission also occur.

Class—*Kinetofragminophorea* :

1. Oral infraciliature only slightly distinct from somatic infraciliature.
2. Cytostome often apical or mid ventral, on surface of body or at bottom of atrium or vestibulum.
3. Prominent cytopharyngeal apparatus.
4. Compound ciliature, oral or somatic typically absent.

Subclass—*Vestibuliferia* :

1. Apical or near vestibulum commonly present.
2. Cilia arises apical from anterior parts of somatic kineties (basal body) and leading to cytostome.
3. Free living or parasitic, in digestive tract of vertebrates and invertebrates.

Order—*Trichostomatida* :

1. Somatic ciliature not reduced.
2. Vestibular without buccal cilia.

Type—*Balantidium* : State distinctive features as described previously.

11. *PARAMOECIUM*

Habit and Habitat

Fresh water animal, abundantly found in ponds, ditches, pools, streams, lake etc., containing decaying vegetable matter. Holozoic, feeds on bacteria and minute protozoans, algae, diatoms etc. Can be easily cultured in hay infusion in laboratory.

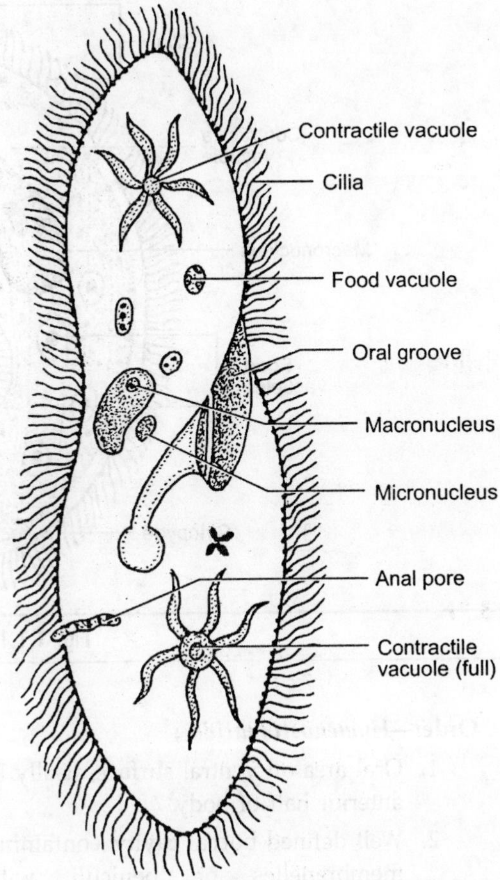

Fig. 1.1.1.11 : *Paramoecium*

Systematic Position

Subkingdom	**Protozoa**
Phylum	**Ciliophora**
Class	**Oligohymenophorea**
Subclass	**Hymenostomatia**
Order	**Hymenostomatida**
Type	***Paramoecium***

Distinctive features

1. Slipper shaped body, anterior end bluntly round, and posterior end pointed.
2. Cilia cover the entire animal.
3. Two contractile vacuoles, one anterior and other posterior with radiating canals.
4. Oral groove or **peristome** and cytopharynx present.
5. Kidney shaped macronucleus and a small micronucleus present.

Identification with reasons

State the characters up to—

Phylum—*Ciliophora* as those of *Balantidium*.

Class—*Oligohymenophorea* :

1. Generally well defined oral apparatus, at least partially in buccal cavity.
2. Oral ciliatures clearly distinct from somatic ciliature consisting of compound organells on left side.
3. Cytostome present at bottom of buccal or infundibular cavity, usually ventral and/or near anterior end.
4. Cyst not uncommon.

Subclass—*Hymenostomatia* :

1. Often with uniform and heavy body ciliation.
2. Buccal cavity, when present, ventral.
3. **Kinetodesmata** regularly present, usually conspicuous.
4. Sessile form and stalk, colony, and cyst formation relatively rare.
5. Fresh water forms predominant.

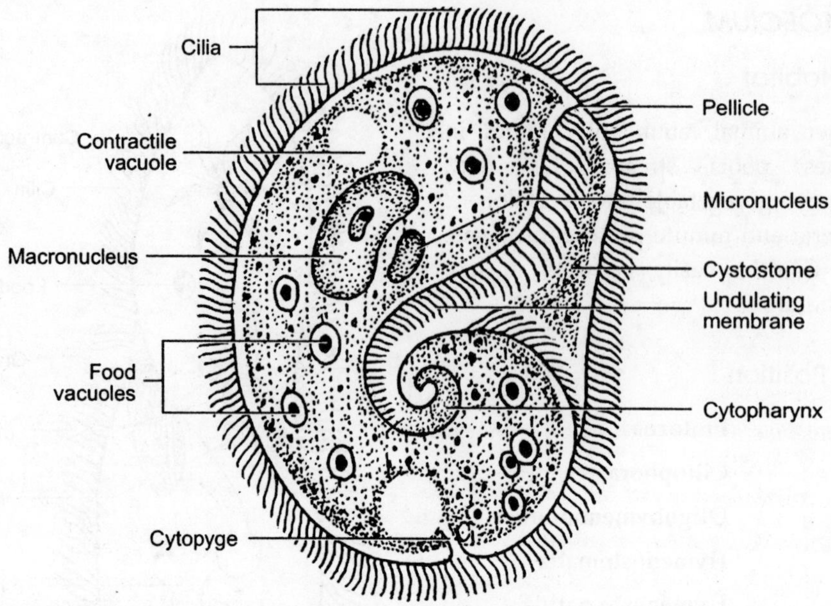

Fig. 1.1.1.12 : *Nyctotherus*

Order—*Hymenostomatida* :

1. Oral area on ventral surface, usally in anterior half of body.
2. Well defined buccal cavity, containing membranelles or peniculi with infraciliary bases typically 3-4 rows of wide kinetosomes.

Type—*Paramoecium* : State the distinctive features as described previously.

12. *NYCTOTHERUS*

Habit and Habitat

Nyctotherus an intestinal commensal of mammals, anuran amphibians, fishes and insects. It is found associated with other parasitic protozoans as *Opalina* and *Balantidium*.

Systematic Position

Subkingdom	**Protozoa**
Phylum	**Ciliophora**
Class	**Polymenophorea**
Subclass	**Spirotrichia**
Order	**Heterotrichida**
Type	***Nyctotherus***

Distinctive features

1. Body kidney shaped and dorso-ventrally flattened with longitudinal rows of equal sized cilia.

2. The cytostome leads into a long, curved, coiled cytopharynx.

3. Permanent **cytopyge** is posterior.

4. Micronucleus and irregular shaped oval macronucleus are in the front of cytopharynx.

5. Cilia are large and wind clockwise in the cytopharynx.

Identification with reasons

State the characters up to—

Phylum—*Ciliophora* as those of *Paramoecium*.

Class—*Polymenophorea* :

1. Well-developed, conspicuous adoral zone of numerous buccal or peristomial organelles, often extending out on to body surface.
2. Somatic ciliature complete or reduced or appearing as cirri.
3. Cytostome at bottom of buccal cavity or infundibulum.
4. **Cytoproct** often absent.
5. Cyst, very common in some groups.

Subclass—*Spirotrichia* :

1. Extensive stair case like zone of adoral memebranelles which spirals towords the cytostome.
2. Usually with reduced body cilia.
3. Buccal ciliature is well developed.

Order—*Hetrotrichida* :

1. Body with heavy **holotrichous ciliation.**
2. Macronucleus oval.
3. Body cilia arranged in rows.
4. Endoparasite.

Type—*Nyctotherus* : State the distinctive features as stated above.

13. *PLASMODIUM*

Habit and Habitat

Plasmodium is an intracellular blood parasites of man and other vertebrates. It causes **Malaria** in man. Man is infected whenever a female mosquito bites him and injected several sporozoites into the blood and thus communicating diseases from man to man. Mosquito here acts as **vector** or carrier of the host.

Economic status

Four species of *Plasmodium* are capable of causing malaria in man. They are *Plasmodium vivax, P.malariae, P.falciparum* and *P.ovale*. Different stages in man produce toxin, which result in fever and anaemia. Malaria is transmitted from man to man by female *Anopheles* mosquito. *P.vivax* is the cause of "tertian" or "benign tertian" malaria. *P. malariae* is the cause of "quartan" malaria. *P. falciparum* is the cause of "malignant tertian" or "subtertian" malaria, now preferably called "falciparum" malaria. *P.ovale* is the cause of "mild tertian" fever.

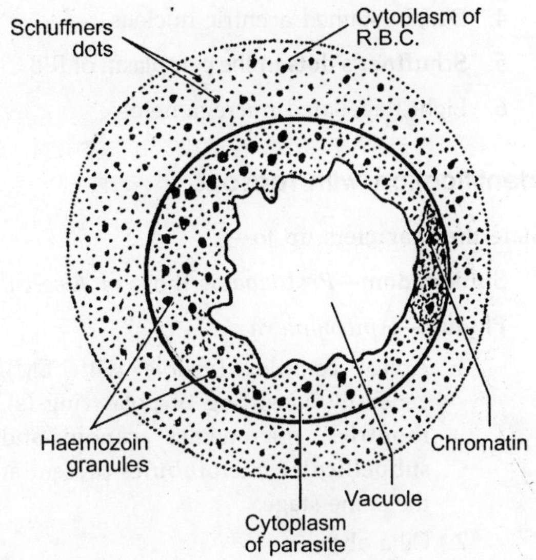

Fig. 1.1.1.13 : *Plasmodium*

Systematic Position

Subkingdom	**Protozoa**
Phylum	**Apicomplexa**
Class	**Sporozoea**
Subclass	**Coccidia**
Order	**Eucoccidiida**
Type	***Plasmodium***

Distinctive features

1. Trophozoite, rounded ring like (signet ring) located within R.B.C.
2. Large central vacuole present. **Chromatin granules** on one side of the vacuole.
3. Cytoplasm scanty in the periphery and contains **haemozoin granules.**

4. Deeply stained acentric nucleus.
5. **Schuffner's dots** in the cytoplasm of R.B.C.
6. Eight flagellated microgametes.

Identification with reasons

State the characters up to—

Subkingdom—*Protozoa* as those of *Amoeba.*

Phylum—*Apicomplexa* :

1. Apical complex (visible with EM), generally consisting of polar ring (s), **rhoptries**, micronemes, **conoid** and **subpellicular microtubules** present at the some stage.
2. Cilia absent.
3. Micropore(s) generally present at the some stage.
4. Sexuality by **syngamy**.
5. All are parasitic.

Class—*Sporozoea* :

1. Conoid if present, forming complete cone.
2. Reproduction generally both asexual and sexual.
3. **Oocysts** generally containing infective **sporozoites** resulting from **sporogony**.
4. Locomotion by body movement, gliding or undulation of longitudinal ridges.
5. Microgametes of some groups are with flagella.
6. Pseudopods ordinarily absent, if present used for feeding, not locomotion.

Subclass—*Coccidia* :

1. Gamonts ordinarily present. Mature gamonts small, typically intracellular, without **mucron** or **epimerite**.
2. **Syzygy** generally absent, if present involves markedly anisogamous gametes.
3. Life cycle consisting of **merogony, gametogony** and **sporogony**.

Order—*Eucoccidiida* :

1. Merogony in vertebrates and sprogony in invertebrates.
2. Macrogamete and microgamont developing independently.
3. No syzygy.
4. Conoid ordinarily absent.
5. Microgamont producing eight flagellated microgametes or many as in *Eimeria.*
6. Zygote motile (**ookinete**) in *Plasmodium;* nonmotile in *Eimeria.*
7. Naked sporozoites, with three-membraned wall.

Type—*Plasmodium* : State the distinctive features as described previously.

14. *EIMERIA*

Habit and Habitat: *Eimeria* is an intracellular pathogenic parasite in the gut epithelium of rabbits, cattle, frog, duck etc. Horse, sheep, goat, pig, dog, guineapig frequently suffer by this protozoan attack.

Economic Status

In animals the pathogenic effect varies with different species and with the severity of the infection. Chickens harbour at least eight species of *Eimeria,* six species are in cattle, six species in rabbit, and one species in pigeons. Of these species *E. tenella* and *E. necatrix* are very harmful. *E. tenella* causes severe injury to the caeca in young chicks, with often fatal bloody diarrhoea (caecal coccidiosis). *E. necatrix,* causes more chronic intestinal disorder in older bird. *E.bovis* in cattle causes red or watery diarrhoea, weakness and emaciation, which is often fatal.

Systematic Position

Subkingdom	**Protozoa**
Phylum	**Apicomplexa**

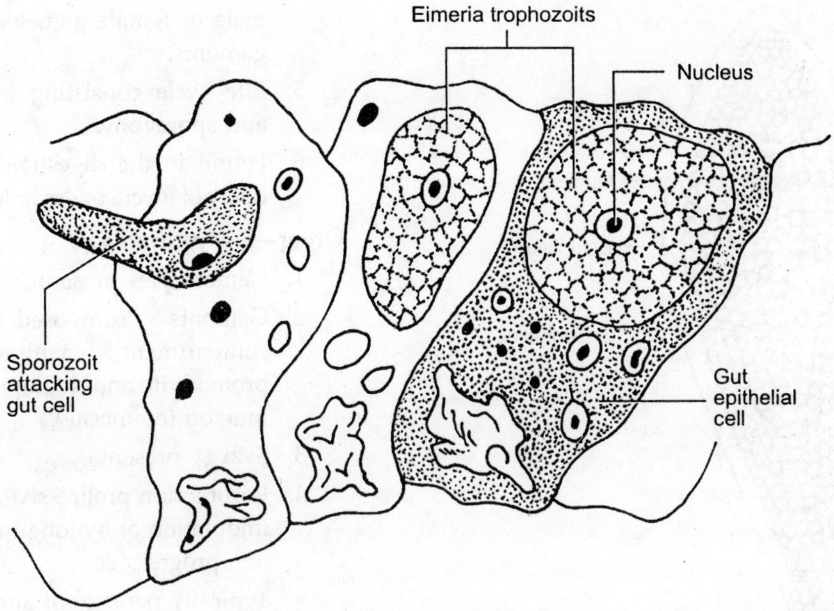

Fig. 1.1.1.14 : *Eimeria*

Class	**Sporozoea**
Subclass	**Coccidia**
Order	**Eucoccidiida**
Type	***Eimeria***

Distinctive features

1. Intracellular parasite, having small, spherical and simple body containing granular endoplasm and a minute nucleus.

2. Sporozoites are encased in sporocyst within oocyst.

3. A microgametocyte produces a number of biflagellate microgametes.

Identification with reasons

State the characters up to—

Order—*Eucoccidiida* as those of *Plasmodium*.

Type—*Eimeria* : State the distinctive features as in above.

15. *MONOCYSTIS*

Habit and Habitat

Monocystis is a non-pathogenic endo parasite in the seminal vesicle of earthworm. The various phases of its life-cycle become complete within the seminal vesicle. It obtains nutrition from seminal fluid.

Systematic Position

Subkingdom	**Protozoa**
Phylum	**Apicomplexa**
Class	**Sporozoea**
Subclass	**Gregarinia**
Order	**Eugregarinida**
Type	***Monocystis*** (Trophozoite stage)

Distinctive features

1. Body elongated, usually flattened and pointed at both ends.

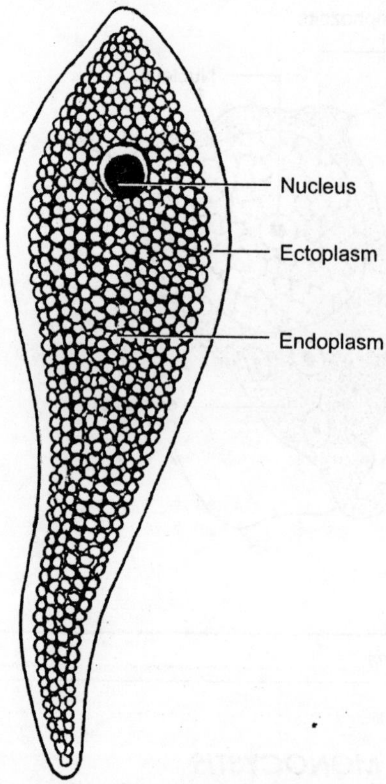

Fig. 1.1.1.15 : *Monocystis.* Trophozoite stage

2. Cytoplasm is differentiated into another non-granular ectoplasm and an inner granular endoplasm.

3. Body is covered with a distinct pellicle.

4. Nucleus anterior to the centre.

Identification with reasons

State the characters up to—

Class—*Sporozoea* as those of *Plasmodium.*

Subclass—*Gregarinia* :

1. Mature gamonts large extracellular.
2. Mucron or epimerite in mature organism and mucron formed from conoid.
3. Generally syzygy present in gamonts.
4. Gametes usually similar (isogamous) or nearly so, with similar numbers of male or female gametes produced by gamonts.
5. Life cycle consisting of gametogony and sporogony.
6. Found in the digestive tract or body cavity of invertebrates or lower chordates.

Order—*Eugregarinida* :

1. Gametocytes present.
2. Gamonts composed of single compartment without definite protomerite and deutomerite, but with mucron (epimerite).
3. Syzygy present.
4. Locomotion progressive, by gliding or undulation or by longitudinal ridge or non-progressive.
5. Typically parasite of annelids.

Type—*Monocystis* : State the distinctive features as described previously.

16. *MYXOBOLUS*

Habit and Habitat: *Myxoblous* is an extracellular spore forming parasite of fresh water fish and aquatic annelids. In fish the parasite lives in the urinary bladder, gall bladder, musculature, gills and tissues.

Systematic Position

Subkingdom	**Protozoa**
Phylum	**Myxozoa**
Class	**Myxosporea**
Order	**Bivaldulida**
Type	***Myxobolus***

Distinctive features

1. Ovoid in shape.

2. Two pyriform polar capsules at the anterior end of the spore, each with a filament twisted into 5 or 6 cells and with a long filament.

3. **Sporoplasm** occupy posterior region.

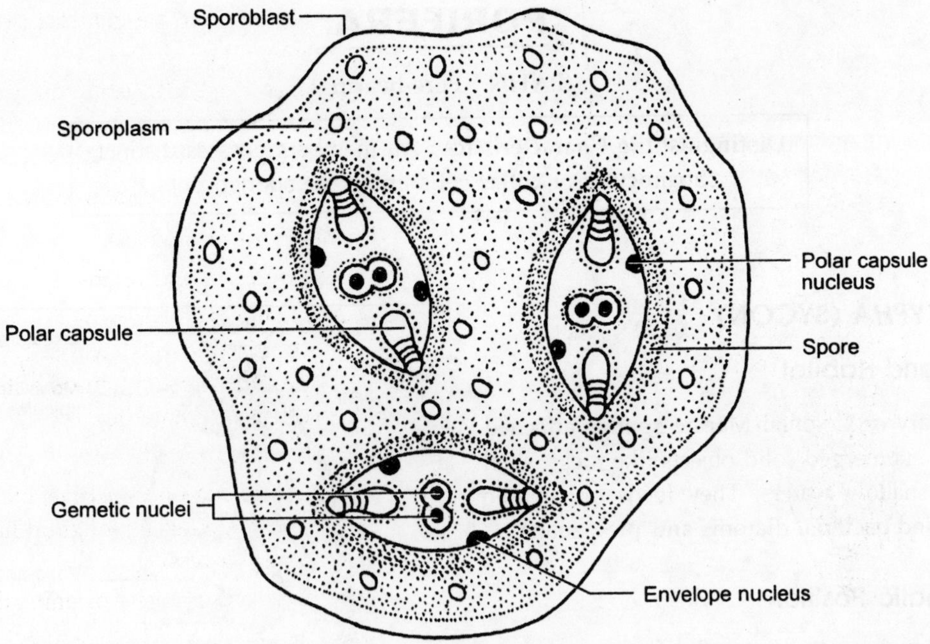

Fig. 1.1.1.16 : *Myxobolus*

Identification with reasons

State the characters up to—

Subkingdom—*Protozoa* as those of *Amoeba*.

Phylum—*Myxozoa* :

1. Multi-cellular origin spores, with one or more polar capsules and sporoplasms.
2. Presence of 1, 2, or 3 (rarely more) valves.
3. All are parasitic.

Class—*Myxosporea* :

1. Spore with one or two sporoplasm and one to six (typically two) polar capsules.
2. Each polar capsule are with coiled polar filament. Filament probably used for anchoring.
3. Generally spore membrane two, occationally upto six valves.
4. Well developed **trophozoite** stage.

Order—*Bivaldulida* :

1. Spore wall consists of two valves.
2. Spore with **two polar** capsules.

Type—*Myxobolus* : State the distinctive features as described previously.

PORIFERA

(LATIN : Pore-bearing)

Distinguishing Characteristics : All members possess distinct **choanocytes** or **collar cells** lining the spongocoel.

1. *SCYPHA* (SYCON)

Habit and Habitat

Solitary or Colonial Marine sponge attached to some submerged solid objects like rocks, shell etc., in shallow waters. They feeds on organic matter and bacteria, diatoms and protozoans.

Systematic Position

Phylum	**Porifera**
Class	**Calcarea**
Subclass	**Calcaronea**
Order	**Sycettida**
Type	***Scypha (Sycon)***

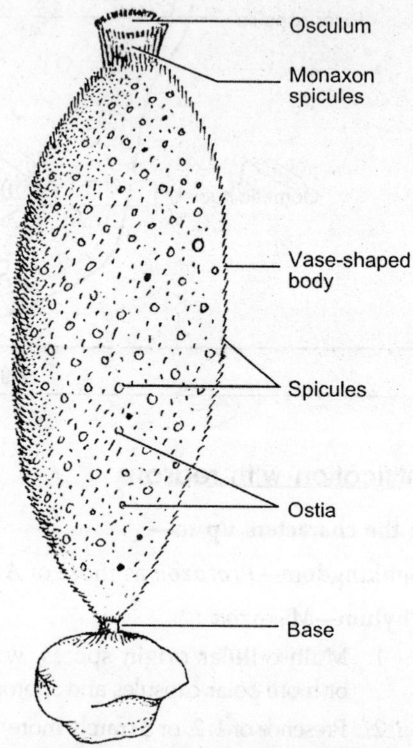

Fig. 1.1.1.17 : *Scypha (Sycon)*

Distinctive Features

1. Vase-shaped sponge with several cylinders attached to common base.
2. Each cylinder is dilated at the middle and opens to the outside by **osculum.**
3. Thick body wall through which spicules project.
4. **Choanocytes** are restricted to the radial canals.

Identification with Reasons

Phylum—*Porifera* :

1. Body with cellular grade of organisation.
2. Body surface perforated by number of many pores or **ostia**.
3. All are usually sessile.
4. Lack of any organ system.
5. Choanocytes or **collar cells** lining the **spongocoel.**
6. Possess canal system for flow of water.
7. Soft structure of sponge body is provided by calcareous or siliceous **spicules**, by organic fibres, or by a combination of fibres and spicules.

Class—*Calcarea* :

1. Skeleton made up of calcareous spicules.

Classification is based on Marshall, A., Williams, W.D.(eds.), 1972.

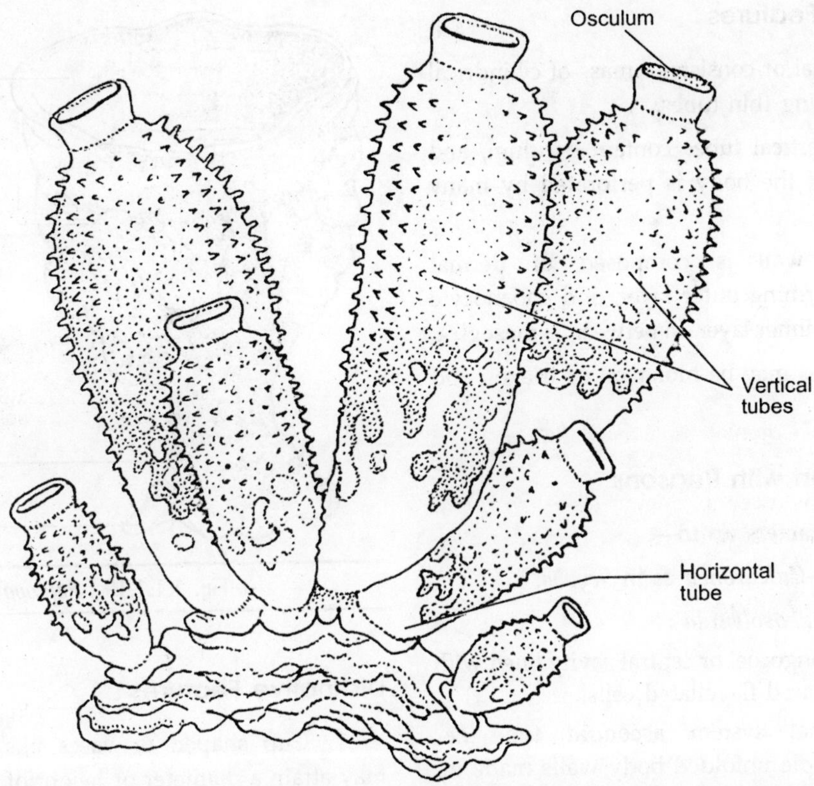

Fig. 1.1.1.18 : *Leucosolenia* (Colony)

2. Marine, usually shallow water species.
3. Spicules are **monaxon** or **triaxon** or **tertraxon**.

Subclass—*Calcaronea* :

1. Choanocyte nucleus is apical in position.
2. The flagellum of each choanocyte arises directly from the nucleus.
3. The tri-radiate spicules usually with one long ray.

Order—*Sycettida* :

1. **Canal system** either **syconoid** or **leuconoid** type i.e. gastral layer is discontinuous, confined to flagellated chambers.
2. Choanocytes are restricted to the **radial canals.**

Type—*Scypha (Sycon)* : State the distinctive features as stated above.

2. *LEUCOSOLENIA*

Habit and Habitat

Marine, Colonial sponge found in shallow water, attached on sea-shore rocks.

Systematic Position

Phylum	**Porifera**
Class	**Calcarea**
Subclass	**Calcaronea**
Order	**Leucosolenida**
Type	***Leucosolenia***

Distinctive Features

1. Colonial or consists of mass of cylindrical or anastomosing thin tubes.

2. The vertical tubes contain osculum, and the surface of the body is perforated by many pores or ostia.

3. Body wall is composed of dermal epithelium forming outer layer, and choanocyte layer forming inner layer. In between is **mesoglea.**

4. Spicules may be monaxon or triaxon and calcareous.

Identification with Reasons

State the characters up to—

Subclass—*Calcaronea* as in *Scypha.*

Order—*Leucosolenida* :

1. Spongocoel or central cavity lined with collared flagellated cells.
2. Canal system **asconoid** type *i.e.,* simple unfolded body walls made up of an outer epidermis, the **pinacocytes** and an inner lining of **choanocytes.**

Type—*Leucosolenia* : State the distinctive features as stated above.

3. *POTERION* (NEPTUNE'S CUP)

Habit and Habitat

Solitary marine, attached to a substratum by stalk. It takes organic matter, bacteria, diatoms and protozoa as food.

Systematic Position

Phylum	**Porifera**
Class	**Demospongiae**
Subclass	**Tetractinomorpha**
Order	**Hydromerida**
Type	***Poterion***

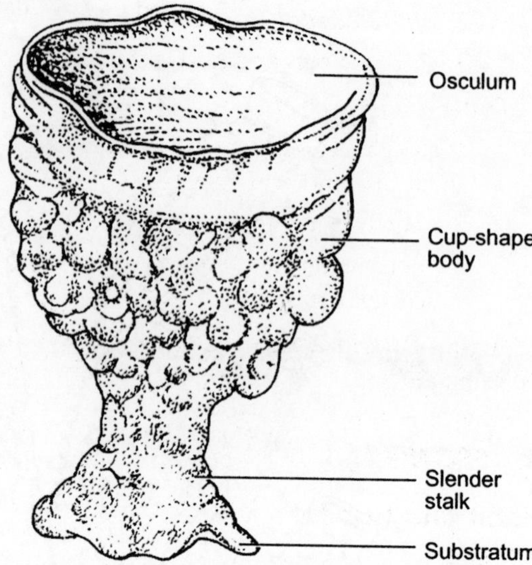

Fig. 1.1.1.19 : *Poterion*

Distinctive Features

1. Cup shaped or vases like body, which may attain a diameter or height of 1 to 2 meters.

2. Incurrent apertures or ostia on the outer surface and excurrent openings on the inner surface of the cup.

3. A large osculum at anterior end.

4. The posterior end has a root like process in the form of a stalk.

Identification with Reasons

State the characters up to—

Phylum—*Porifera* as in *Scypha.*

Class—Demospongiae :

1. Devoid of skeleton or with one to four rayed siliceous spicules, or of **spongin fibres**, or both.
2. The siliceous spicules are not triaxon.
3. The canal system is Leuconoid type.

Subclass—*Tetractinomorpha* :

1. Skeleton with tetraxon spicules.

2. Without spongin fibres.
3. **Megascleres** and **microscleres** distinct.

Order—*Hydromerida* :

1. Megascleres generally monaxon, pointed at one end and knobbed at other.
2. Microscleres in form of asters.
3. Body is typically radiate and a spicule cortex is present.

Type—*Poterion* : State the distinctive features as stated above.

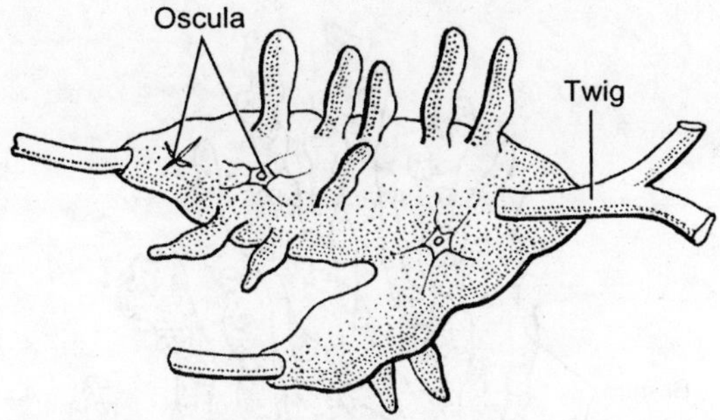

Fig. 1.1.1.20 : *Spongilla*

4. *SPONGILLA*

Habit and Habitat

Most common fresh-water sponge, formed in ponds, lakes and slow stream. Colonial, growing on submerged sticks and plants. Found all over the world. Feeds on particles of organic matter, bacteria and protozoa.

Systematic Position

Phylum	**Porifera**
Class	**Demospongiae**
Subclass	**Ceractinomorpha**
Order	**Haplosclerida**
Type	***Spongilla***

Distinctive Features

1. Body is of irregular shape and size and profusely branched colony.

2. The body wall consists of very thin dermal membrane perforated with dermal pores and several oscula.

3. Found in fresh water.

4. Microscleres are absent.

Identification with Reasons

State the characters up to—

Class—*Demospongiae* as in *Poterion*.

Subclass—*Ceractinomorpha* :

1. Skeleton consists of monaxonid megascleres and sigmoid or chelate microscleres.
2. Spongin may or may not present, when present, in most forms spongin fibres incorporate some spicules or detritus.

Order—*Haplosclerida* :

1. Monaxon megascleres are of one kind, *viz.* diactinal (pointed at both end) and embedded in spongin fibre.
2. Microscleres are absent.
3. Spongin fibres generally present.

Type—*Spongilla* : State the distinctive features as stated above.

5. *HALICLONA (CHALINA)* (Finger Sponge)

Habit and Habitat

Commonly found and widely distributed in all shallow water. Some species found in deep water.

Systematic Position

Phylum	**Porifera**
Class	**Demospongiae**
Subclass	**Ceractinomorpha**
Order	**Haplosclerida**
Type	***Haliclona (Chalina)***

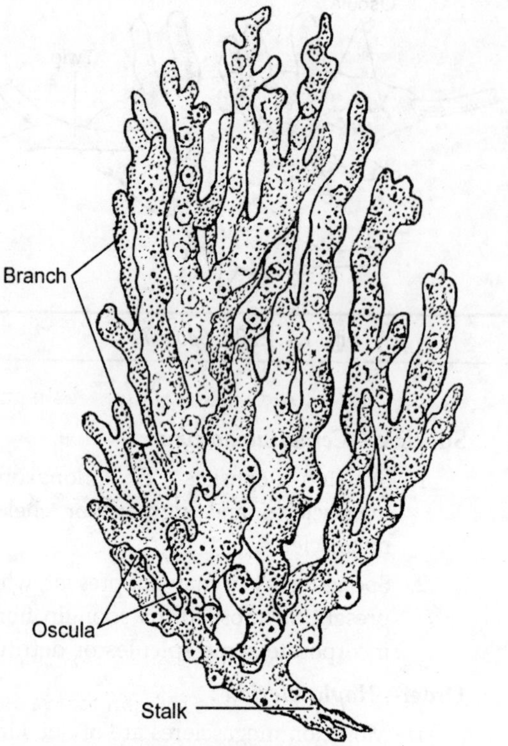

Fig. 1.1.1.21 : *Haliclona*

Distinctive Features

1. Body is more or less flattened and consisting of bunches of fingerlike branches with rounded orifices.

2. The skeletal comprises of spongin and siliceous spicules.

3. Megascleres joined in a network or reticulum.

4. Canal system leucon type.

Identification with Reasons

State the characters up to—

Order—*Haplosclerida* as in *Spongilla*.

Type—*Haliclona* : State the distinctive features as stated above.

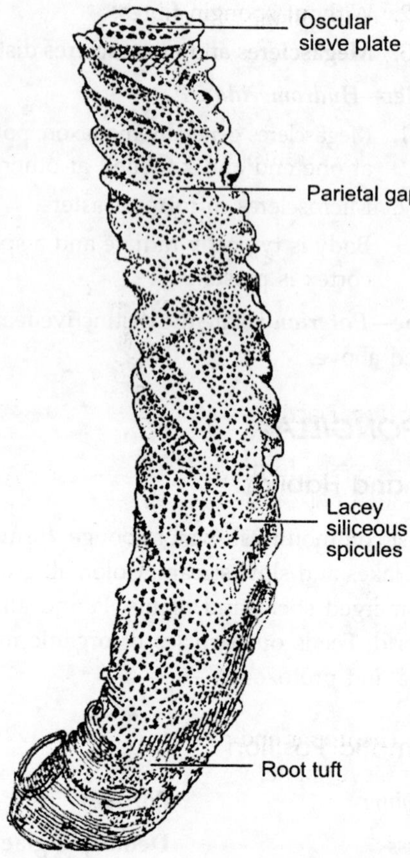

Fig. 1.1.1.22 : *Euplectella*

6. *EUPLECTELLA* (Venus's Flower-Basket)

Habit and Habitat

Euplectella is a solitary animal; found in deep waters at depths ranging from 100 m to over 5,200 m in slow running water.

Special features

It has a great social and mythological value. The skeleton of *Euplectella* is in great demand in Japan for wedding gift. Individual sponges often harbour a single pair of shrimp, one of each sex, that enter the sponge when small and reached reproductive adult hood imprison together within the spongocoel. Glass-sponge with imprisoned pair of shrimp symbolized "life unto death".

Systematic Position

Phylum	**Porifera**
Class	**Hexactinellida**
Subclass	**Hexasterophora**
Type	***Euplectella***

Distinctive Features

1. Deep-sea forms, attached to rooting tufts; size about 15-30 cm in length.

2. The body is long, curved, erect tube with root tuft of spicules.

3. Body is composed of four and six-rayed siliceous spicules, which joined together forming a network.

4. The upper end of the tube is closed by an oscular sieve- plate.

Identification with Reasons

State the characters up to—

Phylum—*Porifera* as those of *Scypha*.

Class—*Hexasterophora* :

1. Six-rayed siliceous spicules which may be separate or united to networks.
2. Spongin is absent.
3. Cellular epidermal is lacking.
4. Choanocytes found only in finger-shaped chambers.

Subclass—*Hexactinellida* : Parenchymal microscleres are typical hexasters.

Type—*Euplectella* : State the distinctive features as stated above.

Economic Importance of Sponges

Sponges are of great advantage to man. Sponges of commerce have formed a sponge industry. The spongin skeleton of certain species of sponge are used in automobile industry.

Sponges are largely used by man for washing and mopping. Sponges are used in surgical operations for absorbing excess fluid, blood etc. They are used as packing materials and used for sound-absorption. Sponges are also used in glass manufacturing industry for cleaning purposes.

CNIDARIA (= Coelenterata)

(Greek : a sting thread) (Greek : hollow gut)

> **Distinguishing Characteristics :**
> Presence of cnidae (nematocyst); **Planula larvae** in the life cycle.

1. *OBELIA*

Habit and Habitat

Marine, colonial, found in shallow water, attached on the surface of seed weed, rocks, shells etc.

Systematic Position

Phylum	**Cnidaria**
Class	**Hydrozoa** = Water animal
Order	**Hydroida**
Suborder	**Leptomedusae (Thecata)**
Type	***Obelia***

Distinctive Features

1. Tree like, branched, colonial from with stem.
2. **Hydrorhiza** is the basal position of the stem for attachment.
3. Vertical stem **hydrocaulus** consisting of outer **perisarc** and inner **coenosarc.**
4. Zooid in hydrocaulus consists of bell-shaped polyp with hydrothecae and tentacle.
5. Near proximal region, cylindrical **blatostyle** enclose **gonothecae**, produced medusa-buds.

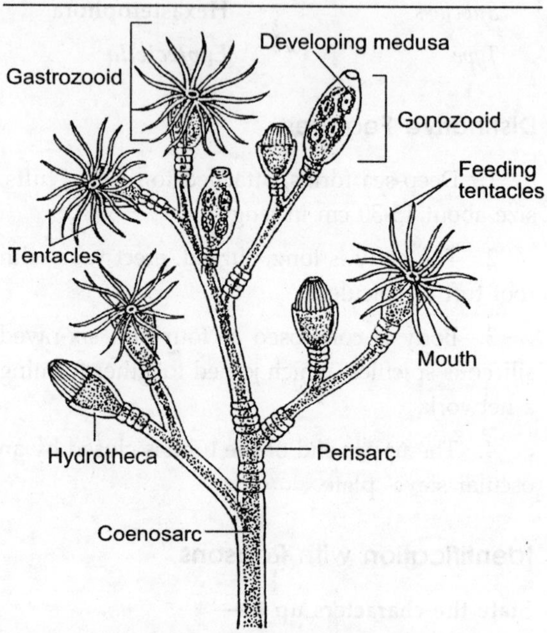

Fig. 1.1.1.23 : *Obelia*

Identification with Reasons

Phylum—*Cnidaria* :

1. Radially symetrical and possess only two layers of living tissue (diploblastic).
2. Single opening of the digestive system- the mouth, surrounded by large number of tentacles.
3. Presence of **cnidoblast cells** with nematocysts.
4. Single gastrovascular cavity or coelenteron.
5. Polypoid and medusoid from present in adult.

Class—*Hydrozoa* :

1. Two types of zooids, polyp and medusa are present.
2. Gastrodermal tissue lacks nematocysts, which are restricted to the epidermis.

Classification is based on Pechenik, J. A., 2002.

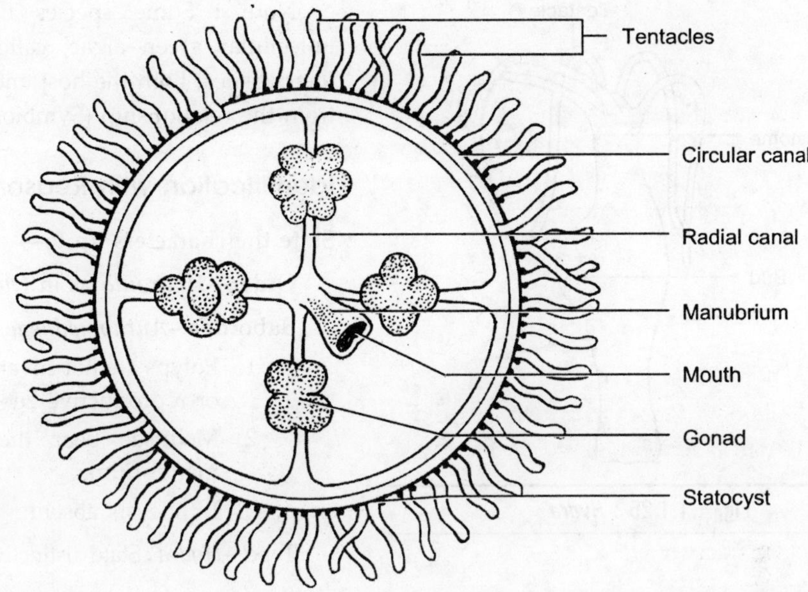

Fig. 1.1.1.24 : Medusa of *obelia.*

3. No cells are found with the **mesoglea.**
4. Gastrovascular cavity undivided.

Order—*Hydroida* :

1. Fixed zoophyte stage.
2. Sense organs are ectodermal in origin.

Suborder—*Leptomedusae* (Thecata) :

1. Hydrotheca in polyp and gonotheca in balstostyle.
2. The medusae with gonads, and they are associated with the radial canal.
3. **Lithocysts** or eye spot present.

Type—*Obelia* : State the distinctive features as stated above.

2. *MEDUSA OF OBELIA*

Habit and Habitat

Marine, free swimming, carnivorous feeds on small living animals or bodies of animals.

Systematic Position

Phylum	**Cnidaria**
Class	**Hydrozoa**
Order	**Hydroida**
Suborder	**Leptomedusae**
Type	**Medusa bud of *Obelia***

Distinctive Features

1. A solitary free-swimming bell or umbrella like with covex ex-umbrella and concave sub-umbrellar surface.
2. Medusa with velum and contain tentacles.
3. Sense organs at the bases of some tentacles.
4. Manubrium hangs from the centre of sub-umbrella, having mouth.
5. Four radial canals each with one gonad and one circular canal present.
6. Distinct radial symmetry.

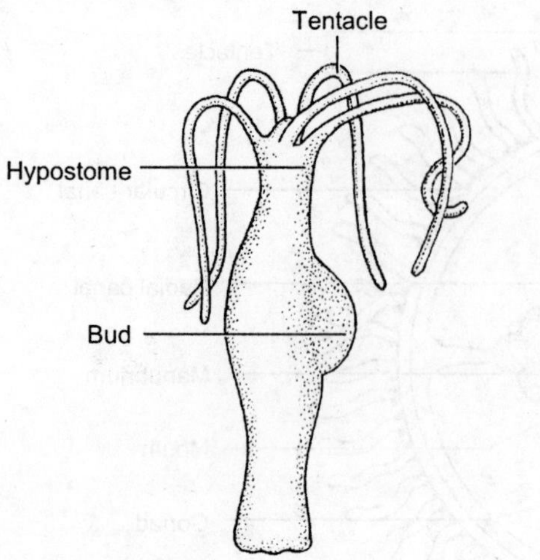

Fig. 1.1.1.25 : *Hydra*

3. *HYDRA*

Habit and Habitat

Solitary, fresh water found attached with some objects in ponds, lakes, streams, ditches and in clear permanent stagnant water. Holozoic; takes *Tubifex*, insects' larvae, crustaceans and their larvae as food.

Systematic Position

Phylum	**Cnidaria**
Class	**Hydrozoa**
Order	**Hydroida**
Suborder	**Anthomedusae (Athecata)**
Type	***Obelia***

Distinctive Features

1. Solitary poly morphic having contractile, tabular and cylindrical body.

2. Posterior end contains pedal or basal disc.

3. Mouth is at the base of conical **hypostome**, which is encircled by 6-10 contractile, slender, hollow tentacles.

4. Sides of hydra bear gonads and buds.

Note : Some species of *Hydra* harbour unicellular green algae, called *zoochlorellae*, in their tissues. Both the host and the algae benefit from the relationship (Symbiosis).

Identification with Reasons

State the characters up to—

Order—*Hydroida* as in *Obelia.*

Suborder—*Anthomedusae* (Athecata) :

1. Polyps are not covered by hydrothecae or reproductive zooid by gonothecae.
2. Medusae bear the gonads on the manubrium.
3. Lithocysts absent.

Type—***Hydra*** : State distinctive features as above.

4. L.S. OF *HYDRA*

Systematic Position

As in *Hydra.*

Distinctive Features

1. Stained microscopic preparation shows tissue grade of organization and diploblastic layers of tissue.

2. Body walls are composed of outer ectoderm and inner endoderm with a non cellular mesoglea layer.

3. Body and tentacles are hollow and covers gastrovascular cavity.

4. Ectoderms are generally provided with epitheliomuscular cells.

5. Endoderm are made of digestive muscular cells and gland cells.

6. Buds and gonads are found on sides.

5. *PORPITA*

Habit and Habitat

Marine, free-floating at the surface; feed on small fishes.

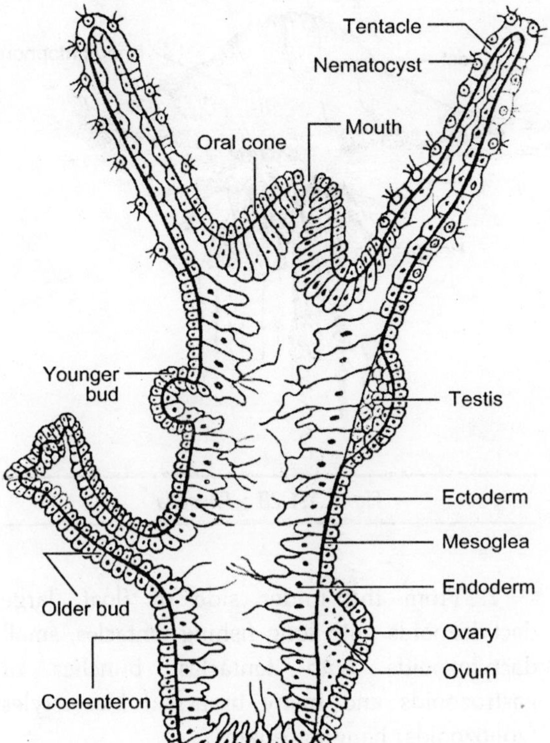

Fig. 1.1.1.26 : L.S. of *Hydra*

Systematic Position

Phylum	**Cnidaria**
Class	**Hydrozoa**
Order	**Hydroida**
Sub-Order	**Chondrophora**
Type	***Porpita***

Distinctive Features

1. Disc-shaped medusa like body, enclosing a chamber, chitinous and porous shell, having air, which acts as a float.

2. A large gastrozooid is present in the centre having mouth, which is encircled by several blastostyles.

3. Edge of the disc is beset with numerous **dactylozooids** or **tentaculozooids**.

4. Polymorphic, colonial form.

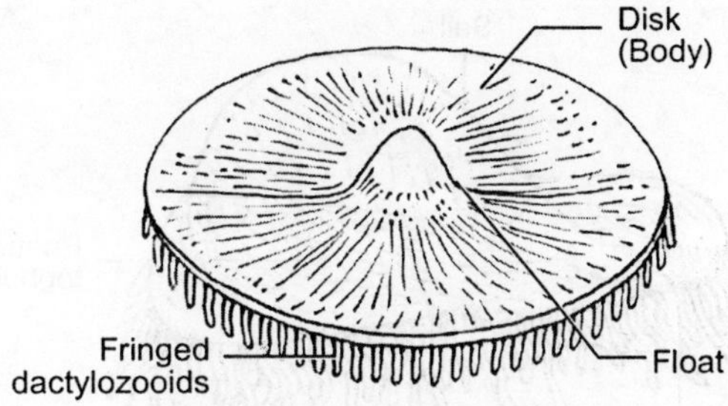

Fig. 1.1.1.27 : *Porpita*

Identification with Reasons

State the characters up to—

*Order—**Hydroida*** as in *Obelia.*

*Sub-Order—**Chondrophora*** :

1. A large polyp with a chitinous, gas-filled floats.
2. Adult is a free sexual medusa.

*Type—**Porpita*** : State distinctive features as in above.

6. *VELELLA*

Habit and Habitat

Marine, Pleagic, feeds on small fishes.

Systematic Position

Phylum	**Cnidaria**
Class	**Hydrozoa**
Order	**Hydroida**
Sub-Order	**Chondrophora**
Type	***Velella***

Distinctive Features

1. A polymorphic, rhomboidal floating colony; on super surface there is a sail or crest.

2. On the dorsal surface there are pneumatophores containing air.

3. On the ventral surface hangs a single large gastrozooids having mouth.

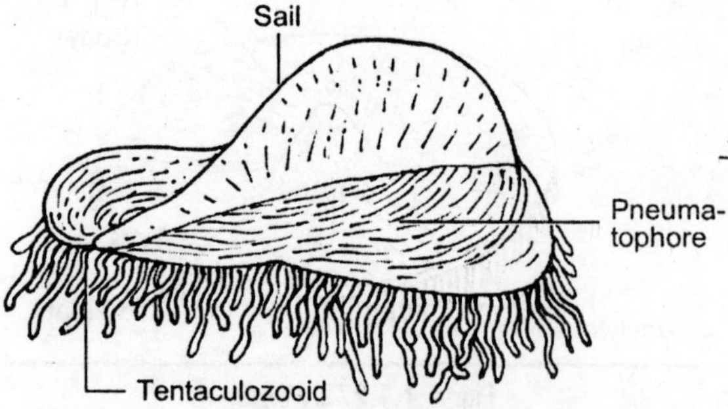

Fig. 1.1.1.28 : *Velella*

4. The edge of the body is beset with numerous dactylozooid or tentaculozooids.

5. Gonozooids are found around the single gastrozooids.

Identification with Reasons

State the characters up to—

Sub-Order—*Chondrophora* as in *Porpita.*

Type—*Velella* : State distinctive characters as in above.

7. PHYSALIA

Habit and Habitat

Marine, free-floating or swimming colony; voracious carnivores.

Systematic Position

Phylum	**Cnidaria**
Class	**Hydrozoa**
Order	**Siphonophora**
Type	***Physalia* (Portuguese man-of -war)**

Distinctive Features

1. On the dorsal surface there is a large crested float or **pneumatophore.**

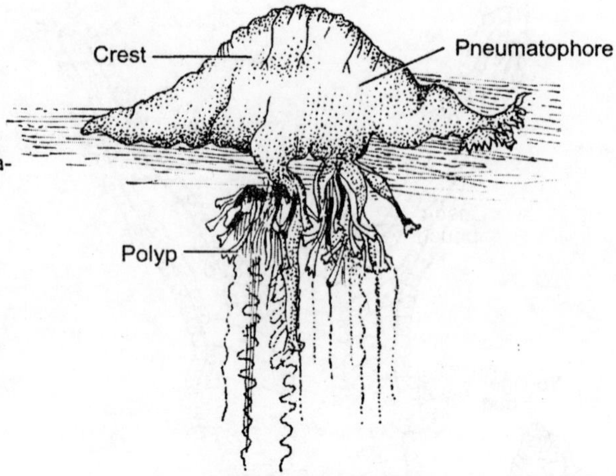

Fig. 1.1.1.29 : *Physalia*

2. From the under side of float, large dactylozooids with long fishing tentacles, small dactylozooids with tentacles, bunches of gastrozooids and many branched blastostyles (gonozooids) hanging down.

Special Features

It is very poisonous. The long trailing tentacle inflicts severe pain, burning sensation, skin lesions, eruptions etc. due to the batteries of nematocyst situated in the tentacle.

Identification with Reasons

State the characters up to—

Class—*Hydrozoa* as in *Obelia.*

Order—Siphonophora :

1. Colonial forms and exhibits extreme polymorphism, in which polypoid and medusoid forms are present.
2. No free sexual medusae.

Type—*Physalia* : State the distinctive features as in above.

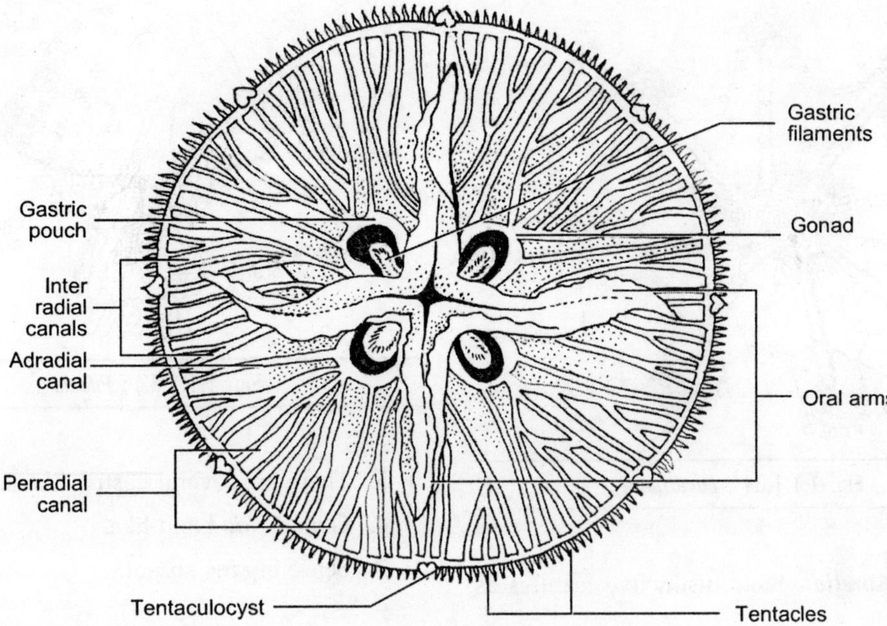

Fig. 1.1.1.30 : *Aurelia*

8. *AURELIA*

Habit and Habitat

Solitary, marine and medusoid jelly fish, swimming close to the surface of the sea. Carnivorous, taking eggs, fishes, small larvae etc. as food.

Systematic Position

Phylum	**Cnidaria**
Class	**Scyphozoa (Cup animal)**
Order	**Semaeostomeae (Discomedusae)**
Type	***Aurelia***

Distinctive Features

1. Body consists of a gelatinous saucer – shaped umbrella; distinguish into convex ex-umbrella and concave sub-umbrella.

2. The disc is surrounded by numerous, delicate marginal tentacles having eight notches.

3. A pendant manubrium hangs down from the centre of sub umbrella and is surrounded by four long oral arms.

4. Highly branched radial canals arising from the centre.

Identification with Reasons

State the characters up to—

Phylum—*Cnidaria* as in obelia.

Class—*Scyphozoa* :

1. Body is gelatinous, transparent, saucer-shaped umbrella.
2. Presence of four horseshoe-shaped gonads near the centre of umbrella.
3. No true velum.

Order—Semaeostomeae (Discomedusae) :

1. Flattened saucer-shaped or disc shaped umbrella.
2. Tentaculocysts usually eight, of which four perradial and four interradial.
3. Square mouth with four long oral arms.

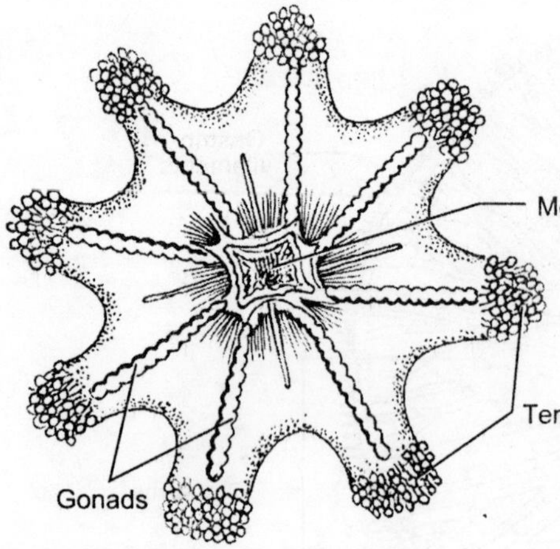

Fig. 1.1.1.31 : *Lucernaria*

Type—*Aurelia* : State distinctive features as in above.

9. *LUCERNARIA*

Habit and Habitat

It is sessile, sedentary, marine animal; carnivorous.

Systematic Position

Phylum	**Cnidaria**
Class	**Scyphozoa**
Order	**Stauromedusae (Lucernariida)**
Type	***Lucernaria***

Distinctive Features

1. Body is trumpet-shaped or conical.

2. Attached temporarily to seaweeds by a peduncle which is developed from the centre of ex-umbrella.

3. The margin of the umbrella is prolonged into eight short hollow adradial arms.

4. Each arm possesses groups of short adhesive tentacles.

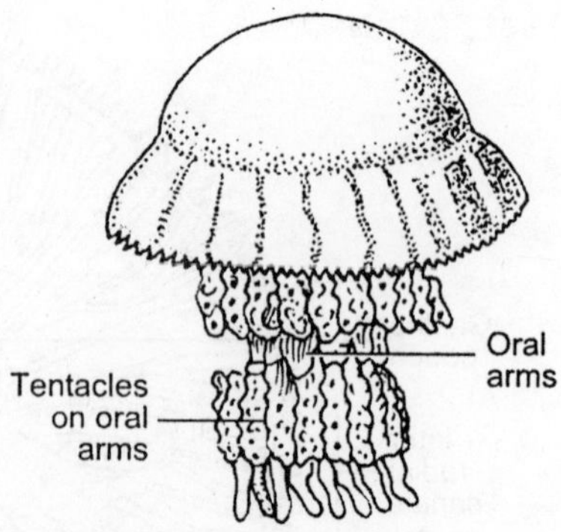

Fig. 1.1.1.32 : *Pilema*

5. There are several gastric filaments.
6. Gonads are band-like.
7. Sense organs absent.

Identification with Reasons

State the characters up to—

Class—*Scyphozoa* as those on *Aurelia*.

Order—*Stauromedusae* :

1. Conical or vase-shaped umbrella.
2. Adult sessile, attached by an ex-umbrella peduncle.
3. No tentaculocysts.

Type—*Lucernaria* : State distinctive features as in above.

10. *PILEMA*

Habit and Habitat

Solitary, marine medusoid jelly fish, carnivorous feeds on small fishes.

Systematic Position

Phylum	**Cnidaria**
Class	**Scyphozoa**
Order	**Rhizostomeae**
Type	***Pilema***

Distinctive Features

1. The body is hemispherical without marginal tentacles.

2. Eight oral arms hang down from the centre of the umbrella are of great length and folded to give a root-like appearance.

3. The mouth is obliterated by the growth of oral lobes across it and is obliterated by a large number of funnel-like apertures, called suctorial mouth.

4. On the outer surface of the mouth near the bell an additional mouth bearing fringed out growth or **scapulets** are found.

5. The suctorial mouth lead into canals in the thickness of the arms and finally open into the stomach.

Identification with Reasons

State the characters up to—

Class—Scyphozoa as those on *Aurelia.*

Order—Rhizostomeae :

1. Oral lobes are fused and thus obliterate the month.
2. Oral lobes possess numerous small mouths which have many canals open by funnel-shaped apertures on the edge of the arms.

Type—*Pilema* : State distinctive features as in above.

11. *TEALIA* (Common Sea anemone)

Habit and Habitat

The most abundant and known off shore animals. Marine; solitary, remains attached to a variety of substratum including rocks, logs, shells, sandy and muddy beds at low water mark, left high and dry by the ebbing tide, carnivorous, commonly feed an molluscs, crustaceans and other invertebrates.

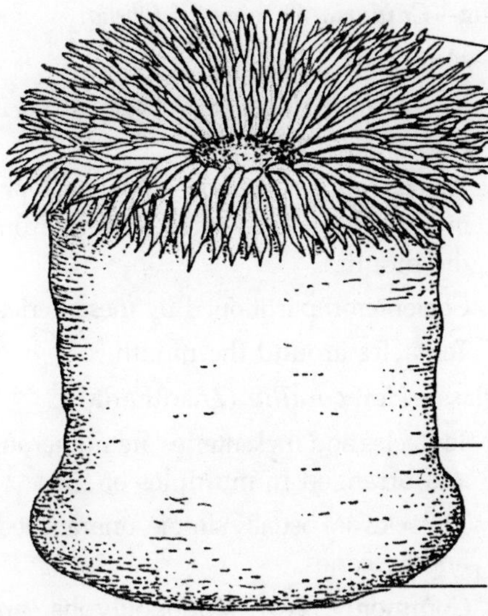

Fig. 1.1.1.33 : *Tealia* (common Sea-anemone)

Systematic Position

Phylum	**Cnidaria**
Class	**Anthozoa (Flower animals)**
Sub class	**Hexacorallia (Zoantharia)**
Order	**Actiniaria**
Type	***Tealia***

Distinctive Features

1. Body short, cylindrical, flower like, and divided into oral disc, column and pedal disc.

2. Oral disc flattened and crowned with whorls of hollow tentacles.

3. Column is cylindrical; pedal disc or base for firm attachment to substratum.

4. Mouth is in the form of a slit and opens into the gastrovascular cavity.

5. Six pairs of **mesenteries** divided the gastrovascular cavity.

Identification with Reasons

State the characters up to—

Phylum—*Cnidaria* as those of *Obelia*.

Class—***Anthozoa*** **:**

1. No medusa stage, only polyp from is present.
2. Presence of ciliated groove (**Siphonoglyph**) in the pharyngeal wall leading from the mouth.
3. Coelenteron partitioned by mesenteries.
4. Tentacles around the mouth.

Sub class—***Hexacorallia*** **(Zoantharia) :**

1. Tentacles and mesenteries are numerous and arranged in multiples of six.
2. Tentacles are usually simple, unbranched, hollow cones.
3. Commonly two siphonoglyphs are present.

Order—***Actiniaria*** **:**

1. Calcareous skeleton absent.
2. One or more siphonoglyphs present.

Type—***Tealia*** **:** State distinctive characters as in above.

12. *METRIDIUM* (Sea-anemone) :

Same as those of *Tealia* as described above except distinctive features.

Type—***Metridium*** **:**

1. Body is short, cylindrical, radially symmetrical and generally longer than broad and divided into oral disc, column and pedal disc.
2. Oral disc is lobed, flat and crowned with numerous and short tentacles.
3. The lower thick wall of the column is smooth and is perforated by small openings called **cinclides**.
4. Pedal or basal disc is broad.

13. *ADAMSIA*

Habit and Habitat

Marine: found attached to gastropods shell occupied by hermit crab, *Eupagurus*.

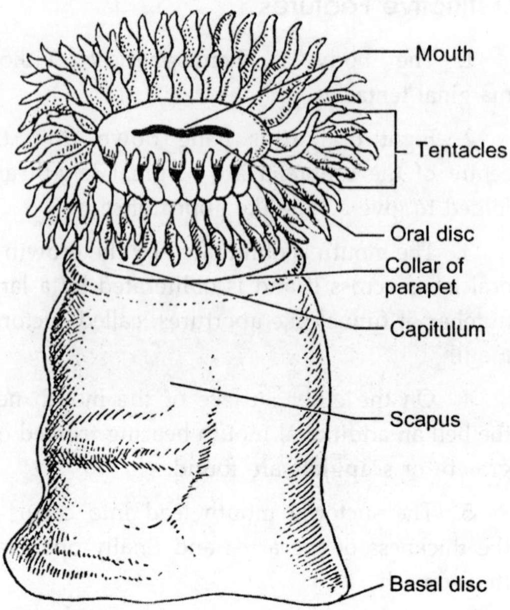

Fig. 1.1.1.34 : *Metridium* (Sea-anemone)

Systematic Position

Phylum	**Cnidaria**
Class	**Anthozoa**
Sub class	**Hexacorallia (Zoantharia)**
Order	**Actiniaria**
Type	***Adamsia***

Distinctive Features

1. Body consist of base, column and oral disc.

2. Oral disc has central mouth and marginal tentacles, bearing nematocysts.

3. Flat, sucker like pedal disc usually remains attached to the molluscan shell especially to the empty gastropod shell.

4. A band of cinclidial tubercles is situated at the base of column.

5. Gullet with siphonoglyphs, gastrovascular cavity contains six pair's mesenteries.

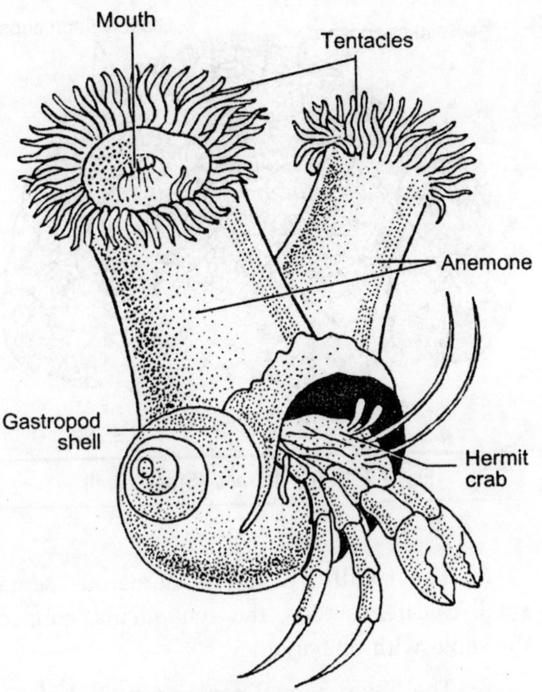

Fig. 1.1.1.35 : *Adamsia* (attached to empty gastropodshell)

Identification with Reasons

State the characters up to—

Order—*Actiniaria* as those of ***Tealia***.

Type—*Adamsia* : State distinctive features as in above.

Special Features

Adamsia shows commensalisms. It lives on gastropod shells inhabited by hermit crab. *Adamsia* is carried from place to place by the hermit crab, and in this way receiving a varied and abundant food supply. On the other hand, the hermit crab is protected from its enemies by retreating into shell, exposing sea anemone which is protected by poisonous stinging capsules. These individual benefit each other but with out any morphological, physiological and organic relationship.

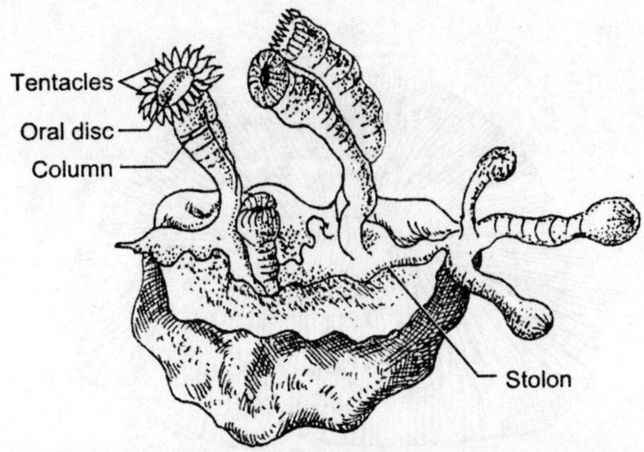

Fig. 1.1.1.36 : *Zoanthus* colony

14. *ZOANTHUS*

Habit and Habitat

Colonial sea anemones arising from an stolen like base. **Epizoic**, grows on other animals, such as sponges, shells, hermit crabs, corals etc.

Systematic Position

Phylum	**Cnidaria**
Class	**Anthozoa**
Sub class	**Hexacorallia (Zoantharia)**
Order	**Zoanthidea**
Type	***Zoanthus***

Distinctive Features

1. Body is divisible into column and oral disc having mouth and oral tentacles.

2. Cylindrical small polyps arise singly from a network of stolen.

3. Tentacles 48 to 60 in numbers, arranged in one or two rows.

4. Single siphonoglyph present.

Identification with Reasons

State the characters up to—

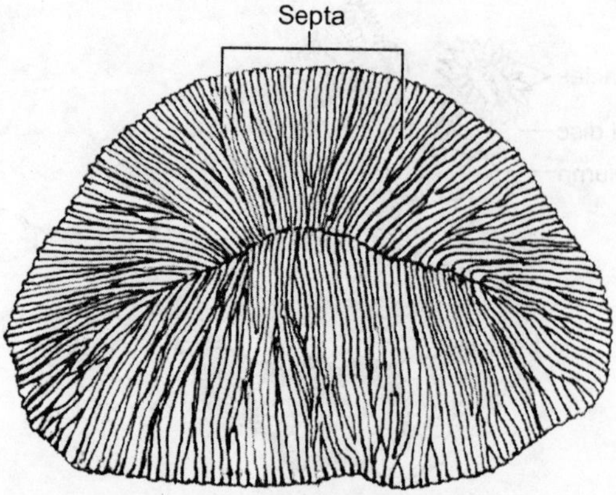

Fig. 1.1.1.37 : *Fungia* (Mushroom coral)

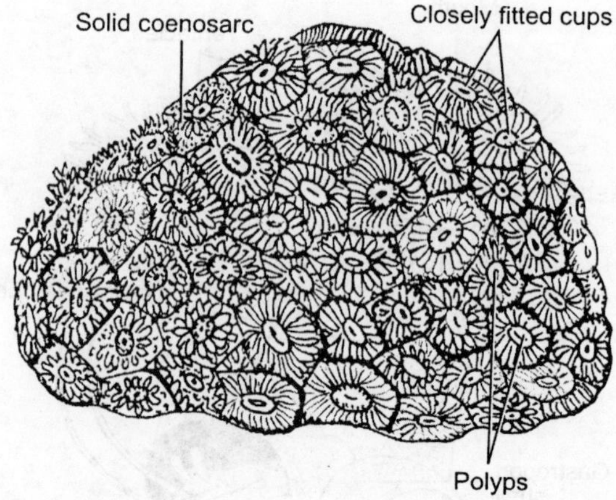

Fig. 1.1.1.38 : *Astraea* (Stony coral)

Subclass—*Hexacorallia* as those of *Tealia.*

Order—*Zoanthidea* :

1. Colonial with one ventral siphonoglyph.
2. Unbranched tentacles.
3. Body wall is usually encrusted with foreign bodies of calcareous nature.

Type—*Zoanthus* : State distinctive features as in above.

15. *FUNGIA* (Mushroom coral)

Habit and Habitat

Marine, solitary coral: attached to hard substratum in shallow waters of warm sea.

Systematic Position

Phylum	**Cnidaria**
Class	**Anthozoa**
Sub class	**Hexacorallia (Zoantharia)**
Order	**Madreporaria (Scleractinia)**
Type	***Fungia***

Distinctive Features

1. Discoidal, mushroom-shaped coral, with upper convex and lower concave surface having thecae to use lower surface.

2. The **corallite** contains numerous septa, small calcareous rods, the synapticula, connect the septa with one another.

3. The living animal contains a single large polyp with numerous tentacles.

Identification with Reasons

State the characters up to—

Subclass—*Hexacorallia* as those of *Tealia.*

Order—*Madreporaria* :

1. Solitary or colonial possess an ectodermal calcareous skeleton.
2. Siphonoglyph absent

Type—*Fungia* : State distinctive features as in above.

16. *ASTRAEA* (Stonycoral)

Habit and Habitat

Marine, Colonial: found in tropical and subtropical waters, attached to hard substratum in shallow and coastal water.

Systematic Position

Phylum	**Cnidaria**
Class	**Anthozoa**

Sub class	**Hexacorallia (Zoantharia)**
Order	**Madreporaria**
Type	**Astraea**

Distinctive Features

1. Stony coral, contains ectodermal, calcareous or calcium carbonate skeleton, called as **corallium.**

2. Compact and massive colony formed by buds, growing more or less parallel with each other.

3. Skeleton takes the form of cups called theca containing polyps.

Identification with Reasons

State the characters up to—

Order—Madreporaria as those of *Fungia.*

Type—*Astraea* : State distinctive characters as in above.

17. *MADREPORA* (Horn coral)

Habit and Habitat

Marine, Colonial, Symbiotic coral: found in shallow waters of warm seas of Indo-Pacific region.

Systematic Position

Phylum	**Cnidaria**
Class	**Anthozoa**
Sub class	**Hexacorallia (Zoantharia)**
Order	**Madreporaria (Scleractinia)**
Type	***Madrepora***

Distinctive Features

1. Branched, partly porous or reticulate colony.

2. Colony consists of cylindrical cups or corallites, in which flower like polyps lodged.

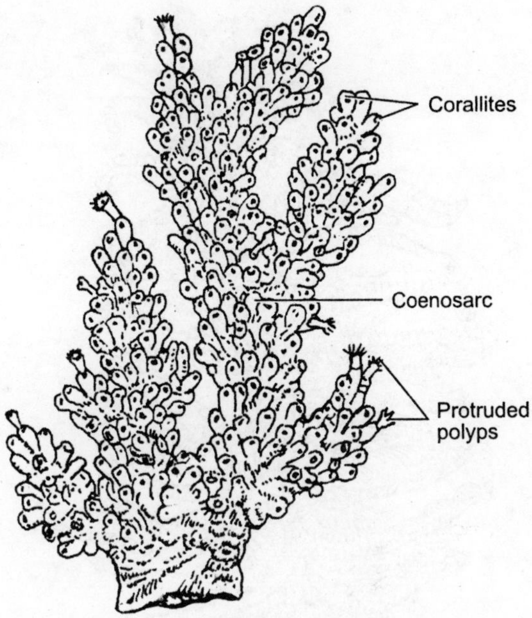

Fig. 1.1.1.39 : *Madrepora* (Horn coral)

3. Terminal polyps bear twelve tentacles.

Identification with Reasons

State the characters up to—

Order—*Madreporaria* as those of *Fungia.*

Type—*Madrepora* : State distinctive features as in above.

Special Features

It plays as important role in coral-reef formation. The colony increases in size by growth and budding of the polyps.

18. *PENNATULA* (Sea Pen)

Habit and Habitat

Sedentary and colonial, remain embedded in muddy or sandy bottom of sea waters by the stalk, found in shallow and costal waters of Atlantic and Pacific oceans.

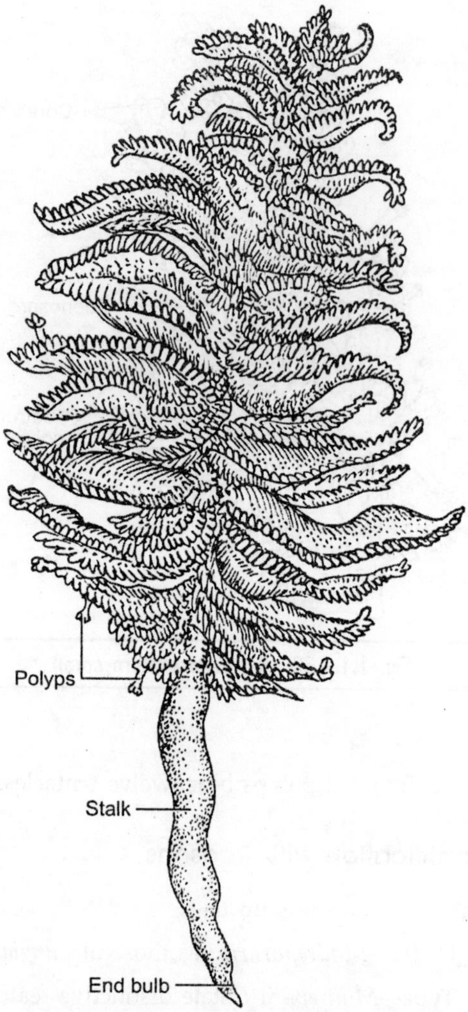

Fig. 1.1.1.40 : *Pennatula*

Systematic Position

Phylum	**Cnidaria**
Class	**Anthozoa**
Sub class	**Octocorallia (Alcyonaria)**
Order	**Pennatulacea**
Type	***Pennatula***

Distinctive Features

1. Colony is feather like, having a single axial polyp and several secondary polyps.

2. Posteriorly the central axis or rachis forms a peduncle with an end bulb.

3. Anteriorly rachis bears lateral branches or pinnules containing gastrozooids with tentacles.

4. Smaller polyps or siphonozooids on the lateral side of the rachis devoid of tentacles.

Identification with Reasons

State the characters up to—

Class—*Anthozoa* as those of *Tealia.*

Subclass—*Octocorallia* :

1. Polyps bear eight tentacles and are subdivided by eight complete mesenteries.
2. Tentacles are pinnate, that is they bear numerous lateral out folding called pinnules.
3. One ventral siphonoglyph is present.

Order—*Pennatulacea* :

1. Elongated colony, one end embedded in the mud, while the opposite end bears the laterally placed polys.
2. The stem is supported by a calcareous or horney skeleton.
3. Polyps are dimorphic.

Type—*Pennatula* : State distinctive features as in above.

19. PTEROEIDES (Sea Pen)

Habit and Habitat

Same as Pennatula. It is found in Europe and in the Bay of Bengal Coast.

Systematic Position

Phylum	**Cnidaria**
Class	**Anthozoa**
Sub class	**Octocorallia**
Order	**Pennatulacea**
Type	***Pteroeides***

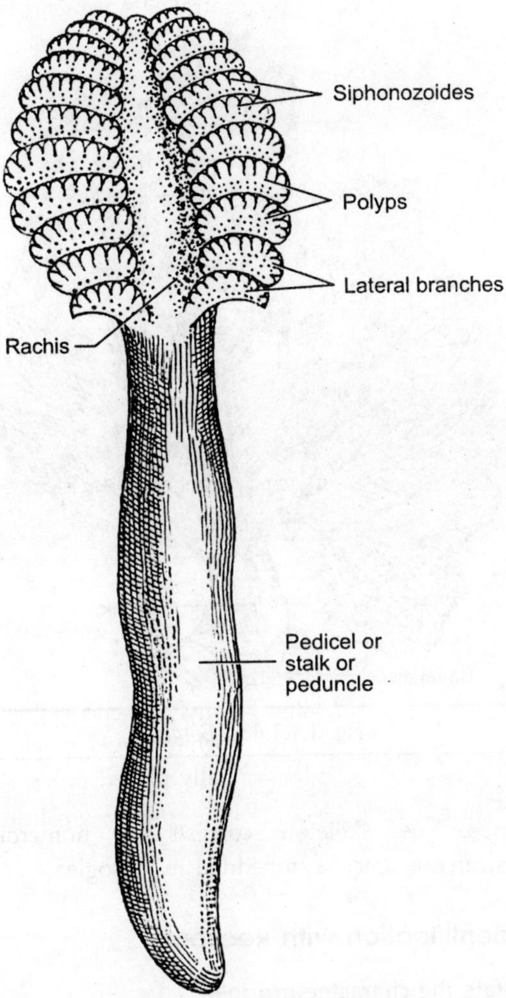

Fig. 1.1.1.41 : *Pteroeides*

Distinctive Features

1. Colony Composed of an elongated primary axial polyp and several secondary polyps originating laterally from it.

2. Lateral branches originate much above the middle of the body and very closely set together.

3. Fan like spicules support the lateral branches.

4. Peduncle or stalk is elongated.

5. Numerous polyps or gastrozooids on the lateral branches.

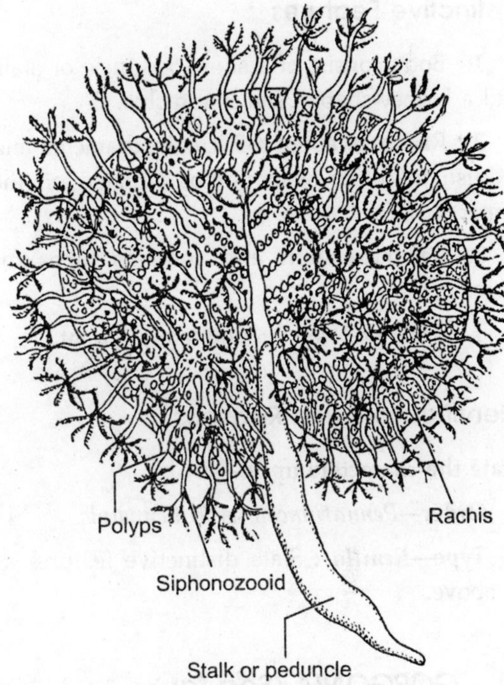

Fig. 1.1.1.42 : *Renilla*

Identification with Reasons

Same as *Pennatula*.

Type—*Pteroeides* : State the distinctive features as in above.

20. *RENILLA* (Sea Pansy)

Habit and Habitat

Marine: benthic, dimorphic colony, found in the shallow water of Carolina coast and West Indies.

Systematic Position

Phylum	**Cnidaria**
Class	**Anthozoa**
Sub class	**Octocorallia (Alcyonaria)**
Order	**Pennatulacea**
Type	***Renilla***

Distinctive Features

1. Body consist of a short peduncle or stalk and a kidney-shaped leaf like rachis.

2. Rachis is about 6-7 cm in diameter and reddish in colour with polyps situated on the upper surface.

3. Peduncle short and contains tubes for water current.

4. Axial skeleton is completely absent.

Identification with Reasons

State the characters up to—

Order—*Pennatulacea* as in *Pennatula.*

Type—*Renilla* : State distinctive features as in above.

21. *GORGONIA* (Sea fan)

Habit and Habitat

Marine: colony remains attached to rock and other hard substratum by basal plate; found in shallow tropical and subtropical water and flourish in Indo-Pacific region.

Systematic Position

Phylum	**Cnidaria**
Class	**Anthozoa**
Sub class	**Octocorallia (Alcyonaria)**
Order	**Gorgonacea**
Type	***Gorgonia***

Distinctive Features

1. Erect, branched and fan-shaped colonies of yellow or red colour.

2. Branches unite so as to form a network.

3. Polyps minute and contain eight pinnate tentacles.

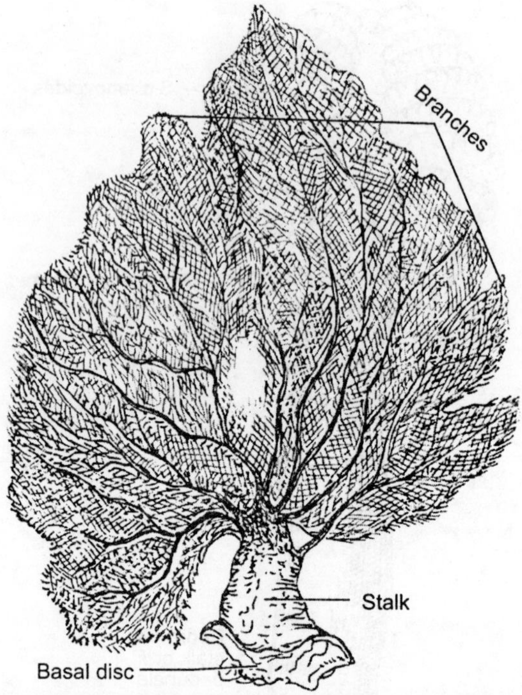

Fig. 1.1.1.43 : *Gorgonia*

4. The skeleton consists of numerous calcareous spicule embedded in mesoglea.

Identification with Reasons

State the characters up to—

Order—*Octocorallia* as in *Pennatula.*

Order—*Gorgonacea* :

1. Calcareous or horny skeleton forming a branched axis throughout the colony.
2. Spicules are present in the mesoglea.
3. Siphonoglyph absent.

Type—*Gorgonia* : State distinctive features as in above.

Special Features

It is very familiar museum specimen due to fan shaped body. The dried skeleton often kept as ornament.

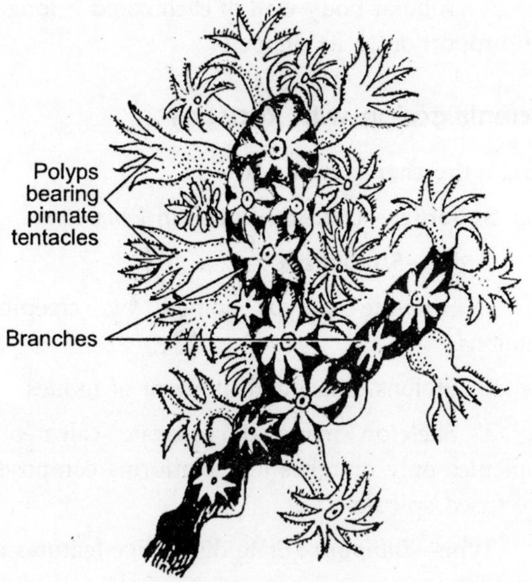

Fig. 1.1.1.44 : *Corallium*

22. *CORALLIUM* (Red coral or Precious Coral)

Habit and Habitat

Marine littoral: colony living on rocky or hard bottoms of warm waters. It is found in Mediterranian, Eastern Atlantic and the coast of Japan, Africa and Italy.

Systematic Position

Phylum	**Cnidaria**
Class	**Anthozoa**
Sub class	**Octocorallia (Alcyonaria)**
Order	**Gorgonacea**
Type	***Corallium***

Distinctive Features

1. Erect, branching colony, with calcareous axis of fused spicules.
2. The skeleton is hard, calcareous, red in colour.
3. A branching axial rod runs through out the colony.
4. Calcareous spicules absent.
5. Dimorphic polyps are delicate, white coloured and retractile.

Identification with Reasons

State the characters up to—

Order—*Gorgonecea* as in *Gorgonia.*

Type—*Corallium* : State distinctive features as in above.

Economic Importance

It is a precious coral and of great commercial value. It is extensively used in jewellery.

23. TUBIPORA (Organ-pipe coral)

Habit and Habitat

Marine: colonial polypoid Anthozoa; widly distributed in coral reefs; found in warm water of Atlantic, Indian and Pacific Ocean.

Systematic Position

Phylum	**Cnidaria**
Class	**Anthozoa**
Sub class	**Octocorallia (Alcyonaria)**
Order	**Stolonifera**
Type	***Tubipora***

Distinctive Features

1. Colony consists of vertical tubes closely fitted together.
2. The tubes are joined together by transverse platforms formed by fused spicules.
3. The tubes are containing brightly coloured polyps.
4. The base of the original polyp forms feathered expansion from which polyps originate.

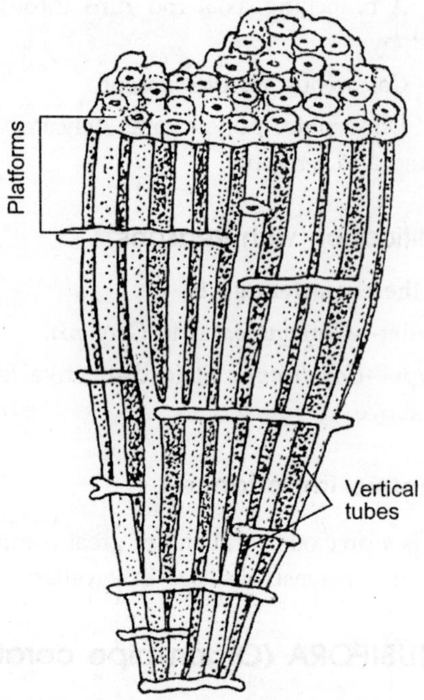

Fig. 1.1.1.45 : *Tubipora*

5. Tubular body wall of each zooid is longer in proportion to its diameter.

Identification with Reasons

State the characters up to—

Suborder—*Octocorallia* as in *Pennatula.*

Order—*Stolonifera* :

1. Separate polyps joined by creeping stolons.

2. Stolons embedded in a mat of tissues.

3. Skeleton made of separate calcareous spicules only or tubes and platforms composed of fused spicule.

Type—*Tubipora* : State distinctive features as in above.

CTENOPHORA

(Greek : Comb bearing)

Distinguishing Characteristics :
Plates of fused cilia arranged in rows; adhesive prey capturing **colloblasts cells** present.

An outline classification of **Phylum-Ctenophora** after Ruppert, E.E. and Barnes, R.D., 1994 (for those types that are included in the text)

Classes

Tentaculata	Nuda
Order—**Cydippida**	**Beroida**—*Order*
Example : *Hormiphora*	Example : *Beroe*

1. *HORMIPHORA*

Habit and Habitat

Marine: planktonic, *i.e.,* weak swimmers, carried about by ocean currents; predators upon larval fish, other ctenophores, small crustaceans and other zoo plankton; found in all seas.

Systematic Position

Phylum	**Ctenophora**
Class	**Tentaculata (Micropharyngea)**
Order	**Cydippidea (Cydippida)***
Type	***Hormiphora***

Distinctive Features

1. Pear-shaped, transparent gelatinous body
2. A slit-like mouth is situated at the centre of the oval pole.
3. Single sense organ on the aboral pole.
4. Two ciliated polar plates are transversely arranged with the depression.

Classification is based on Pechenik, J. A., 2002.

* *Mentioned by Ruppert, E.E. and Barnes, R.D., (1994).*

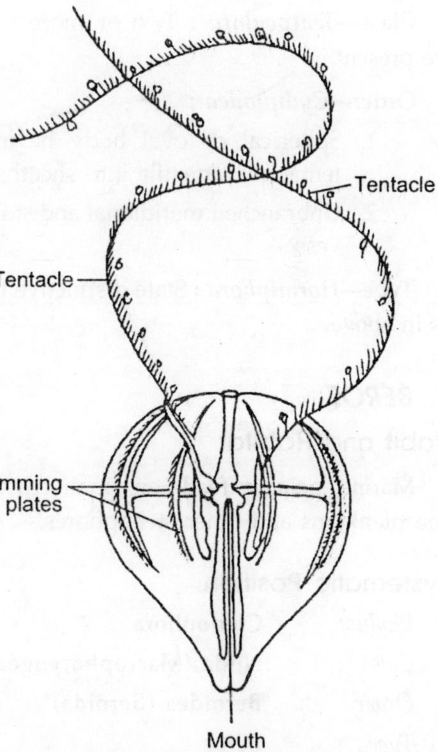

Fig. 1.1.1.46 : *Hormiphora*

5. Surface bears eight equally spaced meridional swimming-plates or comb plates.
6. Two extensible, long and solid tentacles arises on the two sides of the broader end..

Identification with Reasons

Phylum—*Ctenophora* :

1. Biradially symmetrical, triploblastic, spherical or oval in shape.

2. Cilia fused to form comb-like structures which are arranged in eight meridional rows or **swimming plates (costae)**.
3. Tentacle when present, are two in number, having adhesive cells known as colloblasts.
4. Nematocysts are absent.
5. Gastrovascular cavity has a single opening, called mouth.

Class—*Tentaculata* : Two or more tentacles are present.

Order—*Cydippidea* :

1. Spherical or oval body having two tentacles, retractile into sheaths.
2. Unbranched meridional and stomodeal vessel.

Type—*Hormiphora* : State distinctive features as in above.

2. *BEROE*

Habit and Habitat

Marine, pelagic; feeds on small crustaceans, zoo planktons and other ctenophores.

Systematic Position

Phylum	**Ctenophora**
Class	**Nuda (Macropharyngea)**
Order	**Beroidea (Beroida)***
Type	***Beroe***

Distinctive Features

1. Body is conical, laterally compressed and without tentacles.
2. Large wide mouth at the truncated oral end.
3. Rounded aboral end contains sense organs.
4. Large **stomodaeum, infundibulum,** perradial and infundibular canals, aggregated into a small space at the aboral pole.
5. The meridional canals send off branches, which form a complex network of tubes.

* *Mentioned by Ruppert, E.E. and Barnes, R.D., (1994).*

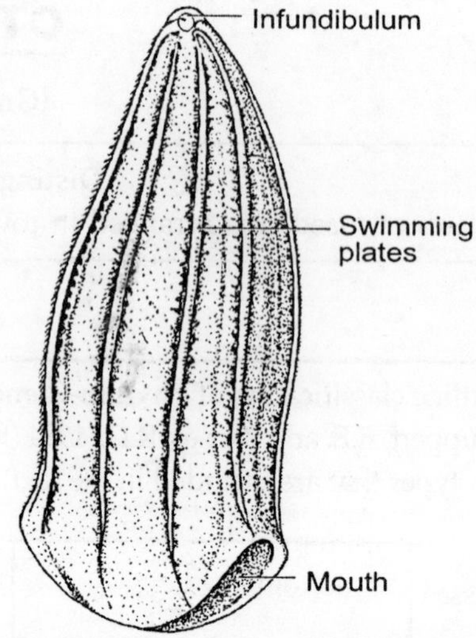

Fig. 1.1.1.47 : *Beroe*

Identification with Reasons

State the characters up to—

Phylum—*Cetnophora* as in *Hormifora.*

Class—*Nuda* : Ctenophora without tentacles.

Order—*Beroidea* :

1. Very wide mouth and the gullet occupies the greater part of the interior of the body.
2. Meridional canals are produced into a complex system of anastomosing branches.

Type—*Beroe* : State distinctive features as in above.

Economic Importance

Ctenophores playing major role in estuarine ecology. Nearly all ctenophores are **bioluminescent**. The adaptive significance of bioluminescence is not known. Some species may use bioluminescence to avoid detection by visual predators, producing light of ambient intensity. Luminescence together with iridescence of comb rows, the get-up and the beauty movement, make living ctenophores among the most glorious of animals to observe.

PLATYHELMINTHES

(Greek : flatworm)

Distinguishing Characteristics :
Dorsoventrally flattened body; anus, circulatory and respiratory system absent.

1. *DUGESIA* (*Planaria*)

Habit and Habitat

Free living common in streams, lake and in the mud at the bottom of ponds; cosmopolitan; carnivorous.

Systematic Position

Phylum	**Platyhelminthes**
Class	**Turbellaria**
Order	**Tricladida**
Type	***Dugesia***

Distractive Features

1. Elongated, dorsoventrally flattened animal.
2. Body divisible into an anterior head and a posterior pointed end.
3. Triangular head with two ear like auricles and two eyes with multicellular retina and pigmented cups.
4. Ventrally situated mouth encircled by proboscis pore.
5. Three main intestinal limbs, one extended upwards and two backwards.

Identification with Reasons

Phylum—*Platyhelminthes* :

1. Body dorsoventrally flattened, unsegmented, bilaterally symmetrical and without coelom.

Classification is based on Ruppert, E.E., Barnes, R.D., 1994.

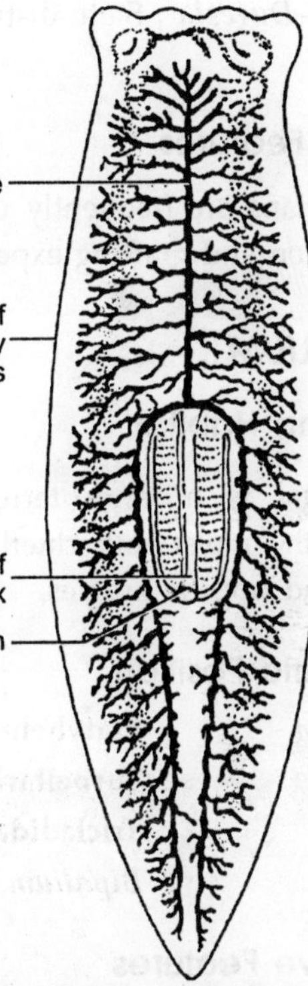

Fig. 1.1.1.48 : *Dugesia*

2. Incomplete digestive system, anus absent.
3. Skeletal, circulatory and respiratory systems are absent.
4. Excretory organs are flame cells.
5. Suckers, hooks or spines usually present.

Class—*Turbellaria* :

1. Non-parasitic form without hooks and suckers.

2. Ciliated epidermis and **rhabdites** always present

Order—*Tricladida* :

1. Free living, gut with pharynx.
2. Three main intestinal limbs, one anterior and two posterior.
3. Posterior mouth.
4. Single genital aperture behind the mouth.

Type—*Dugesia* : State distinctive features as in above.

Special Features

Planerians are frequently used in studies of regeneration and grafting experiments.

2. *BIPALIUM*

Habit and Habitat

Bipalium is a large terrestrial species in habiting the moist soil; chiefly found in green houses and tropical jungles.

Systematic Position

Phylum	**Platyhelminthes**
Class	**Turbellaria**
Order	**Tricladida**
Type	***Bipalium***

Distinctive Features

1. Largest land planarians measuring 20-25 cm in length.

2. Animal with expanded head and long twisted body.

3. Numerous eyes on the margin of the head and sides of the body.

4. Stripes on the dorsal surface and a creeping sole on the ventral surface.

Identification with Reasons

State characters of each taxon as those of *Dugesia*.

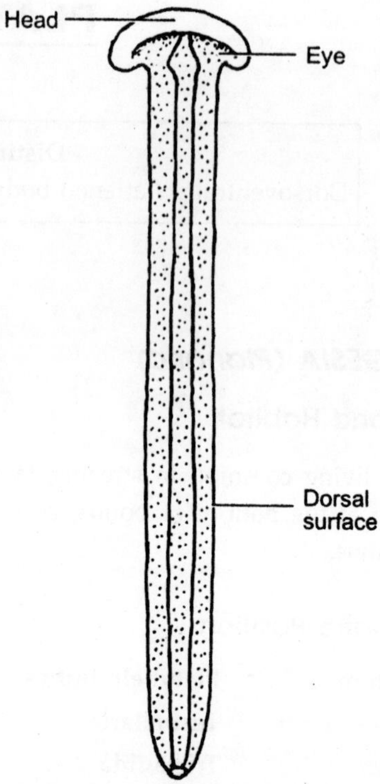

Fig. 1.1.1.49 : *Bipalium*

Type—*Bipalium* : State districtive features as in above.

3. *FASIOLA*

Habit and Habitat

It is commonly known as liver-fluke. Adult fluke is an endoparasite in the bile ducts of sheep. This parasite has a snail as the sole intermediate host.

Systematic Position

Phylum	**Platyhelminthes**
Class	**Trematoda**
Subclass	**Digenia**
Type	***Fasiola***

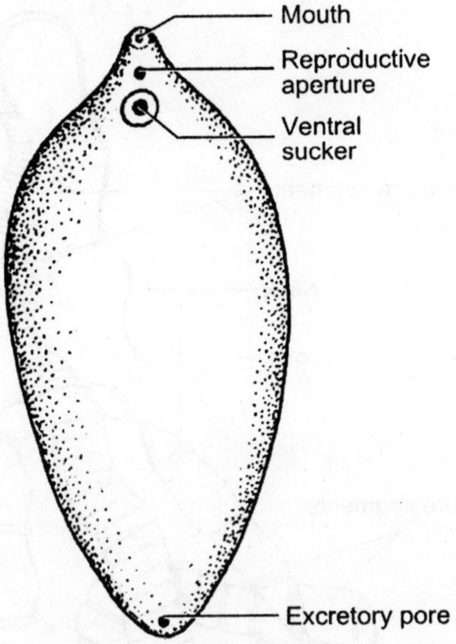

Fig. 1.1.1.50 : *Faciola*

Distinctive Features

1. The body is leaf-like, dorsoventrally flattened.
2. Anterior end terminate into a conical projection called cephalic cone.
3. At the apex of the cephalic cone is a small ventrally placed mouth, surrounded by oral sucker.
4. At the short distance behind the cephalic cone is an adhesive **acetabulum** or ventral sucker.
5. Between oral and ventral suckers is gonopore.
6. Excretory pore at the posterior end.

Identification with Reasons

State the characters up to—

Phylum—*Platyhelminthes* as those of *Dugesia.*

Class—*Trematoda* :

1. Endoparasites, epidermis devoid of cilia and rhabdites.
2. Gut always present and provided with oral sucker, pharynx and bifurcated intestine.
3. Single posterior excretory pore present.

Subclass—*Digenia* :

1. An anterior sucker is usually present, surrounding the month.
2. A simple acetabulate ventral sucker present.
3. Endoparasites with complex life always require one or more interme-diate hosts.
4. The first on a gastropod snail and second one if required is usually an arthropod. The definitive host is vertebrate.

Type—*Fasiola* : State the districtive features as in above.

Economic Importance

The liver-fluke causes **'liver rot'** in sheep. The yield of wool, leather and meat is greatly reduced in infected sheep causing great loss to poultry. Liver-fluke, which is parasite in man, also causes liver diseases, occulation of bile ducts and serious anaemia, diarrhoea, dysentery and ulcers.

4. *SCHISTOSOMA*

Habit and Habitat

It is known as blood-fluke, and lives in the blood of mammals, while larvae are found in snails; the trematode is found in the sub-mucous tissue of the human urinary bladder or rectum.

Systematic Portion

Phylum	**Platyhelminthes**
Class	**Trematoda**
Subclass	**Digenia**
Type	***Schistosoma***

Distinctive Features

1. The body is cylindrical, elongated with oral and highly developed ventral sucker.
2. Dioecious, female is lodged during copulation in the **gynaecophoric canal** of the male.
3. Male is larger than the female and posseses a deep groove called gynaecophoric canal.

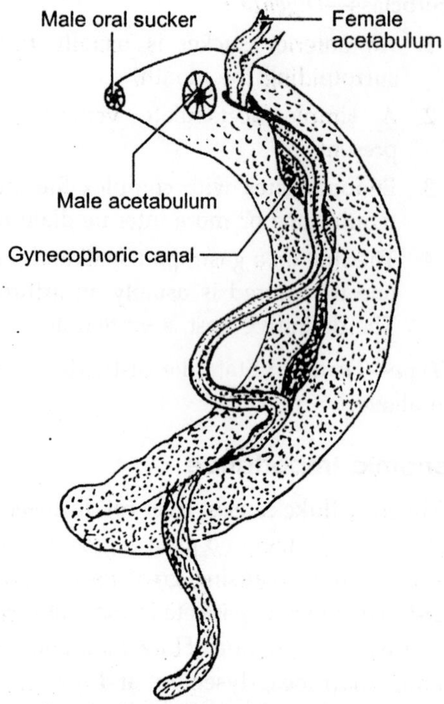

Fig. 1.1.1.51 : *Schistosoma*

4. Genital aperture immediately below the ventral sucker.

Systematic Position

State characters of each taxon as those of *Fasiola.*

Type—*Schistosoma* : State distinctive features as in above.

Economic Importance

The liver-fluke causes **'Schistosomiasis'**. The symptoms of which are: body ache, cough, followed by dysentery and anaemia. Patient become weak and exhausted and ultimately dies.

5. *TAENIA*

Habit and Habitat

It is found as an endoparasite in gut of vertebrate host. Man is the primary host and pig is its secondary host. Scolex lies burried in the intestinal mucosa. **Cystecercus stage** live in the muscles of pig.

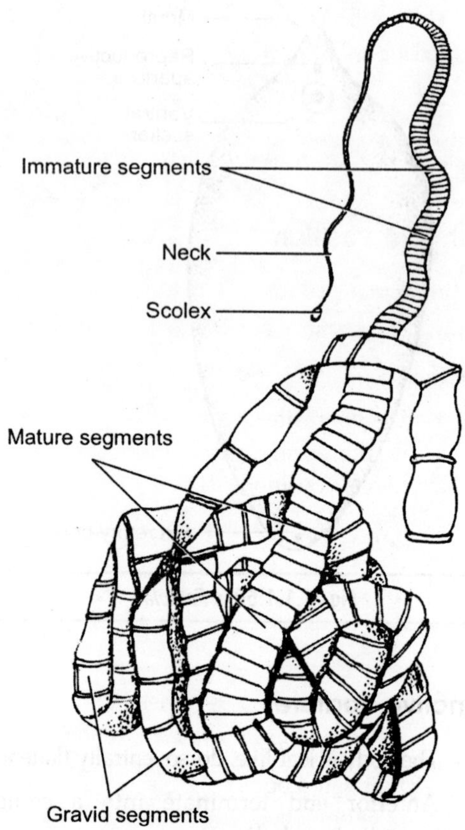

Fig. 1.1.1.52 : *Taenia* (Entire)

Systematic Portion

Phylum	**Platyhelminthes**
Class	**Cestoidea**
Subclass	**Eucestoidea**
Order	**Taeniodea (cyclophyllidea)**
Type	***Taenia***

Distinctive Features

1. Ribbon like body. Body is differentiated into **scolex**, neck and a large number of immature, mature and gravid segments or **strobila**.

2. The mature segments contain fully developed hermaphroditic genital organs.

3. The adhesive scolex contains four suckers and round **rostellum** crowned with double row of hooks.

Systematic Position

State the characters up to—

Phylum—*Platyhelminthes* as these of *Dugesia*.

Class—*Cestoidea* :

1. Endoparasites with attachment organ in the form of suckers, tentacles or hooks confined to the anterior region.
2. Gut completely absent.
3. Body divided into head, or scolex, neck and **proglottids**.
4. Each segments except the head and neck is provided with one or two sets of complete sex organs.

Subclass—*Eucestoidea* :

1. Elongated body divided into a number of segments and covered by cuticle.
2. Expanded scolex bears four suckers.
3. Life cycle involves free-swiming six-hooked larva or intact egg entering gut of intermediate host.

Order—*Taeniodea* (cyclophyllidea) :

1. Scolex consists of four acetabulate suckers usually with rostellum.
2. Two-host life cycle: invertebrate or vertebrate.

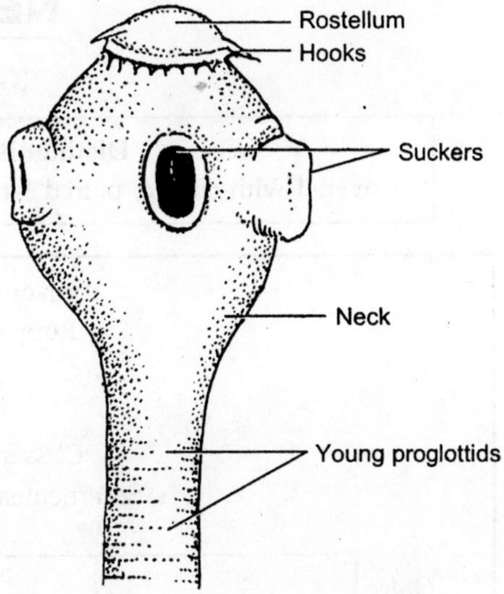

Fig. 1.1.1.52A : Head of *Taenia*

Type—*Taenia* : State distinctive characters as in above.

Economic Importance

The worms are responsible for causing severe symptoms, in children and adults. It causes nausea, abdominal pain and nervous disorder resembling epilepsy, anaemia and eosinophilia. Some times, bladder worm stage may harbour in man, causing a disease known as **cysticercosis**. The disease is very fatal. It infects eyes, brain, spinal cord and other delicate organs.

NEMATODA

(Greek : thread)

Distinguishing Characteristics :
Covered with cuticle; paired lateral sensory organs (amphids) on the head.

An outline classification of **Phylum-Nematoda** after Ruppert, E.E. and Barnes, R.D., 1994 (for those types that are included in the text)

Classes

- Secernentea or Phasmida
  - *Order*
    - **Ascaridida** — Example : *Ascaris*
    - **Strongylida** — Example : *Ancylostoma*
    - **Spirurida** — Example : *Wuchereria*
- Adenophorea or Aphasmida
  - *Order*—**Trichocephalida** — Example : *Beroe*

1. *ASCARIS* (Round worm)

Habit and Habitat

Endoparasite, found in the intestine of man and pig; it is found in all parts of the world.

Systematic Position

Phylum	**Nematoda (=Nemata)**
Class	**Secernentea (Phasmida)**
Subclass	**Rhabditida**
Order	**Ascaridida**
Type	*Ascaris lumbricoides*

Distinctive Features

1. Common round worm, having elongated and cylindrical body.

2. Unsegmented body, tapering on both ends with distinct lateral lines on both sides of the body.

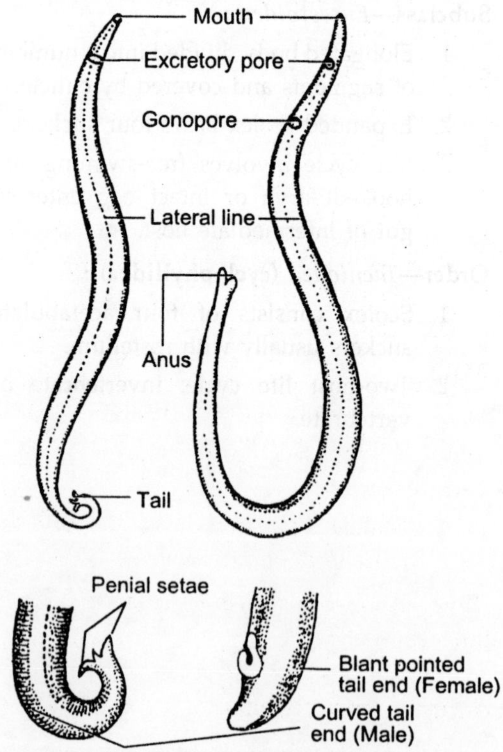

Fig. 1.1.1.53 : *Ascaris*

Classification is based on Parker, S.P. ed. 1982.

3. Female is larger than the male and the tail end is bluntly pointed.

4. The tail end of male is vertically curved containing two copulatory spicules.

5. **Amphids** are found on the head.

Identification with Reasons

Phylum—*Nematoda* :

1. Pseudocoelomate and acoelomate worms, with no external segmentation.
2. Cylindrical body, tapered at both ends and covered with a multilayered cuticle.
3. Presence of muscular pharynx, intestine, rectum and posterior anus.
4. Respiratory and blood vascular system absent.

Class—*Secernentea* (Phasmida) :

1. Nearly all terrestrial, majority are parasites on vertebrates, invertebrates or plants.
2. Epidermal or caudal gland absent.
3. Possess phasmids.

Subclass—*Rhabditida* : Species are free living, parasite or both.

Order—*Ascaridida* :

1. Large parasite with elongated and cylindrical body.
2. Mouth is surrounded by three prominent lips.
3. Teeth bearing cutting plates absent.

Male :

1. The tail end of male is ventrally curved, containing two copulatory spicules.

Female :

1. Female is larger than the male and the tail end is bluntly pointed.

Type—*Ascaris lumbricoides* : State distinctive features as in above.

Economic Importance

Ascaris is pathogenic to man. Larvae cause long inflammation and fatal pneumonia. Sometimes, fever, anaemia, leucocytosis and eosinofilia result. Mature Ascaris may cause, some abdominal discomfort and colic pains with vomitting, diarrhoea, and slight temperature. Appendicitis is also caused by Ascaris. Toxic substances produced by Ascaris cause convulsions, delirium, coma and general nervousness.

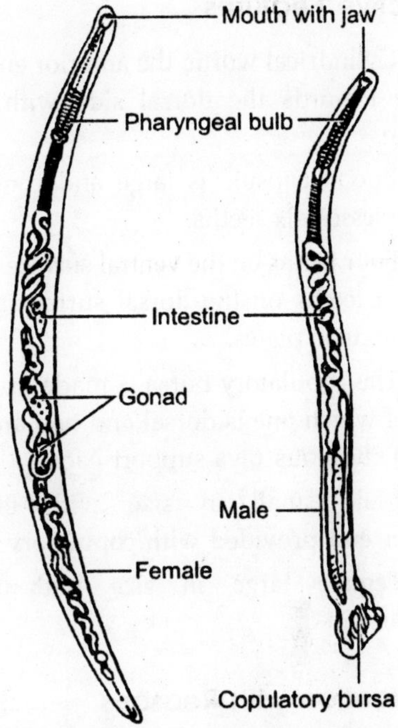

Fig. 1.1.1.54 : *Ancylostoma*

2. *ANCYLOSTOMA* (Hook worm)

Habit and Habitat

Endoparasite in the small intestine of man especially in jejunum, less often in the duodenum and realy in ileum; sucks blood and body fluids. Distributed in tropical and subtropical countries.

Systematic Position

Phylum	**Nematoda (=Nemata)**
Class	**Secernentea (Phasmida)**
Subclass	**Rhabditida**
Order	**Strongylida**
Type	***Ancylostoma duodenale***

Distinctive Features

1. Cylindrical worm, the anterior end is bent slightly towards the dorsal side with an oral aperture.
2. Buccal capsule is large and conspicuous and possesses six teeths.
3. Four teeths on the ventral surface are hook like, two teeths on the dorsal surface are knob like triangular plates.
4. The copulatory bursa is made up of three lobes, of which one is dorsal and two are lateral, thirteen chitinous rays support each.
5. Male small in size with expanded perterior end provided with copulatory bursa.
6. Female large in size with tapering posterior end.

Identification with Reasons

State the characters up to—

Subclass—*Rhabditida* as in *Ascaris.*

Order—*Strongylida* :

1. Teeth-bearing cutting plates present.
2. Mouth without distinct lips.

Male :

1. Smaller in size.
2. Posterior end expanded like a umbrella, possess copulatory bursa.
3. Genital opening posterior with cloaca.

Female :

1. Longer in size.
2. Posterior end tapering with no copulatory bursa.
3. Genital opening at the junction of posterior and middle third of the body.

Type—*Ancylostoma duodenale* : State distinctive features as in above.

Economic Importance

Larvae of Ancylostoma cause lesions in the skin which include **ancylostomal dermatitis** and creeping eruption. Lesions in lungs causes bronchitis and broncho-pneumonia. Adult worm is responsible for a disease known as *ancylostomiasis* or *hook worm diseases* characterised mainly by anaemia.

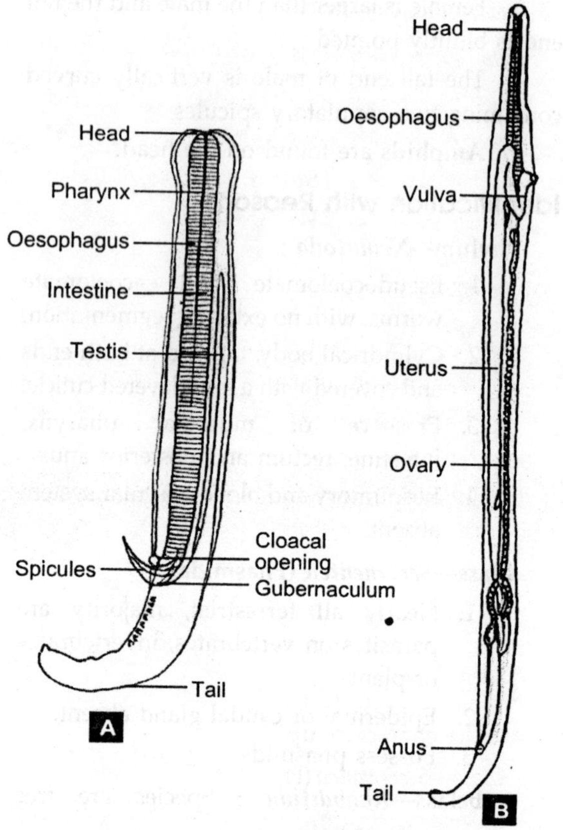

Fig. 1.1.1.55A : *Wuchereria bancrofti.* A. Male, B. Female

3. *WUCHERERIA*

Habit and Habitat

The adult worms live in the lymphatic vessels and lymph nodes of man only; Culex, Anopheles and Aedes mosquito acts as intermediate hosts into which microfilariae undergo further development. Infected stage enter human body during mosquito bite, migrates into lymphatic system and develop into adult.

Systematic Position

Phylum	**Nematoda**
Class	**Secernentea (Phasmida)**
Subclass	**Spiruria**
Order	**Spirurida**
Type	***Wuchereria bancrofti***

Distinctive Features

1. Hair like transparent worms.
2. Filiform and both ends are tapering.
3. Body curved without kinks.
4. The posterior end of male is curved ventrally and with two unequal length spicules.
5. The tail end of female is narrow, unbent and abruptly pointed.
6. Male and females usually remain coiled.

Identification with Reasons

State the characters up to—

Class—*Secernentea* (Phasmida) as in Ancylostoma.

Subclass—*Spiruria* : Parasite on annelids or vertebrates. Arthopods, usually insects serve as intermediate hosts.

Order—*Spirurida* :

1. Filiform worms without lips, buccal capsule and bursa.
2. Pharynx is both muscular and glandular.
3. Vulva is anteriorly situated and with an ovejector.

Type—*Wuchereria bancrofti* : State distinctive features as in above.

Type—*Microfilaria* (*Wuchereria* larva) :

1. Body slightly flat and invested by a delicate sheath which is projecting slightly beyond the extremities.
2. Cephalic space without granules is present at the anterior end.
3. Nerve ring at the anterior region and genital cells and anus at the posterior region.

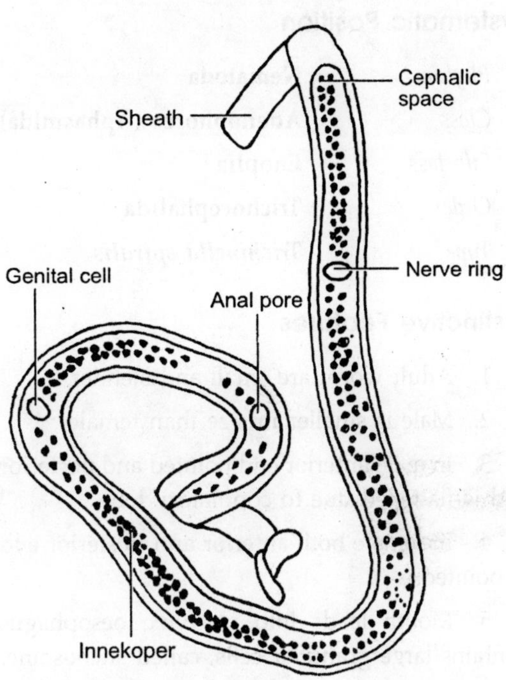

Fig. 1.1.1.55B : Microfilaria *(Wuchereria larvae)*

4. Central axis of the body, composed of granules which extends from anterior to posterior end.

Economic importance

Wuchereria causes *elephantiasis* or *wuchereriasis* or *Bancroft's filariasis*. It effects limbs and genitalia. The effected limbs and genitalia become elongated and produced a tumor like solidity. The symptoms thus produced is the manifestation in acute infective stage. Microfilariae cause filarial fever, mental depression and headache.

4. *TRICHINELLA* (whip worm)

Habit and Habitat

Trichinellla in encysted condition is found in the muscular tissues in various parts of the body. It is found in the intestine of man, pig and other vertebrates. Parasite enter in human body when improperly cooked infected 'Pork' is eaten.

Systematic Position

Phylum	**Nematoda**
Class	**Adenophorea (Aphasmida)**
Subclass	**Enoplia**
Order	**Trichocephalida**
Type	***Trichinella spiralis***

Distinctive Features

1. Adult worm are small and slender.
2. Male is smaller in size than female.
3. In male anterior end pointed and posterior end is flattened due to copulatory lobe.
4. In female both anterior and posterior end is pointed.
5. Mouth leads into pharynx, oesophagus contains large granullar cells, called stichosome.
6. Vulva is anterior.

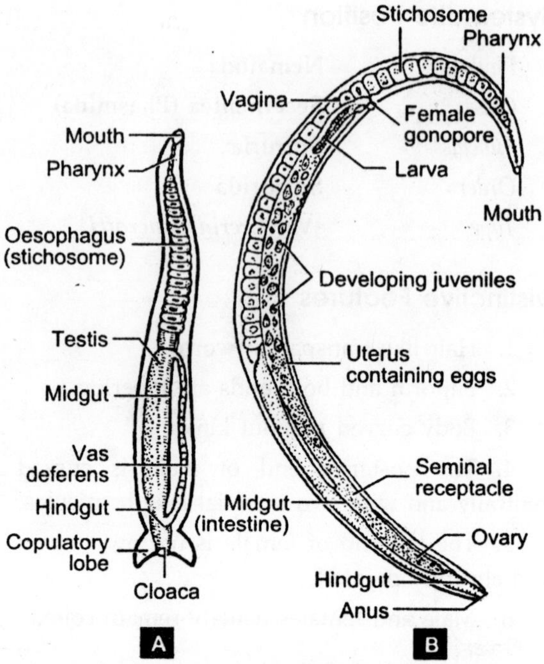

Fig. 1.1.1.56 : ***Trichinella spiralis*. A. Male, B. Female**

Identification with Reasons

State the characters up to—

Phylum—*Nematoda* in Ascaris.

Class—*Adenophorea* (Aphasmida) :

1. Caudal sensory organs or phasmid absent.
2. Caudal glands and epidermal glands present.
3. Excretory system rudimentary or absent.
4. Mostly marine specics.

Subclass—*Enoplia* : Free living and parasitic form.

Order—*Trichocephalida* :

1. Body anteriorly filiform and pharyngeal region is narrow.
2. Mouth without lips.
3. Anterior muscular region short, posterior region long with stichosome.

Type—*Trichinella spiralis* : State distinctive features as in above.

Economic Importance

Trichinella, causes hemorrhage, muscular pain, abdominal pain, diarrhoea, thrombosis, restlessness, renal failure, degeneration of muscles, and swelling of various parts of the body.

Ruppert, E.E., and Barnes, D.B. (1994) have divided Nematoda into Class – ***Secernentea*** *or* ***Phasmida*** *and class –* ***Adenophorea*** *or* ***Aphasmida*** *and 14 ordes.*

ANNELIDA

(Latin, annulus: little ring and G: eidos: form)

> **Distinguishing Characteristics :**
> One or more pairs of chitinous setae, metamerism appears.

An outline classification of **Phylum-Annelid** after Ruppert, E.E. and Barnes, R.D., 1994 and Parker, S. P. (Ed.) 1982 (for those types that are included in the text)

Classes

- Polychaeta
  - *Order*
    - **Phyllodocida** — Example : *Nereis*, *Aphrodite*
    - **Chaetopterida** — Example : *Chaetopterus*
    - **Capitellida** — Example : *Arenicola*
    - **Sabellida** — Example : *Sabella*
    - **Terebellida** — Example : *Terebella*
- Oligochaeta
  - *Order*—**Haplotaxida**
    Example : *Pheretima*, *Tubifex*
- Hirudinea or Hirudinoidea
  - *Order*—**Arnchynchobdellida**
    Example : *Hirudo*

1. *NEREIS* (Ragworm or clam worm)

Habit and Habitat

Marine; live in tubes or burrows hidden under stones or sea weeds while some are commensal with hermit crabs or certain other polychaetes. They are free living, predaceous, nocturnal, carnivorous, often found buried in the intertidal zone, feeds on small insects, molluscas and worms. It has cosmopolitan distribution.

Systematic Position

Phylum	**Annelida**
Class	**Pollychaeta (G: many setae)**
Order	**Phyllodocida**
Type	**Nereis**

Classification is based on Pechenik, J.A., 2002.

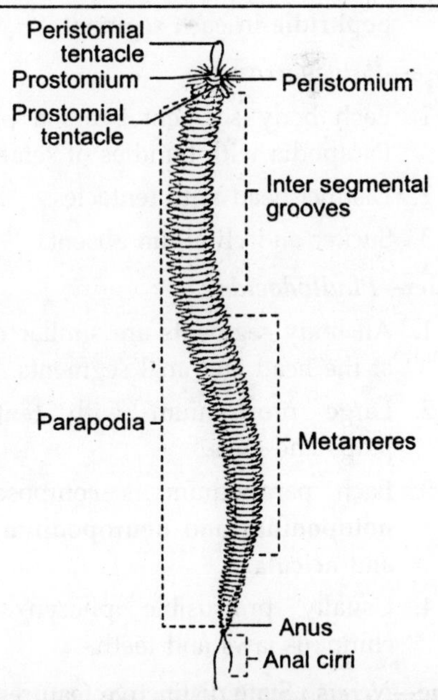

Fig. 1.1.1.57 : *Nereis*

Distinctive Features

1. Body elongated, cylindrical and flattened, not divided into an **atokal** and **epitokal** region.

2. Well developed head with **prostomium** and **peristomium.** Peristomium bears a pair of tentacles, palps and eyes. Peristomium with four pairs of peristomial tentacles.

3. Each segment is provided with a pair of **parapodia**, except head and last segment.

4. Setae and parapodia all similar.

5. The anal segment contains a pair of anal cirri.

Identification with Reasons

Phylum—*Annelida* :

1. Bilaterally symmetrical, body metamerically segmented.
2. Distinct cephalization.
3. Body is covered by a thin and moist cuticle.
4. Unjointed appendages.
5. Mouth and anus terminal.
6. Excretory system consists of a pair of **nephridia** in each segment.

Class—*Polychaeta* :

1. Each body segment bears a pair of Parapodia with bundles of setae.
2. Distinct head with tentacles.
3. Sucker and **clitellum** absent.

Order—*Phyllodocida* :

1. All body segments are similar except at the head and anal segments.
2. Large prostomium with tentacles, palps and eyes.
3. Each parapodium is composed of **notopodium** and **neuropodium, cirri** and **acicula.**
4. Usually protrusible pharynx with chitinous jaws and teeth.

Type—*Nereis* : State distinctive features as in above.

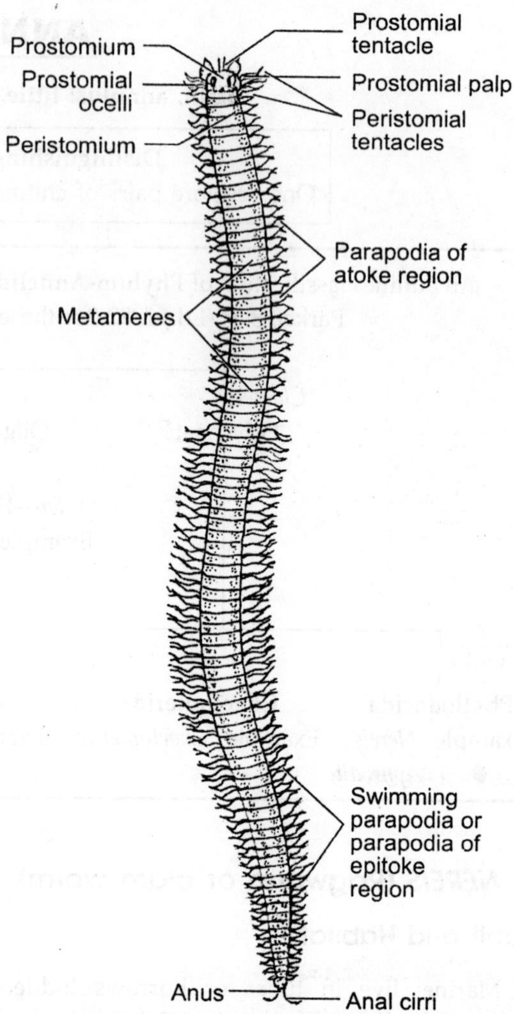

Fig. 1.1.1.58 : *Heteronereis.* Entire

2. *HETERONEREIS* (Sexual stage of Nereis)

Characters and systematic position similar as those of *Nereis.*

Distinctive Features

1. Elongated cylindrical body divided into an anterior asexual **atoke** and a posterior sexual **epitoke.**

2. Epitokal parapodia large foliaceous lobe and setae flattened and fan like.

3. A pair of anal cirri present.

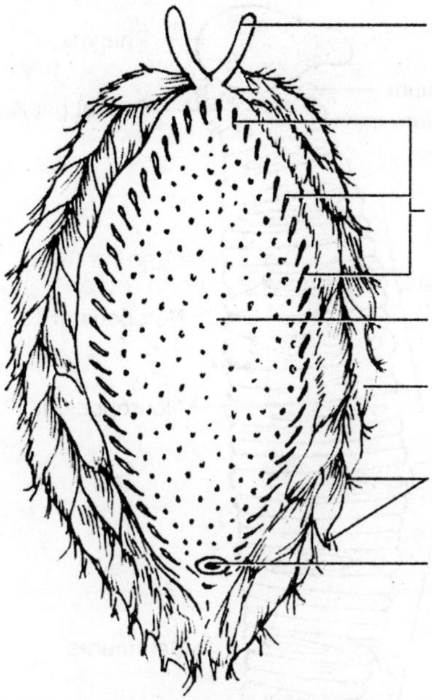

Fig. 1.1.1.59 : *Aphrodite*

Special Feature

Marine; free swimming, swims to sea surface in great number at night for sexual reproduction. They perform nuptial dance.

3. *APHRODITE* (=*Aphrodita*)(Sea-mouse)

Habit and Habitat

Marine: bottom dweller, lives partly buried in fine mud with the posterior end projecting into water, found in shallow waters below low-tide mark; carnivorous, walk of the surface of sea bottom, using thick, tubular, highly muscular parapodia. Cosmopolitan.

Systematic position

Phylum	**Annelida**
Class	**Polychaeta**
Order	**Phyllodocida**
Type	***Aphrodite***

Distinctive Features

1. Oval, dorsoventrally flattened. Dorsal surface convex and ventral surface flat, forming a creeping sole.

2. Body is covered dorsally by felt-like setae, called **elytra**.

3. Notopodia contain, stiff setae, soft setae and iridescent setae.

4. Anus is dorsally situated.

Identification with Reasons

State the characters up to—

Order—*Phyllodocida* as in *Nereis*.

Type—*Aphrodite* : State distractive features as in above.

Special Feature

Beautiful iridescent colour changes from gold to peacock-blue during movement of the animal.

4. *CHAETOPTERUS*

Habit and Habitat

Marine; bottom dweller; found in sandy and muddy bottom; lives in secreted U-shaped leathery tube which is buried in the intertidal zone; sedentary; captures food particles with mucus nets (filter-feeder); feeds on plankton and detritus. World wide distribution.

Systematic Position

Phylum	**Annelida**
Class	**Polychaeta**
Order	**Chaetopterida**
Type	***Chaetopterus***

Distinctive Features

1. Long worm, body is divided into three distinct **tagmata**—anterior, middle and posterior.

2. Anterior flattened region with a funnel-shaped collar having two peristomial palps.

3. Tenth segment has a pair of wing like aliform notopodia.

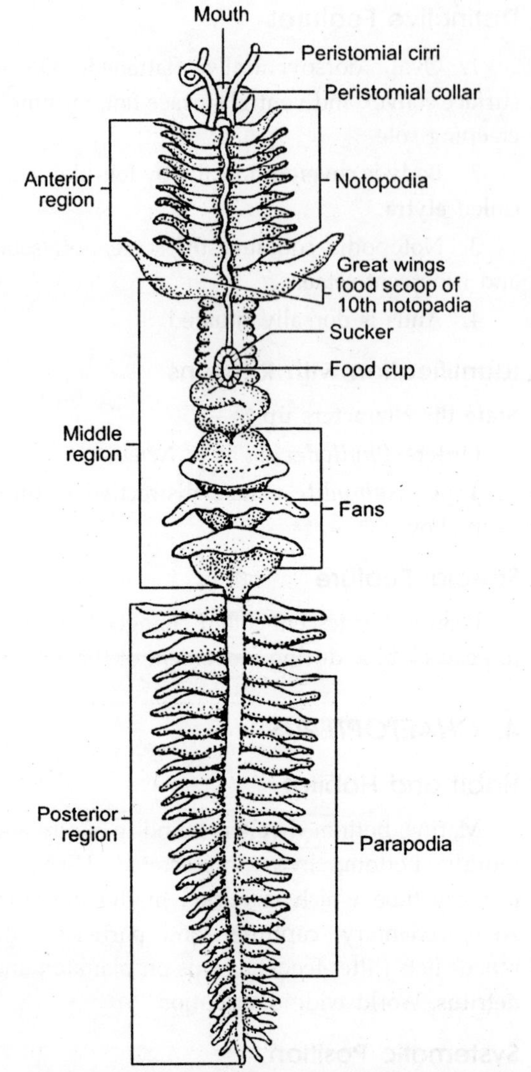

Fig. 1.1.1.60 : *Chaetopterus*

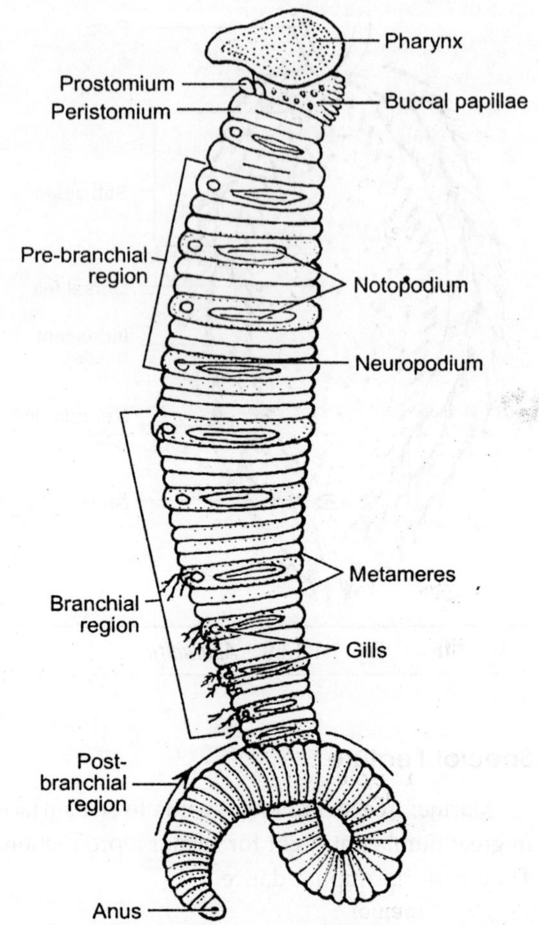

Fig. 1.1.1.61 : *Arenicola*

4. Middle region has three pairs of fans which are modified parapodia.
5. Posterior segments longer, bearing parapodia.

Identification with Reasons

State the characters up to—

Class—*Polychaeta* as in *Nereis.*

Order—*Chaetopterida* :

1. Sedentary, live in secreted leathery tubes.
2. Body divided into three distinct tagmata.
3. Non protrusible pharynx without jaws.
4. Both uniramous and biramous parapodia present; acicula and cirri absent.

Type—*Chaetopterus* : State distinctive features as in above.

5. *ARENICOLA*

Habit and Habitat

Marine: bottom dweller; found in muddy sand in intertidal areas of estuaries and bays with little wave action; lives in a U-shaped burrows made of sand and mucous; deposit feeders; ingesting sediment and digesting the organic components.

Systematic Position

Phylum	**Annelida**
Class	**Polychaeta**
Order	**Capitellida**
Type	***Arenicola***

Distinctive Features

1. Body elongated, cylindrical differentiated into pre-branchial, branchial and post-branchial regions.
2. Body segments divided into annuli.
3. Pre-branchial segments bear parapodia.
4. Branchial segments bear parapodia and gills.
5. Post-branchial segments without parapodia and gills.

Identification with Reasons

State the characters up to—

Class—*Polychaeta* as in *Nereis*.

Order—*Capitellida* :

1. No anterior appendages.
2. Prostomium small and trilobed with no eyes or tentacles.
3. Parapodia biramous with reduced neuropodia.

Type—*Arenicola* : State distractive features as in above.

6. *SABELLA* (Feather-duster or fan worms)

Habit and Habitat

Marine; bottom living; tube dwelling, worms secrete mucous tubes within the burrows, tube made of mud, sand and shell fragments; suspension feeders, world wide distribution.

Systemic Position

Phylum	**Annelida**
Class	**Polychaeta**
Order	**Sabellida**
Type	***Sabella***

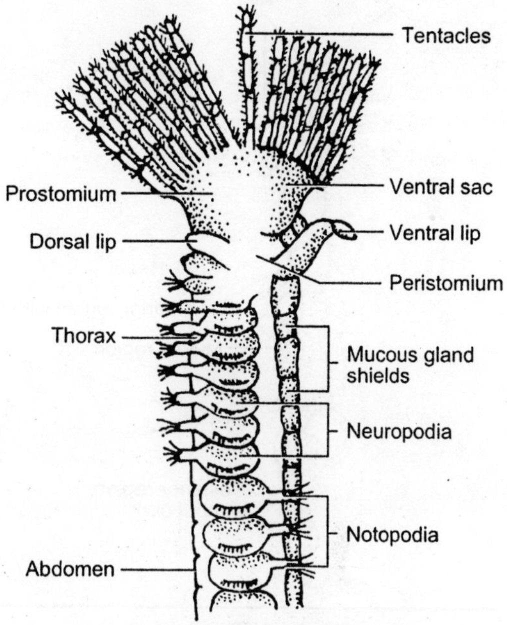

Fig. 1.1.1.62 : *Sabella*

Distinctive Features

1. Body cylindrical, tapering and divided into two grossly unequal division.
2. Prostomium contains several gill filaments forming a branchial crown.
3. Anterior region shorted and bears parapodia with prominent notopodia but no neuropodia.
4. Peristomium forms a collar surrounding the crown.
5. Abdomen is largest having parapodia in each **metamere**.

Identification with Reasons

State the characters up to—

Class—*Polychaeta* as in *Nereis*.

Order—*Sabellida* :

1. Body contains of two unequal regions.
2. Jaws absent, pharynx not eversible.
3. A branchial crown present.
4. Prostomium collar like.

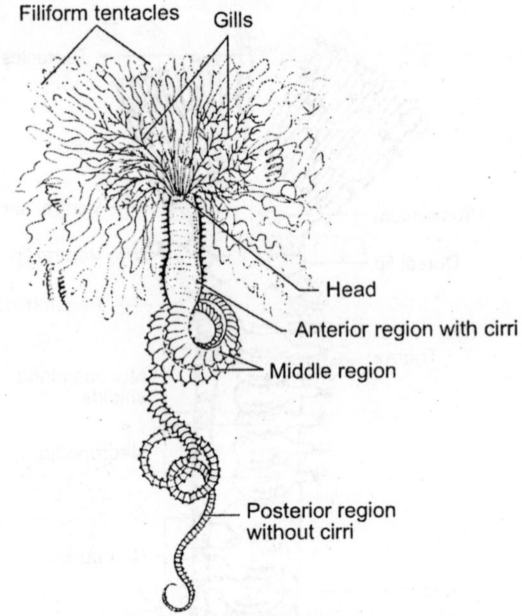

Fig. 1.1.1.63 : *Terebella*

Type—*Sabella* : State distinctive features as in above.

7. *TEREBELLA*

Habit and Habitat

Sedentary; lives in muddy bottom in the intertidal zone within permanent tube; deposite feeder.

Systemic Position

Phylum	**Annelida**
Class	**Polychaeta**
Order	**Terebellida**
Type	***Terebella***

Distinctive Features

1. Elongated, cylindrical body divided into anterior and posterior regions.

2. Horse shoe-shaped head, made up of prostomium and peristomium.

3. Prostomium contains several long filiform tentacles and two pairs of eye spots.

4. Behind the head are branched gills or branchiae.

5. Anterior 1/3 body is thickened, while major parts of posterior body tapers narrowly.

Identification with Reasons

State the characters up to—

Class—*Polychaeta* as in *Nereis.*

Order—*Terebellida* :

1. Permanent tube dwellers.
2. A pair of highly branched gills for gas exchange.
3. Numerous grooved tentacles project form the prostomium.
4. Absence of conspicuous parapodia.

Type—*Terebella* : State distinctive features as in above.

8. *PHERETIMA* (Earthworm)

Habit and Habitat

Terrestrial; they usually live in the upper layers of slightly damp soils, lawns, gardens and up to the depth of 30-45 cm, nocturnal; feeds on dead organic matter which is taken along with soil;

Systemic Position

Phylum	**Annelida**
Class	**Clitellata**
	(Clite1-L: a pack saddle)
Subclass	**Oligochaeta (G: few setae)**
Order	**Haplotaxida**
Type	***Pheretima***

Distinctive Features

1. The body is long, narrow, cylindrical and perfectly adapted for burrowing life.

2. Distinct external and internal segmentation.

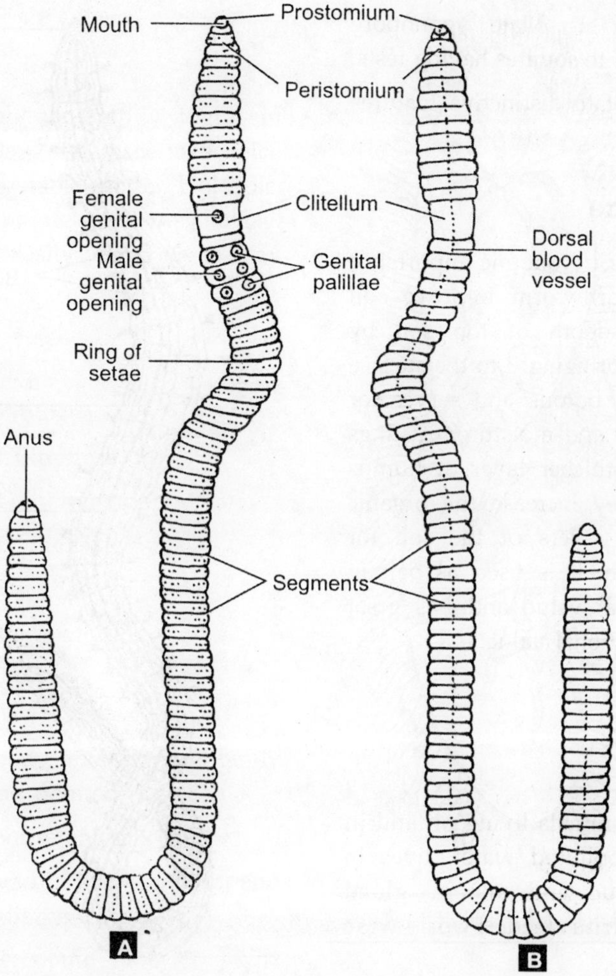

Fig. 1.1.1.64 : *Pheretima.* A–Ventral view, B–Dorsal view.

3. Each segment contains setae in the ventral body wall, except the first, last and cliteller segments.

4. A dark median line of dorsal vessel is found.

5. Spermathecal pores in the inter-segmental groove of 5/6, 6/7, 7/8, and 8/9 segments.

6. Female gonopore on the 14th segments and male gonopore on the 18th segment.

Identification with Reasons

State the characters up to—

Phylum—*Annelida* as in *Nereis*.

Class—*Clitellata* :

1. **Clitellum** used to secrete a cocoon for eggs.
2. Hermaphodite.
3. Development direct without larvae.

Subclass—*Oligochaeta* :

1. Small head with cephalic appendages.
2. Relatively few setae present.
3. Testes anterior to ovaries and with only one or two pairs of male or female gonads.
4. Presence of **seminal receptacles.**

Order—*Haplotaxida* : Male gonopore situated at some distance to somites having tests.

Type—*Pheretima* : State distinctive features as in above.

Economic Importance

Earthworm are of great economic importance to man. Burrows of earthworm improve soil aeration; increase the depth of top soil by burrowing into subsoil, bringing it to the surface as a result soil become porous and serves for better penetration of air and moisture. Castings of earthworm forms a thicker layer of humus surface of the soil. They increase the organic content of the surface layers of the soil for favourable plant growth. It is used as bait in fishing. It has medicinal value and has great experimental and educational value.

9. *TUBIFEX*

Habit and Habitat

Fresh water; small animals living on and in bottom sediments of polluted water, lives in tubes made up of mud and mineral glued together by mucous; detritus feeder; world wide distribution.

Systemic Position

Phylum	**Annelida**
Class	**Clitellata**
Subclass	**Oligochaeta**
Order	**Haplotaxida**
Type	**Tubifex**

Distinctive Features

1. Long, cylindrical segmented body.

2. Four bundles of setae on dorsal and ventral side per segment.

3. Dorsal bundle setae are forked and hair like, while ventral bundle setae only forked.

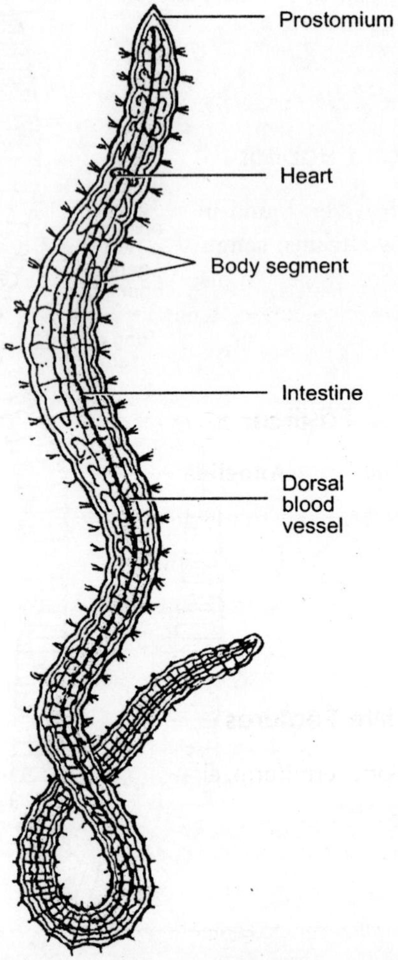

Fig. 1.1.1.65 : *Tubifex*

4. Male genital pore on the 11th segment and female genital pore on the 12th segment.

Identification with Reasons

State the characters up to—

Order—*Haplotaxida* as in *Pheretima*.

Type—*Tubifex* : State distinctive features as in above.

Economic Importance

Used as food for aquarium fishes. It is encouraged to grow in filter beds of sewage disposal plants, probably they have purificatory functions.

10. *HIRUDO* (Leech)

Habit and Habitat

Fresh water; found in ponds, lakes, swamps and slow streams; **sanguivorous**; sucking blood from fishes, frogs and also the domestic animals who used the water in which leeches are present; mainly in tropical distribution.

Systemic Position

Phylum	**Annelida**
Class	**Clitellata**
Subclass	**Hirudinea**
Order	**Arnhynchobdellae**
Type	***Hirudo medicinalis***

Distinctive Features

1. Soft, vermiform, elongated, dorsoventrally flattened body.
2. Moist and slimy skin, olive green dorsally and orange red ventrally.
3. Well developed anterior and posterior suckers.
4. Thirty four body segments and each segment is superficially divided into five annuli.
5. A pair of eyes on each of the first five segments on the dorsal side.

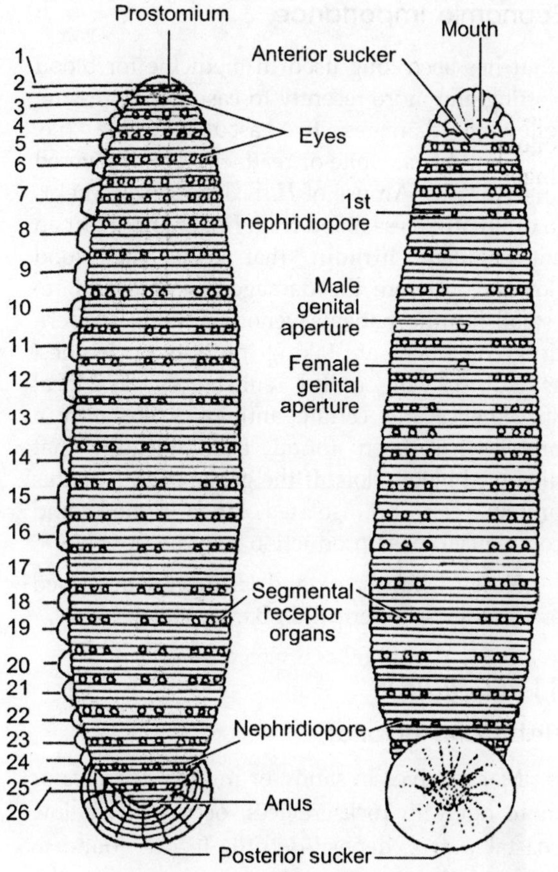

Fig. 1.1.1.66 : *Hirudo medicinalis*
Left–Ventral view, Right–Dorsal view.

Identification with Reasons

State the characters up to—

Class—*Clitallata* as in *Pheretima.*

Subclass—*Hirudinea* :

1. Each body segment is made up of several annuli.
2. Body segment numbers fixed.
3. Absence of parapodia or setae.
4. Internal segmentation obscure.
5. Coelom often filled up with **Botryoidal tissue.**

Order—*Arnhynchobdellae* :

1. Thirty four body segment present.
2. Presence of anterior and posterior suckers.
3. Fertilization is accomplished by inserting penis into the vagina of an individual.
4. Presence of three well developed toothed jaw.

Type—*Hirudo medicinalis* : State distinctive features as in above.

Economic Importance

It has been long used in medicine for blood clotting and more recently to ease fluid pressure following damages to vascular tissue (i.e. following a snake bite or reattachment of severed finger or ear). An use of *Hirudo medicinalis,* helps to drain off excess blood from finger and inject an anticoagulant, **hirudin,** that keeps the blood flowing out from the damaged wound and re-establishes the patients venous system. Modern drug companies are taking interests on leeches, as sources of useful substances as local anaesthetics and certain antibiotics. Recently a protein has been found from hirudin that dissolved blood clots. If the genes coding of this protein can be isolated and cloned and commercial mass production may be possible.

Note : Oligochaeta and Hirudinea described as Class by Ruppert, E.E., Barnes, R.D., 1994.

11. *ECHIURUS*

Habit and Habitat

Marine; lives in sandy or muddy burrows or more rarely in rock crevices; occurs in shallow coastal waters; deposit-detritus feeder; found in warm and temperate seas.

Systemic Position

Phylum	**Annelida**
Class	**Echiura* (G:echis=serpent like)**
Order	**Echiuroinea**
Type	***Echiurus***

Distinctive Features

1. Body divided into a permanently everted, grooved proboscis and a non segmented trunk.
2. Proboscis is flat, narrow and non retractile.
3. A pair of large, hooked, ventral setae projected from the setal sac.
4. Some species have one or two posterior circles of setae.

* *Ruppert, E.E., Barnes R.D., (1994) considered* ***Echiura*** *as a Phylum (lack of segmentation and elongated prostomium in echiurans).*

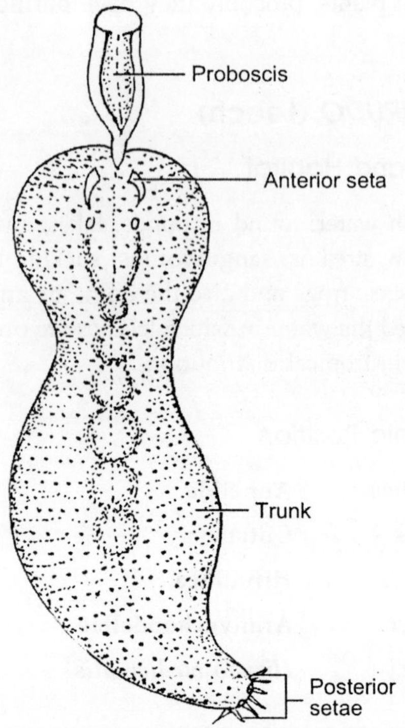

Fig. 1.1.1.67 : *Echiurus*

Identification with Reasons

State the characters up to—

Phylum—*Annelida* as in *Nereis.*

Class—*Echiura* :

1. Body made of an anterior flat, everted proboscis and a posterior, cylindrical wide trunk.
2. Trunk with a pair of antro-ventral seta and posterior anus.
3. One to many metanephridia present.

Order—*Echiuroinea* :

1. Seven or fewer pairs of nephridia present.
2. Show little or no sexual dimorphism.
3. Short proboscis extent upto 2 cm to 4 cm.

Type—*Echiurus* : State distinctive features as in above.

SIPUNCULIDA

(Sipunculus = G: little tube)

Distinguishing Characteristics :
Anterior part of the body forms an eversible and fully retractable introvert, with the mouth at its end.

1. *SIPUNCULUS* (Peanut worms)

Habit and Habitat

Marine: sedentary creatures that burrow in the sand or mud; bores into coral rocks, an annelid tubes or a molluscan shell; deposit or detritus feeders. Cosmopolitan distribution.

Systematic Position

Phylum	**Sipunculida (Sipuncula)**
Class	**Sipunculidea**
Order	**Sipunculiformes**
Type	***Sipunculus***

Distinctive Features

1. The body is divided into an anterior introvert with mouth and a posterior trunk.
2. The mouth is surrounded by tentacles.
3. The digestive tract is spiral and u-shaped.
4. The anus is opening dorsally in the anterior end of the trunk.
5. Retractile muscles are present which help to move the **introvert**.
6. Body contains an enormous coelomic cavity.

Identification with Reasons

Phylum—*Sipunculus* :

1. Body consists of a plump, unsegmented trunk and an anterior introvert.

Classification is based on Parker, T.J., Haswell, W.A., 1972.

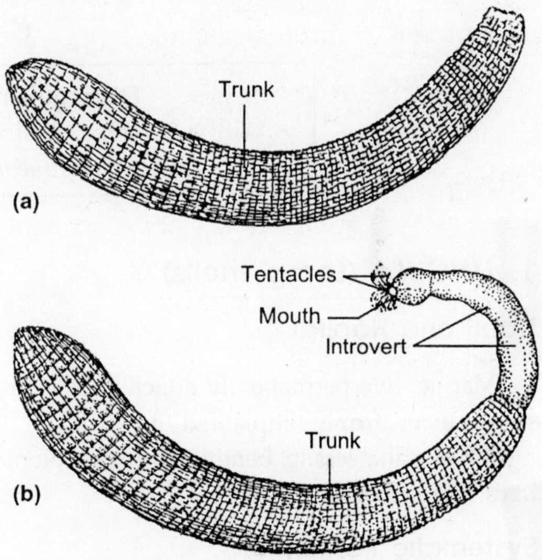

Fig. 1.1.1.68 : *Sipunculus*
(a) Introvert withdrawn and (b) extended

2. Introvert is fully retractable into the body.
3. Introvert containing the mouth and a posterior trunk.
4. The mouth is surrounded by tentacle or a dorsal **nuchal organ**.
5. The gut is spiral and u-shaped.
6. Their is a single or one pair of **metanephridia**.

Class—*Sipunculidea* :

1. Mucous-covered tentacles surrounding mouth.
2. Introvert hooks (when present) simple.

3. Hollow structures usually irregularly distributed.

Order—*Sipunculiformes* :

1. Longitudinal muscles thickened into bonds.
2. Generally coelom extended into epidermis as channels.
3. Metanephridia one pair.

Type—*Sipunculus* : State distinctive features as in above.

BRACHIOPODA

(G: arm foot)

Distinguishing Characteristics :
With a bivalved shall with dorsal and ventral valves, generally attached by a pedicle or stalk.

1. *LINGULA* (Lamp shells)

Habit and Habitat

Marine: live permanently attached to a solid substrate or firmly implanted in sediment by means of stalk; sessile, benthic and non-colonial; lives on microscropic particles.

Systematic Position

Phylum	**Brachiopoda**
Class	**Inarticulata**
Order	**Lingulida**
Type	***Lingula***

Distinctive Features

1. Body is enclosed in a bivalved shell, one dorsal and one ventral.

2. Valves are generally attached by a pedicle or stalk.

3. In the anterior surface of the body **lophophore** is present which surrounds the mouth and is beset with ciliated tentacles.

4. The alimentary canal is usually v-shaped.

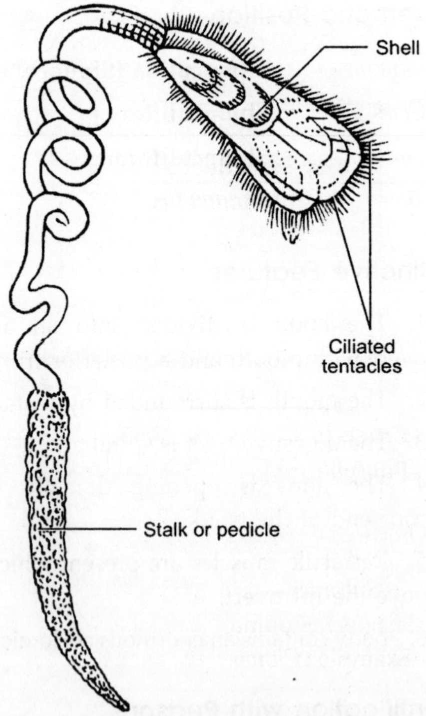

Fig. 1.1.1.69 : *Lingula*

5. The coelom is spacious and continues into the mantle lobes.

6. One or two pairs of metanepridia.

Classification is based on Parker, T.J., Haswell, W.A., 1972.

Identification with Reasons

Phylum—*Brachiopoda* :

1. The body is covered with a bivalved shell with dorsal and ventral valves.
2. Valves are generally attached by a pedicle or stalk.
3. Lophophore with a single row of alternating tentacles leading to a food groove.
4. Lophophore arms are with two mesocoelic channels.
5. One or two pairs of metanephridia which also acts as gonoducts.

Class—*Inarticulata* :

1. Valves are not united by a hinged, but held together by muscles only.
2. Lophophore lacks internal skeletal support.
3. Digestive tract complete with arms.

Order—*Lingulida* :

1. Pedicle valve ventral, but both valves modified to permit pedicle passage.
2. Shell used to form burrows in sediment.
3. The mineral of the shell is calcium phosphate.

Type—*Lingula* : State distinctive features as in above.

MOLLUSCA

(Latin: soft)

Distinguishing Characteristics :
Body is made up of head, visceral mass and locomotory or digging foot; epidermis forming mantle, secretes calcareous spicules or produces one or more shells; chitinous ribbon or **radula** having small teeth in the mouth (absent in bivalves).

An outline classification of **Phylum-Mollusca** after Ruppert, E.E. and Barnes, R.D., 1994
(for those types that are included in the text)

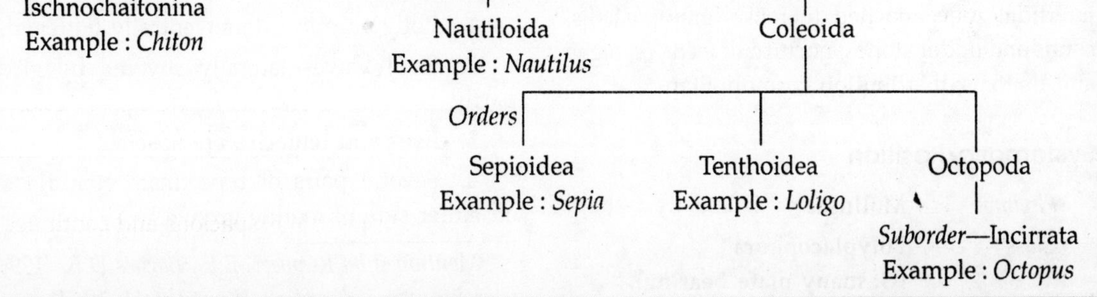

Chart contd. next page

Class—**Gastropoda**

- *Subclasses*
  - Prosobranchia
    - *Orders*
      - Archaeogastropoda (Diotocardia, Asphidobranchia) Example : *Patella, Haliotis*
      - Mesogastropoda Example : *Pila, Cypraea*
      - Neogastropoda Example : *Murex*
  - Opisthobranchia
    - *Orders*
      - Anaspidea Example : *Aplysia*
      - Nudibranchia Example : *Doris*
  - Pulmonata
    - *Orders*
      - Stylommatophora Example : *Achatina, Helix*
      - Basommatophora Example : *Lymnaea, Planorbis*

Class—**Bivalvia**

- *Subclasses*
  - Pteriomorphia
    - *Orders*
      - Mytiloida Example : *Mytilus*
      - Pterioida Example : *Pinctada*
  - Palaeoheterodonta
    - *Orders*—Unionoida Example : *Lamellidens*
  - Heterodonta
    - *Orders*
      - Veneroida Example : *Ensis*
      - Myoida Example : *Teredo*

1. *CHITON*

Habit and Habitat

Marine; sluggish, slow moving; found in the intertidal zone attached to rocks, empty shells, corals and under stones; nocturnal; feeds on algae and diatoms; distribution cosmopolitan.

Systematic Position

Phylum	**Mollusca**
Class	**Polyplacophora (G: many plate bearing)**
Order	**Ischnochitonida *(Neoloricata)****
Type	***Chiton***

Distinctive Features

1. Body elliptical dorsoventrally flattened.
2. Shell valves laterally covered by girdle partly.
3. Eyes and tentacles are absent.
4. Several pairs of bipectinate **ctenidia** are on either side of mantle groove.

---

* *Mentioned by Ruppert, E.E., Barnes D.B. (1994) Classification is based on Hyman, L.H. 1967.*

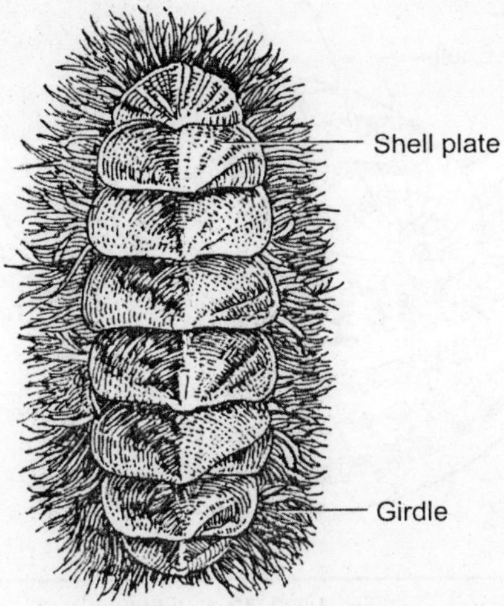

Fig. 1.1.1.70 : *Chiton* (Dorsal view)

Identification with Reasons

Phylum—*Mollusca* :

1. Soft–bodied, unsegmented with distinct head at unjointed appendages.
2. Distinct ventral and muscular foot present.
3. Body is commonly protected by an exoskeletal calcareous shell of one or more pieces

Class—*Polyplacophora* :

1. Body bilaterally symmetrical and dovsoventrally flattened, with a broad creeping ventral foot bordered by a groove.
2. External shell made of longitudinal series of eight transverse shell valves.
3. Ctenidia in numerous pairs.

Order—*Ischnochitonida* :

1. Girdle well developed but not covering plates.
2. Shell valves imbricate and teethed.
3. Shell plates possess a middle calcarious layer.

Type—*Chiton* : State distinctive features as in above.

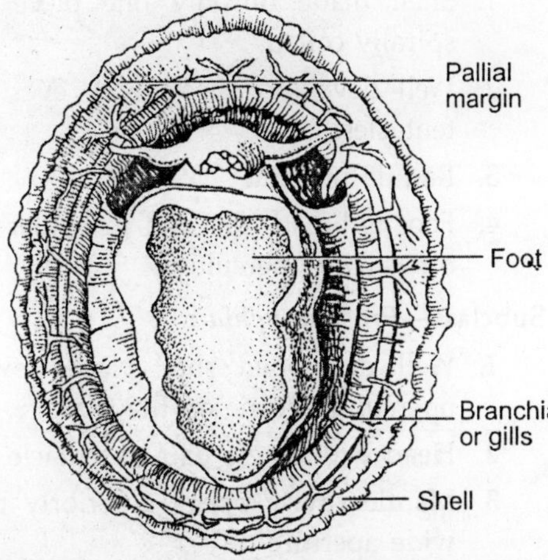

Fig. 1.1.1.71 : *Patella* (Ventral view)

2. *PATELLA* (True limpet)

Habit and Habitat

Marine; found attached to rocks and other hard substrate on intertidal zone; nocturnal; feeds on algae and other marine vegetation; widely distributed except North and South America.

Systematic Position

Phylum	**Mollusca**
Class	**Gastropoda (G: stomach foot)**
Subclass	**Prosobranchia**
Order .	**Archaeogastropoda**
Type	***Patella***

Distinctive Features

1. Shell rounded and oval; operculum absent.
2. Foot large with a broad creeping sole.
3. Ctenidia are replaced by leaf-like secondary gills.

Identification with Reasons

State classification and characters up to—

Phylum—*Mollusca* as in *chiton*.

Class—*Gastropoda* :

1. Shell made of only one piece and spirally coiled.
2. Well developed head with eyes and tentacles.
3. Radula present.
4. Foot is broad, flat and muscular and often with operculum.

Subclass—*Prosobranchia* :

1. Well developed shell closed by an operculum borne on the foot.
2. Head with single pair of tentacle.
3. Mantle cavity opens anteriorly by a wide aperture.
4. Two ctenidia are anterior to heart.

Order—*Archaeogastropoda* :

1. Ctenidia replaced by two rows of bipectinate gills in mantle cavity.
2. Heart with pair of auricles.
3. Siphon, proboscis absent.

Type—*Patella* : State distinctive features as in above.

3. *HALIOTIS* (Ear-shell)

Habit and Habitat

Marine; living on solid substrate, attached very tightly below tidal level in tropical and temperate seas; herbivorous; feeds on algae and seaweeds.

Systematic Position

Phylum	**Mollusca**
Class	**Gastropoda**
Subclass	**Prosobranchia**
Order	**Archaeogastropoda**
Type	***Haliotis***

Distinctive Features

1. Shell spirally twisted and ear – shaped and perforated by rows of small holes near the left margin.

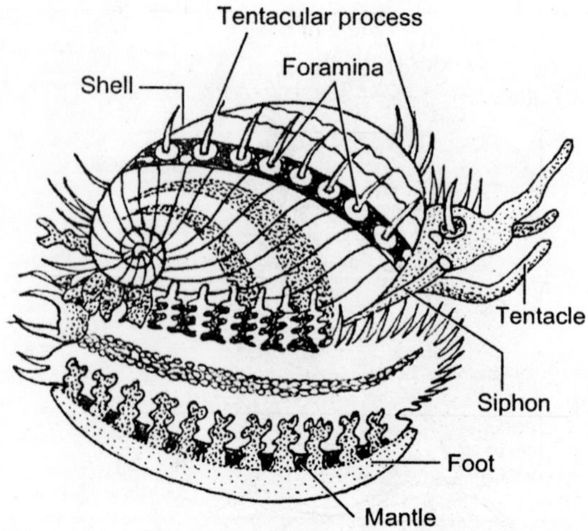

Fig. 1.1.1.72 : *Haliotis*

2. Large foot has a fold of skin called **epipodium**, bearing sensory epipodial tentacles.

3. Head with two pairs of tentacle and a pair of **ommatophores** with eyes at their tips.

Identification with Reasons

State the characters up to—

Order—*Archaeogastropoda* as in *Patella*.

Type—*Haliotis* : State distinctive features as in above.

Economic Importance

Haliotis tissue is eaten by humans. It is collected for food and is named **Abalone**. Shells are of great economic importance, as ornaments, button and buckles.

4. *PILA* (Apple snail)

Habit and Habitat

Fresh water; abundant in ponds, pools, tanks, lakes, marshes, paddy fields and in water having succulent vegetation; herbivorous; feeds on aquatic plant like *Vallisneria*, *Pistia* etc; remain dormant during prolonged drought; amphibious; common in India, U.S.A. and Europe.

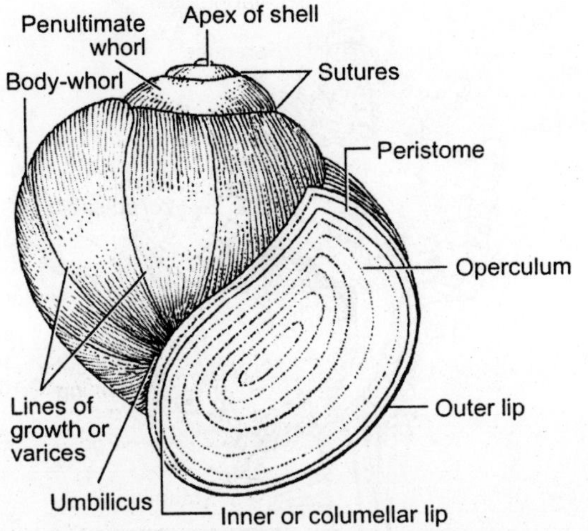

Fig. 1.1.1.73 : *Pila* (Entire animal)

Systematic Position

Phylum	**Mollusca**
Class	**Gastropoda**
Subclass	**Prosobranchia**
Order	**Mesogastropoda**
Type	***Pila***

Distinctive Features

1. Body is covered by thick globular univalve shell comprising of whorl and apex.

2. The shell is spirally coiled round the axis and the surface of the shell is marked by lines of growth.

3. Wide shell aperture covered by operculum.

Identification with Reasons

State the characters up to—

Subclass—*Prosobranchia* in *Haliotis.*

Order—*Mesogastropoda* :

1. Single, **monopectinate** gill, one auricle and one nephridium present.
2. With or without operculum.
3. One sided gill attached for its whole lengths into wall of mantle cavity.

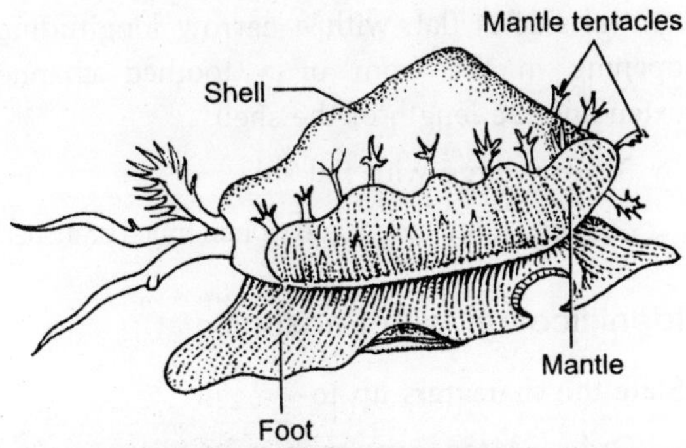

Fig. 1.1.1.74 : *Cypraea* (Cowry)

4. Siphon and proboscis present.
5. Radula usually has seven teeth in each row (tacnioglossate).

Type—*Pila* : State distinctive features as in above.

Economic Importance

Pila has great medicinal, experimental and study value.

5. *CYPRAEA* (Cowry)

Habit and Habitat

Marine; lives on rocky ground in shallow, tropical seas; carnivore; found in Indian and pacific oceans.

Systematic Position

Phylum	**Mollusca**
Class	**Gastropoda**
Subclass	**Prosobranchia**
Order	**Mesogastropoda**
Type	***Cypraea***

Distinctive Features

1. Large, oval shell with single whorl.

2. The dorsal side of the highly polished and often highly coloured shell is convex and the

ventral side is flat, with a narrow longitudinal opening in the from of a toothed channel extending the length of the shell.

3. Foot large with flat sole.
4. Operculum absent and siphon much branched.

Identification with Reasons

State the characters up to—

Order—*Mesogastropoda* as in *Pila*.

Type—*Cypraea* : State distinctive features as in above.

Economic Importance

Cowry shells are used as ornaments, curios and mantle-piece decoration. Shells used for playing an Indian game.

6. *MUREX*

Habit and Habitat

Marine; carnivore; predators; feeds on living and dead animals; found in shallow waters of tropical and temperate seas.

Systematic Position

Phylum	**Mollusca**
Class	**Gastropoda**
Subclass	**Prosobranchia**
Order	**Neogastropoda**
Type	*Murex*

Distinctive Features

1. Spirally, coiled shell is dextral (coiling right handed) and on the shell lays three or more rows of spines.
2. Aperture round, ending below in a long canal.
3. Mantle is drawn out into a cylindrical extension called **siphon**.
4. Large foot with flat creeping sole.

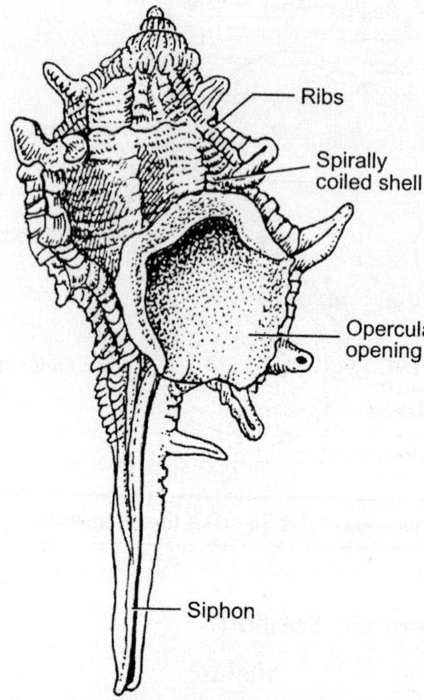

Fig. 1.1.1.75 : *Murex*

Identification with Reasons

State the characters up to—

Subclass—*Prosobranchia* as in *Pila*.

Order—*Neogastropoda* :

1. Single monopectinate ctenidium.
2. Single kidney and heart with one auricle.
3. Well developed **bipectinate osphradium.**
4. Radula has no more than three teeth per row (rachiglossate).

Type—*Murex* : State Distinctive features as in above.

Economic Importance

Some species of *Murex* bore holes through shells of barnacles, gastropods and bivalves. They are destructive to oyster industry. Phoenicians and Romans dyed their ceremonial robs purple obtained from many species of *Murex*. *Murex* has

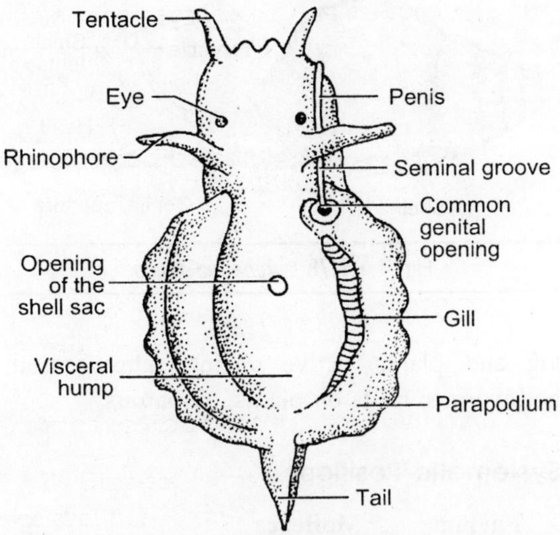

Fig. 1.1.1.76 : *Aplysia*

also ornamental importance and is also used for producing sound during worshipping.

7. *APLYSIA* (Sea-hare)

Habit and Habitat

Marine; free swimming; found crawling among seaweeds; herbivorous; feeds mainly on the sea weeds; found in India, Asia, West Indies.

Systematic Position

Phylum	**Mollusca**
Class	**Gastropoda**
Subclass	**Opisthobranchia**
Order	**Anaspidea (= Aplysiacea)**
Type	***Aplysia***

Distinctive Features

1. Soft, shiny, fleshly body animal with district head, eyes and tentacles.

2. **Rhinophores** resemble rabbit ears.

3. Broad foot possesses a pair of lateral folds called Parapodia.

4. Visceral mass is raised into a prominent hump.

5. Shell is internal and a central shell aperture is on the trunk.

6. Posterior small tail is present.

Identification with Reasons

State the characters up to—

Class—*Gastropoda* as those of *Patella*.

Subclass—*Opisthobranchia* :

1. Shell often reduced or wanting.
2. Nervous system untwisted by **detorsion**.
3. Respiratory structures are secondary gills.
4. Heart with single auricle which is posterior to the ventricle.
5. Visceral hump is partially or completely uncoiled.
6. Hermaphrodite.

Order—*Anaspidea* :

1. A thin internal shell covered by mantle.
2. Rhinophores resemble rabbit ears.
3. Well developed parapodia.
4. Visceral hump attached along foot.

Type—*Aplysia* : State Distinctive features as in above.

8. *DORIS* (Sea lemon)

Habit and Habitat

Marine; found under stones at intertidal zone and between weeds; carnivorous; feeds most commonly on sponges; distribution cosmopolitan.

Systematic Position

Phylum	**Mollusca**
Class	**Gastropoda**
Subclass	**Opisthobranchia**
Order	**Nudibranchia (Acoela)**
Type	***Doris***

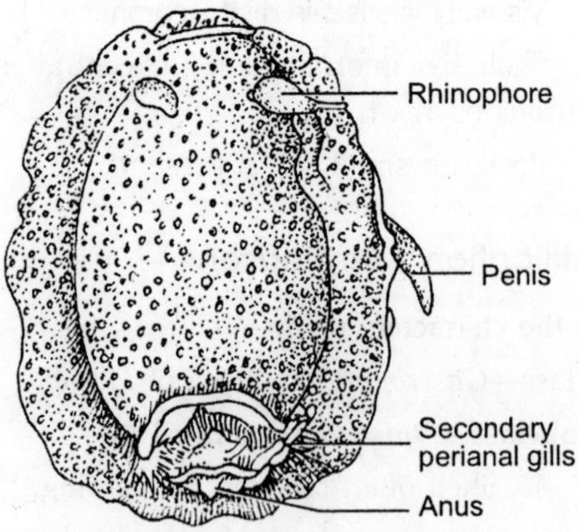

Fig. 1.1.1.77 : *Doris*

Distinctive Features

1. Shell-less body more or less ovoid mass with a convex warty dorsal side.
2. Head with retractile, short tentacles or rhinophores.
3. Anus is surrounded by a circlet of secondary branchiae.
4. Foot has large creeping sole.

Identification with Reasons

State the characters up to—

Subclass—*Opisthobranchia* as in *Aplysia.*

Order—*Nudibranchia* :

1. Adult lacks a shell, ctenidia and mantle cavity.
2. Secondary gills arranged in a circular tuft around the anus.
3. Visceral mass attached along foot.
4. Parapodia absent.

Type—*Doris* : State Distinctive features as in above.

9. *ACHATINA FULICA* (Garden-snail)

Habit and Habitat

Terrestrial; commonly found in garden; nocturnal; spend daytime under stones, wooden log and plants; active during rainy season; herbivorous; feeds on plants; oviparous.

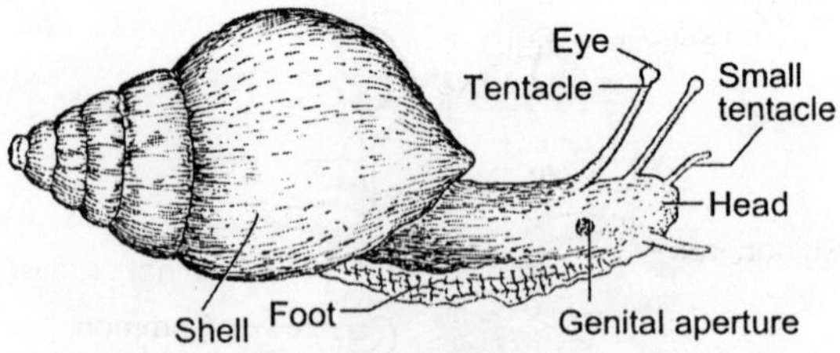

Fig. 1.1.1.78 : *Achatina fulica*

Systematic Position

Phylum	**Mollusca**
Class	**Gastropoda**
Subclass	**Pulmonata**
Order	**Stylommatophora**
Type	***Achatina fulica***

Distinctive Features

1. Body soft and extremely slippery.
2. Ovate–conical, elongated, thick shell covering the body is pale yellowish with dark stripes.
3. Spire well marked, conical and body whorl extensive and capacious.
4. Small head with two pairs of tentacles, posterior pair large with eyes and anterior pair small.
5. Genital aperture is on the right side, posterior to head.
6. Flat foot with large ventral sole.

Identification with Reasons

State the characters up to—

Class—*Gastropoda* as in *Patella.*

Subclass—*Pulmonata* :

1. Shell may or may not be present.
2. Eyes at the tip of the tentacles.

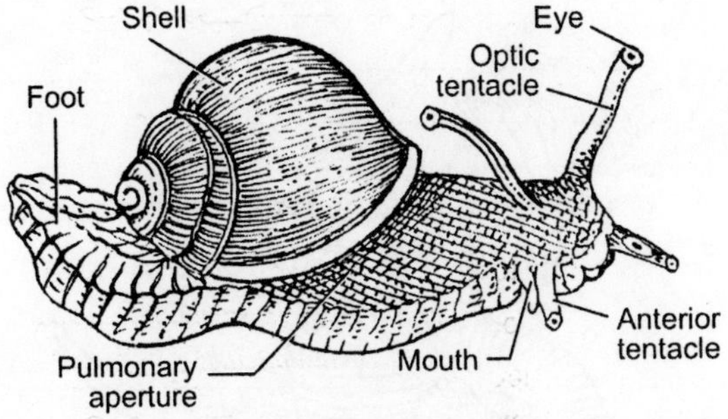

Fig. 1.1.1.79 : *Helix*

3. Ctenidium absent, mantle cavity modified as respiratory sac, used for air or water breathing.
4. Anus directed forward.

Order—*Stylommatophora* :

1. Presence of spiral shell.
2. Two pair of tentacles, with the upper ones each bearing an eye at the tip.
3. Operculum absent.

Type—*Achatina* : State Distinctive features as in above.

Economic Importance

Achatina causes much damage to plants and regarded as agricultural pests. It is also eaten by humans in some parts of the world; the shells are used as utensils and decorations.

10. *HELIX* (Land-snail)

Habit and Habitat

Terrestrial, found in moist and shady places like lawns and gardens; nocturnal; hibernates in the soil during winter; feeds on leafy vegetation.

Systematic Position

Phylum	**Mollusca**
Class	**Gastropoda**
Subclass	**Pulmonata**
Order	**Stylommatophora**
Type	***Helix***

Distinctive Features

1. Shell is thin, external, heliciform with a low conical spire.
2. Shell has smooth aperture and prominent lines of growth.
3. Head with two pairs of tentacles, the posterior pair bear eyes.
4. Radula is well developed with several rows of curved teeth.

Identification with Reasons

State the characters up to—

Order—*Stylommatophora* as in *Achatina*.

Type—*Helix* : State Distinctive features as in above.

11. *LYMNAEA* (Pond snail)

Habit and Habitat

Fresh water; found in swamps, ponds, lakes and streams; distribution cosmopolitan, especially, in temperate regions.

Systematic Position

Phylum	**Mollusca**
Class	**Gastropoda**
Subclass	**Pulmonata**
Order	**Basommatophora**
Type	***Lymnaea***

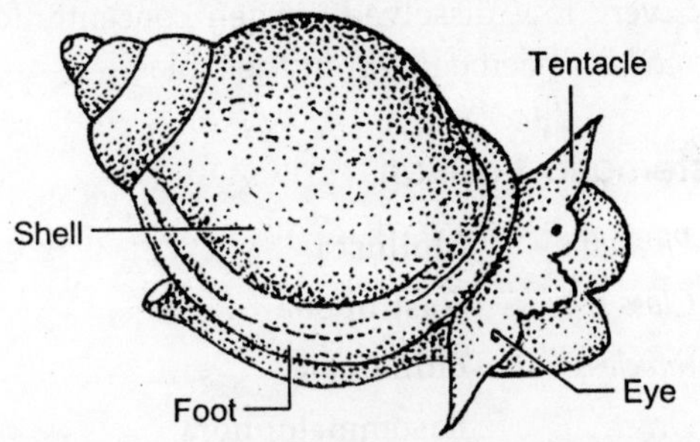

Fig. 1.1.1.80 : *Lymnaea*

Distinctive Features

1. Spiral external shell which is thin, horny and fragile.
2. Body – whorl very large, having an aperture.
3. Foot rounded, tentacles flattened.
4. **Pulmonary sac** covered with an external lobe.

Identification with Reasons

State the characters up to—

Subclass—*Pulmonata* as in *Achatina*.

Order—*Basommatophora* :

1. Snail with spiral external shell.
2. One pair of tentacles.
3. Eyes usually located at the base of the tentacles.

Type—*Lymnaea* : State Distinctive features as in above.

Economic Importance

Lymnaea serves as intermediate hosts of variety of trematodes and cestodes. It is eaten by humans in different parts of the world.

12. *PLANORBIS*

Habit and Habitat

Fresh water; air-breathing; live in ponds and lakes in large numbers; commonly found in temperate regions; they can live in polluted areas with very low dissolved oxygen concentrations due to rich haemoglobin in their blood.

Systematic Position

Phylum	**Mollusca**
Class	**Gastropoda**
Subclass	**Pulmonata**
Order	**Basommatophora**
Type	***Planorbis***

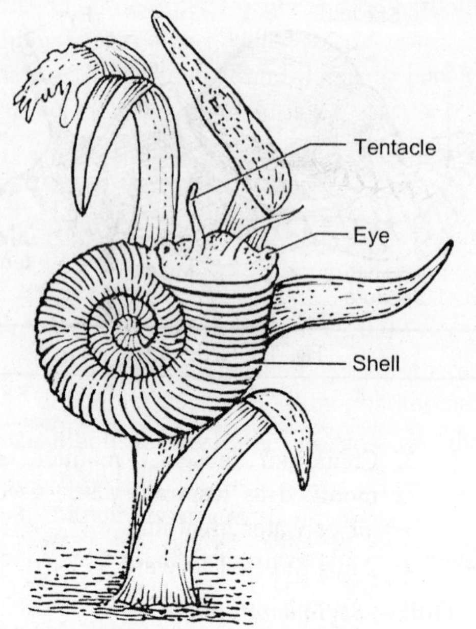

Fig. 1.1.1.81 : *Planorbis*

Distinctive Features

1. Black shell is discoidal and sinistrally spiral.
2. Head with filiform tentacles and a pair of eyes.
3. Pulmonary sac is enclosed by a lobe of the mantle, which serves as a gill.

Identification with Reasons

State the characters up to—

Order—*Basommatophora* as in *Lymnaea*,

Type—*Planorbis* : State Distinctive features as in above.

Economic Importance

Planorbis acts as intermediate hosts in the transmission of **Schistosomiasis**, a disease caused by trematode flatworms.

13. *MYTILUS* (Sea-mussel)

Habit and Habitat

Marine; sedentary; lives attached by byssal threads to solid substrates in marine and

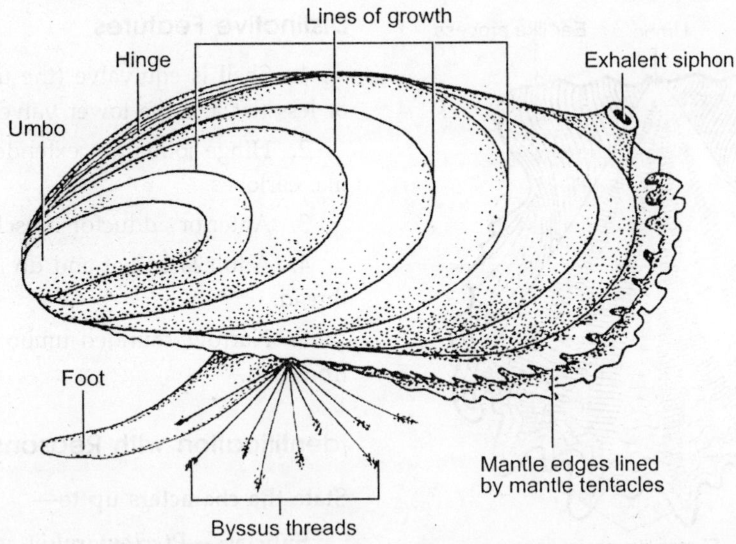

Fig. 1.1.1.82 : *Mytilus*

estuarine habitats; feeds on protozoans, diatoms and detritus; distribution cosmopolitan.

Systematic Position

Phylum	**Mollusca**
Class	**Bivalvia (Pelecypoda)** (L: two valved. G; hatchet foot)
Subclass	**Pteriomorphia**
**Family*	**Mytilidae**
Type	***Mytilus***

Distinctive Features

1. Body is enclosed by shell of two equal valves pointed in front and rounded behind.
2. Adductor muscles two and unequal (**heteromyrian**).
3. The shell is marked with lines of growth.
4. **Umbo** near the anterior end.
5. Cylindrical foot with **byssus threads**.

Identification with Reasons

State the characters up to—

Phylum—Mollusca as in *Chiton.*

Class—*Bivalvia* (Pelecypoda) :

1. Body flattened laterally.
2. Body is enclosed by two-valved shell, hinged together middorsally with distinct lines of growth.
3. Highly muscular tongue-shaped foot.
4. Head rudimentary, without eyes and tentacles.

Subclass—*Pteriomorphia* :

1. Gills usually filibranch and adjacent gill filaments connected by tuffs of interlocking cilia.
2. Adducter muscles unequal (**hetero-myarian**) or equal (**monomyarian**).
3. Foot usually with byssal gland.
4. Umbo at or near the anterior end.

Family—*Mytilidae* :

1. Attached by bysal threads.
2. Umbo near the anterior end.
3. Foot cylindrical without a groove.

Type—*Mytilus* : State Distinctive features as in above.

* *Mentioned as order – Mytiloida by Ruppert, E.E., Barnes R.D., (1994).*

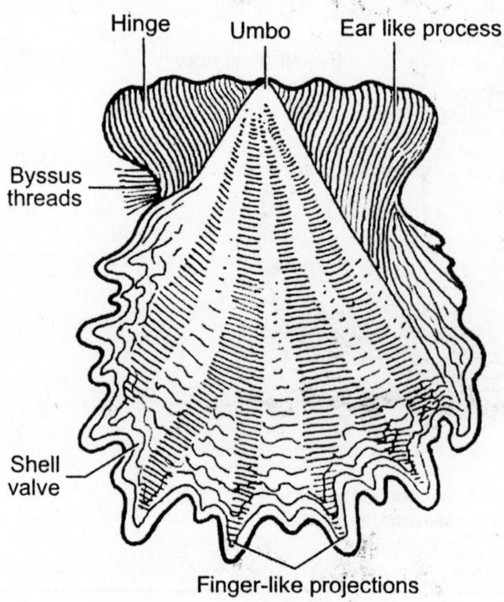

Fig. 1.1.1.83 : *Pinctada*

Economic Importance

Mytilus edulis has become a major pollution **bioindicator species.** It is also taken as food in Europe.

14. *PINCTADA* (Indian Pearl oyster)

Habit and Habitat

Marine; sedentary; attached to rocks, stone, dead corals at about a depth of 18 - 22 meter and at a distances of 19 km. from the shore; filter feeder; feeding on protozoans, diatoms etc; distribution cosmopolitan but common in tropical and sub tropical waters.

Systematic Position

Phylum	**Mollusca**
Class	**Bivalvia (Pelecypoda)**
Subclass	**Pteriomorphia**
**Family*	**Ostreidae**
Type	***Pinctada***

* *Mention as order – **Ostreoida** by Ruppert, E.E., Barnes R.D., (1994).*

Distinctive Features

1. Shell is equivalve (the upper valve more or less flat and the lower valve convex).
2. Hinge joints are extended on either side like earlobes.
3. Anterior adductor muscle absent.
4. Adult lack foot and do not secrete byssal thread.
5. Narrow, rounded umbo and hinge at the front end.

Identification with Reasons

State the characters up to—

Subclass—*Pteriomorphia* as in *Mytilus.*

Family—*Ostreidae* :

1. Adults lack foot and do not secrete byssal thread.
2. Anterior adductor muscle lacking.
3. Shell lacks a **nacreous layer**.
4. Adult firmly cemented to the substrate by the left valve.

Type—*Pinctada* : State Distinctive features as in above.

Economic Importance

Oyster has considerable commercial importance. It's shell secretes pearls. When some foreign particles accidentally enter in the body of oyster concentric **layers of nacre** is secreted and deposited around the foreign particle to give rise to **pearl**. They are cultivated in pearl oyster industry.

15. *LAMELLIDENS MARGINALIS* (Fresh water mussel)

Habit and Habitat

Fresh water; found in muddy bottom of ponds, lakes and streams. Filter feeder; feeds on microscopic organism; commonly found in India, Europe and U.S.A.

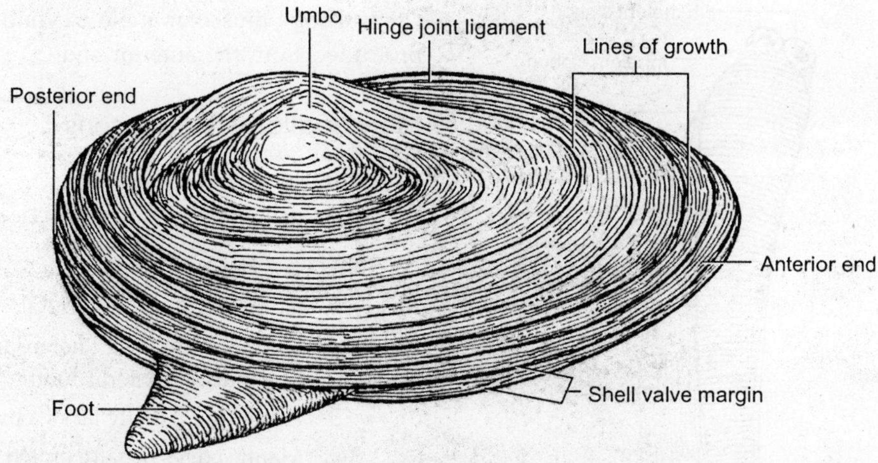

Fig. 1.1.1.84 : *Lamellidens marginalis*

Systematic Position

Phylum	**Mollusca**
Class	**Bivalvia (Pelecypoda)**
Subclass	**Paleoheterodonta**
**Family*	**Unionidae**
Type	***Lamellidens marginalis***

Distinctive Features

1. Unsegmented body, flattened from side to side.

2. Completely enclosed in equal, dark brown, bivalve shells.

3. Two valves are united together along the dorsal side by a straight hinge- line ligament.

4. Aboral side mantle lobes produced inhalent and exhalent siphon side by side.

5. Foot is large, muscular and tongue-shaped.

6. Adductor muscles are of same size.

Identification with Reasons

State the characters up to—

* *Mentioned as order – **Unionoida** by Ruppert, E.E., Barnes R.D., (1994).*

Class—*Bivalvia* as those of *Mytilus*.

Subclass—*Paleoheterodonta* :

1. Shell in equilateral (umbo is not at the centre) and equivalve.
2. Foot without byssus gland.
3. Mantle margin fused posteriorly.

Family—*Unionidae* :

1. Well developed **Periostracum** (inner-margin of mantle).
2. Adductor muscles two, similar (**isomyarian**) and strong.
3. Hinge dentition schizont type.
4. Foot long and compressed with out a groove.
5. Umbo near the anterior end.

Type—*Lamellidens marginalis* : State distinctive features as in above.

16. *ENSIS* (Razor clam)

Habit and Habitat

Marine; estuarine, burrowing animal; live in the sand about low tide mark; suspension feeder; found in India and U.S.A.

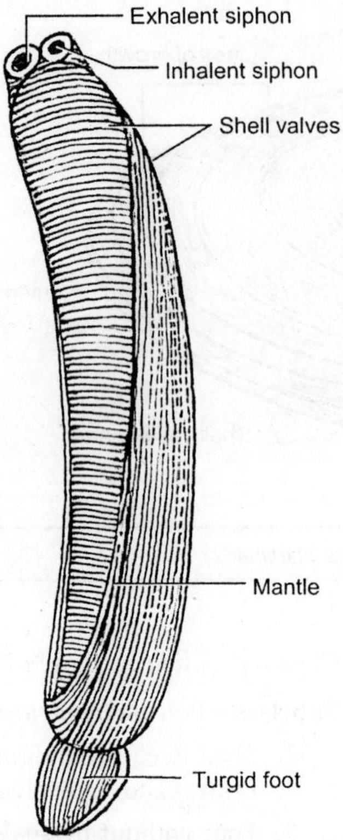

Fig. 1.1.1.85 : *Ensis*

Systematic Position

Phylum	**Mollusca**
Class	**Bivalvia (Pelecypoda)**
Subclass	**Heterodonta**
**Family*	**Cultellidae**
Type	***Ensis* (Rajor clams)**

Distinctive Features

1. Shells long, narrow, thin and equivalve, which are slightly bent in the middle.

2. Shells contain parallel margins and lines of growth and have gaping ends.

3. Umbo at the anterior end.

4. Inhalent and exhalent siphons are short and united.

5. Foot muscular and cylindrical and protruded from the anterior end.

Identification with Reasons

State the characters up to—

Class—*Bivalvia* as in *Mytilus.*

Subclass—***Heterodonta*** **:**

1. Adductor muscles similar.
2. Hinge dentition heterodont (with cardinal and lateral tooth).
3. Adult foot usually lacks a byssal gland.
4. Mantle edges usually united posteriorly, leading to development of siphon.

Family—*Cultellidae* :

1. Shell valve has gape at the anterior end.
2. Umbo at the anterior end.
3. Adductor muscles two.

Type—*Ensis* : State distinctive features as in above.

Note : Solen the other rajor shells having the same features as *Ensis*, except shell curvature.

17. *TEREDO* (Ship worms)

Habit and Habitat

Marine or estuarine; mostly bore into wood, immersed in sea-water; lives in long burrows in wood or clay; eats cellulose and living micro organisms; distribution cosmopolitan.

Systematic Position

Phylum	**Mollusca**
Class	**Bivalvia (Pelecypoda)**
Subclass	**Heterodonta**
**Family*	**Teredinidae**
Type	***Teredo***

Distinctive Features

1. Body elongated and vermiform.

---

* *Mentioned as order –* ***Veneroida*** *by Ruppert, E.E., Barnes R.D., (1994).*

* *Mention as order –* ***Myoida*** *by Ruppert, E.E., Barnes R.D., (1994).*

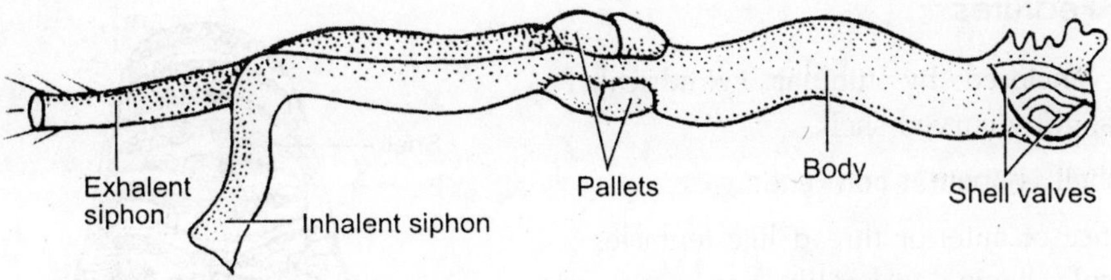

Fig. 1.1.1.86 : *Teredo*

2. Shell valves very small, animal protrudes posteriorsly from the shell.
3. The exhalent and inhalent siphons are large capable of contraction and expansion.
4. On the junction of the body and siphon are pallets which act as operculum.
5. Foot is very small and serves as adhesive organ.

Identification with Reasons

State the characters up to—

Subclass—*Heterodonta* as those of *Ensis*.

Family—*Teredinidae* :

1. Adductor muscles three, anterior, posterior and ventral.
2. Shell small and animal protrudes porteriorly from the shell.
3. Mostly bore into wood.

Type—*Teredo* : State distinctive features as in above.

Economic Importance

Teredo causes considerable economic damage to wooden ships and pilings.

18. *DENTALIUM* (Tusk-shell)

Habit and Habitat

Marine; sedentary; lives in sand or mud substrate mostly in deep waters; detritus feeders and micro-carnivores; found in the entire major ocean.

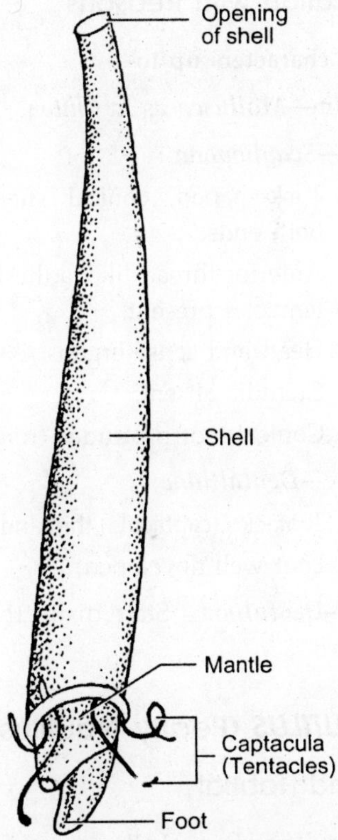

Fig. 1.1.1.87 : *Dentalium*

Systematic Position

Phylum	**Mollusca**
Class	**Scaphopoda (G: Spade foot)**
Family	**Dentallidae**
Type	***Dentalium***

Distinctive Features

1. Body enclosed by tubular cylindrical, arched and slightly curved shell.

2. The shell is open at both ends.

3. Presence of anterior thread like tentacles called **captacula**, having sucker-like tips.

4. Conical foot protrudes from shell from anterior end.

Identification with Reasons

State the characters up to—

Phylum—*Mollusca* as in *Chiton.*

Class—*Scaphopoda* :

1. Tusk-shaped, conical shell open at both ends.
2. Anterior thread like, adhesive feeding tentacles present.
3. Head and sense organs reduced.
4. Ctenidia absent.
5. Conical foot protrudes from shell.

Family—*Dentallidae* :

1. Tentacles (captacula) thin and numerous.
2. Foot well developed.

Type—*Dentalium* : State distinctive features as in above.

19. *NAUTILUS* (Pearly Nautilus)

Habit and Habitat

Marine, found in shallow waters near the shore; bottom dweller; nocturnal; feeds on small crustaceans and dead animal matters; occurs in Indian and Pacific oceans.

Systematic Position

Phylum	**Mollusca**
Class	**Cephalopoda (G: head foot)**
Subclass	**Nautiloidea**
Type	***Nautilus***

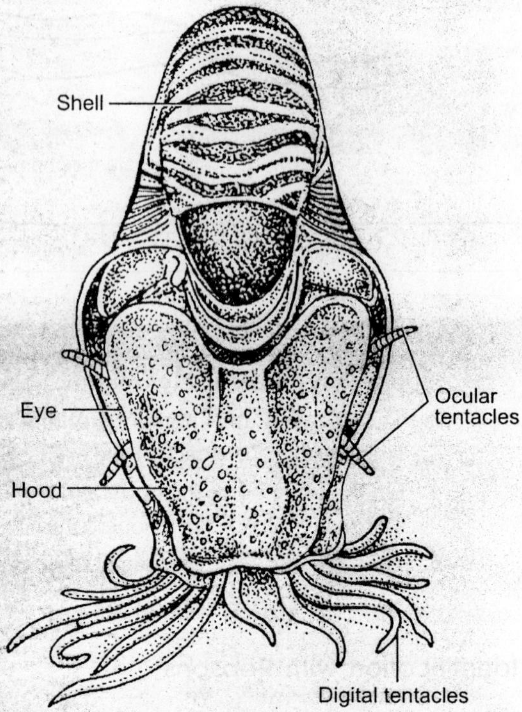

Fig. 1.1.1.88 : *Nautilus*

Distinctive Features

1. Body is distinguishable into a large conical head with mouth, eyes and tentacles.

2. Body enclosed by external, calcareous shell, which is coiled and divided internally by transverse septa.

3. Animal possesses eighty to ninety tentacles without suckers.

4. Shell colour usually dark brown or black with transverse bands of white and yellowish brown alternately.

Identification with Reasons

State the characters up to—

Phylum—*Mollusca* as in *Chiton.*

Class—*Cephalopoda* :

1. Head is well developed and possesses a pair of highly developed eyes and a number of arms or tentacles.
2. Foot modified to from arm-like tentacles generally with suckers.
3. Shell spiral-chambered, some times internal or absent.

Subclass—*Nautiloidea* :

1. True external shell secreted by the mantle.
2. Shell spiral and chambered.
3. Foot modified to from flexible arms and siphon.
4. Eyes are simple and without lens.
5. **Ink gland** absent.
6. Tentacles sucker less.

Type—*Nautilus* : State distinctive features as in above.

Economic Importance

Body used as food and shell used for ornamental and useful purposes.

20. *SEPIA* (Cuttle fish)

Habit and Habitat

Marine; good swimmer; found in shallow waters; carnivorous; feeds on crustaceans, small fishes and other animals; cosmopolitan in distribution.

Systematic Position

Phylum	**Mollusca**
Class	**Cephalopoda**
Subclass	**Coleoida (Dibranchiata)**
Order	**Sepioidea**
Type	***Sepia***

Distinctive Features

1. Body is differentiated into head, neck and visceral hump or trunk.

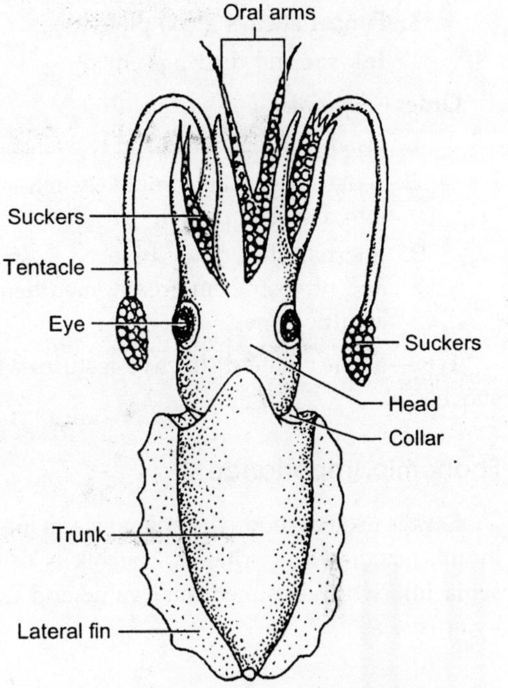

Fig. 1.1.1.89 : *Sepia*

2. Internal calcareous shell, called **cuttle bone** is present.

3. Presence of prominent eyes.

4. Trunk is shield – shaped and bordered by narrow lateral fins on either side.

5. Head with eight arms with suckers and two large tentacles with suckers only at the tips.

6. In male, the left fourth arm is **hectocotylised**.

Identification with Reasons

State the characters up to—

Class—*Cephalopoda* as in *Nautilus*.

Subclass—*Coleoida* :

1. Shell, if present is always internal and is completely surrounded by mantle tissue.
2. Head with eight non retractile arms and in addition may bear pairs of tentacles with suckers.

3. Funnel always a complete tube.
4. **Ink sac** and duct present.

Order—*Sepioidea* :

1. Head with eight arms and two tentacles.
2. Tentacles can be completely retracted into a special pocket.
3. Internal shell with reduced rostrum and pro-ostracum greatly modified as a cuttle bone.

Type—*Sepia* : State distinctive features as in above.

Economic Importance

Sepia is used as food. The animal ejects ink by its ink gland, when irritated. The ink is called **sepia ink**, which has medicinal value and used by artists.

21. *LOLIGO* (Squid)

Habit and Habitat

Marine; fast swimming; found in shallow or deep waters; carnivorous; feeds on crustaceans, small fishes and other squids; world wide in distribution.

Systematic Position

Phylum	**Mollusca**
Class	**Cephalopoda**
Subclass	**Coleoidea (Dibranchiata)**
Order	**Teuthoidea (= Decapoda)**
Type	***Loligo***

Distinctive Features

1. Body elongated, flattened and spindle shaped, with pointed posterior end.

2. Body is differentiated into head, collar or neck and visceral hump or trunk.

3. Fin is triangular with pointed ends covers the posterior end of the trunk.

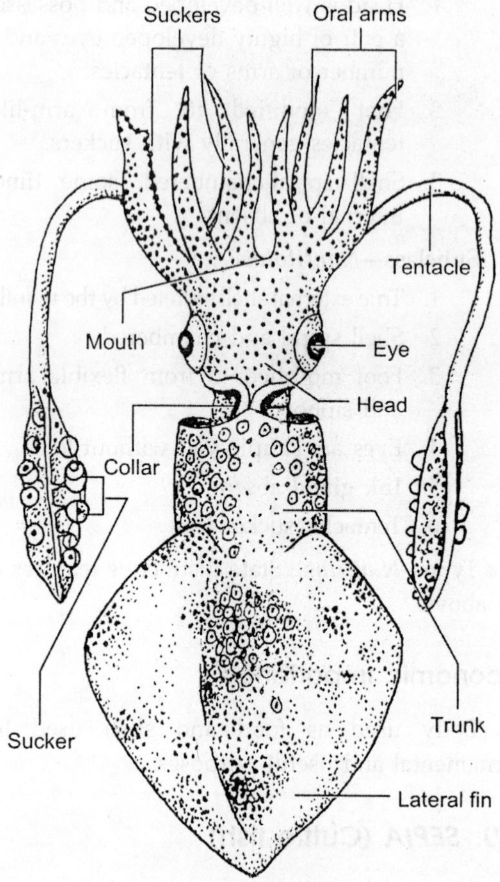

Fig. 1.1.1.90 : *Loligo*

4. Foot is represented by arms and very long tentacles.

5. Two eyes in the head region.

6. In male the left fourth arm is **hectocotylised.**

Identification with Reasons

State the characters up to—

Subclass—*Coleoidea* as in Sepia.

Order—*Teuthoidea* :

1. Head bears eight arms and two tentacles.
2. Tentacles never withdrawn into special pits.

3. Internal shell lacks a rostrum and pro-ostracum in the from of an elongated pen.

Type—*Loligo* : State Distinctive features as in above.

Economic Importance

Loligo has food value. Certain species are being caught for food each year.

22. *OCTOPUS*

Habit and Habitat

Marine; bottom dwellers, non swimmers and individuals tend to live by themselves in small, protective caves; nocturnal; carnivorous; feeds on crustaceans, cephalopods and fishes; distribution cosmopolitan.

Systematic Position

Phylum	**Mollusca**
Class	**Cephalopoda**
Subclass	**Coleoidea**
Order	**Octopoda**
Type	***Octopus***

Distinctive Features

1. Body short and rounded and is differentiated into a head and visceral hump.

2. A pair of eyes at the base of the head.

3. Tentacles larger than the body and with **pedunculate suckers.**

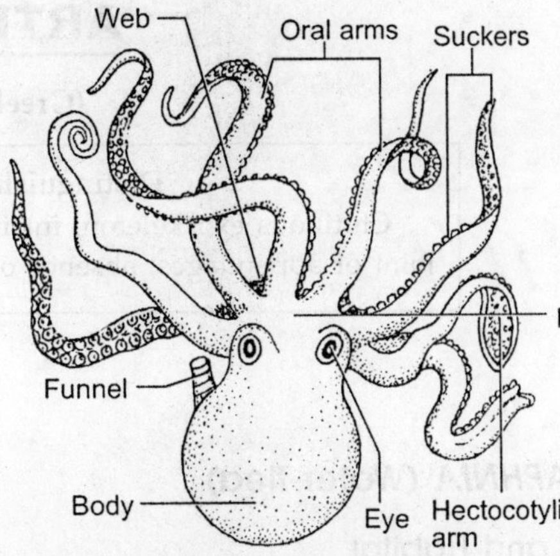

Fig. 1.1.1.91 : *Octopus*

4. Shell lacking.
5. In male right third arm is **hectocotylised.**

Identification with Reasons

State the characters up to—

Subclass—*Coleoidea* as in *Sepia.*

Order—*Octopoda* :

1. Presence of eight arms and no tentacles.
2. Internal shell lacking.
3. Lateral fin absent.

Type—*Octopus* : State distinctive features as in above.

Special Features

Octopus is largely used for studying behavioural sciences. Octopuses have complex behaviour; they can learn and remember.

ARTHROPODA

(Greek: jointed foot)

Distinguishing Characteristics :
Chitinous exoskeleton; intrinsic musculature between individual joint of appendages; absence of motile cilia in adult and larval stages.

1. *DAPHNIA* (Water-flea)

Habit and Habitat

Freshwater animal, abundant in ponds and ditches; feeds on plankton, cosmopolitan.

Systematic Position

Phylum	**Arthropoda**
Subphylum	**Mandibulata**
Class	**Crustacea (L: a crust)**
Subclass	**Branchiopoda**
Order	**Cladocera (Diplostraca)**
Type	**Daphnia**

Distinctive Features

1. Bilaterally compressed body, enclosed in a transparent bivalved carapace, ending into sharp caudal spine and anteriorly into pointed rostrum.

2. Rounded head bears large **biramous antennae.**

3. Large sessile single, huge eye is very distinct.

4. Five pairs leaf like thoracic appendages and abdomen is devoid of appendages.

5. In female, there is a brood pouch near the back.

Identification with Reasons

Phylum—*Arthropoda* :

Classification is based on Marshall A., Williams, W.D. (eds.), 1972.

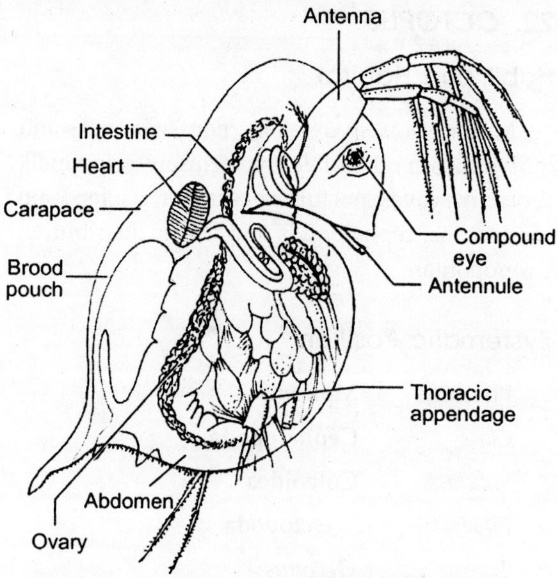

Fig. 1.1.1.92 : *Daphnia* (female)

1. Bilaterally symmetrical and metamerically segmented.
2. Majority of the segments bear a pair of jointed appendages.
3. Body is covered with a thick chitinous exo-skeleton.

Subphylum—*Mandibulata* :

1. Body usually divided into head, thorax and abdomen.
2. One or two pairs of antennae present.
3. One pair of **mandibles** presents.

Class—*Crustacea* :

1. Body consists of head, thorax and abdomen with **telson**. Often thoracic segments fuse with the head to from **cephalothorax**.
2. Compound eyes generally present.
3. Appendages are generally biramous.
4. Antennae are two pairs.

Subclass—*Branchiopoda* :

1. Four pairs thoracic appendages are flat, leaf like, having gills on their margin.
2. Carapace is generally present.

Order—*Cladocera* :

1. Bivalved carapace narrow, laterally compressed, encloses trunk and appendages.
2. Unjointed caudal style claw like.
3. Trunk appendages four to six pairs.

Type—*Daphnia* : State distinctive features as in above.

Economic Importance

Serves as aquarium food for fishes.

2. *TRIOPS* (Tadpole shrimp)

Habit and Habitat

Found in fresh or brackish water; feeds on small animals, plants and bottom sediments; world wide distribution.

Systematic Position

Phylum	**Arthropoda**
Subphylum	**Mandibulata**
Class	**Crustacea**
Subclass	**Branchiopoda**
Order	**Notostraca**
Type	***Triops***

Distinctive Features

1. Elongated body with broad, horse-shoe shaped **carapace**.

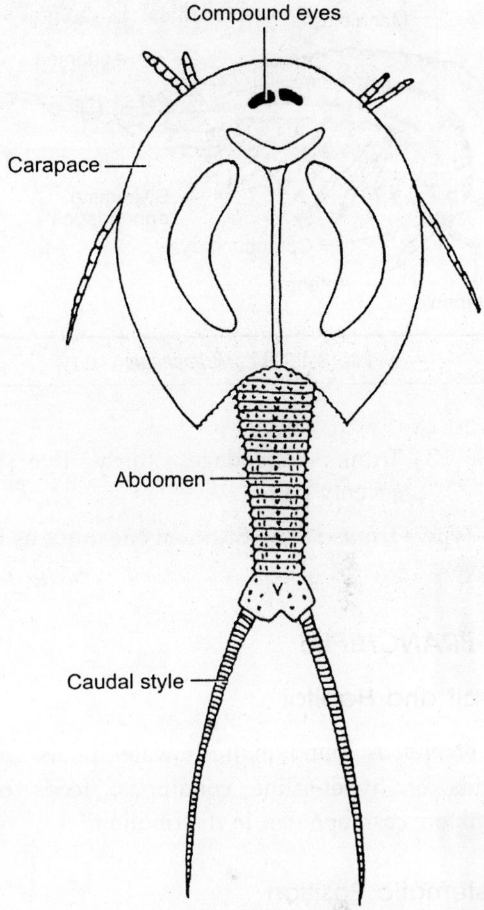

Fig. 1.1.1.93 : *Triops*

2. Dorsal surface of the carapace near its anterior end, are with paired eyes, median eye and dorsal organ.

3. At the caudal region is a forked pair of long caudal style.

Identification with Reasons

State classification and characters up to—

Subclass—*Branchiopoda* as in *Daphnia*.

Order—*Notostraca* :

1. Broad carapace covering the head and anterior part of the trunk.
2. Many jointed caudal style.

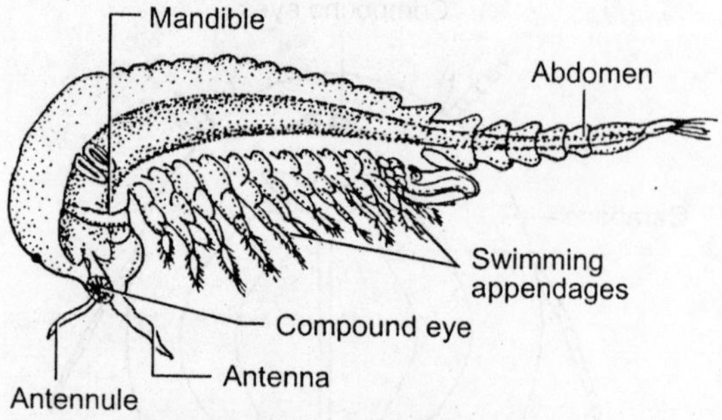

Fig. 1.1.1.94 : *Branchipus*

3. Trunk Appendages thirty five to seventy pairs.

Type—*Triops* : State distinctive features as in above.

3. *BRANCHIPUS*

Habit and Habitat

Branchipus found in fresh water pools and ponds or hyper-saline conditions; feeds on plankton; cosmopolitan in distribution.

Systematic Position

Phylum	**Arthropoda**
Subphylum	**Mandibulata**
Class	**Crustacea**
Subclass	**Branchiopoda**
Order	**Anostraca**
Type	**Branchipus**

Distinctive Features

1. Elongated, transparent body without carapace.
2. Body divisible into head, thorax and abdomen.
3. Head contains downward bent rostrum.
4. Thorax contains several flattened, swimming appendages associated with **branchiae.**
5. Abdomen short and without appendages.
6. Last segments with style.

Identification with Reasons

State characters up to—

Subclass—*Branchiopoda* as in *Daphnia.*

Order—*Anostraca* :

1. Lack carapace.
2. Stalked eyes.
3. Nineteen body segments bear swimming appendages.
4. Unjointed caudal style.

Type—*Branchipus* : State distinctive features as in above.

4. *CYCLOPS* (Water-flea)

Habit and Habitat

Fresh water and brackish water found commonly in ponds and back waters; feed on microscopic organism, mainly diatoms; cosmopolitan in distribution.

Systematic Position

Phylum	**Arthropoda**
Subphylum	**Mandibulata**
Class	**Crustacea**
Subclass	**Copepoda**
Order	**Cyclopoida**
Type	**Cyclops**

Distinctive Features

1. Pear-shaped body with broad anterior and narrow posterior end.
2. The head and first thoracic segment fuse to form cephalothorax.
3. A median **Nauplius eye** is present near anterior end.

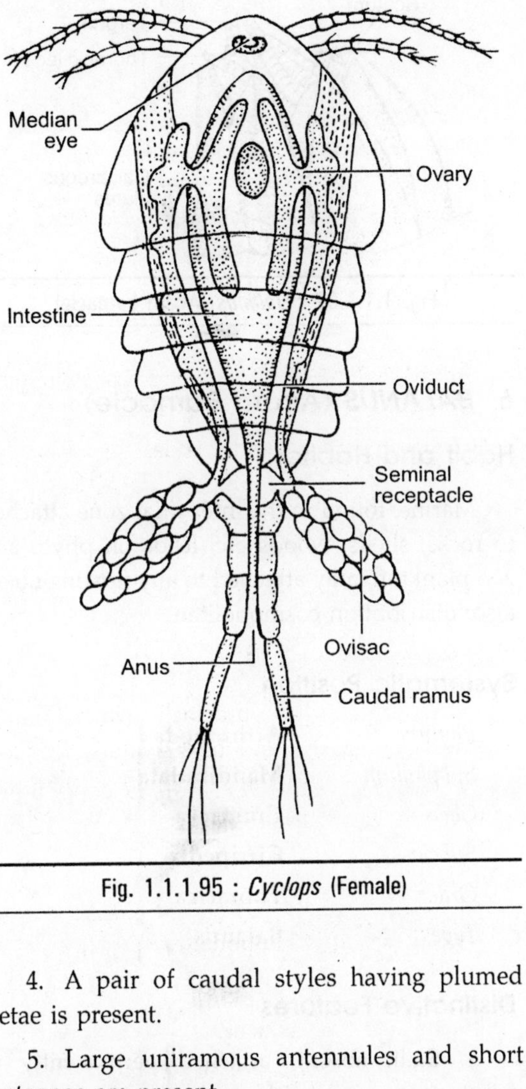

Fig. 1.1.1.95 : *Cyclops* (Female)

4. A pair of caudal styles having plumed setae is present.

5. Large uniramous antennules and short antennae are present.

6. Female carry two lateral egg sacs attached to the abdomen.

Identification with Reasons

State characters up to—

Class—*Crustacea* as in *Daphnia.*

Subclass—*Copepoda* :

1. First thoracic segment is fused with the head and covered by carapace.
2. An anterior, median eye present.
3. Thorax with six-segments; each has a biramous appendages.
4. Abdomen with four segments and without appendage.
5. Caudal style with setae.

Order—*Cyclopoida* :

1. Most species are free- living, but some are commensal, and others are parasitic.
2. Well-developed antennae; the second pair are some time uniramous.
3. The genital aperture is on a special genital segment preceeding abdomen.

Type—*Cyclops* : State the distinctive features as in above.

Economic Importance

Cyclops makes an important link in aquatic food chains because they feed on microscopic organisms and are eaten by macroscopic carnivores.

5. *ARGULUS* (FISH-LICE)

Habit and Habitat

Blood sucking ectoparasite living on gills or body surface of fresh water and marine fishes and a few amphibian, can swim very actively, leaves the host during breeding season; distribution world wide.

Systematic Position

Phylum	**Arthropoda**
Subphylum	**Mandibulata**
Class	**Crustacea**
Subclass	**Branchiura**
Order	**Arguloida**
Type	**Argulus**

Distinctive Features

1. Body consists of an oval flattened cephalothorax and short abdomen.

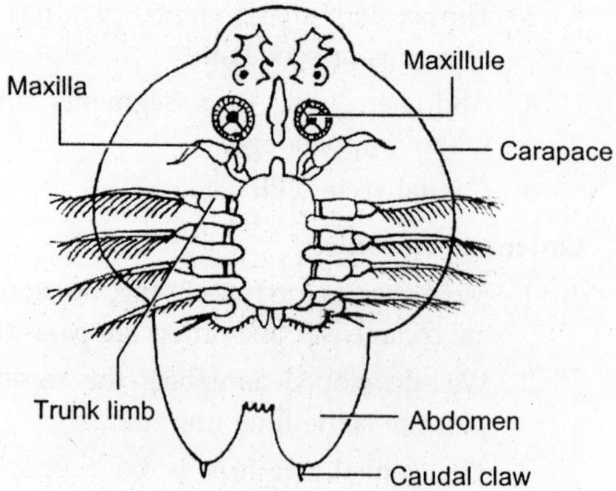

Fig. 1.1.1.96 : *Argulus*

2. Head and most of the thorax are covered by a posteriorly notched carapace.

3. Two pairs of small antennae; the antennules have a large claw used for attachment.

4. Sessile large compound eyes present.

5. Maxillules and maxillae forming suckers with spine for attachment.

6. Four pairs of swimming feet or thoracic legs.

Identification with Reasons

State characters up to—

Class—*Crustacea* as in *Daphnia*.

Subclass—*Branchiura* :

1. Flat, shield like carapace.
2. Four pairs of **cirriform limbs.**
3. Suctorial mouth.
4. Unsegmented abdomen, bilobed and ends with a minute pair of caudal claws.

Order—Arguloida :

1. Head and thorax covered with a large, disc-like carapace.
2. Sessile compound eyes present.
3. Appendages serve as suckers and claws used for attachment.

Type—*Argulus* : State the distinctive features as in above.

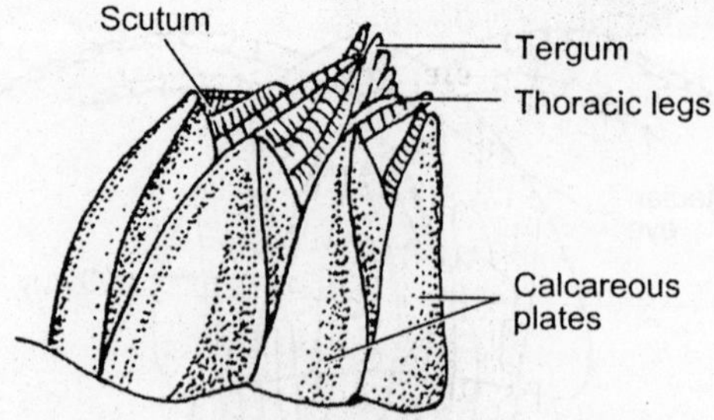

Fig. 1.1.1.97 : *Balanus* (Acorn barnacle)

6. *BALANUS* (Acorn Barnacle)

Habit and Habitat

Marine; found in the inter tidal zone attached to rocks, shells, woods etc; feeds on phyto and zoo plankton; may attached to any floating object also; distribution cosmopolitan.

Systematic Position

Phylum	**Arthropoda**
Subphylum	**Mandibulata**
Class	**Crustacea**
Subclass	**Cirripedia**
Order	**Thoracica**
Type	**Balanus**

Distinctive Features

1. Stalk absent, attached permanently to substrates or rock with shell cover.

2. Individually enclosed by six calcareous plates, united immovably.

3. The opening of shell is provided with a four fold cover two **scuta** and two **terga.**

4. Through the opening six pairs of delicate thoracic legs protrude out.

Identification with Reasons

State characters up to—

Class—*Crustacea* as in *Daphnia*.

Subclass—*Cirripedia* :

1. Marine, either free or parasitic.
2. Saccular carapace forms a pair of folds - the mantle, enclosed the animals and forming a hard shell.
3. Six pairs of cirri form thoracic appendages help in feeding.
4. Abdomen lacking.

Order—*Thoracica* :

1. Non-parasitic sessile animal.
2. Mantle produces calcareous plates.
3. With **fleshy stalk** or without a stalk.

Type—*Balanus* : State the distinctive features as in above.

Economic Importance

Barnacles act as fouler of ship bottom. A moderate encrustation of barnacles can greatly reduce ship speed and fuel efficiency.

7. *LEPAS* (Goose barnacle)

Habit and Habitat

Marine, sessile; found in the open sea attached to floating sea weeds, logs, turtle, and ship; plantonic filter feeder; cosmopolitan in distribution.

Systematic Position

Phylum	**Arthropoda**
Subphylum	**Mandibulata**
Class	**Crustacea**
Subclass	**Cirripedia**
Order	**Thoracica**
Type	***Lepas***

Distinctive Features

1. Body consists of fleshy long stalk and **capitulum.**

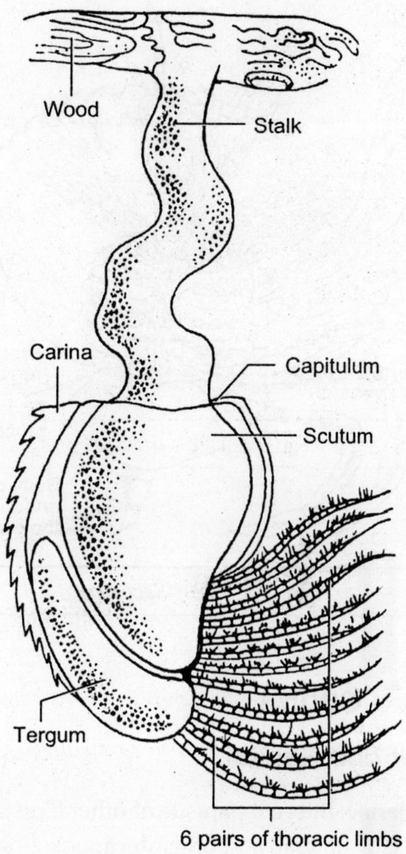

Fig. 1.1.1.98 : *Lapas* (goose barnacle)

2. Capitulum is enclosed by a bivalved carapace in which, a pair of scuta, a pair of small terga and a **carina** are embedded.

3. Long and stout penduncle or stalk acts as a hold-fast organ.

4. Six pairs of cirriform biramous appendages fringed with tuft of setae arise from the thorax.

Identification with Reasons

State characters up to—

Class—*Thoracica* as in *Balanus.*

Type—*Lepas* : State the distinctive features as in above.

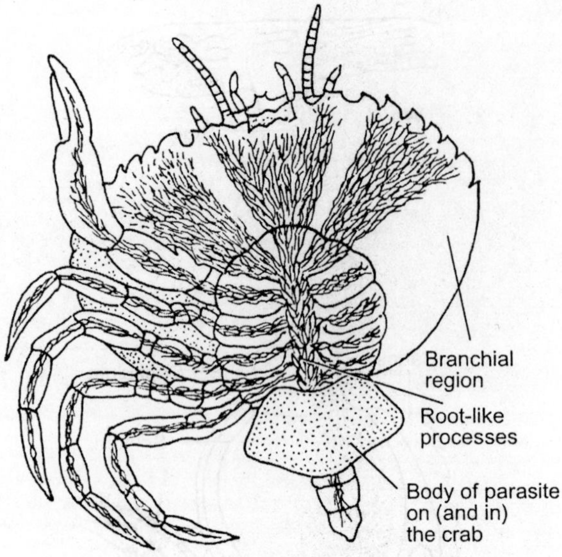

Fig. 1.1.1.99 : *Sacculina*

8. *SACCULINA*

Habit and Habitat

Marine; internal parasite of other Crustaceans, especially on crab and other decapods; first larval stages are free swimming and later stages are, first ectoparasite, then endoparasite and finally ectoparasite; occur from shallow water to depths exceeding 4,000 m.

Systematic Position

Phylum	**Arthropoda**
Subphylum	**Mandibulata**
Class	**Crustacea**
Subclass	**Cirripedia**
Order	**Rhizocephala**
Type	***Sacculina***

Distinctive Features

1. Compressed fleshy tumour like body attached to the abdomen of the crab on the ventral side.

2. Appendages, segmentation, mouth and anus are absent.

3. From the attached end, it sends fine roots in viscera of the host body.

4. Cloacal opening posterior.

Identification with Reasons

State characters up to—

Subclass—*Cirripedia* as in *Balanus*.

Order—Rhizocephala :

1. Parasitic, with body reduced to ramifying **stolon** and gonad sac.
2. Mantle present but calcareous shell absent.
3. Appendages, segmentation, and digestive system absent.

Type—Sacculina : State distinctive features as in above.

Special Features

Interesting animal as it shows extreme case of degeneration due to parasitism. Only the study of the developmental history discovers its real Systematic Position. Moreover the male crab acquires female characters, while in female crab the ovary degenerates–a case of **parasitic castration.**

9. *MYSIS*

Habit and Habitat

Generally marine, sometimes, found in fresh waters, including lakes, wells and caves; most species are heavily preyed on by various fish; usually lives in great swarms, mainly carnivorous.

Systematic Position

Phylum	**Arthropoda**
Subphylum	**Mandibulata**
Class	**Crustacea**
Subclass .	**Malacostraca**

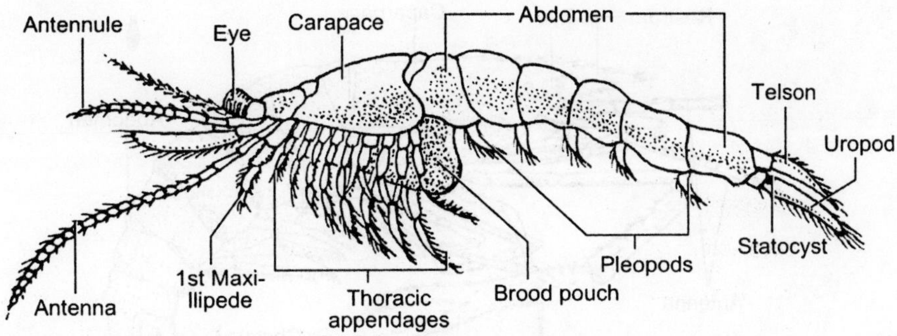

Fig. 1.1.1.100 : *Mysis*

Superorder	**Peracarida**
Order	**Mysidacea**
Type	**Mysis**

Distinctive Features

1. Elongated, shrimp-like and laterally compressed, transparent body.
2. Cylindrical, stalked compound eyes present.
3. Most of the thorax is covered by a carapace.
4. Thoracic limbs bear eight pairs of biramous appendages.
5. First pair of thoracic appendages is small and form maxillipeds.
6. Posterior thoracic limbs bear a brood pouch.
7. Telson without caudal style.
8. **Uropod** bears **statocyst.**

Identification with Reasons

State characters up to—

Subclass—*Crustacea* as in *Daphnia.*

Subclass—*Malacostraca* :

1. Body made up of nineteen segments: five in head, eight in thorax and six in abdomens.
2. Endoskeleton of head united with few or more thoracic segments to form **cephalothoracic carapace.**
3. Appendages, on the sixth abdominal segment are flattened to form uropods.
4. Caudal style absent.
5. Large, stalked compound eyes present.

Super order—*Peracarida* :

1. Carapace when present fused to anterior thoracic segments.
2. Four or more thoracic segments always distinct.
3. Eggs are brooded within a chamber.

Order—*Mysidacea* :

1. Carapace covers most of the thorax.
2. Eyes on movable stalk.
3. A tail fin formed by uropods.

Type—*Mysis* : State distinctive characters as in above.

10. *MACROBRACHIUM* (Fresh water Prawn)

Habit and Habitat

Marine, estuarine and fresh water; bottom dweller; nocturnal; feeds on small animals and detritus; migrate to brackish water during breeding; commonly found in India.

Systematic Position

Phylum	**Arthropoda**
Subphylum	**Mandibulata**

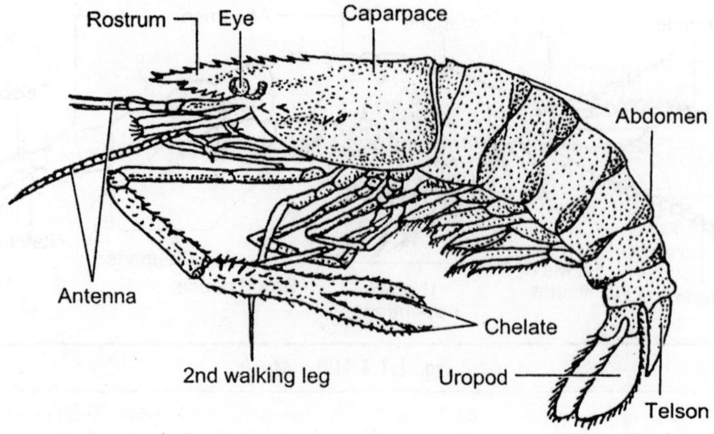

Fig. 1.1.1.101 : *Macrobrachium*

Class	**Crustacea**
Subclass	**Malacostraca**
Superorder	**Eucarida**
Order	**Decapoda**
Type	***Macrobrachium***

Distinctive Features

1. Elongated body and divided into cephalothorax, abdomen and telson.
2. Carapace is anteriorly produced into rostrum.
3. Two prominent eyes on the head.
4. Abdomen dorsally rounded, laterally compressed normally bent under the cephalothorax.
5. Second abdominal segment overlap the first.
6. First and second walking legs possess pincers and are called **chelate legs**.
7. Last three walking legs are non-chelate.
8. Antennale spines and hepatic spines present.
9. Long rostrum serrated on both sides.

Identification with Reasons

State characters up to—

Subclass—*Malacostraca* as in *Mysis*.

Suborder—*Eucarida* :

1. Large carapace, fused with all thoracic segments.
2. Stalked eyes.

Order—*Decapoda* : First three pairs of thoracic appendages are maxillipedes and rest five pairs are walking legs.

Type—*Macrobrachium* : State distinctive features as in above.

11. *PENAEUS*

Habit and Habitat

Found in fresh water streams, rivers, lakes, ponds and estuaries, bottom dweller and nocturnal; feeds on plants and small animals.

Systematic Position

Phylum	**Arthropoda**
Subphylum	**Mandibulata**
Class	**Crustacea**
Subclass	**Malacostraca**
Super order	**Eucarida**
Order	**Decapoda**
Type	***Penaeus monodon***

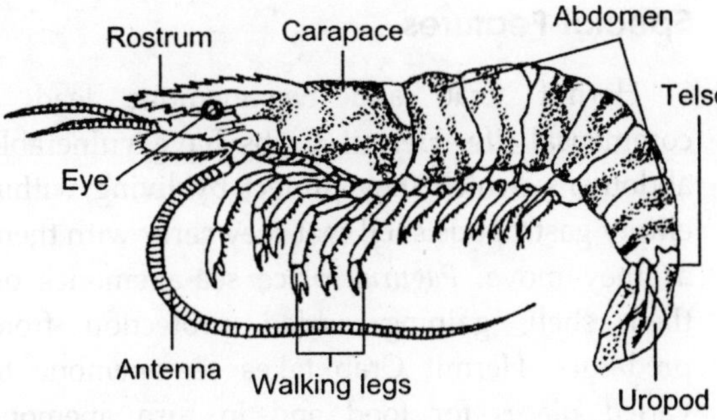

Fig. 1.1.1.102 : *Penaeus*

Distinctive Features

1. Second abdominal segment does not overlap the first.
2. More or less similar walking legs except the fourth or fifth which are comparatively larger.
3. **Rostral crest** triangular in outline.
4. Greenish targa with a dorsal black spot.

Identification with Reasons

State characters up to—

Order—*Decapoda* as in *Macrobrachium.*

Type—*Penaeus monodon* : State distinctive features as in above.

12. *PENAEUS INDICUS*

Habit and Habitat

Same as in *P. monodon.*

Systematic Position

Phylum	**Arthropoda**
Subphylum	**Mandibulata**
Class	**Crustacea**
Subclass	**Malacostraca**
Super order	**Eucarida**
Order	**Decapoda**
Type	***Penaeus indicus***

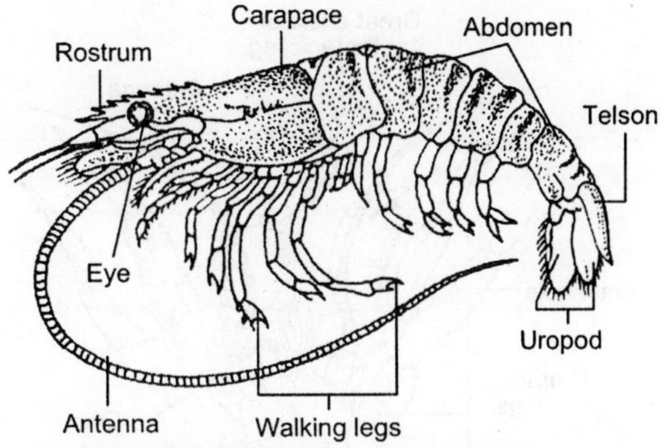

Fig. 1.1.1.102 : *Penaeus indicus*

Distinctive Features

1. Rostrum lacks ventral teeth.
2. Well developed third walking leg.
3. Rostral crest non-triangular in profile.
4. First abdominal segments overlap the posterior border of carapace.
5. Whitish targa lacks black spot.

Identification with Reasons

State characters up to—

Order—Decapoda as in *Macrobrachium.*

Type—*Penaeus indicus* : State distinctive features as in above.

13. *EUPAGURUS* (Hermit Crab)

Habit and Habitat

All marine; leaves inside the empty shell of gastropods; scavenger and detritus feeder.

Systematic Position

Phylum	**Arthropoda**
Subphylum	**Mandibulata**
Class	**Crustacea**
Subclass	**Malacostraca**

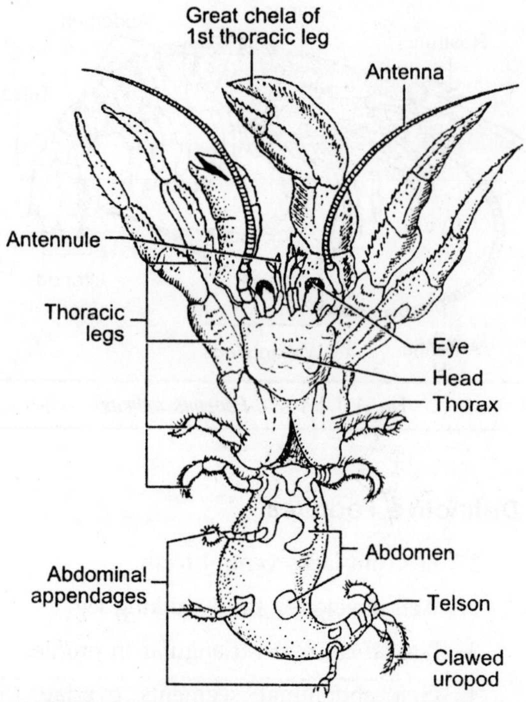

Fig. 1.1.1.104 : *Eupagurus* (Hermit crab)

Super order	**Eucarida**
Order	**Decapoda**
Type	***Eupagurus*** **(Hermit crab)**

Distinctive Features

1. Body consists of flattened cephalothorax and soft, fleshy and elongated abdomen.
2. First thoracic leg is chelate and the right side chelate leg is large and acts as lid.
3. Right side abdominal appendages are absent, while left side appendages are reduced.
4. Uropods are hook like.

Identification with Reasons

State characters up to—

Order—*Decapoda* as in *Macrobrachium.*

Type – *Eupagurus* : State distinctive features as in above.

Special Features

Hermit Crab and Sea-anemone lead a commensal life. Hermit Crabs have vulnerable abdomens, which they protect by living within empty gastropods shell that they carry with them as they move. *Pagurus* place sea-anemones on their shell, gaining added protection from predators. Hermit Crab takes sea-anemone to varied places for food and in turn anemone protects itself as well as the Crab by its sting cells.

14. *HIPPA* (Mole Crab)

Habit and Habitat

Marine; lives in the sand bottom and beaches in soft sand along the wave line, burrows with rapidity; feeds on plankton and detritus; found in tropical and subtropical waters.

Systematic Position

Phylum	**Arthropoda**
Subphylum	**Mandibulata**
Class	**Crustacea**
Subclass	**Malacostraca**
Super order	**Eucarida**
Order	**Decapoda**
Type	***Hippa***

Distinctive Features

1. Carapace oval and strongly convex.
2. Abdomen reduced and ventrally flexed.
3. Long hairy antennae.
4. **Rostrum** reduced and pointed.
5. Well developed fan like **pleopod**.

Identification with Reasons

State characters up to—

Order—*Decapoda* as in *Macrobrachium.*

Type—*Hippa* : State distinctive features as in above.

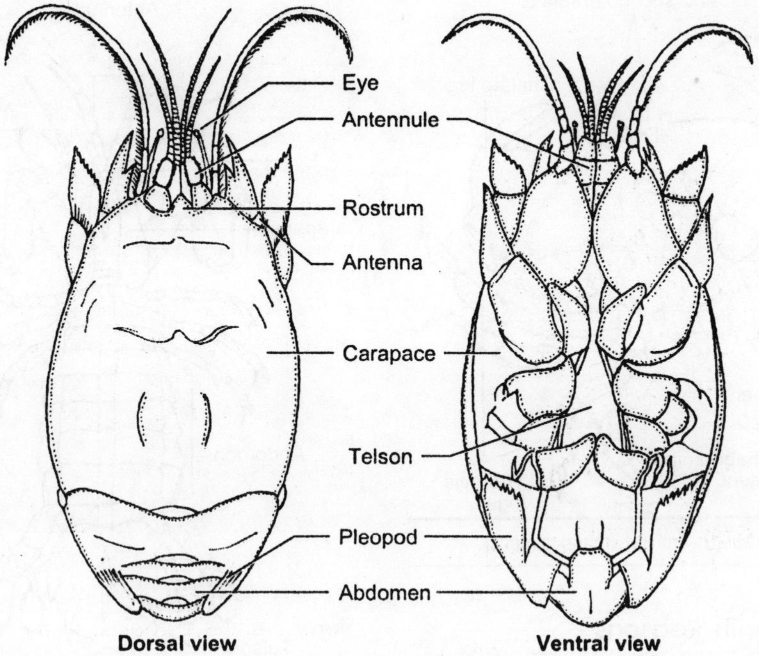

Fig. 1.1.1.105 : *Hippa* (Mole crab)

15. *CRAB*

Habit and Habitat

Terrestrial, freshwater and marine; commonly found buried in rocks and in sand; scavenger and detritus feeder; distribution cosmopolitan.

Systematic Position

Phylum	**Arthropoda**
Subphylum	**Mandibulata**
Class	**Crustacea**
Subclass	**Malacostraca**
Superorder	**Eucaridia**
Order	**Decapoda**
Type	**Crab**

Distinctive Features

1. Carapace oval and broader than long.

2. Small abdomen permanently flexed under the cephalothorax.

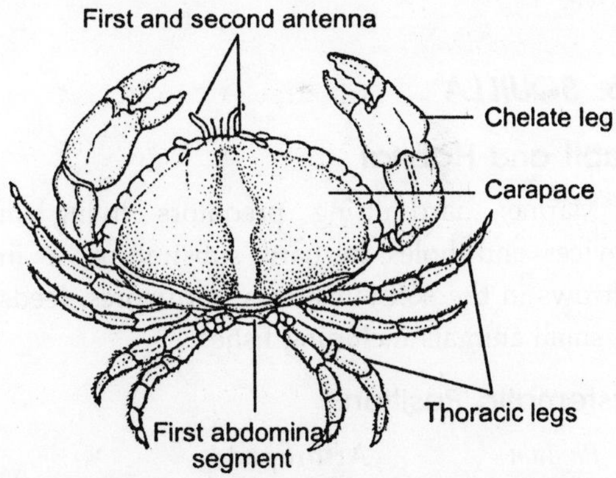

Fig. 1.1.1.106A : *Crab* [Dorsal aspect]

3. Five pairs of thoracic legs are well developed.

4. Rostrum and uropod absent.

5. The abdomen is narrower in male but some what boarder in female.

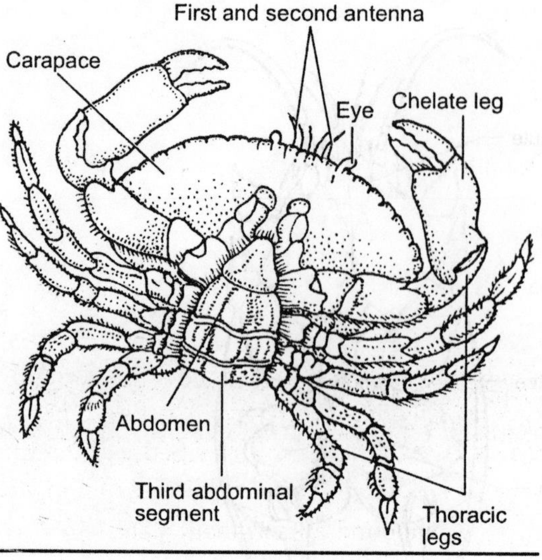

Fig. 1.1.1.106B : *Crab* [B–Ventral (male) aspect]

Identification with Reasons

State classification and characters up to—

Order—*Decapoda* as in *Macrobrachium.*

Type—*Crab* : State distinctive features as in above.

16. *SQUILLA*

Habit and Habitat

Marine; hard-hitting predators living in crevices and holes in hard substratum or in burrows in the soft substrates; carnivores, feeds on small animals including fishes.

Systematic Position

Phylum	**Arthropoda**
Subphylum	**Mandibulata**
Class	**Crustacea**
Subclass	**Malacostraca**
Super order	**Hoplocarida**
Order	**Stomatopoda**
Type	**Squilla**

Distinctive Features

1. Carapace small, flat and covers three of the seven thoracic segments.

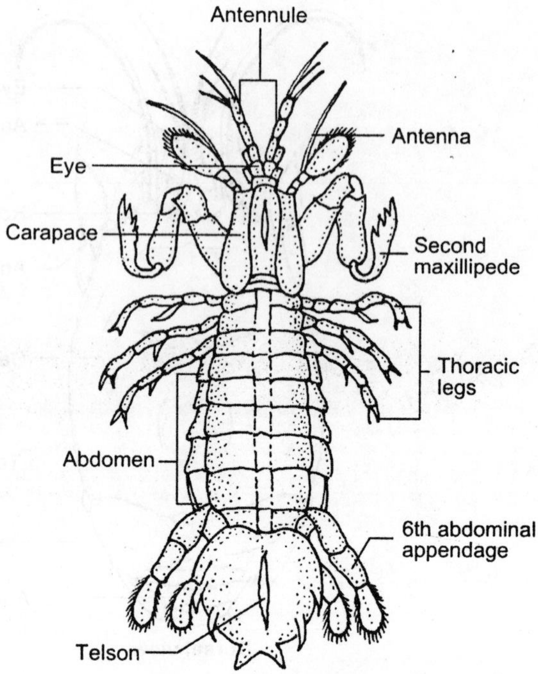

Fig. 1.1.1.107 : *Squilla*

2. Elongated body with large and broad abdomen.

3. The second pair of the thoracic appendages are large and sub-chelate and **raptorial.**

4. Gills or branchiae on the first five pairs of pleopods.

5. Sixth pleopod forms large uropod.

6. Telson very broad.

Identification with Reasons

State classification and characters up to—

Subclass—*Malacostraca* as in *Macrobranchium.*

Superorder—*Hoplocarida* :

1. Small carapace covering three thoracic segments.
2. Tri flagellate antennules.
3. Pleopods with branchiate gills.
4. Large second pair of thoracic appendages are subchelate.

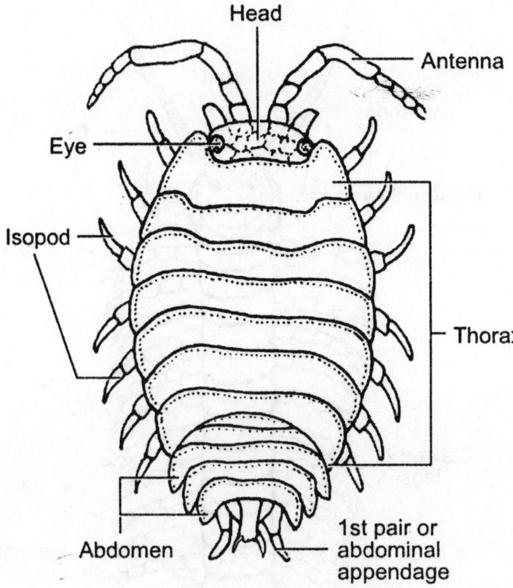

Fig. 1.1.1.108 : *Oniscus*

Order—*Stomatopoda* :

1. Elongated body with broad abdomen.
2. Anterior thoracic limbs as sub-chelate raptorial claw.

Type—*Squilla* : State distinctive features as in above.

17. *ONISCUS* (Wood louse)

Habit and Habitat

Terrestrial; found under stones, leaves, logs, bark and matted vegetation; feeds on decaying vegetation or animal matters.

Systematic Position

Phylum	**Arthropoda**
Subphylum	**Mandibulata**
Class	**Crustacea**
Subclass	**Malacostraca**
Super order	**Paracarida**
Order	**Isopoda**
Type	***Oniscus***

Distinctive Features

1. Body broad, oval and dorsoventrally compressed.
2. First thoracic segment fused with head to from cephalothorax, which is devoid of carapace.
3. Compound eyes without stalk.

Identification with Reasons

State classification and characters up to—

Super order—*Paracarida* as in *Mysis.*

Order—*Isopoda* :

1. Dorsoventrally flattened body.
2. Carapace absent.
3. Eyes without stalk.
4. All thoracic segments are similar with uniform appendages.

Type—*Oniscus* : State distinctive features as in above.

18. *JULUS* (Milliped)

Habit and Habitat

Terrestrial; lives in darkened damp places under stones, or logs or decaying leaves or

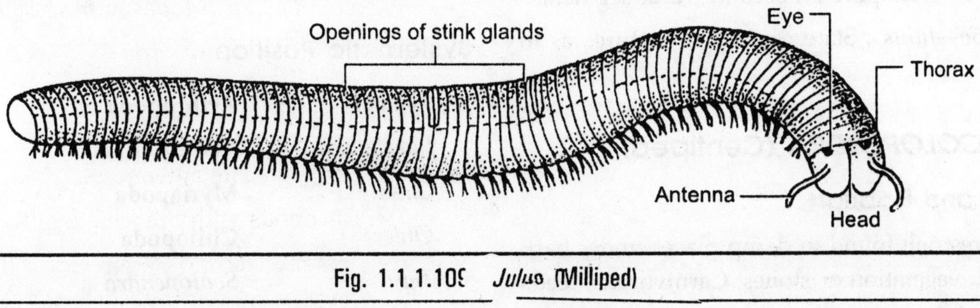

Fig. 1.1.1.109 *Julus* (Milliped)

rotating wood or moss; herbivorous, food consists of dead and decaying vegetable matter, cultivated plants, leaves and grass; when disturbed rolls up into a coil or ball.

Systematic Position

Phylum	**Arthropoda**
Subphylum	**Mandibulata**
Class	**Myriapoda**
Order	**Diplopoda**
Type	***Julus***

Distinctive Features

1. Elongated, cylindrical body.
2. Antenna short, club-shaped.
3. Three thoracic and trunk segment bears two pairs of walking legs.
4. Poison claw absent but sting glands present.

Identification with Reasons

State classification and characters up to—

Subphylum—*Mandibulata* as in *Mysis.*

Class—*Myriapoda* :

1. Possess many legs and all appendages are uniramous (single-branched).
2. Respiration by tracheae.
3. Gonopore anteriorly on body.

Order—*Diplopoda* :

1. Segmented body and each trunk segments possess two pairs of legs.
2. Poison fangs absent but possess a pairs of **poison gland.**
3. Antennae short.
4. Gonopore on second trunk segment.

Type—*Julus* : State distinctive features as in above.

19. *SCOLOPENDRA* (Centiped)

Habit and Habitat

Terrestrial; found in damp places under bark or thick vegetation or stones; Carnivorous, feeds on larvae and adult insects, worms and other small creatures, kills prey by poison of the poison-claws.

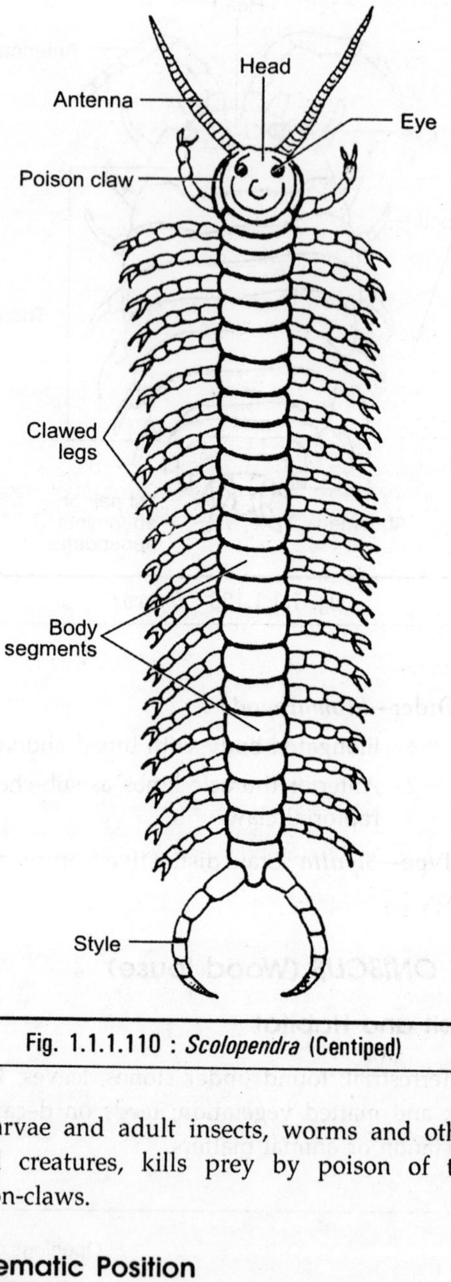

Fig. 1.1.1.110 : *Scolopendra* (Centiped)

Systematic Position

Phylum	**Arthropoda**
Subphylum	**Mandibulata**
Class	**Myriapoda**
Order	**Chilopoda**
Type	***Scolopendra***

Distinctive Features

1. Elongated, dorsoventrally flattened body.
2. Body segments have twenty one pairs of walking legs.
3. Leg is provided with single claw.
4. The first pair of trunk-appendages possess a pair of sharp claws with poison glands.

Identification with Reasons

State classification and characters up to—

Class—*Myriapoda* as in ***Julus.***

Order—*Chilopoda* :

1. First pair of trunk appendages modified into a pair of **poison fangs.**
2. Mandible retractable into head pouch.
3. One pair of walking legs in each segment.

Type—*Scolopendra* : State distinctive features as in above.

Note : Bite of centiped is dangerous to man, but not fatal.

20. *LEPISMA* (Silver fish)

Habit and Habitat

Terrestrial; common household pest, hiding in damp cool places and in books, pictures, clothes and wallpaper etc., feeds on starch, paper, cotton, paste, etc., nocturnal.

Systematic Position

Phylum	**Arthropoda**
Subphylum	**Mandibulata**
Class	**Insecta (= Hexapoda)**
Subclass	**Apterygota**
Order	**Thysanura**
Type	**Lepisma**

Distinctive Features

1. Body covered with shining scales.

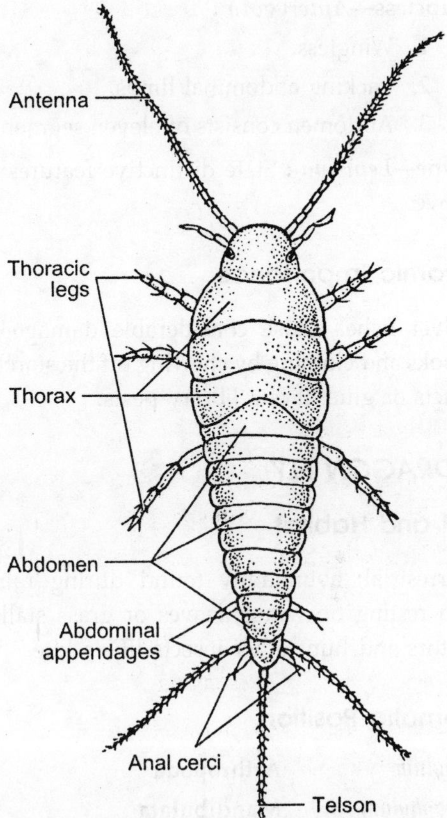

Fig. 1.1.1.111 : *Lepisma* (Silver fish)

2. Antennae are long, filiform and many segmented.
3. Abdomen with **anal cerci** and **caudal filament.**

Identification with Reasons

State classification and characters up to—

Subphylum—*Mandibulata* as in *Daphnia.*

Class—*Insecta* :

1. Body divided into head, thorax, and abdomen.
2. Three pairs of thoracic legs.
3. Antenna one pair.

Subclass—*Apterygota* :

1. Wingless.
2. Lacking abdominal limbs.
3. Abdomen consists of eleven segments.

Type—*Lepisma* : State distinctive features as in above.

Economic Importance

Silver fishes cause considerable damage to the books and clothing by chewing off the starchy products or glue. It is a library pests.

21. *DRAGON FLY*

Habit and Habitat

Terrestrial; abundantly found during rainy season resting on wings, leaves or grass stalks, predators and hunt other insects like a kite.

Systematic Position

Phylum	**Arthropoda**
Subphylum	**Mandibulata**
Class	**Insecta**
Subclass	**Pterygota**
Order	**Odonata**
Type	***Dragon fly***

Distinctive Features

1. Two pairs of transparent, membranous, glossy wings with complex cross-veins.
2. Head with a pair of large compound eyes and small hair like antenna.
3. Elongated abdomen.
4. Mouth parts are of biting and masticatory form.

Identification with Reasons

State classification and characters up to—

Class—*Insecta* as in *Lepisma.*

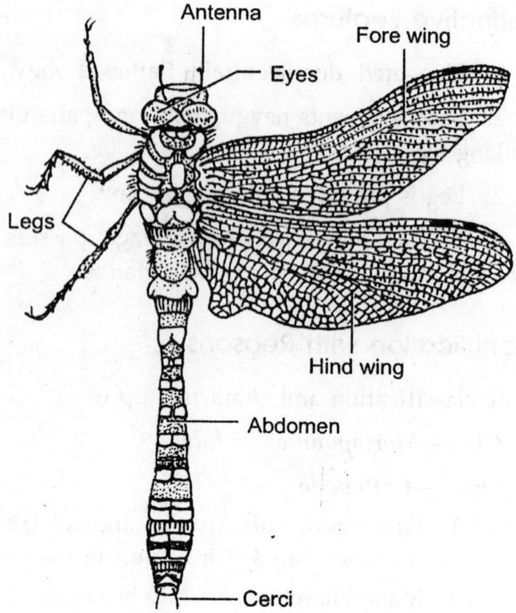

Fig. 1.1.1.112 : *Dragon fly*

Subclass—*Pterygota* : With wings or secondarily wing less.

Order—*Odonata* :

1. Slender body with strong chewing mouth parts.
2. Head with huge compound eyes and small hair like antennae.
3. Two pairs of similar, narrow, net-veined wings held straight out from the body when at rest.
4. Abdomen long and cerci very short.

Type—*Dragon fly* : State distinctive features as in above.

22. *PERIPLANETA* (Cockroach)

Habit and Habitat

Terrestrial; nocturnal, very active runner, avoiding daylight in holes and crevices in the day, and found in worm damp places like latrine, sewers, kitchens, bakeries, godowns, stores etc; omnivorous; found all over the world.

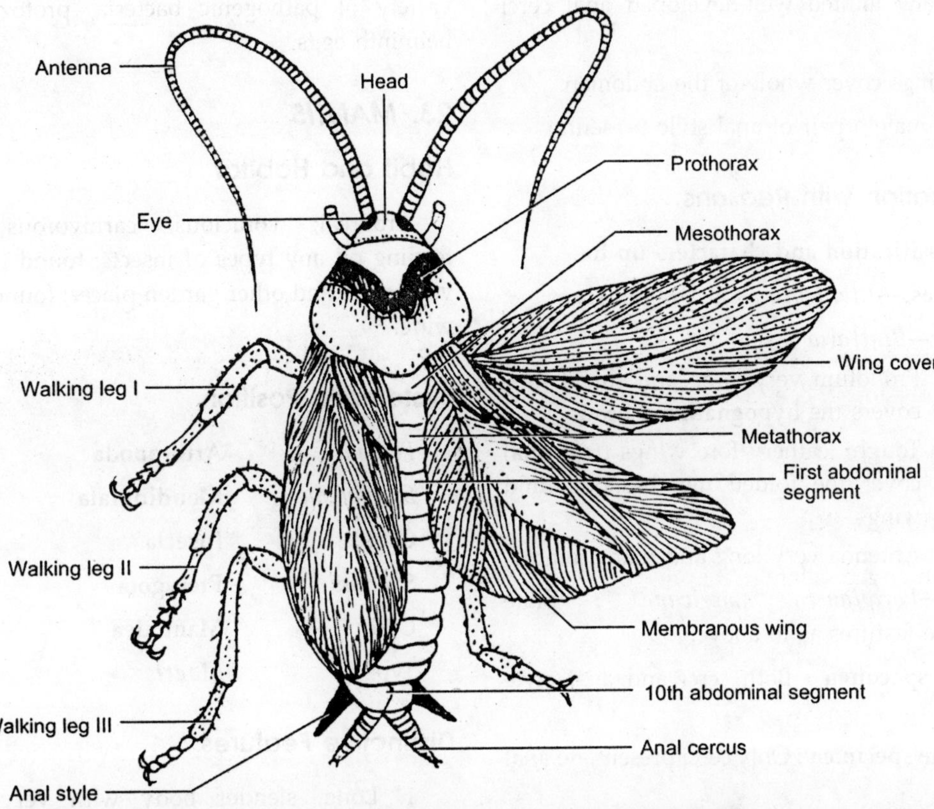

Fig. 1.1.1.113A : ***Periplaneta americana. Dorsal view (Male)***

Systematic Position

Phylum	**Arthropoda**
Subphylum	**Mandibulata**
Class	**Insecta**
Subclass	**Pterygota**
Order	**Blaterria**
Type	***Periplaneta americana***

Distinctive Features

1. Narrow, elongated, dorsoventrally flattened body.

2. Thorax has three pairs of walking legs and two pairs of wings.

3. **Pronotum** almost cover the **hypognathous** head.

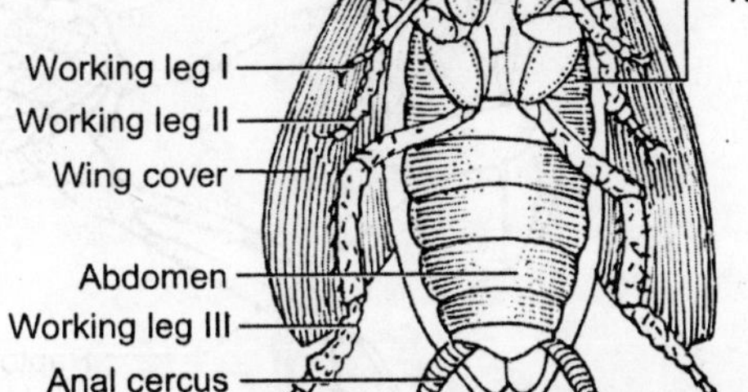

Fig. 1.1.1.113B : ***Periplaneta americana***
Ventral view (female)

4. Many jointed well-developed anal cerci present.

5. Wings cover whole of the abdomen.

6. In male a pair of anal style present.

Identification with Reasons

State classification and characters up to—

Subclass—*Pterygota* as in **Dragon fly.**

Order—*Blattaria* :

1. Pronotum very large and shield like, covers the **hypognathous** head.
2. Tough, leathery fore wings (**tegmina**) cover the folded membranous hind wings.
3. Antenna very long and many jointed.

Type—*Periplaneta americana* : State distinctive features as in above.

Male specimen : Both cerci and anal style present.

Female specimen : Only cerci present; no anal styles.

Economic Importance

Cockroaches are pests of human habitations, destroying food and other things leaving behind a disagreeable odour. They act as a transmitter of variety of pathogenic bacteria, protozoa and helminth eggs.

23. *MANTIS*

Habit and Habitat

Arboreal; voracious, carnivorous form, feeding on any types of insects; found in green vegetation and other garden places; found world wide.

Systematic Position

Phylum	**Arthropoda**
Subphylum	**Mandibulata**
Class	**Insecta**
Subclass	**Pterygota**
Order	**Mantodea**
Type	***Mantis***

Distinctive Features

1. Long, slender body with very much elongated pro-thorax.

2. Pro-thoracic pair of legs are modified for capturing the prey (**Reptorial**).

3. Head is small, triangle and can be turned sidewise.

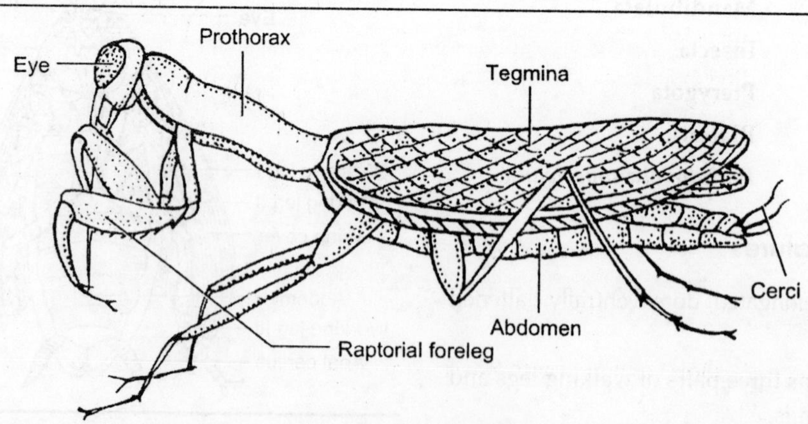

Fig. 1.1.1.114 : *Mantis*

4. Wings are folded, flat and over lapping the sides of the body.

Identification with Reasons

State classification and characters up to—

Subclass—*Pterygota* as in Dragon fly.

Order—*Mantodea* :

1. Pro-thorax is much elongated, pro-thoracic legs are modified for grasping and holding the prey (**raptorial**).
2. Antenna long and filiform.
3. Small triangular head bears large compound eyes.
4. Possess two pairs of wings.

Type—*Mantis* : State distinctive features as in above.

24. STICK INSECT

Habit and Habitat

Arboreal; nocturnal and has great power of mimicry and resembles plant stem; found in tropical forests and thick vegetation; leaf eater.

Systematic Position

Phylum	**Arthropoda**
Subphylum	**Mandibulata**
Class	**Insecta**
Subclass	**Pterygota**
Order	**Phasmida**
Type	**Stick insect**

Distinctive Features

1. Elongated, cylindrical body with long legs and filiform antenna.

2. Pro-thorax is short while meso and metathorax are longer.

3. Tegmina small.

4. Long walking legs are simple and similar.

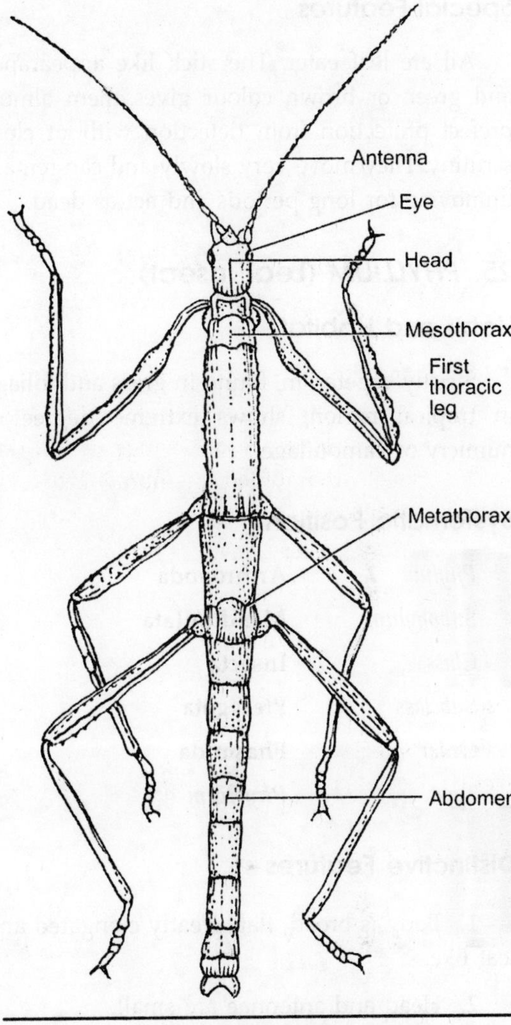

Fig. 1.1.1.115 : Stick insect

Identification with Reasons

State classification and characters up to—

Subclass—*Pterygota* as in Dragon fly.

Order—*Phasmida* :

1. Prothorax shorter than other thoracic segments.
2. Similar long and slender legs.
3. Long filiform antenna.

Type—Stick insect : State distinctive features as in above.

Special Features

All are leaf eater. The stick like appearance and green or brown colour gives them almost prefect protection from detection without close scrutiny. They move very slowly and can remain unmoving for long periods and act as dead.

25. *PHYLLIUM* (Leaf insect)

Habit and Habitat

Strictly vegetarian, found in grass and foliage in tropical region; shows extreme degree of mimicry or camouflage.

Systematic Position

Phylum	**Arthropoda**
Subphylum	**Mandibulata**
Class	**Insecta**
Subclass	**Pterygota**
Order	**Phasmida**
Type	***Phyllium***

Distinctive Features

1. Body is broad, flat, greatly elongated and leaf like.

2. Head and antennae are small.

3. **Tagmen** or first pair of wings is shorter than the other pair.

4. Wings are coloured, marked and veined like leaves.

5. Legs are flattened and leaf coloured.

Identification with Reasons

State classification and characters up to—

Order—*Phasmida* as in Stick insect.

Type—*Phyllium* : State distinctive features as in above.

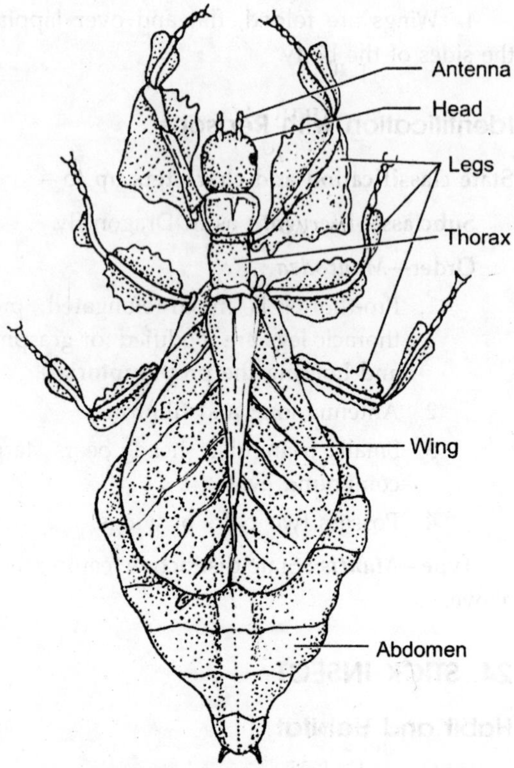

Fig. 1.1.1.116 : *Phyllium* (Leaf insect)

26. *FORFICULA* (Earwig)

Habit and Habitat

Nocturnal, free living herbivorous or carnivorous, active at night feeding actively on insects, caterpillars, slug or occasionally on green plants. During day time hide under bark, in the soil or crevices.

Systematic Position

Phylum	**Arthropoda**
Subphylum	**Mandibulata**
Class	**Insecta**
Subclass	**Pterygota**
Order	**Dermaptera**
Type	***Forficula***

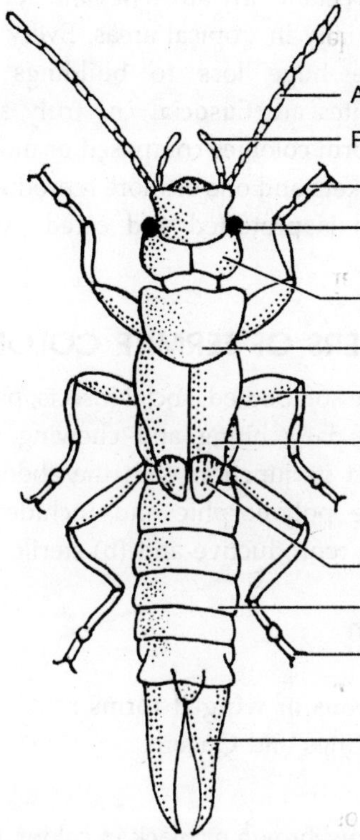

Fig. 1.1.1.117 : *Forficula*

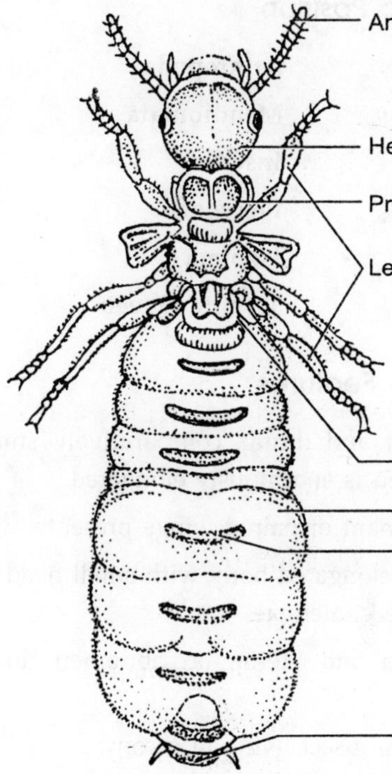

Fig. 1.1.1.118 : *Termite queen*

Distinctive Features

1. Medium-sized insects.
2. Fore wings short, leathery, without veins.
3. Hind wings membranous, semicircular, folded beneath the fore wings at rest.
4. Thorax with three pairs of walking legs.
5. At the end of the abdomen, cerci form forcep-like appendages.

Identification with Reasons

State classification and characters up to—

Subclass—*Pterygota* as in **Dragon fly.**

Order—*Dermaptera* :

1. Elongated body with tough exoskeleton.
2. Antennae long.
3. Fore wing shorter and without vein; hind wing membranous, semicircular and radially veined.
4. Forceps-like appendages on the end of the abdomen.

Economic Importance

Forficula is frequently destructive to ornamented plants. It destroys roses, dahlias and other flowering plants eating off the petals at the base as a result flowers dropdown.

27. TERMITE QUEEN

Habit and Habitat

Subterranean; lives underground in a chamber, the royal cell with in the termite colony; fed by worker termites with regurgitated saliva and fungal hyphae.

Systematic Position

Phylum	**Arthropoda**
Subphylum	**Mandibulata**
Class	**Insecta**
Subclass	**Pterygota**
Order	**Isoptera**
Type	**Termite Queen**

Distinctive Features

1. Head and thorax comparatively smaller but abdomen is enormously elongated.
2. Remnant of paired wings present.
3. Soft elongated body with small head and short, beaded antennae.
4. Targa and Sterna on abdomen do not grow.
5. Social insect living in colony.

Identification with Reasons

State classification and characters up to—

Subclass—*Pterygota* as in Dragon fly.

Order—*Isoptera* :

1. Social forms with soft exoskeleton.
2. Members of the colony differentiated as reproductive members, workers and soldier.
3. With biting mouth parts.
4. Head bears slender, short beaded antennae.
5. Wings when present, similar, with basal fractures at point where they break when shed.

Type—Termite Queen : State distinctive features as in above.

Economic Importance

Termites are remarkably destructive of wooden structures, such as homes, trees, but are extremely important in nutrient and energy recycling especially in tropical areas. Every year termites cause huge loss to buildings and libraries. Termites are **Eusocial** *i.e.,* truly social insects. They form colonies composed or more or less sterile workers and one or more reproductive queens. Queen is protected and cared by her offspring.

27A. MEMBERS OF TERMITE COLONY

Termites are soft bodied, social insects, pale in colour. Mouth parts biting and chewing type. Thorax wide at its junction with the abdomen. Terminates are polymorphic and include two forms, viz., (a) reproductive and (b) sterile.

Identification

1. Macropterous or winged forms : (winged Kings and Queen).

Characters :

1. Yellow, brown or black in colour. Body well sclerotized.
2. Wing two pairs, larger than the body and very well developed. Ultimately, these wings are discharged.
3. Compound eyes are two, large and widely separated.
4. Adapted for a brief aerial life.

Economic Importance : Adapted for a brief aerial life and forms the new colonies.

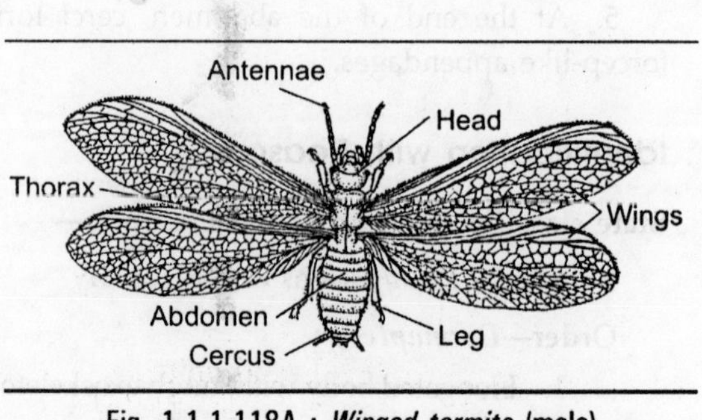

Fig. 1.1.1.118A : *Winged termite* (male)

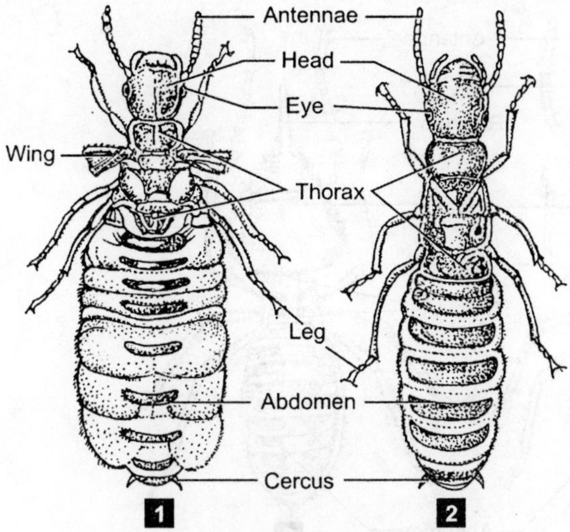

Fig. 1.1.1.118B : *Macropterous form*
1–Dealated queen, 2–King.

2. **Brachypterous or short winged forms :** (Supplementary reproductive forms, consisting both Kings and Queens)

Characters :

1. Body is usually less sclerotized and pigmented then primary reproductives.
2. Straw-coloured or greyish white in colour.
3. Compounds eyes are generally reduced.
4. Two pairs of wings are vestigial, short and stump like.

Economic Importance : Not normally found in colonies. They appear after one or both primary reproductives die and replace them so that the colony can continue.

3. **Apterous or wingless forms :**

Characters :

1. Body is not pigmented.
2. Wigs are completely absent.
3. Compound eyes are vestigial.
4. Ocelli are completely lacking.

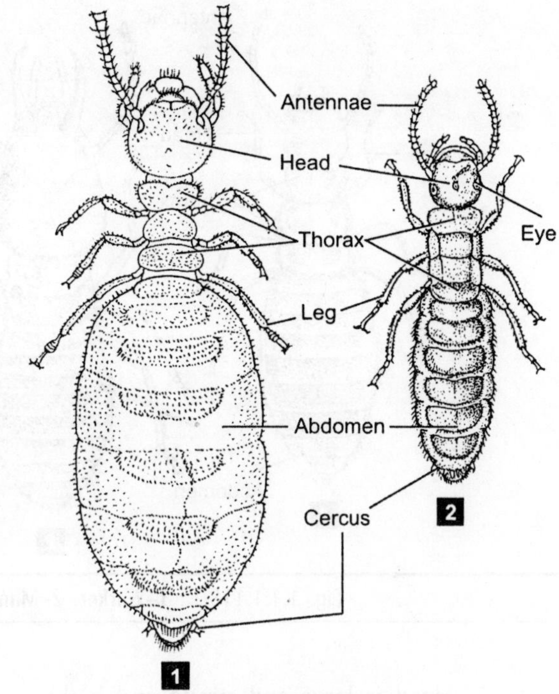

Fig. 1.1.1.118C
1–Apterous queen, 2–Brachypterous form.

4. **Workers (Sterile castes) :**

Characters :

1. Small white insects, without wings.
2. Head is directed down words, with small, broad jaws.
3. Compound eyes are absent.
4. Mandible powerful and adapted for gnawing wood and other vegetable tissue.

Economic Importance : They show marked care for the eggs and young. They also feed and tend the Queens, forage for food, cultivate fungus garden and excavate the **gallaries** and tunnels which serve for the nest.

(5) **Soldiers :**

Characters :

1. Bigger in size than workers.

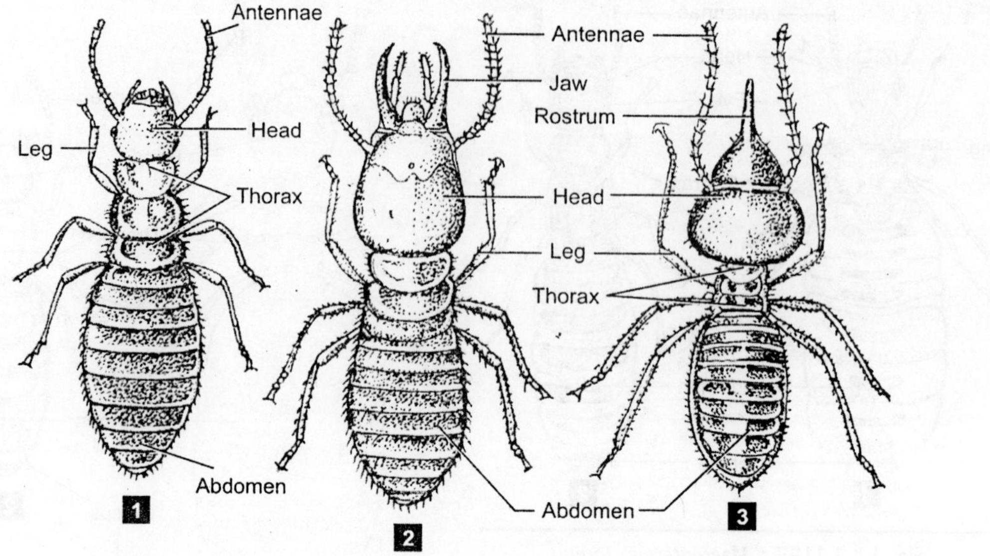

Fig. 1.1.1.118D : 1–Worker, 2–Mandibulate soldier, 3–Nasute soldier.

2. Head is large and strong and highly sclerotized.
3. Pointed jaws are placed ventrally in front of the head.

Two types of soldier

(a) *Mandibulate type* :

Characters : With large and powerful jaws but no frontal rostrum.

(b) *Nasute type* :

Characters: Median frontal rostrum, but jaws are small or vestigial.

Economic Importance

Their work is to defend the colony and attack the intruders into the colony.

28. GRASSHOPPER

Habit and Habitat

Commonly found in green field and in abundant leafy vegetation; feed on leafy vegetation; solitary, serve as pest to agriculture and harmful to mankind.

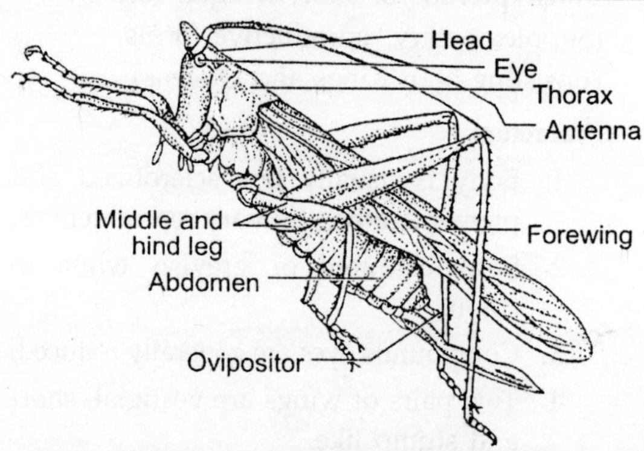

Fig. 1.1.1.119 : *Grasshopper* (female)

Systematic Position

Phylum	**Arthropoda**
Subphylum	**Mandibulata**
Class	**Insecta**
Subclass	**Pterygota**
Order	**Orthoptera**
Type	***Grasshopper***

Distinctive Features

1. Typically with long, jumping hind legs.
2. Narrow opaque fore wings and membranous broad hind wings present.
3. Chewing mouth parts with large eyes.
4. Female has a long ovipositor.

Identification with Reasons

State classification and characters up to—

- **Subclass—*Pterygota* as in Dragon fly.**

Order—*Orthoptera* :

1. Medium to large sized insect with large eyes and chewing or biting mouth parts.
2. Typically with long, saltatorial (jumping) hind legs.
3. Fore wings typically with tegmina covering folded, membranous hind wings.
4. Abdomen with short, unsegmented cerci.

Type—*Grasshopper* : State distinctive features as in above.

Economic Importance

Grasshoppers live on vegetation and cause it a great damage. They also feed on grasses and render heavy damage to range and pasture field. Grasshopper is beneficial too. They used for baiting fishes and also food to birds. Grasshopper also taken by man as food, either roasted or fried.

29. LOCUST

Habit and Habitat

Arboreal, herbivorous, voracious eater of vegetation and crops both by nymphs and adults; polymorphic; exist in two phases; Gregarious (migratory) phase and Solitary phase; individuals migrate long distance causing great damages to crops and vegetation; found in Africa, Arabia,

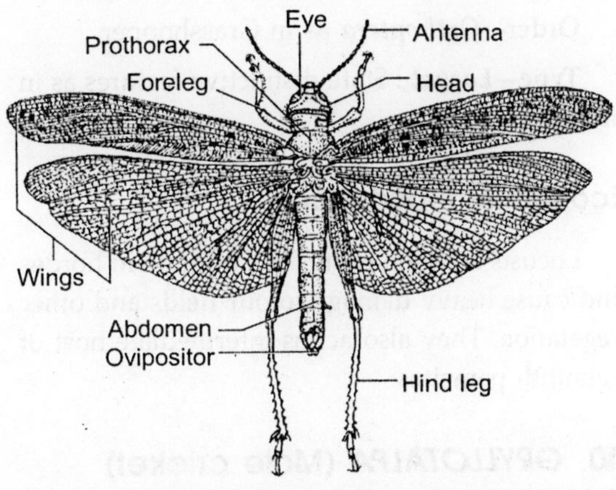

Fig. 1.1.1.120 : *Locust* (Female)

Iraq, Iran, Afghanistan and north eastern India; three kinds of locusts are found; *Locusta migratoria*, *Schistocerca gregaria* and *Patanga succinta*; *P. succinta* (Bombay locust) is confined to India.

Systematic Position

Phylum	**Arthropoda**
Subphylum	**Mandibulata**
Class	**Insecta**
Subclass	**Pterygota**
Order	**Orthoptera**
Type	**Locust**

Distinctive Features

1. Antennae short.
2. Forewing leathery and with many brownish-black spot.
3. Hind legs are largest and strong and adapted for jumping.
4. Female has short ovipositor and male has anal style.

Identification with Reasons

State classification and characters up to—

Order—*Orthoptera* as in Grasshopper.

Type—*Locust* : State distinctive features as in above.

Economic Importance

Locusts are crop pest. They migrate in hordes and cause heavy damage to our fields and other vegetation. They also act as intermediate host of helminth parasites.

30. *GRYLLOTALPA* (Mole cricket)

Habit and Habitat

Nocturnal; subterranean; burrowing insects, found in fields near ponds in soil; carnivorous, feeding on small insects; wide spread over Indian plains, lower hills, warms parts of Asia and Europe.

Systematic Position

Phylum	**Arthropoda**
Subphylum	**Mandibulata**
Class	**Insecta**
Subclass	**Pterygota**
Order	**Orthoptera**
Type	***Gryllotalpa***

Distinctive Features

1. Body covered with short fine hairs.
2. Head with large eyes and short antennules.
3. Mouth parts biting and chewing type.
4. Fore legs broad and provided with stout spines and adapted for digging and are called fossorial legs.
5. Fore wings short, hind wings large, extend backwards and appear as slender processes beyond the tegmina.
6. A pair of cerci at the end of the abdomen.

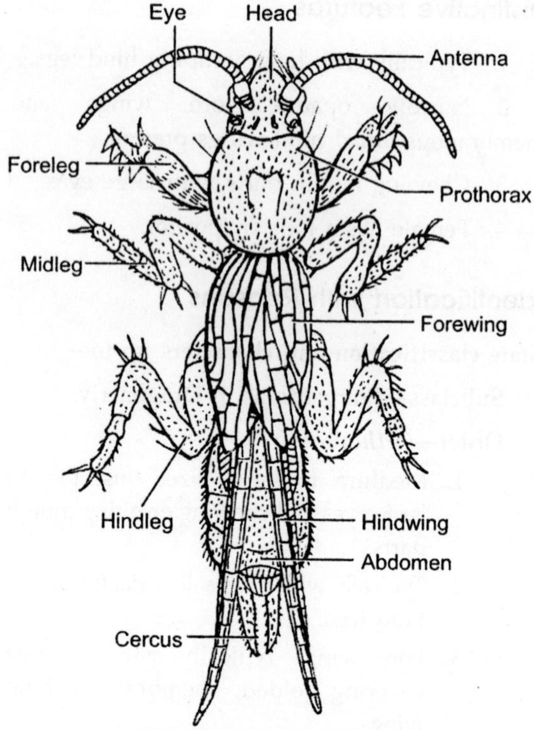

Fig. 1.1.1.121 : *Gryllotalpa*

Identification with Reasons

State classification and characters up to—

Order—*Orthoptera* as in Grasshopper.

Type—***Gryllotalpa*** : State distinctive features as in above.

Economic Importance

Mole crickets are agricultural pests causing damage to the cultivated crops. They burrow through the soil and destroy the roots of the plants.

31. *GRYLLUS* (House-cricket)

Habit and Habitat

Omnivorous; destructive to paper, clothes, fruit, vegetables, etc.; lives in warm places and

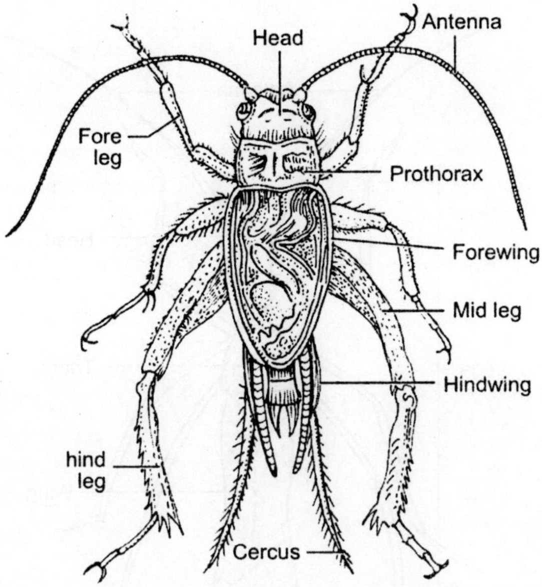

Fig. 1.1.1.122 : *Gryllus* (House cricket)

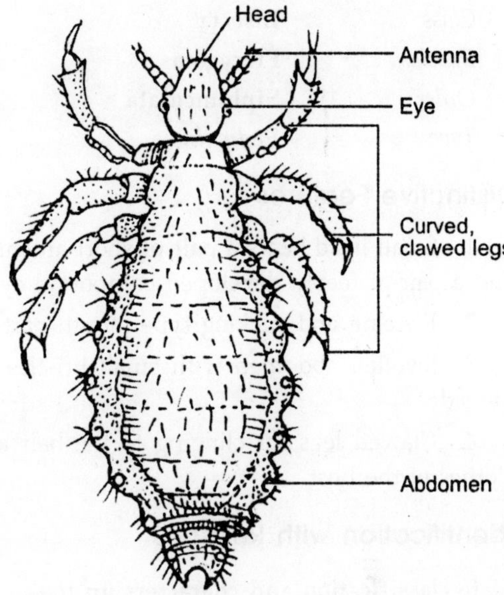

Fig. 1.1.1.123 : *Pediculus* (Human body louse)

hides in books, kitchens, bakeries, hole and crevices; male produce characteristics chirping sound.

Systematic Position

Phylum	**Arthropoda**
Subphylum	**Mandibulata**
Class	**Insecta**
Subclass	**Pterygota**
Order	**Orthoptera**
Type	*Gryllus*

Distinctive features

1. Body broad anteriorly and narrows down posteriorly.
2. Antennae long.
3. Mouth parts biting and chewing type.
4. Fore legs short and meta-thoracic legs large and adapted for leaping purpose.
5. Ovipositor is spear-shaped.

Identification with Reasons

State classification and characters up to—

Order—*Orthoptera* as those of **Grasshopper**.

Type—*Gryllus* : State distinctive features as in above.

Economic Importance

Gryllus are destructive to all sorts of clothes, paper, fruits and vegetables etc.

32. *PEDICULUS* (Human body and head lice)

Habit and Habitat

Blood sucking ectoparasite of monkey, apes and man; lives in the hairy parts of body; found world wide.

Systematic Position

Phylum	**Arthropoda**
Subphylum	**Mandibulata**

Class	**Insecta**
Subclass	**Pterygota**
Order	**Siphunculata**
Type	***Pediculus***

Distinctive Features

1. Small head bears a pair of short antennae and a pair of feebly developed compound eyes.
2. Piercing and sucking type mouth parts.
3. Swollen abdomen with small bristles on the side.
4. Clawed legs for clinging to the hair and clothes of the host.

Identification with Reasons

State classification and characters up to—

Subclass—*Pterygota* as in Dragonfly.

Order—*Siphunculata* :

1. Small wing less insects.
2. Dorsoventrally flattened body with dark marking along the side.
3. Short antennae.
4. Thoracic segments fused and contain clawed legs.
5. Abdomen swollen.
6. Eyes reduced or absent.

Type—*Pediculus* : State distinctive features as in above.

Economic Importance

Primary transmitters of epidemic typhus, trench fever, and relapsing fever. Unsanitary conditions, crowded camps are the ideal places for *Pediculus* to thrive.

33. *LEPTOCORISA* (Rice bug, Paddy bug, Gundhi bug)

Habit and Habitat

Phytophagous; feed on various portions of plants; mouth parts modified for piercing and sucking; pest of rice, active during early morning and dusk; emits a very obnoxious smell and thus called Gundhi bug.

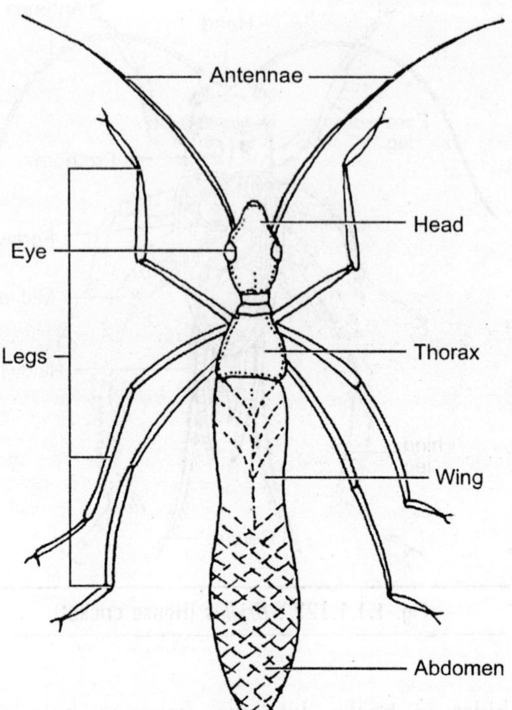

Fig. 1.1.1.124 : *Leptocorisa*

Systematic Position

Phylum	**Arthropoda**
Subphylum	**Mandibulata**
Class	**Insecta**
Subclass	**Pterygota**
Order	**Hemiptera**
Type	***Leptocorisa***

Distinctive Features

1. Slender greenish brown with large compound eyes.
2. Long legs.
3. Four jointed antennae inserted well up on sides of head.

4. Apical part of the fore wing is membranous (Heteroptera)

5. Abdomen without cerci.

Identification with Reasons

State classification and characters up to—

Subclass—*Pterygota* as in **Dragon fly.**

Order—*Hemiptera* :

1. Usually two pairs of wings, fore wing modified as hemi elytra, with membranous apex and leathery base to cover membranous hind wings.
2. Legs absent or adapted for running, jumping, digging, grasping prey or swimming.
3. Piercing and sucking mouth parts.
4. Gradual or incomplete metamorphosis.
5. Herbivorus and predaceous
6. Beak arises from the front of the head and extends ventrally and posteriorly.

Type—*Leptocorisa* : State distinctive features as in above.

Economic Importance

Leptocorisa are major agricultural pests. They are serious pests of paddy, sucking the liquid from developing rice grains when they are in soft milky stage.

34. *BELOSTOMA* (Giant water-bug)

Habit and Habitat

Fresh water; found in ponds and lakes of tropical countries; Carnivorous; feeds on insects, snails, fishes, tadpoles and small frogs etc.

Systematic Position

Phylum	**Arthropoda**
Subphylum	**Mandibulata**
Class	**Insecta**
Subclass	**Pterygota**
Order	**Hemiptera**
Type	***Belostoma***

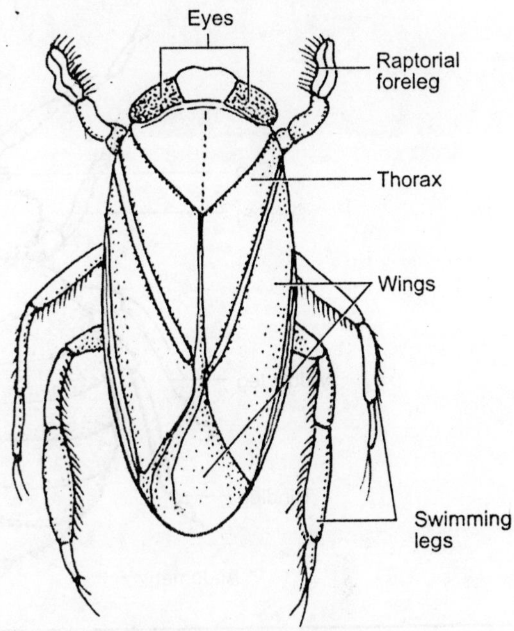

Fig. 1.1.1.125 : *Belostoma*

Distinctive Features

1. Large insect with elongated body covered by leathery brownish exoskeleton.

2. Mouth parts piercing and sucking type.

3. Legs flat, fore legs short and raptorial while posterior legs adopted for swimming and crawling.

4. Thorax broad and some what triangular.

Identification with Reasons

State classification and characters up to—

Order—*Hemiptera* as in ***Leptocorisa.***

Type—*Belostoma* : State distinctive features as in above.

Note : *Belostoma* if handle inflict a painful wound.

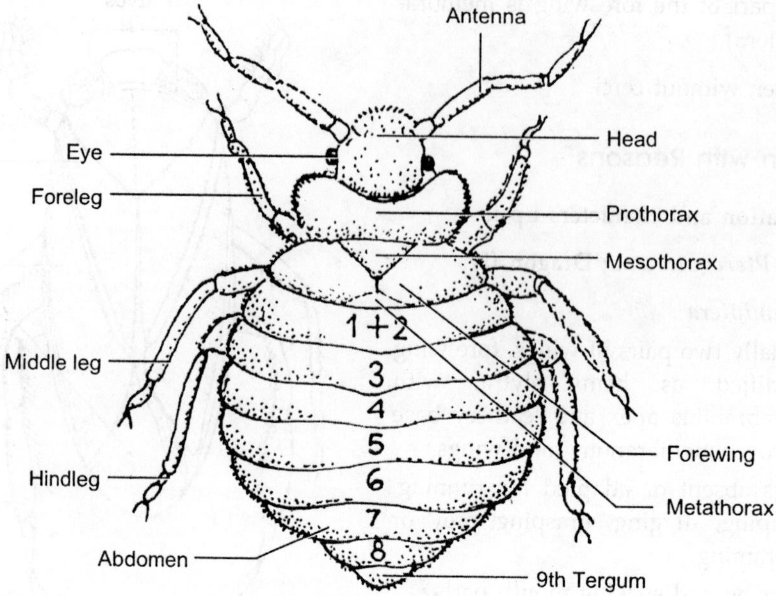

Fig. 1.1.1.126 : *Cimex* (Bed bug)

35. *CIMEX* (Bed-bug)

Habit and Habitat

Blood sucking ectoparasite of birds and mammals; nocturnal; during day time hidden in crevices of walls, floors, furniture's etc., biting causes itching and swelling.

Systematic Position

Phylum	**Arthropoda**
Subphylum	**Mandibulata**
Class	**Insecta**
Subclass	**Pterygota**
Order	**Hemiptera**
Type	***Cimex***

Distinctive Features

1. Flattened, oval body, usually without wings.

2. Fore wing in the form of spiny pad.

3. Head is short, broad and set inside the lateral extension of pro-thorax.

4. Mouth parts piercing and sucking, form a suctorial rostrum.

5. Prominent eyes at the sides of the head.

6. In male the abdomen is pointed at the end, while in female, it is rounded.

Identification with Reasons

State classification and characters up to—

Order—*Hemiptera* as in *Leptocorisa*.

Type—*Cimex* : State distinctive features as in above.

36. *APHIS* (Aphid; plant lice)

Habit and Habitat

Phytophagus; sucking the juice of every kind of plants; severe agricultural pests and vectors of viral, bacterial, and fungal diseases of plant; distribution world wide.

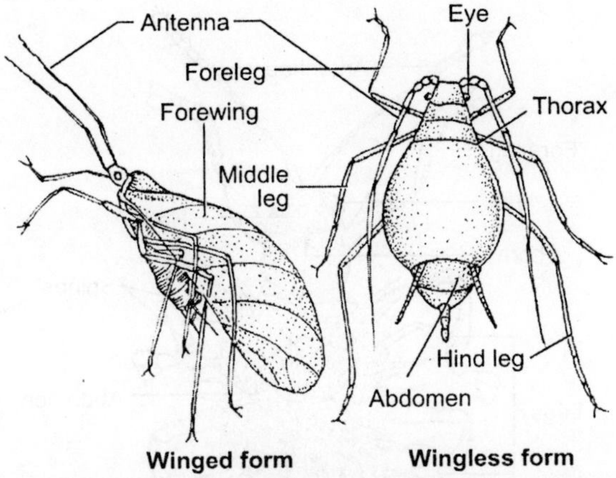

Fig. 1.1.1.127 : *Aphis* (Aphid)

Systematic Position

Phylum	**Arthropoda**
Subphylum	**Mandibulata**
Class	**Insecta**
Subclass	**Pterygota**
Order	**Homoptera**
Type	***Aphis* (Aphid)**

Distinctive Features

1. Small green insect with rounded body form and a pair of Siphon.

2. Head is distinct with long straight antennae.

3. Long thread like sucking mouth parts extended from a short proboscis.

4. Abdomen large and elongated.

5. Both asexual and sexual phases and both winged and wingless forms occur.

Identification with Reasons

State classification and characters up to—

Subclass—*Pterygota* as in Dragon fly.

Order—*Homoptera* :

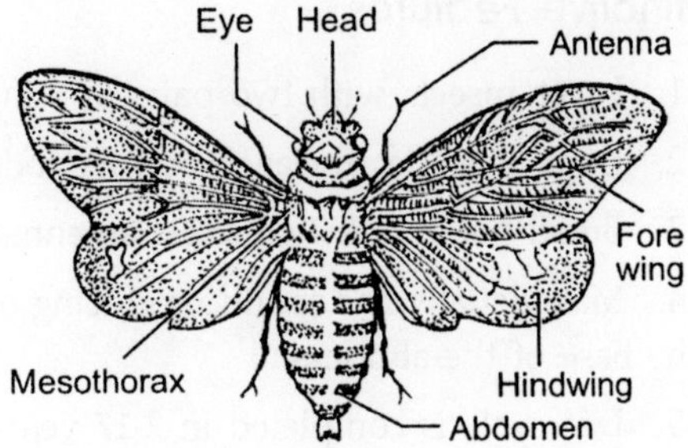

Fig. 1.1.1.128 : *Cicada*

1. Herbivorus and be arises from back of the head.
2. Piercing and sucking mouth parts.
3. Typically with similar, membranous wings.
4. Abdomen without cerci.
5. Incomplete metamorphosis.

Type—*Aphis* (Aphid) : State distinctive features as in above.

Economic Importance

Aphids are well guarded by ants. Certain ant use aphids as **"ant cows"**. Aphids produce a sweet juice or honey called honey dew which is an attractive food for ants.

37. *CICADA*

Habit and Habitat

Nocturnal; occurs most abundantly in forests and well wooded moist localities; nymphs and adults feed on the tree sap; found in tropical, sub-tropical and temperate forests.

Systematic Position

Phylum	**Arthropoda**
Subphylum	**Mandibulata**
Class	**Insecta**
Subclass	**Pterygota**
Order	**Homoptera**
Type	***Cicada***

Distinctive Features

1. Large insects with two pairs of wings.
2. Large wings lie at on angle over the abdomen.
3. Broad head with two short antennae.
4. Male usually has sound producing organs at the base of the abdomen.
5. Life cycle is completed in 2-17 years.

Identification with Reasons

State classification and characters up to—

Order—*Homoptera* as in ***Aphis* (Aphid).**

Type—*Cicada* : State distinctive features as in above.

Economic Importance

Cause serious damages to the twigs and branches of fruit and hard wood trees.

38. *HISPA ARMIGERA* (=*Dicladispa armigera*)

Habit and Habitat

Serious pest of rice; feeds on various parts of plants; cause damage both by larva and adult; distributed throughout India.

Systematic Position

Phylum	**Arthropoda**
Subphylum	**Mandibulata**
Class	**Insecta**
Subclass	**Pterygota**
Super order	**Holometabola**
Order	**Coleoptera**
Type	***Hispa armigera* (=*Dicladispa armigera*)**

Distinctive Features

1. Small bluish black beetle; with numerous short spines on the body.

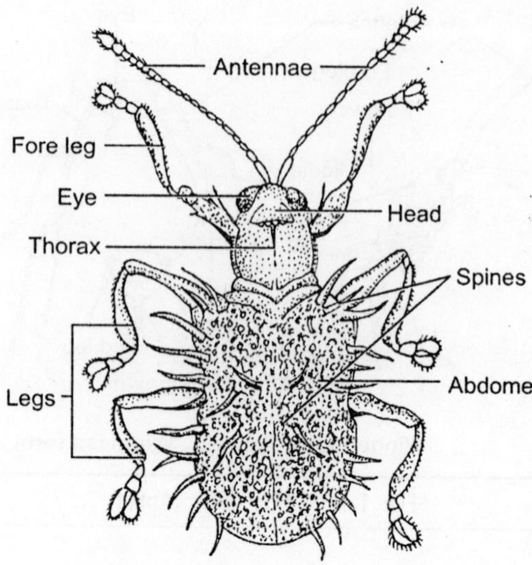

Fig. 1.1.1.129 : *Hispa armigera*

2. Two pairs of wings, front pair serving as a protective sheath for the rear pair.
3. Antennae are short but clubbed at the anterior.

Identification with Reasons

State classification and characters up to—

Subclass—*Pterygota* as in **Dragon fly.**

Superorder—*Holometabola* :

1. Exhibits complete metamorphosis.
2. Wings developed internally.

Order—*Coleoptera* :

1. Insect with hard skeleton.
2. Forewings modified as hardened elytra.
3. Hind wings membranous and completely hidden by elytra when at rest.
4. Biting and chewing mouth parts.
5. Usually ten fragmented abdomen without cerci.
6. Complete metamorphosis.

Type—*Hispa armigera* : State distinctive features as in above.

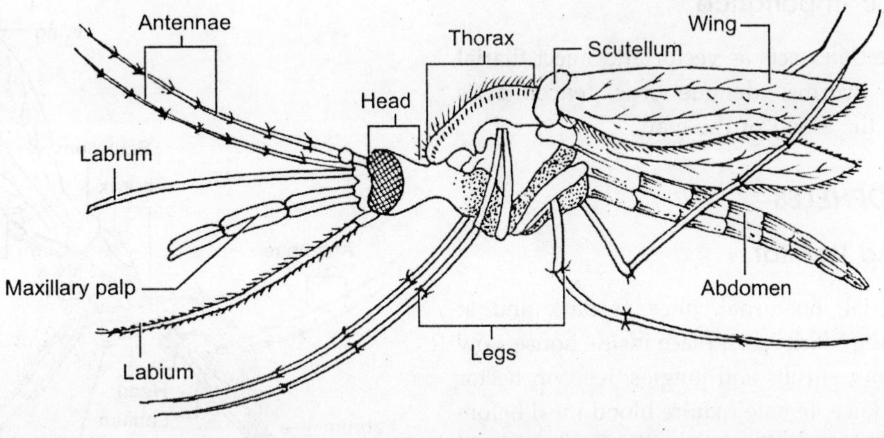

Fig. 1.1.1.130 : *Culex* (Female)

Economic Importance

Major agricultural pest, both larva and adult cause serious damage to paddy plant.

39. *CULEX*

Habit and Habitat

Terrestrial; nocturnal, active at dark and night; live in swarms and prefer to stay in damp places inside houses, cow sheds, shrubs and jungles; feed on nectar and plaint juice; females require blood meal before egg laying; larva are aquatic, distribution world wide.

Systematic Position

Phylum	**Arthropoda**
Subphylum	**Mandibulata**
Class	**Insecta**
Subclass	**Pterygota**
Super order	**Holometabola**
Order	**Diptera**
Type	***Culex***

Distinctive Features

1. Soft, slender body covered with small scales.
2. Trilobed **scutellum** with a tuft of bristles on each lobe.
3. Wing unspotted.
4. In adult female the proboscis is modifying for piercing.
5. Body remain parallel to the surface at rest.

Male :

1. Antennae plumose.
2. Maxillary palps are usually as long as proboscis and are five jointed.

Female :

1. Pilose antennae.
2. Maxillary palps are short and three jointed.

Identification with Reasons

State classification and characters up to—

Superorder—*Holometabota* as in *Hispa armigera.*

Order—*Diptera* :

1. Small sized insects with one pair of membranous flying fore wings.
2. Posterior pair of club shaped wings-halteres, used for balancing during flight.
3. Mouth parts Piercing, sucking or sponging type.
4. Complete metamorphosis.

Type—*Culex* : State distinctive features as in above.

Economic Importance

Female *Culex* acts as vector and inject filarial nematode worms, dengue-virus, encephalitis virus into the blood of the man.

40. *ANOPHELES*

Habit and Habitat

Terrestrial; nocturnal, bites at dark and at night; lives in dark moist place inside house, cow sheds, farms, shrubs and jungles; feed on nectar and plant juice; female require blood meal before egg laying; larva are aquatic; distribution world wide.

Systematic Position

Phylum	**Arthropoda**
Subphylum	**Mandibulata**
Class	**Insecta**
Subclass	**Pterygota**
Super order	**Holometabola**
Order	**Diptera**
Type	***Anopheles***

Distinctive Features

1. Soft, slender body, scale less.
2. Crescentic scutellum with bristles on the posterior margin.
3. Wings are marked with dark spots.
4. In adult female the proboscis is modified for piercing.
5. Body remains at an angle to the surface at rest.

Male :

1. **Plumose** antennae.
2. Maxillary palps are larger than the proboscis and are five jointed and appear club shaped.

Female :

1. **Pilose** antennae.

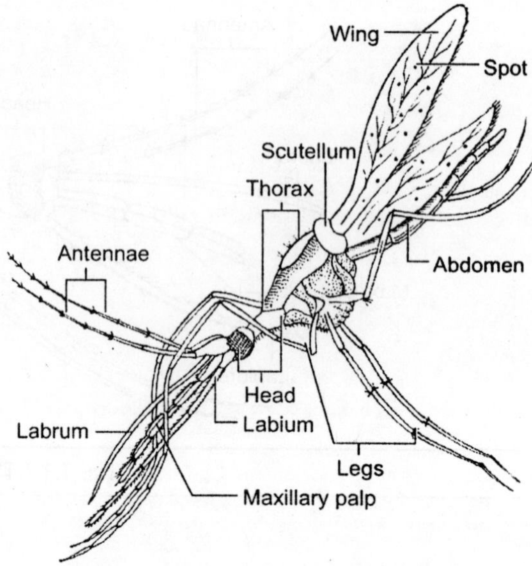

Fig. 1.1.1.131 : *Anopheles* (Female)

2. Maxillary palps usually as long as the proboscis.

Identification with Reasons

State classification and characters up to—

Order—*Diptera* as in *Culex*.

Type—*Anopheles* : State distinctive features as in above.

Economic Importance

Female *Anopheles* are carrier of *Plasmodium*, the protozoan parasite responsible for human malaria.

P.malariae, P.vivax, P.falciparum and *P.ovale* are all passed on to man by the biting of female anopheles.

41. *AEDES*

Habit and Habitat

Terrestrial, domestic, diurnal, biting principally in the morning and late afternoon;

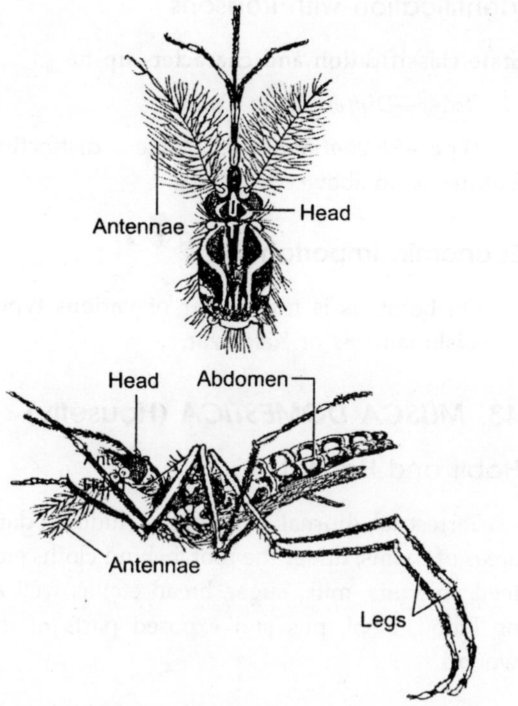

Fig. 1.1.1.132 : *Aedes* (female); head of male *Aedes*

lives hiding behind pictures or under furniture, feed on nectar and plant juice; female lays eggs after a blood meal; important breeding places are drinking water jars, water plants, neglected flower vases, unused toilets and ice box drains, old tins, and holes in trees: distribution world wide.

Systematic Position

Phylum	**Arthropoda**
Sub Phylum	**Mandibulata**
Super class	**Insecta**
Subclass	**Pterygota**
Super order	**Holometabola**
Order	**Diptera**
Type	*Aedes*

Distinctive Features

1. Small black species conspicuously marked with silvery-white on the legs and abdomen.

2. Adult with a lyre-shaped white mark on the thorax.

3. Wings are clear and some what iridescent.

Male :

1. Antennae densely plumose.
2. Maxillary palps in male are as long as the proboscis.

Female :

1. Pilose antennae.
2. Has very short palpi which are white at the tip.

Identification with Reasons

State classification and characters up to—

Order—*Diptera* as in *Culex.*

Type—*Aedes* : State distictive features as in above.

Economic Importance

Female *Aedes* acts as vector and transmit viruses of yellow fever, dengue and miscellaneous viruses by their bites.

42. *PHLEBOTOMUS* (Sand flies)

Habit and Habitat

Nocturnal; during day time hide in cool, damp places, favour dark rooms, thick mud-walled houses; females are blood suckers and biting causes irritation; feeds on man and other mammals; found in all worm and tropical region of the world.

Systematic Position

Phylum	**Arthropoda**
Subphylum	**Mandibulata**
Class	**Insecta**
Subclass	**Pterygota**
Super order	**Holometabola**
Order	**Diptera**
Type	***Phlebotomus***

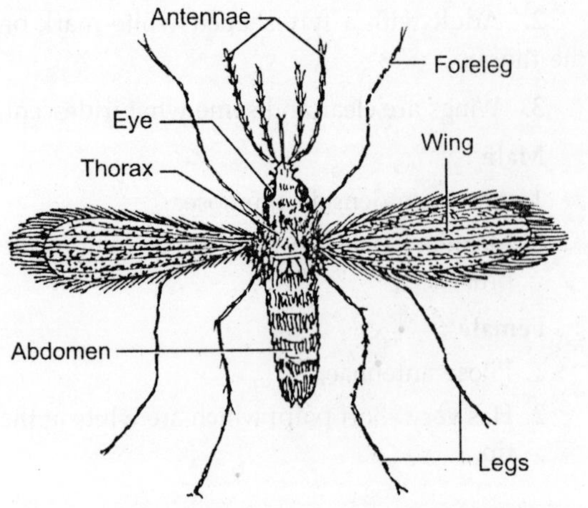

Fig. 1.1.1.133 : *Phlebotomus*

Distinctive Features

1. Small, dull coloured insects with hairy body.
2. Slender antennae, long maxillary palp and proboscis longer than the head.
3. Narrow hairy wings with parallel veins and no cross veins except near the base.
4. Legs, body and wings are covered with long coarse hairs.

Identification with Reasons

State classification and characters up to—

Order—*Diptera* as in *Culex*.

Type—*Phlebotomus* : State distinctive features as in above.

Economic Importance

Phlebotomus is transmitter of various types of **Leishmaniasis** or Kala-azar.

43. *MUSCA DOMESTICA* (Housefly)

Habit and Habitat

Terrestrial; diurnal; at night they hides in dark areas of rooms, under the roof, behind cloths etc.; feed on fruits, milk, sugar, bread etc; as well as on blood, stool, pus and exposed parts of the wound.

Systematic Position

Phylum	**Arthropoda**
Subphylum	**Mandibulata**
Class	**Insecta**

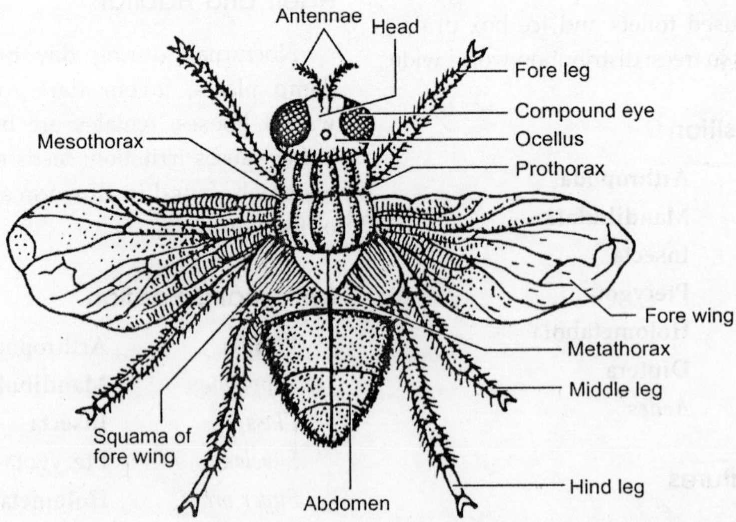

Fig. 1.1.1.134 : *Musca domestica*

Subclass	**Pterygota**
Super order	**Holometabola**
Order	**Diptera**
Type	***Musca domestica***

Distinctive Features

1. Robust and short body.

2. Small head with large compound eyes and short, flexible antennae contain dorsal spiny **arista**.

3. Second thoracic segment large, marked with four longitudinal dark lines.

4. Abdomen segmented, oval and hairy.

Identification with Reasons

State classification and characters up to—

Order—*Diptera* as in ***Culex.***

Type—*Musca domestica* : State distinctive features as in above.

Economic Importance

Housefly co-exists with man. They are the carrier of various well known diseases such as: typhoid, diarrhoea, paratyphoid, amoebic dysentery, tuberculosis, conjunctivitis etc.

44. *XENOPSYLLA* (Rat-flea)

Habit and Habitat

Ectoparasite on worm-blooded mammals; adult frequently jump from one host to another and transferring disease among host; feeds on blood and is capable of free existence for some parts of life.

Systematic Position

Phylum	**Arthropoda**
Subphylum	**Mandibulata**
Class	**Insecta**
Subclass	**Pterygota**

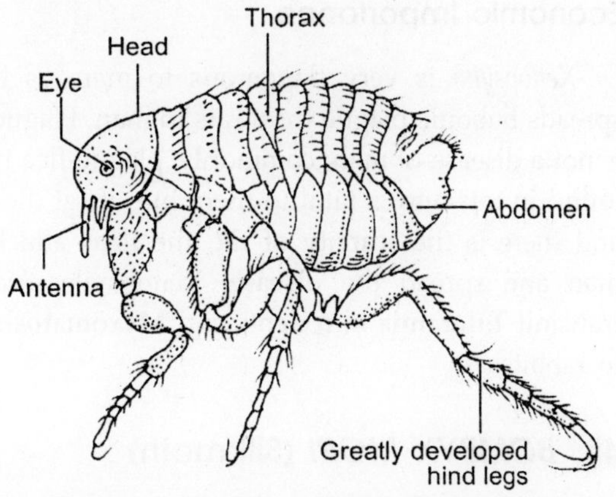

Fig. 1.1.1.135 : *Xenopsylla* (Rat-flea)

Super order	**Holometabola**
Order	**Siphonaptera**
Type	***Xenopsylla***

Distinctive Features

1. Small laterally compressed body furnished with backwardly projecting bristles.

2. Head small, eyes simple.

3. Wings are absent.

4. Biting and sucking type mouth parts.

5. Legs long and stout, adapted for jumping.

Identification with Reasons

State classification and characters up to—

Super order—*Holometabola* as those of ***Hispa.***

Order—*Siphonaptera* :

1. Ectoparasite on birds and mammals.
2. Piercing and sucking mouth parts.
3. Antennae short, eyes simple, no ocelli.
4. Legs long, stout, adapted for jumping.
5. Small, wingless insects.
6. Abdomen without cerci.

Type—*Xenopsylla* : State distinctive features as in above.

Economic Importance

Xenopsylla is very dangerous to man, as it spreads bubonic plague from rats to man. Plague is not a disease of man; commonly, plague-flea is found in rats and is fatal to them but, if rat dies and there is the scarcity of rat, the fleas attack man and spread the disease. Xenopsylla also transmit **Tularamia** in rodents and **Myxomatosis** in rabbits.

45. *BOMBYX MORI* (Silkmoth)

Habit and Habitat

Nocturnal; larvae typically feed on plants, adult moths nonfeeding; domesticform; bred and reared in China, Japan and India.

Systematic Position

Phylum	**Arthropoda**
Subphylum	**Mandibulata**
Class	**Insecta**
Subclass	**Pterygota**
Super order	**Holometabola**
Order	**Lepidoptera**
Type	***Bombyx mori***

Distinctive Features

1. Body stout, creamy white in colour.
2. Antennae short, feathery and pectinate.
3. Hind wings smaller than forewings and wings are horizontally held during resting period.
4. Proboscis absent and labial palp rudimentary.
5. Scales hair-like.

Identification with Reasons

State classification and characters up to—

Super order—*Holometabola* as those of ***Hispa.***

Order—*Lepidoptera* :

1. Variable antennae often clubbed or hooked, or knobbed or feathery.

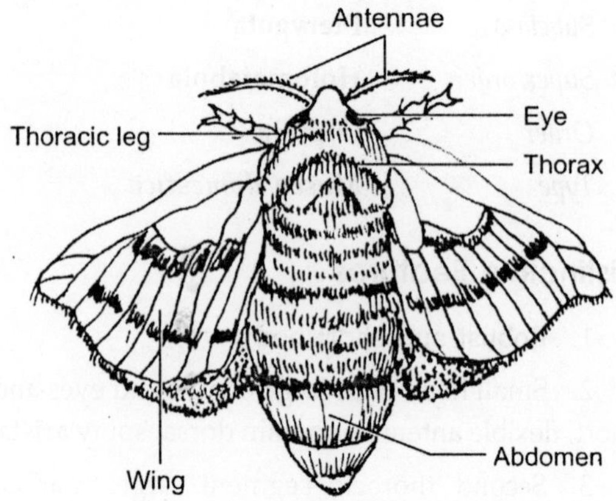

Fig. 1.1.1.136 : *Bombyx mori*

2. Generally with two pairs of membranous wings, often large.
3. Wings with overlapping scales and often hairy.
4. Mouth parts highly modified as coiled proboscis used for sucking.
5. Large compound eyes.
6. Complete metamorphosis.
7. Larvae are caterpillars and usually plant feeder.
8. Adults feed little or not at all.

Type—*Bombyx mori* : State distinctive features as in above.

Economic Importance

Silk moth has a long-standing commercial importance for producing silk of commerce and also played a major role in the development of molecular biology. The mRNA molecules isolated obtain in large quantities are from silk protein of *B.mori.*

46. LARVA OF *B.MORI* (Caterpiller)

Habit and Habitat

Domesticated; Phytophagus; feeds on mulberry leaf.

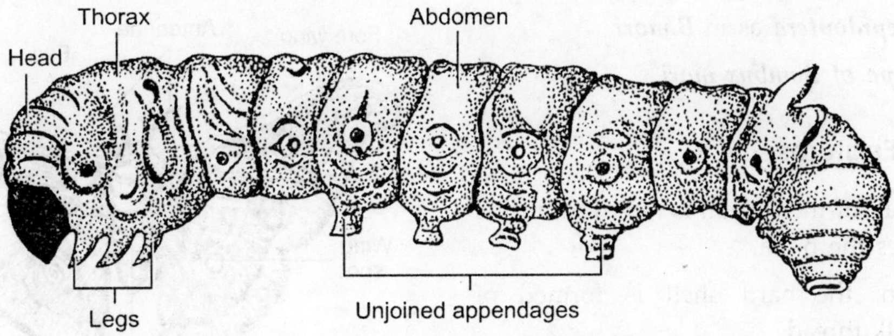

Fig. 1.1.1.136A : Adult larva of *Bombyx mori*

Systematic Position

Phylum	**Arthropoda**
Subphylum	**Mandibulata**
Class	**Insecta**
Subclass	**Pterygota**
Super order	**Holometabola**
Order	**Lepidoptera**
Type	***Bombyx mori***

Identification with Reasons

State classification and characters up to—

Order—*Lepidoptera* as in *B.mori.*

Type—*Larva of Bombyx mori.*

Distinctive Features

1. Body elongated, cylindrical, and divided into prominent head, thorax, and abdomen.
2. Soft body is provided with thick hairy covering.
3. Antennae, compound eyes, and wings are absent.
4. Chewing and biting mouth parts.
5. Thorax with three pairs of jointed legs, the tip of which bears a recurved hook.
6. Ten segmented and bear stumpy unjointed appendages on the third, fourth, fifth, and nineth segments.

47. PUPA OF *B.MORI*

Habit and Habitat

Mature caterpillar become non-feeding and

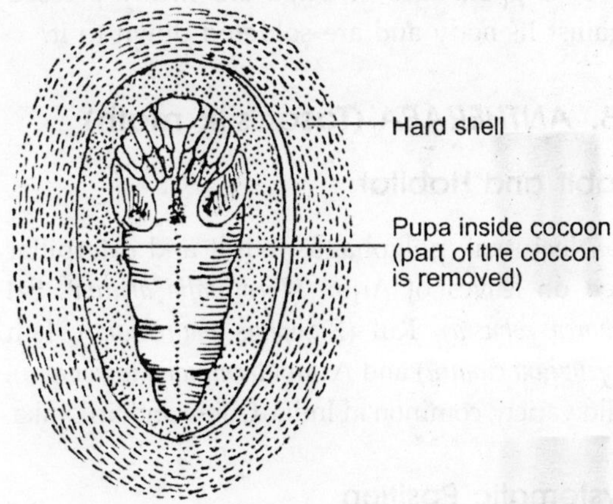

Fig. 1.1.1.136B : Pupa of *Bombyx mori*

secretes fine, long continuous thread of silk from the silk gland which form the Cocoon or pupal case around it.

Systematic Position

Phylum	**Arthropoda**
Subphylum	**Mandibulata**
Class	**Insecta**
Subclass	**Pterygota**
Super order	**Holometabola**
Order	**Lepidoptera**
Type	***Bombyx mori***

Identification with Reasons

State classification and characters up to—

Order—*Lepidoptera* as in *B.mori.*

Type—*Pupa of Bombyx mori.*

Distinctive Features

1. One half of the cocoon is removed; inside the cocoon lies the pupa.

2. Cocoon, the hard shell is formed of continuous silk thread.

3. Within cocoon, pupa transforms into imago.

4. Pupa remain immobile.

5. Appendages of pupa are firmly pressed against its body and are soldered down to it.

48. *ANTHERAEA* (Tasar silk moth)

Habit and Habitat

Nocturnal; polyphagus; adult and caterpillar feed on leaves of Arjun (*Terminalia arjuna*), Sal (*Shorea robusta*), Kul (*Zizyphus mauritiana*), Jam (*Syzygium cumini*) and Asan (*Terminalia tomentosa*); wild variety, common in India, China and SriLanka.

Systematic Position

Phylum	**Arthropoda**
Subphylum	**Mandibulata**
Class	**Insecta**
Subclass	**Pterygota**
Super order	**Holometabola**
Order	**Lepidoptera**
Type	***Antheraea***

Identification with Reasons

State classification and characters up to—

Order—*Lepidoptera* as in *Bombyx*.

Type—*Antheraea*.

Distinctive Features

1. Body stout, bipectinate antennae are short and feathery.

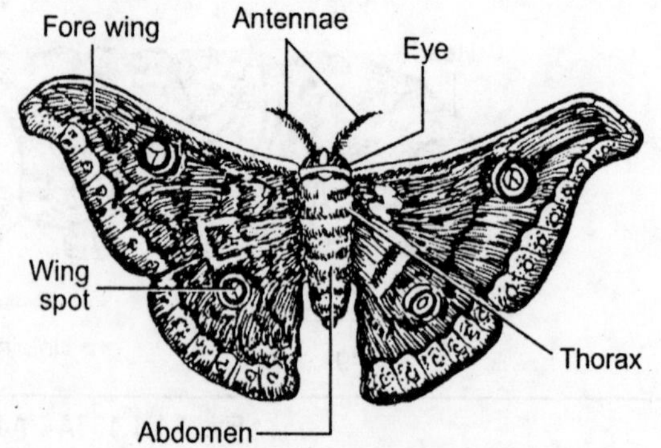

Fig. 1.1.1.137 : *Antheraea*

2. Scales hair like.

3. Proboscis absent; labial palp minute.

4. Forewings larger then hind wing and with transparent eye spot near the centre of each wings.

5. Characteristic colour- patterns are found.

Economic importance

Antheraea produces tasar, a coarse variety of silk which is also of commercial importance.

49. *TRYPORYZA INCERTULAS* (Rice pest)

Habit and Habitat

Nocturnal; much attracted to light; adult non-injurious; larvae penetrate into the paddy plant stem and feed the juice of the phloem.

Systematic Position

Phylum	**Arthropoda**
Subphylum	**Mandibulata**
Class	**Insecta**
Subclass	**Pterygota**
Super order	**Holometabola**
Order	**Lepidoptera**
Type	***Tryporyza incertulus***

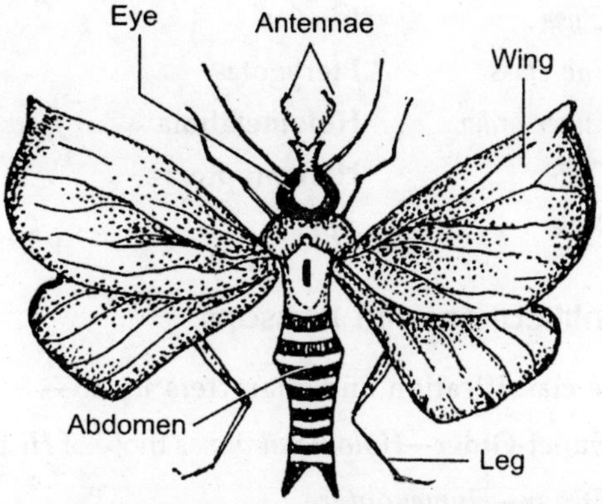

Fig. 1.1.1.138 : *Tryporyza* (Male)

Identification with Reasons

State classification and characters up to—

Order—*Lepidoptera* as in Bombyx.

Type : *Tryporyza incertulus.*

Distinctive Features

1. Body stout, antennae short, filiform and not swollen at tip.
2. Wings yellowish brown in colour.
3. Reduced proboscis and labial palp distinct.
4. Female moth are larger than male and each fore wing bears a central distinct black spot.
5. Male pale whitish yellow, fore wing spot inconspicuous.
6. Female have a prominent tuft of brownish yellow silken hair at the tip of their abdomen.

Economic Importance

Serious major pest of rice, adult lays eggs on stem and leaves of paddy plant. On emergence larva migrate and penetrate into the plant stem and feed, thus causing much damage to the plant.

50. BUTTERFLY

Habit and Habitat

Diurnal; found in the lower slops of hills where vegetation is abundant and varied; feeds on plant juices, nectar, and fruit juices; larvae feeds on various parts of plants, particularly leaves; distribution cosmopolitan.

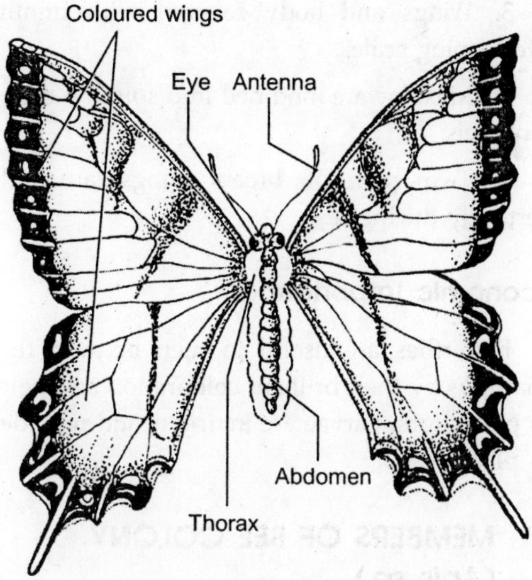

Fig. 1.1.1.139 : Butterfly

Systematic Position

Phylum	**Arthropoda**
Subphylum	**Mandibulata**
Class	**Insecta**
Subclass	**Pterygota**
Super order	**Holometabola**
Order	**Lepidoptera**
Type	***Butterfly***

Identification with Reasons

State classification and characters up to—

Order—*Lepidoptera* as in *Bombyx*.

Type—*Butterfly*.

Distinctive Features

1. Body slender and wings brightly coloured.
2. Head with large compound eyes and clubbed antennae (with swollen tips).

3. Wings and body covered with minute overlapping scales.

4. Maxillae are modified into spirally- coiled proboscis.

5. Two pair of broad wings are held vertically during rest.

Economic Importance

Butterflies are useful to men because they charm us by their brilliant colouration and helps as pollinators. Larvae are injurious and regarded as plant pests.

51. MEMBERS OF BEE COLONY (*Apis sp.*)

Honey bee is a social insect and lives in a permanent colony. *Apis indica* (Indian – bee) is common Honey bee found in the forests and plains throughout India. A bee colony consists of three casts namely, Queen, drones (males) and workers.

Systematic Position

Phylum	**Arthropoda**
Sub Phylum	**Mandibulata**
Class	**Insecta**
Sub class	**Pterygota**
Super order	**Holometabola**
Order	**Hymenoptera**
Type	***Apis***

Identification with Reasons

State classification and characters up to—

Super-Order—*Holometabola* as those of ***Hispa.***

Order—*Hymenoptera*

1. Chewing mouth parts but also modified for lapping or sucking.
2. Winged or wingless species.
3. Wings are transparent with only a few veins.
4. Larvae are caterpillers or are grub like with chewing mouth parts.

Distinctive Features

(1) Queen :

1. Larger in size, abdomen greatly distended and is conical.

2. Wings are short.

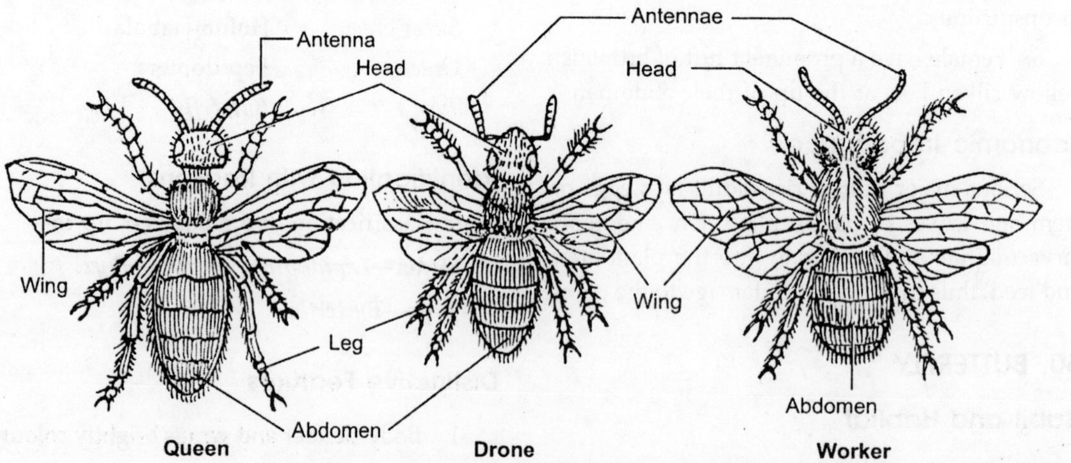

Fig. 1.1.1.140 : Queen, Drone and Worker of *Apis*

3. Well developed legs for walking over the hive.

4. Presence of **ovipositor**, which serves for egg-laying and defensive organ (**sting**).

5. **Pollen collecting baskets** and wax secreting structures are absent.

Economic Importance : The duty of the Queen is to produce eggs and she actually develops the whole colony. She is highly protected in the **Royal chamber**.

(2) Drones (Male) :

1. Body is comparatively broad and abdomen is blunt without any sting.

2. Head is quite round and eyes are very large.

3. Last part of the abdomen is squarish.

4. Wax-glands and pollen-collecting apparatus are absent.

5. They develop from the unfertilized egg parthenogenetically.

Economic Importance : They perform no work and depend for food on the workers. The main work of the drone is to fertilize the virgin female.

(3) Workers (Sterile female) :

1. Small about 12mm.

2. Body is densely covered with hairs.

3. Body colour is usually brown or black.

4. Possess sting with poison sac.

5. **Wax gland** and **pollen basket** present.

6. Mouth sucking type, proboscis well developed.

Economic Importance : Worker bees perform works namely protection, food collection, nursing etc. of the colony. They collect nectar, pollen, propolis and prepare hive, repairs hive and produce honey. They protect the hive and drive away all the drones after mating with the Queen.

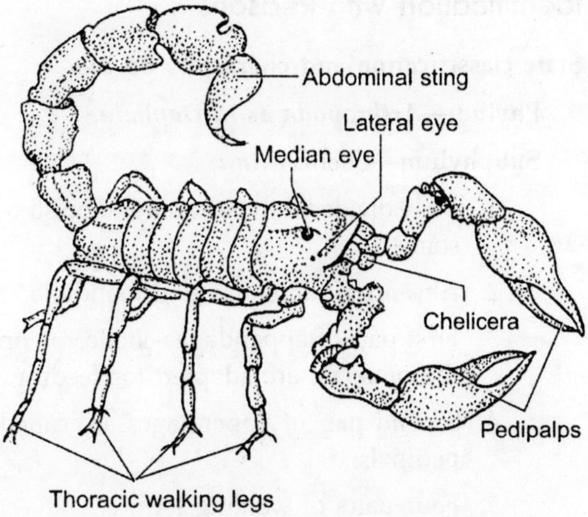

Fig. 1.1.1.141 : *Buthus*

52. *BUTHUS* (Scorpion)

Habit and Habitat

Terrestrial; nocturnal; viviparous; found in sand, crevices, cracks and under wood, stone, loose bark; carnivorous; feeds on insects, spiders and other small animals.

Systematic position

Phylum	**Arthropoda**
Subphylum	**Chelicerata (Chelicera=G:claw)**
Class	**Arachnida**
Order	**Scorpiones**
Type	**Buthus**

Distinctive Features

1. Elongated body, differentiated into **Prosoma**, **metasoma** and **opisthosoma**.

2. Prosoma bears a pair of chelate **chelicerae**, a pair of large chelate **pedipalps** and four pairs of walking legs.

3. The last metasomatic segment is telson containing a sting.

Identification with Reasons

State classification and characters up to—

Phylum—*Arthropoda* as in ***Daphnia*.**

Subphylum—*Chelicerata* :

1. Body divided into prosoma and opisthosoma.
2. Absence of antennae and mandible.
3. First pair of appendages chelicerae on the prosoma are adapted for feeding.
4. Second pair of appendages are called pedipals.
5. Four pairs of walking legs.

Class—*Arachnida* :

1. Prosoma possesses paired chelicerae, pedipalpi and four pairs of walking legs.
2. Compound eyes absent.
3. Abdomen without external gills and locomotor appendages.
4. Epicuticle become waxy.
5. The book gills modified to developed book lungs and trachae.

Order—*Scorpiones* :

1. Prosoma and opisthosoma joined broadly; no pedicel.
2. Three-jointed chelicerae, fangless.
3. Opisthosoma segmented with a terminal poison sting.

Type—*Buthus* : State distinctive features as in above.

Special Features

Harmful to mankind; venom is neurotoxic; it's sting causes extreme pain.

53. *ARANEUS* (Garden spider)

Habit and Habitat

Terrestrial; sedentary; spins web; found in houses and gardens; carnivorous; feeds on spiders, insects and small animals; cosmopolitan distribution.

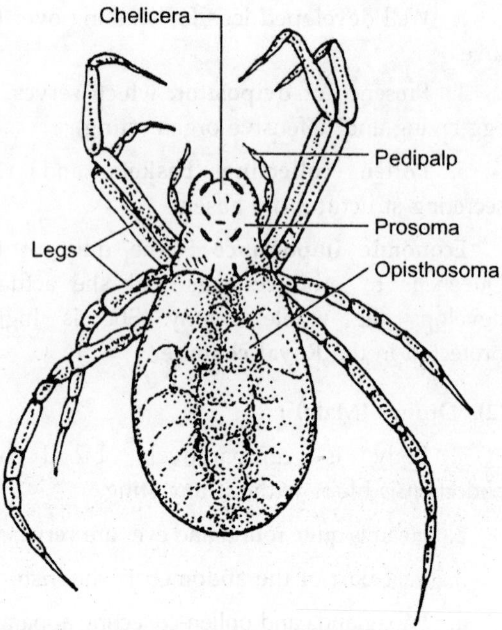

Fig. 1.1.1.142 : *Araneus*

Systematic Position

Phylum	**Arthropoda**
Subphylum	**Chelicerata**
Class	**Arachnida**
Order	**Araneae**
Type	***Araneus***

Distinctive Features

1. Body differentiated into anterior prosoma and posterior opisthosoma.
2. Prosoma and opisthosoma are jointed by narrow predicel.
3. Prosoma bears eight eyes and six pairs of appendages.
4. Chelicerae contain poison glands.
5. Opisthosoma, soft, rounded and without telson.
6. At the tip of abdomen are **spinnerets** which produce silk for making web.

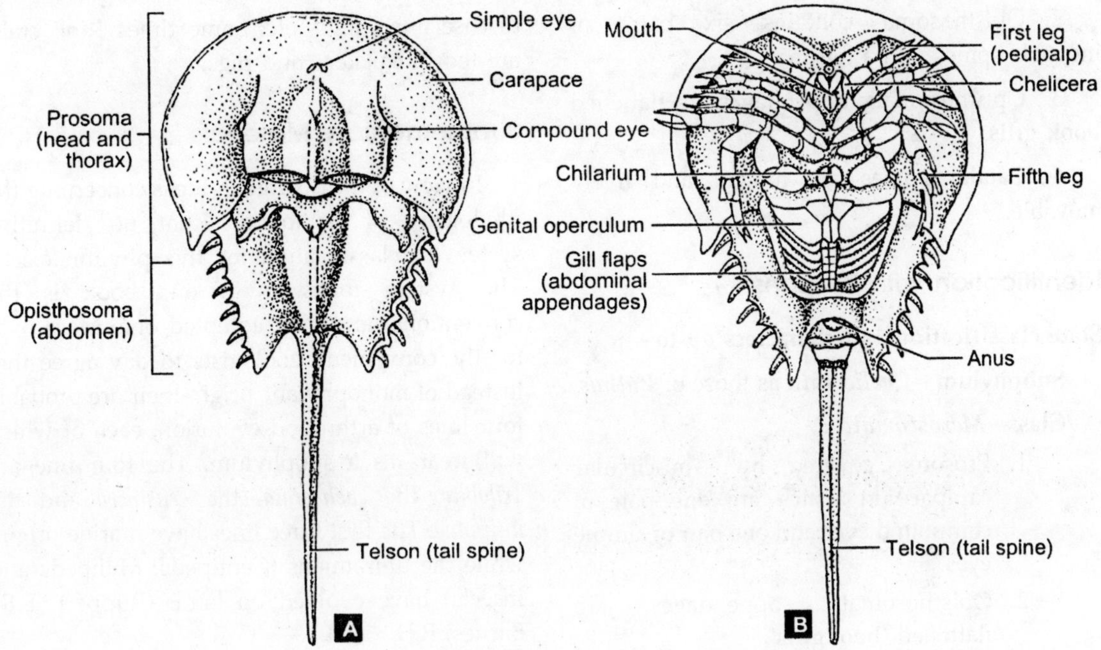

Fig. 1.1.1.143 : *Limulus.* A–Dorsal view; B–Ventral view.

Identification with Reasons

State classification and characters up to—

Class—*Arachnida* as in *Buthus*.

Order—*Araneae* :

1. Opisthosoma attached to prosoma by a narrow pedicel.
2. Chelicerae two-jointed with poison glands.
3. Opisthosoma rounded with terminal spinnerets.
4. A large sternum is on the ventral surface.

Type—*Araneus* : State distinctive features as in above.

Special Features

Garden spider construct webs of various types. Some spiders hunt on the ground and make silk tents only for protection while moulting.

54. *LIMULUS* (King Crab)

Habit and Habitat

Marine; estuarine; found burrowing in the sand or muddy bottom partly buried; lives in shallow water along the shore; feeds on bivalves and small animals; active at night.

Systematic Position

Phylum	**Arthropoda**
Subphylum	**Chelicerata**
Class	**Merostomata**
Order	**Xiphosura**
Type	***Limulus***

Distinctive Features

1. Body is differentiated into a broad, semicircular, horse-shoe-shaped carapace, which is convex above and narrow triangular opisthosoma.

2. Opisthosoma contains six pairs of movable spines and a telson.

3. Opisthosomatic appendages are flattened **book gills.**

4. Caudal spines are hinged and freely movable.

Identification with Reasons

State classification and characters up to—

Subphylum—*Chelicerata* as those of ***Buthus.***

Class—*Merostomata* :

1. Prosoma covered by semi-circular carapace in which are one pair of compound eyes and one pair of simple eyes.
2. Opisthosomatic appendages are flattened **'bookgills'**.
3. Terminal portion of body drawn out into an elongated spine called telson.

Order—*Xiphosura* :

1. Semicircular, horse-shoe-shaped carapace covering the prosoma.
2. Opisthosoma has a dorsal shield with serrated edges.
3. Prosoma and opisthosoma are joined by a transverse hinge.
4. Elongated caudal spine.
5. Prosoma with five pairs of similar appendages.

Type—*Limulus* : State distinctive features as in above.

Economic Importance

Limulus are not true crabs; for its majestic look, it is called King crab. Despite the small number of species and limited geographical distributions, students have profited considerably from studying their biology, specially the principle of vision. King crab blood components are routinely used to test injectable pharmaceutical solutions to screen for venereal disease, meningitis etc. Some times King crabs are fed to chicken and pigs.

Supplementary Materials

In view of the divergent ideas concerning the phylogeny of Arthopoda, that no definitive system of classification for this phylum exists. The system followed in our book is the conventional and long accepted one. More over for the convenient, zoologists to day agree that instead of monophylatic origin their are probable four lines of arthropod evolution, each of which will treats as a subphylum. The four line are *Trilobita*, the *chelicerata*, the *crustacea* and the *uniramia*. The first three lines have marine origin, while the uniramians (Centipeds, Millipeds and Insects) have evolved on land. (Ruppert, E.E., Barnes, R.D., 1994.)

Systematic résumé of the Subphylum Crustacea

1. Daphnia

Class—**Branchiopoda**

Subclass—**Diplostraca** (body enclosed within a laterally compressed carapace)

Order—**Cladocera**

Type—***Daphnia***

2. Triops

Class—**Branchiopoda**

Subclass—**Calamanostraca** (thorax covered by carapace)

Order—**Notostraca**

Type—***Triops***

3. Branchipus

Class—**Branchiopada**

Subclass—**Sarsostraca** (Trunk compossed of 11 to 18 segments with appendages)

Order—**Anostraca**

Type—***Branchipus***

4. Cyclops

Class—**Copepoda**

Order—**Cyclopoida**

Type—***Cyclops***

5. Argulus

Class—**Branchiura**

Type—***Argulus***

6. Balanus

Class—**Cirripedia**

Order—**Thoracica**

Sub-Order—**Balanomorpha** (Symmetrical, sessile barnacles with a wall surmounted by the paired movable terga and scuta.)

Type—***Balanus***

7. Lepas

Class—**Cirripedia**

Order—**Thoracia**

Sub-Order—**Lepadomorpha** (Pedunculate barnacles)

Type—***Lepas***

8. Sacculina

Class—**Cirripedia**

Order—**Rhizocephala**

Type—***Sacculina***

9. Mysis

Class—**Malacostraca**

Subclass—**Eumalacostraca**

Super Order—**Peracarida**

Order—**Mysidacea**

Type—***Mysis***

10. Macrobrachium

11. Penaeus

12. Penaeus indicus

13. Eupagurus

14. Hippa

15. Crab

Class and Subclass same as in Mysis.

Super Order—**Eucarida**

Order—**Decapoda**

Type—***Macrobracium, Penaeus, P. Indicus, Eupagurus, Hippa, Crab***

16. Squilla

Class—**Malacostraca**

Sub-class—**Haplocarida**

Order—**Stomatopoda**

Type—***Squilla***

17. Oniscus

Class, Subclass and *Super order* same as in Mysis.

Order—**Isopoda**

Type—***Oniscus***

Systemic résumé of the Subphylum Uniramia (Antenata)

18. Julus

Subphylum—**Uniramia** (Unbranched appendages, presence of a single pair of antennae, gas exchange organs are tracheae, excretory organs are malpighian tubules.)

Class—**Diplopoda**

Subclass—**Helminthomorpha** (Segments more or less cylindrical or some what flattented in cross section. At least one pair of legs (gonopods) of the seventh segment is modified for sperm tranfer.)

Order—***Julida*** (Trunk composed of 30-90 cylindrical segments. Both pairs of legs of the seventh segment modified as gonopods in male.)

Type—***Julus***

19. Scolopendra

Class—**Chilopoda**

Subclass—**Epimorpha** (Brooed eggs, young possess all segments on hatching. Adult has 21 or more pairs of legs.)

Order—**Scolopendromorpha** (Posses 21 pairs of legs. Targal plates not alternatings in size. With or without eyes.)

Type—***Scolopendra***

20. Silverfish

Class—**Insecta**

Subclass—**Ectognatha** (Mouth parts not sunk into a pouch. Presence of an ovipositor derived from the eighth and ninth abdominal segments.)

Order—**Thysanura** (Three styliform appendages on the abdomen. Small compound eyes)

Type—***Lepisma***

21. Dragonfly

Order—**Odonata**

Type—***Dragonfly***

22. Cockroach

Order—**Orthoptera** (Large headed insects with strong chewing mouth parts and compound eyes. Femur of hind leg enlarged for jumping in many species. Winged and winged less species. Most winged forms have membranous hind wings folded fan like beneath leathery fore wings. Gradual metamorphosis.

Type—***Periplaneta americana***

23. Mantis

Order—**Orthoptera**

Type—***Mantis***

24. Stick insect

Order—**Orthoptera**

Type—***Stick insect***

25. Phyllium (Leaf insect)

Order—**Orthoptera**

Type—***Phyllium***

26. Forficula (Earwig)

Order—**Dermaptera**

Type—***Forficula***

27. Termite Queen

Order—**Isoptera**

Type—***Termite queen***

27A. Members of Termite colony

As above

28. Grasshopper

Order—**Orthoptera**

Type—***Grasshopper***

29. Locust

Order—**Orthoptera**

Type—***Locust***

30. Gryllotalpa (Mole cricket)

Order—**Orthoptera**

Type—***Gryllotalpa***

31. Gryllus (House cricket)

Order—**Orthoptera**

Type—***Gryllus***

32. Pediculus (Human body and head lice)

Order—**Anoplura** (Mouth parts adopted for sucking, small wingless insects, ectoparasite on human, act as vectors of disease.)

Type—***Pediculus***

33. Leptocorisa (Gundi bug)

Order—**Hemiptera**

Type—***Leptocorisa***

34. Belostoma (Giant water-bug)

Order—**Hemiptera**

Type—***Belostoma***

35. Cimex (Bed-bug)

Order—**Hemiptera**

Type—***Cimex***

36. Aphis (Aphid; Plant lice)

Order—**Homoptera**

Type—***Aphis (Aphid)***

37. Cicada

Order—**Homoptera**

Type—***Cicada***

38. Hispa

Order—***Hispa armigera*** (=*Dicladispa armigera*)

39. Culex

Order—**Diptera**

Type—***Culex***

40. Anopheles

Order—**Diptera**

Type—***Anopheles***

41. Aedes

Order—**Diptera**

Type—***Aedes***

42. Phlebotomus (Sandflies)

Order—**Diptera**

Type—***Phlebotomus***

43. Musca (Housefly)

Order—**Diptera**

Type—***Musca domestica***

44. Xenopsylla (Rat-flea)

Order—**Siphonaptera**

Type—***Xenopsylla***

45. Bombyx mori

Order—**Lepidoptera**

Type—***Bombyx mori***

46, 47. Larva and pupa of Bombyx

Same as above.

48. Antheraea

Order—**Lipidoptera**

Type—***Antheraea***

49. Tryporyza (Rice pest)

Order—**Lepidoptera**

Type—***Tryporyza incertulus***

50. Butterfly

Order—**Lepidoptera**

Type—***Butterfly***

51. Apis

Order—**Hymenoptera**

Type—***Apis***

52. Buthus (Scorpion)

Order—**Scorpiones**

Type—***Buthus***

53. Araneus (Garden spider)

Order—**Araneae**

Type—***Araneus***

54. Limulus (King crab)

Order—***Xiphosura***

Type—***Limulus***

This résumé has been designated to meet the requirement of those invertebrae types that are included in the practical syllabus. This account is not intended to be equivalent to the extensive treatment. The characters and full account should be followed from the previously dealt Arthropod section.

ONYCHOPHORA

(Greek: claw bearer)

Distinguishing Characteristics :
Second pair of appendages are highly modified to form jaws surrounding the mouth; third pair of appendages form oral papillae; a pair of clawed, stumpy legs on each trunk segment.

1. *PERIPATUS*

Habit and Habitat

Terrestrial; found in moist habitats in tropical environments and southern temperate region; living in damp forests under leaves or stones, in bark of trees and rock crevices; nocturnal; carnivorous; feeding on various smaller arthropods.

Systematic Position

Phylum	**Onychophora**
Type	***Peripatus***

Distinctive Features

1. Worm like body with homonomous segmentation and enclosed by a thin, flexible chitinous cuticle.
2. A pair of clawed, lobopodial limbs is found on each trunk segment.
3. Second pair of appendages highly modified to form jaws surrounding the mouth.
4. Pair of appendages form stubby oral papillae.

Identification with Reasons

Phylum—*Onychophora* :

1. A pair of clawed, limbs is found on each trunk segment.
2. Second pair of appendages highly modified to form jaws.
3. Slime glands discharge adhesive material through opening on the oral papillae.

Type—*Peripatus* : Characters same as in above.

Special Features

Peripatus is zoologically very interesting because it exhibits both arthropod and annelid characters. The evolutionary history of the Onychophorans have been subject of controversy, as it is impossible to tell whether these animals were more like arthropods or more like annelids. Traditionally they have been regarded as separate phylum transitional between annelids and arthropods.

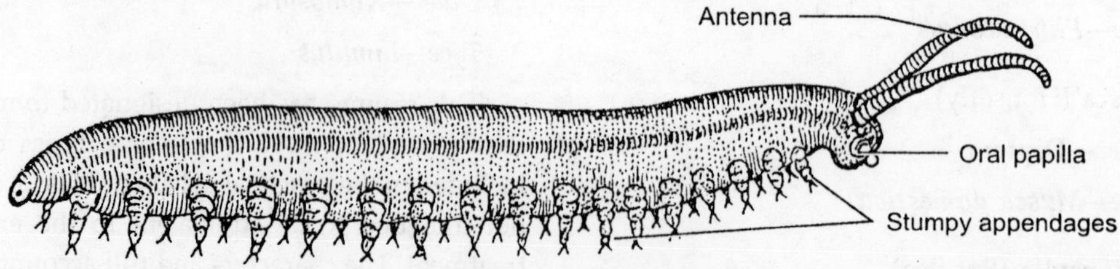

Fig. 1.1.1.144 : *Peripatus*

Classification is based on Pechenik, J.A., 2002.

ECHINODERMATA

(Greek: spine skin)

Distinguishing Characteristics :
A complex series of fluid- filled canals (water-vascular system) derived from pair of coelomic compartments, which leads to thin walled numerous flexible feeding and locomotory structures called **podia** or **tube feet**; adults show a **five-point** (Pentamerous) **radial symmetry**; endoskeleton of calcarious ossicles divided from mesodermal tissue.

An outline classification of **Phylum-Echinodermata** after Ruppert, E.E. and Barnes, R.D., 1994 (for those types that are included in the text)

- *Sub Phylum* **Crinozoa**
  - *Subclass* Articulata
    - *Order*—Comatulida
      Example : *Antedon*
- *Sub Phylum* **Asterozoa**
  - *Classes*
    - Asteroida
      - *Orders*
        - Paxillosida
          Example : *Astropecten*
        - Forcipulata
          Example : *Asterias, Pentoceros*
    - Ophinoroidea
- *Sub Phylum* **Echinozoa**
  - *Classes*
    - Echinodea
      - *Subclass* Euechinoidea
        - *Superorders*
          - Echinacea
            - *Order* Echinoida
              Example : *Echinus*
          - Gnathostomata
            - *Order* Clypeasteroida
              Example : *Clypeaster*
    - Holothuroida
      - *Orders*
        - Dendrochirotida
          Example : *Cucumeria*
        - Aspidochirotida
          Example : *Holothuria*

1. *ANTEDON* (sea-lily)

Habit and Habitat

Marine; found at moderate depths of water; remains attached to rock and stones by cirri; creeps over the sea bottom; Suspension feeders; feeds on diatoms, algae, protozoa and small planktons; worldwide in distribution.

Systematic Position

Phylum	**Echinodermata**
Subphylum	**Crinozoa**
Class	**Crinoidea (G: lily like)**
Order	**Comatulida**
Type	***Antedon***

Distinctive Features

1. Body composed of a central calyx from which slender arms and cirri radiate.
2. Central mouth leads into five **ambulacral grooves** which give branches into each pinnules and contain tube feet.
3. Ten arms with numerous pinnules.

Identification with Reasons

Phylum—*Echinodermata* :

1. Body radially symmetrical and pentamerous.
2. Head absent; distinct **oral surface** and **aboral surface** present.
3. Surface with spines and ossicles.
4. Body is distinguishable into oral and aboral surfaces.

Subphylum—*Crinozoa* :

1. An aboral cup shaped (calyx) body.
2. Upwardly directed pentamerous arms.
3. Ambulacra, as radially divergent systems, restricted to the aboral surface.

Classification is based on Marshall, A.J., Williams, W.D., (eds.), 1972.

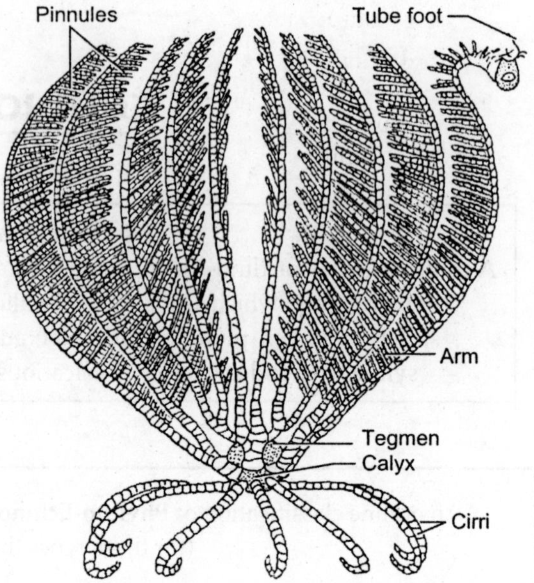

Fig. 1.1.1.145 : *Antedon* (Sea-lily)

4. Attached to the substrate by aboral surface or by a stalk.

Class—*Crinoidea* :

1. Body is supported above the substrate by a long stalk or by cirri.
2. Calyx equipped with arms bearing pinnules.
3. Oral surface directed upward.
4. Mouth and anus both on the oral surface.

Order—*Comatulida* :

1. Calyx is dicyclic *i.e.*, formed of basals and infrabasals.
2. Arms bear pinnules and branched.
3. Stalk absent; cirri present.

Type—*Antedon* : State distinctive features as in above.

2. *OPHIURA* (Brittle-star)

Habit and Habitat

Marine; found in all oceans at all depths; hide under rocks or in crevices during daytime, active

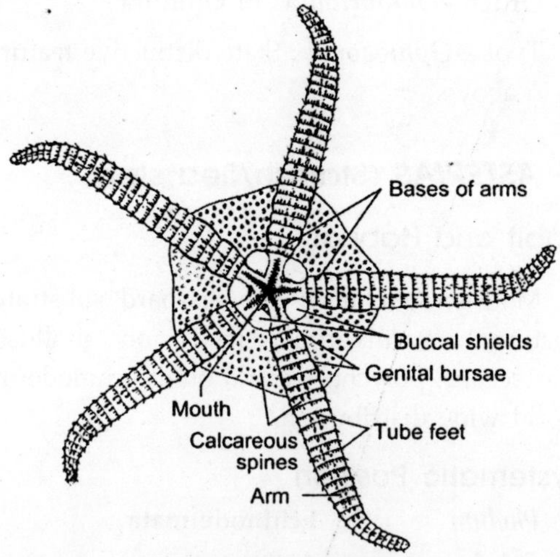

Fig. 1.1.1.146 : *Ophiura*

only at night; deposit feeders, ingesting sediments and assimilating the organic fraction.

Systematic Position

Phylum	**Echinodermata**
Subphylum	**Asterozoa**
**Class*	**Stelleroidea**
Subclass	**Ophiuroidea (G: Snake-like)**
Order	**Ophiurida**
Type	***Ophiura***

Distinctive Features

1. Presence of oral and aboral surfaces with distinct pentagonal **central disc** and five highly flexible, fragile, snake-like arms.
2. Arms many times longer than the central disc.
3. **Madreporite** is oral.
4. Presence of two slits in each genital bursa.

Identification with Reasons

State characters up to—

Phylum—*Echinodermata* as in *Antedon*.

* *Subclass describe as class by Ruppert, E.E., Barnes D.B., (1994).*

Sub-phylum—*Asterozoa* :

1. Star shaped body without stem.
2. Pentamerous arms directed horizontally.
3. Mouth on the oral and anus on the aboral surface.
4. Ambulacral ossicles are biserial and the basal one are dislocated from the calyx.

Class—*Stelleroidea* :

1. Long arms arising from a small central disc.
2. Madreporite present.
3. Tube feet on lateral branches of radial water canal.

Subclass—*Ophiuroidea* :

1. Flattened body with pentagonal or rounded central disc.
2. Arms usually five, long, slender, smooth or spiny.
3. Slender arms distinctly demarcated from central disc.
4. The ambulacral plates fuse together in pairs to form vertebrae.
5. Suctorial tube feet absent.
6. Anus lost.

Order—*Ophiurida* :

1. Disc and arms completely covered in dorsal and ventral plates.
2. Flexible arms extends from a central body, are built of jointed, vertebral ossicles (vertebrae)
3. Spines on arms are borne laterally and directed outwardly.

Type—*Ophiura* : State distinctive features as in above.

3. *OPHIOCOMA*

Habit and Habitat

Marine; found commonly in shallow waters; remain hide during daytime beneath stones, vegetations etc; active at night; deposit feeder; world wide distribution; common Indian specimen.

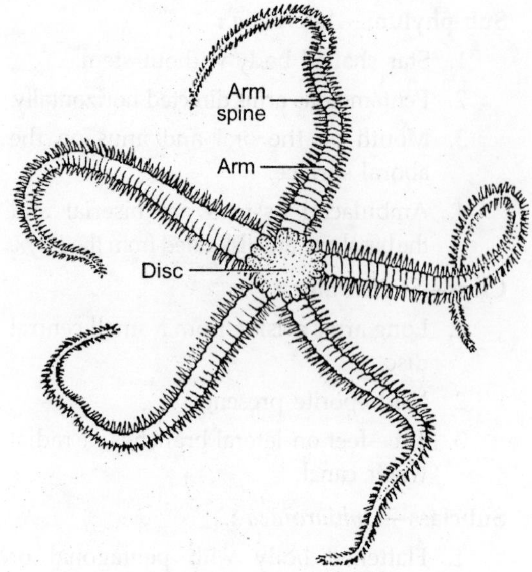

Fig. 1.1.1.147 : *Ophiocoma*

Systematic Position

Phylum	**Echinodermata**
Subphylum	**Asterozoa**
**Class*	**Stelleroidea**
Subclass	**Ophiuroidea**
Order	**Ophiurida**
Type	***Ophiocoma***

Distinctive Features

1. Body consists of a small, rounded central disc.
2. Five-long, slender tapering arms inserted on the under-surfaces of the disc.
3. Presence of ten elongated slits, two on either side of the base of the arm, called genital bursae.
4. Single row of short **tube feet** projects on each side between lower and lateral plates.
5. Madreporite indistinct.

Identification with Reasons

State classification and characters up to—

Order—*Ophiurida* as in **Ophiura**.

Type—*Ophiocoma* : State distinctive features as in above.

4. *ASTERIAS* (Starfish/Sea star)

Habit and Habitat

Marine; sedentary; live on hard substrates; nocturnal; carnivorous; feeds on mollusca, crustaceans, polychaetes and other echinoderms; world wide distribution.

Systematic Position

Phylum	**Echinodermata**
Subphylum	**Asterozoa**
**Class*	**Stelleroidea**
Subclass	**Asteroidea**
Order	**Forcipulata**
Type	***Asterias***

Distinctive Features

1. Star shaped flattened body with a central disc and five-radiating arms.
2. Aboral surface contains spines all over.
3. Arms and Madreporite on aboral surface.
4. Madreporite in inter-radial position.
5. Oral surface contains a mouth, from which five ambulacral groove with tube feet extend in each arm.
6. Arms with terminal tentacles.

Identification with Reasons

State classification and characters up to—

Class—*Stelleroidea* as in **Ophiura.**

Subclass—*Asteroidea* **:**

1. Pentagonal or star-shaped body with a central disc.
2. Distinct oral and aboral surface.
3. Usually five arms radiating from the disc.

* *Subclass describe as class by Ruppert, E.E., Barnes D.B., (1994).*

* *Subclass describe as class by Ruppert, E.E., Barnes D.B., (1994).*

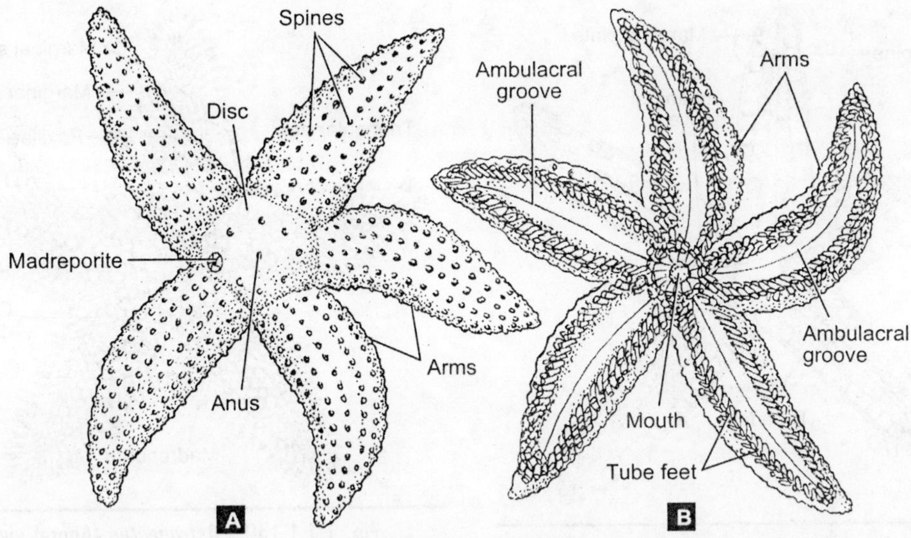

Fig. 1.1.1.148 : *Asterias*, A–Aboral view, B–Oral view

4. Ambulacral grooves are open with rows of tube feet.
5. Ambulacra are restricted to the oral surface.
6. Central disc and arms fused.

Order—*Asteroidea* :

1. Forcipulate pedicellariae (jaw like projection of body wall).
2. Papulae on oral and aboral surface.
3. Marginal plate inconspicuous.
4. Suctorial tube feet usually arranged in four rows.

Type—*Asterias* : State distinctive features as in above.

5. *PENTOCEROS* (Sea Pentagon)

Habit and Habitat

Marine; found in sea from shallow water to deep water; carnivorous; feeds on molluscs, crustaceans, polychaets; commonly found in Indian Seas.

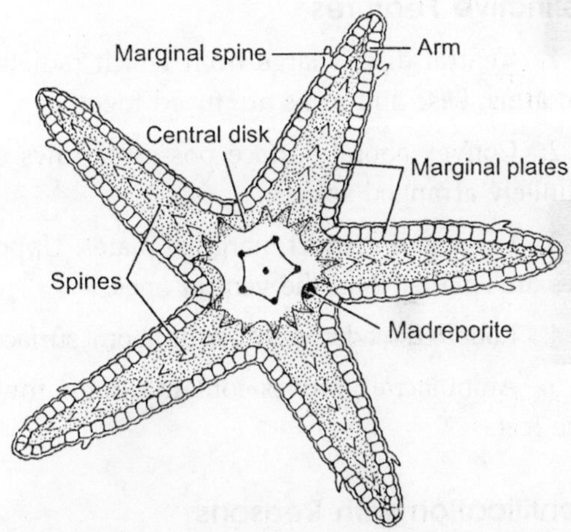

Fig. 1.1.1.149 : *Pentaceros* (Aboral view)

Systematic Position

Phylum	**Echinodermata**
Subphylum	**Asterozoa**
**Class*	**Stelleroidea**
Subclass	**Asteroidea**
Order	**Forcipulata**
Type	***Pentoceros***

* *Subclass describe as class by Ruppert, E.E., Barnes D.B., (1994).*

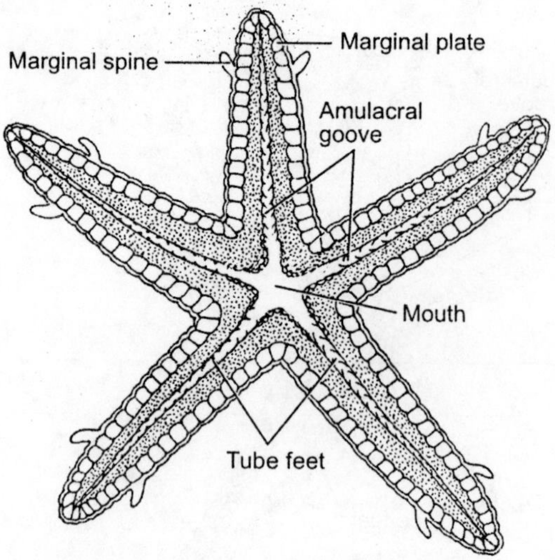

Fig. 1.1.1.149A : *Pentaceros* (Oral view)

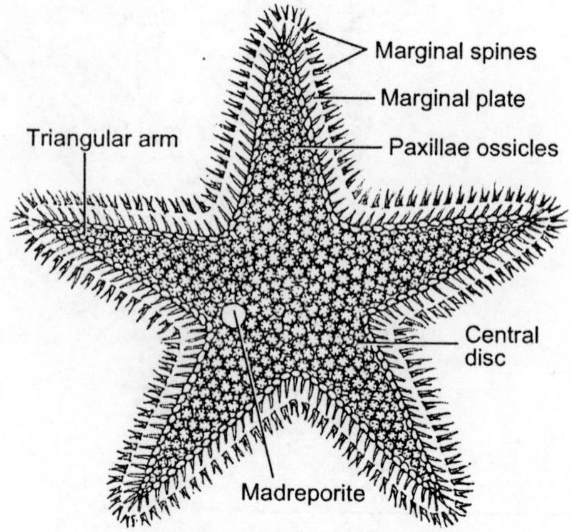

Fig. 1.1.1.150 : *Astropecten* (Aboral view)

Distinctive Features

1. Central disc is large from which radiates five arms. Disc and arms are fused together.

2. Convex aboral surface possesses rows of definitely arranged spines.

3. Large and distinct marginal plates. Upper ones are smaller than the ventral ones.

4. Rounded madreporite on the aboral surface.

5. Ambulacral grooves contain double rows tube feet.

Identification with Reasons

State characters up to—

Order—*Forcipulata* as in *Asterias*.

Type—*Pentoceros* : State distinctive features as in above.

6. *ASTROPECTEN* (Mud star)

Habit and Habitat

Marine; bottom dwelling; living in or on sediment; found in intertidal zone lies partly buried in sand with central part of its body raised above the sand; feeds on bivalves, molluscs, snails, crustaceans and polychaets; distributed in tropical and subtropical waters.

Systematic Position

Phylum	**Echinodermata**
Subphylum	**Asterozoa**
**Class*	**Stelleroidea**
Subclass	**Asteroidea**
Order	**Paxillosida**
Type	***Astropecten***

Distinctive Features

1. Star shaped body having five pointed central disc with five short flexible triangular arms with tapering tips.

2. Ambulacral grooves with double rows of tube feet.

3. Arms and disc bordered by plates and spines.

4. Aboral surface granular; contains blunt spines, branchiae and pedicellariae.

---

* *Subclass describe as class by Ruppert, E.E., Barnes D.B., (1994).*

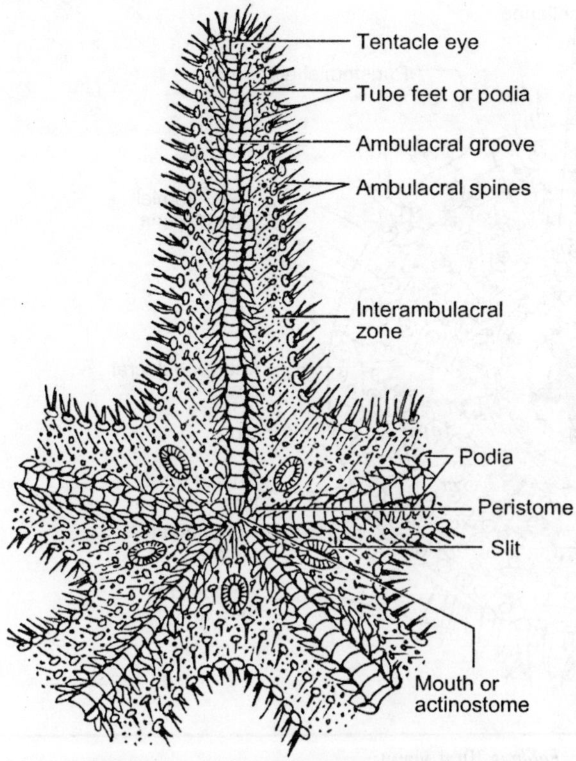

Fig. 1.1.1.150A : *Astropecten* (Oral view)

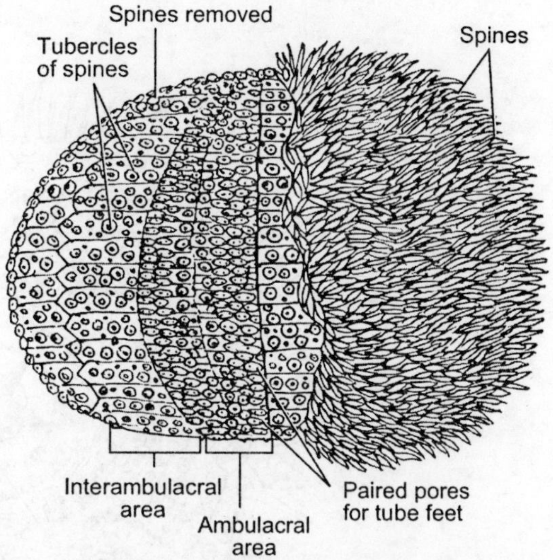

Fig. 1.1.1.151 : *Echinus*
(Spine removed from a portion of the test)

Identification with Reasons

State characters up to—

Subclass—*Asteroidea* as in **Ophiura.**

Order--*Paxillosida* :

1. Tube feet lack suckers and arranged in two rows.
2. Pedicellaria spine-like.
3. Marginal plate distinct and double series.

Type—*Astropecten* : State distinctive features as in above.

7. *ECHINUS* (Sea-Urchin)

Habit and Habitat

Marine; bottom living; common in intertidal zone; omnivorous; suspension-feeder; feeds on sea-weeds and small animals; distribution world wide.

Systematic Position

Phylum	**Echinodermata**
Subphylum	**Echinozoa**
Class	**Echinoidea (G: Spine-like)**
Order	**Echinoida**
Type	***Echinus***

Distinctive Features

1. Body enclosed in a globular test, which is covered with numerous movable spines.
2. Oral surface contains **Aristotle's lantern** (jaw apparatus).
3. Test distinguishes into ambulacral and inter-ambulacral zones.
4. Tube feet arranged in rows in the ambulacral areas.
5. External branchiae present.

Identification with Reasons

State characters up to—

Phylum—*Echinodermata* as in ***Antedon.***

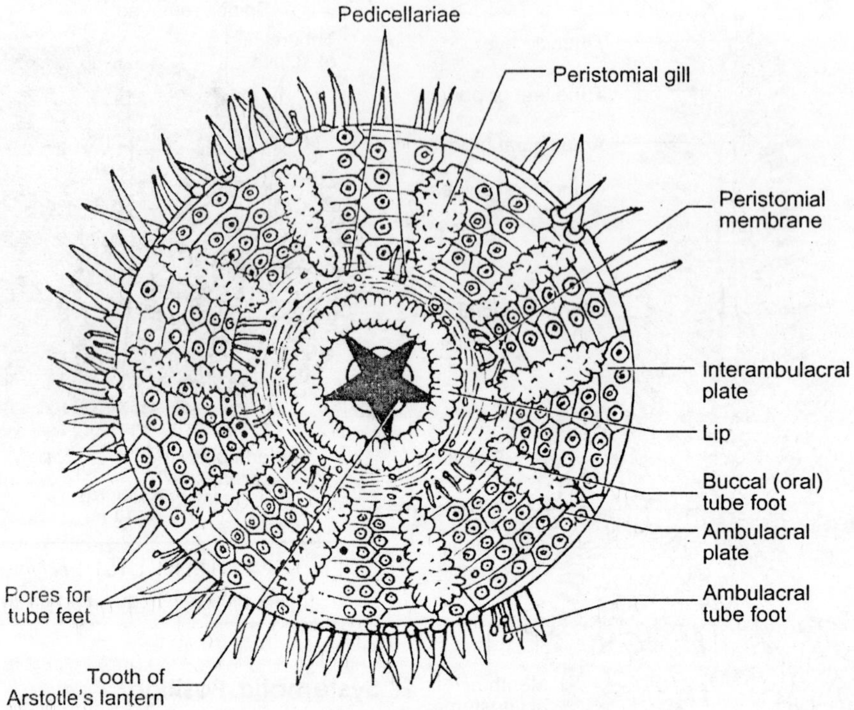

Fig. 1.1.1.151A : *Echinus* (Oral view)

Subphylum—*Echinozoa* :

1. Globoid or circular test with no arms and attachment organs.
2. Mouth and anus lie at opposite poles of the body.
3. Five meridional ambulacra present.

Class—*Echinoidea* :

1. Body spherical, disc shaped or heart shaped enclosed in the shell of calcareous plates.
2. Body surface is provided with articulating movable spines.
3. Mouth generally with elaborate jaw apparatus (Aristotle's lantern).
4. Tube feet meridionally arranged, primitively suctorial.

Order—*Echinoida* :

1. Ossicles are joined to form a rigid test, which lacks sculpture.
2. No petaloid ambulacra.
3. Ambulacral and interambulacral plates, each in two rows.

Type—*Echinus* : State distinctive features as in above.

8. *CLYPEASTER* (Sand-dollars/Sea biscuits)

Habit and Habitat

Marine; lives on sandy bottom of shallow waters; deposit feeder; feeds on sand and other organic inclusion; found in tropical and subtropical waters; distribution cosmopolitan.

Systematic Position

Phylum	**Echinodermata**
Subphylum	**Echinozoa**

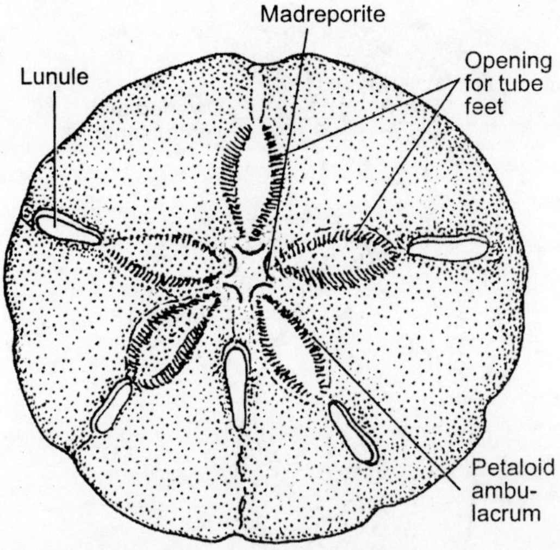

Fig. 1.1.1.152 : *Clypeaster* (Aboral view)

Class	**Echinoidea**
Order	**Clypeasteroida**
Type	***Clypeaster***

Distinctive Features

1. Body is circular, disc-like and dorsoventrally flattened.

2. Aboral surface is entirely covered by small spines.

3. Double rows of tube feet form five petaloid ambulacra on the aboral surface.

4. Madreporite on the aboral surface.

Identification with Reasons

State characters up to—

Class—*Echinoidea* as in *Echinus.*

Order—*Clypeasteroida* :

1. Possess numerous but very short spines.

2. Oval or circular flat test.

3. Petal-shaped ambulacral system on aboral surface.

Type—*Clypeaster* : State distinctive features as in above.

9. *CUCUMARIA* (Sea-Cucumber)

Habit and Habitat

Marine; bottom dweller; found in shallow water; well represented in temperate and cold water; detritus feeder; ingesting sediment and assimilating the organic fraction; distribution world wide.

Systematic Position

Phylum	**Echinodermata**
Subphylum	**Echinozoa**
Class	**Holothuroidea**
Order	**Dendrochirotida**
Type	***Cucumeria***

Distinctive Features

1. Lack a test, the body being more or less naked.

2. The oral end contains ten bushy tentacles.

3. Tube feet is five meridional rows on the bodywall.

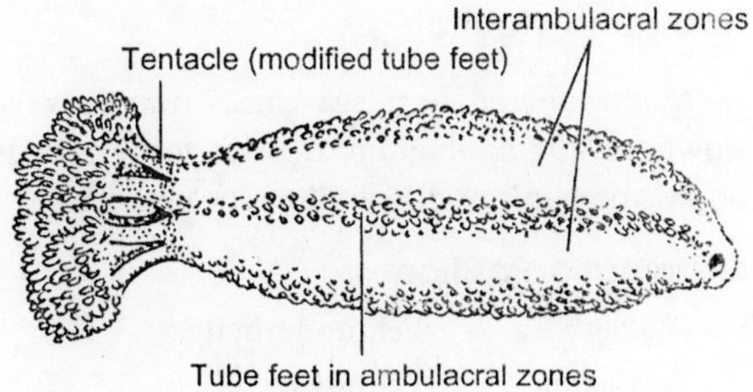

Fig. 1.1.1.153 : *Cucumeria*

* Ruppert, E.E., and Barnes D.B., (1994) mentioned that **Echinus** and **clypeaster** are under *Subclass*—**Euechinoidea**. Echinus is placed under *Super order*—**Echinacea** (seaurchin with rigid test and solid spines, lantern present) and in a separate *Order*—**Echinoida**. On the other hand clypeaster is placed under *super order*—**Gnathostoma** (mouth is in centre of oral surface but anus has shifted out of a apical centre, latern prsent) and put under *Order*—**Clypeasteroida**.

4. Presence of prominent collar or introvert.
5. Highly branched respiratory tree present.

Identification with Reasons

State characters up to—

Subphylum—*Echinozoa* as in ***Echinus*.**

Class—*Holothuroidea* :

1. Worm-shaped body with greatly elongated oral and aboral axis.
2. Skin soft, thin and without spines and **pedicellariae.**
3. Circlet of much branched tentacles at the oral end.
4. Reduced calcareous ossicles are embedded in the body wall.
5. Tube feet usually present.

Order—*Dendrochirotida* :

1. Tentacles are greatly branched and elongated.
2. Thick walled but flexible body contains numerous small and separate calcareous ossicles.

Type—***Cucumeria*** : State distinctive features as in above.

10. *HOLOTHURIA*

Habit and Habitat

Marine; found near sea coasts; moves very slowly on the sea-bottom; deposit feeder; feeds on microorganism; distribution cosmopolitan.

Systematic Position

Phylum	**Echinodermata**
Subphylum	**Echinozoa**
Class	**Holothuroidea**
Order	**Aspidochirotida**
Type	***Holothuria***

Distinctive Features

1. Vermiform animal with hard, stiff and leathery skin.

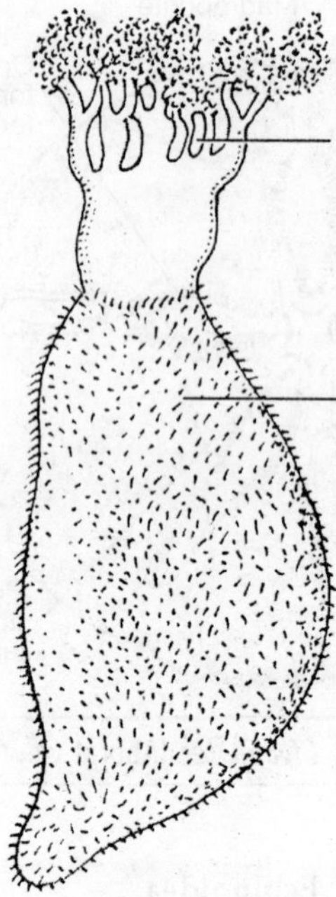

Fig. 1.1.1.154 : *Holothuria*

2. Tentacles more than ten.
3. Various calcareous ossicles are embedded in the body wall.
4. The retractile tube feet are densely scattered all over the body surface.
5. **Respiratory tree** present.

Identification with Reasons

State characters up to—

Class—*Holothuroidea* as in ***Cucumeria*.**

Order—*Aspidochirotida* :

1. Peltate oral tentacles.
2. Podia many.
3. Respiratory tree present.

Type—*Holothuria* : State distinctive features as in above.

Economic Importance

Commercially important as food source in the Far East. Only the body wall is eaten.

HEMICORDATA

(G: Hemi=half, chord=string)

Distinguishing Characteristic :
A conspicuous dorsal extension of the pharynx forms an anterior buccal tube or stomochord.

1. *BALANOGLOSSUS* (Acornworm)

Habit and Habitat

Marine; inhabitants of shallow water forming mucous lined burrows in sandy or muddy sediments: suspension feeders; planktonic organisms and detritus adhere to mucous on the proboscis and are conveyed along ciliated tracts to the mouth; distribution world wide.

Systematic Position

Phylum	**Hemichordata**
Class	**Enteropneusta** (G. entero = gut; Pneustikos = for breathing; or gut breathing)
Type	***Balanoglossus***

Distinctive Features

1. Worm shaped, soft body, divisible into, anterior proboscis (prosoma), middle collar (mesosoma) and posterior trunk (metasoma).
2. Short blunt proboscis.
3. Collar funnel shaped.
4. Trunk cylindrical and much elongated.
5. Mouth and anus terminal.
6. Lateral genital wings present.
7. Two rows of hepatic caeca present in the middle of the body.

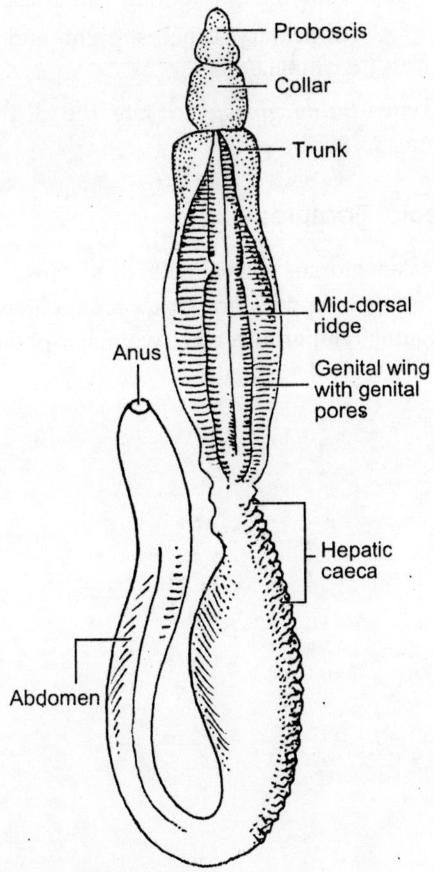

Fig. 1.1.1.155 : *Balanoglossus*

Identification with Reasons

Phylum—*Hemichordata* :

1. Worm like, marine, soft bodied animal.
2. Body is divisible into proboscis, collar and trunk.

Classification is based on Pechenik J.A., 2002.

3. Buccal diverticulum or **stomochord** is present, which is sometimes considered as notochord.
4. Gill slits present or absent, when present numerous in number.
5. Sexes separate.

Class—*Enteropneusta* :

1. Acorn worm (conical shape of the anterior region) having vermiform body.
2. Numerous paired gill slits present.
3. Alimentary canal straight and anus terminal.

Type—*Balanoglossus* : State the districtive features as in above.

Special Features

Balanoglossus and allied forms have phylogenetic importance. It has an apparent and intriguing relationship with two major phyla: the Echinodermata and the Chordata. Lack of notochord exclude them from phylum Chordata, still they posses other chordates characteristics: pharyngeal gill slits, and a dorsal, hollow nerve chord. 18S rDNA sequence analysis suggests a closer link between hemichordates and echinoderms than between hemichordates and chordates. Some other studies suggest a closer relationship between vertebrate chordates or other lophophorates.

Balanoglossus is still a debatable issue, it is better to put it and its allies in a separate phylum Hemichordata.

Note : *Saccoglossus* is another common enteropneusts found in Indian coasts. The classification, distribution and habit and habitat are same as those of *Balanoglossus.* It can be distinguished from Balanoglossus by having exceptionally elongated proboscis, branchial synapticula and perihaemal diverticula absent.

Chapter 1.1.2

Chordates

CHORDATA

(G: String)

Dorsal and hollow nerve cord; at least in some stage of development notochord is formed from the roof of the archenteron and running along the length of the animal ventral to the nerve cord, from anterior to posterior; pharyngeal gill slits (stigmata).

1. *ASCIDIA* (*Sea squirts*)

Habit and Habitats

Marine; sedentary, inhabiting shallow and deep water; ingesting microscopic prey and show ciliary mode of feeding; cosmopolitan in distribution.

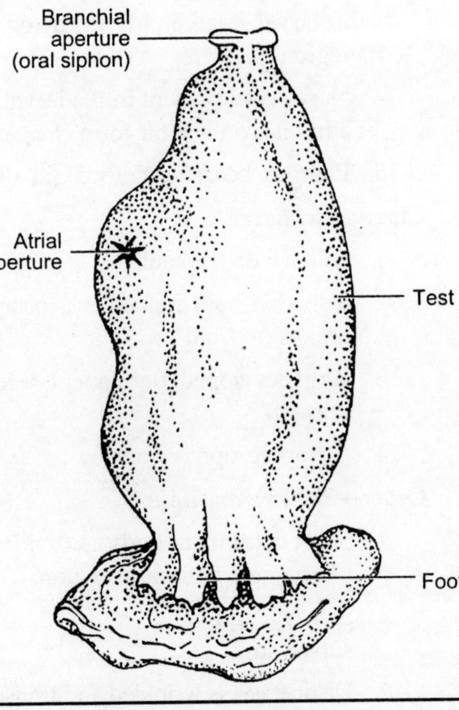

Fig. 1.1.2.1 : Ascidia

Systematic Position

Phylum	**Chordata**
Subphylum	**Urochordata**
Class	**Ascidiacea**
Order	**Phlebobranchia**
Type	***Ascidia***

Distinctive Features

1. Body short and cylindrical with a broad base.

2. Test translucent, wrinkled and brownish.

3. Anterior end bears large branchial and small atrial aperture, borne on siphon.

Classification is based on that by J. Z. Young, 1981.

4. Atrial aperture lies behind the branchial aperture.

5. Presence of pharyngeal gill slits(stigmata)

Identification with Reasons

Phylum—*Chordata* :

1. Bilaterally symmetrical, triploblastic, coelomate animal.
2. Presence of dorsal, hollow, tubular nerve cord.
3. At least in some stages of development notochord is present.
4. Presence of pharyngeal gill slits (stigmata).

Subphylum—*Urochordata* :

1. Solitary or colonial, fixed or free swimming.
2. Body covered by translucent tunic or test.
3. Notochord absent in adult found only in larval stage and is restricted to tail region.
4. Nerve cord present in the larva, but in adult it is only in the form of a ganglion.
5. Pharynx bears numerous gill slits.

Class—*Ascidiacea* :

1. Solitary or colonial.
2. Adult without notochord, locomotary organs and tail.
3. Well developed permanent thick test present.
4. Atriopore opens dorsally.

Order—*Phlebobranchia* :

1. Short cylindrical body is divided into pharyngeal and basal region.
2. Solitary and adult attached to solid substratum.
3. Distinct test is wrinkled and translucent.

Type—*Ascidia* : State distinctive features as in above.

Special Features

Ascidia show **retrogressive metamorphosis.** The adults become usually sessile after undergoing metamorphosis, when the notochord, nerve cord and tail are lost. The larval stage is known as a **tadpole** because of its superficial resemblance to the developmental stages of frogs. Metamorphosis includes the resorption of the larval notochord and tail. The only chordate feature of ascidians is the presence of the pharyngeal gill slits (stigmata).

2. *CIONA*

Habit and Habitat

Marine; solitary and sedentary, found in shallow water; attached with rocks; microphagous, feeds on microscopic animals and plants; found in western sea waters.

Systematic Position

Phylum	**Chordata**
Subphylum	**Urochordata**
Class	**Ascidiacea**
Order	**Phlebobranchia**
Type	***Ciona***

Distinctive Features

1. Body is cylindrical, tabular, transparent and longer than breadth.

2. Eight lobed branchial opening and six lobed artrial opening situated anteriorly.

3. Body covered by a transparent test through which longitudinal muscle bands are clearly seen.

4. Single gonad lies behind the intestinal loop.

Identification with Reasons

Same as *Ascidia* except the distinctive features.

Type—*Ciona* : State distinctive features as in above.

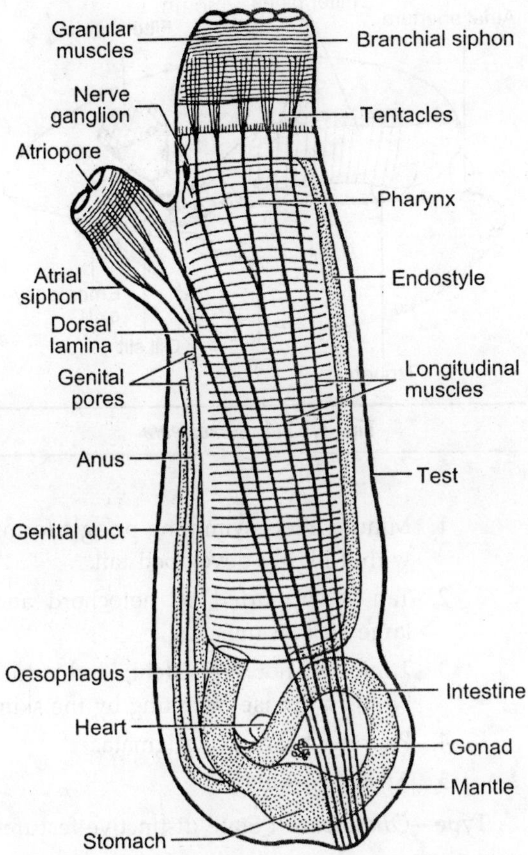

Fig. 1.1.2.2 : *Ciona*

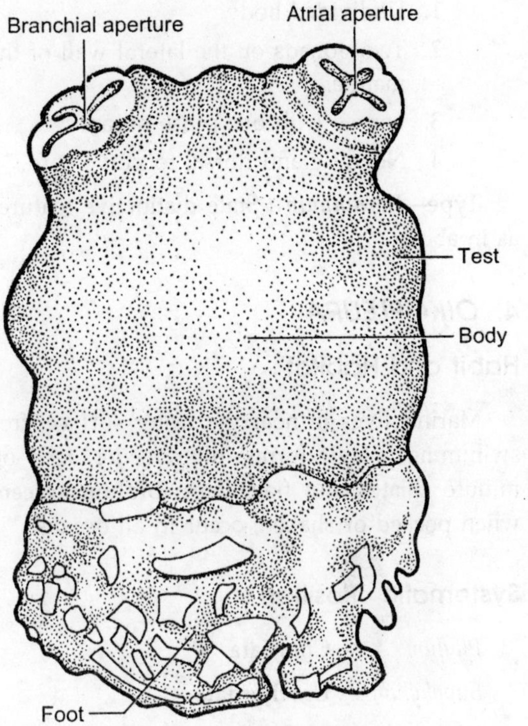

Fig. 1.1.2.3 : *Herdmania*

3. *HERDMANIA* (Monoascidian)

Habit and Habitat

Marine; solitary; adult is sessile and remain attached to the substratum: microphagous; feeds on microscopic animals and plants; provides shelters to hydroids, anemones and gastropods.

Systematic Position

Phylum	**Chordata**
Subphylum	**Urochordata**
Class	**Ascidiacea**
Order	**Stolidobranchia**
Type	***Herdmania***

Distinctive Feature

1. Oblong bag like body.
2. Body regionated into body proper and foot.
3. Foot lacking when attached to rock or molluscan shell.
4. Body enclosed in a thick tough, transparent test.
5. Branchial aperture is on the anterior end and opposite end is the posterior end.
6. Atrial aperture is on the dorsal side and opposite end is the ventral side.

Identification with Reasons

State classification and characters upto—

Class—*Ascidiacea* as in *Ascidia.*

Order—*Stolidobranchia* :

1. Undivided body.
2. Two gonads on the lateral wall of the mantle.
3. Larvae with one sense organ.
4. Neural gland dorsal to nerve cord.

Type—*Herdmania* : State distinctive features as in above.

4. *OIKOPLEURA*

Habit and Habitat

Marine; lives in near the water surface; free swimming pelagic forms; filter feeder; feed on minute ciliates and flagellates; **bioluminescent** when pocked or shaken; occur in all oceans.

Systematic Position

Phylum	**Chordata**
Subphylum	**Urochordata**
Class	**Larvacea (= Appendicularia)**
Type	**Oikopleura.**

Distinctive Features

1. Minute, microscopic, larva like in appearance with rounded body and a tail like appendage.

2. Test absent and each individual makes a 'house' out of its test, called as **oikoplastic epithelium**.

3. Broad tail is supported by notochord and large striped muscles.

4. Pharynx bears two stigmata which open directly to the exterior.

5. Atrium lacking.

Identification with Reasons

State classification and characters upto—

Subphylum—*Urochordata* as in *Ascidia*.

Class—*Larvacea* :

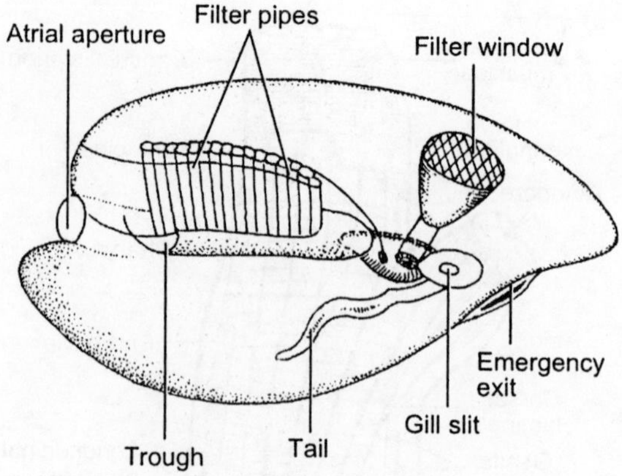

Fig. 1.1.2.4 : *Oikopleura*

1. Minute, Free-swimming pelagic forms with a highly developed tail.
2. Tail is supported by notochord and large striped muscles.
3. Tunic is not persistent; individual builds a '**house**' secreting by the skin.
4. Pharynx bears two stigmata.
5. Atrium lacking.

Type—*Oikopleura* :_State distinctive features as in above.

Special Features

The general structural organization resemblances with a typical ascidian tadpole and it may have evolved from ascidians ancestors through the process of **Neoteny**. The animals become sexually mature while retaining the larval forms. The former adult characters are deleted from the life cycle, and the new adult character gradually become better adapted to its new mode of life through Natural selection.

5. *PYROSOMA*

Habit and Habitat

Marine; pelagic, colonial: microphagous, feed on plankton, exhibit striking bioluminescense; found in all oceans from the surface to a considerable depth.

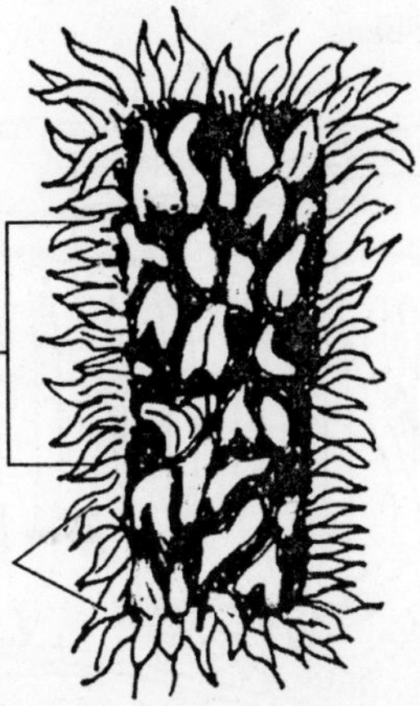

Fig. 1.1.2.5 : *Pyrosoma*

Systematic Position

Phylum	**Chordata**
Subphylum	**Urochordata**
Class	**Thaliacea**
Order	**Pyrosomida**
Type	***Pyrosoma***

Distinctive Features

1. Colony consists of many zooids or individuals embedded in a common test.

2. One end of the colony is closed and the other end open.

3. Branchial and atrial openings are at opposite ends.

4. Branchial openings open to the hollow of the colony.

5. A buccal appendage close to the mouth is present.

Identification with Reasons

State classification and characters upto—

Subphylum—*Urochordata* as in *Ascidia*.

Class—*Thaliacea* :

1. Free-swimming pelagic forms and colonial in at least part of the life cycle.
2. Permanent test bears circular muscle bands.
3. Pharynx bears two large or many small stigmata.
4. Adult devoid of notochord and tail.

Order—*Pyrosomida* :

1. Compact tubular colony.
2. Muscle bands at the end of the body.
3. No larval forms in the life history.
4. Exhibit striking bioluminescence.

Type—*Pyrosoma* : State distinctive features as in above.

Special Features

All individuals exhibit **bioluminescence** probably through the chemical activities of bacterial symbionts. Animal produces light or stimulation and when all colonies produce light, the entire sea surface is illuminated. The light is produced by bioluminescent cells, which contain curved inclusions.

6. *DOLIOLUM*

Habit and Habitat

Marine; pelagic free swimming and solitary: microphagous; feeds on plankton; found in all oceans and most common in warmer waters.

Systematic Position

Phylum	**Chordata**
Subphylum	**Urochordata**
Class	**Thaliacea**
Order	**Doliolida**
Type	***Doliolum***

Distinctive Features

1. Barrel-shaped body with complete muscle bands forming eight regular rings.

2. Pharynx bears many variable gill slits.

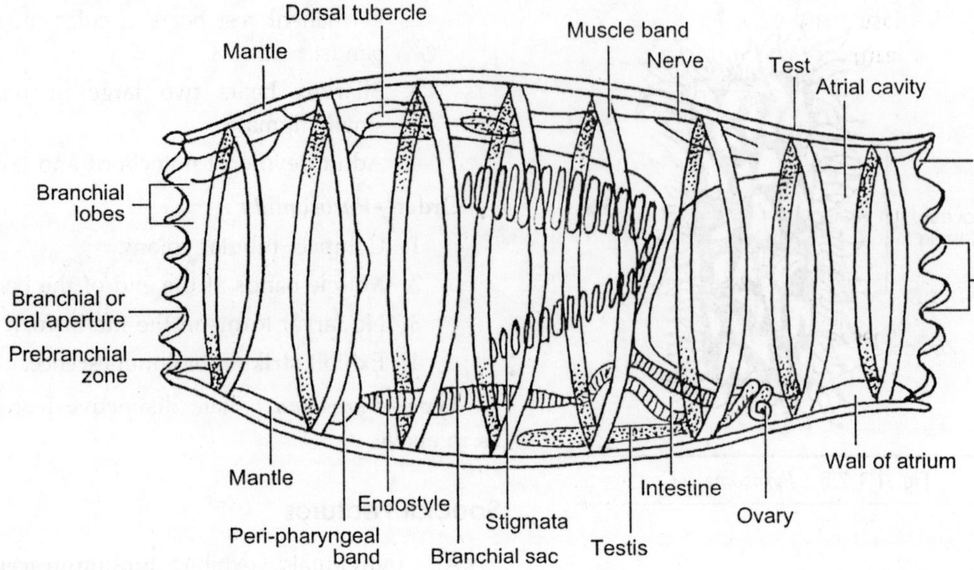

Fig. 1.1.2.6 : *Doliolum*

3. The open ends of the barrel are fringed with lobes.

Identification with Reasons

State classification and characters up to—

Class—*Thaliacea* as in *Pyrosoma.*

Order—*Doliolida* :

1. Barrel-shaped body.
2. Muscles bands in the form of eight complete rings.
3. Branchial and atrial aperture in opposite end.
4. Variable number of gill-slits.

Type—*Doliolum* : State distinctive features as in above.

7. *SALPA*

Habit and Habitat

Marine; pelagic, free swimming, found on surface and down to depth of 1,500m: feed on plankton; live in all the world's oceans and are specially abundant in warm waters.

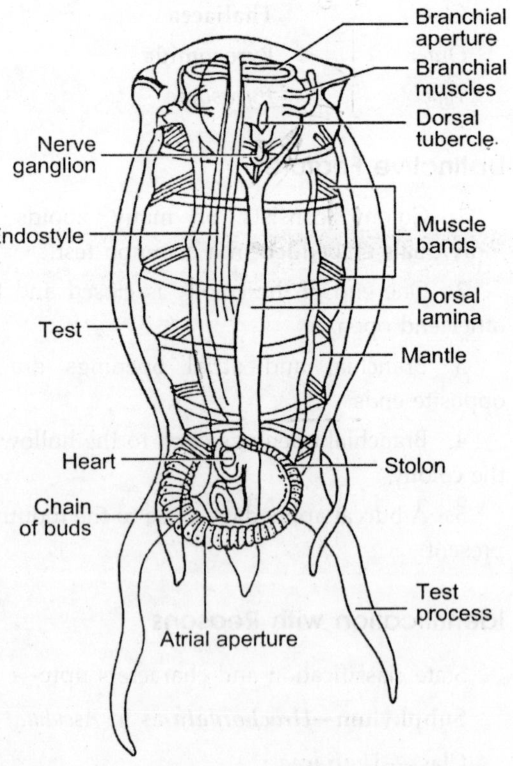

Fig. 1.1.2.7 : *Salpa*

Systematic Position

Phylum	**Chordata**
Subphylum	**Urochordata**
Class	**Thaliacea**
Order	**Salpida**
Type	***Salpa***

Distinctive Features

1. Cylindrical body covered by a thin transparent test with two long processes.

2. Seven incomplete muscles bands surround the body.

3. Presence of stolon which gives rise to a chain of buds.

4. Single tentacle on the dorsal surface.

Identification with Reasons

State classification and characters upto—

Class—*Thaliacea* as in *Pyrosoma*.

Order—*Salpida* :

1. Cylindrical body covered with transparent test.
2. Muscles bands incomplete ventrally.
3. No larval stage.

Type—*Salpa* : State distinctive features as in above.

Special Features

Salpa exhibits polymorphism and exits in two phases – asexual phase or **oozooid** and sexual phase or **blastozooid.** Sexual and asexual generations alternate. The sexual form produces a single egg which develops within the mother and is nourished by a diffused placenta.

8. *BRANCHIOSTOMA* (Commonly called as Lancelet)

Habit and Habitat

Marine; occur in shallow water and live partially buried in sandy substrates living only its anterior end protected: microphagous: food consists of protozoas, algae, diatoms and other organic particles suspended in sea-waters; obtain the food by filtering the sea-water: cosmopolitan in distribution.

Systematic Position

Phylum	**Chordata**
Subphylum	**Cephalochordata (= Acrania. G: without caranium) (G: head string)**
Type	***Branchiostoma lanceolatus***

Note : Yarell (1836) gave the name *Amphioxus lanceolatus*. Costa (1834) suggested the genus

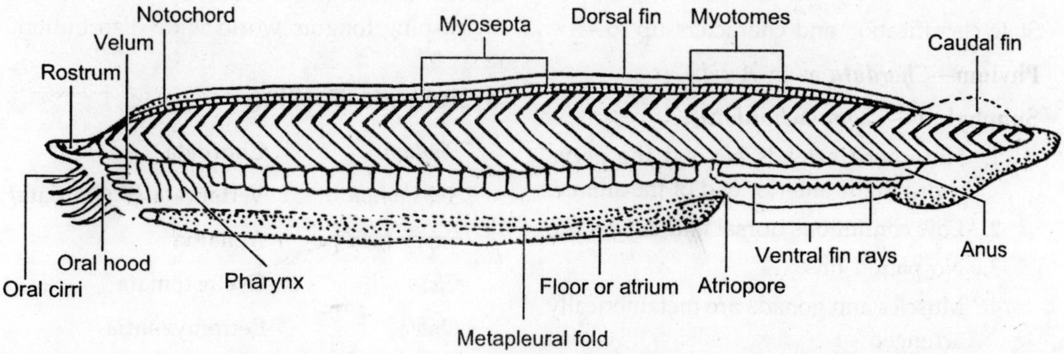

Fig. 1.1.2.8 : *Branchiostoma*

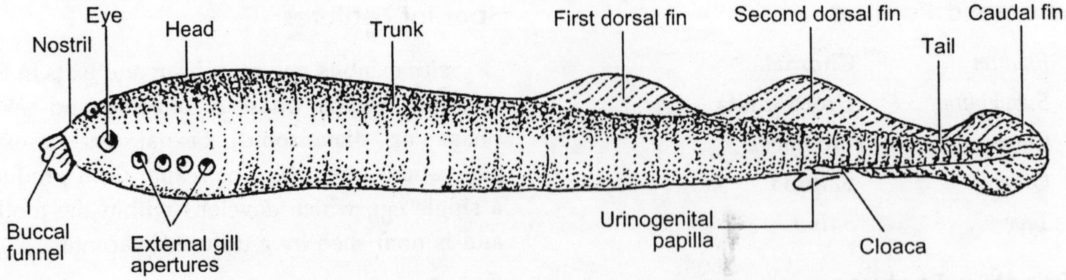

Fig. 1.1.2.9 : *Petromyzon*

name as *Branchiostoma*. According to the law of priority the name of the genus has been established as *Branchiostoma*.

Distinctive Features

1. Body is elongated, flattened and pointed at both ends.
2. Dorsal, ventral and caudal fins are low and continuous.
3. Two lateral fins or **metapleural folds** present.
4. Mouth is ventral and guarded by oval hood bearing oral cirri.
5. V- shaped **myotomes** are arranged on both sides of the body.
6. Huge pharynx with numerous gill slits open through median and ventral atriopore.

Identification with Reasons

State classification and characters up to—

Phylum—***Chordata*** as in *Ascidia*.

Subphylum—***Cephalochordata*** :

1. Notochord extends beyond the nerve chord to the anterior end of the animal.
2. Low continuous dorsal and caudal fin.
3. No paired fins.
4. Muscles and gonads are metamerically arranged.
5. Large Pharynx with numerous gill slits.

Type—*Branchiostoma lanceolatus* : State distinctive features as in above.

Special Importance

Lancelets are eaten by people in some parts of China. *Branchiostoma* is of special zoological interest, because they may tell us about our own vertebrate origin and about the underlying genetic changes that resulted in vertebrate evolution. Comparisons of gene sequences coding for 18S rRNA suggest that cephalochodates are our closest non-vertebrate relative.

9. *PETROMYZON* (commonly called as lamprey)

Habit and Habitat

Marine; and fresh water: ectoparasite on other fishes and remains attached on the ventral surface of fish: carnivorous; eat the flesh of fishes by rasping tongue: world wide distribution.

Systematic Position

Phylum	**Chordata**
Subphylum	**Vertebrata (= Craniata)**
Super class	**Agnatha**
Class	**Cyclostomata**
Order	**Petromyzontia**
Type	***Petromyzon***

Distinctive Features

1. Eel-like cylindrical body.

2. Body differentiated into head, trunk and tail with two dorsal and one caudal fin.

3. Circular mouth with horny teeth and rasping tongue.

4. The upper lip is greatly developed to form a suctorial buccal funnel.

5. Well developed branchial basket.

Identification with Reasons

State classification and characters up to—

Phylum—*Chordata* as in *Ascidia.*

Subphylum—*Vertebrata* :

1. Notochord is replaced by **vertebral column.**
2. Distinct cranium or brain case is present.
3. Presence of well developed head.
4. Post anal tail is usually present.

Subclass—*Agnatha* :

1. Devoid of functional jaws.
2. Paired fins absent.

Class—*Cyclostomata* :

1. Mouth is circular and suctorial.
2. Naked and glandular skin.
3. Single median nostril.
4. Continuous and persistent notochord.

Order—*Petromyzontia* :

1. Dorsal fin well developed.
2. Well developed branchial basket.
3. Seven pairs of gill slits.
4. Gill pouches open into a respiratory tube below the oesophagus.
5. Two semi-circular canals present.
6. Metamorphosis present in the life cycle.

Type—*Petromyzon* : State distinctive features as in above.

Special importance

Cyclostomes represent the most primitive craniates that are living today. The cyclostomes are represented by two distinct groups, lampreys and hagfishes. Their nearest relatives are the ancient ostracoderms.

Economic Importance

Lamprey injures and destroys fishes by eating the flesh and sucking the blood thus causing secondary infection. It is used as bait for commercial fishing.

10. *AMMOCOETES LARVA*

Habit and Habitat

Fresh water: remain buried in the mud: microphagous; devours microorganism: ciliary mode of feeding: distribution world wide.

Systematic Position

Phylum	**Chordata**
Subphylum	**Vertebrata**
Super class	**Agnatha**
Class	**Cyclostomata**
Order	**Petromyzontia**
Type	***Ammocoetes larva***

Distinctive Features

1. Cylindrical, small, transparent eel-like body.

2. Median dorsal fin is continuous with the caudal fin.

3. Suctorial buccal funnel is absent.

4. Mouth is bounded by an oral hood.

5. Several branched buccal tentacles present.

6. Seven pairs of branchial sacs present.

7. Teeth lacking.

8. Photoreceptors in the tail.

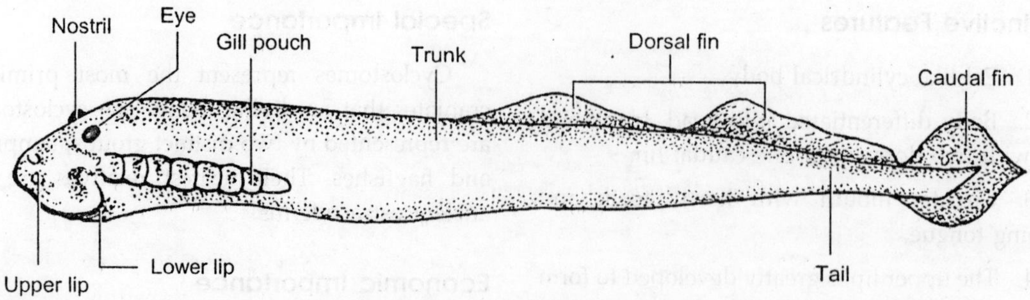

Fig. 1.1.2.10 : *Ammocoetes larvae*

Identification with Reasons

State classification and characters up to—

Order—*Petromyzontia* as those of Lamprey.

Type—*Ammocoetes larva* : State distinctive features as in above.

Remark

It is larval stage of the lamprey. Structurally this larval stage shows an intermediate form between cephalochordates and vertebrates.

11. *MYXINE* (Commonly called as Hagfish)

Habit and Habitat

Purely marine; live in muddy sea bottom: sluggish: saprophagous; feed on the flesh and viscera of dead fishes: nocturnal.

Systematic Position

Phylum	**Chordata**
Subphylum	**Vertebrata**
Super class	**Agnatha**
Class	**Cyclostomata**
Order	**Myxinoidea**
Type	***Myxine***

Distinctive Features

1. Slender eel-like soft body without scales.
2. Four pairs of barbels or tentacles on the head.
3. Buccal funnels and jaws are absent.
4. Single inconspicuous median dorsal fin extends up to caudal fin.
5. Six pairs of gill pouches open to the exterior by one pair of gill slits.

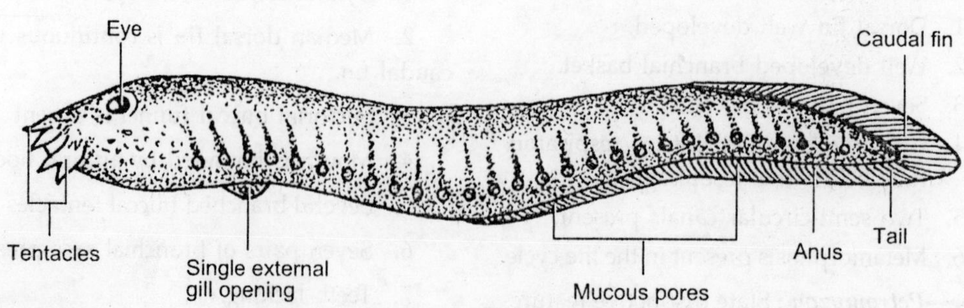

Fig. 1.1.1.11 : *Myxine*

Identification with Reasons

State classification and characters up to—

Class—*Cyclostomata* as those of Lamprey.

Order—*Myxinoidea* :

1. Dorsal fin inconspicuous.
2. Branchial basket is ill-developed.
3. Gill pouches six to fifteen pairs.
4. Gill slits one to fifteen pairs.
5. Only one semicircular caudal fin is present.
6. No metamorphosis.

Type—*Myxine* : State distinctive features as in above.

Economic Importance

Hagfishes are injurious to fish industry. They damage fish caught in nets. Some times hagfishes enter into the body of fishes and eat entire soft part, leaving only the skin and bones.

First successful class of the chordates

FISHES

Fishes are purely aquatic vertebrates, which exploit every corner of the territory of water. Fishes have under gone extensive adaptive modifications and dominate the water home. They are conspicuous for their astronomical number and varieties of form, habit and habitats.

The classification of fishes is a very difficult task and there is no unanimously accepted scheme regarding fishes. Fishes of present day world are broadly grouped into seven classes: **Pterichthys, Coccostei, Acanthodii, Elasmobranchii, Holocephali, Teleostomi** and **Dipnoi**. The first three classes include the extinct fishes and are collectively called as the **Placoderms**, which were the progenitors of the modern fishes. The classes Elasmobranchii and Holocephali include the **cartilegínous fishes**, while Telestomi and Dipnoi include **bony fishes**. The bony fishes represent the highest developed forms. This group includes the **crossopterygians**, from which the amphibians have evolved.

The fishes include large varieties of forms and all are built on a fundamental plan. Fishes are purely aquatic, breathing oxygen dissolved in water through gills. They are provided with fins. The skin is covered with **scales** and is abundantly coated with **mucus**. They have **paired** and **unpaired fins**, which are supported by **dermal fin rays**. Organs of smell, sight and hearing are found on the head. Eyes are devoid of eyelids. Most fishes lay eggs, which hatch directly into larvae.

Series—PISCES

12. *SCOLIDON* (Commonly called as Dogfish/Shark)

Habit and Habitat

Marine; living in shallow tropical and warm temperate seas and some are fresh water; Carnivorous; mainly fish eater; predaceous; attacked prey by its sharp teeth; active swimmers; distribution worldwide.

Systematic Position

Phylum	**Chordata**
Subphylum	**Vertebrata**
Superclass	**Gnathostomata**
Class	**Elasmobranchii (plate gills)**
Subclass	**Selachii**
Super order	**Selachoidea (Pleurotremata)**
Order	**Lamniformes**
Type	***Scoliodon sorrakowah***

Classification is based on Berg. 1943.

Distinctive Features

1. Spindle shaped body divisible into head, trunk and tail.
2. Dorsoventrally compressed head flattened into snout.
3. Wide slit-like mouth opening on the ventral side.
4. Well defined heterocercal tail.
5. In male claspers present.
6. In female claspers absent.
7. Spiracle present.

Identification with Reasons

State classification and characters up to—

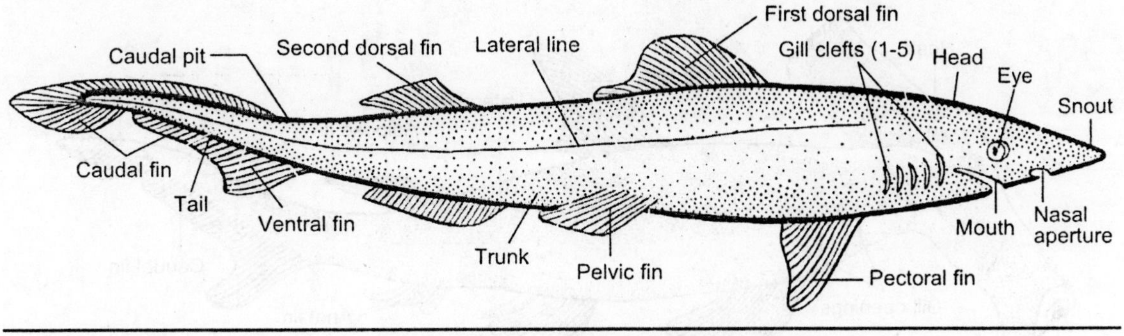

Fig. 1.1.2.12 : *Scoliodon sorrakowah*

Subphylum—*Vertebrata* as in ***Petromyzon.***

Superclass—*Gnathostomata* :

1. Mouth is bounded by upper and lower jaw.
2. Paired appendages present.

Class—*Elasmobranchii* (Chondrichthys) :

1. Body is covered with minute placoid scales.
2. Presence of paired and unpaired fins supported by fin rays.
3. Endoskeleton cartilagenous.
4. Ventrally placed mouth with rows of teeth.
5. Cloaca present.
6. Tail is heterocercal.
7. Male with claspers.

Subclass—*Selachii* :

1. Five to seven pairs of gill slits.
2. Gill placed in a separate cleft and not covered by operculum.
3. Spiracle when present, situated behind each eye.
4. Spiral valve is present in the intestine.

Superorder—*Selacoidea* :

1. Gills slits on lateral side.
2. Anterior margin of pectoral fins are free, not united with the head.
3. Tail heterocercal.

Order—Lamniformes :

1. Both dorsal fins not supported by spines.
2. Anal fin present.
3. Five gill arches.

Type—*Scoliodon sorrakowah* : State distinctive features as in above.

Note : According to law of priority *Scolidon* (Bleeker, 1853) is used as *Carcharinus* (Ruppell, 1837).

Economic Importance

Scolidon has great experimental value because of its size and availability. It is popular for type study and dissecting specimen in classes. Scoliodon is widely used for food, oil and skin.

13. *SPHYRNA* (Commonly called as hammer-headed Shark)

Habit and Habitat

Marine: found in tropical seas; predaceous: consume other fishes; ferocious in nature and lash at their prey with their head: distribution world-wide.

Systematic Position

Phylum	**Chordata**
Sub Phylum	**Vertebrata**
Superclass	**Gnathostomata**

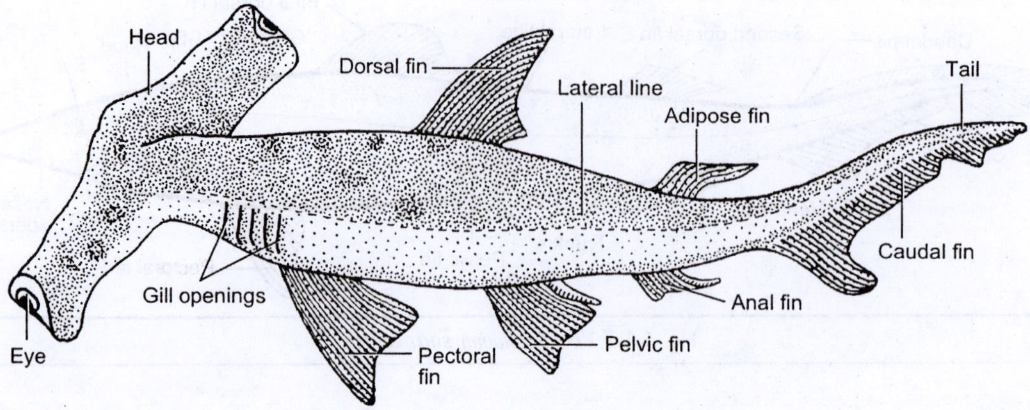

Fig. 1.1.2.13 : *Sphyrna*

Class	**Elasmobranchii (plate gills)**
Subclass	**Selachii**
Superorder	**Selachoidea**
Order	**Lamniformes**
Type	***Sphyrna***

Distinctive Features

(*Sphyrana* synonymous with *Zygaena*)

1. Head is flattened in front and expanded side ways, thus resembling a hammer.
2. Eye is located at the tip of each lateral extension of the head.
3. Spiracle absent.
4. Nostrils lie ventrally at the base of the lateral lobes.
5. Gill slits five pair and lateral in position.
6. Male with clasper and in female clasper absent.

Identification with Reason

State the classification and characters up to—

Order—*Lamniformes* as in *Scoliodon*.

Type—*Sphyrna* : State distinctive features as in above.

Economic Importance

Sphyrna is used in extracting liver oil.

Special Features

Sphyrna is viviparous with **yolksac placenta.**

14. *TRYGON* (commonly called as Sting-Ray)

Habit and Habitat

Marine; found lying quietly on the bottom of the ocean in shallow waters ; carnivorous; catch small fishes, arthropods, gastropods etc; distribution world wide.

Systematic Position

Phylum	**Chordata**
Subphylum	**Vertebrata**
Superclass	**Gnathostomata**
Class	**Elasmobranchii**
Subclass	**Selachii**
Super order	**Batoidea (Hypotremata)**
Order	**Rajiformes**
Type	***Trygon***

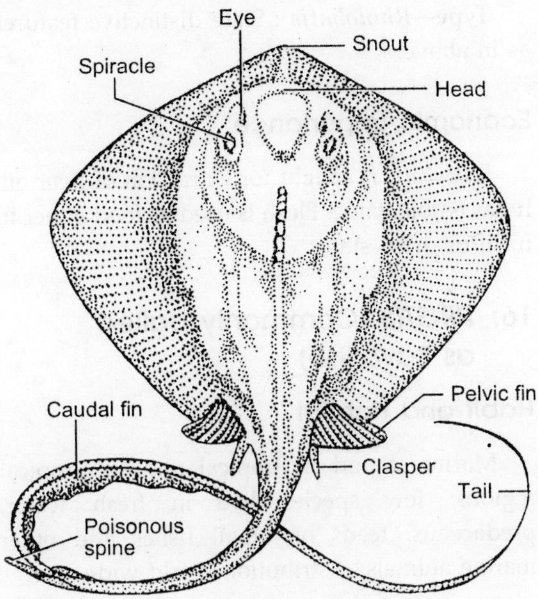

Fig. 1.1.2.14 : *Trygon* (Male)

Distinctive Features

1. Body flat disc-shaped.
2. Head, trunk and pectoral fin form a rounded outline.
3. Dorsal fin absent.
4. Tail is reduced, slender and whip like.
5. Tail terminates in a small caudal fin and contains a large mid-dorsal poisonous sting.
6. Male with clasper and in female clasper absent.

Identification with Reasons

State classification and characters up to—

Subclass—*Selachii* as in *Scolidon.*

Super order—*Batoidea* :

1. Gill opening ventrally placed.
2. Anterior margin of the pectoral fin fused with side of the body or head.
3. Teeth flattened.
4. Dorsal fin, if present placed far behind.
5. Spiracle present behind the eye.

Order—*Rajiformes* :

1. Body dorso-ventrally flattened.
2. No electric organs between head and pectoral fin.

Type—*Trygon* : State distinctive features as in above.

Economic Importance

Trygon is valuable in commercial fisheris. Their flesh is edible and liver yields a fair amount of oil.

Special Features

Poisonous sting inflicts wounds on the victim. Wounds sometimes complicated by gangrene on bathers fishermen.

15. *RHINOBATIS* (= RHINOBATUS) (Commonly called as guitarfish)

Habit and Habitat

Marine; bottom living; viviparous; carnivorous; feed mainly on invertebrates and small fishes; mostly distributed in tropical and sub-tropical seas.

Systematic Position

Phylum	**Chordata**
Subphylum	**Vertebrata**
Superclass	**Gnathostomata**
Class	**Elasmobranchii**
Subclass	**Selachii**
Super order	**Batoidea (Hypotremata)**
Order	**Rajiformes**
Type	***Rhinobatis***

Distinctive Features

1. Body assumes sub-rhombic shape and edge of the fins are free.

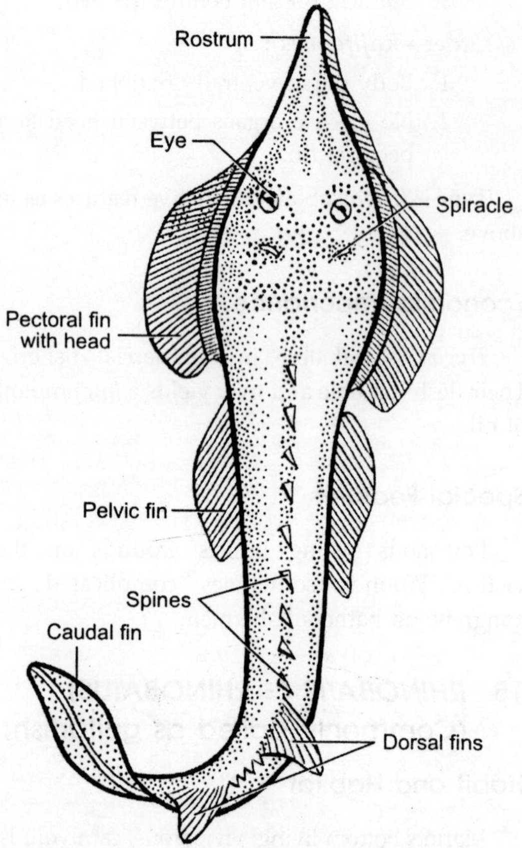

Fig. 1.1.2.15 : *Rhinobatis*

2. Sub-rhombic head tapers into triangular snout and an elongated caudal region, resembling guitar-like shape.

3. Pectoral fins are greatly enlarged along the head and trunk.

4. Tail is elongated and strong with two dorsal fins and a caudal fin.

5. Entire back contains median row of minute denticles .

6. No electric organs.

Identification with Reasons

State classifications and characters up to—

Order—*Rajiformes* as in *Trygon*.

Type—*Rhinobatis* : State distinctive features as in above.

Economic Importance

Rhinobatis is caught for extraction of liver oil. It has edible value. Flesh is used as food either in fresh or dried state.

16. *PRISTIS* (Commonly called as Saw-Fish)

Habit and Habitat

Marine; found in tropical and sub-tropical regions; few species live in fresh water; predaceous; feeds on small fishes and other marine animals; distribution world wide.

Systematic Position

Phylum	**Chordata**
Subphylum	**Vertebrata**
Superclass	**Gnathostomata**
Class	**Elasmobranchii**
Subclass	**Selachii**
Super order	**Batoidea**
Order	**Pristiformes**
Type	***Pristis***

Distinctive Features

1. Body is elongated and slightly broadened laterally.

2. Head is drawn out into a long rostrum armed with denticles giving the appearance of a saw.

3. Two dorsal fins and first dorsal fin is opposite to pelvic fin.

4. Well developed tail ends in a heterocercal caudal fin.

5. No electric organ.

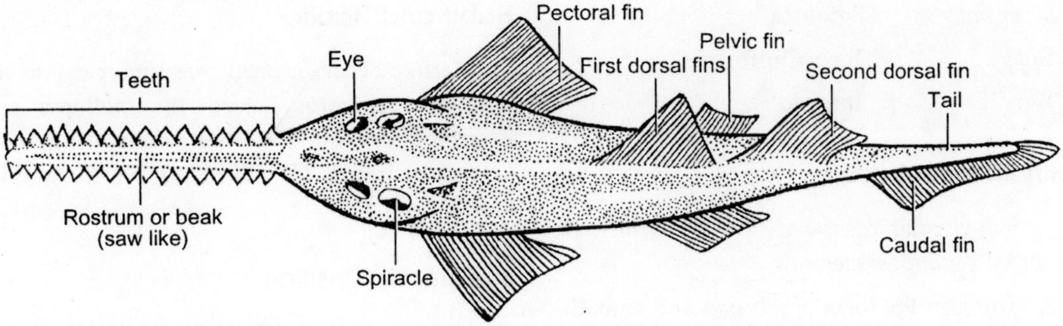

Fig. 1.1.2.16 : *Pristis*

Identification with Reasons

State classification and characters up to—

Super order—*Batoidea* as in *Trygon*.

Order—*Pristiformes* :

1. Dorsal fin two and first dorsal fin is opposite to pelvic fin.
2. Presence of elongated, toothed saw-like rostrum.
3. Tail ends in a heterocercal caudal fin.
4. Gill opening laterally.
5. No electric organs.

Type—*Pristis* : State the distinctive features as in above.

Economic Importance

Pristis is invaluable for liver oil of high vitamin value and its skin for making scale boards.

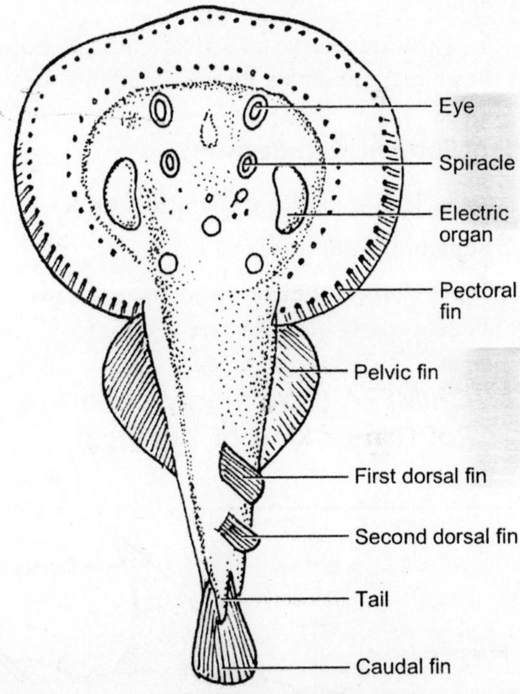

Fig. 1.1.2.17 : *Torpedo*

17. *TORPEDO* (Commonly called as Electric-ray)

Habit and Habitat

Marine: lives in the flat, sandy or muddy bottom of the ocean in shallow water; Carnivorous: feed mainly on invertebrates and small fishes; distribution worldwide.

Systematic Position

Phylum	**Chordata**
Subphylum	**Vertebrata**
Superclass	**Gnathostomata**
Class	**Elasmobranchii**
Subclass	**Selachii**

Super order	**Batoidea**
Order	**Torpediniformes**
Type	*Torpedo*

Distinctive Features

1. Sub-circular disc shaped body, which is supported by endoskeleton.

2. Trunk in the form of a broad and smooth disc contain electric organ.

3. Skin smooth and without scales.

4. Dorsal fin two and spineless caudal fin present.

5. Fins extended so far forward that the front of the animal presents a rounded outline.

Identification with Reasons

State classification and characters up to—

Superorder—*Batoidea* as in *Trygon.*

Type—*Torpedo* : State distinctive features as in above.

18. *CHIMERA* (Commonly called as Rat Fish or King of Herrings)

Habit and Habitat

Marine; occur in deep seas; live close to the bottom; carnivorous; feed on molluscs and invertebrates; found in the coasts of Europe, Japan, Australia, New Zealand, North America and Africa.

Systematic Position

Phylum	**Chordata**
Subphylum	**Vertebrata**
Superclass	**Gnathostomata**
Class	**Holocephali (only bone)**
Subclass	**Chimaerae**
Order	**Chimaeriformes**
Type	***Chimaera***

Distinctive Features

1. Small fleshy lipped mouth with tooth plates attached to jaw.

2. Fan like large pectoral fins present.

3. Tail is long, thin and whip like.

4. Single external branchial opening protected by operculum.

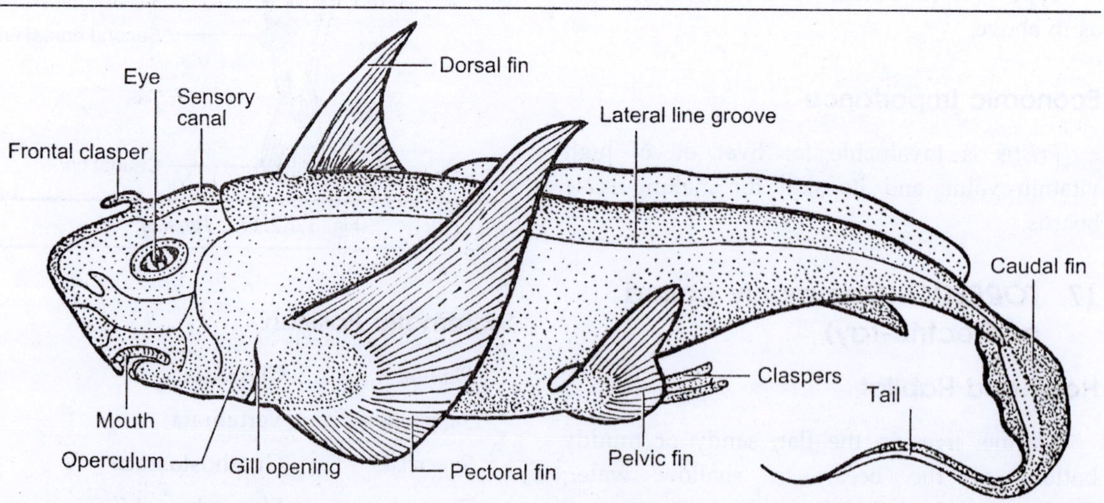

Fig. 1.1.2.18 : *Chimaera*

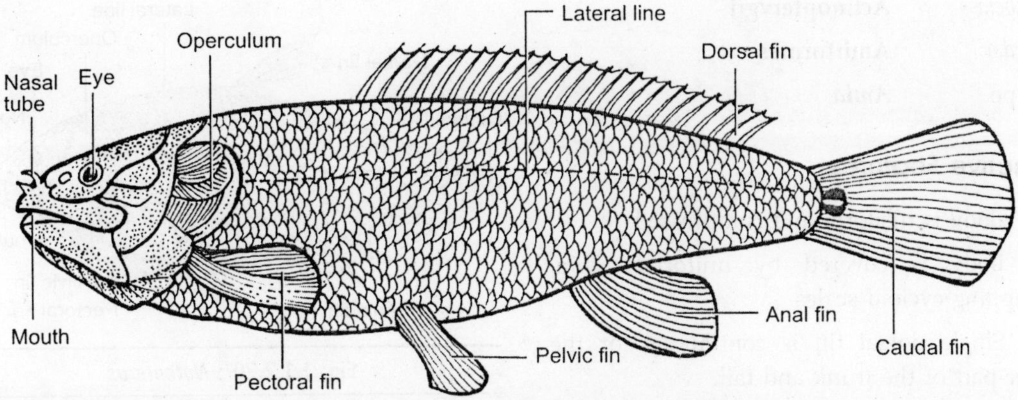

Fig. 1.1.2.19 : *Amia*

5. Erectile spine in front of the dorsal fin.

6. A pair of claspers near pelvic fins and a extra pair in front of pelvic fins.

7. Male with extra claspers, the **cephalic claspers** on the head.

8. Spiracle and cloaca are absent.

Identification with Reasons

State classification and characters up to—

Superclass—*Gnathostomata* as in *Scoliodon.*

Class—*Holocephali* :

1. Operculum covers the gills.
2. Small fleshy lipped mouth with tooth plates attached to jaw.
3. Spiracle absent.

Subclass—*Chimaerae* :

1. Cloaca absent and urinogenital aperture behind anus.
2. Male with extraclaspers, the cephalic claspers on the head.

Order—*Chimaeriformes* :

1. An erectile spine in front of the dorsal fin.
2. Extra claspers in front of usual pelvic fin.

Type—*Chimaera* : State distinctive features as in above.

Remarks

Chimaera is an interesting fish, having relation with typical cartilage fishes(Elasmobranch) and bony fishes(Teleost). They preserve today some features of elasmobranch life in palaeozoic times though in other respects they are aberrant. The affinities of the Holocephali and their relative are still debated.

19. *AMIA* (Commonly known as Bow Fish)

Habit and Habitat

Aquatic; found in fresh water, rivers and lakes; voracious; carnivorous; feeding upon insects, crustaceans and other insects; found in Central and South America.

Systematic Position

Phylum	**Chordata**
Subphylum	**Vertebrata**
Superclass	**Gnathostomata**
Class	**Teleostomi (osteichthys) (bonyfish)**

Subclass	**Actinopterygii**
Order	**Amiiformes**
Type	*Amia*

Distinctive Features

1. Body is fusiform and compressed.
2. Body is covered by uniformly thin, overlapping cycloid scales.
3. Single dorsal fin is continuous for the greater part of the trunk and tail.
4. Anal fin short.
5. Tail is homocercal, but internally still heterocercal.
6. Two comb like structures are found on the throat.

Identification with Reasons

State classification and characters up to—

Superclass—*Gnathostomata* as those of *Scoliodon.*

Class—*Teleostomi* :

1. Mainly bony endoskeleton.
2. Gills covered by movable operculum.
3. Tail homocercal.
4. Scales lack ganoine layer.
5. Presence of paired and unpaired fins with fin rays.
6. Cloaca absent.
7. Claspers are absent in male.
8. Swim bladder present.

Subclass—*Actinopterygii* :

1. The paired fins lack a central axis.
2. Nostrils do not open into the mouth.
3. Basal of the fins extend beyond the body limit.

Order—*Amiiformes* :

1. Thin cycloid scales without ganoine layer.
2. The tail fin is externally homocercal.
3. Swim bladder is modified into bilobed dorsal lung.

Type—*Amia* : State distinctive features as in above.

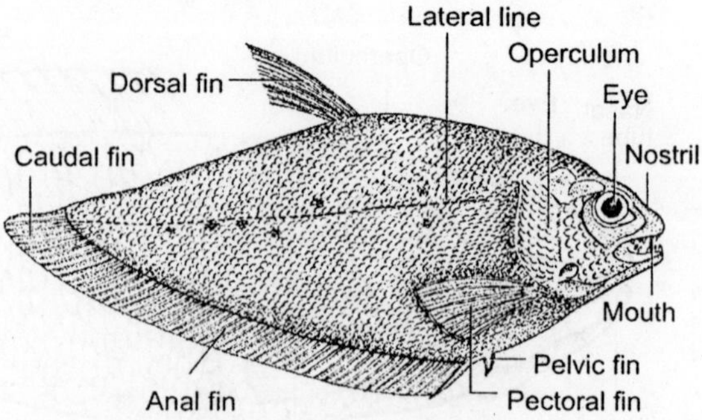

Fig. 1.1.2.20 : *Notopterus*

20. *NOTOPTERUS*

Habit and Habitat

Aquatic inhabits in fresh water or brackish water; carnivorous, predaceous: feed on insects, crustaceans and other small fishes: found in India, Myanmar, Malaya, and West Africa.

Systematic Position

Phylum	**Chordata**
Subphylum	**Vertebrata**
Superclass	**Gnathostomata**
Class	**Teleostomi**
Subclass	**Actinopterygii**
Order	**Clupeiformes**
Type	***Notopterus***

Distinctive Features

1. Laterally compressed body with short precaudal region.
2. Body silvery white with fins grey spots along the back.

3. A single dorsal fin is present.

4. The anal fin is very much elongated and meeting with the caudal fin.

Identification with Reasons

State classification and character up to—

Sub class—*Actinopterygii* as in *Amia*.

Order—*Clupeiformes* :

1. Scales are well developed and cycloid type.
2. A single dorsal fin is present.
3. Caudal fin is homoceral.
4. Anal fin is very much elongated and meeting with the caudal fin.

Type—*Notopterus notopterus* : State distractive features as in above.

Economic Importance

Notopterus regarded as food fish. The flesh of the fish is very rich in nutrition and well flavoured.

21. *GADUSIA* (Commonly known as Khoria)

Habit and Habitat

Brackish water: omnivorous: feed on suspended organic particles and larvae of different invertebrates: found in India, Bangladesh and Myanmar.

Systematic Position

Phylum	**Chordata**
Sub phylum	**Vertebrata**
Super class	**Gnathostomata**
Class	**Teleostomi**
Sub class	**Actinopterygii**
Order	**Clupeiformes**
Type	***Gadusia***

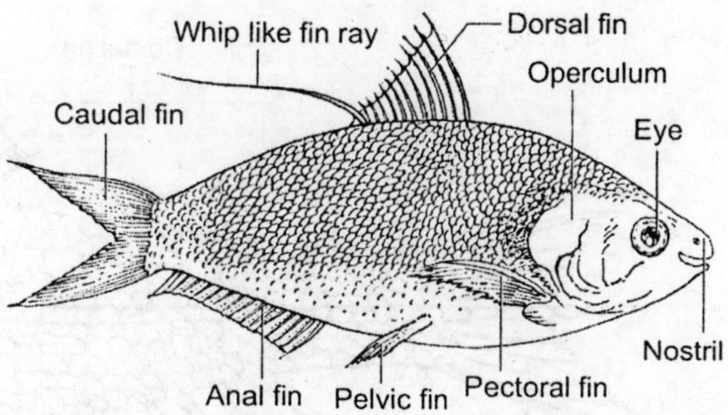

Fig. 1.1.2.21 : *Gadusia*

Distinctive Features

1. Body laterally compressed, oblong with ventral profile more convex than dorsal profile.

2. Silvery white body covered with shiny cycloid scales.

3. Dorsal fin single and the last fin ray is elongated into a whip like organ reaching almost the caudal fin.

4. Lower jaw prominent and eyes are surrounded by broad adipose tissues.

Identification with Reasons

State classification and characters up to—

Order—*Clupeiformes* as in Notopterus.

Type—*Gadusia* : State distinctive features as in above.

Economic Importance

Gadusia is a food fish with less economic value.

22. *HARPODON* (Commonly known as Bombay Duck)

Habit and Habitat

Marine: Shows trends of migration for feeding purposes: found abundantly in the costal waters of Arabian Sea along the Bombay coast: also found on the eastern coasts as far as Orissa and West Bengal: carnivorous: feeds on small fishes.

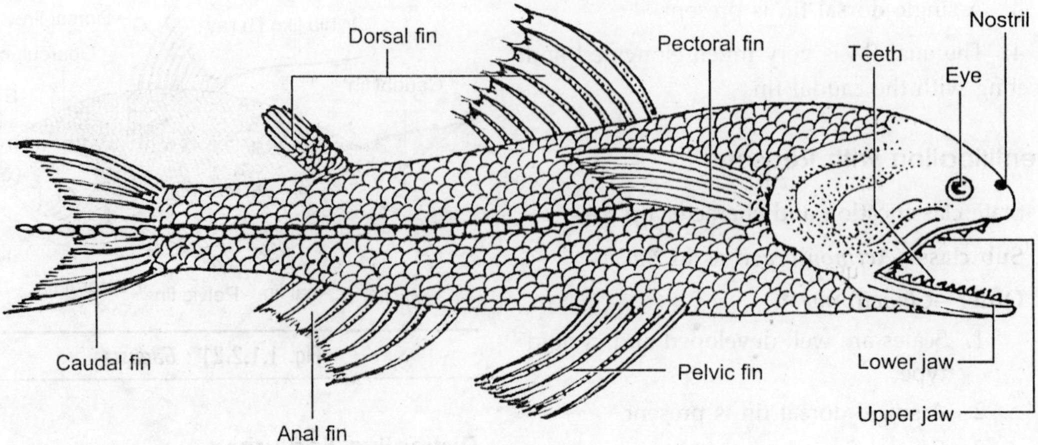

Fig. 1.1.2.22 : *Harpodon*

Systematic Position

Phylum	**Chordata**
Sub phylum	**Gnathostomata**
Class	**Teleostomi**
Sub class	**Actinopterygii**
Order	**Scopeliformes**
Type	***Harpodon neherius***

Distinctive Features

1. Pulpy body is elongated, compressed, with rounded abdomen.

2. Snout short, bluntly rounded and lower jaw slightly longer than upper jaw.

3. Pelvic fin longer than pectoral fin.

4. Dorsal fin two, second one is very small.

5. Fins are with adipose tissue.

Identification with Reasons

State classification and characters up to—

Subclass—*Actinopterygii* as in *Amia.*

Order—*Scopeliformes* :

1. Large mouth and jaws sub equal with numerous teeth.
2. Dorsal and anal fins are without spines.
3. Adipose fins are present.
4. Swim bladder is usually absent.
5. Possess phosphorescent organs.

Type—*Harpodon neherius* : State distinctive features as in above.

Economic Importance

Harpodon is a well-known commercial fish in India. Sun-dried fishes are consumed as food.

23. *LABEO* (Commonly known as Carp)

Habit and Habitat

Fresh water: found in ponds and rivers: herbivorous: column feeder: feed on algae, decayaing plants, debris, and phytoplankton: widely distributed in tropical and temperate regions, specially found in India and Myanmar.

Systematic Position

Phylum	**Chordata**
Sub phylum	**Vertebrata**
Super class	**Gnathostomata**

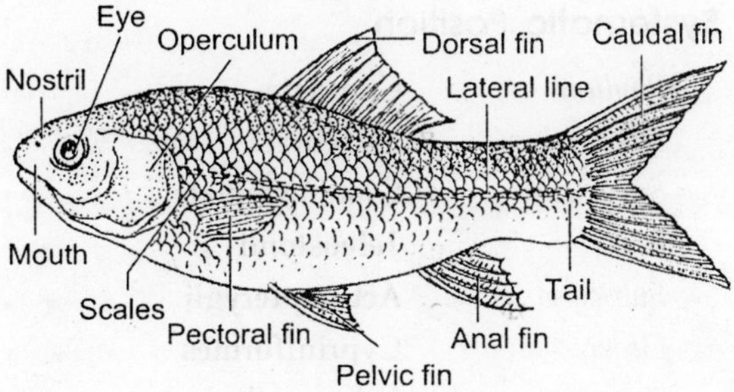

Fig. 1.1.2.23 : *Labeo*

Class	**Teleostomi**
Sub class	**Actinopterygii**
Order	**Cypriniformes**
Type	***Labeo rohita***

Distinctive Features

1. Body is compressed, fusiform and covered by large over lapping cycloid scales.

2. Colour of the body is orange to reddish.

3. Head prominent with blunt snout projecting beyond the narrow mouth.

4. Lips thick, a pair of filamentous short barbels arises from upper lip.

5. Caudal fin deeply forked.

Identification with Reasons

State classification and characters up to—

Sub class—*Actinopterygii* as in *Amia*.

Order—*Cypriniformes* :

1. Scales are absent on the head.
2. No teeth in the jaws.
3. Barbels are very short or absent.
4. Single large dorsal fin.
5. Fins are supported by soft fin rays.
6. Lateral line is distinct.
7. Ventral fin is abdominal in position.

Type—*Lobeo rohita* : State distinctive features as in above.

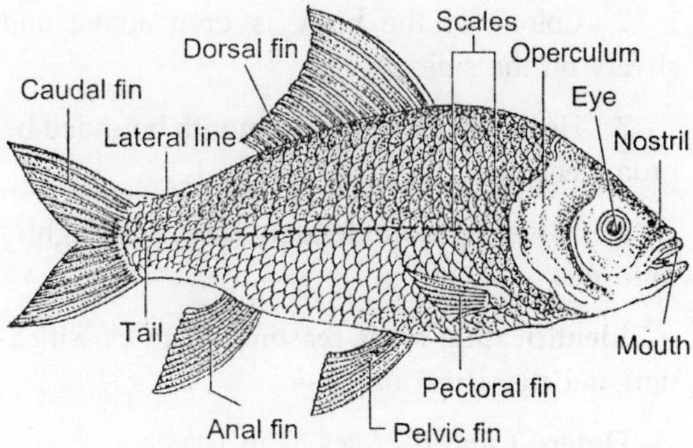

Fig. 1.1.2.24 : *Catla*

Economic Importance

Labeo has a great food value. They are consumed as food fish all over India and hence are of high commercial value.

24. *CATLA* (Commonly known as largest Carp)

Habit and Habitat

Fresh water: found in ponds, rivers, jheels, bheels and pools: omnivorous: surface feeder: feed on crustaceans, algae, organic debris and zoo plankton: found all over India.

Systematic Position

Phylum	**Chordata**
Sub phylum	**Vertebrata**
Super class	**Gnathostomata**
Class	**Teleostomi**
Sub class	**Actinopterygii**
Order	**Cypriniformes**
Type	***Catla catla***

Distinctive Features

1. Deep compressed body with abruptly broad trunk.

2. Colour of the body is grey above and silvery on the sides.

3. Head large with broad mouth bounded by prominent lips.

4. Lower lip is moderately thick and slightly upturned

Identification with reasons: State classification and characters up to—

Order—*Cypriniformes* as in *Labeo.*

Type—*Catla catla* : State distinctive features as in above.

Economic Importance

Catla is cultured and used as food fish all over India. *Catla* is the largest and fastest growing carp.

24A. *CALBASU*

Habit and Habitat

Found only in fresh water. Like rohu they also live in the middle layer of ponds or fresh water reservoirs. They are omnivorous or herbivorous.

Systematic Position

Phylum	**Chordata**
Sub phylum	**Vertebrata**
Super class	**Gnathostomata**
Class	**Teleostomi**
Sub class	**Actinopterygii**
Order	**Cypriniformes**
Type	***Labeo calbasu***

Distinctive Features

1. Snout truncated or rounded.
2. Lips thick, fleshy and forming a labial fold.
3. Barbels four.
4. Anal fin short with seven to eight rays.
5. Caudal fin deeply forked.
6. Dorsal fin with sixteen rays.
7. Blackish body colour.

Identification with reasons : State classification and characters upto—

Order—*Cypriniformes* as in ***Labeo.***

Type—*Labeo calbasu.*

Economic Importance

Calbasu is cultured as food fish and has thus commercial value.

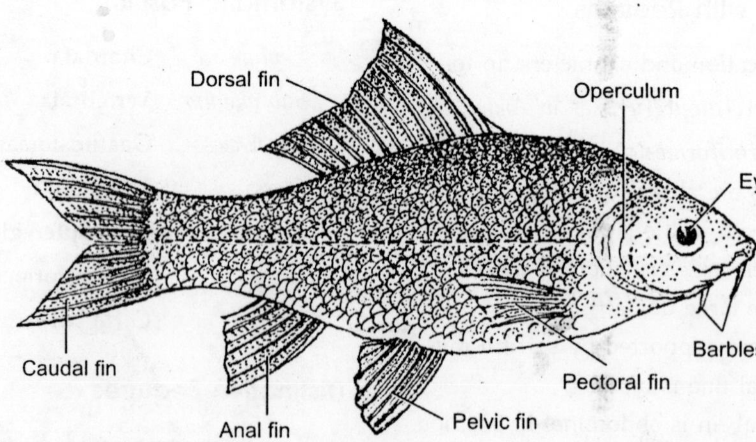

Fig. 1.1.2.24A : *Labeo calbasu*

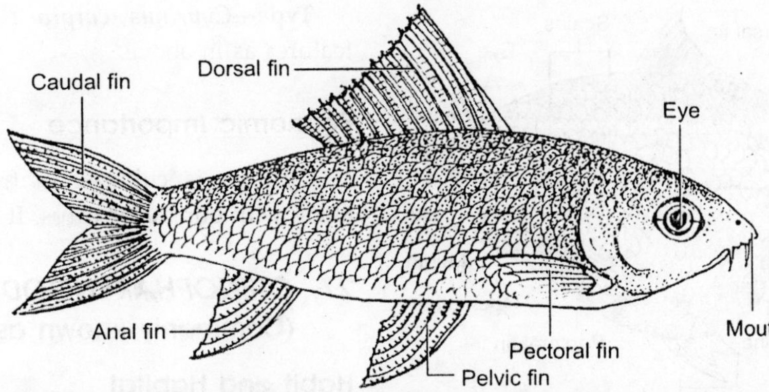

Fig. 1.1.2.25 : *Cirrhinus*

25. *CIRRHINUS*

Habit and Habitat

Fresh water: live at the bottom of the ponds and reservoirs: omnivorous: bottom feeder: feed on decomposed organic matter: found in India, Bangladesh and Myanmar.

Systematic Position

Phylum	**Chordata**
Sub phylum	**Vertebrata**
Super class	**Gnathostomata**
Class	**Teleostomi**
Sub class	**Actinopterygii**
Order	**Cypriniformes**
Type	***Cirrhinus mrigala***

Distinctive Features

1. Moderately elongated, slender body and sub-cylindrical in shape.

2. Head small, mouth wider, lips thinner and with horny sheath.

3. Upper lip entire.

4. Colour of the body is silvery, dark grey along the back, sometimes with coppery tinge.

5. Barbel two and small.

6. Pectoral, pelvic and anal fins are orange in colour with marginal black tinge.

Identification with Reasons

State classification and characters up to—

Order—*Cypriniformes* as in *Labeo*.

Type—*Cirrhinus mrigala* : State distinctive features as in above.

Economic Importance

Cirrhinus is cultivated by pond culturists together with *Labeo* and *Catla*. They attain a good size and used as food fish hence of commercial value.

26. CYPRINUS (Commonly called common European Carp, American Rohu)

Habit and Habitat

Fresh water: cultured in pond along with our major carps: in young stage they feed on zooplankton: adult herbivorous: feed on phytoplankton, decomposed aquatic plants and organic food mixed up with soil: bottom feeder: found in Europe : brought in our country from Bangladesh in 1957.

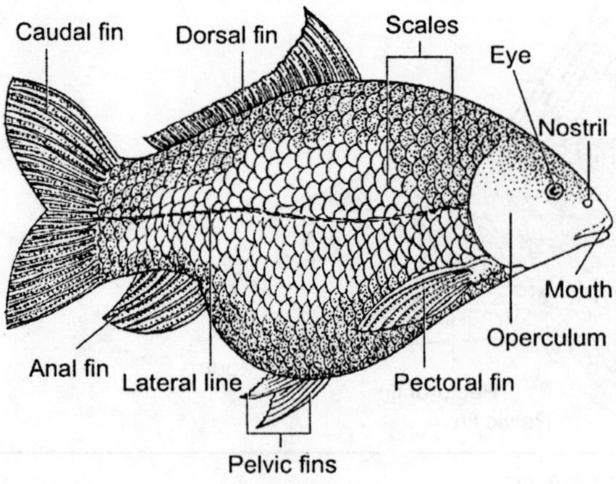

Fig. 1.1.2.26 : *Cyprinus*

Systematic Position

Phylum	**Chordata**
Sub phylum	**Vertebrata**
Super class	**Gnathostomata**
Class	**Teleostomi**
Sub class	**Actinopterygii**
Order	**Cypriniformes**
Type	***Cyprinus carpio***

Distinctive Features

1. Elongated body, robust anteriorly and with broad rounded abdomen.

2. Body completely covered by large cycloid scales which are irregular in shape.

3. Body colour orange to red.

4. Mouth bounded by prominent fleshy lips.

5. Barbel two.

6. Anal fin short with three spines and five rays.

Identification with Reasons

State classification and characters up to—**Order**—***Cypriniformes*** as in *Labeo*.

Type—***Cyprinus carpio*** : State distinctive features as in above.

Economic Importance

Cyprinus is cultured in fresh water found along with our major carps. It is a food fish.

27. *CTENOPHARYNGODON* (Commonly known as Grass Carp)

Habit and Habitat

Fresh water: river carp of china: herbivorous: column feeder: feed on phytoplankton: cultured in our fresh water pond along with our major carp: brought in India from Japan in 1959.

Systematic Position

Phylum	**Chordata**
Sub phylum	**Vertebrata**
Super class	**Gnathostomata**
Class	**Teleostomi**
Sub class	**Actinopterygii**
Order	**Cypriniformes**
Type	***Ctenopharyngodon idella***

Distinctive Features

1. Body sub-cylindrical anteriorly with rounded abdomen and compressed posteriorly.

2. Mouth terminal, upper jaw slightly elongated than lower jaw.

3. Barbels absent.

4. Lateral line continuous slightly decurved.

5. Anal fin short with ten rays.

6. Colour of the body is backish grey on back and silvery dark below.

Identification with Reasons

State classification and characters up to—

Order—***Cypriniformes*** as in *Labeo*.

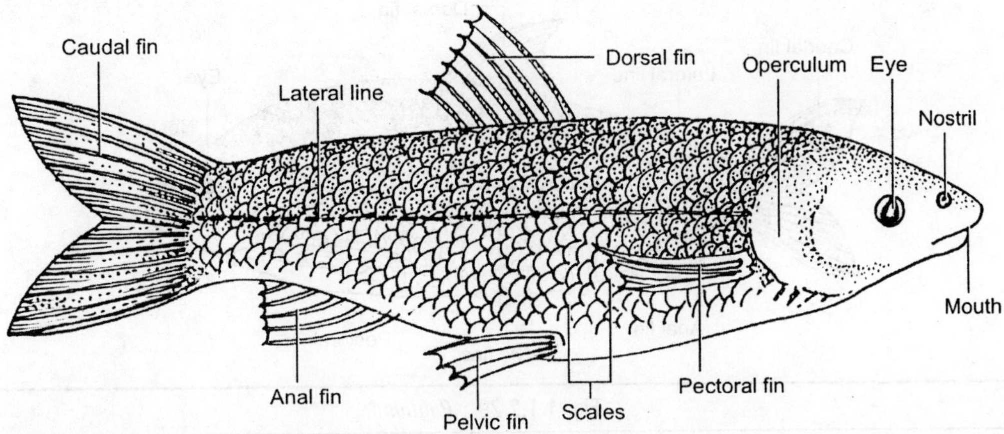

Fig. 1.1.2.27 : *Ctenopharyngodon*

Type—*Ctenopharyngodon idella*: State distinctive features as in above.

Economic Importance

Ctenopharyngodon is cultured in fresh water pond along with our major carps. They control the unwanted aquatic weed of pond.

28. *HYPOPTHALMICHTHYS* (Commonly known as Silver Carp)

Habit and Habitat

Fresh water river carp of China: cultured in fresh water ponds: in young stage feed on unicellular algae and in adult stage feed on phytoplankton: surface feeder: brought to our country from Hong Kong in 1959.

Systematic Position

Phylum	**Chordata**
Sub phylum	**Vertebrata**
Super class	**Gnathostomata**
Class	**Teleostomi**
Sub class	**Actinopterygii**
Order	**Cypriniformes**
Type	***Hypopthalmichthys molitrix***

Distinctive Features

1. Laterally compressed body with short keel form breast to vent.

2. Colour of the body is silvery white and fins are reddish brown.

3. Mouth rounded, upper jaw protruded upward and is little larger than the lower jaw.

4. Barbels absent.

5. Scales minute and lateral line with hundred ten to hundred fifteen scales.

6. Dorsal fin with ten rays and anal fin with fourteen to seventeen rays.

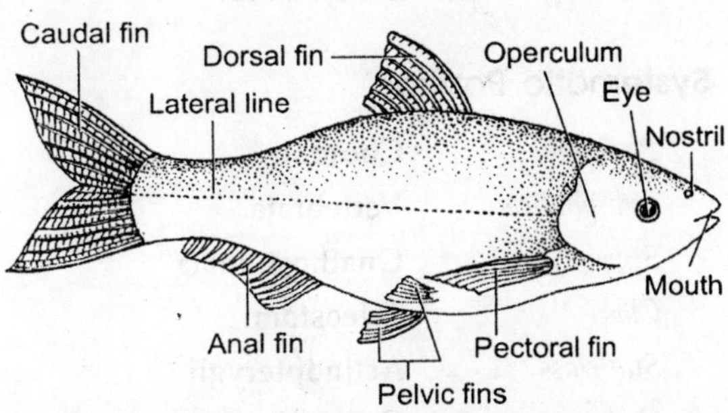

Fig. 1.1.2.28 : *Hypopthalmichthys*

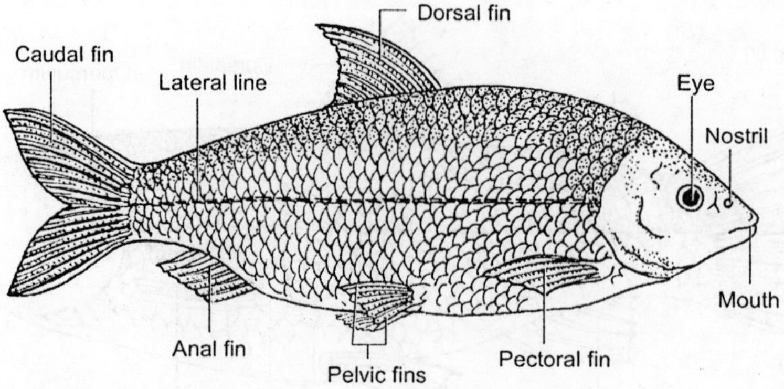

Fig. 1.1.2.29 : *Puntius*

Identification with Reasons

State classification and characters up to—

Order—*Cypriniformes* as in *Labeo.*

Type—*Hypopthalmichthys molitrix* : State distinctive features as in above.

Economic Importance

Cultured along with major carp. It is a good fish food and it clears the unwanted aquatic weeds from the ponds.

29. *PUNTIUS*

Habit and Habitat

Fresh water: inhabiting in rivers, ponds and lakes: herbivorous: surface feeder: distributed in India, Bangladesh and Myanmar.

Systematic Position

Phylum	**Chordata**
Sub phylum	**Vertebrata**
Super class	**Gnathostomata**
Class	**Teleostomi**
Sub class	**Actinopterygii**
Order	**Cypriniformes**
Type	***Puntius***

Distinctive Features

1. Body short to moderately elongated and silvery white in colour.
2. Mouth is arched, lips simple and operculum is short.
3. Fins are whitish or yellowish.
4. Barbels are smaller and four in number.
5. Dorsal fin short and nine to thirteen rays with or without spine.
6. Anal fin short with seven to nine rays.

Identification with Reasons

State classification and characters up to—

Order—*Cypriniformes* as in *Labeo.*

Type—*Puntius* : State distinctive features as in above.

Economic Importance

Various species of *Puntius* are taken as food fish but have less commercial value.

30. *LABEO BATA* (Commonly known as Bata Fish)

Habit and Habitat

Fresh water: found in Ponds, lakes and ditch:

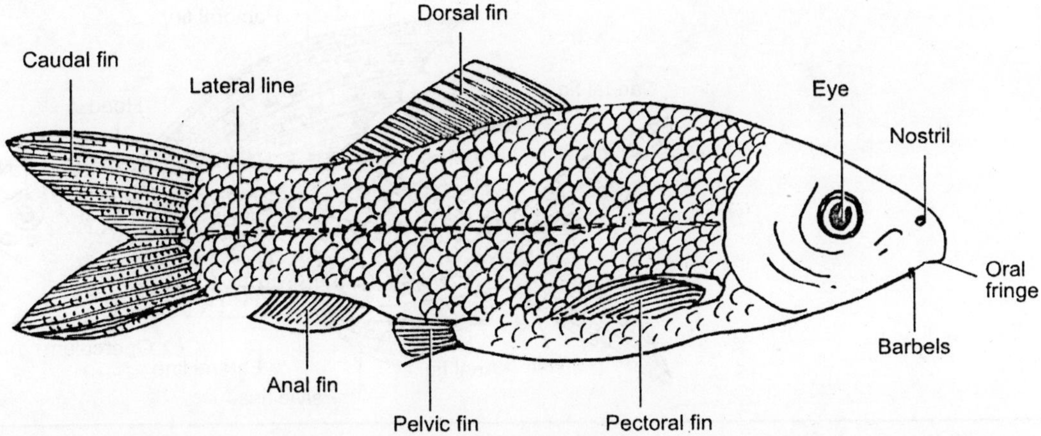

Fig. 1.1.2.30 : *Labeo bata*

omnivorous: feed on phytoplankton and zoo planton: distributed in India and Bangladesh.

Systematic Position

Phylum	**Chordata**
Sub phylum	**Vertebrata**
Super class	**Gnathostomata**
Class	**Teleostomi**
Sub class	**Actinopterygii**
Order	**Cypriniformes**
Type	***Labeo bata***

Distinctive Features

1. Body short to moderately long, dorsal profile more convex than ventral one.
2. Colour of the body is silvery white.
3. Scales small and shiny.
4. Lips thin and continuous.
5. A pair of very short maxillary barbels present.
6. Dorsal fin with eleven to twelve rays.

Identification with Reasons

State classification and characters up to—

Order—*Cypriniformes* as in *Labeo rohita.*

Type—*Labeo bata* : State distinctive features as in above.

Economic Importance

Labeo bata is regarded as food fish but has less commercial value.

31. *EXOCOETUS* (Commonly known as Flying Fish)

Habit and Habitat

Marine: found in seas: pelagic: often skittering near boats: carnivorous: feeding on prawn and young fishes and their eggs: distributed in tropical and warmer Atlantic and India oceans and abundance in the seas of South India.

Systematic Position

Phylum	**Chordata**
Sub phylum	**Vertebrata**
Super class	**Gnathostomata**
Class	**Teleostomi**
Sub class	**Actinopterygii**
Order	**Beloniformes**
Type	***Exocoetus***

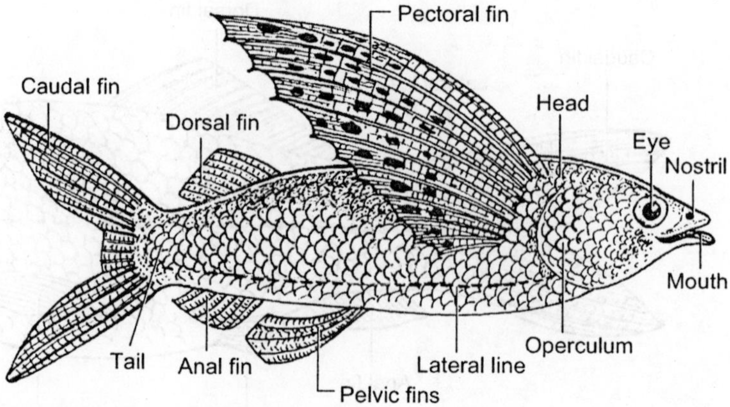

Fig. 1.1.2.31 : *Exocoetus*

Distinctive Features

1. Body is elongated and the sides of the body are silvery white.

2. Pectoral fins are greatly elongated to form wing-like structures.

3. Head is blunt and eyes large.

4. Lateral line is located low on the body.

5. Enlarged pectoral fin with black spots.

6. Tail is hypoblastic. Ventral lobe of tail fin large.

Identification with Reasons

State classification and characters up to—

Sub class—*Actinopterygii* as those of *Amia*.

Order—*Beloniformes* :

1. Body is elongated spindle shaped.
2. Large pectoral fin shifted dorsally and adapted for gliding.
3. Pelvic fin abdominal and fins with out spines.

Type—*Exocoetus* : State distinctive features as in above.

32. *HIPPOCAMPUS* (Commonly known as Sea-Horse)

Habit and Habitat

Marine: swim vertically by their dorsal fin :can anchor to any object by the **prehensile tail**: carnivorous:feeding on small fishes: cosmopolitan distribution: found in almost all warm sea waters.

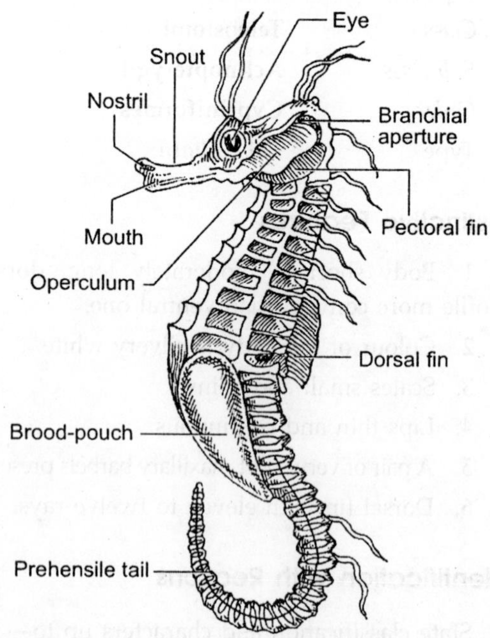

Fig. 1.1.2.32 : *Hippocampus*

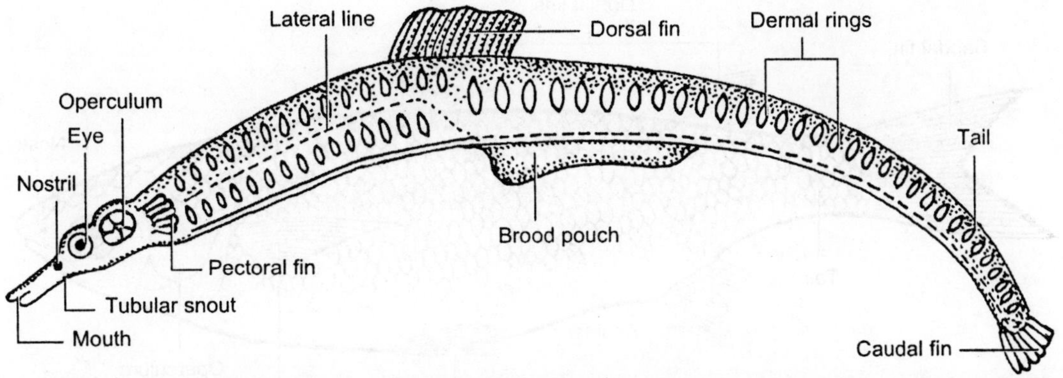

Fig. 1.1.2.33 : *Syngnathus*

Systematic Position

Phylum	**Chordata**
Sub Phylum	**Vertebrata**
Super class	**Gnathostomata**
Class	**Teleostomi**
Sub class	**Actinopterygii**
Order	**Syngnathiformes**
Type	***Hippocampus***

Distinctive Features

1. Shape of the body is like horse neck and head with trunk and tail.
2. Mouth is situated at the tip of the tubular snout.
3. Trunk with ten to twelve ring like exoskeleton.
4. Branchial aperture small, placed dorsally.
5. Gills are covered by opercula.
6. Dorsalfin is single, ventral and caudal fins are absent.
7. Tail prehensile and curved.
8. Male contains brood pouch.

Identification with Reasons

State classification and character up to—

Subclass—*Actinopterygii* as in *Amia*.

Order—*Syngnathiformes* :

1. Body is covered with ring-like exoskeleton of enlarged soft plate.
2. Mouth is terminated and suctorial.
3. Snout cylindrical.
4. Tail long and pectoral fin well developed.

Type—*Hippocampus* : State distinctive features as in above.

Special Features

Sea horse is a very interesting and most familiar fish. They strongly deviated from fish-like appearance. Male shows parental care.

33. *SYNGNATHUS* (Commonly known as Pipe Fish)

Habit and Habitat

Marine: a few fresh water form: found in tropical and subtropical arears: carnivorous: feed on small crustaceans: foods are sucked by pipette-like snout: distributed all over the world.

Systematic Position

Phylum	**Chordata**
Sub Phylum	**Vertebrata**
Super class	**Gnathostomata**

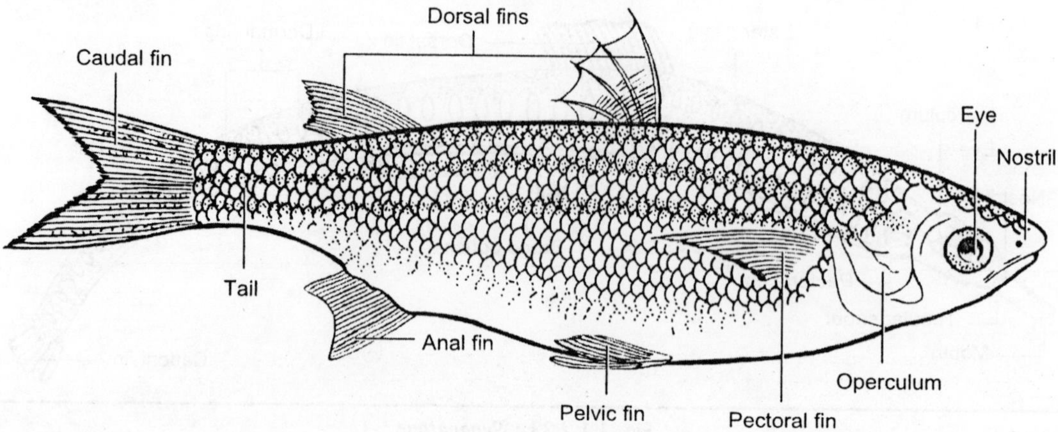

Fig. 1.1.2.34 : *Liza parsia*

Class	**Teleostomi**
Sub class	**Actinopterygii**
Order	**Syngnathiformes**
Type	***Syngnathus***

Distinctive Features

1. Body is elongated and covered by ring like exoskeletal rings.

2. Head is produced anteriorly in the from of snout.

3. Mouth is terminal, suctorial, toothless and lies at the extremety of the snout.

4. Gills are reduced and gill slits is reduced to a small opening.

5. Single dorsal fin and ventral fin absent.

6. Long tail is non prehensile and with small caudal fin.

7. Male with brood pouch on the ventral side.

Identification with Reasons

State classification and characters up to—

Order—*Syngnathiformes* as in *Hippocampus.*

Type—*Syngnathus* : state distinctive features as in above.

34. *LIZA*

Habit and Habitat

Typically estuarine species: carnivorous: feed on eggs and larvae of fishes mixed up in the soil and mud: found all along the coast of India, but is more common in the Bay of Bengal.

Systematic position

Phylum	**Chordata**
Sub Phylum	**Vertebrata**
Super class	**Gnathostomata**
Class	**Teleostomi**
Sub class	**Actinopterygii**
Order	**Mugiliformes**
Type	***Liza parsia***

Distinctive Features

1. Medium sized, average length is about 12-15 cm.

2. Colour of the body is greenish above with three to six stripes along the body.

3. Body compressed with rounded abdomen.

5. Body and head covered by cycloid scales.

6. Eyes with adipose lids.

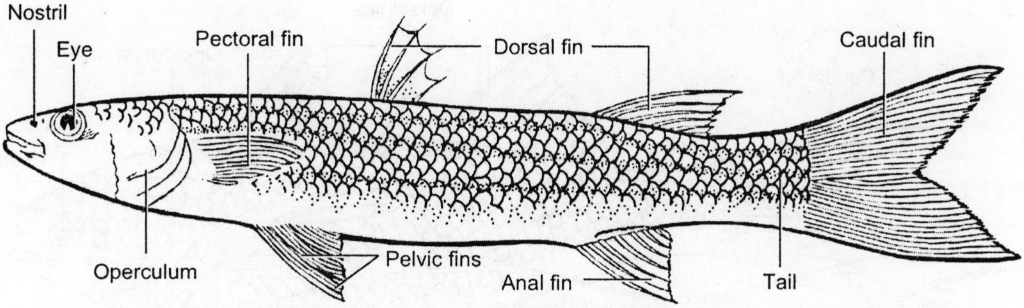

Fig. 1.1.2.35 : *Mugil tade (Bhangan)*

7. First dorsal fin with four soft rays.

8. Anal fin originates advance of second dorsal fin.

Identification with Reasons

State classification and characters up to—

Subclass—*Actinopterygii* as in *Labeo.*

Order—*Mugiliformes* :

1. Body and head are covered by cycloid and ctenoid scales.
2. Pelvic is abdominal.

Type—*Liza parsia* : State distinctive features as in above

Economic Importance

Liza constitutes an important food in east coast of India.

35. *MUGIL* (Commonly known as Bhangan)

Habit and Habitat

Estuarine: carnivorous: feed on eggs and larvae of fishes and invertebrates along with organic substances along with mud and soil: migrating into estuarines and rivers: distributed along the coastal waters of India.

Systematic Position

Phylum	**Chordata**
Sub Phylum	**Vertebrata**
Super class	**Gnathostomata**
Class	**Teleostomi**
Sub class	**Actinopterygii**
Order	**Mugiliformes**
Type	***Mugil tade***

Distinctive Features

1. Body elongated cylindrical with rounded abdomen.

2. Largest, Mullets attains a length of about 90 cm.

3. Head long and inter orbital space flat.

4. Scales cycloid in young, ctenoid in adult.

5. Origin of second dorsal and anal fin are more or less on the same line.

7. Eyes with adipose tissue.

8. Three to four strips along the body.

Identification with Reasons

State classification and characters up to—

Order—*Mugiliformes* as in *Liza.*

Type—*Mugil tade* : State distinctive features as in above.

Remark

Mugil is a food fish.

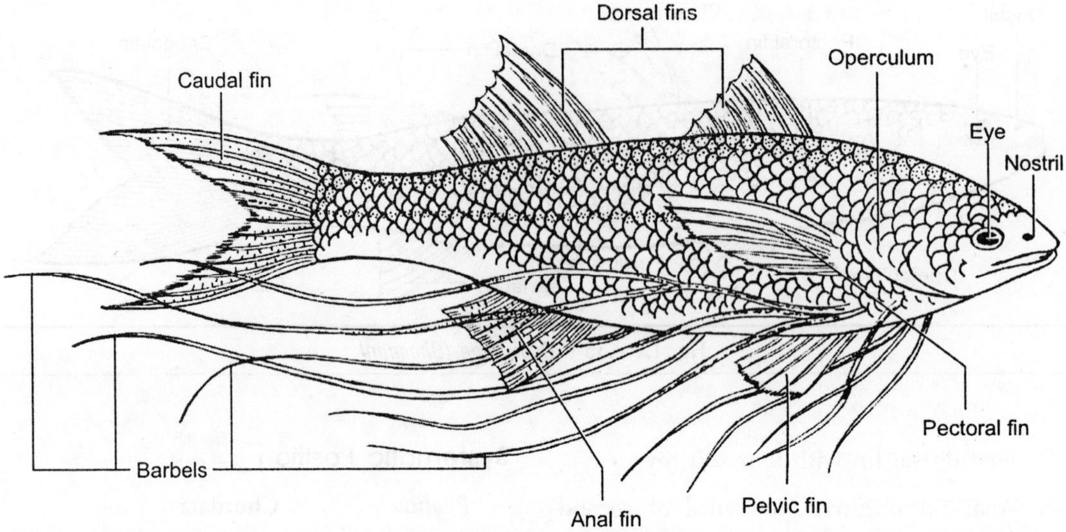

Fig. 1.1.2.36 : *Polynemus*

36. *POLYNEMUS* (Commonly known as Tapse)

Habit and Habitat

Estuarine: migrating into fresh water rivers: feed on organic matter suspended in waters and present in soil and mud.

Systematic Position

Phylum	**Chordata**
Sub Phylum	**Vertebrata**
Super class	**Gnathostomata**
Class	**Teleostomi**
Sub class	**Actinopterygii**
Order	**Polynemiformes**
Type	***Polynemus***

Distinctive Features

1. Body compressed and abdomen rounded.
2. Head short and maxilla extends forward to form the snout.
3. Bilobed pectoral fin with seven long fin rays acting as barbeles. The largest fin ray is about twice the length of the body.
4. Dorsal fin two, first dorsal fin rays are spiny.
5. Body covered by ctenoid scales.
6. Caudal fin deeply forked, with long pointed lobes.

Identification with Reasons

State classification and characters upto—

Sub class—*Actinopterygii* as in *Labeo.*

Order—*Polynemiformes* :

1. Large mouth lies below the snout.
2. Two dorsal fins are present.
3. Bilobed pectoral fins.
4. Body covered by ctenoid scales.

Type—*Polynemus* : State distinctive features as in above.

Remark

Polynemus is a food fish.

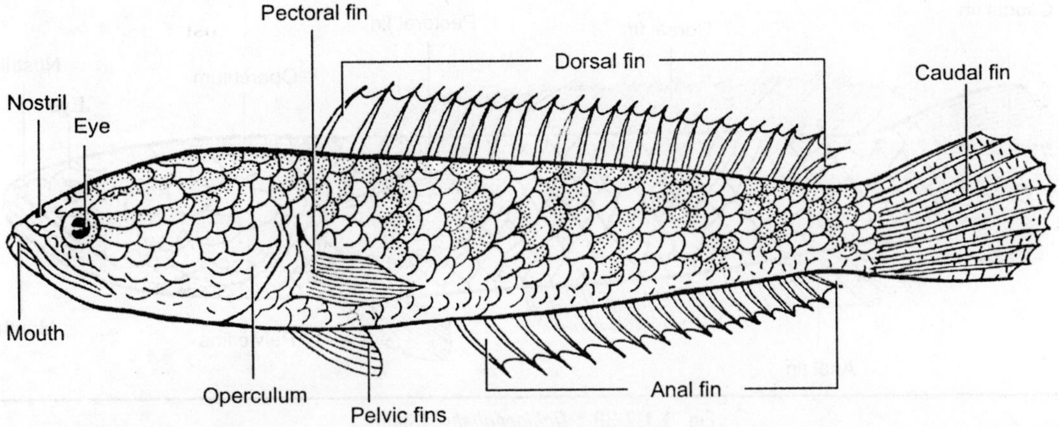

Fig. 1.1.2.37 : *Channa* (Lata)

37. *CHANNA* (Commonly known as Snake-heads and are called livefish or jiol mach)

Habit and Habitat

Fresh water: found in ditches and ponds: prefers stagnant muddy waters: carnivorous: feeds on small fishes and their spawns, insects and tadpoles: builds crude nest to rear their young ones: air-breathing habit: can survive out of water for few hours as they possess **supra branchial organ**: abundant in India, Mayanmar, Sri Lankha, Malay peninsula, Thailand, China and Africa.

Systematic Position

Phylum	**Chordata**
Sub Phylum	**Vertebrata**
Super class	**Gnathstomata**
Class	**Teleostomi**
Sub class	**Actinopterygii**
Order	**Ophiocephaliformes**
Type	***Channa* (Ophiocephalus) *punctatus***

Distinctive Features

1. Smaller is size, head is triangular and tapers into a pointed snout and resembles that of a snake.
2. Head is covered with large scales.
3. Dorsal and anal fins single, long and without spine.
4. Body elongated, anteriorly cylindrical, posteriorly compressed.
5. Body colour variable, usually greenish brown above, yellow below with coloured spots and patches.
6. Caudal fin rounded.

Identification with Reasons

State classification and characters up to—

Sub class—*Actinopterygii* as those of *Amia*.

Order—*Ophiocephaliformes* :

1. Head is depressed and covered by large plate-like scales.
2. Scale cycloid or pseudoctenoid.
3. Fins are devoid of spines.
4. Accessory respiratory organ present.
5. Caudal fin rounded.

Type—*Channa punctatus* : State distinctive features as in above.

Economic Importance

Useful as food

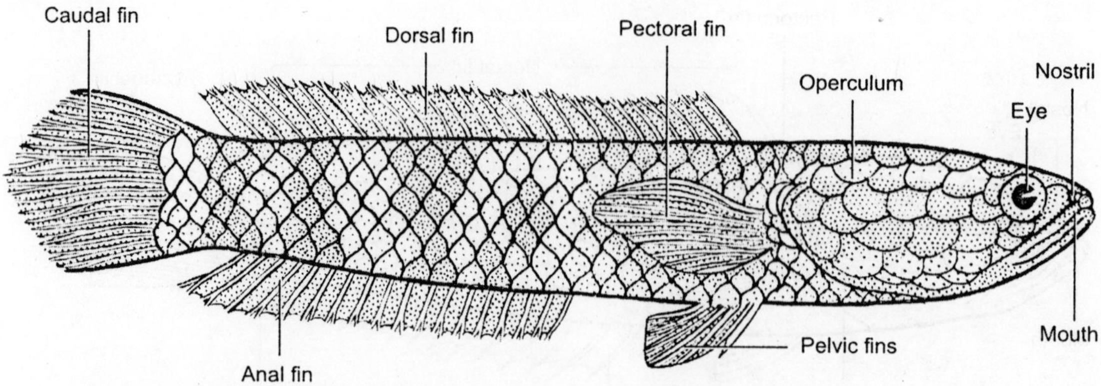

Fig. 1.1.2.38 : *Ophiocephalus* (Shol)

38. *OPHIOCEPHALUS* (Commonly known as Shol)

Habit and Habitat

Same as *Channa*

Systematic Position

Phylum	**Chordata**
Sub Phylum	**Vertebrata**
Super class	**Gnathostomata**
Class	**Teleostomi**
Sub class	**Actinopterygii**
Order	**Ophiocephaliformes**
Type	*Ophiocephalus striatus*

Distinctive Features

1. Medium-sized, adult may be 1-1·5 feet in length.
2. Body dark brown or black above, yellowish or orange below.
3. Fins are pigmented with yellow dots in rows.
4. Jaws with sharp teeth.

Identification with Reasons

State classification and characters up to—

Order—*Ophiocephaliformes* as in *Channa*

Type—*Ophiocephalus striatus* : State distinctive features as in above

Economic Importance

Useful as food.

Remark

Fish belong to the Order Ophiocephaliformes are commonly knows as Snake-heads. Snake-heads or Murrels being typical live fishes, having soft flesh devoid of fats, are considered to have medicinal value and provide nutritious food.

39. *AMPHIPNUOUS*

Habit and Habitat

Found in fresh and brakish water: amphibious: generally live in mud holes or in the grassy sides of ponds: carnivorous: feed on insect larvae, shrimps and other small fishes: distributed in India, Bangladesh and Myanmar.

Systematic Position

Phylum	**Chordata**
Sub Phylum	**Vertebrata**
Super class	**Gnathostomata**

Class	**Teleostomi**
Sub class	**Actinopterygii**
Order	**Symbranchiformes**
Type	***Amphipnous cuchia***

Distinctive Features

1. Long, snake like body devoid of scales.
2. Colour of the body is brownish with numerous black spots.
3. Dorsal and anal fins are devoid of fin rays.
4. Paired fins are absent.
5. Gill rudimentary, gill slit triangular.
6. Scales small, present in the tail only.
7. Lateral line prominent.
8. Pharynx with accessory sacs for air breathing.

Identification with Reasons

State classification and characters upto—

Sub class—***Actinopterygii*** as those of *Amia*.

Order—***Symbranchiformes*** :

1. Body elongated, compressed and snake like.
2. Scales small and arranged in longitudinal in rows.
3. Dorsal, caudal and anal fin are continuous.
4. Gill slit single, transverse and ventral.
5. Pneumatic sac may or may not be present.

Type—***Amphipnous cuchia*** : State distinctive features as in above

Remark

Amphipnuous has a respiratory sac. Gills are rudimentary and they lost all its power of aquatic respiration. It frequents the surface of water at regular intervals to inhale air, similar to those of frog.

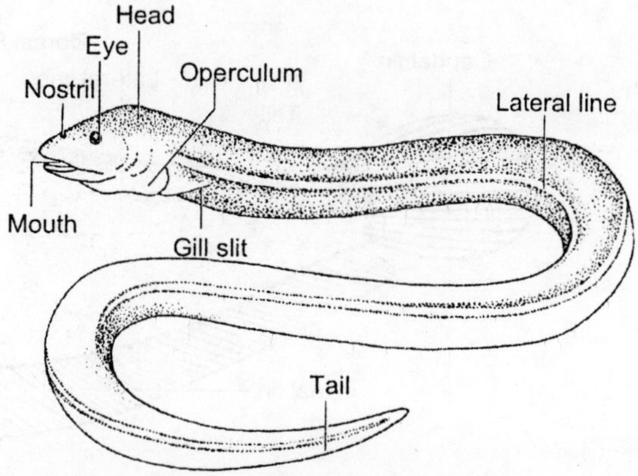

Fig. 1.1.2.39 : *Amphipnuous*

40. *HETEROPNEUSTES* (Commonly called as Catfish; known as Singhi)

Habit and Habitat

Found in fresh water and estuaries: bottom dwellers: omnivorous: feed on snails, insect larvae, crustaceans and small fishes: act as scavengers by living in dirty ponds and muddy water: distributed in India, Bangladesh, Myanmar, Sri Lanka and Malaya.

Systematic Position

Phylum	**Chordata**
Sub Phylum	**Vertebrata**
Super class	**Gnathostomata**
Class	**Teleostomi**
Sub class	**Actinopterygii**
Order	**Siluriformes**
Type	***Heteropneustes fossilis***

Distinctive Features

1. Elongated, slender body not covered by scales.
2. Head flattened and laterally compressed.
3. Dorsal fin short and without spine.

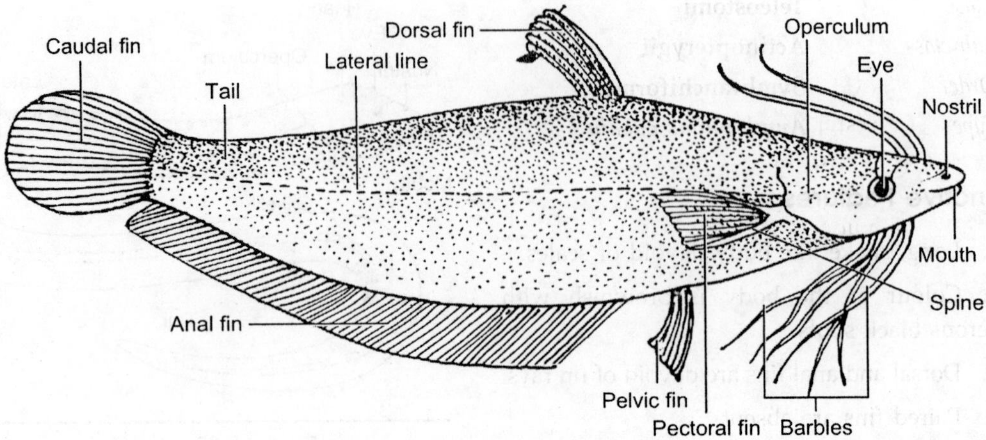

Fig. 1.1.2.40 : *Heteropneustes* (Singhi)

4. Anal fin elongated and is separated from caudal fin by a notch.

5. Pectoral fin strong with poison spine.

6. Barbels long, four pairs.

7. First fin ray of dorsal and pectoral fins are modified into a sharply pointed spine.

8. Presence of an accessory respiratory air tube in addition to the gills.

Identification with Reasons

State classification and characters upto—

Sub class—*Actinopterygii* as in *Amia*.

Order—*Siluriformes* :

1. Skin naked or with bony plate but never with scales.
2. Barbels one to four pairs.
3. Pectoral fin with a spine.

Type—*Heteropneustes fossilis* : State distinctive features as in above

Economic Importance

Heteropneustes is popular as food. They are considered to have medicinal value and provide nutritious food particularly to the children and sick persons. They can easily be cultured in cages.

41. *CLARIAS* (Commonly called as Catfish; known as Magur)

Habit and Habitat

Same as Heteropneustes.

Systematic Position

Phylum	**Chordata**
Sub Phylum	**Vertebrata**
Super class	**Gnathostomata**
Class	**Teleostomi**
Sub class	**Actinopterygii**
Order	**Siluriformes**
Type	***Clarias batrachus***

Distinctive Features

1. Elongated body with a broad depressed head with top and sides covered with bony plates.

2. Barbels four pairs around the snout of which one pair is larger and rest are smaller.

3. Colour of the body uniform brown or grayish black.

4. Dorsal fin without spine and extending from neck to the caudal fin.

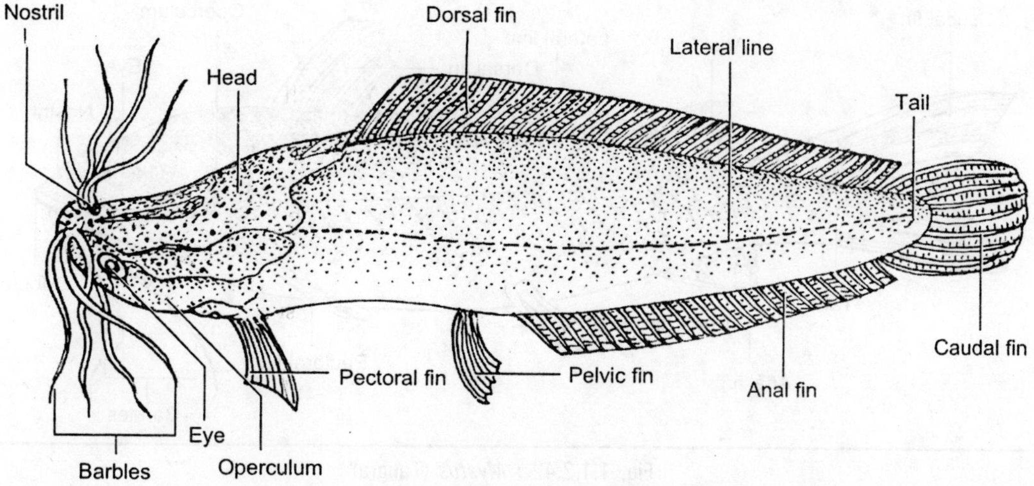

Fig. 1.1.2.41 : *Clarias* (Magur)

5. Pectoral fins are with non poisonous spines.

6. Well developed accessory respiratory organs are located dorsal to the gills.

Identification with Reasons

State classification and characters upto—

Order—*Siluriformes* as in *Heteropneustes.*

Type—*Clarias batrachus* : State distinctive features as in above.

Economic Importance

Same as *Heteropneustes*

42. *MYSTUS*

Habit and Habitat

Brackish water and fresh water fishes: bottom deweller: bottom feeder omnivorous: predatory in habits: attack small carps and prawns: widely distributed in India, Bangladesh, Myanmar and Sri Lanka.

Systematic Position

Phylum	**Chordata**
Sub Phylum	**Vertebrata**
Super class	**Gnathostomata**
Class	**Teleostomi**
Sub class	**Actinopterygii**
Order	**Siluriformes**
Type	***Mystus***

Distinctive Features

1. Body is brownish above with silvery sides.
2. Well developed adipose dorsal fin.
3. Circular black spot on the adipose fin.
4. Snout is long with barbels.
5. The first fin ray of the pectoral fins and dorsal fins are modified into a large stout spine.

Identification with Reasons

State classification and characters upto—

Order—*Siluriformes* as in *Heteropneustes.*

Type—*Mystus* : State distinctive features as in above.

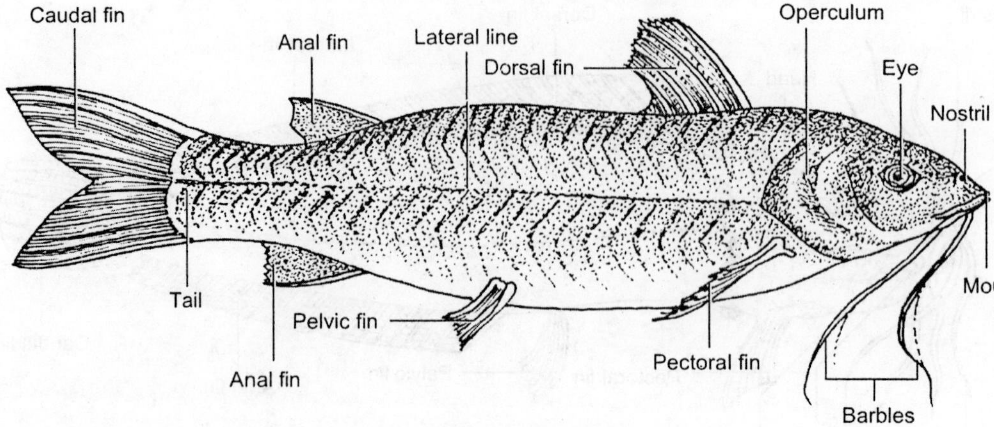

Fig. 1.1.2.42 : *Mystus* (Tangra)

Economic Importance

Mystus is regarded as foodfish.

43. *LATES*

Habit and Habitat

Marine: in habits the saline estuaries: occasionally found in fresh water streams: carnivorous: feed on small fry of fishes and crustaceans larvae: found in the seas from India to China, especially in the estuaries of large rivers on the eastern coasts.

Systematic Position

Phylum	**Chordata**
Sub Phylum	**Vertebrata**
Super class	**Gnathostomata**
Class	**Teleostomi**
Sub class	**Actinopterygii**
Order	**Perciformes**
Type	***Lates calcarifer***

Distinctive Features

1. Body is spindle-shaped and perfectly stream-lined.

2. Colour of the body is grey, dark green along back and silvery below.

3. Mouth is wide and terminal, bounded by jaws with teeth.

4. Operculum spiny.

5. First dorsal fin with seven spiny rays of which third one is the largest.

6. Second dorsal fin with twelve rays which the first one is spiny, rest is soft.

7. Pelvic fin with five rays of which the first one is spinous rod.

Identification with Reasons

State classification and characters up to—

Sub class—*Actinopterygii* as in *Amia*.

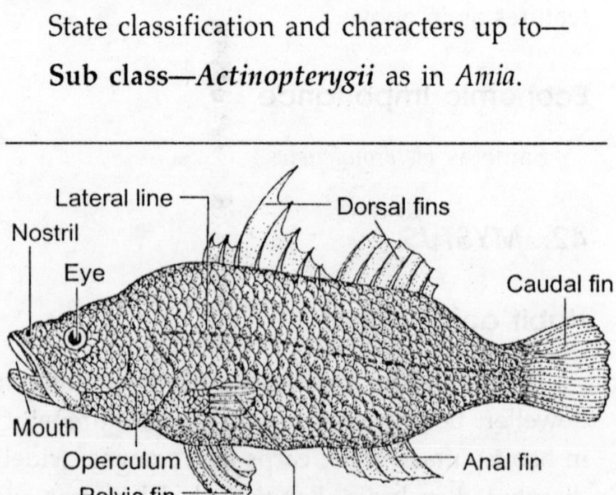

Fig. 1.1.2.43 : *Lates*

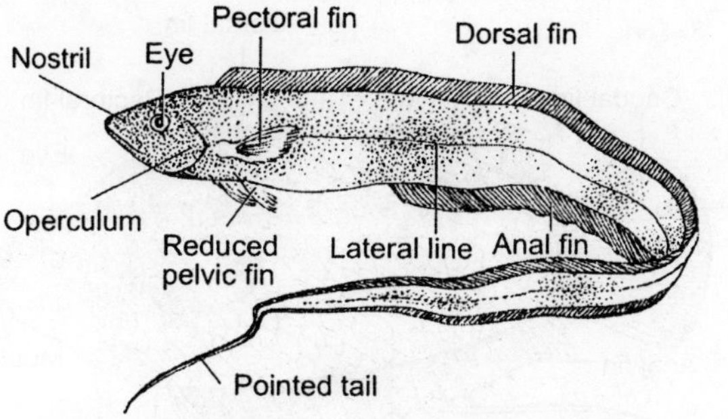

Fig. 1.1.2.44 : *Trichiurus* (Ribbon fish)

Order—*Perciformes* :

1. Body covered with ctenoid scales.
2. Head depressed covered with large scale or plates.
3. Dorsal and anal fins are provided with both Spiny and soft fin rays.

Type—*Lates calcarifer* : State distinctive features as in above.

Economic Importance

Lates is a good game fish and is highly esteemed as high quality food.

44. *TRICHIURUS* (Commonly called as Ribbon fish)

Habit and Habitat

Marine: enters estuaries: carnivorous: voracious feeder: feed on small fishes, crustaceans etc: common in Arabian Sea, Bay of Bengal and extends on to the Indian ocean and beyond.

Systematic Position

Phylum	**Chordata**
Sub Phylum	**Vertebrata**
Super class	**Gnathostomata**
Class	**Teleostomi**
Sub class	**Actinopterygii**
Order	**Perciformes**
Type	***Trichiurus***

Distinctive Features

1. Ribbon-shaped fish, having a very thin body flattened from side to side.
2. Dorsal and anal fins narrow with numerous soft rays.
3. Caudal fin absent.
4. Pelvic fin very reduced.
5. Trunk tapers to a long filamentous tail.

Identification with Reasons

State classification and characters up to—

Order—*Perciformes* as in *Lates.*

Type—*Trichiurus* : State distinctive features as in above.

Economic Importance

Ribbon fishes are eaten in the fresh condition but they were mostly sun-dried and sold as cured fish.

45. *ANABAS* (Commonly called as Climbing Perch)

Habit and Habitat

Both fresh water and brackish water: carnivorous: feed on small fishes and spawns: posses **supra-branchial organ** by which they can live out of water for long period: migrate from pond to pond during rainy season walking on their serrated operculum: distributed in India, Bangladesh, Myanmar, Sri Lanka and Malaya.

Systematic Position

Phylum	**Chordata**
Sub Phylum	**Vertebrata**
Super class	**Gnathostomata**
Class	**Teleostomi**
Sub class	**Actinopterygii**
Order	**Perciformes**
Type	***Anabas***

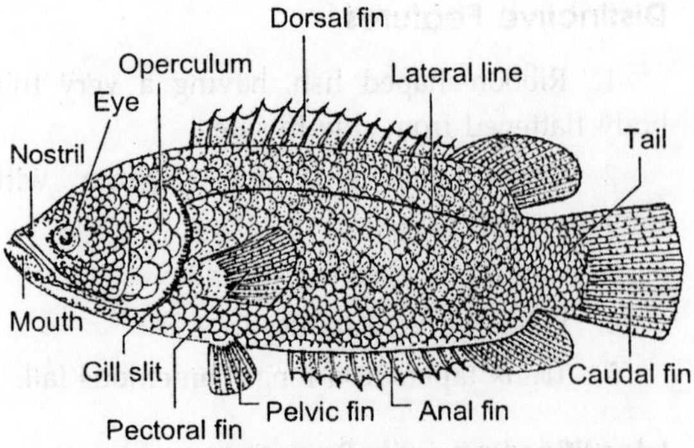

Fig. 1.1.2.45 : *Anabas*

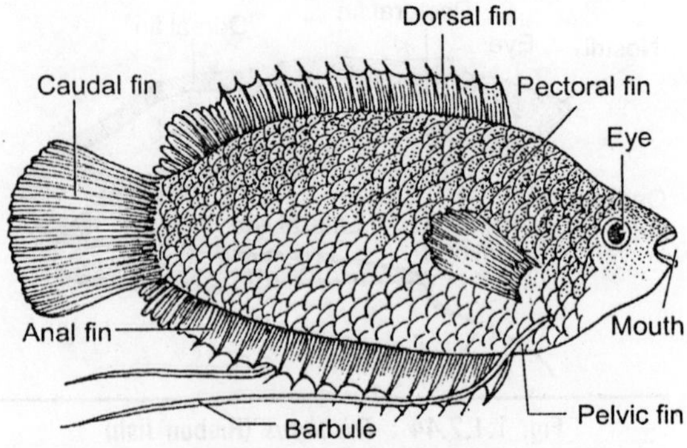

Fig. 1.1.2.46 : *Trichogaster*

Distinctive Features

1. Body oblong, laterally compressed and covered with ctenoid scales.

2. Dorsal and anal fins are elongated and have two distinct lobes. Each anterior lobe is spiny and posterior lobe soft rayed.

3. Operculum with backwardly directed spines.

4. Pectoral and pelvic fin is spine less.

Identification with Reasons

State classification and characters up to—

Order—*Perciformes* as in *Lates*.

Type—*Anabas* : State distinctive features as in above.

Economic Importance

Anabas is regarded as food fish.

46. TRICHOGASTER (Known as Kholisa Fish)

Habit and Habitat

Same as Anabas.

Systematic position

Phylum	**Chordata**
Sub Phylum	**Vertebrata**
Super class	**Gnathostomata**
Class	**Teleostomi**
Sub class	**Actinopterygii**
Order	**Perciformes**
Type	***Trichogaster***

Distinctive features

1. Body laterally compressed, head small, eyes large.

2. Continuous dorsal fin with anterior part supported by fifteen to eighteen spine and posterior part by seven to thirteen rays.

3. Continuous anal fin having fifteen to twenty spines in anterior part and eleven to nineteen rays in posterior part.

4. Mouth small, upturned.

5. Pelvic fin elongated into filliform barbule.

6. Colour bands present on the body.

Identification with Reasons

State classification and characters upto—

Order—*Perciformes* as in *Lates*.

Type—*Trichogaster* : State distinctive features as in above.

Economic Importance

Trichogaster is regarded as food fish.

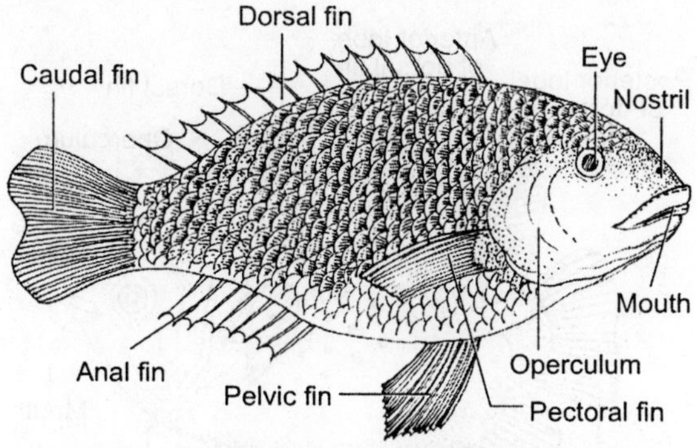

Fig. 1.1.2.47 : *Tilapia*

47. *TILAPIA*

Habit and Habitat

Exotic species introduced into India in 1952: original home is Africa: fresh water: found in ponds and lakes: thrives under cultivation: carnivorous: feed on other small fishes and mosquito larvae: prolific breeder: mouth brooders.

Systematic Position

Phylum	**Chordata**
Sub Phylum	**Vertebrata**
Super class	**Gnathostomata**
Class	**Teleostomi**
Sub class	**Actinopterygii**
Order	**Perciformes**
Type	***Tilapia/Oreochromis***

Distinctive Features

1. Thick body, laterally compressed.

2. Body with many coloured patches with blackish dorsal side.

3. Mouth opening large and lower jaw is far upraised.

4. Single dorsal fin is supported by twelve to eighteen fin rays.

5. Pectoral, pelvic and anal fins are well developed.

Identification with Reasons

State classification and characters up to—

Order—*Perciformes* as in *Lates.*

Type—***Tilapia/Oreochromis*** : State distinctive features as in above.

Economic Importance

Food fish and a valuable source of protein. Can be used to control mosquito.

48. *PAMPUS* (Commonly known as Grey or Silver-pomfret)

Habit and Habitat

Marine: carnivorous: feed on small fishes and zoo planktons: found through out the east and west coasts of India also in the northern parts of Bay of Bengal.

Systematic Position

Phylum	**Chordata**
Sub Phylum	**Vertebrata**
Super class	**Gnathostomata**
Class	**Teleostomi**
Sub class	**Actinopterygii**
Order	**Perciformes**
Type	***Pampus***

Distinctive Features

1. Body ovate, compressed from side to side and covered with thin, small scales.

2. Upper part of the body silvery grey, while head and body with fine black spot over a silvery back ground.

3. Dorsal fin with a distinct posterior portion, which is hidden in the skin.

4. Caudal fin deeply forked and the ventral lobe is larger then the dorsal lobe.

5. Pelvic fin well developed.

6. Mouth terminal, jaws equal, lips thin.

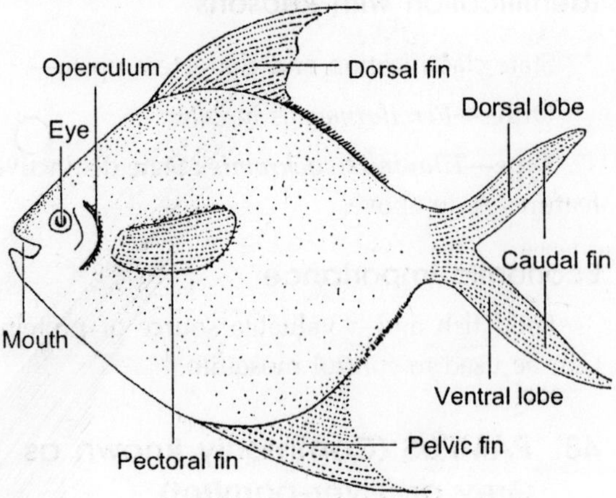

Fig. 1.1.2.48 : *Pampus*

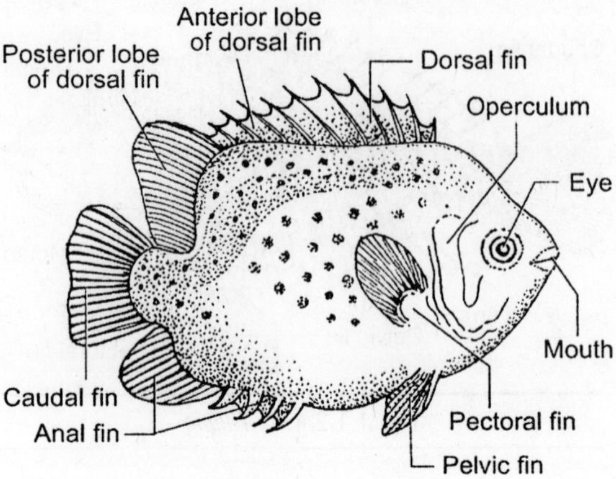

Fig. 1.1.2.49 : *Scatophagus*

Identification with Reasons

State classification and characters upto—

Order—*Perciformes* as in *Lates.*

Type—*Pampus* : State distinctive features as in above.

Economic Importance

Commercially important fish on account of their soft flesh, delicate flavours and absence of hard bones, they are regarded as good fish food.

49. *SCATOPHAGUS* (Commonly known as Chanda)

Habit and Habitat

Marine: found in brackish water: carnivorous: feed on very small fishes, crustaceans etc: occurs through out Indo-pacific region.

Systematic Position

Phylum	**Chordata**
Sub Phylum	**Vertebrata**
Super class	**Gnathostomata**
Class	**Teleostomi**
Sub class	**Actinopterygii**
Order	**Perciformes**
Type	***Scatophagus***

Distinctive Features

1. Body ovoid, laterally compressed, covered with small scales.
2. Dorsal fin two, anterior one spiny and posterior one fan-shaped.
3. Anterior dorsal fin with eleven spiny fin rays.
4. First fin ray of pelvic fin is spiny.
5. Anal fin bilobed, anterior with four spines.
6. Dorsal side of the body with black spots, ventral side greyish white.

Identification with Reasons

State classification and characters up to—

Order—*Perciformes* as in *Lates.*

Type—*Scatophagus* : State distinctive features as in above.

Economic Importance

Scatophagus is a food fish.

50. *NANDUS* (Commonly known as Nados)

Habit and Habitat

Fresh water: carnivorous: feed on small fishes and spawns: distributed in India, Bangladesh.

Systematic Position

Phylum	**Chordata**
Sub Phylum	**Vertebrata**
Super class	**Gnathostomata**
Class	**Teleostomi**
Sub class	**Actinopterygii**
Order	**Perciformes**
Type	*Nandus*

Distinctive Features

1. Small to medium sized, brilliantly coloured patches is on the dorsal side, ventral surface is whitish.

2. Dorsal fin divided into anterior and posterior parts, anterior part bear twelve spines, posterior parts is soft and with seven coloured stripes.

3. Caudal fin has six coloured stripes and twelve soft fin rays.

4. Pelvic fin with six fin rays of which the first one is short and spiny and with six black stripes.

5. Anal fin bilobed, anterior with three spines and posterior with six black stripes.

6. Circular black patch on either side of the caudal peduncle.

7. Both jaws with very small teeth.

Identification with Reasons

State classification and characters up to—

Order—*Perçiformes* as in *Lates*.

Type—*Nandus* : State distinctive features as in above.

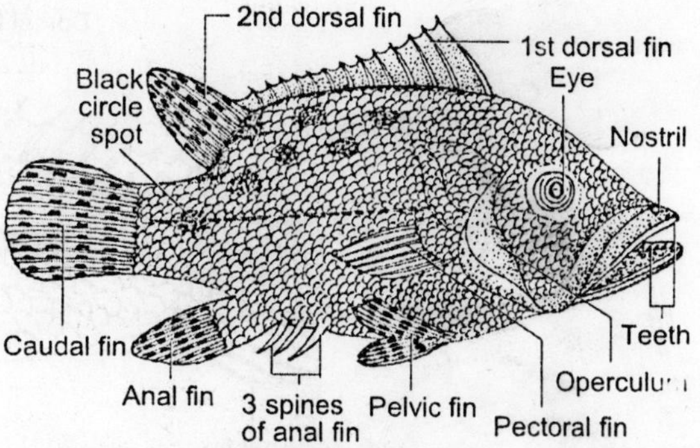

Fig. 1.1.2.50 : *Nandus*

Economic Importance

Nandus is a food fish.

51. *BARILIUS* (Commonly known as Bhola)

Habit and Habitat

Brakish water: some marine: carnivorous: eat variety of meat, larvae etc: distributed in India and Bangladesh.

Systematic Position

Phylum	**Chordata**
Sub Phylum	**Vertebrata**
Super class	**Gnathostomata**
Class	**Teleostomi**

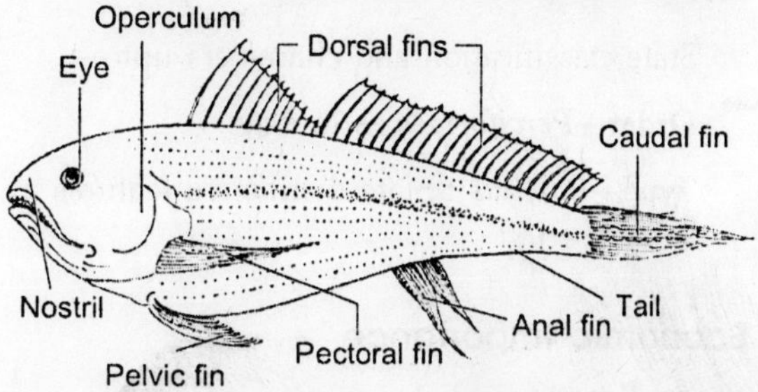

Fig. 1.1.2.51 : *Barilius*

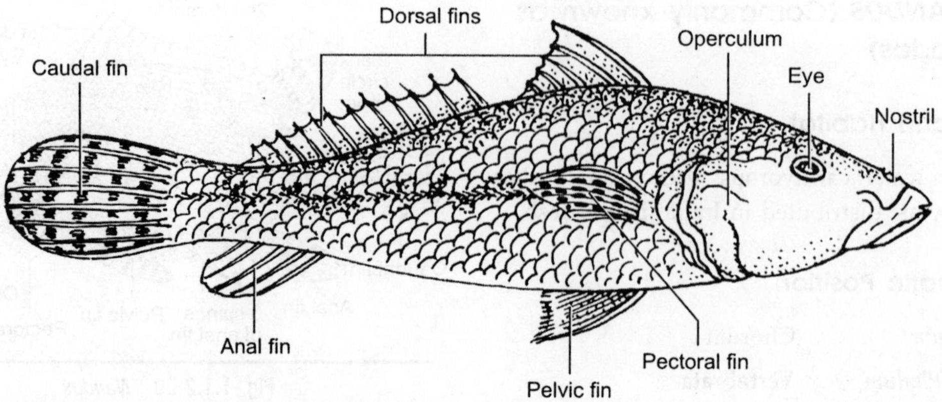

Fig. 1.1.2.52 : *Glossogobius*

Sub class	**Actinopterygii**
Order	**Perciformes**
Type	***Barilius***

Distinctive Features

1. Body is fusiform, maximum breadth just behind the operculum, trunk abruptly tapers.

2. Colour of the body is light orange with clusters of dots.

3. Mouth large, teeth on both jaws and a knob above symphysis of lower jaw.

4. Dorsal fins two, first dorsal fin with eleven to twelve fin rays, and second with twenty one to twenty two fin rays.

5. Body with vertical bars.

Identification with Reasons

State classification and characters upto—

Order—*Perciformes* as in *Lates.*

Type—*Barilius* : State distinctive features as in above.

Economic Importance

Barilius is food fish.

52. *GLOSSOGOBIUS* (Commonly called as Bele in Bengali)

Habit and Habitat

Freshwater: some times found in brackish water: omnivorous: voracious eater and eat mosquito larvae: distributed in plains of India.

Systematic Position

Phylum	**Chordata**
Sub Phylum	**Vertebrata**
Super class	**Gnathostomata**
Class	**Teleostomi**
Sub class	**Actinopterygii**
Order	**Perciformes**
Type	***Glossogobius***

Distinctive Features

1. Body elongated, cylindrical with compressed rounded abdomen.

2. Males are brown and females are yellowish grey markings with black spot.

3. Dorsal fin separated into two lobes.

4. Eyes large and are situated almost side by side on top of the head.

5. First lobe of the dorsal fin with six spines.

6. Pelvic fin oblong united and situated just below the pectoral fin.

7. Caudal fin oblong.

8. Lower jaw elongated.

Identification with Reasons

State classification and characters upto—

Order—*Perciformes* as in *Lates.*

Type—*Glossogobius* : State distinctive features as in above.

Economic Importance

Glossogobius is food fish. It can be cultured for mosquito control.

53. *CYNOGLOSSUS* (Commonly called as Flat-fish or Tongue-fish)

Habit and Habitat

Marine: bottom-living in coastal waters lying on side: carnivorous: feed on molluscs: is common on the west coast of India.

Systematic Position

Phylum	**Chordata**
Sub Phylum	**Vertebrata**
Super class	**Gnathostomata**
Class	**Teleostomi**
Sub class	**Actinopterygii**
Order	**Pleuronectiformes**
Type	***Cynoglossus***

Distinctive Features

1. Highly compressed body, upper side which is exposed is coloured, the lower side is white.

2. Both eyes are situated on the upper side.

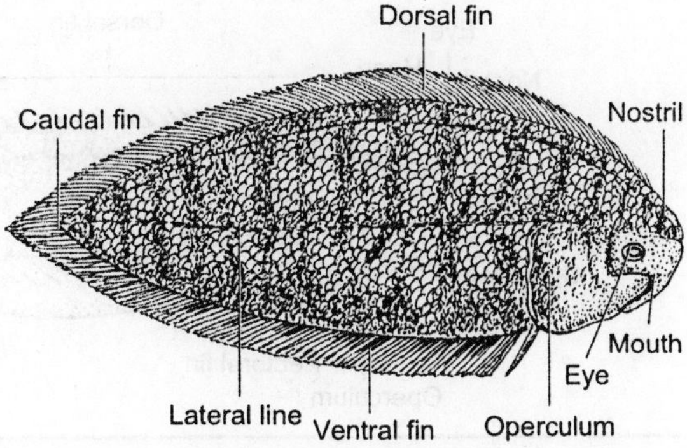

Fig. 1.1.2.53 : *Cynoglossus*

3. Dorsal, caudal and ventral fins are continuous structures giving the body a spear-head shape.

4. Pectoral is absent.

5. Pelvic fin is only on the upper side.

6. Dorsal fin originates beyond the limit of the eyes.

Identification with Reasons

State classification and characters up to—

Sub class—***Actinopterygii*** as in *Amia.*

Order—***Pleuronectiformes*** :

1. Highly compressed body and the eyes are shifted to one side of the head.
2. Head is asymmetrical.
3. Fins lack spines.
4. Dorsal side coloured, ventral side is white.
5. Dorsal and ventral fins are long and continuous with the caudal fin.

Type—***Cynoglossus*** : State distinctive features as in above

Economic Importance

Flat fish is an important food fish in west coast of India and is sought commercially to a great extent.

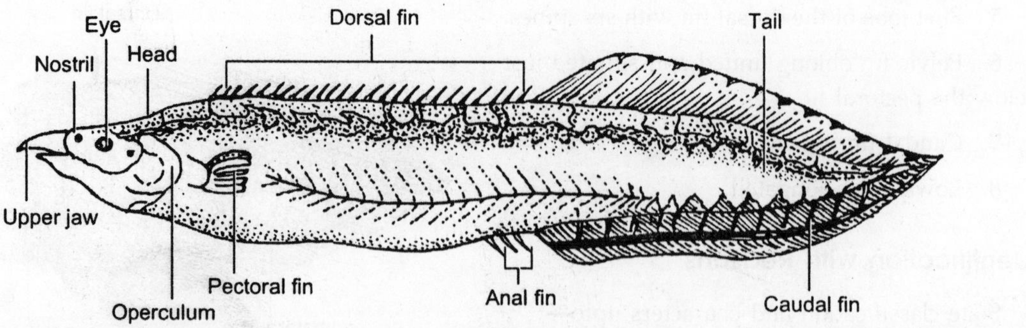

Fig. 1.1.2.54 : *Mastacembelus*

54. *MASTACEMBELUS* (Commonly called as Spiny-eel)

Habit and Habitat

Freshwater: found in ponds, ditches and in shallow rivers: omnivorous: feed mainly on decomposed organic matters:

Systematic Position

Phylum	**Chordata**
Sub Phylum	**Vertebrata**
Super class	**Gnathostomata**
Class	**Teleostei**
Sub class	**Actinopterygii**
Order	**Mastacembeliformes**
Type	***Mastacembelus***

Distinctive Features

1. Eel like body with tubular nostril situated at the end of the snout.

2. Both side of the body are pointed.

3. Upper jaw is elongated and pointed and hangs like a beak.

4. Dorsal fin long, proceeded by seven to forty free spines.

5. Anal fin long and with three spines.

6. Caudal fin short and united with dorsal and anal fin.

Identification with Reasons

State classification and characters up to—

Sub class—*Actinopterygii* as in *Amia*

Order—*Mastacembeliformes* :

1. Dorsal, caudal and anal fins become united to form a continuous fin.
2. Dorsal fin is preceded by some free spines.
3. Anterior nostrils tubular.
4. Pelvic fin absent.

Type—*Mastacembelus* : State distinctive features as in above.

55. *ECHENEIS* (Commonly called as Sucker-fish)

Habit and Habitat

Marine: attaches itself by means of adhesive disk to other fishes, whales, turtles and other floating object: carnivorous: feeding on small fishes: distributed all over the tropical and warm seas: dwell in the surface area of the sea.

Systematic Position

Phylum	**Chordata**
Sub Phylum	**Vertebrata**
Super class	**Gnathostomata**
Class	**Teleostei**

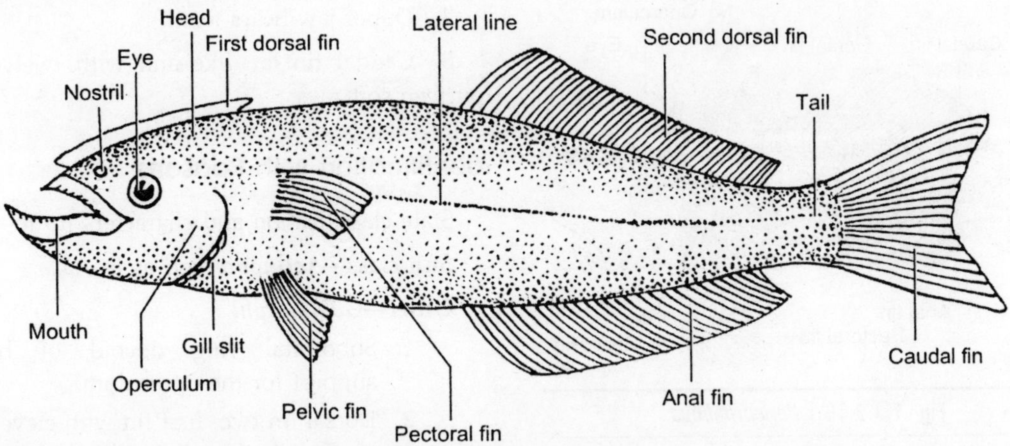

Fig. 1.1.2.55 : *Echeneis*

Sub class	**Actinopterygii**
Order	**Echeneiformes**
Type	***Echeneis***

Distinctive Features

1. Body is elongated and covered with cycloid scales.

2. The first spinous dorsal fin is modified into an adhesive disk or sucker.

3. Second dorsal fin and anal fin apposed to each other and without any spine.

4. Lower jaw slightly extended than the upper jaw.

5. Ventral fin with a spine and five rays.

6. Air bladder absent.

Identification with Reasons

State classification and characters up to—

Sub class—*Actinopterygii* as in *Amia*

Order—*Echeneiformes* :

1. First dorsal fin is modified into a flat oval adhesive disc on the upper surface of the head.
2. The other dorsal and anal fins are spineless.
3. Body covered by cycloid scales.
4. Airbladder absent.

Type—*Echeneis* : State distinctive features as in above.

56. *PERIOPTHALMUS* (Commonly known as mud skipper)

Habit and Habitat

Marine: very often found near river mouth and mangrove vegetation: carnivorous: feed on very small fishes, spawns and shrimps: skip swiftly over mud with great speed: during high tide, take rest on branches of trees and plants.

Systematic Position

Phylum	**Chordata**
Sub Phylum	**Vertebrata**
Super class	**Gnathostomata**
Class	**Teleostei**
Sub class	**Actinopterygii**
Order	**Gobimorphi**
Type	***Periopthalmus***

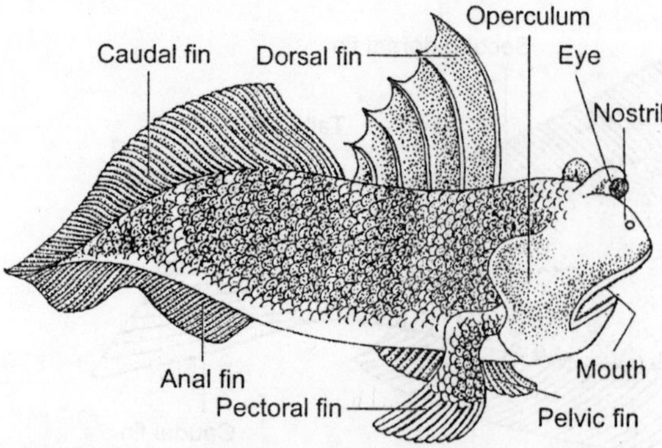

Fig. 1.1.2.56 : *Periopthalmus*

Distinctive Features

1. Basal portion of the pectoral fin is stumpy and are united to form a disc.

2. Second dorsal and anal fin originate from the same place at opposite end of the body.

3. Projected eyes are large and placed side by side on the top of the head.

4. Upper jaw bears teeth.

5. Caudal fin fan like and with twelve to fourteen soft rays.

Identification with Reasons

State classification and characters up to—

Sub class—*Actinopterygii* as in *Amia*

Order—*Gobimorphi* :

1. Suborbital ring devoid of body support for the operculum.
2. Dorsal fin two, first fin with eleven to twelve fin rays of which the first one much longer.
3. Pelvic fins located below the pectorals and each pectoral fin has one spine and five to six soft rays.
4. Gills are four and the gill membrane is attached to isthmus.
5. Swim bladder absent.

Type—*Periopthalmus* : State distinctive features as in above.

FIRST TERRESTRIAL VERTEBRATES: AMPHIBIA

During the later part of the Devonian period same osteolapid fishes lived in the fresh water gave rise to the terrestrial population, the **Amphibia**. Amphibia are the first of the great group of land vertebrate the **tetrapoda**. While coming on land, they underwent tremendous biological changes to become land-dwelling animals. The modern Amphibia are quite different from their remote palaeozoic ancesters. They have lost their scales to use their moist skin for excretion carbon dioxide. They are capable of respiration on land by lungs while larvae breathes in water by gills. The **axial** and **appendicular** skeletal frame work becomes altered to a new method of locomotion on land.

Class Amphibia is divided into the three Sub classes (i) **Labyrinthodontia**, (ii) **Lepospondyli** and (iii) **Liss amphibia**. Modern Amphibia are placed in a separate subclass Liss amphibia with three orders: **Urodela** (newts and salamanders) retain original long body and larval characteristics and exhibits **paedogenesis** and **Neoteny**; the **Anura** (frogs and toads) have lost their post anal tail; and the **Apoda** are limbless.

1. *ICHTHYOPHIS*
(Commonly known as Cecilian)

Habit and Habitat

Terrestrial: burrowing; occurs in moist and damp areas under stones and rocks: carnivorous; feed on earthworms, insects and small vertebrates: restricted throughout tropics and found in eastern Himalaya, South India, Sri Lanka, and S.E. Asia.

Systematic Position

Phylum	**Chordata**
Sub Phylum	**Vertebrata**
Super class	**Gnathostomata**
Class	**Amphibia** [Gk. amphi = both; bios = life]
Subclass	**Liss amphibia** [Gk. Lissos = smooth]
Order	**Apoda** (= Gymnophiona = Ceacelia) [Gk, a = without;podos=foot; Gymnos = naked]
Type	***Ichthysophis***

Classification is based on J. Z. Young, 1981.

Distinctive Features

1. Worm like, elongated, limbless animal.
2. Body is covered with minute scales which are hidden in skin.
3. Skin smooth with transverse grooves.
4. Lidless eyes covered by the skin.
5. Mouth subtermimal.
6. Conical, flap like protrusible sensory tentacles in between eyes and nostrils.
7. Tail short, vestigial and conical.

Identification with Reasons

State classification and characters up to—

Super class—*Gnathostomata* as in *Scolidon*.

Class—*Amphibia* :

1. Naked, moist, glandular, rough or smooth skin.
2. Minute dermal scales if present deeply embedded in the skin.
3. Body is ectothermic (poikilothermus).
4. Two pairs of limbs usually with four digits and hind limbs with five digits.

Subclass—*Liss amphibia* : Skin naked, moist and smooth.

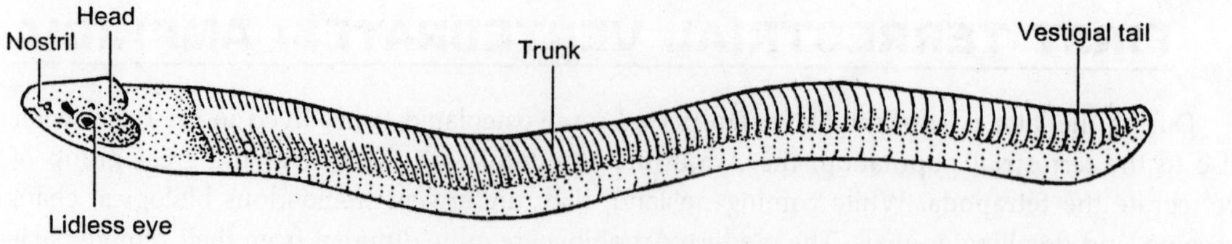

Fig. 1.1.2.57 : *Ichthyophis*

Order—*Apoda* :

1. Vermiform without limbs and limb girdles.
2. Tail if present is short and conical.
3. Male with single median protrusible couplatory organ.

Type—*Ichthyophis* : State distinctive features as in above.

2. *URAEOTYPHLUS*

Habit and Habitat

Same as those of *Ichthyophis.*

Systematic Position

Phylum	**Chordata**
Sub Phylum	**Vertebrata**
Super class	**Gnathostomata**
Class	**Amphibia**
Subclass	**Liss amphibia**
Order	**Apoda**
Type	***Uraeotyphlus***

Distinctive Features

1. Elongated, vermiform cylindrical body.
2. Skin is transversely grooved containing minute scales.
3. Small conical head bears lidless eyes covered by skin.
4. Mouth ventral.
5. Tentacular opening is below the nostril.

Identification with Reasons

State classification and characteristics up to—

Order—*Apoda* as in *Ichthyophis.*

Type—*Uraeotyphlus* : State distinctive features as in above.

3. *CRYPTOBRANCHUS* (Commonly called Hell-bender)

Habit and Habitat

Aquatic: lives in mountain rivers; carnivorous; feed on insects, worms and crustacean; found in United States and in Eastern Asia.

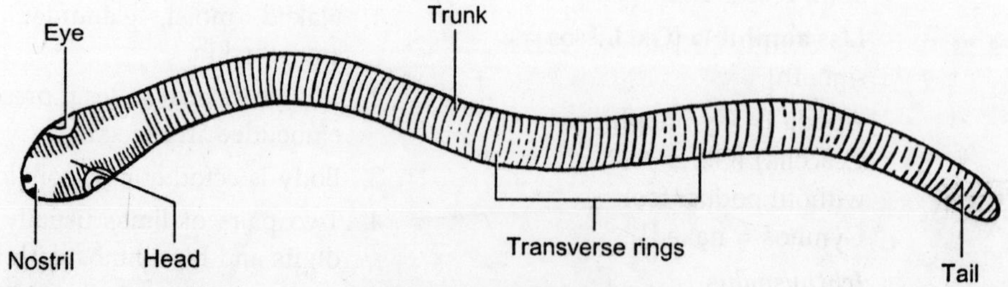

Fig. 1.1.2.58 : *Uraeotyphlus*

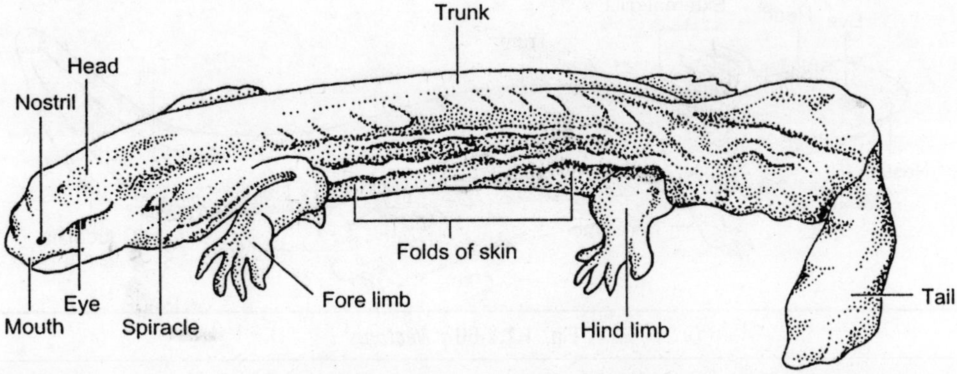

Fig. 1.1.2.59 : *Cryptobranchus*

Systematic Position

Phylum	**Chordata**
Sub phylum	**Vertebrata**
Super class	**Gnathostomata**
Class	**Amphibia**
Subclass	**Liss amphibia**
Order	**Urodela (=Caudata)**
Type	***Cryptobranchus***

Distinctive Features

1. Body is elongated and depressed.
2. Skin is soft and flabby with fold of skin in the lateral side of the body.
3. Limbs are of equal size.
4. Laterally compressed tail.
5. No gills in adults; spiracle present.
6. Eye lids absent.
7. Vomerine teeth are parallel to jaw.

Identification with Reasons

State classification and characters up to—

Super class—*Liss amphibia* as in *Ichthyophis*.

Order-*Urodela* :

1. Lizard like body with distinct head, trunk and well developed tail.
2. Limbs are of equal size.
3. Eyes are small and lidless.
4. Skin devoid of scales.
5. Teeth may or may not be present.

Type—*Cryptobranchus* : State distinctive features as above.

4. *NECTURUS* (Commonly called as Mudpuppy)

Habit and Habitat

Aquatic: lives in the fresh water river and lakes of Eastern North America; crawling on the bottom: Carnivorous; feed on worms, fishes and small frogs.

Systematic Position

Phylum	**Chordata**
Subphylum	**Vertebrata**
Super class	**Gnathostomata**
Class	**Amphibia**
Subclass	**Liss amphibia**
Order	**Urodela (=Caudata)**
Type	***Necturus***

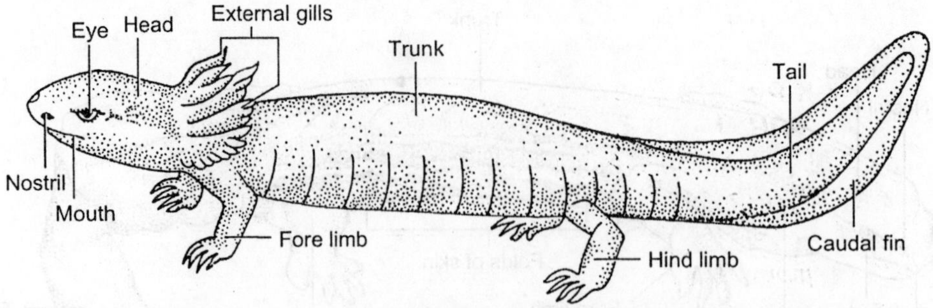

Fig. 1.1.2.60 : *Necturus*

Distinctive Features

1. Aquatic permanent larva.
2. Small eyes with out eye lids.
3. Three pairs of external bushy gills and two gill opening on each side.
4. Mouth sub terminal.
5. Limbs are short and weak and provided with four digits.
6. Laterally compressed tail provided with tail fin.

Identification with Reasons

State the classification and characters up to—

Order—*Urodela* as in *Cryptobranchus*.

Type—*Necturus* : State distinctive features as in above.

Special features

It is very interesting animals. Metamorphorsis is incomplete. Adults are permanently **neotenic larva**.

5. *PROTEUS* (Commonly called Olm or Cave Salamander)

Habit and Habitat

Aquatic: inhabits in the water of deep cave under darkness in the underground streams: feeds on aquatic crustaceans and small fish: distributed in Eastern Europe, Austria and Germany.

Systematic Position

Phylum	**Chordata**
Subphylum	**Vertebrata**
Super class	**Gnathostomata**
Class	**Amphibia**
Subclass	**Liss amphibia**
Order	**Urodela (=Caudata)**
Type	***Proteus***

Distinctive Features

1. Eel-like body with laterally compressed trail.

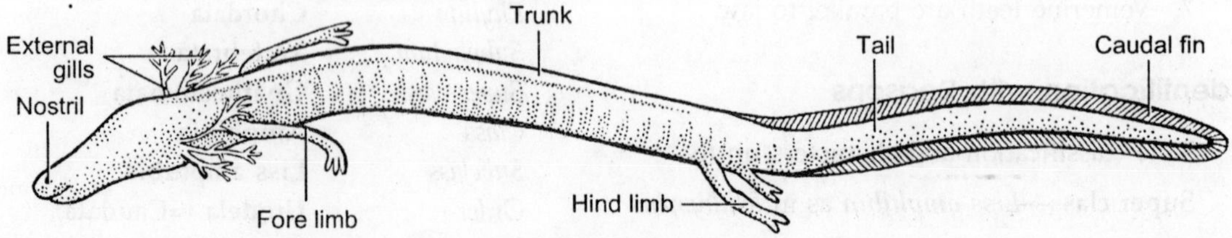

Fig. 1.1.2.61 : *Proteus*

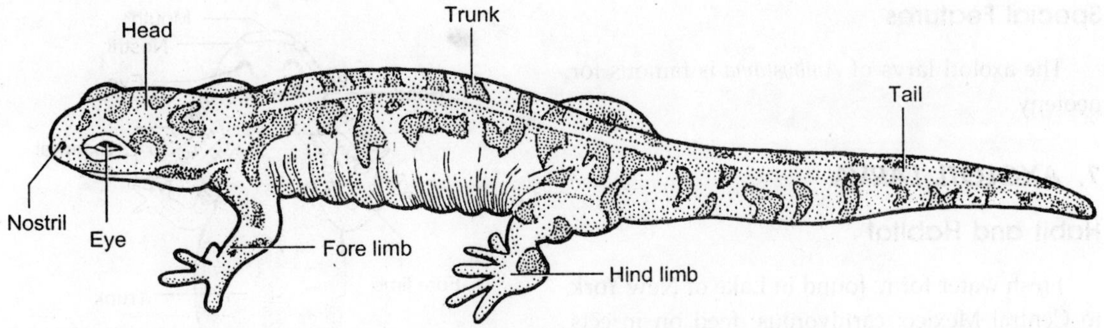

Fig. 1.1.2.62 : *Ambystoma*

2. Broad head with a blunt snout.

3. Rudimentary eyes sunken deep in the head are without eyelids.

4. Three pairs of feathery gills present.

5. Limbs are ill – developed, hind limb with two toes and forelimbs with three toes.

6. Both jaws have teeth.

Identification with Reasons

State classification and characters upto—

Order—*Urodela* as in *Cryptobranchus.*

Type-*Proteus* : State distinctive features as in above.

Special Features

It is a permanent neotenic larva with three pairs of gills.

6. *AMBYSTOMA* (Commonly called Tiger Salamander or Spotted Salamander)

Habit and Habitat

Terrestrial: found in North America, Central Mexico: carnivorous: feed on worm, snail, insects and crustacean: lay eggs near pond.

Systematic Position

Phylum	**Chordata**
Subphylum	**Vertebrata**
Super class	**Gnathostomata**
Class	**Amphibia**
Subclass	**Liss amphibia**
Order	**Urodela (=Caudata)**
Type	***Ambystoma***

Distinctive Features

1. Lizard- like body, with traverse grooves.

2. Depressed head with mouth.

3. Head contain eyes and a pair of poison gland.

4. Eyelids present.

5. External gills and tail fin are absent.

6. A series of palatine teeth is present.

7. Limbs well developed; fore limbs with four digits and hind limbs with five digits.

Identification with Reasons

State classification and characters up to—

Order—*Urodela* as in *Cryptobranchus.*

Type—*Ambystoma* : State the distinctive features as in above.

Special Features

The axolotl larva of *Ambystoma* is famous for neoteny.

7. *AXOLOTL LARVA*

Habit and Habitat

Fresh water form; found in Lake of New York to Central Mexico: carnivorous; feed on insects larva and worms.

Systematic Position

Phylum	**Chordata**
Subphylum	**Vertebrata**
Super class	**Gnathostomata**
Class	**Amphibia**
Subclass	**Liss amphibia**
Order	**Urodela (=Caudata)**
Type	**Axolotl larva**

Distinctive Features

1. Body is divided into head, trunk and tail with broad terminal mouth.
2. Three pair of external gills and four pair of open gill clefts.
3. Weakly developed limbs.
4. Ill developed eyes with eyelids.
5. Tail with caudal fin.
6. Neotenic larva of *Ambystoma.*

Identification with Reasons

State classification and characters up to—

Order—*Urodela* as in *Crytobranclus.*

Type—*Axototl larva* : State the distinctive characters as in above.

Special Features

Larva is sexually mature and lays eggs. This phenomenon is called as **Neoteny or**

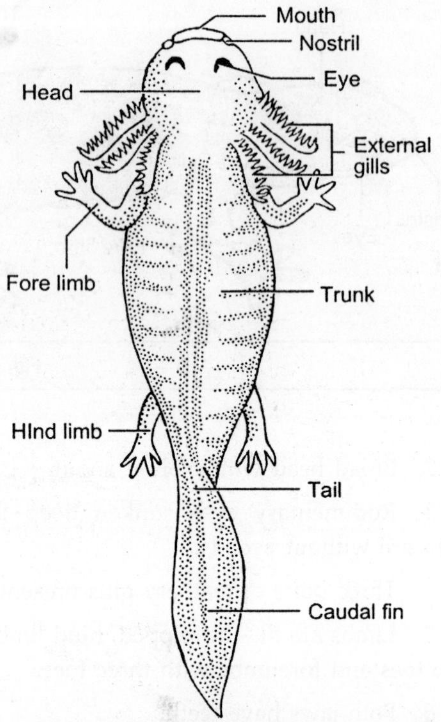

Fig. 1.1.2.63 : *Axolotl larva*

Paedogensis. Metamorphosis can be induced by injecting thyroid into axolotl larva.

8. *TRYLOTOTRITON SP.*

Habit and Habitat

Aquatic: found in pools and bogs of Darjeeling, Sikkim, Manipur and Arunachal Pradesh: carnivorous: feed on insects larva, earthworms and small fishes.

Systematic Position

Phylum	**Chordata**
Subphylum	**Vertebrata**
SuperClass	**Gnathostomata**
Class	**Amphibia**
Subclass	**Liss amphibia**
Order	**Urodela (=Caudata)**
Type	***Tylototriton***

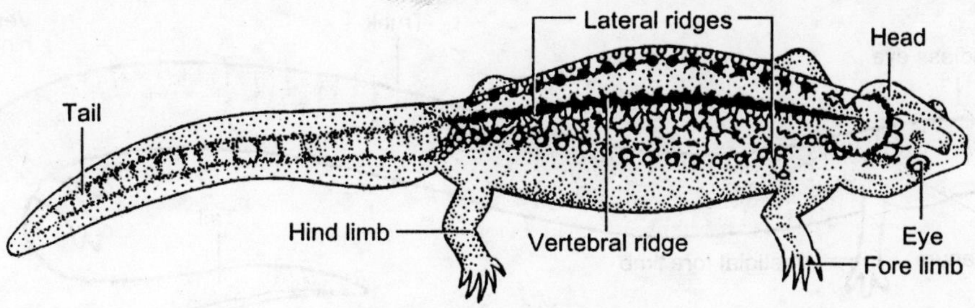

Fig. 1.1.2.64 : *Tylototriton*

Distinctive Features

1. Lizard like body with laterally compressed tail.

2. Back of the body bears a median vertebral ridges and two laterally placed beaded lines.

3. Flattened head bears two ridges on the dorso-lateral side.

4. Head contains prominent eyes.

Identification with Reasons

State classification and characters up to—

Order—*Urodela* as in *Cryptobranchus*.

Type—*Tylototriton* : State distinctive features as in above.

Remark

It is an Indian salamander and commonly called Darjeeling salamander. This is the only salamander found in the Eastern Himalayas of India.

9. *SALAMANDRA* (Commonly called Fire Salamandra)

Habit and Habitat

Terrestrial: inhabits under logs, stones, cracks and crevices: sluggish and feed on worms, snails, insects, and same crustaceans: distributed in Europe, Eastern Asia and North America.

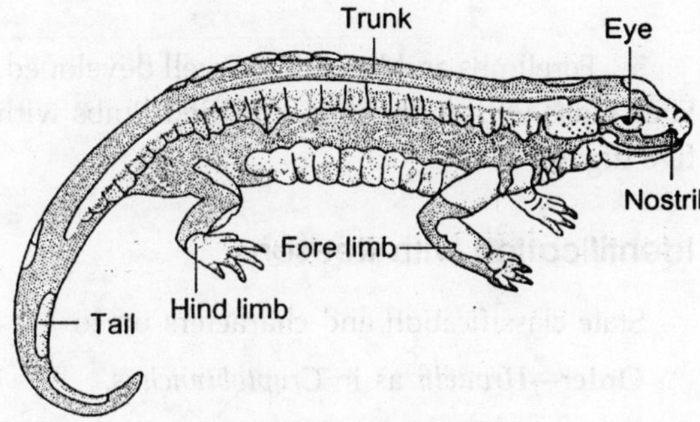

Fig. 1.1.2.65 : *Salamandra*

Systematic Position

Phylum	**Chordata**
Sub phylum	**Vertebrata**
Super class	**Gnathostomata**
Class	**Amphibia**
Sub class	**Liss amphibia**
Order	**Urodela (=Caudata)**
Type	***Salamandra***

Distinctive Features

1. Lizard like animal, female is larger than male.

2. Body is black with brilliant yellow orange or some times red patches.

3. Head contains prominent eyes and nostrils.

4. Tail without tail fin.

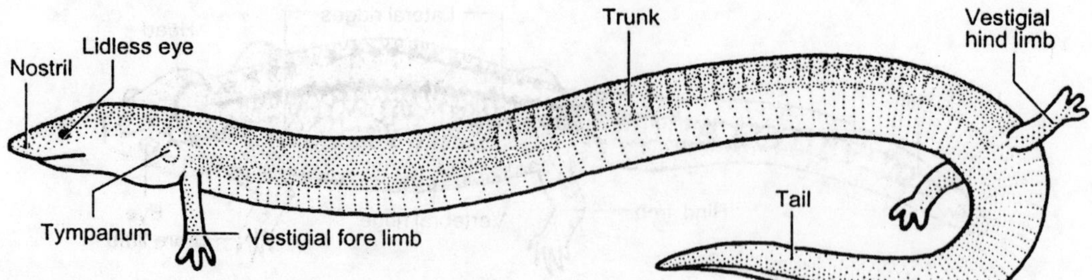

Fig. 1.1.2.66 : *Amphiuma*

5. Forelimbs and hind limbs well developed. Four limbs with four digits and hind limbs with five digits.

Identification with Reasons

State classification and characters up to—

Order—*Urodela* as in *Cryptobranchus.*

Type—*Salamandra* : State distinctive features as in above.

Special Features

Viviparous, eggs develop in oviduct. Hibernate in soil and breeds in running water.

10. *AMPHIUMA* (Commonly called Congo Eel)

Habit and Habitat

Found in swampy meadows and in rice fields: Carnivorous; feed on molluscs, crayfishes and small fishes: distributed in North America and South America.

Systematic Position

Phylum	**Chordata**
Sub phylum	**Vertebrata**
Super class	**Gnathostomata**
Class	**Amphibia**
Sub class	**Liss amphibia**
Order	**Urodela (=Caudata)**
Type	***Amphiuma***

Distinctive Features

1. Eel like body covered with smooth and black skin.
2. Eyes without eyelids.
3. Limbs poorly developed and rudimentary. Toes two or three.
4. No external gills. Gill clefts are two in number.

Identification with Reasons

State the classification and characters up to—

Order—*Urodela* as in Cryptobranchus.

Type—*Amphiuma* : State distinctive features as in above.

Special Features

Male constructs a nest in which female lay eggs.

11. *TRITURUS* (Commonly called British Newt)

Habit and Habitat

Terrestrial: feed on worms, snail, insects and insects larvae from water: distributed in Asia , North Africa, Europe and North America.

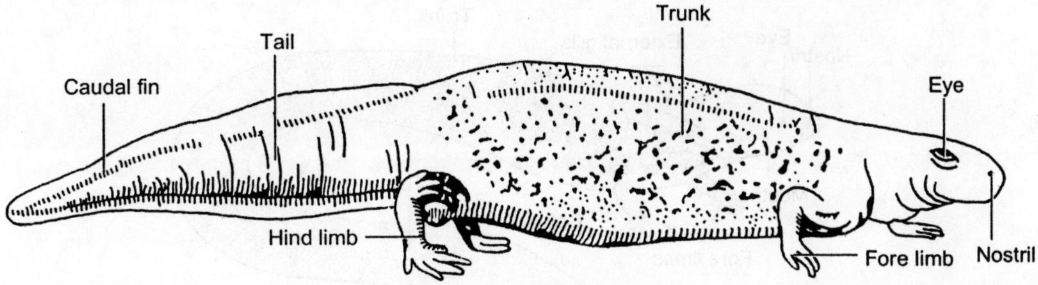

Fig. 1.1.2.67 : *Triturus* (Female)

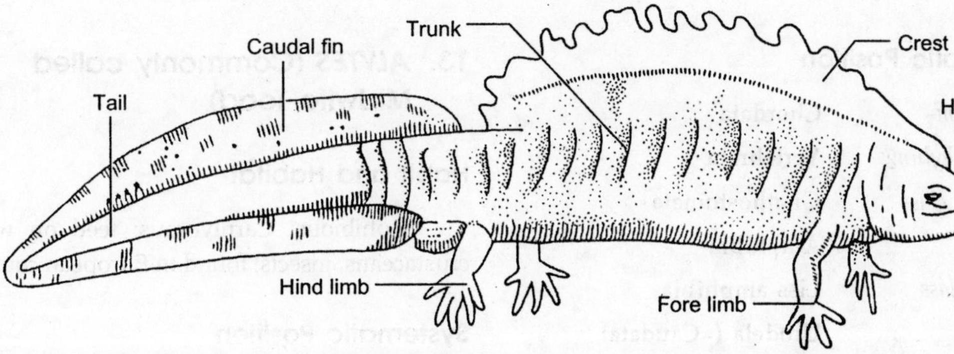

Fig. 1.1.2.67A : *Triturus* (Male)

Systematic Position

Phylum	**Chordata**
Sub phylum	**Vertebrata**
Super class	**Gnathostomata**
Class	**Amphibia**
Sub class	**Liss amphibia**
Order	**Urodela (=Caudata)**
Type	***Triturus***

Distinctive Features

1. Elongated and cylindrical body.

2. Laterally compressed tail provided with caudal fin.

3. Fore limbs with four digits and hind limbs with five digits.

4. Eyes with eyelids.

5. Male develops a crest during breeding season.

6. Head is conical and compressed, containing wide mouth and small nostrils.

Identification with Reasons

State classification and characters up to—

Order—*Urodela* as in Cryptobranchus.

Type—*Triturus* : State distinctive features as in above.

12. *SIREN* (Commonly called Mud Eel)

Habit and Habitat

Aquatic: live in ponds and muddy ditches: Carnivorous: feed on small fishes and invertebrates: found in U.S.A.

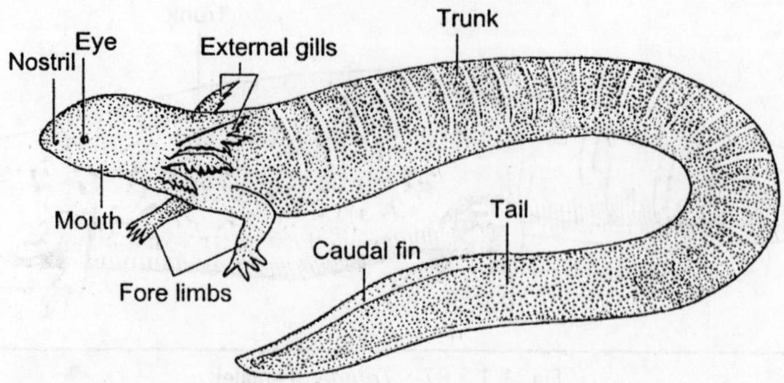

Fig. 1.1.2.68 : *Siren*

Systematic Position

Phylum	**Chordata**
Sub phylum	**Vertebrata**
Super class	**Gnathostomata**
Class	**Amphibia**
Sub class	**Liss amphibia**
Order	**Urodela (=Caudata)**
Type	***Siren***

Distinctive Features

1. Animal is slender, elongated, eel-like without hind limbs.
2. Forelimbs small containing four digits.
3. Head is conical with small eyes and nostrils.
4. Three pairs of external gills which are persistent.
5. Eyes are with out eyelids.
6. Tail with small caudal fin.

Identification with Reasons

State the classification and characters up to—

Order—*Urodela* as in *Cryptobranchus.*

Type—*Siren* : State distinctive features as in above.

13. *ALYTES* (Commonly called Midwife Toad)

Habit and Habitat

Amphibious: Carnivorous: feed on worms, crustaceans, insects; found in European countries.

Systematic Position

Phylum	**Chordata**
Sub phylum	**Vertebrata**
Super class	**Gnathostomata**
Class	**Amphibia**
Sub class	**Liss amphibia**
Order	**Anura** [Anura. Gk. an = without. oura = tail]
Type	***Alytes***

Distinctive Features

1. Body has warty surface, grey, brown, green, or red coloured.
2. Upper jaw and vomer are beset with teeth.
3. Eyelids present.
4. Tympanum is large and distinct.
5. Non protrusible tongue is in the form of a round disc and remains attached at its base.
6. Male with out vocal sac.

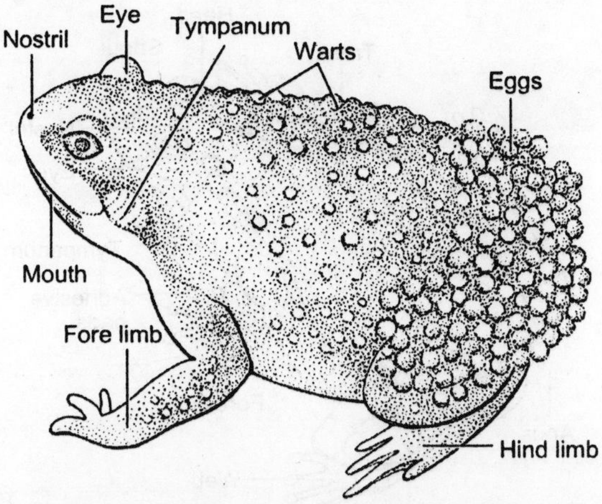

Fig. 1.1.2.69 : *Alytes*

Identification with Reason

State classification and characters up to—

Subclass—*Liss amphibia* as in *Cryptobranclus.*

Order—*Anura* :

1. Body short, broad and trunk fused.
2. Neck and tail absent.
3. Wide mouth and large eyes with well developed eyelids.
4. Tongue may be present or absent.
5. Fore limbs short, hind limbs long and help in long leap.
6. No traces of external gills or gill slits.
7. Tympanum distinct.

Type—*Alytes* : State distinctive features as in above.

Special Features

Males exhibited **parental care**. The female lays egg in strings and the male winds the strings around posterior body and thighs and goes in moist soil. When eggs develop to tadpole stage, the toad comes in water and larvae hatch.

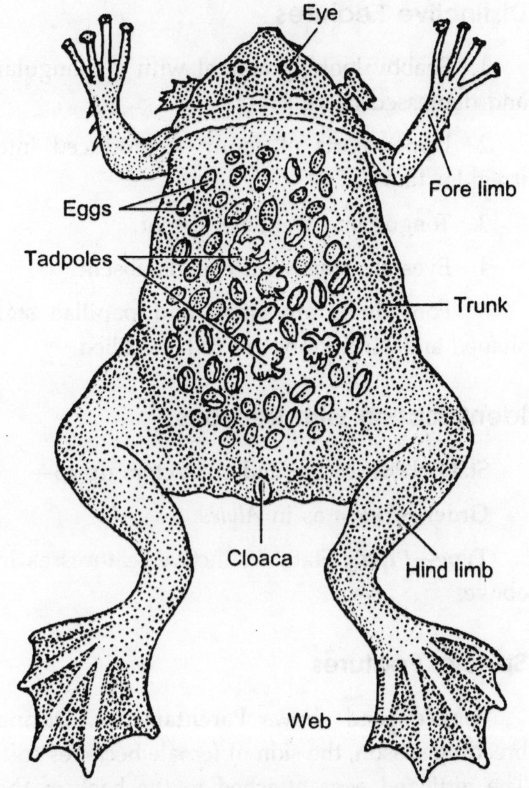

Fig. 1.1.2.70 : *Pipa*

14. *PIPA* (Commonly known as Surinam Toad)

Habit and Habitat

Aquatic: Carnivorous: feed on worms, snails, crustaceans and insects larvae: found in South Africa and America.

Systematic Position

Phylum	**Chordata**
Sub phylum	**Vertebrata**
Super class	**Gnathostomata**
Class	**Amphibia**
Sub class	**Liss amphibia**
Order	**Anura**
Type	***Pipa***

Distinctive Features

1. Shabby looking animal with a triangular and depressed head.

2. Upper parts of snout in produced into irregular flaps.

3. Tongue and teeth are absent.

4. Eyes are small and eyelids absent.

5. Fore limbs are small and papillae star shaped and hind limbs are fully webbed.

Identification with Reasons

State classification and characters up to—

Order—*Anura* as in *Alytes.*

Type—*Pipa* : State distinctive features as in above.

Special Features

Female Toad shows **Parental care.** During breeding season, the skin of female becames soft. The fertilized eggs attached to the back of the female and young's are carried in pits in the back of female.

15. *HYLA* (Commmonly called Tree Frog)

Habit and Habitat

Arboreal: living on trees and rocks: feed on insects: nocturnal; distributed in India, China, Japan, Africa, Central and South America.

Systematic Position

Phylum	**Chordata**
Sub phylum	**Vertebrata**
Super class	**Gnathostomata**
Class	**Amphibia**
Sub class	**Liss amphibia**
Order	**Anura**
Type	***Hyla***

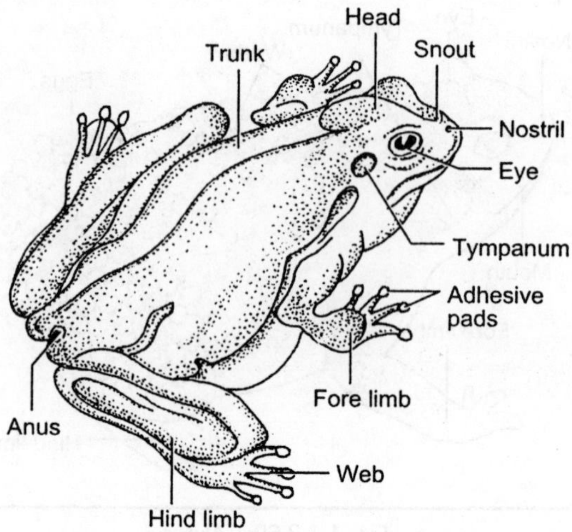

Fig. 1.1.2.71 : *Hyla*

Distinctive Features

1. Smooth skin on the dorsal surface, while skin on the ventral surface bears papillae.

2. Terminal phalanges are claw-like and expanded to form adhesive pads, which are used to climb trees or rocks.

3. Upper jaw is toothed, while the lower jaw is tooth less.

4. Eyes well developed with horizontal pupil.

5. Tympanum is distinct.

Identification with Reasons

State the classification and characters up to—

Order—*Anura* as in *Alytes.*

Type—*Hyla* : State distinctive features as in above.

Special Features

Adapted from amphibious to arboreal life. They also change their colour according to their environment. They exhibit parental care.

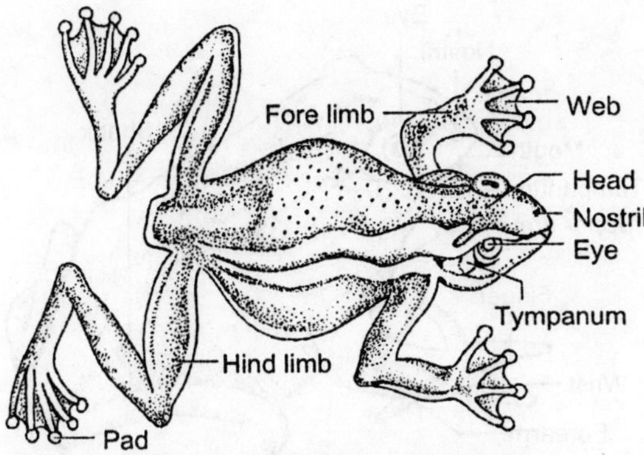

Fig. 1.1.2.72 : *Rhacophorous*

16. *RHACOPHOROUS* (Commonly called Flying Frog)

Habit and Habitat

Terrestrial: arboreal in habit, attached themselves to branches and trunk of trees: can glide passively from tree by their webbed feet: feed mainly on insects: found in Africa, South Eastern India, Japan and Madagascar.

Systematic Position

Phylum	**Chordata**
Sub phylym	**Vertebrata**
Super class	**Gnathostomata**
Class	**Amphibia**
Sub class	**Liss amphibia**
Order	**Anura**
Type	***Rhacophorus***

Distinctive Features

1. Slender body with broad, conical head.
2. Limbs are elongated and thin.
3. Well developed webs on fore and hind limbs. Membranous webs acts as parachutes.
4. Fingers and toes are provided with adhesive disc like pads.
5. Eyes with well developed eyelids.
6. Females are larger than males.

Identification with Reasons

State classifications and characters up to—

Order—*Anura* as in *Alytes.*

Type—*Rhacophorous* : State distinctive features as in above.

Special Features

Rhacophorous shows extreme degree of adaptive radiation. They attached themselves to branches and trunk of trees. Some species can glide passively from tree to tree or parachuting while alighting from trees. They built foamy nest on aquatic plants above water in which female lays eggs. Both male and female offered parental care. They can change colour rapidly.

17. *BUFO* (Commonly called True Toad)

Habit and Habitat

Terrestrial: remain in dark and moist places, preferably near water hiding under logs and stones: nocturnal: enters water only to breed and spawn: carnivorous: catches insects by its sticky tongue: distribution world wide, except in Australia and Madagascar.

Systematic Position

Phylum	**Chordata**
Sub phylum	**Vertebrata**
Super class	**Gnathostomata**
Class	**Amphibia**
Sub class	**Liss amphibia**
Order	**Anura**
Type	***Bufo***

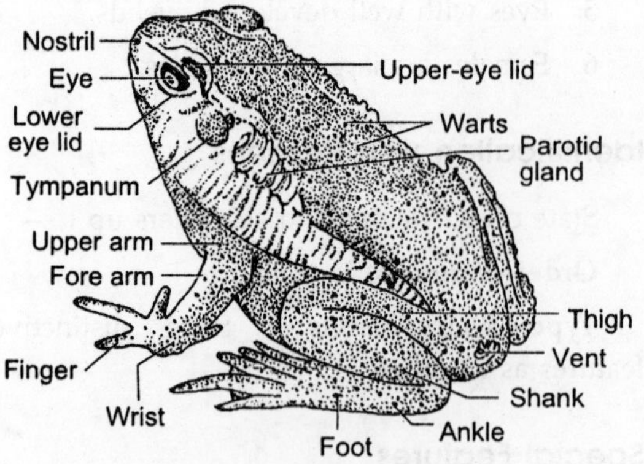

Fig. 1.1.2.73 : *Bufo* (Toad)

Distinctive Features

1. Body divisible into head and trunk.
2. Skin rough with warts on the dorsal surface.
3. Behind tympanum there is a pair of large parotid gland.
4. Eyes are large and prominent with eye lids.
5. Fore limbs are smaller than the hind limbs bear poorly developed webs.
6. Both upper and lower jaws are tooth less.
7. In male a thumb pad develops during breeding seasons at the basal part of the inner finger.

Identification with Reasons

State classifications and characters up to—

Order—*Aunura* as in *Alytes*.

Type—*Bufo* : State distinctive features as in above.

Special Features

Bufo is cold-blooded or **Poikilothermous** or **exothermic** (body temperature depends on the external temperature of environment). During winter they undergo **hibernation** or winter sleep.

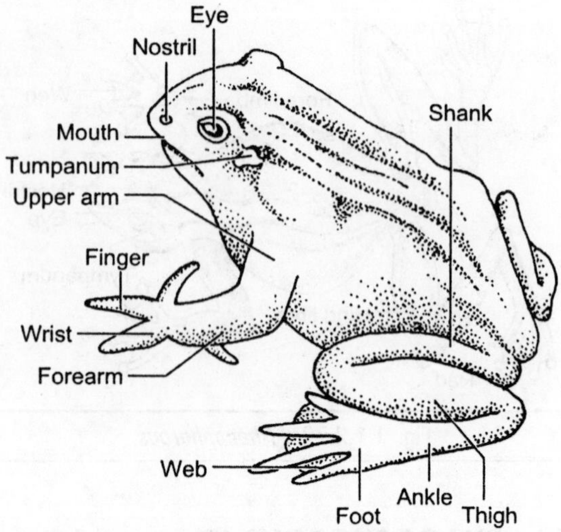

Fig. 1.1.2.74 : *Rana* (Frog)

18. RANA (Commonly called true frog)

Habit and Habitat

Aquatic or semi-aquatic: catching flies and insects with there tongue: cosmopolitan; except in Australia and South Africa.

Systematic Position

Phylum	**Chordata**
Sub phylum	**Vertebrata**
Super class	**Gnathostomata**
Class	**Amphibia**
Sub class	**Liss amphibia**
Order	**Anura**
Type	***Rana***

Distinctive Features

1. Body divisible into head and trunk.
2. Smooth slimy skin which is decorated with colour patches.
3. Triangular head with blunt snout.
4. Upper jaws beset with small conical teeth.

5. Hind legs are long and bears poorly developed webs.

Identification with Reasons

State classification and characters up to—
Order—*Anura* as in *Alytes*.

Type—*Rana* : State the distinctive features as in above.

19. *TADPOLE LARVAE*

Habit and Habitat

Aquatic: attaches it self to the aquatic weeds for few days till mouth is formed: free-swiming tadpole larvae ingests aquatic weeds.

Systematic Position

Same as those of **Toad** or **Frog**.

Distinctive features of developing larva

1. Limbless body divided into an ovoid head, a short trunk and a slender tail.
2. An adhesive sucker is present on the ventral side of the head in freshly hatched larva.
3. Elongated tail with dorsal and ventral fin.
4. A pair of horny jaws surround the small mouth appeared in advance stage.
5. Spirally coiled intestine.
6. Three pairs of branched external gills.

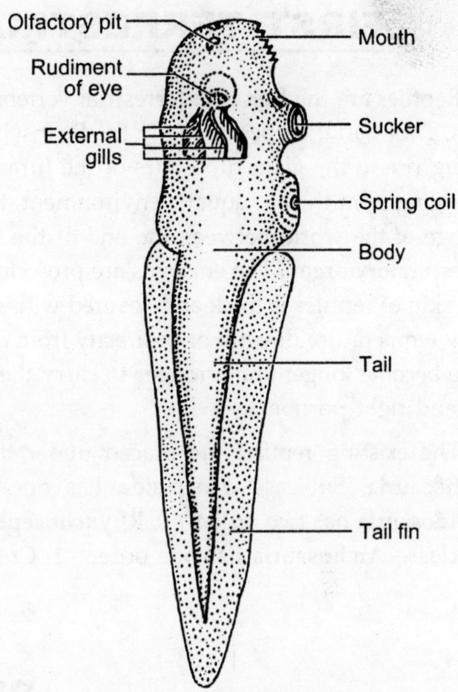

Fig. 1.1.2.75 : *Tadpole larva of toad*

Advance tadepole larvae

Distinctive features of advance tadpole larva

1. External gills disappear and internal gills are formed.
2. Gills and gill slits are covered by operculum.
3. Hind limbs appearing.
4. Fore limbs remain hidden under the operculum and subsequently emerge through it.
5. Gradually tail becomes absorbed into the body.

FIRST TERRESTRIAL AMNIOTA: REPTILES

Reptiles are the first true terrestrial vertebrates that give up the practice to go back to water to lay eggs. They had their beginning in the Pennsylvanian period and dominated the earth during Mesozoic giving rise to the giant dinosaurs of the Jurassic and Cretaceous and then they decline in Coenozoic. Emerging from the old aquatic environment, the reptiles become successfully adapted to the profound change of the world between the end of the Cretaceous and the Eocene. They have developed lungs as respiratory organ. The embryos are protected by the development of **chorion, amnion and allantois.** The skin of reptiles is thick and covered with scales. They are **ectothermic**, thus they can regulate their body temperature directly or indirectly from out side energy (solar). For locomotion on land the limbs have become longer and stronger to carry the weight of the body. The ventricle is partly divided into left and right portion.

The existing reptiles are placed under three Subclasses : 1. **Anapsida**, 2. **Lepidosauria** and 3. **Archosauria**. Subclass—Anapisida has one order : 1. **Chelonia**, tortoise and turtles. Subclass—**Lepidosauria** has two orders : 1. **Rhynchocephalia**, the *Sphenodon*; 2. **Squamata**, the lizards and snakes. Subclass—**Archosauria** has one order : 1. **Crocodilia**, the crocodiles.

REPTILIA

1. *CHELONIA MYDAS* (Commonly called Green Turtle)

Habit and Habitat

Marine: found in the costal water: lay eggs at night on the sea beaches: herbivorous: feed on sea grass and algae: distributed in tropical and sub tropical regions: found in Indian, Pacific and Atlantic oceans and coasts of United States.

Systematic Position

Phylum	**Chordata**
Subphylum	**Vertebrata**
Super class	**Gnathostomata**
Class	**Reptilia**
Sub class	**Anapsida**
Order	**Chelonia**
Type	***Chelonia mydas***

Classification is based on J. Z. Young, 1981.

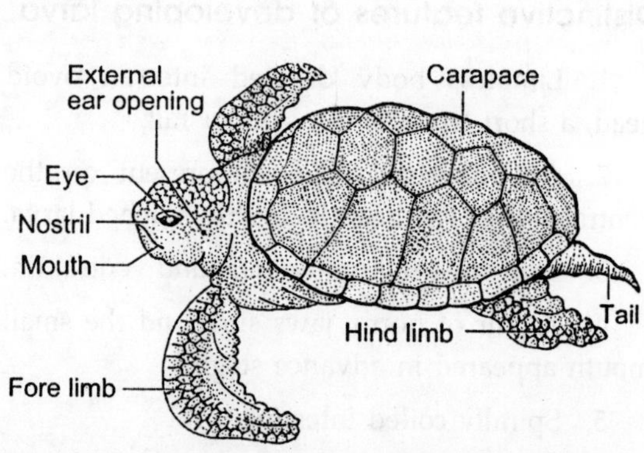

Fig. 1.1.2.76 : *Chelonia*

Distinctive Features

1. The carapace is depressed, heart shaped and covered with smooth bony shields.

2. The head is covered by single pair of prefrontal shields.

3. Well developed eyes, provide with eyelids and nictitating membrane.

4. Limbs flipper like and single clawed.
5. Fore limbs form wing like paddles.
6. Teeth absent and the jaws are covered by horny pads.

Identification with Reasons

State classification and characters upto—

Super class—*Gnathostomata* as in *Scoliodon*.

Class—*Reptilia* :

1. Skin dry, cornified, usually covered with epidermal scales or scutes.
2. Single external nasal opening on the scales.
3. Two pairs of pentadactyl clawed limbs in most classes.
4. A post anal tail is usually present.
5. Cloacal aperture either transverse or longitudinal.

Sub class—*Anapsida* :

1. Devoid of temporal vacuity.
2. Roof of the skull is solid.

Order—*Chelonia* :

1. Body is enclosed by dorsal convex **carapace** and a flat ventral plate, **plasteron**, which remains joined at side.
2. The head, neck and limbs retractile inside the enclosed body in non marine form.
3. Jaws covered with horny sheaths.
4. Pentadactyl limbs are modified into paddle in marine form or webbed in fresh water form.
5. Tail is short.

Type—*Chelonia mydas* : State systematic features as in above.

Economic Importance

Sea turtles armour is used for various purposes. Their flesh is very delicious and edible. *Chelonia* is in the verge of extinction because of mass killing.

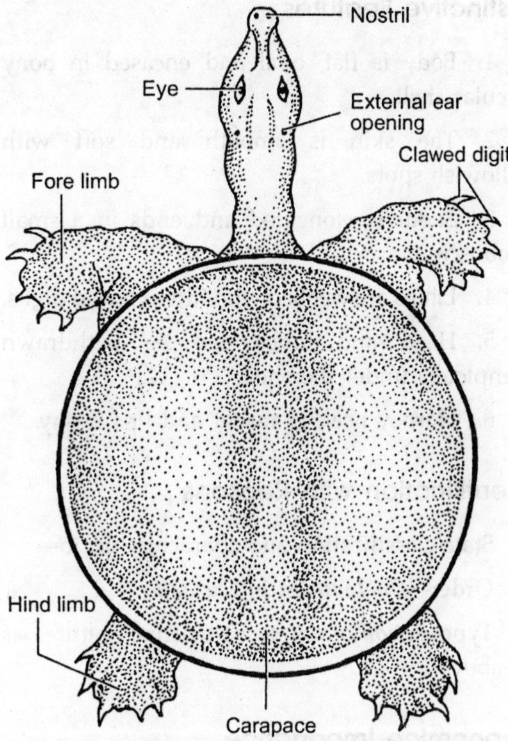

Fig. 1.1.2.77 : *Trionyx*

2. *TRIONYX* (Commonly called as Soft-shelled Terrapin or Tortoise)

Habit and Habitat

Aquatic: found in fresh water rivers and ponds: carnivorus: feed on fishes, frogs and mollusks: distributed in india, North America, Africa.

Systematic Position

Phylum	**Chordata**
Subphylum	**Vertebrata**
Super class	**Gnathostomata**
Class	**Reptilia**
Sub class	**Anapsida**
Order	**Chelonia**
Type	***Trionyx***

Distinctive Features

1. Body is flat, oval and encased in bony circular shell.

2. The skin is smooth and soft with yellowish spots.

3. Snout is elongated and ends in a small proboscis.

4. Limbs are webbed with three clawed digits.

5. Head and limbs can be withdrawn completely under the shell.

6. Horney beak is absent and lips fleshy.

Identification with Reasons

State classification and characters up to—

Order—*Chelonia* as in *Chelonia.*

Type—*Trionyx* : State distructive features as in above.

Economice Importance

Flesh is edible and sold in the market for food.

3. *TESTUDO* (Known as Giant Land Tortoise)

Habit and Habitat

Terrestrial or aquatic: diurnal : hibernate during winter season: herbivorous: sometimes feed on small worms and insects: distributed in tropical and temperate places of Africa, Europe and Asia.

Systematic Position

Phylum	**Chordata**
Subphylum	**Vertebrata**
Super class	**Gnathostomata**
Class	**Reptilia**
Sub class	**Anapsida**
Order	**Chelonia**
Type	*Testudo*

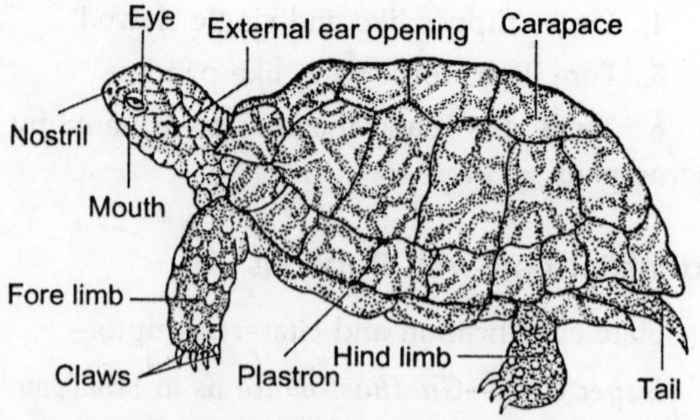

Fig. 1.1.2.78 : *Testudo*

Districtive Features

1. Body short, broad and provided with horney epidermal scales.

2. Domed-shell, carapace is very convex, while plasteron is concave

3. Bent neck can be with drawn into shell along with head.

4. Teeth absent, jaws with horney sheaths having sharp cutting edges.

5. Limbs are short, broad and digits webless.

Identification with Reasons

State classification and characters upto—

Order—*Chelonia* as in *Chelonia.*

Type—*Testudo* : State distinctive features as in above.

4. *CHRYSEMYS* (Commonly known as Painted Terrapin)

Habit and Habitat

Aquatic: common in fresh water but is adapted for amphibious life: carnivorous: feed on insects, worms and other animals: commonly found in India, Sri Lanka, United States, Europe, Asia and Africa.

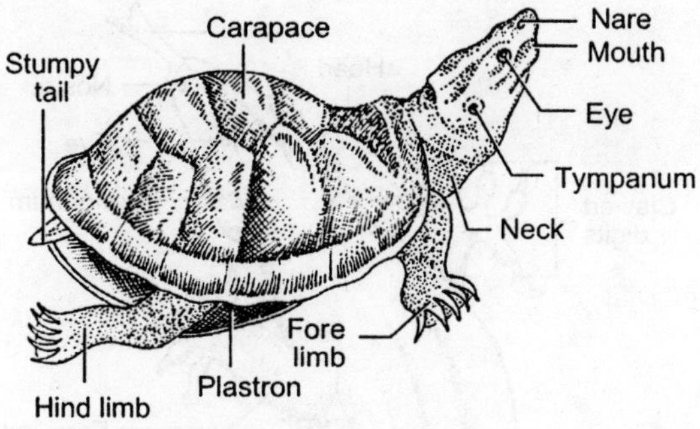

Fig. 1.1.2.79 : *Chrysemys*

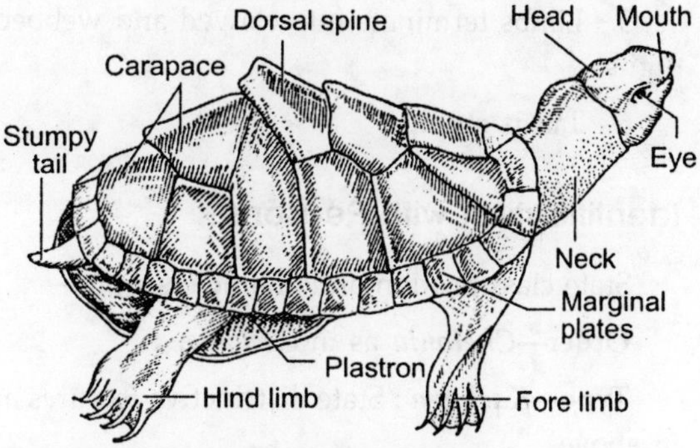

Fig. 1.1.2.80 : *Kachuga*

Systematic Position

Phylum	**Chordata**
Sub phylum	**Vertebrata**
Super class	**Gnathostomata**
Class	**Reptililia**
Sub class	**Anapsida**
Order	**Chelonia**
Type	***Chrysemys***

Distinctive Features

1. Body is short, broad and is enclosed in bony armour of dorsal carapace and ventral plasteron.

2. Carapace is convex and dome shaped.

3. Head is triangular and there is a fold of skin behind the head.

4. Well developed fore limbs and backwardly directed hind limbs.

5. Digits with three claws and toes partly webbed.

6. Neck is retractile and tail stumpy.

Identification with Reasons

State classification and characters up to—

Order—*Chelonia* as in *Chelonia*.

Type—*Chrysemys* : State distinctive features as in above.

5. *KACHUGA* (Commonly known as Tortoise)

Habit and Habitat

Aquatic: commonly found in rivers, especially concentrate at bathing places: omnivorous: food consists of both plants and animals: distributed in Indian rivers, especially Ganges.

Systematic Position

Phylum	**Chordata**
Sub phylum	**Vertebrata**
Super class	**Gnathostomata**
Class	**Reptilia**
Sub class	**Anapsida**
Order	**Chelonia**
Type	***Kachuga***

Distinctive Features

1. Body is enclosed in carapace-plasteron box.

2. The first three neural plates of the carapace are produced into backwardly directed spines.

3. Head is retractile into carapace and covered with smooth skin.

4. Teeth absent but jaws are covered with horney beak, having sharp, cutting edges.

5. Limbs terminate into clawed and webbed digits.

6. Tail is stumpy.

Identification with Reasons

State classification and characters upto—

Order—*Chelonia* as in *Chelonia.*

Type—*Kachuga* : State distrinctive features as in above.

6. *HEMIDACTYLUS* (Commonly known as Wall Lizard)

Habit and Habitat

Terrestrial: nocturnal: hide on the back of photoframes on the wall, crevices and other dark places: carnivorous: feed on insects, flies and spiders: clinging to walls and ceilings by adhesive pads: distribution world-wide.

Systematic Position

Phylum	**Chordata**
Sub phylum	**Vertebrata**
Super class	**Gnathostomata**
Class	**Reptilia**
Sub class	**Lepidosauria** [Gk. Lepis = scales; saures=lizard]
Order	**Squamata** [L. Squamatus = scaly]
Sub order	**Lacertilia** [L. Lacerta = a lizard]
Type	***Hemidactylus***

Distinctive Features

1. Body is slender and covered with minute small scales.

2. Head is broad, flat, eyes lack movable eyelids.

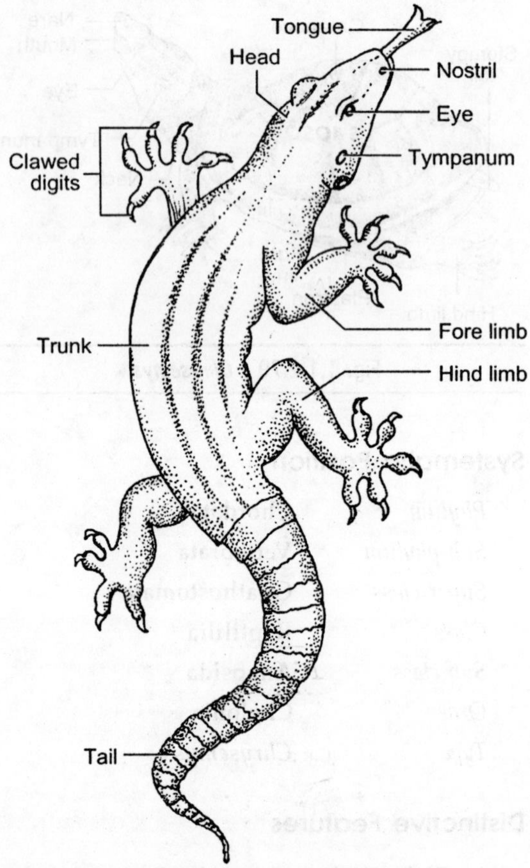

Fig. 1.1.2.81 : *Hemidactylus*

3. Clawed digits with transverse rows of adhesive lamellae.

4. Protrusible tongue is short, sticky and notched.

5. Tail breaks off when chased by enemy, later on it regenerates.

Identification with Reasons

State classification and characters up to—

Class—*Reptilia* as in *Chelonia.*

Sub class—*Lepidosauria* :

1. Scaly lizard
2. Two temporal vacuities in the skull.

Order—*Squamata* :

1. Skull with superior temporal fossa.
2. Cloacal aperture is a transverse slit.
3. Quadrate is movable.

Sub order—*Lacertilia* :

1. Elongated body with two pairs of limbs except limbless lizards.
2. Usually movable upper and lower eye lids are present.
3. Tongue is broad usually entire.

Type—***Hemidactylus*** : State distinctive features as in above.

Special Features

Tail is swollen at the base, **Autotomy** is found in these lizards, it breaks off its tail, when chased by enemy. Caudal vertebrae are loosely attached. There is interesting unossified zone between vertebrae which are easily broken and then regenerates. The wall lizard is adapted to walk on walls, roofs and on smooth surfaces.

7. *GEKKO* (Commonly called as House-Gecko)

Habit and Habitat

Terrestrial: arboreal: lives in trees of the jungles and also in houses: insectivorous: distribution cosmopolitan but specially found in warm climate of Asian and European countries.

Systematic Position

Phylum	**Chordata**
Sub phylum	**Vertebrata**
Super class	**Gnathostomata**
Class	**Reptilia**
Sub class	**Lepidosauria**
Order	**Squamata**
Sub order	**Lacertilia**
Type	***Gekko***

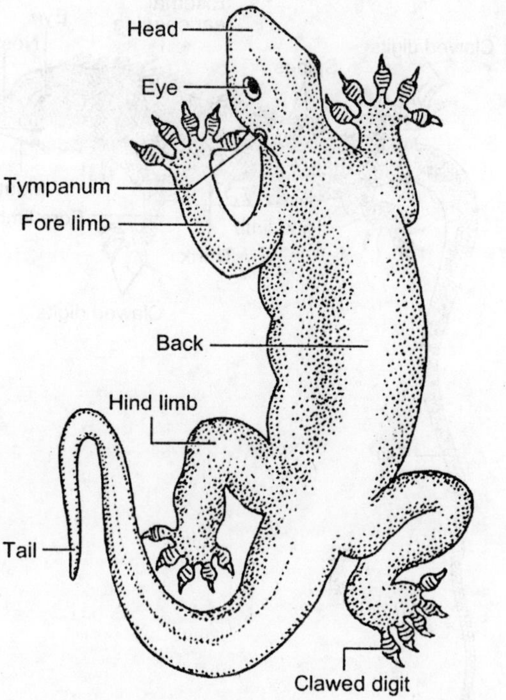

Fig. 1.1.2.82 : *Gekko*

Distinctive Features

1. Body is covered by minute grandular scales.
2. Head is small, body and tail are elongated.
3. Tongue is entire and protusible.
4. Toes with ridged pads furnished with minute bristles.
5. Fore and hind limbs are adapted for walking on walls and smooth surfaces.
6. Tail regenerates if broken.

Identification with Reasons

State classification and characters up to—

Sub order—*Lacertilia* as in *Hemidactylus.*

Type—*Gekko* : State distinctive features as in above.

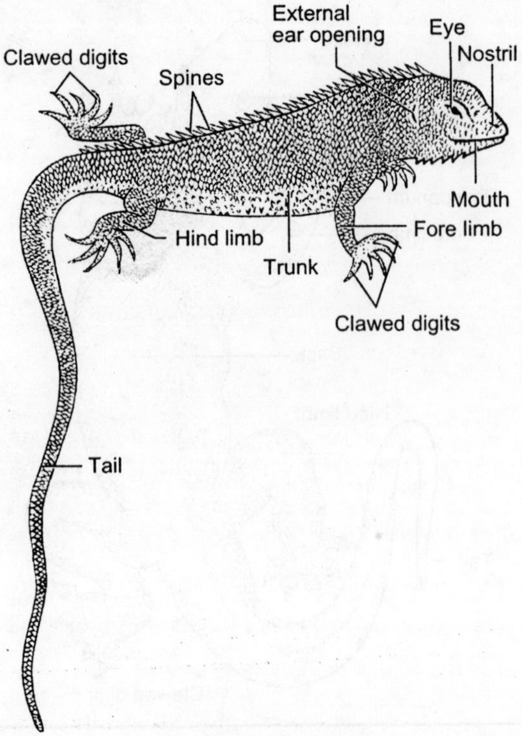

Fig. 1.1.2.83 : *Calotes*

Remark

Geckos emit a peculiar clicking and cheeping sounds.

8. *CALOTES VERSICOLAR*

(Commonly known as blood suckers, so called because of the red colour of the throat)

Habit and Habitat

Terrestrial: arboreal: found in the bushes, hedges and house gardens of India, Malaysia and China: insectivorous: swift runner.

Systematic Position

Phylum	**Chordata**
Sub phylum	**Vertebrata**
Super class	**Gnathostomata**
Class	**Reptilia**
Sub class	**Lepidosauria**
Order	**Squamata**
Sub order	**Lacertilia**
Type	***Calotes versicolor***

Distinctive Features

1. Body is elongated, slender and covered with overlapping, epidermal rough scales.
2. A crest of spine in mid-dorsal line of the body is present.
3. Movable broad head bears one or two spines on the back.
4. Eye lids are movable.
5. Long cylindrical tail does not break.
6. Limbs normal with clawed digits.

Identification with Reasons

State classification and characters up to—

Sub order—*Lacertilia* as in *Hemidactylus.*

Type—*Calotes versicolor* : State distinctive features as in above

Remark

Calotes is famous for colour changes. It is largely governed by temperature, environment and hormone from pituitary.

9. *DRACO* (Commonly known as Flying Lizard)

Habit and Habitat

Terrestrial: arboreal: living on trees: can glide from tree to tree with the help of **patagium**: insectivorous: feed on small insects: found in India, Malaysia, Europe, Africa, Asia and Australia.

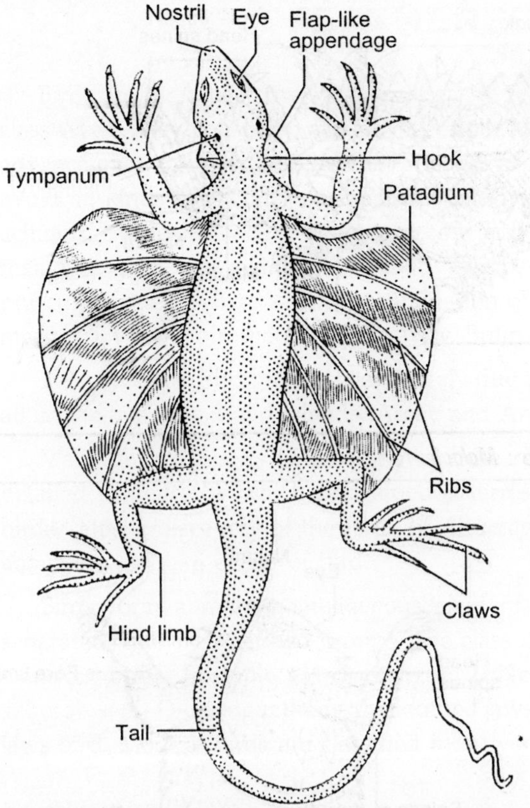

Fig. 1.1.2.84 : *Draco*

Systematic Position

Phylum	**Chordata**
Sub phylum	**Vertebrata**
Super class	**Gnathostomata**
Class	**Reptilia**
Sub class	**Lepidosauria**
Order	**Squamata**
Sub order	**Lacertilia**
Type	***Draco***

Distinctive Features

1. Small lizard with long, slender, whip like tail.

2. Large parachute web on each side of the body is supported by four or five ribs.

3. The neck contains three hooks.

4. Below the neck there are sacs like structures known as **gular pouches**.

5. Eyes with movable eye lids.

Identification with Reasons

State classification and characters up to—

Suborder—***Lacertilia*** as in *Hemidactylus*.

Type—***Draco*** : State distinctive features as in above.

Special Features

Draco shows extreme adaptation for flying life.

10. *MOLOCH HORRIDUS* (Commonly known as Devil Lizard)

Habit and Habitat

Terrestrial: found in deserts of Australia: insectivorous: body adapted for dry condition.

Systematic Position

Phylum	**Chordata**
Sub phylum	**Vertebrata**
Super class	**Gnathostomata**
Class	**Reptilia**
Sub class	**Lepidosauria**
Order	**Squamata**
Sub order	**Lacertilia**
Type	***Moloch horridus***

Distinctive Features

1. Lizard like body is covered with heavy armour of scaly spines.

2. Head spines are largest, tail and limb spines are smaller in size.

3. Skin is rough and much hygroscopic.

4. Limbs are clawed.

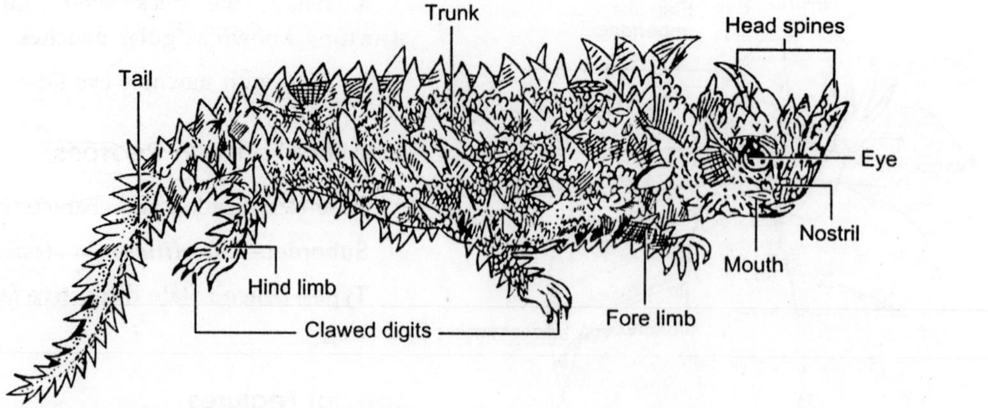

Fig. 1.1.2.85 : *Moloch*

Identification with Reasons

State clasiification and characters up to—

Sub order—*Lacertalia* as in *Hemidactylus.*

Type—*Moloch horridus* : State distinctive features as in above.

11. *PHRYNOSOMA* (Commonly called Horned Toad)

Habit and Habitat

Terrestrial: adapted for desert life: lives on dry ground and sandy places: insectivorous: found in the desert of Mexico and South-Western United States.

Systematic Position

Phylum	**Chordata**
Sub phylum	**Vertebrata**
Super class	**Gnathostomata**
Class	**Reptilia**
Sub class	**Lepidosauria**
Order	**Squamata**
Sub order	**Lacertilia**
Type	***Phrynosoma***

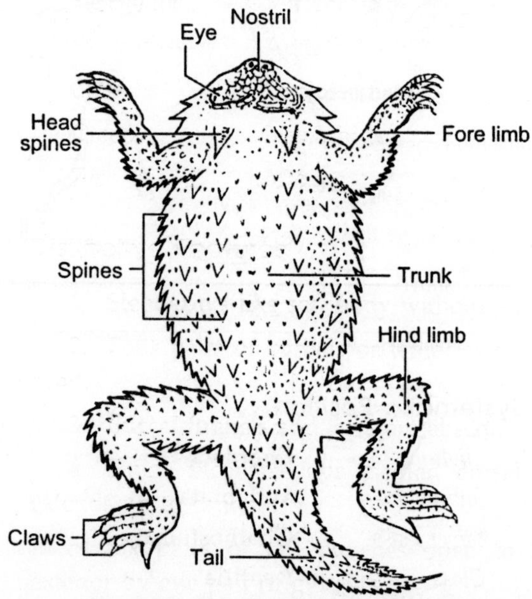

Fig. 1.1.2.86 : *Phrynosoma*

Distinctive Features

1. Broad, flat and spiny body with short spiny tail.
2. Dorsal side of the body is covered with spiny scales and ventral side with keeled scales.
3. Head with five spikes on each side.
4. Tongue is fleshy and non-protrusible.

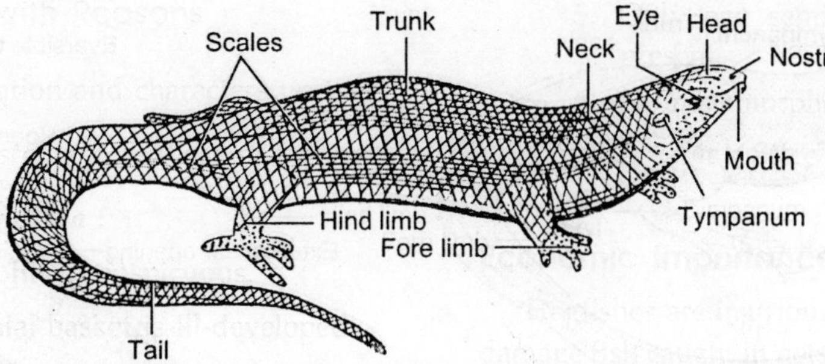

Fig. 1.1.2.87 : *Mabuia*

Identification with Reasons

State classification and characters up to—

Sub order—*Lacertilia* as in *Hemidactylus.*

Type—*Phrynosoma* : State distinctive features as in above.

Special Features

Adapted for desert environment, ovoviviparous, give birth to the youngs.

12. *MABUIA* (Commonly called as Skink)

Habit and Habitat

Terrestrial: adapted for burrowing life: live in dry abundant places : insectivorous: found in India, Myanmar, Bangladesh and China.

Systematic Position

Phylum	**Chordata**
Sub phylum	**Vertebrata**
Super class	**Gnathostomata**
Class	**Reptilia**
Sub class	**Lepidosauria**
Order	**Squamata**
Sub order	**Lacertilia**
Type	***Mabuia***

Distinctive Features

1. Body is covered with overlapping smooth shiny scales.
2. Eyes are small and with movable eyelids.
3. Flat tongue with scale like papillae.
4. Limbs are weak and slender.

Identification with Reasons

State classification and characters up to—

Sub order—*Lacertilia* as in *Hemidactylus.*

Type—*Mabuia* : State distinctive features as in above.

13. *VARANUS* (Commonly called as Monitor Lizard)

Habit and Habitat

Terrestrial semi aquatic: live in dry places under stones and rocks: leads a burrowing life: carnivorous: feed on rats, mices, tortoise, squirrels and dead bodies of animals: nocturnal: distributed in India, Sri Lanka, Malaysia, Southern Asia, South-east Australia and Africa.

Systematic Position

Phylum	**Chordata**
Sub phylum	**Vertebrata**
Super class	**Gnathostomata**

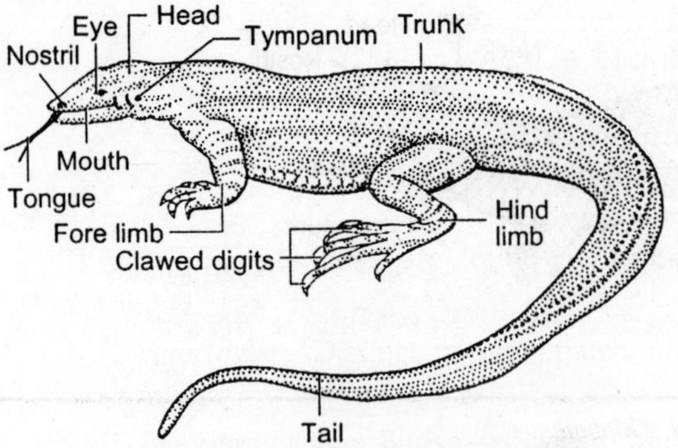

Fig. 1.1.2.88 : *Varanus*

Class	**Reptilia**
Sub class	**Lepidosauria**
Order	**Lacertilia**
Type	***Varanus***

Distinctive Features

1. Elongated body with movable head, long neck and large trunk.
2. Body is covered with bead like scales.
3. Long tail is compressed.
4. Tongue is long and deeply bifurcated and protrusible.
5. Short limbs are with powerful claws.

Identification with Reasons

State classification and characters up to—

Sub order—*Lacertilia* as in *Hemidactylus.*

Type—*Varanus* : State distinctive features as in above.

Economic Importance

Varanus are beneficial to man. Meat is used as food and skin is used for making valuable bags and slippers.

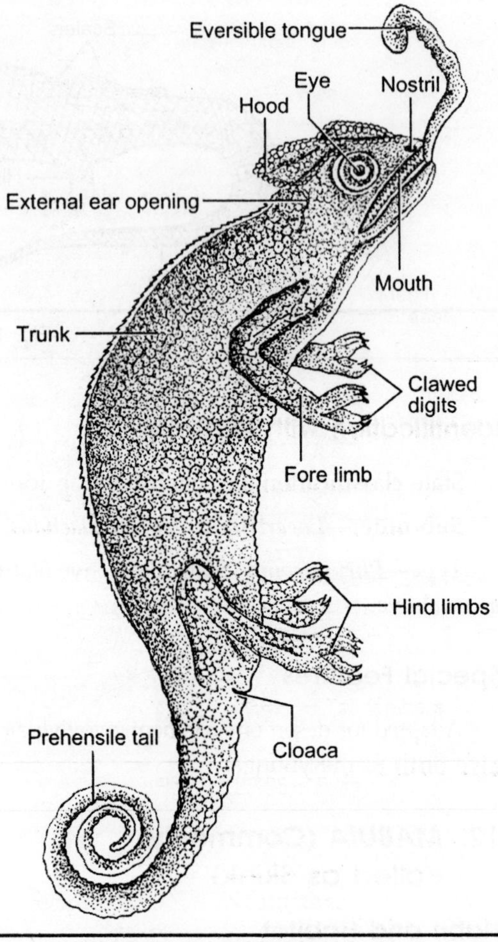

Fig. 1.1.2.89 : *Chemaeleo*

14. *CHEMAELEO*

Habit and Habitat

Arboreal: limbs and tails are modified for grasping: insectivorous: caught insects by their very long tongue: change colour very often : found in India, Sri Lanka, Asia, Africa, Southern Arabia, Spain, Europe.

Systematic Position

Phylum	**Chordata**
Sub phylum	**Vertebrata**

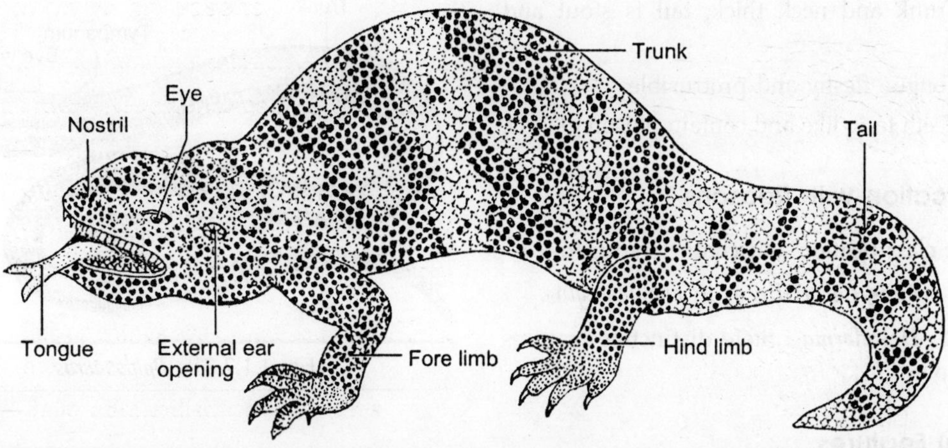

Fig. 1.1.2.90 : *Heloderma*

Super class	**Gnathostomata**
Class	**Reptilia**
Sub class	**Lepidosauria**
Order	**Squamata**
Sub order	**Lacertilia**
Type	***Chemaeleo***

Distinctive Features

1. Laterally compressed body with helmet-shaped head with prominent crest above.

2. Eyes are large and can move independently.

3. Digits are arranged in groups of two and three for grasping

4. Sticky club shaped tongue is protrusible, long and spoon like.

5. Tail is prehensile.

Identification with Reasons

State classification and characters up to—

Suborder—*Lacertilia* as in *Hemidactylus*.

Type—*Chemaeleo* : State distinctive features as in above.

15. *HELODERMA* (Commonly called as Gila Monster or Beaded Lizard)

Habit and Habitat

Terrestrial: lives in desert under rocks and in burrows: sluggish animal: carnivorous: feed on snake, eggs, birds, lizards, and young mammals: found in desert of Mexico and USA.

Systematic Position

Phylum	**Chordata**
Sub phylum	**Vertebrata**
Super class	**Gnathostomata**
Class	**Reptilia**
Sub class	**Lepidosauria**
Order	**Squamata**
Sub order	**Lacertilia**
Type	***Heloderma***

Distinctive Features

1. Heavy body is brightly black coloured with irregular rosy patches.

2. The limbs are short, powerful and well developed.

3. Trunk and neck thick, tail is stout and rounded.

4. Tongue fleshy and protrusible.

5. Teeth fang like and contain poison glands.

Identification with Reasons

State classification and characters up to—

Sub order—*Lacertilia* as in *Hemidactylus*.

Type—*Heloderma* : State distinctive features as in above.

Special Features

Heloderma is the only poisonous lizard. The poision is very effective on small animals and rarely to man.

16. *OPHIOSAURUS* (Limbless lizard) (Commonly called as glass snake)

Habit and Habitat

Terrestrial: found in humid places of grass lands and forests in burrows : nocturnal: restricted to South-Eastern Europe, North America, South-West Asia, Africa and India.: carnivorous: feed on worms, caterpillars, slugs and insects.

Systematic Position

Phylum	**Chordata**
Sub phylum	**Vertebrata**
Super class	**Gnathostomata**
Class	**Reptilia**
Sub class	**Lepidosauria**
Order	**Squamata**
Sub order	**Lacertilia**
Type	***Ophiosaurus***

Distinctive Features

1. Snake like body is covered with epidermal scales which are rectangular and arranged in circular and longitudinal rows.

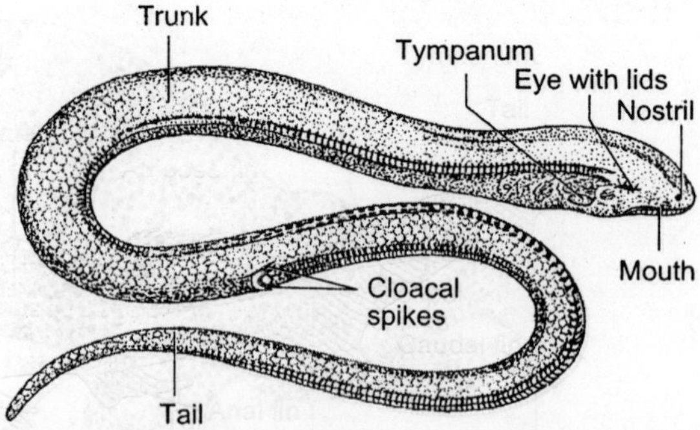

Fig. 1.1.2.91 : *Ophiosaurus*

2. Eyes reduced with movable eyelids

4. Tympanum and forked tongue present.

5. Lateral longitudinal fold of skin terminates into a locomotory spike near the cloaca.

6. Limbs are absent.

7. Tail breaks off and quickly regenerates.

Identification with Reasons

State classification and characters up to—

Sub order—*Lacertilia* as in *Hemidactylus*.

Type—*Ophiosaurus* : State distinctive features as in above.

17. *TYPHLOPS* (Commonly known as Blind Snake)

Habit and Habitat

Terrestrial : burrower : live in under logs, leaves, wet-forests and gardens: carnivorous : feed on earth worm, ants, termites and small insects: found in tropical and sub tropical countries of the world: non-poisonous harmless snakes.

Systematic Position

Phylum	**Chordata**
Sub phylum	**Vertebrata**
Super class	**Gnathostomata**
Class	**Reptilia**

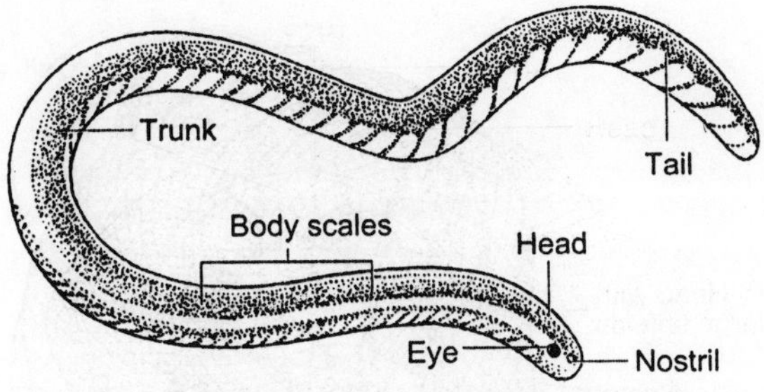

Fig. 1.1.2.92 : *Typhlops*

Sub class	**Lepidosauria**
Order	**Squamata**
Sub order	**Ophidia [L. serpents]**
Type	***Typhlops***

Distinctive Features

1. Earthworm like body is elongated, cylindrical and covered with overlapping shinny scales.
2. Head and tail blunt and eyes indistinct.
3. Tongue slender, bifid and protrusible.
4. Limbs and ear openings are absent.
5. Lower jaws without teeth.

Identification with Reasons

State classification and characters up to—

Order—*Squamata* as in *Hemidactylus.*

Sub order—***Ophidia*** :

1. Worm like, elongated, cylindrical animal, devoid of limbs and limb girdles (except Boas and Pythons)
2. Devoid of movable eyelids and eyes are protected by scales.
3. Slender, bifid, protrusible tongue is present.
4. Temporal fossa absent.
5. Vomeronasal organ or **organ of Jacobsons** is well developed.

Type—***Typhlops*** : State distinctive features as in above.

18. *NATRIX* (=*Tropidonotus*) (Commonly called as Grass Snake or Water Snake)

Habit and Habitat

Aquatic : found near tank, ponds, rivers and paddy fields: hidden under water vegetation or grasses: diurnal: carnivorous: feed on frogs , toads, fishes: found in India and USA: non poisonous.

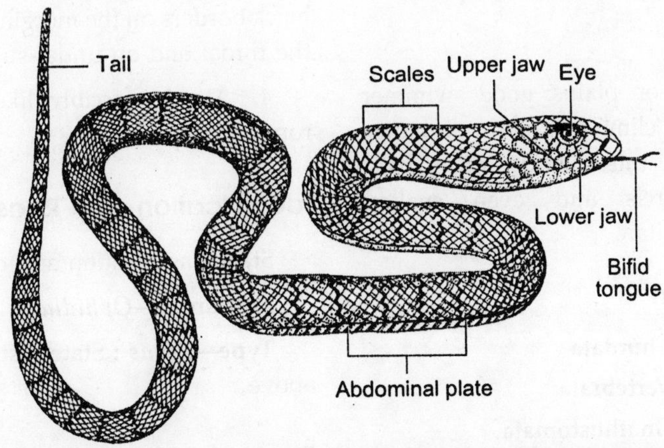

Fig. 1.1.2.93 : *Natrix*

Systematic Position

Phylum	**Chordata**
Sub phylum	**Vertebrata**
Super class	**Gnathostomata**
Class	**Reptilia**
Sub class	**Lepidosauria**
Order	**Squamata**
Sub order	**Ophidia**
Type	***Natrix* (=*Tropidonotus*)**

Distinctive Features

1. Short and stout body covered with scales with dark blotches on the back.
2. Head is distinct and contains bulging eyes.
3. Head shield are large and belly scales are plate like.
4. Tail is long.

Identification with Reasons

State classification and characters up to—

Sub order—*Ophidia* as in *Typhlops*.

Type—*Natrix* : State distinctive features as in above.

19. *PTYAS* (Commonly called as Rat Snake or Dhaman)

Habit and Hatitat

Terrestrial: found on plains: good swimmer and a fast runner: can climb tree: non-poisonous: carnivorous: feed on rats, lizards, mice, bats, toads, frogs, squirrels and even snakes: distribution cosmopolitan.

Systematic Position

Phylum	**Chordata**
Sub phylum	**Vertebrata**
Super class	**Gnathostomata**
Class	**Reptilia**
Sub class	**Lepidosauria**
Order	**Squamata**
Sub order	**Ophidia**
Type	***Ptyas***

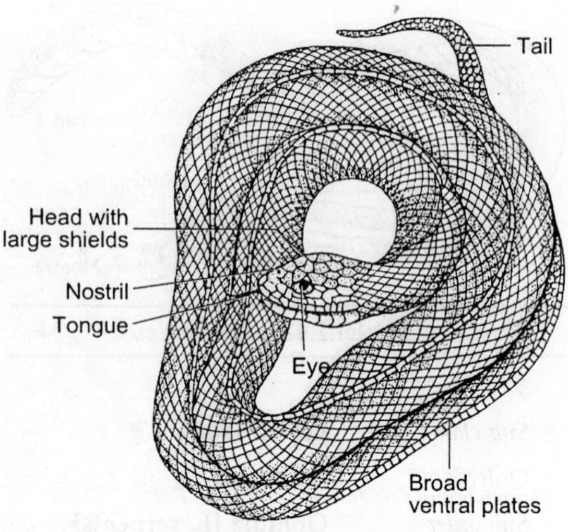

Fig. 1.1.2.94 : *Ptyas*

Distinctive Features

1. Body is long, robust with long tail, no limbs and about three metres in length.
2. Head with large shields is differentiated from neck and with large eyes and nostrils.
3. Colour is yellowish with characteristic black borders on the margin of lips, on the side of the throat and on under-surface to tail.
4. Attacks forcibly like whip hence called rope snake.

Identification with Reasons

State classification and characters up to—

Sub order—*Ophidia* as in *Typhlops*

Type—*Ptyas* : State distinctive features as in above.

Economic Importance

Ptyas controls rodents in the field.

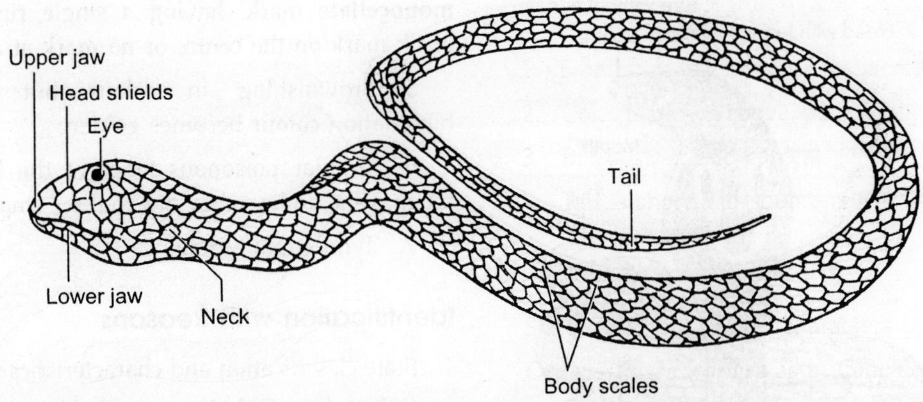

Fig. 1.1.2.95 : *Dryophis*

20. *DRYOPHIS* (Commonly called as Green Whip Snake)

Habit and Habitat

Terrestrial; arboreal: non poisonous: inhabits high grasses, bushes or trees: can leap or spring: remain suspended by its long thin tail: carnivorous: feed on frogs, birds, lizards and some insects: found in Asia, South America, Australia and all over India.

Systematic Position

Phylum	**Chordata**
Sub phylum	**Vertebrata**
Super class	**Gnathostomata**
Class	**Reptilia**
Sub class	**Lepidosauria**
Order	**Squamata**
Sub order	**Ophidia**
Type	***Dryophis***

Distinctive Features

1. Body is long and slender with a long tail.

2. Elongated head has a dorsal groove medially and elongated rostral shield.

3. Colour is parrot green with yellow stripes along the side.

4. Large eyes with horizontal and elongated pupil.

Identification with Reasons

State classification and characters up to—

Sub order—*Ophidia* as in *Typhlops*.

Type—*Dryophis* : State distinctive features as in above.

21. *NAJA NAJA* (Commonly called as Cobra)

Habit and Habitat

Terrestrial: generally nocturnal : found in jungles, bushes, open fields and near human habitation, living in holes, under stones and mud walls: carnivorous: feed on frogs, toads, rats, lizards, birds and other snakes: oviparous; hibernate during winter: **highly poisonous**: venom is neurotoxic: found in India, Africa, China, Australia, Philippines, Tasmania, and Egypt.

Systematic Position

Phylum	**Chordata**
Sub phylum	**Vertebrata**

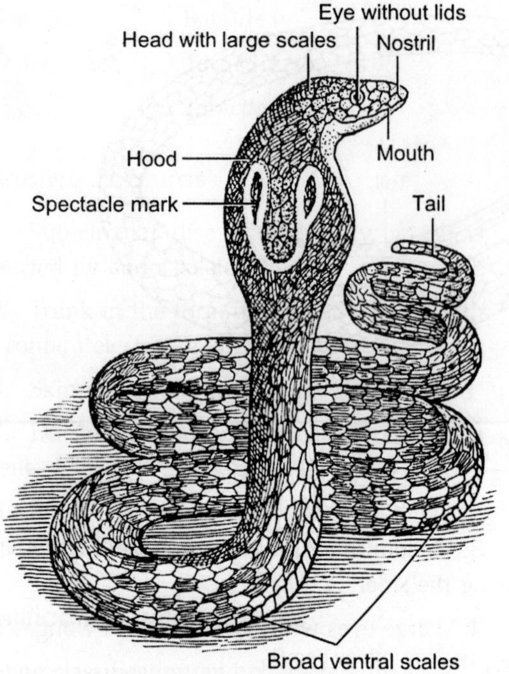

Fig. 1.1.2.96 : *Naja naja* (Cobra)

Super class	**Gnathostomata**
Class	**Reptilia**
Sub class	**Lepidosauria**
Order	**Squamata**
Sub order	**Ophidia**
Type	***Naja naja***

Distinctive Features

1. Body is covered by smooth oblique scales and with out limbs or limb girdles. The scales on the neck are twenty five to thirty five, in the mid body twenty one to twenty five and at vent fifteen to seventeen.

2. The caudal scales are paired.

3. Head is small, neck is dilatable and cervical ribs are elongated.

4. Neck region expands and form hood, which contains **binocellate** mark on the dorsal surface with a white band around the mark or **monocellate** mark, having a single ring with black mark on the centre or no mark at all.

5. Brownishing in colour but during hibernation colour becomes golden.

6. Tubular poisonous fangs at the front of the maxilla , followed by one to three small teeth.

7. Tympanum absent.

Identification with Reasons

State classification and characteristics up to—**Sub order—*Ophidia*** as in *Typhlops.*

Type—*Naja naja* : State distinctive features as in above.

Then examine the hood marking of the specimen, there you will see either of the three following markings. State accordingly.

1. Binocellate marking on the dorsal surface with a white band in the centre, the specimen is *Naja naja.*
2. Monocellate marking having a single ring with a black mark in the centre, the specimen is *Naja naja kaouthia.*
3. No mark in the hood at all, the specimen is *Naja naja oxiana.*

22. *OPHIOPHAGUS (NAJA) HANNAH* (Commonly called as King Cobra or Hamadrayad)

Habit and Habitat

Terrestrial: build nests on the trees and take care of the young ones: found in fairly dense forests, particularly at a higher altitude: carnivorous: feed on other snakes both poisnous and non poisonous: found in the Mangrove forests of Sunderban of Bengal, Orissa, Assam, Nilgiris, Andamans, China and Malaya.

Systematic Position

Phylum	**Chordata**
Sub phylum	**Vertebrata**

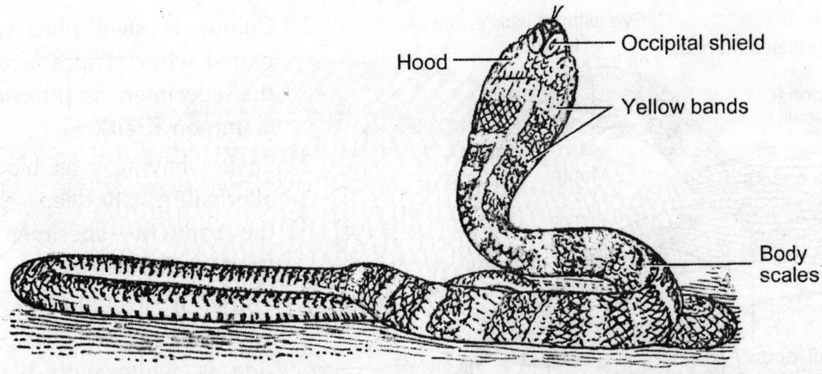

Fig. 1.1.2.97 : *Ophiophagus* (King cobra)

Super class	**Gnathostomata**
Class	**Reptilia**
Sub class	**Lepidosauria**
Order	**Squamata**
Sub order	**Ophidia**
Type	***Ophiophagaus* (=*Naja*) *hannah***

Distinctive Features

1. Body is covered with smooth oblique scales and with out limbs or limb girdles.

2. Scales on the neck are only seventeen to nineteen, the ventral scales being two hundred forty to two hundred fifty four.

3. Head is small, neck is dilatable and cervical ribs are elongated.

4. Neck region expands and form hood, which is less wide, but longer then that of cobra.

5. The colour of the young snake is olive or light black with thick faint yellow bands all over the body but in older snake the bands becomes faint or absent.

Identification with Reasons

State classification and characteristics up to—

Sub order—*Ophidia* as in *Typhlops*.

Type—*Ophiophagus* (*Naja*) *hannah* : State distinctive features as in above.

Remark

Cobras are deadly poisonous snakes. The venom is neurotoxic; death is due to respiratory failure. Mantras, herbs and charms can not cure the patient, only remedy is the antivenin injection.

23. *BUNGARUS* (Commonly called as Krait)

Habit and Habitat

Terrestrial: nocturnal: live in cracks in walls or in heaps of stones or bricks or logs: carnivorous: feed on toads, mice and other snake: found in South East Asia, all over India and Malaya.

Systematic Position

Phylum	**Chordata**
Sub phylum	**Vertebrata**
Super class	**Gnathostomata**
Class	**Reptilia**
Sub class	**Lepidosauria**
Order	**Squamata**
Sub order	**Ophidia**
Type	***Bungarus***

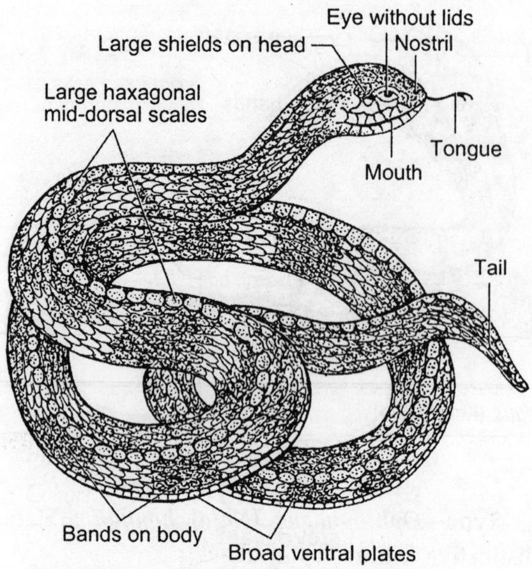

Fig. 1.1.2.98 : *Bungarus* (Krait)

Distinctive Features

1. Body is elongated, cylindrical and covered by smooth polished scales. Scales are in fifteen to seventeen rows.

2. Colour is steel blue. Steel blue patches alternate with rows of the paired white stripes across the body or jet black cross strips alternating with a deep yellow strips on the body or steel blue or black and is devoid of any white or yellow marking.

3. Body contains enlarged chain of hexagonal scales on the dorsal side.

4. Ventral scales beyond the anal region in a single row or scales beyond the vent is continuous.

Identification with Reasons

State classification and characteristics up to—

Sub order—*Ophidia* as in *Typhlops*.

Type—*Bungarus* : State distinctive features as in above.

Then examine the specimen, there you will see either of the three following body colours. State accordingly.

1. Colour is steel blue with rows of paired white stripes across the body–the specimen is *Bungarus caeruleus* (common Krait).
2. Snakes having a jet black cross strip alternating with deep yellow strips on the body–the specimen is *Bungarus fasciatus* (banded Krait).
3. Colour is steel-blue or black and is devoid of any white strips. The ventral side is white with black spots–the specimen is *Bungarus niger*.

Remark

Kraits are deadly poisonous. The venom is neurotoxic. The victim feel sleepy and immediate antivenom is the remedy to save the patient from death.

24. *VIPERA* (Commonly called as Chandrabora)

Habit and Habitat

Terrestrial: found in rocky and bushy region where the body colour harmonizes with the surroundings: remain coiled with head in the centre of the coil: carnivorous: feed mice, lizards and birds: found in Europe, Asia, Sri Lanka, Myanmar and all over India.

Systematic Position

Phylum	**Chordata**
Sub phylum	**Vertebrata**
Super class	**Gnathostomata**
Class	**Reptilia**
Sub class	**Lepidosauria**
Order	**Squamata**
Sub order	**Ophidia**
Type	***Vipera russelli***

Distinctive Features

1. Body is elongated, cylindrical, without limbs and limb girdles.

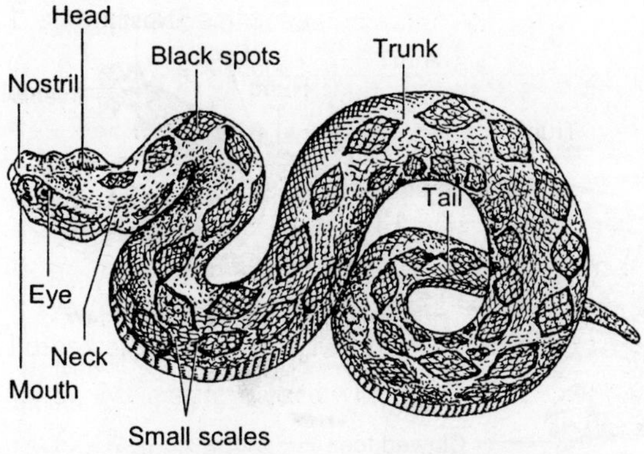

Fig. 1.1.2.99 : *Vipera* (Chondrabora)

2. Colour is brown with deep elliptical patches in three rows.

3. Head is triangular and head scales on the top are very small, imbricate and keeled.

4. Paired erectile fangs in front of upper jaws.

5. No pit between nostrils and eye (Pitless viper).

Identification with Reasons

State classification and characters up to—

Sub order—*Ophidia* as in *Typhlops*.

Type—*Vipera russelli* : State distinctive features as in above.

Special Features

Poisonous snake, it can strike but not bite. Strike gives rise to swelling and pain. Can be cured by treatment. The snake hisses loudly and continuously.

25. *HYDROPHIS* (Commonly known as Sea Snake)

Habit and Habitat

Aquatic: live in water: very good swimmer: carnivorous: feed on fishes, eels and on sea food: oviparous: found in India, along the Pacific coast from Southern Mexico to Northern South America, in the Bay of Bengal and Malayan regions.

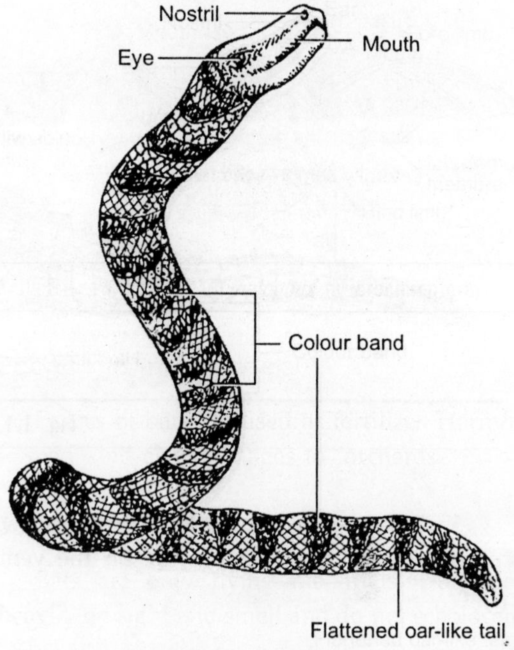

Fig. 1.1.2.100 : *Hydrophis*

Systematic Position

Phylum	**Chordata**
Sub phylum	**Vertebrata**
Super class	**Gnathostomata**
Class	**Reptilia**
Sub class	**Lepidosauria**
Order	**Squamata**
Sub order	**Ophidia**
Type	***Hydrophis***

Distinctive Features

1. Body is compressed posteriorly, tail is flat and shaped like a paddle.

2. Body is rounded and covered by small scales which are not overlapping.

3. Head is indistinct and covered by large shields.

4. Nostril is on top of the head.

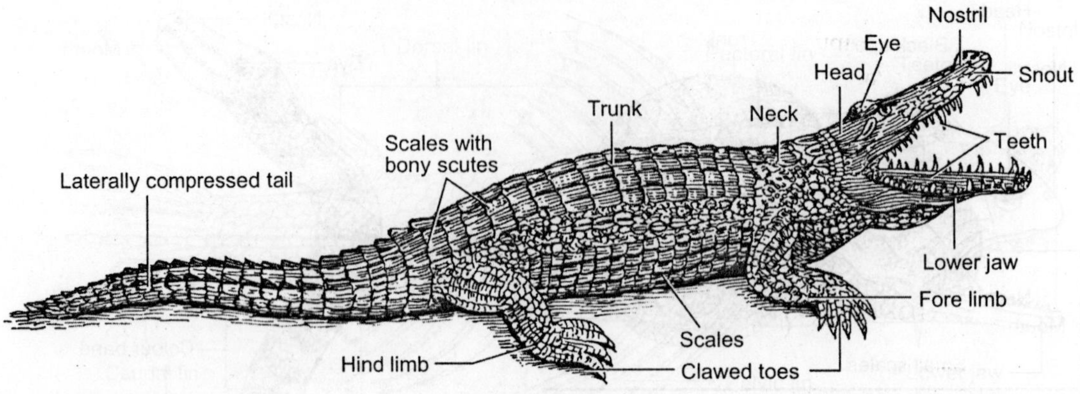

Fig. 1.1.2.101 : *Crocodylus*

5. Body colour is black with a yellowish crescent mark above and white on the ventral side.

Identification with Reasons

State classification and characters up to—

Sub order—*Ophidia* as in *Typhlops*

Type—*Hydrophis* : State distinctive features as in above.

Remark

Poisonous snake. Poison is neurotoxic. Bite is seldom due to the position of the fangs which is bit behind.

26. *CROCODYLUS* (Commonly called as Mugger or Marsh Crocodile)

Habit and Habitat

Amphibious: spending mainly in water and part on dry land : live in rivers, lakes, ponds, marsh lands, swamps, logs, brackish waters and estuaries: on land live in forests, grass lands, deserts and river banks: carnivorous: feed on frogs, snakes, fish, birds and mammals: may attack human beings: found in central America, Africa, Southern Asia, Malaya, North Australia, East Indies and India.

Systematic Position

Phylum	**Chordata**
Sub phylum	**Vertebrata**
Super Class	**Gnathostomata**
Class	**Reptilia**
Sub class	**Archosauria**
Order	**Crocodilia**
Type	***Crocodylus***

Distinctive Features

1. Body is elongated with triangular head, which is prolonged into a stout snout.

2. Teeth unequal in size, the fourth mandibular teeth fits in to the notch in the upper jaw.

3. Fore and hind limbs short and ending in clawed toes with webs between.

4. Long tail laterally compressed.

5. Body is covered by leathery armour.

Identification with Reasons

State classification and characters up to—

Class—*Reptilia* as in *Typhlops*.

Sub class—*Archosauria* :

1. Skull diapsid type (closed upper temporal vacuity) with antorbital vacuities behind the nostrils and palate.
2. The lower jaw with vacuities between dentary and angular.

Order—*Crocodilia* :

1. Elongated body is covered by epidermal scales which are supported by dermal bones or scales.
2. Long, laterally compressed tail is provided with large crests.
3. Nostrils are at the top of the snout.
4. Fore limbs with five digits and hind limbs with four digits with webs between.
5. Cloacal opening longitudinal.
6. Heart four chambered.

Type—*Crocodylus* : State distinctive features as in above.

Remark

Crocodiles are the top consumer in rivers, lakes, ponds, marsh lands, swamps, logs, brackish waters and estuaries. Crocodiles are dangerous to mankind.

27. *GAVIALIS* (Commonly called as Gharial)

Habit and Habitat

Amphibious: fresh water animal: live in Ganges and other large rivers: carnivorous: feed on fishes only: found in India, Myanmar, and Malaysia.

Systematic Position

Phylum	**Chordata**
Sub phylum	**Vertebrata**
Super class	**Gnathostomata**
Class	**Reptilia**
Sub class	**Archosauria**
Order	**Crocodilia**
Type	***Gavialis gangeticus***

Distinctive Features

1. Largest crocodile attains a length of over six metre.

2. Head is large, rimmed with various conical teeth.

3. Snout is extremely long and slender.

4. Teeth sub equal, while first and fourth lower teeth fit into the groove of the upper jaw.

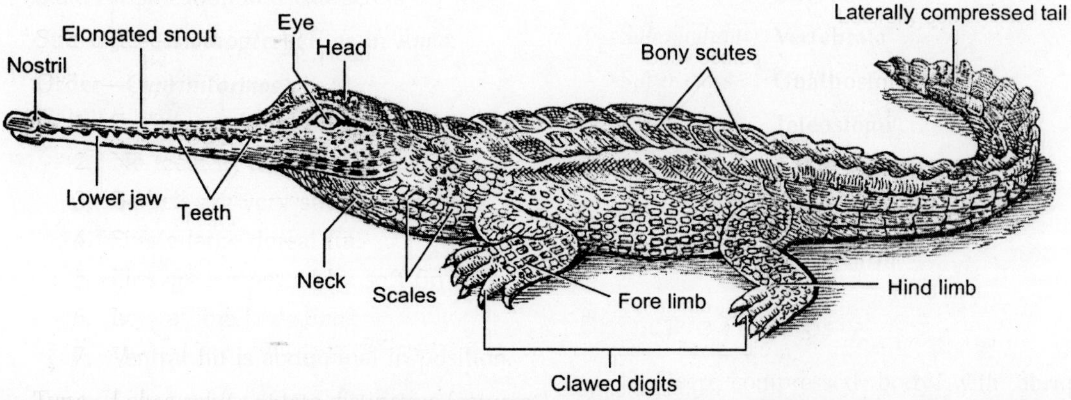

Fig. 1.1.2.102 : *Gavialis*

5. Tail is laterally compressed, and very powerful.

6. Fore and hind limbs are short and ending in clawed toes with webs between.

Identification with Reasons

State classification and charactertes up to—

Order—*Crocodilia* as in *Crocodylus*.

Type—***Gavialis gangeticus*** : State distinctive features as in above.

Remark

Gavialis gangeticus is the only species found in India. Harmless to man.

FEATHERED BIPED BIRDS

Birds are the best known and most easily recognized of all animals which are included under the class **Aves.** They constitute a highly specialized group of vertebrates which have made a significant advance over all lower vertebrates. Flight is of course the characteristic features by which they easily avoid all kinds of enemies on land and find new habitats. The light body and strength for flight are achieved by the presence of pneumatic bone and air sacs. Teeth and heavy jaws are lost. They have high metabolic rate, can regulate their body temperature and have high performance respiratory and muscular system. There is little accumulation of fat and the young develop outside the body, by such means they maintain the specific gravity. Brain development of them is also high.

The success of bird life has been largely due to the great variety and habitat selection. Bird occupies all continents, ocean and islands. Arctic and Antarctic and lives from sea level to high mountains.

Many characteristics of bird shows close resemblance to those of reptiles. Though we have no detailed evidence by which cold blooded terrestrial reptiles gave rise to warm blooded biped flying birds, but the discovery of the fossil of *Archaeopteryx* from the upper Jurassic rocks of Bavaria speaks about the reptilian origin of bird.

Birds form a more homogeneous class of vertebrates than reptiles, therefore, they cannot be separated into well defined groups. The class Aves is divided into two subclasses: 1. **Archeornithes** - ancient birds. Example : *Archaeopteryx.* 2. **Neornithes:** new birds. Neornithes is divided into two suborders: (1) **Odontognathae** – the toothed jaws bird. Example : *Hesperonis.* (2) **Neognathae** – the new jaws bird. Modern birds are classified into twenty eight orders.

AVES

1. *COLUMBA LIVIA* (Commonly called as Pigeon)

Habit and Habitat

Terrestrial; tree living; build nests in buildings, houses, ware houses, sheds, railway stations; seed and grain eater. Young are nourished by milk secreted by the crop; distribution worldwide.

Systematic Position

Phylum	**Chordata**
Subphylum	**Vertebrata**
Superclass	**Gnathostomata**
Class	**Aves [L. Av = a bird]**
Subclass	**Neornithes [G. Neos = modern]**
Superorder	**Neognathae [G. gnathos = jaw]**
Order	**Columbiformes**
Type	***Columba livia*** **[L. columbi = dove]**

Classification is based on J. Z. Young, 1981.

Distinctive Features

1. Body is divided into head, neck, trunk and tail.

2. Entire body except feet are covered by feathers.

3. Two dark bars on wing and a band across the end of the tail.

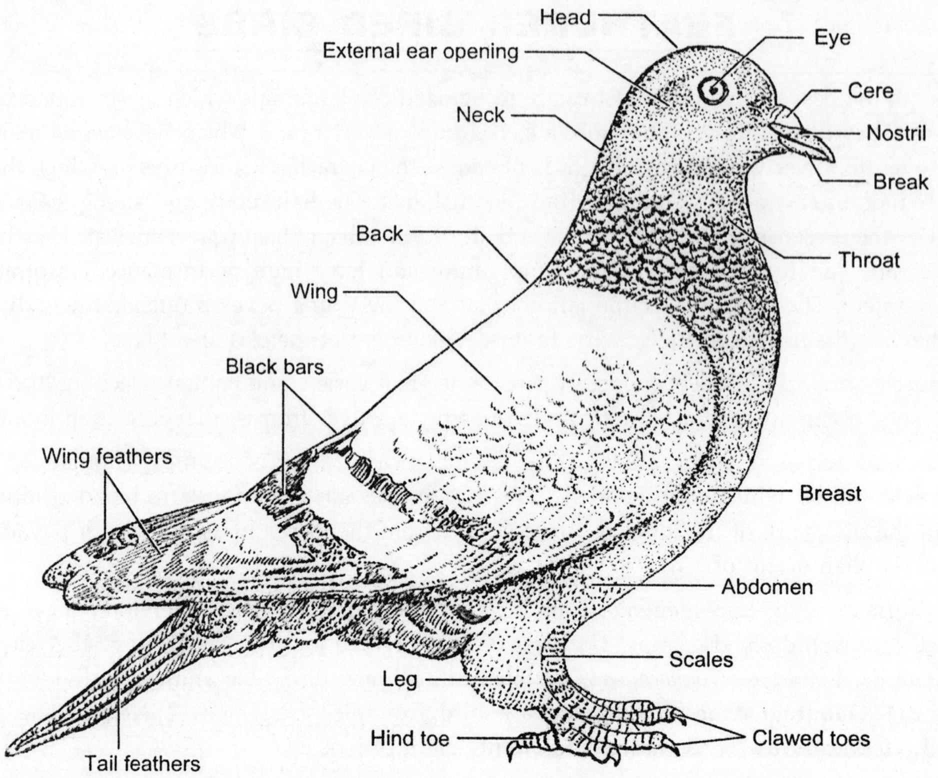

Fig. 1.1.2.103 : *Columba livia*

4. Head contains large eyes, slit like nasal opening and a short and slender beak.

5. Presence of cere at the base of the upper beak.

Identification with Reasons

State classification and characters upto—

Super class—*Gnathostomata* as in *Scoliodon*.

Class—*Aves* :

1. Endothermic, derive energy from metabolism for temperature regulation.
2. Body is covered by special integumentary derivative called feathers.
3. Head is small, fairly long movable neck and mouth is provided with specialized derivative called beaks.
4. Fore limbs are modified as wings.
5. Hind limbs are with clawed digits and scales.

Subclass—*Neornithes* :

1. Tail short with a terminal pygostyle.
2. In most members, the teeth are absent.
3. The retrices are arranged in a semicircle in the pygostyle.
4. Well developed sternum bears keel or carina.

Super order—*Neognathae* : Vomer slender, palatine protruded posteriorly and movably articulated with pterygoids(neognathus condition).

Order—*Columbiformes* :

1. Head and neck are small with short, slender beak.
2. Each foot with four toes, three are forwardly directed and one backwardly.
3. Skin soft and thick, wings long and pointed.

Type—*Columba livia* : State distinctive features as in above.

Special Features

Pigeons are common domestic birds. It lives where ever human habitations are found, therefore they are a varity called *domesticus*. They subvert the beauty of the building and statues with their fecal matter.

2. *PSITTACULA* (Common name Parakeet or Parrot)

Habit and Habitat

Arboreal: lives in flocks: commonly found in the fruit trees, ripe crops, orchards and cultivation; feed on fruits crops and seeds; built nests in crevices and hollow trees; found in India, Pakistan, Myanmar, Sri Lanka and Bangladesh.

Systematic Position

Phylum	**Chordata**
Subphylum	**Vertebrata**
Superclass	**Gnathostomata**
Class	**Aves**
Subclass	**Neornithes**
Superorder	**Neognathae**
Order	**Psittaciformes**
Type	***Psittacula***

Distinctive Features

1. Brilliant blue-green, yellow or red plumage with deeply hooked red bill.

2. Beak strong, narrow, sharp edge and hooked at the tip.

3. Feet suitable for grasping, two toes in front and two behind.

4. Tail feathers elongated.

5. Female lacks the black and rose-pink collar of male.

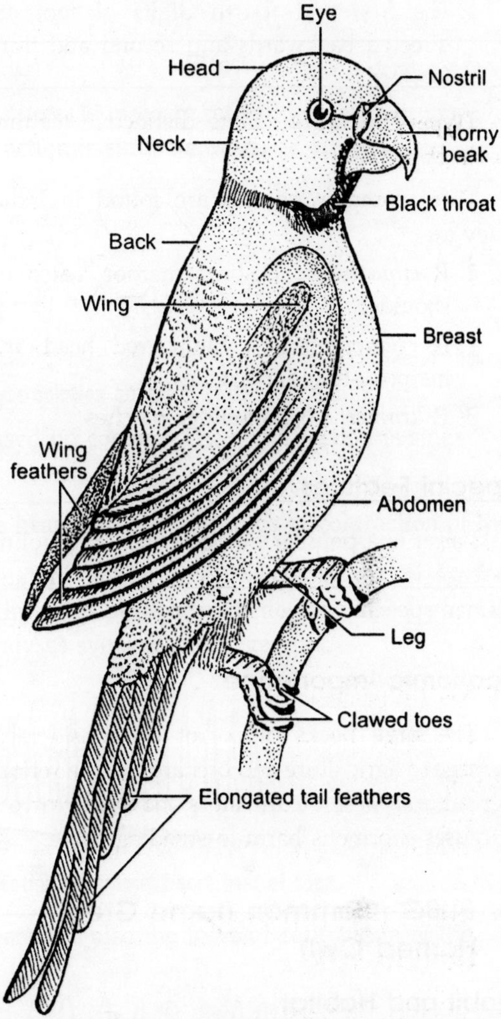

Fig. 1.1.2.104 : *Psittacula*

Identification with Reasons

State classification and characters up to—

Super order—*Neognathae* as in *Columba*.

Order—*Psittaciformes* :

1. Adapted for arboreal life and can also climb.
2. Beaks are used for climbing and feet for grasping, climbing and holding the food.
3. Upper mandible movable on frontal bone of skull.

4. The first and fourth digits of foot are directed backwards and second and third forward.

Type—*Psittacula* : State distinctive features as in above.

Three common species are found in India. They are :

1. *P. eupatoria* : Distinct maroon patch on shoulder.
2. *P. cyanocephala* : Bluish-red head and maroon shoulder patches.
3. *P. krameri* : No shoulder patches.

Special Features

Parrot is a popular domesticated bird, found in cage in almost every home and it imitates human speech.

Economic Importance

The large flocks of parrot do considerable damage to agriculture and orchards. It is a serious agricultural pest to cultivators and food growers. It causes enormous harm to standing crops.

3. *BUBO* (Common name Great Horned Owl)

Habit and Habitat

Nocturnal; hide in daytime in secluded leafy branches or tree holes or woody places;live in plains and foothills near human habitations; carnivorous; feeds on insects, rats, rabbits, mice, birds and other animals; world wide distributions specially found in India, Pakistan and Myanmar.

Systematic Position

Phylum	**Chordata**
Subphylum	**Vertebrata**
Superclass	**Gnathostomata**
Class	**Aves**
Subclass	**Neornithes**
Superorder	**Neognathae**
Order	**Strigiformes**
Type	***Bubo***

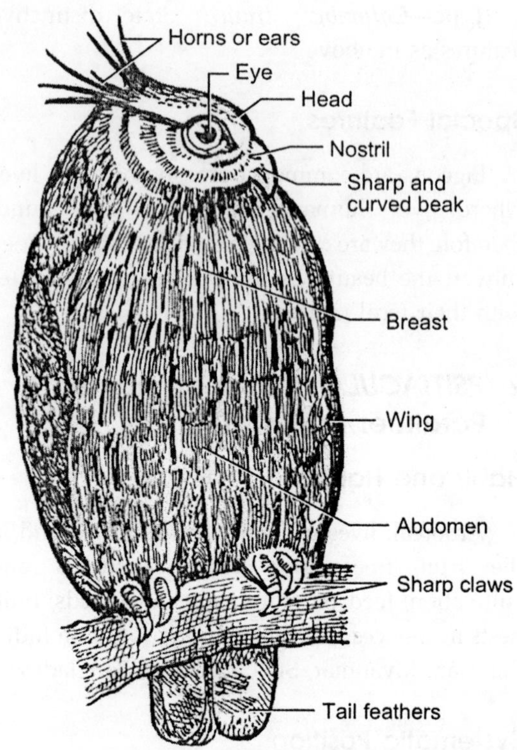

Fig. 1.1.2.105 : *Bubo*

Distinctive Features

1. Large, heavy and dark brown owl with monkey shaped head, which is large and round.
2. Contour feathers are loosely arranged and extended up to the digits.
3. Eyes are large, rounded and forwardly directed.
4. Ears opening large with flap-like cover.
5. Short beak and strongly built.
6. Feet adapted for grasping; claws sharp.

Identification with Reasons

State classification and characters up to—

Super order—*Neognathae* as in *Columba.*

Order—*Strigiformes* :

1. Contour feathers are loosely arranged and extend up to the digits.
2. Eyes are large and round and directed frontally.
3. Ear openings are on the lateral side of the head.

Type—*Bubo* : State distinctive features as in above.

Economic Importance

Bubo is of great economic value. By destroying the harmful animals like rats, mice, lizards, birds etc. they became useful to men.

4. *ALCEDO* (Common name Kingfisher)

Habit and Habitat

Found near the tanks, streams, lakes and the sea shore; prey upon aquatic insects, tadpoles and small fishes by their long bill; dive into the water to catch fish; nests are builds up in the tree holes; distribution cosmopolitan; found throughout India.

Systematic Position

Phylum	**Chordata**
Subphylum	**Vertebrata**
Super class	**Gnathostomata**
Class	**Aves**
Sub class	**Neornithes**
Super order	**Neognathae**
Order	**Coraciformes**
Type	***Alcedo***

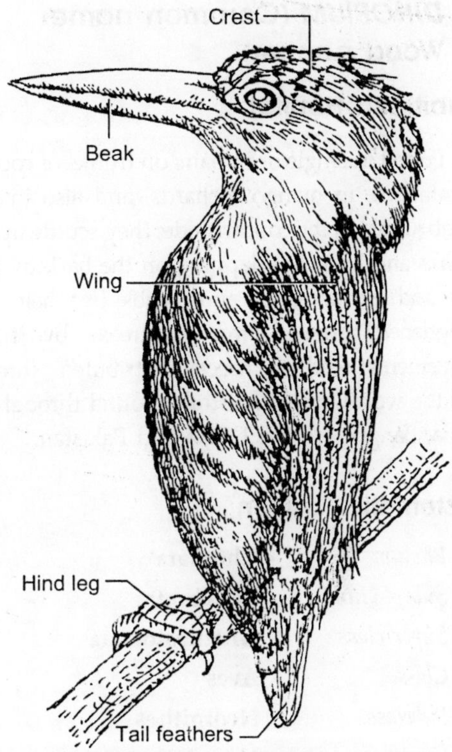

Fig. 1.1.2.106 : *Alcedo*

Distinctive Features

1. Small in size and body covered with blue and green feathers.
2. Wings and legs are short.
3. Tail short stumpy.
4. Straight long beak.

Identification with Reasons

State classifications and characters up to—

Super order—*Neognathae* as in *Columba.*

Order—*Coraciformes* :

1. Little bird and the body is brilliantly colored.
2. Beak is long.
3. Three anterior toes are united.

Type—*Alcedo* : State the distinctive features as in above.

5. *DINOPIUM* (Common name Wood pecker)

Habit and Habitat

Found in single or in pairs on trunks of coconut plantations, in mango orchards, and also in light scrub jungle in the country side; they scuttle up tree trunks and branches, lapping on the bark by their bill and extract ants and grubs by their long roughened tongue; climb up trees by jerking movements; feed on insects; distributed through out the world except Australia; found through out India, Bangladesh, Sri Lanka and Pakistan.

Systematic Position

Phylum	**Chordata**
Sub phylum	**Vertebrata**
Super class	**Gnathostomata**
Class	**Aves**
Subclass	**Neornithes**
Super order	**Neognathae**
Order	**Piciformes**
Type	***Dinopium***

Distinctive Features

1. Dorsal surface is golden yellow with black and white spots.
2. Crown and occipital crest crimson.
3. Tongue is roughened or has barbs near the tip and protrusible.
4. Tail is stiff and wedge shaped.

Identification with Reasons

State classification and characters up to—

Super order—*Neognathae* as in *Columba.*

Order—*Piciformes* :

1. Beak is long, stout and pointed.
2. Second and third toes are directed forward and first and fourth toes are directed backward. (**Zygodactylous** toes).

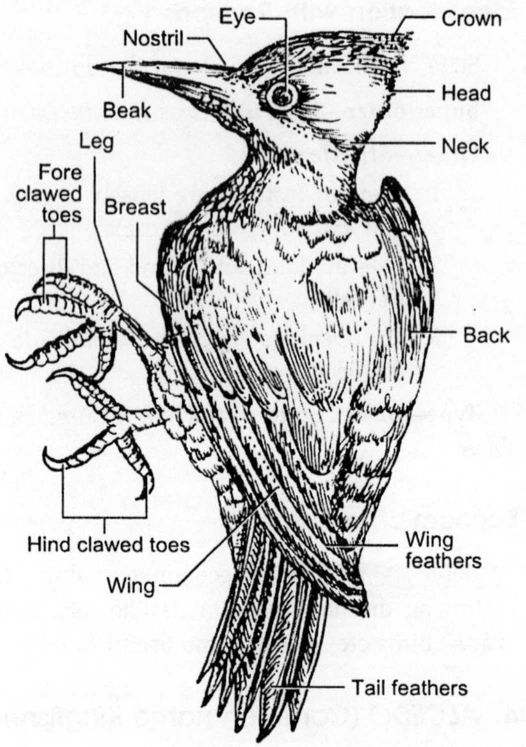

Fig. 1.1.2.107 : *Dinopium*

3. Tail feathers stiff with pointed tips.

Type—***Dinopium*** : State distinctive features as in above.

6. *PASSER* (Common name House Sparrow)

Habit and Habitat

Are seen almost with the human habitation; move freely and nest in the houses; omnivorous; feeds on insects, grains, flower nectar and kitchen left off; flock damage crops and gardens: found through out the world.

Systematic Position

Phylum	**Chordata**
Sub phylum	**Vertebrata**
Super class	**Gnathostomata**

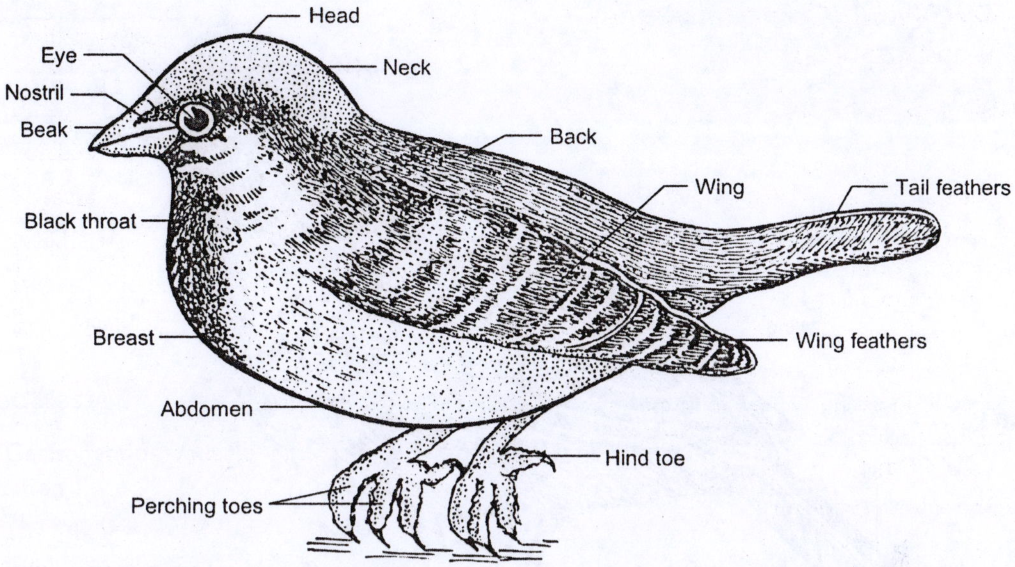

Fig. 1.1.2.108 : *Passer (Male)*

Class	**Aves**
Sub class	**Neornithes**
Super order	**Neognathae**
Order	**Passeriformes**
Type	***Passer domesticus***

Distinctive Features

1. Small, fragile body.

2. Inconspicuous coloured, female is ashy-grey-brown and male is earthy- gray- brown with black spots; abdomen whitish in both sexes.

3. Beak short, hard and curved.

4. Male with black patch on the ventral side of the neck.

Identification with Reasons

State classification and characters up to—

Superorder—*Neognathae* as in *Columba.*

Order—*Passeriformes* :

1. Small birds, head more or less rounded.
2. Beak is strong, hard and curved.
3. Toes three in front and one behind.

Type—*Passer domesticus* : State distinctive features as in above.

Economic Importance

Sparrows are useful and harmful to man kind as flock damage crops and gardens and destroy several agricultural pests. Ignored by men, as it is commensal with them.

7. *PYCNONOTUS* (Common name Bul-Bul)

Habit and Habitat

Found in scrubs jungles and in gardens; remain near or away from human habitation;feed on banyan and peepul figs, termites, moth caterpillars and other insects; found in Indian sub continent and in Bangladesh, Srilanka and Mayanmar.

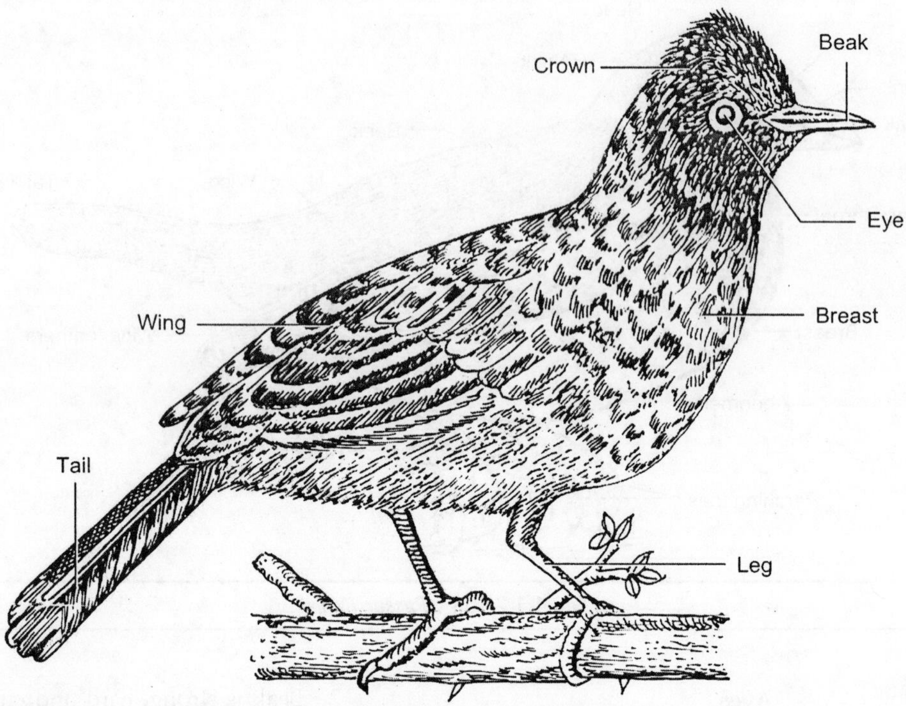

Fig. 1.1.2.109 : *Pycnonotus*

Systematic Position

Phylum	**Chordata**
Subphylum	**Vertebrata**
Superclass	**Gnathostomata**
Class	**Aves**
Subclass	**Neornithes**
Superorder	**Neognathae**
Order	**Passeriformes**
Type	*Pycnonotus*

Distinctive Features

1. Small in size and colour smoke brown.
2. Mop crested black head bird.
3. Scale like markings on the breast and crimson patch at the root of the tail.
4. Bill and legs are red in colour.

Identification with Reasons

State classification and characteristics up to—

Order—*Passeriformes* as in *Passer.*

Type—*Pycnonotus* : State distinctive features as in above.

8. *PLOCEUS* (Common name Weaver Bird or Baya)

Habit and Habitat

Gregarious; lives in flocks, flocks seen about open cultivation; construct nests, which is bottle shaped; feed on paddy grains, other cereals and also insects; found in Indian subcontinent, Bangladesh, Pakistan, Sri Lanka and Myanmar.

Systematic Position

Phylum	**Chordata**

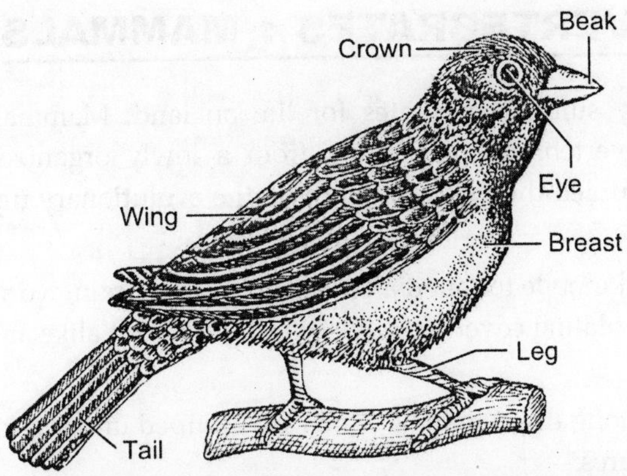

Fig. 1.1.2.110 : *Ploceus*

Sub Phylum **Vertebrata**
Superclass **Gnathostomata**
Class **Aves**
Subclass **Neornithes**
Superorder **Neognathae**
Order **Passeriformes**
Ploceus

Distinctive Features

1. Small bird, size that of a sparrow.
2. Color is brown above, plain whitish below.
3. Bill is stout and conical.
4. Tail is short and square cut.

Identification with Reasons

State classification and characters up to—

Order—*Passeriformes* as in *Columba*.

Type—*Ploceus* : State distinctive features as in above.

Special Features

Ploceus is known as weaver bird. They build bottle shaped nests with a long vertical entrance tube. Nest is pendent and woven out of paddy leaves and rough - edge grasses, hanging in clusters from date palms or any other trees usually over water. Flock damages cereal crops.

ENDOTHERM TERRESTRIAL VERTEBRATES : MAMMALS

The mammals and the birds are the most fully suited vertebrates for life on land. Mammals constitute a highly specialized tetrapod group of vertebrates originating from a lowly organized reptilian stalk during the Mesozoic era, and have occupied the top most place in the evolutionary tree by virtue of their super organized brain.

Mammals may be defined as highly percipient and mobile tetrapod animals, with large brain, worm blooded, left aortic arch and a water proofed hairy insulating covering, whose young are born alike and are nourished by milk from mammae.

A wide variety of structure and form is present among the mammals which are grouped under class mammalia. They have the following distinctive features:

1. Body covered by hairs.
2. Presence of sweat and sebaceous glands.
3. Presence of mammary glands.
4. Four chambered heart with left aortic arch.
5. Presence of double occipital condyle.
6. Seven cervical vertebrae.
7. Non nucleated RBC.

Living members of mammals are divided into two sub classes: **Prototheria** and **Theria**. Prototheria includes one order—**Monotremata**. **Theria** comprises of two infra classes—**Metatheria** and **Eutheria**. Metatheria has one order—**Marsupialia** and **Eutheria** (= Placentalia) consists of sixteen living orders under four main cohorts on the basis of the foot structure. Mammals have conquered all available habitats and show extensive adaptive radiation. Parental care has reached the extreme stage of development.

MAMMALIA

1. *SUNCUS* (Commonly called as House Shrew or Ground Shrew)

Habit and Habitat

Terrestrial: nocturnal: lives in burrows: some times comes out to our house in search of food: feeds on insects, small invertebrates and rodents: found in Asia, Africa, Europe and North America.

Systematic Position

Phylum	**Chordata**
Sub phylum	**Vertebrata**
Super class	**Gnathostomata**
Class	**Mammalia**
Sub class	**Theria** [G.Therion = beast]
Infra class	**Eutheria** [G. eu = true]
Cohort	**Unguiculata** [L. unguiculus = little nail]
Order	**Insectivora** [L. inseta = insect; vora = to eat]
Sub order	**Lipotyphla**
Type	***Suncus***

Classification is based on J. Z. Young, 1981.

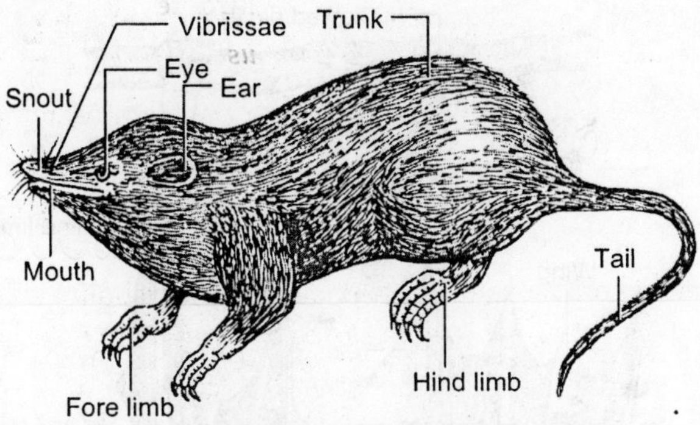

Fig. 1.1.2.111 : *Suncus*

Distinctive Features

1. Small size looks like a rat with rounded **pinnae.**

2. Prolonged snout formed proboscis. Proboscis bears numerous long vibrissae.

3. Tail is short more than half of the body length.

4. Scent gland present which imparts a light musky odour.

5. Body is blackish brown or dark brown.

Identification with Reasons

State classification and characters up to—

Super class—*Gnathostomata* as in *Scoliodon.*

Class—*Mammalia* :

1. Endothermal homoeotherm animal.
2. Generally body is covered with epidermal hairs.
3. External fleshy pinnae is present.
4. Mammary glands are present which are well developed in female.
5. Presence of sweat and sebaceous glands in the skin.
6. Dentition **thecodont**, **heterodont** and **diphyodont**.
7. A muscular diaphragm is present between the thoracic and abdominal cavity.

Sub class—*Theria* :

1. Mammary glands are provided with nipples or teats.
2. Teeth present both in young and adults.
3. Cloaca absent.
4. Testis in the scrotal sac in most cases.
5. Give birth to young ones.

Infra class—*Eutheria* :

1. The anal and urinogenital apertures are separate.
2. Presence of allantoic placenta.

Cohort—*Uniguiculata* :

1. Possess nails or claws.

Order—*Insectivora* :

1. Nocturnal insectivores with a snout.
2. Mammary glands are many and are distributed on the ventral surface.
3. Scrotum absent, testes are internal.
4. Teeth have sharp molar cusps.
5. Dental formula 3. 1. 4. 3/3. 1. 4. 3.

Sub order—*Lipotyphla* :

1. Five clawed digits in each fleshy limb.
2. **Plantigrade** foot posture.
3. Incomplete bony palate.

Type—*Suncus* : State distinctive fertures as in above.

2. *TALPA* (Commonly called as Mole)

Habit and Habitat

Subterranean : highly speacilised for burrowing: lives in tunnels : feeds mainly on earthworm and small insects: found through out the Holarctic and Oriental region:

Systematic Position

Phylum	**Chordata**
Sub phylum	**Vertebrata**
Super class	**Gnathostomata**
Class	**Mammalia**
Sub class	**Theria**

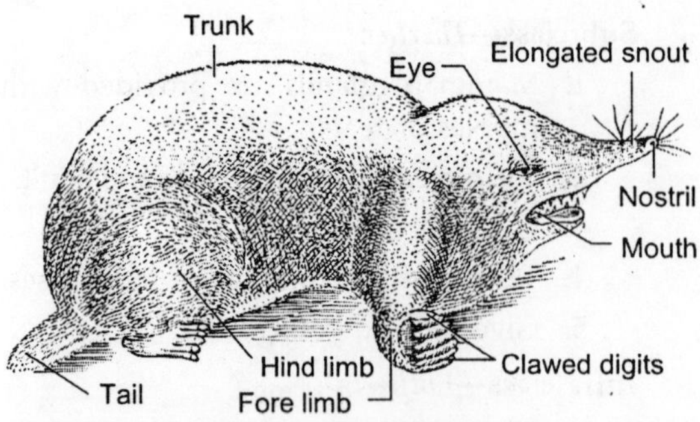

Fig. 1.1.2.112 : *Talpa*

Infra class	**Eutheria**
Cohort	**Unguiculata**
Order	**Insectivora**
Sub order	**Lipotyphla**
Type	***Talpa***

Distinctive Features

1. Cylindrical body is covered with yellowish-brown smooth fur.
2. Neck is very short and cervical vertebrae fused.
3. Eyes are very small and covered opaque skin.
4. External ears absent.
5. Fore limbs with broad digging claws and fore feet turned out wards.
6. Tail very short.

Identification with Reasons

State classification and characters up to—

Sub order—*Lipotyphla* as in *Suncus*.

Type—*Talpa* : State distinctive features as in above.

3. *PTEROPUS* (Commonly called as Fruit Bats or Flying Fox)

Habit and Habitat

Arboreal: nocturnal: many roost in colonies in trees: during day time they live in caves, crevices,

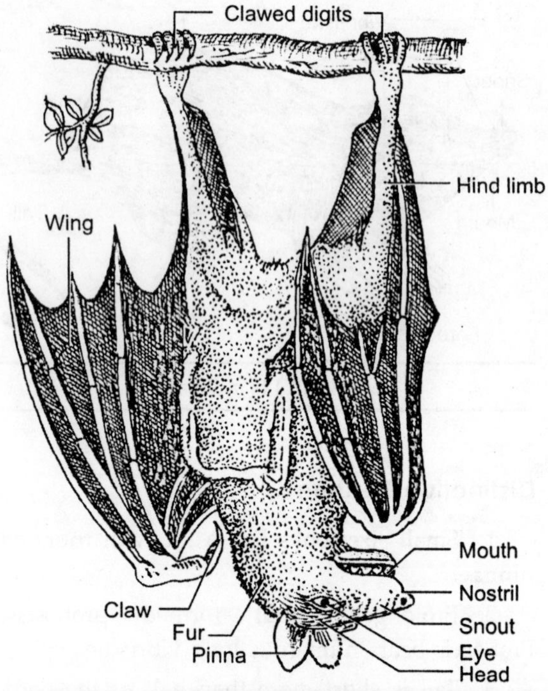

Fig. 1.1.2.113 : *Pteropus* (Megachiroptera)

old abandoned buildings, sleeps during day time hanging head down wards by the help of hind claws: frugivorous: eats various fruits but mainly feed on juicy sweet fruits: young remains cling to mothers body: found in parts of Asia, the Pacific, Australia, Africa, Sri lanka and in India.

Systematic Position

Phylum	**Chordata**
Sub phylum	**Vertebrata**
Super class	**Gnathostomata**
Class	**Mammalia**
Sub class	**Theria**
Infra class	**Eutheria**
Cohort	**Unguiculata**
Order	**Chiroptera [L.Cheir=hand: pteron=wing]**
Sub order	**Megachiroptera**
Type	***Pteropus***

Distinctive Features

1. Body is dark brown coloured and a large size bat.
2. Mainly fruit eaters with flattened grinding teeth.
3. Snout elongated without nose leaf.
4. The first and second digits of the fore limbs bear claw.
5. Molars are marked with longitudinal grooves.
6. Tail is short.
7. Foot with five clawed digits.

Identification with Reasons

State classification and characters up to—

Cohort—*Unguiculata* as in *Suncus*.

Order—*Chiroptera* :

1. Fore limbs are modified to form wings.
2. Fore limb bones and all the finger, except the pollex are elongated to support the patagium.
3. Interfemoral membrane is present between the femurs.
4. Pinna is well developed and complicated foliaceous nasal folds around the nose is many.
5. Dental formula in general is 3.1.4.3/ 3.1.4.3 but undergo modification
6. Testes abdominal.

Sub order—*Megachiroptera* :

1. Large in size and frugivorous.
2. Large eyes with elongated snout.
3. Tail without internal membrane.
4. Pinna is simple and without foliaceous nasal folds.
5. A second digit of the fore limbs is provided with claws.
6. Molars without sharp cusps but are with longitudinal grooves.

Type—*Pteropus* : State distinctive features as in above.

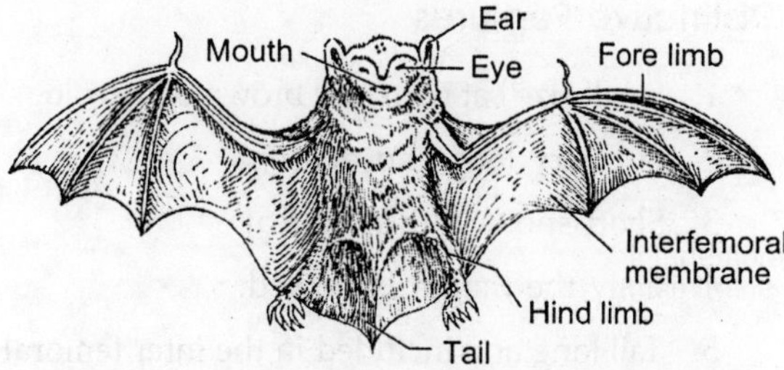

Fig. 1.1.2.114 : *Rhinolophus* (Microchiroptera)

Economic Importance

Faeces of bats are used as fertilizer. Harmful to man kind as it damages to orchards.

Special Features

Bats are only flying mammals. Depended heavily on sight and smell and do not echolocate.

4. *RHINOLOPHUS* (Commonly called as Horse-Shoe Bat)

Habit and Habitat

Gregarious: some solitary: nocturnal: found over water, near abandoned houses, warehouses and caves: hang upside down by their hind limb claws: mostly insectivorous but may be carnivorous: catch insects by echolocation: found all over the world, especially in North-East Africa, Asia and India.

Systematic Position

Phylum	**Chordata**
Sub phylum	**Vertebrata**
Super class	**Gnathostomata**
Class	**Mammalia**
Sub class	**Theria**
Infra class	**Eutheria**
Cohort	**Unguiculata**
Order	**Chiroptera**
Suborder	**Microchiroptera**
Type	***Rhinolophus***

Distinctive Features

1. Small size bat with dull brown body colour.
2. Fur very long and dense.
3. Short snout with nose leaf.
4. Only the thumb is clawed.
5. Tail long and included in the inter femoral membrane.
6. Molar with cusps and with transverse grooves.

Identification with Reasons

State classification and characters up to—

Order—*Chiroptera* as in *Pteropus.*

Sub order—*Microchiroptera* :

1. Insectivorous bats and small size.
2. First digit of the fore limb is clawed.
3. Tail when present is enclosed in the interfemoral membrane.
4. Large pinna with extra lobe inside called **tragus** and auricular appendage is present.
5. Eye small and snout is short.
6. Molars are with sharp cusps and transverse grooves.

Type—*Rhinolophus* : State distinctive features as in above.

Economic Importance

Useful as insect destroyer. Faecal matters are used as valuable fertilizer.

Special Features

Microchiropterans have **echolocation** apparatus. While flying they constantly send out ultrasonic sound waves, which strike on the object or wire reflected back to bat. Sonic sounds are emitted through the nose or mouth and received by the ear.

5. *MANIS* (Commonly called as Scaly ant Eaters or Pangolin)

Habit and Habitat

Nocturnal: powerful diggers: found in the plain and lower slopes of hill of India, Sri Lanka, Assam, Eastern Himalayas, Myanmar, South China and South Eastern Asia: feeds on ants and termites.

Systematic Position

Phylum	**Chordata**
Sub phylum	**Vertebrata**
Super class	**Gnathostomata**
Class	**Mammalia**

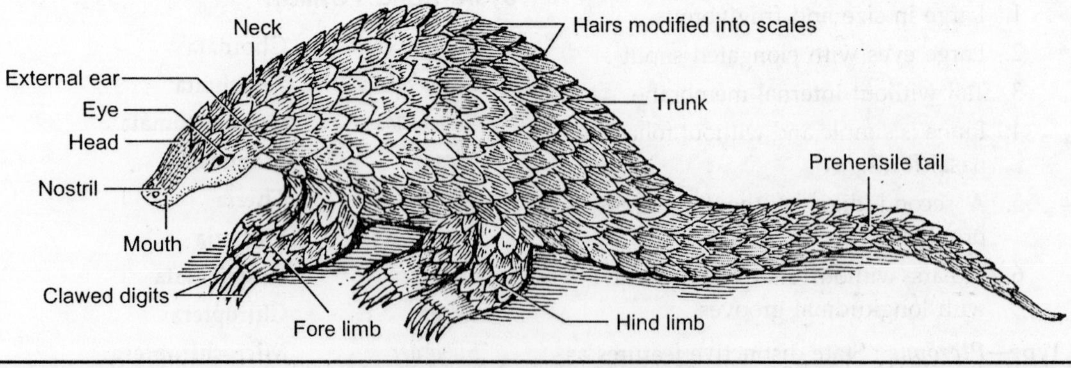

Fig. 1.1.2.115 : *Manis* (Pangolin)

Sub class	**Theria**
Infra class	**Eutheria**
Cohort	**Unguiculata**
Order	**Pholidota** (G. Pholis = scale)
Type	***Manis***

Distinctive Features

1. Large in size, measured about ninety centimetre.
2. Body is divided into head, neck, body and tail.
3. On the dorsal surface of head, body and tail horny scales are present in imbricate fashion but ventral surface covered with hairs.
4. Eyes small and pinnae are ill developed.
5. Tongue is long, sticky and protrusible.
6. Teeth are absent.
7. Limbs short but powerful. Move slowly with arched back and the fore limb and anterior surface in contact with the ground.

Identification with Reasons

State classification and characters up to—

Cohort—*Unguiculata* as in *Suncus*.

Order—Pholidota :

1. On the dorsal surface of head, body and tail horny scales are present in an imbricated fashion.
2. Eyes are small and pinna is ill developed.
3. The tail is long and tapering.
4. Limbs are short, powerful and five digits are provided with curved, sharp claws.
5. Tongue is long, sticky and protrusible.
6. Teeth are absent.
7. Facial part is prolonged to form a short muzzle.

Special Features

When disturbed it rolls its body into a ball. In its pyloric stomach, small pebbles or stones are found and hence is it called as **stone eater** or **bajrakit**.

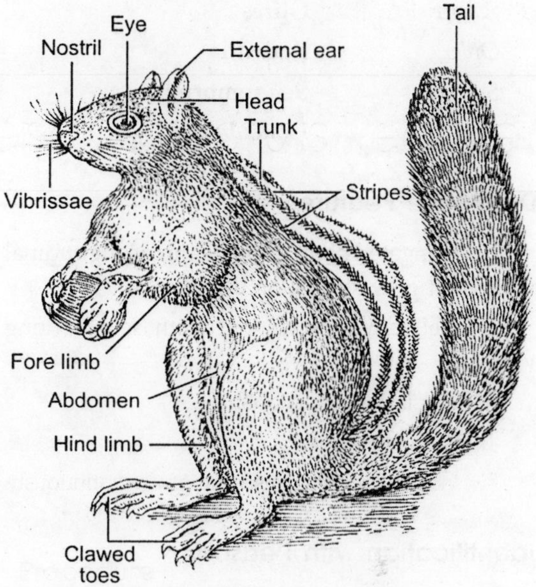

Fig. 1.1.2.116 : *Funambulus*

6. *FUNAMBULUS* (Commonly called as Squirrel)

Habit and Habitat

Lives on trees as well as on grounds: adapted for arboreal life: fast runner: active at day time: herbivorous: feed on fruits, seeds and nuts: world wide distribution: found on all continents and islands.

Systematic Position

Phylum	**Chordata**
Sub phylum	**Vertebrata**
Super class	**Gnathostomata**
Class	**Mammalia**
Sub class	**Theria**
Infra class	**Eutheria**

Cohort	**Glires**
Order	**Rodentia**
Sub order	**Sciuromorpha**
Type	***Funambulus***

Distinctive Features

1. Elongated body contains three longitudinal white and grey stripes on dorsal side.
2. Well developed snout with several long vibrissae.
3. Tail is long and bushy.
4. Eyes are large.
5. Incisors exposed, chisel like, grow continuously.

Identification with Reasons

State classification and characters up to—

Infra class—*Eutheria* as in *Suncus*.

Cohort—*Glires* :

1. Gnawing teeth mammals.

Order—*Rodentia* :

1. Body is covered with soft hairs.
2. Eyes large or small.
3. Pinna is well developed.
4. Fore limbs are smaller than hind limbs and with blunt claws.
5. Single pair of upper incisors which are chisel like.
6. Presence of a gap between the incisors and the cheek teeth called **diastema.**
7. Tail is short or long.
8. Canine and anterior premolars are absent.
9. Testes are abdominal or inguinal in position.

Sub order—*Sciuromorpha* :

1. Two upper premolars and one lower premolars are present.
2. Infraorbital foramen is not elongated.

Type—*Funambulus* : State distinctive features as in above.

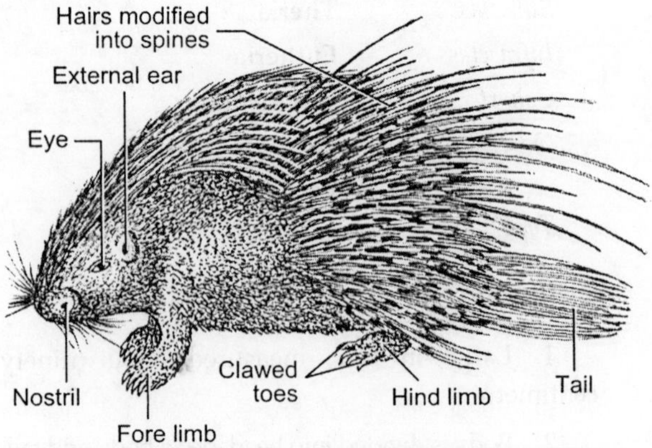

Fig. 1.1.2.117 : *Hystrix* (Porcupine)

7. *HYSTRIX* (Commonly called as Porcupine)

Habit and Habitat

Burrower; inhabits river banks, nallahs and bunds: nocturnal: comes only after sunset: herbivorous: live on plant roots, bulbs and fallen fruits; found in Asia and Africa.

Systematic Position

Phylum	**Chordata**
Sub phylum	**Vertebrata**
Super class	**Gnathostomata**
Class	**Mammalia**
Sub class	**Theria**
Infra class	**Eutheria**
Cohort	**Glires**
Order	**Rodentia**
Sub order	**Hystricomorpha**
Type	***Hystrix***

Distinctive Features

1. Body of the animal measured about ninety centemetre in length.

2. Hairs on the body are modified to form black and white quills on the body and tail and can be erected like spine by muscles.

3. Snout is covered with short and stiff bristles.

4. Head contains crest of black bristles.

Identification with Reasons

State classification and characters up to—

Order—*Rodentia* as in *Funambulus*.

Sub order—Hystricomorpha :

1. Infra orbital is elongated.
2. In some forms hairs are modified to form quills.
3. Body quills are solid and tail quills hollow.

Type—*Hystrix* : State distinctive features as in above.

Economic Importance

Hystrix is harmful to crop of potato, carrots and cabbage. Dangerous to dog. Attack its enemies by erecting its spines which penetrates into the body of enemy.

8. *CAVIA* (Commonly called as Guineapig)

Habit and Habitat

Terrestrial; lead a subterranean life; active during day time: herbivorous; **coprophagous** taking their soft nocturnal faeces; distribution world wide.

Systematic Position

Phylum	**Chordata**
Sub phylum	**Vertebrata**
Super class	**Gnathostomata**
Class	**Mammalia**
Sub class	**Theria**

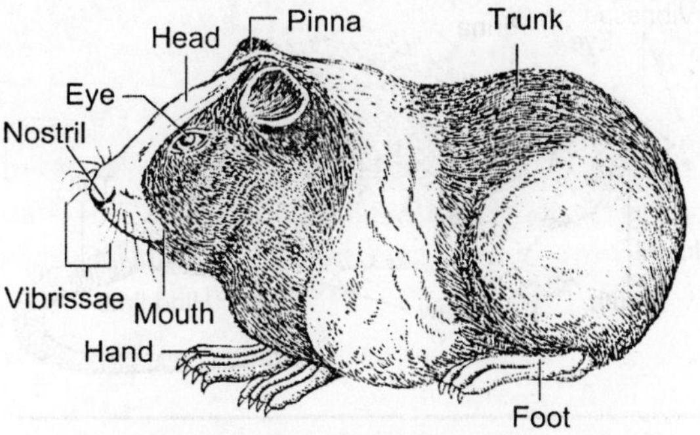

Fig. 1.1.2.118 : *Cavia* (Guineapig)

Infra class	**Eutheria**
Cohort	**Glires**
Order	**Rodentia**
Sub order	**Hystricomorpha**
Type	***Cavia porcellus***

Distinctive Features

1. Body is covered with soft hairs.

2. On the side of the snout vibrissae present.

3. Tail shortened and only tail bone is perceptable.

4. Upper lip has a cleft in the middle.

5. Round and large eyes are present.

6. Fore limb with four digits and hind limb with three digits.

7. Abdominal teats or nipple in female.

Identification with Reasons

State classification and characters up to—

Sub order—*Hystricomropha* as in *Histrix*.

Type—*Cavia porcellus* : State distinctive features as in above.

Remark

Guineapigs are much used in the laboratory as a dissecting animals and for various experiments.

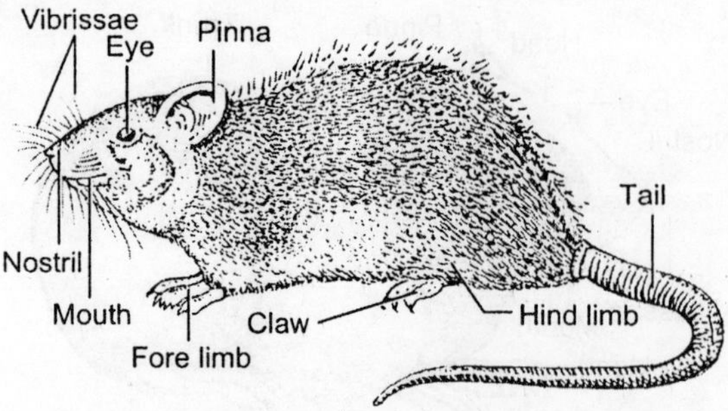

Fig. 1.1.2.119 : *Rattus*

9. *RATTUS* (Commonly called as Black Rats)

Habit and Habitat

Terrestrial: subterrestrial: arboreal: live in the ground burrows, on the trees holes , in the dwelling house and in cultivated fields: omnivorous : feed on variety of plant and animal substances and stored grains: found all parts of the world: prefers warmer and dried conditions.

Systematic Position

Phylum	**Chordata**
Sub phylum	**Vertebrata**
Super class	**Gnathostomata**
Class	**Mammalia**
Sub class	**Theria**
Infra class	**Eutheria**
Cohort	**Glires**
Order	**Rodentia**
Sub order	**Myomorpha**
Type	***Rattus***

Distinctive Features

1. Head and body are covered with dull grey soft hairs.

2. Tail is elongated, tapering and covered by scales.

3. Snout long with well developed vibrissae.

4. Teats five to six pairs, two to third pairs inguinal and three pairs pectoral.

Identification with Reasons

State classification and characters up to—

Order—*Rodentia* as in *Funambulus*.

Sub order—*Myomorpha* :

1. Infra orbital opening is large.
2. A great elongation of the metatarsal of the three central digits.

Type—*Rattus* : State the distinctive features as in above.

Economic Importance

Rats are great menace to mankind. They may be considered as parasites of man. Rat destroy crop and stored grains. They spoil more than they eat. Rats act as carrier of several disease-germs viz. cholera, typhus, plague, rat-bite fever. Rat is largely used in various biophysical and biochemical studies.

10. *BANDICOTA* (Commonly called as Lesser Bandicoot)

Habitat and Habitat

Terrestrial: fossorial: nocturnal: live in the burrows on the dykes of the field: omnivorous: feed on creales vegetables, stored grains and other agricultural products: when food became scares they take shelters in domestic houses: found in India, Myanmar, Bangladesh, Thailand, Sri lanka, Sumatra, Java, Vietnam.

Systematic Position

Phylum	**Chordata**
Sub phylum	**Vertebrata**
Super class	**Gnathostomata**
Class	**Mammalia**

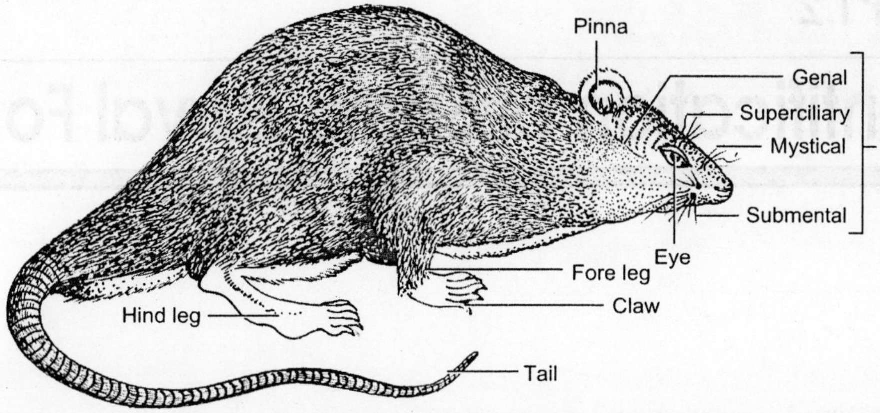

Fig. 1.1.2.120 : *Bandicota*

Sub class	**Theria**
Infra class	**Eutheria**
Cohort	**Glires**
Order	**Rodentia**
Sub order	**Myomorpha**
Type	***Bandicota bengalensis***

Distinctive Features

1. Body covered by coarse hairs.
2. Slightly pointed snout, vibrissal long, stiff and many in number.
3. Pinna semicircular.
4. Tail longer then the trunk and covered by scales.
5. Mammary glands eight pairs.

Identification with Reasons

State the classification and characters up to—

Suborder—*Myomorpha* as in *Rattus.*

Type—*Bandicota bengalensis* : State distinctive features as in above.

Economic Importance

Bandicota is regarded as pest. They destroy stored grains and eat domestic food. It spreads Plague and typhus fever.

Chapter 1.2

Identification of Some Larval Forms

External fertilization usually leads into a free living, swimming larva, that grows and differentiates in the water. Larval forms are also found in species that have internal fertilization. In indirect development there is a larval stage, which has little resemblance with the adult of their own species, and acquiring adult form through metamorphosis. The larva of most invertebrate species are ciliated. The cilia helps in locomotion and food collection. The larva can be defined as, **a free living developmental stage in the life history of many invertebrates that has little resemblance with the adults of their own species** and acquiring adults form through **metamorphosis.**

1. *EPHYRA LARVA*

Characters

1. It is a medusoid stage with tetramerous symmetry.

2. It is umbrella shaped with eight bifid long arms four perradial and four interradial.

3. Each arm contains two marginal lappets and short tentaculocyst.

4. Entric cavity opens into mouth which is four sided and possesses parradial, interradial and adradial canals.

5. Ephyra larva develops into adults Aurelia.

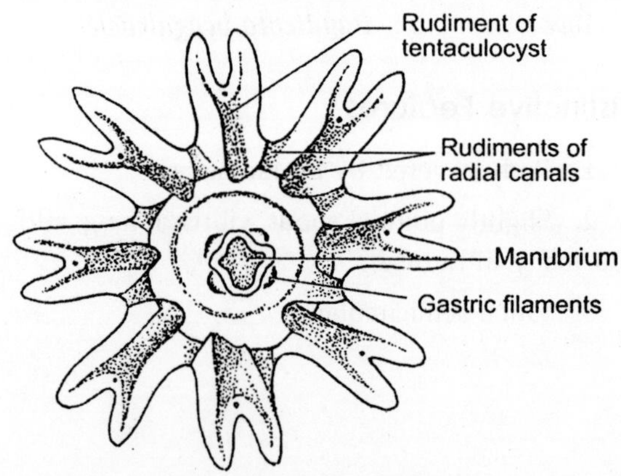

Fig. 1.2.1 : *Ephyra larva.*

2. *MIRACIDIUM LARVA*

Characters

1. Microscopic, conical, dorsoventrally flattened and covered with ciliated epidermal plate.

2. The broad anterior end contains apical papilla at the base of which an apical gland presents.

3. Body wall consists of muscle layers, plate like glandular epithelium and fluid filled mesenchyme cells.

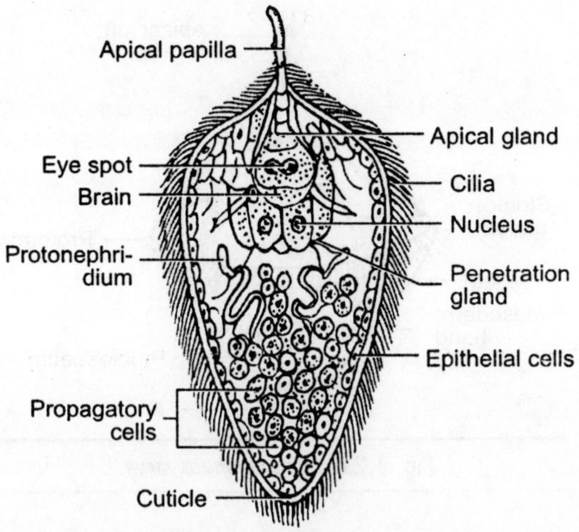

Fig. 1.2.2 : *Miracidium larva.*

4. Internally penetration glands, a pair of eye spots, brain, mass of propagatory cells and a pair of flame cells called as protonerphridia are present.

5. It changes into next or second-stage larva sporocyst of *Fasciola hepatica.*

3. *SPOROCYST LARVA*

Characters

1. The body is rounded, sac like structure having germ balls and four flame cells.

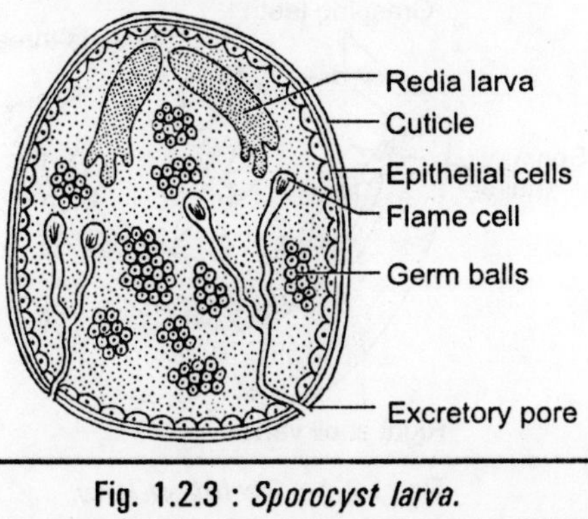

Fig. 1.2.3 : *Sporocyst larva.*

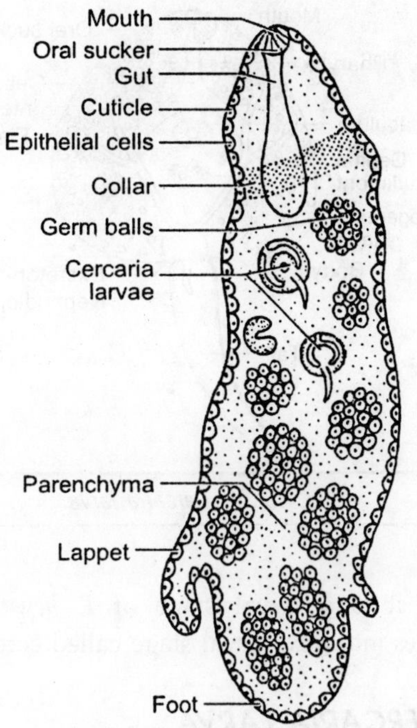

Fig. 1.2.4 : *Redia larva.*

2. The body wall layers are cuticle, epithelial cells, muscles layers and mesenchyme.

3. It is a non feeding stage larva of *F.hepatica* and found in the body of snail.

4. It changes with next stage larva *redia.*

5. It is the sporocyst larva of *F.hepatica.*

4. *REDIA LARVA*

Characters

1. The body is elongated and cylindrical and anteriorly, there is a muscular ring like collar and birth pore.

2. Posterioly, larva contains two lappets near foot.

3. The body wall is composed of cuticle, epithelial layer and delicate mesenchyme.

4. Larva contains mouth, pharynx and elongated gut.

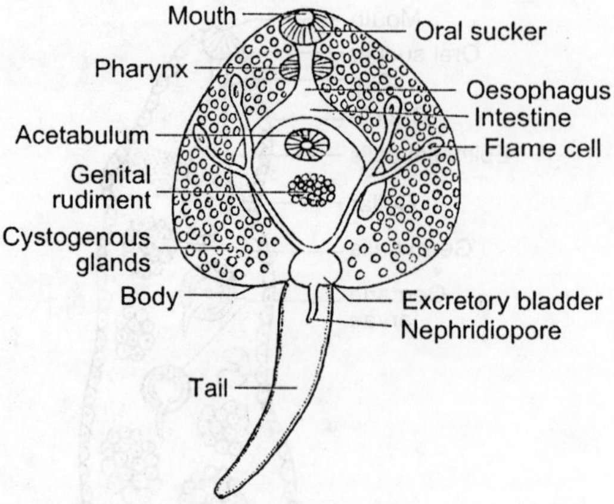

Fig. 1.2.5 : *Cercaria larva.*

5. It is a larval stage of *F. hepatica* and changes into next larval stage called cercaria.

5. *CERCARIA LARVA*

Characters

1. Oval shaped body with a long simple tail which is covered with cuticular spines.

2. Body wall is composed of cuticle, circular, longitudinal and diagonal muscle fibres and mesenchyme.

3. Beneath the muscle layer are numerous unicellular **cystogenous glands.**

4. Digestive system consists of mouth, oral suckers, pharynx, oesophagus and forked intestine.

5. Acetabulum is just below the forked intestine.

6. It is a free swimming larva of *F.hepatica.*

6. *TROCHOPHORE LARVA*

Characters

1. Body is transparent, oval or pear shaped larva with broad anterior and narrow posterior end.

2. In the broader anterior end an apical sensory plate with a small tuft of cilia is present.

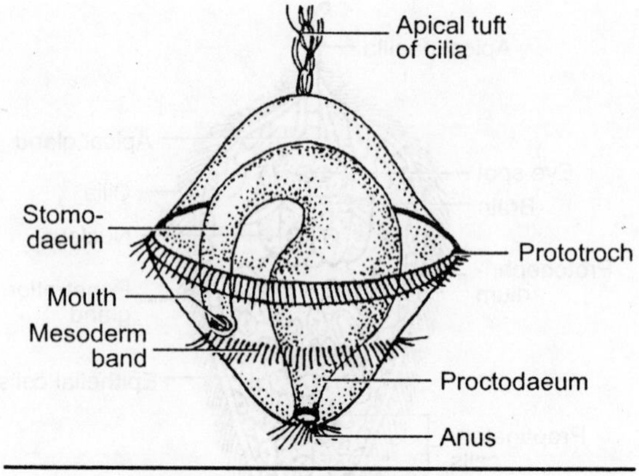

Fig. 1.2.6 : *Trochophore larva.*

3. Mouth is on the lateral side of stomodium, which leads into a sac like stomach, a short intestine.

4. Intestine opens through anus at the terminal part of narrow posterior end.

5. Around the body at the equatorial plane, a double circlet of cilia, the **prototroch** is present.

6. Large blastocoel and a pair of protonephridia are present.

7. It is the larva of polychaeta.

7. *GLOCHIDIUM LARVA*

Characters

1. Microscopic bivalved larva consists of conical shell containing curved hook bearing spines or teeth ventrally and mantle.

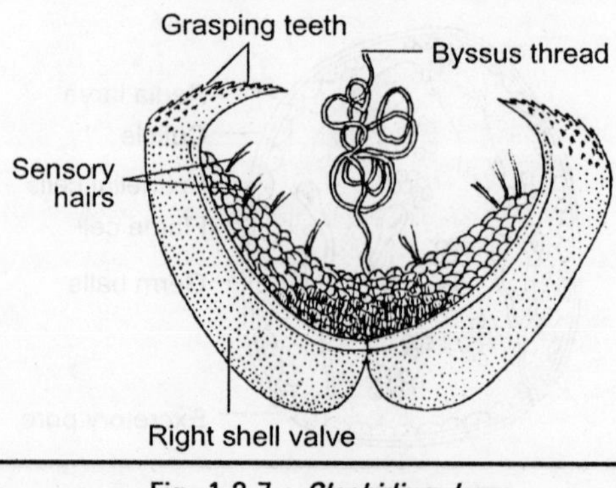

Fig. 1.2.7 : *Glochidium larva.*

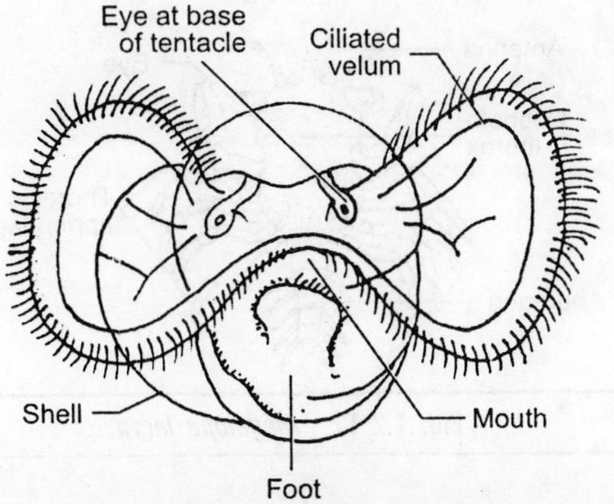

Fig. 1.2.8 : *Veliger larva.*

2. Shell consists of two triangular valves united dorsally and free ventrally.

3. Shell valves operated by large adductor muscle situated at the base.

4. Small mantle lobes bear brush-like sensory hairs.

5. **Byssus gland** present near adductive muscles which gives rise to a long filamentous sticky byssus thread.

6. Glochidium is the larva of *Unio* and leads an ectoparasite life, attaches to fish skin.

8. *VELIGER LARVA*

Characters

1. It is a modified **trochosphere** or **trochophore.**

2. Presence of ciliated organ, curved gut and larval nephridia.

3. Shell assumes spiral shape due to unequal growth.

4. The **protroch** of trochosphere expands to form a ciliated disc called velum.

5. Velum is bilobed.

6. Foot is ventral and between the mouth and anus.

7. It is the larva of snail and calm.

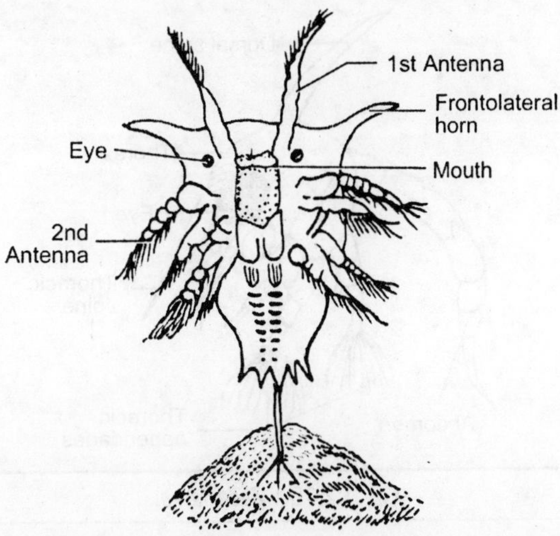

Fig. 1.2.9 : *Nauplius*

9. *NAUPLIUS LARVA*

Characters

1. Minute, conical and microscopic with board anterior and narrow posterior end.

2. Body is oval and divided into indistinct head, trunk and anal region.

3. It contains single median eye and frontal and lateral horns.

4. Three pairs of appendages namely uniramous antennules, biramous antennae and biramous mandibles.

5. Larva contains mouth and gut.

6. It is the larva of crustacea.

10. *ZOEA LARVA*

Characters

1. Body is differentiated into unsegmented cephalothorax and long abdomen.

2. Cephalothorax produced into a large median dorsal spine, a venteral thoracic spine and two small lateral spines.

3. Six pairs of thoracic appendages develop as buds.

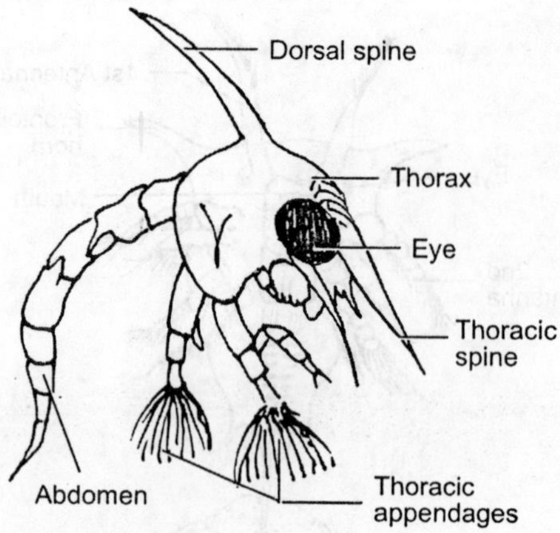

Fig. 1.2.10 : *Zoea larva.*

4. Long and slender six segmented abdomen contains a biforked **telson**.

5. The head contains large compound eyes, antennules, antennae, mandible, maxillae and well developed maxillipeds.

6. It is the larva of crab.

11. *MYSIS* LARVA

Characters

1. Elongated, transparent body is differentiated into cephalothorax and six-segmented abdomen with a telson.

2. Biramous thoracic appendages are present.

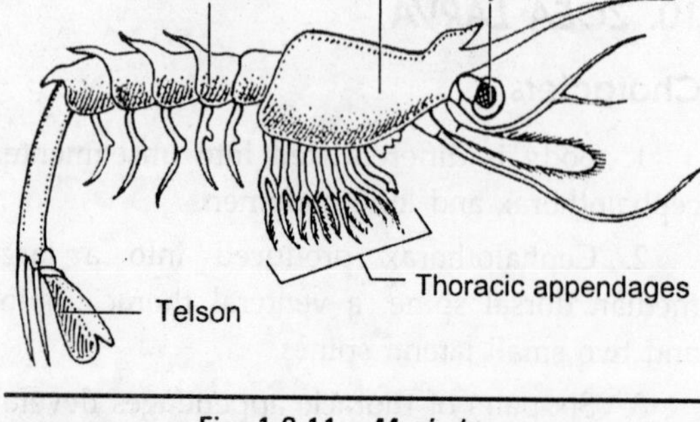

Fig. 1.2.11 : *Mysis larva.*

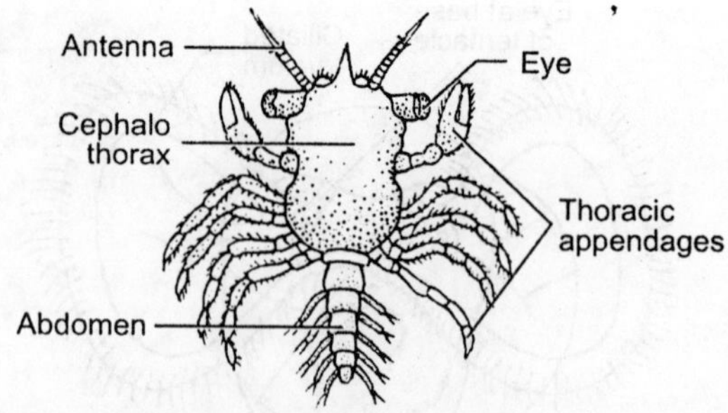

Fig. 1.2.12 : *Megalopa larva.*

3. Presence of a pair of stalked compound eyes on the head.

4. Parts of the thorax is covered by a carapace.

5. It is the larva of *Penaeus.*

12. *MEGALOPA* LARVA

Characters

1. The body consists of crab-like cephalothorax and a large abdomen.

2. Abdomen is well-developed and possesses biramous pleopods.

3. The abdomen has a pair of telsons.

4. Five pairs of thoracic appendages are well-developed.

5. Large stalked eyes are distinctly visible.

6. It is the larval stage of crabs.

13. *BIPINNARIA* LARVA

Characters

1. Transparent, bilateral symmetrical, elongated body possesses two ciliated bands, the pre-oral and post-oral.

2. The pre-oral loop encircles around mouth while post-oral ciliated band forms a complete ring between the mouth and anus.

3. Body has series of prolongation called arms which are provided with muscles.

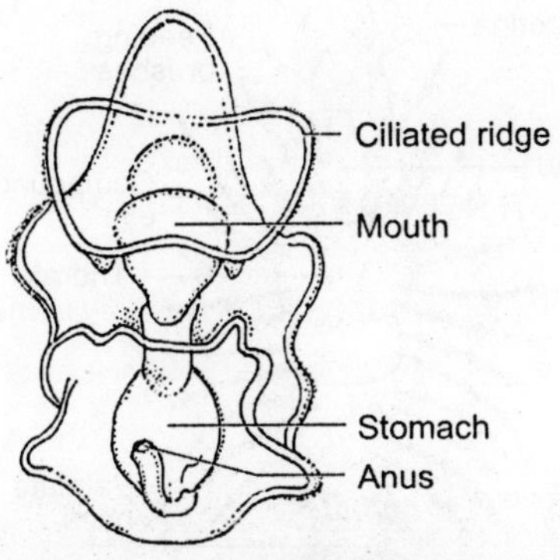

Fig. 1.2.13 : *Bipinnaria larva.*

4. Larval mouth, alimentary canal and anus well developed.

5. Bipinnaria is the free swimming larval stage of *Asterias sp.*

14. *PLUTEUS*

Characters

1. Transparent body with long, five to six pairs arms supported by calcareous rods with ciliated bands at the margin.
2. Pre-oral lobe small.
3. Post-anal part well developed.
4. Two types of Pluteus larvae are present: *Echinopluteus* and *Ophiopluteus.*

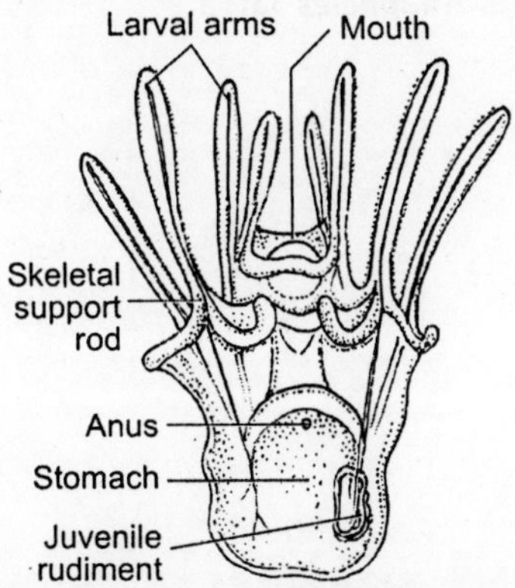

Fig. 1.2.14 : *Echinopluteus larva.*

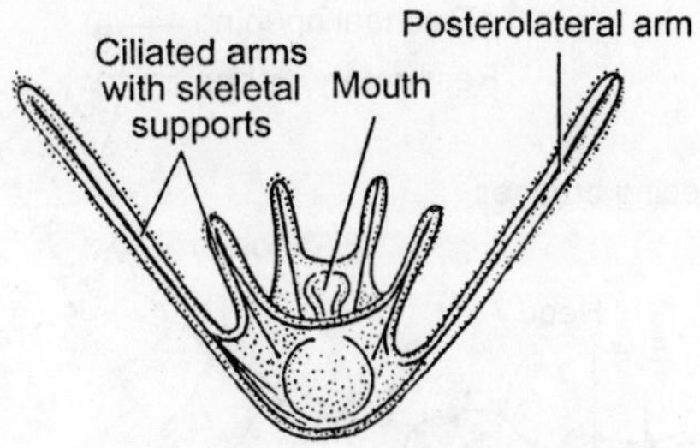

Fig. 1.2.15 : *Ophiopluteus larva.*

A. Echinopluteus

Characters :

1. Arms five to six pairs, pigmented and supported by calcareous rods.
2. Postero-lateral arms are very short and projected backward.
3. Pre-oral and antero-dorsal arms present,
4. Mouth and anus well developed.
5. Echinopluteus is the free swimming larval forms of **Echinoidia.**

B. Ophiopluteus

Characters :

1. Arms four pairs, slender and supported by calcareous rods.
2. Postero-lateral arms largest and directed forwards giving larva V-shaped appearance.
3. Ciliated bands are at the edge of arms.
4. Mouth and anus well developed.
5. Ophiopluteus is free swimming larval forms of **Ophiuroidea.**

15. *CULEX LARVA*

Characters

1. Transparent larva differentiated into head, thorax and abdomen.
2. Head contains compound eyes, simple antennae, mandibles, maxillae and feeding brushes.
3. Broad unsegmented thorax contains tufts of hairs.

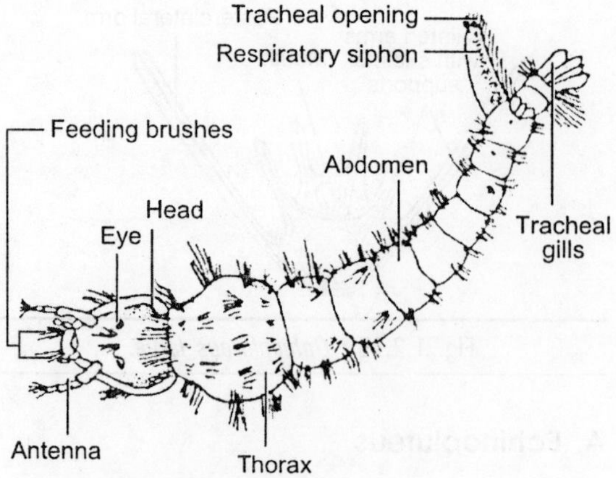

Fig. 1.2.16 : *Culex larva.*

4. Abdomen contains nine segments without palmate hairs.

5. Eight abdominal segment has a long **respiratory siphon**.

6. Last abdominal segment with **tracheal gill**.

7. Head hangs downwards at an angle with the water surface during respiration.

8. It is *Culex larva.*

16. *ANOPHELES LARVA*

Characters

1. Transparent larva is differentiated into head, thorax and abdomen.

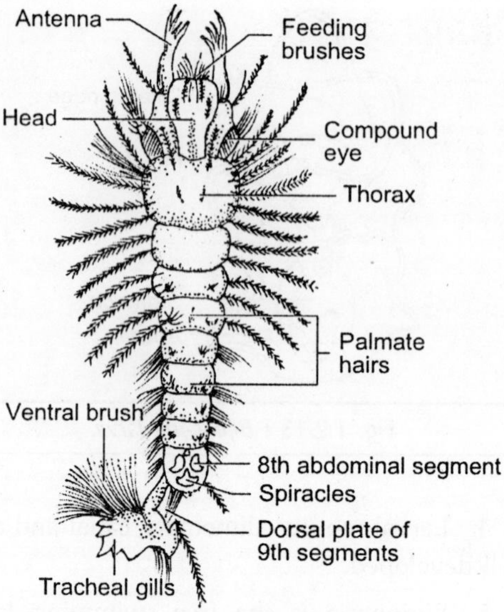

Fig. 1.2.17 : *Anopheles larva.*

2. Head contains compound eyes, simple antennae, mandibles, maxillae and feeding brush.

3. Broad unsegmented thorax contains tufts of hairs.

4. Abdomen contains nine segments with palmate hairs.

5. Last abdominal segments contain tracheal gills with ventral brush.

6. Head remain parallel with the water surface during respiration.

7. It is ***Anopheles* larva.**

Chapter 1.3

Identification of life history stages of some species

Sexual reproduction in the metazoa usually involves the fusion of gametes, the sperm and the ovum. The study of eggs and embryos produced by sexual reproduction reveals many different patterns of development and life cycle. The difference between direct and indirect development is a familiar features of life histories. In direct development the adult stage is attained by progressive growth and differentiation, whereas in indirect development there is a larval stage, differing both in its structure and its habit from the adult, and attaining adult from through a radical and sometimes sudden metamorphosis. Thus life histories are composed of a series of stages of progressive differentiation.

1. *CULEX* MOSQUITO

See chapter 1.1.1. Non chordate: Arthropod, item no. 39 and also chapter 1.5. item no. 16 and 17.

2. *ANOPHELES* MOSQUITO

See chapter 1.1.1. Non chordate: Arthropod, item no. 40 and also chapter 1.5., item no. 16 and 17.

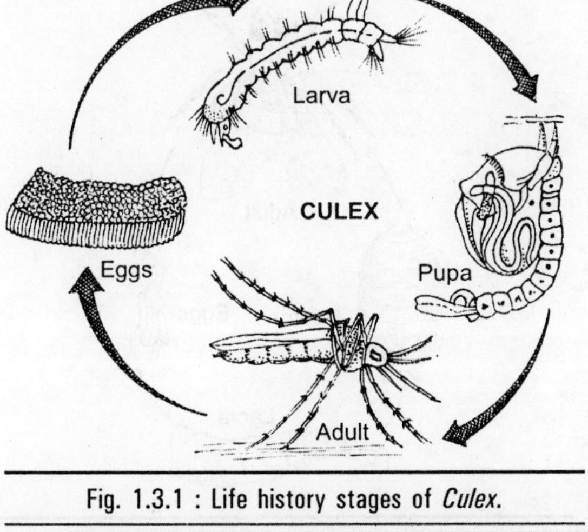

Fig. 1.3.1 : Life history stages of *Culex*.

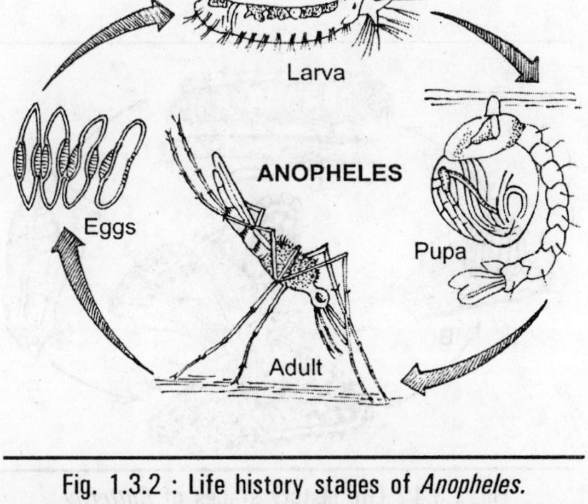

Fig. 1.3.2 : Life history stages of *Anopheles*.

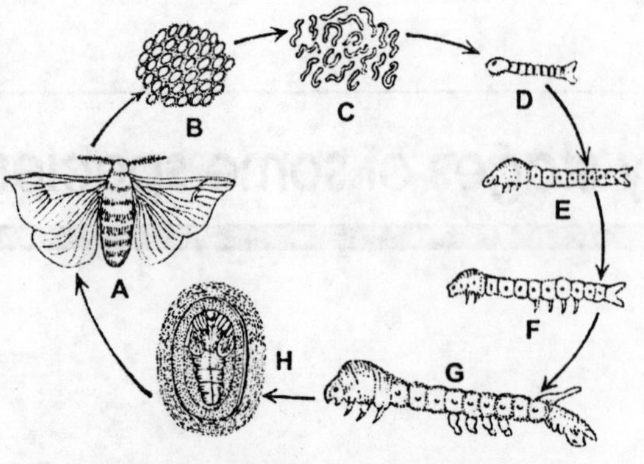

Fig. 1.3.3 : Life history stages of *Bombyx*.
A. Adult moth, B. Eggs, C. Newly born caterpillers. D-G. Different stages of larva, H. Pupa inside cocoon (Part of the cocoon is removed.).

3. *BOMBYX* (SILK MOTH)

See chapter 1.1.1. Non chordate: Arthropod, item no. 44, 45 and 46.

4. *BUTTERFLY*

See chapter 1.1.1. Non chordate: Arthropod, item no 49.

Larva :

1. Soft body is provided with thick hairy covering.
2. The larvae are called **caterpillars**.
3. Presence of well- defined segmentation.
4. Powerful mandibles.
5. Three pairs of jointed thoracic legs and four to five pairs of unjointed abdominal legs.
6. Less developed antennae.

Pupa :

1. It is resting inactive instar, incapable of feeding.
2. Presence of non-articulated mandible.
3. Appendages of pupa are firmly pressed against its body.
4. Pupa is called as **chrysalis** and is suspended by its caudal extremity by a slender stalk.
5. Pupa construct cocoon by silk.

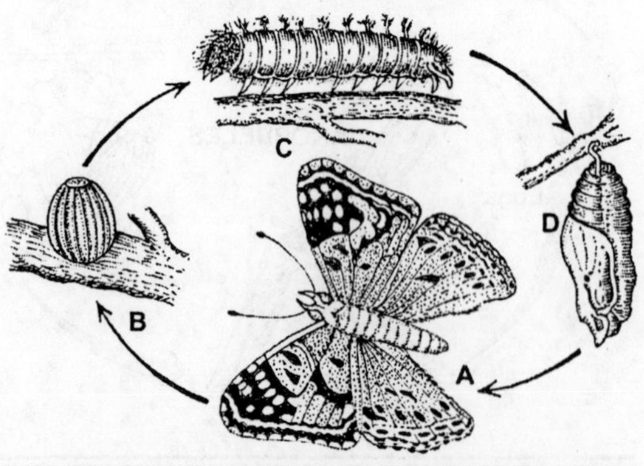

Fig. 1.3.4 : Life history stages of *Butterfly*.
A. Adult moth, B. Eggs, C. Caterpillar, D. Pupa.

5. *HOUSE FLY*

Habit and Habitat

Day time visitors of human habitation, favouring residing and breeding places are open drains, garbages, kitchen and pantries; sits over uncovered foods: carrying germs of various diseases.

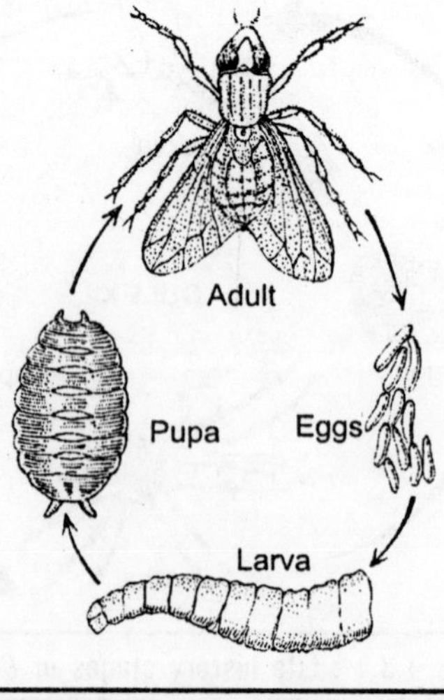

Fig. 1.3.5 : Life history stages of *Housefly*.

Systematic Position

Phylum	**Arthropoda**
Sub Phylum	**Mandibulata**
Class	**Insecta**
Subclass	**Petrygota**
Super order	**Holometabola**
Order	**Diptera**
Type	*Musca*

Distinctive Features

1. Head small and semicircular in outline.
2. Paired antennae are short and flexible, the distal segment largest and contains a brush like **arista.**
3. Spongin type well-developed mouth parts.
4. Mandible absent.
5. Numerous hairs and bristles present on legs surface.
6. Curved claws on the terminal end of each leg.
7. Anterior pair of wing is broad, transparent, hairy and triangular.
8. Abdomen oval, hairy and segmented.

Identification with Reasons

State classification and characters up to—

Order—**Diptera** as in *Culex* in chapter 1.1.1.Non chordate: Arthropod. Item no 39.

Type—**Musca** : State distinctive features as in above.

Larva :

1. The larva is called **maggot**, white and 2mm in length.
2. The anterior end is narrower than the posterior end.
3. Larvae are apodus and no sign of head.
4. Body is divisible into thirteen segments; the anterior rudimentary segment contains two oral lobes.
5. Mouth is in between the two oral lobes.
6. A hook like mandibular sclerite is on the anterior end of the mouth.
7. Sixth to seventh segments possess seminiferous pads on its ventral side.

Pupa :

1. It is resting inactive instar, incapable of feeding.
2. Cocoon is wanting and the pupa is enclosed in a puparium.
3. Spiracles projected out wards from the hard covering.

6. *RANA* (FROG)

See chapter 1.1.2. Chordate: Amphibia, item no 18 and 19.

Chapter 1.4

Major and Minor Dissection of Invertebrates and Vertebrates

Dissection constitutes very significant and important practical work. The object of dissection of animals in the laboratory is to make students familiar with structures studied in the theoretical classes. Therefore the significance of the study is to learn the various internal structures and their correlation with body.

BASIS OF DISSECTIONS

- While dissecting the animal, a set of instruments required for dissection of specimens and handling materials are necessary.
- All the instruments in the dissecting box should be clean and sharp. Students should always have drawing papers, lead pencils and an eraser.
- One should study about the internal structures of the animal to be dissected before hand.
- Student will be supplied with properly killed or anaesthetised specimens before dissection.
- The animal should be washed before dissection to remove excess of formalin or other fixative, if it is not freshly killed.
- All invertebrates are dissected from the dorsal side since the double ventral nerve cord goes along the mid-ventral region. On the other hand vertebrates are always dissected from the ventral side, because the hard vertebral column is placed along the mid dorsal line.
- The animal is fixed in the dissecting tray properly with pins and and the pins are inserted obliquely to avoid obstruction in the handling of the instruments while dissecting. The tray is submerged with clean water and incision is given along the structure. While opening the animal one should never make deep incisions. The body wall layers are removed by cutting with scissors in such a manner that all the internal organs remain intact.
- Dissection should be kept submerged in water in the dissecting tray. One should keep small fragmented black paper below the nerves and nerve cord of the dissection. Flag label may be made only when asked for. It can be done by cutting small arrow head flags and pined to the wax, keeping the pointed ends just touching the organ.
- The dissection should be made tidy by removing the unwanted organs and tissues and should be displayed nicely on the tray.
- A good number of materials are processed in the laboratory for

observation. Materials may be macroscopic or microscopic. Macroscopic materials may be left in 90% glycerol and covered with another watch glass placed in an inverted position. In case of microscopic animals, stain or unstained the material is placed at the centre of a glass slide and mounted with glycerine. A piece of paper is placed below the watch glass or glass slide and a clear note on the identification of the object should be written in the paper.

A clean, labelled diagram of the dissection or preparation should be drawn in one's practical note book and date is to be recorded. One should also follow carefully the instructions given by the teacher in the practical class.

Dissections INVERTEBRATES

1. EARTHWORM (*Pheretima posthuma*)

Pheretima is a common Indian earthworm. It is very delicate animal. Place the specimens in a Petri dish and pour 50 per cent alcohol. The specimens are removed after stopping of movements and placed in a large beaker containing water.

The earthworms are put on a dissecting tray, keeping the dorsal surface up wards and the animal is pinned down straight at its two ends in the tray. A longitudinal mid-dorsal incision is made along the entire length of the worm by a pair of fine scissors through the skin above the dossal vessel. The cut should not deep with the scissors to avoid injury to internal organs. The worm is next fixed with few pins aside the loose flaps of the body wall. During this process the skin is held with a pair of fine forceps and the septa is torn with a fine needle to loosen the body wall.

(a) External features :

Before dissecting the earthworm examined the external morphology of the worm. The following parts are observed and drawn separately dorsal and ventral surface of the animal.

(1) **Shape :** Elongated, cylindrical body with pointed anterior and blunt posterior end. There is

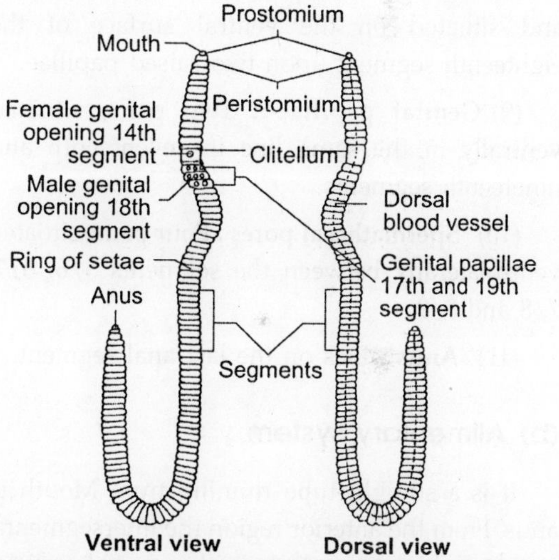

Fig. 1.4.1 : *External features of an earth worm (Dorsal and ventral view) (Pheretima posthuma)*

dark median line running through out the length of the body. This line is due to dorsal blood-vessel.

(2) **Segmentation :** The whole body is divided into a number of ring like grooves or series of segments called **metameres.**

(3) **Number of segments :** The number is about a hundred to hundred – twenty.

(4) **Mouth :** Mouth is small cresentic opening, lies in the first segment, called **peristome** and shifted ventrally by fleshy lobe, called **prostomium.**

(5) **Clitellum :** In the mature worm, in the 14-16 segments, there is a prominent band or girdle, called **clitellum.**

(6) **Setae :** Ventrally, middle of each segment bears a ring of fine rod – shaped **setae** or **chaetae** which can be observed by a hand lens. These are absent in the first and last segments as in the clitellum.

(7) **Female genital pore or opening :** It lies on the ventral surface of the clitellum in the fourteenth segment.

(8) **Male genital aperture :** Two in number and situated on the ventral surface of the eighteenth segment upon two raised papillae.

(9) **Genital papillae :** Two pairs, situated ventrally in the same line in seventeenth and nineteenth segments.

(10) **Spermathecal pores :** Four pairs, situated ventrolaterally between the segments 5/6, 6/7, 7/8 and 8/9.

(11) **Anus :** Lies on the last anal segment.

(b) Alimentary system :

It is a straight tube running from Mouth to anus. From the anterior region the intersegmental septa, musculature along with seminal vesicles are removed to fully exposed the alimentary canal. The following parts are observed :

(1) **Mouth :** Small cresentic opening, lies in the first segment, ventral and posterior to the prostomium.

(2) **Buccal cavity :** It is a sac like structure and posterior to the Mouth.

(3) **Pharynx :** Found in segments three and four, it is a pear shaped structure and lies posterior to the buccal cavity.

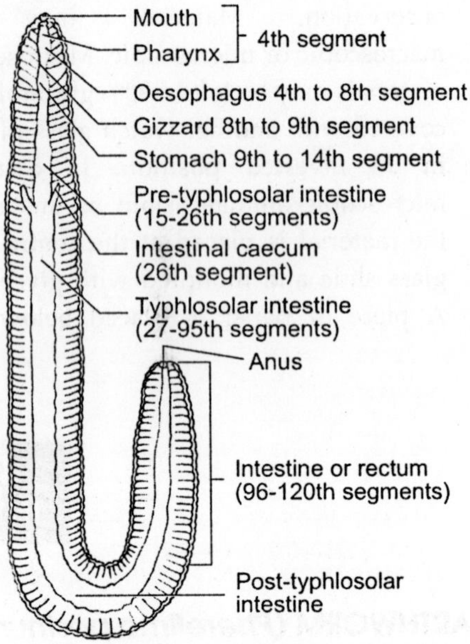

Fig. 1.4.2 : *Digestive system of an earth worm (Pheretima posthuma)*

(4) **Oesophagus :** It is a narrow, tubular portion lies behind the pharynx and extends up to the eighth segments.

(5) **Gizzard :** It is hard, bulbous and thick walled muscular almost round organ, situated in the eighth and nineth segments.

(6) **Stomach :** It is a narrow tube, lies segments nine to fourteenth, glandular and vascular structure.

(7) **Intestine :** It starts from fifteenth segment and continues up to anus. In the twentysixth segments, intestine gives a pair of in the **intestinal caeca.** Intestine divided into three regions:

(a) **Pre-typhlosolar region :** Extends from the fifteenth to nineteenth segments, no median fold.

(b) **Typhlosolar region :** Continues from twentysixth to ninetyfifth segments having **typhlosole** or **internal fold.**

(c) **Post-typhlosolar region :** Continues from ninetyfifth to hundredtwentiath segments. It is called **rectum.**

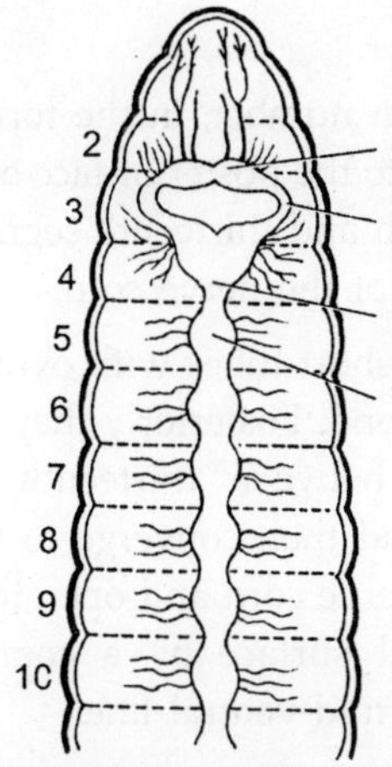

Fig. 1.4.3 : *Nervous system of an earth worm (Pheretima postuma) (Anterior & Dorsal view)*

(c) Nervous System :

The animal is dissected as in case of alimentary system taking care that the nerve ring, which is dorsoventral in orientation, is not damaged. The Oesophagus is cut and the entire alimentary canal is carefully removed by detaching it from the body wall to which it is attached. The nerve rings is found encircling the buccal chamber and the ventral nerve cord is clearkly seen. The buccal chamber is carefully removed leaving nerve ring. Small pieces of the black paper are placed under nerve cord and the following parts are studied.

(1) **Supra pharyngeal or cerebral ganglia :** Bilobed mass, lying dorsal to the pharynx, fused to from the dorsal part of the ring or so: called brain.

(2) **Peri-pharyngeal connective :** From cerebral ganglia two peri pharyngeal connectives pass one on either side of the pharynx down ward and back wards to join the subpharyngeal ganglia.

(3) **Sub-pharyngeal ganglia :** The two subpharyngeal ganglia situated below the pharynx, are fused together as a bilobed mass from which a ventral nerve cord runs back wards.

(4) **Ventral nerve cord :** Found ventral to alimentary tube. It starts from the sub-pharyngeal ganglia and extends upto the posterior end of the body. In each segments it forms a swollen ganglion. Two pairs of small nerves arise from each ganglion.

(d) Reproductive System :

The earthworm is **hermaphrodite** i.e., both male and female reproductive organs are found in the same individual.

The animal is dissected as in the case of alimentary system. The gut is removed from the body wall starting from about twenty five segment and the cut end is held with a pair of forceps in one's left hand and detached with a fine needle from septa connecting the gut, lifting

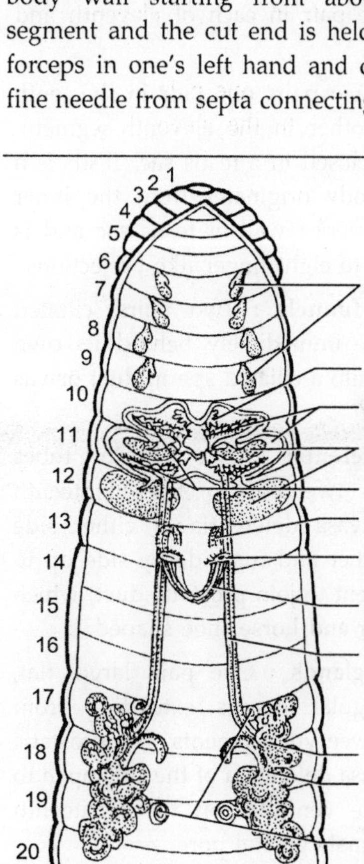

Fig. 1.4.4 : *Reproductive system of an earth worm (Pheretima postuma)*

a little, care have taken from about fifteen segments in removing the gut because ovaries are attached to the posterior face of 12/13 segments below the gut. The gut is removed. First male reproductive system is exposed. The testes sacs are teased to expose enclosed structures. The following parts are observed.

(i) Male system :

(1) **Testis sac :** Two pairs, situated below the oesophagus, one pairs in the tenth and another in the eleventh segment. White or yellow in colour.

(2) **Seminal vesicles :** Two pairs, white or yellowish white situated on both side of oesophagus. One pair in each of eleventh and twelvth segment.

(3) **Testes :** Two pairs one pair in the tenth segment and another in the eleventh segment. Each testis is enclosed in a testis sac. Testis is a minute white body originates from the inner surface of the anterior wall of testis sac and is made up of five to eight finger like projections.

(4) **Seminal funnels :** Two pairs, ciliated structure and lie immediately behind its own testis and open into a ciliated sperm duct or vas deferens.

(5) **Vasa deferentia :** Two slender tubes extending from twelveth to the eighteenth segments. Two vasa deferentia of either side come close together and run side by side up to eighteenth segment to join prostatic duct, which is thick muscular and horse shoe shaped.

(6) **Prostate glands :** One pair, large, flat, white and irregular organs extending from seventeenth to twentyth segments. The prostatic duct with two vasa deferentia of the side open to the anterior side ventrally on the eighteenth segment by the male genital pore.

(7) **Accessory gland :** One pair in the seventeenth and another pair in the nineteenth segment on the ventral body wall, one or each side of the nerve cord. The gland looks like rounded or coiled circles.

(ii) Female system :

(1) **Ovaries :** Two in number, in the form of white masses attached to the posterior face of the septum between twelfth and thirteenth segment, situated on either side of the nerve cord.

(2) **Oviducts :** Two short tubes with oviducal funnels at the anterior end. Posteriorly they pass through the septum between thirteenth and fourteenth segments and then converge to meet each other below the nerve cord and open to the outside on the ventral surface by a common genital aperture in the mid-ventral line.

(e) Septal nephridia :

See chapter 1.5 item no 4.

2. LEECH (*Hirudinaria* sp.)

It is the common Indian cattle leech. It is commonly found in fresh water ponds, ditches, tanks, swamps, lakes and sluggish stream. They are ectoparasite and **sanguivorous**, as they feed on the blood of cattle.

(a) External features :

A freshly killed specimen is taken and the following external features are studied.

(1) **Shape and size :** Slimy, vermiform, elongated and dorsoventrally flattened body. Size ten to fifteen centimetre in length.

(2) **Colour :** The body is greenish brown and paler below than above.

(3) **Segmentation :** Metameres or somites are usually thirty three in number. Each external segment is further divided into five annular rings or annuli.

(4) **Division of the body :** Whole body is divided into six regions:

(a) *Cephalic :* Consists of segment from one to five.

(b) *Preclitellar :* Segments from six to eight.

(c) *Clitellar :* From nine to eleven.

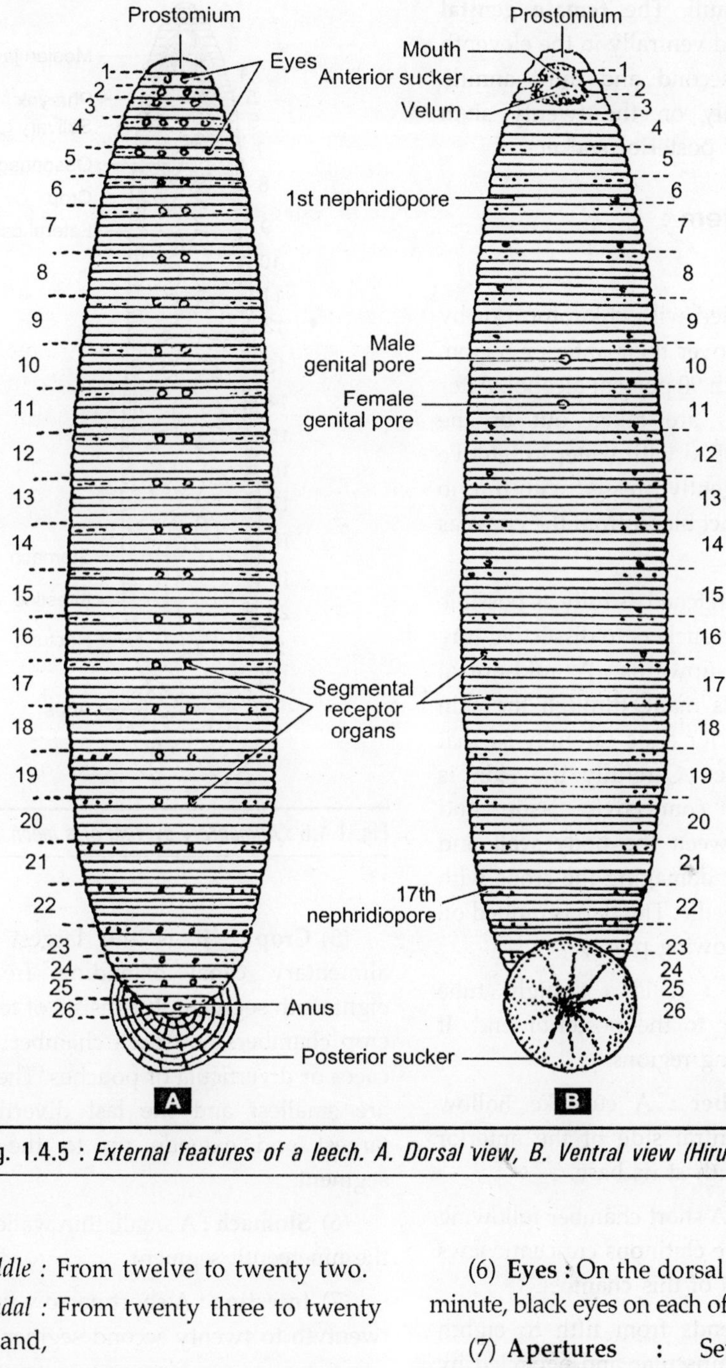

Fig. 1.4.5 : *External features of a leech. A. Dorsal view, B. Ventral view (Hirudinaria sp.)*

(d) *Middle :* From twelve to twenty two.

(e) *Caudal :* From twenty three to twenty six and,

(f) *Posterior sucker :* From twenty seven to thirty three segments.

(5) **Sucker :** Each end of the body contains a sucker namely anterior and posterior sucker.

(6) **Eyes :** On the dorsal side, there is a pair of minute, black eyes on each of the first five segments.

(7) **Apertures :** Seventeen pairs of **nephridiopores**. Segments six to twenty two have each a pair of minute nephridiopores.

The **male genital aperture** is situated mid ventrally in a grove of tenth somite between second and third annuli. The **female genital aperture** is situated mid ventrally in the eleventh somite between the second and third annuli. **Anus** lies mid-dorsally on the twenty sixth segment at the base of posterior sucker.

(b) Alimentary system :

Procedure :

The leeches are killed with chloroform or by pouring common salt over their anterior region. They may be killed with 90 percent alcohol. After killing the specimens are taken out of the container and it is washed with water to remove slim and the body is gently pressed with thumb and first finger to extract blood from the crops as far as possible.

The leech is fully stretched as fully as possible and pinned on both the suckers to dissecting tray with dorsal surface upwards. A mid-dorsal longitudinal incision is made through the skin from anterior to posterior ends carefully so that the gut remains intact. Carefully the skin is separated from the connective tissue and **botryoidal tissue** between the body wall and digestive tube without damaging the crops with a fine arrow headed needle. The flap is pinned on sides and sees the following points.

Alimentary canal : It is a straight tube running from anterior to the posterior end. It consists of the following regions.

(1) **Pre-oral chamber :** A cup-like hollow depression on the ventral side of the anterior sucker containing mouth at its base.

(2) **Buccal cavity :** A short chamber following the mouth. Three white chitinous crescentic jaws are present in the wall of this chamber.

(3) **Pharynx :** Extends from fifth to eighth segment. It is highly muscular and encircled by salivary glands.

(4) **Oesophagus :** A very short, narrow tube leads into crop.

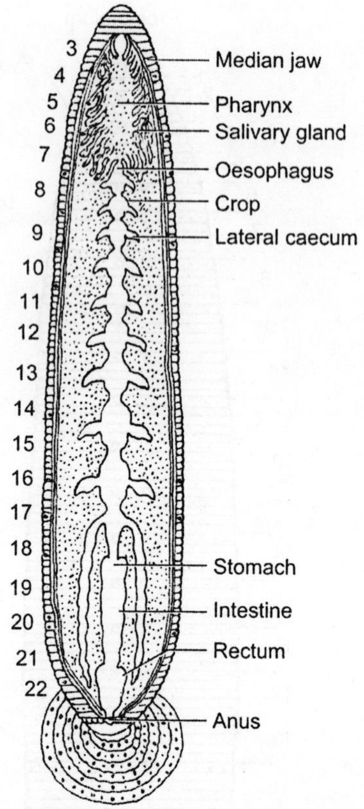

Fig. 1.4.6 : *Alimentary system of a leech. (Hirudinaria sp.)*

(5) **Crop :** It is the largest part of the alimentary canal extending from ninth to eighteenth segment. It consists of ten thin-walled crop chambers and each chamber has a pair of caeca or diverticula or pouches. The anterior pair are smallest and the last diverticula are the largest and extende up to the twenty two segment.

(6) **Stomach :** A small, thin walled chamber in the nineteenth segment.

(7) **Intestine :** A short narrow tube runs from twentyth to twenty second segment.

(8) **Rectum :** The last part of the gut narrows down posteriorly and runs from twenty second to twenty sixth segment. It opens to the out side by the anus.

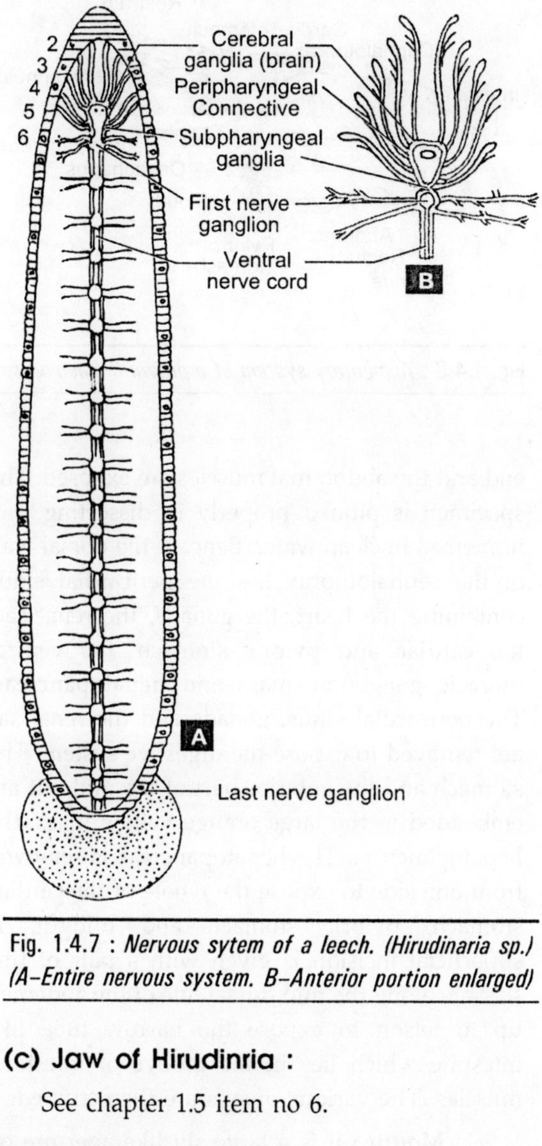

Fig. 1.4.7 : *Nervous sytem of a leech. (Hirudinaria sp.) (A–Entire nervous system. B–Anterior portion enlarged)*

(c) Jaw of Hirudinria :

See chapter 1.5 item no 6.

(d) Nervous system :

Procedure :

Same as in earthworm. Nervous system consists of the following.

(1) **Suprapharyngeal or cerebral ganglia :** A pair of fused ganglia situated above the pharynx.

(2) **Peripharygeal connectives :** Two in number, run on each side of the pharynx connecting the cerebral ganglia with sub-pharyngeal ganglia.

(3) **Subpharyngeal ganglia :** Two ganglia fused together and are situated below the pharynx.

(4) **Ventral nerve cord :** From the subpharyngeal ganglia two ventral nerve cord runs back word from sixth to twenty sixth segment. In each segment nerve cords bear a ganglion and from each ganglion two pairs of nerves arise. The terminal ganglionic mass is ovoid.

3. PRAWN (*Palaemon sp.*)

Palaemon commonly called prawn is common Indian fresh water prawn. They are nocturnal and come out in the search of food.

(a) External features :

A fresh prawn is taken and the following features are studied.

(1) **Shape and size :** The body is elongated, bilaterally symmetrical and spindle shaped. The size of the adult prawn varies from genus to genus.

(2) **Colour :** Adult animal is pale blue or greenish. Formalin preserved specimen becomes orange-pink.

(3) **Exoskeleton :** The entire body including the appendages are covered by an exoskeleton which forms a ring around the segment. On dorsal ride, exoskeleton is convex, called **tergum** and the ventral narrow part is called as **sternum.** The down word projection of tergum is called as **pleuron.** Exoskeleton is not continuous throughout the body, but is localized so as to from harden plates or **sclerites.** Between the sclerites are situated **arthropodial membrane.**

(4) **Divisions of the body :** The body is composed of nineteen appendage bearing segments. The body is regionated into anterior cephalothorax and posterior abdomen.

(a) **Cephalothorax :** It is large, immovable, unjointed and more or less cylindrical in shape. It is formed by five appendage bearing head segments. The cephalothorax

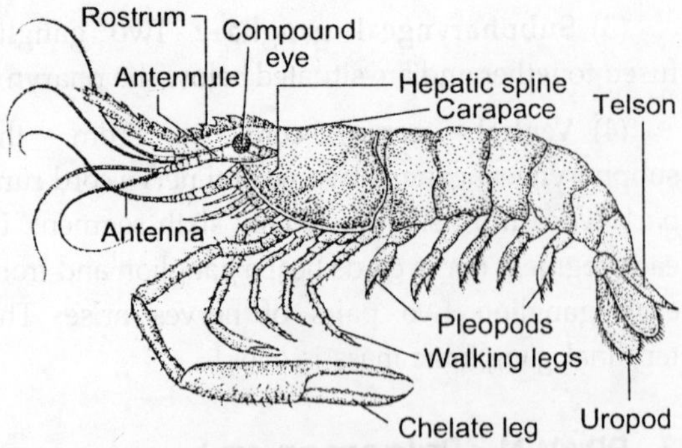

Fig. 1.4.8 : *External features of a prawn (Palaemon sp.)*

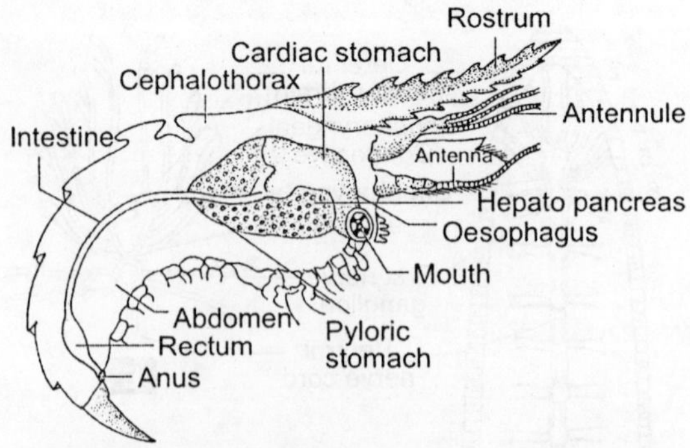

Fig. 1.4.9 : *Alimentary system of a prawn (Palaemon sp.)*

is covered by a large chitinous shield, called the **carapace**, which is dorsally and anteriorly extended into a serrated **rostrum**. Both side of the carapace bears convex flaps called gill covers or **branchiostegites.** At the base of the rostrum, on either side, there is movable jointed stalk large compound eye. Behind and below the eye are situated on either side a pair of spine like projections, the anterior one is antennal spine and the posterior is hepatic spine.

(b) **Abdomen** : It consists of six posterior movable segments and a terminal conical plate or **telson.** Abdomen is rounded dorsally and compressed laterally, and flexed or bent in wards under cephalo-thorax. Abdomen bears six pairs of appendages and is called **pleopods** or **swimmeretes.**

(b) Alimentary system :

Procedures :

A specimen is taken in hand and the carapace is lifted from lateral side. It is cut loose at its anterior margin and inner lining membrane to remove it completely. All the abdominal terga and pleura are removed by giving two parallel incisions along the ventrolateral lines of the abdomen staring from the anterior to posterior end and the abdominal muscles are exposed. The specimen is pinned properly in dissecting tray immersed in clean water. Beneath the dorsal wall of the cephalothorax lies the pericardial sinus containing the heart, the gonads, the renal sac, the cardiac and pyloric stomach, the ventral thoracic ganglionic mass and **hepatopancreas.** The pericardial sinus, gonads and the renal sac are removed to expose the digestive system. The stomach and the anterior part of the mid gut are embedded in the large orange yellow mass, the hepatopancreas. The hepatopancreas is removed from one side to expose the whole of the cardiac stomach, pyloric stomach and midgut. A superficial incision is given with a pair of fine scissors along the mid dorsal line of the abdomen up to telson to expose the narrow tube like intestine which lies in the groove of intestinal muscles. The various organs are then studied.

(1) **Mouth :** It is a large slit like aperture on the ventral side of the head. It is surrounded by labrum in front and laterally by plate like incisor processes of mandibles and behind by labium.

(2) **Buccal cavity :** A short, anteroposteriorly compressed cavity next to mouth.

(3) **Oesophagus :** A short wide vertical tube joined with the buccal cavity and stomach.

(4) **Stomach :** A large, wide sac like chamber consisting of cardiac and pyloric parts. Cardiac

stomach is longer and bag like. Pyloric stomach is small narrow chamber situated below the posterior end of cardiac stomach.

(5) **Intestine :** It is a long, narrow, more or less straight tube which ascends gradually between two lobes of hepatopancreas and runs back to the sixth abdominal segment to end in rectum.

(6) **Rectum :** It is the terminal part of the hind gut. It is a small chamber, wider anteriorly and narrows down posteriorly and opens to the outside by anus.

(7) **Hepatopancreas :** A large, orange red, bilobed dense and glandular mass, occupies major portion of the cephalothorax. Two ducts of the two lobes of hepatopancreas open separately into the ventral chamber of the pyloric stomach.

(c) Hastate plate :

See chapter 1.5 item no 9.

(d) Nervous system :

Procedure :

The specimen is dissected in the same way as in the case of alimentary canal. Brain is situated at the base of rostrum in front of oesophagus. The endopharyngeal skeletal plates found in the ventral side of the thorax are removed to exposed the under lying nerves till the brain is seen. The oesophagus is cut and pulled out with the stomach to expose the nerves arising from the brain. Two nerve cords that arise from the posterior end of the brain and run backwards around the oesophagus and unite with the suboesophageal ganglionic mass are traced.

A cut is made medianly with the scalpel between the large flexor muscles to expose the ventral nerve cord in the abdominal region. These muscles are pressed laterally and are pinned in the dissecting tray. As the muscles are stretched, the nerve cord with its ganglia is very clearly seen. The various parts are studied.

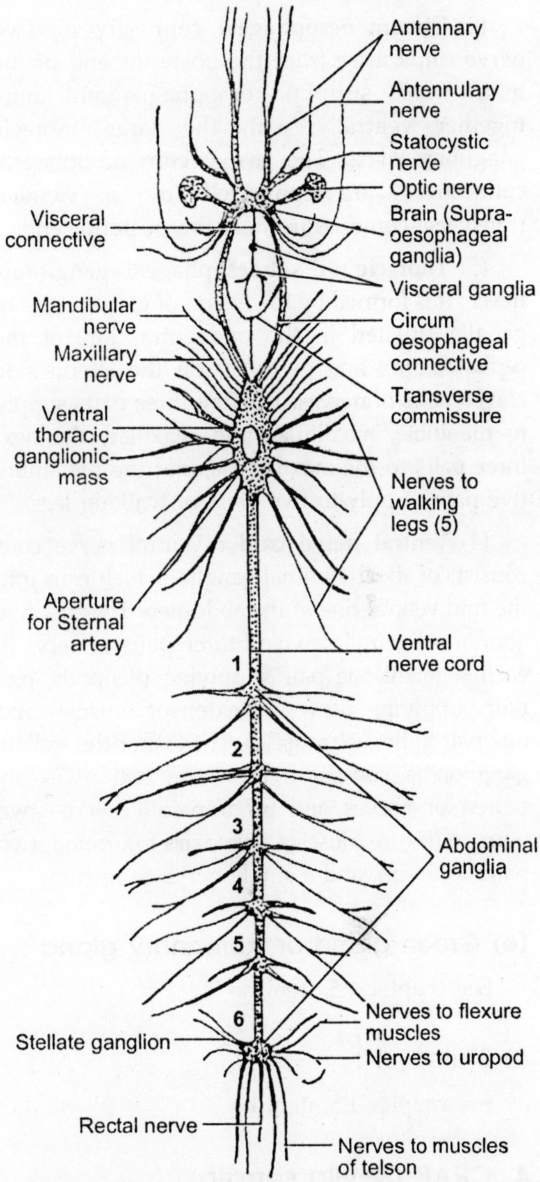

Fig. 1.4.10 : *Nervous system of a prawn (Palaemon sp.)*

(1) **Brain or supra oesophageal ganglia :** It is a bilobed structure situated at the base of rostrum in front of oesophagus. Brain gives, a pair of optic nerves to eye stalk, a pair of antennular nerves to antennules, a pair of antennary nerves to antennae, a pair of ophthalmic nerves to eye muscles and a pair of tegumental nerves to labrum.

(2) **Circum oesophageal connectives :** Two nerve cords arise from the posterior end of the brain which surround oesophagus and unite together ventrally with the large thoracic ganglionic mass. The two circum oesophageal connectives are connected by a slender, transverse commissure near the posterior end.

(3) **Thoracic or suboesophageal ganglionic mass :** It is formed by the fusion of eleven pairs of ganglia situated in the midventral line of the cephalothorax and gives off on the lateral side eleven pairs of nerves. The first three pairs supply to mandible, maxillulae and maxillae, the next three pairs to the maxillipede, and the remaining five pairs supply to five pairs of walking legs.

(4) **Ventral nerve cord :** Ventral nerve cord consists of six abdominal ganglia, which runs into the mid ventral line of the abdomen. The first five abdominal ganglia give off three pairs of nerve in each segment-one pair supplying pleopods, one pair supplying nerves to extensor muscles and one pair to flexure muscles. The last or the stellate ganglion is comparatively large and gives six paired of nerves and an unpaired nerve—two pairs to flexure muscles, two pairs to uropod, two pairs to telson, and a single nerve to rectum.

(e) Green gland or Antennary gland :

See chapter 1.5, item no 7.

(f) Statocyst :

See chapter 1.5, item no 8.

4. CRAB (Scylla serrata)

The crabs are of several types and vary in size, shape and colouration. They are adapted for aquatic respiration; it may live for a considerable time on land.

(a) External features :

A live crab is held, after clipping the last one segment of the first pair of walking legs. The following features are studied.

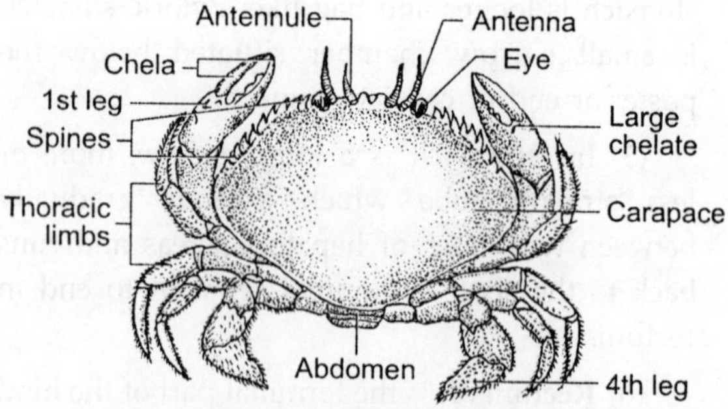

Fig. 1.4.11 : *Extrenal features of a crab. (Dorsal view–Male) (Seylla serrata)*

(1) **Shape :** Body is more or less oval in out line and the width is greater then the length.

(2) **Colour :** Greenish dorsally and pale orange-red ventrally.

(3) **Divisions of the body :** The body is divisible into cephalothorax and abdomen.

(a) **Cephalothorax :** It is the most prominent part in the crab and the exoskeleton of cephalothorax is covered dorsally by a flat, broad and oval carapace. The anterior and lateral margins of carapace are semicircular and serrated. The ventral side of the cephalothorax is covered by **sternites**. The exoskeleton which covers the two sides of cephalothorax is called as **pleuron**. Cephalothorax contains a mouth, one pair of stalked compound eyes and

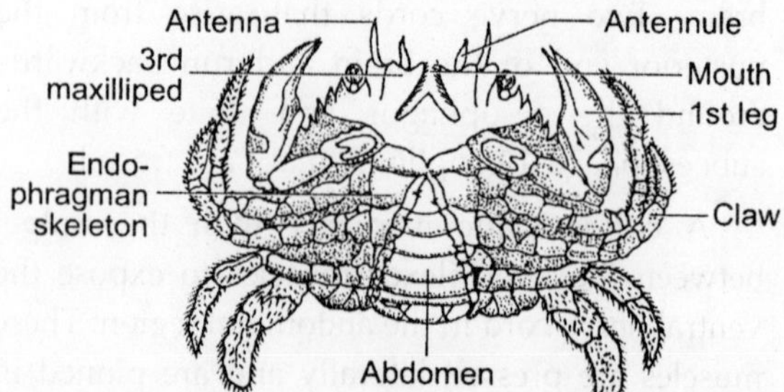

Fig. 1.4.11A : *Extrenal featu... a crab. (Ventral view–Male) (Seyll... ata)*

cephalothoracic appendages. Cephalic appendages are one pair antennule, one pair antenna, one pair mandible, one pair first maxillae and one pair second maxillae. Thoracic appendages are three pairs of maxillipeds and five pairs of walking legs. The first pair of walking leg is with powerful **chela** and the rest are clawed.

(b) **Abdomen :** It is six segmented, flap like and remain flexed in the ventral side of the cephalothorax. Each abdominal segment bears a pair of rudimentary **pleopods**. The abdomen of female crab is broad.

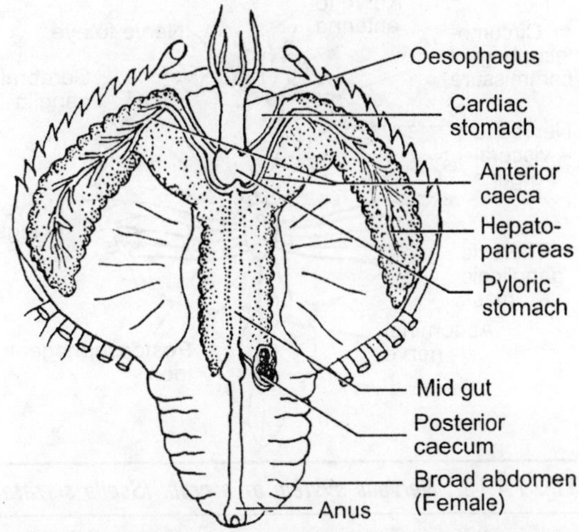

Fig. 1.4.12 : *Alimentary system of a crab. (Scella serrata)*

(b) Alimentary system :

Procedure :

The crab is held with left hand and the portions of legs are clipped with a stout scissors. The crab is then cut along the margin of the dorsal carapace with a stout scissors or a bone cutter and carefully the carapace is removed from the linning membrane below it. The specimen is fixed on a dissecting tray with help of pins through the lateral side of the abdominal sternites and through the walking legs, keeping the dorsal surface of the crab up wards. The flexed abdomen is released from the ventral side of the cephalothorax, stretched and pinned on the tray. The tergites of the abdomen are cut along the margin starting from the junction of the cephalothorax and the abdomen. The tergites are removed. The organs in the cavities are exposed. The fore gut and anterior part of the mid gut is exposed by partly removing the orange yellow organ the hepatopancreas. The pericardial sinus with the heart is completely removed, the gonads, the renal sac are also removed to expose the major part of the alimentary canal. The followings parts are observed.

(1) **Mouth :** It is a square opening on the ventral side of the anterior end, guarded by two mandibles.

(2) **Oesophagus :** It is short and open in stomach.

(3) **Stomach :** A large oval, muscular sac found just beneath the carapace. It consisting of large anterior cardiac and small posterior pyloric parts.

(4) **Intestine :** A narrow tube runs through out the abdomen and ends in rectum.

(5) **Anterior or superior caeca :** A pair of narrow, blind diverticula arises from the junction of the pyloric chamber and intestine.

(6) **Posterior or inferior caeca :** An elongated, coiled, blind tube arises from the intestine just before it enter in the abdomen.

(7) **Rectum :** Wider tube extends down posteriorly to open in the anus.

(8) **Digestive gland :** A large, yellow – orange mass occupies major portion of cephalothorax, called as **hepatopancreas**. Its ducts open in the pyloric chamber.

(c) Nervous system :

Procedure :

The specimen is dissected as in case of alimentary system. The endopharyngeal skeletal plates found on the ventral side of the thorax are

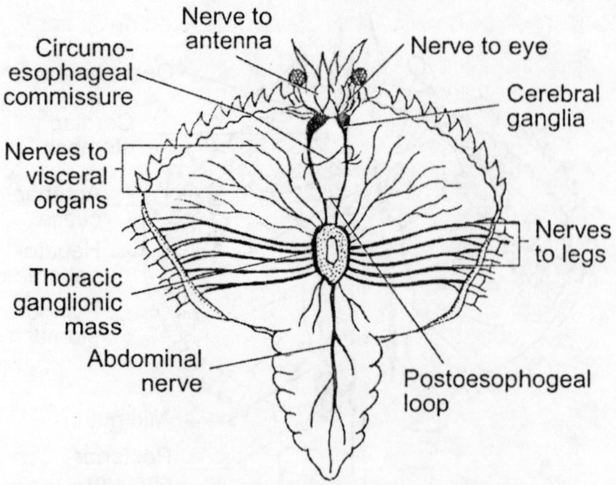

Fig. 1.4.13 : *Nervous system of a crab. (Scella serrata)*

removed and the underlying nerves are exposed till the brain is exposed. Two nerve cords arise from the lateral side of the brain are traced that runs backward and downward round the oesophagus and cardiac stomach and unite with the ventral thoracic ganglionic mass.

As the circumoesophageal commissure is found encircling oesophagus, the oesophagus is cut and removed carefully leaving commissure. The digestive system along with hepatopancreas is removed along with the other organs that lie dorsal to the nervous system to expose it. The various parts are studied.

(1) **Brain or supra oesophageal ganglia :** It is placed at the anterior end and on the dorsal side, formed by the fusion of two cerebral ganglia. From the brain following paired nerves arise.

(a) **optic** to eye,

(b) **ophthalmic** to eye muscles,

(c) **antennulary** to antennules, and

(d) **antennary** to antennae.

(2) **Circumoesophageal connectives:** From the side of each cerebral ganglion arises are oesophageal connective. Two connectives run backward and downward encircling the oesophagus and unite with the ventral thoracic ganglionic mass in the middle of thorax.

(3) **Circumoesophageal commissure :** Circumoesophageal connective are joined by a fine commissue encircling the oesophagus.

(4) **Postoesophageal loop :** Circumoesophageal connectives are joined transversely by a connective before they join the ventral thoracic ganglionic mass.

(5) **Ventral thoracic ganglionic mass :** A large mass is formed by the fusion of suboesophageal, thoracic abdominal ganglia and major part of the entire ventral nerve cord. This mass supply nerves to maxillulae, maxillipeds, thoracic limbs and abdominal muscles.

(6) **Ventral nerve cord :** It runs porteriorly from the ventral thoracic ganglionic mass to the posterior tip of the abdomen and gives off a few branches at the posterior end.

(d) Reproductive system :

Procedure :

The specimen is dissected as in case of alimentary system. The carapace is carefully removed, and also hepato pancreas and posterior portions of gills. In both sexes the reproductive organs are united in middle and are extended laterally with the fold of the carapace.

(i) Male reproductive system :

(1) **Testes :** It is paired, tubular structure united in the middle by a transverse testicular isthmus and extened laterally with in the fold of the carapace.

(2) **Vas deferens :** It is paired; long, narrow convoluted tubes arise behind the isthmus and ran back word and have anterior convoluted parts, less convoluted median part and straight posterior part.

(3) **Genital pore :** It is present in last leg (fifth walking leg) and a flexible process of the coax.

(ii) Female reproductive system :

(1) **Ovaries :** It is paired structure united in

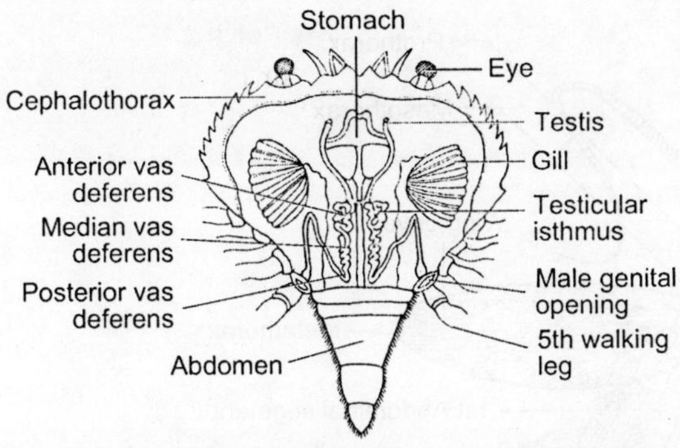

Fig. 1.4.14 : *Reproductive system of a Male crab. (Scylla serrata)*

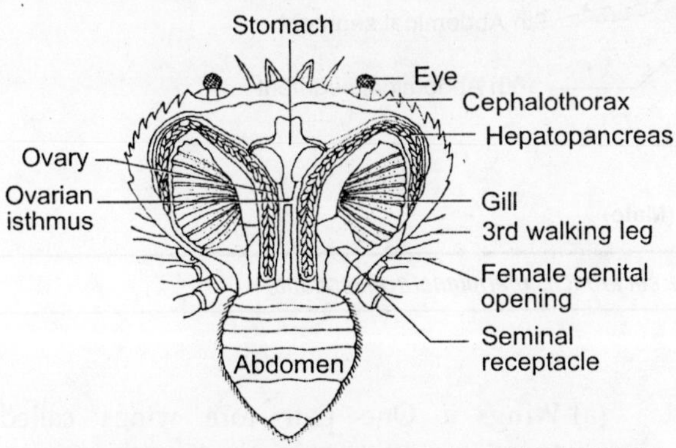

Fig. 1.4.15 : *Reproductive system of a Female crab. (Scylla serrata)*

the middle by a transverse ovarian isthmus and extend laterally within the fold of carapace.

(2) **Oviducts :** Each ovary give rise to one duct from the ventral side, posterior to ovarian isthmus and runs back ward.

(3) **Spermatheca or seminal receptacles :** It is the dilated chamber of the oviducts and opens in sternum.

(4) **Genital pore :** A small round opening at the third walking legs.

5. COCKROACH (*Periplaneta americana*)

It is common Indian cockroach and most suited for laboratory study. Nocturnal, omnivorous house hold pest. It prefers dark places of kitchen, godown, drain pipes, bakeries etc.

(a) External features :

A fresh cockroach is taken, which has been killed with chloroform or by drowning in water. The following features are studied.

(1) **Shape and colouration :** Body is dorsoventrally flattened, elongated and reddish brown in colour with a yellowish area near pronotum and two dark patches over it.

(2) **Exoskeleton :** Entire body is covered by a chitinous cuticle which contains several sclerites. The skeletal plates of thorax and abdomen are dorsal **terga**, ventral **sterna** and lateral **pleura**.

(3) **Segmentation and divisions :** Body is metamerically segmented and the segments and regionated into three distinct parts each is called **tagma**. The tagmata are head, thorax and abdomen.

(i) Head :

It is small body, more or less triangular in shape and is formed by fusion of six segments such as **epicarnium, frons, clypeus, labrum** and **genae**. Head contains **compound eyes, antennae** and mouth parts or **trophi.**

(a) **Compound eyes :** One pair prominent eyes on sides. Eyes are large in males and small in females and made up of several visual units or **ommatidia.**

(b) **Antennae :** A pair of thread-likes, many jointed and filiform antennae is attached in front of the eyes.

(c) **Ocellus :** Paired circular areas one on each side of the eye, near the base of antenna.

(d) **Mouth and Mouth parts :** Placed at the anterior end of the head and is provided with several appendages—the mouth parts or **trophi**. Mouth parts are composed of **labrum, mandibles, maxillae, labium** and **hypopharynx.** Head is connected to thorax by short, slender neck or cervicum.

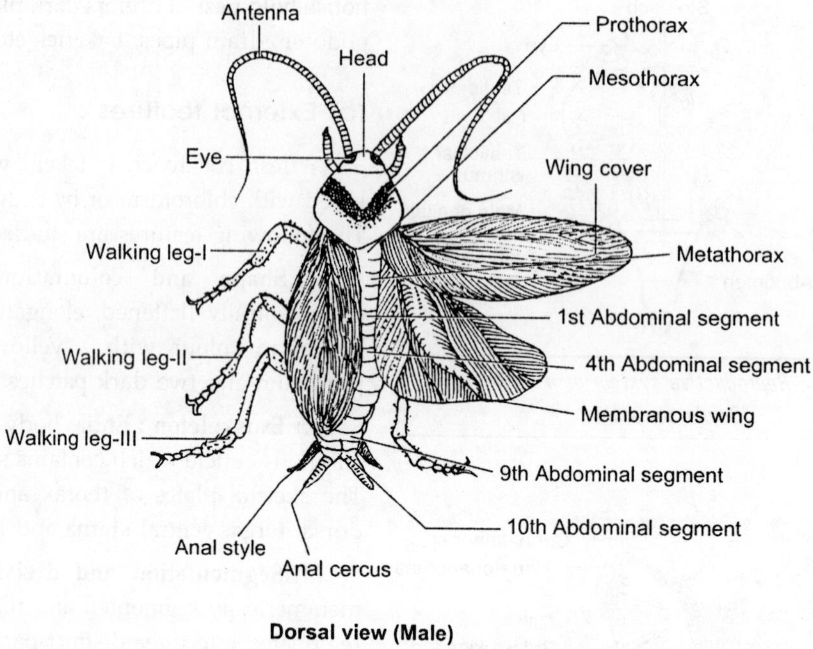

Fig. 1.4.16 : *External features of a cockroach. (Periplaneta americana)*

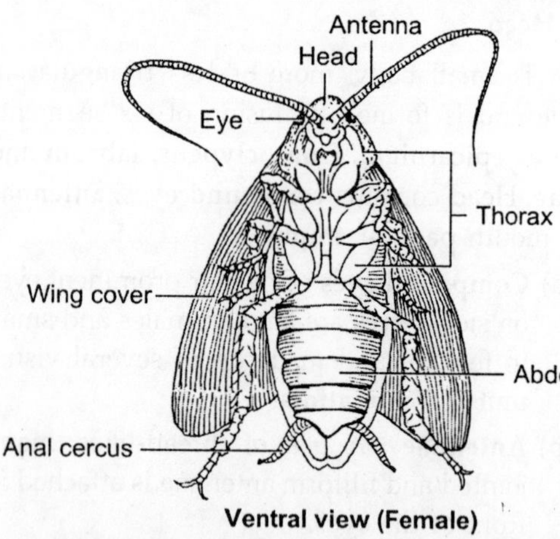

Fig. 1.4.16A : *External features of a cockroach. (Periplaneta americana)*

(ii) Thorax :

It is composed of three segments namely Prothorax, mesothorax and metathorax. A large sclerite covers the prothorax dorsally. Thorax contains :

(a) **Wings** : One pair fore wings called **tegmina** or **elytra** orginate from the tergum of mesothorax. It is dark, opaque, leathery and protective. One pair hind wings or true wings originate from the tergum of metathorax. It is thin, membranous and used in flight.

(b) **Spiracle** : Two pairs of spiracles one pair which is the largest present on the mesothorax and one pair on the metathorax.

(c) **Walking legs** : Three pairs, each thoracic segment carries a pair of walking legs. All the legs are of similar shape and structure and made up of **coxa**, **trochanter**, **femur**, **tibia, tarsus, claws** and **arolium** or **pulvillus**.

(iii) Abdomen :

Ten segments. Both sexes bears on the tergites of the tenth abdominal segment, a pair of ventrolateral, many jointed **cerci**. In addition to anal cerci the posterior end of the abdomen in a male cockroach bears a pair of unjointed stylets on the ninth segment. Abdomen bears eight pairs of **spiracles** or **stigmata**, one pair each on soft cuticle of terga and sterna. The last part of the abdomen in both sexes is provided with certain genital appendages.

Digestive System

Procedure :

A freshly killed cockroach is held with the left hand and the wings are removed, antennae and legs are cut off close to their bases. The lateral membranes between terga and sterna are cut of the thorax up to the antherior edge of pronotum with a pair of fine scissors. The specimen is pinned in abdominal sterna and coxa of legs in a dorsal position on a dissecting tray. The head is fixed by pinning between mandibles.

A transverse incision is made along the posterior edge of the ninth segment and other segments are gently removed and very carefully proceed forward up to the anterior end of the thorax, so that the under lying organs and tissues are not disturbed. Clear water is put in the tray, fat bodies, tracheae and other muscles are removed to expose internal organs. Intestine is uncoiled and alimentary canal is stretched on one side and the following parts are studied.

The alimentary system includes alimentary canal and digestive glands. Alimentary canal is divisible into, foregut, midgut or mesenteron and hindgut.

(1) **Fore gut** : It includes mouth, pharynx, oesophagus, crop and gizzard.

(a) **Mouth :** Ventral in position and lies at the anterior end of the head.

(b) **Buccal cavity** : A small chamber between the mandibles and maxillae on either side anteriorly bounded by labium and posteriorly by labrum. On the floor of the buccal cavity lies the hypopharynx or tongue.

(c) **Pharynx** : A small tube which leads into the oesophagus.

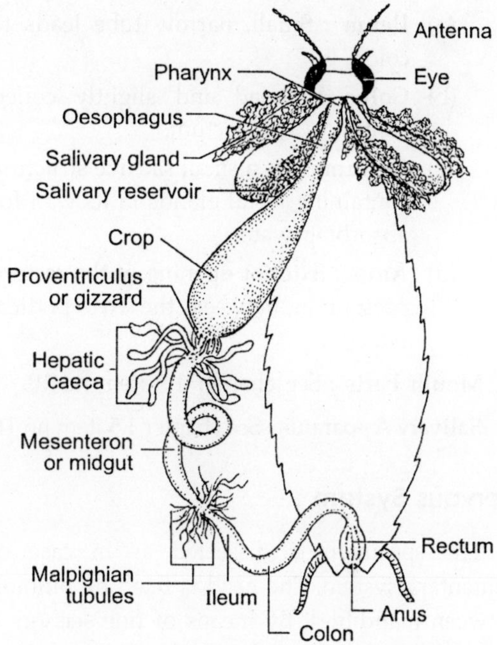

Fig. 1.4.17 : *Alimentary system of a cockroach. (Periplaneta americana)*

(d) **Oesophagus** : A short tube extends posteriorly up to the prothorax.

(e) **Crop** : A dilated sac like structure extends within the abdominal cavity.

(f) **Gizzard or proventriculus** : A short rounded thick wall structure posterior to the crop is divided into anterior and posterior part. Anterior part contains sixchitinous teeth in its inner wall and posterior part possesses two circular hairy cushions.

(2) **Midgut or mesenteron** : A slender tube running from gizzard to the hindgut. Eight hollow slender tubes called **hepatic caeca** or **digestive diverticula** are present near the junction of the fore and mid gut.

(3) **Hind gut or proctodaeum :** A small narrow tube divisible into, ileum, colon, rectum and anus. At the beginning of ileum, is surrounded by numerous (sixty to seventy) yellowish treads malpighian tubules which are excretory organs.

(a) **Ileum :** Small, narrow tube leads to colon.

(b) **Colon :** Broad and slightly coiled continues into rectum.

(c) **Rectum :** It is a small sac like structure containing rectal glands in its wall for absorbing water.

(d) **Anus :** Exterior opening of the rectum present in between the two podical plates.

Mouth Parts : See chapter 1.5 item no 10.

Salivary Apparatus : See chapter 1.5 item no 11.

Nervous System :

The specimen is dissected as in case of alimentary system. The head is fixed by pinning between mandibles. By means of fine scissors a rectangular cut is made in the head around clypeus and anterior epicranial plate. The plate is removed carefully and exposes the cerebral ganglia. The pharynx is cut and pulled out with oesophagus. Fat bodies, tracheae and viscera are removed to expose the ventral nerve cord. The following parts are studied.

(1) Two fused **cerebral ganglia** or brain lies dorsally on head. Nerves arise from it supply to eyes, antennae and adjacent structures.

(2) Short **circumesophageal connectives** arising from brain run along the oesophagus to meet the sub oesophageal ganglion.

(3) Single **suboesophageal ganglion** which is formed by the fusion of two ganglia lies on lower or ventral side of head. Nerves arise from it supply to labrum, mandible and both pairs of maxillae.

(4) Double **ventral nerve cord** starts from suboesophageal ganglion and run along the mid-ventral line and extends upto posterior end. It is connected by pairs of fused ganglia, three thoracic and six abdominal. The three thoracic ganglia and the last abdominal ganglion are large in size. Nerves arising from the thoracic ganglia supply the muscles of the thoracic segments. First five abdominal ganglia send out pair of nerves to supply different parts of abdominal segments. The sixth abdominal ganglion sends out seven to eight pairs of nerve to different structures present in the last four abdominal segments.

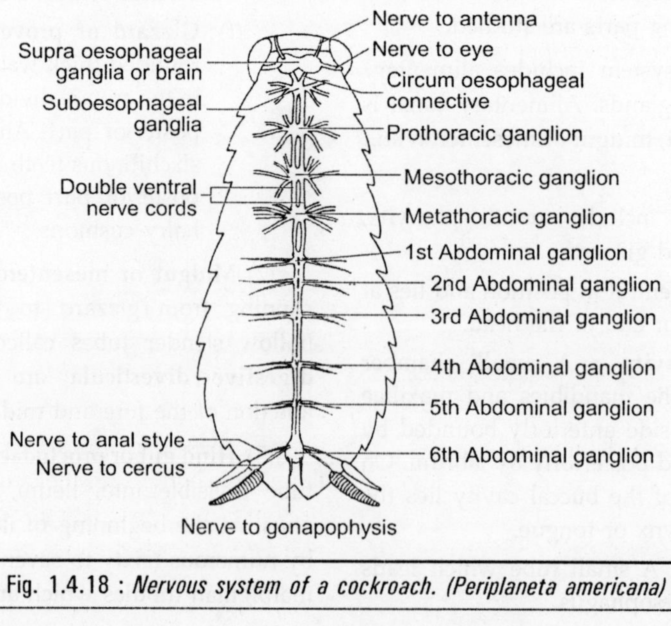

Fig. 1.4.18 : *Nervous system of a cockroach. (Periplaneta americana)*

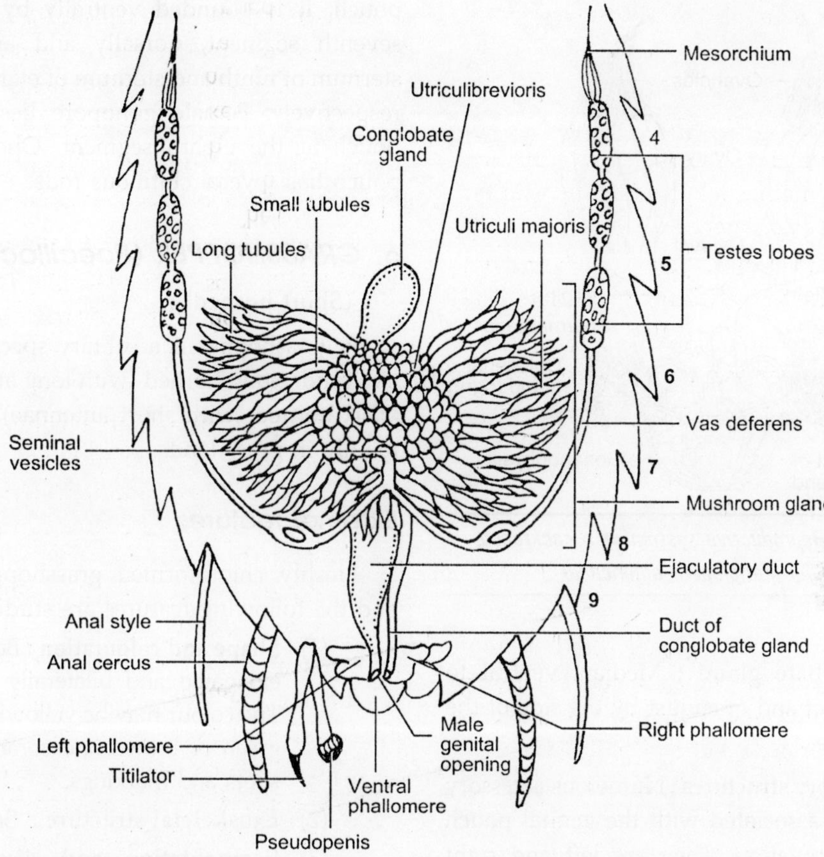

Fig. 1.4.19 : *Reproductive system of a cockroach. (Male) (Periplaneta americana)*

Reproductive System :

The specimen is dissected as in case of alimentary system but care is taken while removing the terga. Paired testes are present in the lateral side of the abdominal cavity under the terga of fourth and fifth abdominal segments. It is difficult to distinguish testes from the surrounding fat as they are of the same colour. The testes are carefully located in the lateral side of fourth and fifth abdominal segments and then proceed to trace the vasa deferentia.

Male reproductive system :

(1) **Testes :** Paired, small in size, rounded, consists of 3 or 4 lobed.

(2) **Vas deferens :** Two slender ducts originating from the testes run posteriorly and open with in sac called **seminal vesicle.**

(3) **Ejaculatory duct :** It is a common duct to which the two seminal vesicles open. The duct runs up to reproductive pouch and opens through male gonopore.

(4) **Mushroom gland :** Numerous blind slender and thread live structure are given out from the outer wall of seminal vesicle which give a mush room like appearance. Central region of the gland contains smaller diverticula is **utriculi breviores** and two lateral region with slender vesicle is **utriculi majores.**

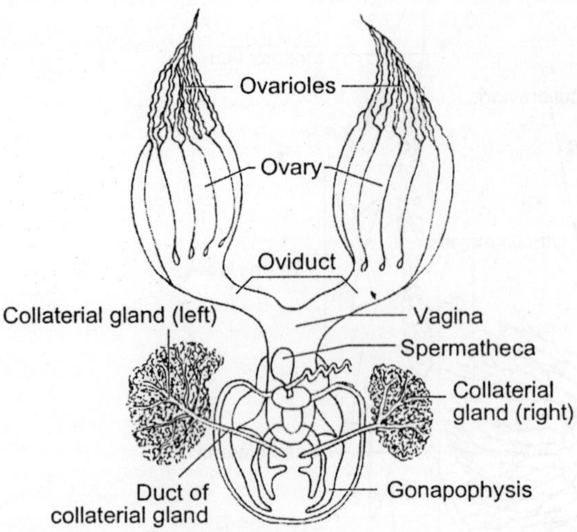

Fig. 1.4.19 : *Reproductive system of a cockroach. (Female) (Periplaneta americana)*

(5) **Conglobate gland :** Median, ventral to ejaculatory duct and open just by the side of the male gonopore.

(6) **Accessory structures :** Numerous accessory structures are associated with the genital pouch to help in copulation. They are left and right **phallomere**, **pseudopenis**, ventral **phallomere** and **titilator**.

Female reproductive system :

(1) **Ovaries :** Paired, placed in the lateral and posterior end of the abdominal cavity under the fourth and eighth terga. Each ovary has eight beaded tube like **ovarioles**, which are held together.

(2) **Oviducts :** Two, each ovary give rise to one oviduct which is formed by the fusion of ovarioles at the posterior end. Two ducts unite in the median position and from a short vagina.

(3) **Spermathecae :** Paired unequal organs placed in between the oviducts. Two spermathecae open into the genital pouch on the ninth segment.

(4) **Genital pouch :** Seventh, eighth and ninth abdominal segments modified to from genital pouch. It is bounded ventrally by sternum of seventh segment, dorsally and anteriorly by sternum of ninth and sternum of eighth segments respectively. Female gonopore lies within the pouch at the eighth segment. Opening of the pouch has several chitinous rods.

6. *GRASSHOPPER (Poecillocerus sp.)*

(Short horned)

A grasshopper is a solitary species and is of two kinds: long horned (with long antennae) and short horned (with short antennae). They occur mainly in grass lands.

External features :

Freshly chloroformed grasshopper is taken and the following features are studied.

(1) **Shape and colouration :** Body is slender elongated and bilaterally symmetrical. The colour may be yellowish green, leaf green or brown and some with varied spots and markings.

(2) **Exoskeletal structure :** See cockroach.

(3) **Segmentation and division :** See cockroach.

(I) **Head :** See cockroach.

(II) **Thorax :** See cockroach. In male, the third segment of the leg bears rows of chitinous hooks

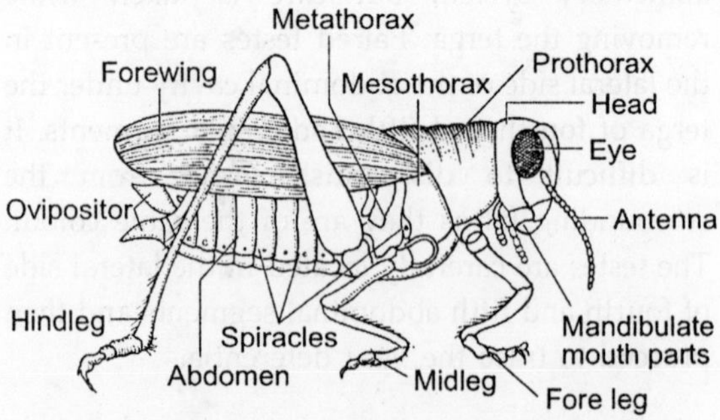

Fig. 1.4.21 : *External features of a grass hopper. (Poecillocerus sp.)*

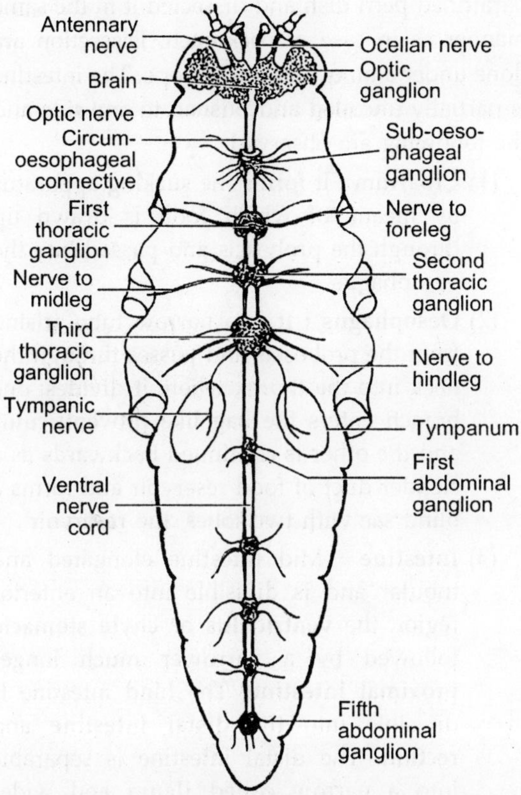

Fig. 1.4.22 : *Nervous system of a grass hopper. (Poecillocerus sp.)*

called **stridulatary organ**. By pressing the hooks along the hard outer surface of the wing on prothorax, they produce sound.

(III) **Abdomen** : See cockroach.

Nervous System :

Procedure : The specimen is dissected as in case of cockroach. The features are studied as in case of cockroach. Here the ventral nerve cord along its path bears three thoracic and five abdominal ganglia.

7. *BLUE BOTTLE FLY (Calliphora sp.)*

Blue bottle flies are day time visitors of human habitation. The fly gives frequent flying visits to the house and sits over uncovered foods. They can be easily collected from fruit markets or confectionary. They can be killed with chloroform.

External features :

A freshly killed specimen is taken and the following are observed.

Medium sized bluish grey coloured fly with red eyes. The body of the fly consists of three tagmata head, thorax and abdomen.

(1) **Head** : Relatively large size, on extensive portion of head is occupied by a pair of much prominent compound eyes containing nearly 4000 ommatidia and three small simple eyes are present. Head contains paired, short and flexible antennae. It also contains mouth parts which are composed of labrum, maxillae, labium and hypopharynx. The Mouth parts are well developed and contains two portions **proboscis** and **food channel**. The labium forms the proboscis which is usually expanded distally to from a pair of fleshy lobes or **labella** which is traversed by numerous channels called **pseudotracheae** opening externally along the margin. The proboscis consists of a proximal **rostrum** and a distal **haustellum**. Rostrum carries anteriorly the maxillary palps. The haustellum carries the labrum and the hypopharynx on its

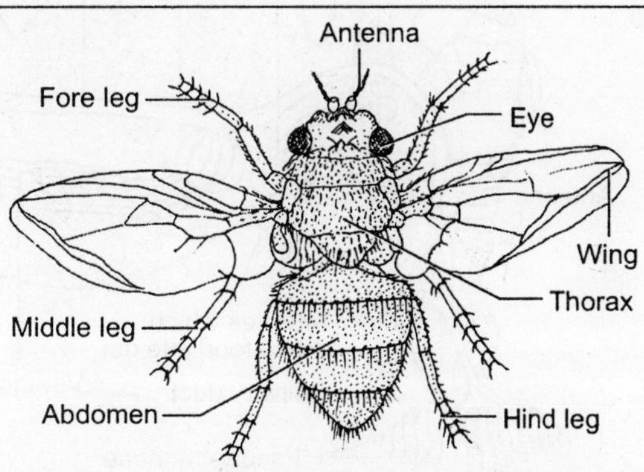

Fig. 1.4.23 : *External system of a blue bottle fly. (Calliphora sp.)*

dorsal face. The food channel is formed by labrum, epipharynx and hypopharynx.

(2) **Thorax** : Three segments are clearly demarcated and are covered by bristles. Thorax bears three pairs of legs and two pairs of wings. Three pairs of legs have insect like divisions. Terminal end of each leg is provided with a pair of carved claws. In between the claws lie two pad like pulvilli. Numerous hairs and bristles occur through out the surface of the legs. Wings two pairs, anterior pair is prominent, broad, transparent, hairy and triangular and the posterior pair is modified into **halters**. Thorax contains two pair **spiracles.**

(3) **Abdomen :** It is oval, hairy and segmented. In female the visible segments are the second to the fifth, while the sixth to tenth segments form the retractile ovipositor. In male the ninth and tenth segments are with the genitalia from the **hypopygium**. Abdomen contains eight pairs of spiracles.

The Alimentary system :

Procedure :

The killed specimens are pinned in a paraffined petri dish and dissected it in the same manner as in case of cockroach. Dissection are done under a binocular microscope. The intestine is partially uncoiled and pushed to one side and the following are observed.

(1) **Cibarium :** It forms the sucking apparatus by means of which food is drawn up through the proboscis and passed into the oesophagus.

(2) **Oesophagus :** It is a narrow tube arising from the proboscis and passes through the neck into the thorax where it divides: one branch enters the bag like proventriculus and the other is continued backwards as a slender duct of food reservoir and forms a blind sac with two lobes, the **reservoir.**

(3) **Intestine :** Mid intestine elongated and tubular and is divisible into an anterior region the **ventriculus** or **chyle stomach**, followed by a narrower much longer **proximal intestine**. The hind intestine is divisible into the **distal intestine** and **rectum**. The distal intestine is separable into a narrow coiled **ileum** and wider **colon**. The rectum is pyriform and

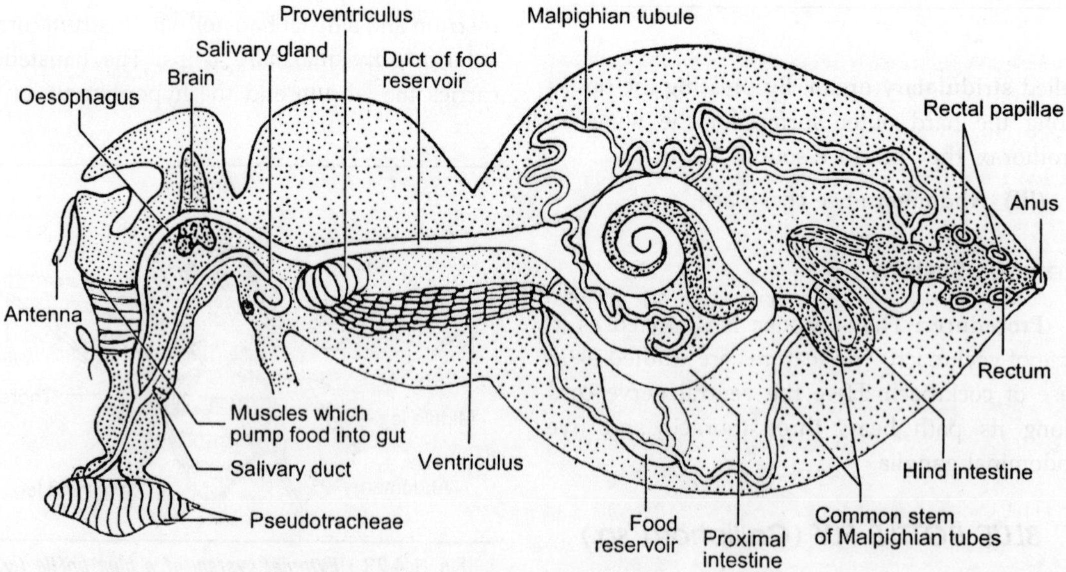

Fig. 1.4.24 : *Alimentary system of a blue bottle fly. (Calliphora sp.)*

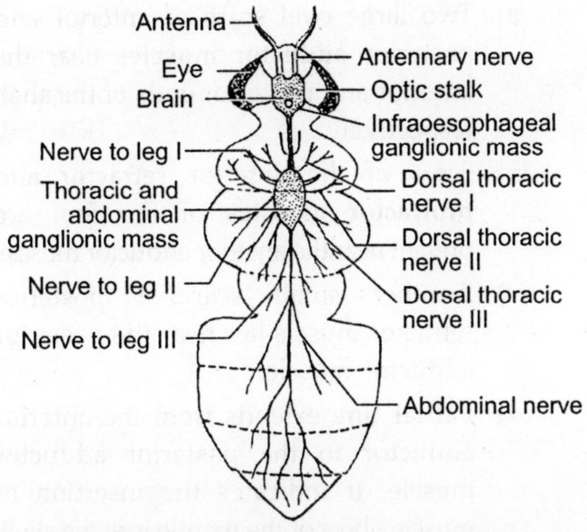

Fig. 1.4.25 : *Nervous sytem of a blue bottle fly. (Calliphora sp.)*

muscular chamber provided with four papillae. Two pairs of malpighians tubules open in the hind intestine. Rectum opens through rounded aperture called **anus.**

Salivary glands : Two elongated and tubular glands are situated in the thorax and the head. The common salivary duct passes to the base of hypopharynx, where it expels the secretion down the salivary groove.

Labial glands : They are spherical aggregations of gland cells present on the proboscis at the base of labella.

The Nervous System :

Procedure :

Same as in case of alimentary system. The head is fixed and the epicranial plate of the head is carefully removed to expose the cerebral ganglia. The oesophagus is cut anterior to the brain and the entire alimentary canal is removed to expose the nervous system. The following are observed.

(1) **Cerebral ganglia :** Two cerebral and the infraoesophageal ganglia are fused to from the brain. It sends nerves to antennae, mouth parts and eyes.

(2) **Thoracic and abdominal ganglionic mass :** Fused thoracic and abdominal ganglia situated in the anterior part of the thorax.

(3) **Ventral nerve cord :** A short double chain ventral nerve cord connects the thoracic and abdominal ganglionic mass with the brain.

(4) **Abdominal nerves :** A large abdominal and a few small nerves arise from the posterior end of the thoracic and abdominal ganglionic mass. It supplies nerves to abdomen.

8. *FRESH WATER MUSSEL (Lamellidens marginalis)*

Lamellidens is an aquatic form and found abundantly in fresh water and in the bottom of lakes, ponds and rivers. Generally they are found half embedded in mud or sand at the bottom.

External features :

A live specimen or a preserved specimen is taken and a scalpel is inserted between the two halves of the shell near the posterior end and carefully separate the mantle from one of the valve of the shell keeping the blade of the scalpel against the inner surface of the shell and the posterior adductor muscle is cut. The handle of another scalpel is inserted in this gap and the shell is raised upward to release the valve. In the

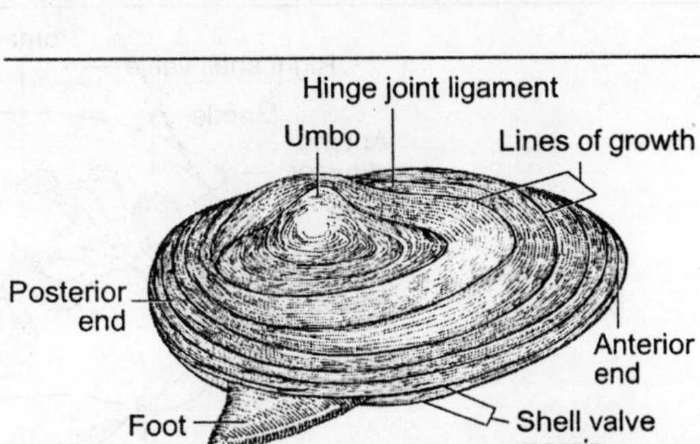

Fig. 1.4.26 : *External features of a fresh water mussel. (Lamellidens marginalis)*

same way the blade of a scalpel is inserted near the anterior end and the anterior adductor muscle is cut. The shell is pulled backwards till the hinge is broken. The followings are observed:

(1) **Size** : From five to ten centemetre in length.

(2) **Colour** : Outer surface is rough and greyish brown, while the inner surface white, smooth and lustrous.

(3) **Shape** : Anterior side is roughly oval in out line and the posterior end is slightly narrower. The body is compressed and enclosed by shell.

(4) **Shell** : It consists of two equal lateral valves united dorsally along a hinge line by elastic hinge-ligament. Beneath the shell, a delicate layer called **mantle** envelops the visceral mass. Two lateral halves of mantle called mantle lobes. At the aboral side mantle lobes produced two short tubes the **inhalant** and **exhalent siphons**. A tongue-shaped foot protrudes between the two valves towards the oval end.

(5) **Umbo** : It is a swollen, whitish, knob like structure represents the anterior side of the shell. Series of concentric **lines of growth** are found all around the surface of the shell which starts from umbo.

(6) **Macroscopic structure of Shell** : Some markings are also observed on the inner side of the shell.

(a) Two large oval scars of anterior and posterior **adductor muscles** near the anterior and posterior ends of the shell respectively.

(b) Scars of the anterior **retractor** and **protractor muscles** of the foot are present near the anterior adductor muscle.

(c) Another small scar of posterior retractor muscle lies near the posterior adductor muscle.

(d) **Pallial line** extends from the anterior adductor to the posterior adductor muscle. It indicates the insertion of muscle fibres of the mantle into the shell.

Alimentary system :

Procedure :

The specimen is dissected as in case of external features. The other valve of the shell is separated in same way. Alimentary system is best dissected in a preserved *Lamellidens.*

The animal is pinned on a dissecting tray with the unturned mantle fold and through the edge of the foot. The gills of the sides are removed. The mouth below the anterior adductor muscles are located which is bounded by two pairs of labial palps. A little incision is made on posterodorsally to the mouth on the skin. The skin is held with pair of forceps and it is carefully

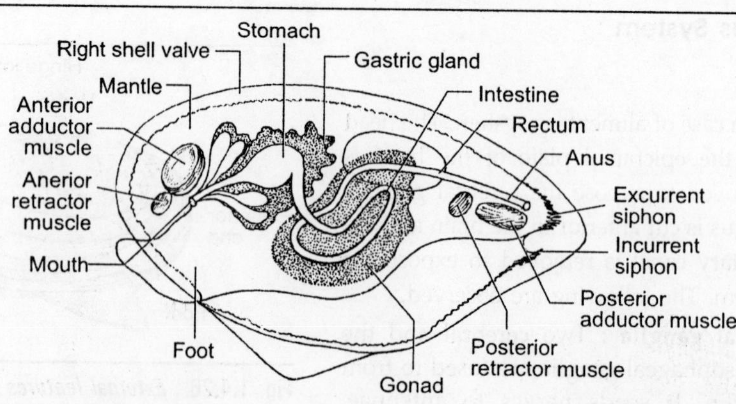

Fig. 1.4.27 : *Alimentary system of a fresh water mussel. (Lamellidens marginalis)*

separated from the underlying irregular digestive gland, with an arrow headed needle. A part of digestive gland is removed to exposed the oesophagus and the stomach.

From the posterior end of stomach starts the intestine which enters into the visceral mass dorsal to the foot and turns several turns and continues into the rectum. The gonad and other tissue is carefully removed and the coiled intestine is exposed.

The pericardial sinus with the heart is removed to exposed the junction of intestine with rectum. The **suprabranchial chamber** is cut open to expose the rectum. The following parts are observed.

(1) **Mouth** : Transverse slit, bounded by two pairs of labial palps.

(2) **Oesophagus** : Short, wide tube opens into stomach.

(3) **Stomach** : Wide, thick walled rounded spacious sac surrounded by digestive glands.

(4) **Intestine** : Coiled, slender tube. Leaving stomach, intestine has several turns in the visceral mass, forms two incomplete loops then goes up and takes the level of the stomach.

(5) **Rectum** : Terminal part of gut runs posteriorly through the ventricle of the heart, which is actually wrapped around it and open through the anus.

(6) **Digestive gland** : A pair of large, dark brown and irregular in shape. They open in the stomach by small ducts.

Nervous system :

Procedure :

Same as in alimentary system. The specimen is pinned with the mantle fold with the foot facing upwards. Carefully gill lamellae of both the sides are removed. The posterior adductor muscle is located, at the base of which lies the **visceral ganglia**. Carefully the peritoneal covering is removed to observe the visceral

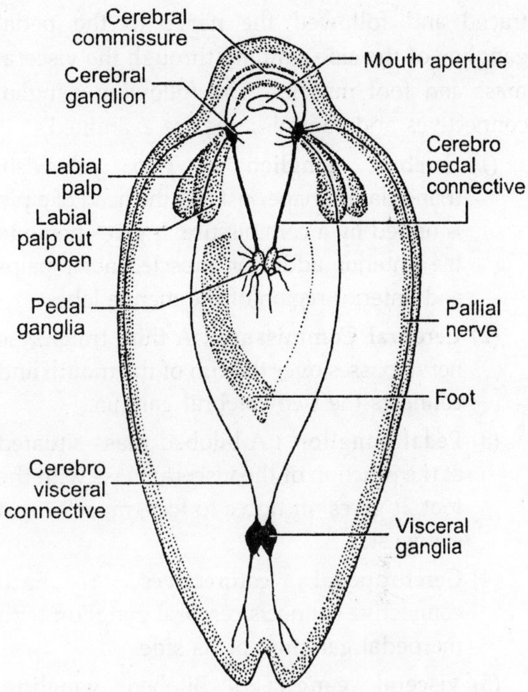

Fig. 1.4.28 : *Nervous system of a fresh water mussel. (Lamellidens marginalis)*

ganglia. While removing the covering two **cerebrovisceral connectives** became visible. The cerebrovisceral connective runs through the visceral mass to join the cerebral ganglion of the side.

Just below the anterior retractor muscle, at the base of labial palps a small, yellowish, triangular **cerebro pleural ganglion** or **cerebral gangilon** is found. The ganglion of either side is located with the help of a fine needle and also located the **cerebral commissure** passes over the upper rim of the mouth.

The muscular part of the foot is cut. A vertical incision is given along the middle line of the cut border of the foot and the two halves is slowly separated with a stout forceps located the **pedal ganglia** in anterior one-third foot.

The cerebropedal connectives which orginate at the posterior border of cerebral ganglion is traced and followed the nerve to the pedal ganglion of the side running through the visceral mass and foot muscles. The following ganglia, connectives and commissures are examined.

(1) **Cerebral ganglion** : Two yellowish, triangular in shape, just like the head of a pin is united by a commissure. It gives nerve to the anterior adductor muscle, labial palps and anterior region of the mantle lobes.

(2) **Cerebral Commissure :** A thin, transverse nerve passes over the rim of the mouth and connects the two cerebral ganglia.

(3) **Pedal ganglion :** A bilobed mass situated at the junction of the visceral mass with the foot. It gives off nerve to foot, muscles and statocyst.

(4) **Cerebropedal connectives** : Each connective connects cerebral ganglion with the pedal ganglion of its side.

(5) **Visceral ganglia** : Bilobed ganglion, flattened triangular in shape, situated mid ventrally on the posterior adductor muscles. It supplies nerves to the posterior part of the mantle, kidney, gill and posterior adductor muscles by its branches.

(6) **Cerebrovisceral connectives :** Two, each connect the anterior end of the visceral ganglia with the cerebral ganglion of the side.

9. *APPLE SNAIL (Pila globosa)*

Pila is quite abundant in the fresh water ponds, tanks, lakes and rice fields. They are **amphibious** i.e. can live in water and also on land.

External features :

Pila is killed with hot water. Care is taken so that the flesh is not cooked. It can also be killed by keeping then in one percent chloralhydrate solution overnight. (See **narcotization**)

Shell and Operculum : A killed *Pila* is held and the followings are observed :

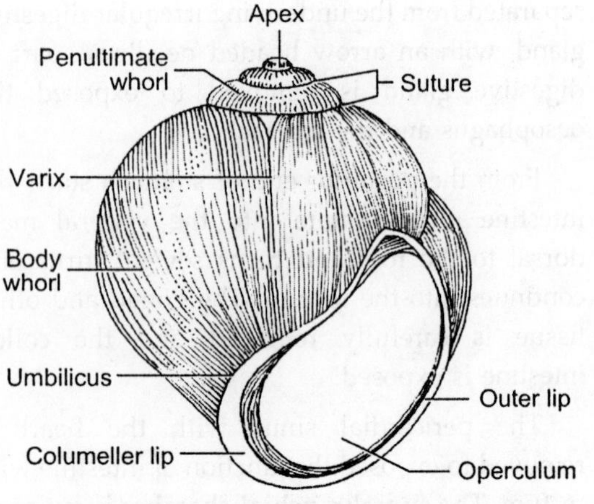

Fig. 1.4.29 : *External features of an apple snail. (Pila globosa) (Oral view)*

(1) **Shell** : Univalve globose type, closely coiled in a spiral manner around a central axis, the **columella.** Shell consists of whorls starting from the top and gradually increases in size. Immediately after apex is **penultimate whorl** and then **body whorl.** Whorls are demarcated by sutures. The first part of the shell at the top of the spiral is the **apex.** The surface of the shell is marked by numerous lines of growth, called **varices.** The shell opens outside by wide mouth or aperture. The smooth and continuous margin of the aperture is called **peristome,** it outer margin is **outer** lip and inner margin is **columellar lip.**

(2) **Operculum :** The aperture of the shell is closed by **operculum,** which is attached dorsally tightly to the hinder part of the foot.

Soft parts :

The shell is broken piece by piece with the blunt end of a scalpel or with a bone cutter, then the shell is completely removed while keeping operculum on the lower side and the following are observed.

After the shell is removed, the body of *Pila* is divisible into 3 parts.

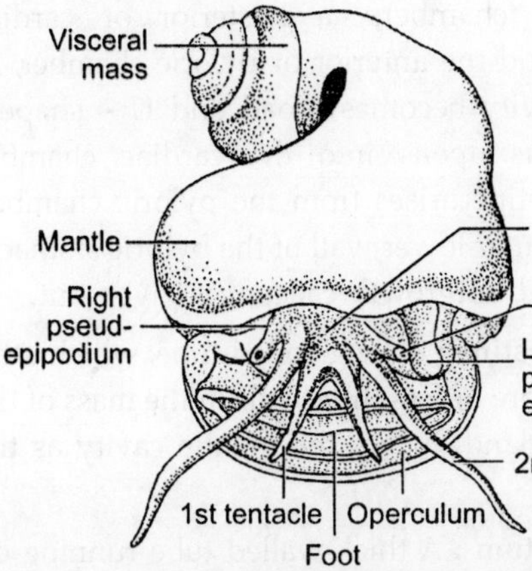

Fig. 1.4.29A : *External features of an apple snail. (After removing the shell) (Pila globosa)*

(1) **Head :** Anterior fleshy part of the body over hanging foot. It is prolonged into a contractile snout, which bears mouth and two pairs of tentacles. First pair is smaller while the second pair is long and contractile. At the base of each second pair of tentacles is a **stalked eye** or **ommatophore.**

(2) **Foot :** It is large, highly muscular and is roughly triangular in outline. The dorsal surface of the foot bears **operculum**, the smooth ventral surface is **sole.**

(3) **Visceral mass :** It forms the dorsal lump and occupies the greater part of the body – whorl and other inner whorls of the shell. It is coiled like the shell. The skin of the visceral mass is the **mantle** or **pallium.** Mantle encloses a large cavity in dorsolateral position is called **mantle cavity** or **pallial cavity.** On the right and left side of the head, over the foot, the mantle is prolonged into right and left **nuchal lobes** or **pseudepipodium.**

The Alimentary system :

Procedure :

The body whorl is broken up to suture as in case of external features. The operculum is removed with a sharp scalpel. The collumellar

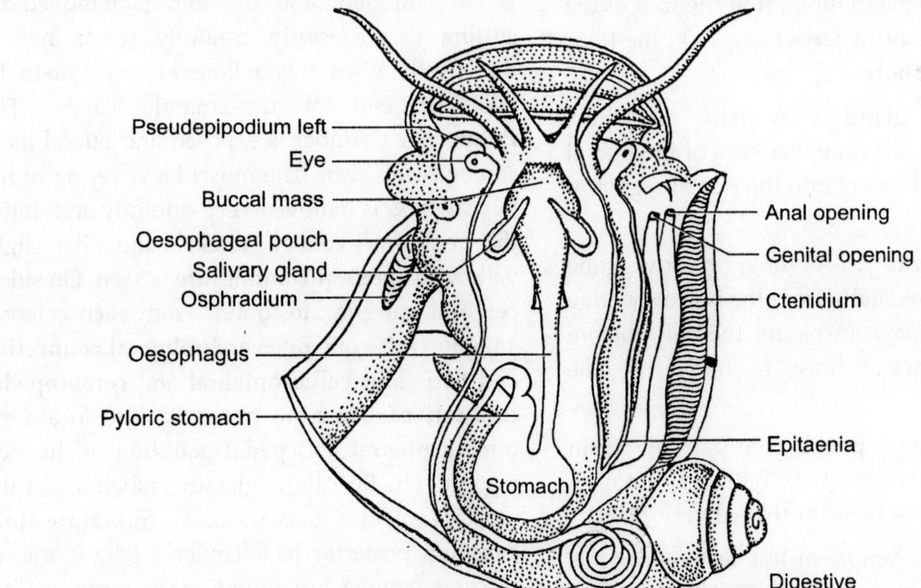

Fig. 1.4.30 : *Alimentary system of an apple snail. (Pila globosa)*

muscle are detached form the collumella and the *Pila* is freed from the collumella by slowly uncoiling the body in a direction opposite to the twist of the collumella. The snail is placed in a dissecting tray and a longitudinal incision is made from the edge of the mantle on the right side and left to the **ctenidium** up to the posterior limit. The mantle flap is reflected to the left and pinned.

A longitudinal cut is given along the middle of the floor of the mantle cavity and preceded anteriorly up to the head to expose the foregut. The flap is turned and the buccal mass is located. The oesophagus is traced next to the buccal mass. The oesophagus is followed and the stomach is located embedded in the digestive gland. The tissue of digestive gland is removed to expose the coiled intestine. The rectum is located on the right side of the mantle running on the left side of the ctenidium. The following are observed:

(1) **Mouth :** A median vertical slit at the tip of the snout.

(2) **Buccal mass :** It is a strong, thick walled structure next to the mouth. From the roof of the buccal cavity a pair of **jaws** hangs. On the floor lays the **odontophore**.

(3) **Salivary gland :** A pair of white, branching gland one on either side of the buccal mass. Each gland enters into the dorsal region of the buccal cavity.

(4) **Oesophagus :** A long, narrow tube entering posteromedially from the posterior edge of the buccal mass turns to the left, before entering the visceral mass to open into the stomach.

(5) **Oesophageal pouch :** A pair of cream coloured sac – like structure, lying one on each side of oesophagus beneath the salivary gland.

(6) **Stomach :** A more or less rectangular sac, dark red colour embedded partially in the digestive gland. A transverse arched ridge divides the cavity of stomach into two inter connective chambers, a posterior or cardiac chamber and the anterior or pyloric chamber, so that its cavity becomes broad and U – shaped. Oesophagus opens into the cardiac chamber while intestine arises from the pyloric chamber. From the outer lower wall of the pyloric stomach, small, blind outgrowth **caecum** is given out.

(7) **Intestine :** A long coiled tube, which after forming more or less three coils in the mass of the digestive gland, enters the mantle cavity as the rectum.

(8) **Rectum :** A thick walled tube running on the left side of the ctenidium and opens through anus.

(9) **Digestive gland :** A long, spirally coiled bilobed, somewhat brownish or dull green mass. Two main ducts from two lobes unite to form a single duct before opening into the stomach.

The Nervous system :

Procedure :

Same as in alimentary canal. The mantle fold is cut obliquely and the fold is removed by cutting it posteriorly carefully so as not to damage the visceral ganglion as it lies close to the posterior end of the mantle cavity. The Pulmonary chamber is exposed and buccal mass can be easily seen. The muscular covering of the buccal mass is removed very carefully and under the muscular coat a thick, band like, light yellowish **cerebral commissure** is seen. On sides, **cerebral ganglia** are found. From each cerebral ganglion arise one fine **cerebrobuccal connective** and two stout **cerebropleural** and **cerebropedal connectives**. They are traced back to locate the **buccal**, **pleural** and **pedal ganglion** of the side respectively. Buccal ganglia are embedded in the muscles of the buccal mass. **Supraintestinal ganglion** posterior to left nuchal lobe is traced. Visceral ganglia are found in posterior region. The very thin supra intestinal nerve passing over the oesophagus and going deep to join the right

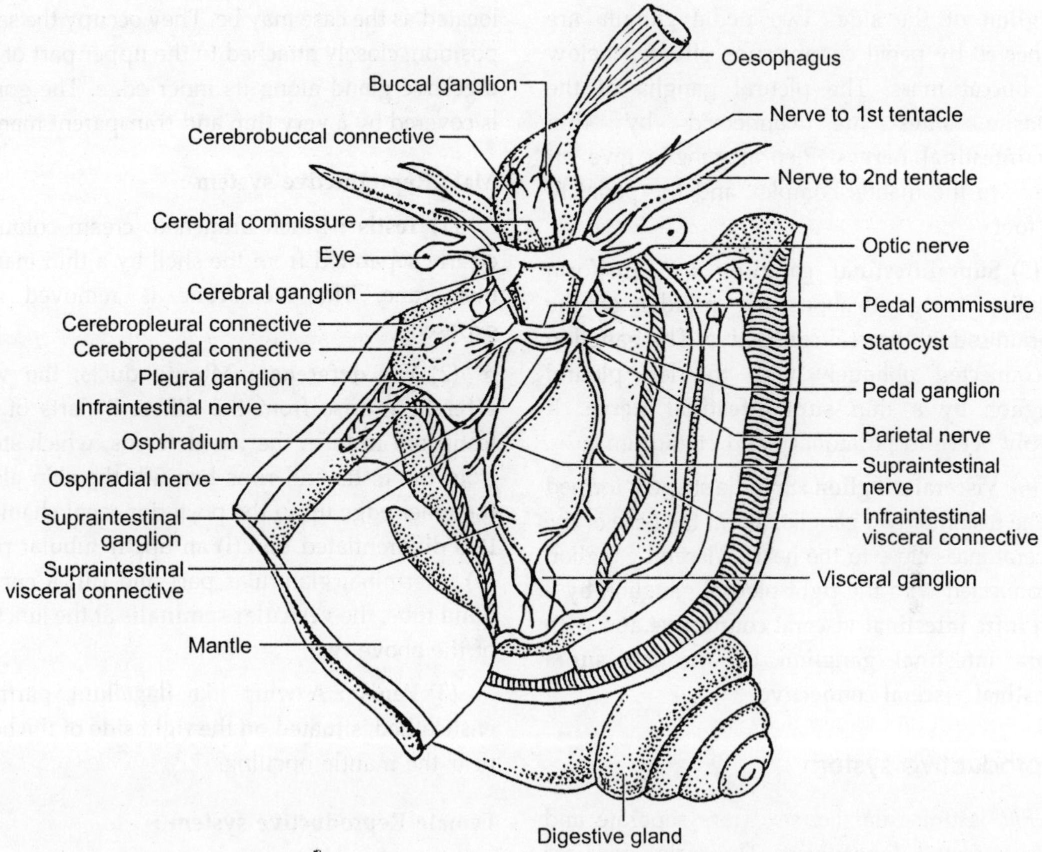

Fig. 1.4.31 : *Nervous system of an apple snail. (Pila globosa)*

pleuropedal ganglionic mass is exposed. The oesophagus is cut at its base and the major part is removed to expose the stout **pedal commissure.** The **infraintestinal nerve** that lies close to the **pedal commissure** is located. The **infraintestinal** and **supraintestinal visceral connective** are traced from the right and left pleuropedal ganglia to the visceral ganglion. The following are observed:

(1) **Cerebral ganglia :** Two, triangular, situated anteriorly on the dorsolateral side of the buccal mass. Each ganglion sends nerves to the snout, tentacles, buccal mass, eyes and statocyst. Both the ganglia are connected dorsal to the buccal mass by a thick, band like cerebral commissure.

(2) **Buccal ganglia :** Paired, situated on the dorsolateral side of the buccal mass at its junction with the oesophagus. Two buccal ganglia are connected with each other by a **buccal commissure**, located ventral to the oesophagus. Each cerebral ganglion is connected by **cerebropleural** and **cerebropedal connectives** with pleural and pedal ganglia of its side respectively. Each buccal ganglion supplies nerves to the buccal mass and radular sac.

(3) **Cerebrobuccal connectives :** Nerve connecting the cerebrals with the buccal ganglia.

(4) **Pleuropedal ganglia :** A pair of ganglionic mass, each is situated ventrolaterally to the buccal mass on each side. Each ganglionic mass is formed by the fusion of **pleural** and **pedal**

ganglion of the side. Two pedal ganglia are connected by pedal commissure, situated below the buccal mass. The pleural ganglia of the opposite sides are connected by thin **infraintestinal nerves**. Pleural ganglia give off nerves to the mantle complex and the pedal to the foot.

(5) **Supraintestinal ganglion :** A fusiform ganglion lying at about the middle of the supraintestinal visceral connective. The ganglion is connected obliquely with the left pleural ganglion by a thin **supraintestinal nerve.** It supply nerve to pulmonary and ctenidium.

(6) **Visceral ganglion :** Spindle shaped, formed by the fusion of two ganglia, lies at the base of the visceral mass close to the heart. Visceral ganglion is connected with the right pleural ganglion by a thin **infra intestinal visceral connective** and with **supra intestinal ganglion** by the left **supra intestinal visceral connective.**

Reproductive system :

Pila is unisexual i.e. sexes are separate and there is sexual dimorphism. The male shells are generally smaller than those of female shell.

Procedure : The snail is dissected as done previously. The spirally coiled digestive gland carefully uncoiled and the testis or ovary is located as the case may be. They occupy the same position, closely attached to the upper part of the digestive gland along its inner edge. The gonad is covered by a very thin and transparent mantle.

Male Reproductive system :

(1) **Testis :** Much branched, cream coloured clearly separated from the shell by a thin mantle membrane. The membrane is removed and exposed.

(2) **Vas deferens :** Minute ducts, the **vasa efferentia,** arise from the different parts of the testis join to form the vas deferens, which starts from the testis and runs beneath the skin along the inner edge up to the posterior renal chamber. It is differentiated into (i) an upper tubular part, (ii) a terminal glandular part and (iii) a curved blind tube, the **vesicular seminalis** at the junction of the above two.

(3) **Penis :** A whip like flagellum, partially ensheathed, situated on the right side of the body near the mantle opening.

Female Reproductive system :

(1) **Ovary :** Branched, orange coloured, occupies the same position as in case of testis.

(2) **Oviduct :** Formed by the joining minute ducts coming from the ovary. The tubular part of

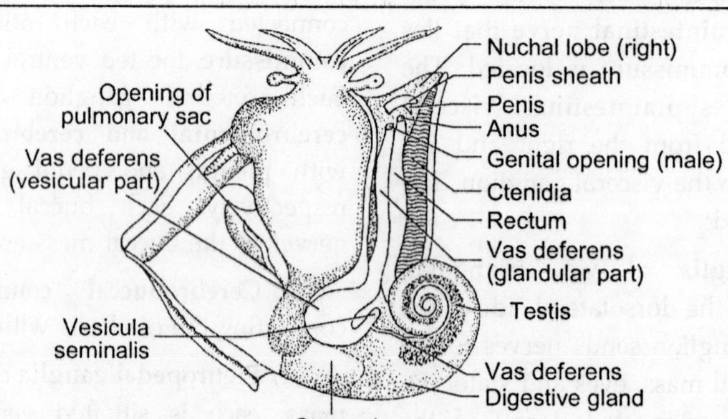

Fig. 1.4.32 : *Male reproductive system of an apple snail (Pila globosa)*

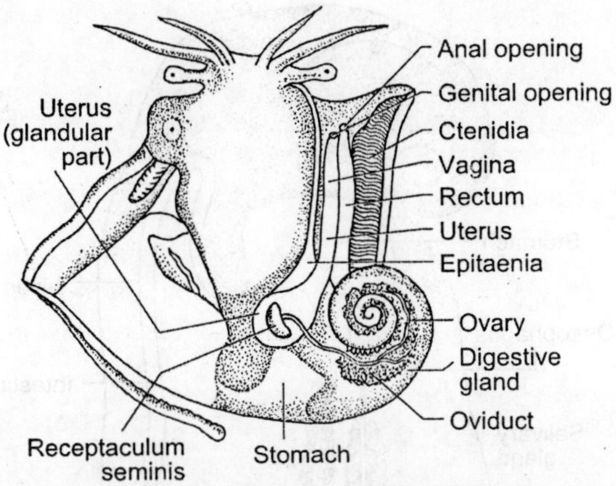

Fig. 1.4.33 : *Female reproductive system of an apple snail (Pila globosa)*

duct runs downward along the margin of the digestive gland and opens into a bean shaped **receptaculum seminis.** The lower glandular part of the duct lies on the floor of the mantle cavity parallel to the rectum and divisible into a large, pear shaped **uterus,** which lies below the intestine to the right of the renal chamber and a narrow, cream coloured tube **vagina** running forward and open into the mantle cavity a little behind the anus.

10. *LAND SNAIL (Achatina fulica)*

Achatina is terrestrial, found in open grasslands, gardens and hedges. They are found to climb vertical walls and trees.

External features :

Killing method is same as in case of *Pila.* It also can be killed by prolonged drowning in water. Land snails are put in a large glass jar, filled with water to the brim leaving no air space. A lid is placed and some heavy weight is put on the lid. Snails are killed in this way and they drop down at the bottom of the jar. Snails are washed thoroughly and the mucus on their body are removed.

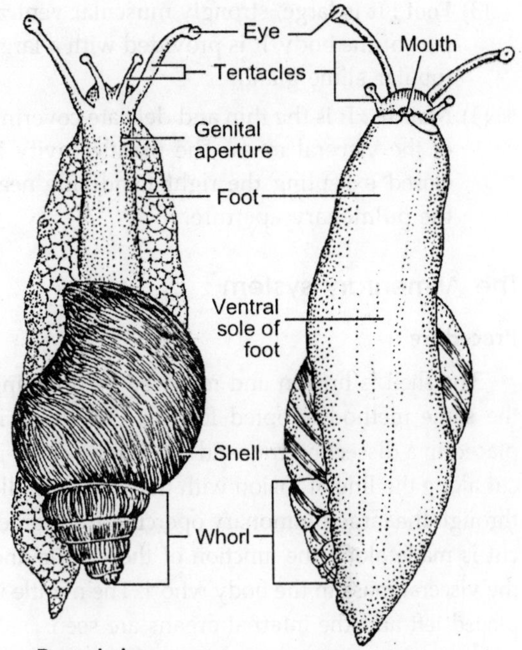

Fig. 1.4.34 : *External features of a land snail. (Achatina fulica)*

The following are observed :

(1) **Shell :** Univalve globose type, coiled in a spiral manner around a central axis. Shell with a high conical spire and made up of five to seven whorls. Colour of the shell is yellowish – brown, with broad bands. The body coils of the shell are known as whorls and they are in open communication. Note, operculum is absent.

Now, the body whorls are broken and the shell is removed completely and the soft parts are observed.

(2) **Head :** Anterior fleshy soft part of the body overhanging foot. It is long and can be projected to a great extent. It is produced anteriorly into a snout, which bears semicircular mouth and two pairs of tentacles. First pair is smaller, while second pair is longer. Eyes are two and are placed at the tip of the larger tentacles. Genital aperture is on the right side of posterior to head.

(3) **Foot :** It is large, strongly muscular ventral part of the body. It is provided with a large tubular slime gland.

(4) **Mantle :** It is the thin and delicate covering of the visceral mass. The mantle cavity is closed excepting the right hand side near the pulmonary aperture.

The Alimentary system :

Procedure :

The shell is broken and removed by applying the same method adopted for *Pila*. The snail is placed in a dissecting tray and the mantle collar is cut along the line of fusion with the visceral stalk, through the large pulmonary operculum. Another cut is made along the junction of the mantle and the visceral mass in the body whorl. The mantle is placed left and the internal organs are seen.

The specimen is pinned through the foot and the flap of the mantle. The organs in the visceral stalk and the perivisceral region are exposed by giving an incision on the middle of the dorsal wall of the visceral stalk anteriorly up to the head and posteriorly to the meeting point of the body whorl and the visceral mass. The tissue of digestive gland is removed to expose partially coiled intestine and stomach. The following are observed :

(1) **Mouth :** Semicircular ventral, ventrally placed at the anterior of the snout.

(2) **Buccal mass :** Large, highly muscular body and is the anterior part of the alimentary canal. It is guarded by semicircular cartilagenous jaw on the dorsal surface.

(3) **Salivary gland :** Paired, elongated, cream white. The ducts open in the buccal cavity at the base of the oesophagus.

(4) **Oesophagus :** Narrow tube, originating from the posterior ends of the buccal mass and dilates to form crop.

(5) **Crop :** A thin walled elongated sac divided into anterior large and posterior small chamber. A digestive duct opens to the posterior chamber.

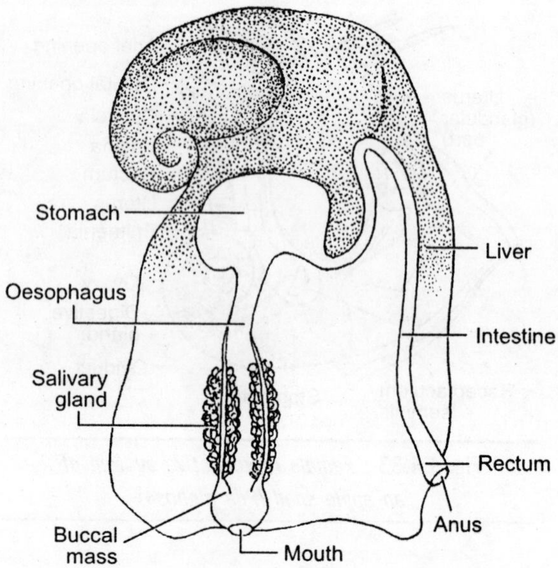

Fig. 1.4.35 : *Alimentary system of a land snail. (Achatina fulica)*

(6) **Stomach :** A thick walled spherical chamber remains embedded in the digestive gland. It receives ducts from the digestive gland.

(7) **Intestine :** A fairly long, thin walled, coiled tube, forms a loop in the first part and passes through the substance of the digestive gland.

(8) **Rectum :** A straight tube, attach to the posterior wall of the mantle cavity and opens to the exterior through anus in the pulmonary aperture.

(9) **Digestive gland :** A large, spirally coiled bilobed, blackish brown mass occupying major portion of the visceral mass.

The Nervous system :

Procedure :

The procedure described previously is followed. The nervous system of *Achatina* is greatly condensed, where the cerebral, pleural,

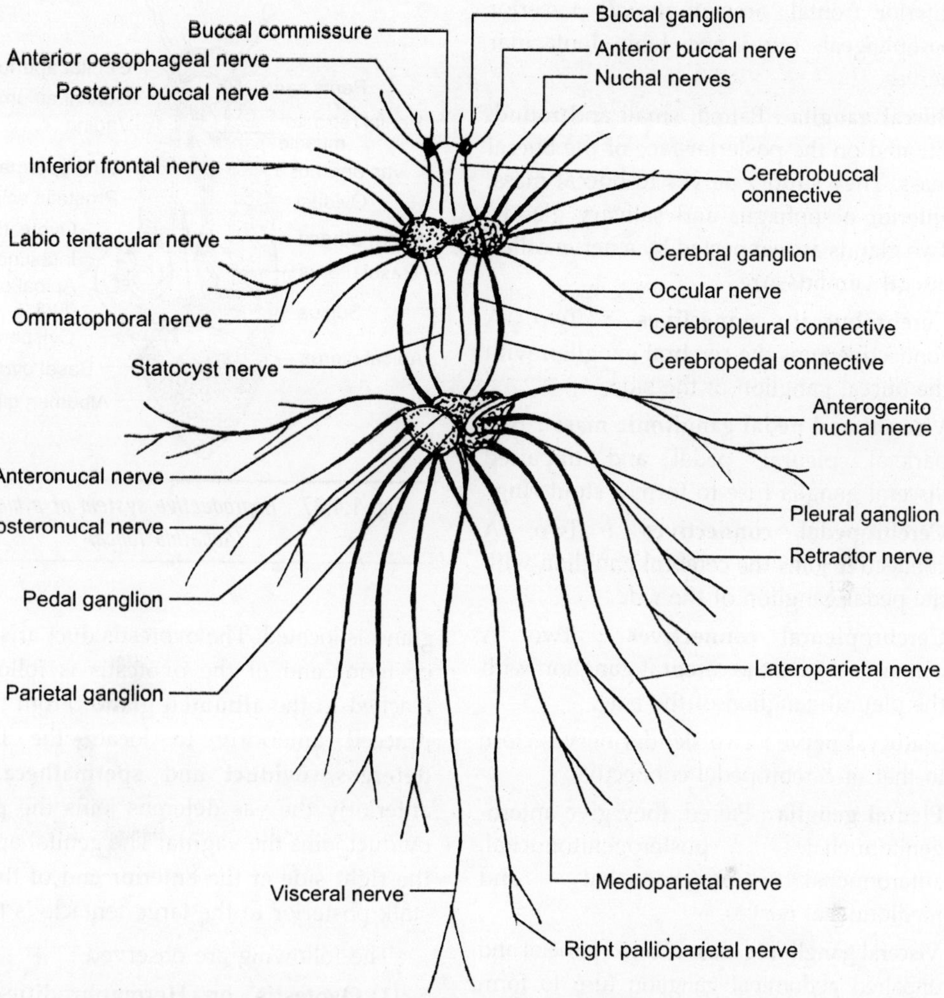

Fig. 1.4.36 : ***Nervous system of a land snail. (Achatina fulica)***

pedal and visceral ganglia are concentrated in the head region due to shortening of nerves between them. All these ganglia are paired excepting the visceral that is unpaired.

The thin membrane covering the cerebral ganglia is cleared. The oesophagus is cut near its origin and removed to expose the ventrally situated pleural, visceral and pedal ganglionic ring. All these ganglia are located around the oesophagus and thus form a **circumenteric nerve ring**. The buccal ganglia on the posterior side of the buccal mass are exposed. The commissure and connectives are traced. The statocyst nerve and cerebropedal connective of the side are ensheathed. The sheath is carefully removed and the fine statocyst and cerebropedal connective of all the nerves are separated, the visceral nerve is the longest and runs posteriorly.

The following are observed :

(1) **Cerebral ganglia** : Paired, almost round, yellowish red in colour, fused at the medial side. They send off ocular, superior frontal,

inferior frontal, ommatophoral, posterior oesophageal, nucal and labio tentacular nerves.

(2) **Buccal ganglia :** Paired, small and round, situated on the posterior face of the buccal mass. They supply nerves to buccal mass, anterior oesophagus and salivary glands. Two glands are connected by a nerve called **buccal commissure.**

(3) **Cerebrobuccal connectives :** Two. A connective joins the cerebral ganglion with the buccal ganglion of the side.

(4) **Visceral and pedal ganglionic mass :** Two parietal, pleural, pedal and unpaired visceral ganglia fuse to form a stout ring.

(5) **Cerebropedal connectives :** Two. A connective joins the cerebral ganglion with the pedal ganglion of the side.

(6) **Cerebropleural connectives :** Two. A connective joins the cerebral ganglion with the pleural ganglion of the side.

(7) **Statocyst nerve :** Two slender nerves close to that of cerebropedel connective.

(8) **Pleural ganglia :** Paired, they give antero-genitonuchal, posterogenitonuchal, anteronuchal, posteronuchal and medionuchal nerves.

(9) **Visceral ganglionic mass :** Paired parietal and unpaired abdominal ganglion fuse to form this ganglionic mass. It gives retractors, lateroparietals, medioparietals, pallioparietals, and one longest visceral nerve.

(10) **Pedal ganglia :** Paired, they give rise to nine pairs of nerves which form regular plexuses in the sole of the foot along the borders.

The Reproductive system :

Procedure :

Achatina is hermaphrodite. The specimen is dissected as in case of alimentary system. Carefully uncoiling the digestive gland and the **ovotestis** or **hermaphroditic gland** which is embedded in the posterior lobe of the digestive

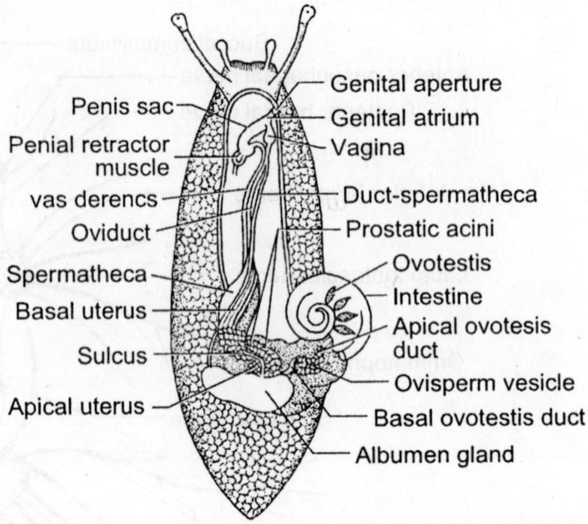

Fig. 1.4.37 : *Reproductive system of a land snail. (Achatina fulica)*

gland is located. The ovotestis duct arising at the posterior end of the ovotestis is followed and reached to the **albumen gland.** From this gland proceed anteriorly to locate the large **vas deferens**, **oviduct** and **spermatheca**. Further anteriorly the vas deferens joins the **penis** and oviduct joins the **vagina**. The genital aperture on the right side at the anterior end of the visceral stalk posterior to the large tentacle is traced.

The following are observed :

(1) **Ovotestis or Hermaphroditic gland :** Finger like lobes covered by transparent membrane.

(2) **Ovotestis duct :** It has three regions; the first region is **apical ovotestis duct**, the second region is enlarged and much folded **ovisperm vesicle** and the basal region is the **short ovotestis duct.**

(3) **Albumen gland :** Slightly curved, cream white gland.

(4) **Spermoviduct :** It is a fairly long sac—like structure formed the male portion of the **genital duct (sulcus)** and **uterus.** The uterus has two regions, one is the cream white **apical uterus** and the other is the light yellow **basal uterus.**

(5) **Oviduct :** The duct joints the basal uterus posteriorly and vagina anteriorly.

(6) **Spermatheca :** A thin walled, tubular adhere to the anterior end of the spermoviduct. The long duct of spermatheca joins the oviduct.

(7) **Vagina :** A muscular chamber. The anterior wider end received the oviduct. Posterior end is narrow.

(8) **Vas deferens :** A long, narrow tube joins the penis.

(9) **Penis :** A curved, muscular structure covered by a sheath.

(10) **Genital atrium :** A small chamber connected with the penis and vagina. Genital atrium opens through common genital aperture situated on the right side posterior to the head.

11. *CUTTLE FISH (Sepia sp.)*

Sepia is a marine mollusc found usually near the surface of water, Sepia is popularly known as **cuttlefish**. Dissection is done in specimens preserved in formaldehyde.

External Features :

The specimen is thoroughly washed in water and the following are observed :

(1) **Shape and size :** The body of sepia is cylindrical, dorsoventrally flattened and twenty centemetre in length.

(2) **Colour :** It is usually brown with white marks.

(3) **Body :** The body is divisible into a distinct head and dorsoventrally flattened **trunk** separated by narrow **neck**.

(a) **Head :** Head bears two large eyes and ten oral arms. Eight of these arms are short and stout and possesses on their inner surface four longitudinal rows of suckers. The fourth pair of the arms is comparatively longer and is distinguish as tentacles. In male Sepia, the tentacle of the left side is modified into a **hectocotylized arm** and act as intromittent organ. Mouth is situated in the middle of arms surrounding by a fleshy circular lip.

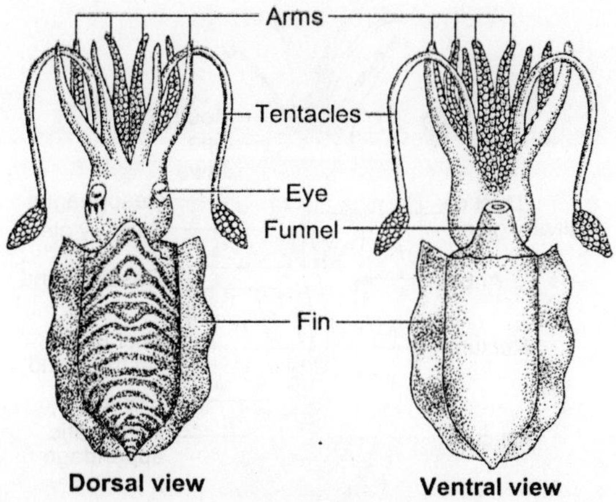

Fig. 1.4.38 : *External features of a cuttle fish. (Sepia officinalis)*

(b) **Trunk :** It is shield shaped covered by a thick muscular mantle having a rounded lobe in front. Trunk is bordered by a fin on either side.

(4) **Shell :** It is a flat ovoidal structure, situated in the upper side of the mantle. It is internal and known as **cuttle-bone.**

(5) **Funnel :** It is a large, conical, muscular funnel-like tube present on the mid-dorsal side of the oral end.

The Alimentary system :

Procedure :

A formalin fixed preserved specimen is taken. Thoroughly washed with water and the specimen is dissected dorsally. The *Sepia* is held on the palm of the left hand, and a deep transverse cut is given along the neck, passing below the ridge formed by mantle. Now from the neck region two lateral longitudinal cut one on each side on the mantle up to the aboral end

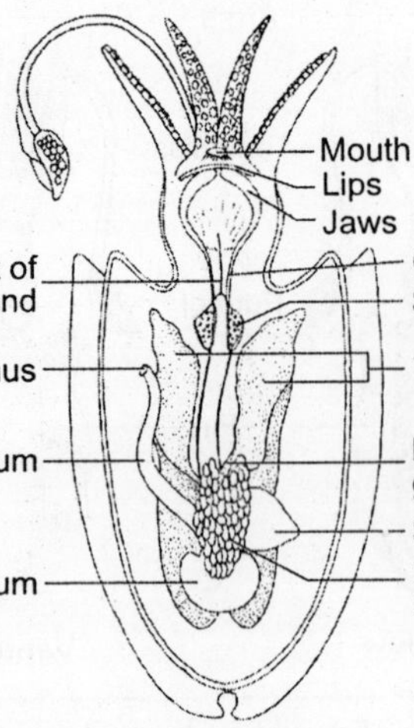

Fig. 1.4.39 : *Alimentary system of a cuttle fish. (Sepia officinalis)*

is made and carefully the mantle is removed by separating it from the tissues to expose the pallial complex. The specimen is placed on a dissecting tray and pinned along the fins. To expose the mouth and the foregut a medial incision is given in the head from neck up to the base of the arms. The flaps are stretched and pinned. The dorsal part of the cranium is removed and the cerebral ganglia to fully expose the oesophagus.

The following are observed :

(1) **Mouth :** Anterior, surrounded by the bases of the ten long oral arms. The lip is fleshy, circular and provided with many **papillae.** Lip contents two **jaws.**

(2) **Buccal mass :** Thick walled, muscular, containing **odontophore.**

(3) **Oesophagus :** A narrow straight tube running to wards the aboral end to open into stomach.

(4) **Stomach :** Thick walled, muscular has two inter connected chambers. The first chamber is oval **gizzard** and the second chamber is spirally coiled **caecum.**

(5) **Intestine :** A short round tube arises from caecum and runs anteriorly towards the oral end as rectum.

(6) **Rectum :** A round tube runs forward, parallel and ventral to oesophagus, which opens into the mantle cavity through **anus.**

(7) **Salivary gland :** A pair of glands situated close to the oesophagus in front of digestive gland. Each is connected with a duct which unites. The common duct opens into the buccal cavity.

(9) **Digestive gland :** Near the salivary glands up to aboral posterior end of the body extends on each side a large brown digestive gland. There are two sets of digestive glands: the larger consist of a fused pair of brownish glandular lobes called **liver**, whose ducts enter the stomach near the origin of caecum. The other set are smaller, cream coloured and surround the liver ducts and constitute the **pancreas.**

Dissections VERTEBRATES

1. *LATA FISH* (*Channa punctatus*)

Lata is a bony fish inhabits the fresh water ponds and ditches: Live fishes are collected from the market. They are killed in hot water (not boiling).

The live fishes are washed with tap water and placed in a container. Sufficient amount of hot water is poured in the container and covered with a lid. The fishes are removed when their movement stops and quickly the fishes are put in cold water.

External features :

Procedure :

A lata fish is procured and the following external features are examined.

(i) **Shape** : Elongated, cylindrical body.

(ii) **Colouration :** The colour of fish varies with water. Back is greenish and yellow on sides. Abdomen is with dark stripes.

(iii) **Divisions** : Body is differentiated into head, trunk and tail.

(a) **Head** : It is triangular and tapers into a pointed snout. It is depressed and covered by large plate-like scales resembling the head shields of snakes. Head contains the following structures.

(i) **Mouth** : It is terminal, and bounded by thick lips having folds. Teeth are present on jaw and palate.

(ii) **Nostrils** : Paired nostrils are situated dorsally in front of the eyes.

(iii) **Eyes** : Two large circular eyes situated laterally on the head.

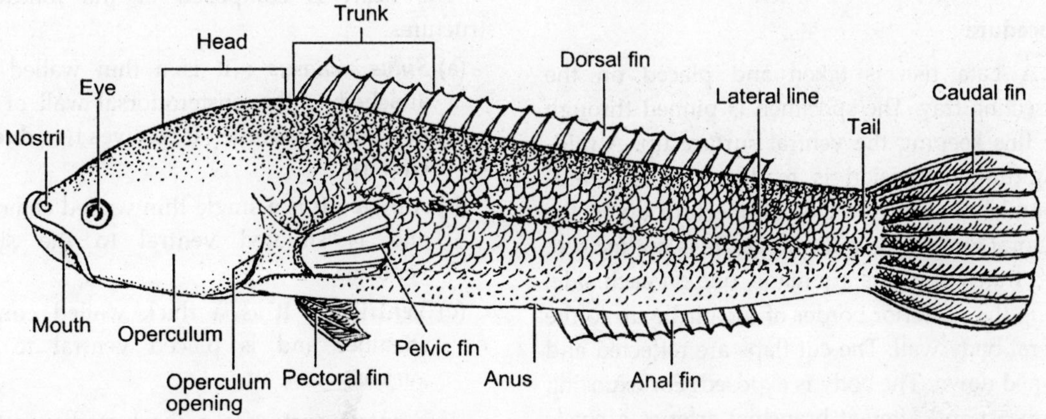

Fig. 1.4.40 : *External features of a lata fish. (Channa punctatus)*

(iv) **Operculum** : In the posterior region of head, on each side a gill cover or operculum is present. Situated between the **operculum** and side of the body there is large gill opening.

(b) **Trunk** : It is the region which extends from the operculum to the base of the tail. Both trunk and tail are covered by **ctenoid scales.** Lateral line is slightly curved and runs on the lateral side of the body. It is interrupted at places.

Trunk contains both paired and unpaired or median fins.

(i) *Unpaired fins* : Dorsal fin is single and extends up to tail. It is supported by spinous bony fin rays. Anal fin or ventral fin is situated just behind the anus and extends up to tail. It is also supported by spinous bony fin rays.

(ii) *Paired fins* : Pectoral fins are situated at the anterolateral side of the trunk behind the operculum and provided with soft rays. Pelvic fins are located on the ventral side behind the pectorals and also provided with soft rays.

(iii) *Tail* : Trunk is followed by tail. Tail is **homocercal.** The caudal fin is situated at the tip of the tail and is fan-shaped. It is supported by rays.

Afferent branchial system :

Procedure :

A Lata fish is taken and placed on the dissecting tray. The specimen is pinned through the fins keeping the ventral surface up. A mid-longitudinal incision is made along the mid-ventral line from the cloacal aperture up to the pectoral girdle. The pectoral girdle is not incised. Two transverse cuts are made, one on each side, along the posterior border of the girdle up to the lateral body wall. The cut flaps are reflected and pinned down. The body is exposed. For exposing the heart and afferent branchial arteries a cut is given through the pectoral girdle and the dental

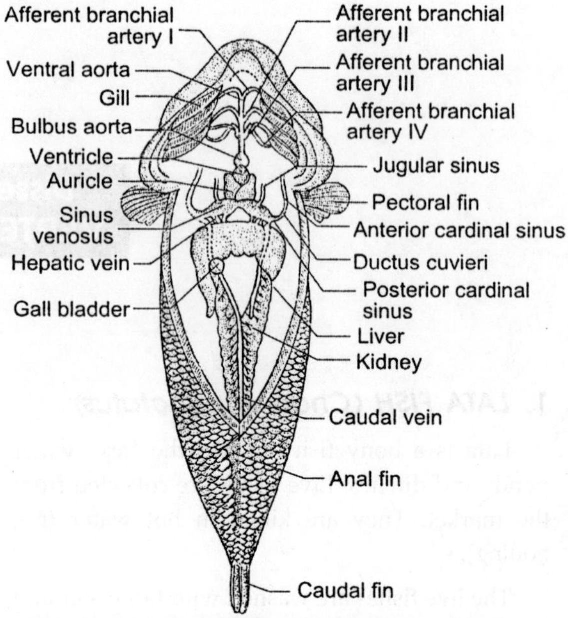

Fig. 1.4.41 : ***Afferent branchial system of a lata fish. (Channa punctatus)***

plates on the roof the pharynx are removed. Care has taken not to injure the heart and underlying blood vessel. The pericardium is removed and the following are observed.

(1) **Heart** : It is located ventral to the oesophagus in the pericardial chamber. It is enclosed with in a thin pericardium.

The heart is composed of the following structures.

(a) ***Sinus venosus*** : It is a thin walled sac attached to the posterodorsal wall of the pericardial chamber. It receives two **ductus cuvieri.**

(b) ***Auricle*** : It is a single thin walled chamber and is situated ventral to the sinus venosus.

(c) **Ventricle** : It is a thick walled conical chamber and is placed ventral to the auricle.

(2) **Ventral aorta** : Is a short median artery that arises from the base of the ventricle, the

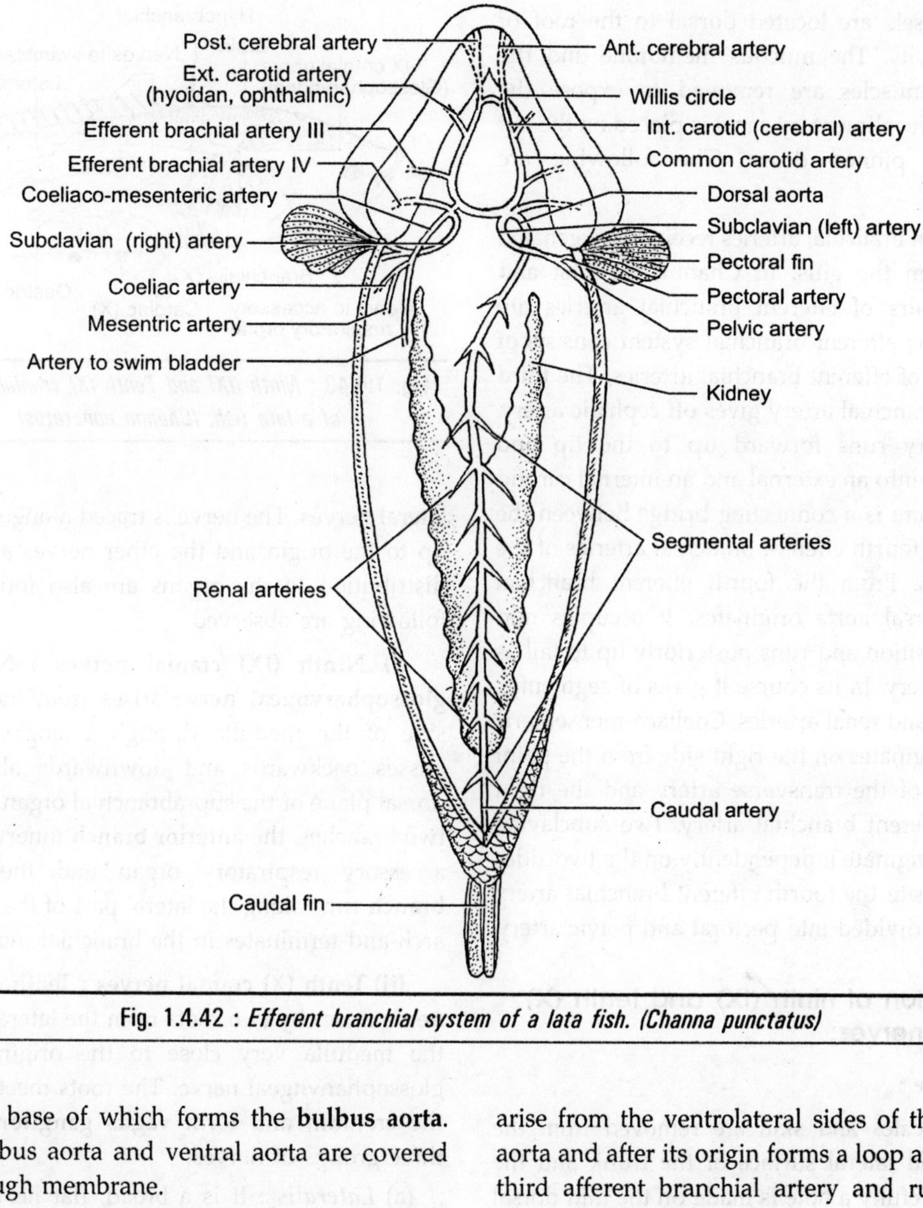

Fig. 1.4.42 : ***Efferent branchial system of a lata fish. (Channa punctatus)***

swollen base of which forms the **bulbus aorta.** The bulbus aorta and ventral aorta are covered by a tough membrane.

(3) **Afferent branchial arteries** : From the ventral aorta two pair of afferent branchial arteries emerges out to supply to gills. All the four pairs have independent origin from the ventral aorta.

The first and second pairs of arteries are straight arteries leading to the corresponding pair of gills. The third pair are lateral in origin and run to the third pair of gills. The fourth pairs arise from the ventrolateral sides of the ventral aorta and after its origin forms a loop around the third afferent branchial artery and run to the fourth pair of gills.

Efferent branchial system :

Procedure :

The procedure for dissection is same as in afferent branchial system. The efferent branchial arteries are located in the roof of the buccal cavity and pharynx. The fish is dissected from the roof of pharynx and the flaps are pinned down. The blood vessels are located dorsal to the roof of buccal cavity. The mucous membrane and the adjacent muscles are removed to expose the arteries. The alimentary canal is placed on the left side and pinned down. The following are observed.

Efferent branchial arteries received oxygenated blood from the gills. In Channa the first and second pairs of efferent branchial arteries are absent. The efferent branchial system consists of two pairs of efferent branchial arteries. The third efferent branchial artery gives off cephalic artery. This artery runs forward up to the tip and bifurcates into an external and an internal carotid artery. There is a connecting bridge between the third and fourth efferent branchial arteries of the two sides. From the fourth efferent branchial artery dorsal aorta originates. It occupies mid dorsal position and runs posteriorly up to tail as caudal artery. In its course it gives of segmental, gonadial and renal arteries. Coeliaco-mensenteric artery originates on the right side from the point of union of the transverse artery and the right fourth efferent branchial artery. Two subclavian arteries originate independently on the two sides just opposite the fourth efferent branchial artery and each divided into pectoral and pelvic artery.

Distribution of ninth (IX) and tenth (X) cranial nerves:

Procedure :

The scales and skin are removed from the dorsal and lateral surface of the trunk and the head. Carefully a hole is made on the mid dorsal region of the cranium with the help of a sharp scalpel. With a pair of stout forceps the cranium is broken by pushing it through the hole. The dorsal and lateral walls of the cranium are removed. Care is taken not to damage the brain and the roots of the cranial nerves. To expose brain the covering sheath is removed with a brush. A longitudinal incision is made with a sharp scalpel along the lateral line to expose the

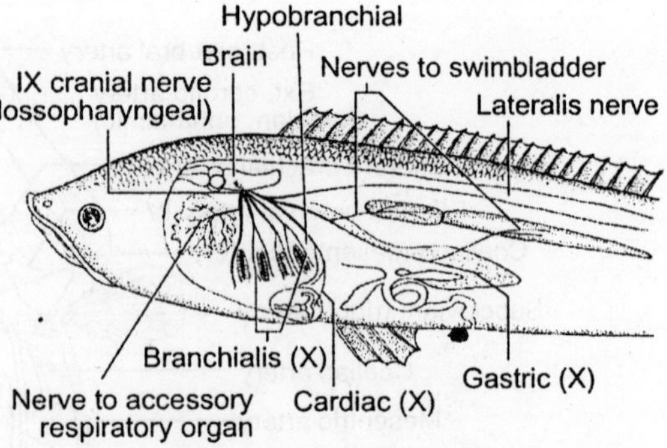

Fig. 1.4.43 : *Ninth (IX) and Tenth (X) cranial nerves of a lata fish. (Channa punctatus)*

lateral nerves. The nerve is traced along it course up to the origin and the other nerves and their distribution to the organs are also found. The following are observed.

(i) Ninth (IX) cranial nerves : Ninth or **glossopharyngeal nerve** arises from the lateral side of the medulla through a single root. It passes backwards and downwards along the dorsal plane of the suprabranchial organ. It gives two branches, the **anterior branch** innervates the accessory respiratory organ and the **lateral branch** runs along the lateral part of the first gill arch and terminates in the branchial muscles.

(ii) Tenth (X) cranial nerves : Tenth of **vagus nerve** arises by two roots from the lateral side of the medulla very close to the origin of the glossophanyngeal nerve. The roots meet outside the cranium and form **vagus ganglion**. It has three groups of nerves.

(a) ***Lateralis*** : It is a broad, flat nerve arises from the lateral part of vagus ganglion. It enters deep into the muscles to reach the vertebral column and proceeds straight up to the tail.

(b) ***Visceralis*** : This nerve originates from the vagus ganglion and passes down ward and gives two branches. The **dorsal branch** supplies nerves to the dorsal muscle. The **middle branch** after runing

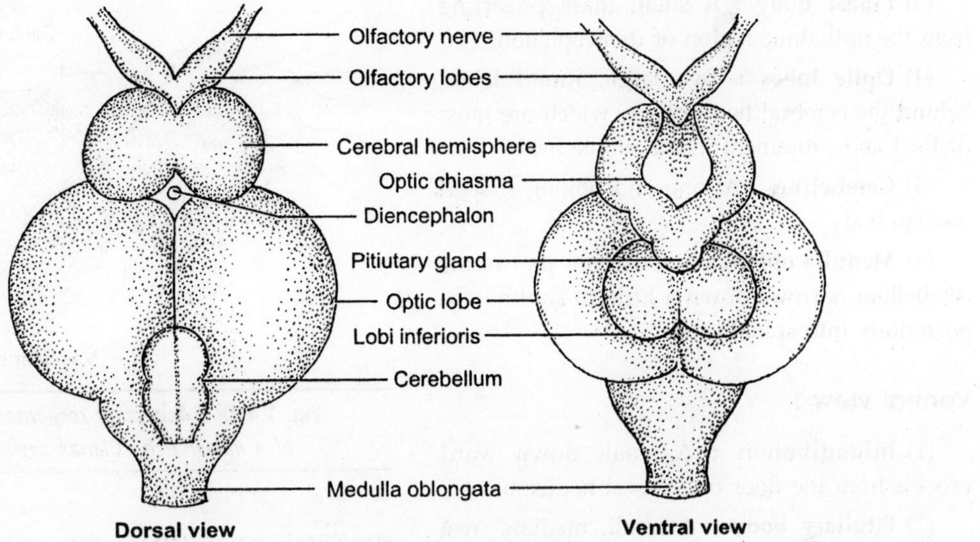

Fig. 1.4.44 : *Brain of a lata fish. (Channa punctatus)*

some time bifurcates. The **dorsal branch** innervates the kidney and airbladder, while the ventral branch, **the gastric** innervates the stomach. A thin and most ventral branch, **the cardiac** innervates the heart.

(c) ***Branchialis*** : Four groups of nerves. The **first branchialis** originated from the vagus ganglion moves and bifurcates into a thinner **pre-trematic** and a stouter **post-trematic**. Pre-trematic innervates the first gill arch while the post-trematic to second gill arch. The **second branchialis** arises from the vagus ganglion, moves and bifurcate as pre and post-trematic branches innervate second and third gill arch respectively. The third and fourth branchialis arise at the base of the visceralis from a common origin. The pre and post-trematic branches of the third branchialis supplies to third and fourth gill arches respectively. The **fourth branchialis** innervates the fourth gill arch by a fine pre-trematic branch.

The brain :

Procedure :

To expose the brain the same procedure is adapted as in case of dissection of cranial nerves. Bones from the anterior and posterior border of the cranium are removed. The roots of all the cranial nerves arising from the brain are cut and the anterior portion of medulla oblongata. A transverse cut is given at the posterior end of the medulla and carefully the brain is lifted with a pair of fine forceps and this separates the brain from the cranium. The brain is placed in a watch glass containing water. Different parts are observed.

The **brain** consists of **forebrain** or **prosencephalon**, the **midbrain** or **mesencephalon** and the **hindbrain** or **rhombencephalon**.

Dorsal view :

(1) **Olfactory lobe** : Two well developed round structure in front of the two cerebral hemisphere.

(2) **Cerebrum** : Two well developed large, semicircular bodies, posterior to the olfactory lobe. Individual lobe is called cerebral hemisphere.

(3) **Pineal body** : A small mass projecting from the opthalmic region of diencephalon.

(4) **Optic lobes** : Two large, round lobes, behind the cerebral hemisphere, which are most distinct and prominent in the whole brain.

(5) **Cerebellum** : A long and oblong shaped median body.

(6) **Medulla oblongata** : Posterior part of the cerebellum narrows down a little to continue as posteriorly into spinal card.

Ventral view :

(1) **Infundibulum** : A small down ward process from the floor of cerebral hemisphere.

(2) **Pituitary body** : A small, median, oval body to the anterior end of infundibulum.

(3) **Lobi inferiors** : Two, some what large bean-shaped bodies situated on either side of the pituitary body.

(4) **Optic nerves** : Two nerves of the optic lobes simply cross one another.

(5) **Saccus vasculous** : A narrow, median, highly vascular structure at the tip of infundibulum.

2. *MAGUR FISH (Clarias batrachus)*

Accessory respiratory organ :

Procedure :

The live specimen is killed by striking the head with a hard object. The operculum is cut along the border of the branchial chamber and the suprabranchial chamber which is enclosed in a highly vascularised membrane is located. The membrane is cut, the specimen is fixed on a dissecting tray in a lateral position. Now the cut membrane of the supra branchial chamber is pinned down and observed the accessory respiratory organ.

Structural organisation :

The accessory respiratory organs are two, one in each supra branchial chamber. It is a complex structural organization and consists of the following.

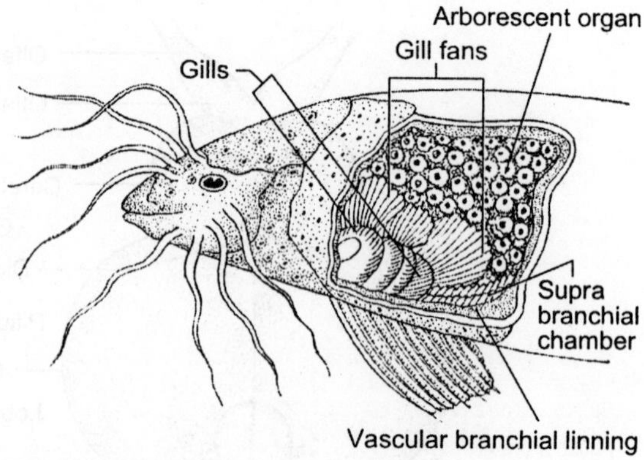

Fig. 1.4.45 : *Accessory respiratory organ of a magur fish. (Clarias batrachus)*

(1) An elaborate tree-like, red colour dendritic structure growing from the upper end of the second and fourth gill arches of either side.

(2) A pair of highly vascularized suprabranchial chambers which house the tree like structure.

(3) The entrance of the suprabranchial chamber is guarded by fan-like structure called gill-fans.

3. *SINGI FISH (Heteropneutes fossilis)*

Accessory Respiratory Organ :

Procedure :

The branchial chamber is exposed as in case of *Clarias*. With the help of a sharp scalpel a deep cut is given along the lateral line up to the tail. Care is taken not to damage the air tube. The muscles are separated along the course of incision to expose the air tube. The air tube is traced from posterior upto its anterior end in branchial chamber. The organ is observed.

Pneumatic sac :

(1) A pair of tubular pneumatic sacs, one on each side of the body arises as the out growth

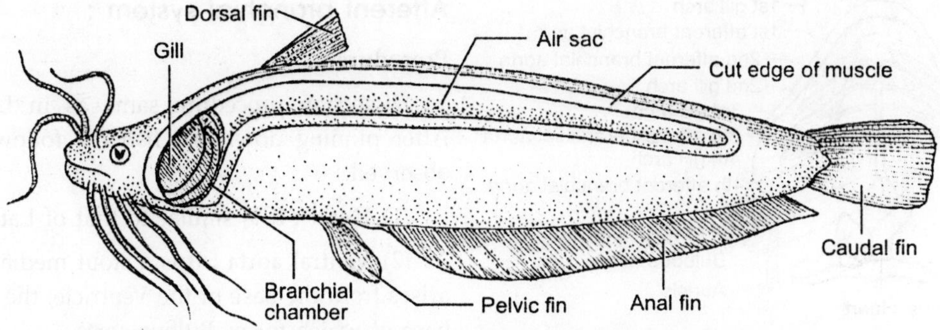

Fig. 1.4.46 : *Accessory respiratory organ of a singi fish. (Heteropneutes fossilis)*

from the branchial chamber and extends almost up to the tail.

(2) Anteriorly the tubes communicate by a large opening with the branchial chamber.

(3) The narrow posterior end is closed.

4. *ROHU FISH (Labeo rohita)*

Rohu is a bony fish inhabits rivers, beels, lakes, ponds etc. Fishes are collected from market and taken for dissection.

External features : (see Fig. 1.1.2.23)

Procedure :

Collected fish is taken and the following external features are examined :

(1) **Shape** : The body is roundish and elongated with a small head and a convex abdominal profile.

(2) **Coloration :** The body colour is brownish gray or bluish on the back.

(3) **Division** : Body is differentiated into head, trunk and tail.

(a) **Head :** Head is long and extends from the snout up to the posterior margin of operculum. Snout is depressed and is continued beyond the jaws. Head contains the following structures.

(i) **Mouth :** Mouth is inferior crecentric and not terminal. It is bounded by thick fringed lips having a distinct fold above and below. One pair of thin, sharp tiny barbels are situated on the maxilla.

(ii) **Nostrils :** Parried nostrils are situated dorsally in front of the eyes and open into the olfactory sacs.

(iii) **Eyes** : Eyes are large but lid less. Cornea is covered by a transparent layer of skin.

(iv) **Operculum :** In the posterior region of head, on each side a gill cover or operculum is present, situated between the operculum and side of the body there is a large gill opening.

(b) **Trunk :** Trunk extends from the operculum to the base of the tail. Both trunk and tail are covered by thin rounded **cycloid scales.** The lateral line runs on the lateral side of the body. Trunk contains both parried and unpaired or median fins.

(i) **Unpaired fins :** Dorsal fin is single and lies midway between snout and caudal base. It has fifteen to sixteen fin rays. The anal or ventral fin is situated just behind the anus. The anal fin possesses seven fin rays. Caudal fin is **homocercal**-two lobes are of equal size. The fin has nineteen fin rays.

(ii) **Paired fins :** Pectoral and pelvic fins are paired. Pectoral fins are situated at the anteroateral side of trunk behind the operculum. Each has seventeen to

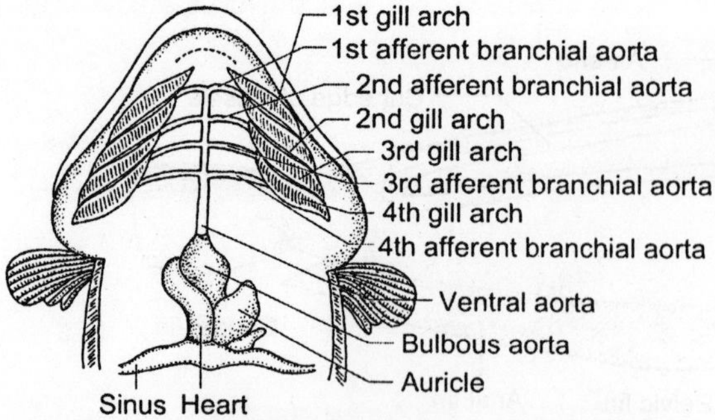

Fig. 1.4.47 : *Afferent branchial system of a rohu fish. (Labeo rohita)*

nineteen fin rays. Pelvic fins are located on the ventral behind the pectorals. Each has nine fin rays.

(c) **Tail :** Trunk is followed by a **homocercal tail.**

Afferent branchial system :

Procedure :

Dissection procedure same as in Lata fish After pinning down the fish the following are observed.

(1) **Heart :** It is similar to that of Lata fish.

(2) **Ventral aorta :** It is a stout median artery arises from the base of the Ventricle, the swollen base of which forms Bulbus aorta.

(3) **Afferent branchial arteries :** Ventral aorta runs forward along the ventral surface of the pharynx up to the posterior end of the hyoid. From the ventral aorta four pairs of afferent branchial arteries emerge out to supply to the respective gill. The first afferent originates by bifurcation of the ventral aorta itself, while the second, third and fourth pairs of the afferent

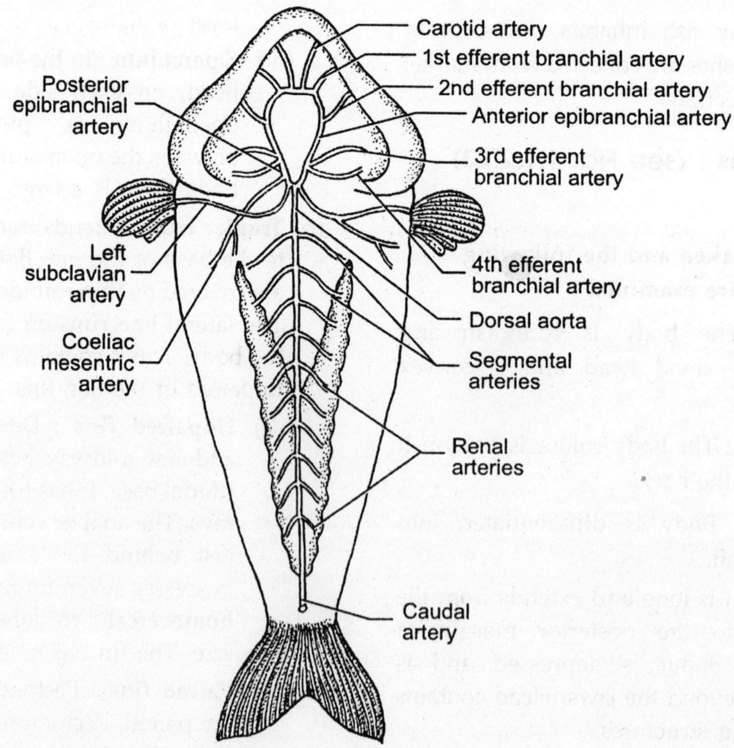

Fig. 1.4.48 : *Efferent branchial system of a rohu fish. (Labeo rohita)*

branchial arteries have separate and independent origins from the ventral aorta.

Efferent branchial system :

Procedure :

Dissection procedure same as in Afferent branchial system. The fish is dissected from the roof of the pharynx and the flaps are pin down. The blood vessels are located dorsal to the roof of buccal cavity. The alimentary canal is placed on the left side and pin down. The following are observed:

Efferent branchial arteries received oxygenated blood from the gills. There are four pairs of efferent branchial arteries corresponding to the number of gills. The first and second efferent branchial arteries on each side unite and form anterior **epibranchial artery,** then they unite posteriorly along the middorsal line to form dorsal aorta. The third and fourth efferent branchial arteries on each side unite to form posterior epibranchial artery. First efferent branchial artery gives rise to **cephalic artery, external carotid artery** and **internal carotid artery**. The dorsal aorta extends backward, mid dorsal in position and continue as caudal artery in the tail.

The dorsal aorta gives the following arteries :

(1) **Coeliaco-mesentric artery :** Supplies the swim bladder, stomach, intestine, liver spleen etc.

(2) **A pair of subclavian arteries :** One branch supplying pectoral fins and the other branch to pelvic fin.

(3) The rest of **dorsal artery** gives numbers of paired branches such as **segmental arteries renal arteries** and **genital arteries.**

Fifth (Vth) and Seventh (VIIth) cranial nerves of a Rohu fish :

Procedure :

The same technique is followed as in case of Lata fish. The brain is exposed and the nerves are traced from their root and their courses are followed.

The fifth and seventh cranial nerves arises from the lateral side of the medulla oblongata and form a trigemino facial ganglionic complex.

1. **Fifth (Vth) cranial nerve : Trigeminal** or fifth (Vth) after its origin devides into three branches : (a) **superficial opthalmicus,** (b) **maxillary** and (c) **mandibular.**

(a) **Superficial opthalmicus :** A thin ventral branch after its origin runs beside the opthalmicus branch of the seventh cranial nerve to innervate the orbital surface bordering its rim and the skin.

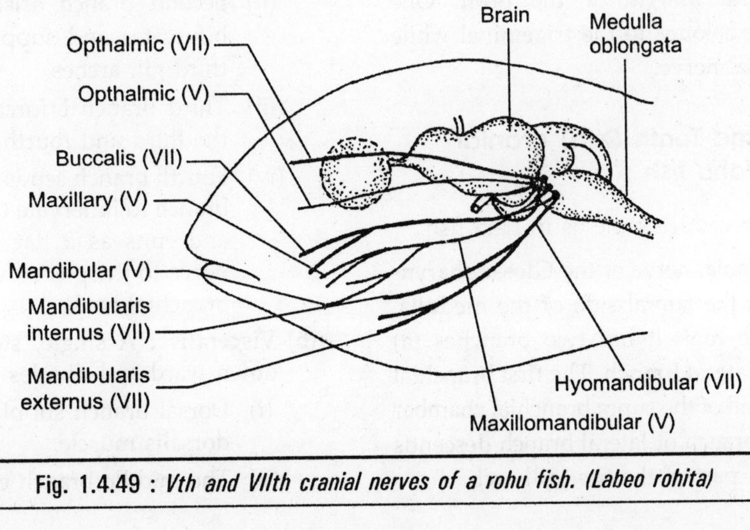

Fig. 1.4.49 : *Vth and VIIth cranial nerves of a rohu fish. (Labeo rohita)*

(b) **Maxillary** : A flattened nerve moves downward and forwarded along the posteroventral part of orbit to innervate the upper lip.

(c) **Mandibular** : It is posterior to the maxillary and turns backward making a sharp angle at the maxillomandibular junction and innervate the mandibular muscles.

2. **Seventh (VIIth) cranial nerve** : The seventh or **facial nerve** divides into three branches : (a) **opthalmic superfacialis,** (b) **buccalis** and (c) **mandibularis.**

(a) **Opthalmic superfacialis** : A thick dorsal branch running straight forward to innervate the snout area.

(b) **Buccalis** : It emerges from the base of the infra orbital trunk and moves outward and then downward to innervates the posterior part of infra orbital sensory canal.

(c) **Mandibularis** : Leaving the cranium through a foramen of the hyomandibular bone, it divides into two branches : **mandibularis externus** and mandibularis internus. Mandibular branch innervates lower jaw. Posteriorly mandibular branch gives off an opercular nerve which goes to the operculum.

Note : Opthalmic nerves are two nerves going across the dorsal margin of the orbit. One opthalmic nerve belongs to the trigeminal while the other to facial nerve.

Ninth (IXth) and Tenth (Xth) cranial nerves of a Rohu fish :

Dissection procedure same as in Lata fish.

Ninth (IX) cranial nerve or the **Glossopharyngeal** arises form the lateral side of the medulla, through a single root. It has two branches (a) **arterior** and (b) **lateral branch.** The first branchial innervetes the wall of the supra branchial chamber and the second branch or lateral branch descends along the lateral part of the first gill arch.

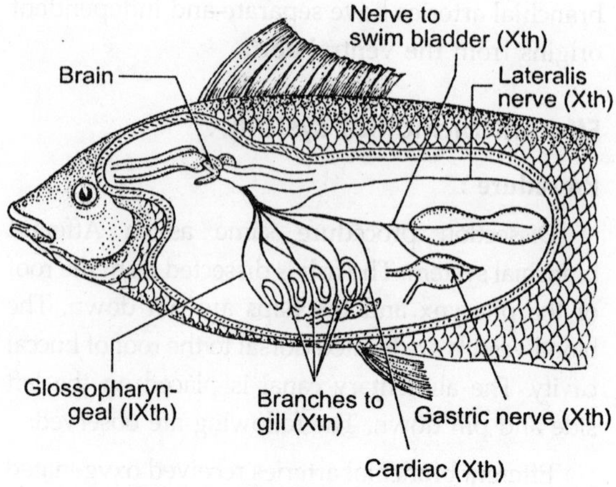

Fig. 1.4.49A : *IXth and Xth cranial nerves of a rohu fish. (Labeo rohita)*

The **tenth** or **Vagus** arises by two roots from the lateral side of the medulla oblongata posterior to the original of the ninth (IX) nerve. The two roots meet to form **Vagus ganglion.** Tenth has three branches : (a) **Branachials,** (b) **Viscreralis** and (c) **Lateralis.**

(a) **Branchialis :** A group of four branchial nerves.

(i) First branchialis move out ward and bifurcates. One branch innervates the first gill arch and the second branch innervates the second gill arch.

(ii) Second branch arises independently, bifurcates and supplying second and third gill arches.

(iii) Third branch bifurcate and innervate the third and fourth gill arches.

(iv) Fourth branch sends a fine pretrematic branch to innervate the fourth gill arch and runs as a flat nerves to supply nerve to the posteroventral part of the branchial basket.

(b) **Visceralis** : A single, stout nerve moves down ward and divides into branches.

(i) Dorsal branch supplying the retractor dorsalis muscle.

(ii) The middle branch enters the visceral

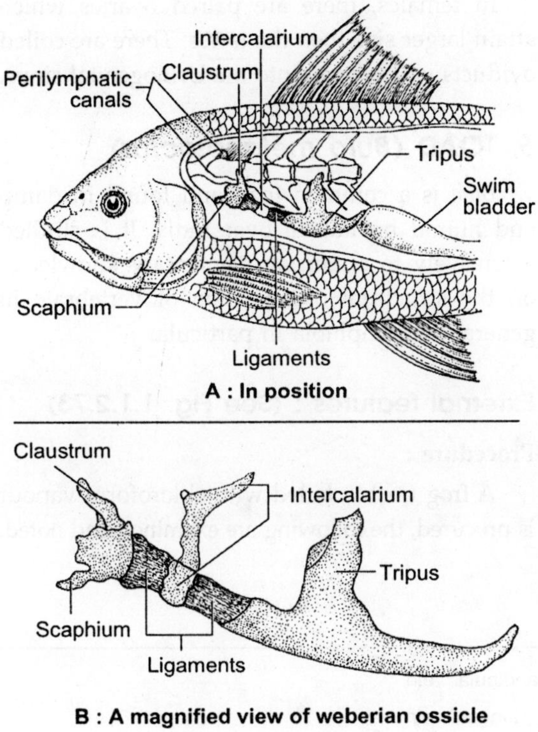

Fig. 1.4.49B : *Weberian ossicles of a rohu fish. (Labeo rohita)*

cavity and bifurcates. Dorsal branch innervate the air bladder. The ventral branch give branches to stomach as gastric, and to heart as cardiac.

(c) **Lateralis :** It arises from the lateral part of vagus ganglion. It innervates the lateral line of trunk.

Weberian ossicle (Auditory apparatus) :

Procedure :

The scales are removed from the lateral side of the fish and posterior portion of the operculum is cut with a pair of stout scissors. A deep incision is given with a sharp scalpel along the lateral line starting from the cut end of operculum up to the dorsal fin. The muscles are separated and the vertebral column and the air bladder is exposed. Care is taken not to damage the swim bladder.

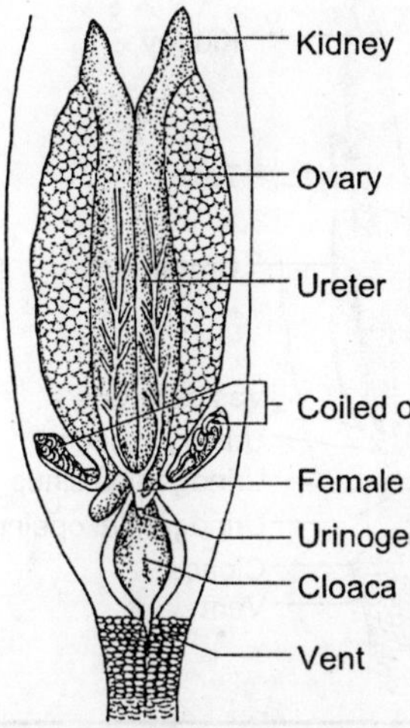

Fig. 1.4.50A : *Urinogenital system of a Tilapia fish (female) (Tilapia / Oreochromis sp.)*

The anterior end of the swim bladder is located and the chain of ossicles called weberian ossicles is followed, which are connected with it.

Auditory apparatus :

The perilymphatic sac and the anterior end of the swim bladder are connected by a series of four ossicles which are articulated. They are **tripus, intercalarium** and **scaphium** form a chain, while the fourth one, **claustrum** lies dorsal to the scaphium and lies with the perilymphatic sac.

5. URINOGENITAL SYSTEM OF *TILAPIA*

Dissection procedure same as in Lata fish.

The kidneys are extremely elongated running along the whole length of the visceral cavity. They are partly fused in the middle region. The ureters one from each kidney unite at the posterior and to from a common ureter and they open into a thin walled Urinary bladder, situated ventral to the cloaca. The Urinary bladder opens into the Urinogental sinus.

In males the paired testes extend the whole length of the abdominal cavity. Each testes has vas deferens which opens into the Urinogenital sinus.

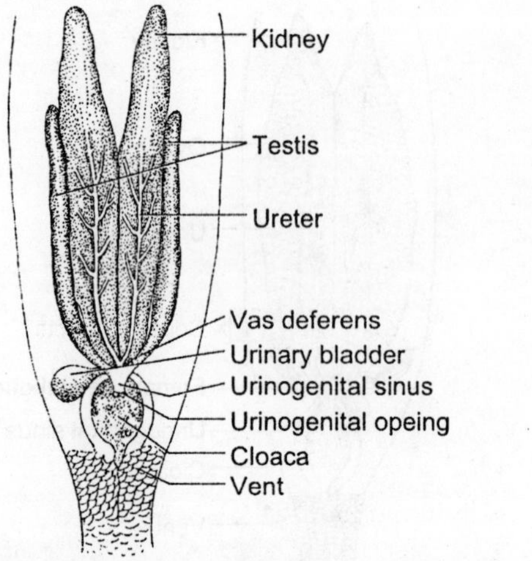

Fig. 1.4.50B : *Urinogenital system of a Tilapia fish (male) (Tilapia / Oreochromis sp.)*

In females, there are paired ovaries which attain larger size then the testes. There are coiled oviducts which open into the Urinogenital sinus.

6. TOAD (*Bufo melanostictus*)

Bufo is a common toad and found in damp and humid habitat all over India. It is studied extensively to obtain an elementary knowledge on the structural organization of vertebrate in general and amphibia in particular.

External features : (See Fig. 1.1.2.73)

Procedure :

A frog freshly killed with chloroform vapour is procured, the following are examined and noted.

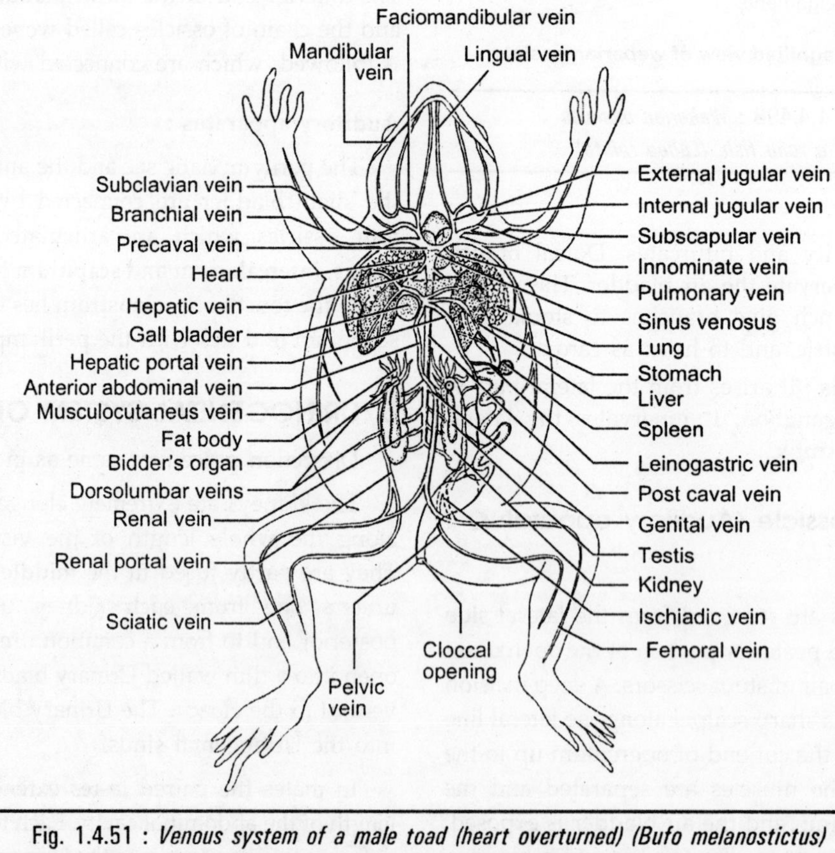

Fig. 1.4.51 : *Venous system of a male toad (heart overturned) (Bufo melanostictus)*

(1) **Shape :** The body is stream-lined, adapted for swimming.

(2) **Texture :** The skin is naked and has a rough texture. Ennumerable warts are present on the dorsal side of the body.

(3) **Colouration :** The dorsal skin is blackish grey while ventral is yellowish grey.

(4) **Divisions :** The body is divisible into head and trunk.

(a) **Head :** It is semi circular in outline. It is broad depressed and having blunt snout. At the terminal end of the head there is a wide opening mouth is located. The following structures are found in the head.

- ***Nostrils or external nares*** : Two small opening lying at the anterior dorsal side of the head.
- ***Eyes :*** Eyes are large, very prominent and protruding. They are situated on either side of the head. Each eye has a thick **upper eyelid** and a small mobile **lower eyelid**. A transparent **nictitating** membrane is present.
- ***Tympanum or eardrum :*** Behind each eye, there is a circular area, eardrum is present.
- ***Parotid gland :*** Just behind each tympanum there is an elongated elevation called parotid gland.
- ***Vocal sacs :*** Found in male only under the head.

(b) **Trunk** : The trunk is broad, short and flattened. Trunk contains anterior fore limbs and posterior hind limbs.

- ***Fore limbs :*** These are smaller than the hind limbs. Each comprises of **brachium** (upper arm), **antebrachium** (fore arm) and a **manus** (hand) containing **wrist, palm** and **four digits**. In male, a cushion-like thumb pad develops during breeding season at the basal part of index finger.
- ***Hind limbs :*** They are strongly built and are much longer the fore limbs. Each comprises of **femur** (thigh), **crus** (shank) and **pes** (foot). Foot is made up of **ankle, sole** and **five elongated slender digits**. The digits are united by webs.
- ***Vent or cloacal aperture :*** It is located on the dorsal side of the posterior end of the body and between the hind limbs.

Venous system :

Procedure :

A chloroformed killed toad is obtained and laid on its back in a dissecting tray. The limbs are pinned and pressed on the tray with the handle of the forceps. The tray is filled with water so that the toad is completely submerged. The abdominal skin is lifted with the forceps and a mid longitudinal cut is made from cloacal region extending anteriorly up to the lower jaw. The skin of the limbs is cut along the length with lateral incisions. The skin is separated from the underlying muscles and the flaps of the skin are pinned down.

An incision is given on the abdominal muscles at about its middle either on the left or right side of the mid ventral line. It is to be continued for a short length both anteriorly and posteriorly. The cut end of the abdominal muscles is lifted with a pair of forceps and the anterior abdominal vein is separated from the dorsal surface of the wall. The mid-longitudinal incision is continued up to the pectoral girdle and preceded backward up to the cloacal region. The flaps of abdominal wall are pinned down. Care is taken not be pull down unduly skin covering of the thorax to the outside so as not to damage **musculo-cutaneous vein**. Alimentary canal is separated from neighbouring organs by cutting mesenteries. The veins in the dissecting position lie above the arteries.

The pericardium is removed and the heart studied. The heart is partly venous and partly arterial and is the major organ both in venous and arterial system. Carefully the heart is turned up wards and the pulmonary veins, pre-caval, post-caval vessels and a thin walled sinus venosus are located. The pre-caval, post caval, hepatic portal and renal portal systems are traced. The two pelvic veins are separated from the thigh muscles by gently pulling the anterior abdominal vein forwards. The muscle lying on the other side of the thigh are removed to expose femoral vein. The pubic symphysis is cut through lying between the two thighs by a sharp scalpel to expose the sciatic veins coming from the inner side of the hind limbs.

Deoxygenated blood from all parts of the body is collected by the following veins except pulmonary vein and bring back to heart.

(1) ***Pulmonary vein*** : Two veins from two lungs collect oxygenated blood from lungs and unite to a common vein just before opening into left auricle.

(2) ***Caval veins*** : Two large caval veins or venae cava open into the sinus venosus. They are :

(a) **Pre-caval vein** : Two pre-cavals collect blood from anterior part of the body. Each pre-caval is formed by the union of three branches. They are :

(i) ***External jugular*** : It is formed by two veins, a *lingual* collects blood from tongue and a **faciomandibular** from snout and lower jaw.

(ii) ***Innominate*** : It is also formed by the union of two veins. An **internal jugular** collects blood from brain and orbit and a **sub-scapular** from shoulder.

(iii) ***Subclavian*** : It has two branches. A **brachial vein** collects blood from the fore limb and a **musculocutaneous vein** collect blood from muscles and skin.

(b) **Post-caval vein** : It is a large median vein receives blood from legs, kidneys, gonads and liver. This vein ascends to enter into the sinus venosus. It comprises of the following vessels.

(i) ***Femoral vein*** : Collects blood from outside of the leg. It gives out a branch, the **pelvic** which moves forward and joins its fellow of the other side to from the **anterior abdominal vein** which opens into the left lobes of liver.

(ii) ***Sciatic vein*** : Collects blood from the inner side of the femur.

(iii) ***Renal portal system*** : Each femoral vein divides into a pelvic and the renal portal vein. The sciatic vein joins with the renal portal vein, which runs by the side of the corresponding kidney and enters into it to break into capillaries. Four or five **renal veins** emerge from kidney. The renal veins of two sides unite with the post-caval vein. Each renal portal vein receives two or three **dorso lumber veins** carrying blood from the body wall.

(iv) ***Genital vein*** : A pair from the gonads join with the post-caval vein.

(v) ***Hepatic*** : A pair, from two lobes of the liver, join with the post-caval vein.

(vi) ***Hepatic portal system*** : This system comprises of **hepatic portal vein** and **anterior abdominal vein.** Hepatic portal vein formed by the union of **gastric vein** collects blood from stomach, an **intestinal vein** from intestine, a **duodenal vein** from duodenum, a **pancreatic vein** from pancreas and a **splenic vein** from spleen.

The hepatic portal and the anterior abdominal vein join together and the common vein enters into the two lobes of liver.

The arterial system :

Procedure :

The procedure adapted in case of dissection of venous system is followed. The veins are removed. All the mesenteries are kept intact.

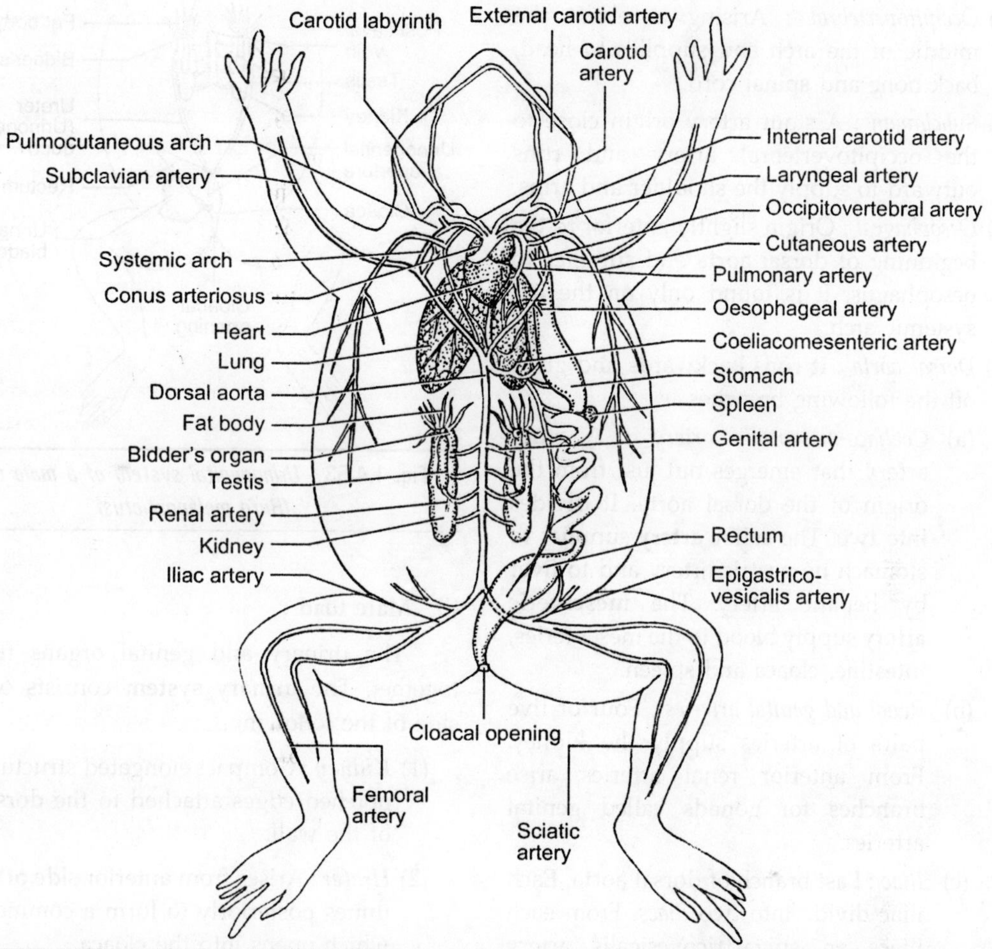

Fig. 1.4.52 : *Arterial system of a male toad. (Bufo melanostictus)*

The pubic symphysis is cut between the two thighs in order to expose the iliac arteries. The muscle lying on the outer and inner side of the thigh are removed to expose femoral and sciatic artery.

The heart is exposed and the **conus arteriosus** is located arising from the base of the ventricle at its right side and followed to trace the **truncus arteriosus** and its branches.

Truncus arteriosus :

It divides into two branches, each of which again gives three arches.

(1) *Carotid arch* : It is the anterior most one. It runs anteriorly, forms a swelling, the carotid labyrinth and divides into two—

(i) *External carotid* : Inner branch, supplies to tongue and hyoid.

(ii) *Internal carotid* : Outer branch, supplies to orbit and brain.

(2) *Systemic arch* : Two systemic arches encircle around oesophagus and join together dorsally to the dorsal aorta. Each systemic arch gives the following branches.

(i) *Laryngeal* : A very short branch supplies the laryngotracheal chamber.

(ii) *Occipitovertebral* : Arising at about the middle of the arch and supplies to head, back bone and spinal cord.

(iii) *Subclavian* : A stout artery origin close to the occipitovertebral artery and runs outward to supply the shoulder and arms.

(iv) *Oesophageal* : Origin slightly anterior to the beginning of dorsal aorta and supplies to oesophagus. It is found only on the left systemic arch.

(v) *Dorsal aorta* : It runs backwards and gives off the following branches.

(a) *Coeliaco-mesenteric artery* : A stout artery that emerges out just from the origin of the dorsal aorta. It divides into two. The **celiac artery** supplies to stomach by gastric artery and to liver by hepatic artery. The **mesenteric artery** supply blood to the mesenteries, intestine, cloaca and spleen.

(b) *Renal and genital arteries* : Four or five pairs of arteries supply the kidney. From anterior renal arteries arise branches for gonads called genital arteries.

(c) *Iliac* : Last branch of dorsal aorta. Each iliac divide into two *iliacs*. From each iliacs an epigastricovesicalis artery arise to supply the urinary bladder and ventral body wall. Iliac then enters into the hind limb and divides into a femoral artery supplying to the outer side of the leg, and a sciatic artery supplying to inner side of the leg.

Urinogenital system :

Procedure :

The toad is dissected as usual. The alimentary canal is separated from mesenteries and the oesophagus and the anterior part of the rectum are incised. These are removed and the pubic symphysis is cut to expose the cloaca and also the circulatory system, tearing only the posterior part of the dorsal aorta. Observed the following.

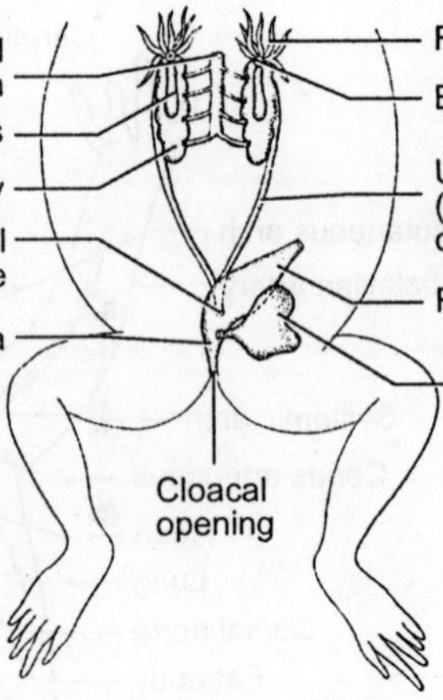

Fig. 1.4.53 : *Urinogenital system of a male toad. (Bufo melanostictus)*

(i) Male toad :

The urinary and genital organs function together. The **urinary system** consists on each side of the following.

(1) ***Kidney*** : Compact elongated structure with notched edges attached to the dorsal side of the wall.

(2) ***Ureter*** : Arises from anterior side of kidney, unites posteriorly to form a common duct which opens into the cloaca.

(3) ***Urinary bladder*** : It is a bilobed and thin walled sac, opening into the ventral side of the cloaca. The **genital system** comprises of :

(1) *Testes* : Two elongated white bodies, attached with the ventral side of the kidney by *mesorchium.*

(2) *Vasa efferentia* : **Semiferous tubules** of each testis communicate with collecting tubules of kidney by fine ducts **vasa efferentia**, which opens into the ureter. In male this duct is known as **urinogenital duct.**

(ii) Female toad :

The urinary and genital system functions independently. The Urinary system is like that of

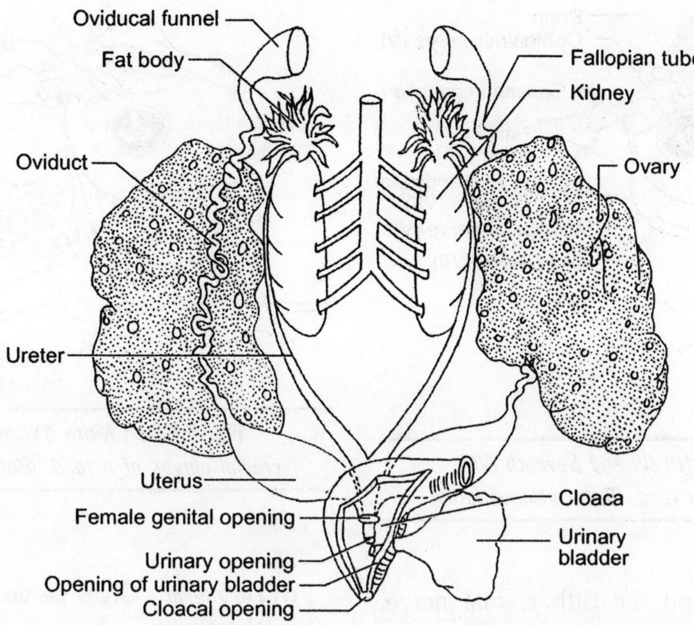

Fig. 1.4.54 : *Urinogenital system of a female toad. (Bufo melanostictus)*

male toad. The genital system consists on each side of the following :

(1) *Ovary :* Each is an irregular much folded sac attached with the ventral side of the kidney by *mesovarium*.

(2) *Oviduct :* Two very long coiled tubes placed on the lateral side of the body. It has got three parts:

(i) *Oviducal funnel* : It is the funnel-like anterior part of the oviduct which opens directly into coelom.

(ii) *Median portion* : Extensively coiled and glandular.

(iii) *Uterus* (ovisac) : The dilated thin walled part. Two uteri unite to from a common median tube which opens to the dorsal side of the cloacal chamber.

Cranial nerves :

Distribution of fifth (V) and seventh (VII) cranial nerves :

Procedure : The same technique used in case of *Channa* is followed and the brain and the cranial nerve roots are exposed. Nerves from the origin are traced and followed their courses.

(a) **Fifth (V) cranial nerve :** The fifth (V) or **trigeminal nerve** arises from the lateral side of the medulla oblongata. After its origin it gives off a swelling the **gasserian glanglion (prootic ganglion)**. It has three branches :

(i) *Opthalmic* : After its origin it runs across the dorsal of the orbit and sends nerve fibres to the upper eyelid, snout and mucous membrane of the nose.

(ii) *Maxillary* : It runs along the postero-ventral border of the orbit and gives nerve fibres to the upper lip, upper jaw and the lower eyelid.

(iii) *Mandibular* : It is posterior to the maxillary and gives nerves fibres to the muscles of the lower jaw.

(b) **Seventh (VII) cranial nerve :** The seventh (VII) or **facial nerve** arises from the medulla

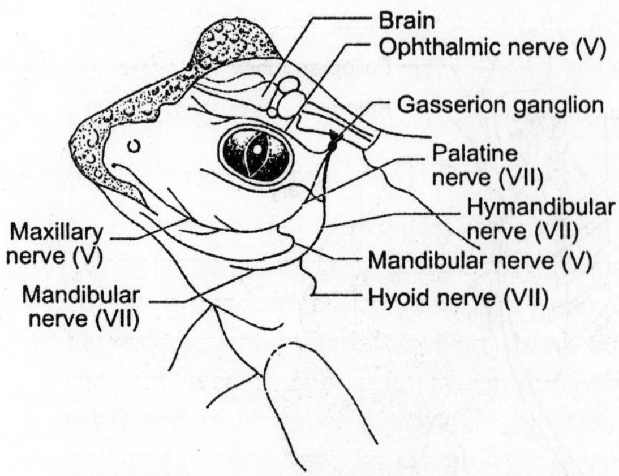

Fig. 1.4.55 : *Fifth (V) and Seventh (VII) cranial nerves of a toad. (Bufo melanostictus)*

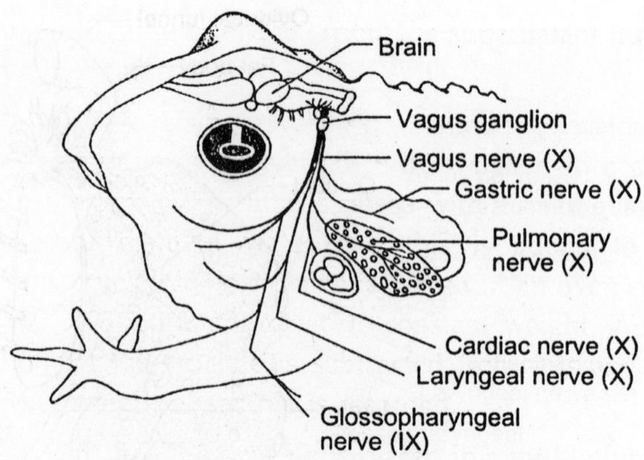

Fig. 1.4.56 : *Ninth (IX) and Tenth (X) cranial nerves of a toad. (Bufo melanostictus)*

oblongata just behind the fifth cranial nerve. After its origin from the gasserian ganglion as a separate nerve along with the fifth cranial nerve it divides into two branches.

(i) *Palatine* : It runs along the floor of the orbit and gives nerves to the roof of buccal cavity.

(ii) *Hyomandibular* : Runs outward and back ward around the auditory capsule, divides at the angle of the jaws and supplies nerve to tongue, lower jaw muscles and face.

• **Distribution of ninth (IX) and tenth (X) cranial nerves :**

(a) **Ninth (IX) cranial nerve :** The ninth (IX) or **glossopharyngeal** nerve arises from the lateral side of medulla oblongata. Immediately after origin it runs into the vagus ganglion and emerges from it as on independent nerve. It is also joined with the hyomandibular branch of seventh nerve just behind the tympanum. It innervates the tongue and pharyngeal wall.

(b) **Tenth (X) cranial nerve :** The tenth (X) or **vagus** or **pneumogastric** nerve arises from the medulla oblongata by many roots. After its origin it runs into the vagus ganglion. It gives off four branches which run postero-ventrally.

(i) *Laryngeal* : Gives nerve to laryngotracheal chamber.

(ii) *Cardiac* : Gives nerve to the heart.

(iii) *Pulmonary* : Innervates the lung.

(iv) *Gastric* : Gives nerve to alimentary canal.

7. GARDEN LIZARD (*Calotes versicolor*)

Calotes is a common Indian garden lizard. *Calotes* passes much of its time on branches and twigs. It is often seen running swiftly on the ground.

External features :

Procedure :

A chloroform killed lizard is obtained and examined and the following are noted.

(1) **Shape :** Typically lizard like body.

(2) **Texture :** The skin is highly modified for living on land. Body is covered by rough epidermal scales and the skin glands are absent.

(3) **Colouration :** Body is brown or greyish on the dorsal surface and dirty whitish ventrally. The colour of the body is changeable.

(4) **Division :** Body is divisible into head, neck, trunk and tail. A crest of spines (frills) is found in mid-dorsal line of the body.

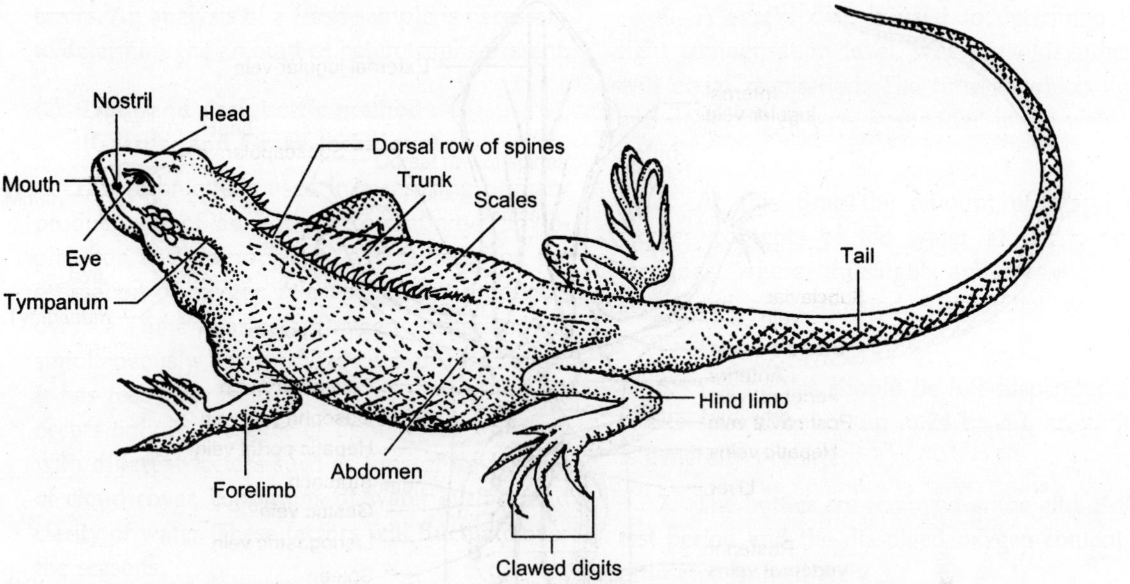

Fig. 1.4.57 : *External features of a garden lizard. (Calotes versicolor)*

(a) **Head :** It is more or less triangular with a short, pointed snout and contains the following structures.

(i) **Nostrils :** The tip of the snout contains a pair of large nostrils.

(ii) **Mouth :** It is a transverse aperture, terminal in position and without lips.

(iii) **Eyes :** Eyes have two movable eyelids and a movable nictitating membrane.

(iv) **Tympanum or eardrum :** Behind each eyes there is a circular area, the eardrum is present.

(b) **Neck :** It is elongated, small and narrow.

(c) **Trunk :** It is elongated, subcylindrical and forming major part of the body. It contains two pairs of limbs. (1) **Fore** and (2) **hind limbs**. The fore limb is divisible into **brachium, antibrachium** and **manus**. The hind limb is divisible into **femur**, **crus** and **pes**. The limbs end in five digits and each digit contains a sharply pointed claws. (3) **Cloacal aperture** or **vent**: It is a transverse opening and located in the posterior-ventral part of trunk at the base of the tail. A large cloacal plate is situated in front of the cloacal aperture.

(d) **Tail :** It is long, slender and tapering. It measures as long as the head and body.

Venous system :

Procedure :

To expose the venous system the procedure adopted as in *Bufo*. Pericardium is removed and the heart is observed. The heart is reddish brown, broadly conical, muscular structure situated ventral to the alimentary canal. The heart consists of a **sinus venosus**, two **auricles** and a **ventricle**. Carefully turned the heart upwards and the pulmonary veins, pre-caval, post-caval vessels and a thin walled sinus venosus, which is restricted to the dorsal side of the right auricle are located. The pre-caval, post caval, hepatic and renal portal systems are traced.

De-oxygenated blood from all parts of the body is collected by the following veins (except pulmonary vein) and brings back to heart.

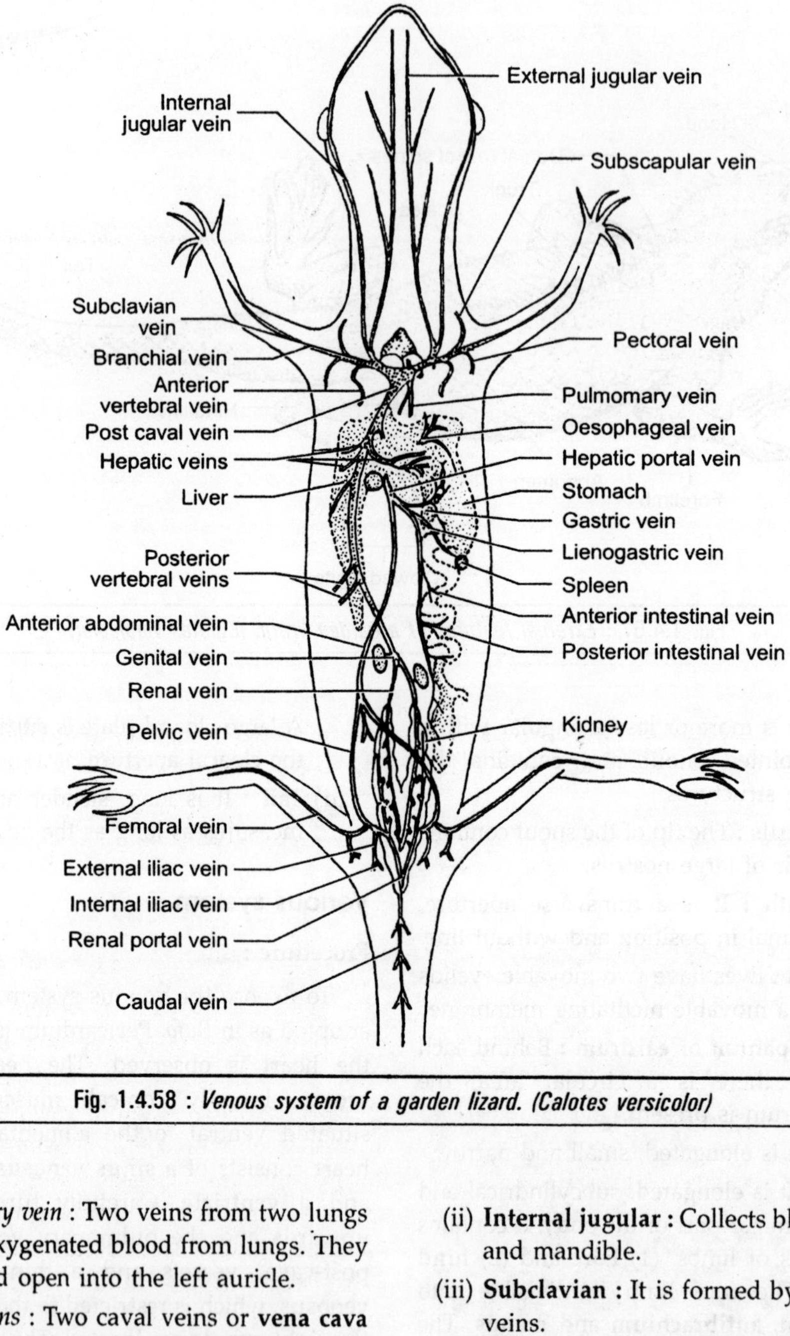

Fig. 1.4.58 : ***Venous system of a garden lizard. (Calotes versicolor)***

(1) *Pulmonary vein* : Two veins from two lungs collect oxygenated blood from lungs. They unite and open into the left auricle.

(2) *Caval veins* : Two caval veins or **vena cava** open into the sinus venosus. They are:

A. ***Pre-caval vein*** : Each pre-caval vein receives the following veins.

(i) **External jugular :** Collects blood from neck and mouth.

(ii) **Internal jugular :** Collects blood from head and mandible.

(iii) **Subclavian :** It is formed by the following veins.

(a) *Brachial* : Collects blood from the fore limb.

(b) *Subscapular* : Draws blood from the scapular region and muscle of the back.

(c) *Pectoral* : Collects blood from the pectoral region.

(d) *Vertebral* : Draws blood from the anterior portion of the vertebral column.

B. ***Post-caval vein*** : It is a large vein situated a little to the right of the middle line. Major part of it is embedded in the right lobe of the liver. Two efferent renal trunks from kidneys extend forward and unite together below gonad and form the **post-caval vein**. The post-caval vein after receiving a **hepatic vein** from the liver, continued up wards and open into the sinus venosus. The post-caval vein collects blood from the posterior region and receives the following veins.

(i) **Caudal vein** : Caudal vein draws blood from tail, which on entering the kidney divides into two capillaries and from a renal portal vein.

(ii) **Renal portal vein** : Received veins from the cloaca and rectum. Two renal portal veins with their branches form the **renal portal system**. Portal veins end in the kidney.

(iii) **Femoral vein** : Femoral from each hind limb enters the pelvic region and goes forward as pelvic vein.

(iv) **Internal iliac** : This vein from the pelvic region joins the pelvic vein.

(v) **External iliac** : A short vein connecting the pelvic with the renal portal vein.

(vi) **Anterior abdominal vein** : The pelvic veins unite to form a median anterior abdominal vein which opens into the left lobe of the liver.

(vii) **Renal** : Paired, arises from the kidneys and the two forms narrow loop at the posterior end.

(viii) **Genital** : Two, arise from the gonads and join the post-caval vein.

(ix) **Oesophageal** : Collect blood from the oesophagus.

(x) **Hepatic portal system** : It is formed by the union of the following veins and joins the abdominal vein and enters the liver.

(a) *Gastric* from stomach,

(b) *Intestinal* from intestine,

(c) *Duodenal* from duodenum,

(d) *Splenic* from spleen,

(e) *Pancreatic* from pancreas and

(f) *Rectal* from rectum.

Arterial system :

Procedure :

Same as in case of venous system. The veins which are not required are removed. The heart is located and the pericardium is removed and three aortic arches arising from the ventricle are observed and their branches along their courses are traced.

(1) **Pulmonary aorta** : It originates from the right side of the ventricle and divides into two pulmonary arteries, each going to lung. They carry de-oxygenated blood to lungs.

(2) **Systemic aorta** : The truncus divides into right and left systemic arches. Right originates from the left side of the ventricle and left arch from the right side.

(a) ***Left systemic aorta*** : Originates from the right side of the ventricle, moves forward, then curves round the heart and meet the right systemic aorta. It sends only the *ductus caroticus* and *oesophageal arteries.*

(b) ***Right systematic aorta*** : Originates from the left side of the ventricle, moves forward, then curves the heart and meet the left systematic aorta to from *dorsal aorta.* It gives various arteries.

(i) *Innominate artery (common carotid)* : It originates from the apex of the curvature of the right systemic arch, which divides into the right carotid and left carotid arteries. Each carotid divides into an internal carotid and external carotid arteries.

(ii) *Ductus caroticus* : It connects the internal carotid artery with systemic arch of its side.

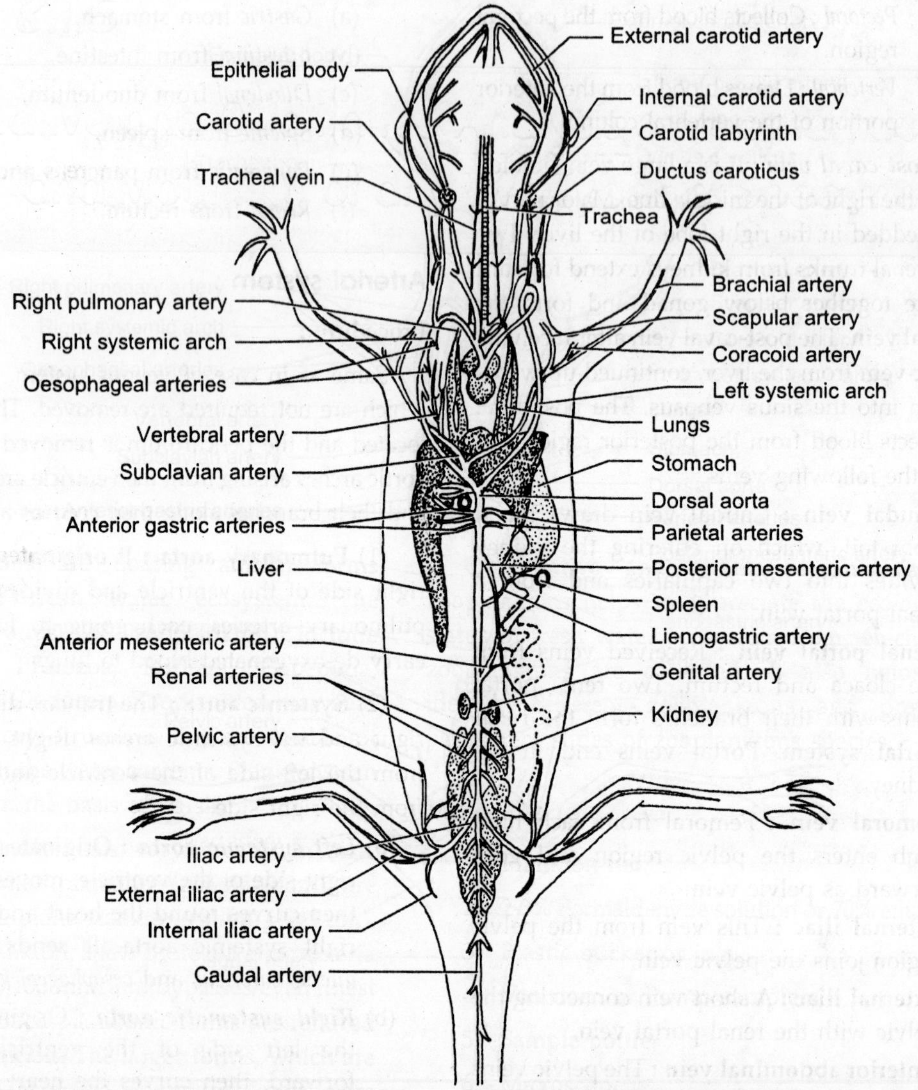

Fig. 1.4.59 : *Arterial system of a garden lizard. (Calotes versicolor)*

(iii) *Oesophageals* : Three or four arteries originate as small vessels posterior to the root of ductus caroticus.

(iv) *Subclavian* : Originates from the right systemic aorta and bifurcates into right and left sub clavian arteries which supply to fore limbs.

(v) *Vertebral* : A branch from right systemic aorta supplies to vertebral.

(c) ***Dorsal aorta*** : The right and left systemic arches unite together forming dorsal aorta, which runs posteriorly and gives several arteries to viscera.

(i) *Anterior mesenteric* : Single, supplying the ileum.

(ii) *Parietals* : Paired, small vessels supplying the trunk vertebrae, thoracic ribs and costal muscles.

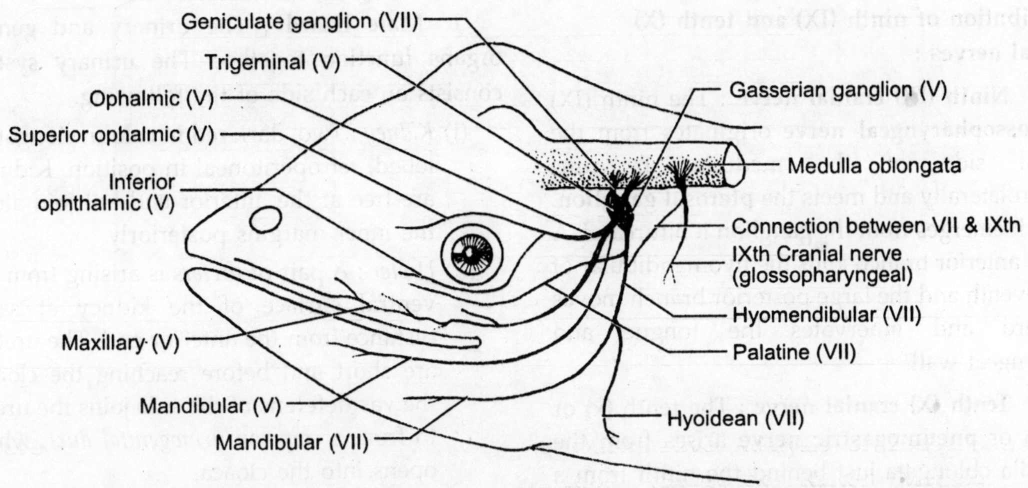

Fig. 1.4.60 : *Fifth (V) and Seventh (VII) cranial nerves of a garden lizard. (Calotes versicolor)*

(iii) *Gastric* : Small branches supplying the stomach.

(iv) *Posterior mesenteric* : Large vessels, supplies the caecum and colon.

(v) *Lienogestric* : Arises behind the posterior mesenteric, supplies spleen, intestine and liver.

(vi) *Genital* : One pair, supplies gonad.

(vii) *Renal* : Two to four pairs, supplying the kidneys.

(viii) *Iliac* : One pair, supplies legs.

(ix) *Caudal* : To tail.

Cranial nerves :

Distribution of fifth (V) and seventh (VII) cranial nerves :

Procedure : The same technique is followed as in case of *Channa*. The brain is exposed and the nerves are traced from their root and their courses are followed.

1. **Fifth (V) cranial nerve** : The fifth or *trigeminal* arises from the ventrolateral side of medulla oblongata at its anterior end. Immediately after its origin it gives off a swelling, the **gasserian ganglion**. It has three branches.

(1) *Opthalmic* : After its origin it bifurcates and runs the dorsal and ventral border of the orbit as *superior* and *inferior* opthamic nerves. They innervate the upper eyelid, snout and mucous membrane of the nose.

(2) *Maxillary* : It runs forward along floor of the orbit and gives nerve fibres to the upper lip, upper jaw and the lower eyelid.

(3) *Mandibular :* It is posterior to the maxillary and gives nerve fibres to the muscles of the lower jaw.

2. **Seventh (VII) cranial nerve** : The seventh (VII) or *facial nerve* arises from the side of the medulla oblongata behind the fifth from a swelling, the **geniculate ganglion**. It divides into two branches;

(i) *Palatine* : Slightly displaced the eye ball from the root and observe the branch. It is a slender branch runs along the floor of the orbit and gives nerve to the root of buccal cavity.

(ii) *Hyomandibular* : The stout branch runs outward and backward around the auditory capsule divides at the angle of the jaws as *anterior mandibular* and *posterior hyodian*. Hyomandibular suppler nerve to tongue, lower jaw muscles and face.

Distribution of ninth (IX) and tenth (X) cranial nerves :

1. **Ninth (IX) cranial nerve :** The ninth (IX) or **glossopharyngeal nerve** originates from the lateral side of the medulla. It runs posterolaterally and meets the **pterosal ganglion.** After it emerges from the ganglion it bifurcates. A short, anterior branch joins the hyomandibular of the seventh and the large posterior branch moves forward and innervates the tongue and pharyngeal wall.

2. **Tenth (X) cranial nerve :** The tenth (x) or **vagus** or **pneumogastric nerve** arises from the medulla oblongata just behind the ninth from a separate ganglion called *vagus ganglion.* It gives off six branches.

(i) *Laryngeal* : Gives nerve to larynx.

(ii) *Oesophageal* : Gives nerve to oesophagus.

(iii) *Cardiac* : To heart.

(iv) *Pulmonary* : To lung.

(v) *Gastric* : To stomach.

(vi) *Vagosympathetic* : To vertebral column.

Urinogenital system :

Procedure :

The same techinique used as in case of *Bufo.*

1. **Male lizard :** The urinary and genital organs function together. The urinary system consists on each side of the following.

(i) *Kidney* : Two, dark red in colour, elongated, lobed, retroperitoneal in position. Kidneys are free at the anterior end but join along the inner margins posteriorly.

(ii) *Ureter* : A pair of *ureters* is arising from the ventral surface of the kidney at same distance from the anterior end. The ureters are short and before reaching the cloaca, the vas deferens of the side joins the ureter to from a *common urinogenital duct,* which opens into the cloaca.

(iii) *Urinary bladder* : A small, single, thin walled sac arises from the lateral wall of the cloaca on its ventral surface.

The genital system comprises of :

(i) *Testes* : Paired, white oval bodies situated in the abdominal cavity in front of the kidneys. The right one slightly anterior than the left one. They remain suspended by **mesorchium.**

(ii) *Vasa deferentia* : Two, elongated convoluted tubes proceeds posteriorly open along with ureter into the cloaca separately by small papillae. Anteriorly each vas deferens

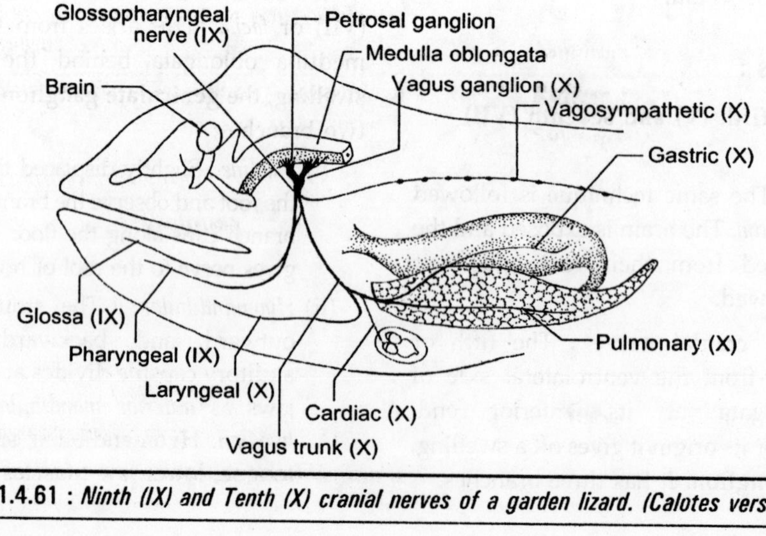

Fig. 1.4.61 : *Ninth (IX) and Tenth (X) cranial nerves of a garden lizard. (Calotes versicolor)*

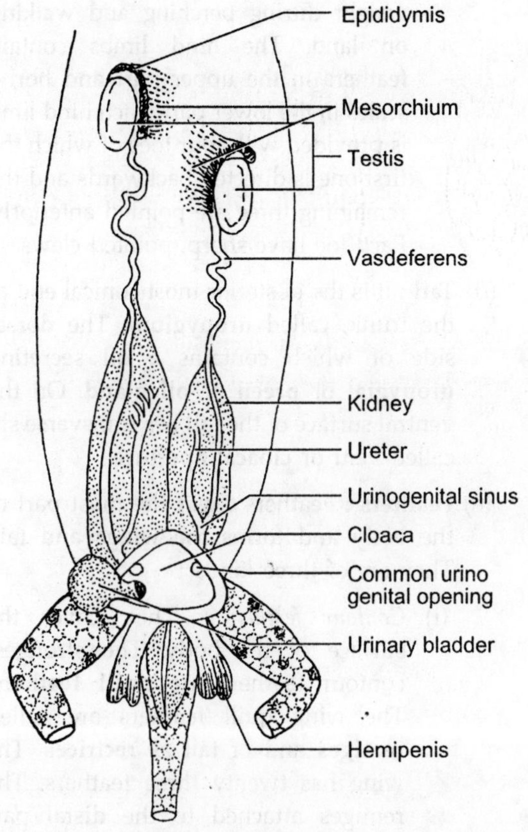

Fig. 1.4.62 : *Urinogenital system of a garden lizard (Male). (Calotes versicolor)*

receives a club-shaped epididymis which connects the testis by narrow **vasa efferentia**.

2. **Female lizard** : The urinary and genital system functions independently. The urinary system is like that of male lizard. The genital system consists on each side of the following.

(i) *Ovary* : Each is of clusters of round cells of different size, fixed to the dorsal side by mesentery called **mesovarium**.

(ii) *Oviduct* : Two, large folded ducts attached with the body wall by fold of peritoneum called broad ligament. The anterior end of the oviduct is wide; funnel shaped and is situated near the corresponding ovary. The posterior side of the oviducts are wide and

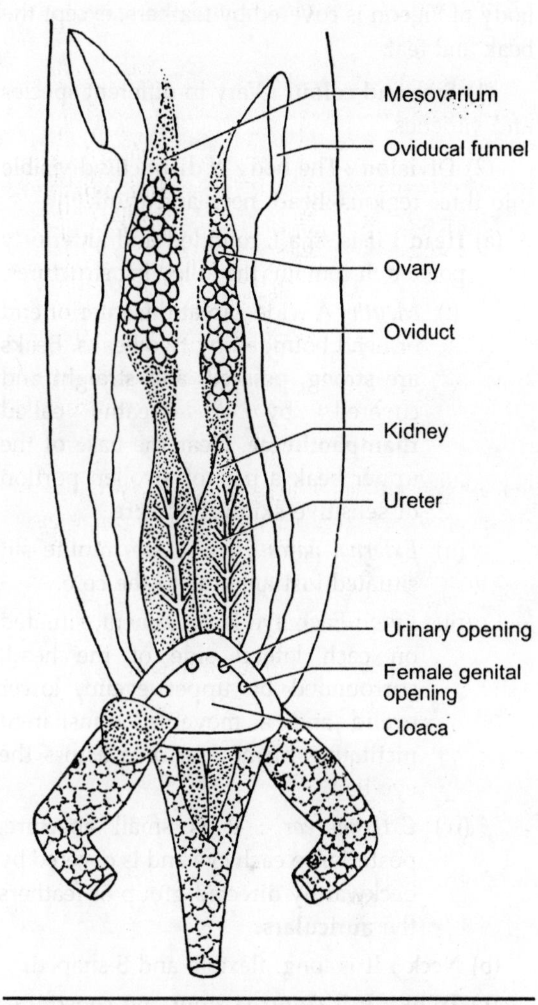

Fig. 1.4.63 : *Urogenital system of a garden lizard (Female). (Calotes versicolor)*

open independently into the cloaca on its dorsal surface.

8. PIGEON (*Columba livia domestica*)

The pigeons are seen almost throughout the world. They exhibit most of the peculiar characters of birds and are available in large numbers.

External features : (See Fig. 1.1.2.103)

A chloroformed killed pigeon is taken and examined the external morphology. The entire body of Pigeon is covered by feathers, except the beak and feat.

(1) **Size and colour :** Vary in different species and varieties.

(2) **Division :** The body is distinctly divisible into three regions, head, neck and trunk.

(a) **Head :** It is small, rounded and anteriorly pointed. It contains the following structures.

(i) *Mouth* : A wide gap at the anterior end of head bounded by two beaks. Beaks are strong, pointed, and straight and covered by a sheath called **rhamphothecae.** Near the base of the upper beak a pair of swollen portion of sensitive skin called **cere.**

(ii) *External nares* : These are minute slit situated just anterior to the cere.

(iii) *Eyes* : Each eye is prominent, situated on each lateral side of the head, surrounded by upper eyelid, lower eyelid and a movable, transparent nictitating membrane runs across the eye ball.

(iv) *External ear* : It is small aperture, posterior to each eye and is covered by backwardly directed group of feathers the **auriculars.**

(b) **Neck :** It is long, flexible and S-shaped.

(c) **Trunk :** It is compact stout and constitutes major part of body. It is divided into large thorax and a small abdomen. The keel is seen and felt as a prominent ridge in the thorax. The trunk contains two pairs of limbs.

(i) *Fore limbs :* These are modified as wings. Each wing has three divisions, **upper arm, fore arm** and **hand**. There are two folds of skin, one between upper arm and fore arm, called **pre-patagium** and the other between upper arm and trunk, called **post-patagium.**

(ii) *Hind limbs :* Paired hind limbs are modified to carry the entire body weight during perching and walking on land. The hind limbs contain feathers in the upper part and horny scales in the lower part. Each hind limb is provided with four toes of which the first one is directed backwards and the remaining three are pointed anteriorly. Each toe have sharp, pointed claws.

(d) **Tail :** It is the posterior most conical end of the trunk, called **uropygium**. The dorsal side of which contains a oil secreting **uropygial** or **preen** or **oil gland**. On the ventral surface of the tail lies transverse slit called **vent** or **cloacal aperture**.

(e) **Feathers :** Feathers cover the most part of the body and forms the wings and tail. They are of three kinds.

(i) *Contour feathers* : They cover the surface of the body. The strongest contour feathers are **quill feathers.** The wing quill feathers are called **remiges** and of tail as **rectrices**. The wing has twenty three feathers. The remiges attached to the distal part **(carpometacarpus)** of hand are **primaries** and to radius and ulna are **secondaries** and those to the proximal part **(humerus)** are **tertiaries**. Feathers attached to the first pre axial digit called **alaspuria** or **bastard wing**. A typical feather in divided into naked hollow, cylindrical basal **calamus** or **quill** and **upper rachis** or **shaft** surrounded by **vexillum** or vane. The vane contains delicate obliquely arranged **barbs** on each side of the rachis. The distal edge of each barb carries two sets of **barbules**-proximal and distal. Barbules of adjacent barbs are fastened together by **hooks** and **hooklets.**

(ii) *Filoplumes or Hair feathers* : These are very thin, hair-like degenerated feathers without interlocking apparatus, situated at the base of the contour feathers.

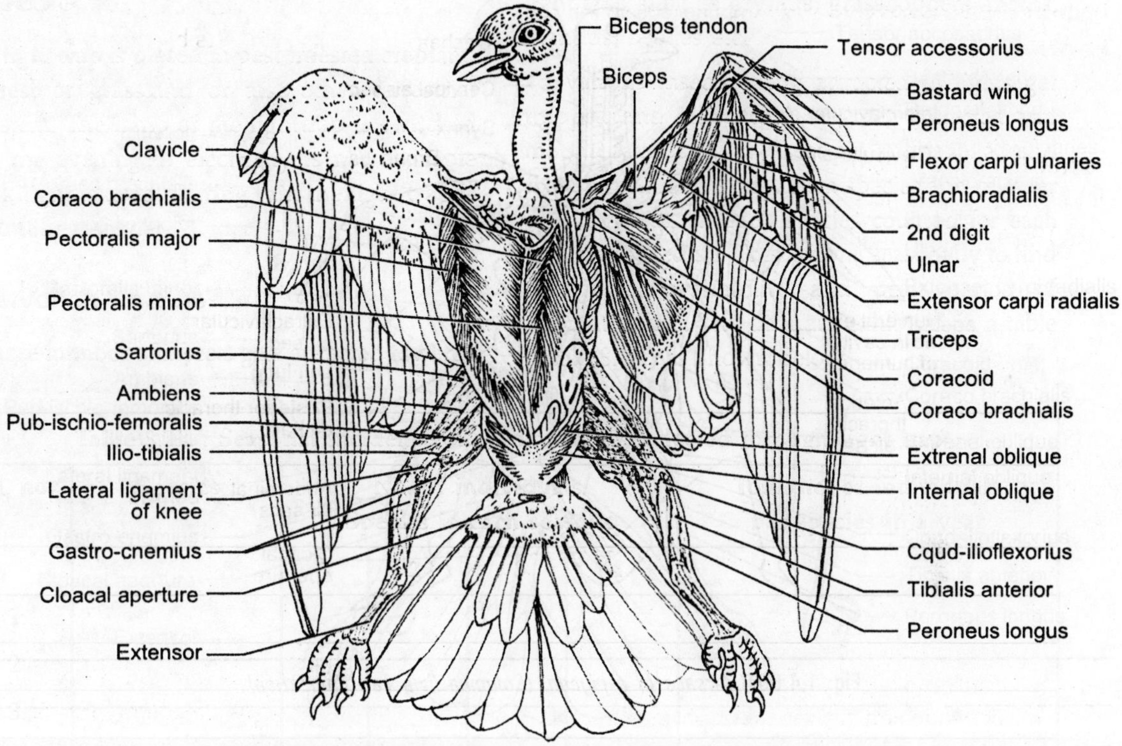

Fig. 1.4.64 : ***Flight muscles of a pigeon. (Columba livia var. domestica)***

(iii) *Down feathers or plumules* : They are provided with a short quill or stem and barbs are arranged in a circle at the top of the stem. They are found below contour feathers and are with out interlocking system (**Fig. 1.5.37**).

Flight muscles :

The muscles which are concerned with the activity of the wings are called **flight muscles**.

Procedure :

The contour feathers are plucked from the ventral surface. After complete defeathering the bird is placed on a dissecting tray and the pigeon is pinned with ventral surface upward through wings and limbs. A longitudinal incision is made in the skin with a pair of scissors from the cloacal aperture up to the lower beak along the mid ventral line. A lateral incision is given in the skin along the wings and limbs. The flaps of skin are pinned after separating it from the under lying muscles. After removal of the skin the muscles exposed between the body and the keel is occupied by two pectoralis major. With a sharp scalpel cut is given through the origin of pectorlis major by the side of the keel, the posterior edge of sternum and the clavicle are freed from the underlying muscles. It is turned up ward and the following flight muscles are studied.

(1) **Pectoralis major** : Paired muscles of immense size, arises from whole of the keel from the posterior part of the sternum and clavicles. There flat tendons are inserted on the great trochanter of humerus. They are the main depresser muscles of the wings which lower the wings during flight.

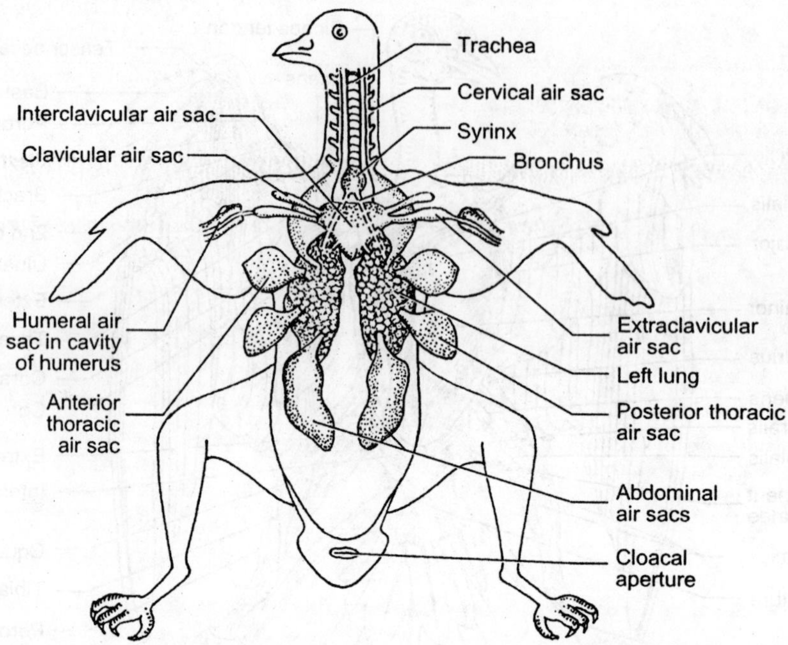

Fig. 1.4.65 : *Airsacs of a pigeon. (Columba livia var. domestica)*

(2) **Pectoralis minor (or subcalvius or supracoracoideus)** : These paired muscles are situated on the dorsal side of the pectoralis major and on the anterior side of the sternum proper. The tendon, formed by its fibres passes through the *foramen triosseum* (aperture between furcula, scapula and coracoid) and is inserted on the anterodorsal surface of humerus. They are the main elevator muscles of the wings (rises the wing).

(3) **Coracobrachialis longus and Coracobrachialis brevis** : They are small muscles originate from the coracoid and scapula and are attached with the humerus. It helps in lowering of the wing.

Air sacs :

Procedure:

The pigeon is dissected as in case of flight muscles. The ventral abdominal wall from the cloaca, prior to sternum is cut longitudinally. By a strong scissors or bone cutter a cut is made through the side of sternum along its attachment to the ribs up to the anterior border of the coelom and the sternum is lifted with the fingers, so as not to injure the internal organs. The coracoids are cut from the sternum and also the furcula are cut at the two clavicles. The sternum is removed to expose fully the underlying organs. The lungs and air sacs are dorsal to the visceral organs. The visceral organs are removed with care, so as not to puncture the air sacs.

A blow pipe is inserted in to the trachea that inflates the air sacs to make out the outline of them. A thread is tied in trachea and the following air sacs are examined.

There are nine major air sacs. They are—

(a) *Cervical air sacs* : Paired, formed at the base of the neck.

(b) *Interclavicular air sacs* : Single, median, found between the two clavicles.

(c) *Anterior thoracic air sacs* : Paired, anterior to the posterior thoracic and located to wards the anterior parts between the lungs and the ribs.

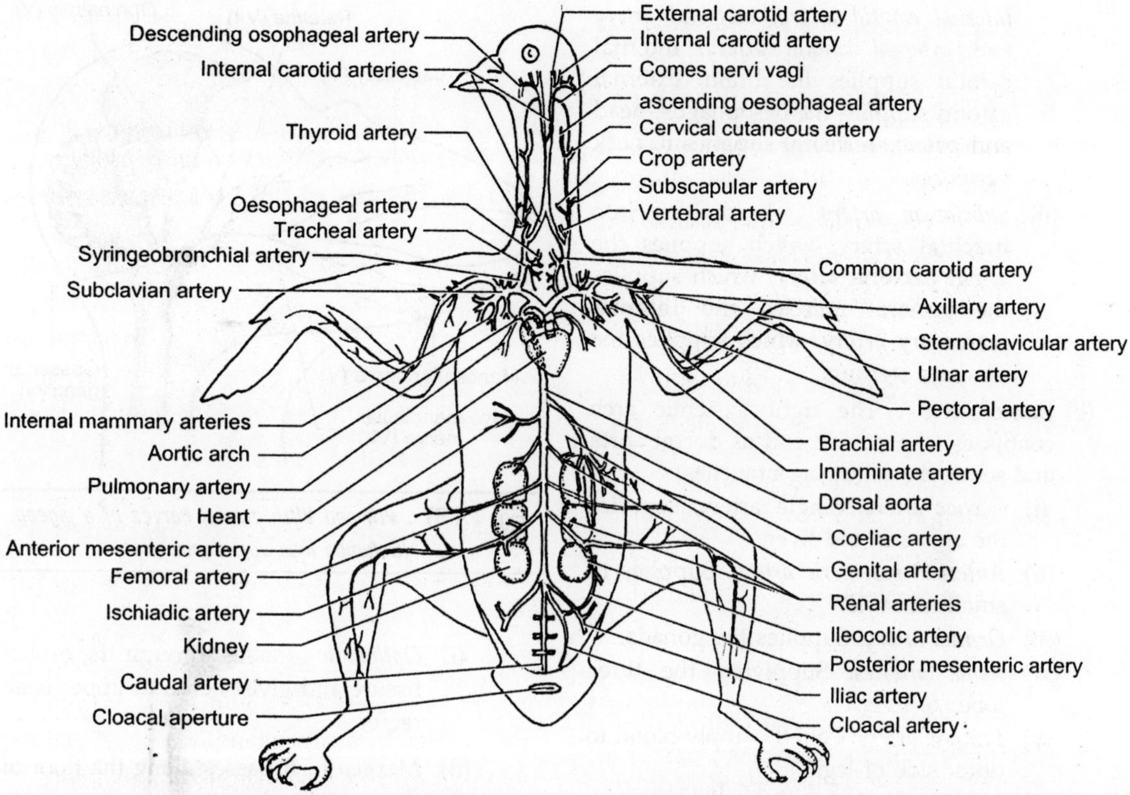

Fig. 1.4.66 : ***Arterial system of a pigeon. (Columba livia var domestica)***

(d) *Posterior thoracic air sacs* : Paired, placed on the posterior parts of the lungs.

(e) *Abdominal air sacs* : Paired, these are the largest air sacs and lie among the coils of intestine.

Arterial system :

Procedure :

The specimen is dissected in the same way as in case of air sacs. The pericardium is removed along with the fat and connective tissues at the base of the heart and blood vessels. The pulmonary aorta arising from the right ventricle and the aortic arch arising from the left ventricle are identified. The arteries are traced along their courses. The following aortae and their branches comprises the arterial system of pigeon:

- **Pulmonary arch and right aortic arch :**

The following arteries supply blood to various parts of body.

(1) *Pulmonary arch* : The pulmonary arch arises from the right ventricle and divides into two pulmonary arteries which convey deoxygenated blood to lungs, one on each side.

(2) *Right aortic arch* : Only right systemic arch persist in adult. It originates from the left ventricle and then curves over the right bronchus and give rise to two **innominate arteries.**

(a) *Innominate arteries* : One on each side gives rise to **common carotid** and **subclavian artery.**

(i) *Common carotid artery* : It extends upwards along the ventrolateral side of the neck region and divides into *internal carotid artery, vertebral artery* and *external carotid artery.* Internal carotid supplies the brain, external carotid supplies the oesophagus, head and palate, vertebral supplies to neck vertebrae.

(ii) *Subclavian artery* : It divides into **brachial artery**, which supplies the wing, **pectoral artery**, which supplies the pectoral muscles and **internal mammary artery**, which supplies the ribs and sternum.

(b) *Dorsal aorta* : The right systemic arch continues to posterior end as dorsal aorta and sends the following branches.

(i) *Coeliac artery*: Single artery, supplies the stomach and liver.

(ii) *Anterior mesenteric artery*: Supplies to small intestine.

(iii) *Genital artery*: Supplies the gonads.

(iv) *Renal arteries*: Supplying the three lobes of liver.

(v) *Femoral artery*: Paired, supply blood to outer side of legs.

(vi) *Ischiatic artery:* Supply to the posterior part of legs.

(vii) *Internal iliac artery*: Supply to the hip region.

(viii) *Posterior mesenteric*: Szupplies the mesenteries of the posterior side.

(ix) *Caudal artery*: Single, supplies to the tail.

Cranial nerves :

(i) Distribution of fifth (V) and seventh (VII) cranial nerves :

Procedure : The technique adopted for *Channa* is followed. The brain is exposed and the nerves are traced from their root and their courses are followed.

• **Fifth (V) cranial nerve :** The fifth (V) or **trigeminal** originates by many roots from the lateral part of medulla oblongata ventral to optic lobe. It has three branches.

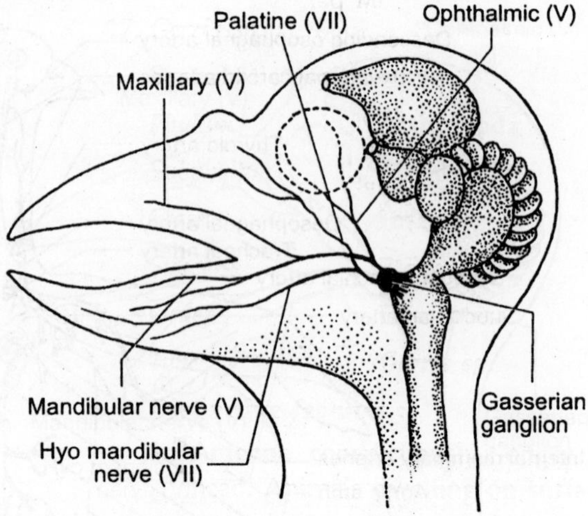

Fig. 1.4.67 : *Vth and VIIth cranial nerves of a pigeon. (Columba livia var. domestica)*

(i) *Opthalmic* : Passes through the orbital fissure and gives nerve to upper beak region.

(ii) *Maxillary* : It passes along the floor of the orbit and innervates the upper mandible.

(iii) *Mandibular* : It is the largest branch enters into the orbit and running forward and supplying temporal muscles and lower mandible.

• **Seventh (VII) cranial nerve :** The seventh (VII) or **facial** arises just back of the trigeminal on the later side of the medulla and runs back wards into two branches.

(i) *Palatine* : A slender branch runs along the floor of the orbit and innervates the roof of buccal cavity.

(ii) *Hyomandibular* : It is a stout branch, runs posterior to maxillary and forward round the tympanic region and gives nerve fibres to the post tympanic area and lower jaw.

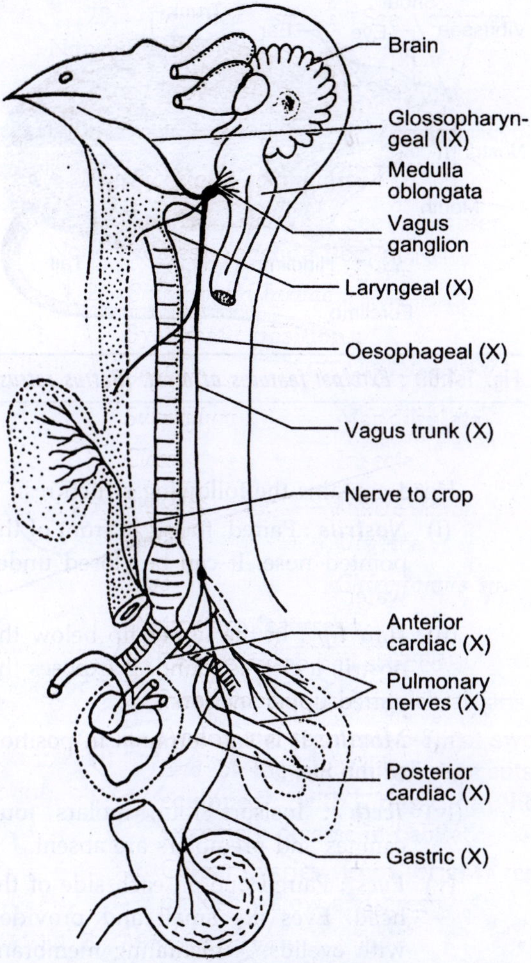

Fig. 1.4.68 : *IXth and Xth cranial nerves of a pigeon. (Columba livia var. domestica)*

(ii) Distribution of ninth (IX) and tenth (X) cranial nerves :

• **Ninth (IX) cranial nerve :** The ninth (IX) or **glossopharyngeal nerve** arises on the side of the medulla close to the auditory and the tenth nerve. After it has emerged from the cranium it bifurcates, the upper short branch joins the hyomandibular of seventh and the large posterior branch innervates the tongue and pharynx.

• **Tenth (X) cranial nerve :** The tenth (X) or **pneumogastric** or **vagus nerve** arises close to the glossopharyngeal and runs straight back along the side of the neck to the body cavity, where it sends four major branches.

(i) *Laryngeal*: Gives nerve to syrinx.

(ii) *Pulmonary*: Gives nerve to lungs.

(iii) *Cardiac*: Supplies the heart.

(iv) *Gastric*: To crop, stomach and gizzard.

9. FOWL (*Gallus sp.*)

Distribution of fifth (Vth) and Seventh (VIIth) Cranial nerves :

Procedure :

Same as adapted for *Channa*.

• **Fifth (V) cranial nerve :** The fifth (V) or **trigeminal** arises on the side of the medulla oblongata beneath the optic lobe and passes forward to the Gasserian ganglion where it divide into three branches.

(i) **Opthalmic :** Runs through the floor of the skull into the orbit and gives two branches, the **superficial opthalmic** and **inferior opthalmic.**

(ii) **Maxillary :** It passes along the floor of the orbit and innervates the upper mandible.

(iii) **Mandibular :** It is the largest branch enters into the orbit and running forward and supply temporal muscles and lower mandible.

• **Seventh (VII) cranial nerves :** The seventh (VII) or **facial** arises just back of the trigeminal on the lateral side of the medulla. It has two branches.

(i) **Palatine :** A slender branch runs along the floor of the orbit and innervates the roof of buccal cavity.

(ii) **Hyomandibular :** It runs posterior to mandibular and gives nerve fibers to the post tympanic area and lower jaw.

10. RAT (*Rattus sp.*)

The rat is a typical mammal due to the presence of mammary gland and hairs. In India

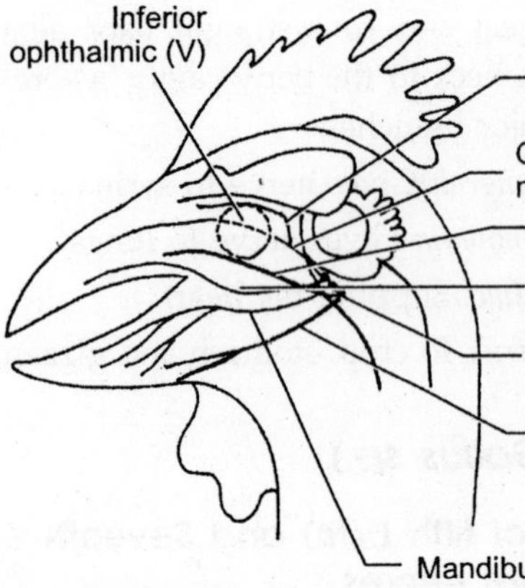

Fig. 1.4.68A : *Vth and VIIth cranial nerves of a Fowl. (Gallus sp.)*

there are nearly seventy species of rats of which, the most commonly found domestic rats are *Rattus rattus, Mus musculus* and *Rattus norvegicus. Rattus norvegicus albinus* or white rats are reared and bred for use in laboratories. *Rattus rattus* are medium-sized brownish rats common in domestic houses.

External features :

Procedure :

The rat is first killed with chloroform. As the domestic house rats often carry parasites, it is immersed in a solution of a suitable disinfectant (viz., lysol) or any other antiseptic solution and is thoroughly washed with water before use. The rat is put in a dissecting tray and the following structures are noted.

(1) **Size :** About twenty five centimetre including tail.

(2) **Shape and colour :** Elongated and brownish to blackish in colour.

(3) **Divisions :** Regionated into the head, trunk and tail. Soft hairs cover and protect the entire body.

(a) ***Head*** : It is elongated, some what conical and pointed anteriorly to terminal nose.

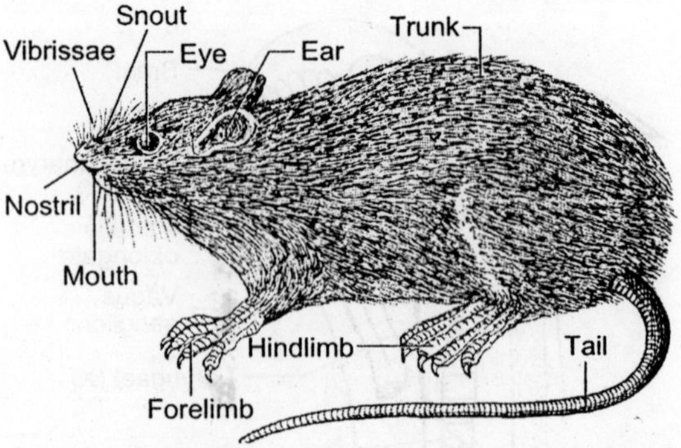

Fig. 1.4.69 : *External features of a rat. (Rattus rattus)*

Head contains the following structures.

(i) ***Nostrils*** : Paired, found in front of the pointed nose. It can be closed under water.

(ii) ***Hair lip*** : In the upper lip below the nostril, a cleft is found. It exposes the paired upper incisors.

(iii) ***Mouth*** : It is sub terminal in position of the head.

(iv) ***Teeth*** : Insisors four, molars four, canines and premolars are absent.

(v) ***Eyes*** : Paired one on each side of the head. Eyes are small and provided with eyelids. A nictitating membrane is attached to anterior cornea of each eye. External eyelid has lachrymal glands.

(vi) ***Pinnae or external ear*** : Found on the postero-lateral area of the head. They are large and hair less.

(b) ***Neck*** : It is short, joining the head to the trunk.

(c) ***Trunk*** : It is regionated into **thoracic**, **abdominal** and **pelvic area**. On the ventral surface of the female, one pair of **thoracic nipples**, two pair's **axillary nipples**, one pair of **abdominal nipples** and two pairs of **inguinal nipples** are seen. The pelvic region contains the urethral opening, vulval opening and anus. Male is provided with two large testes in the **scrotal sacs.** The

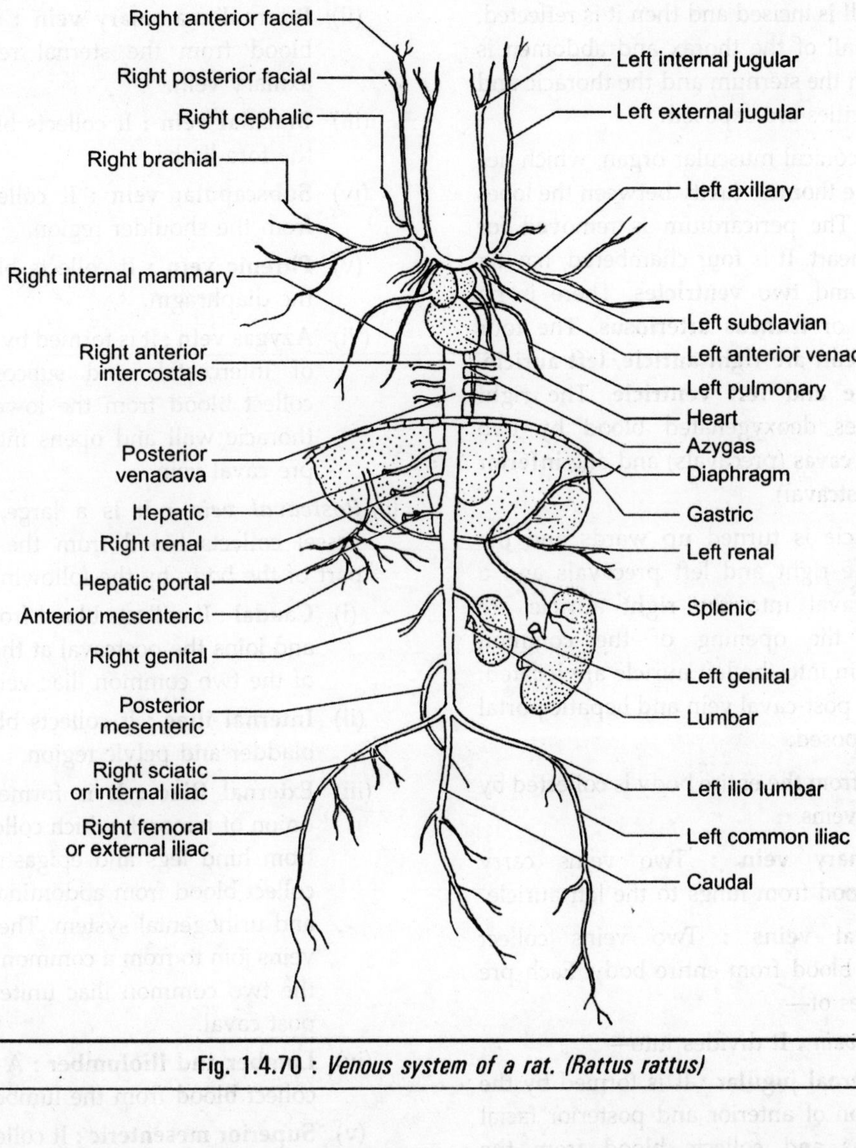

Fig. 1.4.70 : *Venous system of a rat. (Rattus rattus)*

trunk contains fore limbs and hind limbs.

(d) ***Tail*** : It is larger in comparison to head and trunk. It contains small over lapping vestigial scales.

Venous system :

Procedure :

The rat is killed with chloroform, disinfected with a suitable disinfectant (viz., Lysol) or any other antiseptic solution. The specimen is then put on its back on a dissecting tray. The limbs are fixed with pin. A longitudinal cut is made in the skin from the pubic symphysis to the tip of the snout. A transverse incision is given along the length of the limbs. The skin is separated from the underlying muscles and then is fixed with pin.

A transverse incision is made in the abdominal wall in pubic symphysis region. A parallel incision is made to the ribs on both sides laterally and anteriorly up to the anterior border of the thoracic cavity. The clavicle and muscles are cut. The attachment of the diaphragm to the abdominal wall is incised and then it is reflected. The ventral wall of the thorax and abdomen is separated from the sternum and the thoracic and abdominal cavities are exposed.

Heart is a conical muscular organ, which lies obliquely in the thoracic cavity between the lobes of the lungs. The pericardium is removed for studying the heart. It is four chambered, having **two auricles** and **two ventricles**. There is no **sinus venous** or **truncus arteriosus**. The four chambers of heart are **right auricle, left auricle, right ventricle** and **left ventricle**. The right auricle receives deoxygenated blood by two **superior vena cavas** (precavals) and one **inferior vena cava** (postcaval).

The ventricle is turned up wards and the opening of the right and left precavals and a single post caval into the right auricle are located, and the opening of the common pulmonary vein into the left auricle are located. The pre-caval, post-caval vein and hepatic portal system are exposed.

The blood from the entire body is collected by the following veins.

(1) **Pulmonary vein** : Two veins carry oxygenated blood from lungs to the left auricle.

(2) **Pre-caval veins** : Two veins collect deoxygenated blood from entire body. Each pre caval comprises of—

(a) ***Jugular vein*** : It divides into—

(i) **External jugular** : It is formed by the union of anterior and posterior facial veins and collects blood from the muscles of head, lower jaw, tongue and salivary glands.

(ii) **Internal jugular** : It collects blood from the brain and joins with the external jugular.

(b) ***Subclavian vein*** : It is a short vein and opens into the pre-caval vein. It receives:

(i) **Anterior or superior intercostal vein** : It collects blood from the intercostal spaces and opens into the right precaval vein.

(ii) **Internal mammary vein** : It collects blood from the sternal region and axillary vein.

(iii) **Brachial vein** : It collects blood from the fore limbs.

(iv) **Subscapular vein** : It collects blood from the shoulder region.

(v) **Phrenic vein** : It collects blood from the diaphragm.

(vi) **Azygas vein** : It is formed by the union of intercostals and subcostals and collect blood from the lower part of thoracic wall and opens into the left pre caval vein.

(c) ***Postcaval vein*** : It is a large, medium vessel collects blood from the posterior part of the body by the following veins.

(i) **Caudal** : It collects blood from the tail and joins the postcaval at the junction of the two common iliac veins.

(ii) **Internal iliac** : It collects blood from bladder and pelvic region.

(iii) **External iliac** : It is formed by the union of femoral, which collects blood from hind legs and epigastric, which collect blood from abdominal muscles and urinogental system. The two iliac veins join to from a common iliac, and the two common iliac unite with the post caval.

(iv) **Lumber and iliolumber** : A few pairs collect blood from the lumber region.

(v) **Superior mesenteric** : It collects blood from the ilium.

(vi) **Inferior mesenteric** : It collects blood from the large intestine.

(vii) **Genital** : Paired, collects blood from gonads.

(viii) **Renal** : Paired, collects blood from kidneys.

(ix) **Hepatic** : Collects blood from the lobes of liver.

(d) ***Hepatic portal vein*** : It contains the following veins:

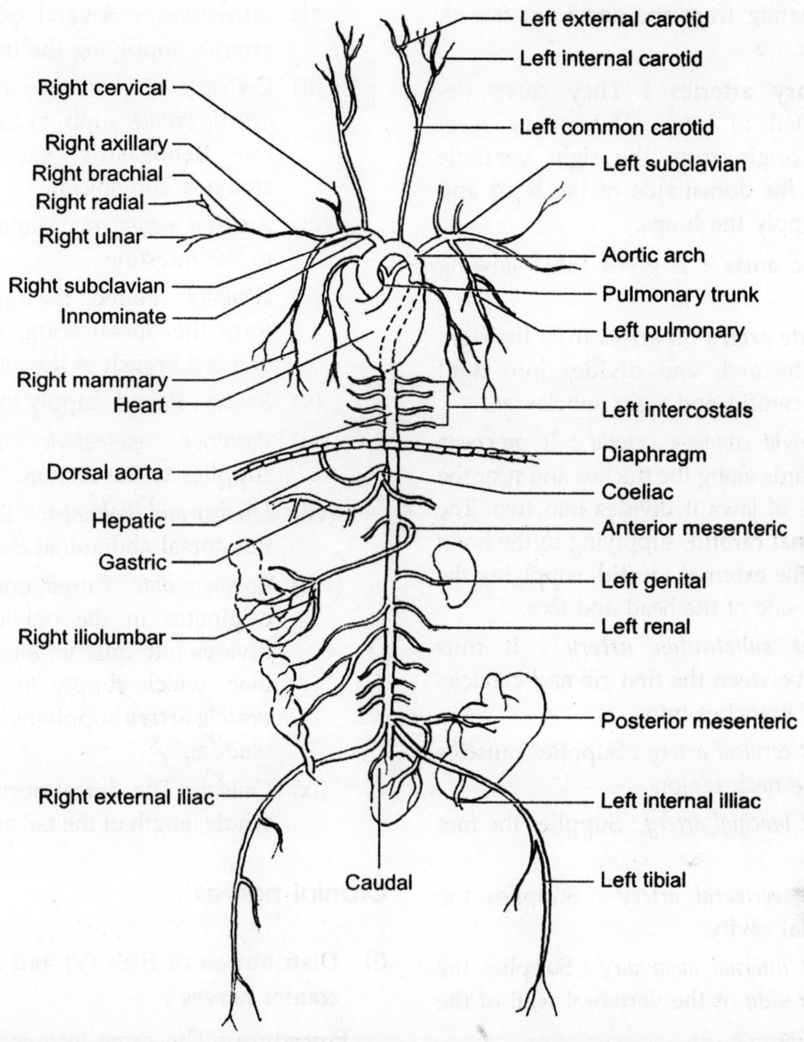

Fig. 1.4.71 : *Arterial system of a rat. (Rattus rattus)*

(i) **Lienogastric** : Returning blood from stomach, spleen and pancreas.

(ii) **Superior** (anterior) **mesenteric** : It collects blood from ileum, caecum and colon.

(iii) **Inferior** (posterior) **mesenteric** : It collects blood from rectum.

(iv) **Duodenal** : Collects blood from duodenum. Hepatic portal veins opening in the liver.

Arterial system :

Procedure :

Same as in case of venous system. The veins which are not required are removed. The heart is located and the pericardium is removed and the systemic aorta arising from the base of the left ventricle is observed. It passes dorsal to the pulmonary arch and curves round the dorsal side of the heart and runs postero-dorsally and medially as dorsal aorta. The courses of the branches originating from the aorta are traced. The arteries are.

(1) **Pulmonary arteries :** They carry de-oxygenated blood to lungs. Pulmonary aorta soon after its origin from the right ventricle curves over on the dorsal side of the heart and bifurcates to supply the lungs.

(2) **Systemic aorta :** It gives the following branches:

(a) ***Innominate artery*** : It arises from the right side of the arch and divides into right common carotid and right subclavian.

(i) *The right common carotid* : It proceeds forwards along the trachea and near the angle of jaws it divides into two. The **internal carotid**, supplying to the brain and the **external carotid**, supplying the right side of the head and face.

(b) ***The right subclavian artery*** : It runs outward between the first rib and clavicle. It further branches into:

(i) *Right cervical artery* : Supplies muscles of the neck region.

(ii) *Right brachial artery* : Supplies the fore limb.

(iii) *Right vertebral artery* : Supplies the cranial cavity.

(iv) *Right internal mammary* : Supplies the inner side of the vertebral wall of the thorax.

(v) *Right subscapular* : Supplies upper arm, scapula and thorax.

(vi) *Right costocervical* : Supplies neck and upper part of the thorax.

(c) ***The left common carotid artery*** : Originates directly from the systemic arch and has internal and external branches.

(d) ***The left subclavian artery*** : Arises directly from the aortic arch and supplies to shoulder girdle and arms. Branches are similar to those of the right one.

(3) **Dorsal aorta :** It comprises the following arteries :

(i) *Intercostals* : Several pairs of small arteries supplying the thoracic wall.

(ii) *Coeliac* : It gives two branches. The *hepatic branch* supplies to the liver and the **lienogastric** supplies to the stomach and spleen.

(iii) *Anterior mesenteric* : Unpaired, supplies to the intestine.

(iv) *Genitals* : Paired, the right one arises form the dorsal aorta, while the left one is a branch of the left renal artery.

(v) *Renals* : Paired, supply to two kidneys.

(vi) *Posterior mesenteric* : Unpaired, supplies to the rectum.

(vii) *Lumbar and iliolumbar* : They supply to the dorsal abdominal wall.

(viii) *Common iliac* : Paired, one on each side originates in the pelvic region and divides into internal iliac and external iliac, which supply to the legs, and ***vesicle artery*** supplying to the urinary bladder.

(ix) *Caudal* : The dorsal aorta extends the whole length of the tail as *caudal artery*.

Cranial nerves :

(i) Distribution of fifth (V) and seventh (VII) cranial nerves :

Procedure : The same technique used as in case of *Channa*. The brain is exposed and roots of the cranial nerves and their courses followed.

• **Fifth (V) cranial nerve :** The fifth or trigeminal arises from the lateral region of ***Pons varolli***. It has three branches.

(1) **Ophthalmic :** After its origin it divides into three branches and runs forward as *lachrymal*, *frontal* and *nasociliary*. Lachrymal innervates the lacrymal gland and the conjunctiva. Frontal traverses the orbit sends branches to the skin of the fore head. Nasociliary or anterior ethmoidal nerve enters the cranial

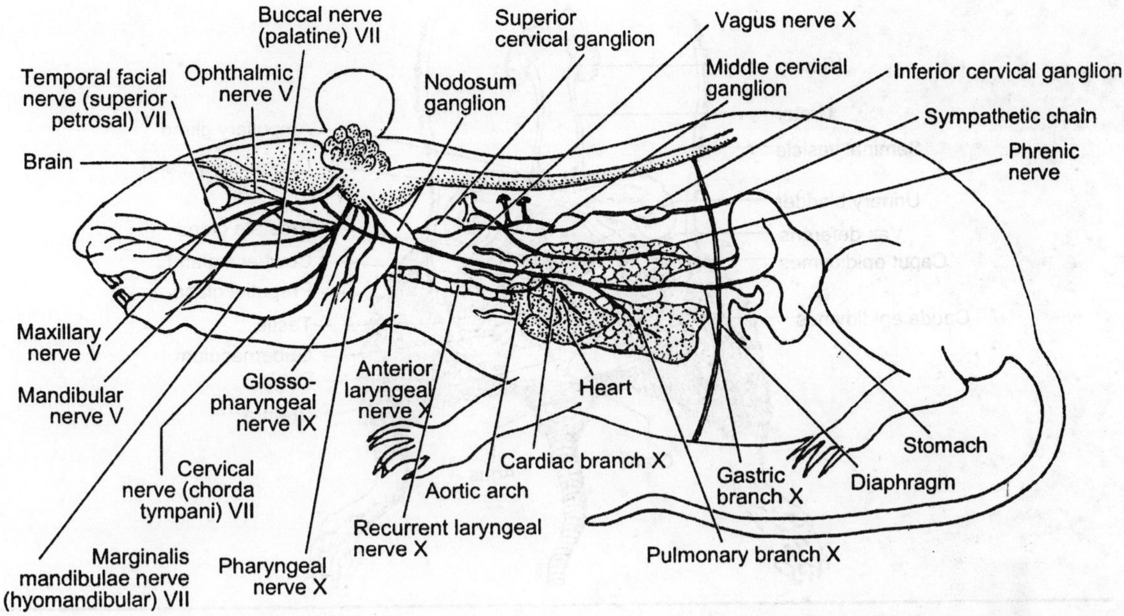

Fig. 1.4.72 : *Brain and V, VII, IX and X cranial nerves of a rat. (Rattus rattus)*

cavity and sends branches to mucous membrane and tip of the nose.

(2) **Maxillary :** It runs forward along the floor of the orbit and gives nerve fibres to the side of the nose, the upper and lower lips.

(3) **Mandibular :** It is posterior to the maxillary, terns down ward and gives nerve fibres to the durameter, mandible, pinna, mucous membrane of the mouth, tongue, teeth and skin of the lower jaw.

• **Seventh (VII) cranial nerves :** The seventh or facial nerve arises from the anterolateral border of the medulla oblongata behind the fifth and enters the facial canal. It divides into four branches.

(i) ***Superfacial pectoral*** : It runs through the middle ear and innervates soft palate and lacrymal gland.

(ii) ***Palatine :*** It passes back ward and upward between the pinna and supplies palate and nose.

(iii) ***Hyomandibular*** : It runs anteriorly and dorsally over the zygomatic arch and supplies the muscles of the lower jaw and lip.

(iv) ***Chorda tympani*** : It enters the tympanic cavity and supplies the snout, tongue and salivary glands.

(ii) Distribution of ninth (IX) and tenth (X) cranial nerves :

• **Ninth (IX) Cranial nerve :** The ninth (IX) or glossopharyngeal nerve originates from the postolateral side of the medulla oblongata and runs anteriorly between the external and internal carotid arteries. After its origin it divides into two. A small, *pharyngeal* joins the pharyngeal nerve of tenth and forms the pharyngeal plexus. The *anterior glossal* moves forward and supply the tongue and pharyngeal wall.

• **Tenth (X) cranial nerve :** The tenth (X) or vagus nerve originates from the posteroventral

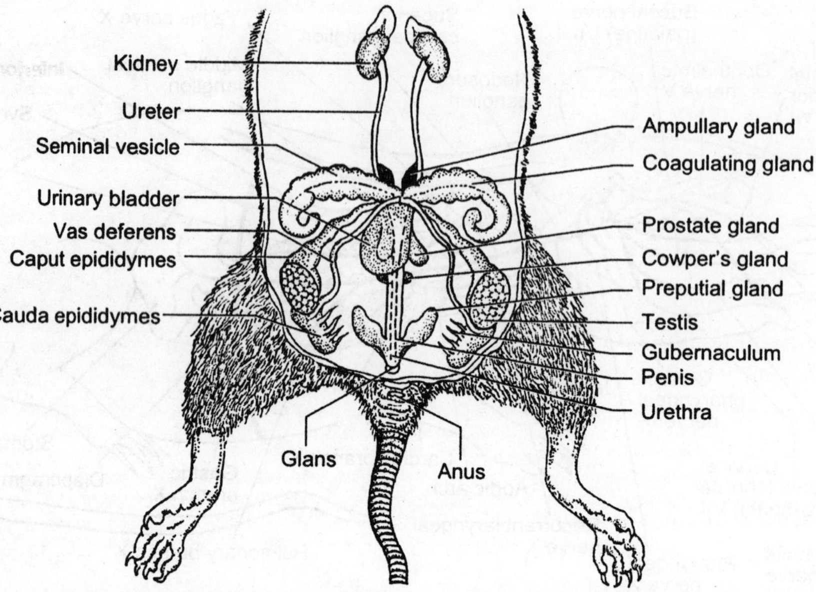

Fig. 1.4.73 : *Urinogenital system of a male rat. (Rattus rattus)*

region of the medulla oblongata and runs back wards outside the carotid artery and possesses **nodosum ganglion**, just before it gives off the pharyngeal branch. The vagus runs backward through the neck and enters abdomen. Two nerve arise from the nodosum ganglion :

(i) *Pharyngeal* : It arises from the anterior end of ganglion and joins the pharyngeal branch of ninth.

(ii) *Superior anterior laryngeal* : It arises from the posterior end of the ganglion and supplies the larynx and thymus gland.

(iii) *Recurrent laryngeal* : It originates near thorax from the main trunk of the vagus. In the left side it curves around aorta and around subclavian in the right side, and finally turns anteriorly to supply the larynx.

(iv) *Cardiac* : Supplies the heart.

(v) *Pulmonary* : Supplies the lung of the side.

(vi) *Gastric* : The main trunk runs backward and sends branch to oesophagus, liver, and other visceral organs.

Urinogenital system :

The urinogenital system consists of two separate systems, called urinary and genital system. In rat they are not completely separate. The two systems are inter dependence and there fore, are usually described together under the heading of urinogenital system.

Procedure :

The rat is dissected as usual. The alimantary system is separated from mesenteries and the oesophagus and the anterior part of the rectum are incised and removed. The sides of the pubic symphysis are cut and the cut portion is removed to expose the posterior part of urinogenital system. The fat covering is removed to exposed the testes.

The Urinary System :

(1) **Kidneys** : Two dark red, bean-shaped bodies, having on the median inner side a notch, the hilus. The right kidney is some what more

anteriorly situated than the left.

(2) **Ureters** : From hilus arises two narrow, long, thin, whitish tube runs backwards along the dorsal abdominal wall and open posteriorly into the urinary bladder.

(3) **Urinary bladder** : A small, pear-shaped muscular sac, having transparent wall, opens into the urethra.

(4) **Urethra** : The bladder continues posteriorly as urethra. In male it is long and muscular and opens to outside through an opening located at the tip of penis. In females the urethra is short and independent of the genital ducts and open through an aperture called **urinary aperture** at the tip of **clitoris**.

• **Male genital system :**

Procedure : Testes are found in the scrotal sac, which is present in front of the anus and between the hind legs. The scrotum is cut open to expose the scrotal sac. The testes are found in the muscular portion of the sac.

(1) ***Testes*** : Paired, white, oval bodies, attached to the hinder end of the scrotal sac by a band of tissues called **gubernaculum**.

(2) ***Epididymes*** : Irregular and convoluted tube found along the inner edge of testes. At anterior end it forms **caput epididymes** and at posterior end **cauda epididymes**. Cuda epididymes gives rise to vas deferens.

(3) ***Vas deferens*** : Paired tubes. Each vas deferens coils round the ureter from the outside before opening into the urethra behind the urinary bladder.

(4) ***Urethra*** : A common passage for urine and sperms opening into the penis.

(5) ***Penis*** : A long, median muscular organ hangs in front of anus from the ventral wall of the abdomen, covered distally by a loose fold of skin, called prepuce.

(6) ***Seminal vesicles*** : Paired, large curved flabby organs associated with vasa deferentia.

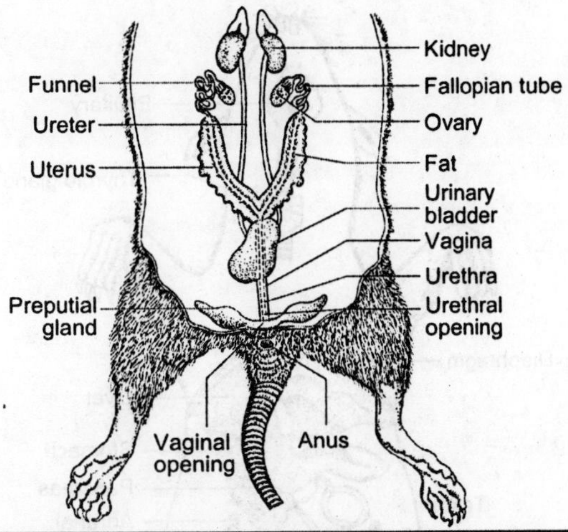

Fig. 1.4.74 : ***Urinogenital system of a female rat. (Rattus rattus)***

(7) ***Coagulating glands*** : Paired glands attached to the seminal vesicles.

(8) ***Prostate glands*** : Two small lobed bodies attached to the junction of bladder and open into urethra by small ducts.

(9) ***Cowpar's glands*** : Two ovoid glands, situated behind the prostate gland on the dorsal wall of the urethra and open by short ducts into it.

(10) ***Preputial glands*** : Paired flat elongated bodies situated at the sides of the penis.These glands secrete odorous secretion.

• **Female genital system :**

(1) ***Ovaries*** : Paired small yellowish, compact bodies on outer sides of the kidneys and are suspended by *mesovarium*.

(2) ***Oviducts*** : Two oviducts corresponding to the ovaries. They are regionated into two areas.

(a) *Fallopian tube* : Coil and convoluted anterior part of the oviduct. Each duct opens in the abdominal cavity by a wide funnel shaped fallopian funnel, facing just outside the ovary.

(b) *Uterus* : The posterior dilated, thickend, broad ends of the fallopian tubes.

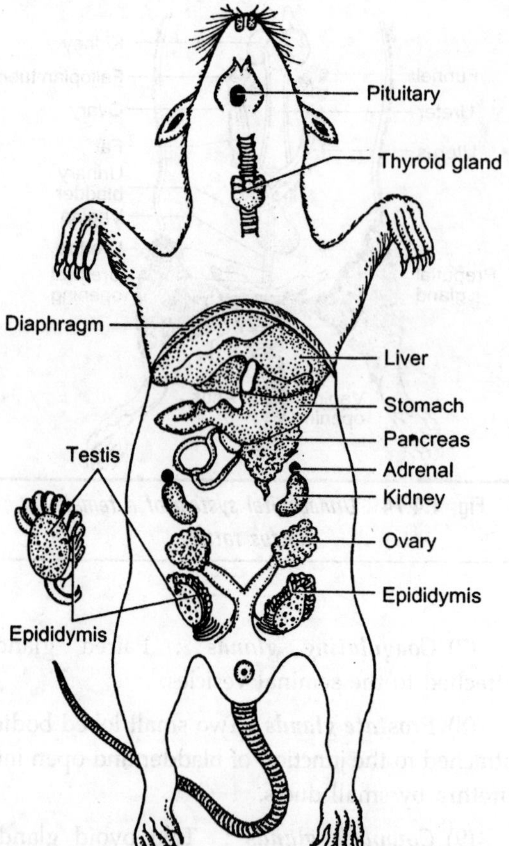

Fig. 1.4.75 : *Diagram showing the position of endocrine gland of a male and female rat. (Rattus rattus)*

(c) Vagina : The common chamber formed by the union of the two uteri. It opens to the exterior by a slit like opening the vulva.

(d) Preputial glands : Like male.

(e) Clitoris : A rod-like structure, found anterior to vulva.

DISSECTION AND MOUNTING OF ENDOCRINE GLANDS OF A RAT

Procedure :

Chloroformed killed rats are placed in a dissecting tray and the limbs are fixed with pin. Dissection procedure is same as done previously. After exposing the body cavity the various endocrine glands are absorbed. Endocrine glands produce secretions, which are called **hormones.** Dissection and mounting procedure is same as described in Chapter 1.5.

1. **Thyroid :** It is a bilobed gland, situated just below the larynx. It produces a hormone, called thyroxin which regulates metabolism. The gland is dissected out and placed in a clean slide and mounted in glycerine for observation.

2. **Adrenals or Suprarenals :** Paired, small ovoid glands situated close to the kidneys. It produces adrenalin or medulla-cortex. The former regulates the carbohydrate metabolism. The gland is dissected out and placed in a clean slide and mounted in glycerine for observation.

3. **Pancreas :** It is suspended by delicate mesenteries between the stomach and duodenum. It produces insulin from islets of langerhans. The hormone insulin is responsible for conversion of excess glycogen into glucose. The gland is dissected out and placed in a clean slide and mounted in glycerine for observation.

4. **Testis :** Paired, elongated, oval bodies found in the scrotal sac. Scrotal sacs are a pair of pouches in front of the anus and between the hind legs. Testes are exposed by cutting the scrotal sac. Testes produce hormones and sex cells. Testis is dissected out and is placed in a watch glass containing normal saline solution for observation.

5. **Ovary :** Paired, small, yellowish, structures lies on outer sides of the kidneys and suspended in the body cavity by mesovarium. Ovary is dissected out and is placed in a watch glass containing normal saline solution for observation.

6. **Pituitary :** The gland is found attached to the ventral side of the diencephalon of the brain and lies in the depression on the floor of the cranium. It has anterior and posterior lobes. The

anterior lobes originate from the roof of the pharynx as an out growth, while the posterior lobe develops from the floor of the brain as a down growth.

OBSERVATION OF THE MOVEMENT OF SPERM IN RAT

Procedure :

Method I : In rat sperm are obtained from the epididymis. Along the side of each testis there is a mass of twisted tubules called epididymis. With a sharp instrument the epididymis is cut across and the cut surface is drawn over a clean slide. Few drops of physiological saline solution is poured over the material to observe the movement of sperm.

Method II : The Epididymis is exposed and opened and then a drop of fluid is placed to a clean slide containing normal saline solution. The material is spread with the edge of second slide for observation.

Chapter 1.5

Mounting and Display of Dissected Parts

Mounting is a process in which part(s) of a specimen dissected out and placed in a suitable medium for observation under microscope or a dissecting binocular or powerful lens.

Before mounting of any organ or its part, one has to study thoroughly about the structure which is to be stained and mounted. The animal is dissected to take out the structure to be mounted and the methods as described in the following pages are followed. After mounting and study, the slides prepared are drawn.

A. NON CHORDATA

1. Parapodium of *Neanthes (Nereis) sp.* :

Parapodia are paired, lateral out growth of the body and present in all segments except head and the last segment. A biramous structure with a dorsal **notopodium** and a ventral **neuropodium**. Presence of a dorsal cirrus in notopodium and a ventral cirrus in neuropodium. Both of the them bear bundles of fine **setae** and **acicula** in **setigerous sacs**.

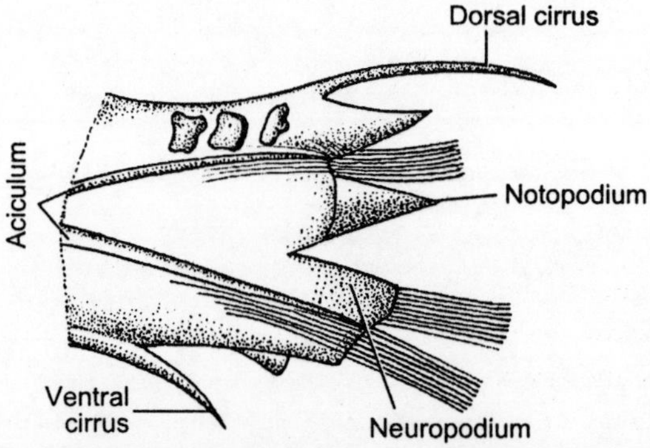

Fig. 1.5.1 : Parapodium of *Neanthes sp.*

Preparation :

1. A portion of the body wall with the parapodium is separated by giving a V-shaped cut on the lateral wall of the body close to the base of parapodium by two incisions.

2. The body wall is separated from the parapodium with a pair of fine forceps taking care not to damage the acicula.

3. The parapodium is placed in a watch glass and washed thoroughly with water.

4. Passed through thirty, fifty and seventy percent alcohol and stained in Borax carmine following the single stain procedure.

5. Dehydrated, cleared and mounted in Canada balsam.

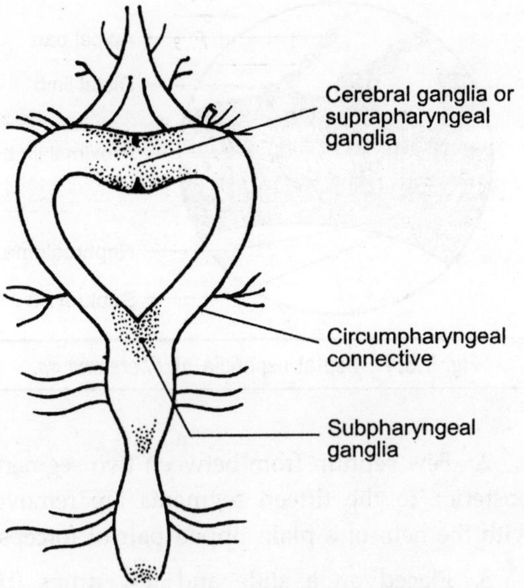

Fig. 1.5.2 : Nerve ring of *Phretima sp.*

2. Nerve ring of *Pheretima sp.* :

It consists of a pair of **suprapharyngeal** or **cerebral ganglia** or brain, which is a fused swelling situated in the third segment above the pharynx. From these ganglia two **circumpharyngeal connectives** pass one on either side of the pharynx downwards and back wards to join the **subpharyngeal** or **suboesophageal ganglia**. These are situated below the pharynx partly in the third and partly in the fourth segment of the body. The cerebral ganglia, circumpharyngeal connectives and subpharyngeal ganglia form a ring of nerves and ganglia, commonly called as nerve ring.

Preparation :

1. Earthworm is dissected from the dorsal side and alimentary canal is removed from posterior region up to the fourth segment leaving the pharynx intact. Nerve will be visible.

2. Few drops of acetoalcohol are poured on the nervous system and one has to wait for 5 – 10 minutes.

3. The muscles and connective tissue of the anterior side are cleared off. Nerve ring will be visible.

4. The connective tissue and the pharynx are removed.

5. The nerve ring is separated and the entire nerve ring with two to three ganglia of the nerve cord is dissected out.

6. Placed on a glass slide, mounted in glycerine and covered with a cover glass. Observed under a simple microscope.

3. Setae of *Pheretima sp.* :

Setae are chitinous slightly bent fine rod – shaped body present in all segments of earth worm, except the first and last segments as well as in the clitellum. Each seta is enclosed in a small pouch, called setal sac, which is attached to the muscles. Muscles help in protrusion and retraction of the setae. Two third of each seta remain with in the body and one third projects out. A seta consists of base, middle part and free end or neck.

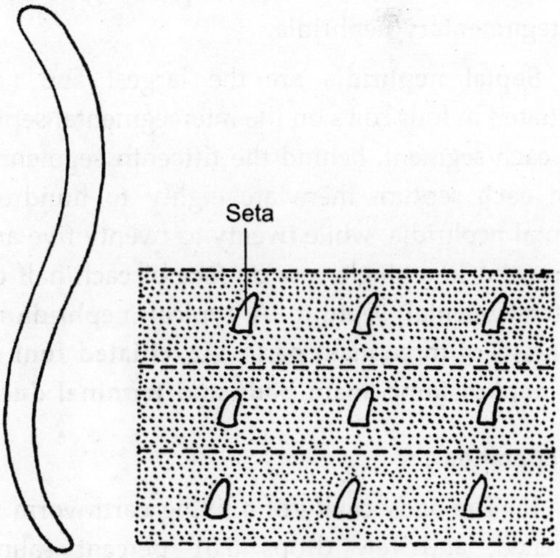

Fig. 1.5.3 : Setae of *Pheretima sp.*
A–Isolated seta, B–Setae in situe.

Preparation :

1. Earthworm is dissected, the nerve cord and connective tissues are removed.

2. A rectangular small piece of skin of the ventral surface behind the clitellar region is incised. This cut piece is put in a test tube containing ten percent caustic potash (KOH) solution.

3. The contents of test tube are boiled till the cut piece become transparent. Care is taken not to dissolve the muscle.

4. The contents of the tube are poured in a watch glass and washed thoroughly in water to remove KOH.

5. The transparent piece is placed on a clear glass slide and fully stretch, with the help of two needles.

6. Mounted in glycerine and observed under microscope.

4. Septal nephridia of *Pheretima sp.* :

Nephridia are distinguished into three types according to their arrangement and position in the body. These are **septal, pharyngeal** and **integumentary** nephridia.

Septal nephridia are the largest and are situated in four rows on the intersegmental septa in each segment, behind the fifteenth segments. On each septum there are eighty to hundred septal nephridia, while twenty to twenty five are arranged in a single row on face of each half of intersegmental septum. A typical nephridium consists of three main parts, viz, **ciliated funnel** or **nephrostome, main body** and **terminal duct**.

Preparation :

1. The body cavity of a living earthworm is opened, and few drops 0.01 percent saline solution is poured on it. Septa connecting the alimentary canal with the body wall is seen. The nephridia are arranged on both the faces of a septum.

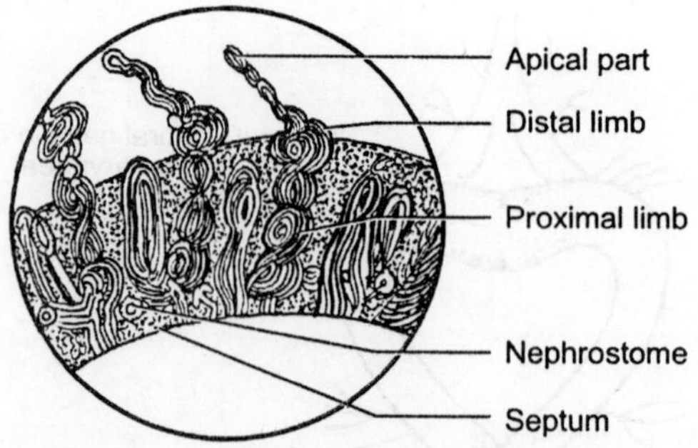

Fig. 1.5.4 : Septal nephridia of *Pheretima sp.*

2. Few septum from between two segments posterior to the fifteen segments are removed with the help of a plain tipped pair of forceps.

3. Placed on a slide and few drops 0.01 percent saline water is poured on it. The water is removed and it is mounted in glycerine.

4. Examined under microscope and nephridia is observed on both the faces of a septum.

5. Ovary of *Pheretima sp.* :

Below the alimentary canal are on either side of the nerve cord, there is a pair of small, white masses in the thirteenth segments attached to posterior face of the septum between twelveth and thirteenth segments. Ovary consists of a number of filaments fused proximally, while free

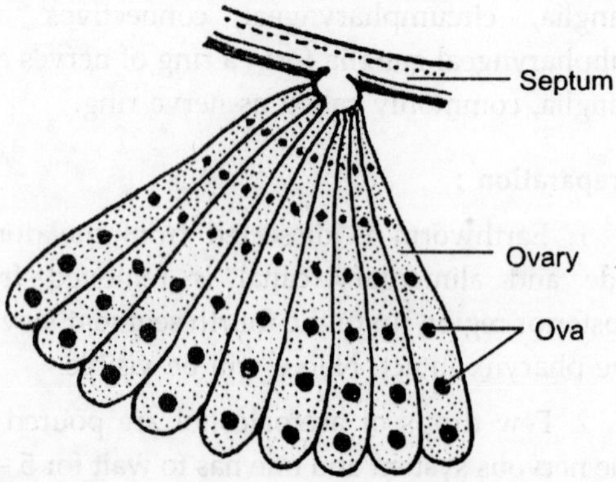

Fig. 1.5.5 : Ovary of *Pheretima sp.*

distally. Distal parts of the filaments are beaded in appearance and possess ova.

Preparation :

1. The earthworm is dissected out and the gut is detached from the body wall starting from thirteeth segments, the gut is cut and the cut end is held with a pair of forceps, it is lifted a little and the septa attached to it is separated from the body wall with a fine needle. Few drops of Bouin's fluid is poured, the ovary will take yellow colour. On the hinder part of thirteenth segments lies a pair of ovaries.

2. Carefully the intestine is removed proceeding up to the anterior end.

3. Carefully one ovary is removed with a pair of fine forceps and placed on a slide.

4. Mounted in glycerine and observed under microscope the digited structure which contains ova.

6. Jaw of *Hirudinea sp.* :

On the ventral side of the anterior sucker, is a hollow depression, the oral chamber, at the base of which there is a narrow tri – radiate mouth, which leads into a very short chamber, the buccal cavity. Within the buccal cavity lie the three white chitinous **crescentric jaws**. One jaw is mid-dorsal and the other two are ventrolateral in position. Along the cut border lies **antagonistic** muscles and the free end is provided with **teeth** or **denticles**, which are arranged in a single rows. On each side of the jaw, there are several projections, called **salivary papillae**.

Preparation :

1. A previously killed leech is held and the buccal cavity is opened, the three chitinous triangular plates are seen.

2. A jaw is removed by cutting it muscles of attachment with a fine scissors and placed on a watch glass containing ten percent KOH solution for twenty five to thirty minutes.

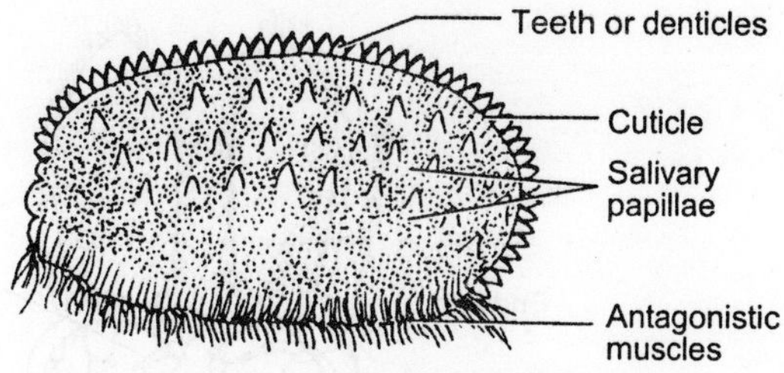

Fig. 1.5.6 : Jaw of *Hirudinea sp.*

3. Muscles are removed by a pair of forceps and scalpel as far as practicable.

4. Washed thoroughly with water and the jaw is placed on a clean slide.

5. Mounted in glycerine, examined under microscope and the semicircular jaw off leech is seen.

7. Green gland or Antennary gland of *Macrobrachium sp.* :

Antennary gland is situated in prawn within the coxal podomeres of the second antennae. Each gland is just like a Pea seed. It consists of three parts: (i) **end sac.** (ii) **labyrinth** and (iii) **bladder.**

The end sac is small bean shaped structure, situated in the anterior part of the gland. The labyrinth is made up of a mass of coiled tubules, joined on one end with the end sac and, on the other, with the bladder. The urinary bladder is situated on the inner side of the end sac.

Preparation :

1. One second antenna is removed from the cephalothorax of prawn by cutting it at its base.

2. The ventral wall of coxa and basis, close to the outer border is cut off.

3. A transverse incision is given along the distal border of the basis and the flap is turned inside out. The antennary gland is exposed.

4. The entire gland is removed carefully from the antenna after cutting the ureter close to renal pore.

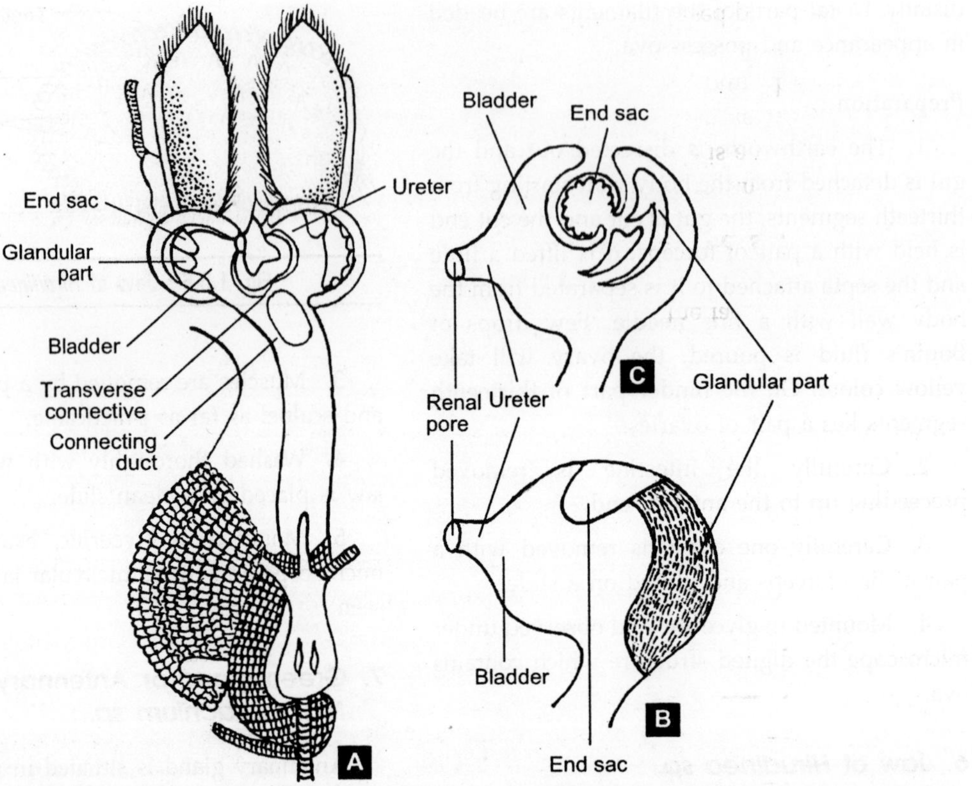

Fig. 1.5.7 : Green gland or Antennary gland of *Macrobrachium sp.* A–In situ, B–Gland dissected out, C–Sectional view.

5. It is placed in a watch glass with water and a thick section is cut with a sharp razor.

6. Washed and examined under a powerful dissecting binocular.

8. Statocyst of *Macrobrachium sp.* :

Statocysts are the balance organs of prawn. These are pair of small, white, hollow flask shaped structure, situated inside the precoxa of the antennule attached to its dorsal wall. It is open to the outside through a narrow pore. The cavity of the statocyst is filled with sand grains. On removing the sand grains it is found that there is an oval ring of elongated, delicate **receptor setae**, which is attached to the inner wall of the statocyst and are arranged in an oval outline. Each seta has a swollen base and a pointed shaft.

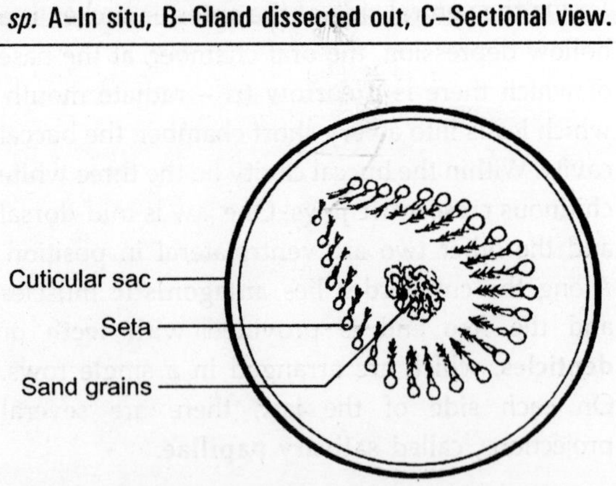

Fig. 1.5.8 : Statocyst of *Macrobrachium sp.*

Preparation :

1. One antennule from the prawn is removed by cutting it at its base.

2. The cavity of the precoxa region is removed by cutting along its inner border and

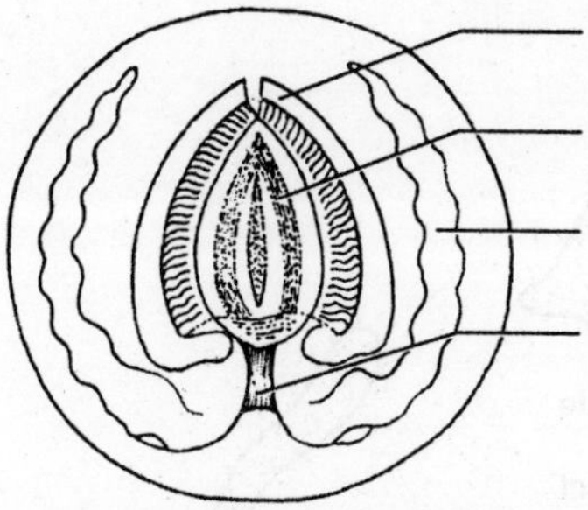

Fig. 1.5.9 : Hastate of *Macrobrachium sp.*

also transversely at its distal end. The statocyst is exposed.

3. The statocyst is cut with a sharp scalpel close to the base and separated from the precoxa.

4. It is placed in water in a watch glass. Sand grains with a fine, soft brush are washed.

5. Mounted in glycerine and observed under microscope.

6. For permanent preparation, stained in borax carmine, dehydrated and mounted.

9. Hastate plate of *Macrobrachium sp.* :

Hastate plate is a large and triangular just like the head of spear found on the floor of the cardiac stomach with its upper surface covered with a thick growth of delicate setae. Its posterior triangular border forms the anterior valve of **cardiopyloric aperture.**

Preparation :

1. A prawn is dissected out and the cardiac stomach is taken out by cutting the oesophagus and the junction of cardiac and pyloric stomachs.

2. The hastate plate is exposed, by giving an incision on the roof of the cardiac stomach. The hastate plate is seen occupying the floor of the cardiac stomach which looks like the head of a spear.

3. The hastate plate is removed from the stomach by cutting along its border with a pair of fine scissors.

4. It is placed on a watch glass containing drops of water, it is then cleared with a fine brush and mounted in glycerine.

10. Mouth parts of *Periplaneta sp.* :

Mouth parts are situated ventrally on the head. The mouth parts of cockroach are composed of paired appendages like **mandibles, maxillae, hypopharynx, labrum and labium.**

Preparation :

1. The head of killed cockroach is held in between the thumb and the first finger of left hand, ventral surface facing up wards.

2. The mouth parts are separated with the help of a pair of fine forceps one by one and arranged in a glass slide in their natural order.

3. Mounted in glycerine and observed under a lens.

11. Salivary apparatus of *Periplaneta sp.* :

On each side of the thoracic cavity of cockroach lies a pair of **salivary gland.** Each gland consist of two leaf like lobes and a reservoir. The secretory lobes, reservoirs and their ducts together constitute the **salivary apparatus.**

Preparation :

1. A cockroach is pinned down in a dorsal position on a dissecting tray.

2. A dorsal incision is given up to thorax and either end of the cut body wall is pinned down. The alimentary canal is visible along with a pair of salivary gland with their ducts and a receptacle, on either side of oesophagus.

3. All the tracheae and fat in the region are carefully removed where the salivary apparatus is lodge.

4. The apparatus is separated from the crop and traced the ducts anteriorly running from the glands and receptacle along the sides of the crop and then ventral to the oesophagus.

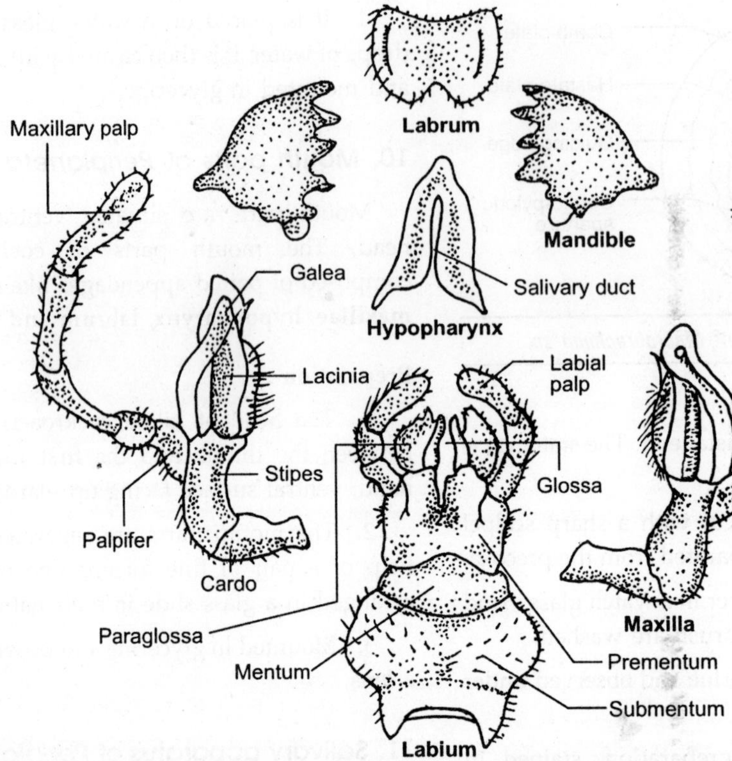

Fig. 1.5.10 : Mouth parts of *Periplaneta sp.*

5. The head of cockroach is pinned down with ventral surface up wards.

6. The glands and receptacles are detached from the crop and the common salivary duct up to hypo pharynx is traced.

7. The hypopharynx is separated from the specimen by cutting it at the base and the entire apparatus is brought out by carefully lifting it with a spatula and placed it on a glass slide.

8. Each part is spread equally and mounted in glycerine.

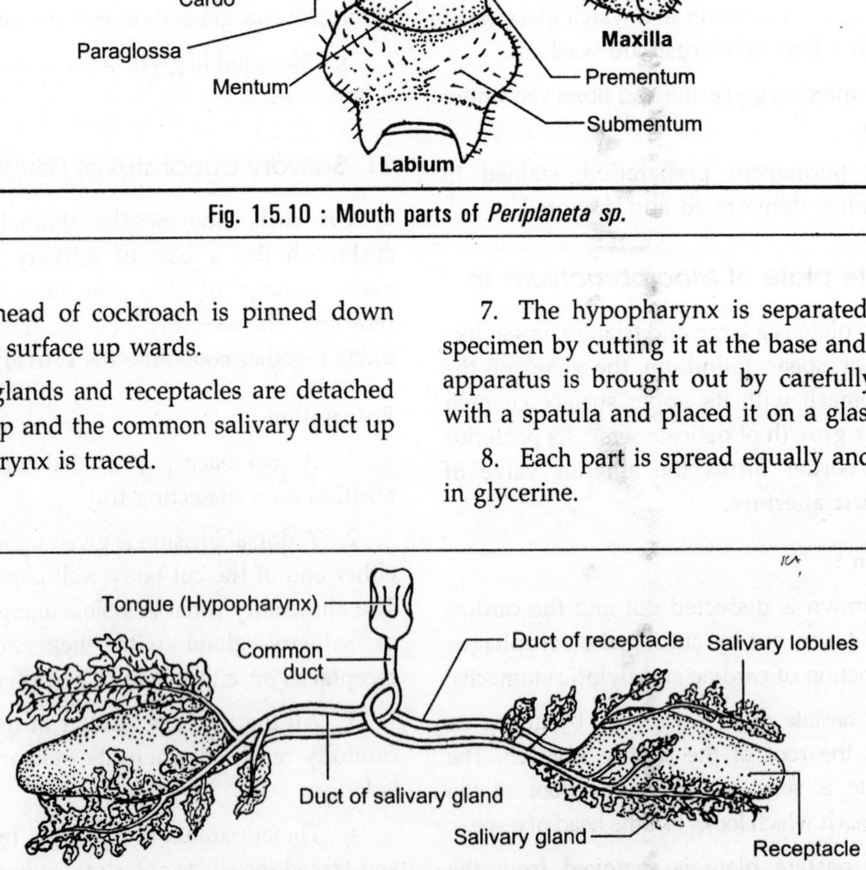

Fig. 1.5.11 : Salivary apparatus of *Periplaneta sp.*

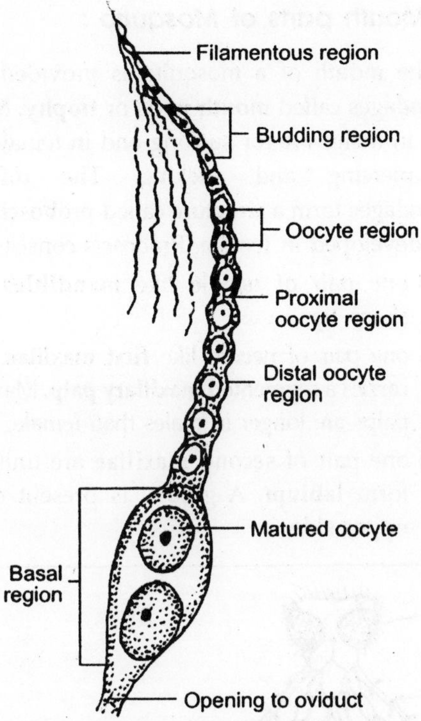

Fig. 1.5.12 : Ovariole of *Periplaneta sp.*

12. Ovariole of *Periplaneta sp.* :

A pair of ovaries will be seen on the posterolateral abdominal cavity of female cockroach. Each ovary is made up of eight beaded tube like ovarioles.

Preparation :

1. The female reproductive system of a cockroach is dissected out.

2. The ovary is located and one of the ovariole is dissected out and mounted in glycerine on a clean glass slide.

3. Examined under microscope and the linear arrangement of eggs at different stage of maturity with the ovariole is observed.

13. Mouth parts of *Musca sp.* :

It is convenient to use the whole head of a fly to study the mouth parts. The head of a house fly

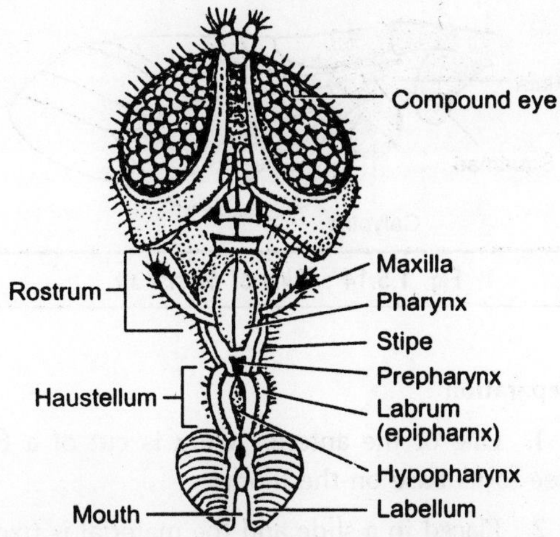

Fig. 1.5.13 : Mouth parts of *Musca sp.*

is small and semi circular in out line. The mouth parts are well developed and contain two portions – (a) **proboscis** and (b) **food channel**.

Proboscis is formed by fleshy labium which may be separated into – (i) **rostrum**, (ii) **haustellum**, and (iii) **labellum**. Rostrum carries in front a pair of maxillary palp. The food channel is formed by **labrum-epipharynx** and **hypopharynx**. Mandible is absent.

Preparation :

1. The head of a fly is severed with a fine scissor. The head is placed in a hard test tube containing five to ten percent KOH solution and boiled for five to ten minutes.

2. The solution cooled and decanted and the material is washed thoroughly in water and the last trace of KOH is removed.

3. The head is placed on a clean glass slide, poured few drops of glycerine and mounted to study under microscope.

14. Wing of *Musca sp.* :

Thorax of a fly contains two pairs of wings only the anterior pair is prominent, broad, transparent, hairy and triangular. The posterior pair is modified into stick like **halters**.

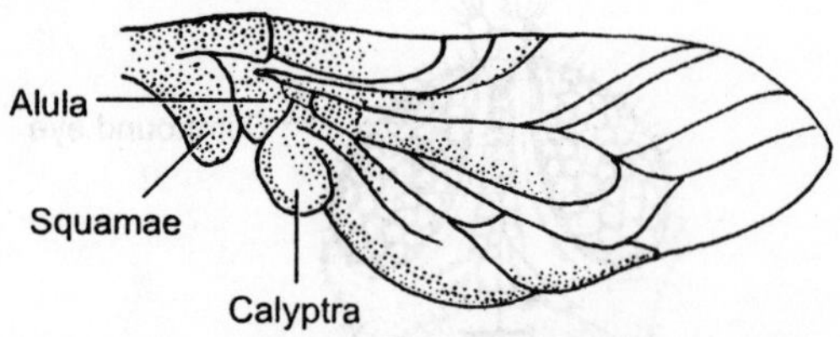

Fig. 1.5.14 : Wing of *Musca sp.*

Preparation :

1. One of the anterior wing is cut of a fly close to its base on the thorax.

2. Placed in a slide and the material is fixed in seventy per cent alcohol and mounted in glycerine.

15. Mouth parts of *Mosquito* :

The mouth of a mosquito is provided with appendages called **mouth parts** or **trophy**. Mouth parts in males are for sucking and in females are for piercing and sucking. The different appendages form a structure called **proboscis**. It is well developed in female. Proboscis consists of—

(a) one pair of needle like **mandibles** with sharp tips,

(b) one pair of needle like **first maxillae**. Each carries a segmented **maxillary palp**. Maxillary palps are longer in males than female,

(c) one pair of **second maxillae** are united to form **labium**. A groove is present on its upper side,

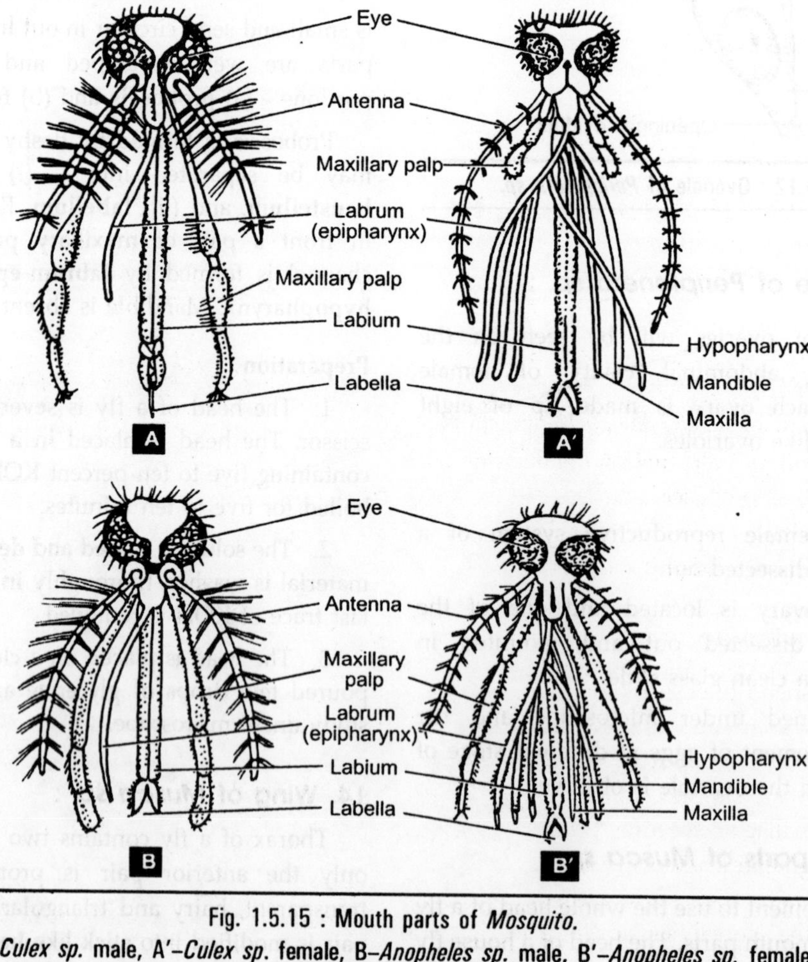

Fig. 1.5.15 : Mouth parts of *Mosquito.*
A–*Culex sp.* male, A′–*Culex sp.* female, B–*Anopheles sp.* male, B′–*Anopheles sp.* female.

(d) the **labrum** and **epiphanynx** are fused to form a single structure with sharp free end,

(e) A chitinous blade like **hypopharynx** with pointed tip. Hypopharynx and epipharynx unite to form a tube through which blood is sucked.

Preparation :

Same as in *Musca* mouth parts. Observed under low magnification of the microscope and the following characters are identified.

- Mouth parts of male ***Culex*** **mosquito** :
  - (i) Maxillary palp three segmented but as long as proboscis, last segment pointed.
  - (ii) Antennary segments with dense bushy bristles.
- **Mouth parts of female** ***Culex*** **mosquito :**
  - (i) Maxillary palp three segmented but short in length.
  - (ii) Antennary segments bear scanty hairs or bristles.
- **Mouth parts of male** ***Anopheles*** **mosquito :**
  - (i) Maxillary palp five segmented and longer than the proboscis and last segment club shaped.
  - (ii) Bushy antennary bristles but comparatively long.
- **Mouth parts of female** ***Anopheles*** **mosquito :**
  - (i) Maxillary palp five segmented and as long as proboscis and last segment is pointed.
  - (ii) Antennary bristles are scanty.

16. Mosquito larvae :

Mosquito spend their larva life in water. The larva is transparent and is differentiated into **head, thorax** and **abdomen**. Head contains compound eyes, simple antennae, chewing mandibles, maxillae and feeding brushes. Thorax is broad, unsegmented and contains tufts of hairs. Abdomen contains nine segments and tracheal opening on the dorsal surface of the eighth segments. Tracheal gills are found in ninth segments. The head of *Culex* larva hangs down

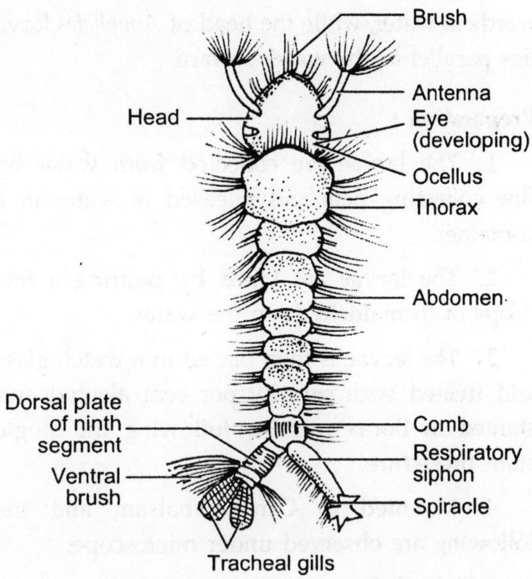

Fig. 1.5.16 : Larva of *Culex sp.*

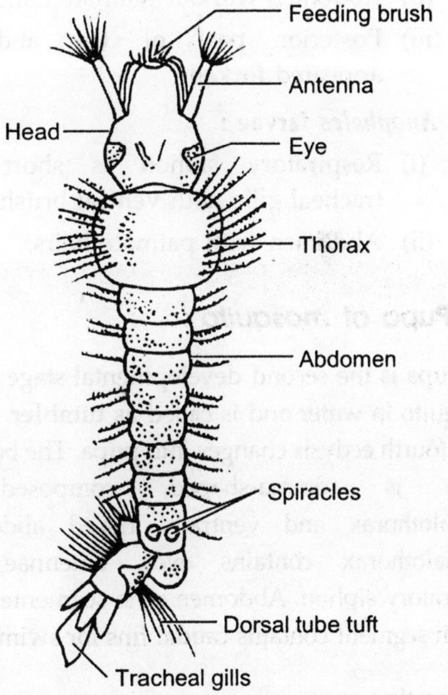

Fig. 1.5.17 : Larva of *Anopheles sp.*

wards in water, while the head of *Anopheles* larva lies parallel to the water surface.

Preparation :

1. The larvae are collected from water by fine collecting nets and released in water in a container.

2. The larvae are killed by pouring a few drops of formaldehyde in the water.

3. The larvae is then placed in a watch glass and treated with seventy per cent alcohol and stained in Borax carmine following the single stain procedure.

4. Mounted in Canada balsam and the following are observed under microscope.

- ***Culex* larvae :**
  - (i) Respiratory siphon in the eighth segment and long.
  - (ii) Abdomen without palmate hairs.
  - (iii) Posterior part of the abdomen appeared forked.
- ***Anopheles* larvae :**
  - (i) Respiratory siphon is short and tracheal gills with ventral brush.
  - (ii) Abdomen with palmate hairs.

17. Pupa of *mosquito* :

Pupa is the second developmental stage of the mosquito in water and is called as **tumbler.** Larva after fourth ecdysis changes into pupa. The body of pupa is comma-shaped, composed of cephalothorax and ventrally flexed abdomen. Cephalothorax contains eyes, antennae and respiratory siphon. Abdomen nine segmented and eighth segment contains caudal fins for swimming.

Preparation :

Same as in mosquito larvae. Observed under microscope and the following characters are identified.

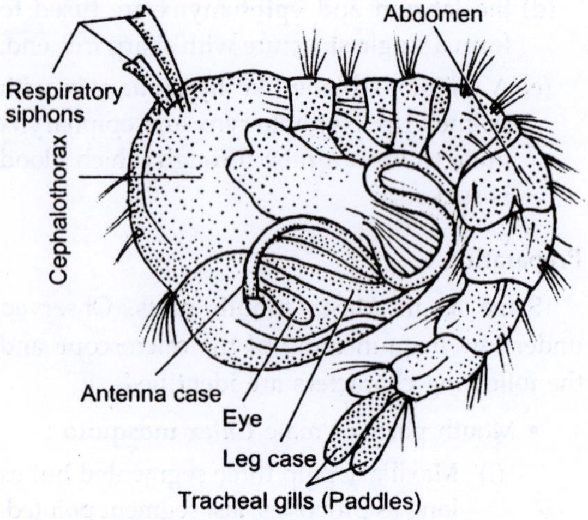

Fig. 1.5.18 : Pupa of *Culex sp.*

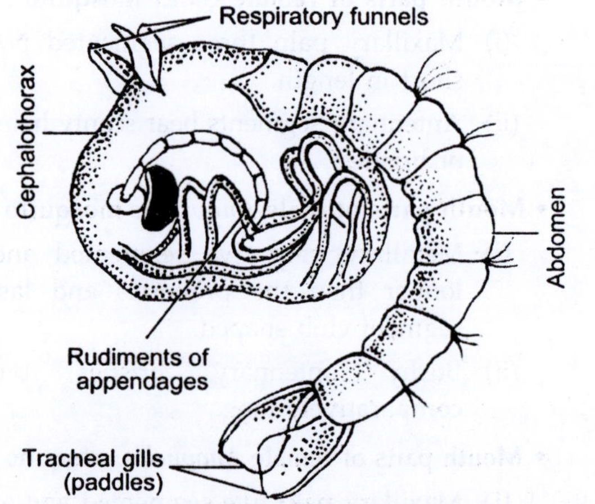

Fig. 1.5.19 : Pupa of *Anopheles sp.*

- ***Culex* Pupa :**
  - (i) The pupa is colour less.
  - (ii) Respiratory siphons are long.
  - (iii) Palmate hairs are absent.
- ***Anopheles* Pupa :**
  - (i) Green coloured pupa.
  - (ii) Respiratory siphon is short and funnel-shaped.
  - (iii) Palmate hairs are present.

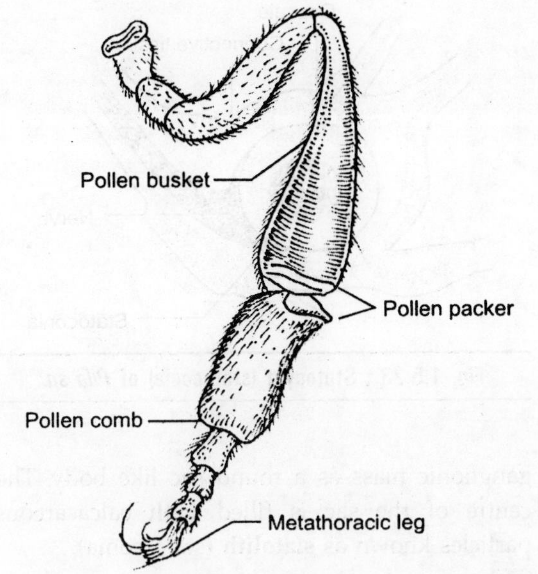

Fig. 1.5.20 : Pollen basket and Pollen packer of *Apis sp.*

18. Pollen basket and Pollen packer of *Apis sp.* :

Both pollen basket and pollen packer are present in each hind leg or metathoracic legs of *Apis sp*. The Pollen basket is a depression which is situated on the outer surface of the tibia and the basket is covered by rows of long, curved bristles. The first tarsal segments bears at its inner part **pollen combs.** Near the junction of tibia and first tarsal segment there is a structure called **pollen packer**. The pollen packer is found by a spiny **pecten** and a notched plate called **auricle** on the first tarsal segment. Worker bee collects pollen on its body – hairs, which are combed by pollen combs. The Pecten scrapes the pollen from the comb, while auricle pushes up the pollen mass in pollen basket.

Preparation :

1. The hind leg of worker bee previously killed by chloroform is dissected out.

2. The leg is placed on a glass slide and examined with a high power hand lens or under a simple microscope and noted the pollen basket and pollen packer.

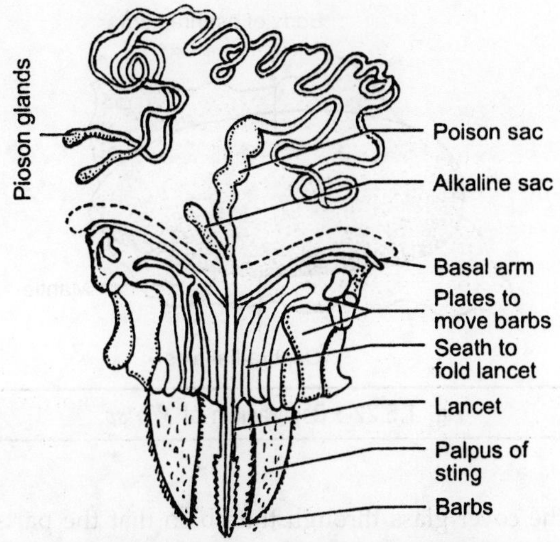

Fig. 1.5.21 : Sting apparatus of of *Apis sp.*

19. Sting apparatus of *Apis sp.* :

In Queens and worker honey bee the posterior most end of the abdomen is provided with a sting which is connected to the internal **poison gland**. The sting has a dorsal **stylet sheath** and a pair of ventral **lancets.** The Proximal end of sting has a swollen bulb, a pair of bifid arms, one on each side of it. The distal tip is beset with spines called **barbs. Oblong, fulcral** and **quadrate** are three pairs of plate remain associated with sting to act as lever. From each oblong plate arises an elongated palp to enclose the sting.

Preparation :

1. The ovipositor of a chloroform killed worker honey bee is dissected out and placed in ten percent KOH solution in watch glass for ten minutes.

2. Washed thoroughly in water and the material is placed on a glass slide.

3. Few drops of glycerine are poured and a cover glass is placed on it. Pressed gently with

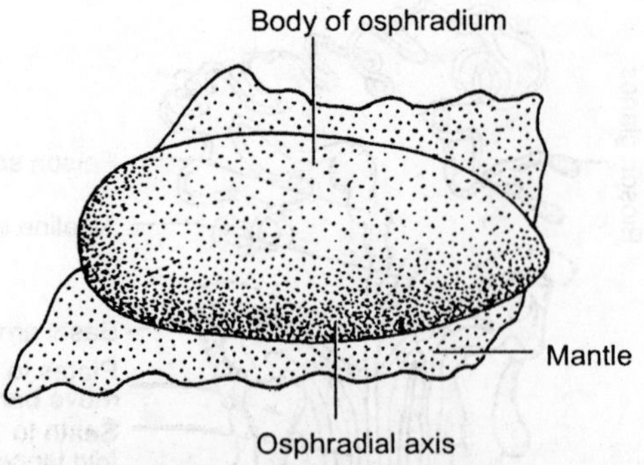

Fig. 1.5.22 : Osphradium of *Pila sp.*

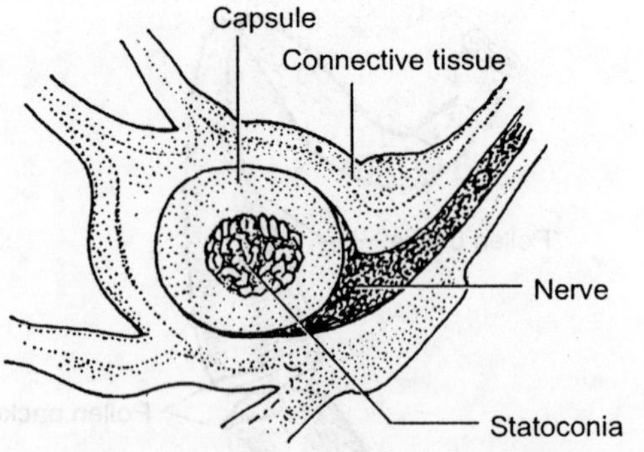

Fig. 1.5.23 : Statocyst (statoconia) of *Pila sp.*

the cover glass through thumb so that the parts spread uniformly.

4. Observed under microscope the various parts of poison sting.

20. Osphradium of *Pila sp.* :

The mantle cavity of *Pila* is imperfectly divided into left and right chamber. The right chamber contains **ctenidium, rectum** and the **genital duct** and the left chamber the **pulmonary sac**. The floor of the opening of the left mantle chamber gives out a spoon shaped **nuchal lobe**. Close to the left nuchal lobe, is a comb like organ of taste, called **osphradium**.

Preparation :

1. The shell of *Pila* is removed and the mantle cavity is dissected out and the nuchal lobes are located.
2. The osphradium is held with a pair of fine forceps and separated from the mantle with a pair of fine scissors.
3. Washed with water and mounted in glycerine on a slide.
4. Observed under microscope.

21. Statocyst of *Pila sp.* :

Statocysts of Pila are two in number and situated one on each side of the pleuropedal ganglionic mass as a round sac like body. The centre of the sac is filled with calcacareous particles known as **statolith** (stato conia).

Preparation :

1. The round sac is located and it is incised with a pair of fine scissors.
2. It is removed and mounted in glycerine on a glass slide.
3. Observed under microscope.

22. Radula of *Pila sp.* :

Covering the floor of the buccal cavity of *Pila* a chitinous ribbon like structure is found. This is known as radula. It is an elongated structure bearing transverse rows of serrations. Each transverse row contains about seven teeth. One is **central** or **rachidian** and three **lateral**. At the posterior part of radula lies the radular sac.

Preparation :

1. The mantle cavity is dissected out as before and buccal mass is located. The buccal mass is cut longitudinally and carefully the radula is separated by separating the muscle of attachment.
2. With the help of pair of forceps and a needle the radula is taken out and mounted in glycerine on a glass slide.
3. Observed under microscope.

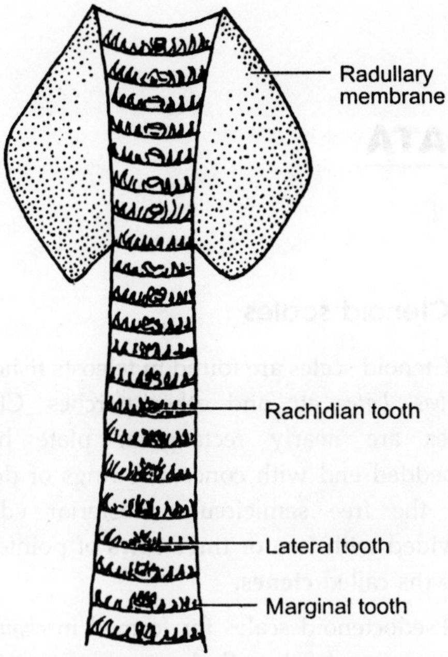

Fig. 1.5.24 : Radula of *Pila sp.*

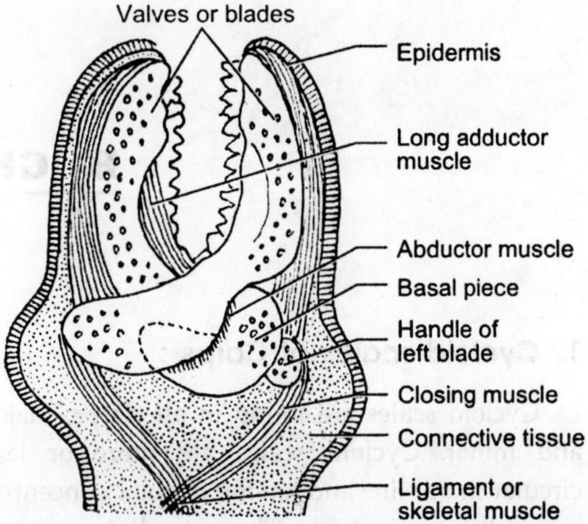

Fig. 1.5.25 : Pedicellaria (Crossed type) of *Asterias sp.*

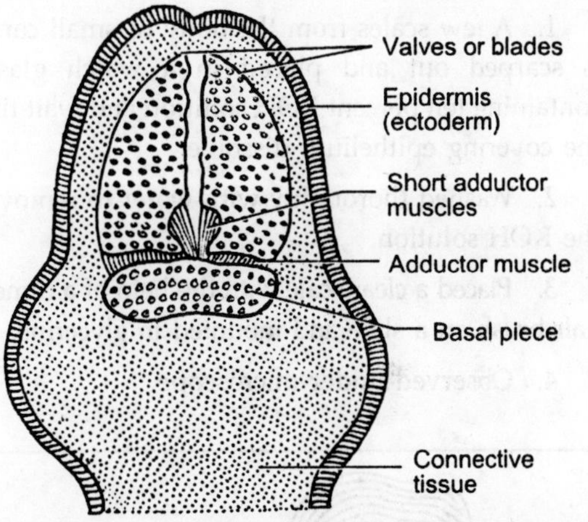

Fig. 1.5.26 : Pedicellaria (Straight type) of *Asterias sp.*

23. Pedicellariae of *Asterias sp.* :

Scattered all over the body surface of starfish there are many microscopic bodies called **pedicellariae**. A pedicellariae consists of a **flexible stalk** and three calcareous pieces one **basilar** and two **jaws** or **valves.** Jaws are attached to the basilar plate. Pedicellariae are of two types—straight or crossed. Where two jaws are more or less straight on the basilar piece, are called **straight pedicellariae**. When the basal portion of the jaws are curved and crossed each other, are called **crossed pedicellariae.**

Preparation :

1. The very small bodies in between the spines of the oral and aboral surfaces of Starfish are located and a few of them are separated, both of small and large sizes.
2. Placed on slide and mounted in glycerine.
3. Observed under a microscope.

B. CHORDATA

1. Cycloid scales of Carps :

Cycloid scales are found in all Carps, major and minor. Cycloid scales are more or less circular in out line and having circular concentric rings all over the body. The outer lining area is smooth, unbroken and not spiny.

Preparation :

1. A few scales from the body of small carp is scarped out and placed in a watch glass containing ten percent KOH solution and wait till the covering epithelium dissolve.

2. Washed thoroughly with water to remove the KOH solution.

3. Placed a clear scale with the help of a camel hair brush on a slide and mounted in glycerine.

4. Observed under microscope.

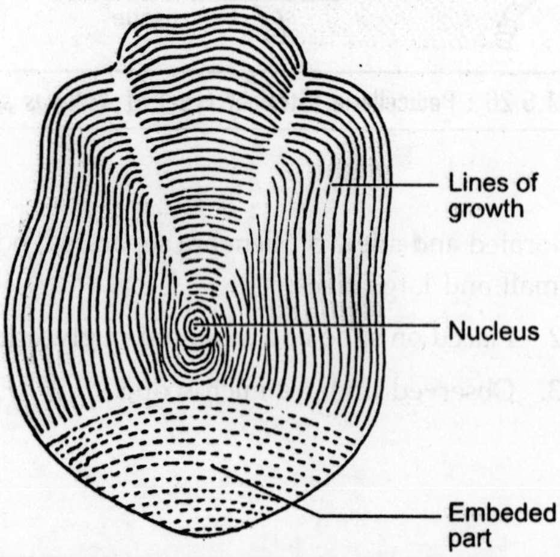

Fig. 1.5.27 : Cycloid scale of carp.

2. Ctenoid scales :

Ctenoid scales are found in teleosts fishes viz. *Anabas, Lates* etc and other perches. Ctenoid scales are nearly rectangular plate having embedded end with concentric rings of dentine and the free semicircular posterior edge is provided with two or three rows of pointed out growths called **ctenes.**

Psedoctenoid scales for formed in *channa sp.* and in sal, sol fishes. Scales are like ctenoid but difference is that spines do not project beyond the limit of the scales.

Preparation :

Same as in case of cycloid scale.

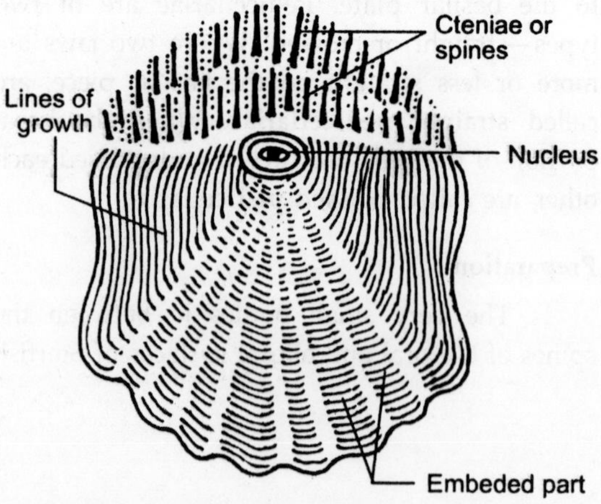

Fig. 1.5.28 : Ctenoid scale of *Anabas*.

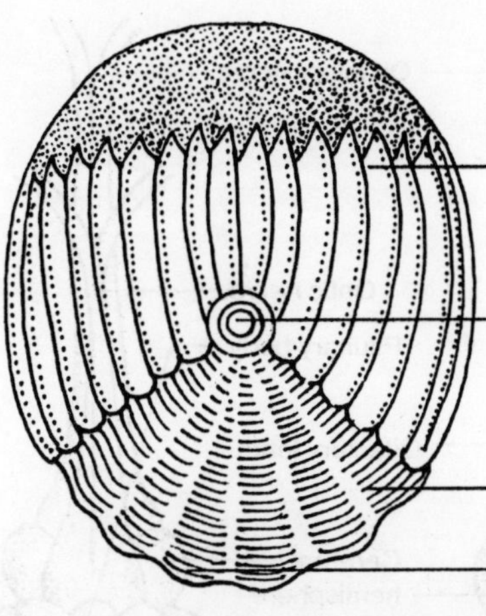

Fig. 1.5.29 : Pseudoctenoid scale.

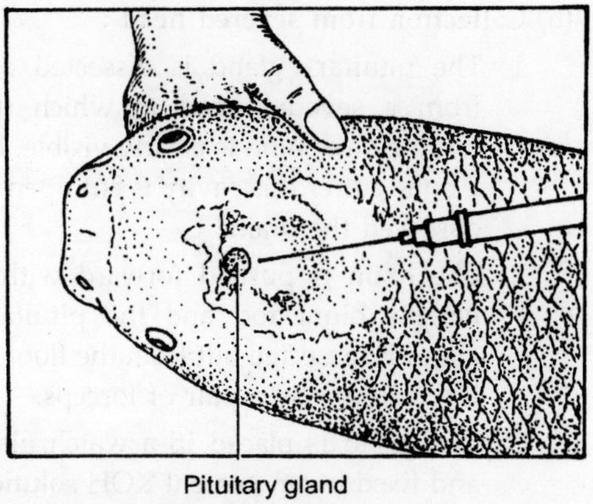

Fig. 1.5.30 : Position of Pituitary gland of *Labeo sp.*

3. Pituitary gland of a carp :

In carps, the pituitary gland or hypophysis is a small oval body, lying ventrally just behind the crossing of optic nerves. It is connected to the brain by infundibulum and covered by a membrane.

Preparation :

(a) **Collection from whole fish :**

1. The head of a small carp is held with left hand in a dorsoventral position and dissected out the bones of the roof of the cranium by a bone cutter or by a stout scalpel to exposed the brain.
2. The fatty matters around the brain is washed away with clean water with the help of a fine brush.
3. With the help of forces the brain is slightly pused aside to locate the small round body which is lodged in a depression and covered by a membrane.
4. The covering of the gland is gently removed and taken out the gland with a bend sharp needle.
5. The gland is placed in a watch glass containing ten per cent formaline for fixation.
6. The gland is washed with water and observed.

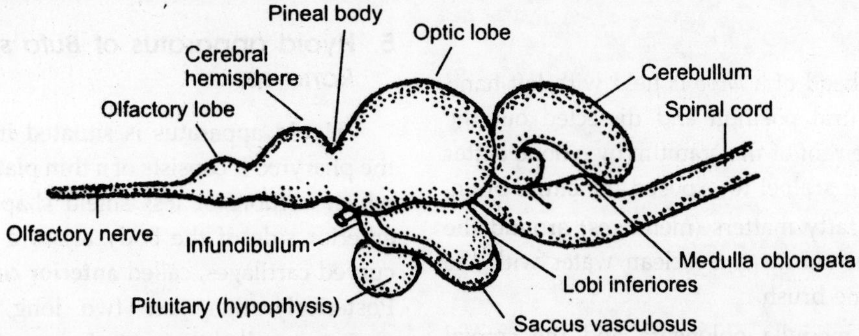

Fig. 1.5.30A : Pituitary gland of *Labeo sp.* (fully exposed)

(b) **Collection from severed head :**

1. The pituitary gland is dissected out from a severed head in which the foramen magnum is clearly visible. By a bone cutter this cavity is cut open to observed the brain.
2. The brain is pushed forward with a narrow blunt rod and the pituitary gland is dissected out from the floor of the cranium by a pair of forceps.
3. The gland is placed in a watch glass and fixed in ten percent KOH solution for observation.

4. Brain of *Labeo sp.*

The brain of *Labeo rohita* is more specialised than that of scoliodon. The **cerebral hemispheres** are not separate. Pallium or roof of the prosencephalon is thin and non-nervous and the floor is raised so as to form a prominent thickening called **corpora striata.** In front of the cerebral hemispheres are moderately developed **olfactory lobes.**

The reduced diencephalon is provided with a dorsal **pineal body** and a ventral **pituitary body**. Behind the cerebral hemispheres on the dorsal side are large optic lobes with two large **lobi inferioris** on the ventral surface. Optic chiasma is absent but optic nerves simply cross one another. The **cerebellum** is prominent and prolongs anteriorly to form **valvula cerebelli**, which is the characteristic of a teleostean fish.

Preparation :

1. The head of a *labeo* is held with left hand in dorsoventral position and dissected out the bones of the roof of the cranium by a bone cutter or by a stout scalpel to exposed the entire brain.

2. The fatty matters (meninges) around the brain is washed away by clean water with the help of a fine brush.

3. The medulla oblongata and the cranial nerves are cut close to their origin and the brain is lifted up from the anterior end, with a pair of fine forceps and taken out from the cranium.

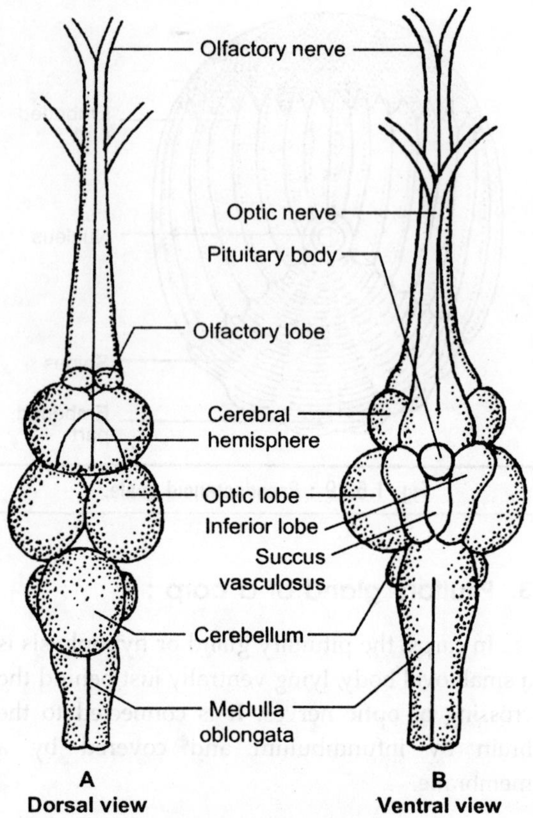

Fig. 1.5.30B : Brain of *Labeo sp.*

4. The brain is placed in a watch glass and mount in glycerin for observation.

5. Hyoid apparatus of *Bufo sp.* or *Rana sp.* :

Hyoid apparatus is situated in the floor of the pharynx. It consists of a thin plate of cartilage, which is more or less shield shaped. From the anterior side of the body arises a pair of long, curved cartilages, called **anterior cornua** (horns). Posteriorly, there are two long, stout, bony processes, called the **posterior cornua**. Anteriorly, in front of the hyoid is a pair of large flat

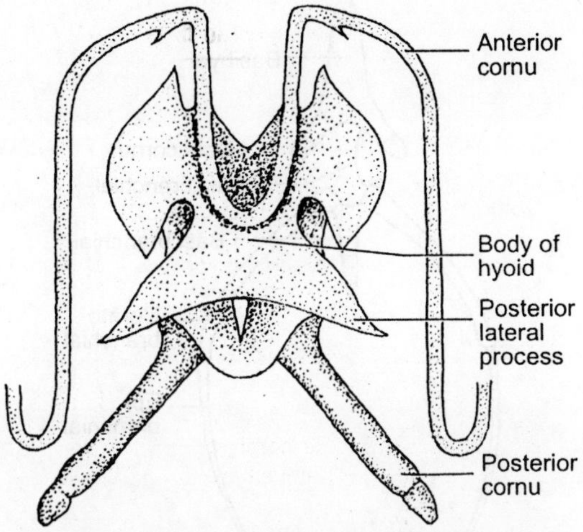

Fig. 1.5.31 : Hyoid apparatus of *Bufo sp.*

processes called the **alar process**. Similarly, there is a pair of smaller posterior processes.

Preparation :

1. A frog or toad previously killed by chloroform is pinned down in a dissecting tray.

2. The skin is removed from the ventral surface of the buccal floor and the anterior portion of the throat.

3. First and second finger of one's left hand is placed below the ventral muscle of the buccal floor to feel the body of the hyoid which is situated just dorsal to the ventral muscle.

4. The muscles and connective tissue of the floor are removed with a pair of forceps and the body of hyoid is exposed.

5. The muscles are removed to trace anterolaterally directed anterior cornua up to their point of connection with the auditory capsule.

6. Similary the posterior cornua are traced posterolaterally.

7. The hyoid apparatus is removed and placed in a watch glass containing ten percent KOH solution for ten minutes to facilitate clearing of the muscles.

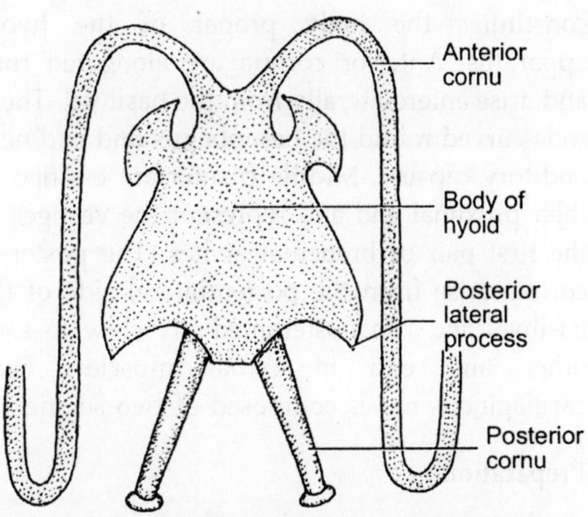

Fig. 1.5.32 : Hyoid apparatus of *Rana sp.*

8. Washed thoroughly with water several times to removed the traces of KOH.

9. Mounted in glycerine in a watch glass.

6. Hyoid apparatus of *Calotes sp.* :

The hyoid apparatus of *Calotes* occupies the floor of the buccal cavity. Hyoid apparatus consists of the **basihyal, anterior cornua, middle cornua** and **posterior cornua.** The basihyal

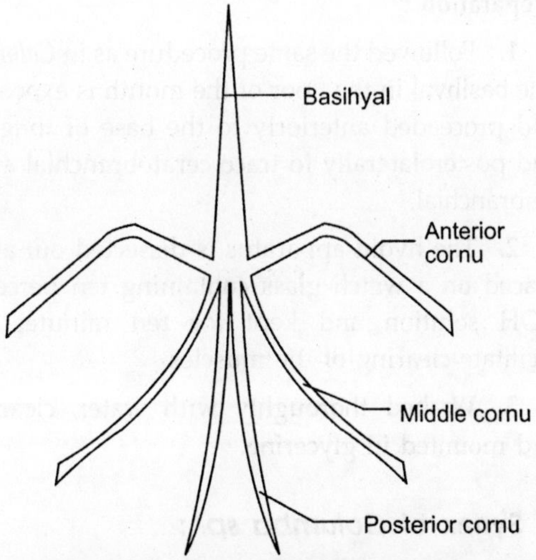

Fig. 1.5.33 : Hyoid apparatus of *Calotes sp.*

constitutes the body proper of the hyoid apparatus. Anterior cornua are elongated rods and arise anterolaterally from the basihyal. These rods curved round the oesophagus and end near auditory capsule. Middle cornua are ossified at their proximal end and represent the vestiges of the first pair of branchial arches. The posterior cornua arise from the posterolateral side of the basihyal and run posteromedially close to each other and end in throat muscles. Each cartilaginous rod is composed of two segments.

Preparation :

Procedure same as in case of frog or toad. Dissected out and placed in a watch glass containing ten percent KOH solution. Cleared, mounted in glycerine.

7. Hyoid apparatus of *Columba sp.* :

The hyoid apparatus occupies the same relative position as in *Calotes*. The hyoid apparatus of bird consists of an arrow shaped **basihyal** from which two short cornua originate. Basihyal continued as a slender **basibranchial** from which originates a pair of posterior cornua. Each cornu represented by **ceratobranchial** and **epibranchial** segments.

Preparation :

1. Followed the same procedure as in *Calotes*. The basihyal in the floor of the mouth is exposed and proceeded anteriorly to the base of tongue and posterolaterally to trace ceratobranchial and epibranchial.

2. The hyoid apparatus is dissected out and placed on a watch glass containing ten percent KOH solution and kept for ten minutes to facilitate clearing of the muscles.

3. Washed thoroughly with water, cleared and mounted in glycerine.

8. Syrinx of *Columba sp.* :

Syrinx or voice box is the characteristic of birds which is responsible for sound – production. Larynx leads into the long, flexible **trachea**, which is situated ventral to the oesophagus anteriorly. Before it entire into the coelomic cavity, trachea divides to from the **right** and **left bronchi**. Near the junction of neck and trunk, the trachea is swollen into a chamber called the **Syrinx**. Syrinx is small cartilaginous

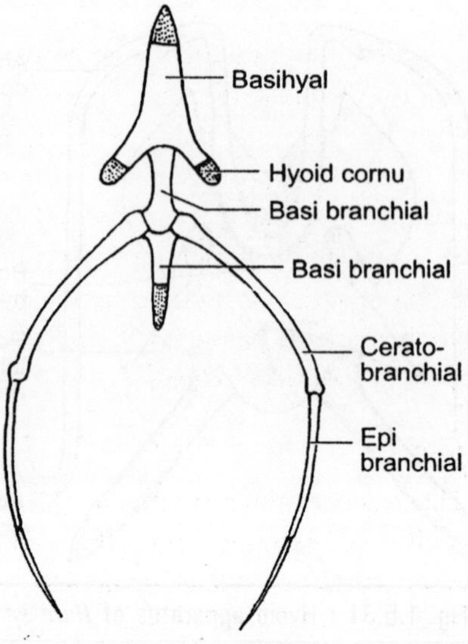

Fig. 1.5.34 : Hyoid apparatus of *Columba sp.*

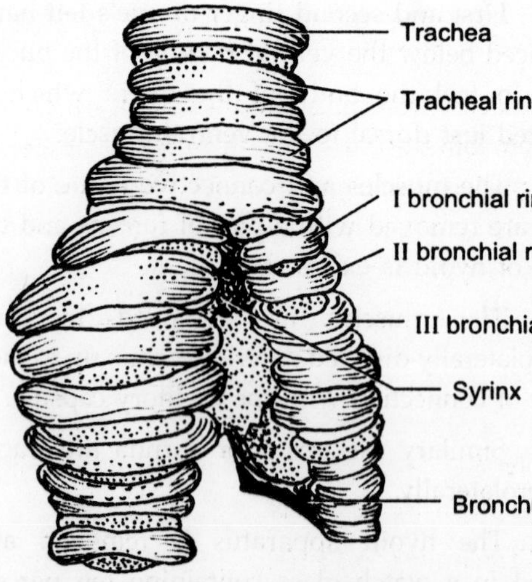

Fig. 1.5.35 : Syrinx apparatus of *Columba sp.* (with trachea and branchi)

chamber having membranous wall. A bar of cartilage called the **pessulus** is present at the junction of two bronchi. It extends dorsoventrally inside the chamber and holds a small fold of mucous membrane called the **membrane semilunaris.** The sound is produced by the vibration of the membrana semilunaris. Modification of sound is effected by the action of the **hypobronchial muscles**, the **intrinsic syringeals** and **sterno-tracheals.**

Preparation :

1. The trachea along with two bronchi in body cavity is exposed and the Syrinx is observed, a slightly dilated chamber at the junction of trachea and two bronchi.

2. The trachea is cut at about 5-6 mm anterior to the syrinx and bronchi close to the lungs.

3. The syrinx is carefully removed and the external muscles are detached and it is put in a watch glass.

4. Washed with water and mounted in glycerine.

5. To examine the internal structure of syrinx, the syrinx is cut longitudinally by putting it in water in a paraffin lined petridish and the structure is observed.

9. Pecten of *Columba sp.* :

In Pigeon projecting into the vitreous body of eye from blind spot, a characteristics structure called **pecten** is present. It is composed of a dark pigmented plate. The plate is folded fanwise and assume the from of a **comb.**

Preparation :

1. One of the eye balls is removed from the socket of a pigeon and the eye ball is cut into two equal halves by giving an equatorial incision with a pair of fine scissors.

2. The upper half along with lens is removed and then push from below the lower half of the remaining eye by one's ring finger.

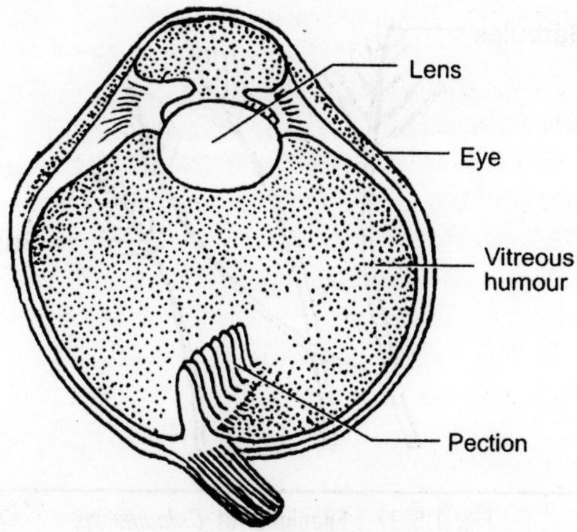

Position of pecten

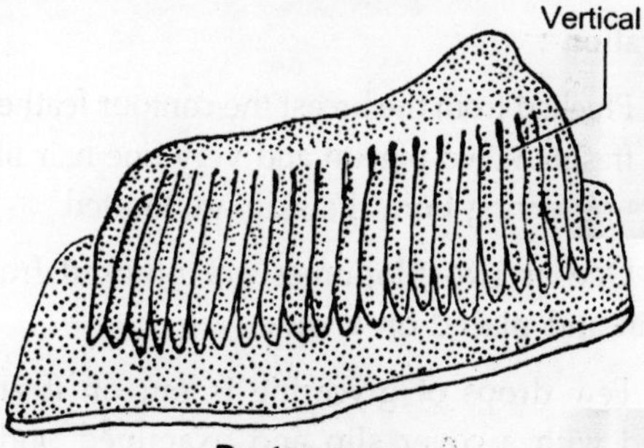

Pecten proper

Fig. 1.5.36 : Pecten of *Columba sp.* (above, position of pecten)

3. At the top of the evaginated portion the black, longitudinal folded comb like pecten at the point of entry of the optic nerve in the eye ball is observed.

4. The entire pecten is separated from the eye ball with a pair of fine scissors and placed in a glass slide.

5. A drop of glycerine is put in it and mounted.

10. Filoplume of *Columba sp.* :

Filoplumes are seen in plucked birds. These are minute hair like feathers, consisting of a long axis and a few barbs at the distal end.

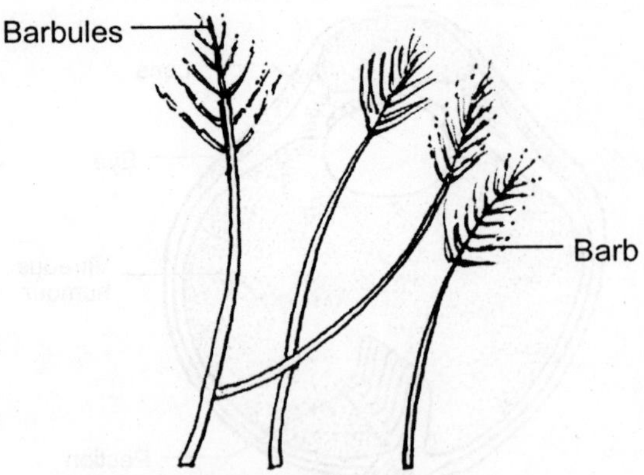

Fig. 1.5.37 : Filoplume of *Columba sp.*

Preparation :

1. Plucked from the breast the contour feathers from a freshly killed pigeon and very fine hair like feathers projecting from the skin are located.

2. One or two filopumes are plucked from the skin and placed on a slide.

3. Few drops of glycerine is put on it and covered with a cover slip and examined under low power of microscope.

11. Hyoid apparatus of *Gallus sp.* :

The hyoid apparatus of *gallus sp.* occupies the same relative position as in *calotes sp.*

The hyoid is made up of a median portion, which consists of three bones placed end to end and two pairs of *cornua*. The three median bones are : (a) *ceratohyal* or *entoglossal*, which is arrow shaped, situated in the tongue with short process extending laterally. These cartilages are : anterior *cornua* or *horns*. (b) *basihyal*, which is a bony piece uniting posteriorly with long *horns* or *cornua*, and (c) *urohyal* or *basibranchial* projects from posterior cornua.

A second posterior pair of *cornua* or *horns*, situated at the point of the junction of *basihyal* or *urohyal*, is long united process directed backwards. The first part or *cerotobranchial* run

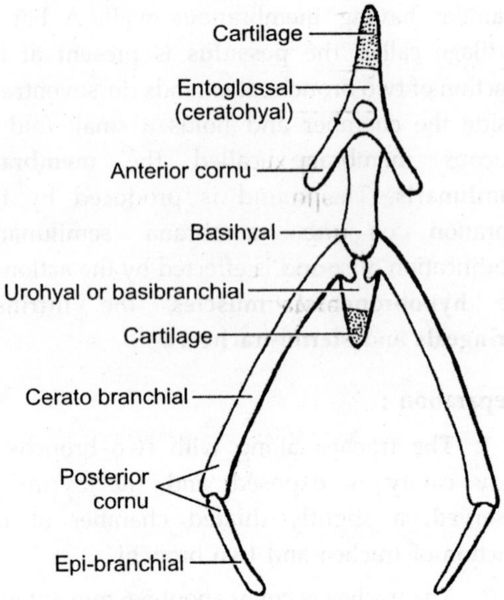

Fig. 1.5.38 : Hyoid appratus of *Gallus sp.*

posteriorly from the junction of basihyal and the second part is *epibranchial*, continuous with the cerato branchials, curved posteriorly and gradually narrow down.

Preparation :

Follow the same procedure as in *calotes*.

12. Brain of *Gallus sp.* :

The brain of *gallus sp.* consists of very small *olfactory lobes*. The *cerebral hemispheres* are large and smooth. The cerebral hemispheres are well developed posteriorly and as such they conceal all the *diencephalon* expect the *pineal body* or *epiphysis*. The mid brain consists of a pair of spherical bodies the *optic lobes* which is situated at the sides of the brain between the *cerebrum* and *cerebellum*.

The cerebellum is large and consists of a large median transversely grooved area, called *vermis* and two small lateral lobes or *flocculi*. Between the cerebral hemisphere and the cerebellum is situated ventrally an *infundibulum*. The

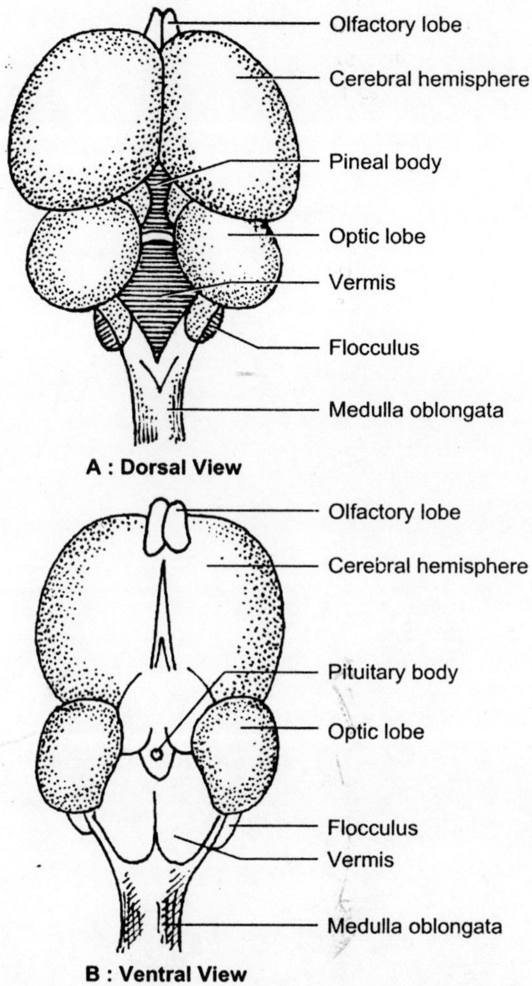

Fig. 1.5.39 : Brain of *Gallus sp.*

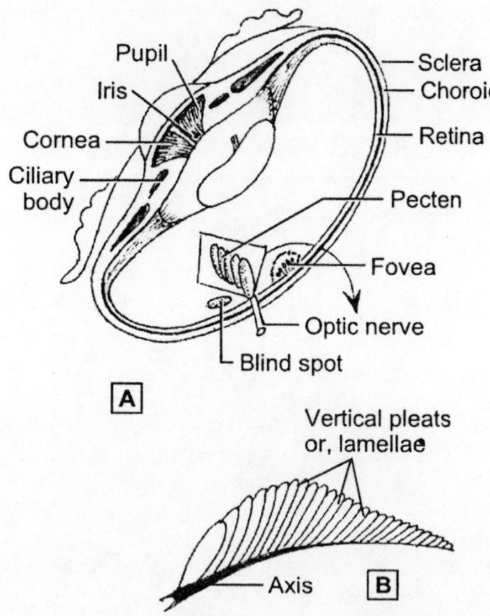

Fig. 1.5.40 : Pecten of *Gallus sp.*
A. Pecten inposition; B. Structure of pecten.

infundibulum together with hypophysis projecting from it constitutes the *pituitary gland.*

Preparation :

The head of the gallus is held with the left hand in dorsoventral position and the skin is removed. The rest procedure is same as in *labeo sp.*

13. Pecten of *Gallus sp.*

Structure and preparation same as in columba.

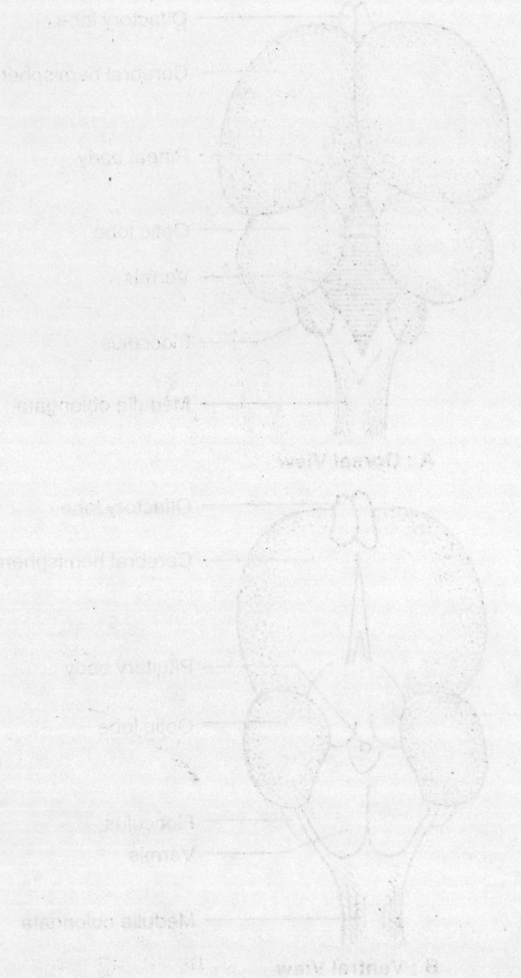

Section 2

OSTEOLOGY

SECTION AGENDA

The support structure of the vertebrate body is a marvel of complex engineering designed to give maximum strength and mobility. Each bone is a particular shape because it has a specific role. Bones (L=OS) make up the Skeleton.

The term **skeleton** refers to the framework of animal body. In vertebrates it is composed of cartilage or bone or combination of two and serves for the support, attachment of muscles and protection for certain delicate vital organs, which it more or less completely surrounds. These supporting bones are designated as the **endoskeleton**.

Dermal skeleton consists of bony structures, scale, plates that originate in the dermis of the skin as a result mesenchymal origin.

Visceral (pharyngeal) skeleton includes the gill arches and its derivatives. The **axial skeleton** includes the visceral skeleton, cranium, sternum, vertebral column and ribs. The paired limb skeletons and the girdles (pectoral and pelvic) make up the **appendicular skeletons.**

Each of the different parts of skeleton is designed to do a particular job. The Skull or Cranium protects the brain and also eyes and ears. The back bone or spine is made up of a chain of small bones, called vertebrae, and forms the central axis of the skeleton. Girdles, gives extra stability, while limb bones carryout grasping and complicated movements.

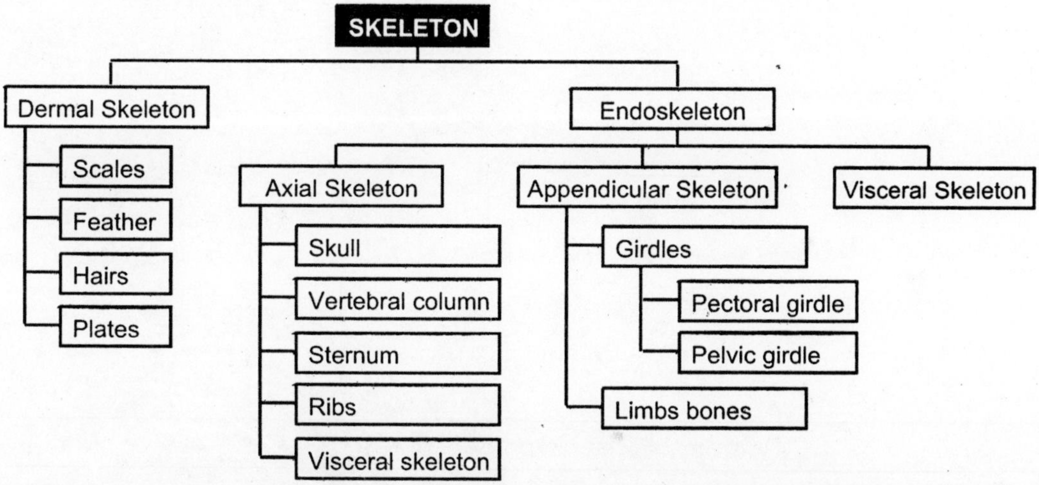

Figure 2.1 : Scheme for classification of skeletal elements of the vertebrate.

Chapter 2.1

Identification of Endoskeletal Elements of Some Vertebrate

The variation in structure and nature of skeleton in different classes of vertebrates is so varied that individual study may draw confusion.

I. SKULL

The skeletal framework of vertebrate head is known as the skull. The portion of the skull which surround and protect the brain is called as the cranium.The skull of most vertebrate are derived from three embryonic components viz.,

(i) **Chondrocranium :** Composed of cartilage and which contributes chiefly to the base of the skull includes sense capsules.

(ii) **Dermatocranium :** Made from membrane bones which roof over the chondrocranium or closely applied to the lower surface.

(iii) **Splanchnocranium :** Derived from visceral skeleton and is primitively cartilaginous although some may be invested or covered by membranous bone.

The comparative anatomy of the skull is very complex and homologies are often difficult to assign.

In order to remove this confusion the skeletal system may be studied in a comparative manner. We have given a chart for ready identification of tetrapod skull in the next page.

Identifying characters of vertebrate skull :

1. It is a compound structure, made of cartilages, bones or bones and cartilages.
2. Presence of cranium which has occipital, parietal and frontal rings.
3. Presence of **foramen of magnum** at the posterior end.
4. Presence of **occipital condyle**(s) at the back of the skull.
5. It bears three **sensory capsules** – olfactory, optic and otic.

AMPHIBIAN SKULL

A. Class Amphibia :

Characters :

1. Skull is triangular in shape, broad and dorsoventrally flattened.
2. Bony cranium is smaller and dorsally placed orbits.
3. Two occipital condyles present.
4. Supra and basioccipitals are absent.
5. Frontal and parietal fused.
6. An unpaired sphenethmoid constitute posterior wall of olfactory chamber.

SKULL

Dicondylic (Two occipital condyles) **Amphibia**	**Monocondylic** (Single occipital condyle) **Reptilia**
1. Cranium is smaller and dorsaly – placed orbits. 2. Supra and basi – occipitals are absent. 3. Vomers and vomarine teeth present. (Dentition is absent in Toad) 4. Tympanic bulla is absent. 5. Jaw suspension autostylic ie. Upper jaw fused with Cranium.	1. The skull has considerable weight. 2. Beak absent. Jaw bones possess teeth. 3. Skull bones are several and with distinct sutures. 4. Secondary palate absent.
Mammalia	**Aves**
1. Cranial part smaller but facial part is elongated and orbits are laterally placed. 2. Presence of supra and basic – occipitals. 3. Vomers and vomarine teeth are absent. Premaxilla bear thecodont and diphyodont teeth, dentition heterodont. 4. Tympanic bulla present. 5. Jaw suspension Craniostylic ie. lower jaw directly articulates with membrane bone.	1. Skull is very light due to spongy bones. 2. Skull with a long pointed beak without teeth. 3. Skull closely fused and with obliterated sutures. 4. Secondary palate present.

7. Dagger – shaped parasphenoid, located on the floor of brain box.
8. Squamosal hammer – shaped.
9. Palatine long, delicate and rod – shaped.

(a) ***Bufo sp.* :** State the vertebrate skull characters and amphibian skull characters. Next write the following characters :

Characters :

1. Anterior end of skull is parabolic.
2. Frontoparietals are broad and flat.
3. Supraorbital ridge well developed.
4. Upper and lower jaws are toothless.
5. Lower jaw is composed of two rami and united anteriorly by ligament.
6. Each half of lower jaw is developed from **Meckel's cartilage** and consists of **dentary, angulosplenial** and **mentomeckelian cartilage.**

(b) ***Rana sp.* :**

Characters :

1. Anterior end of skull slightly pointed.
2. Nasal and frontoparietals thin and narrow.
3. Supraorbital ridge ill developed.
4. Presence of conical teeth on vomer, premaxilla and maxilla.
5. Lower jaw same as in *Bufo sp.*

REPTILIAN SKULL

B. Class Reptilia :

Characters :

1. Well ossified, heavy and box like skull.
2. Occipital condyle single.
3. Presence of parietal and coronary sutures.
4. **Temporal vacuity** may or may not be present.
5. Basioccipital and basisphenoid large.
6. Presphenoid narrow and elongated.
7. Conical and **homodont teeth** may or may not be present.

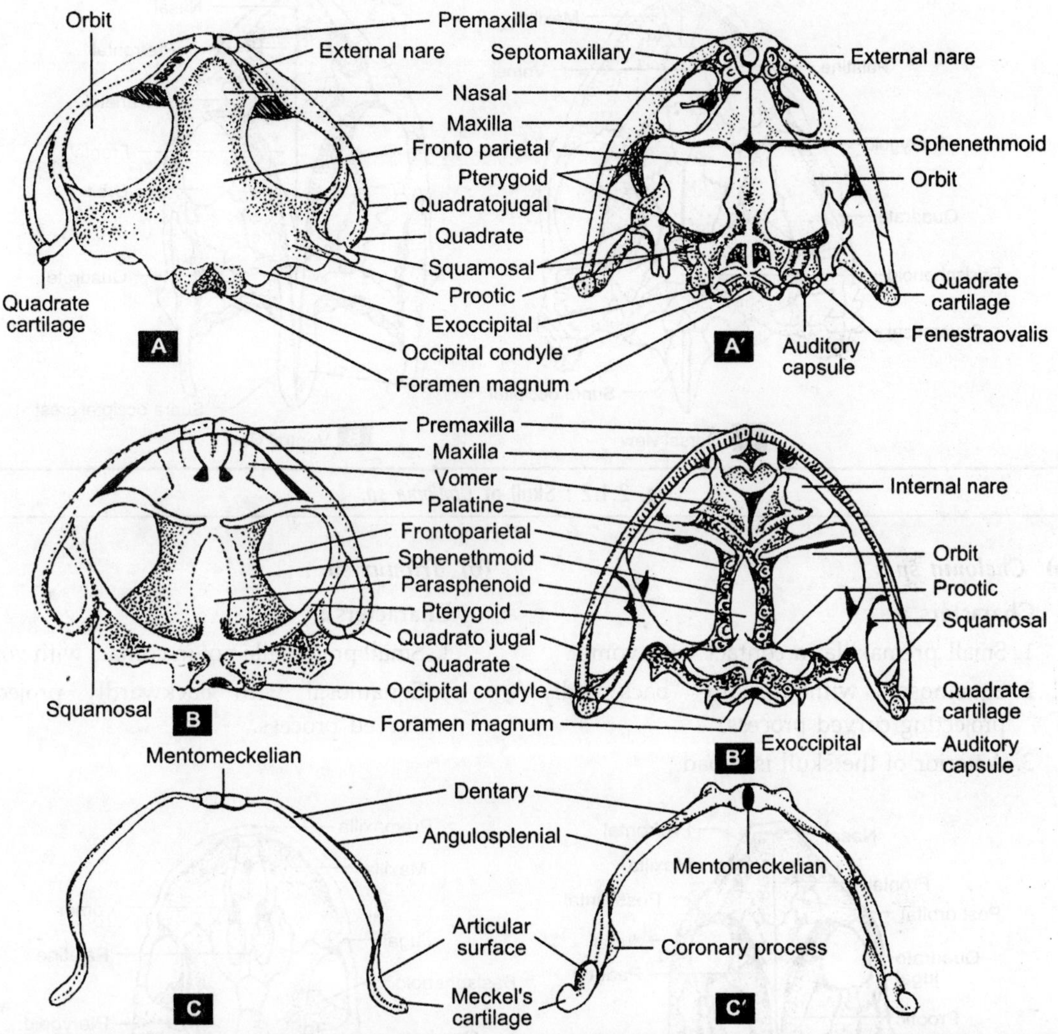

Fig. 2.1.1 : Skull and mandibule. A–C : *Bufq sp.*, A′–C′ : *Rana sp.* A–A′ : Dorsal view, B–B′ : Ventral view.

(a) Skull of *Chelonia sp.* and *Trionyx sp.* :

State the vertebrate skull characters and reptilian skull characters. Next write the following characters :

Sub class—*Anapsida* :

Characters :

1. Skull bones immovably articulated.
2. Skull roof is solid.
3. Complete bony ring in the temporal region.
4. Temporal vacuity absent.

Order—*Chelonia* :

1. Occipital condyle is tripartite.
2. Parietal extended vertically downward.
3. Quadrate is immovably articulated to skull.
4. Single nasal opening.
5. A prominent supraoccipital crest extended backwardly.
6. Nasal triangular, flat and fused in the middle.
7. Jaw is without any teeth but provided with horny sheath.

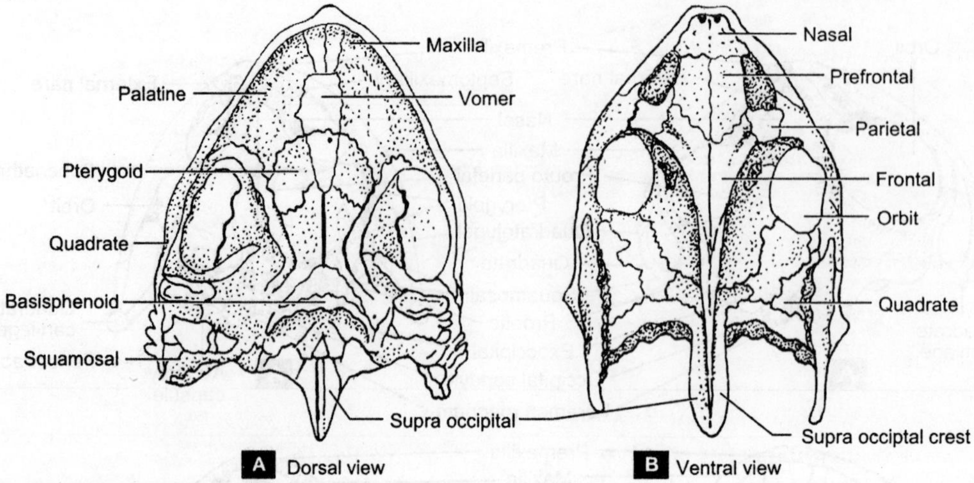

Fig. 2.1.2 : Skull of *Chelonia sp.*

(a) *Chelonia sp.* :

Characters :

1. Small premaxilla in contact with vomer.
2. Squamosal without any backwardly projecting curved process.
3. Anterior of the skull is broad.

(b) *Trionyx sp.* :

Characters :

1. Small premaxilla not in contact with vomer.
2. Squamosal with backwardly projecting curved process.

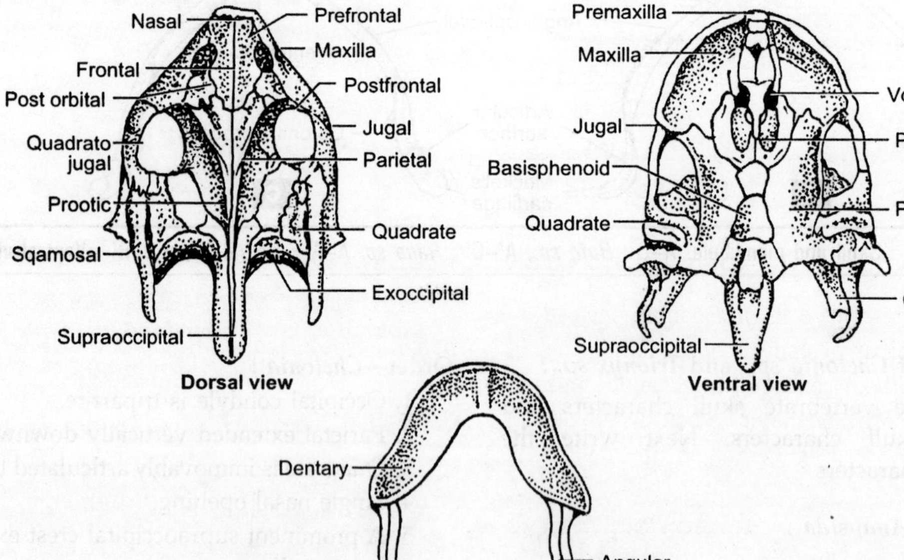

Fig. 2.1.3 : Skull and mandible of *Trionyx sp.*

3. Anterior end of the skull is narrow.
4. Two rami of the mandibule united anteriorly.
5. Lower jaw composed of dentary, angular, supra-angular and articular.
6. Articular bears a facet at the base.

(b) Skull of *Varanus sp.* and *Calotes sp.* :

State the vertebrate skull characters and the reptilian skull characters. Next write the following characters.

Sub Class—*Lepidosauria* :

Characters :

1. Skull with supra temporal and infra temporal vacuities.
2. Temporal arches two – supra and infra temporal.
3. Antero-orbital vacuity absent.
4. Secondary palate absent.

Order—*Squamata* :

Characters :

1. Elongated skull.
2. Infra-temporal vacuity is secondarily lost.
3. Entire occipital condyle.
4. Jaws with teeth.
5. External nostrils are two.

Sub order—*Lacertilia* :

Characters :

1. Quadrate is movably articulated with the ex–occipital.
2. Well developed supratemporal arch, infra-temporal absent.
3. Teeth **pleurodont** or **acrodont**.
4. Parietals are fused and bear a median parietal foramen.
5. Presence of epipterygoid.

(a) *Varanus sp.* :

Characters :

1. Elongated skull has narrow anterior end with elongated facial part.
2. Skull bones are several and with distinct sutures.
3. Premaxilla and maxilla possess pleurodont teeth.
4. Teeth are **polyphyodont** and **homodont** type.
5. Presence of basioccipital, exoccipital and rectangular supraoccipital.
6. Tympanic cavity is small containing a single ear oscicle, the columella.
7. **Mandible** consist of six bones – dentary, angular, supra-angular, coronoid, splenial and articular.
8. Articular bears a cartilage at the base for articulation with Quadrate.

(b) *Calotes sp.* :

Characters :

1. Skull small and slightly triangular.
2. Cranium small and facial part short.
3. Presence of cartilagenous inter-orbital.
4. Squamosal is sickle – shaped and is attached to parietals.
5. Pterygoid form a bridge between quadrate (posteriorly) and palatine (anteriorly).
6. Premaxilla and **maxilla** bear acrodont teeth.
7. Vomer very small and articulates with maxilla.
8. Mandible consists of six bones – dentary, angular, supra–angular, coronoid, splenial and articular.
9. Articular bears a facet at the base for articulation with quadrate.

(c) Skull of a Poisonous and non Poisonous Snake :

State the skull characters up to—

Order—**Squamata** as those of *Varanus sp.* then state the following characters.

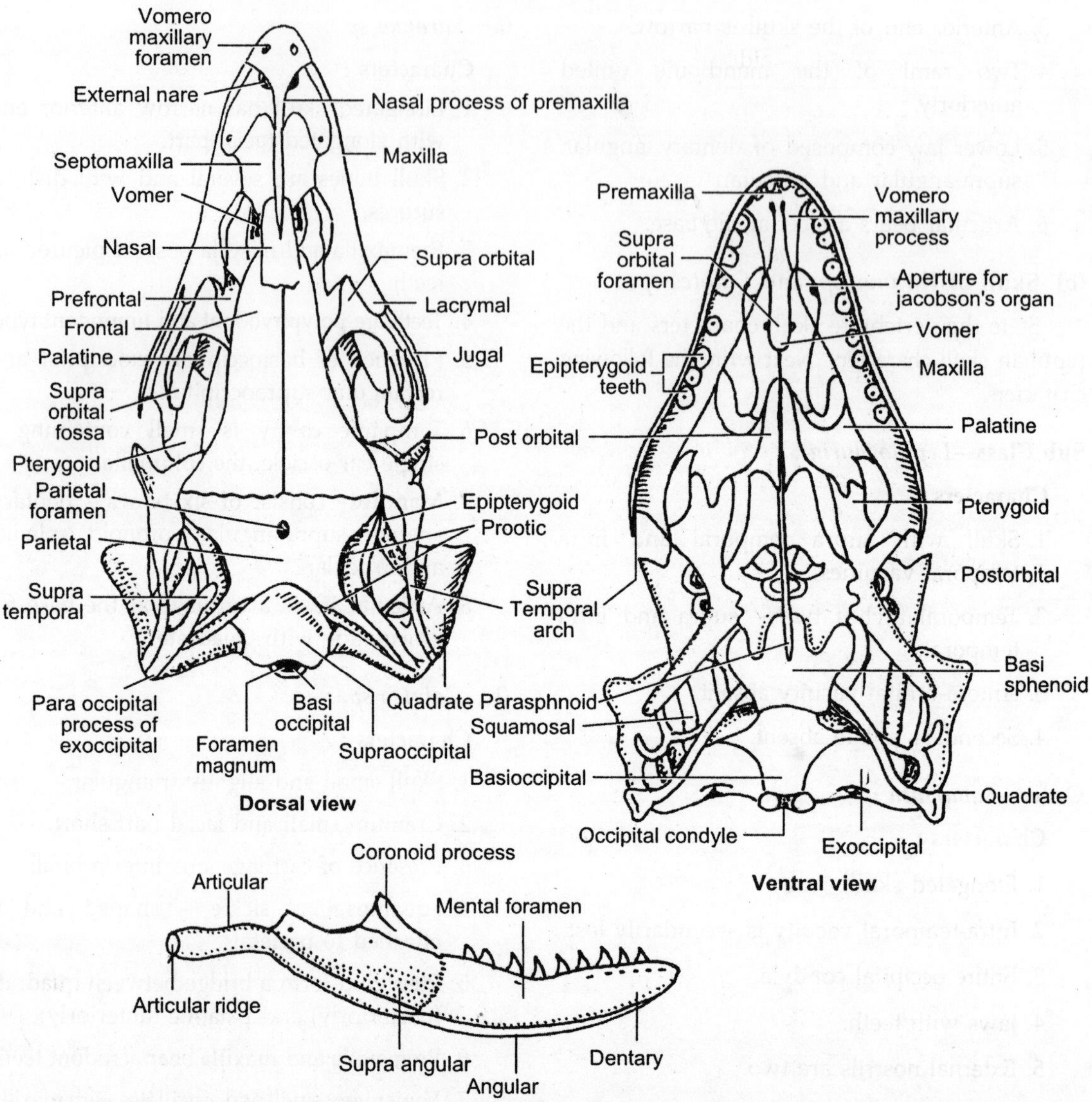

Fig. 2.1.4 : Skull and mandible of *Varanus sp.*

Sub order—*Ophidia* :

Characters :

1. Elongated cranial cavity.
2. Quadrate is highly movable and articulates with squamosal.
3. Occipital condyle distinctly tripartite.
4. Temporal foramen and fossae are absent.
5. Basisphenoid is prolonged into rostrum.
6. Maxilla, palatine and pterygoid is movable.
7. Premaxilla fused to form a single piece.
8. Inter-orbital septum is absent.
9. Teeth pleurodont type and recurved and present a maxilla, pterygoid, palatine and dentary.
10. Horizontal squamosal extends beyond the posterior limit of the skull.

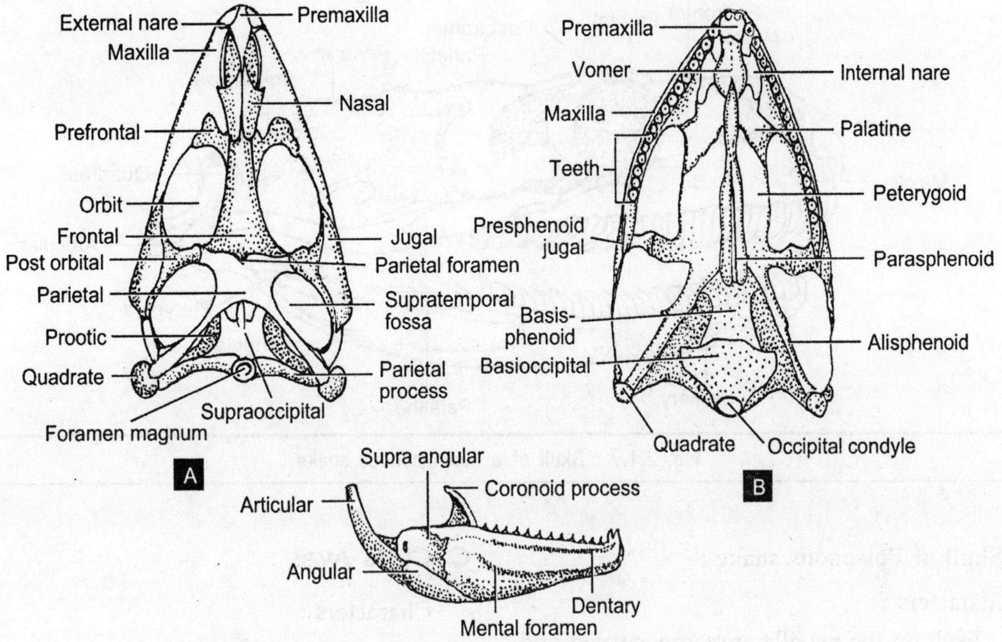

Fig. 2.1.5 : Skull and mandible of *Calotes sp.* A–Dorsal view, B–Ventral view.

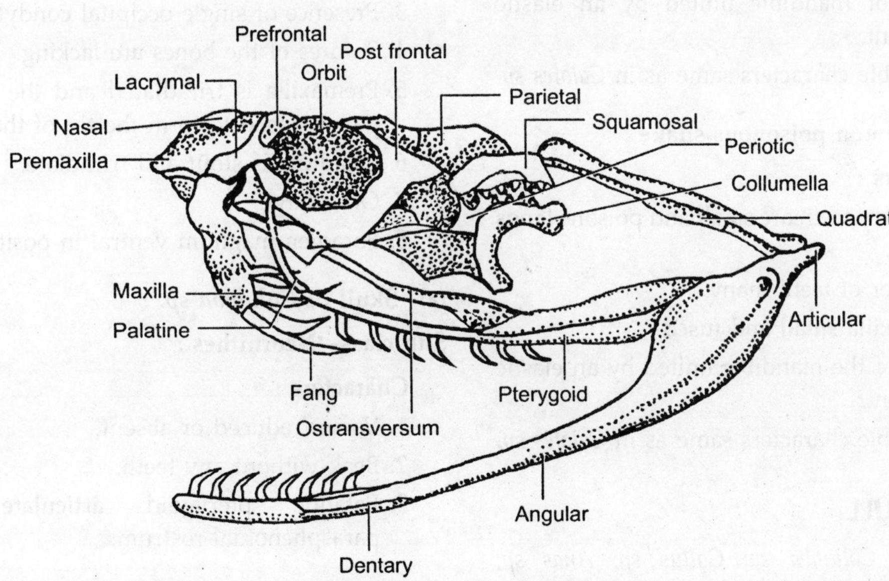

Fig. 2.1.6 : Skull of a poisonous snake.

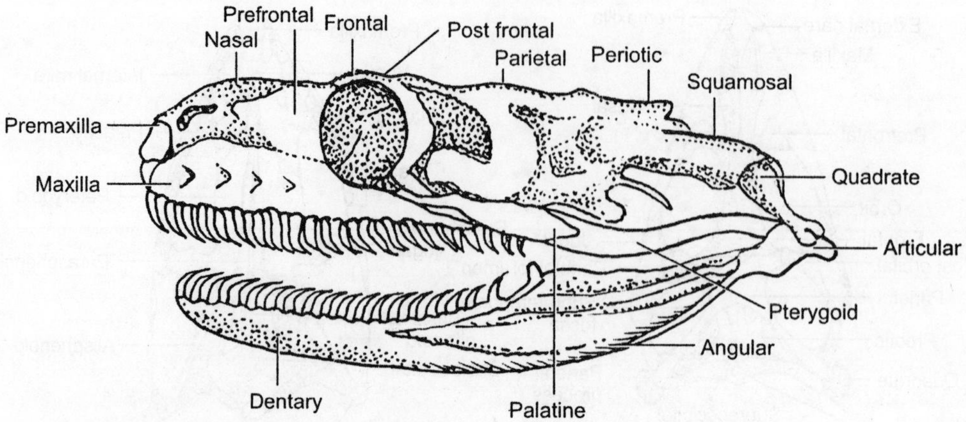

Fig. 2.1.7 : Skull of a nonpoisonous snake.

(a) Skull of Poisonous snake :

Characters :

1. Teeth on the maxilla are large, curved and grooved, known as poison **fangs.**
2. Premaxilla fused and without any tooth.
3. Rami of mandible united by an elastic ligament.
4. Mandible characters same as in *Calotes sp.*

(b) Skull of non poisonous snake :

Characters :

1. Maxillary teeth are small and poisons fangs absent.
2. Number of teeth many.
3. Premaxilla small and fused.
4. Rami of the mandible united by an elastic ligament.
5. Mandible characters same as in *Calotes sp.*

AVIAN SKULL

Skull of *Columba sp; Gallus sp; Anas sp; Psittacula sp; Corvus sp.*

State the skull characters as before and then the following characters are to be stated.

C. Class Aves :

Characters :

1. Cranium is enlarged and rounded.
2. Bones of the skull are thin, light and smooth.
3. Presence of single occipital condyle.
4. Sutures of the bones are lacking.
5. Premaxilla is triradiated and the anterior part is fused to form the tip of the beak.
6. Quadrate is stout and triradiate.
7. Orbit large.
8. Foramen magnum ventral in position.

(a) Skull of *Columba sp.* :

Sub class—Neornithes :

Characters :

1. Vomer reduced or absent.
2. Beak without any teeth.
3. Palatine pterygoid articulate with parasphenoidal rostrum.

Super order—Neognathae :

Characters :

1. Slender vomer separates the palatines imperfectly.

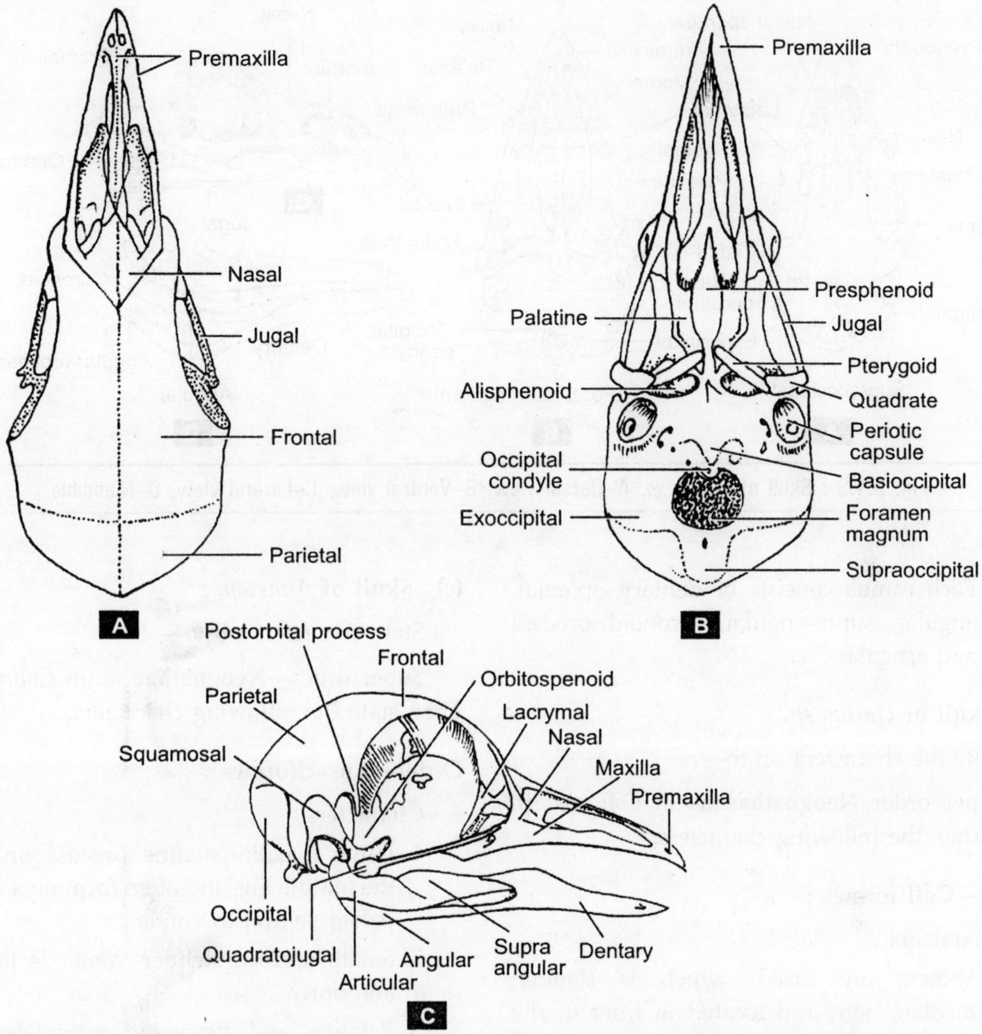

Fig. 2.1.8 : Skull of *Columba sp.* A–Dorsal view, B–Ventral view, C–Lateral view.

2. Palatine extends backwards up to the base of the cranium.
3. Palatine protruded posteriorly and movably articulated with pterygoid.
4. Double headed quadrate.

Order—Columbiformes :

Characters :

1. Maxillo–palatine process slender and do not unite with one another.
2. External nostrils are elongated, ovoid and slit like.
3. Anterior end of beak is slightly arched adapted for picking up grains.

(a) ***Columba sp.*** **:**

Characters :

1. Vomer absent.
2. Beak slightly pointed, narrow and weak.
3. Laterally flattened rami uniting in front and expanded at their hind end.

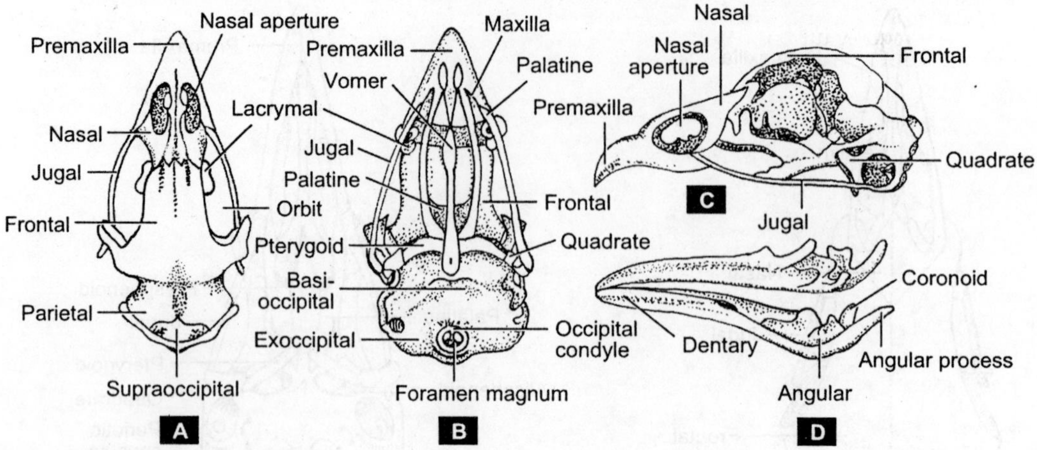

Fig. 2.1.9 : Skull of *Gallus sp.* A–Dorsal view, B–Ventral view, C–Lateral view, D–Mandible.

4. Each ramus consists of dentary, splenial, angular, supra-angular, coronoid process and articular.

(b) Skull of *Gallus sp.* :

State the characters up to—

Super order **Neognathae** as in *Columba sp.* Then state the following characters.

Order—Galliformes :

Characters :

1. Vomers are fused, which is slender, median, bony rod located in front of the rostrum.
2. Maxillo-palatine process slender, spongy and not united with one another.
3. Pterygoid short, stout and backwardly directed bones.

(b) *Gallus sp.* :

Characters :

1. Beak short, strongly build and lightly arched at the anterior end.
2. External nostrils are large and oval.
3. Write the characters 3 and 4 as in *Columba sp.*

(c) Skull of *Anas sp.* :

State characters up to—

Super order—**Neognathae** as in *Columba sp.* Then state the following characters.

Order—Anseriformes :

Characters :

1. Large maxillo-palatine process united in the middle line and often forming a spongy palate ventral to vomer.
2. Small, narrow, slender vomer is tapering anteriorly.
3. Palatine and pterygoid articulates with rostrum.
4. Basipterygoid processes are lacking.

(c) *Anas. sp.* :

Characters :

1. Beak broad enlongated and anterior end rounded.
2. Absence of frontal hinge between upper beak and skull.
3. External nostrils large and ovoid.
4. Internasal septum incomplete.
5. Write the characters 3 and 4 as in *Columba sp.*

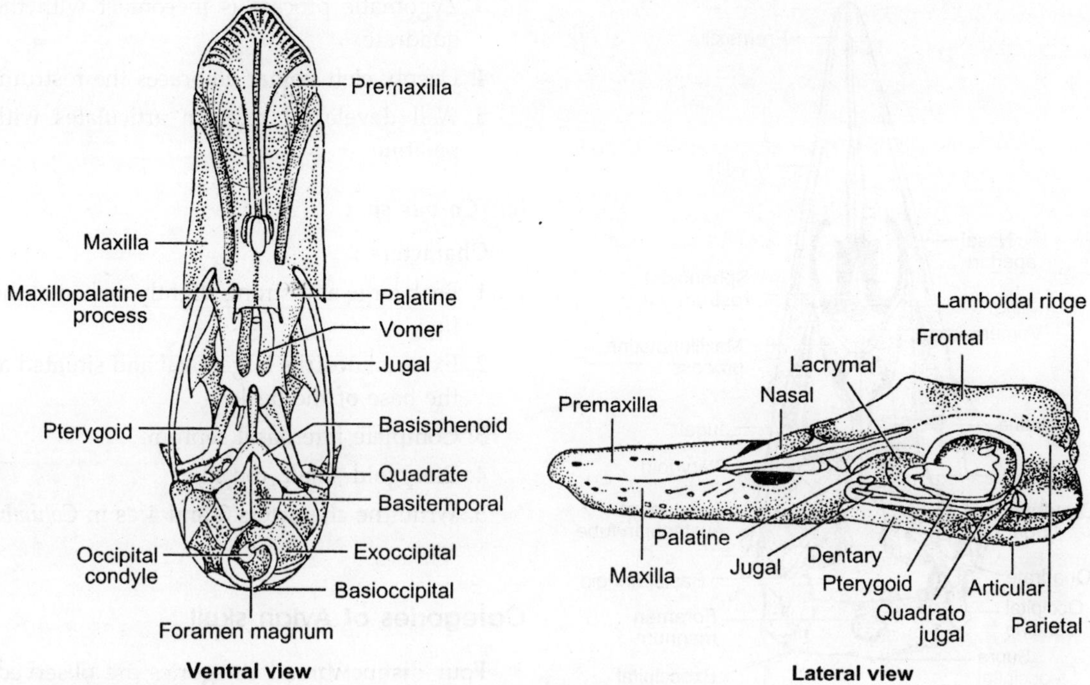

Fig. 2.1.10 : Skull of *Anas sp.*

(d) Skull of *Psittacula sp.* :

State characters up to—

Super order **Neognathae** as in *Columba sp.* Then state the following characters.

Order—*Psittaciformes* :

Characters :

1. Maxilla small and premaxilla well developed.
2. Vomar reduced.
3. Basipterygoid process absent.
4. Maxillo palatine process large, united and spongy.

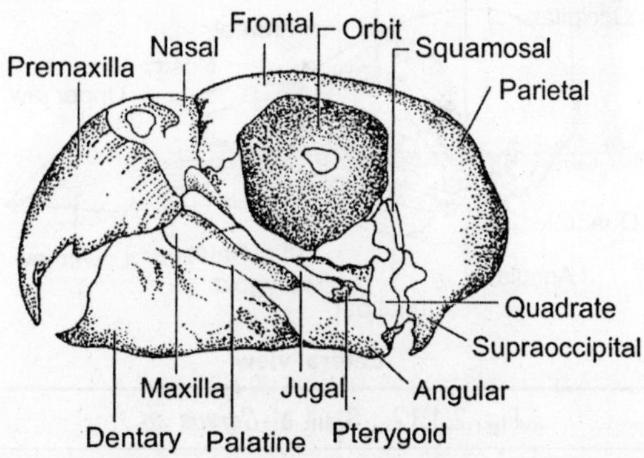

Fig. 2.1.11 : Skull of *Psittacula sp.*

(d) *Psittacula sp.* :

Characters :

1. Presence of frontal hinge between upper jaw and skull.
2. Beak strong, board, narrow and hooked anteriorly.
3. External nostrils small, rounded and internasal septum complete.
4. Write the characters 3 and 4 as in *Columba sp.*

(e) Skull of *Corvus sp.* :

State characters up to—

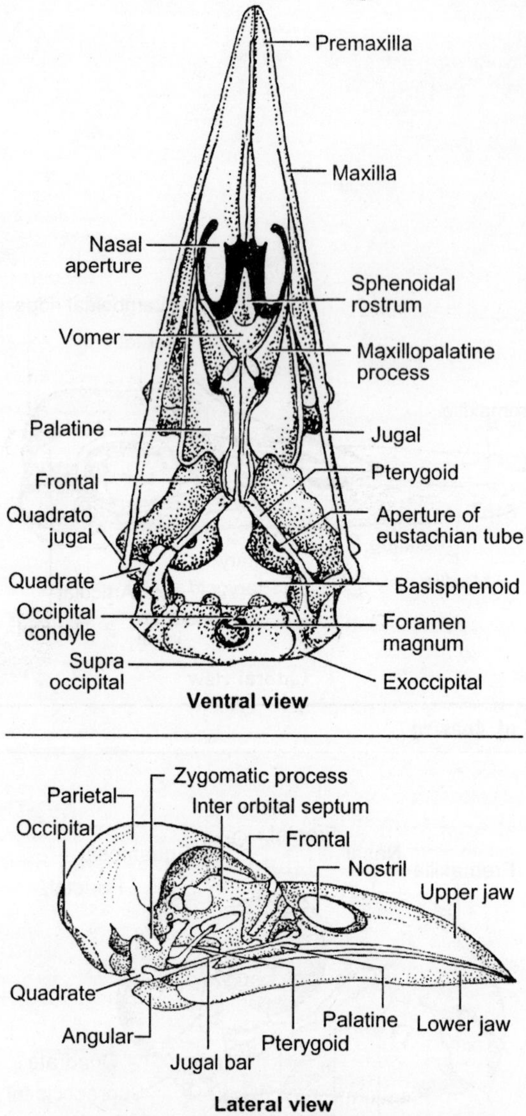

Fig. 2.1.12 : Skull of *Corvus sp.*

Super order **Neognathae** as in *Columba sp.* Then state the following characters.

Order—Passeriformes :

Characters :

1. Vomar is broad and truncated in front.
2. Maxillo palatine process slender but not united with one another.
3. Zygomatic process is in contact with the quadrate.
4. Deeply cleft vomar embraces the rostrum.
5. Well developed rostrum articulates with palatine.

(e) *Corvus sp.* **:**

Characters :

1. Beak large and strong slightly arched at the tip.
2. External nostrils large, oval and situated at the base of the beak.
3. Complete internasal septum.
4. Pterygoid small, rod like.
5. Write the characters 3 and 4 as in *Columba sp.*

Categories of Avian skull :

Four distinct broad categories are observed. These are :

1. Dromoeognathous skull (Seen in ratites, *Tinamus sp.*) :

Characters :

1. Basipterygoid processes are large which arises from the basis sphenoid to articulate with the posterior ends of the pterygoids.
2. Vomer large and broad.
3. Palatine does not articulate with rostrum.
4. Maxillo–palatine processes small and do not unite with one another.

2. Schizognathous Skull : (Seen in *Gallus sp; Columba sp.* etc) :

Characters :

1. Vomer small and pointed in front. (absent in *Columba sp.*).
2. Basipterygoid processes either absent or spring from the base of rostrum.
3. Maxillo–palatine do not unite with one another or with vomer.
4. Palatine and pterygoids articulate with parasphenoidal rostrum.

3. **Aegithognathous skull** (Found in passerines, *Corvus sp.* **etc). :**

Characters :

1. Broad vomer truncated in front.
2. Maxillo–palatine process slender but not united with one another.
3. Basipterygoid process absent.
4. Well developed rostrum articulates with palatines.

4. **Desmognathous Skull :** (Found in duck, anus, parrots etc.) :

1. Vomer small.
2. Maxillo–palatines are large and spongy and unite with one another in the middle line ventral to vomer.
3. True cranio-facial hinge is present between the skull and upper jaw in parrots.

MAMMALIAN SKULL

State the skull characters and then write the following :

D Class—Mammalia :

Characters :

1. Cranium large.
2. Presence of well developed sutures between firmly united bones.
3. Presence of two **occipital condyles**.
4. Presence of **tympanic bulla**.
5. **Zygomatic arch** well developed.
6. Quadrate absent.
7. Bony secondary palate formed by premaxilla, maxilla and palatine separates the nasal passage.
8. Lower jaw is made up of single bone, dentary and articulates to skull through squamosal.
9. Teeth, **theocodont** and **heterodont** type.

Sub class—Theria :

Characters :

1. Tympanic bone is ring like and forms a tympanic bulla.

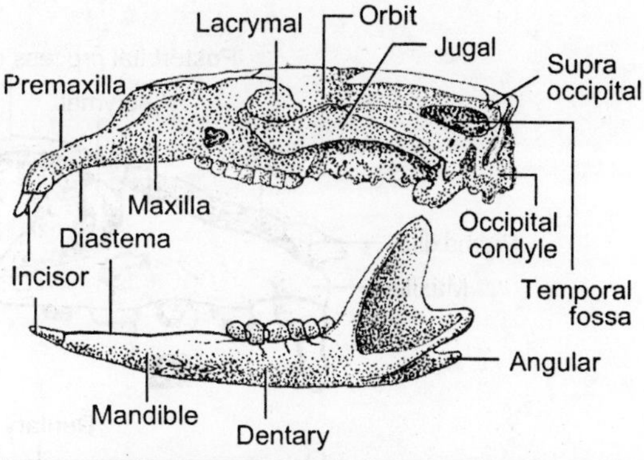

Fig. 2.1.13 : Skull and Mandible of *Macropus sp.*

2. Orbital and temporal fossae are confluent.
3. Palate is fenestrated.
4. Base of zygomatic arch is not perforated.
5. Pterygoid small.

(a) Skull of *Macropus sp.* :

State the above characters and then write the followings :

Infra class—**Metatheria** and

Order—**Marsupialia** :

Characters :

1. Orbit and temporal fossa remain fully confluent.
2. No post orbital bar.
3. Base of the zygomatic arch is perforated.
4. Bony palate is incomplete posteriorly.
5. Jugal bones participate in the formation of glenoid articulation.
6. Tympanic bulla is partly formed by alisphenoid.

Macropus sp. :

Characters :

1. Small skull with a flat top.
2. Orbit and temporal fossa remain fully confluent.

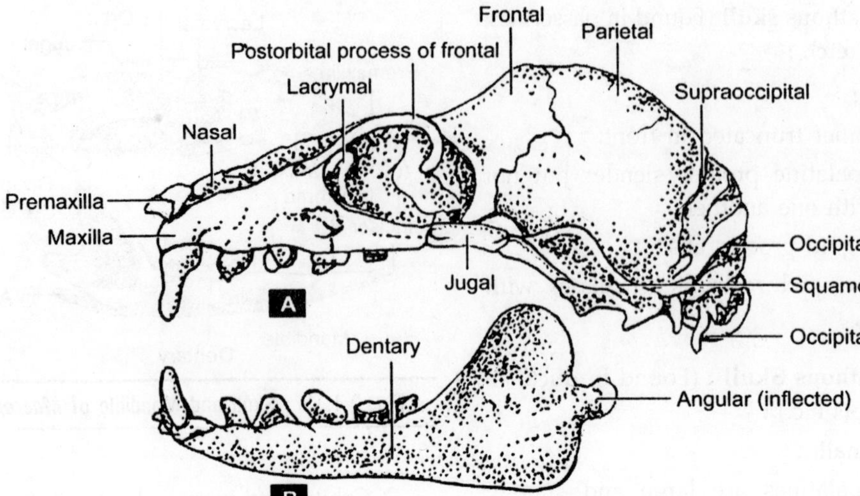

Fig. 2.1.14 : Skull and Mandible of Megachiroptera.

3. Lower jaw with inflected inner angle.
4. Dentition is heterodont, thecodont and **monophyodont** (except last premolar).
5. Canine absent and presence of **diastema**.
6. Dental formula 3.1.2.4/1.0.2.4=34.

(b) Skull of Megachiroptera and Microchiroptera :

State characters up to—

Sub class—**Theria** as those of *Macropus sp.* then write the following :

Infra class—Eutheria :

Characters :

1. Tympanic bone is ring like and forms tympanic bulla.
2. Secondary palate without large posterior vacuites.
3. Jugal dose not participate in the formation of **glenoid cavity**.
4. Alisphenoid is never associated with the bulla.
5. Jaws with teeth, teeth **thecodont, diphyodont** and **heterodont** type.
6. Dentition never exceeds 3.1.4.3/3.1.4.3 = 44.

Order—*Chiroptera* :

Characters :

1. Bones light and sutures of the skull are obliterated.
2. Post orbital process of the frontal is well developed.
3. A **lacrymal foramen** is present out side the orbit.
4. Orbit and temporal fossae are confluent.
5. Zygomatic arch long and cylindrical.
6. Orbit incomplete.
7. Ill developed tympanic bulla.

(i) Megachiroptera :

Characters :

1. Elongated snout.
2. Molar teeth without cusps and crown with longitudinal ridges.
3. Bony palate extended beyond first molar.
4. Angle of the lower jaw is inflected.

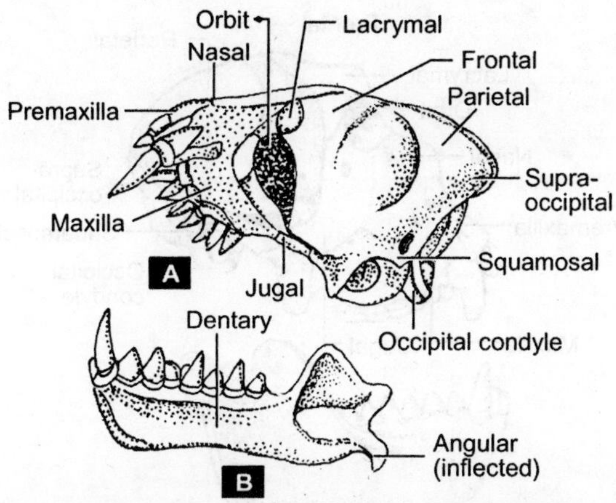

Fig. 2.1.15 : Skull and Mandible of *Microchiroptera*.

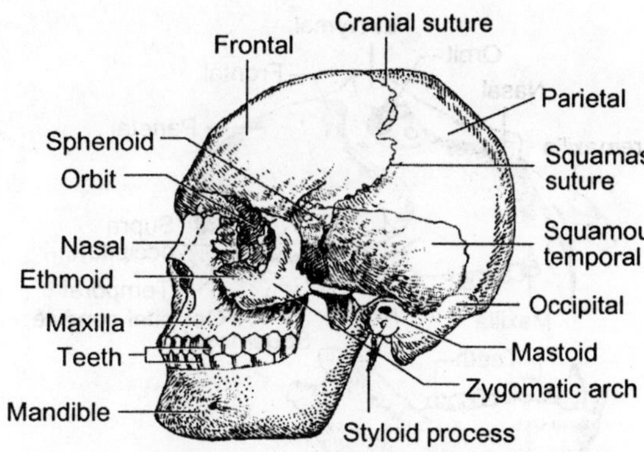

Fig. 2.1.16 : Skull and Mandible of *Homo sapiens sapiens.*

(ii) Microchiroptera :

Characters :

1. Snout is short.
2. Molar with cusps and bear transverse ridges.
3. Bony palate dose not extend beyond last molar.
4. Lower jaw with inflected inner angle.

(c) Skull of Human :

State characters up to—

Infra class—**Eutheria** as those in *Megachiroptera sp*. Then state the following.

Order—Primate :

Characters :

1. Cranium almost round.
2. Orbital ring complete and the eye socket confluent or separated with temporal fossae.
3. Number of teeth reduced.
4. Presence of bony eye socket.
5. Zygomatic arch is complete.
6. Lacrymal foramen extra orbital or infra orbital.
7. Incissor teeth two only.

Sub order—*Anthropoidea* :

Characters :

1. Enlarge Skull cap. (Calvaria).
2. Orbit sockets are forwardly directed and closely placed.
3. Temporal and orbital fossae are separated by a bony partition.
4. **Foramen magnum** broad and faces downward.
5. Lacrymal foramen infra orbital.
6. Mandibular symphysis strongly developed.

Homo sapiens sapiens :

Characters :

1. Brain box large.
2. Skull vault arched and voluminous.
3. Fore head high.
4. Nuchal or sagittal crest and supra orbital torus absent.
5. Nasal apertures long, pyriform or oval.
6. Internasal septum narrow.
7. Dental formula 2.1.2.3/2.1.2.3=32.

(d) Skull of Platyrrhine monkey :

State characters up to—

Sub order—**Anthropoidea** as in above. Then state the following.

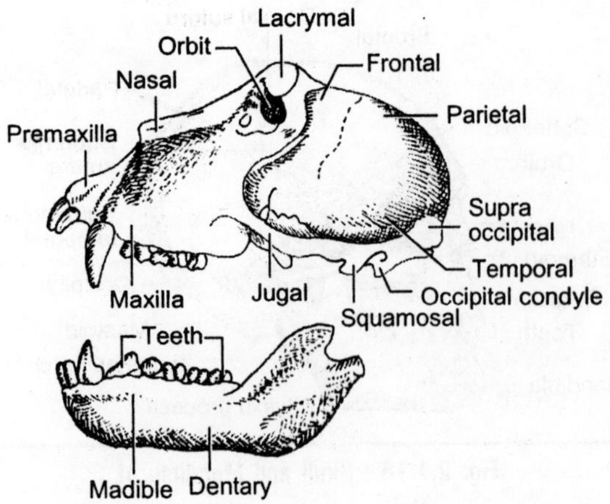

Fig. 2.1.17 : Skull and Mandible of Platyrrhine monkey.

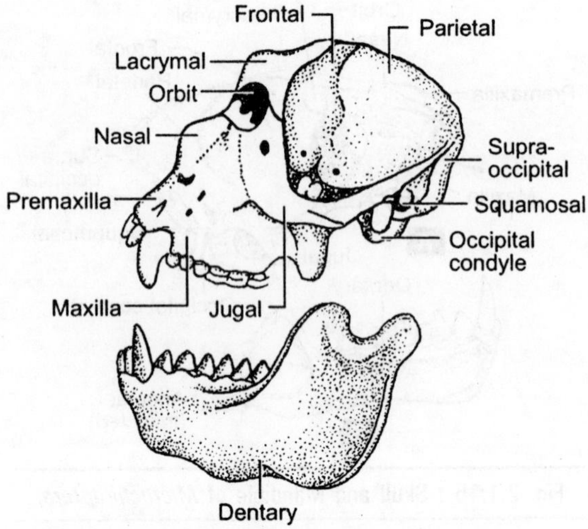

Fig. 2.1.18 : Skull and Mandible of Catarrhine monkey.

Platyrrhine monkey :

Characters :

1. Frontal bone is triangular and wedged between the parietals.
2. Orbito-temporal foramen is large.
3. Fore head rounded and inter-narial septum is broad and triangular.
4. Tympanic bulla well developed.
5. Palate short and number of premolar is three.
6. Dental formula 2.1.3.2/2.1.3.2 = 32 or 2.1.3.3/2.1.3.3 = 36.

(e) Skull of Catarrhine monkey :

State characters up to—

Sub order—*Anthropoidea* as in above. Then write the followings :

Catarrhine monkey :

Characters :

1. Frontal is not wedge between the parietals.
2. Nasal aperture narrow.
3. Orbito-temporal foramen small.
4. Tympanic bulla ill developed.
5. Upper molar carry four **bilophodont cusps.**
6. Palate long and premolar always 2.
7. Dental formula 2.1.2.3/2.1.2.3=32.

(f) Skull of Guineapig, Rat and Rabbit :

State characters up to—

Infra class—*Eutheria* as those of Chiroptera. Then write the followings.

Order—Rodentia :

Characters :

1. Skull small and orbit incomplete.
2. Slender zygomatic arch.
3. Orbital and temporal fossae confluent.
4. Premaxilla and maxilla large.
5. Large tympanic bulla flask–shaped.
6. Presence of large, sharp chisel – shaped incisor.
7. Diastema well developed.
8. Canine absent.
9. Symphysis of the lower jaw absent.
10. Angular process long.

(i) *Cavia sp.* :

Characters :

1. Anterior premolar absent.

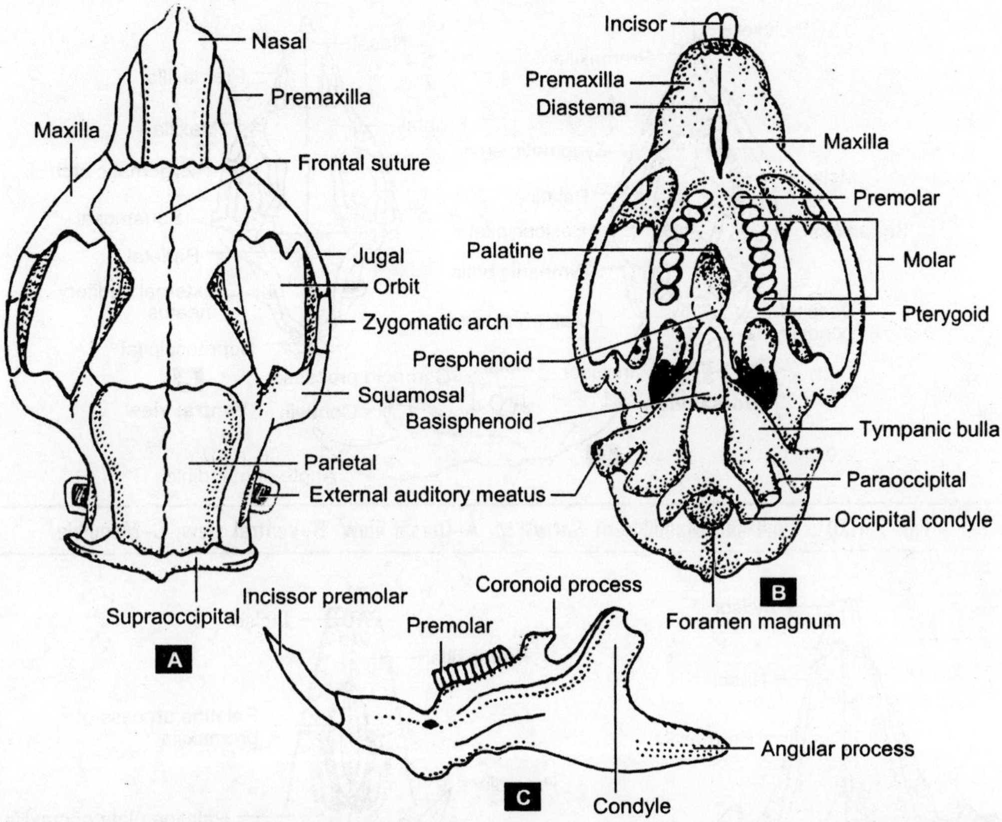

Fig. 2.1.19 : Skull and Mandible of *Cavia sp.* A–Dorsal view, B–Ventral view, C–Mandible.

2. One pair of incisor in each jaw.
3. Infraorbital foramen large.
4. Dental formula 1.0.1.3/1.0.1.3 = 20.

(ii) *Rattus sp.* :

Characters :

1. Narrow and elongated facial portion.
2. One pair of incisor in each jaw.
3. Premolar absent.
4. Dental formula 1.0.0.3/1.0.0.3=16.

(iii) Rabbit skull :

State characters up to—

Infra class—**Eutheria** as those of *Megachiroptera sp*. Then write the followings.

Order—*Lagomorpha* :

Characters :

1. Skull small.
2. Orbit small.
3. Squamosal has large glenoid fossa.
4. Maxilla laterally fenestrated.
5. Two pairs of incisor in the upper jaw and one pair in lower jaw.
6. Lower jaw articulates with squamosal.
7. Presence of three premolars in the upper jaw and two in the lower jaw.
8. Anterior symphysis of the lower jaw present.

***Oryctolagus sp.* :**

Characters :

1. Cranial part of the skull is smaller but facial part elongated.

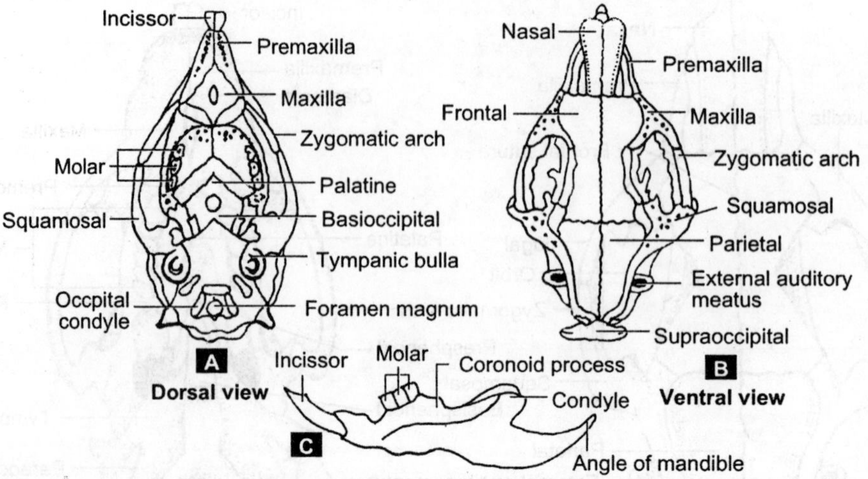

Fig. 2.1.20 : Skull and Mandible of *Rattus sp.* A–Dorsal view, B–Ventral view, C–Mandible.

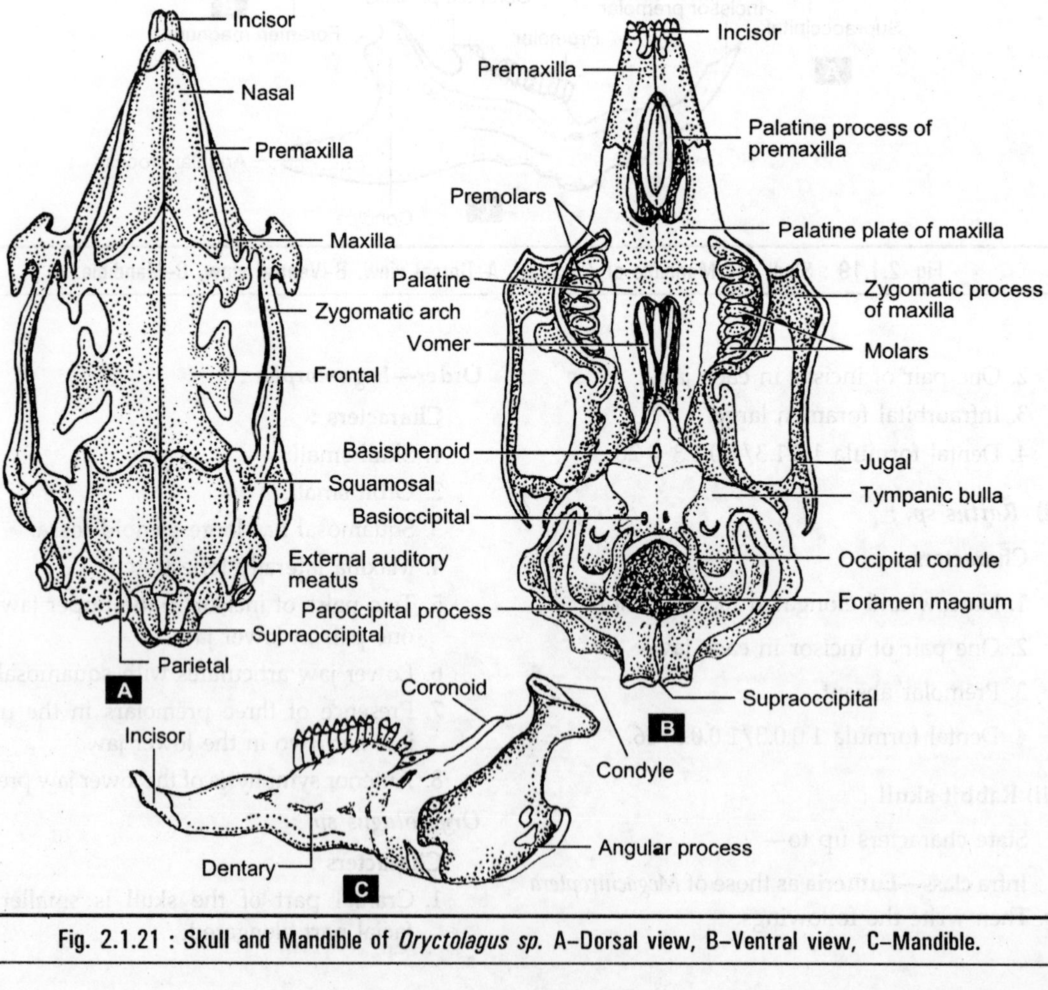

Fig. 2.1.21 : Skull and Mandible of *Oryctolagus sp.* A–Dorsal view, B–Ventral view, C–Mandible.

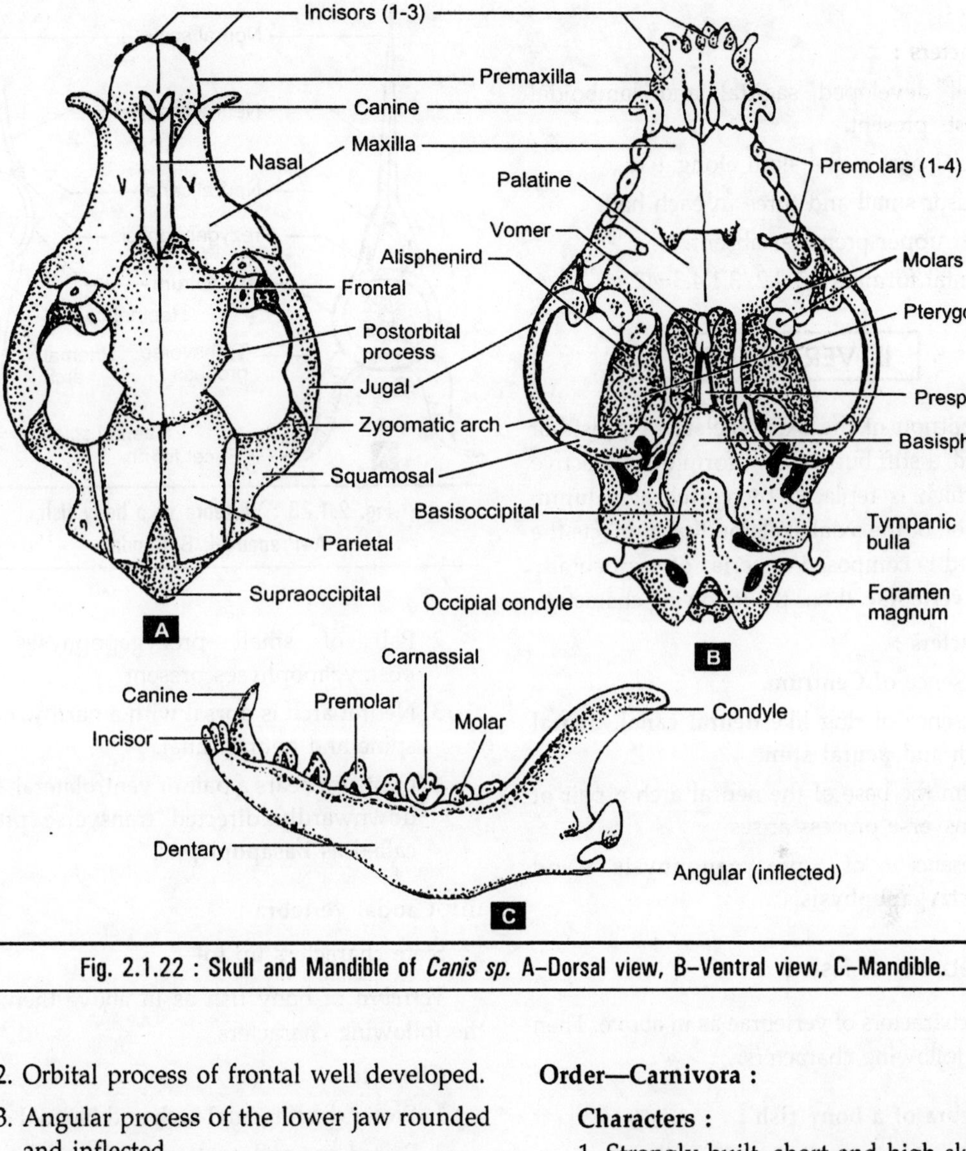

Fig. 2.1.22 : Skull and Mandible of *Canis sp.* A–Dorsal view, B–Ventral view, C–Mandible.

2. Orbital process of frontal well developed.
3. Angular process of the lower jaw rounded and inflected.
4. Vacuities present in the dentary and anterior to the condyle.
5. Dental formula 2.0.3.3/1.0.2.3=28.

(g) Skull of dog :

State characters up to—

Infra class—***Eutheria*** as those of *Megachiroptera sp.* Then write the following.

Order—Carnivora :

Characters :

1. Strongly built, short and high skull.
2. Zygomatic arch strongly built and outwardly curved.
3. Canine large, pointedly sharp and recurved slightly.
4. Well developed tympanic bulla.
5. Last upper premolar and first lower molar coincide to form **carnassial set of teeth** pattern.
6. Condylar process of dentary large and rounded.

Canis sp. :

Characters :

1. Well developed sagittal and lamboidal crests present.
2. Facial part of the skull elongated.
3. Incisor small and three in each half.
4. Last upper premolar absent.
5. Dental formula 3.1.4.2/3.1.4.3=42.

II. VERTEBRAE

The portion of the axial skeleton consists of **notochord**, a stiff but resilient form of connective tissue, which is replaced by **vertebral column.** The portion of the axial skeleton that protects the spinal cord is composed of series of segmentally arranged **vertebrae.** It has the following characters.

Characters :

1. Presence of **Centrum.**
2. Presence of ring like **neural canal, neural arch** and **neural spine.**
3. Form the base of the neural arch a pair of transverse process arises.
4. Presence of **prezygapophysis** and **postzygapophysis.**

A. Vertebrae of Fish :

State characters of vertebrae as in above. Then state the following characters.

(a) Vertebra of a bony fish :

Characters :

1. Centrum **amphicoelous** i.e., both ends of the centrum have concavities (biconcave).
2. Presence of long neural arch and backwardly directed neural spine.

(i) Precaudal vertebra (trunk vertebra) :

Characters :

1. Centrum bears a pair of short outwardly directed transverse processes for articulation of ribs.

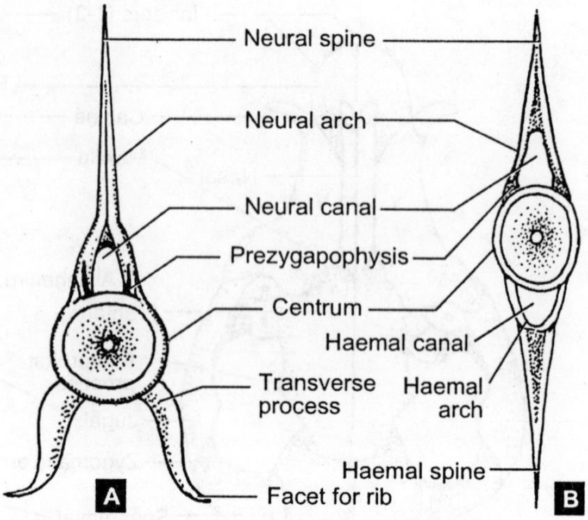

Fig. 2.1.23 : Vertebra of a bony fish. A–Precaudal, B–Caudal.

2. Pair of small prezygapophyses and postzygapophyses present.
3. Neural arch is dorsal with a narrow neural spine and neural canal.
4. Centrum bears a pair of ventrolateral, short downwardly directed transverse process called as **basapophysis.**

(ii) Caudal vertebra :

State characters up to—

Vertebra of bony fish as in above then state the following characters.

Characters :

1. Presence of **haemal arch** and **haemal canal.**
2. Paired pre and postzygapophyses present.
3. Neural arch and neural spine present.
4. Presence of a pointed and backwardly directed median haemal spine.

B. Vertebrae of Amphibia :

Characters :

1. Centrum **procoelous** i.e., Anterior end of the centrum is concave, while the posterior surface is convex.

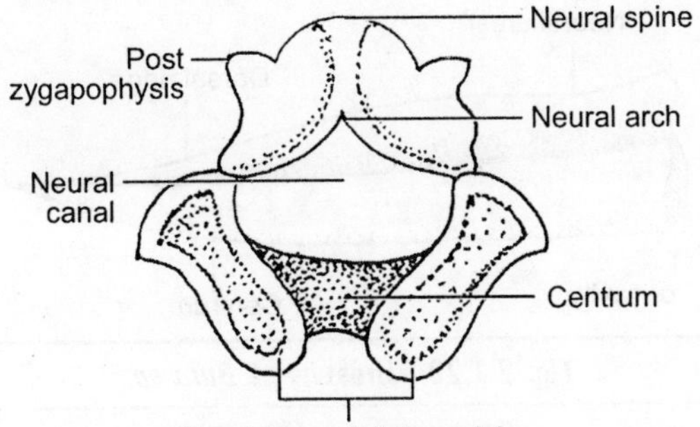

Fig. 2.1.24 : Atlas of *Bufo sp.*

2. Neural spine short (neural spine well developed in *Rana sp.*).
3. Transverse process laterally directed.
4. Vertebrae **notocentrous** type.

(a) Atlas (*Bufo sp.*) :

State the characters of vertebra and the characters of amphibian vertebra as in above. Then write the following.

Characters :

1. Small, ring like vertebra.
2. Neural arch enclosed a large neural canal.
3. Reduced Centrum possesses two concave facets on its anterior side for articulating with the occipitals condyle of the skull.
4. Transverse process and prezygaphyses are absent.
5. Postzygapophyses present.
6. It is the first vertebra and is the only cervical vertebra of amphibia.

(b) Typical vertebra of *Bufo sp.* :

State characters up to amphibian vertebra as in above. Then state the following :

Characters :

1. Short, solid, cylindrical centrum.
2. Centrum procoelous.
3. Neural arch well developed, neurel spine short and backwarldly directed.
4. Short, straight, rod like transverse processes are laterally directed.

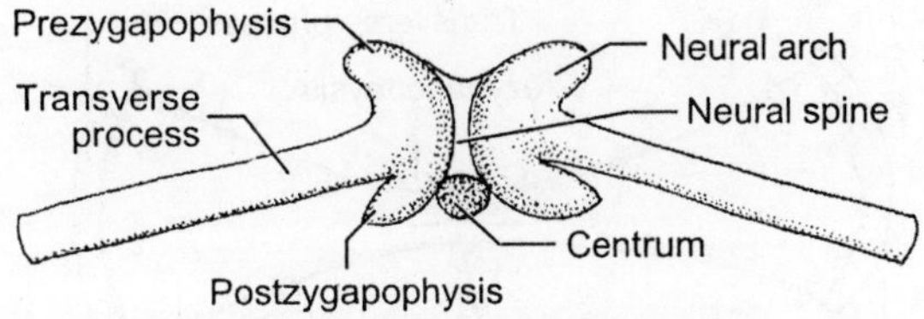

Fig. 2.1.25 : Typical vertebra axis (2nd vertebra) of *Bufo sp.*

5. Pre and postzygapophyses are well developed.

Note : Form second to eighth vertebra all are typical vertebra. The second vertebra has broad, wing like, well developed transverses processes, while the third and fourth have thick, transverse processes. The fifth, sixth, seventh and eighth vertebrae possess slender transverse processes.

(c) Eighth vertebrah of *Bufo sp.* :

Characters :

1. Centrum **amphicoelous.** i.e. concave both in front and behind (biconcave). Rest characters are as those of typical vertebra .

(d) Ninth vertebra of *Bufo sp.* :

State characters up to amphibian vertebra as in above. Then state the following characters.

Characters :

1. Centrum procoelous.
2. Transverse process fan-shaped with cartilaginous margins.
3. Two rounded condyles on the posterior margin of centrum.

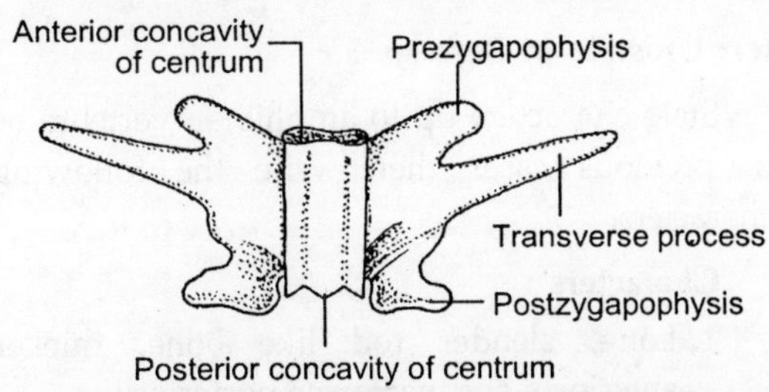

Fig. 2.1.26 : Eighth vertebra of *Bufo sp.*

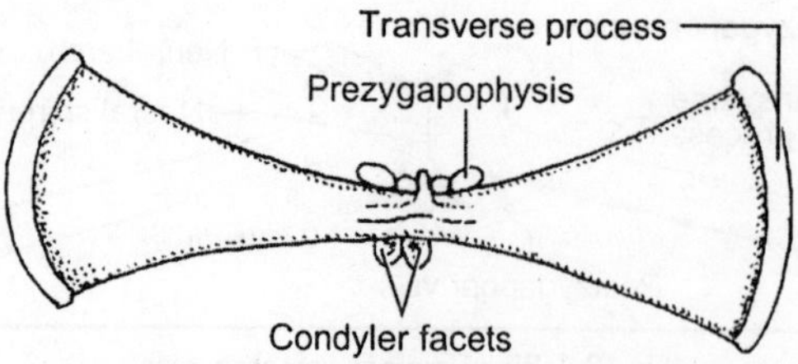

Fig. 2.1.27 : Ninth vertebra of *Bufo sp.*

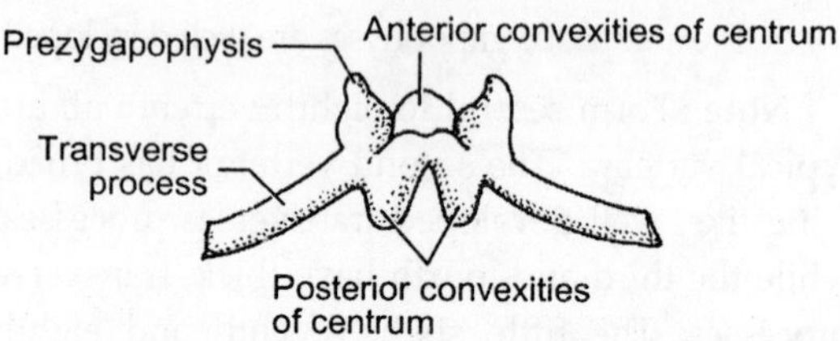

Fig. 2.1.28 : Ninth vertebra of *Rana sp.*

4. Prezygapophyses well developed.
5. Post zygapophyses absent.

(d) Ninth Vertebra of *Rana sp.* :

Characters :

1. Centrum biconvex (convex both anteriorly and posteriorly).
2. Transverse processes are strong, broad and cylindrical.
3. Two small knob like convexities at the posterior end of centrum.
4. Prezygapophyses well developed.
5. Postzygapophyses absent.

Note : It is the only sacral vertebra in frogs and toads.

(e) Urostyle of *Bufo sp.* :

State characters up to amphibian vertebra as in previous cases then write the following characters.

Characters :

1. Long, slender rod like bone, thicker anteriorly and narrower posteriorly.

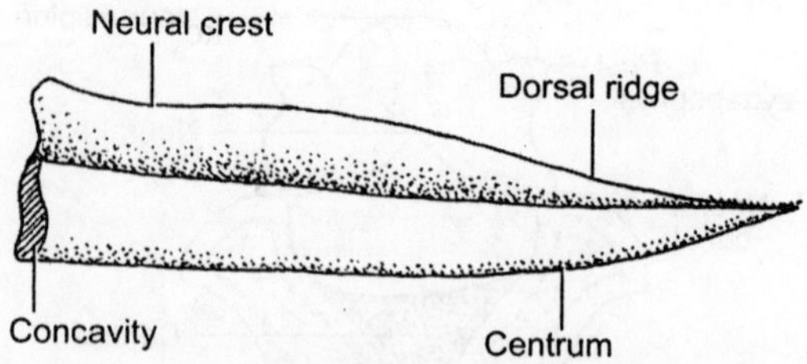

Fig. 2.1.29 : Urostyle of *Bufo sp.*

2. Presence of two con-cavities at the anterior end of centrum.
3. Presence of a mid-dorsal crest.
4. A median narrow short canal runs throughout the urostyle.

C. Vertebrae of Reptilia :

State characters up to vertebra as in amphibia then state the following characters.

Class Reptilia :

Characters :

1. **Procoelous** centrum.
2. Neural spine long and transverse processes short.
3. Thoracic vertebrae bear ribes.

(i) VERTEBRAE OF *CALOTES SP.* :

(a) **Atlas of *Calotes sp.* :**

State the above characters and then write.

Characters :

1. Small ring like bone.
2. Centrum inconspicuous.
3. Incomplete neural arch.
4. A pair of prezygapophyses is present.
5. Neural spine and post zygapophyses absent.
6. A pair of posterolateral processes below the postzygapophyses.
7. Neural canal is divided by a transverse ligament into two upper neural canals and below **Odontoid process.**

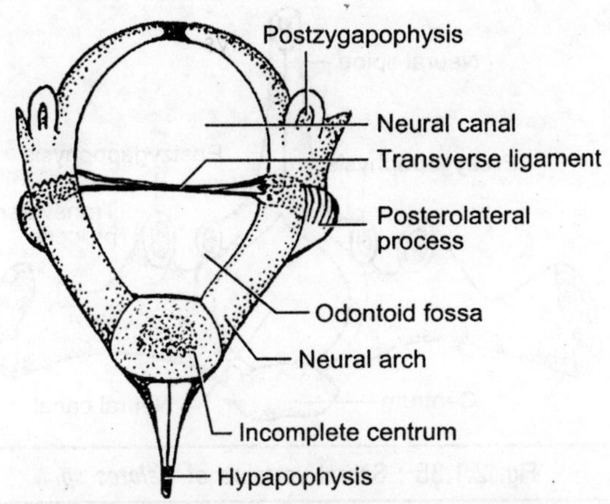

Fig. 2.1.30 : Atlas (first vertebra) of *Calotes sp.*

(b) Axis of *Calotes sp.* : State the characters up to—

Class—***Reptilia*** as in above then state the following characters.

Characters :

1. Centrum **procoelous.**
2. It bears a small peg like process called **Odontoid process** on its anterior side.
3. The mid ventral line is provided with a ridge called **hypapophysis.**
4. Presence of well developed backwardly directed neural spine.

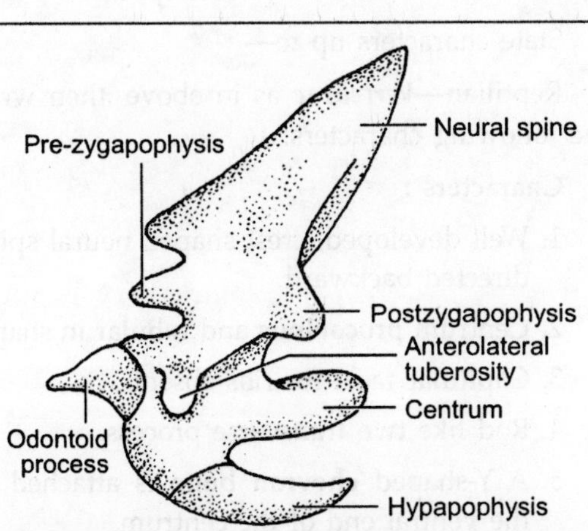

Fig. 2.1.31 : Axis (second vertebra) of *Calotes sp.*

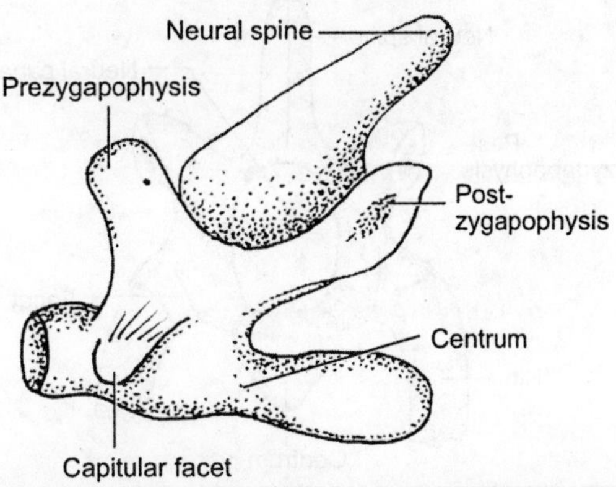

Fig. 2.1.32 : Typical cervical vertebra of *Calotes sp.*

5. Transverse process absent.

(c) Typical cervical vertebra of *Calotes sp.* :

State characters up to—

Reptilian—***Vertebra*** as in above .then write the following characters.

Characters :

1. Centrum **procoelous** and well developed.
2. Neural arch enclosing a neural canal.
3. Well developed backwardly directed crest like neural spine present.
4. A pair of **capitular facet**, one an each side of the centrum.
5. Transverse process absent.
6. Pre and postzygapophyses present.

Note : In *Calotes sp.*cervical region composed of eight cervical vertebrae of which first one is atlas, second one is axis, and the rest are more of less identical.

(d) Typical thoracic vertebra of *Calotes sp.* :

State characters up to—

Reptelian—***Vertebra*** as in above and then write the following characters.

Characters :

1. Centrum **procoelous.**
2. Well developed narrow neural spine backwardly directed.

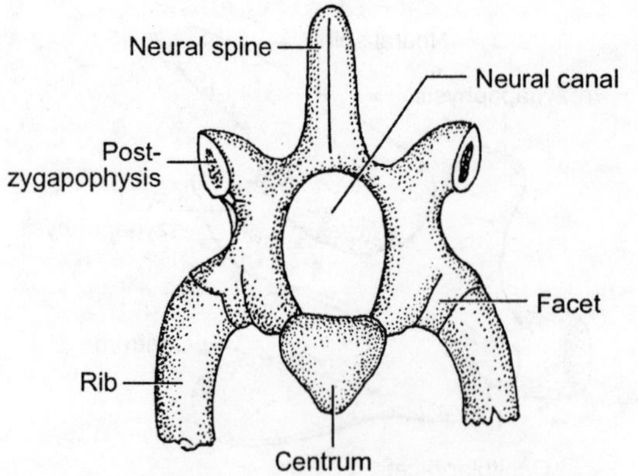

Fig. 2.1.33 : Typical thoracic vertebra of *Calotes sp.*

3. Hypapophysis absent.
4. Vertebra bears ribs which reach the sternum.
5. Pre and postzygapophyses well developed.
6. Well developed facets at the junction of neural arch and centrum below prezygapophyses.
7. Transverse process absent.

Note : There are four thoracic vertebrae in *Calotes sp.*

(e) Lumber vertebra of *Calotes sp.* :

Characters same as those of thoracic vertebrae. Only the ribs are short and do not reach sternum.

Note : There are eleven lumber vertebrae in *Calotes sp.*

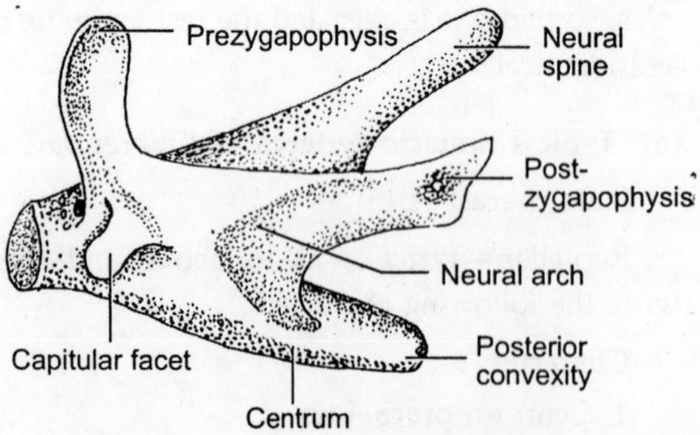

Fig. 2.1.34 : Lumber vertebra of *Calotes sp.*

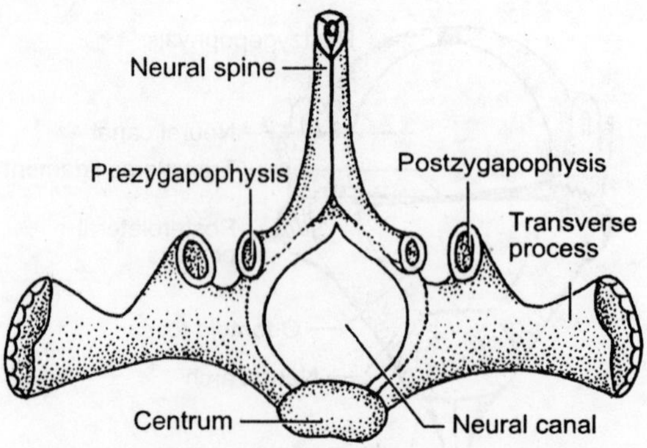

Fig. 2.1.35 : Sacral vertebra of *Calotes sp.*

(f) Sacral vertebra of *Calotes sp.* :

State characters up to reptilian vertebrae as in above. Then write the following.

Characters :

1. Well developed **procoelous** centrum.
2. Well developed crest shaped neural spine, backwardly directed.
3. Transverse process well developed.
4. Pre and postzygapophyses well developed.
5. Capitular facet absent.

Note : There are only two sacral vertebrae in *Calotes sp.*

(g) Caudal vertebra of *Calotes sp.* :

State characters up to—

Reptilian—***Vertebrae*** as in above .then write the following characters.

Characters :

1. Well developed, crest shaped neural spine directed backward.
2. Centrum **procoelous** and tubular in shape.
3. Capitular facet for ribs absent.
4. Rod like two transverse process.
5. A Y-shaped **chevron bone** is attached to the ventral end of the centrum.
6. Pre and postzygapophyses well developed.

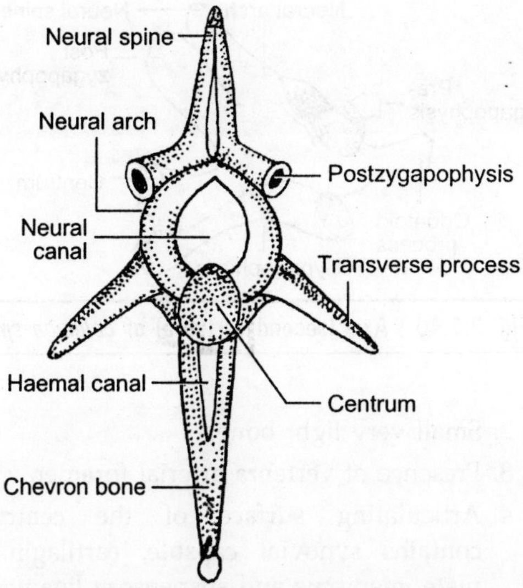

Fig. 2.1.36 : Caudal vertebra of *Calotes sp.*

Note : There are about fifty caudal vertebrae of which first fifteen are called **anterior** and rest are **posterior** caudal vertebrae. Posterior caudal vertebrae possess all the characters of anterior one, except that transverse process and chevron bones which are lacking in then.

(ii) VERTEBRAE OF SNAKE :

(a) Typical pre caudal or trunk vertebra of snake :

State characters up to—

Reptilian—*Vertebra* as before. Then write the followings.

Characters :

1. Centrum **procoelous** and well developed.
2. Neural spine thick, flat and crest like.
3. Neural arch encloses a neural canal.
4. Pre and post zygapophyses well developed.
5. Two extra articular surface develop from the anterior and posterior side of the neural arch, **zygosphene** at the anterior and **zygantrum** at the posterior face.
6. Presence of ventral hypapophysis .
7. Transverse processes small and short.
8. Presence of capitular facets.

(b) Caudal vertebra of snake :

State characters up to—

Precaudal—*Vertebra* (from point 1-5) as in above, then write the following characters.

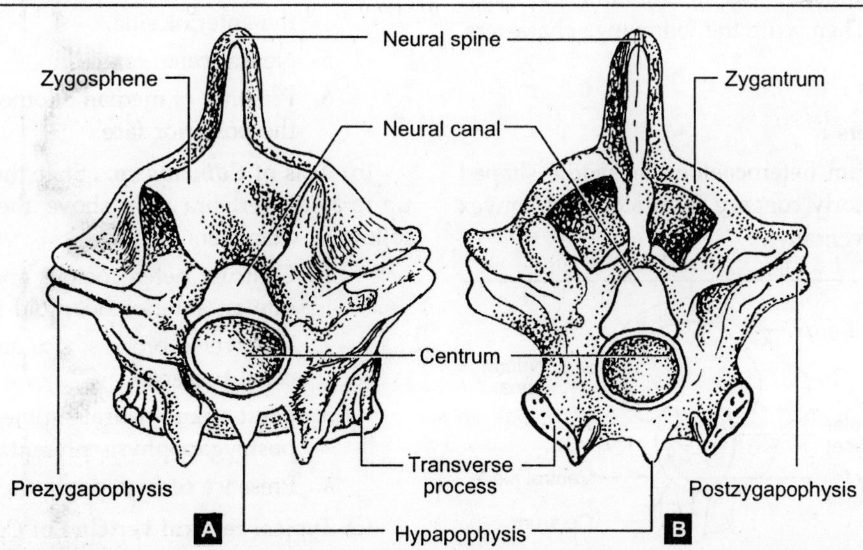

Fig. 2.1.37 : Precaudal or trunk vertebra of a snake. A–Anterior view, B–Posterior view.

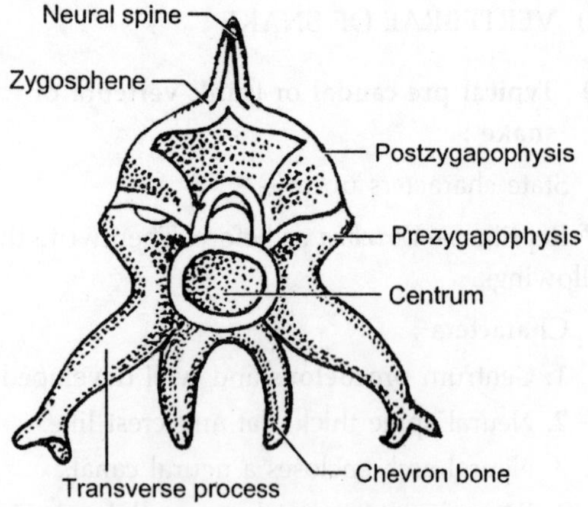

Fig. 2.1.38 : Caudal vertebra of a snake.

1. Hypapophysis absent.
2. Capitular facets absent.
3. Transverse process rod shaped and short.
4. **Chevron bone** present.

Note : Snakes possess large number of vertebrae except atlas and axis, all vertebrae bear ribs.

D. Vertebrae of Bird :

State characters up to vertebra as stated previously, Then write the followings characters.

Class—Aves :

Characters :

1. **Centrum heterocoelous** or saddle shaped (anteriorly concave from side and convex dorsovenrally).

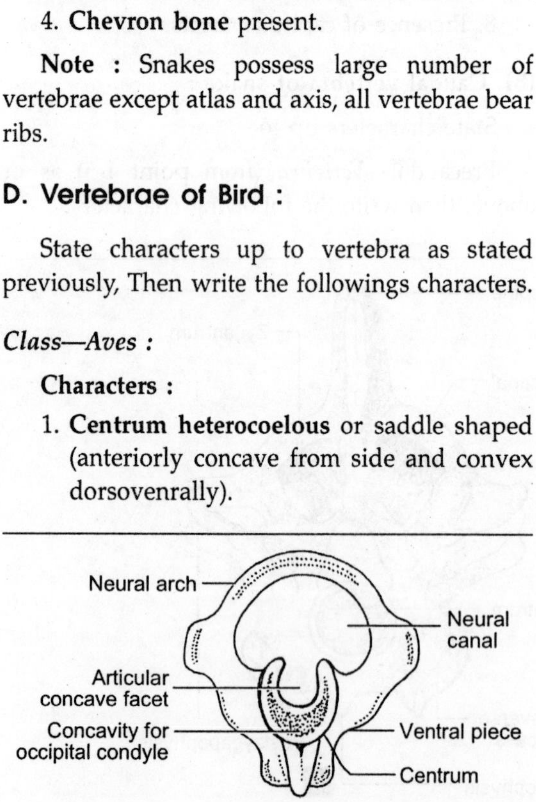

Fig. 2.1.39 : Atlas of *columba sp.*

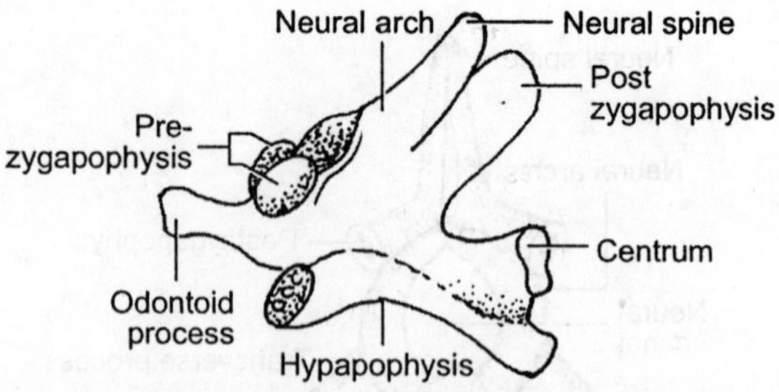

Fig. 2.1.40 : Axis (second vertebra) of *columba sp.*

2. Small very light bone.
3. Presence of vertebra arterial foramen.
4. Articulating surface of the centrum contains synovial capsule, cartilaginous plate, meniscus and suspensory ligament.

(a) **Atlas of *Columba sp.* :** State the characters up to avian vertebra as in above, then write the following characters.

Characters :

1. Light, ring like composed of two bones but sutures absent.
2. Centrum, neural spine, ribs, transverse process and prezygapophysis absent.
3. Postzygapophysis present.
4. A concavity for occipital condyle on the anterior side.
5. Neural canal large.
6. Presence of median odontoid notch on the posterior face.

(b) **Axis of *Columba sp.* :** State the characters up to avian vertebra as in above, then write the following characters.

1. Centrum heterocoelous and produced into an anterior odontoid process.
2. Transverse process, ribs and vertebral canal present.
3. Crest like neural spine, pre and postzygapophysis present.
4. Presence of hypapophysis.

(c) Typical cervical vertebra of *Columba sp.* : State the characters up to avian vertebra as above, then write the following characters.

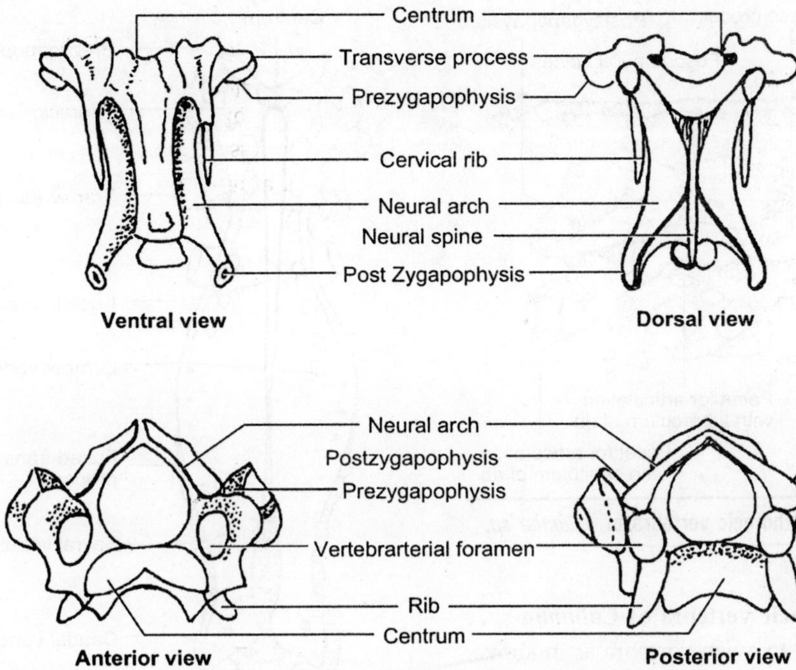

Fig. 2.1.41 : Typical cervical vertebra of *Columba sp.*

Characters :

1. Centrum heterocoelous.
2. Neural arch short and neural spine rudimentary.
3. Transverse process short and irregular and fused with thin backwardly directed reduced cervical ribs.
4. Presence of vertebra arterial foramen.
5. Prezygapophyses flat.
6. Postzygapophyses project downward and outwards.

Note : Pigeon possess fourteen cervical vertebrae of which first is atlas second is axis and rest twelve are typical cervical vertebrae.

(d) Free thoracic vertebra of *Columba sp.* : State characters up to avian vertebra as in above, then write the following characters.

Characters :

1. Short and stout heterocoelous centrum.
2. Presence of hypapophyses on the ventral side of centrum.

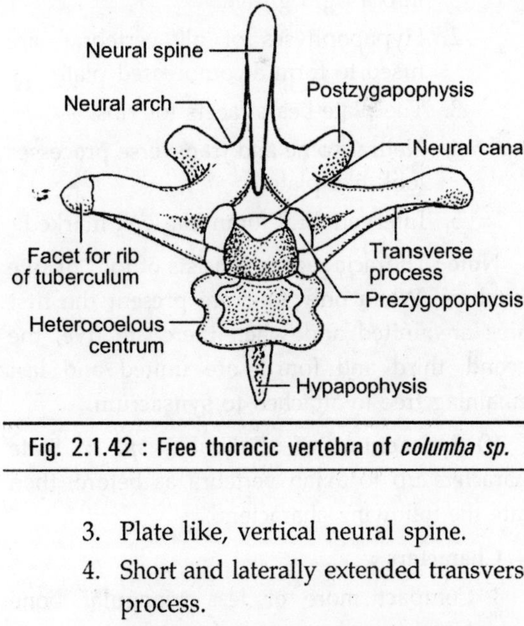

Fig. 2.1.42 : Free thoracic vertebra of *columba sp.*

3. Plate like, vertical neural spine.
4. Short and laterally extended transverse process.
5. Well developed pre and postzygapophyses.
6. Centrum and transverse process bear facets for thoracic ribs.

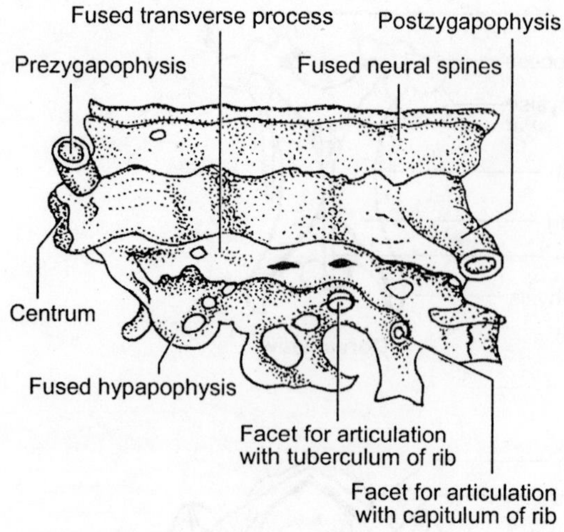

Fig. 2.1.43 : Fused thoracic vertebra of *columba sp.*

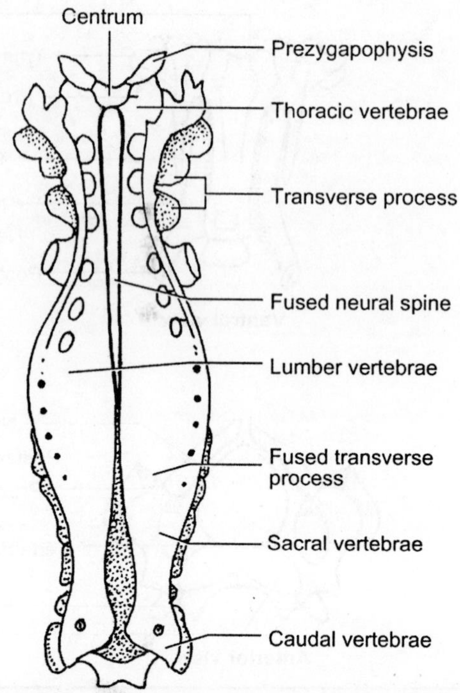

Fig. 2.1.44 : Synsacrum of *columba sp.*

(e) **Fused thoracic vertebra of *Columba sp.* :** State characters up to avian vertebra as in above. Then write the following characters.

Characters :

1. Four or five thoracic vertebrae fused into a rigid girder.
2. Hypapophyses of all vertebrae are fused to form a compressed plate.
3. The plate bears facets for ribs.
4. Neural spine and transverse processes look like plate.
5. Intervertebral foramina well marked.

Note : Thoracic region consists of four or five vertebrae. When only four are present the first three are united and when there are five, the second, third and fourth are united and last remaining free to attached to **synsacrum.**

(f) Synsacrum of *Columba sp.* : State characters up to avian vertebra as before, then state the following characters.

Characters :

1. Compact more or less triangular bone formed by the fusion of vertebrae.
2. Compond bone is formed by the fusion of thirteen vertebrae, the vertebrae are, one thoracic, five lumber, two sacral, and five anterior caudal vertebrae.
3. The heterocoleus centrum and prezygapophyses are distinguished anteriorly.
4. Tranverse processes of first three vertebrae are free and rest (lumber and sacral) are fused to from a compact lateral plate on either side.
5. Crest like neural spine is well developed on anterior side.
6. Anterior vertebrae possesses facets for ribs and intervertebral foramina.

(g) **Pygostyle of *Columba sp.*** : State characters up to avian vertebra as before then state the following characters.

Characters :

1. Compressed skeletal structure formed by the fusion of four hindermost caudal vertebrae.

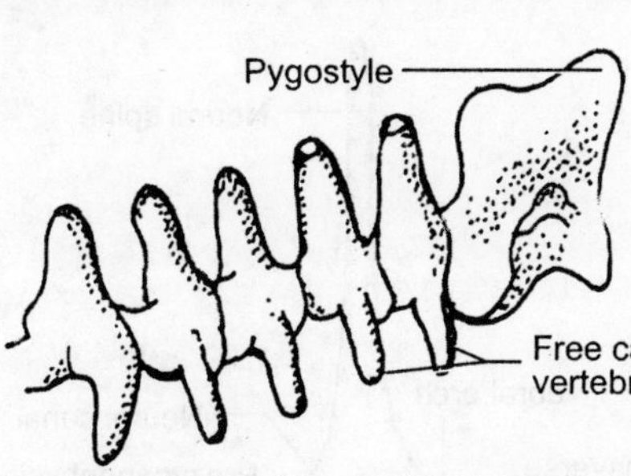

Fig. 2.1.45 : Pygostyle of *columba sp.*

2. Centra and neural spines not distinguishable.
3. First four vertebrae are with well marked transverse processes.
4. The last vertebra is very thin, flat and is turned upward (**Ploughshare bone**).

E. Vertebrae of Mammalia :

State characters up to vertebra as stated previously then write the following characters.

Class—Mammalia :

Characters :

1. Shout bony element with **acoelous centrum** i.e. flattened centrum both anteriorly and posteriorly.
2. Pre and postzygapophyses are mounted on specially developed **metapophysis** and **anapophysis.**
3. Presence of intervertebral foramen.

(a) Atlas of *Cavia sp.* : State characters as in above, then write the following characters.

Characters :

1. Ring like bone, centrum insignificant.
2. Neural canal large, neural spine very small.
3. Transverse process small and flat perforated at the base by anteroposterior vertebraterial foramen.
4. Zygapophyses are absent.
5. Anterior surface is provided with a pair of facets for articulation with occipital condyle.

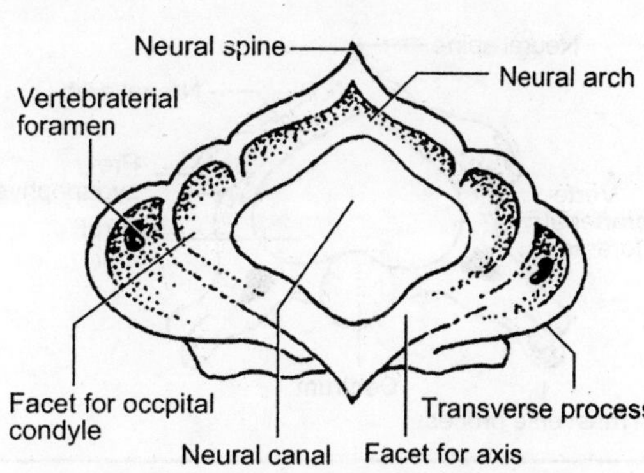

Fig. 2.1.46 : Atlas of *cavia sp.*

(b) Axis of *Cavia sp.* :

Characters :

1. Centrum is broad and drawn anteriorly into a peg like odontoid process.
2. Large thin crest like neural spine extends anteroposteriorly.
3. Small rod like transverse process also directed backwardly.
4. Vertebraterial foramen present.
5. Pre and postzygapophyses are present.

(c) **Typical cervical vertebra of *Cavia sp.* :** State characters up to—

Class—*Mammalia* as in above. Then state the following characters.

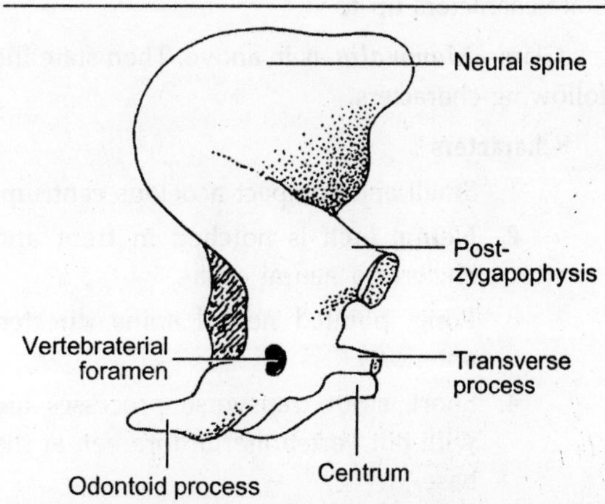

Fig. 2.1.47 : Axis of *cavia sp.*

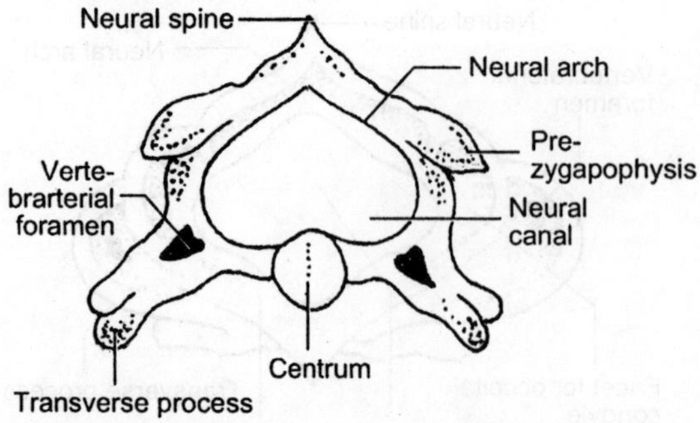

Fig. 2.1.48 : Typical cervical vertebra of *cavia sp.*

Characters :

1. Centrum small but broad from side to side.
2. Short neural spine directed upward.
3. Neural arch with upper lamina and a notched pedicel at lower end.
4. Bifid transverse process stubby directed outwards and downwards and is perforated by vertebraterial foramen at the base.
5. Pre and post zygapophyses are well developed.

Note : Cervical vertebrae are seven in number in mammals. The first is atlas and second is axis and rest are typical vertebrae.

(d) Anterior thoracic vetebra of *Cavia sp.* : State characters up to—

Class—***Mammalia*** as in above. Then state the following characters.

Characters :

1. Small and compact acoelous centrum.
2. Neural arch is notched in front and encloses a neural canal.
3. Long, pointed neural spine directed anteriorly.
4. Short, stout transverse processes are with out vertebraterial foramen at the base.
5. Concave capitular and tubercular facets for articulation of ribs on

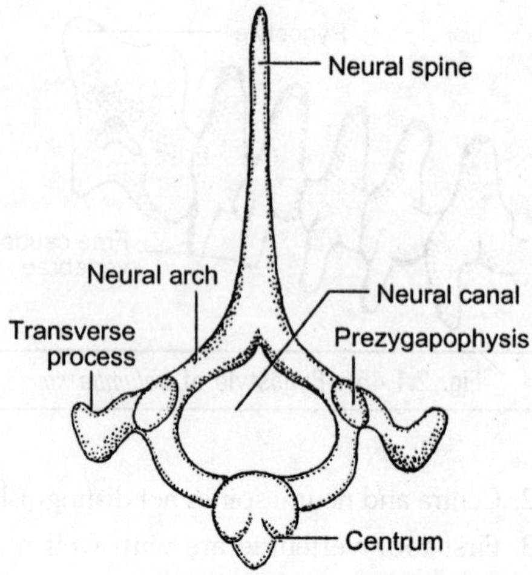

Fig. 2.1.49 : Anterior thoracic vertebra of *cavia sp.*

centrum and transverse processes respectively.

6. Pre and postzygapophyses present.

(e) Posterior thoracic vertebra of *Cavia sp.* : State characters up to—

Class—***Mammalia*** as in above. Then write the following characters.

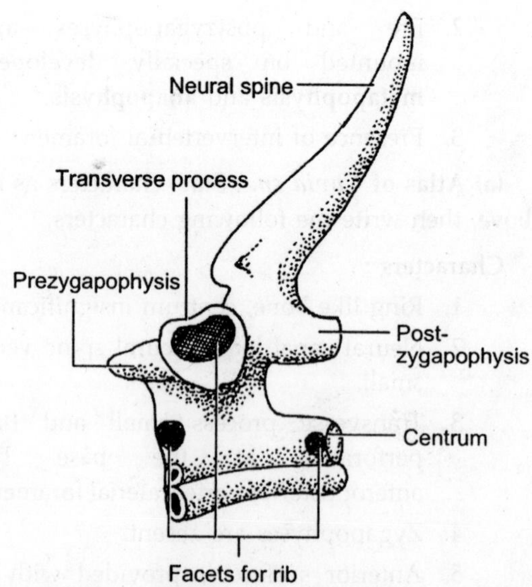

Fig. 2.1.50 : Posterior thoracic vertebra of *cavia sp.*

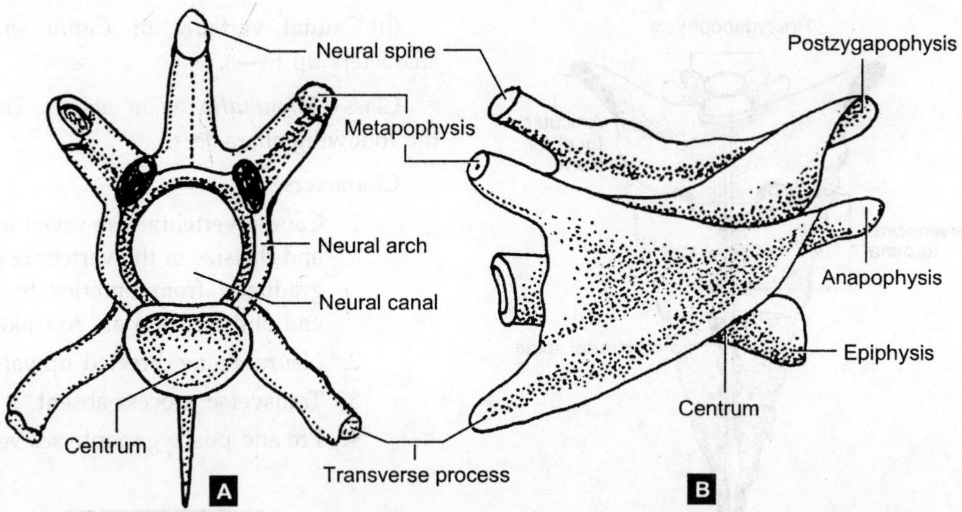

Fig. 2.1.51 : Lumber vertebra of *cavia sp.* A–Anterior view, B–Lateral view.

Characters :

1. Small acoelous centrum having lateral facets for articulating ribs.
2. Neural arch with notches in front and behind and encloses a small neural canal.
3. Neural spine flat side wise and directed forward.
4. Small transverse process lacks facet for rib articulation.
5. Pre and postzygapophyses well developed.

Note : *Cavia* has twelve thoracic vertebrae of which the last three are called **posterior thoracic** and first nine are called **anterior thoracic** vertebrae.

(f) **Lumber vertebra of *Cavia sp.* :** State characters up to—

Class—***Mammalia*** as in above. Then write the following.

Characters :

1. Large, stout, flat neural spine directed anteriorly.
2. Transverse processes are large and expanded, directed forward and downward.
3. Zygapophyses, metapophysis and anapophysis well developed.
4. Epiphysis well developed and a ventromedian ridge, hypapophysis present.

Note : Total number of lumber vertebrae in *Cavia* are seven only.

(g) Sacrum of *Cavia sp.* : State characters up to—

Class—***Mammalia*** as in above. Then write the following characters.

Characters :

1. Three to four sacral vertebrae are fused to from a compound triangular plate like structure having broad anterior and narrow posterior end.
2. Neural arch encloses a narrow neural canal.
3. Well developed neural spine is flattened and bridge like.
4. Presence of intervertebrarterial foramen.
5. The first sacral vertebra has well developed prezygapophyses.
6. Transverse process flat and laterally expanded has in its margin articular facet for ilium.

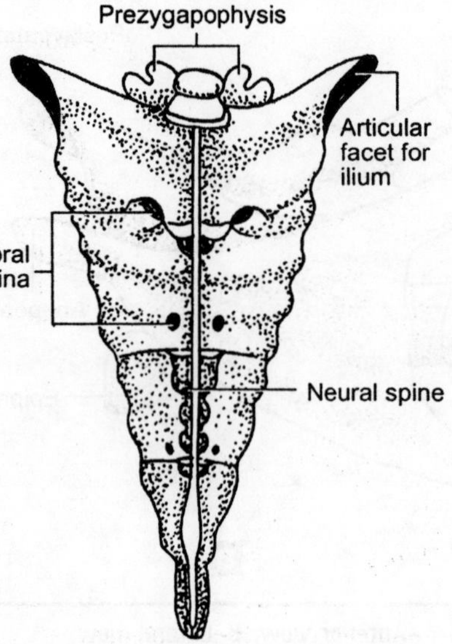

Fig. 2.1.52 : Sacrum of *cavia sp.* (Dorsal view)

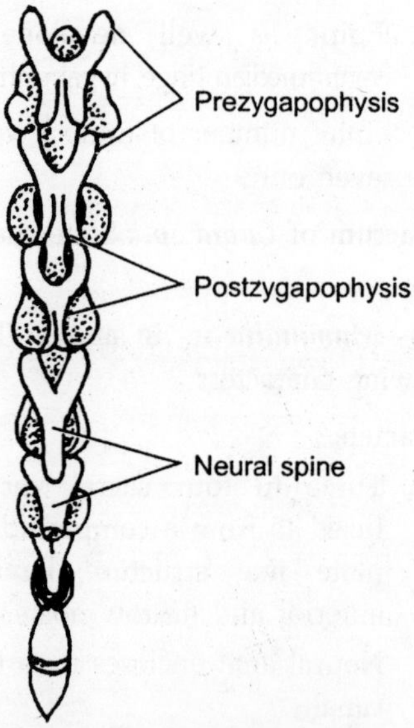

Fig. 2.1.53 : Caudal vertebrae of *cavia sp.*

7. Anapophysis absent, metapophysis is much reduced.

Note : Male *Cavia* has four sacral while female has three sacral vertebrae.

(h) **Caudal vertebra` of *Cavia sp.* :** State characters up to—

Class—***Mammalia*** as in above. Then write the following characters.

Characters :

1. Caudal vertebrae are seven in number and the size of the vertebrae decreases gradually from anterior to posterior end and last two are rod like.
2. Neural spine directed upward.
3. Transverse process absent.
4. Pre and postzygapophyses very small.

III. GIRDLES

The appendicular skeletons in all tetrapods show a basic similarity. There is a shoulder, or pectoral girdle to which the forelimbs are attached and a hip or pelvic girdle supporting the hind limbs. The pectoral girdle usually does not have any association with the vertebral column except through some muscles and ligaments. The pelvic girdles however attaches directly with the vertebral column. Each girdle is composed of two halves and each half is composed of three bones. In the pectoral girdle of the tetrapods the number of membrane bones has been reduced from the condition in fishes and just the clavicle and interclavicle may persist. The bones of the pectoral girdle typically consists of ventral coracoid, which meets the sternum radially, a scapula extending dorsally and a clavicle which lies on the ventral side between scapula and sternum and anterior to the coracoid. The pelvic girdle is composed of a ventral **ischium**, a dorsal **ilium** and an anterior **pubis**.

A. Pectoral Girdles :

(a) Pectoral girdles of *Bufo sp.* :

Characters :

1. Compound bony structure formed of bones and cartilage and consists of two similar

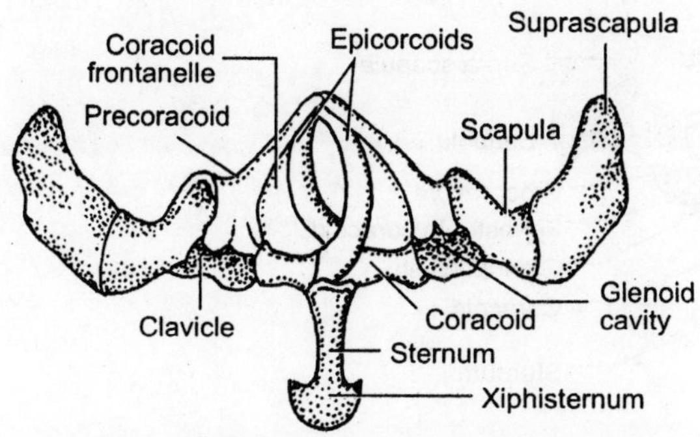

Fig. 2.1.54 : Pectoral girdle of *Bufo sp.* (Ventral view)

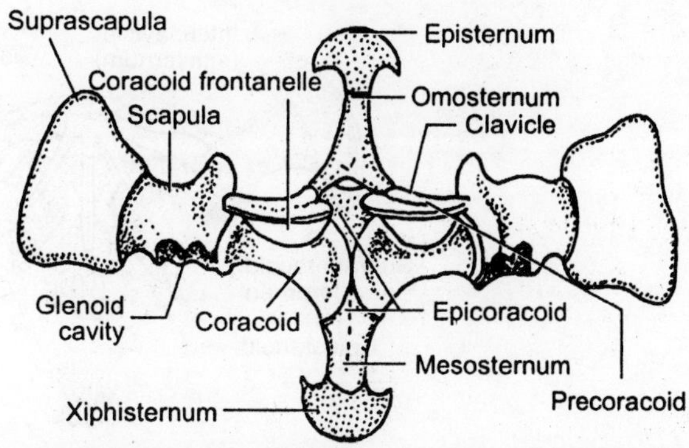

Fig. 2.1.55 : Pectoral girdle of *Rana sp.* (Ventral view)

halves unite at the mid-ventral plane but separated in the dorsal side.

2. Each half consists of dorsal **suprascapula** lateral **scapula**, ventral **coracoid** and ventral **clavicle.**
3. Suprascapula, a cartilaginous broad plate, attached to outer end of scapula, is present.
4. Scapula is a short plate like bone, slightly narrow in the middle but expanded at both ends.
5. Coracoid a short, stout and flat bone attached to scapula is present.
6. Clavicle is thin rod like, attached with scapula and **epicoracoid.**
7. Between scapula and coracoid **gleneoid cavity** is situated.
8. Presence of a thin strip of cartilaginous precoracoid which is separated from the coracoid by **coracoid frontanelle.**
9. At the junction of coracoid and epicoracoid, a small cartilaginous plate **sternum** is present.
10. At the terminal part of sternum is the **xiphisternum.**

(b) Pectoral girdle of *Rana sp.* :

Characters : State all the above characters except point eight, nine, and ten. Then write the followings.

1. Precoracoids do not over lap.
2. A Y-shaped **omosternum** and flattened cartilaginous episternum are present on the anterior part of sternum.
3. A thick bone bordered by the cartilage, **mesosternum** is present on the posterior part of sternum.
4. Below mesosternum long, flat cartilaginous **xiphisternum** is present.

(c) Pectoral girdle of *Calotes sp.* :

Characters :

1. Compound structure made of cartilage and bones consists of two equal halves, united at the mid ventral plane and firmly attached with T- shaped inter clavicle.
2. Each half is composed of supra scapula, scapula, coracoid, interclavicular and clavicle.
3. Suprascapula a flattened, cartilaginous plate articulating ventrally with scapula, while dorsal margin is free and curved.
4. Scapula is a flattened bony plate attached to suprascapula.
5. Coracoid is a flat bone partly ossified and partly cartilaginous having coracoid fenestra on the anterior side.
6. On the anterior portion of coracoid, a cartilaginous precoracoid and on the inner side of coracoid, epicoracoid are present.

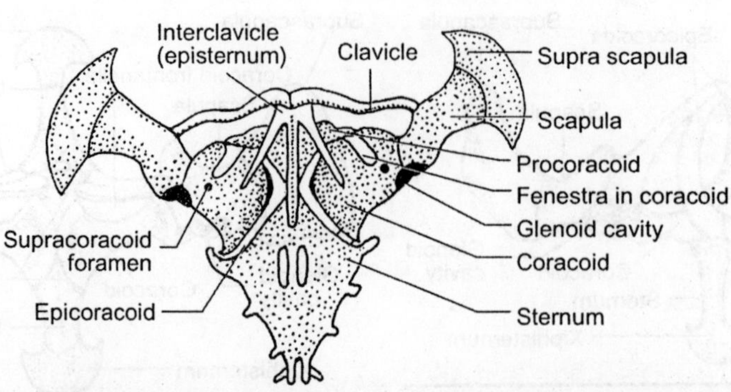

Fig. 2.1.56 : Pectoral girdle of *Calotes sp.* (Ventral view)

7. At the junction of scapula and coracoid, glenoid cavity is present.
8. Interclavicle is T- shaped investing bone between two halves of pectoral girdle.
9. The clavicle is a short and curved bone articulating with interclavicle and scapula.
10. Sternum a median rhomboidal cartilaginous plate bears four pairs of sternal ribs, pair of sternal foramen and posterior xiphisternum.

(d) Pectoral girdle of *Columba sp.* :

Pectoral girdle consists of ossified two similar half and is connected by a median **furcula.** Each half is constituted by **coracoid, scapula, jointed clavicle; interclavicle** and a huge **sternum.**

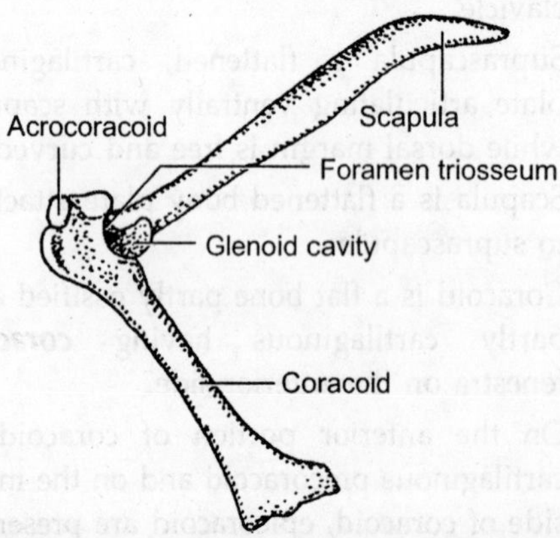

Fig. 2.1.57 : Half of pectoral girdle of *Columba sp.*

(e) Coracoid, scapula, clavicle of *Columba sp.* :

Characters :

1. Coracoid is a large, stout and pillar shaped bone, having a broad bone to articulate with the sternum.
2. At the distal end of the coracoid is a small **acrocoracoid process.**
3. Scapula is flat, sabre shaped bone, anteriorly it contains a cup shaped depression and on inner side it has acromian process.
4. The glenoid cavity is formed by scapula and coracoid.
5. On the dorsal side of the glenoid cavity, the acrocoracoid, acromion process and articular facet of furcula (clavicle) formed a small cavity called **foramen triosseum** is present.
6. Clavicle fused to form V- shaped furcula or **merry thought or wish bone.**
7. Clavicles is slender, thin and slightly curved bone.
8. Ventrally the clavicles unite to from interclavicle or **Hypocleidium** (episternum).
9. Each clavicle bears an articular facet on its free end.

(f) Furcula of *Columba sp.* :

Characters :

1. Very light, v-shaped bone.

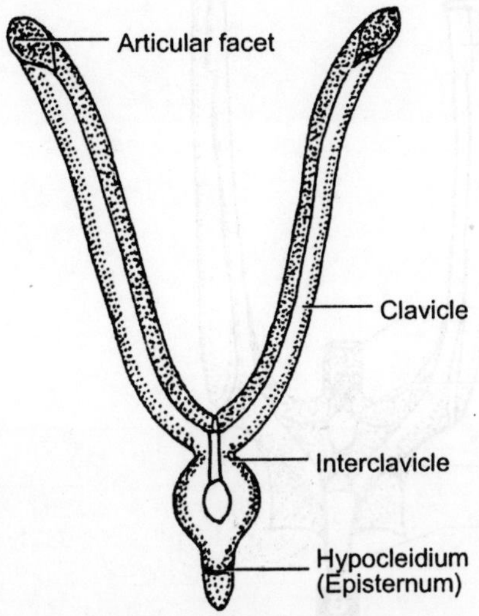

Fig. 2.1.58 : Furcula of *columba sp.*

2. Each arm of 'v' represents the slender, thin, and slightly curved clavicle which fused with its partner at the ventral side only to form inter clavicle.
3. Each clavicle bears an articular facet on its free end.

(g) Sternum of *Columba sp.* :

Characters :

1. Oblong bony structure having a dorsal plate known as **metasternum** and a ventral boat shaped plate called **keel or carina.**

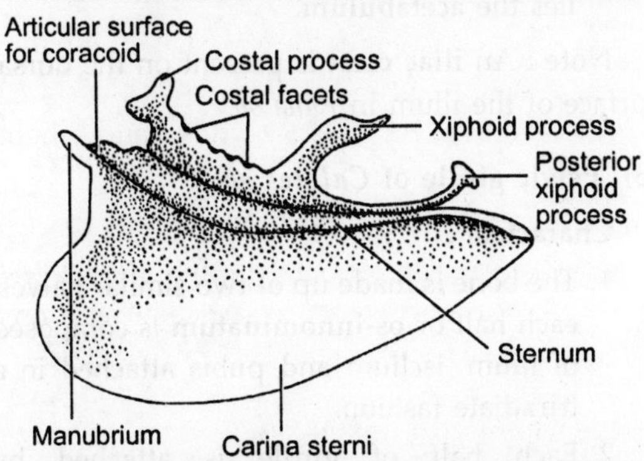

Fig. 2.1.59 : Sternum of *columba sp.*

2. Carina is broad anteriorly and gradually tapers posteriorly. At its anterior end it bears a process called **rostrum.**
3. Metasternum is concave-dorsally and convex ventrally and bears a pair of laterally placed grooves called **coracoid grooves** situated just above the rostrum.
4. Anteriorly metasternum bears a process called anterior metasternum process.
5. Metasternum also bears paried costal processes, five pairs of costal facets, a backwardly directed paired **xiphoid process.**

(h) Pectoral girdle of *Cavia sp.*

Characters :

1. Compound, thin, flat, plate like bone, each half of the girdle consists of suprascapula, spine and glenoid cavity.
2. Scapula is large and triangular plate, narrow posteriorly, broad and flat distally.
3. Suprascapula is present as a strip of cartilage along the broad margin of scapula.
4. Scapular spine lies on the dorsomedian side of the scapula.
5. At the proximal end of spine inner acromion and outer metacromion processes present.

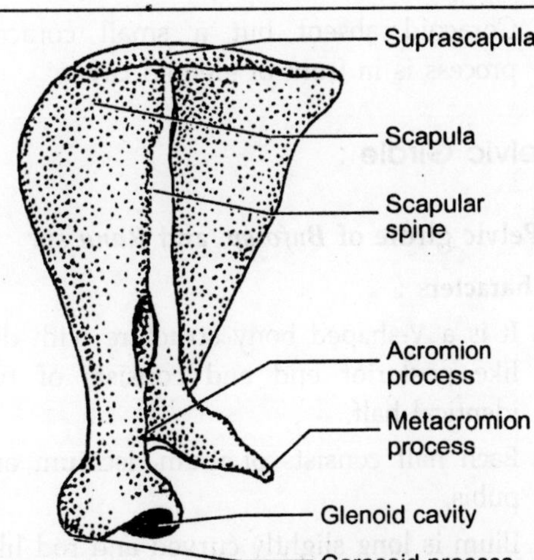

Fig. 2.1.60 : Pectoral girdle of *cavia sp.*

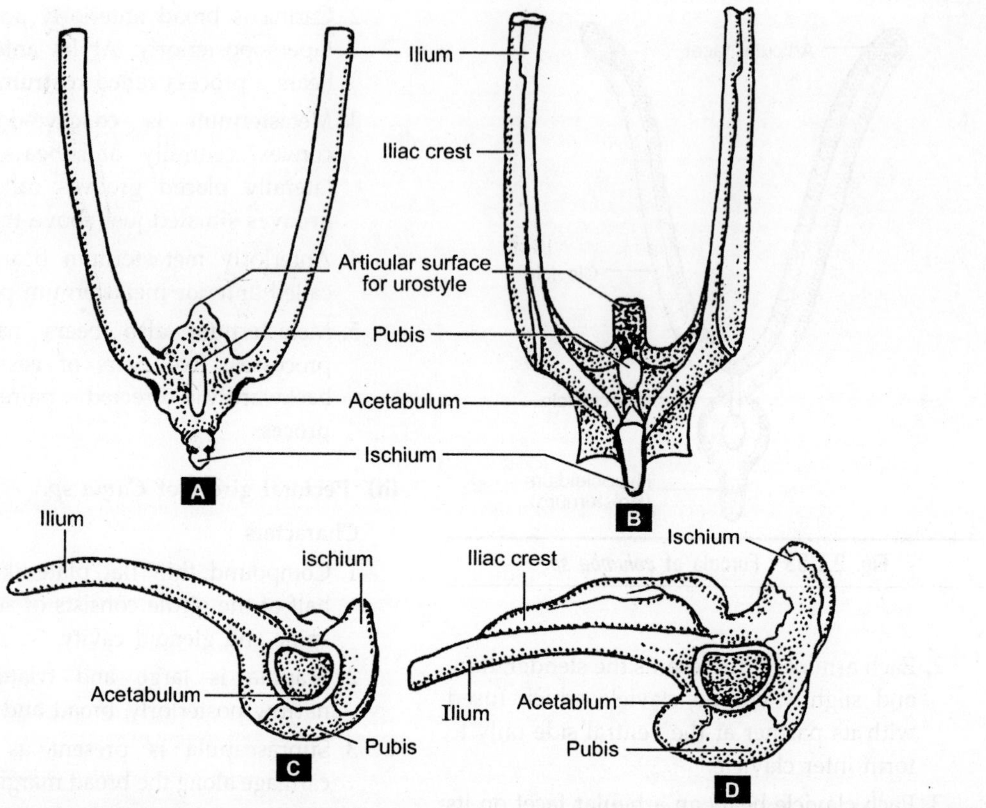

Fig. 2.1.61 : Pelvic girdle of *Bufo sp.* and *Rana sp.* AB–Ventral view, CD–Left side view, AC–*Bufo sp.*, BD–*Rana sp.*

6. At the apex of the scapula is the glenoid cavity.
7. Coracoid absent but a small coracoid process is in front of glenoid cavity.

B. Pelvic Girdle :

(a) Pelvic girdle of *Bufo sp.* and *Rana sp.*

Characters :

1. It is a V-shaped bony structure with disc like posterior end and consists of two identical half.
2. Each half consists of ilium, ischium and pubis.
3. Ilium is long slightly curved and rod like.
4. Ischium small, two ischia form the posterior part of the disc.
5. The flat posterior end of two ilia forms the anterior and dorsal part of the disc.
6. Pubis is small and cartilaginous, two pubis form the ventral part of the disc.
7. At the junction of ilium, ischium and pubis lies the **acetabulum.**

Note : An **iliac crest** is present on the dorsal surface of the ilium in *Rana sp.*

(b) Pelvic girdle of *Calotes sp.*

Characters :

1. The bone is made up of two similar halves, each half or **os-innominatum** is composed of ilium, isclium and pubis attached in a triradiate fashion.
2. Each half of girdle is attached by **ischiopubis symphysis.**

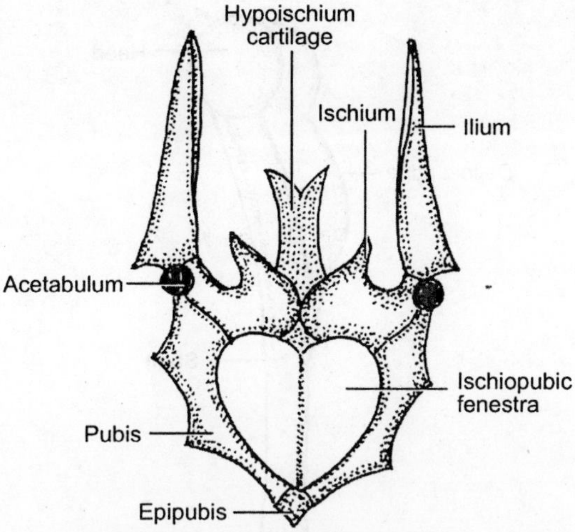

Fig. 2.1.62 : Pelvic girdle of *calotes sp.*

3. At the tip of pubis is cartilaginous epipubic and a pubic prominence at the other tip.
4. Between two is situated the hypoischium cartilage.
5. Acetabulum is situated close to the pubic prominence.
6. **Obturator foramen** is incomplete.

(c) Pelvic girdle of *Columba sp.*

Characters :

1. Compact bony element composed of ilium, ischium and pubis.
2. Ilium broad, thin, large and flat, concave anteriorly and convex posteriorly.
3. Ischium thin, broad plate fused with the ilium at the posterior end.

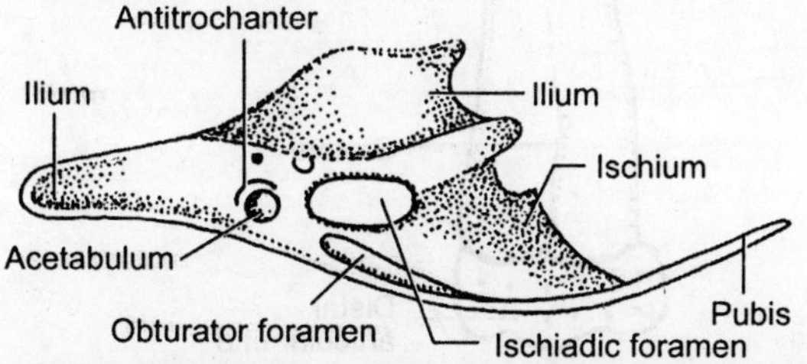

Fig. 2.1.63 : Half pelvic girdle of *columba sp.* (Lateral view)

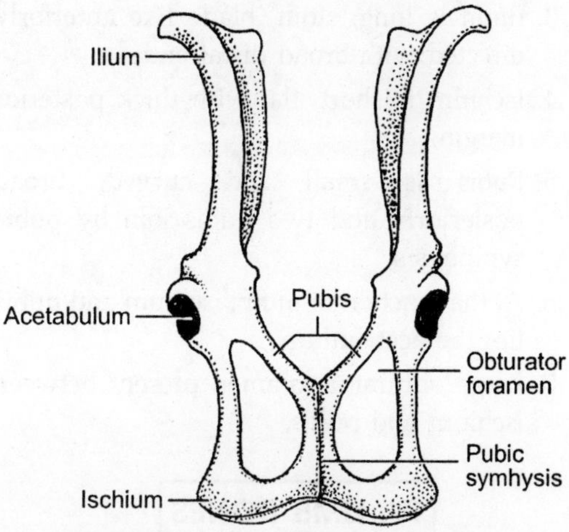

Fig. 2.1.64 : Pelvic girdle of *cavia sp.*

4. **Ilio-ischiatic foramen** separates ilium and ischium.
5. Pubis is slender, curved and long, parallel to the ischium and separated by an obturator foramen.
6. At the junction of ilium, ischium and pubis lies the acetabulum.

(d) Pelvic girdle of *Cavia sp.*

Characters :

1. Each half of girdle is known **os-innominatum** and consists of ilium, ischium and pubis.
2. Pubic symphysis joint each half.

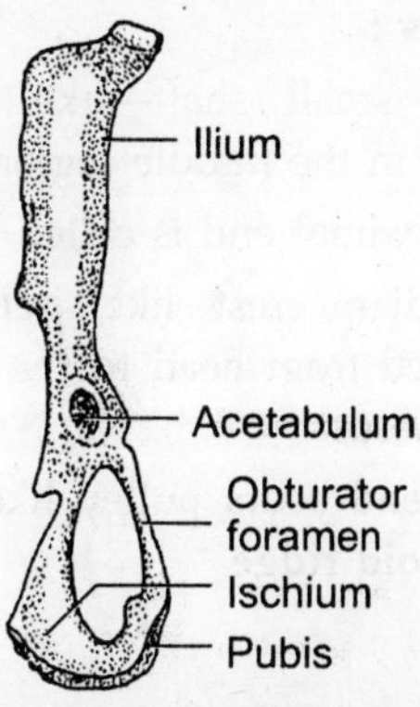

Fig. 2.1.65 : Os-innominatum of *cavia sp.*

3. Ilium is long, stout, blade like anteriorly directed with broad distal end.
4. Ischium is short, flat with thick posterior margin.
5. Pubis is small and curved, broad posteriorly and two pubis join by pubic symphysis.
6. At the junction of ilium, ischium and pubis lies the acetabulum.
7. Large obturator foramen present between ischium and pubis.

IV. LIMB BONES

The limbs include both the fore and the hind limbs are arranged on the same general pattern. In the forelimbs there is a single bone, the **humerus**, in the upper arm. The head of the humerus joins the pectoral girdle at the glenoid fossa. Two bones arranged parallel to each other are found in the fore arm. They are the lateral **radius** and median **ulna**. Corresponding bones of the hind limbs are the **femur**, which joins the acetabulum, **tibia** and **fibula**, comparable to the radius and ulna.

• Fore Limbs

A. Fore limb bones of *Bufo sp.*

(a) Humerus :

Characters :

1. Single, small shaft—like bone, slightly curved in the middle region.
2. The proximal end is called head.
3. A median crest like **deltoid ridge** is extended from head to the middle region of humerus.
4. Distal end bears pulley like **trochlea** and **condyloid ridge.**

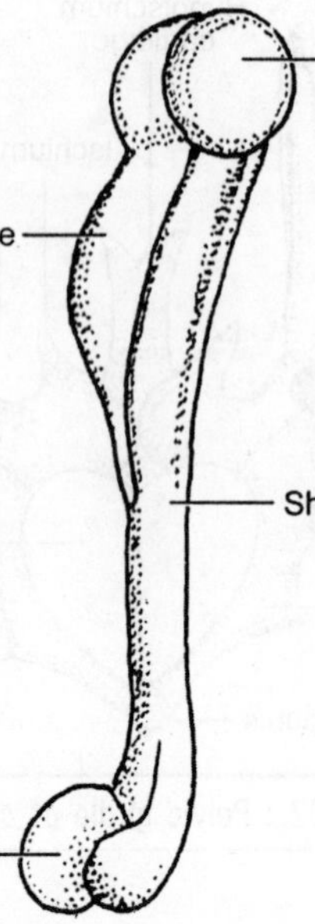

Fig. 2.1.66 : Humerus of *Bufo sp.*

(b) Radio—Ulna :

Characters :

1. It is formed by fusion of two separate bones radio and ulna.

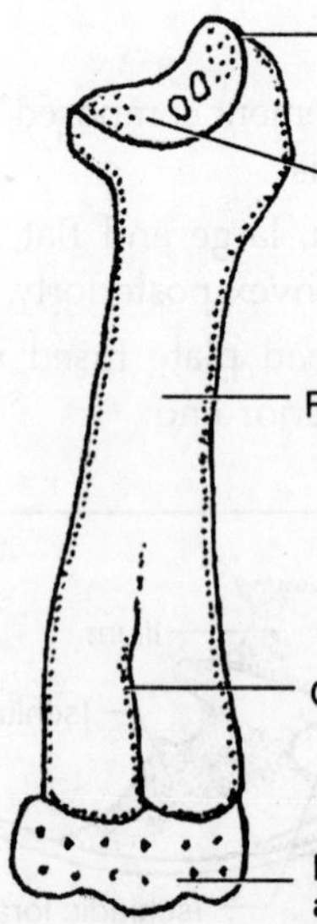

Fig. 2.1.67 : Radio-ulna of *Bufo sp.*

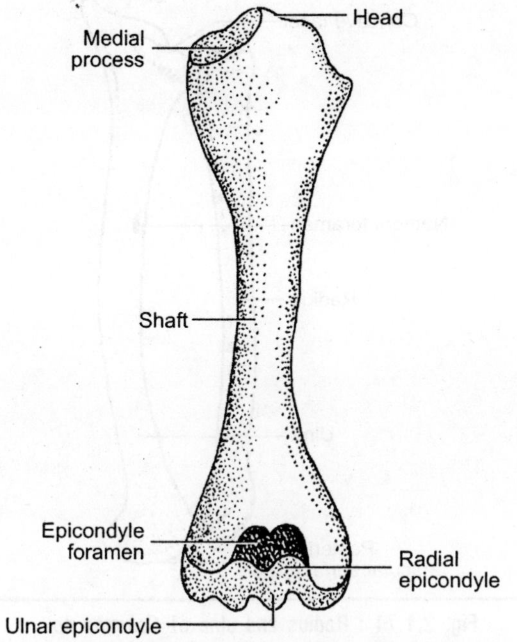

Fig. 2.1.68 : Humerus of *Calotes sp.*

2. Presence of a distal groove indicates the line of fusion.
3. The anterior end of radio – ulna bones a cup like concavity for accommodating the trochlea of humerus.
4. The anterior end of the cup is drawn out into an **olecranon process.**
5. Posterior articular end is flattened to give attachment of six **carpal bones.**

B. Fore limb bones of *Calotes sp.*

(a) Humerus :

Characters :

1. Slender and slightly curved bone with both ends slightly swollen.
2. Proximal end with a head and a tuberosity.
3. Distal end bears a pulley like trochlea with two articular condyles.
4. Deltoid ridge is absent.
5. Epicondylar foramen at the distal end.

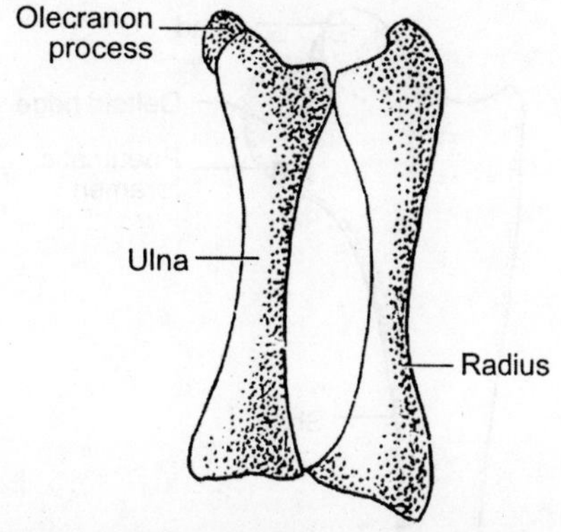

Fig. 2.1.69 : Radio-ulna of *Calotes sp.*

(b) Radio and Ulna :

Characters :

1. Two narrow shaft like elongated bones are approximated at the anterior and posterior ends but widely separated in the middle.
2. Anterior terminal end of stout ulna produces short olecranon process.
3. Both ulna and radius have convex articular facet at their distal ends.
4. From the distal end of preaxial radius **styloid process** is produced.

C. Fore limb bones of *Columba sp.*

(a) Humerus :

Characters :

1. Single, narrow, shaft–like strongly built bone with broad head.
2. Head contains one smaller pre-axial tuberosity or deltoid ridge and the postaxial tuberosity, which contains a large **pneumatic foramen.**
3. Distal end contains two trochlear articular surfaces for radius and ulna and separated by an intercondylar foramen.

(b) Radius and Ulna :

Characters :

1. Two long separated bones **radius** and **ulna**

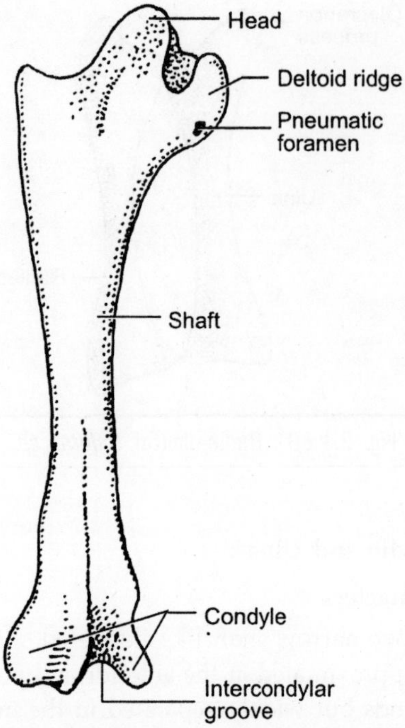

Fig. 2.1.70 : Humerus of *Columba sp.*

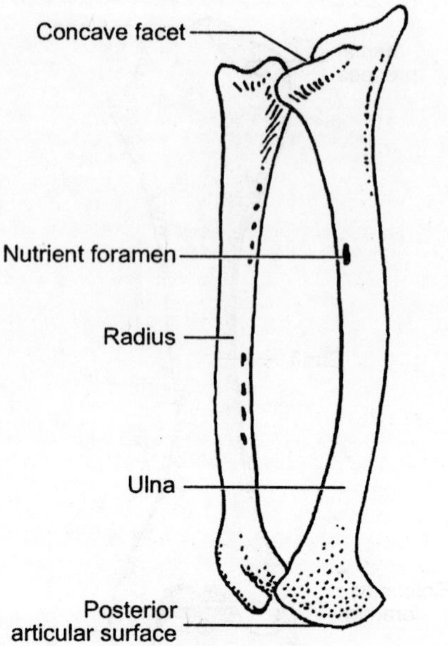

Fig. 2.1.71 : Radius and ulna of *Columba sp.*

join at their ends but separated widely in the middle.

2. Radius is straight and slender.
3. Proximal end of radius posseses articular surface.
4. **Ulna** is larger, stout and slightly curved.
5. Proximal end of ulna contains semilunar notch which is drawn into an olecranon process.
6. Radius posses a styloid process at the distal end.
7. Ulna contains a convex articular surface at the distal end.
8. Presence of nutrient foramen.

(c) Carpo-metacarpus :

Characters :

1. Three metacarpals fuse with distal row of carpals to form **Carpo–metacarpus.**

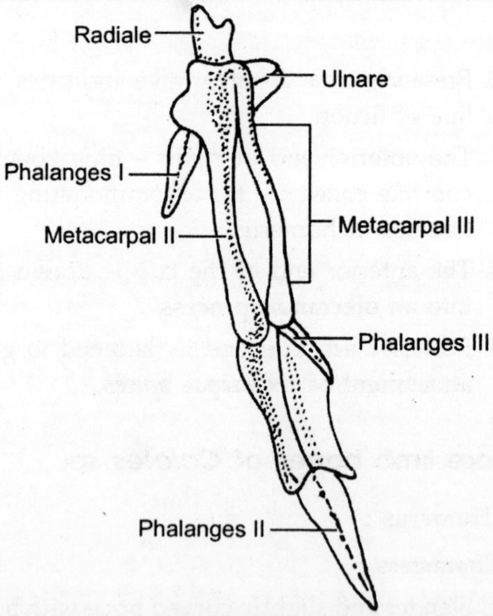

Fig. 2.1.72 : Carpometacarpus of *Columba sp.*

2. The first metacarpal is lump like projection while the other two are rod like.
3. Articulating with first metacarpal is a single pointed phalanx I.

4. The second metacarpal is straight and stout.
5. The third metacarpal is slender and slightly curved outwardly.
6. Proximal end bears fused radiale and ulnare bones.
7. Distal end bears articular surface to articulate **phalanges.**

D. Fore limb bones of *Cavia sp.*

(a) Humerus :

Characters :

1. Long shaft-like, rounded bone, the proximal end of which contains a round head.
2. Head contains greater and lesser tuberosities and a bicipital groove.
3. Deltoid ridge present below the head.
4. Distally humerus contains a pulley-like trochlea to articulate with ulna.
5. Just above trochlea a supratrochear foramen is present.
6. Presence of olecranon and coronoid fossae anterior to the trochlea.

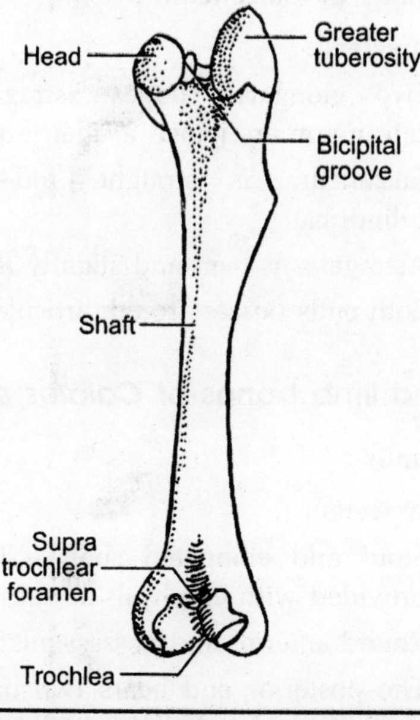

Fig. 2.1.73 : Humerus of *Cavia sp.*

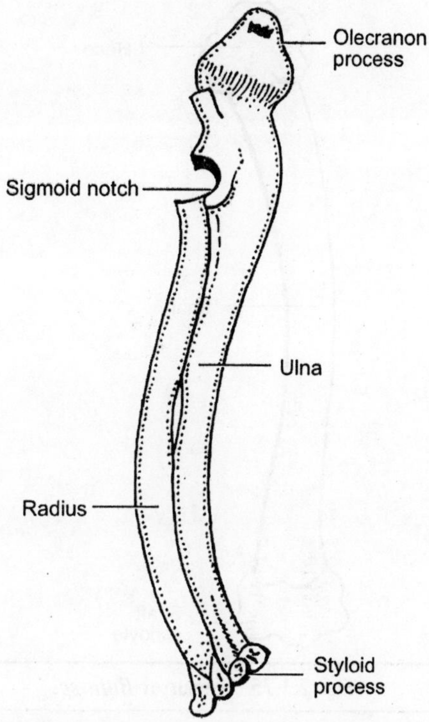

Fig. 2.1.74 : Radius and ulna of *Cavia sp.*

(b) Radius and Ulna :

Characters :

1. Radius and ulna, two long slender bones lying close to each other and united firmly at both ends.
2. The radius is small and slightly curved at the middle has a concave facet at the proximal end and a styloid process at the distal end.
3. Ulna is longer and slightly curved, having an olecranon process at the anterior end.
4. Ulna possesses a sigmoid notch at the neck.
5. The flat distal end of ulna possesses a styloid process.

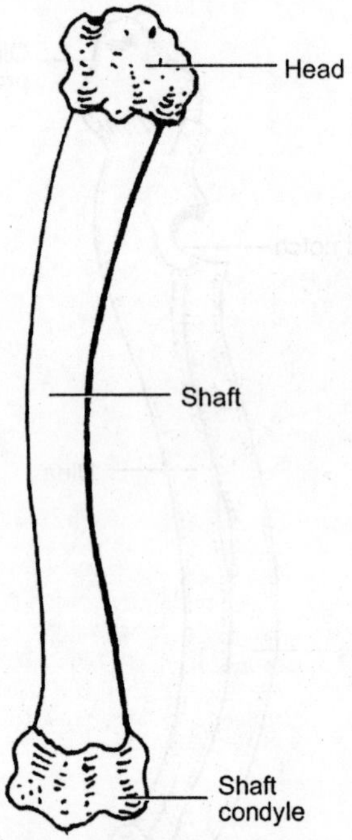

Fig. 2.1.75 : Femur of *Bufo sp.*

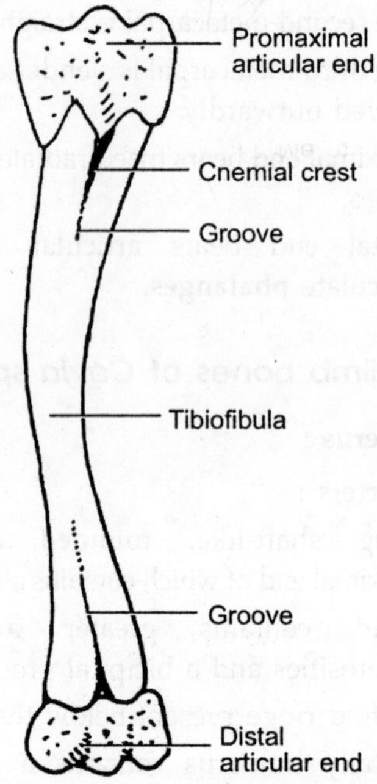

Fig. 2.1.76 : Tibiofibula of *Bufo sp.*

• HIND LIMBS

A. Hind limb bones of *Bufo sp.*

(a) Femur :

Characters :

1. Single, long shaft-like bone gently curved at the middle.
2. Both anterior and posterior ends of the bone are slightly flattened.
3. Proximal end possess a rounded head.
4. Slightly flattened distal end possess condyle with a concave articular facet.

(b) Tibio–Fibula :

Characters :

1. Two almost equal sized long bones, **tibia** and **fibula** fused to form a single shaft.
2. Proximal end possess a **cnemial crest**.
3. Both Proximal and distal end of the bone are expanded and possess small grooves indicate the fusion.
4. Both ends of tibia and fibula are expanded to articulate with the condyle.

(c) Astragalas Calcaneum :

Characters :

1. Two elongated bones **astragalas** and **calcaneum** are joined at their both ends.
2. Calcaneum is straight, rod–like and cylindrical.
3. Astragalas is bent and slightly flat.
4. Both ends possess rough articular surface.

B. Hind limb bones of *Calotes sp.*

(a) Femur :

Characters :

1. Stout and elongated shaft – like bone, provided with epiphysis at both ends.
2. Round anterior end possess a lateral head.
3. The posterior end bears two ball-shaped condyles and a shallow patellar grooves.

4. The anterior side of femur is provided with a **trochantar** and trochanteric groove.

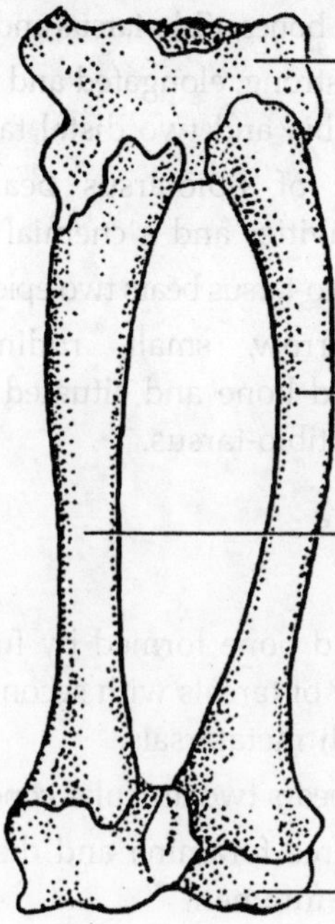

Fig. 2.1.77 : Astragalas calcaneum of *Bufo sp.*

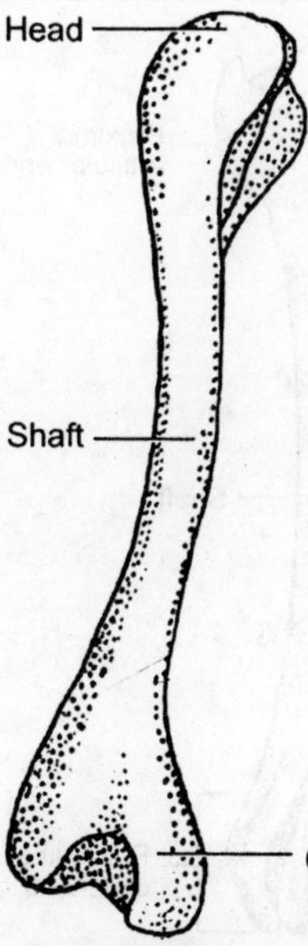

Fig. 2.1.78 : Femur of *Calotes sp.*

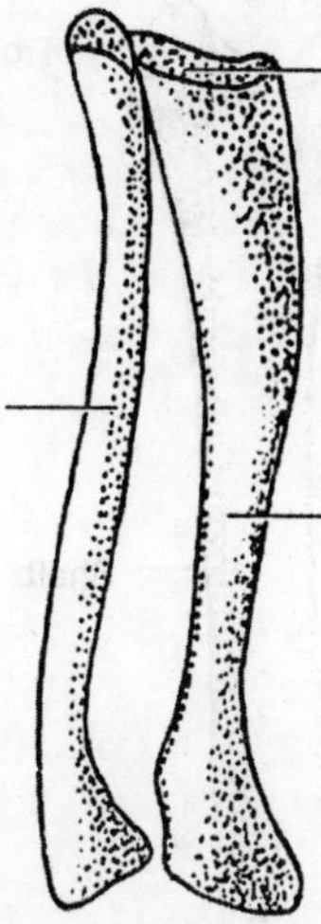

Fig. 2.1.79 : Tibia and Fibula of *Calotes sp.*

(b) Tibia – Fibula :

Characters :

1. Two slender, elongated bones, tibia and fibula separated but placed very close to each other.
2. Pre-axial tibia, straight and stouter bears a cnemial crest at it's anterior end and two concavities at the posterior.
3. Slender fibula bears single concavity at the anterior end.
4. Distal end of fibula is flat.

C. Hind limb bones of *Columba sp.*

(a) Femur :

Characters :

1. Single, cylindrical, short bone with broad ends.
2. Proximal end contains a rounded head and outwardly a greater trochonter with flat articular surface.
3. Distal ends bears two pulley like condyle separated by inter condylar fossa.

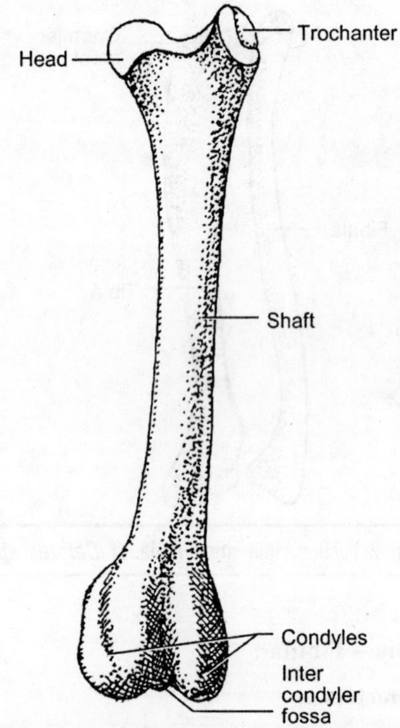

Fig. 2.1.80 : Femur of *Columba sp.*

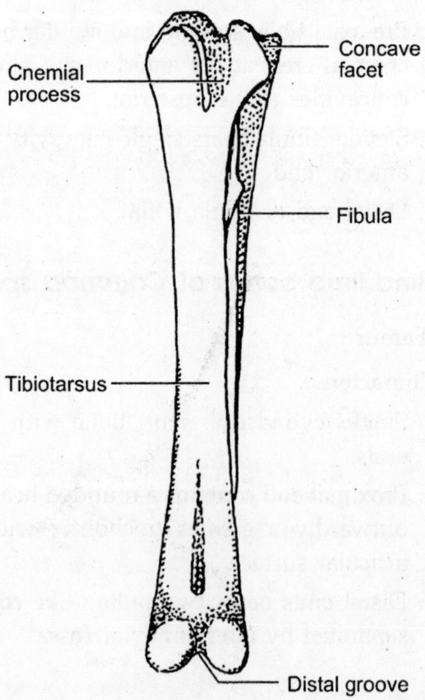

Fig. 2.1.81 : Tibiotarsas and Fibula of *Columba sp.*

(b) Tibio-tarsus and Fibula :

Characters :

1. Compound structure made of two partly fused elongated bones, **tibio-tarsus** and **fibula.**
2. Tibio-tarsus is strong, elongated and formed by fusion of tibia and two distal tarsals.
3. Proximal end of tibiotarsus bears two articular concavities and a cnemial crest.
4. Distal end of tibio-tarsus bears two epicondyle.
5. Fibula is narrow, small, rudimentary distally pointed bone and situated on the lateral side of tibio-tarsus.

(c) Tarso-metatarsus :

Characters :

1. Compound bone formed by fusion of distal row of tarsals with second, third and fourth metatarsals.
2. Proximal end bears two articular concavities.
3. Neck bears three foramina and distal end bears a single foramen.
4. Ventral side of the proximal end bears a crest.

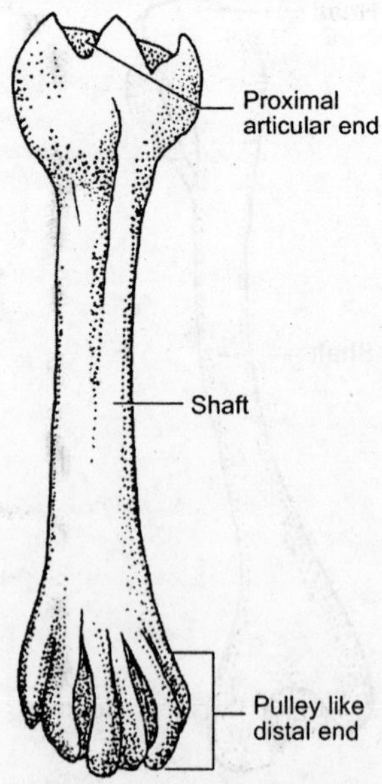

Fig. 2.1.82 : Tarsometatarsus of *Columba sp.*

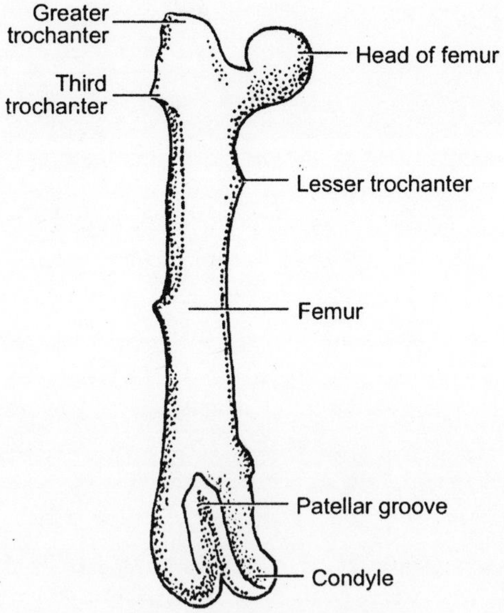

Fig. 2.1.83 : Femur of *Cavia sp.*

5. Three bones are separated by pulley like depression at their distal ends and each free end is itself pulley like.

D. Hind limb bones of Cavia sp.

(a) Femur :

Characters :

1. Single, elongated stout bone slighly curved at both ends.
2. Proximal end possess a prominent round head.
3. Three elevated areas, greater tochanter, lesser trochanter and third trochanter are present near the neck.
4. Presence of a lateral crest.
5. Distal end bears two round condyle separated by condylar notch or patellar groove.

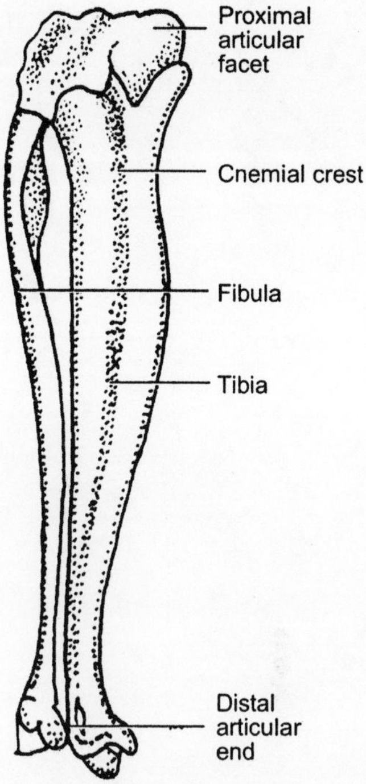

Fig. 2.1.84 : Tibia and Fibula of *Cavia sp.*

(b) Tibia and Fibula :

Characters :

1. Compound bony structure composed of two bones, **tibia** and **fibula** lying close together and attached at the ends only.
2. Tibia is stout, slightly bent, bears two articular facets at the proximal end and two irregular articular facets at it's distal ends.
3. Proximal ends of tibia bear a cnemial or tibial crest.
4. Fibula is slender and narrow toward the distal end and fused with fibula.
5. Distal ends of fibula bears articular surface.

Section 3

DEVELOPMENTAL BIOLOGY

SECTION AGENDA

The actual meaning of **embryology** is the study of embryos. The term embryo denotes the juvenile stage of an animal while it is within the egg membrane or in the body of mother. It would be artificial to limit the studies of the juvenile stages of animal life to the period before the animal is hatch or born. Therefore it world be proper to study the life history of an animal as a whole. Accordingly, the science of embryology can be interpreted as **the study of the development of animals.**

Every one of the higher animals starts life as a single cell, the fertilized ovum or **zygote**, which initiates the life of a new individual. The first phase in an individual's complete life history is called **ontogenic development**. Ontogenic development includes **embryogenesis** which is used to denote the development from the egg or **blastogenesis** which is used for the development of new individuals by means of asexual reproduction. Thus, the embryology or the developmental biology deals with the study of ontogenic development viz. the **embryogenetic development** or **blastogenetic development** of an individual (Balinsky, 1970). It is more usual, however, to regard embryology as the phase of development before **metamorphosis** in the amphibia, before hatching in birds, and before birth in mammals.

There are several trends of thought in embryology, in this edition we will be dealing only with the experimental embryology and descriptive embryology.

Experimental embryology uses experiment as method of investigation which seeks to find out why a process occurs at that specific time and in just that particular way.

The term **descriptive embryology** is applied to the method of embryological study concerned with the direct observation and description of embryological development.

Chapter 3.1

Experimental Embryology

(a) Observation of Live sperm under a microscope :

Live and healthy gametes are very essential for the process of fertilization and further development of the offsprings. Examination of the live gametes is very vital for the determination of the mortality rate, number of the gametes and morphology etc.

Materials required :

1. Phosphate buffered saline.
2. Neutral red/or other vital dye.
3. Watch glass
4. Petridish
5. Pasteur pipette
6. Forceps
7. Scissors
8. Bovine or goat serum
9. Male and female matured albino rats

Method :

1. Male rats are properly anaesthetized with anaesthetic ether. The scrotum is opened. The epididymis is incised out from the testis. Epididymis is kept in PBS at 37°. 10% Bovine Fetal calf serum/Goat serum is added into it to keep it healthier. Next with the help of a 25 G needle fitted on a hypodermic syringe inner part of the epididymis is flashed out to procure maximum number of sperm.

2. In case of female gamete, a female rat is dissected out, the ovary is collected including the fallopian tube and kept in PBS. Next with the help of a Pasteur pipette PBS is flashed so that the mature eggs may be disassociated.

3. After collection of both the gametes only clear part of the PBS is obtained in two different test tubes, i.e. visibly junk materials should be avoided.. This is taken in a centrifuge tube to prevent clumping the gametes are centrifuged. A minimum volume for centrifugation is necessary so that a good number of gametes could be observed under a microscope. After the centrifugation is over desired volume of PBS is left in the tube and the supernatant is discarded. The suspension is further mixed properly and a drop of it is taken on a clean glass slides separately for each suspension containing male and female gametes.

4. Vital dye can also be added to the suspension to distinguish the gametes. The gametes are observed under a compound microscope.

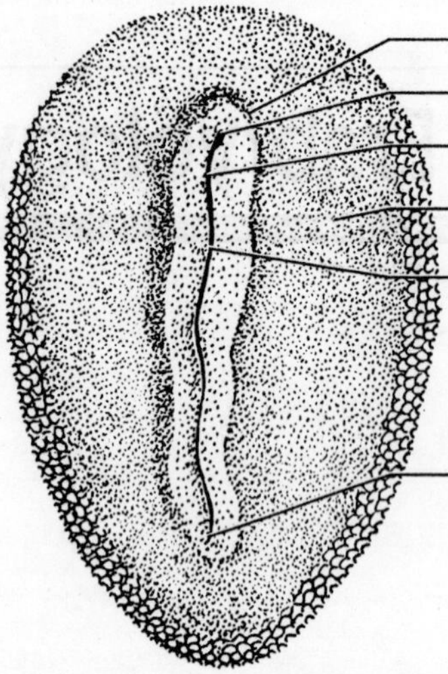

Fig. 3.1.1 : 16-hours stage of entire chick embryo.

(b) Staining and mounting of chick embryo :

Method :

1. A fertilized egg of chick is taken in the early hours of incubation, especially in the 16- 20 hours stage.

2. The shell membrane of the egg is broken into two halves through the middle striking it gently with the handle of a scalpel. Next the two ends of the egg are held and separated the egg by pressing with thumb.

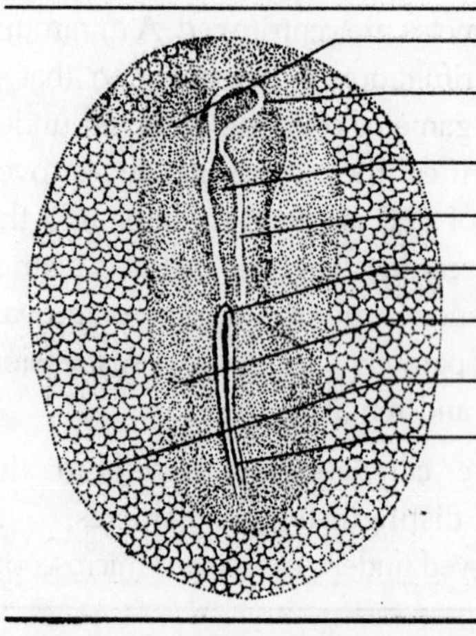

Fig. 3.1.2 : 19-hours stage of entire chick embryo.

3. The whole materials of the egg are released without damaging **the vitelline membrane** around the egg in a clean Petri dish containing Ringers' solution. The embryo appears as a small white body on the surface of the yolk at the centre.

4. The vitelline membrane is held with a pair of forceps and the entire **germinal disc** is dissected with a fine scissor and the germinal disc along with the growing embryo is removed from the yolk by directing a stream of saline solution between the embryo and the yolk with a small pipette. The embryo is then placed in a clean watch glass containing Ringers' solution. The embryo should be lying in the watch glass with the yolky side down. One has to be very careful while removing embryo of less than 18 hours of duration.

5. The solution is decanted and the embryo is fixed in aqueous Bouins solution. The fixation time for embryo up to 18 hours of incubation is four hours. Fixation is done by dropping the fixative over the **blastoderm** gently using a pipette. The blastoderm is floated by injecting the fixative below it.

6. The embryo should be thoroughly washed in distilled water (three changes for one hour duration). The yellow colour of the picric acid in the Bouins can be removed by adding lithium carbonate in alcohol series used for dehydration. Alternatively addition of 3 ml of ammonium hydroxide to 100 ml of 70% alcohol produces a solution.

7. Cedar wood oil is used for clearing the embryo. It requires about twenty four hours for the embryo to become translucent.

8. The oil is removed by transferring the embryo to xylol. Three changes of forty minutes each in xylol are normally enough. The embryo can know be mounted in DPX or Canada balsam.

9. When the embryos are required to be stained, the stains used can either be Haematoxylin or Borax carmine.

- **For staining with Haematoxylin** : After washing the embryos in water and removing the picric acid, it is washed in water for two hours.. Next the embryos are put in 4% iron alum for twelve hours. Washed briefly in water for ten minutes and stained in Heidenhain's Haematoxylin for six hours or more. If overstained it can be destained with 2% iron alum. When there is a uniform greyish colour remains, the embryo should be removed from iron alum and should be kept in slowly running tap water. This will stop the destaining process.

  For staining with Borax carmine, the dehydration process has to be started. After reaching the 70% alcohol, the embryo has to decolourised with ammonium hydroxide. After the yellow colur is removed, the embryo is transferred to 70% alcohol (for ten minutes) and then into Borax carmine (for one hour).

- The embryo is then dehydrated through 70%, 80%, 90%, 95% and absolute alcohol (leave for five minutes in each change).

10. Cleared and mounted in DPX or Canada balsam.

(c) Preparing sections of chick embryos :

The procedure for removing and fixing the blastoderm from the egg is same as section 3.1 b. Procedure carried out uptill clearing in cedar wood oil, next embedded in paraffin wax. Sectioning is done at 10μ thickness. Fixing and staining is done as point (5-9) of section 3.1.b. (see also section 4.5.12).

Chapter 3.2.1

Identification of Embryonic Stages of Frog

(a) Staining and mounting of toad/frog embryo :

Toad/frogs breeds during rainy season. The eggs are deposited either as compact masses (frog) or in strings (toads). The eggs are laid in water in gelatinous aggregations. Each egg has a black and white hemisphere. The former is directed upwards and surrounded by a sphere of jelly.

Method :

1. The fertilized eggs are collected from water.

2. These collected eggs are brought to the laboratory and placed on a clean tray.

3. The jelly coat is divested and the eggs are cultured in Holtfreter's saline.

4. Some gastrula are collected and fixed in aqueous Bouin's fluid for study.

5. Rest of the eggs are kept in fresh Holtfreter's saline and cultured further until they are hatched. After fifteen days of the zygote formation the eggs hatch into larvae.

6. The hatched out embryo are transferred in fresh tap water for further culture.

7. These larvae are fed liberally with boiled spinach and Basela leaves. Water was changed twice to check cannibalism.

8. Different stages of embryo and larvae are collected from the culture and fixed them in aqueous Bouin's fluid for study.

9. For whole mount study the embryo or larvae is put in a clear watch glass and washed throughly with 70% alcohol to remove picric acid.

10. Hydrated with distilled water and stained in light haematoxylin if required, dehydrated and mounted. (See chapter 4.5)

11. To study the histological differentiation, the usual procedure is followed for microtome sectioning at 8 microns thickness.

12. The section are stained with dilute haematoxylin,dehydrated and mounted in Canada Balsam or DPX. (See chapter 4.5)

(b) Whole mount of frog embryo at different stages :

1. **Early gastrula :**
   - (i) Embryo becomes spherical.
   - (ii) It is covered externally by several layers of small, black cells, constitute

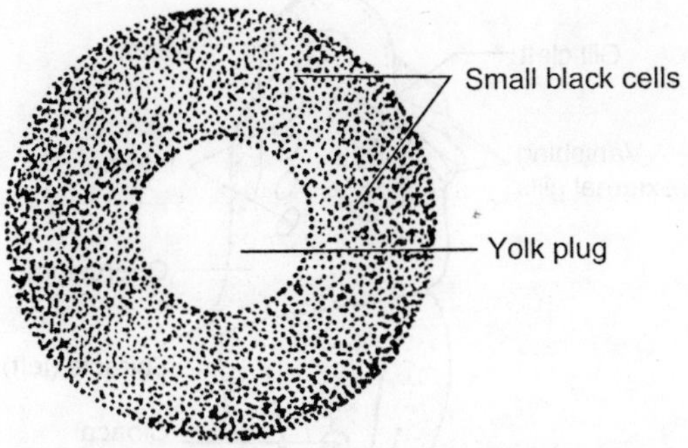

Fig. 3.2.1.1 : Early gastrula of frog.

ectoderm and containing **endo dermal** and **mesodermal** components internally.

(iii) **Blastopore** is complete.

(iv) **Yolk plug** is observed.

2. **Late gastrula :**

(1) Blastopore is very small.

(ii) Neural plate area becomes evident as neural folds begin their elevation.

3. **Neurula :**

(i) Embryo become oval shaped.

(ii) Blastopore has closed.

(iii) Neural folds are well developed.

(iv) **Neurenteric passage** way between neural folds and dorsal evagination of archenteric space is visible.

(v) New caudal opening is forming.

(vi) Tail rudiment elevation is indicated.

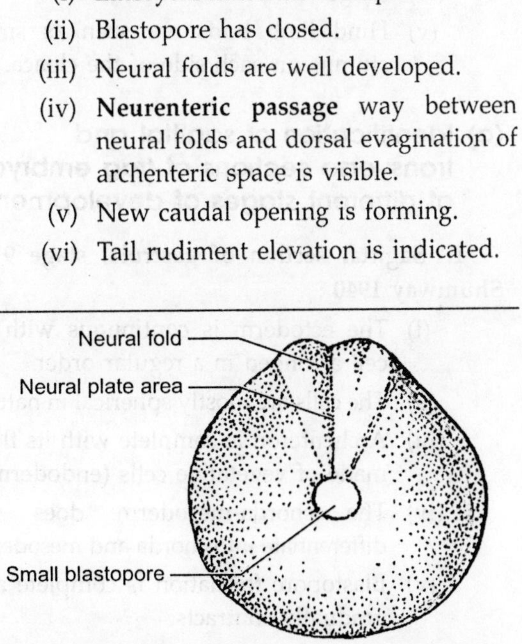

Fig. 3.2.1.2 : Late gastrula of frog.

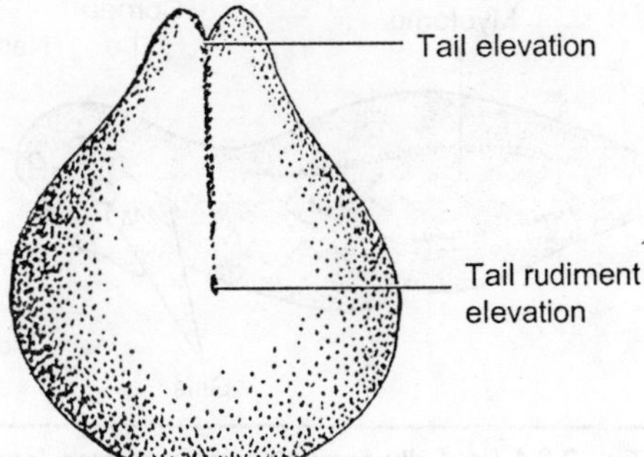

Fig. 3.2.1.3 : Neurula of frog.

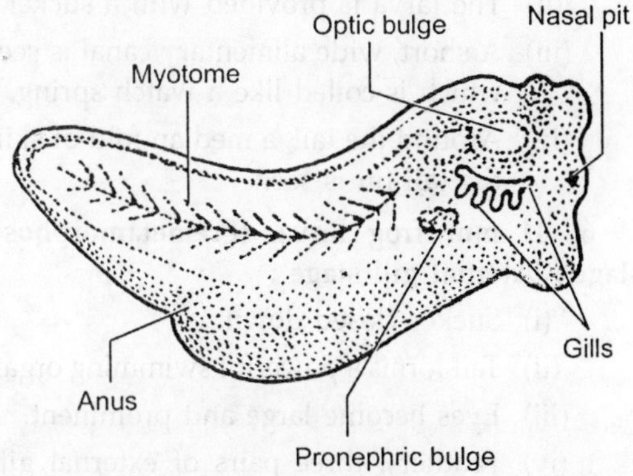

Fig. 3.2.1.4 : Tail bud stage of frog embryo. (4 mm embryo stage)

(vii) Proctodaeal opening and tail rudiment are seen.

4. **4 mm. Frog embryo, stage 19 (Shumway, 1940). Tail bud stage :**

(i) Dark coloured larva is with out limb and mouth.

(ii) Larva is provided with a prominent head and a fusiformed body.

(iii) Tail is short and compressed.

(iv) Six vertical ridges on each side of the head and neck are seen.

5. **7mm. Frog larva stage (Shumway, 1940). Fully formed tadpole :**

(i) Fish like young tadpole with a mouth surrounded by horny jaws.

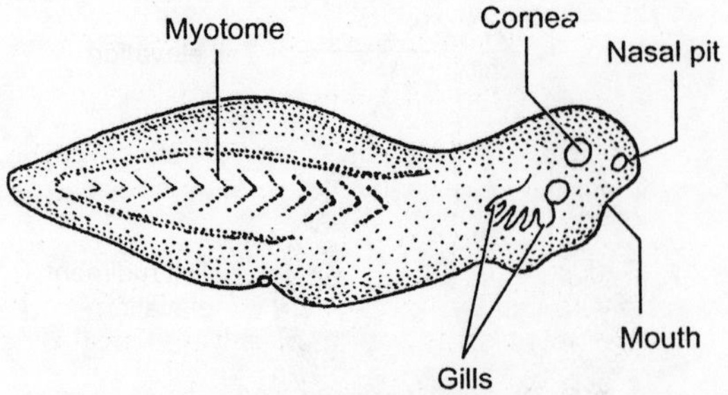

Fig. 3.2.1.5 : Fully formed tadpole of frog larva. (7 mm larval stage)

(ii) The larva is provided with a sucker.
(iii) A short, wide alimentary canal is seen which is coiled like a watch spring.
(iv) Around the tail, a median fold of skin, the tail fin is seen.

6. **11 mm frog larva. (Premetamorphosis stage). External gill stage :**
   (i) Sucker closed down.
   (ii) Tail forms a powerful swimming organ.
   (iii) Eyes become large and prominent.
   (iv) Feathery, three pairs of external gills are seen.
   (v) Eye rudiment is visible.

7. **Hind limb stage :**
   (i) Sucker disappeared.
   (ii) External gills are lost and internal gills are appearing.
   (iii) Operculum is seen on each side of the head.

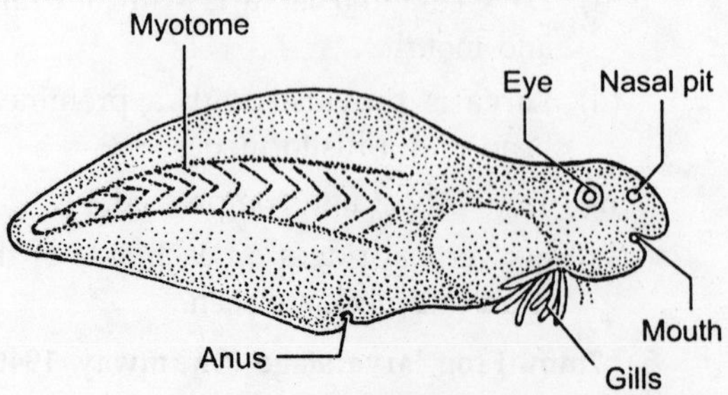

Fig. 3.2.1.6 : External gill stage of frog. (11 mm long tadpole larva)

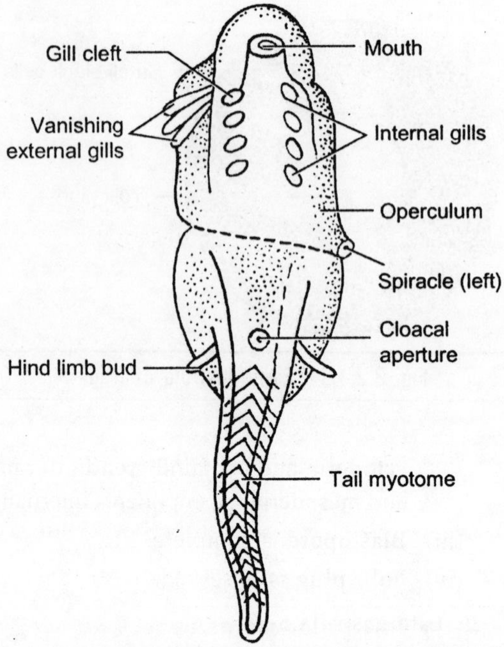

Fig. 3.2.1.7 : Hind limb stage of frog tadpole. (15 mm long larva)

   (iv) Large and prominent eyes.
   (v) Hind limb buds are seen as small stumps on each side of the cloaca.

(c) Identification of sagittal and transverse sections of frog embryo at different stages of development :

1. **Sagittal section of gastrula, stage 9 of Shumway 1940 :**
   (i) The ectoderm is continuous with its cell arranged in a regular order.
   (ii) The cells are mostly spherical in nature.
   (iii) Archenteron is complete with its floor made of vegetative cells (endoderm).
   (iv) The chorda-mesoderm does not differentiate into chorda and mesoderm.
   (v) Blastopore formation is complete and gradually contracts.
   (vi) The yolk plug is seen.

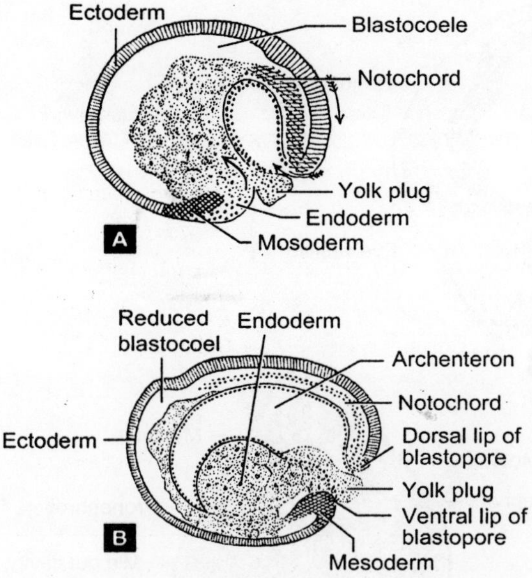

Fig. 3.2.1.8 : Sagittal section of gastrula of frog showing the cell arrangement. A–Early gastrula, B–Late gastrula.

2. **T.S of neurala, stage 16 of Shumway 1940 :**
   (i) The neural tube is established.
   (ii) The neural tube contains a compact mass of cells particularly surrounds the **neurocoel.**
   (iii) The ectoderm is continuous and is still maintains contact with the neural tube.
   (iv) Optic vesicle is formed.
   (v) Optic vesicle comes in contact with the outer ectoderm to from the lens placodes.
   (vi) Chorda-mesoderm is differentiated into notochord, somites and lateral plate mesoderm.
   (vii) Gut formation is complete.
   (viii) Yolk platelets are also seen.
3. **T.S of tail bud embryo, stage 19 of Shumway, 1940 :**
   (i) The neural tube is with a large neural cavity.

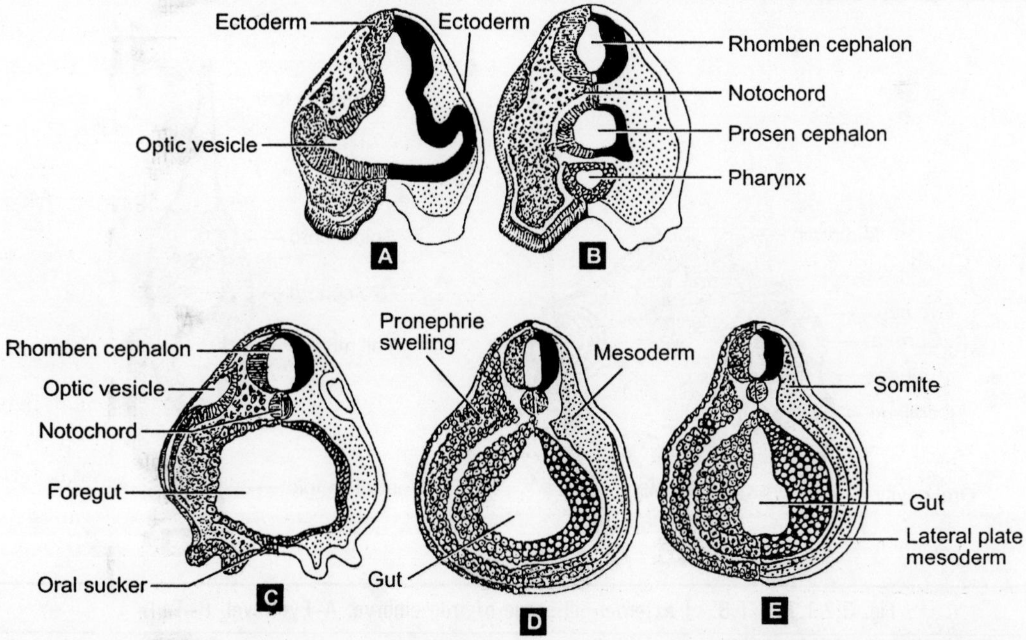

Fig. 3.2.1.9 : T.S. of neurula of frog. A-E–Head to trunk region.

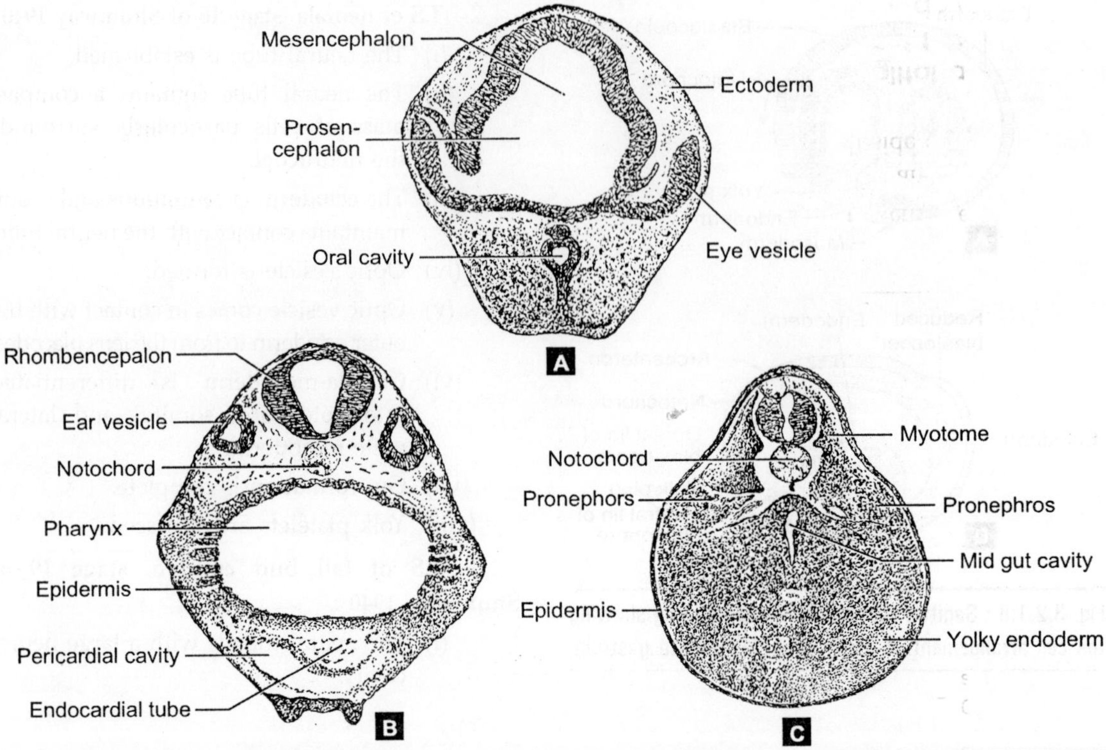

Fig. 3.2.1.10 : T.S. of tail bud stage of frog embryo. A–Eye rudiment, B–Ear vesicle and heart, C–Pronephros.

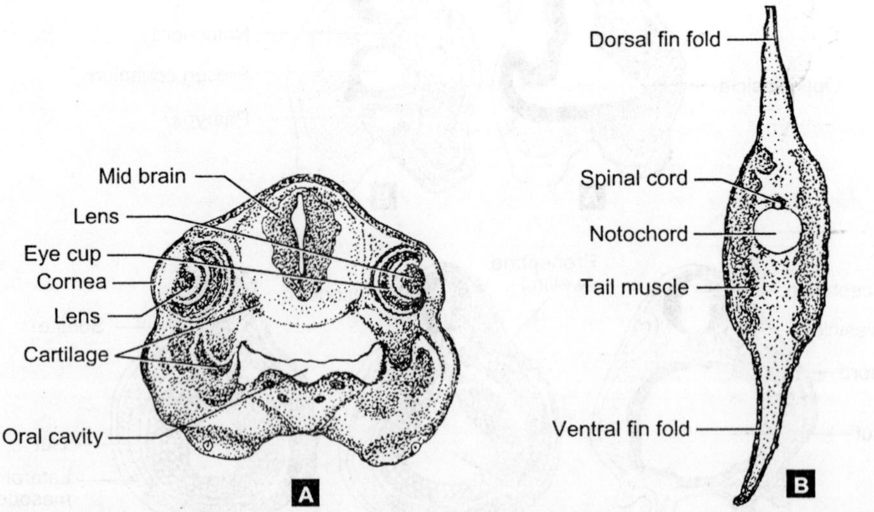

Fig. 3.2.1.11 : T.S. of external gill stage of frog embryo. A–Eye lavel, B–Tail.

(ii) Neural tube is separated from the ectoderm.

(iii) Lens is formed and is separated from the ectodem.

(iv) A large number of mesenchyme cells are seen round the neural tube.

(v) Auditory placodes are seen.

(vi) The development of nephric tubules and gut has taken place.

4. **T.S. of external gill stage :**

(i) In the neural tube the peripheral mantle and the germinal layers are established.

(ii) Eye cup and lens are seen.

(iii) Complete gut.

(iv) The notochord is large, conspicuous and vacuolated.

(v) The heart becomes divided into chambers.

(vi) The nephric tubules and ducts are very prominent.

(vii) In the tail, the formation of the neural tube is complete and spinal chord is seen.

(viii) Presence of dorsal and ventral fin fold.

Chapter 3.2.2

Identification of Embryonic Stages of Chick

A. Structure of the hen's egg

The fully formed egg contains a large amount of yolk and it is called **polylecithal** (**macrolecithal**) and **telolecithal**. The active cytoplasm at the animal pole is very little and is in the form of a small disc with a zygote nucleus and is called **germinal** or **blastodisc**. Most of the space of animal pole and vegetal pole of the egg is occupied by the **yolk**. The yolk and disc are bounded by a plasma membrane.

After fertilization the ovum is surrounded by various membranes, added to the delicate **vitelline membrane**. The fully formed and laid egg is surrounded on the outer side by a **calcareous shell**. Beneath the shell, there are two layers of shell membrane, an **outer shell membrane** and an **inner shell membrane**. In between the shell membranes are air space. Beneath the shell membrane is the **albumen** or white of egg. In it lie two spirally twisted cords or **chalazae**, one chalazae are situated on the broad side of the egg while another lie at the narrow end of the egg. Cleavage starts in the germinal disc. The cleaved celles are called **blastomeres**. The central mass of blastomeres lying above the subgerminal cavity and is called

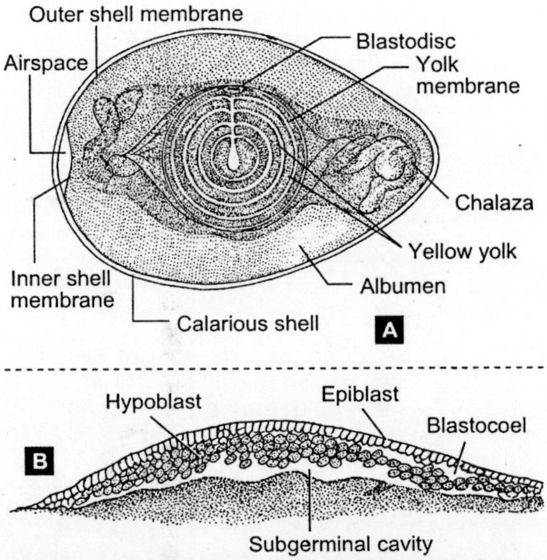

Fig. 3.2.2.1 : The egg of hen and its membrane. A–Structure of hen's egg, B–L.S. through blastoderms.

blastoderm, contains formative or embryonic cells which later on form the embryo proper. Cleavage converts the germinal disc into a disc shaped blastula, called **discoblastula** which floats atop the yolk mass. The discoblastula has an anterior **epiblast** and inner **hypoblast**, both enclosing a blastocoel in the central area. The

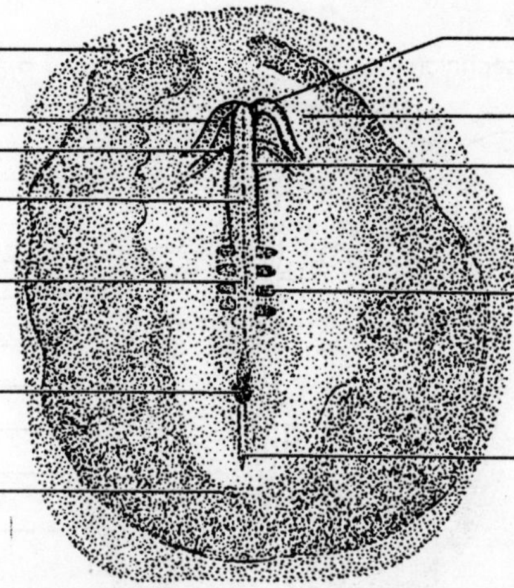

Fig. 3.2.2.2 : 24 hours stage of entire chick embryo.

central area of discoblastula consists of a translucent area, called **area pelucida** and the peripheral area of discoblastula is opaque, called **area opaca.**

A. Whole mounts of Chick embryo at different stages :

1. **24 hour (First day) stage :**

(i) The area opaca is divided into **area vasculosa** and **area vittelina.**

(ii) Abundant blood islands within the area vasculosa.

(iii) The space between the head process and the pro-amnion is subcephalic pocket.

(iv) Foregut is established and the anterior intestinal portal the opening of the foregut also appears.

(v) Neural folds are very prominent.

(vi) In the posterior region there is a faint **primitive streak.**

(vii) Between the neural pore there is the anterior **neuropore.**

(viii) Notochord is present.

(ix) Distinct head fold situated above the pro-ammion.

(x) Somites are about four pairs and well-formed.

2. **33 hour Stage :**

(i) Neural folds unite dorsally and the neural tube is formed.

(ii) At the anterior end there is a little ventral flexion of the head process.

(iii) Optic vesicles are well developed.

(iv) The vesicles of the fore brain, mid-brain and hind-brain are formed.

(v) Distinct notochord is present.

(vi) **Heart** is thin wall and 'S' shaped.

(vii) There are two large **omphalomesenteric veins.**

(viii) Indication of the development of the aortic arches is evident.

(ix) Pellucida, opaca, vasculosa and vitellina areas are easily distinguishable.

(x) Primitive streak is visible at the posterior end.

(xi) Twelve pairs of somites are present.

3. **48 hour stage :**

(i) Embryo shows cranial flexure and twisting of head over the right side.

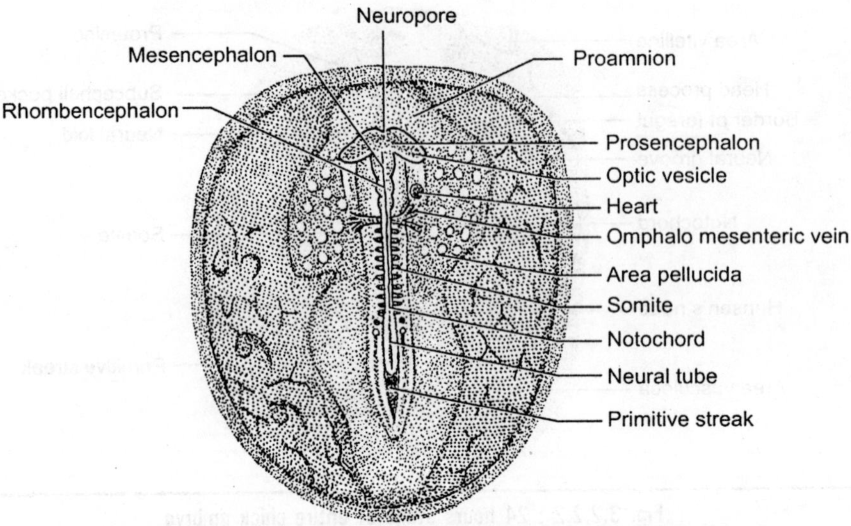

Fig. 3.2.2.3 : 33 hours stage of entire chick embryo.

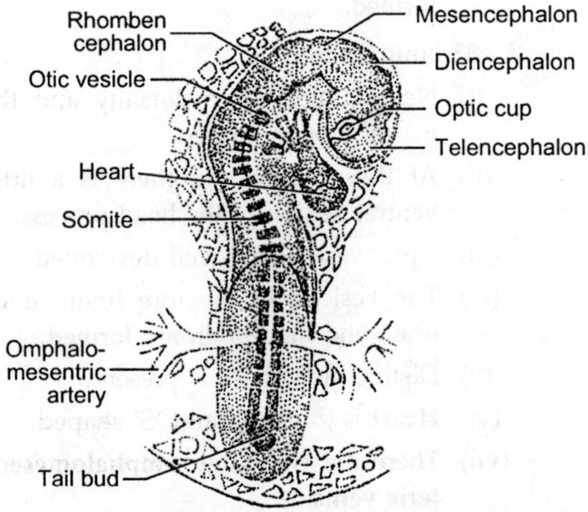

Fig. 3.2.2.4 : 48 hours stage of entire chick embryo.

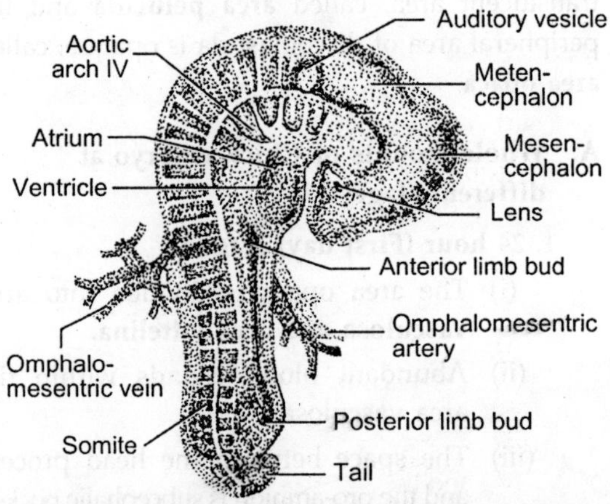

Fig. 3.2.2.5 : 72 hours stage of entire chick embryo.

(ii) Fore, mid and hind brain distinct and fore brain takes the anterior most places.

(iii) Heart is differentiated into ventricular, arterial and sinus region.

(iv) Second aortic arches develop.

(v) Vitelline veins and the arteries are distinct.

(vi) **Somites** are 26 pairs.

4. **72 hour stage :**

(i) Torsion as well as flexure is increased thus bilateral symmetry is lost.

(ii) Lumbar flexure and caudal flexure in the tail region.

(iii) A four pair of pharyngeal pouches have appeared.

(iv) First four aortic arches visible.

(v) Area vasculosa considerably expanded

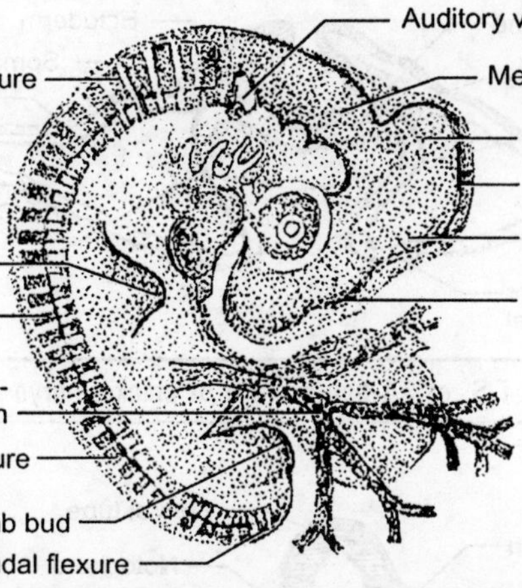

Fig. 3.2.2.6 : 96 hours stage of entire chick embryo.

and vitelline arteries and veins are well developed.

(vi) Eye develops *lens*.

(vii) Mesodermal somites are thirty five pairs.

(viii) Limb buds are visible.

5. **96 hour stage :**

(i) Flexure and torsion are much more complete and gives the embryo a 'C' shaped appearance.

(ii) **Head, trunk** and **tail** are clearly demarcated.

(iii) Very large vitelline veins and arteries are visible.

(iv) Wing buds and leg buds appear.

(v) Somites are continued up to tail process and there number is forty two pairs.

B. Identification of transverse sections of chick embryo at different stages of development

1. **T.S at 24 hour stage :**

(a) **T.S through head :**

(i) T.S shows ecto, endo and mesoderm.

(ii) **Splanchnopleure** of the either side bends towards the middle line along the lateral margin of the intestinal portal.

(iii) Neural plate ectoderm folding to form a neural groove.

(iv) Notochord is situated below the neural plate.

(v) Gut is immediately caudal to anterior intestinal portal.

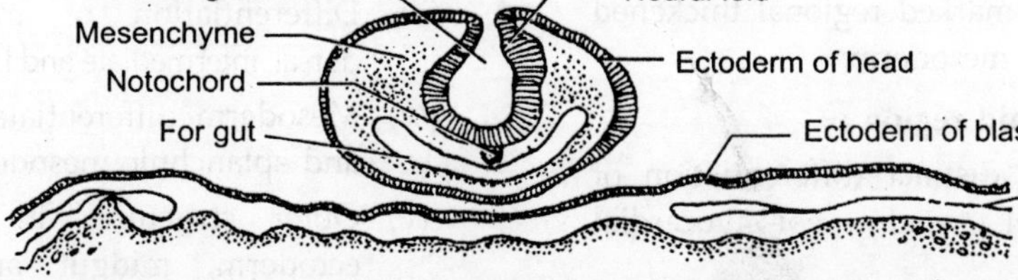

Fig. 3.2.2.7 : T.S. of 24 hours stage of chick embryo through head region.

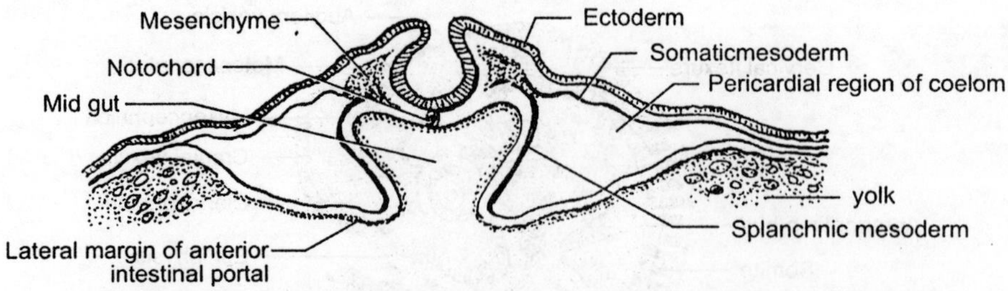

Fig. 3.2.2.8 : T.S. of 24 hours stage of chick embryo through mid region.

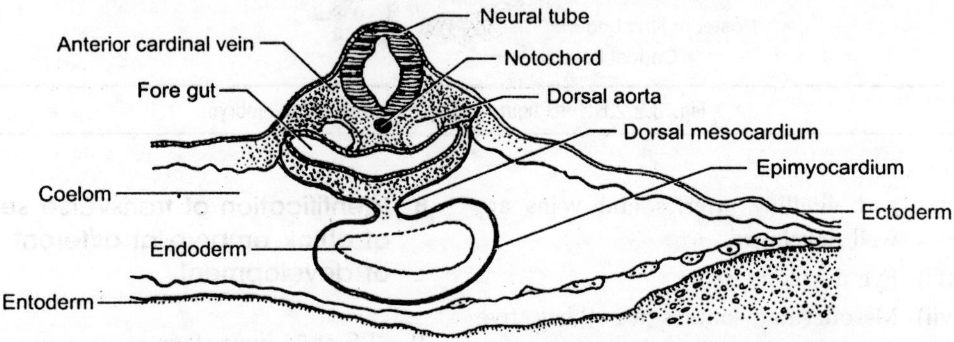

Fig. 3.2.2.9 : T.S. of 33 hours stage of chick embryo through mid body region.

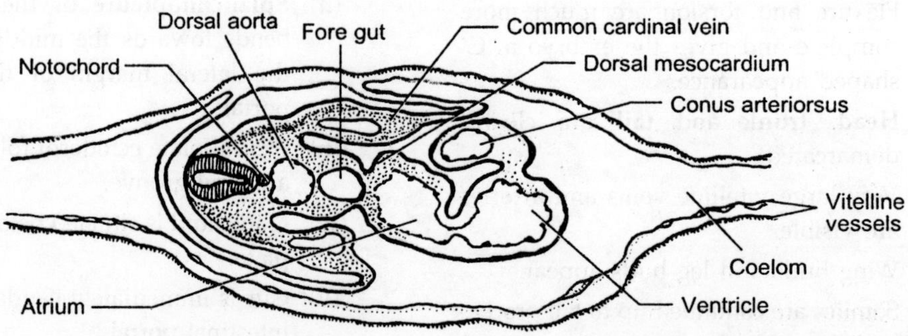

Fig. 3.2.2.10 : T.S. of 48 hours stage of chick embryo through mid body region.

(vi) There is a marked regional thickened splanchnic mesoderm.

(b) **T.S through mid region :**

(i) T.S shows distinct differentiation of endocardial and epi-myocardial primordial.

(ii) Neural groove is closing to form neural canal.

(iii) Differentiation of mesoderm into dorsal, intermediate and lateral mesoderm.

(iv) Mesoderm differentiates into somatic and splanchnic mesoderm.

(v) Outer structures in the T.S. are ectoderm, midgut and pericardial region of the coelom.

2. T.S. through mid body region at 33 hours stage :

(i) T.S shows formation of neural tube is complete.

(ii) Notochord is situated below the neural tube and an aorta on either side of the notochord is visible.

(iii) Dorsal mesocardium, epi-myocardium and endocardium is visible.

(iv) Vitelline vein is also seen.

3. T.S. through mid body region at 48 hours stage :

(i) T.S. shows complete formation of notochord.

(ii) On the ventral border of the notochord aortae are visible.

(iii) Cardinal and Vitelline veins are present.

(iv) Coelom is present.

(v) Sinus venosus and bulbus arteriosus are also seen.

Section 4

Histology and Histochemical Procedures

SECTION AGENDA

The basic histological study comprises the examination of routine paraffin sections, on the slides, stained with haematoxylin and Eosin. They are observed under light microscope. Now the subject matter of histology goes far beyond the above knowledge. Indeed, it is the object of histology to understand not only the micro anatomy of cells, tissues, and organs but also to know about their function in structural terms.

Histology also depends on **histochemistry** and **autoradiography** for examining tissues such as cell, tissues and organ culture and on specialized microscopic techniques to observe the specimen. Histology also includes specific chemical procedures to tissues. These procedures give more detailed knowledge about a tissue. These chemical procedures are enzymes, antibodies and radioactive isotopes and are included under the heading of **histochemistry** and **cytochemistry**.

Chemical substances in tissues and cells can be indentified by chemical reactions that produce insoluble colour compounds. In addition to chemical reactions in the tissue, other methods are frequently used. They are **interference microscopy**, which determines the mass of cells or tissues; and **microspectophotometry**, which determines the localization and quantitation of DNA and RNA in the cells, by means of ultraviolet light.

Thus it is the object of histology to understand not only the microanatomy of cells, tissues or organs, but also to learn as much as possible about their function in structural terms.

Chapter 4.1

Basic Histological Techniques

In view of the characteristics of the light microscope and the principles of image formation; it is evident that material to be observed with such a system must satisfy certain specifications. Under the optical microscope, tissues are examined by **transillumination**. Since tissues and organs are usually too thick for transillumination, they must be relatively thin. With few exceptions, most tissues are colourless, so that observing them unstained in the light microscope is difficult. So they should be coloured. Tissue must be transparent if they are to be observed by transillumination.

Types of preparation

Living materials, such as protozoa, invertebrate embryos, larvae, flatworms, nematodes, rotifers, gastrotrichs etc. and vertebrate and invertebrate blood, spermatozoa, oocytes and isolated tissue cells, may be studied for a limited time with microscope. Parasitic forms and tissue cells are examined in normal saline or **Ringer's solution** unstained or stained. For staining, the dyes used are, vital or **supravital dyes**. The most frequently used are **janus green** and **neutral red**.

Wet whole mounts are limited to organism or parts of organisms that are small enough to be mounted entire and studied by transmitted light. Such material is dehydrated, with ethyl alcohol, which is then replaced by a clearing agent, such as xylene. Finally the material is placed in a mounting medium and the object is mounted on a slide under a cover glass. Before or during dehydration such preparations are stained. Protozoa, small insects and other arthropods, embryos, larvae and many parasites are prepared as wet whole mount.

Smears are prepared by spreading a thin film of material on a slide or cover glass. Protozoa, blood, lymph, spermatozoa, contents of digestive tract, faeces, vaginal epithelium are the materials that are commonly prepared by such techniques.

Squashes are prepared by keeping the material in a fluid between slide and cover glass and tapping gently on the cover glass until the material separates. It is used exclusively for the study of Drosophila salivary gland cells chromosomes, testes cells chromosome, etc.

Sectioning provides a method which retains the structural unit in its natural condition so that it's interrelationship to other structural units and the whole may be observed. The first step in preparing a tissue sample for examination with light microscope is **fixation.**

Fixatives are chemicals that preserve the physical structure of the cell and convert the cell's contents into a stable compound. Numerous chemicals and mixtures of chemicals are used as fixatives. Chemicals employed as fixers are, acetic acid, formalin (40% formaldehyde), alcohol, picric acid, mercuric chloride, potassium dichromate etc. The most commonly used mixtures are Bouin's fluid, composed of picric acid, formalin, acetic acid and, water; and Zenker's formalin (Helly's fluid), containg formalin, potassium dichromate, mercuric chloride and water. A simple fixative most commonly used is formaldehyde – a ten percent solution of formalin in saline and a two to six percent solution of buffered glutaraldehyde. Formaldehyde and glutaraldehyde are known to react with amino groups of protein and thus act as good fixative.

Fixatives, such as formaldehyde, have the possibility to further react with any staining procedure which may be applied afterward in the process. As a result, any remaining fixative is washed out by placing the blocks in running water overnight or by consecutive changes of water and/or a buffer. There are many means of washing the tissues (using temperature, pH and osmotically controlled buffers), but generally simple washing in tap water is sufficient.

If the tissues are to be embedded in paraffin or plastic, all traces of water must be removed: water and paraffin are immiscible. The removal of water is **dehydration**. The dehydration process is carried on by passing the tissue through a series of increasing gradient of alcohol concentrations. The blocks of tissue are transferred sequentially to 30%, 50%, 70%, 80%, 90%, 95%, and 100% alcohols for about two hours each. The blocks are then placed in a second 100% ethanol solution to ensure that all water is removed. Note that ethanol is hydroscopic and absorbs water vapor from the air. Absolute ethanol is only absolute if steps are taken to ensure that no water has been absorbed. The alcohol is then replaced by a lipid solvent, such as xylene. As the tissue becomes impregnated with the solvent, it usually becomes transparent. After dehydration, the tissues can be embedded in paraffin, nitrocellulose or various formulations of plastics. Paraffin is the least expensive and therefore the most commonly used material. More recently, plastics have come into increased use, primarily because they allow thinner sections (about 1.5 microns compared to 5—7 microns for paraffin). This process is called **infiltration** or **embeddings.** For **paraffin embedding**, first clear the tissues. **Clearing** refers to the utilization of an intermediary fluid that is miscible with ethanol and paraffin, as these two compounds are immiscible. Benzene, chloroform, toluene or xylene are the mostly used clearing agents, though some histologists favour mixtures of various oils (cedar wood oil, methyl salicylate, creosote, clove oil, amyl acetate or Cellosolve).

Be cautious of all organic solvents. Most are implicated as carcinogenic agents. Heed all precautions for the proper use of these compounds

The most frequently used clearing agent is toluene. It is used by moving the blocks into a 50:50 mixture of absolute ethanol: toluene for two hours. The blocks are next placed into pure toluene and then into a mixture of toluene and paraffin (also 50:50). They are then placed in an oven at 56-58°C (the melting temperature of paraffin).

The blocks are transferred to pure paraffin in the oven for one hour and then into a second pot of melted paraffin for an additional two-three hours. During this time the tissue block is completely infiltrated with melted paraffin.

Subsequent to infiltration, the tissue is placed into an embedding mold and melted paraffin is poured into the mould to form a block. The blocks are allowed to cool and they harden thus are then ready for sectioning. The hardened paraffin containing the tissue is trimmed, usually into block form. The small blocks are then sectioned by a cutting machine called a **microtome**. The process is called as **sectioning.** Microtome is rather a simple device consists of a stationary knife holder/blade and a specimen holder which advances by pre-set intervals with each rotation of the flywheel mounted on the right hand side. A control knob adjusts internal clams which advance the paraffin block with each stroke. It is relatively easy to section paraffin at 10 microns but requires a lot of skill and practice to cut at 5 microns. Since each section comes off of the block serially, it is possible to align all of the sections on a microscope slide and produce a serial section from one end of a tissue to the other. The sections are placed on a slide which is previously smeared by a small amount of **Mayer's albumin** serving as an adhesive. The paraffin from the tissue is dissolved out with xylene and the tissue is rehydrated by being passed through a graded series of ethanol to water.

While virtually anyone can cut a section within minutes of being introduced to the microtome, proper use of the microtome is an art form and requires practice and inventiveness. Many a cell biology research project has depended on the skills inherent in the use of this instrument. A microscope is nearly useless without a good thin, flat, and undistorted section from properly fixed, dehydrated and embedded tissue.

Staining

Most tissues are colourless, and it is very difficult to observe them unstained in the light microscope. Therefore methods of staining tissue have been devised. Most dyes used in histological study behave like acidic or basic compounds and, have a tendency to form electrostatic linkages with ionisable radicals of the tissue. A **basic dye** is a dye molecule that carries a positive charge (charges) on the coloured part of the molecule. In contrary **acidic dye** molecule carries a negative charge on the coloured portion. Any tissue component that reacts with a basic dye is said to be **basophilic**. Heterochromatin and nucleoli of the nucleus, ergastoplasm and matrix of cartilage are the basophilic tissue components. Tissue components that stain with acidic dyes are said to be **acidophilic**. Such tissue components are most cytoplasmic filaments and extra cellular matrix fibres.

Toluidine blue and methylene blue are basic dyes. Haematoxylin is not a basic dye. It is used with a mordant, and, it is due to the mordant that haematoxylin closely resemble those of a basic dye. The linkages in the tissue – mordant – haematoxylin complex are not a simple salt linkages and when placed in water, haematoxylin dose not dissociate from the tissue. Because of this character, haematoxylin is followed by eosin or other acidic dyes. True basic dyes are not generally used where it is followed by an acid dye because, the basic dye tends to dissociate. Acid dyes bind primarily to tissue components by electrostatic linkages in a manner similar to but opposite that of basic dye. Acid dyes such as, orange G, eosin, acid fuchsin stain mostly the basic components present in cytoplasmic protein.

Of all dyes haematoxylin and eosin are the most commonly used dyes in histology. Many other dyes are used in different histological procedures.

Chapter 4.2

Basic Histochemical Techniques

The chemistry of tissues and the cells is studied by both microscopy and chemical analysis.Chemical substances in tissues and cells can be identified by chemical reactions that produce insoluble coloured compounds, observed with the light microscope or electron scaterring precipitates that can be observed with the electron microscopy

For a histochemical reaction to be recognised as valid and meaningful the following are the basic requirements:

1. **The substances being analyzed must not diffuse out of their original sites:** This can be done by use of proper fixative used to that preserve the cell structure and prevent diffusion of compounds to be assessed.

2. **The product of the reaction should be insoluble and coloured or electron scattering:** This property of the product helps the study of the product through electron and light microscopy.

3. **The method engaged should be specific for the substance or chemical groups being studied.**

4. **The procedure must not block or denature reactive groups.**

There are two well – known applications of the Schiff reagent in histochemistry. One is the **Feulguen's reaction** to study DNA and the other is the **Periodic acid – schiff (PAS) reaction** for polysaccharides and oligosaccharides.

Feulgen reaction depends on the treatment of fixed tissues by mild acid hydrolysis with hydrogen chloride (N – HCl at 60^0C). This process separates purine groups from the deoxypentose sugar of DNA, promoting the formation of aldehyde group in deoxyribose. These in turn react with Schiff's reagent (Fuchsin – sulphurous acid) to produce a purple dye in the nuclear chromatin alone. The reaction is specific for DNA. RNA fails to react in this procedure because it lacks deoxyribose sugar. RNA has great affinity for basic stains – it reacts intensely with toluidine or methylene bule. Feulgen reaction, together with a system of **histophotometry**, it is possible to estimate the amount of DNA in the nuclei of cells.

Periodic acid (HIO_4) is an oxidant. The PAS reaction based on the oxidative action of Periodic acid on 1:2 – glycol groups (CHOH—CHOH) present in the glucose residues. Periodic acid breaks the bond (C—C) between these adjacent carbon atoms and converted into aldehyde

Table : Substances of biological interest and histochemical methods

Compounds/Substances	Histochemical method
Iron (Fe^{3+})	Reacted with a mixture of potassium ferricyanide ,the ferric ions in the tissue can be detected through the formation of blue ppt. of ferric ferrocyanide (Perl's reaction)
Phosphates	Demonstrated by the reaction with silver nitrate forming a black ppt. of reduced silver.
Lipids	Demonstrated by staining with dyes such as Sudan IV and Sudan black.
DNA	Demostrated through Feulgen's reaction.
RNA	Stained through basic stains (affinity for basic stain or **basophilia**) viz., toluidine or methylene blue.
Proteins	Demonstated by Millon reaction and Sakaguchi reaction
Polysaccharides	Polysaccharides such as glycogen can be demonstrated through PAS reaction
Oligosaccharides	Oligosaccharides of glycoproteins containing sulphate or carboxyl groups react strongly with alcian blue. Neutral glycoproteins can also be demonstrated through PAS reaction.
Acid phosphatase	Enzyme acid phosphatase can be demonstated through Gomori method
Dehydrogenases	Demonstrated through the reaction with tetrazole (H^+ acceptor) converting the substrate to **formazan** which can be visualised as darkly stained ppt.

groups. These aldehyde groups react with Schiff's reagent give a substituted dye which is red in colour. This can be seen in the light microscope, and such substances are PAS – positive. Since many tissue components give a PAS- positive reaction, the specificity of this reaction depends on pretreatment with salivary amylase—an enzyme that specifically digest glycogen. PAS – positive materials which give no reaction following pretreatment with amylase are considered to contain glycogen. Using this method glycogen can be observed in normal liver tissue and striated muscles tissue. The identification of basement membrane material and reticular fibres which contains PAS – positive sugar groups can be identified by PAS reaction.

Many histochemical methods are used to reveal and identify enzymes. One of the most useful enzyme methods is that for **acid phosphatase**. This method permits the localization of acid phosphatase enzymė's activity and to identify lysosomes, cell organelles that contain acid phosphatase. The formalin - fixed tissue sections are incubated at 37°C for four hours in a solution containing sodium glycerolphosphate and lead nitrate buffered to pH5.0. The enzyme hydrolyzes the glycerophosphate and liberate phosphate ion that reacts with lead nitrate to produce an insoluble, colourless precipitate of lead phosphate at the site of enzyme activity. After washing briefly, the slides are immersed in dilute yellow ammonium sulphide that reacts with the colourless lead phosphate to produce a black precipitate of lead sulphide. Histochemical methods are well established for ATPase, alkaline phosphatase and several enzymes.

MICROSCOPY

1. Light Microscope :

The most fundamental requirements for the students in the histology laboratory are the ability to use the light microscope properly and to understand its capacities and limitations. The microscope used by the students in the laboratory is referred to as **light microscope, bright field microscope** or **compound microscope**. The microscope is composed of both mechanical and optical parts. Optical components consist of three systems of lenses – the **substage condenser, objective** and **ocular**. The rest parts of the microscope are mechanical components.

Sub stage condenser includes a set of lenses mounted between the mirror and the microscope stage. Condenser serves as a critical light – focusing system i.e. it projects a cone of light to illuminate the object placed on the stage to be observed. The objective lens situated at the bottom of the tube next to the material, enlarges the object and projects its image in the direction of ocular lens. The microscope is usually provided with three objectives arranged on a rotating nosepiece. The ocular lens situated at the top of the tube, next to the eye, further magnifies the image and project it on to the retina of eye, but it dose not increase resolution. The degree of total magnification is obtained by multiplying the magnifying power of the objective and ocular lens.

The usefulness of the light microscope depends on the ability to magnify and ability to resolve structural details. **Resolution** is the ability to distinguish two closely particles as being distinct from one another. The degree of resolution depends not only on the optical system, but also on the wave length of the light source and specimen thickness. Resolution may be improved by using short wave length light / or by slowing down the light. Short wave length light may be obtained by using a blue ("day light") bulb or a blue filter in front of a white bulb. Resolution is further improved by the use of immersion oils between the objective and the material. The resolving power of the good light microscope is approximately 0.2μm.

The resolving power of an objective is dependent upon what portion of the light cone of the objective can utilized or pick up as light is transmitted through the material. The quality of image depends on the resolution power of the microscope. **Magnification** is independent of its resolving power and in itself accomplishes little, since increasing size may not give any new detail of structure and is only of value when accompanied by a high resolution capacity.

The inherent capacity of an objective together in light is indicated by the "**numerical aperture**" (N.A.) of the objective. The numerical aperture of a given object is calculated on the basis of the refractive index of the medium in which it is to be used. Resolution is function of N.A. and of the light wave length used. Therefore, the higher the N.A. of the objective, the greater should be the angle of light emerging from the material. The angle of light is dependent upon the adjustment of the substage iris diaphragm. The greater the N.A. of the objective, the opening of the sub stage iris must be wider.

The compound microscope form an inverted image of the object while the simple microscope, or magnifying glass dose not invert the image. An objective lens system compound microscope has several numbers written on it. The first number engraved on the left in the first line denotes the magnification; to its right is N.A. The number on the left in the second line is the tube length in millimetres; to its right denotes the thickness in millimetres of the cover slip to be used in dry field examination.

2. Phase Contrast Microscope :

Phase contrast microscope is used to examine unstained fresh tissue or living cells. Phase contrast microscopy is based on the fact that light passing through regions of different refractive indexes slows down and changes direction. Different organelles of a living cell exhibit different refractive indexes and consequently alter the light passing through them. This creates a phase differences between two adjoining regions. With the phase contrast microscope these phase differences are transformed into differences of light intensity so that the object can be visualized with a useful amount of contrast. Here the light waves initially slows down by the specific tissue component have been furthered diminished by a set of rings in the special optical system.

Interference microscope and the **differential interference microscope** are the modifications of phase contrast microscope. The first one also allows for quantitation of tissue mass and second one is specially useful for assessing surface properties of cells and biological specimens.

Chapter 4.3

Some Important Histochemical Staining

The present chapter is primarily concerned with the preparation of the reagent, methods of staining the tissue and the result obtained therein.

1. Feulgen Reaction

- **Preparation of Schiff's reagent :**

1. 1 g of basic fuchsin is dissolved in 200 ml. of boiling distilled water. Shaken for 5 minutes and cooled to exactly 50°.

2. Filtered and added to the filtrate 20 ml of N—HCl.

3. Cooled to 25° and added 1 g of sodium or potassium metabisulphite ($Na_2S_2O_5$).

4. The solution is allowed to stand in the dark for 14-24 hours.

5. Added 2 g of activated charcoal and shaken for one minute.

6. Filtered and the filter is kept in the dark at 0-4°. Before use the reagent is allowed to reach room temperature.

- **Method :**

1. Formalin fixed paraffin sections are brought to water.

2. Rinsed briefly in cold N—HCl.

3. Placed in N—HCl at 60°, kept for 10 minutes for hydrolysis.

4. Rinsed in cold N—HCl and then in distilled water.

5. The slide is transferred to Schiff's solution and kept for one hour.

6. Drained and rinsed with freshly prepared bisulphite solution. (5ml. 10% $K_2S_2O_5$, 5ml. N—HCl and made to 100ml with water) thrice.

7. Rinsed in water.

8. Dehydrated through graded alcohol.

9. Cleared in xylene and mounted in balsam or DPX.

- **Result :**

DNA (chromatin) stains reddish-purple.

[See Fig. 4.3.1 colour plates (Section 4).]

2. Determination of RNA (Pyronin y)

Methyl-Green-Pyronin Y method for RNA (Kurnick, 1955)

- **Preparation of stain :**

1. Two percent aqueous solution of Pyromin Y extracted with $CHCl_3$ is prepared by shaking in

a separating funnel until the chloroform layer becomes colour-less.

2. Two percent aqueous solution of methyl green is prepared and extracted with $CHCl_3$ as above until the chloroform layer is no longer violet coloured.

3. Mixed 12.5 ml Pyronin Y solution and 7.5 ml of methyl green with 30ml distilled water during use.

- **Procedure :**

1. The section is brought to water.

2. The slide is kept in methyl green-pyronin for 6 minutes.

3. Blotted with filter paper.

4. Immersed in two changes of 4-butyl alcohol, 5 minutes in each.

5. The slide is kept in xylene for 5 minutes.

6. The slide is kept in cedar wood oil for 5 minutes.

7. Mounted in DPX.

- **Result :**

Chromatin clear green, nucleoli bright red, cytoplasmic RNA bright red.

[See Fig. 4.3.2 colour plates (Section 4).]

3. The Periodic Acid-Schiff Technique : (Hotch Kiss, 1948)

- **Preparation of the solution :**

1. **Periodic acid :**

(i) 0.4 gm of Periodic acid (HIO_4, $2H_2O$) is dissolved in 35ml of ethyl alcohol and in it added 5ml of 1/5(M) sodium acetate (27.2 g of the hydrated salt in 1000 ml) and 10 ml of distilled water.

(ii) The solution is kept in dark at 17°-22° and used at this temperature.

(iii) The solution is discarded if it turns brown.

2. **Schiff's reagent :** For preparation see Fuelgen.

- **Method :**

1. The paraffin sections are brought to water.

2. Immersed in Periodic solution 5 minutes for oxidation.

3. Washed in distilled water.

4. Immersed in Schiff's reagent for 10-15 minutes.

5. Washed in running water for 5-10 minutes.

6. Dehydrated through graded alcohol, cleared in xylene and mounted in balsam or in DPX.

- **Result :**

Carbohydrate – containing proteins stain in various shades of purplish – red. Glycogen stains deeply.

[See Fig. 4.3.3 colour plates (Section 4).]

4. Lead Nitrate Method for Acid Phosphatase : (after Gomori, 1950)

- **Method :**

1. Formalin fixed frozen sections are cut 10-15 μm thick and mounted in slides without any adhesive and dried for 1-2 hours.

2. Incubated at 37° for 4 hours in 0.01 (M) sodium β–glycerophosphate in 0.05 (M) acetate buffer (pH5.0) containing 0.004(M) lead nitrate.

3. Washed briefly and immersed in dilute yellow ammonium sulphide, 1-2 minutes.

4. Washed and counter stained with 1 percent aquous eosin for 5 minutes.

5. Washed thoroughly and mounted in glycerine jelly.

- **Result :**

1. Black precipitate of lead sulphide indicates the presence of acid phosphatase.

[See Fig. 4.3.4 colour plates (Section 4).]

5. Modified Gomori Method for Alkaline Phosphatase : (after Fredricsson, 1956)

- **Method :**

1. Small blocks of tissue are fixed in 90 percent alcohol for 24 hours at 22° (two changes).

2. The tissue is transferred to 96 percent alcohol and keeps for 1 hour.

3. Transfered to absolute alcohol, and two changes are made of 15 minutes each.

4. Transfered to benzene, and two changes are made 30 minutes each.

5. The tissue is embedded in paraffin at 56°, given three changes, 20 minutes each.

6. Cut sections of 3-10 μm and mounted on albumenised slides without any floating medium.

7. The paraffin are removed by dipping the slides in xylene and passing through absolute acetone, and then into 40 percent acetone to water.

8. Incubated in a closed coplin jar at 37° for 30 minutes in the following mixture:

2 percent sodium b-glycerophosphate—	25 ml
2 percent sodium veronal—	25 ml
2 percent calcium nitrate—	25 ml
0.8 percent Magnesium chloride—	25 ml
Acetone—	40 ml

9. Rinsed in 40 percent acetone.

10. Treated the slide with 2 percent cobalt nitrate in 40 percent acetone for 5 minutes.

11. Rinsed again in 40 percent acetone.

12. Treated with dilute yellow ammonium sulphide in 40 percent acetone for 3 minutes.

13. Rinsed in 40 percent acetone, dehydrated in alcohol, cleared in xylene and mounted in mounting medium.

- **Result :**

Black deposit of cobalt sulphide indicate sites of enzyme activity.

[See Fig. 4.3.5 colour plates (Section 4).]

6. Millon Reaction :

- **Preparation of the Reagent :**

1. Added 10 g $HgSO_4$ to 100ml of 10 percent H_2SO_4 and heated until dissolved.

2. This is made up to 200ml.

3. Added 0.5 ml of 0.25 percent $NaNO_2$.

- **Method :**

1. After deparaffinization the formalin fixed tissue sections are passed through 50 percent alcohol to water.

2. Sections are placed in small beaker containing the above reagent and boiled gently.

3. After gentle boiling the solution is allowed to come to room temperature.

4. The slides are removed from the beaker and washed thrice in distilled water (two minutes in each wash).

5. Dehydrated, cleared in xylene and mounted in balsam or DPX.

- **Result :**

Tyrosine – containing proteins are stained red, pink or yellowish-red.

7. Metachromatic Methods :

Metachromasia is a phenomenon where by a dye (such as toluidine blue) changes colour after reacting with a tissue component. It is due to the presence of numerous acidic groups in the structure amenable to metachromatic staining.

Toluidine Blue Method :

1. After deparaffinization the sections are brought to water.

2. Stained in 0.1 percent toluidine blue in 30 percent ethanol for 5-20 minutes.

3. Rinsed in 95 percent alcohol.

4. Dehydrated in absolute alcohol.

5. Cleared in xylene and mounted in DPX or balsam.

- **Result :**

Metechromatic substances red or pink (γ-meta-chromasia) or purple (β-metachromasia). Nuclei blue.

[See Fig. 4.3.6 colour plates (Section 4).]

Chapter 4.4

Identification of Some Vertebrate Tissue Sections

Tissues are groups of cells having the same structure and functions. The concept of tissues provides a basis for understanding and recognizing the many cell types within the body and how they interrelate. Cell aggregations are classified under four tissue types. These basic tissues are given in the table below.

Each of these fundamental tissues is defined by a set of general morphological characteristics. In this chapter we are presenting only some of the tissue types.

I. EPITHELIAL TISSUE :

The epithelial tissue are made up of close aggregations of cells that are in apposition over a larger part of their surface with very little intercellular materials. In its simplest form the epithelial tissue consists of single continuous

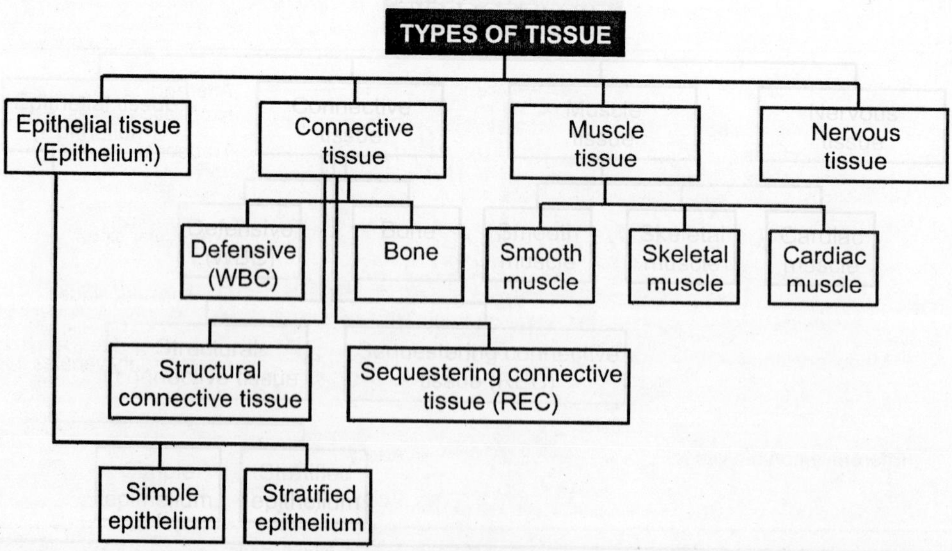

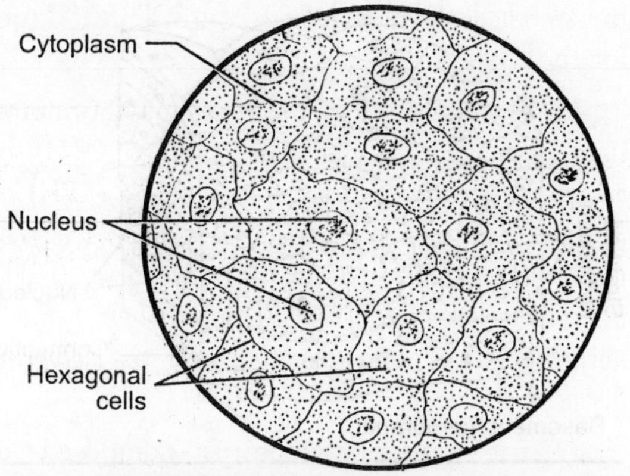

Fig. 4.4.1 : Squamous epithelium.

layer of cell of similar type. It covers the free body surfaces either externally or internally.The different component parts of the epithelial tissue are epithelial cells, intercellular substances and basement membrane. The epithelial tissue can be simple or stratified. Some important epithelial tissues are described below:

(A) Squamous epithelium :

(1) These are simple epithelial tissue, consist of very thin and flattened, roughly hexagonal cell. They are closely fitted to give a mosaic appearance.

(2) The width and depth of the cell is greater than its height.

(3) Each cell consists of a distinct, rounded nucleus in the centre. The cells has faintly stained granular cytoplasm.

(4) The basal surface of the tissue adheres to the **basement membrane.**

(5) It is found in the lining of the lungs, Bowman's capsule and blood vessels.

(B) Ciliated epithelium :

(1) These are simple epithelial tissue .Single layer cells are columnar but are broader than long.

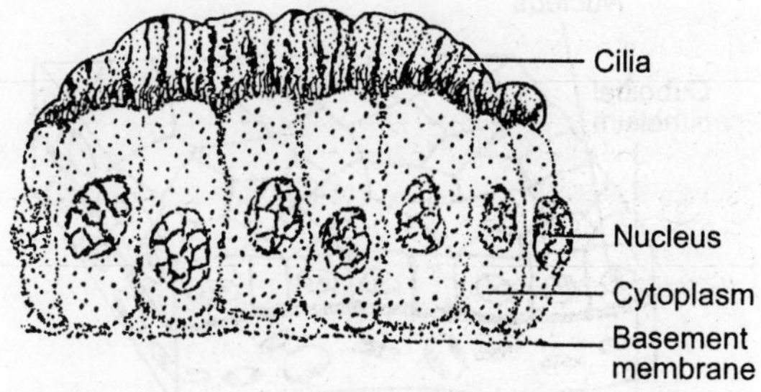

Fig. 4.4.2 : Ciliated epithelium.

(2) Each cell possesses at its free end a bunch of cilia which may be motile or non-motile.

(3) The basal surface of the tissue adheres to the basement membrane.

(4) It is found in the mucous membrane of nasopharynx, alveoli, uterine tube and endometrium of uterus, efferent ductile of testes etc.

(C) Columnar epithelium :

(1) These are simple epithelial tissue .Single layer of elongated cells, situated side by side like columns and lie at right angles to the surface.

(2) The height of cell is greater than their width.

(3) Cells are tapered towards their attached ends.

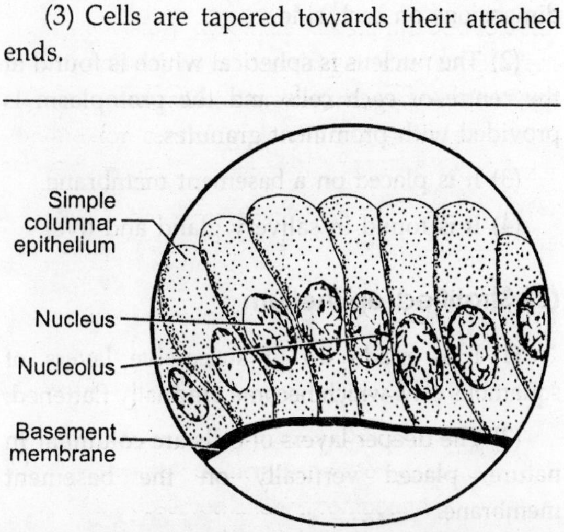

Fig. 4.4.3 : Columnar epithelium cells.

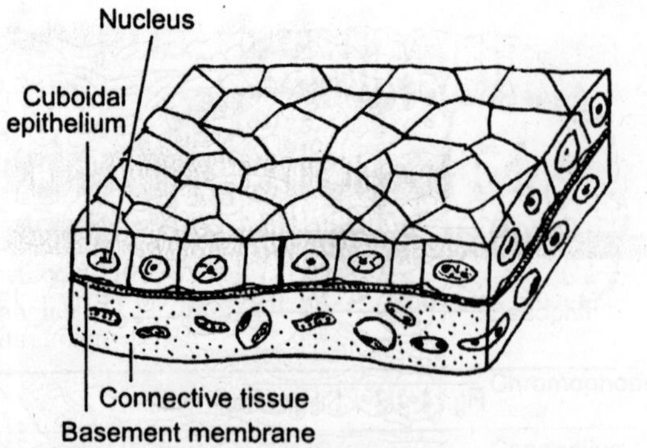

Fig. 4.4.4 : Cuboidal epithelium.

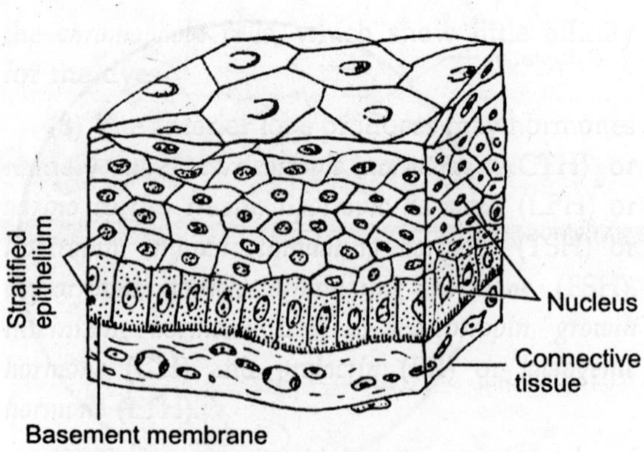

Fig. 4.4.5 : Stratified epithelium.

(4) Each cell is with prominent ovoid nuclei at the basal part.

(5) Free surface of the cell is covered by a thick straight border.

(6) The basal surface of the tissue adheres to the basement membrane.

(7) It is found in the mucous membrane of gastro-intestinal tract, greater part of the male urethra and greater vestibular glands.

(D) Cuboidal epithelium :

(1) These are simple epithelial tissue.The cuboidal cells are isodiametric having the same dimensions on each side

(2) The nucleus is spherical which is found at the centre of each cells and the protoplasm is provided with prominent granules.

(3) It is placed on a basement membrane.

(4) It is found in salivery gland and liver.

(E) Stratified epithelium :

(1) It is made up of successive layers of squamous cells which become gradually flattened.

(2) The deeper layers of cells are columnar in nature, placed vertically on the basement membrane.

(3) The successive layers of cells consist of polyhedral cells and most superficial cells are flattened.

(4) These tissues are found as keratinized epithelium in skin, hairs, nails, scales, hoofs, horns, enamel of teeth etc. In the non-keratinized form they are found in the buccal cavity, tongue, lower part of pharynx, oesophagus, cornea, cervix of uterus and vocal cord.

II. CONNECTIVE TISSUE :

The connective tissue originates from widely scattered free mesenchyme cells. It is named so because it forms basis of connection between cells and tissues. It binds together functionally more active tissue of the body. The connective tissue proper is made up of cells and extracellular *fibres* in an amorphous *ground substance* containing the *tissue fluid*. Connective tissue can be classified as embryonal connective tissue and adult connective tissue. Adult connective tissue can again be classified into **connective tissue proper, fluid connective tissue** and **sclerous connective tissue.**

(A) Areolar tissue :

(1) On the brown ground substance the wavy bundles of white fibres, fine elastic fibres singly and curled.

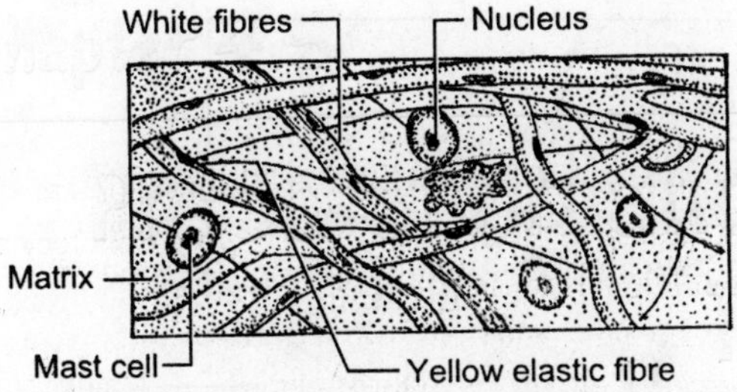

Fig. 4.4.6 : Areolar or loose connective tissue.

(2) The elastic fibres are fine single that may be straight or wavy.

(3) The lamellar cells or the fibroblasts are elongated in nature with projections having an ovoid nucleus and one or two nuclei.

(4) Masts cells are present singly or often in groups along the blood vessels.

(5) Plasma cells are small having eccentrically placed nucleus with condensed chromatin.

(6) Large and small lymphocytes are rounded cells which have a condensed chromatin clumps but no nucleoli.

(7) Occasional pigment cells are present.

(8) Adipose cells are also noticed with a narrow rim of cytoplasm and eccentrically placed cytoplasm.

(9) It is found in the subcutaneous tissue in eyelid, scrotum, penis etc. It is also found in the sub-mucous coat of alimentary tract. On the sub-serous coat between muscles, vessels and nerves.

(B) Yellow elastic fibres :

(1) These are large straight fibres that usually run singly but may also form loose bundles in which single fibre can be identified.

(2) Individual fibre freely branch and anastomose with each other to create a network.

(3) When divide the fibres quickly curl up due to elastic recoil at the broken ends.

(4) There is absence of any nuclei.

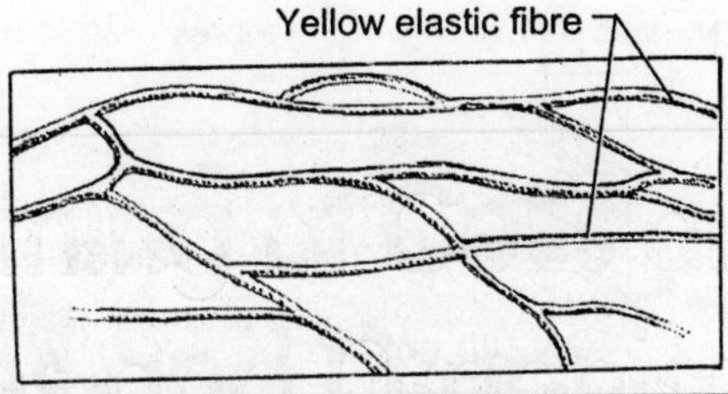

Fig. 4.4.7 : Yellow elastic fibre.

(5) These are found in the mucous membrane of trachea, bronchi and lung alveoli. They form the connective tissue framework of the lung. They form the capsule of spleen. Found in the vocal cords and coats of large blood vessels.

(C) White fibrous tissue :

(1) The fine, wavy or irregular bundles of white fibres faintly stained red and lie parallel to each other in different directions.

(2) Presence of deeply stained tendon cells which are angular-stellate type arranged in single rows are typically found in the surface of the transverse section of white fibres.

(3) The nuclei placed opposite to one another.

(4) The eccentric nuclei are more deeply stained.

(5) Matrix or ground substances are of fewer amounts.

(6) These are found in the tendon and **aponeurosis** of muscles (which connects muscles with bones). They are also found in the ligaments.

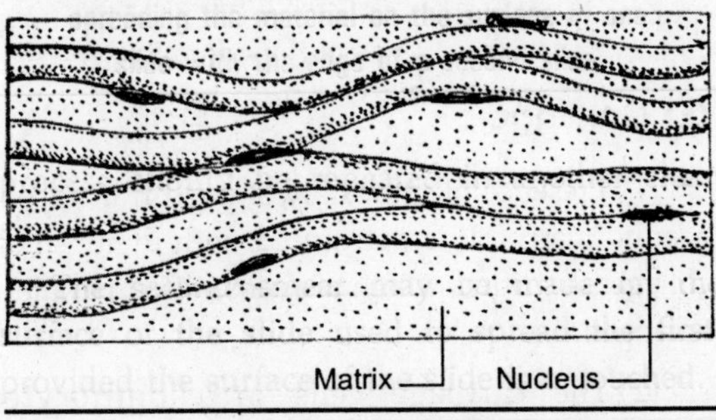

Fig. 4.4.8 : White fibre.

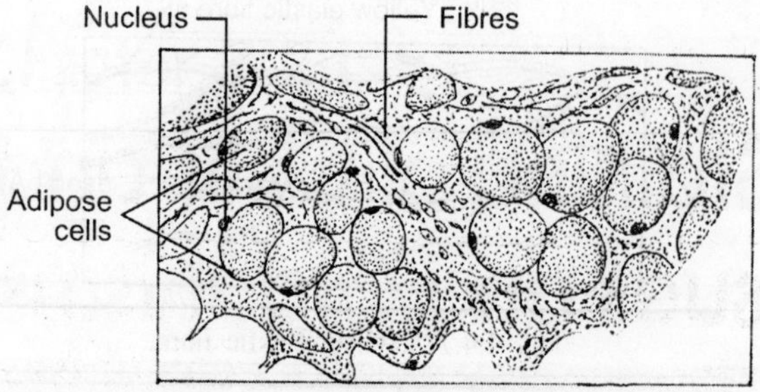

Fig. 4.4.9 : Adipose tissue.

These are also found in the **periosteum**, **perichondrium** and fibro-cartilage.

(D) Adipose tissue :

(1) The adipose cells have unstained protoplasm having solution of fat during tissue preparation and deeply stained nuclei that is pushed to one side of the cell and compressed in the peripheral rim of the cytoplasm.

(2) The adipose cells are close to each other separated by small amounts of connective tissue in which **fibroblasts** are compressed.

(3) The preponderance of the adipose cells are alone and are around the blood vessels.

(4) The orange stained fat globules occupy major part of the cell.

(5) The adipose tissue is widely distributed in all parts of the body as in greater omentum, mesenteries and in renal fascia of the abdomen, subcutaneous tissue, subpericardial tissue, lactating mammary gland, in synovial membrane, bone marrow etc.

(E) Cartilage :

(1) Presence of uniformly stained translucent ground matrix. It is permeated with many small collagenous fibres which are most often parallely arranged.

(2) The cartilage cells or **chondrocytes** two, three, or four in number are sorrounded

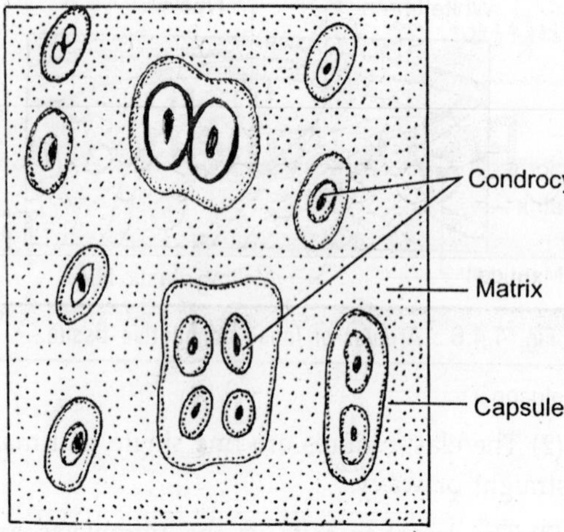

Fig. 4.4.10 : Hyaline cartilage.

by a capsule some of which are in dividing condition.

(3) Small chondrocytes in lacunae usually lie in rows within the fibrous matrix.

(4) It is found in the epiphyseal cartilage, articular cartilage, cartilages of nose, cartilages of larynx, cartilages of trachea and bronchi, symphysis of pubis, external ear, external auditory meatus, epiglottis etc.

(F) Bone tissue :

(1) Compact sections of the bone are formed of **Haversian system** or **osteon**. Each osteon consists of a number of concentric or tubular lamellae of bone matrix surrounding a central or Haversian canal which contains reticular connective tissue and blood vessels may be observed in transverse sections.

(2) **Haversian canals** are in parallel rows intercommunicating with each other and also communicating with the external periosteal surface of bone and internal medullary surface of bones.

(3) Cross-sections anastomoses between central or Haversian canal system.

(4) The bone matrix is composed of more

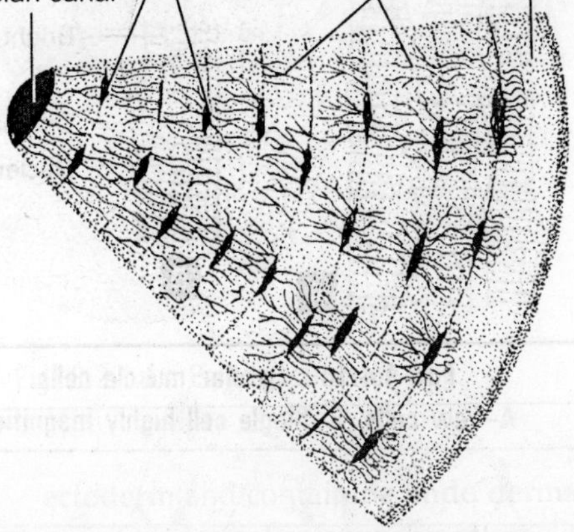

Fig. 4.4.11 : Part of Haversian system or osteon.

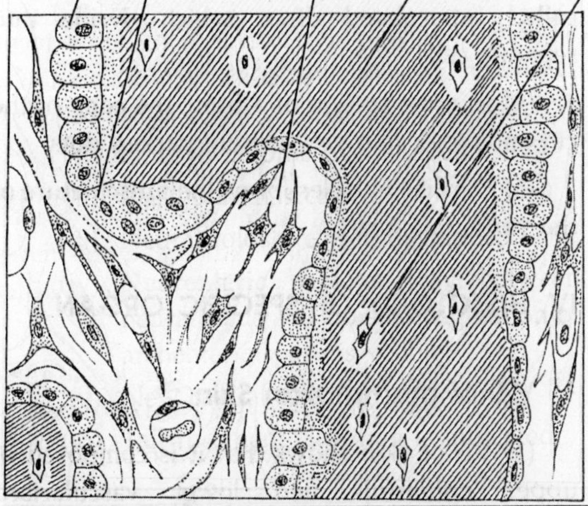

Fig. 4.4.11A : Section showing intramembranous ossification.

collagen fibres and less cement substances. The mesenchyme cells forms **osteoblasts** which change into **osteocytes.** The undifferentiated bone marrow cells develop **osteoclasts**.

(5) The bones are found in the skeletal parts of the body which is made up of bone cells and intercellular ground substance known as the matrix.

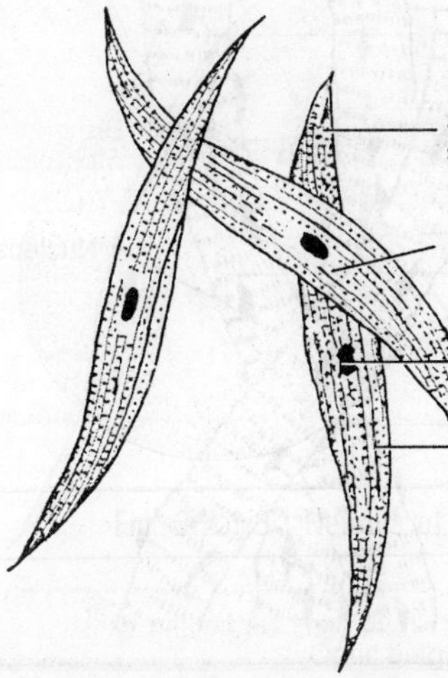

Fig. 4.4.12 : Smooth muscle fibre.

(G) Defensive connective tissue :

Comprising of various white blood cells are described in the Haematology section

III. MUSCLE TISSUE :

(A) Smooth (involuntary) muscle fibre :

(1) Arrangement of muscle fibres may be in groups or bundles, form a network or scattered singly.

(2) Spindle-shaped cells i.e. tapering at both ends with numerous fine longitudinal **myofibrillae**. There is a network of fine connective tissue in between the bundles and envelop the muscle fibre.

(3) Large oval nucleus in the centre, enclose in a mass of **sarcoplasm**.

(4) Each fibre is bounded by a delicate, transparent, thin membrane, called **sarcolemma**.

(5) Absence of striation.

(6) These muscles are found in the hollow tubular and saccular viscera, ducts of exocrine glands, blood vessels, ureter, urethra, urinary bladder and genital ducts. It is found in the walls of the arteries, veins and larger lymphatic channels; nipple of mammary glands, prostrate gland, muscle of iris, sweat glands of skin etc.

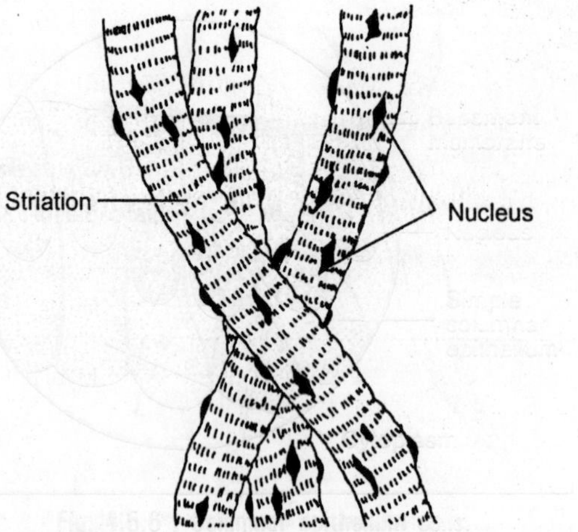

Fig. 4.4.13 : Striated skeletal muscle.

(B) Sketetal (voluntary or striated) muscle :

(1) Muscle tissue is extremely elongated and is cylindrical and tapering at both the ends, which are arranged in bundles, called **fasciculi.**

(2) Each fibre is surrounded by an **endomysium** and each fasciculous by a **perimysium.**

(3) The muscle fibre consists of **sarcoplasm,** a peripherally placed nucleus and covered by a thin **sarcolemma.**

(4) The centre of the fibre is made up of longitudinal bundles, the myofibrillae.

(5) The myofibrils are striated in a regular manner and show alternate light and dark pattern.

(6) At cross section myofibril looks like a series of closely packed dots across the entire width.

(7) These muscles are attached to bones and these includes the biceps, triceps, cuff muscle etc.

(C) Cardiac muscle :

(1) Cardiac muscle tissue is short, cylindrical and exhibits alternate light and dark bands with faint striation.

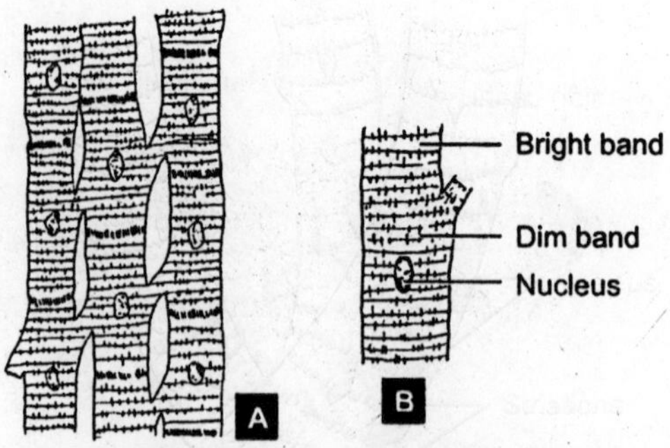

Fig. 4.4.14 : Cardiac muscle cells.
A–Few cells, B–Single cell highly magnified.

(2) Each fibre possesses a central nucleus, enclosed by sarcoplasm and ensheathed by a thin less distinct sarcolemma. The fibres are branched in nature that forms a **syncytium.**

(3) Myofibrillae possesses transverse marking called intercalated discs, which divide into separate cells.

(4) In cross section of a fibre, the edge of the cut ends look like very closely packed dots.

(5) Cardiac muscles forms the musculature of the vertebrate heart.

IV. HISTOLOGY OF SPECIFIC ORGAN

1. V.S of mammalian Skin :

(1) V.S of skin shows that it is composed of upper **epidermis** and lower vascularised fibroelastic connective tissue, **dermis.**

(2) Epidermis consists of stratified squamous epithelium the cells of which are arranged in strata. It has four layers :

(a) *Stratum Corneum* : Outermost keratinized layer of flat cells without nucleus comprising of amorphous matrix with prekeratin filaments.

(b) *Stratum lucidum* : Next layer consists of dead cells without granules.

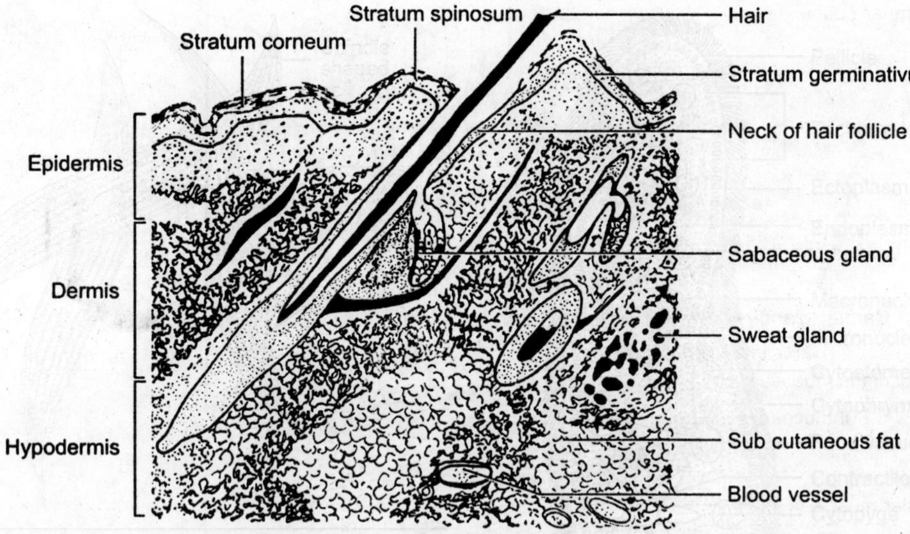

Fig. 4.4.15 : V.S. of mammalian skin.

(c) *Stratum granulosum* : Next to stratum lucidum and contains **keratohyalin granules**.

(d) *Stratum malpighii* : The inner most layer, consisting of germinal layer it produces new cells

(3) Hair shaft arises from hair papilla.

(4) Hair shaft project obliquely on the body surface.

(5) Beneath the epidermis is dermis, which is formed of connective tissue.

(6) Outer part of dermis is dense and bears papillae projecting into the epidermis.

(7) The dermis contains the nerve fibres, tactile corpuscles, blood vessels and various glands such as **sweat glands** and **sebaceous glands.**

(8) Coiled sweat gland opens on the epidermis.

2. T.S. of mammalian Stomach :

(1) The T.S. is composed of from outside **serous coat** or (serosa), **muscular layer, submucosa, muscularis mucosa and mucosa.**

(2) The serous coat forms outer layer and is composed of flat squamous epithelium covering loose connective tissue.

(3) Muscular layer (muscularis externa) consists of smooth muscles arranged in outer longitudinal fibres, middle circular fibres and inner longitudinal and oblique fibres. The oblique fibres are best developed in cardiac stomach.

(4) Submucosa is a layer of loose connective tissue, containing blood vessels and lymphatics, mucous secreting glands and elastic fibres forming cores of the mucosal folds.

(5) Mucosa (luminal mucosa) is the inner most layer, and thrown into folds. Its inner surface is covered by columnar epithelial cells resting on basement membrane and **lamnia propia**. Each folds contain gastric glands, **zymogen** or pepsin – secreting cells and HCl producing cells.

(6) In between mucosa and submucosa lies muscularis mucosa and it consists of an inner circular and outer longitudinal fibres.

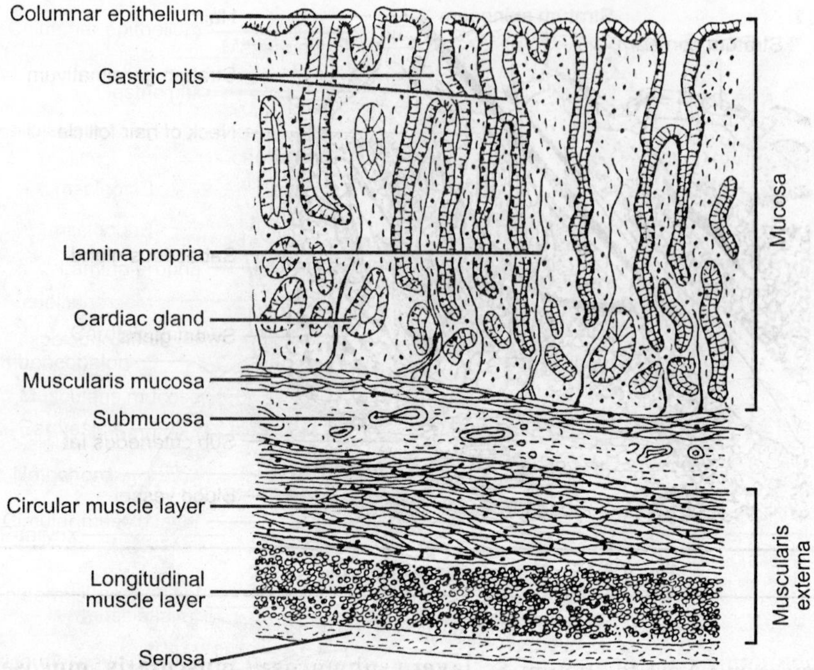

Fig. 4.4.16 : T.S. of mammalian stomach (cardiac region).

3. T.S. of mammalian small intestine :

(1) The T.S shows that it is composed of outer **serosa, muscular coat, submucosa, muscularis mucosa** and **mucosa.**

(2) Serosa forms outer covering. It is derived from visceral peritoneum and is made up of flat squamous epithelial cells (mesothelium).

(3) Muscular coat consists of external longitudinal and internal circular smooth muscle fibres.

(4) Submucosa is composed of loose connective tissue containing blood vessels, nerve plexus, adipose tissue and ganglionated plexus.

(5) Musculais mucosa is very much reduced.

(6) Mucosa and lamina propria (stroma) together are thrown up into villi or folds composed of single layered epithelial and goblet cells. Epithelial cells contain tubular glands or **crypts of Lieberkuhn** and **Brunner's glands. Paneth cells** with **zymogen granules** are seen at the base of crypts of Lieberkuhn.

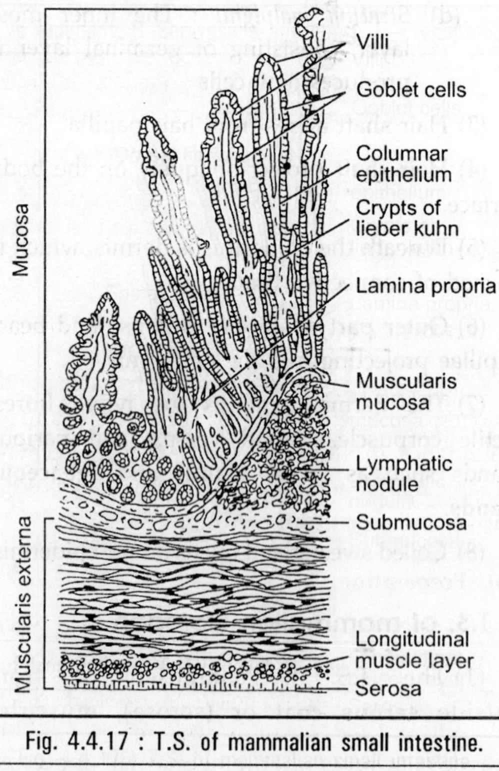

Fig. 4.4.17 : T.S. of mammalian small intestine.

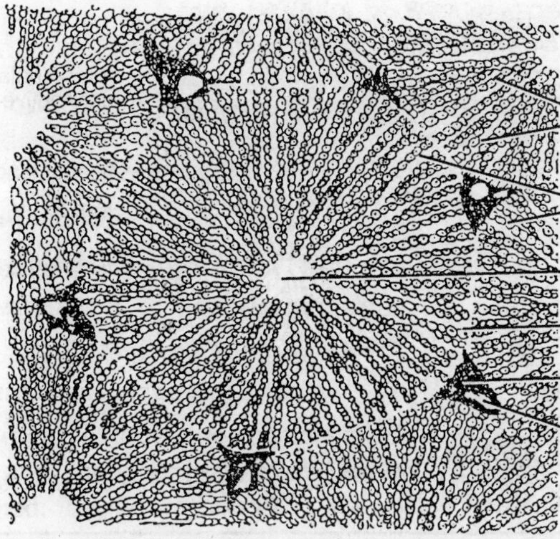

Fig. 4.4.18 : T.S. of mammalian liver.

4. T.S. of mammalian liver :

(1) T.S shows hepatic lobules, bile ducts, blood vessels and connective tissue surrounding the hepatic lobules, forming **Glisson's capsules**.

(2) Nucleated polyhedral hepatic cells arranged radially in each lobule.

(3) In between cells are bile capillaries.

(4) An intralobular vein present in the middle of each lobule (indicating the central axis of the lobule) and in between the lobules is the interlobular vein and hepatic artery. Peripheral to each lobule occurs incomplete ring of regions made up of veins,arteries,ductules and lymphatics that are designated as **portal** areas.

(5) The cords of the polyhedral cells are seen to be separated by slit like sinusoid lined by endothelial cell. Phagocytic **Kupffer's cells** or **stellate cells** in sinusoids.

(6) Polyhedral cells forms trabeculae that are exocrine in nature.

5. T.S of mammalian Pancreas :

(1) T.S. shows that it is composed of large number of alveoli or **acini** (pancreatic lobules). It is a composed tubulo–alveolar racemose gland consisting of both exocrine and endocrine parts. Many tightly packed acini forms a lobule. Lobules are separated from each other by thin septa this comprises the parenchymatous tissue of the pancreas. Acinus is formed of single row of pyramidal cells lying on delicate reticulate membrane. Cytoplasm of each acinar cell is basophilic in nature. An acinar cell bears a spherical nucleus with one or two prominent nucleoli.

(2) Each lobule contains bunch of secretory cells and a lumen connected by loose connective tissue.

(3) In between the lobules rounded mass of cells with unstained cytoplasm appear forming **Islets of Langarhans**, which is richly vascularized and they contain four types of cells :

(a) α-cells which are coarsely granular, secrete **glucagon** hormone.

(b) β-cells containing small granules, secrete another hormone, **insulin**.

(c) δ-cells contain small granules and are less in number, secretes **somatostatin**.

(4) Each acinus guide to an intercalated duct lined by flat cuboidal epithelium.Acinus ducts

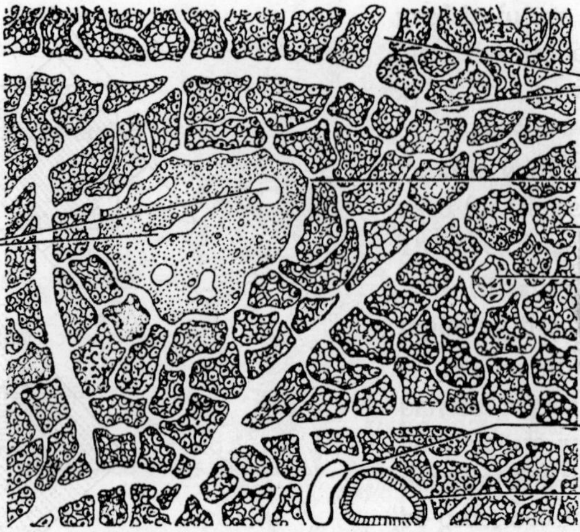

Fig. 4.4.19 : T.S. of mammalian pancreas.

open into inconspicuous intralobular ducts which forms tributaries of more conspicuous interlobular ducts traversing in between the lobules.

6. T.S. of mammalian spleen :

(1) T.S. shows outer cortical and inner medullary part.

(2) It is composed of muscular and fibrous coat, trabeculae, red pulp and while pulp.

(3) The capsule or Splenic coat extends in ward as **trabeculae**, which form a network in the substance of the gland. The trabeculae bring in blood vessels, nerves and gives substantial support to the gland. The capsule is made up of dense ordinary connective tissue made from collagen and elastic fibres and few smooth muscles covered by squamous mesothelium.

(4) In the interstices of the frame work lies a soft pulpy substance, called **splenic pulp** which may be red or white.

(5) The **white pulp** generally forms rounded aggregations of lymphocytes distributed among the red pulp as spherical islands. This forms the

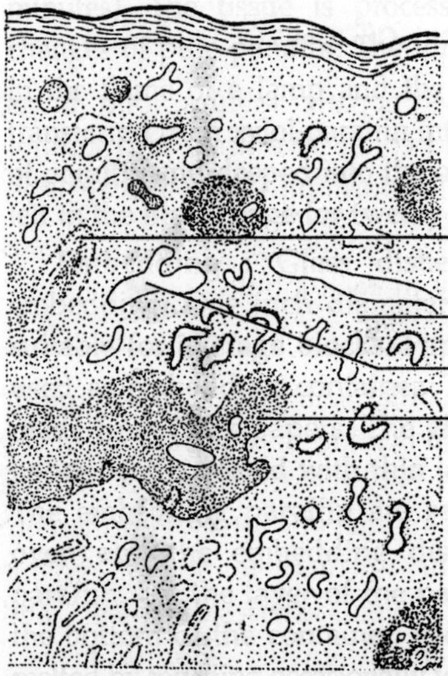

Fig. 4.4.20 : T.S. of mammalian spleen.

lymphatic sheath and **lymphatic nodules.** These nodules are arranged long small arteries emerging from the trabeculae.

(6) The **red pulp** forming the bulk of pulpy substance gives red colour to spleen. Red pulp is carried by network of reticular fibres and enormous number of blood cells found free in the interstices of the network.

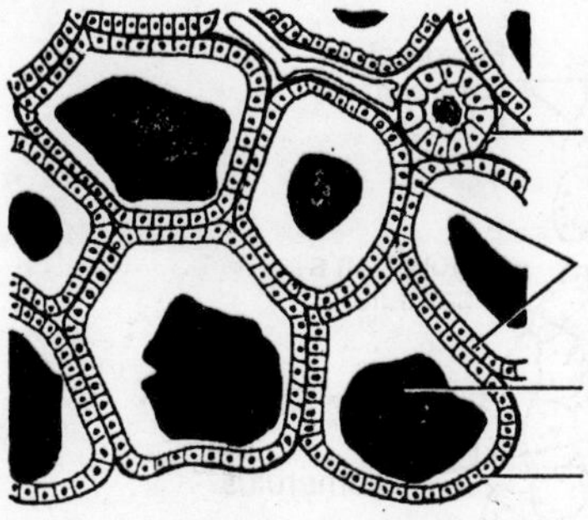

Fig. 4.4.21 : T.S. of mammalian thyroid gland.

7. T.S. of mammalian Thyroid gland :

(1) The T.S shows parenchymatous tissue divided into poorly defined lobules through fine septa arising from the fibro-elastic connective tissue capsule.

(2) Each lobule is made up of a large number of **thyroid follicles** (acinus) of various sizes lined with a single layer of cuboidal epithelium cells lying on a prominent basement membrane. The follicles are enclosed within a meshwork of connective tissue that supports a close network of anatomosing capillaries.

(3) The thyroid follicles are separated from one another by connective tissue strands.

(4) The cavities of the vesicles are filled with a viscous liquid, called thyroid **colloid,** containing iodine in the form of **thyroglobulin** having active component called **thyroxine.**

(5) In the walls of the follicles there are **parafollicular cells** which are found singly that do not reach the the central lumen.

8. T.S. of mammalian Adrenal gland :

(1) T.S. Shows that it's composed of outer cortex and inner medulla.

(2) The cortex is covered by a fibro-elastic connective tissue capsule and is divides into three zones:

Fig. 4.4.22 : T.S. of mammalian adrenal gland.

(a) An outer layer of **zona glomerulosa,** consists of columnar cells.

(b) A middle layer of **zona fasciculata,** consists of larger vacuolated cells that are arranged as one or two cell thick radially oriented columns having fenestrated capillaries in between.

(c) An inner layer of **zona reticularis,** consists of anastomosing network of pigmented reticular cells having pycnotic nucleus.

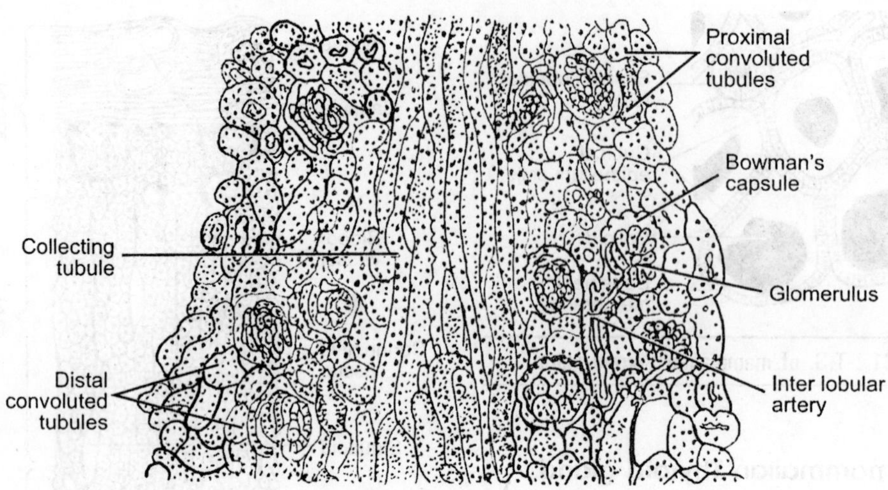

Fig. 4.4.23 : T.S. of mammalian kidney.

(3) The medulla contains elastic fibres, sinusoids and **chromaffin cells** containing granules.

(4) Medulla also bears post-ganglionic **sympathetic ganglionic cells** with prominent nuclei that occur in groups or alone.

9. T.S. of mammalian Kidney :

(1) T.S. shows a tough capsule made of dense connective tissue at the outermost part.

(2) The basic tissue has two distinct portions or lobes– outer granular **cortex** and inner radially striated **medulla.**

(3) Each lobe is made up of conical medullary pyramid and cortical tissue cap which is further sub-divided into many lobules. These lobules are mark out by inter-lobular arteries and bear a central core of medullary rays.

(4) Closely packed uriniferous tubules are present both in cortex and medulla.

(5) In the cortex there are sac like enlargements, **Bowman's capsule** that encloses an encircling network of blood vessels called **glomerulus** that together forms the **renal corpuscles** are visible.

(6) The tubule leaves the Bowman's capsule – by neck and it forms **proximal convoluted tubule**(lined by pyramidal cells with large spherical nuclei and brush border along the open surface), ascending limb, descending limb and **loop of Henle** (having narrow lumen lined by simple squamous epithelial cells). There are also **distal convulated tubule** with wider lumen. These are lined by smaller cells with spherical nuclei.

(7) The medulla is made up of **collecting tubules**, the main ducts which encloses a wider lumen that are lined by cuboidal or columnar epithelium.

10. T.S. of mammalian lung :

(1) T.S. Shows a large number of **alveoli** or air cells lined by cylindrical epithelial cells, **bronchi**, alveolar duct, blood vessels and nerve.

(2) The bronchi are lined by ciliated epithelium resting on a basement membrane.

(3) External to epithelial cells is a layer of mucous membrane containing large number of longitudinal elastic fibres and some lymphoid tissue.

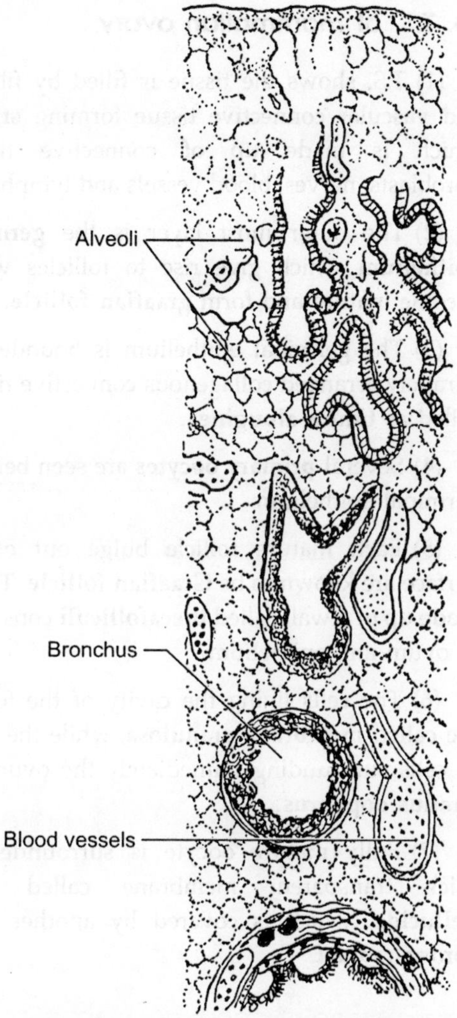

Fig. 4.4.24 : T.S. of mammalian lung.

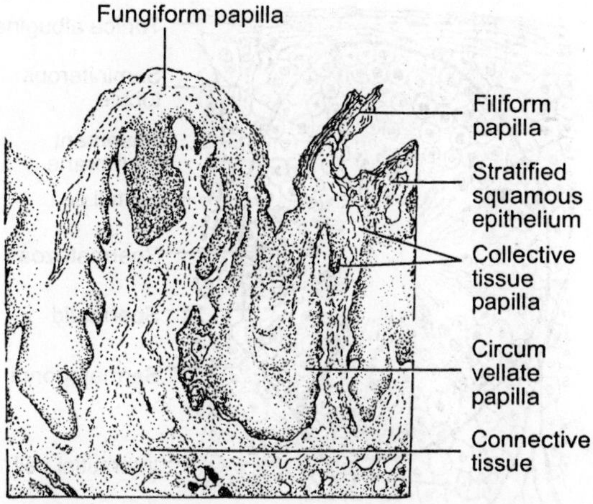

Fig. 4.4.25 : T.S. of mammalian tongue.

11. T.S. of mammalian Tongue :

(1) T.S. Shows stratified squamous epithelium which is not cornified.

(2) Epithelium contain three types of papillae – **filiform, fungiform** and **circumvallate**.

(3) Filiform papillae consists of a connective tissue core covered by stratified squamous keratinized epithelium.

(4) Fungiform papillae are mushroom shaped projections contain a connective tissue core, a stratified squamous surface and **taste bud** on their dorsal surface.

(5) Circumvallate papilla consists of connective tissue and is covered with stratified squamous epithelium.

(6) Papillae contain **serous gland** and **mucous gland.**

(7) Muscularis mucosa and serosa are absent.

(8) Submucosa are not well developed.

12. T.S. of mammalian Testis :

(1) T.S shows an outer thick capsule made from dense connectinve tissue known as **tunica albuginea**. This sends fibrous septa sub-dividing the internal tissue into incomplete lobules.

(2) Each lobule is made up of a large number of rounded or oval **seminiferous tubules, interstitial cells** and blood vessels.

(3) Seminiferous tubule is lined by **germinal epithelium**, the cells of which undergo **spematogenesis** to give sperm.

(4) In the seminiferous tubules there are some long nutritive and supportive cells, **sertoli cells.**

(5) Sertoli cells are slender pillar like cells with large pale staining nucleus.

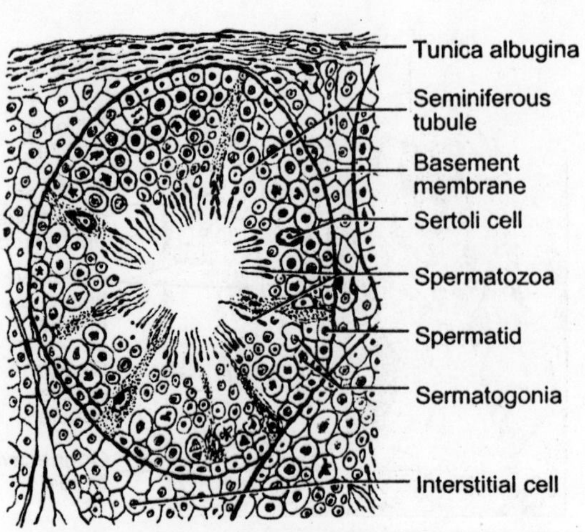

Fig. 4.4.26 : T.S. of mammalian testis.

(6) Each seminiferous tubule contains **spermatogonia, primary spermatocytes, secondary spermatocytes, spermatids** and **sperms** representating various stages of **spematogenesis.**

(7) The interseminiferous tubular tissue is formed by connective tissue which, hold the tubule together and contains blood vessels. Between connective tissue are groups of rounded interstitial cells which produce male hormone **testosterone**, which is responsible for developing secondary sexual characters.

13. T.S. of mammalian ovary :

(1) T.S. shows the tissue is filled by fibrous and vascular connective tissue forming **stroma** which is made up of connective tissue, fibroblasts, nerves, blood vessels and lymphatics.

(2) The outer most layer is the **germinal epithelium** which give rise to follicles which become mature and form **graafian follicle.**

(3) The germinal epithelium is bounded by parallely arranged collagenous connective tissue, called as **tunica albuginea.**

(4) Several **primary oocytes** are seen beneath germinal epithelium.

(5) Each mature follicle bulge out of the surface are known to be **Graafian follicle**. This is made up of a wall called **thecafolliculi** consisting of ovum and epithelium.

(6) The cells lining the cavity of the follicle are called **membrana granulosa**, while the mass of cells surrounding immediately the ovum are **cumlus oophorus.**

(7) Fully mature oocyte is surrounded by thick transparent membrane called **zona pellucida**, which is covered by another layer **corona radiata.**

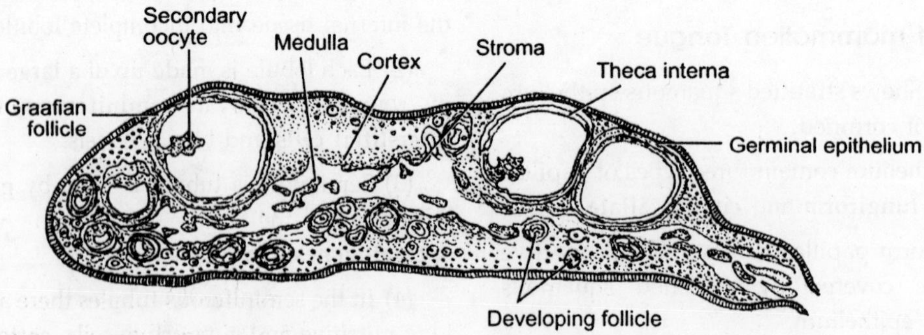

Fig. 4.4.27A : T.S. of mammalian ovary.

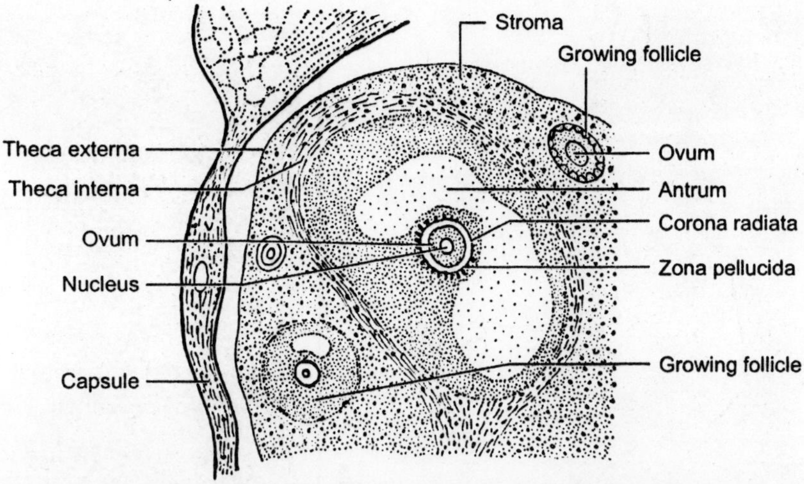

Fig. 4.4.27B : T.S. of mammalian ovary showing a mature follicle.

(8) After the extrusion of ova, a yellowish mass is left, called **corpus luteum**, which produces hormone **progesterone**. It facilitates the implantation of fertilized ovum in the uterus. Corpus luteum further regresses to form the **corpus albicans.**

(9) The medulla is highly vascular but compara-tively small with freely arranged connective tissue, a quantity of elastic fibres and muscles.

14. T.S. of mammalian lymph node :

(1) T.S. shows a tough *capsule* made of dense connective tissue which surrounds the node.

(2) The basic tissue has two distinct portions—*outer cortical* and *inner medullary* portion.

(3) The *trabecule* extend from the capsule into the substance of the node, also comprised of dense connective tissue.

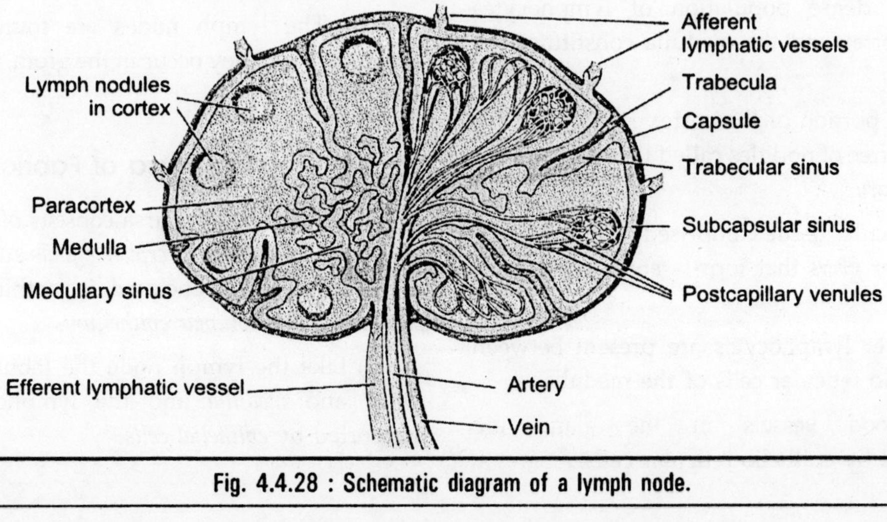

Fig. 4.4.28 : Schematic diagram of a lymph node.

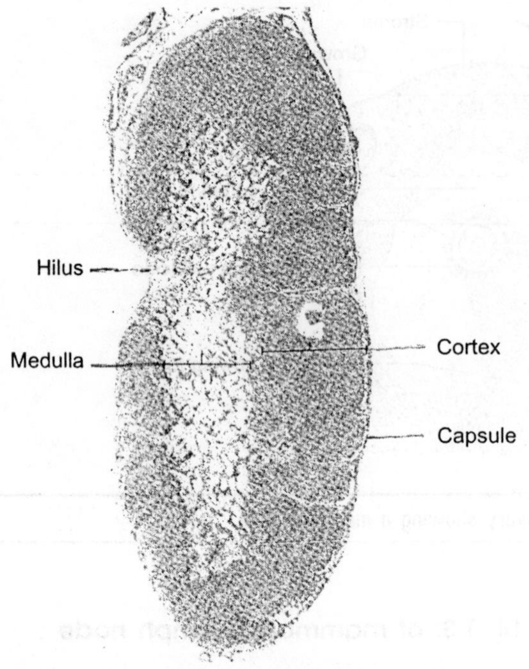

Fig. 4.4.28A : T.S. of mammalian lymph node.

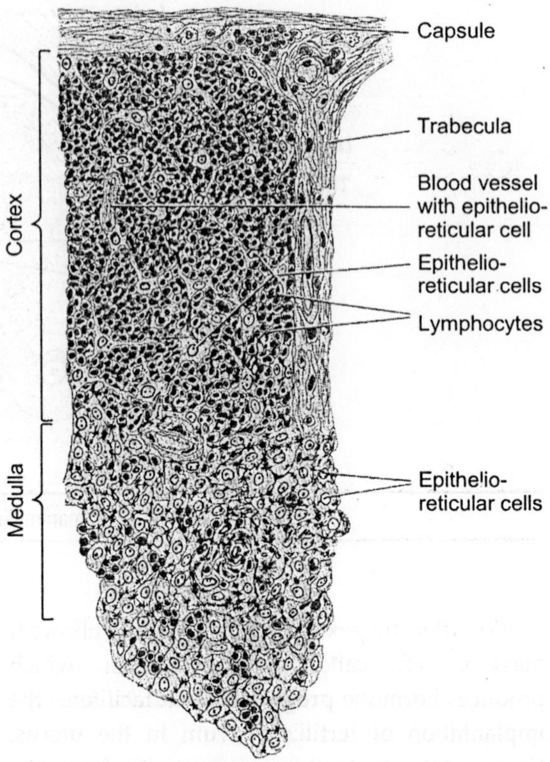

Fig. 4.4.28B : Diagram showing a portion of lymph node.

(4) The *cortex* contains numerous spherical or oval aggregates of lymphocytes called *lymph nodules.*

(5) The *medulla* is the inner most region of the lymph node and extends as far as the *hilus* of the node.

(6) The dense population of lymphocytes between cortex and the medulla constitutes the *para cortex.*

(7) The portion of the cortex adjacent to the medulla is free of nodules called the *deep cortex* or *juxtamedullary.*

(8) Reticular tissue comprised of *reticular cells* and *reticular fibers* that form a supporting mesh work.

(9) Fewer lymphocytes are present between the epithelio reticular cells of the medulla.

(10) Blood vessels in the gland are surrounded by epithelio-reticular cells.

Note : Lymph node sometimes referred to as a lymph gland. It is from here that the white blood cells known as lymphocytes circulate around the body in both the blood and lymph vessels. The lymph nodes are found around major arteries. They occur in the groin, the armpit and the neck.

15. T.S. of Avian Bursa of Fabricius :

(1) T.S. shows the bursa consists of a number of *lymphoid lobules* and *crpt like folds* surrounding a *lumen,* which is enclosed in a thin layer of *pseudostratified columar epithelium.*

(2) Like the lymph node the lobules have a *cortex* and *medulla* and the lymphocytes are supported by *epithelial cells.*

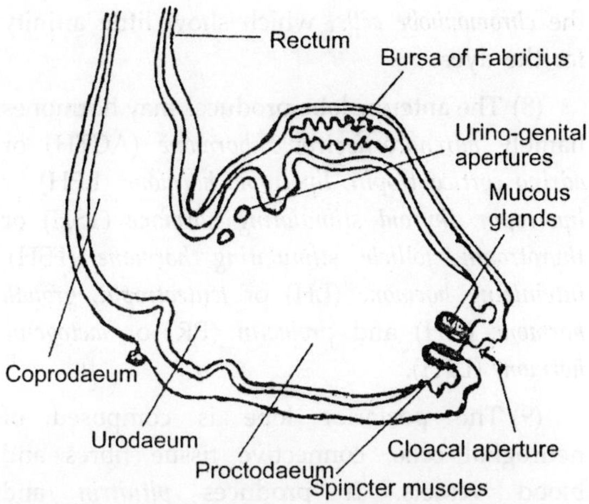

Fig. 4.4.29 : Diagram showing the position of Avian Bursa of Fabricius (interior view).

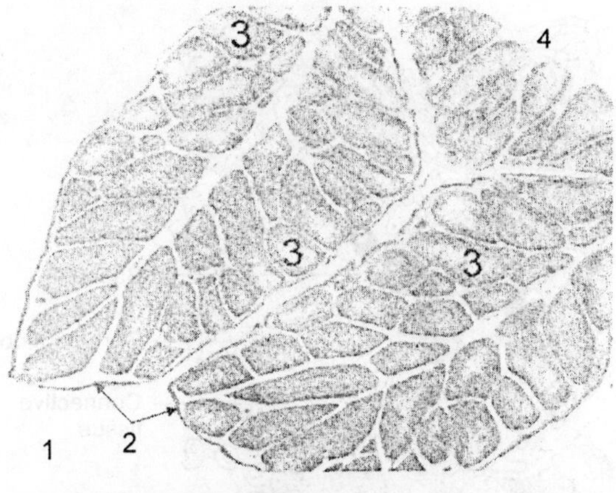

1–Lumen, 2–Pseudostratified columar epithelium
3–Lobules (follicles), 4–Muscularis

Fig. 4.4.29A : T.S. of Avian Bursa of Fabricius (25X).

(3) The folds of the bursa are provided with two distinct types of epithelial cells—*follicular epithelial cells* with few surface *microvilli* and *inter follicular epithelial cells* profusely covered with microvilli.

(4) The epithelium bordering the tunica propria was composed of two layers of cuboidal cells.

(5) The upper layer of the surface epithelium was dorsoventrally flattened and stained paler than the cells of the lower layer.

(6) Blood vessels are present in the cortex of the follicle.

Note : Heironymus Fabricius first described this obscure gland. The cloacal bursa produces lymphocytes and antibodies which have a defensive role in the body. The gland also participates in the production of WBC.

16. T.S. of mammalian pituitary gland :

(1) T.S. shows the gland is more or less glandular in shape and divided into the *adenohypophysis* and a neural portion, the *neurohypophysis*.

(2) Adenohypophysis is further subdivided into—(a) *pars distalis* (anterior lobe), (b) *pars tuberalis* and (c) *pars intermedia*.

(3) Neurohypophysis (posterior lobe) is again divided into—(a) *pars nervosa* and (b) *infundibulum*.

(4) The anterior lobe or pars distalis (constitutes about 75 percent of the hypophysis) is composed of *chromophobes* and *chromophils* cells on the basis of staining properties.

(5) Usually on the outerside are the *basophil cells* which are stained by basic dyes.

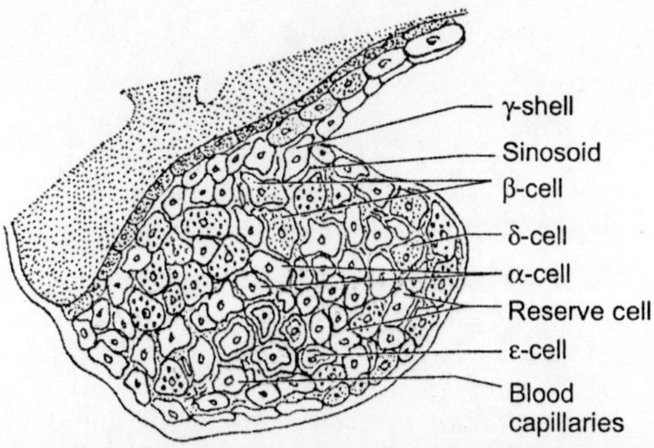

Fig. 4.4.30 : T.S. of mammalian pituitary gland.

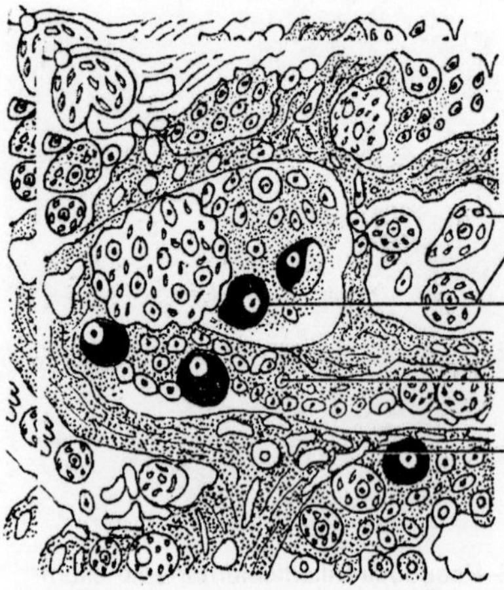

Fig. 4.4.30A : V.L.S. of anterior lobe of mammalian pituitary gland.

(6) In the centre are *acidophil* or *oxyphil* cells which bound acidic dyes.

(7) Scattered throughout the anterior lobe are the *chromophobe cells,* which show little affinity for the dyes.

(8) The anterior lobe produces may hormones namely *adrenocorticotropic hormone* (ACTH) or *adrino cortico tropin, lipotropic hormone* (LPH) or *lipotropin, thyroid stimulating hormone* (TSH) or *thyrotropin, follicle stimulating hormone* (FSH), *luteinizing hormone* (LH) or *leuteotropin, growth hormone* (GH) and *prolactin* (PR) or *lactogenic hormone* (LTH).

(9) The posterior lobe is composed of neuroglial cells, connective tissue fibres and blood vessels. It produces *pituitrin* and *antidiuratic hormone* (ADH) or *vesopressin* hormones.

(10) The pituitary gland or hypophysis is an endocrine gland of utmost importance to organisms.

(Please see colour plates : Section 4)

Chapter 4.5

Preparation and Staining of Cells and Tissues and Organisms

1. Staining and mounting of blood film :

Preparation :

(a) Two clean grease free dry slides are taken.

(b) One of the finger tips is cleaned with absolute alcohol and punctured with a disposable needle. Blood will flow out from the wound.

(c) The slide is kept on a horizontal table and a drop of blood not larger than a pins head is taken on this slide close to its edge. The slide is held firmly with one's left hand.

(d) The ground edge of the other slide is placed at an angle 45° in contact with the drop of blood and then lowered to an angle of 30°. Blood will flow all along the edge and now the slide is pushed gently forward, till the blood is exhausted. As the blood is exhausted a thin blood film (tongue-shaped) is formed over the slide. The entire process must be carried out before the blood clots.

(e) The film is allowed to dry immediately, by waving the slide quickly through the air, to prevent shrinkage of blood cells. A pink coloured thin film of blood will be seen on the slide. If the film is not thin or uniform, the process should be repeated in another clean slide.

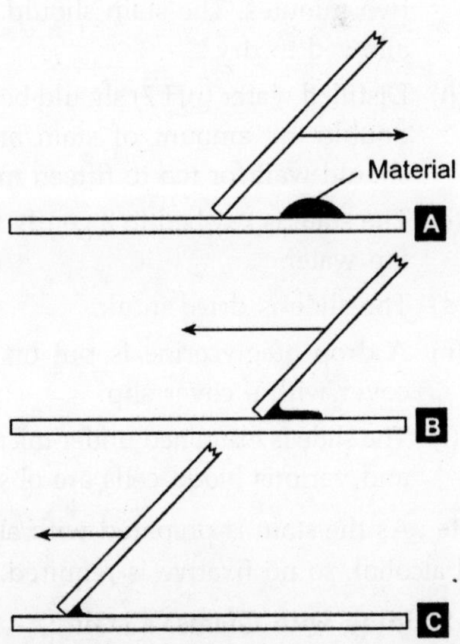

Fig. 4.5.1 : Preparing blood film. A–C–Method of spreading the material on the surface of one slide with the edge of a second slide.

The second smear may be made on the surface of the slide used to spread the first, provided the surface of the slide is untouched.

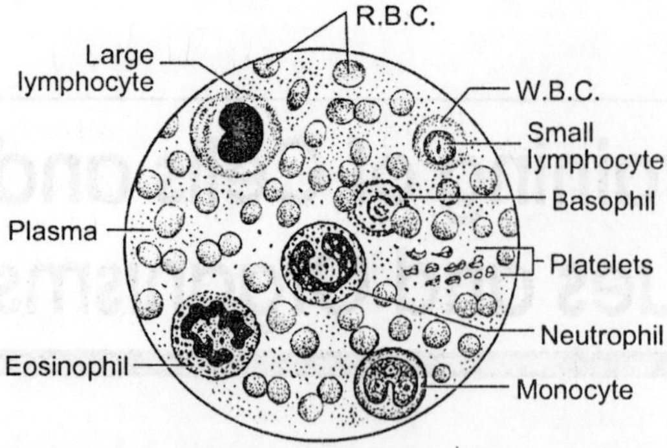

Fig. 4.5.2 : Mammalian blood cells under microscope.

(i) **Staining with Leishman's strain :**

(a) 3-4 drops of leishman's stain is poured over the dried film and kept for one to two minutes. The stain should not be allowed to dry.

(b) Distilled water (pH 7) should be added double the amount of stain and one should wait for ten to fifteen minutes.

(c) The stain is washed in a gently flow of tap water.

(d) The slide is dried in air.

(e) A drop of glycerine is put on it and cover with a cover slip.

(f) The slide is examined under microscope and various blood cells are observed.

Note : As the stain is prepared with absolute methyl alcohol, so no fixative is required.

(ii) **Staining with Giemsa's stain :**

(a) Absolute methyl or ethyl alcohol is poured over the dry film and waited for three to five minutes. The slide is allowed to dry.

(b) The Giemsa stain is diluted by adding 1 ml of distilled water to each one drop of stain.

(c) Diluted stain is poured over the film and waited for twenty to thirty minutes. The slides may be covered to reduce evaporation of the stain.

(d) The slide is washed in a gently flowing tap water. The water is shaken off and the slide is allowed to dry in air.

(e) A drop of glycerine is put on it and covered by a cover slip.

(f) The slide is examined under microscope and the nuclei are observed as deep blue or purple and cytoplasm as pink.

2. Squamous epithelium of Frog/Toad :

Preparation :

(a) A frog/toad is put in a small trough containing sufficient water to submerge the frog/ toad. After some time small flakes of epidermal cells discarded by the Frog/toad will be seen floating.

(b) Few of them are collected with a fine brush over a slide and then fixed with 90 percent alcohol.

(c) Stained in borax carmine.

(d) Dehydrated, cleared and mounted.

(e) Observed under microscope the large polygonal cells with centrally placed blue stained nucleus.

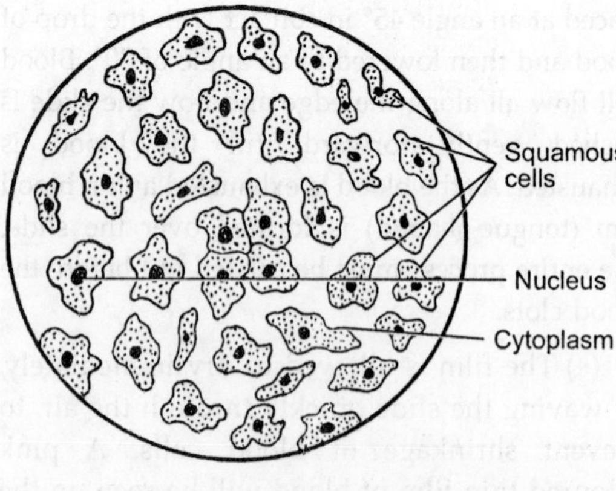

Fig. 4.5.3 : Squamous epithelium of frog.

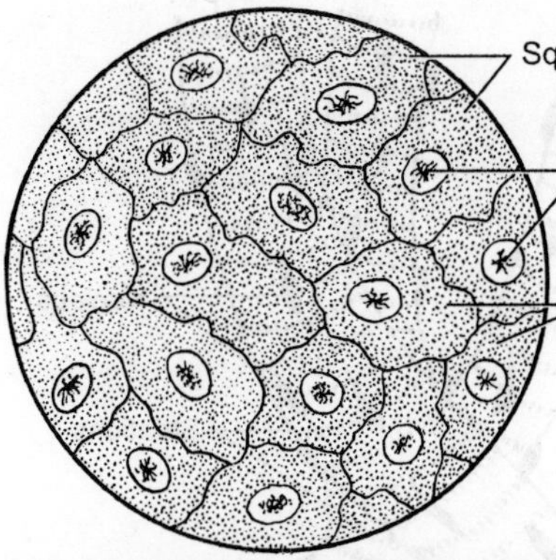

Fig. 4.5.4 : Squamous epithelium of human.

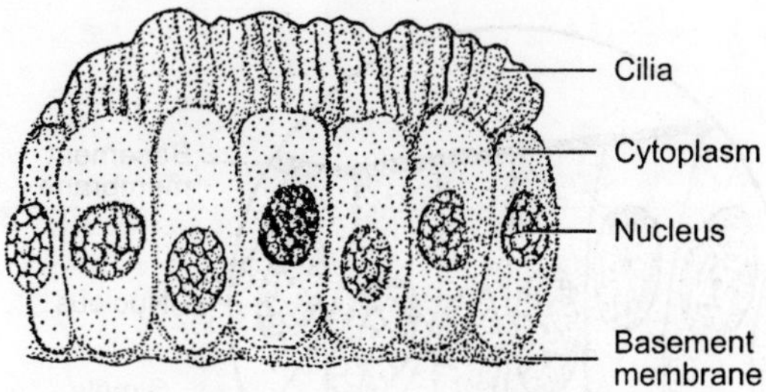

Fig. 4.5.5 : Ciliated epithelium cells.

3. Squamous epithelium of human :

Preparation :

(i) **First method :**

(a) The mucous membrane is scraped of the inner surface of the candidate's lower lip with the help of a clean cover glass.

(b) The scraped material is transfered over a clean slide and moisten it with a drop of normal saline.

(c) The scraping is uniformly spread by pressing a cover glass over the slide. One drop of methylene blue is poured at the edge of cover slip and the stain is drawn inside by putting a blotting paper on the other edge of the coverslip. One should wait for two minutes.

(ii) **Second method :**

(a) The mucous membrane is scraped as before.

(b) A film is drawn on a clean slide with the scraping and allowed to dry for five minutes.

(c) Two to three drops of methylene blue is poured and kept for two to three minutes. Care should be taken so that the stain may not get dry.

(d) The slide is washed in gently flowing tap water.

(e) A drop of glycerine is poured on the slide and then the slide is covered with a cover slip.

The slide is examined under microscope and large polygonal cells arranged like tiles on a floor with centrally place blue stained nucleus with faintly stained granular cytoplasm is observed.

4. Ciliated epithelium :

Preparation :

(a) The roof of the mouth or the mucous membrane of the pharynx of a toad/frog is superficially and gently scraped with a blunt end of a scalpel or with a bent spatula.

(b) The scraping material is transferred over a slide and spread uniformly.

(c) Few drops of 90 percent alcohol are poured on it and waited for one to two minutes.

(d) The slide stained in borax carmine.

(e) Dehydrated, cleared and mounted.

(f) Observed under microscope. Columnar cells with a bunch of fine tapering cilia on the outer surface of the cells are noticed.

5. Columnar epithelial cells :

Preparation :

(a) A small piece of intestine from a freshly killed frog/toad is taken out. This is then put in 70 percent alcohol for nearly twenty-four hours.

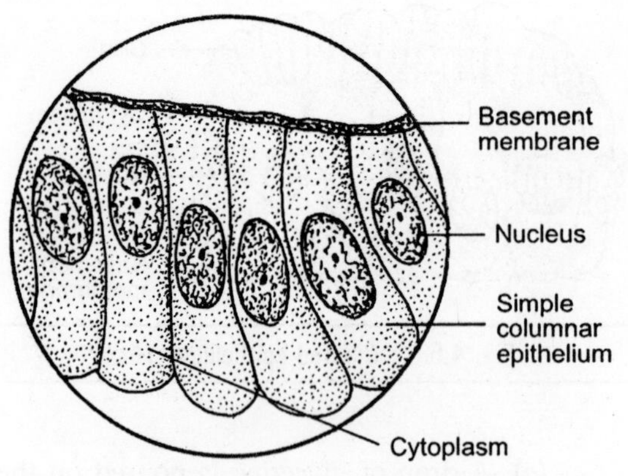

Fig. 4.5.6 : Columnar epithelium cells.

(b) A longitudinal incision is made to open it. The contents of the gut removed with a fine brush and the internal epithelial lining is carefully scraped by a fine scalpel.

(c) The scrapings are transferred in a slide and little water is dropped on it and the material is spread uniformly.

(d) 90 percent alcohol is poured and kept for two to three minutes.

(e) Stained in borax carmine.

(f) Dehydrated, cleared and mounted.

(g) Observed under microscope. Prismatic column like tall cells, closely arranged side by side with prominent usually ovoid nuclei at the basal region.

6. Skeletal muscle at the basal region :

(a) Striated muscle fibre of Toad/Frog :

Preparation :

(a) A toad/frog is killed by pithing on the under surface of the cranium with the tip of stout needle by rotating it clock and anti-clock wise to smash the brain.

(b) A small portion of thigh muscle is cut and placed in a petridish. **Corrosive sublimate** (100 cc mercuric chloride and 5 cc glacial acetic acid v/v) is poured and kept for three to ten minutes.

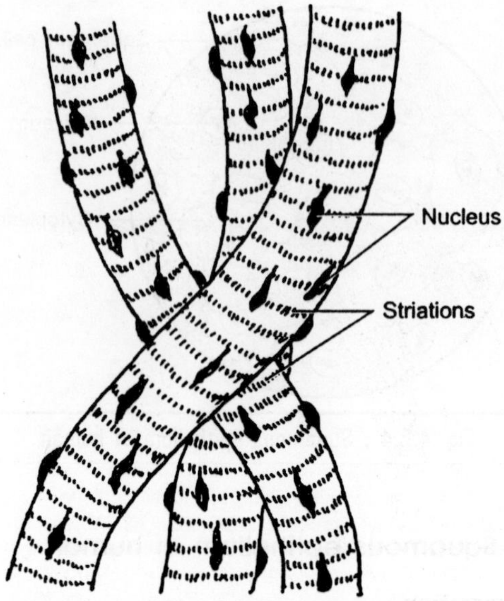

Fig. 4.5.7 : Skeletal muscle fibre (striated) of toad.

(c) Thoroughly washed in 70 percent alcohol to remove fixative.

(e) Stained in borax carmine.

(f) Dehydrated and cleared.

(g) The individual muscle fibre is teased with two fine needles and the fibres are arranged parallel to one another on a slide.

(h) A drop of Canada balsam is put over and covered it with a cover glass.

(i) Examined under microscope and multinucleated striated with dark and light bands are noticed.

(b) Smooth muscle fibre of Toad/Frog :

Preparation :

(a) A small portion of urinary bladder of killed toad/frog is taken in small petridish.

(b) Corrosive sublimate (100 cc mercuric chloride and 5 cc glacial acetic acid v/v) is poured and kept for five to ten minutes.

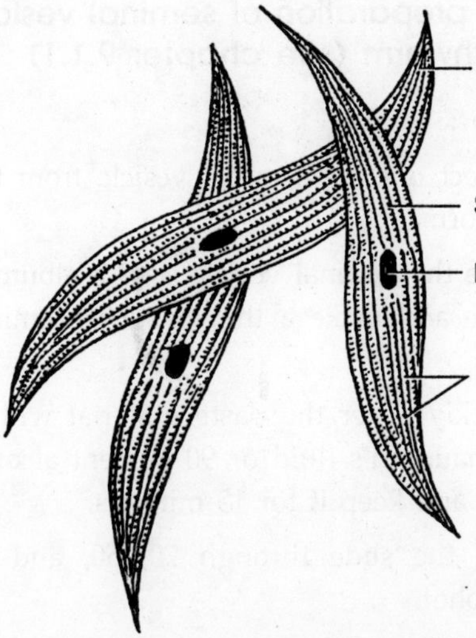

Fig. 4.5.8 : Smooth muscle fibre of toad.

(c) Washed thoroughly in 70 percent alcohol to remove the fixative.

(d) Stained in borax carmine.

(e) Dehydrated and cleared.

(f) The individual fibres are teased with fine needle and arranged on a slide.

(g) A drop of canada balsam is put over it and covered it with a cover glass.

(h) Examined under microscope and uninucleated cells without striations are noticed.

7. Paramoceium sp. :

Paramoceium is common in fresh water ponds, ditches, lake etc. and is specially found in abundance in stagnant ponds rich in decaying organic matter and in sewage water. *Paramoecium* thrive well in culture medium.

(a) Preparation :

(1) Submerged fresh and decomposing weeds from a ditch or pond are collected and placed in a glass jar containing distilled water. Covered and allowed the weeds to decompose. Within a few days swarms of Paramoecia will appear.

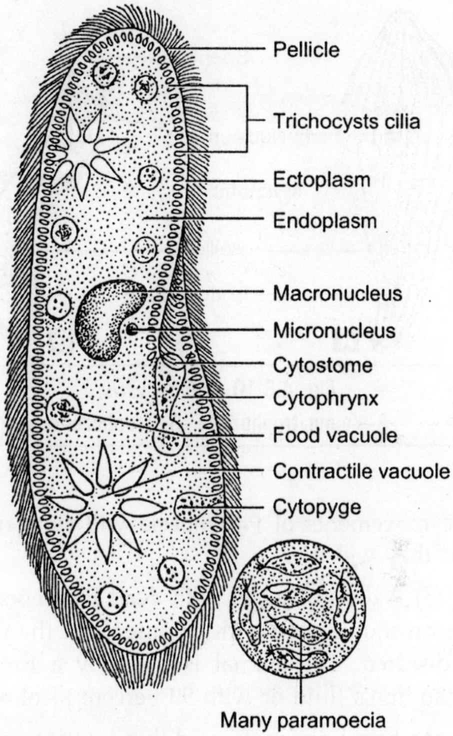

Fig. 4.5.9 : *Paramoecium* and a magnified view showing many *paramoecia* in hay culture.

(2) ***Hay culture :***

(a) Few wheat grains are boiled along with few cut hay stems in 500 ml of distilled water in a conical flask for about fifteen minutes.

(b) The flask is kept in a cool place for about four days.

(c) Paramoecia from the preparation are transferred into this flask by a micropipette. Within a few days many Paramoecia will grow in the calture.

(b) Staining :

(1) About 50 ml culture medium is filtered and a drop of formaldehyde solution is added into it.

(2) A measured amount of culture medium is taken in a watch glass and equal amount of formalinated culture medium is added into it.

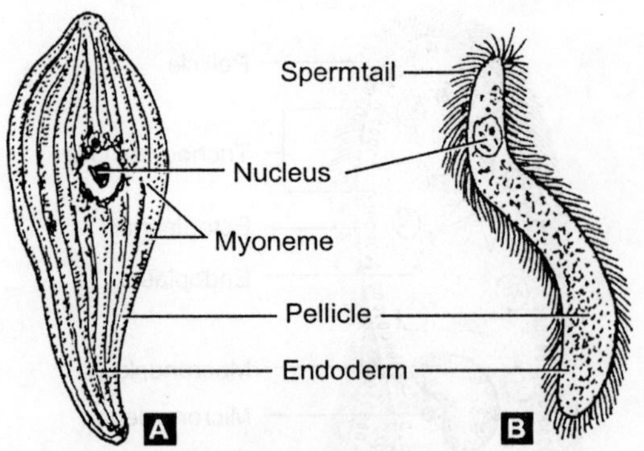

Fig. 4.5.10 : *Monocystis.*
A–Young trophozoite, B–Trophozoite.

The movements of Paramoecia will slow down and they will die.

(3) A drop of water with dead Paramoecia is put on slide coated with albumen and the water is discared. The animal is fixed by a drop of schaudinn's fluid or with 90 percent alcohol.

(4) The slides are passed through descending grades of alcohol (90–70–50–30 percent) and distilled water.

(5) Stained first with haematoxylin and then in eosin following the double staining method (see section 13 of chapter 4.4 a & b) or stain with borax carmine following the single staining method.

(6) Dehydrated, cleared and mounted in Canada balsam or any other suitable mountant.

(7) The stained slides are examined under microscope and the following characters of *Paramoecium sp.* are observed.

(c) Observation :

(a) Slipper shaped animal with round anterior end and a pointed posterior end.

(b) Body covered with cilia arranged in rows.

(c) One large macro and small micronucleus is present.

(d) Two contractile vacuoles at apposite poles with radial canals.

(f) Oral groove and cytopharynx is present.

8. Smear preparation of seminal vesicle of earthworm (see chapter 9.1.1) :

Preparation :

(a) Dissect out the seminal vesicle from the live earthworm.

(b) Place the seminal vesicle over a albumen coated slide and make a thin paste of seminal vesicle.

(c) Quickly cover the paste material with a layer of schaudinn's fluid or 90 percent alcohol for fixation and keep it for 15 minutes.

(d) Pass the slide through 70, 50, and 30 percent alcohol.

(e) Stain in haematoxylin and eosin following the double staning method.

(f) Dehydrate, clear and mount in balsam.

(g) Examine under microscope and if infected observe trophozoites, gametocyst and sprozoite stages of monocystis alone with spermatogonia and spermatozoa of earth worm.

9. Smear preparation of gut parasite of cockroach (see chapter 9.1.2) :

Preparation :

(a) Dissect the live cockroach and open the hind gut by giving a logitudiual incision with a fine scissor.

(b) Pour 0.78 percent normal saline solution in the open gut.

(c) Smear the inner wall of the gut, discarding the faecal matter over an albumen coated slide.

(d) Quickly place the slide into schaudinn's fluid or 90 percent alcohol and wait for 15 minutes.

(e) Pass the slide through 70, 50 and 30 percent alcohol and then to distilled water.

(f) Stain in haematoxylin, wash in distilled water.

(g) Dehydrate and mount in balsam.

(i) Examine under microscope and observe **parasitic Balantidium** and **Nycto therus :**

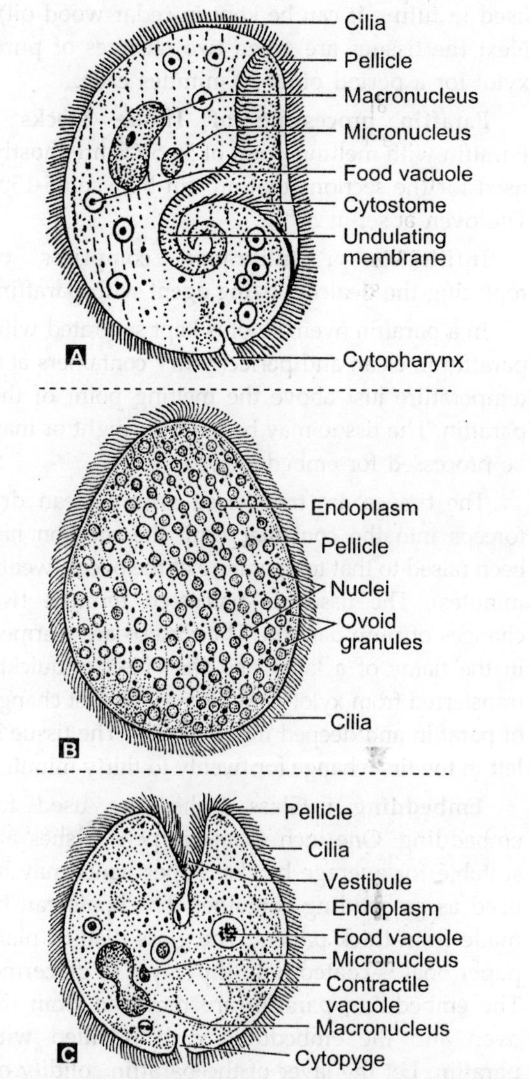

Fig. 4.5.11 : A–*Nyctotherus*, B–*Opalina*, C–*Balantidium*.

10. Smear preparation of rectal contains of Toad/Frog (see chapter 9.1.3) :

Preparation :

(a) Dissect the toad/frog and open the rectum by giving a longitudinal incision with a fine scissor and put 0.78 percent normal saline solution.

(b) Smear a thin film of the rectal contain over an albumen coated slide.

(c) The rest procedure as before.

(d) Examine under microscope and observe **Nyctotherus, Opalina** and **Balantidium.**

11. Preparation of Blood smears :

(a) Place a small drop of blood on the centre of a clean, square cover glass. Immediately drop a second cover glass on the first, positioning it so that the corners protrude beyond the sides of the first as shown in the figure (Fig. 4.5.12).

(b) The blood will spread between the two cover slips which are then slipped apart by pulling in opposite direction in the plane of their surface.

(c) Wave the cover slips through the air so that the film dries.

(d) Process the smears as described in blood film preparation.

12. Microtomy and preparation and staining of tissue :

Microtomy is the process in which *microtome*, a machine, is used for cutting the organs and tissues into very thin sections for histological studies.

The target organs or tissues are fixed, dehydrated, cleaned and embedded in paraffin. The paraffin blocks are then made of embedded material. Such blocks are then mounted on a block holder which is fixed to the microtome. The sections obtained by this process as fine, thin and uniform and useful for various histological studies.

Preparation of tissue :

Fixation : The animal concerned for tissue fixation is first chloroformed, dissected and target organ is taken out. The tissue thus removed must be placed in aqueous Bouin's fluid as soon as possible. If the material is large then after half an hour the material is cut into two-three pieces and again placed in the fixative (Bouin's fluid) for 12 hours to several days. Alcoholic Bouin's fluid may be used as fixative in place of aqueous Bouin's fluid. It is very essential to fix the material in the fixative for the minimum specific time but should be taken out before the maximum specific time is over. Fixative renders

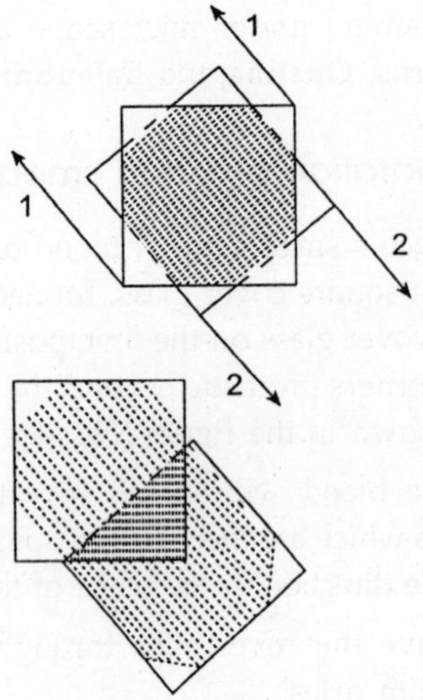

Fig. 4.5.12 : Preparing smears on cover glasses. Corners 1·1 and 2·2 are held in the finger tips of opposite hands.

hardness to the material and also coagulates and render the elements of the material insoluble.

Washing of fixative from the material : After the fixation is completed the material is taken out and placed in a beaker containing distilled water. The material then washed in distilled water, changing water several time as long as yellow colour comes out. In place of distilled water running tap water may also be used for removing the fixative.

Infiltration and embedding : After fixation the tissue has to be infiltrated with paraffin which is necessary for further tissue processing and sectioning.

As paraffin is insoluble in water the tissue must be perfectly *dehydrated* before infiltration with paraffin. Dehydration is done through the use of increasing concentration gradient of alcohol. The dehydration is done by passing the tissue through 30% → 50% → 70% → 90% and two changes in 100% alcohol (twenty to thirty minutes in each)

As alcohol is immiscible in paraffin it must be removed through *dealcoholization*.The tissue has to be transferred from 100% alcohol to 1:1 mixture of 100% alcohol and xylol. (If the tissues are to be used in future it can be kept in cedar wood oil). Next the tissues are given two changes of pure xylol for a period of 20-30 minutes each.

Paraffin processing of Tissue blocks : Paraffin with melting point of 56-58°C are mostly used for the sections that are cut between 5-15μ. The oven is set at 58°C.

Infiltration : Infiltration comprises of replacing the dealcoholizing agent with paraffin.

In a paraffin oven, xylol is kept saturated with paraffin in clean and perfectly dry containers at a temperature just above the melting point of the paraffin. The tissue may be left over night or may be processed for embedding.

The tissues are transferred with a clean dry forceps into the container until the solution has been raised to that temperature of the oven (twenty minutes), The tissue is processed through two changes of pure paraffin, The forceps are warmed in the flame of a lamp and the object is quickly transferred from xylol-paraffin into the first change of paraffin and deeped into the oven. The tissue is left in the first change for twenty to thirty minutes.

Embedding : Glass dishes are used for embedding. One inch diameter petri dishes are suitable for average blocks. Paper boats may be used as embedding container. The boats can be made from thick papers. The glass dish or plain paper boat is coated with thin layer of glycerine. The embedding paraffin are removed from the oven and the embedding dish is filled with paraffin. Let the layer of the paraffin solidify on the bottom of the container, keeping the surface melted by touching occasionally with warm forceps.

The forceps are warmed and the tissue is immediately transferred from the oven into the embedding boat/glass dish. Any trapped air bubbles in the paraffin should be carefully removed during the process. The Embedding dish is then carefully transferred to the cool water and lowered until the surface is just above water. The blocks are left in water for ten to forty minutes.

Trimming : The paraffin block must cautiously trimmed before attachment to the microtome peg.. The block is trimmed with a clean, sharp single edged razor blade. I mm of paraffin is left on the face of the block. Enough paraffin (3-4 mm) must be left on the base of the block.

The paraffin block must be firmly attached to the microtome peg. A spatula is heated and placed on the top of the peg and on the bottom of the paraffin block. The block is immediately pressed against the peg. Some pressure must be applied with the finger on the top of the block so that the block fits uniformly on the peg.

Sectioning : After the tissue block has been attached to the peg of the microtome it is ready for sectioning. The sections should be of known and uniform thickness and free from scratches, breaks or folds. Wheel of the microtome is turned. If uniform sections are produced and the paraffin ribbon is lengthening, the ribbon has to be supported with a soft brush held in the left hand. A ribbon of required length is cut. The ribbon is removed from the knife edge by lifting the ribbon from the knife edge with another brush.

The sections are stored in a tray in order, in which they are cut. It is noticed that one surface of the ribbon is dull and the other is shiny.

Affixation : A small drop of Meyer's albumin solution (affixative) taken on the upper centre of a clean slide. With clean finger tip the affixative is rubbed over the area where section will be affixed. The slide is placed on a flat surface and few drops of water are added to the centre of the slide. The paraffin ribbon is subdivided into desired length with a scalpel.

The sections are transferred to the drop of floating medium on the slides. The section should be placed in the centre of the slide. The slides should be transferred to the thermostatically controlled hotplate whose temperature should be below the melting point of paraffin. Let the sections spread, until they are completely flattened. When the sections are completely flattened, the slides are removed from the hot plate. The excess water is drained away. The sections are oriented to the centre of the slide with the help of a needle tip.

(a) STAINING (Double staining) :

Section of tissue are usually stained with two dyes (double staining) to bring contrast between different histological structures the process is known as double staining.

(a) The paraffin from the dry sections are removed by putting the slides in xylol and leave for five minutes.

(b) The xylol is removed by placing the slide in absolute alcohol through two changes for five minutes in each solution.

(c) The materials are hydrated by placing through 90, 70, 50 and 30 percent alcohol and into distilled water. The slide is left in each solution for five minutes.

(d) The material is stained in haematoxylin (Delafield's). The material may be left for more or less one minute in undiluted stain, for one to five minutes in stain diluted with distilled water.

(e) The slide is rinsed in distilled water and examined under microscope and one should be careful not to let it dry out. The nucleus should be blue and the cytoplasm colourless. It the stain is to dark, destain in acid water (0.5 to 1.0 percent hydrochloric acid in distilled water), agitating it slightly; the slide is removed and rinsed well in distilled water to stop distaining.

(f) Place the slide on the microscope stage and the material is examined. If the stain is too light, the slide may be processed to tap water and restained.

(g) The destained slide is rinsed in running tap water for thirty minutes to one hour. This will remove the iron alum of the stain, clearing the cytoplasm and will avoid later fading of the stain after mounting.

(h) Dehydrated through 30, 50, 70 and 90 percent alcohol; left for two or three minutes in each solution.

(i) Counter stain in eosin. Stained for ten to fifteen seconds. Rinsed quickly through two changes of 90 percent alcohol and the slide is placed in absolute alcohol and then the slide is examined under microscope. One should be careful not to let it dry out. The cytoplasm should be pink. If necessary return to the stain.

(j) Dehydration is completed through two changes of absolute alcohol, left for five minutes in each change.

(k) Dealcoholize and the material is cleaned through two changes of xylol; left for five minutes in each change.

(l) The slide is removed from xylol, and without draining the slide, a drop of balsam or any mountant is placed to one side of the material at about the position that one end of the cover slip will occupy when it is in place.

(m) A cover slip is held with a pair of forceps and brought to the opposite end in contact with the drop of balsam or any other mountant; the cover glass is held at an angle of about 45° until the drop of mountant has flowed out along the edge of the cover glass. Slowly the cover slip is lowered on the material and one should be careful not to allow any air bubble between the cover slip and the slide. The entire process must be carried out before the tissue can dry out.

(n) With a clean cloth the bottom and upper ends of the slide is wiped. If required the cloth is to be slightly moistened with xylol and wipe the slide clean.

(o) The slide is kept flat and allowed to harden the balsam or any other mountant.

Chart for double staining (Permanent preparation) :

Deparaffinze the tissue in xylol I and II → Remove xylol in 100 percent alcohol I and II → Hydrate by running through 90 percent alcohol → 70 percent alcohol → 50 percent alcohol → 30 percent alcohol → Rinse with distilled water (through two changes) → Stain in haematoxylin → Rinse with distilled water → Destain in acid water (if required) → Dehydrate through → 30 percent alcohol → 50 percent alcohol → 70 percent alcohol → 90 percent alcohol → Stain in eosin → Rinse with two changes of 90 percent alcohol → Complete dehydration through 100 percent alcohol I and II → Dealcoholize and clear in xylol (two changes) → Mount in balsam or any other mountant.

Time limit : Time five to fifteen minutes in each reagent.

Summary of Haematoxylin & Eosin staining technique :

Cell and cellular component	Stain reaction
Nucleus	Blue
Cytoplasm	Pink

(b) STAINING (single staining with borax carmine) :

Preparation and staining of wet whole mount : Specimens contracting or reacting suddenly in chemicals are narcotized in water mixed with menthol crystals or magnesium sulphate. Narcotizing liquid is decanted and the animals are then fixed by adding drop by drop formalin. They are preserved in 70 percent alcohol or 5 percent formalin solution.

(i) *Hydra sp.* :

Preparation :

(a) Some water is collected from a pond with some aquatic plants. This is kept in a glass jar and left to stand for a few hours. To naked eyes one can see, whitish threads attached to the leaves and stem of plant and the walls of the jar.

(b) The animals are narcotised by sprinkle a few crystals of menthol or magnesium sulphate on water surface.

(c) Few narcotized hydra are carefully collected with a fine brush or point of fine forceps and placed on a slide.

(d) Animals are fixed by adding drop by drop formalin. The materials are washed through two to three changes of water till the fixative is washed out.

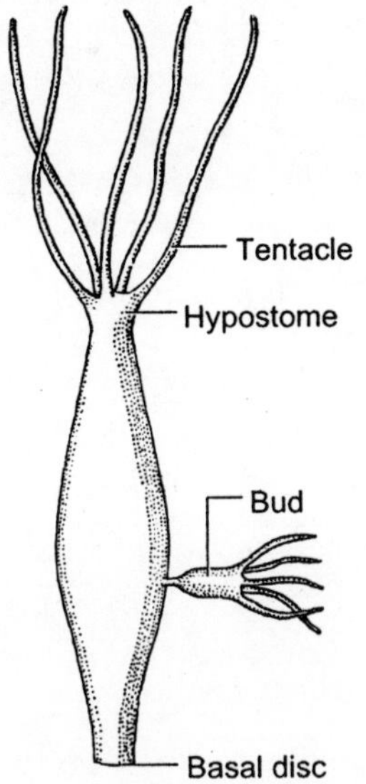

Fig. 4.5.13 : *Hydra.*

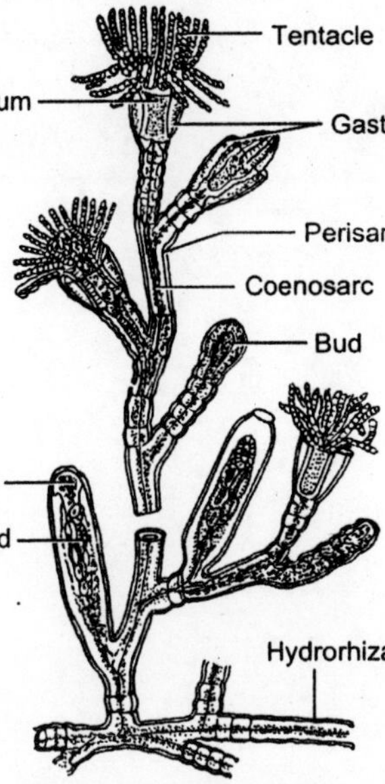

Fig. 4.5.14 : *Obelia colony.*

(e) The material are dehydrated by transferring it through 30, 50 and 70 percent alcohol for five minutes in each solution.

Staining :

(a) The specimen is left in 70 percent alcohol and stained in borax carmine. It is next washed in 70 percent alcohol.

(b) Dehydrated, cleared and mounted in balsam or any other appropriate mountant (see chart for single staining procedure).

(c) Examined under microscope.

Observation :

(a) Almost cylindrical body.

(b) Presence of flat basal disc and a conical hypostome.

(c) Hypostome sorrounded by a circlet of tentacles.

(ii) *Obelia sp.* :

Preparation :

(a) A portion of formalin fixed *Obelia sp.* containing hydranth and blastostyle is cut and placed an a watch glass.

(b) Washed thoroughly with distilled water and stained with borax carmine as described previously.

(c) Dehydrated, cleared and placed in a slide and mounted in canada balsam or any other appropriate mountant (see chart for single staining procedure).

(d) Examine under microscope. and observed.

Observation : The hydranth and blastostyle of obelia colony.

Chart for single staining preparation) :
The material is fixed → Washed thoroughly in tap water → Dehydrated with → 30 percent alcohol → 50 percent alcohol → 70 percent alcohol → Stained in Borax carmine → Destained in acid alcohol (if required) → Dehydrated with 90 percent alcohol → 100 percent alcohol (through two changes) → Dealcoholized and cleared in xylol (two changes) → Mounted in Canada balsam. **Time limit :** Five to fifteen minutes in each reagent.

Section 5

CYTOLOGY

SECTION AGENDA

Chapter 5.1

Preparation and Identification of Salivary Gland Polytene Chromosomes from *Drosophila sp.* and *Chironomus sp.* larva

Introduction

In secretory tissues such as rectum, malphigian tubules, gut, footpads and salivary glands of the dipterans the chromosomes apparently replicate their genetic material many times than the actual separation of the distinct chromatids, through the process known as **endomitosis**. The bundle of multiple replicas is known as the **polytene chromosomes**. In *Drosophila* there are 4 haploid set of chromosomes, for some reason, homologues are tightly paired. Again all four are joined at the **chromocenter**. This is represented as coalesence of heterochromatic areas around the centromeres of all chromosome pairs. These giant polytene chromosomes are found in the salivary glands of dipteran genera such as *Drosophila sp., Chirononius sp.* and *Sciara sp.*, thus are known as the **Salivary gland chromosomes**. Along the length of the chromosomes characteristics stripes called **bands,** which differ in width and morphology, can be recognized and observed. Further there are regions that appear as swollen (**puffs**) or greatly distended (**Balbiani rings**). These regions are related to regions of genetic activity.

(a) From *Drososphila sp.* (Fruit fly) larva :

Materials required :

1. Fruit fly larva (wild type and tandem duplication mutants)
2. Ringers insect saline
3. Fine forceps and probe
4. **Aceto-orcein (2-3%)** : 2-3 g of orcein is dissolved in 100 mL of 45 % glacial acetic acid. The mixture is boiled gently for 5-10 minutes followed by cooling and filtering.
5. **Aceto-lacto-orcein** : To freshly prepared aceto-orcein equal volume of lactic acid is added at room temperature. Mixed gently and left overnight under refrigeration. Filtered immediately before use.
6. 45% acetc acid.
7. **Aceto-methanol/Aceto-ethanol solution** : One part of glacial acetic acid is added to three parts of methanol/ethanol (v/v). Solution should be prepared fresh.
8. Dissecting and regular compound microscopes.
9. Slides.

10. Cover slips.

11. Paint brush.

12. Melted paraffin or Nail polish.

Procedure :

1. A third instar larva is selected, for which the cuticle has not yet hardened, from a wild-type culture of *Drosophila sp*. It is placed into a drop of Ringer's insect saline solution on a slide.

2. The slide is placed on the stage of a dissecting microscope (20x-30x) and the larva is viewed against a dark background. The anterior end of the larva is grasped with a fine point forceps and the posterior portion is pinned down with a probe. Gently the head is pulled off and the tail of the larva is discarded.

3. The salivary glands and their attached fat bodies are located. The glands are semitransparent and attached by ducts to the digestive system (Figure 5.1.1a). The fat bodies are white and opaque. The fat bodies are teased away and discarded.

4. The salivary glands are fixed with a few drops aceto-methanol/ethanol (1:3 v/v) over a slide.

5. The fixative is removed by tilting down the slide. One to three drops of aceto-orcein is placed on fixed tissue over the slide next.

6. A Petri dish is placed over the preparation and allowed to stand for five to ten minutes (it will take a few trials to obtain properly stained chromosomes). The stained slide is washed with 45% acetic acid for thirty to forty five seconds. Excess acetic acid is tilted off. One drop of aceto-lacto-orcein is put over the slide. The stained salivary gland is then covered with a cover slip.

7. The gland preparation is gently squashed in the following manner :

- The slide is placed between several layers of paper towelling.
- The thumb should be placed on the top of the towel immediately over the cover slip and the thumb is gently rolled while exerting a small amount of pressure (as though one is making a finger print). One should not move one's thumb back and forth. One gentle roll is sufficient.
- The slide is removed from the towels, and seal the edges of the cover slip are sealed by using a paintbrush dipped in melted paraffin or nail polish.

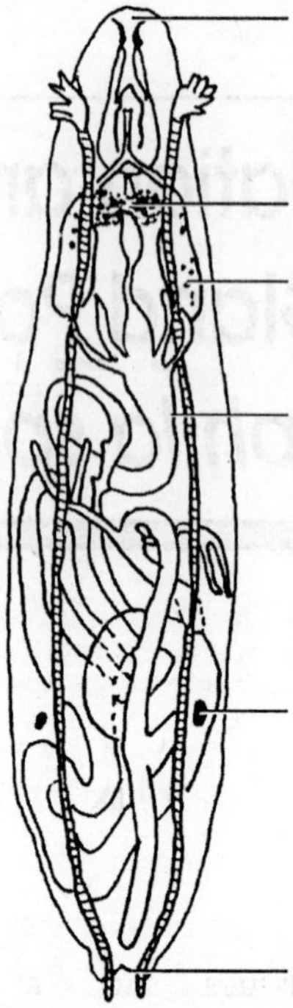

Fig. 5.1.1(a) : Anatomy of 3rd instar larvae of *Drosophila sp.*

8. The slide is examined under the a compound microscope and diagram of the salivary gland chromosome of *Drosophila* sp. (Figure 5.1.1b) is drawn along with the banding patterns that are observed.

Observations :

1. It is well known that genes responsible for different characters (wild type and mutant) are borne by various chromosomes. Locations of these genes are also precisely pointed out in the different bands of the salivary gland

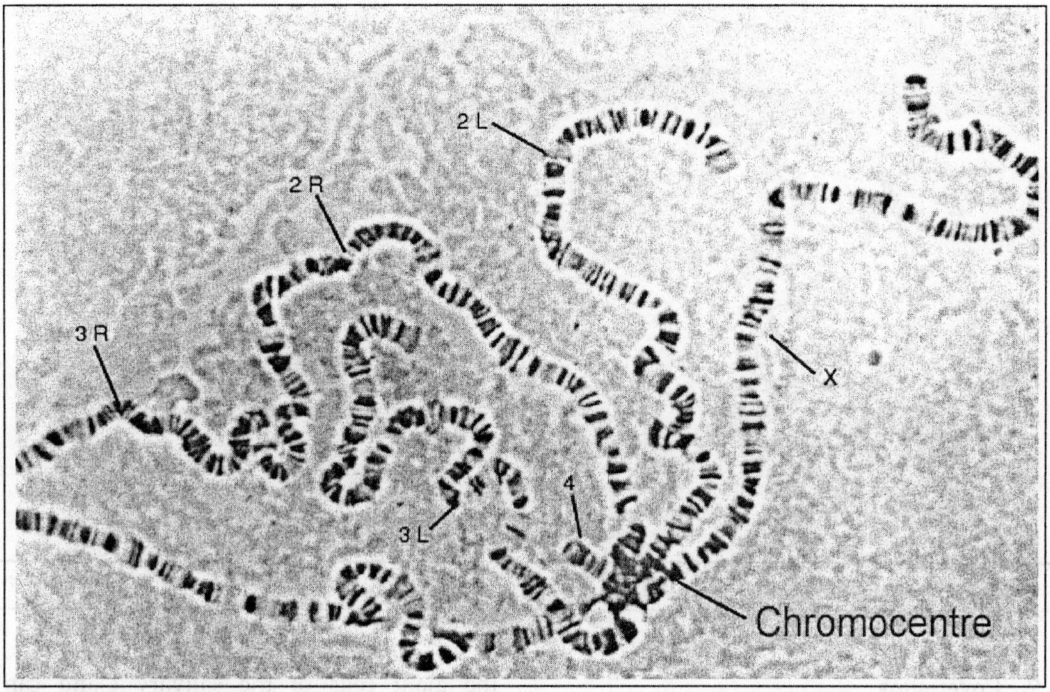

Fig. 5.1.1(b) : Diagram of the salivary gland chromosome.

chromosomes of *Drosophila sp*. Close observation of the salivary gland chromosome morphology reveals the following regions :

(a) The individual chromosome appears as six long arms that radiate out in both male and females. Five long arms radiate from a common chromocentre which is heterochromatic in nature.

(b) Each arm is basically composed of two homologous chromosomes as revealed by split appearance of each arm.

(c) The chromosomes exhibit alternating dark bands and interbands. Bands that are deeply stained with aceto-orcein. These may be single or double. Interband regions are very little stained. In *Drosophila sp*. there are about 5000 bands and 5000 interbands.

(d) Puffs (some swelled out regions) on the different chromosomes are very lightly stained.

2. One can also compare one's squash preparation to that of the prepared slides.

3. The squash technique can also be repeated using larva from a genetic variant known to be the result of a deletion and/or tandem duplication. One has to determine the location of the deleted or duplicated bands on the chromosomes. This feature can be employed to physically map genes on chromosomes.

(b) From *Chironomus sp.* larva :

Materials required :

1. *Chironomus sp.* larvae (Figure 5.1.2).
2. 1 N HCl.
3. Aceto-orcein (see preparation in 5.1a).
4. Aceto-ethanol (Absolute ethyl alcohol : Acetic acid 3:1 v/v).
5. Fine forceps and needle.
6. Glass slides.

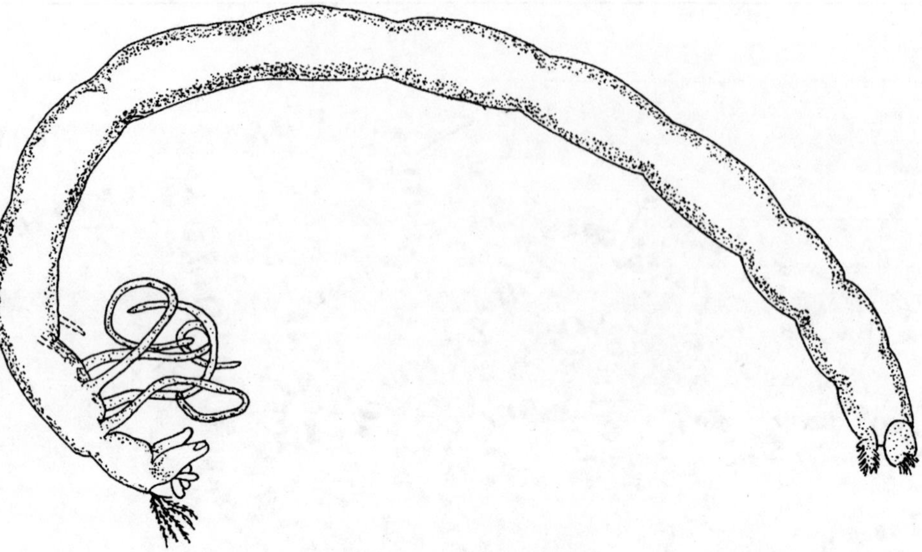

Fig. 5.1.2 : Chironomus larvae.

7. Cover slips.

8. Dissecting and regular compound microscopes.

9. Paint brush (as per 5.1a).

10. Melted paraffin or Nail polish (as per 5.1a).

Procedure :

1. The anterior portion of the larvae is held with the help of fine forceps or a needle. The adjacent body is stretched carefully.

2. The salivary gland is located.

3. Under a dissecting binocular microscope the giant cells of the salivary gland can be detected in a row in the outer circumference of the gland.

4. The glands are fixed in aceto-alcohol for 5 minutes.

5. Both the glands are hydrolysed in 1 N HCl at 57°C for one to two minutes.

6. Stain and Squash technique is applied as done for *Drosophila sp.* salivary gland chromosome (see section 5.1a).

Observation :

Bands and puffs of the polytene chromosomes can be visualised under a compound microscope.

Chapter 5.2

Preparation and Identification of Mitotic Stages

Introduction

An ordinate division of chromosomes routed through consecutive changes and resulting in increase of cell number with equal chromosomal content is termed as **Mitosis**. As chromosome number remains the same from mother cell to the daughter cell it is also termed as **equational division**. The process includes chromosome replication and distribution of these replicated sets of chromosomes into two daughter nuclei. Vohn Mohl and Virchow first described cell division in 1850. Schneider in 1873 gave the first account of mitosis. In 1882, W. Flemming coined the term mitosis. Howard and Pele (1953) denoted four main phases of the cell cycle G_1, S, G_2 and M. M. Cockraum and Mac Canley (1960) explained the biochemical aspects of the process. The onion (*Allium cepa*) root tip remain as the standard introduction to the study of mitosis. The onion has easily observable chromosomes. The mitosis can be divided into interphase, prophase, metaphase, anaphase and telophase.

(a) Interphase :

In earlier research emphasis was placed on the movement of the chromosomes that the cell was considered to be "stationary" when not in mitosis. This term stationary is rarely used today, and the term **interphase** is sufficient for all activities between two mitotic divisions. The cell is highly active during interphase and most of the metabolic and genetic functions of the cell are reduced during the physical division of the nuclear and cellular materials (mitosis).

Interphase is divided into three sub-phases viz., G_1, S and G_2. The basis for this division is the synthesis of DNA. Note also that while the entire cycle may be as long as 24 hours, mitosis is normally less than one hour in length.

Since of the synthesis of DNA in interphase, the quantity of DNA per nucleus is different depending on which sub-phase of interphase the cell is in. The basic amount of DNA in a haploid nucleus is given the value 'C'. A diploid nucleus would be 2C. A triploid and tetraploid cell would be 3C and 4C respectively.

The period within interphase and during which DNA is synthesized is termed the S phase (for Synthesis). The period of interphase preceding the S phase is the G_1 phase (for 1st Growth Phase), while the period subsequent to the S phase is the G_2 phase (for the 2nd Growth Phase). During the G_1 phase, the cell is usually increasing in size and protein content. During S phase, the cell replicates the chromosomes and synthesizes DNA. During G_2 phase, it continues to increase in size, but also begins to build a significant pool of ATP and other high energy phosphates, which are believed to be a significant part of the triggering mechanism for the subsequent karyokinetic and cytokinetic events of mitosis.

Mitosis returns the cells to the 2C state. Meanwhile the number of chromosomes (designated with the letter N) is also changing. For a diploid cell, the number of chromosomes is twice that of a haploid, or 2N. During mitosis, a diploid cell would go from one 2N cell to two 2N cells. Since the daughter cells have the same chromosome number as the parent, mitosis is also referred to as equational division.

(b) Prophase :

The first phase of mitosis is marked by the early condensation of the chromosomes into visible structures. At first, the chromatids are barely visible, but as they continue to coil, the chromosomes become thicker and shorter. The nuclear envelope is still present during this stage, as are any nucleolar structures. The centrioles are moving to the poles of the cell and spindle fibers are just beginning to form.

(c) Metaphase :

During the middle phase of karyokinesis, the chromosomes line up in the centre of the cell, and form a metaphase plate. Viewed on edge, the chromosomes appear to be aligned across the entire cell, but viewed from 90° they appear to be spread throughout the entire cell (visualize a plate from its edge or from above). Each chromosome has a clear primary constriction, the centromere, and attached to each is a definitive spindle fibre. The spindle apparatus is completely formed, and the centrioles have reached their respective poles. The nucleolus and the nuclear envelope have disappeared.

(d) Anaphase :

The movement phase begins precisely as the two halves of a chromosome, the chromatids, separate and begin moving to the opposite poles. The centromere will lead the way in this process, and the chromatids form a V with the centromeres pointing toward the respective poles.

(e) Telophase :

The last phase is identified by the aggregation of the chromatids (now known as chromosomes) at the respective poles. During this phase, the chromosomes uncoil, the nuclear envelope is resynthesized, the spindle apparatus is dismantled and the nucleolus begins to appear.

Procedure I

Materials required :

1. Onion (*Allium cepa*) bulb.
2. Aceto-orcein (1%, 2%) : Mode of preparation same as chapter 5.1.
3. Acetic acid (45%).
4. Aceto-alcohol (v/v) : One part of glacial acetic acid is added to one part of absolute methanol/ethanol.
5. HCl.
6. Glass slides.
7. Coverslips.
8. Compound microscope.
9. Camera lucida.

Method :

1. Roots are grown from the bulb of onion in aerated distilled water at room temperature.

Vigorously growing 2-3 days old healthy roots of 1 cm. in length can be used.

2. Fresh young roots of onion about 2-3 cm long are cut and washed in running water tap water to remove dirt.

3. About 5 mm of the tip portion of the root is cut out.

4. The root tips are fixed in aceto-alcohol i.e. acetic acid : absolute methanol ethanol (1:1 v/v) in specimen tube for 1 hour.

5. The root tips are transferred into a mixture of 2% Aceto-orcein and normal HCl (9%:1 v/v).

6. Gently heated for a few seconds without boiling.

7. The roots are kept in the mixture for 15 minutes to half an hour.

8. The root tips are transferred into a clean grease free glass slide.

9. About 2 mm of the exact portion of the root tip is cut out.

10. About a drop of 1% aceto-orcein solution is added over the root tips and covered with a cover slip.

11. Excess stain is blotted away.

12. The squashing (in 45% acetic acid) is done by applying uniform vertical pressure on the cover slip with a thumb or by tapping the contents under a coverslip with the aid of the flat end of a brush.

13. The squashed preparations are observed under compound microscope. If preparations are of desired nature. The coverslip is covered with paraffin wax temporarily.

Procedure II

Materials required :

1. Onion (*Allium cepa*) bulb.
2. Para-dichlorobenzene.
3. Aceto-carmine (1.5%).
4. Ferric chloride.
5. Carnoy's fixative.
6. Absolute Butanol.
7. Acetic acid (45%).
8. Euparal.
9. Glass phial.
10. Glass slides.
11. Coverslips.
12. Compound microscope.
13. Camera lucida.

Method :

1. Roots are grown from the bulb of onion in aerated distilled water at room temperature. Vigorously growing 2-3 days old healthy roots of 1 cm. in length can be used.

2. Fresh young roots of onion about 2-3 mm long are cut and washed in running water tap water to remove dirt.

3. The washed root tips are treated with saturated aqueous solutions of para-dichlorobenzene for 2.5 hours at 14-16°C.

4. Pre-treated root tips are cleansed and transferred to Carnoy's fixative in which traces of ferric Chloride could be added. The fixation is done for 12 hours.

5. The root tips are kept in a glass phial containing 1.5% aceto-carmine solution. It is warmed for 20-30 minutes.

6. The stained root tips are taken out of the stain and then placed in a grease free clean slide. Root caps are removed and only the meristems are squashed in a drop of acetic acid (45%). The squashing is done by applying uniform vertical pressure on the cover slip with a thumb or by tapping the contents under a coverslip with the aid of the flat end of a brush.

7. The squashed preparations are observed under compound microscope. If preparations are of desired nature. The coverslip is covered with paraffin wax temporarily.

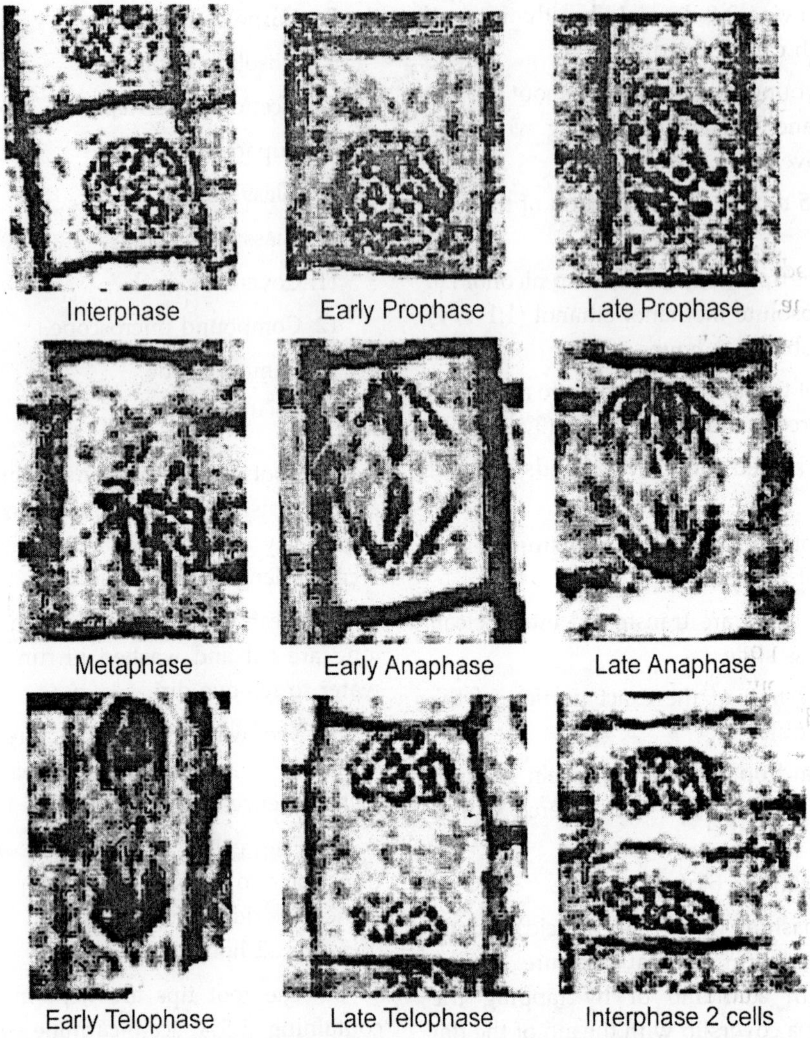

Fig. 5.2.1 : Different mitotic stages of *Allium cepa* root tip.

8. Permanent slides can be prepared in the following way :

(a) Wax is removed from the slide and the cover slip by scrapping it through a blade.

(b) The slide is kept in a petridish including 45% acetic acid for separation of cover slip and the slide from one other.

(c) These are then transferred to a mixture of 45% acetic acid and butanol (1:1v/v) for about 1 minute.

(d) Next they are transferred to absolute butanol for 30 seconds

(e) The slides and the coverglass are mounted in euparal.

Observations :

The slides are observed under a compound microscope and stages are drawn through a camera lucida and are used as a guide to identify, draw and label each of the following stages of mitosis (Figure 5.2.1).

Recording :

Using middle power on the compound microscope, the apical meristem of the onion root tip is focused. This is the area just behind the root cap. Switched to high power. The apical meristem is carefully examined and a sample of at least 50 cells is chosen to classify. Actively dividing group of cells is looked for. The cells will appear to be in rows, so it should be easy to keep track of them. High power examination of individual cells. The stages identified under the microscope can be drawn through a camera lucida fitted to the microscope.

For each of the cells in the sample, the stage of mitosis is identified, and a tally mark is placed in the Tally marks column of the data table, next to the appropriate stage. The tallies for each stage are counted, and the count column of the data table is also filled.

The percentage of cells found in each stage is calculated, and the figures under percent column is noted in the **data table** (Table 5.2.1).

Percentage of cells found in each stage =

$$\frac{\text{Number of stage}}{\text{Total number of samples}} \times 100$$

Table 5.2.1 : Data table: Observations of mitosis in *Allium cepa* root tip

Stage	**Tally Marks**	**Count**	**Percent %**
Prophase			
Metaphase			
Anaphase			
Telophase			

Chapter 5.3

Preparation and Identification of Meiotic Stages

Introduction

Every species have specific number of chromosomes. One of the most important aspects of sexual reproduction is **fertilization** i.e., fusion of male and female gametes that are haploid (n) in nature. Thus production of gametes (haploid) i.e., sperm and ovum from diploid sperm mother cells and oocyte respectively is prime for such sexual reproduction. Thus this requires a special kind of division designated as **Meiosis** (Gk. *meion*, smaller), which, in contrast to fertilization causes halving of chromosome number and consequential production of the haploid gametes. The cell that undergoes meiosis are known as **Meiocytes.** Weismann in 1887 postulated the fact that there is reduction in chromosome content of diploid germ cells of plants and animals to produce haploid gametes.

Meiotic division differs from mitosis in that there are two division cycles instead on one. This division comprises of two successive divisions (designated as meiosis I and meiosis II) with a single doubling of the quantity of chromosome or DNA.

Male grasshoppers are used as a model animal to study the meiosis process in the testis of the specimens. Short horned grasshopper (family Acrididae) under different subfamily such as Acridinae, Catantopinae, Oedipodinae and Pyrogomorphinae are generally used for this purpose.In this experiment *Gesonula punctifrons* under Subfamily Catantopinae is used for temporary squash preparation of meiotic chromosomes of the tesis of male specimens.

Systematic position of short horned Grasshopper :

Phylum	**Arthropoda**
Subphylum	**Mandibulata**
Class	**Insecta**
Subclass	**Pterygota**
Order	**Orthoptera**
Family	**Acrididae (3-4 jointed short antennae)**
Subfamily	**Catantopinae**
Genus	***Gesonula***
Specimen	***Gesonula punctifrons***

Sexual dimorphism :

Male	Female
1. Free end of the abdomen is rounded.	1. Free end of the abdomen is pointed.
2. Absence of ovipositor.	2. Ovipositor present at the end of the abdomen.
3. Subgenital plate found on the 9th sternum.	3. Two pairs of lobes terminate the 9th sternum.

Materials required :

1. Live male short horned grasshopper, *Gesonula punctifrons* (Figure 5.3.1).
2. 45% Acetic acid.
3. **Aceto-carmine stain (2%) :** 2 g of carmine powder is mixed with 45 % acetic acid. Boiled gently for 5 minutes in a reflux condenser. Shaken well and then the solution is filtered after cooling. 1-2 drops of acetic acid saturated with iron acetate is added to it. The stain is kept in exposed condition for its maturation.
4. **Lacto-aceto-carmine :** To freshly prepared 2-3% aceto-carmine equal volume of lactic acid is added at room temperature. It is gently mixed and left standing overnight under refrigerated condition. It is filtered immediately before use.
5. 0.42% potassium chloride (KCl) solution.
6. 0.9% Trisodium citrate solution.
7. Scissors.
8. Forceps.
9. Needles.
10. Watch glass or cavity block.
11. Glacial acetic acid.
12. Absolute alcohol.
13. **Aceto-alcohol :** 10 mL of glacial acetic acid is mixed with 30 mL of absolute ethyl or methyl alcohol).
14. One, 10 mL measuring cylinder.
15. Slides.
16. Cover slips.

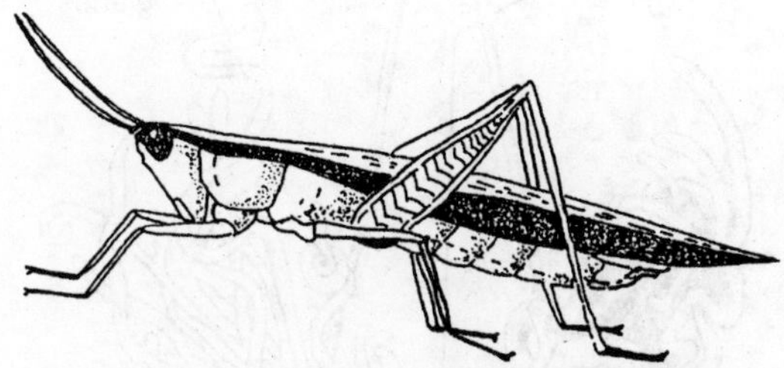

Fig. 5.3.1 : Male short horned grasshopper, *Gesonula punctifrons.*

17. Compound microscope.
18. Camera lucida.
19. Nail polish.
20. Anaesthetic ether.
21. 70% ethyl alcohol.

Procedure :

1. Male grasshopper can be collected from nature, and brought to the laboratory.
2. The insect is made motionless by exposing it to anaesthetic ether for 10-20 seconds.
3. An incision is made on the dorsal side longitudinally along the whole length of abdomen, starting with the last segment to the first segment near the thoracic end; body cavity is opened in water.
4. After separating the alimentary canal, in between the first three segments of abdomen, a cream coloured finger-like compact mass is seen surrounded by fat bodies (see Figure 5.3.2) which is the testis.

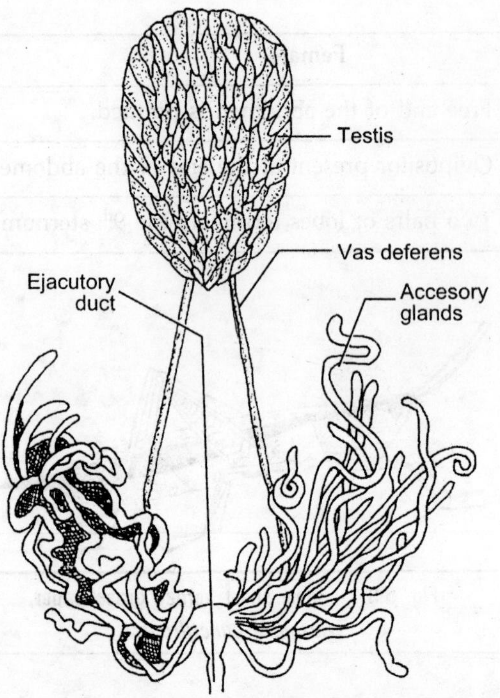

Fig. 5.3.2 : Testis of male *Gesonula punctiforns.*

5. The testis, is removed from the body and kept in 0.42% potassium chloride solution for 15-20 minutes. Potassium chloride solution can be replaced by 0.9% tri-sodium citrate solution. This solution also gives good results.

6. Tissue is then fixed in the freshly prepared aceto-alcohol fixative (which is made by mixing one part of glacial acetic acid and three parts of absolute alcohol). Time of fixation varies in different species or in other tissues but it has been observed that fixing the testicular material for 30 minutes at room temperature gives the best chromosome preparations in grasshopper.

7. Testis is then transferred in 70% ethyl alcohol and kept at low temperature. This material can be used for three-four months.

8. A small portion of testis is separated and placed in cavity block or watch glass containing 2% aceto-carmine stain and incubated at 58-60°C for thirty minutes (but time may vary on the quality of stain used).Excess stain is removed by flooding the stained material with 45% acetic acid.

9. Next, 2-4 testis lobes are put in the centre of a clean grease free slide with a drop of lacto-aceto-carmine and a cover slip is mounted on it, uniform pressure is applied on the cover slip with thumb and the testis lobes are squashed. The cover slip is then ringed with paraffin wax or nail polish to avoid penetration by air; by this method the slide can be observed under microscope for cytological study up to three to four days.

Observation :

1. The slide is placed on the microscope under low power to identify the apical end of the testis and the region where the testis joins with the vas deferens. The apical end is round and packed with cells, while the opposite end is a more open cavity lined with formed sperm.

2. The individual compartments of the testis, known as cysts are to be identified. Connective tissue walls or septa separate the cysts.

3. At the apical end, the cells are grouped into presumptive germ cells known as spermatogonia. These cells are undergoing mitosis and are giving rise to all of the remaining germ cells in the testis. Moving from the spermatogonia in the apical end to the vas deferens at the opposite end of the testis, the cells mature as a group. That is, all of the cells within a given cyst will be in the same approximate stage of meiosis. A good longitudinal section will have nearly all of the stages of meiosis displayed on one section, but one may have to examine several slides to observe all of the stages. Diagram of the different stages can be drawn under a camera lucida.

4. Using Figure 5.3.3 as a guide, each of the different stages are identified, drawn and labeled.

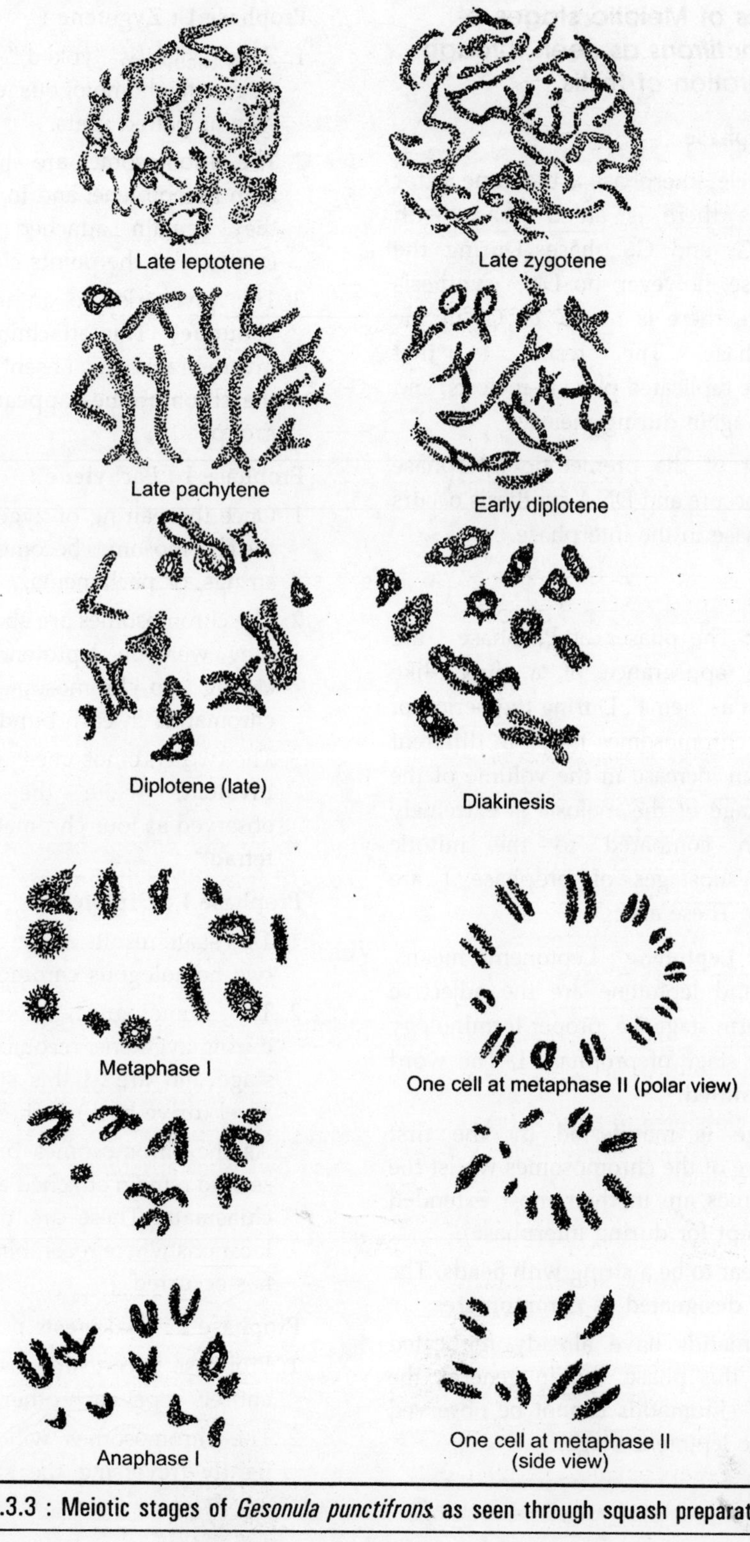

Fig. 5.3.3 : Meiotic stages of *Gesonula punctifrons* as seen through squash preparation.

Characteristics of Meiotic stages of *Gesonula punctifrons* as seen through squash preparation of testis

Premeiotic interphase :

In the first cycle, interphase is the same as for mitosis. That is there is an S phase with corresponding G_1 and G_2 phase. During the second interphase, however, no DNA synthesis occurs. Therefore there is no G_1 or G_2 in the second interphase. The result is that chromosomes are replicated prior to meiosis, and do not replicate again during meiosis.

During most of the premeiotic interphase RNA synthesis occurs and DNA synthesis occurs only in the S phase in the interphase.

Meiosis I :

Prophase I : The phases of Prophase I are named for the appearance of a thread-like structure, known as 'nema'. During the period of Prophase I, the chromosomes move in different ways. There is an increase in the volume of the nucleus. This stage of the meiosis is extremely extended when compared to the mitotic prophase. The substages of prophase I are described below. These are :

Prophase I : Leptotene : Leptonema means, "thin thread" and leptotene are the adjective applied to the term stage, i.e. proper terminology is the leptotene stage of prophase I. The word stage is often omitted.

1. This stage is manifested by the first appearance of the chromosomes whilst the chromosomes are in their most extended form (except for during interphase).
2. They appear to be a string with beads. The beads are designated as **chromomeres.**
3. The chromatids have already replicated earlier to this phase, but in general, the replicated chromatids cannot be observed during the leptotene stage.

Prophase I : Zygotene :

1. Zygos signify, "yoked," and during this stage, the homologous chromosomes are seen as paired units.
2. The chromosomes are shorter and thicker than in leptotene, and in a number of cells they remain attached to the nuclear envelope at the points close to the aster.
3. This gives rise to an image named the "**bouquet**" This attachment is unusual in invertebrates and absent in plants, where the chromosomes appear to be a tangled group.

Prophase I : Pachytene :

1. Once the pairing of zygotene is complete, the chromosomes become visible as "thick" strings, or pachynema.
2. The chromosomes are about 1/4 the length they were in leptotene, and there are clearly two chromosomes, each with two chromatids in each bundle.
3. The two chromosomes are referred to as a **bivalent**, while the same structure observed as four chromatids is known as a **tetrad.**

Prophase I : Diplotene :

1. This stage results as the space between the two homologous chromosomes expand.
2. The homologues have already paired during zygotene, recombined at pachytene stage and are at this stage beginning to repel (drive back) each other.
3. As the chromosomes break up, they are seen to remain attached at points known as **chiasmata.** These are thought to be the locations where recombination of the genes has occurred.

Prophase I : Diakinesis :

1. Prophase I conclude as the homologues entirely repel each other.
2. The chromosomes will carry on to coil tightly (reversing the slight uncoiling of

the diplotene) and will attain their greatest state of contraction.

3. As diakinesis carries on, chiasmata appear to move toward the ends of the chromosomes, a process known as **terminalization.**
4. As this stage is the conclusion of prophase, the nucleolus generally disappears, along with the nuclear envelope.

Metaphase I :

1. The tetrads travel toward the centre and array on a metaphase plate.
2. The nuclear envelope fully disappears.
3. Since the tetrads arrange in a line themselves in the middle of the cell, they fix to spindle fibres in a distinctive manner. The centromeres of a given homologue will attach to the spindles from only one pole.

Anaphase I :

1. The unique event happening at this phase is the partition of the homologues.
2. On the contrary to mitotic anaphase, the centromeres of a given homologue do not divide, and as a result each homologue progresses toward opposite poles.
3. This results in a halving of the integer (number) of chromosomes, and is the root of the reduction division that characterizes meiosis.

Telophase I :

1. Telophase in meiosis is comparable to that of mitosis, apart from that in many species, the chromosomes do not entirely uncoil.
2. If the chromosomes do uncoil and go into a brief interphase, there is absence of replication of the chromatids. It should be remembered that the chromatids have already been replicated prior to Prophase I.

Meiosis II :

Prophase II to Telophase II :

1. These phases are basically the same in meiotic and mitotic division (see Chapter 5.2); the only difference is that the chromosome number is half of the number the cell had before meiosis.
2. Each chromosome (homologue) is composed of two chromatids, and during Anaphase II, the two chromatids of each chromosome spread out and become separate chromosomes.

Chapter 5.4

Preparation and Identification of Barr Body

Introduction

In 1949, Murray Barr and E.G. Bertram found in the nerve cells as well as in most of the body cells of a female cat a small dark body. They recognized this later on as sex-chromatin (a sex-chromosome in shrunk condition). This sex-chromatin is also called the Barr-body after its discoverer. Later on, it was also found in the cells of other mammals but it was further noticed that such structures are absent in the male counterpart and only found in females. Eventually it was Susomo Ohno who proved in 1959 that the Barr-body is one of the two sex-chromosomes in a female. The observations of Ohno soon were confirmed for other female mammals including humans. Though one may be expecting to find two sex-chromosomes in all body cells of a female mammal, only one is to be found and the other is the so called Barr-body. In humans this body is easily detected in the female's buccal mucosal cells and fibroblast but not in the males. This is a highly condensed structure with a diameter of about 1µm and lies against the nuclear envelope. It is Feulgen positive and hence has DNA in it. Later it has been proved that this body is an inactivated X chromosome.

Materials Required :

1. **2% Aceto-orcein stain :** To 100 mL of 60% glacial acetic acid, 2 g of Orcein is mixed and heated to 80°C for about 30-40 minutes. The solution is filtered through Whatman (No.1) filter paper. The stain is stored at room temperature. Shelf life is 4 weeks.
2. Glass slide: Grease free clean and air dried.
3. Coverslip.
4. Compound microscope with oil immersion lenses.
5. Whatman (No. 1) filter paper.
6. Women volunteer.
7. Metal spatula.

Procedure :

1. A women volunteer is asked to rinse her mouth thoroughly. The buccal cavity is scraped through a metal spatula and scraping is spread on a grease free clean air-dried glass slide.

2. This slide is flooded with 4-5 drops of stain. A cover slip is placed over the slide. Let it stand for 5 minutes. The excess stain is removed from the slide.

Observation :

The slide is observed under oil immersion lense of the compound microscope to find out the deeply stained small body i.e. Barr body attached to the nuclear membrane (Figure 5.4.1).

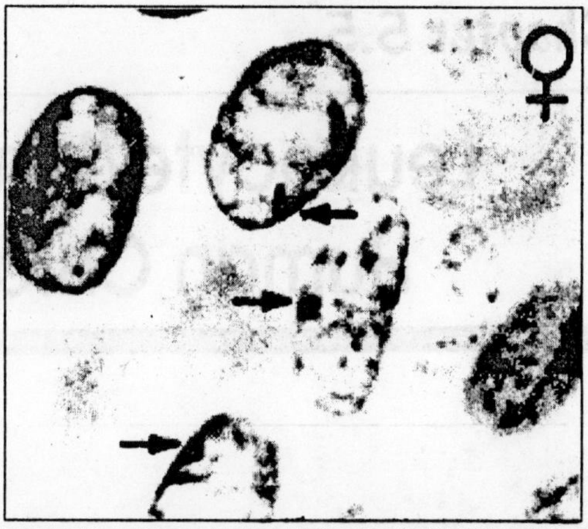

Fig. 5.4.1 : Human female epidermal cells showing the position of the Barr body (indicated by an arrow).

Chapter 5.5

Leukocyte/Lymphocyte Culture and Human Chromosome Karyotyping

Introduction

Karyotyping is an important research tool used to find out the chromosome complement within cultured cells. It is essential to keep in mind that karyotypes evolve with continued culture.

The most excellent mitotic stage for chromosome analysis is prometaphase or metaphase. A representative metaphase chromosome consists of two arms separated by a primary constriction or centromere. Each of the two sister-chromatids has a highly coiled double helix of DNA. Often the sister chromatids are so close to each other that the whole chromosome appears as a single rod-like structure. A chromosome may be distinguished by its total length and the location of its centromere. A chromosome with the centromere at or near the middle is called metacentric. A submetacentric chromosome has a centromere rather shifted from the middle point. If the centromere is clearly off center (e.g., halfway between the middle and the tip of the chromosome), the term very submetacentric may perhaps be used. Acrocentric chromosomes contain their centromeres very near to one end. Telocentric chromosomes, which are absent in human cells, have their centromeres at the very tip of one end. The short chromosome arm is designated p (petite) and the long arm q (one letter after p). Some human chromosomes may also include a secondary constriction, which appears as an unstained gap (also termed a satellite stalk) near the tip. The chromosomal segment distal to this gap emerges as a satellite. Chromosome numbers 13-15 (D group) and 21-22 (G group) have satellites.

A band is defined as that part of a chromosome which is obviously discernible from its adjacent segments by appearing darker or lighter. The chromosomes are visualized as consisting of a continuous series of light and dark bands. Several different technical procedures have been reported that produce banding patterns on metaphase chromosomes.

The term karyotype refers to a display of the chromosomes of a cell by lining them up, beginning with the largest and with the short arm oriented toward the top of the karyotyping sheet.

In humans seven (A–G) groups of autosomes are recognized. Sex chromosomes (X,Y) are placed last. A diagram of the karyotype based on chromosome measurements in many cells is called an **idiogram.**

Human chromosome nomenclature systems were developed during six international conferences on chromosome nomenclature.

The karyotype formula starts on with the total number of chromosomes in the cell, followed by the notation of the sex chromosomes (first the Xs and then the Ys).

The following is a partial list of symbols for chromosome nomenclature :

A–G The chromosome groups

1–22 The autosome numbers

X, Y The sex chromosomes

del Deletion

dup Duplication

inv Inversion

p Short arm of chromosome

q Long arm of chromosome

s Satellite

t Translocation

+/" Placed before the chromosome number, it indicates addition and after it indicates deletion

(+) or deletion (") of the whole chromosome (e.g., +21 denotes an extra chromosome 21, as in Down syndrome). Placed after the chromosome number, the symbol indicates deletion of the chromosome part (e.g., 5p" indicates loss of part of the short arm of chromosome 5, as in the Cri du chat syndrome).

Description of Human Chromosomes :

Chromosome 1 **(Group A)** : The longest chromosome, metacentric; p arm fading at distal end, with at least two dark bands above centromere; q arm with a dark area just below centromere, a light band below, followed by four dark bands of which the middle two are often more intense.

Chromosome 2 **(Group A)** : Almost as long as chromosome 1, submetacentric; p arm without light distal end, with four dark bands; q arm with more than four dark bands.

Chromosome 3 **(Group A)** : Metacentric; both p and q arms with a large light band in the middle; dark area (consisting of two or three close dark bands) distal to the light band of q arm thicker than its counterpart in p arm.

Chromosome 4 **(Group B)** : Very submetacentric; p arm with a thick dark band which may sometimes be resolved as double; q arm with a distinct, prominent dark band below centromere.

Chromosome 5 **(Group B)** : Very submetacentric; p arm with a single dark band; q arm without a prominent dark band below centromere, with three close dark bands in the middle.

Chromosome 6 **(Group C)** : Submetacentric; p arm with a distinct, large light band in the middle; q arm with several dark bands.

Chromosome 7 **(Group C)** : Submetacentric; p arm with a prominent flat dark band at the top; q arm with two distinct dark bands followed by a much less intense dark band.

Chromosome 8 **(Group C)** : Very submetacentric; p arm with two dark bands (sometimes unclear); q arm with two (sometimes three) dark bands, the distal dark band characteristically more intense.

Chromosome 9 **(Group C)** : Varying from slightly submetacentric to very submetacentric; p arm with a broad (often square in shape) dark band (which may be resolved as double); q arm with a variable, proximal light area (from as long as the p arm to almost non-existent) followed by two distinct, equally intense broad dark bands with a light band between them; these three bands (two dark and one light) are equal in thickness.

Chromosome 10 **(Group C) :** Submetacentric; p arm with a dark band; q arm with three distinct dark bands, the proximal dark band most intense.

Chromosome 11 **(Group C) :** Submetacentric; p arm with one or occasionally two dark bands; q arm with a very large light area followed by a dark band which may be resolved into two dark bands.

Chromosome 12 **(Group C) :** Very submetacentric; p arm with a dark band; q arm with a proximal light band (which is relatively thinner than the counterpart of chromosome 11) followed by a dark area in the middle which may be resolved into two, or more often three, bands.

Chromosome 13 **(Group D) :** Acrocentric; p arm with satellites (visible or invisible); q arm with several dark bands.

Chromosome 14 **(Group D) :** Acrocentric; p arm with satellites (visible or invisible); q arm with a distal dark band which is more intense than other dark bands.

Chromosome 15 **(Group D) :** Acrocentric; p arm with satellites (visible or invisible); q arm with dark proximal half and light distal half.

Chromosome 16 **(Group E) :** Metacentric or occasionally submetacentric; p arm light or slightly dark; q arm with a prominent dark band just below centromere, followed by one or two lesser dark bands.

Chromosome 17 **(Group E) :** Submetacentric or very submetacentric; p arm with a thin dark band (may be invisible); centromere dark followed by a distinct, large light band occupying a large part of q arm, q arm with one or two distinct dark bands near the terminal end.

Chromosome 18 **(Group E) :** Submetacentric or very submetacentric; p arm often uniformly dark; q arm with two distinct dark bands, one proximal and one distal; the light band between these two dark bands slightly smaller than the counterpart of chromosome 17.

Chromosome 19 **(Group F) :** Metacentric; p and q arms both light; centromere dark.

Chromosome 20 (Group F) : Metacentric; p arm dark; q arm light, with two narrow, less intense dark bands.

Chromosome 21 **(Group G) :** Acrocentric; p arm with satellites (visible or invisible); q arm with intense dark area proximally, fading out toward distal end.

Chromosome 22 **(Group G) :** Acrocentric; p arm with satellites (visible or invisible); centromere dark; q arm with a narrow, less intense dark band.

X Chromosome **(sex chromosome) :** Submetacentric; p arm with distinct, strong dark band at the middle; q arm with a proximal dark band about the same distance from the centromere as the p arm dark band, with two less intense dark bands near distal end.

Y Chromosome **(sex chromosome) :** Very submetacentric; p arm very short or invisible; q arm dark throughout.

The method starts with the short term human peripheral lymphocyte culture. The basic steps are obtaining blood from a vein keeping it uncloted with help of an anti-coagulant (heparin) and then adding required aliquots to sterile medium + Fetal calf serum (FCS) + Phytohaemagglutinin (PHA) in a sterile screw capped vial. These vials are tightly capped and incubated in a humid incubator at 37°C. Cell divisions starts at about 24 hours. Colchicine or colcemid is then added to arrest the division of the cells at mitosis. At least 3-5 hours before harvesting the cells. The harvested cells are killed and fixed and spread out in dry clean glass slides. The slides are then stained with Giemsa technique and the metaphase chromosomes are photographed.

The photograph is enlarged and cut up into individual chromosomes. The homologous pairs are identified and arranged in order by size (with the exception of the sex chromosomes; these appear last). These tests are typically done on a sample of blood, although any body cell could be used. The cell must be undergoing mitosis – preferably in metaphase – so that the chromosomes are replicated, condensed, and visible under a microscope.

(a) Culture of Peripheral blood leucocyte

Materials required :

1. **Glassware :**

(a) 15 mL or 30 mL Tarson brand sample/culture tubes with caps (Autoclaved).

(b) Disposable plastic syringes 10 mL, 5 mL, 2 mL and 1 mL syringes with No. 22G, 21G, 23G needles in large quantity.

(c) Pasteur pipettes with rubber droppers (Autoclaved).

(d) Centrifuge tubes CORNING (Autoclaved).

(e) Pipettes 1-50 mL CORNING (Autoclaved).

(f) Reagent bottles (Autoclaved).

(g) Glucose bottles empty (500 mL).

(h) Millipore filters with spare filter pads of size 200μm or 220μm fitting in the attachments for syringes should be used after sterilization (Autoclaved).

Note : Autoclaving should be done at 15 lbs/sq. inch for 30 minutes.

2. **Chemicals :**

(a) Culture medium (Medium 199) : Prepared according to manufacturer's specifications. One has to use triple distilled water for dissolving the medium. For immediate use, sodium bicarbonate solution is added till it shows pH 7.2-7.4 revealed by characteristic red/pink colour of the medium owing to indicator phenol red. The medium is sterilized by passing through millipore filter in autoclaved bottle washed as per instructions.

One should also add :

Penicillin 100-200 units/mL.

Streptomycin 50-100 mg/mL.

After checking the sterility by keeping at 37°C for 48 hours. Stored frozen :

(b) Phytohaemagglutinin P. Used as per specifications. Stored frozen.

(c) Heparin sodium (or Laqueniur). To be prepared at a concentration of 1000 units/mL. To be stored at 2°C.

(d) Ethanol (absolute).

(e) Ammonium hydroxide.

(f) 4.5% Sodium bicarbonate ($NaHCO_3$) solution in ampoules of 10 mL or larger vials.

(g) Foetal Calf Serum (FCS).

(h) 0.1 N HCl.

3. **Humid incubator :**

Procedure :

1. Aseptically drawn 5 mL of venous blood into a 10 mL/5 mL syringe containing 0.1 mL of heparin (1000 units/mL) solution and mix thoroughly by inverting the syringe.
2. The needle guard tightly and quickly replaced in order to avoid contamination and leakage.
3. The syringe is kept upright at room temperature for 30 minutes to 1 hour to permit a gravity separation of the erythrocytes from 4 mL of leucocyte rich plasma on the top and white buffy coat of leucocytes in the middle.
4. The needle is bend with the needle guard to 45° and the leucocytes rich plasma is squeezed into a culture tubes, which contains 6 ml of culture medium, supplemented with 2 mL of foetal calf serum (FCS) and 0.2 mL of diluted working solution of Phytohaemagglutinin(PHA).
5. The culture tubes are capped with a sterile rubber stopper or cap and incubated at 37°C (if there are any pH changes, the pH may be raised or lowered by adding $NaHCO_3$ or 0.1 N HCl respectively.) for 72-90 hours.

(b) Basic process of Karyotyping

Materials Required :

1. Colcemid stored at 2°C: Stock solution 1 mg/ml is prepared. Added 1ml stock to 9 ml distilled water i.e. effective strength 100mg/ml. This is used as a working solution.

2. Fixative: Prepared freshly from methyl alcohol 3 parts, Glacial acetic acid 1 part.

3. Hypotonic KCl solution: Prepared from 0.56% KCl, which is 0.075 M in, distilled water, stored at 37°C.

4. Phosphate buffer: 0.067M KH_2PO_4 (9.08 g/L distilled water); 0.067M $Na_2HPO_4 \cdot 2H_2O$ (11.88 g/L distilled water).

5. Centrifuge tubes (Corning).

6. Sorensen's Phosphate buffer pH 6.8: Phosphate buffer: 0.067M KH_2PO_4 (9.08 g/L dist. water); 0.067M $Na_2HPO_4 \cdot 2H_2O$ (11.88 g/L distilled water).

7. 5 % Stock Giemsa solution in phosphate buffer (pH 6.8): 500 mg of Giemsa powder is mixed with 45 mL glycerol and 5 mL methanol. Shaken well and mixed. The remaining glycerol is added and mixed. Remaining 45 mL methanol is also added and mixed. The solution is allowed to rest overnight at 30-50°C. The solution is filtered and kept in a dark screw capped bottle. Working Giemsa solution (10 mL) is prepared by mixing 10 mL of phosphate buffer and 0.5 mL stock Giemsa stain.

8. Distilled water.

9. Xylene, 100 mL.

10. DPX or Euparol.

11. Cover slips.

12. Immersion oil.

13. Compound microscope with high power and oil immersion lens and 10x, 15x eyepieces.

14. Camera to fit the monocular eye pieces.

15. Coplin jars with cover.

16. Vortex mixer.

17. Pasteur pipette.

18. Clinical centrifuge.

19. Humid incubator.

20. Refrigerator.

21. 100 ASA B/W film.

Procedure :

1. ***Harvesting of leukocyte culture* (see chapter 5.5 section (a) for leukocyte culture) :**

(a) 0.1 mL of colcemid (10 mg/mL) should be added to leukocyte culture. The culture tubes are kept at 37°C for 3 hours.

(b) After 3 hours the cells are suspended in the culture medium and transferred to 15 mL conical centrifuge tubes and spun for 10 min at 1000 rpm.

(c) The supernatant is removed and then 2 mL of hypotonic KCl (0.075M) solution is added to the pellet and allowed to stand for 20-25 minutes keeping in an incubator at 37°C and the cells are mixed through pipetting and again centrifuged for 10 minutes at 1000 rpm. The supernatant is discarded

(d) To the pellet 1 mL of ice cold fixative is added drop by drop while shaking gently. More fixative is added until the volume is made up to 10 mL. Mixed well. The colour of the cells becomes blackish.

(e) Tubes are kept in refrigerator at 4°C for 30 minutes.

(f) Again spun at 1000 rpm for 5 minutes. Supernatant is discarded.

(g) The tube is held on a vortex mixer while adding 5 mL of fresh chilled fixative. It is kept on ice for 5 minutes.

(h) Steps (f) and (g) are repeated until cells become colourless.

(i) To the final pellet 1.5 mL of fixative is added.

(j) The cells are then taken in a Pasteur pipette and placed on a slide and observed under a microscope. This can be done in the following way. A clean slide is held in one hand inclined at an angle to the desk top and the pipette is raised with the other hand and the drop is let fall by 12-18 inches upon the raised end of the slide. Due to such an action i.e. due to the impact of the fall the fragile swollen cells will burst and the metaphase chromosome will stick to the glass slide surface. The slides are air dried.

2. **Staining of slides :**

(a) Six coplin jars are arranged with label viz. tap water, DWI, DWII, DWIII, DWIV and Xylene. Note 'DW' stands for distilled water.

(b) The slides are horizontally placed in a Petri dish and the cells are covered with one to two drops of Giemsa stain for 2 minutes.

(c) The slides are placed inside the coplin jars of tap water, under running condition. Let the slide be washed in running water until precipitated and excess stain granules removed.

(d) The slides are passed through four distilled water (DWI to DWIV) jars.

(e) The slides are then transferred to the xylene jar for 10 seconds. The slides are then removed and it is allowed to drip into the jar.

(f) One drop of DPX is put on the cells and covered by a fine rectangular coverslip (00).

(g) The slides are placed in drawers and kept undisturbed.

(h) The slides are then placed under microscope and the lymphocytes (leukocytes) are scanned which are arrested in the metaphase stage.

(i) Photographs are taken of good metaphase spreads (Figure 5.5.1) using 100 ASA B/W film with 20-30 seconds exposure.

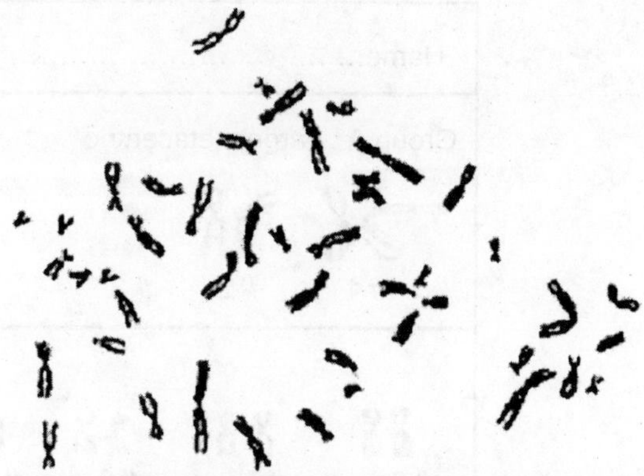

"Spread" of human chromosomes from a single cell

Fig. 5.5.1 : Metaphase spread of Human chromosomes after staining.

Observations :

The slides are examined and chromosomes are drawn under a camera lucida or photographs are taken and Karyotype is prepared through the use karyotype form (Figure 5.5.2).

Analysis :

1. Does the karyotype have an abnormal chromosome number? This may be since chromosomes were missing or cut incorrectly. It is also likely that the karyotype is of a person with a chromosomal abnormality. How can one come to a decision which is the case? *Clue*: These chromosomes were sampled from a living adult. Humans cannot stay alive to adulthood if they are missing any autosomes. Some females may survive with just a single X chromosome (XO). Humans can live with an extra X or Y chromosome (XXX, XXY and XYY). They may also live to adulthood with an extra chromosome 21 (*Down Syndrome*).

2. The number of chromosomes present on the karyotype and the gender of the subject are

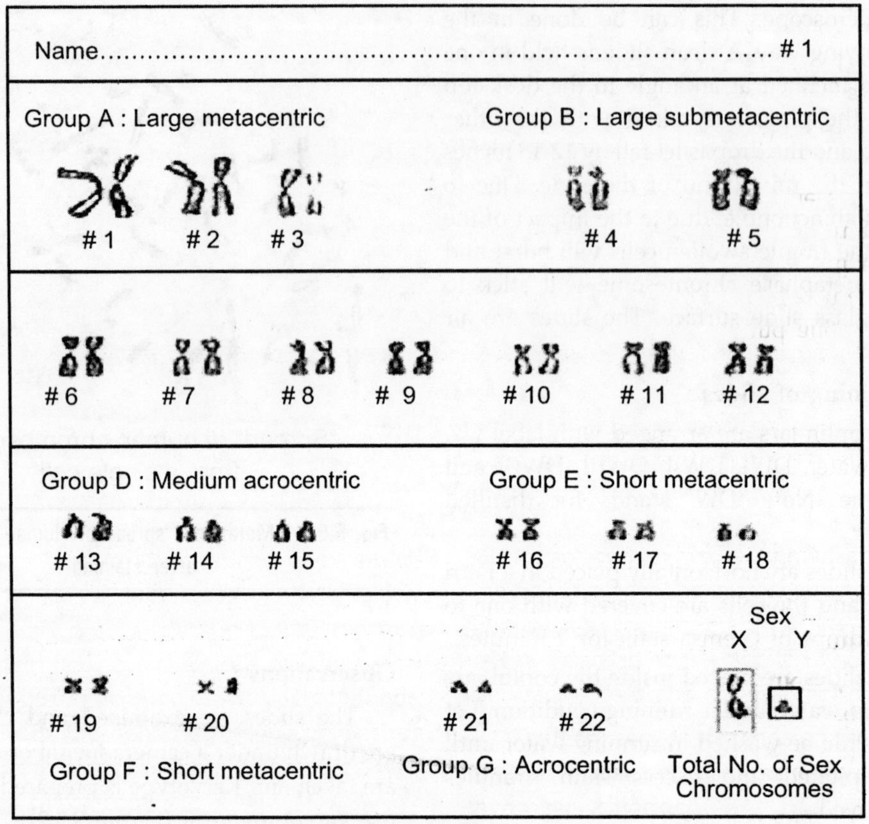

Karyotype Analysis

Fig. 5.5.2 : Karyotype analysis using a Karyotype form with different chromosome arranged according to different groups.

to be recorded. If there is an abnormal number of chromosomes (more or less than 46), note which ones are missing or in excess. The subject may have one of the abnormalities listed earlier. Which abnormality he or she has? Some of the karyotypes exhibit structural abnormalities in individual chromosomes. Cytogeneticists can find even smaller changes than these - down to slight differences in individual bands.

Section 6

GENETICS

SECTION AGENDA

Chapter 6.1

Drosophila Genetics

6.1.1 DROSOPHILA HANDLING, CULTURE AND EXPERIMENTATION

Introduction

Gregor Johann Mendel (1822 – 1884) is famous today but was rather unknown outside Czechoslovakia in his lifetime. He was the first scientist to deduce clear and rational laws which could explain the process of inheritance and it turns out that the rules which Mendel deduced from studies of peas are equally applicable to human inheritance. It is convenient to follow his train of logic beginning with characteristics determined by a single gene and moving on to the complications introduced by multiple genes.

Mendel began by collecting varieties of pea which differed from each other in clearly defined ways. The pea flower has anthers and a stamen which are very close together. It will self fertilise in normal circumstances. It is possible to remove the anthers before they are ready to produce pollen and to cross fertilise the pea plant by bringing pollen from another plant on a paint-brush. Mendel allowed his plants to self fertilise for a number of generations until he was certain that they were **true breeding**, i.e. that the offspring always resembled the parent for the characteristics under consideration. Then he began his experiments.

Through his work Mendel came to two most important principles pertaining to heredity as proposed by him is **Principle of Segregation** and **Principle of Independent Assortment.**

The Principle of Segregation : The phenomenon could be explained if it were assumed that each plant had two copies of the factor influencing the trait. One now calls the factor responsible a **gene**, and it is said that more than one form of the gene can exist and these alternative forms can be called as **alleles**. Mendel explained the results by suggesting that each plant contained two alleles which *did not blend together* but which remained unchanged. In the next generation the plants passed one or other allele at random into a gamete to be combined with a gamete from the other parent. The non-blending followed by separation into the next generation is the *Rule of Segregation*. When a mono-hybrid cross is conducted between two homozygous (for that trait) organism one exhibiting a dominant phenotype and another expressing recessive phenotype. In second filial generation the phenotypic ratios would be 3:1 between the dominant and the recessive traits.

Principle of Independent Assortment : It states that the segregation of one factor pair occurs independently of any other factor pair. But we know that this is only true for loci on non-homologous chromosomes. In case of Independent assortment the phenotypic ratios in the second filial generation is 9:3:3:1.

Drosophila melanogaster or **fruit fly** is used as one of the most important experimental models of classical as well as cytogenetics. Verification of Mendelian Principles was well made by crossing various mutants of the fruit fly.

The fruit fly *Drosophila melanogaster* has 3 pairs of autosomal chromosomes and an X and Y chromosome. Each autosome has two arms that are simply referred to as left (L) and right (R). Each chromosome arm is numbered as follows: X (1-20), 2L (21-40), 2R (41-60), 3L (61-80), 3R (81-100), and chromosome 4 (101-102). Each chromosome arm is also numbered by recombination units, thus allowing one to know the expected recombination frequency between two genes located on the same chromosome arm. The chromosomal locations of individual genes are identified either by numerical location or by recombination units. Sex determination in *Drosophila* is based on the ratio of X chromosome to autosomal sets. Therefore, females which have two X chromosomes have a ratio of 1 whereas; males which have only 1 X chromosome have a ratio of 0.5. In 1910, T.H. Morgan discovered white eye mutation in fruit fly which corresponded with the initiation of *Drosophila* genetics. Since then various mutations have been discovered in *Drosophila* that added up to the knowledge of *Drosophila* genetics.

Material required

1. Different stocks of *Drophila melanogaster.*
2. *Drosphila sp.* collecting bottles or tube.
3. Anaesthetic ether.
4. Etherising bottle fitted with a plastic funnel.
5. Brush.
6. Fly morgue.
7. Culture bottle or food vials (25 ml).
8. Stock bottles. See Figure 6.1.1.1.
9. Ingredients of culture medium with cotton pad.
10. Autoclave or pressure cooker.
11. Glass cover slip.
12. Cavity slides.
13. 100 ml, 250 ml and 500 ml beakers.
14. Glass plate with white back drop.
15. BOD incubator.

Methodology

(a) **Collection of *Drosophila sp.*** : Fruit flies can be easily collected by keeping a cut ripe fruit such as apple, guava etc. in a petri dish in an open apace. The flies get attracted towards fruits and feed on them. From there these flies can easily be collected through a brush. The flies are collected in a wide mouth test tube or glass stoppered bottle (collecting bottle).The flies are brought in to a laboratory anaesthetised segregated into males and females, studied under a microscope and may be cultured.

Virgin female are always utilized for genetic crosses. One of the methods to obtain virgin is to employ **monosex culture**: male and female pupa are separated on the basis of sex combs on tarsi of the first thoracic legs of the male pupa. Individuals from each sex are kept in separate food vials. Thus vials containing the virgin female pupae will contain all virgin females.

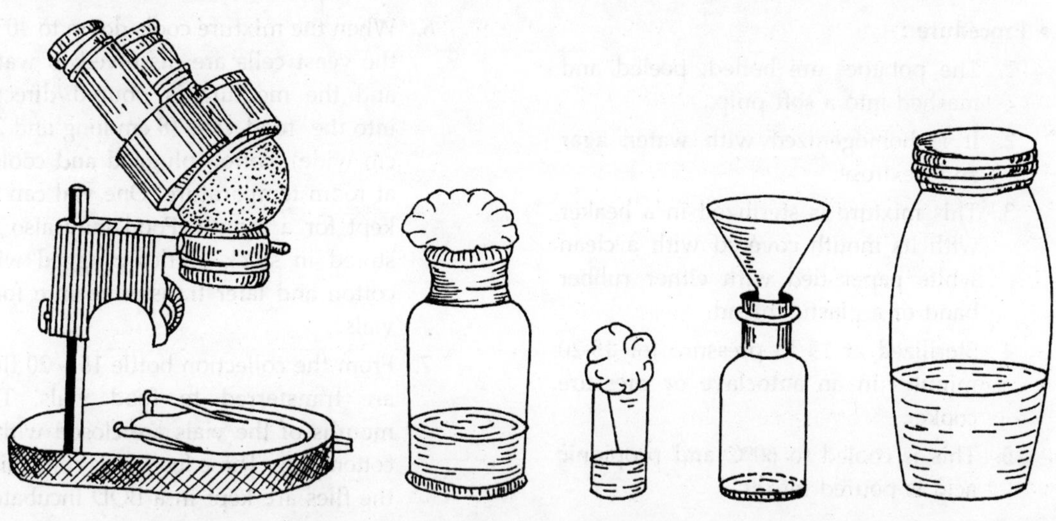

Fig. 6.1.1.1 : Materials required for *Drosophila sp.* culture in a laboratory– (a) Stereo-microscope, (b) Stock bottle, (c) Small food vial, (d) Etherizer, (e) Morgue.

(b) **Anaesthetisation of *Drosophila sp*. :** For counting and subsequent examination of flies warrants, the flies should be anaesthetised through a light dose of anaesthetic ether using a special etherising bottle fitted with a plastic funnel. The plastic funnel has a small cotton pad attached to it.

(c) **Set up for crosses :** Male and female are kept in equal ratio inside food vials having sufficient amount of freshly prepared culture medium. Generally, at 22ºC larvae are located in the top layer of the food in 5 days. Parents from successful culture should be discarded or transferred to fresh food vials for subculture.

(d) **Enumeration of the flies :** The flies are enumerated according to sex and different phenotypic classes within in each sex. Before counting the flies are anaesthetized and placed on glass plates (with a white backdrop). Observed carefully under magnification. Males from the females are separated, making a pile of each sex. Each sex are studied and looked for mutations. Common mutations are eye colour or shape, wing size or shape, or body colour etc. The mutation are observed and described and a name and genetic symbol is assigned and this is noted in laboratory notebook. It should be noted whether the mutation is associated with the males or females. Vial that contains male and female flies, which are the parental generation for the study is referred to as P_1.Every next generation is designated as the filial generation viz., F_1, F_2, etc.

(e) **Culture method :**

1. Bottles of 20ml capacity and test tubes to be used are sterilised. Heat sterilization (at 160ºC) of these bottles may be performed. Before use, these bottles should be washed with sterile water.
2. Preparation of the Culture medium :

- **Ingredients :**

(a) Potato – 100g

(b) Agar – 1g

(c) Propionic acid – 0.8mL

(d) Yeast – 0.5g

(e) Dextrose – 1.0g

(f) Water – 100mL

- **Procedure :**
  1. The potatoes are boiled, peeled and mashed into a soft pulp.
  2. It is homogenized with water, agar and dextrose.
  3. This mixture is sterilized in a beaker with its mouth covered with a clean white paper tied with either rubber band or a plastic thread.
  4. Sterilized at 15 lb pressure for 15-20 minutes in an autoclave or pressure cooker.
  5. This is cooled to 60°C and propionic acid is poured into it.
  6. When the mixture cools down to 40° C the yeast cells are dissolved in water and the medium is poured directly into the food vials (8 cm long and 2.5 cm wide), cotton plugged and cooled at room temparature. One vial can be kept for a month. Food may also be stored in stock bottles plugged with cotton and later transferred into food vials.
  7. From the collection bottle 15 – 20 flies are transferred to food vials. The mouths of the vials are closed with a cotton plug. These food vials containing the flies are kept in a BOD incubator.

6.1.2 OBSERVATIONS OF LIFE CYCLE OF DROSOPHILA

Different stages of the life cycle are collected from the food vials on different days and are fixed in corrosive sublimate (for 5-10 minutes). After they are being fixed the stages are kept in 3% chlorax solution (3 g Sodium hypochlorite dissolved in 100 mL dist. water.). The developmental stages are dehydrated in 50 %, 70%, 90% and then in absolute ethanol cleared in xylol and mounted in DPX in cavity slides for thicker specimens such as larval stages, pupal stages or adult.

Following is the life cycle of *Drosophila melanogaster* :

It is important to realize that a number of factors determine the length of time of each stage in the life cycle. Of these factors, temperature is the most important. At room temperature (about 25°C), the complete cycle takes 10 to 12 days.

The eggs : The eggs are small and ellipsoid and have two filaments at antero-dorsal part of it. Eggs measure 0.5 mm in length and 0.2 mm in breadth. Anterior portion of the egg has an opening called as the micropyle (Figure 6.1.2.1a). They are usually laid on the surface of the culture medium and, with practice, can be seen with the naked eyes. The eggs hatch into larvae after about a day.

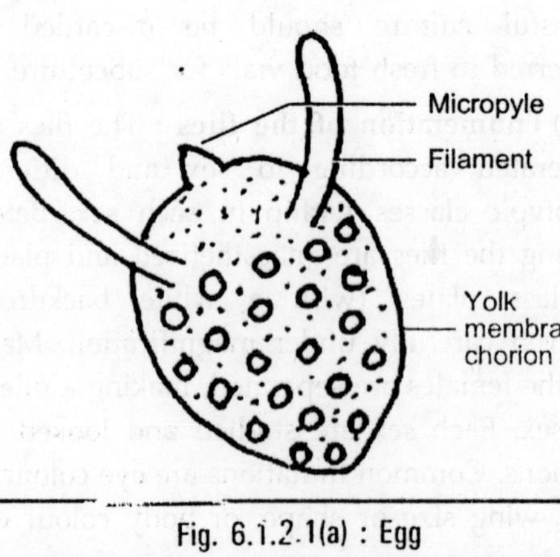

Fig. 6.1.2.1(a) : Egg

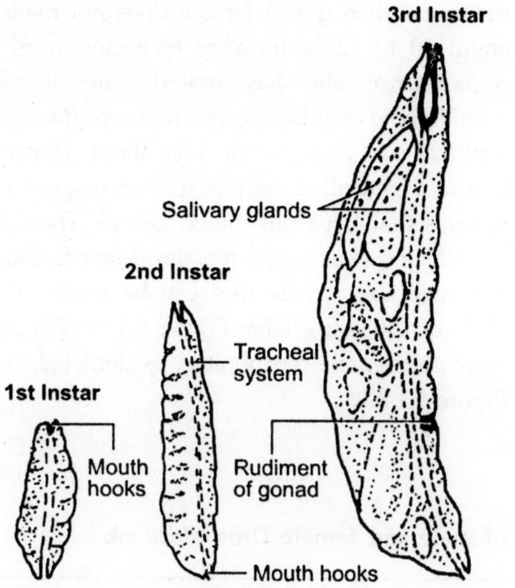

Fig. 6.1.2.1(b) : Larvae.

The larval stage : The worm like larva eats almost continuously, and its black mouthparts can easily be seen moving back and forth even when the larva itself is less distinct. First instar larva is characterized by having body segments and mouth hooks anteriorly. After 24 hrs it molts into the second instar larva. Second instar larva has a cylindrical body. The larva has increased in size and is characterized by mouth hooks and tracheal system. On the fifth day the second instar larva converts into third instar larva. It has a length of about 4-5 mm having a narrow anterior and broad posterior region. Larvae channel through the culture medium while eating; thus channels are a good indication of the successful growth in a culture. The larva sheds its skin twice as it increases in size. In the last of the three larval stages, the cells of the salivary glands contain giant chromosomes, which may be seen readily under low-power magnification after proper staining. The larval stage has got mouth hooks tracheal system and gonad rudiments. The larval stage lasts about 6 days (Figure 6.1.2.1b).

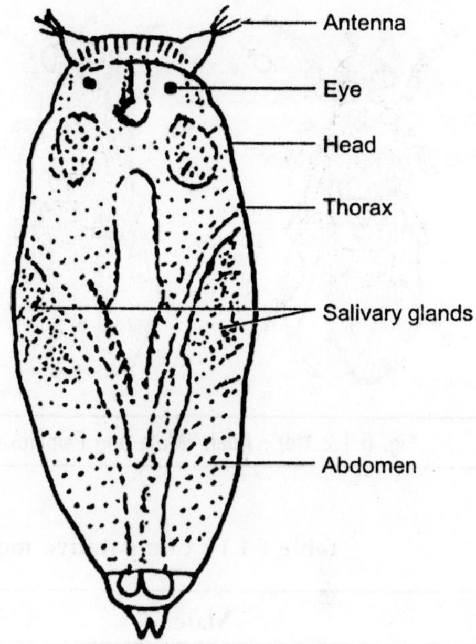

Fig. 6.1.2.1(c) : Pupa.

The pupal stage. On the sixth day the larvae converts into pupa. When a mature larva in a laboratory culture is about to become a pupa, it usually climbs up the side of the culture bottle or onto the paper strip provided in the culture. The last larval covering then becomes harder and darker, forming the pupa case (puparium).In the puparium the body is divisible into head, thorax and abdomen. Head bears antenna and compound eyes. Abdominal segments are also clearly visible. The pupa contains folded impression of the wings and appendages. At first the puparium is white but after 2 hours it changes into yellow then into dark brown (Figure 6.1.2.1c). The pupal stage lasts approximately for 4 days.

The adult stage. When metamorphosis is complete, the adult flies emerge from the pupa case. They are fragile and light in colour and their wings are not fully expanded. These flies darken in a few hours and take on the normal appearance of the adult fly. They live a month or

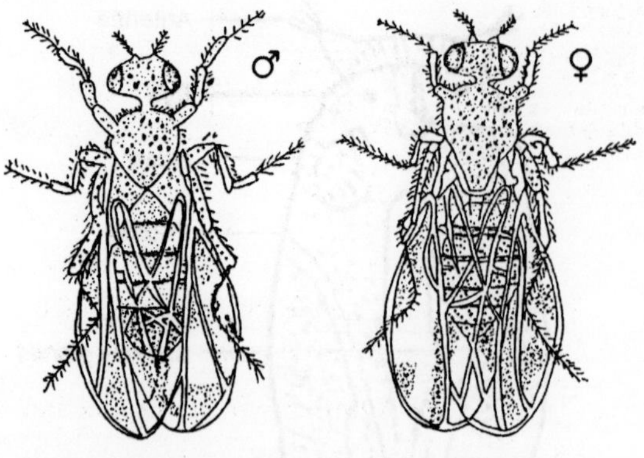

Fig. 6.1.2.1(d) : Adult (Male and Female).

more and then die. A female does not mate for about 10 to 12 hours after emerging from the pupa. Once she has mated, she stores a considerable quantity of sperm in receptacles and fertilizes her eggs as she lays them. Hence, to ensure a controlled mating, it is necessary to use females that have not mated before. These flies are referred to as virgin females. Identification of the male and female flies can be made on the basis of following table (Table 6.1.1). The adult male and female morphology is depicted in the Figure 6.1.2.1d.

Table 6.1.1 : Comparative morphology of male and female *Drosophila sp.*

Male	Female
• Smaller in size.	• Larger in size.
• Body divisible into head, thorax and abdomen. Head contains large compound eyes and a pair of antennae. Antennae bear setae or bristles. Head joined to thorax by a narrow neck.	• Body divisible into head, thorax and abdomen. Head contains large compound eyes and a pair of antennae. Antennae bears bristle. Head joined to thorax by a narrow neck.
• Thorax contains paired three thoracic legs in each of the three thoracic segments and two pairs of wings. Legs bear large number of bristles.	• Thorax contains paired three thoracic legs in each of the three thoracic segments and two pairs of wings. Legs bear large number of bristles.
• First tarsal segment of the first thoracic legs bears a sex comb.	• Sex comb absent.
• Thorax contains wings that extend beyond abdomen.	• Hind wing in the folded condition remains more extended that of the males.
• Posterior abdominal segment is black and rounded.	• Posterior abdominal segment is transparent and with pointed protruding plate.

6.1.3 STUDY OF VARIOUS TYPES OF DROSOPHILA MUTANTS

Introduction

In different organisms trait variation occurs that do not easily reflect the most common relationships between a gene and a trait. Thus geneticists have to pick up cases of genetic variation expressed as mutants to find those with simple correspondence between genotype and phenotype. *Drosophila* stands as good example for mutant study exemplifying the genotype phenotype relationship. All mutants are not useful for genetic analysis only the "Rank 1" mutants are best suited for the purpose. Identification of some important mutants of Drosophila is dealt in this section.

Identification of wild type and mutants

Wild Type :

(a) Body can be divided into head, thorax and abdomen. Head bears antennae and compound eyes.

(b) Eyes bear 800-1000 facets.

(c) Colour grey or brown.

(d) Eyes brick red.

(e) Thorax contains wings, halters and secretellum.

[See Fig. 6.1.3.1 colour plates (section 6)]

Mutant Type :

1. **Yellow body (y) :**
   (a) Colour of the adult yellow.Hair and bristles yellow.
   (b) Body can be divided into head, thorax and abdomen. Head bears antennae and compound eyes.
   (c) Located in 1st chromosome; I-0.0.
   (d) Recessive mutation.

[See Fig. 6.1.3.2(a) colour plates (section 6)]

2. **Curly wings (Cy) :**
   (a) Hind wings are curly they are curled upwards.
   (b) Body can be divided into head, thorax and abdomen. Head bears antennae and compound eyes.
   (c) Located in 2nd chromosome; II-6.1.
   (d) Dominant mutation.

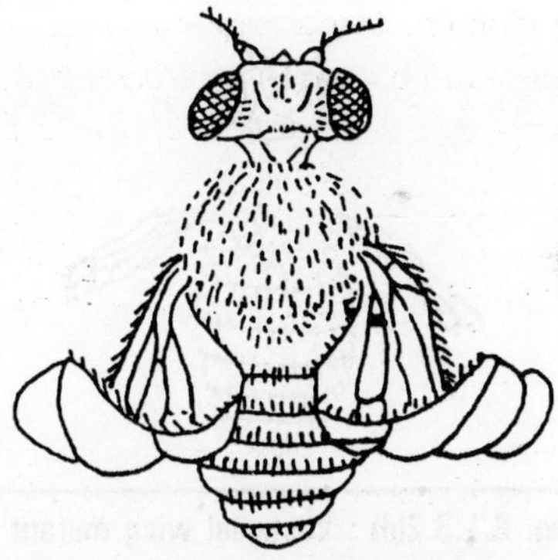

Fig. 6.1.3.2(b) : Curly wing mutant of *Drosophila melanogaster.*

3. **Scalloped wings (sd) :**
   (a) Wings have abnormal edges i.e. scalloped & veins thickened.
   (b) Body can be divided into head, thorax and abdomen. Head bears antennae and compound eyes.
   (c) Located in 1st chromosome; I-51.5.
   (d) Recessive mutation.

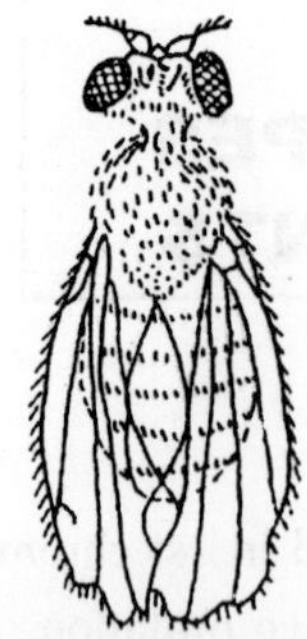

Fig. 6.1.3.2(c) : Scalloped wing mutant of *Drosophila melanogaster.*

4. **Vestigeal wings (vg) :**
   (a) Wings reduced.
   (b) Body can be divided into head, thorax and abdomen. Head bears antennae and compound eyes.
   (c) Located in 2nd chromosome; II-67.0.
   (d) Recessive mutation.

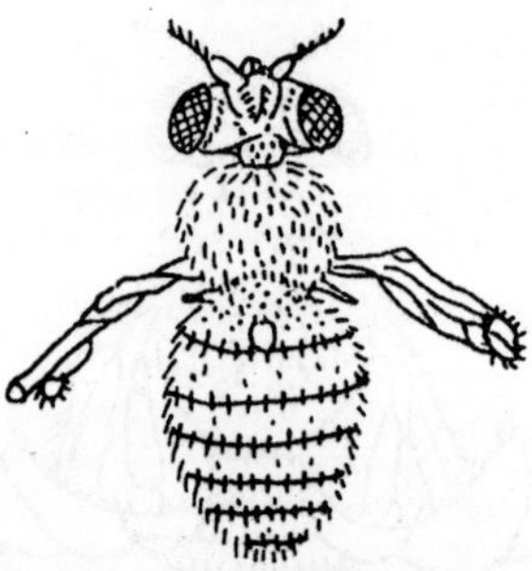

Fig. 6.1.3.2(d) : Vestigeal wing mutant of *Drosophila melanogaster.*

5. **Apterous wings (ap) :**
   (a) Wings are absent hence termed as apterous.
   (b) Bristles found all over the body.
   (c) Body can be divided into head, thorax and abdomen. Head bears antennae and compound eyes.
   (d) Located in 2nd chromosome; II-55.4.
   (e) Recessive mutation.

Fig. 6.1.3.2(e) : Apterous mutant of *Drosophila melanogaster.*

6. **Dumpy wings (dp) :**
   (a) Wings truncated.
   (b) Bristles present all over the body.
   (c) Body can be divided into head, thorax and abdomen. Head bears antennae and compound eyes.
   (d) Located in 2nd chromosome; II-13.0.
   (e) Recessive mutation.

Fig. 6.1.3.2(f) : Dumpy wing mutant of *Drosophila melanogaster*

7. **Curved wings (c) :**
   (a) Wings curved.
   (b) Bristles present all over the body.

(c) Body can be divided into head, thorax and abdomen. Head bears antennae and compound eyes.

(d) Located in 2nd chromosome; II-75.5.

(e) Recessive mutation.

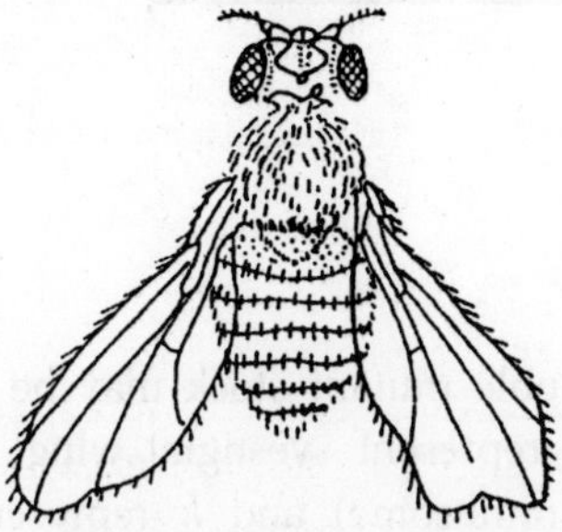

Fig. 6.1.3.2(g) : Curved wing mutant of *Drosophila melanogaster.*

8. **Shaven bristles (sv) :**

(a) Bristles are shaven, reduced.

(d) Located in 4th chromosome; IV-3.0.

(c) Trichogen irregularly displaced.

(d) Recessive mutation.

[See Fig. 6.1.3.2(h) colour plates (section 6)]

9. **Bar eye (B) :**

(a) Eye shape oblong or bar like, with facets less than 300. Smaller than normal eyes.

(b) Other characters similar to normal flies.

(c) Located in 1st chromosome; I-57.0.

(d) Dominant mutation.

[See Fig. 6.1.3.2(i) colour plates (section 6)]

10. **White eyed (w) :**

(a) These flies have white eyes. In these flies, the white gene is totally defective: it produces no red pigment at all.

(b) Located in 1st chromosome; I-1.5.

(c) Recessive mutation.

[See Fig. 6.1.3.2(j) colour plates (section 6)]

11. **Vermillion eye (v) :**

(a) These flies have vermillion eyes.

(b) Located in 1st chromosome; I-33.

(c) Recessive mutation.

[See Fig. 6.1.3.2(k) colour plates (section 6)]

6.1.4 MONOHYBRID AND DIHYBRID CROSS OF DROSOPHILA

Procedure

(i) Monohybrid cross :

If a cross involves only one trait then it is called a monohybrid cross.

1. 10-15 virgin females from a homologous stock for vestigial wings (vg: 2-67.0) (an autosomal recessive gene in) are collected.

2. 10-15 freshly eclosed males from the wild type (vg^+) stock are collected.

3. These flies are put inside a culture bottles. The flies are allowed to mate for 2 days. The parental flies are finally discarded or are subcultured in separate bottles.

4. The culture bottles are kept in a constant temperature in BOD incubator (22°C) for 10 days or until flies emerge.

5. F_1 flies are collected and are examined after appropriate anaesthetized. The observations are recorded.

6. Cross between flies of the F_1 generation is set up.

7. F_2 flies are collected, anaesthetized and examined. The observations are recorded.

(ii) Dihybrid cross :

For carrying out a dihybrid cross the protocol is basically similar apart from for the stocks. For such a case virgin females are collected from a double mutant stock so that the mutant alleles occupy different chromosomes (i.e. unlinked). Such a double mutant stock may be *vg/vg; h/h*, where *vg* represent vestigial wing(located in second chromosome) and *h* represents hairy body (located in third chromosome; 3-26.5).

Observations

(i) Results of monohybrid cross :

The data are to be recorded on the following theoretical explanations :

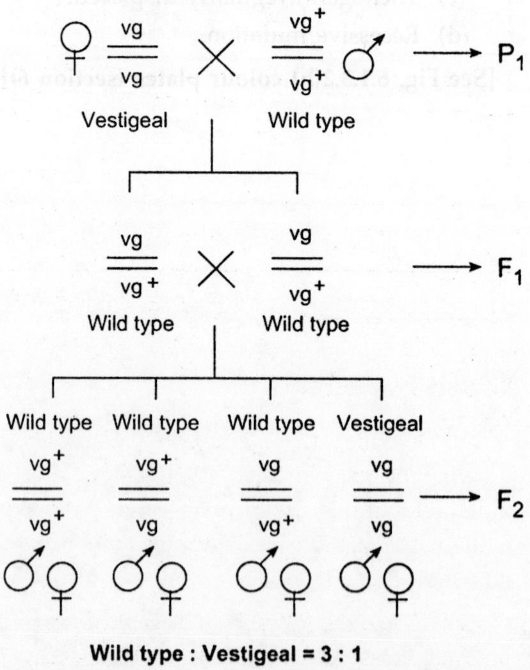

Fig. 6.1.4.1 : Results of monohybrid cross.

(ii) Results of dihybrid cross :

The data are to be recorded on the following theoretical explanations :

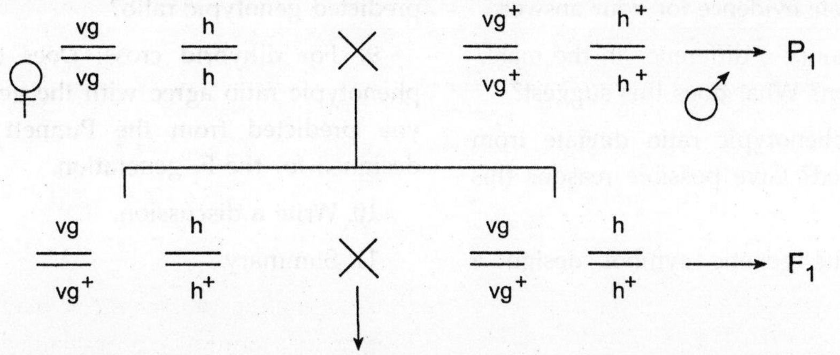

♀ \ ♂ → F_2	vg^+ h^+	vg^+ h	vg h^+	vg h
vg^+ h^+	vg^+/vg^+ h^+/h^+ (Wild)	vg^+/vg^+ h^+/h^+ (Wild)	vg/vg^+ h^+/h^+ (Wild)	vg/vg^+ h/h^+ (Wild)
vg^+ h	vg^+/vg^+ h^+/h (Wild)	vg^+/vg^+ h/h (Hairy)	vg/vg^+ h^+/h (Wild)	vg/vg^+ h/h (Hairy)
vg h^+	vg^+/vg h^+/h^+ (Wild)	vg^+/vg h/h^+ (Wild)	vg/vg h^+/h^+ (Vestigeal)	vg/vg h/h^+ (Vestigeal)
vg h	vg^+/vg h^+/h (Wild)	vg^+/vg h/h (Hairy)	vg/vg h^+/h (Vestigeal)	vg/vg h/h (Vestigeal Hairy)

Wild type: Hairy: Vestigeal : Vestigeal Hairy = 9:3:3:1

Fig. 6.1.4.2 : Results of dihybrid cross through Punnett Square.

Report Preparation

1. Describe the mutation? Complete a drawing of your mutant fly or flies. Was the mutation found in the male or female flies? What is the name of the observed mutation? Why did you choose this name?

2. Record the following dates: the date of the original cross and the date that the parental generation was sent to the morgue. What is the significance of the date the parental generation was sent to the morgue?

3. Record the following data for the F_1/F_2 generation for monohybrid and dihybrid cross respectively :
   - the date of the count
   - the number of wild-type male flies
   - the number of mutant male flies
   - the number of wild-type female flies
   - the number of mutant female flies

4. Tally your data from the F_1/F_2 generation for monohybrid and dihybrid cross respectively. What is your phenotypic ratio?

5. Compare your observed phenotypic ratio to your calculated genotypic ratios. Is your mutation(s) dominant or recessive? Is it sex-linked or autosomal ? State evidence for your answer.

6. Did you notice a difference in the male/female distribution? What does this suggest?

7. Did the phenotypic ratio deviate from what was expected? Give possible reasons this might happen.

8. Using your genetic symbol design a Punnett square for your F_1 cross. Use proper notations for dominant/recessive and sex-linked/autosomal mutations. What is the predicted genotypic ratio?

9. For dihybrid cross, Does the observed phenotypic ratio agree with the genotypic ratio you predicted from the Punnett square you designed for the F_1 generation.

10. Write a discussion.

11. Summary.

Chapter 6.2

Pedigree Analysis

A **pedigree** is a chart or diagrammatic representation of a family containing a systematic list of the ancestors of a given individual. Diagrams or charts are usually constructed for analyzing a particular trait that was happened in the family. The charts are made to symbolize individuals and illustrate relationships among them. From the analysis of the chart one can determine how the trait is inherited.

Pedigree graphic representation :

Pedigrees are represented graphically using certain conventions to symbolize individuals, matings, offspring-to-parent relationships, affected individuals, etc. (Fig. 6.2.1).

General rules :

1. Offsprings or sibs are listed in order of birth from left to right and are labeled with Arabic numerals 1, 2, 3, ... and so on.

2. Each generation is indicated by roman numerals I, II, III, ... and so on.

3. The individual whose case is to be considered is called **proband** and is marked by an arrow pointing to him/her. In case of male he is termed as **propositus** and in case of female she is termed as **proposita.**

Basic principles for pedigree analysis :

Any pedigree analysis follows some basic principles which is specific for that particular family tree—

A. **For dominant trait :**

1. It typically shows mendelian principles.
2. It does not skip generations.
3. An individual with dominant trait must have one of the parent affected with that trait.
4. Unaffected individuals have unaffected parents.

Example : Huntington's disease, brown eyes.

B. **For recessive trait :**

1. A recessive trait generally skips one or more generations.
2. Affected individuals have two affected parents.
3. In case of consanguineous mating i.e., marriage between close relatives, recessive trait is frequently found.
4. Individuals of affected and unaffected parents are of two types—(i) all normal or (ii) 50% normal and 50% affected.
5. Two unaffected parents (both are carriers) may produce individuals with typical mendelian ratio 3: 1 (normal : affected).

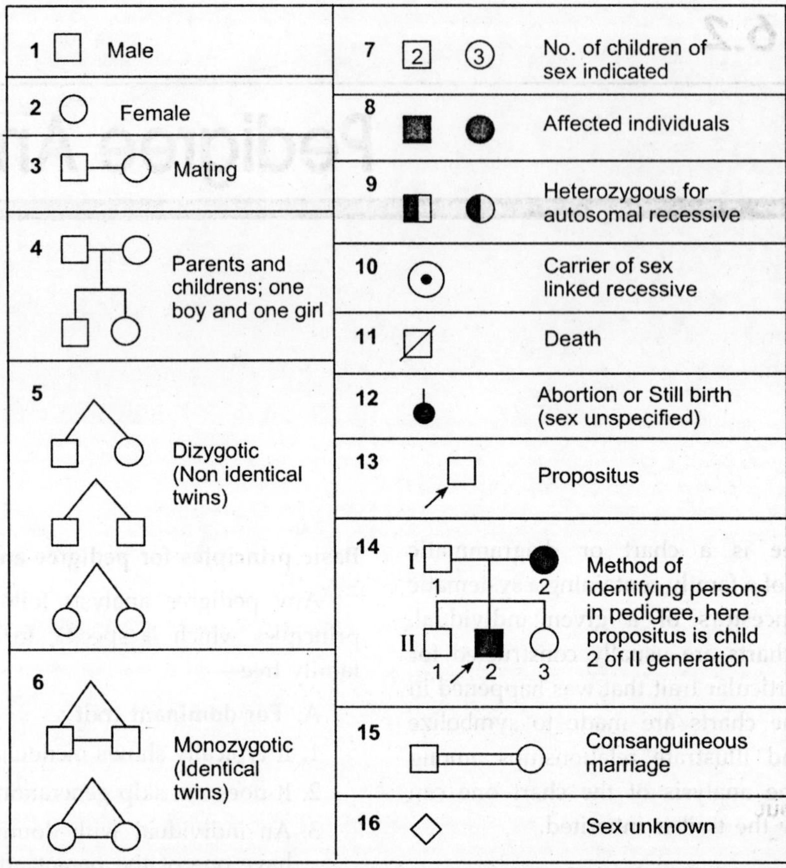

Fig. 6.2.1 : Symbols used in pedigree analysis.

Examples : Albinism, Cystic fibrosis and certain types of haemophilia, Tay-Sachs disease, PKU, blue eyes.

C. **For sex linked or X-linked inheritance :**

1. X-linked trait never passes from father to son.
2. Female offspring of affected father and unaffected mother will be affected.
3. Female offspring of normal father will be normal.
4. Sons of affected female will be affected.
5. All sons and daughters will be affected if their parents are affected.
6. When mother is normal and father is affected the pattern of their progeny will be—(i) all offsprings will be normal, (ii) 50% of their son and 50% of their daughters will be affected.
7. Generally, the trait will be found mostly in males in case of sex linked recessive condition.
8. The trait will occur slightly more often in females in case of sex linked dominant condition.

Examples : Red-green colour blindness, certain types of Haemophilia.

D. **For autosomal inheritance :**

1. An autosomal trait passes from father to son as well as to daughters.
2. Effected father and normal mother will produce offsprings in the following

manner—(i) all sons and daughter will be normal, (ii) 50% of the sons and 50% of the daughters will be affected.

3. In case of recessive autosomal trait approximately same number of sons and daughters will be affected.

   Example of Autosomal recessive includes Albinism, Phenylketonuria, Cystic fibrosis.

   Examples of Autososmal dominant are Neurofibromatosis, Quality of ear wax (dry or wet), Tuberous sclerosis etc.

Problem 1 :

Example of sex linked recessive.

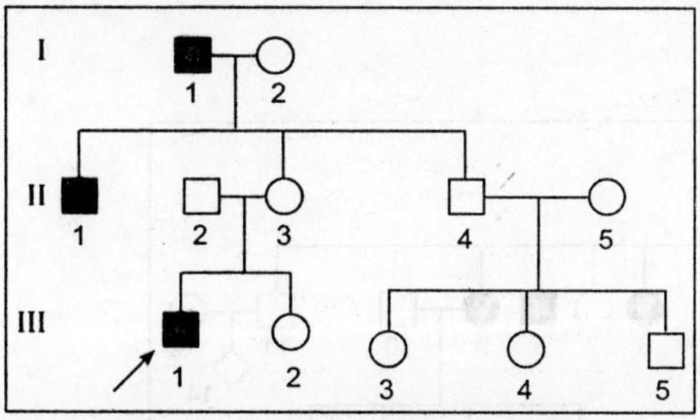

Fig. 6.2.2(a) : Typical pedigree representing the transmission of sex linked recessive trait.

1. **Characterization of the symbols :**

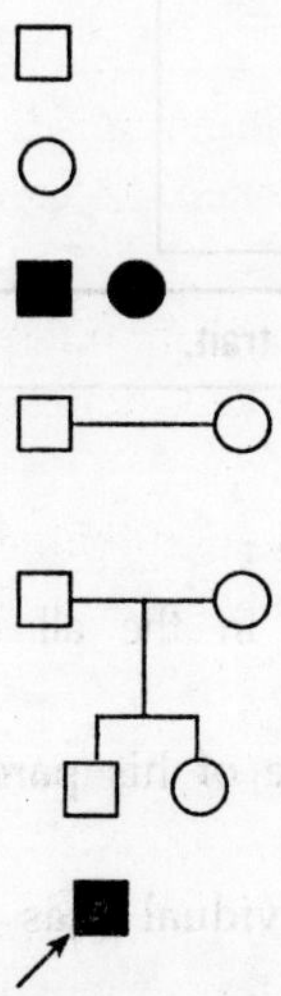

Unaffected male

Unaffected female

Affected individuals

Mating

Vertical line offers their offspring and the horizontal line between them are termed as sibship line. The offspring's are known as sibs.

Male Proband–Propositus; whose case is to be considered.

2. **Statement of proband :**

   (a) According to the proband (III-1) there are three generations and he is of the third generation and is affected.

   (b) The proband has a younger sister (III-2), two cousin sisters (III-3,4) and one cousin brother (III-5).

   (c) In second generation, he has a maternal uncle (II-1) who is affected, unaffected father (II-2) and mother (III-3), one maternal unaffected uncle (II-4) and one unaffected aunt (II-5).

   (d) The proband has an affected grandfather (I-1) and an unaffected grandmother (I-2).

3. **Members to be considered for their clinical importance :**

   The affected members for the trait in the pedigree chart are :

   Proband (III-1)

   His maternal uncle (II-1)

   His grandfather (I-1)

   Member to be considered for their genetical importance for the trait :

   Proband's grandfather (I-1).

4. **Analysis of the pedigree :**

   (a) The trait has shown in all three generations.

   (b) The proband has unaffected parents.

   (c) His cousin brother and sister also have unaffected parents.

   (d) In the second generation his affected grandfather and unaffected grandmother produce both affected and unaffected offspring.

   Comments : Since the unaffected parents of the proband produce 50% affected and 50% unaffected offspring, the trait is recessive rather than dominant.

   (e) One of the sons of his grandfather is affected.

(f) Effected grandfather and unaffected grandmother produce 50% affected son and unaffected daughter.

Comments : The trait is either autosomal recessive or sex-linked recessive

Since the males are only affected in the family, the trait is therefore sex linked recessive rather the autosomal recessive.

5. **Probable genotype of the family member :**

Generations	Members				
	1	2	3	4	5
I	r||↾	r||+			
II	r||↾	+||↾	r||	+||↾	+||+
III	r||↾	r||+ *or* +||+	+||+	+||+	+||↾

(Note : r is designated for recessive)

Problem 2 :

Example of sex linked dominant.

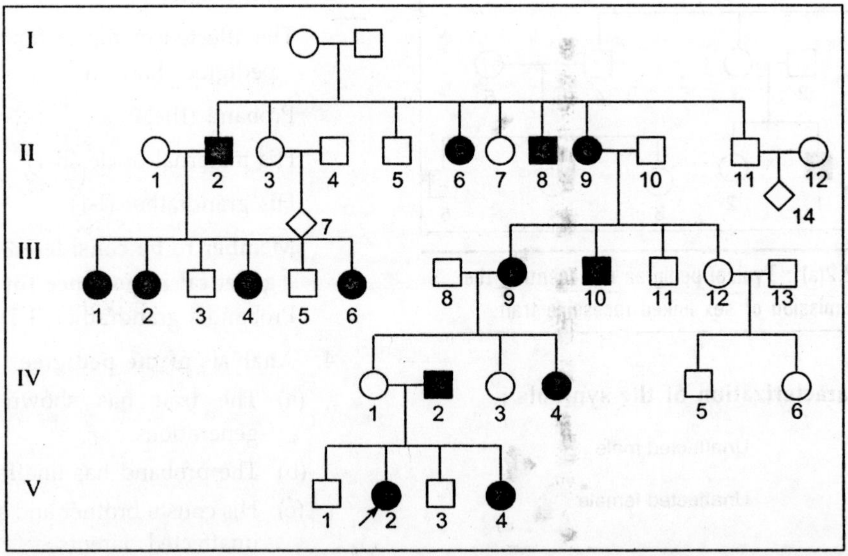

Fig. 6.2.2(b) : Typical pedigree representing the transmission of sex linked dominant trait.

1. **Characterization of the symbols :**

↗● Proband is proposita (because of being female) whose case is to be considered.

Try yourself the explanation of other characters as per rules.

2. **Statement of proband :** Try yourself as per rules.

3. **Analysis of the pedigree :**

(a) The trait has shown in the all five generations.

(b) The proband has one of his parents affected.

(c) Every affected individual has an affected parent.

(d) Normal individuals have normal parents.

(e) Affected male produce affected daughters and unaffected son.

Comments : So the trait appears to be X-linked.

(f) Affected females produced both affected males and females.

Comments : Since the trait do not skip generations and unaffected individuals have unaffected parents and affected individuals have one of the parent affected and females are mostly affected in the pedigree the trait is **dominant.**

Hence the trait is X-linked or **sex linked dominant.**

4. **Probable genotype of the family members :**

<table>
<tr><th rowspan="2">Generations</th><th colspan="14">Members</th></tr>
<tr><th>1</th><th>2</th><th>3</th><th>4</th><th>5</th><th>6</th><th>7</th><th>8</th><th>9</th><th>10</th><th>11</th><th>12</th><th>13</th><th>14</th></tr>
<tr><td>I</td><td>D||+</td><td>+|↾</td><td></td><td></td><td></td><td></td><td></td><td></td><td></td><td></td><td></td><td></td><td></td><td></td></tr>
<tr><td>II</td><td>+||+</td><td>D|↾</td><td>+||+</td><td>+|↾</td><td>+|↾</td><td>D||+</td><td>+||+</td><td>D|↾</td><td>D||+</td><td>+|↾</td><td>+|↾</td><td>+||+</td><td></td><td></td></tr>
<tr><td>III</td><td>D||+</td><td>D||+</td><td>+|↾</td><td>D||+</td><td>+|↾</td><td>D||+</td><td>+|↾
or
+||+
*</td><td>+|↾</td><td>D||+</td><td>D|↾</td><td>+|↾</td><td>+||+</td><td>+|↾</td><td>+|↾
or
+||+
*</td></tr>
<tr><td>IV</td><td>+||+</td><td>D|↾</td><td>+||+</td><td>D||+</td><td>+|↾</td><td>+||+</td><td></td><td></td><td></td><td></td><td></td><td></td><td></td><td></td></tr>
<tr><td>V</td><td>+|↾</td><td>D||+</td><td>+|↾</td><td>D||+</td><td></td><td></td><td></td><td></td><td></td><td></td><td></td><td></td><td></td><td></td></tr>
</table>

(Note : 'D' is designated for dominant) * Sex is not known (III-7, III-14)

Problem 3 :

Example of autosomal recessive.

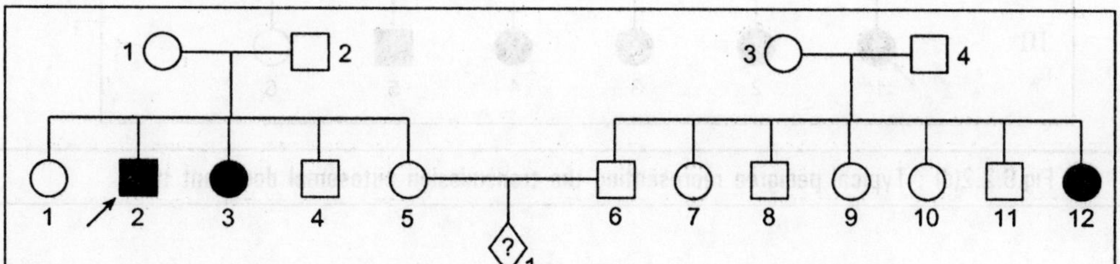

Fig. 6.2.2(c) : Typical pedigree representing inheritance of autosomal recessive trait.

1. **Characterization of the symbols :** Do yourself as per rules.
2. **Statement of proband :** Do yourself as per rules.

3. **Analysis of the pedigree :**
   (a) The trait has shown in only one generation.
   (b) Affected individuals have unaffected parents and unaffected individual (the phenotype/sex is unknown) has unaffected parents.
   Comments : The trait is recessive.
   (c) Approximately same number of males and females are affected for the trait.
   Comments : The trait is ***autosomal recessive.***

4. **Probable genotype of family members :**

Generations	Members											
	1	2	3	4	5	6	7	8	9	10	11	12
I	a||+	a||+	a||+	+||a								
II	a||+	a||a	a||a	+||+	+||+	+||+	+||+	+||+	+||+	+||+	+||+	a||a
III	+||+ *											

* Sex is not known.

Problem 4 :

Example of autosomal dominant.

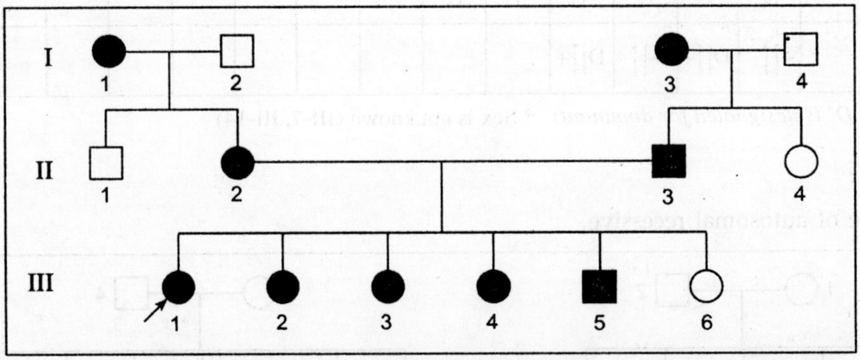

Fig.6.2.2(d) : Typical pedigree representing the transmission autosomal dominant trait.

1. **Characterization of the symbols :** Do yourself as per rules.
2. **Statement of proband :** Do yourself as per rules.
3. **Analysis of the pedigree :**
   (a) The trait has shown in every generation.
   (b) Every affected individual have at least one affected parent. *Comments* : The trait is ***dominant.***
   (c) In the third generation, both affected father and mother produce an unaffected female offspring (III-6). *Comments* : Hence the trait is **autosomal dominant**.

4. **Probable genotype of family members :**

Generations	Members					
	1	2	3	4	5	6
I	D||+	+||+	D||+	+||+		
II	+||+	D||+	D||+	+||+		
III	D||+	D||+	D||+	D||+	D||+	+||+

(Note : D is designated for dominant)

Practice Problems

Some problems to be carried out

1. Describe the pedigree. What is the pattern of inheritance bests accounts for the pedigree?

1.

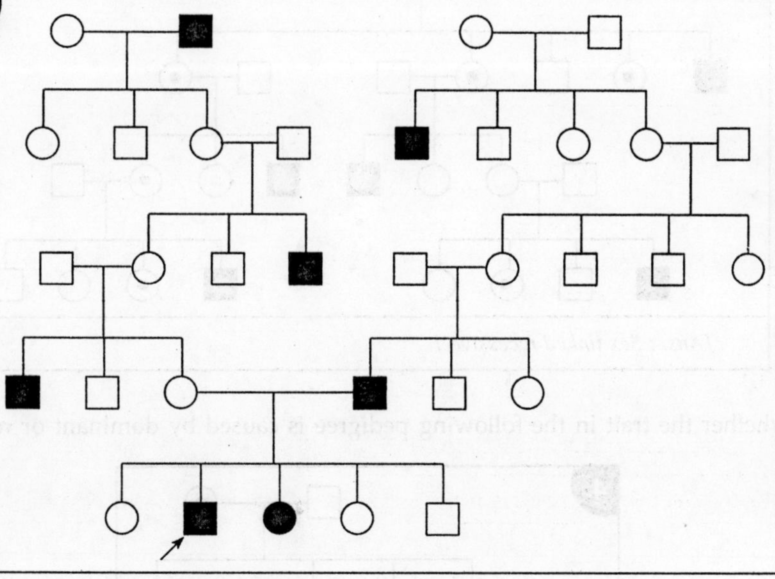

[Ans. : Sex linked recessive.]

2. Write the statement of proband. Analyse the pedigree. What is the mode of inheritance?

2.

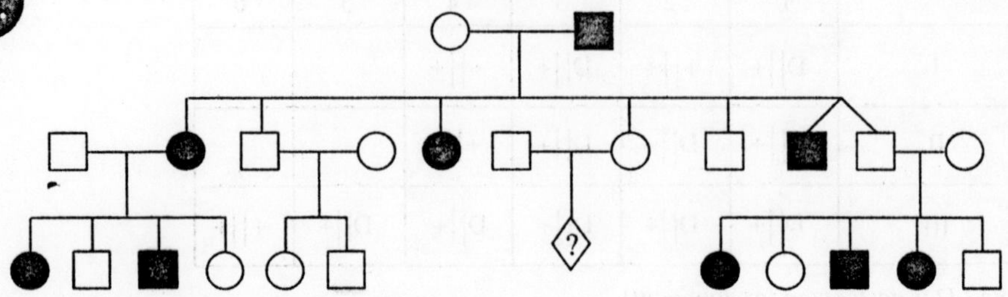

[Ans. : Autosomal dominant.]

3. A trait 'A' found in individuals represented by shaded symbols. Tell what pattern of inheritance is followed by trait 'A'.

3.

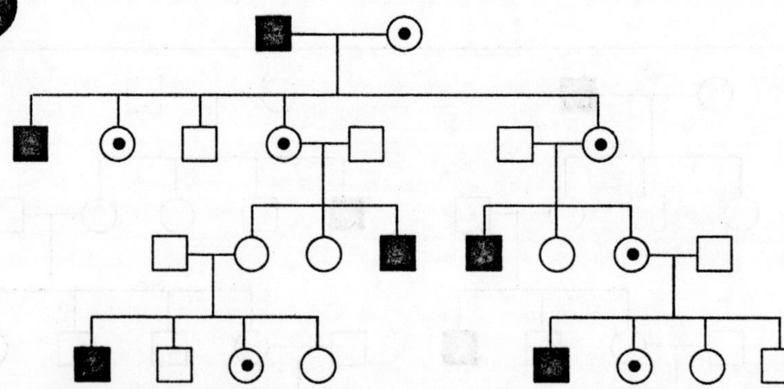

[Ans. : Sex linked recessive.]

4. Indicate whether the trait in the following pedigree is caused by dominant or recessive gene.

4.

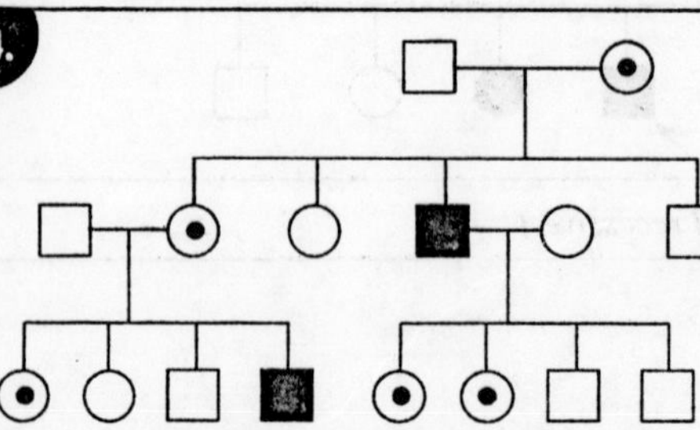

[Ans. : Sex linked recessive.]

5. Analyse the mode of inheritance. Give genotypes of all the members of the pedigree.

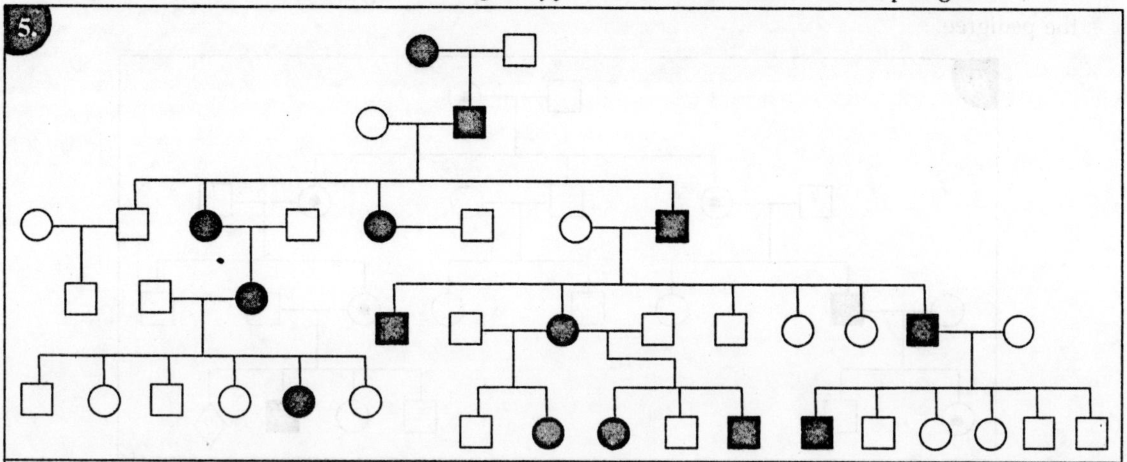

[Ans. : Autosomal dominant.]

6. Describe the pedigree. Explain the mode of inheritance and genotypes of all the members of the pedigree.

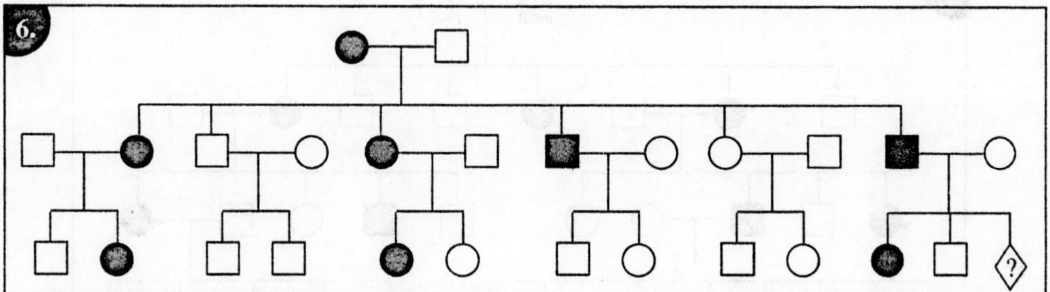

[Ans. : Autosomal dominant.]

7. Describe the pedigree. Analyse the mode of inheritance. State the probable genotypes of all the individuals of the pedigree.

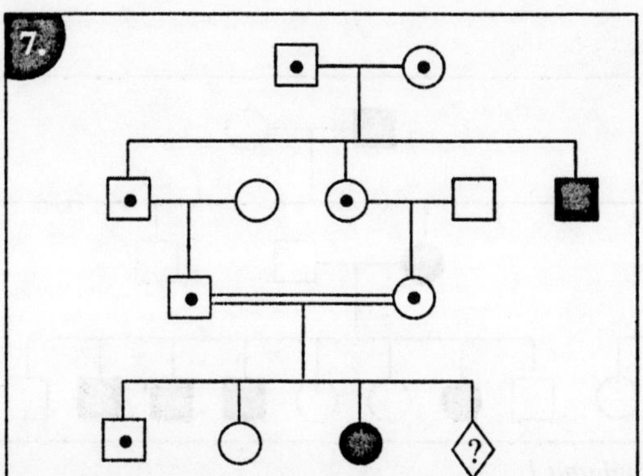

[Ans. : Autosomal recessive.]

8. What is the mode of inheritance of the trait. Write the probable genotypes of all the individuals in the pedigree.

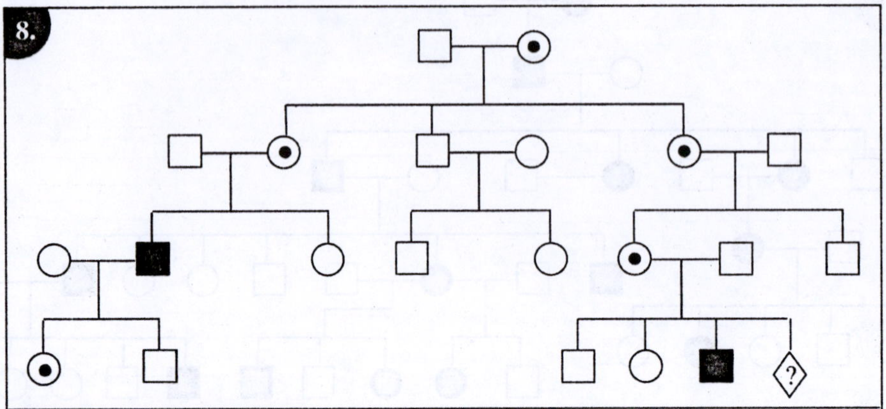

[Ans. : Sex linked recessive.]

9. How can you confirm that this pedigree represents a sex linked dominant trait?

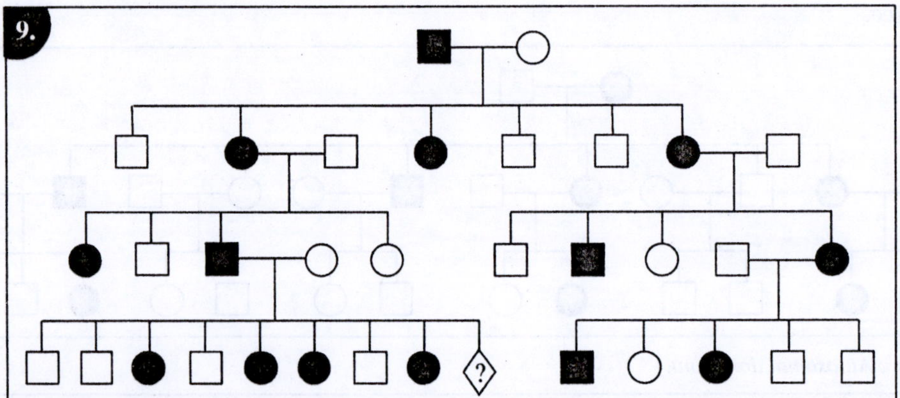

[Ans. : Sex linked dominant.]

10. Describe the pedigree. Explain the mode of inheritance and genotypes of all the members of the pedigree.

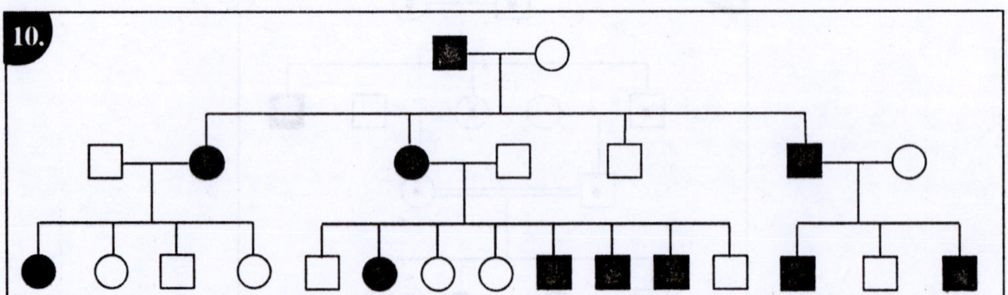

[Ans. : Autosomal dominant.]

11. Explain how this pedigree shows a holandric mode of transmission?

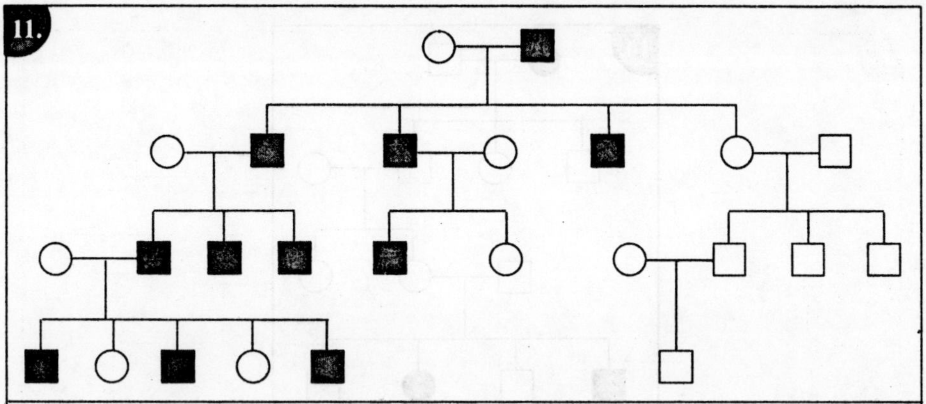

[Ans. : Y-linked.]

12. Write the statement of proband. Analyse the pedigree. What is the mode of inheritance?

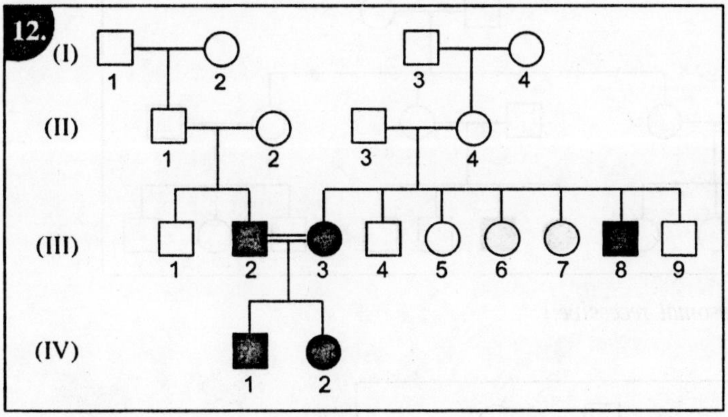

13. Which if any of the following patterns of inheritance is consistent with the pedigree?

(i) Autosomal dominant (ii) Autosomal recessive (iii) Sex linked recessive and (iv) Sex linked dominant.

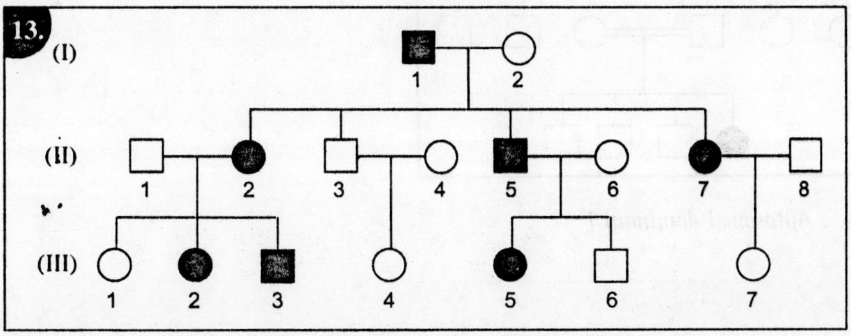

14. How can you prove that the pedigree shows transmission of an Autosomal recessive trait?

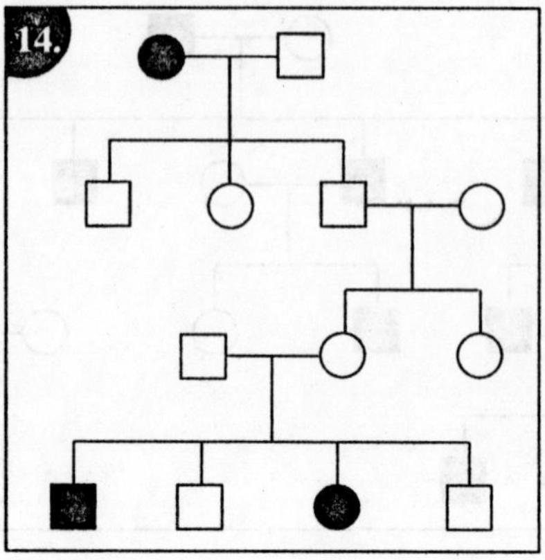

15. Determine the possible mode of inheritance for each trait in pedigree a–c.

(a)

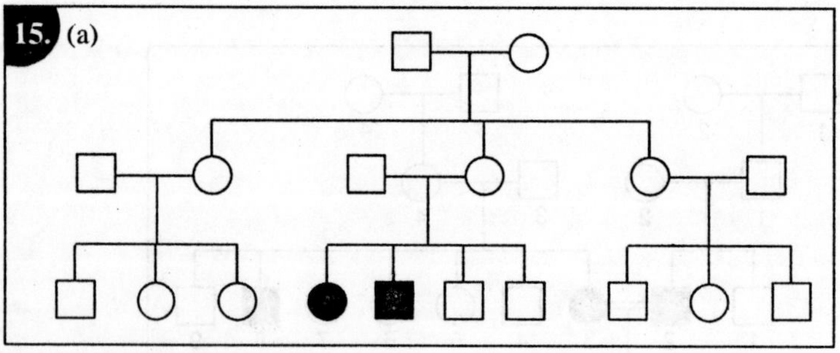

[Ans. : Autosomal recessive.]

(b)

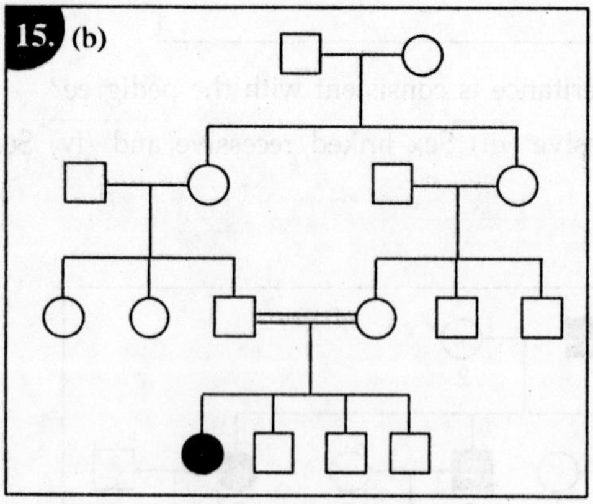

[Ans. : Autosomal dominant.]

(c)

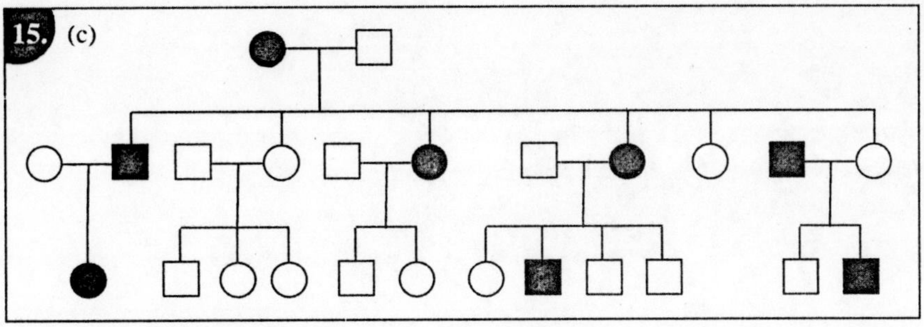

[Ans. : Autosomal dominant.]

16. A pedigree of a family with the disease is shown below, with affected individuals indicated in black. in addition, the blood type of family members is given.

The two alleles at the ALK locus will be denoted ALK^+ and ALK^-. The three alleles at the ABO blood group locus will be denoted IA, IB (which are co-dominant) and I (which is recessive to IA and IB).

(a) What is the genotype of individual 1 at the ALK and ABO loci?

(b) What is the genotype of individual 2 at the ALK and ABO loci?

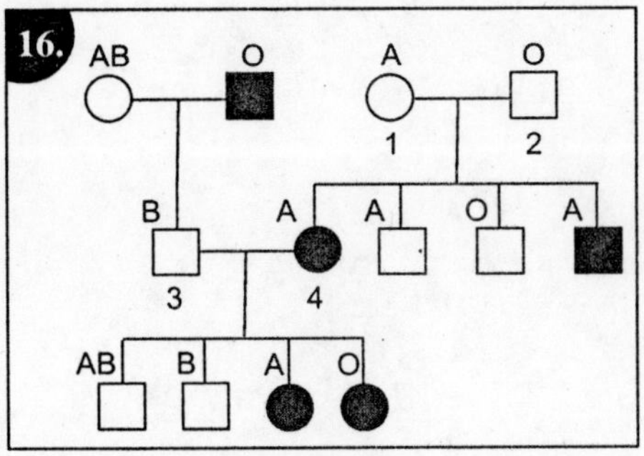

(c) What is the genotype of individual 3 at the ALK and ABO loci? Which alleles of each gene are carried on the chromosome he inherited from his father and which alleles are carried on the chromosome he inherited from his mother?

(d) Individuals 3 and 4 are expecting their fifth child. A physician draws a prenatal blood sample and determines that the child has blood type B. What is the probability that the child will have alkaptonuria? Explain your answer.

Section 7

BIOCHEMISTRY AND PHYSIOLOGY

SECTION AGENDA

Chapter 7.1.1

Qualitative Analysis of Carbohydrates

Introduction

Carbohydrates are widely found in the plant and the animal Kingdom. Carbohydrates exists in the various forms viz., mono-, di-, oligo-, and polysaccharides. Carbohydrates bear active groups such as aldehydes (–CHO) and ketones (=CO). Carbohydrates also bear various hydroxyl (–OH) groups. Chemical properties of the carbohydrates depend upon the number of hydroxyl groups and the presence and the absence of the ketone or aldehyde groups.

1. ***Simple carbohydrates*** : Simplest form which includes monosaccharides or simple sugars.They are classified as :

(a) *On nature of reducing groups :*

(i) *Aldoses* : They possess aldehyde groups, eg. glucose, galactose and ribose.

(ii) *Ketoses* : They possess ketone groups, eg. fructose, erythrulose and sedoheptulose

(b) *According to number of carbon atom* : Dioses (2-C), trioses (3-C; dihydroxyacetone), tetroses (4-C; erythrulose), Pentoses (5-C; ribose), Hexoses (6-C; glucoses, fructoses, levulose, galactose, mannose etc.).

2. ***Compound Carbohydrates*** : They are formed by union of two or more monosaccharides.

(a) *Oligosaccharides* : They are composed of two to nine monosccharides unit linked through **glycosidic linkage**.

(i) *Disaccharides* : They are formed by union of two monosaccharides molecules, eg. sucrose, maltose, lactose and trehalose.

(ii) *Trisaccharides* : They are formed by union of three monosaccharides molecules, eg. raffinose, gentianose, melezitose.

(iii) *Tetrasaccharides* : They are formed by union of four monosaccharides molecules, eg. stachyose.

(b) *Polysaccharides* : They are composed of more than nine monosccharides unit linked through glycosidic linkage, eg. starch, dextrin, insulin, chitin, agar agar, pectin etc.

These carbohydrates give rise to a wide variety of colour reactions. Some simple tests are described below.

(I) GENERAL TESTS FOR CARBOHYDRATES

(i) Benedict's Test for reducing sugars :

Principle : Alkaline solution of copper hydroxide when heated form black cupric oxide.

$$Cu(OH)_2 \longrightarrow CuO + H_2O$$

But in presence of reducing substance; rust brown cuprous oxide is precipitated.

Carbohydrates bearing free or potentially free ketone or aldehyde group, exhibits reducing properties in an alkaline solution.

Materials required :

(i) *Benedict's reagent* :

Solution A : 173 g of sodium citrate and 100 g of sodium carbonate is dissolved in 800 mL of warm water. The solution is filtered and the volume is made up to 850 mL of water.

Solution B : 17.3 g of copper sulphate is dissolved in 100 mL of distilled water.

Solution B is poured to Solution A and the volume is made upto 1 litre with distilled water in a 2 litre breaker.

(ii) *Glucose solution* : Glucose solution is prepared in concentration of 100 g/liters.

Experiment : 5 drops of glucose solution is added to 2 mL of Benedict's solution. The solution is boiled for 2-5 minutes in a water bath and cooled slowly.

Observation : Formation of red, yellow or yellowish orange or green colour precipitate.

Inference : Depending upon the sugar concentration yellow to green precipitate is formed.

(ii) Fehlings Test For Reducing Sugars :

Principle : Blue alkaline cupric hydroxide present in the Fehlings reagent when heated in presence of reducing sugars gets reduced to yellow or red cupric oxide and gets precipitated.

Material required :

(1) *Fehlings solution I (Copper Sulphate solution)* : 34.65 g of copper sulphate is dissolved in 500 mL of distilled water.

(2) *Fehlings solution II (alkaline Solution)* : It is prepared by dissolving 125 g of caustic soda (KOH) and 173 g of Rochelle salt (sodium potassium tartrate) in 500 mL of distilled water.

Experiment : Equal volume (2mL) of Fehling II and I solutions are taken in a test tube and five drops of test solution (reducing sugar solution) is added. Mixture is boiled for five minutes in a water bath; solution is cooled.

Observation : Yellow to brownish red precipitate of cuprous oxide is formed and may become red if particles becomes coarser.

Inference : The yellow or red precipitate of cuprous oxide indicates presence of reducing sugar.

(iii) Molisch's Test :

Principle : Carbohydrates under dehydrating action of the strong mineral acids (viz. hydrochloric and sulphuric acid) form furfural, which then gives coloured compounds in presence of alpha-napthol or thymol. All carbohydrates along with furfural yielding compounds such as glyceraldehydes, glycoaldehyde, acetone, glucoronic acid etc. give the test. The test is also positive for protein having carbohydrate radicals.

Materials required :

1. *Molisch Reagent* : 5% of α-Napthol or thymol is dissolved in 95% ethanol.
2. *Furfural solution* : 1 mL of furfural is dissolved in 100 mL of distilled water.
3. Concentrated H_2SO_4.
4. Carbohydrate sample provided.

Experiment : 3mL of carbohydrate solution is taken in a test tube, 2 drops of 5% Molisch's reagent is added to the carbohydrate solution.

Concentrated H_2SO_4 is taken in another test tube and 2 mL of it is poured along the side of the test tube containing the carbohydrate sample. The contents are allowed to settle.

Observation : A red-cum-violet or purple colour ring appears at the junction of the two liquids.

Inference : The given sample is carbohydrate.

(iv) Trommer's Test for Reducing Sugars :

Principle : Caustic soda reagent enolizes the sugar and also forms cupric hydroxide in presence of copper salts.

In presence of copper salts :

$$CuSO_4 + 2NaOH = Cu(OH)_2 + Na_2SO_4$$

$$2Cu(OH)_2 + R.CHO = Cu_2(OH)_2 + R.COOH + H_2O$$

$$Cu(OH)_2 = Cu_2O + H_2O$$

[Cuprous oxide (red precipitate)]

In absence of reducing salts :

$$Cu(OH)_2 \longrightarrow CuO + H_2O$$

Cupric hydroxide Cupric oxide

Materials required :

1. Copper sulphate solution (0.25%) : 2.5g of copper sulphate is mixed in 1 litre of distilled water.
2. Carbohydrate sample provided.
3. 5% caustic soda.

Experiments : 1mL of 0.25% of copper sulphate solution and 3mL of glucose solution (as a carbohydrate) is taken in 50mL of conical flask. In the mixture 5% caustic soda (NaOH) is added in drops. As a result of which precipitate of cupric hydroxide is formed and later gets re-dissolved.

Observation : A clear blue solution is formed, when boiled due to reduction red or yellow precipitate of cuprous oxide, which forms at the base. In absence of reduction, black precipitate of cupric oxide is formed.

Inference : The red or yellow precipitate indicates the presence of reducing sugars.

(V) Barfoed's Test for Reducing Sugar :

Principle : This test helps in the detection of the monosaccharides. Copper acetate in dilute acid medium is reduced by monosaccharide (reducing sugars). Reduced copper combines with hydroxyl group of water similar to Fehlings test and forms cupric hydroxide (yellow in colour). Further, on heating cupric hydroxide is converted into cuprous oxide (red precipitate) and water.

Materials required :

1. *Barfored's Reagent :* 13.3g of copper acetate is dissolved in 200mL of distilled water. To this mixture 1.8mL glacial acetic acid is added.
2. Carbohydrate sample provided.

Experiments : 1 mL of carbohydrate solution is mixed with 3 mL of Barfored reagent in a test tube. Boiled for a minute and then the mixture is cooled.

Observation : Orange to brick red granules of cuprous oxide is formed and adhere to the side of the test tube or remain suspended in the fluid..

Inference : This indicates the presence of reducing sugar.

(vi) Seliwanoff's Test :

Principle : Ketohexose is converted into the furfural derivative, which combines with resorcinol to give red colour. 4-hydroxymethyl furfural is formed by the action of HCl on ketohexoses. Aldoses do not give furfurals when boiled with dilute HCl.

Fructose (Ketohexose) + dil.HCl $\longrightarrow$ Furfural + Resorcinol $\longrightarrow$ A complex of furfural with resorcinol

Materials required :

1. *Seliwanoff's Reagent :* 0.05g of resorcinol is dissolved in 100mL of dilute (1:2) HCl.
2. Carbohydrate (ketose) sample provided.

Experiments : 1 mL of carbohydrate sample is mixed with 3mL of Seliwanoff's reagent. This mixture is boiled for 25 seconds and then allowed to cool.

Observation : Cherry red to deep red colour appears and later a reddish brown precipitate appears when cooled.

Inference : Cherry red to deep red colour appers if ketohexose (such as fructose) is present.

(vii) Osazones Test :

Principle : This particular tests helps in the detection of reducing sugars.Aldoses and Ketoses when heated with excess amount of phenylhydrazine produces hydrazones and which exhibits osazones of definite crystalline shape.

Materials required :

1. Phenylhydrazine hydrochloride.
2. Sodium acetate.
3. Glacial acetic acid.
4. Carbohydrate sample provided.

Experiments : To about 0.5g of Phenylhydrazine hydrochloride 0.1g of sodium acetate is added. In this about 10 drops of glacial acetic acid is added. Next, to this 5mL of provided carbohydrate sample added. The mixture is shaken well and heated over a flame until a solution is achieved. The test tube is then placed in a boiling water bath for an hour and then after removing from the flame allowed to cool. Crystal's formed are examined under low power microscope.

Observation : Monosaccharide crystals separate out due to heating. Disaccharide separate out after cooling.

Inference : Glucose and fructose and mannose give the same osazone called as 'glucosazone', which is fan, sleeves, needle or cross shape. Where as galactose gives strips and plate like 'galactosazone' where as maltose gives close tufts of short fine crystals refered as 'malatosazone'.

(viii) Moore's Test :

Principle : Glucose condenses reacting with caustic soda and forms caramel.

Materials required :

1. Concentrated caustic soda (NaOH)
2. Concentrated Sulphuric acid (H_2SO_4)
3. Carbohydrate solution provided

Experiment : 2mL of carbohydrate solution is taken in a test tube 1mL of concentrated caustic soda solution is added in to it. The mixture is heated and then few drops of concentrated sulphuric acid are added into it.

Observation : The solution turns yellow changing to orange and then into dark brown, which smells like sugar or caramel.

Inference : The test indicates the presence of reducing sugar, which do not have aldehyde or ketone group in their structure.

(ix) Hydrolysis Test or Inversion Test (for detection of non reducing carbohydrates such as sucrose) :

Principle : Various non-oxidizing acids either hydrolyse or dehydrate carbohydrates, depending on the concentration of the acid and type of carbohydrate. Weak acids hydrolyse polysaccharides to monosaccharides where as strong acids react with carbohydrates producing furfurals.

Sugar is hydrolysed by dilute acids that further produce a distinct reduction to cuprous hydrate and cuprous oxide due to the production of so called **Invert sugar** (mixture of glucose and fructose or Levulose). Polysaccharides are hydrolyzed into glucose by action of dilute acids.

Materials required :

1. Concentrated Sulphuric acid.
2. Sodium carbonate.
3. Carbohydrate solution provided.
4. Benedicts qualitative reagent (described in Benedicts test).

Experiment : 6 mL of carbohydrate sample (non reducing type) solution is taken in a test tube. Few drops of conc. sulphuric acid is added to it. It is boiled for 3 minutes. The solution is cooled and neutralized with solid sodium carbonate until effervescence stops. Benedicts Test is performed with the above solution.

Observation : A red precipitate becomes visible in the mixture.

Inference : The sample contains disaccharide such as sucrose or cane sugar.

(x) Iodine Test :

Principle : This test is useful in the detection of polysaccharides viz., starch, glycogen, dextrin, etc. Colour is formed due to polysaccharide iodine complex which can be removed by alkali and transformed into iodate or iodide from which iodine is further released in addition of acid.

Materials required :

1. *Iodine solution :* A good amount of iodine is added to 2% solution of potassium iodide in distilled water producing a yellow coloured solution.
2. Carbohydrate sample solution provided.
3. 10% NaOH solution.
4. Mineral acid.

Experiment : To about 3mL of carbohydrate solution, 1-2 drops of iodine solution is added.

Observation : Starch gives blue colour, where commercial dextrin or glycogen gives reddish brown colour. The colour is discharged on adding few drops of 10% NaOH but reappears when neutralized with acids. The clear solution indicates presence of dextrin's, where as hazy solution indicates presence of glycogen.

Inference : Colour is formed due to formation polysaccharide iodine complex.

(xi) Precipitation Tests :

These tests are useful in the detection of polysaccharides like starch dextrin, glycogen etc.

(a) Precipitation with Tannic Acid or Basic Lead Acetate Test :

This test is useful for the detection of glycogen and starch.

Materials required :

1. *Basic Lead Acetate Solution :* 180g of lead acetate is dissolved in 700mL of boiling water. Added 110g of lead oxide to it. Boiled for 30 minutes in stirring condition. Cooled and filtered. Volume is made up to 1 litre with distilled water.
2. *20% Tannic acid :* 20g of Tannic acid in 100mL of distilled water.
3. Carbohydrate sample provided.

Experiment : In about 3mL of provided carbohydrate sample solution, about 6mL of 20% Tannic acid or 3mL of basic lead acetate is mixed.

Observation : Precipitation occurs.

Inference : Starch or glycogen detected. After about 5 minutes the solution is filtered and Molisch's test is performed with it. It gives negative result.

(b) Half Saturation with Ammonium sulphate :

The test helps in detection of starch.

Materials required :

1. Half saturated carbohydrate sample solution.
2. Saturated ammonium sulphate solution.

Experiment : To 3mL of half saturated carbohydrate sample solution 3mL of saturated ammonium sulphate is added. The mixture is shaken vigorously and kept undisturbed for 6 minutes.

Observation : A precipitation occurs.

Inference : Starch is detected.

(II) TEST FOR MONOSACCHARIDE

(a) Glucose and Fructose :

Following test are performed :

(i) *Molisch's Test* : A purple or violet ring appears due to furfural formation.

(ii) *Iodine Test* : Negative.

(iii) *Fehling's Test* : At first yellow and then red predicate.

(iv) *Benedict's Test* : A red, yellow or green precipitate will be formed depending upon presence of glucose or fructose.

(v) *Barfoed's Test* : Red granules settles at bottom of the test tube.

(a) For confirming the presence of 'Glucose' carries on Seliwanoff's test and Osazone test. If Seliwanoff's Test is negative, Osazone Test gives yellow crystalline 'Glucosazone' is formed thus glucose is confirmed.

(b) For confirming the presence of fructose one should carry of Seliwanoff's test and Osazone test. If Seliwanoff's Test is positive giving a cherry red colour and later reddish brown precipitate on cooling and with Osazone Test needle like yellow crystals of 'fructosazone' is formed thus fructose is confirmed.

(III) TEST FOR DISACCHARIDES

A. Maltose and Lactose :

The following tests are performed

(i) *Molisch's Test* : A purple or violet ring becomes visible due to formation furfural.

(ii) *Trommer's Test* : At first a yellow and then a red precipitate is visible due to formation of cuprous oxide.

(iii) *Fehling's Test* : At first a yellow and then a red precipitate is visible due to formation of cuprous oxide.

(iv) *Benedict's Test* : A red, yellow or orange precipitate is formed varying on the amount of maltose or lactose present. This is due to the formation of cuprous oxide and cuprous hydroxide.

(v) *Moore's Test* : A yellowish colour results which then changes to a dark brown colour giving out a smell of burnt sugar or caramel.

(vi) *Osazone Test* : Close lifts of short fine crystals of 'maltosazone' are obtained. Circular clusters of fine needles radiating from a dense centre (Hedge Hog Crystal) of 'lactosazone' is formed.

B. Sucrose :

The following tests are performed :

(i) *Molisch's Test* : A purple or violet ring becomes visible due to furfural formation.

(ii) *Inversion Test*: A red precipitate is formed due to the reduction to cuprous hydrate and cuprous oxide.

(iii) *Seliwanoff's Test*: A red colour is obtained. On prolonged boiling due to inversion phenomenon of sucrose, 5-hydroxy-methyl furfural is formed by the action of hydrochloric acid on ketohexose.

(iv) *Osazone Test* : Sucrose does not give Osazone reaction. But when boiled continuously with acid and then neutralized with alkali, due to hydrolysis forms 'glucosazone'.

(IV) TEST FOR POLYSACCHARIDES

A. Test for Starch :

The following tests are performed :

(i) *Molisch's Test* : A purple or violet ring becomes visible due to formation furfural.

(ii) *Iodine Test* : A deep blue colour appears due formation of the starch iodine complex.

(iii) *Tannic acid test* : This is performed for starch, which gives a precipitate, get dissolves when heated.

(iv) *Half saturation test with Ammonium Sulphate* : A precipitate occurs.

(v) *Saturation with Ammonium Sulphate* : A complete precipitation occurs.

(vi) *Hydrolysis Test* : Blue solution turns dark red and ultimately a red precipitate is formed at the bottom. When starch is hydrolyzed with hydrochloric acid it forms glucose, which test positive with Benedicts test.

B. Test for Glycogen :

(i) *Molisch's Test* : A purple or violet ring becomes visible due to formation furfural.

(ii) *Iodine Test* : A reddish-brown or port-wine colour develops as hazy solution due to formation of glycogen-iodine absorption complex.

(iii) *Saturation with Ammonium sulphate* : A complete precipitation occurs.

(iv) *Hydrolysis Test* : A yellow or red colour appears after performing Benedict's test.

Chapter 7.1.2

Qualitative Analysis of Proteins

INTRODUCTION

Proteins are polymers of the bifunctional monomer, **amino acids.** The twenty common naturally occurring amino acids each contain a carbon, an amino group, a carboxylic acid group, and a side chain or side group. These side chains (or R groups) may be nonpolar or polar and uncharged, or charged, depending on the pH and pKa of the ionizable group. Amino acids form polymers through a **nucleophilic attack** by the amino group of an amino acid at the electrophilic carbonyl carbon of the carboxyl group of another amino acid. The carboxyl group of the amino acid must first be activated to provide a better leaving group than OH^-. The resulting link between the amino acids is an **amide linkage** which biochemists call a **peptide bond**. In this reaction, water is released. In a reverse reaction, the peptide bond can be cleaved by water (hydrolysis).

Proteins are polymers of twenty naturally occurring amino acids. Both the sequence of a protein and it's total length make a distinction of one protein from another. The amino acids are all chiral, with the exception of glycine, whose side chain is H. All naturally occuring proteins from all living organisms consist of L-amino acids. The absolute stereochemistry is related to L-glyceraldehyde, as was the case for triacylglycerides and phospholipids. Most naturally occurring chiral amino acids are S, with the exception of cysteine.

Proteins can be classified on the basis of solubility of proteins.

1. *Simple proteins :* These have only α-amino acids or their derivatives and made up of mainly polypeptide chains. These are of few types:
   (a) Fibrous proteins : These are the collagens, elastins, keratins etc.
   (b) Globular proteins : These are Albumins (serum albulmins, lactoalbumins), globulins (serum globulins, ovaglobulins etc.), plant proteins (glutelins, prolamines etc), basic proteins (protamines, histones, globins etc).
   (c) Denatured proteins : These are the acid metaproteins and alkali metaproteins.

2. *Conjugated proteins :* They consist of simple protein molecules attached with some non-protein prosthetic group, eg. nucleoprotein, phosphoprotein, lecithoprotein, lipoproteins, glycoproteins, etc.

3. *Derived protein :* These are the intermediate product of protein hydrolysis through the action of heat, alkali, acid and enzymes.

(a) Primary protein derivative : These are the substances formed from some of the simple and conjugated proteins on denaturation. They are proteans, metaproteins.

(b) Secondary protein derivatives : They are substances formed during the hydrolysis of protein molecules, eg. proteoses, peptones, peptides and amino acids.

GENERAL TEST FOR PROTEINS AND THEIR DERIVATIVES

Proteins are identified by three groups of reactions.

(A) Colour Reactions :

Colour reaction occurs due to reaction between one or more of the constituent's radicals or groups of complex protein molecules and the chemical reagents used in any particular test.

(i) Biuret Test :

Principle : This test depends up on two or more peptide linkages of protein molecule and presence of two ($CONH_2$) groups linked to one another or to same nitrogen or carbon atom.

Materials required

1) Protein sample solution.
2) 0.1% Copper sulphate solution.
3) 10% sodium hydroxide.

Experiment : About 2-3mL of protein sample solution is taken in a test tube, added one or two drops of 0.1 % cupric sulphate solution and 10% NaOH into it and then mixed thoroughly.

Observation : A distinct violet colour or pink colour is noticed. All higher protein gives purple-violet; gelatin gives blue-violet, whereas lower protein (peptones/proteoses) gives rose pink colours.

Inference : Protein or peptone or proteoses present.

(ii) Xanthoproteic Test :

Principle : Most of all proteins give Xanthoproteic reactions. Nitration of phenyl ring of phenylalanine, tyrosine and tryptophan occurs due to reaction with concentrated nitric acid that forms a yellow colour. The nitrobenzene compound freely ionizes in alkaline solution increasing the intensity of the colour.

Materials required :

1) Sample protein solution.
2) Concentrated nitric acid (HNO_3).
3) Concentrated ammonium hydroxide or 2% sodium hydroxide solution.

Experiment : The sample protein solution is treated with few drops of concentrated nitric acid. A white precipitate is formed. It is boiled for a minute and then cooled under tap water and made alkaline with concentrated ammonium hydroxide or 2% sodium hydroxide.

Observation : White precipitated is formed in presence of higher proteins. On heating this precipitate turns yellow and later dissolves to give a yellow solution. Lower proteins such as peptones and proteoses give yellow solution without giving a white precipitate, while after cooling and making the solution alkaline the yellow colour deepens into orange.

Inference : Presence of higher or lower proteins.

N.B. : Gelatin fails to give Xanthoproteic test due to low content of aromatic amino acid.

(iii) Millon's Test :

Principle : Mercury reacting with nitric acid forms mercuric nitrate and nitrite. Phenol group of tyrosine in a tyrosine containing protein (except gelatin) reacts with Millon's reagent to forms red mercury complex with tyrosine.

Materials required :

1. *Millon's Reagent :* Concentrated nitric acid (sp.gr.1.42) is added to mercury (2:1v/v).

This resultant solution is diluted with 2 volumes of water.

2. Sample protein solution.

Experiment : 2 to 3 drops of Millon's reagent is added to 3 mL of protein solution. The mixture is heated over a low flame and brought to boiling very gradually.

Observation : A white precipitate is formed in case of higher protein, which turns brick red upon heating. Lower protein such as peptones and proteoses produce a very small amount of precipitate which get dissolved upon heating thus imparting a red colour to solution.

Inference : Presence of protein, peptones or proteoses with tyrosine residues.

(iv) Ninhydrin Test :

Principle :

When ninhydrin is heated with α–amino acids or proteins or their derivatives made up of α-amino acid group a blue colour is formed. The α-amino acids react with ninhydrin to give carbon monoxide, ammonia and an aldehyde.

The ammonia produced reacts with ninhydrin and produces hydrantin, which is blue in colour. When proteins with amino acids e.g. proline, hydoxyproline when heated with ninhydrin, carbon dioxide evolves forming a red colour substance.

Materials required :

1. Ninhydrin reagent : 0.04 g of ninhydrin (Triketo hydrindene hydrate) is dissolved in 20 mL of distilled water. The reagent should be prepared fresh before use.
2. Protein sample solution provided.

Experiment : 10 drops of freshly prepared 0.1% ninhydrin solution is added to neutral or slightly acidic protein solution. It is boiled for one minute and cooled.

Observation : A violet, purple to blue colour develops.

Inference : This purple colour is due to formation of complex (Rheumans purple) by ninhydrin and ammonia.

(v) Sakaguchi Reaction :

Principle : When protein in alkaline medium in presence of α-napthol and sodium hypochloride or bromine form a red coloration due to formation of coloured compound by the guanidine group of arginine in the protein.

Materials required :

1. 20% NaOH.
2. 1 % α-napthol in alcohol.
3. 2% sodium hypchlorite.
4. Bromine water.
5. Protein sample solution provided.

Experiment : Added 5 drops of 20% NaOH, 3 drops of 1% α-napthol solution and 5 drops of 2% sodium hypochlorite or 2 drops bromine water in 3mL protein solution and mixed thoroughly.

Observation : Intense red colour is formed.

Inference : This red coloured compound is due to arginine rich proteins.

(B) Precipitation Reactions :

1. Precipitation by Heavy Metals :

Principle : The precipitation of protein by heavy metals is due to absorption of metallic cations by negatively charged protein ions.

Materials required :

1. Ferric chloride.
2. Lead acetate.
3. Silver nitrate.
4. Copper Sulphate.
5. Mercuric Chloride (dilute).
6. Hydrochloride acid (dilute).
7. Saturated sodium chloride solution.
8. Protein sample solution provided.

Experiment : To the 3 mL of protein sample provided, dilute mercuric chloride is added in drops.

Observation : A heavy white precipitate is formed which dissolves in an surplus reagent and also on adding up of 2-3mL of saturated sodium chloride solution and is re-precipitated by addition of few drops of dilute hydrochloride acid.

Again with :

1. Ferric chloride : Yellow precipitate is formed which is soluble in excess reagent.
2. Lead acetate : Gives white precipitate.
3. Silver nitrate : Gives white precipitate soluble in excess of reagent.
4. Copper sulphate gives blue precipitate.

2. **Precipitation by Alkaloidal reagents :**

(i) **Precipitation by Esbach's Reagent :**

Principle : When protein reacts with Esbach's reagent insoluble salt protein-picrate is formed due to absorption of negative ions of alkaloidal reagents by positively charged protein ions in solution acidic to protein isoelectric pH.

Materials required :

1. Protein sample solution provided.
2. *Esbach's Reagent :* 10g of picric acid along with 20g of citric acid is dissolved in 1 litre of distilled water.

Experiment : Equal volume of Esbach's reagent is added to equal volume of protein solution.

Observation : Albumin gives yellowish precipitate, which disappears on heating but does not reappears on cooling.

Gelatin gives yellow precipitate, which disappears on heating but does not reappear on cooling.

Inference : Presence of albumin or gelatin depending on the observation.

(ii) **Precipitation by Tannic acid :**

Principle : Insoluble salt protein-tannate is formed when protein reacts with tannic acid giving a white or brown precipitated.

Materials required :

1. Protein sample solution provided.
2. 10 % freshly prepared tannic acid solution.

Experiment : To about 3 mL of protein solution equal volume of freshly prepared 10% tannic acid is added.

Observation : A white or brown precipitate is formed.

Inference : Protein is present in the sample provided.

(iii)**Precipitation by Trichloroacetic acid (TCA) :**

Principle : The protein molecule binds to two TCA molecules and the protein-TCA molecular complex (salt) being insoluble are precipitated out.

Materials required :

1. Protein sample solution provided.
2. 10% Trichloroacetic acid solution.

Experiment : 10% Trichloroacetic acid solution is added to few mL of protein solution.

Observation : A white precipitate is formed.

Inference : Protein present in the sample:

(iv)**Precipitation by Ammonium Sulphate :**

Principle : Proteins are precipitated out in presence of high salt concentration, which is known as **'Precipitation out reaction'**.

Materials required :

1. Proteins sample solution provided.
2. Ammonium sulphate (saturated solution).
3. Solid ammonium sulphate

(a) **Experiment 1 (Full saturation) :** 5mL of protein solution is saturated with solid ammonium sulphate and the contents are shaken vigorously.

Observation : A precipitation occurs which gets dissolved in water.

Inference : All higher and lower proteins except peptones are positive to these tests.

(b) **Experiment 2 (Half saturation)** : 3mL of saturated solution of ammonium suphate is added to an equal volume of protein solution.

Observation : A precipitation occurs.

Inference : Globulins, caseinogen and other higher proteins are precipitated out except albumin.

(C) Coagulation Reactions :

The reactions are due to the colloidal nature and higher molecular weight of the protein macromolecule.

1. Heat Coagulation Test :

Principle : Heating denatures protein such as albumin and globulin. These denatured proteins flocculate near their isoelectric point.

Materials required :

1. Protein samples solution provided.
2. Acetic acid.

Experiment : About 5mL of protein solution is boiled inside a test tube. An opalescence solution is formed. The solution is cooled and acidified with few drops of acetic acid and the mixture is boiled.

Observation : White flocculent coagulam is formed.

Inference : Albumin and globulin gives positive reaction.

2. Heller's Cold Nitric Acid Test :

Principle : Concentrated nitric acid when react with protein forms metaprotein after being coagulated. A ring is formed at the zone of contact of protein solution and nitric acid

Materials required :

1. Protein sample solution provided.
2. Concentrated nitric acid.

Experiment : Concentrated nitric acid is taken in a test tube; few drops of protein are added into it, very slowly through a pipette.

Observation : A sharp white opaque ring is found at the juncture of the two liquids.

Inference : Albumin and globulin is present.

Chapter 7.1.3

Qualitative Analysis of Lipids

Introduction

Lipids are biological molecules, which are soluble in organic solvents, such as chloroform/ methanol, and are sparingly soluble in different aqueous solutions. Chemically lipids are esters of fatty acids or substances capable of forming esters. There are two major classes of lipids, **saponifiable** and **nonsaponifiable**, based on their reactivity with strong bases. The nonsaponifiable classes include the "fat-soluble" vitamins (A, E) and **cholesterol**. The saponifiable lipids contain long chain carboxylic acids, or **fatty acids**, esterified to a backbone molecule, which is either **glycerol** or **spingosine**. The major saponifiable lipids are triacylglycerides, glycerophospholipids, and the sphingolipids. Lipids can be classified as follows :

1. **Simple lipids :** These are the esters of fatty acids with various alcohol. They can be of following types :

(a) *Glycerides or neutral lipids :* These are esters of fatty acids and glycerol, they include :

(i) *Fats :* These are solid at 20°C, eg. Lard, butter, cocoa butter, coconut oil.

(ii) *Oils :* They are liquid at room temperature, eg. Fish oil, liver oil, rape seed oil, castor oil etc.

(b) *Waxes :* These are esters of fatty acids with higher monohydric alcohol other than glycerol, eg. True waxes (wool wax, bee wax), cholesterol ester, vitamin A, vitamin D esters.

2. **Compound or conjugated Lipids :** These are esters of fatty acids which on hydrolysis yield other substances in addition to fatty acids and an alcohol. These are of following types:

(a) *Phospholipids or Phosphatids :* Lipids on hydrolysis yields fatty acids, phosphoric acid and sometimes glycerol and nitrogenous base, eg. Phosphatidic acid, cephalins, lecithins, spingomyelins, plasmalogens.

(b) *Glycolipids :* These are complex lipids containing fatty acid, carbohydrate in combination with long chain aliphatic alcohols or acids and nitrogenous base but devoid of phosphoric acid, eg. cerebrosides, gangliosides, cytolipins.

(c) *Sulpholipids :* Lipids containing carbohydrate, cerebronic acid,sulphuric acid, potassium and nitrogenous base sphingosine.

(d) *Liposproteins :* These are lipid materials bound to proteins. The lipid component mainly consists of cholesterol esters and phospholipids containing stearic acid, palmitic acid and oleic acid.

3. **Derived lipids :** These are compounds formed in the hydrolysis of simple or compound lipids. These includes :

(a) *Fatty acids :* They are straight chain acids of higher series of usually contain an even number of carbon atoms. These may be saturated (acetic acid, propionic acid,palmitic acid, stearic acid etc.) or unsaturated(oleic acid, arachidonic acid etc.).

(b) *Alcohols :* These are straight chain fatty acids containing 16-30 carbon atoms. These are aliphatic alcohols containing b-inone ring. These include cetyl, stearyl alcohols etc.

(c) *Hydrocarbons :* These are unsaturated compound without any carboxyl or alcohol groups containing isoprene rings which cannot be saponified, eg. carotenoids and squalene.

(d) *Fat soluble vitamins :* They are mainly the vitamin A, D, E and K.

(e) *Stererols :* These are esters of higher fatty acids with cyclic alcohol containing perhydrocyclopentanophenathrene ring, eg. cholesterol, ergosterol, sitosterol etc.

(f) *Steroids :* These are compounds having similar structure to sterols containing cyclopentanophenathrene ring, eg. adrenal corticosteroids, sex steroids, some alkaloids etc.

(I) General Test for Lipids :

Fatty materials are generally a mixture of neutral fats, sterols and phospholipids. So there are some general tests for these lipids. These are :

(i) **Solubility Test :** Only a few drops of liquefied fat or oil are allowed to fall in a test tube enclosing distilled water. The test tube is then shaken well.

It is observed that fat particles float in it and do not dissolve.

But when test is done with ether or chloroform or alcohol, the fat particles dissolves in it and they go into the solution easily.

This is due to solubility of fatty substances in non-polar hydrophobic substances such as ether, chloroform or alcohol.

(ii) **Grease Spot Test :** Few drops of oil are put on a strip of paper. A translucent grease spot is observed.

(iii) **Test for presence of free fatty acids :** An alkaline solution of NaOH is mixed to a little alcoholic solution of phenolphthalein. This gives a pink colouration.

To this solution, ethereal solution of fat is added in drops, pink colour of the original solution is discharged.

Thus it can be inferred there is presence of free fatty acid in the fat sample.

(iv) **Saponification through Alkali :** When fats are reacted with basic hydroxides they yield basic salts of acids and glycerol. These reactions are called **saponification**. The salts which are formed are known as **soaps**.

To carry on a saponification reaction, a small amount of oil is taken in a test tube with equal volume of alcoholic potash solution (25%),it is boiled for 30 minutes. Three separate layers are formed. The bottom most layer contains alkali and glycerol, middle one solid soap and upper most layer contain oil. The solid soap may be salted out in excess alkali.

(v) **Emulsification of Fat :** Emulsification of fat is the physical process in which fats are suspended in form of its heaviest particles.

In the process, few drops of oil are put into a solution of soap and strongly shaken. Minute fat globules spread out in the solution. Thus oil is emulsified.

Emulsification is also a physiological process which can be experimented as follows :

Few drops of coconut oil is added to 2 mL of bile salt solution and shaken vigorously. Finely divided oil particles remain in suspension. In this case emulsion is permanent and bile salts acts as emulsifying agent.

(II) Test for Glycerol :

Chemically glycerol is a tri-hydric alcohol, which occurs in conjugation with fatty acids forming esters.

(a) Acrolein Tests :

Principle : Glycerol when heated with dehydrating agents like solid potassium bisulphate, it yields unsaturated aldehyde called as **acrolein.**

Materials required :

1. Glycerol.
2. Crystals of acid potassium metasulphate.

Experiment : 2 drops of glycerol is taken in a dry test tube, in it few crystals of solid potassium bisulphate is added, heated cautiously at first then more strongly.

Observation : Acid vapour with pungent odour of acrolein emerges which blackens a piece of filter paper moistened with ammonium silver nitrate.

Inference : Glycerol present.

(b) Copper Hydroxide Test :

Principle : When copper sulphate and sodium hydroxide suspension is heated normally there is precipitation of cupric hydroxide. But when boiled with glycerol, such precipitation is inhibited by glycerol.

Materials required :

1. Cupric hydroxide suspension: 2.5% copper sulphate solution in 3mL of 5% sodium hydroxide.
2. Glycerol.

Experiment : In cupric hydroxide suspension few drops of glycerol is added.

Observation : A blue colour is obtained. No change is noticed on boiling the solution.

Inference : Glycerol present.

(III) Test for Cholesterol :

(a) Solubilty test :

Principle : It is based on solubility of cholesterol in non polar solvents (alcohol or ether) and insolubility in polar solvent (water).

Materials required :

1. Non-polar solvent (ether or alcohol).
2. Polar solvent (water).
3. Cholesterol crystals.

Experiment : Cholesterol is mixed with ether or alcohol and water.

Observation : Cholesterol dissolved in non polar solvents, but remains undissolved in water.

Inference : Presence of Cholesterol.

(b) Salkowski's Test :

Principle : Cholesterol is oxidized by acid in solution and coloured unstable intermediate compounds are formed.

Materials required :

1. Cholesterol crystals.
2. Chloroform.
3. Concentrated sulphuric acid.

Experiment : A few crystals of cholesterol is dissolved in 1 mL of chloroform and poured in a dry test tube. To such a mixture 3mL of conc. sulphuric acid is added, shaken and allowed to stand.

Observation : The upper chloroform layer turns red, purple in the middle layer and lower most acid layer shows a green fluorescence. This colour of cholesterol gets discharged on adding water, but again returns on addition of concentrated sulphuric acid.

Inference : Cholesterol is present.

(c) Lipschutz Reaction :

Principle : Benzoyl peroxide oxidizes cholesterol to oxy-cholesterol, which forms coloured derivatives when reacting with concentrated sulphuric acid.

Materials required :

1. Cholesterol.
2. Glacial acetic acid.
3. Concentrated sulphuric acid
4. Benzoyl peroxide.

Experiment : To 3mL of a solution of cholesterol in glacial acetic acid, 10 drops of benzoyl peroxide is added and it is boiled for 3 minutes. The mixture is cooled and 5 drops of concentrated sulphuric acid is added in to it and mixed well.

Observation : A red colouration, changing through blue to green develops.

Inference : Cholesterol present.

Chapter 7.2.1

Estimation of Percentage Quantity of Glucose (Benedict's method)

Principle

Glucose is estimated by the quantitative reduction of Benedict's quantitative reagent. Glucose, due to its reducing properties, reduces copper sulphate in hot alkaline medium. Because of the presence of potassium thiocyanate in the Benedict's quantitative reagent a white precipitate of cuprous thiocyanate, instead of the typical red precipitate of cuprous oxide, is produced on reduction of Cu^{++} by glucose. The little amount of potassium ferrocyanide present in the reagent keeps cuprous oxide in solution. Sodium citrate present in the reagent keeps $CuCO_3$ (formed by the interaction of $CuSO_4$ with Na_2CO_3) in solution. The alkali used Na_2CO_3 which is likely to cause less destruction of sugar.

Materials required

1. Glucose solution (supplied).
2. Distilled water.
3. Benedict's Quantitative Reagent (Commercially available).
4. Anhydrous Na_2CO_3.

Procedure

1. 10mL of supplied glucose solution is taken in a 100mL volumetric flask and the volume is made up to the mark with distilled water i.e. it has a dilution of 1:10.

2. Mixed thoroughly and the diluted solution is transferred to a 25mL burette. The initial burette reading is noted.

3. 10mL of Benedict's quantitative reagent is pipetted and transferred carefully to a 100mL conical flask. About 1g of anhydrous Na_2CO_3 along with few glass beads (to avoid bumping) is added and the conical flask is placed over a tripod stand over a wire gauge.

4. The reagent is boiled gently on a low flame and while boiling diluted glucose solution is added drop by drop from a burette.

5. The disappearance of blue colour and appearance of pale white precipitate (of cuprous thiocyanate) indicates the end point of titration. Final burette reading is noted.

Results

No. of observations	Burette reading (mL)		Volume of dilute glucose required (mL)	Mean volume of dilute glucose required (mL)
	Initial	Final		
1.				
2.				
3.				

Calculation

10 mL Benedict's Quantitative Reagent is reduced by 20 mg of glucose.

'x' mL of diluted glucose solution is required to reduce 10 mL of the reagent.

If the mean titration reading is 'x' mL.

'x' mL of diluted glucose solution contains 20 mg of glucose.

Therefore, 1mL of diluted glucose solution contains $20/x$ mg of glucose.

Therfore, 100mL of diluted Glucose solution contains $20 \times 100/x$ mg of glucose.

As the solution is diluted 10 times hence the percentage of original glucose solution is ($20 \times 100 \times 10/x \times 1000$) g or $20/x$ g%.

Chapter 7.2.2

Estimation of Total Carbohydrate through Anthrone Method

Principle

Carbohydrates act as key components of storage and constitutional substance of the plants and animals. They are found as free sugars or as polysaccharides. It is recognized that the fundamental element of the carbohydrates are the monosaccharides. The carbohydrate content can be assessed by acid hydrolysing the polysaccharides into simple sugars and thus measuring the resultant monosaccharides.

Carbohydrates are at initially hydrolysed into simple sugar with dilute hydrochloric acid. In the hot acidic medium glucose is dehydrated to hydroxymethylfurfural. This compound reacts with anthrone, and a green coloured product with absorption maxima at 630 nm is formed.

Materials required

1. 2.5N HCl.
2. Anthrone reagent : 200mg anthrone is dissolved in 100mL of ice cold 95% H_2SO_4. This has to be prepared fresh before use.
3. Standard glucose : Stock – 10mg in 100 mL water. Working standard – 10mL of stock diluted to 100mL with distilled water. Stored in refrigerator after adding few drops of toluene.
4. Test sample provided.
5. Solid sodium carbonate.

Procedure

1. 100mg of sample is weighed into a boiling tube
2. This is hydrolysed in a boiling water-bath for three hours with 5mL of 2.5N HCl and then cooled at room temperature.
3. This is neutralised with solid sodium carbonate until effervescence ceases.
4. The volume is made up to 100mL and then centrifuged.
5. The supernatant is collected and 0.5 and 1mL aliquots are taken for analysis.
6. Standards are prepared by taking 0, 0.2, 0.4, 0.6, 0.8 and 1mL of working standard. '0' is used as a blank.
7. The volume is made up to 1mL in all the tubes including the sample tube by adding distilled water.
8. Next 4mL of anthrone is added to each tube.
9. The tubes are heated for eight minutes in a boiling water bath.

10. Cooled rapidly.

11. This green to dark green colour is read at 630 nm.

12. A standard curve is drawn by plotting concentration of the standard glucose on X- axis versus absorbance in the Y – axis.

13. From this graph the amount of carbohydrate present in the sample is calculated.

Chapter 7.2.3

Assay of Protein

(a) Folin-phenol method :

Principle :

Protein reacts with Folin-ciocalteaous phenol reagent and form a coloured complex. The colour is formed due to reduction of alkaline copper with protein as in the biurette reaction and also reduction of phosphomolybdate by tyrosine and trytophan present in protein.

Materials required :

1. Alkaline sodium carbonate solution (20g/L Na_2CO_3 in 0.1mol/L NaOH).

2. 0.5% Copper sulphate in 1% Sodium Potsssium tartarate solution (5g/L $CuSO_4.5H_2O$ in 10g/L Na, K tartarate). Prepared fresh by mixing the stock solutions.

3. *Alkaline solution (Reagent A) :* Prepared fresh on day of use by mixing 50mL of (1) and 1mL of (2).

4. *Folin ciocalteau's reagent* (Reagent B) is commercially available and is made up of sodium tungstate, sodium molybdate together with phosphoric acid or concentrated HCl. Before using this reagent it is diluted twice with distilled water (25mL reagent + 25mL distilled water; i.e. 1:1 v/v).

5. Standard protein solution (Bovine serum albumin; 1 mg/mL).

6. Unknown test protein sample solution.

7. Spectrophotometer or colorimeter with cuvettes.

8. Test tubes with rack.

Procedure :

Three test tubes are taken, one for blank and two for sample. In each test tube the following reagents one added.

(a) *Blank Tube :*

(i) 1mL of distilled water.

(ii) 5mL of Reagent A (alkaline copper sulphate solution).

(b) *Sample tube (3 nos.) :*

(i) 1st Tube : 0.1mL of test solution + 0.9mL distilled water

2nd Tube : 0.2mL of test solution + 0.8 mL distilled water

3rd Tube : 0.4mL of test solution + 0.6 mL distilled water

(ii) 5mL of Reagent A (alkaline copper sulphate solution).

This is mixed for ten minutes. Then 0.5mL of diluted Ciocalteau's reagent (Reagent B) is added to both the tubes. After thirty minutes blue colour developed O.D is determined against 660 nm filter.

Preparation of standard curve : 100mg of Bovine serum Albumin (BSA) Fraction IV is dissolved in 100 mL of distilled water so that a stock solution of BSA is prepared having a concentration of 1μg/μL. From this stock solution (may be further diluted) 0.2, 0.4, 0.6, 0.8, 1mL is pipetted out into series of test tubes. The volume is made upto 1mL with distilled water. In each 5mL of Reagent A is added. Allowed to stand for 10 miñutes.

In this 0.5mL of diluted Reagent B is added. After 30 minutes O.D against each concentration was noted and a standard curve was obtained by plotting O.D in Y-axis against the respective concentrations in X-axis of a graph paper.

Estimation of concentration of the test solution : Concentration of protein in test solution is determined from standard curve by plotting it's O.D ('x') in the standard curve and knowing the corresponding concentration of the protein in the test solution(s).

Conclusion :

(I) When the above O.D were plotted against the concentration factor a straight line curve was obtained that gives us the standard curve for BSA at given concentrations.

(II) The O.D '*x*' obtained from the colorimeter is compared with the standard curve and the concentration of protein solution determined to be '*y*' mg/mL.

Note : The Lowry method depends on the presence of tyrosine within the protein to be measured. The standard protein must contain *approximately* the same number of tyrosine residues as the sample, or the procedure will be inaccurate. If there are no tyrosine residues in the sample to be measured, the Lowry method of protein determination is useless, and use should be made of the Bradford assay. In general, the Bradford assay is the method of choice for protein determinations.

(b) Bradford method :

Principle :

Bradford method of protein estimation uses a concept of the protein's capacity to bind a dye, quantitatively. This type of protein assay depends on the ability of the proteins to bind to Coomasie brilliant blue G 250 dye thus forming a complex whose extinction coefficient is much greater than that of the free dye.

Materials required :

1. Bovine serum albumin (BSA) fraction IV.
2. Unknown protein sample.
3. Coomasie Brilliant Blue (100mg of Coomassie brilliant blue G250 in 50mL of 95% ethanol).100mL of concentrated. Orthophosphoric acid is added into it. Volume is made up to 200mL by addition of distilled water. Stored refrigerated in amber bottle.Before using the concentrated dye is diluted 4 times in distilled water.
4. Phosphate buffered saline(PBS).
5. Spectrophotometer and tubes.
6. Micropipettes.
7. Test tubes.
8. Vortex.

Procedure :

1. A series of protein standards are prepared using BSA, diluted with PBS to final concentrations of 0 (blank; PBS only), 250, 500, 750 and 1500μg BSA/mL. Also prepared serial dilutions of the unknown sample in PBS to be measured.

2. 100μL of each of the above are added to a separate test tube (or spectrophotometer tube).

3. 5.0mL of Coomasie Blue is next added to each tube and mixed by vortex, or inversion.

4. The spectrophotometer is adjusted to a wavelength of 595 nm, and blank using the tube from step 3 which contains 0μg BSA.

5. After 5 minutes each of the standards and each of the samples are read at 595nm wavelength.

6. Absorbance of the standards v/s their concentrations are plotted to produce a standard curve. The extinction coefficients are computed and the concentrations of the unknown samples are calculated using the standard curve.

Note : This procedure can be used for standard assay, 20-150μg protein; 200-1500μg/ml. The dye exists in two forms (blue and orange) in acid solution. The proteins better binds with the blue form. One should check the absorption of the working dye at 550nm is 1.18. If required one has to adjust either the powder or the water.

(c) By Biuret method :

Principle :

Proteins are made up of amino acids joined through peptide bonds. Two or more peptide bonds react with copper sulphate and result in violet coloured product. Further the Cu atoms form complex with four of the nitrogen atoms of the peptide chain and develop the colour. The intensity of the colour is governed by the number of peptide bonds present.

Materials required :

1. *Biuret Reagent* : 3g of $CuSO_4.5H_2O$ and 9g of sodium potassium tartarate is dissolved in 500mL of 0.2N of NaOH solution. 5g of Potassium iodide is added and the volume is made up to 1 L by adding 0.2N NaOH solution.
2. Bovine serum albumin (BSA) fraction IV.
3. Spectrophotometer and tubes.
4. Water bath at 37°C.
5. Unknown protein sample.

Procedure :

1. Standard dilutions of BSA containing 1, 2.5, 5.0, 7.5 and 10mg/mL protein are prepared. Serial dilutions of the unknown samples are also prepared.

2. 1.0mL of each of the standards, each unknown sample, and 1.0mL of distilled water are added into to separate tubes. 4.0mL of Biuret reagent are then added to each tube. Mixed by vortexing.

3. All of the tubes are incubated at 37°C for 20 minutes in a water bath.

4. Spectrophotometer is turned on and adjusted to read at a wavelength of 540 nm.

5. The tubes from Step 3 are cooled, the spectrophotometer is blanked and all of the standards and samples are read at 540 nm.

6. The absorbance of the standards versus their concentration are plotted to produce a standard graph. The extinction coefficient is computed and the concentrations of the unknown samples are calculated.

Note : The Biuret reaction is one of the important process for the determination of protein concentration. It remains as a rapid determination method, but is not very accurate. It is useful during protein separation procedures since there are fewer salt interference reactions than with the Bradford or Lowry techniques. The colour formed is stable for about 1-2 hours and consequently all spectrophotometer readings must be made as soon as possible after the incubation step.

Chapter 7.2.4 (a)

Estimation of Amino-acids by Formol Titration Method

Principle

A solution containing amino acid is almost neutral in reaction. On adding up of formaldehyde, it reacts with the amino groups of amino acid and so the basic property of this amino acid is lost. The carboxyl group of the amino acids can be titrated with a standard alkali by means of phenolphthalein indicator. The acidity showed as a result of the titration is the gauge for amino acid nitrogen. Ammonia reacts with formaldehyde in a similar fashion, and thus if ammonia is spread in the sample, it must be removed first or else the result is flawed.

Materials required

1. N/10 NaOH solution.
2. Formaldehyde (Formol) solution.
3. Phenolphthalein solution.
4. Sample solution provided.
5. Conical flask (10mL).
6. Pipette (10mL).
7. Volumetric flask (100mL).
8. Burette (25mL).

Procedure

1. The sample is transferred quantitatively, pursued by its washings with distilled water to a 100mL volumetric flask and the volume is made equal to 100mL with distilled water.
2. 10mL of this dilute sample is taken through a pipette into a conical flask and its reaction is tested by adding a drop of phenolphthalein.
3. If acidic, it is cautiously neutralized with N/10 NaOH.
4. The neutralized formol solution is then added and mixed up with neutralized diluted sample.
5. The mixture is then titrated with N/10 NaOH being taken in a burette until the first permanent pink tinge indicates the end-point.

Calculation

1mL of N/10 NaOH = 0.0075g of glycine

'x'mL of N/10 NaOH = 'x' × 0.0075g of glycine.

10mL of diluted amino acid contains 'x' × 0.0075 g of glycine.

100mL diluted amino acid contains

$$\frac{x \times 0.0075 \times 100}{10} \text{ g of glycine}$$

∴ Total quantity of glycine present = 'x' × 0.075g of glycine.

Chapter 7.2.4(b)

Estimation of Amino-nitrogen by Formol Titration Method

Principle

This is generally done by Formol's titration method. The method consists of adding neutralized formol to the sample. The formaldehyde methylates the amino groups of amino acid, as a result, then by allowing the –COOH group to act as acid so as to be titrated. The acidic forms can be titrated with N/10 NaOH and the amount of nitrogen can thus be estimated. As, ammonia and amino group of amino acid reacts with formaldehyde similarly.

Materials required

1. N/10 NaOH solution.
2. Formaldehyde solution.
3. Phenolphthalein solution : Dissolved 1g phenolophthalein in 100mL of 95% ethyl alcohol.
4. Sample solution provided.
5. Distilled water.
6. Volumetric flask (100mL).
7. Conical flask (10mL).
8. Pipette (10mL).
9. Burette (25mL).

Reactions

$4NH_4Cl + 6CH_2O = N_4(CH_2)_6 + 6H_2O + 4HCl$

Procedure

1. 10 ml. of the sample solution is taken in a volumetric flask and the volume is made up to 100 mL with the addition of distilled water.
2. 10 ml of this diluted sample is taken by the pipette into the conical flask and its reaction is tested by adding a drop of phenolphthalein.
3. If acidic, it is then carefully neutralised by adding N/10 NaOH solution.
4. The neutralized formol solution is then added and mixed with the neutralized diluted sample solution.
5. The mixture is then titrated with N/10 NaOH solution, which is being taken in a burette, until the first permanent pink tinge colour indicates the end point.

Calculation

1 mL of N/10 NaOH = 1.4mg of amino-nitrogen.

'x' mL of N/10 NaOH = $1.4 \times x$ mg of amino-nitrogen.

10 mL of diluted solution contains $1.4 \times x$ mg of amino-nitrogen.

100 mL of diluted solution contains $\frac{1.4 \times x \times 100}{10}$ mg of amino-nitrogen.

10mL of original solution contains $1.4 \times x \times 10$ mg of amino-nitrogen.

100mL of original solution contains $\frac{1.4 \times x \times 10 \times 100}{10}$ mg of amino-nitrogen.

Hence, the percentage of the amino-nitrogen in the given solution is $140 \times x$ mg% or, $0.14 \times x$ g%.

Note : The average daily output of amino nitrogen in the urine of an adult on a mixed diet is more or less 0.7g, corresponding to about 500mL of 0.1N base per day, and representing about 2.5 to 4.5% of the total nitrogen. It rises in amount by the ingestion of acids or acid forming foods and in the **acidosis** of starvation or **diabetes**, and falls by the ingestion of alkalies or base-forming foods and in **alkalosis.** The ammonia of the urine is liberated enzymatically in the kidneys primarily from glutamine but also from amino acids and then possible precursors.

Chapter 7.2.5

Isolation of DNA from Goat Liver (Mamur's method)

Principle

Every cell contains DNA, though the amount of DNA in some tissues is very small so they are not suitable source for DNA isolation. Further some tissues have high DNase activity as a result DNA is broken down into smaller fragments. Liver is a very good source of DNA.

Extraction of DNA is carried out in a number of ways. The efficiency of DNA extraction depends upon the sample material, ionic conditions of the extraction medium etc. The procedure stated below is the Murmur's method. The principle includes rupturing of tissue, cell membrane, nuclear membrane followed by deproteinization and precipitation of the nucleic acid using ethyl alcohol.

Materials required

1. *Saline EDTA :* 0.15M NaCl + 0.1M EDTA, pH 8.0. The EDTA and high pH inhibits DNase activity.

2. *Sodium laurylsulfate* (SLS) *or sodium dodecylsulfate* (SDS), 25% solution : This anionic detergent ($NaC_{12}H_{26}SO_4$) lyses most non-metabolising cells, also inhibits enzyme activity and denatures many proteins.

3. *Sodium perchlorate* (5M solution) : The high salt concentration provided by the perchlorate helps to dissociate proteins from DNA.

4. *Chloroform-isoamyl alcohol mixture* (24: 1 v/v) : Chloroform causes surface denaturation of proteins while isoamyl alcohol reduces foaming and helps in separation, retain stability of the layers of the centrifuged deproteinized solution.

5. *Ethyl alcohol* (85%) : It precipitates nucleic acids after deproteinization; denatured ethanol can also be used as an alternative.

6. *Saline citrate* : 0.15M NaCl + 0.015M trisodium citrate, pH 8.0. It maintains ionic strength of dissolved DNA and also chelates divalent ions.

7. *Dilute saline citrate* : 0.015M NaCl + 0.0015M trisodium citrate. In this solution DNA dissolves more readily because of low ionic concentration. DNA should never be dissolved in pure water, other wise denaturation occurs owing to the absence of ions to maintain the double helical structure.

8. *Concentrated saline citrate* : 1.5M NaCl + 0.15 M trisodium citrate. This is employed to bring the dilute saline citrate (7) in which DNA is dissolved up to saline citrate concentration (6) approximately.

9. *Ribonuclease* : 0.2% (w/v, crystalline form) in 0.15M NaCl, pH5.0. This solution is heated at 80°C for 10 minutes to inactivate any contamination of DNase. The RNase digests RNA and facilitates purification of DNA.

10. *Acetate-EDTA* : 3.0M sodium acetate + 0.001M EDTA, pH 7.0. It provides the proper ionic environment in the isopropanol step required for the separation of DNA from RNA or its digestion products.

11. *Isopropanol* : Precipitates DNA selectively. RNA remains held in the solution. In some cases it precipitates and separates DNA from polysaccharides also.

12. Goat Liver.

13. Refrigerated preparative centrifuge.

Procedure

1. Liver homogenate prepared in 2.5mL of buffer (0.25M sucrose + 0.05M Tris + 0.005M $MgCl_2$, pH 7.6)

2. Centrifuged at 2000 rpm × 10 minutes

3. Supernatant removed, pellet extracted which contains the nuclear fraction.

4. Washed in the same buffer.

5. Pellet suspended in 10-15mL of saline EDTA, added SLS or SDS to a final concentration of 1%, mixture is heated at 65°C for 5 minutes and then cooled to room temperature.

6. Next added sodium perchlorate to a final concentration of 1M. To it added an equal volume of Chloroform: Isoamyl alcohol mixture.

7. Shaken gently for 30 minutes in a round glass stoppered flask.

8. Centrifuged at 5000 rpm for 15 minutes. Upper aqueous phase bears the DNA and proteins at the interphase. Final deproteinized aqueous phase is collected in a beaker. Through the side of the beaker 2 volumes of chilled alcohol is added with constant gentle stirring with a glass rod. DNA fibres spooled around through a glass rod.

9. DNA is dissolved in dilute saline citrate.

10. Next added concentrated saline citrate solution to bring final concentration equivalent to saline citrate concentration DNA in solution.

N.B. : If absolute purity of DNA is required, then following steps are performed :

DNA solution → RNase treated → Acetate EDTA added
↓
Pure DNA precipitated ← Centrifuged ← Isopropanol added
↓
Supernatant with RNA

Note : During the course of action vigorous shaking and narrow mouthed pipettes should be strictly avoided so that there is no DNA denaturation.

Chapter 7.2.6

Paper and Thin Layer Chromatography

Introduction

Chromatography is one of the separation methods based on differential migration of components of a mixture. The technique of chromatography was introduced as early as 1906 by a Russian scientist Michael Tswett.

In general, if two solvent phases are in contact with one another and if a solute is added, it will distribute itself among the two phases. The ratio of the concentrations of the solute in the two phases is called **partition coefficient**. Amongst the two phases one can be static and the other mobile. The sample mixture, introduced into the mobile phase goes through repetitive interactions (partitions) between the stationary and mobile phase while being carried all the way through the system by the mobile phase. Various components of the sample mixture interact with the two phases differentially on the basis of little dissimilarity in their physicochemical properties.

While these different rates of interactions direct the migration of sample components through the system, each one of the components migrate at a different rate. The compound that interacts more with the mobile phase and least with the stationary phase migrates quickly. The components showing least interactions with the mobile phase while interacting strongly with the stationary phase migrates slowly (retarded).

The distribution of the solute between the stationary and the mobile phases is a consequence of the stability of forces between the solute molecules and the molecules of each phase. Partition coefficient, as a result, reflects the relative attraction or repulsion that the molecules of the two phases show for the solute molecules and for themselves. These attractive or repulsive interactions go with by a release or intake of energy. Partition chromatography is usually carried out as **Paper chromatography or Thin layer chromatography (TLC).**

7.2.6(A) SEPARATION OF AMINO ACIDS AND SUGARS USING PAPER CHROMATOGRAPHY PRINCIPLE

Cellulose in the form of paper sheets makes for an ideal support medium where water is adsorbed between the cellulose fibres and acts stationary hydrophilic phase. Whatman filter paper or commercially available cellulose plates are used for the purpose. A small drop of the mixture amino acids is spotted on to the paper, and then dried. The chromatogram is prepared by placing the filter paper in a tank containing a suitable solvent and allowed to flow along the sheet. The solvent front is marked and then after drying the paper the positions of the compounds thus separated are visualised and identified by a suitable staining reaction.

The separation of the amino acids (solutes) is based on the liquid - liquid partitioning of these amino acids in paper chromatography. The partitioning happens between the water molecules (static phase) adsorbed to the cellulosic matter of the paper and the organic (mobile) phase.

The separation, identification and (semi) quantification of amino acids by means of paper chromatography is described below. The same methodology can be used to separate other smaller molecules for instance, sugars, organic acids etc. by changing the mobile phase and detection by a suitable staining reaction using (spray) agents.

(i) Separation of amino acids :

Materials required :

1. Whatman No.1 filter paper as chromatographic sheet.
2. Chromatography chamber.
3. Hair dryer or spot lamp.
4. Atomiser, Microsyringe (Hamilton syringe) or micropipette.
5. *Mobile phase (solvent system) :* n-butanol, glacial acetic acid and water are mixed in the ratio of 4:1:5 in a separating funnel and made to stand to equilibrate for 30 minutes. The lower aqueous phase is drained into a beaker and placed inside to saturate the chromatography chamber. The upper organic phase is saved and used for developing the chromatogram.
6. Different individual amino acid are dissolved in distilled water at a concentration of 1mg/mL. Very dilute, 0.05N HCl is used to dissolve the free amino acids tyrosine and phenylalanine. Tryptophan is dissolved in very dilute, 0.05N NaOH.
7. *Extraction of sample* : A known quantity of the sample material (dry/wet) is ground in a pestle and mortar with 10-fold volume of 70% ethanol. The contents are shaken at 55°C for 30 minutes and centrifuged at 10,000 rpm for 10 minutes. The supernatant is collected. The extraction is repeated at 55°C at least twice. The supernatants are pooled and shaken vigorously. The petroleum ether layer is discarded. The alcohol fraction is evaporated to dryness under vacuum using either a water pump or rotary evaporator at 40-45°C. The residue is dissolved in a known volume of absolute ethanol or water for analysis.
8. *Ninhydrin Reagent* : 100mg ninhydrin is dissolved in 100 mL acetone.
9. *Elution Mixture* : Prepared 1% copper sulphate solution. Ethanol and copper sulphate solution are mixed in the ratio 80 : 20 (v/v).
10. Water pump or rotatory vacuum pump.
11. Hot air oven.
12. Table centrifuge.

Procedure :

1. Chromatography sheet is cut carefully to a convenient size (40 × 24 cm). A line is drawn with pencil across the sheet about 5 cm away from one end. A number of points at intervals of 3 cm are marked.
2. A small volume (say 25 μL) of each amino acid is applied as a separate small spot using a microsyringe. A stream of hot air from a hair dryer aids in fast drying of spot. The spot should be as small as possible for better resolution.
3. Similarly, different known aliquots of sample extract are also spotted.
4. After spotting, the sheet is placed in a stainless steel trough in the chromatography chamber firmly held by placing a long steel support rod over the sheet. The spot end of the sheet should be in the trough as in descending chromatography. Or else, the sheet may be rolled

as a cylinder tied together with fine thread or with clip and placed upright with the spots on the bottom in a large petridish for upward movement of solvent (ascending chromatography); (see Figure 7.2.6.1).

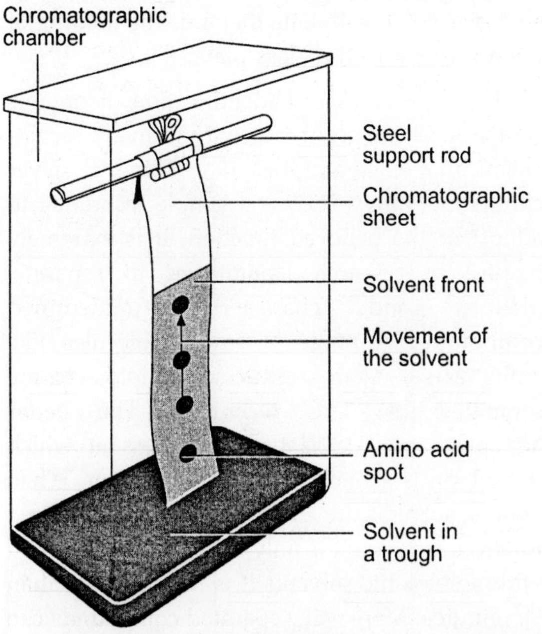

Figure 7.2.6.1 : Paper chromatography set up

5. The organic (phase) solvent is added to the trough/petridish and the chamber closed airtight by covering through a tight glass lid. The chromatogram is developed preferably overnight, until the solvent moves more or less to the other end.

6. The solvent front is noted and the chromatogram is dried to free of solvent in a fume chamber.

7. The chromatogram is sprayed with the ninhydrin reagent using an atomiser. The paper is dried for about 5 minutes at room temperature followed by at 100°C in an oven for 2-3 minutes.

Amino acids emerge as purple spots; hydroxyproline and proline give yellow coloured spots are marked and their R_f values are calculated by the formula.

R_f = Distance (cm) moved by the solute from the origin/Distance (cm) moved by the solvent from the origin.

Comparing the R_f values with that of the authentic amino acids, co-chromatographed, helps in the identification of the amino acids present in the sample.

8. For quantitative estimation, each spot is cut into several small bits and transferred to the bottom of the test tube. 3mL of elution mixture is added. The tubes are shaken vigorously for 15 minutes. The liquid is decanted and the pieces are eluted with another 2mL of elution mixture. The elutions are repeated with small aliquots until the bits are colourless. The eluate is combined and cleared by centrifuging at 10000 rpm for 10 minutes. The intensity of purple colour is read at 570nm in a colorimeter. The spot of leucine (50µg) run as standard is used for comparison.

(ii) Separation of Sugars :

Materials Required :

1. Sugars : D-Glucose, D-Xylose and Lactose and mixture of all three.

2. Solvent system: Ethyl acetate: Pryidine: water :: 14:6:5v/v.

3. Staining reagent :
   - m-phenyldiamine - 0.5g
   - Stannous chloride - 12g
   - Acetic acid - 20mL
   - Ethanol 80mL
4. Chromatographic chamber.
5. Hamilton syringe.
6. Whatman No. 1 filter paper.
7. Hair dryer.

Procedure :

1. 25mL of the solvent is taken in a glass jar and covered with a lid.

2. On Whatman No.1 filter paper 4 drops (25 µL) of one for each solution of sugars is placed through a Hamilton syringe.

3. The paper is curled into a cylindrical form and is placed into chromatographic chamber so that the paper does not touch the sides of the chamber.

4. The chromatogram is run for 3 hours.

5. The paper is taken out and the solvent front is marked with the help of a pencil.

6. The paper is dried; a hair dryer may be used.

7. The papers are then dipped in staining reagent solution.

8. The paper is put in a hot air oven at 100°-105° C for 5-6 minutes. Different spots of the sugars are observed from dark yellow to different forms of colour.

9. The R_f value are recorded and compared.

Notes :

1. The chromatography sheet is handled very carefully until developed, as otherwise amino acids from fingers will contaminate. The chromatographic paper is held between a fold of filter paper piece.

2. As the R_f value differs from run to run, due to solvent system, paper, room temperature, size of the chromatography chamber etc., it is desirable to co-chromatograph the standards every time.

3. Similarly, investigation of organic acid phenolic compounds can also be made by paper chromatography (for solvent systems, spraying agents etc., see under thin layer chromatography).

7.2.6(B). THIN LAYER CHROMATOGRAPHY

Principle :

The general principle concerned in TLC is comparable to that of column chromatography that is adsorption chromatography. In the adsorption process, the solute contends with the solvent for the surface sites of the adsorbent. Basing on the distribution coefficients, the compounds are distributed over the surface of the adsorbent. Obviously, in TLC the partition effect in the separation is also not excluded. The adsorbent usually used have a binding agent such as calcium sulphate that aids the holding of the adsorbent to the glass plate.

The separation and identification of organic compounds is a regular activity in many service chemical laboratories. Thin layer chromatography (TLC) is a simple technique to adopt for the believed function. It is extremely helpful in research laboratories to separate, identify and characterize unidentified compounds. A range of small molecules like amino acids, sugars organic acids, lipids etc. are separated by TLC technique. The better advantage of TLC is the speediness at which separation is achieved in under one hour. When volatile solvents are employed, the time required to effect separation is only about 30 minutes and with nonvolatile solvents it is rarely longer than 90 minutes. Moreover, separated compounds can be spotted by corrosive sprays and elevated temperatures with some thin layer materials which of course is not possible through paper chromatography.

Materials required :

1. Glass plate (20 × 20 cm or 20 × 10 cm).
2. Glass Tank with lid.
3. Spreader/applicator.
4. Developing solvents.
5. Adsorbent silica Gel-G/Aluminate salt.
6. Sample (should be extracted following the procedures specified for each group of compounds, for e.g., extraction with 80% alcohol for amino acids and sugars).
7. Standards.
8. Spraying agent (differs as for the group of compounds of interest).
9. Hot air oven.

Procedure :

A. **Preparation of plates :** Dry, clean glass plates (5 numbers, 20 × 20 cm) are placed on the plastic base plate above a plane surface.

1. Slurry is prepared of the adsorbent in water (sometimes buffer) in the ratio 1:2 (w/v).
2. The slurry is stirred carefully for 1-2 minutes and dispensed into the applicator placed on the head glass plate.
3. The slurry is coated above the glass plates at a thickness of 0.25mm for qualitative analysis by moving the applicator at a uniform speed from one end towards the other. (One has to attain some skill by practising to prepare uniformly coated plates). The plates are left to dry at room temperature for 15-30 minutes.
4. The plates are heated in an oven at 100-120°C for about 1-2 hours to eliminate the moisture and to activate the adsorbent on the plate. The dried plates are placed in a rack and can be stored in a desiccator over silica gel to avoid moisture absorption.

B. **Sample Application :**

1. From one end of the glass plate, 2.5 cm is left and no less than an equal distance from the edges.
2. The sample and standards are applied through a micropipette or microsyringe as small spots. All spots must be placed at equal distance from one end of the plate. It is to be seen that the adsorbent does not peel off at the sample application point. (calculated volumes are applied for quantitative analysis).
3. The sample is allowed to dry to facilitate spotting that can be done continually for a more concentrated sample spot.

C. **Developing Chromatogram :**

1. The developing solvent is poured into the tank to a depth of 1.5cm and allowed to stand for no less than an hour with a cover plate over the top of the tank to make sure that the environment within the tank becomes saturated with solvent vapour. This is called **equilibration.**
2. After equilibration, the cover plate is removed and the thin layer plate (sample applied) is positioned vertically in the tank so that it stands in the solvent with the spotted end dipping in the solvent.
3. The cover plate is restored. The separation of the compounds occurs as the solvent moves upward. The chromatogram is developed at constant temperature in order to avoid anomalous solvent running effects.
4. Once the solvent get to the top of the plate, it is removed from the tank, dried and proceeded for the identification of the separated compounds.

D. **Identification :**

1. If the commercially available adsorbent has a fluorescent dye, the separated compounds will come as blue, green or black spots when viewed under UV light. These areas can be scrapped, eluted with a suitable solvent, and quantitative assessment of the separated compounds can be performed.
2. When such a dye is not used in the adsorbent, one of the following methods of identification can be followed.
   (a) 50% sulphuric acid is sprayed and heated. This will result in most compounds get charred and show up as brown spots.
   (b) The plates are visualised under UV light that will show the positions of UV absorbing or fluorescent compounds.
   (c) Exposed to iodine vapour for unsaturated compounds.

(d) Ninhydrin is sprayed for amino acids.

(e) If the compounds are radioactive, the plates may be subjected to autoradiography or can be scanned by a radiochromatogram scanner.

(f) Table 7.2.6.1 given at the end of this chapter may be referred for the detection of specific compounds.

Notes :

1. Cellulose powder is also used as adsorbent in TLC in some cases. This connects adsorption and partition effects. Here the plates have to be air-dried.

2. The adsorbent, solvent system and the method of identification for certain group of compounds are listed below.

Table 7.2.6.1

Comound	Adsorbent	Solvent system	Identification mode	Colour developed
Mono and di-saccharides	Silica gel G impregnated with 0.2 M sodium borate buffer; pH 8.0 or 0.02M sodium acetate or boric acid	Benzene :glacial acetic acid: methanol:: 20:20:60 (v/v)	Sprayed with 0.2% napthoresocinol in ethanol and 2% aq. TCA (1:1). These has to be mixed just before use. Heated at 100° C for 5-10 minutes.	• Fructose, sorbose, raffinoe and sucrose produce red colour. • Pentoses and uronic acids produces blue colur on standing.
Neutral lipids	Silica gel	Petroleum ether : diethyl ether: acetone::90:10:1 (v/v) or Isopropyl ether : acetic acid :: 24:1 (v/v)	1. Sprayed with ethanolic 0.2% 2, 7 dichlorofluorescein.Has to be seen under UV light. 2. Sprayed with 0.5% ethanolic rhodamine B or 6G	• Light green fluorescence. • Yellow or blue violet spots on a pink background appear.
Amino acids	Silica gel G	96% ethanol: water ::70:30 (v/v); Butan-1-ol:acetic acid: water:: 80:20:20 (v/v)	Sprayed with ninhydrin (0.1%) in acetone and the paltes are heated at 100-110°C for 5 minutes.	Pink or purple.

Chapter 7.2.7(a)

Estimation of DNA by the Diphenylamine Reaction

Principle

While DNA is treated with diphenylamine in an acidic environment, a blue compound is produced having sharp absorption maxima at 595 nm. This reaction is explicit of two deoxypentoses in common and is not specific for DNA. Under acidic condition the straight chain form of a deoxypentose is transformed to highly reactive β-hydroxylevulinaldehyde that reacts with diphenylamine to produce blue complex. In DNA only the purine nucleotides react. Hence the value obtained represents half of the total deoxyribose present.

Materials required

1. DNA (commercial sample) : 0.5mg/mL.
2. Isolated DNA : 10mg.
3. Saline citrate solution (0.15M NaCl; 0.015M sodium citrate, pH 7) : 500mL.
4. *Diphenylamine reagent* : 5g of pure diphenylamine in 500mL of glacial acetic acid and added 13.75mL of concentrated sulphuric acid. This solution should be prepared fresh.
5. Water bath.
6. vis-Spectrophotometer with tubes.

Procedure

1. Separate marked tubes containing 1mL, 2mL and 3mL aliquots of isolated DNA dissolved in standard saline citrate and similar aliquots of 0.5 mg DNA/mL standard were prepared.
2. All the sample tubes along with the blank is made up to 3mL with H_2O.
3. Next added 6mL of diphenylamine reagent to each tube and after mixing, the tubes are heated in boiling water bath for 10 minutes.
4. The tubes are cooled.
5. The absorbance (OD) of the blue solution is read at 600nm against the blank in a spectrophotometer.
6. The standard curve is then constructed with OD versus concentrations of standard DNA solution.
7. The concentration of DNA (isolated) dissolved in the saline citrate solution is calculated from the standard curve.

Chapter 7.2.7(b)

Estimation of RNA by the Means of Oricinol Reaction

Principle

This is mainly a universal reaction for the pentoses and depends on the formation of furfural when the pentose (ribose)s is heated in the company of concentrated hydrochloric acid. Orcinol reacts with the furfural in the presence of Ferric chloride as a catalyst to produce a green colour. Only purine nucleotides give such reaction.

Materials required

1. Standard RNA solution.
2. Sample RNA solution.
3. *Orcinol acid reagent* : 2mL of 10% solution w/v of Ferric chloride. $6H_2O$ is added to 400mL of concentrated HCl.
4. *6% Alcohol Orcinol* : Dissolved 6g orcinol in 100mL of 95% ethanol. This is refrigerated in a dark opaque bottle. It would remain stable for 2 months.
5. Water bath.
6. vis-Spectrophotometer with tubes.

Procedure

1. Standard RNA (50µg RNA/mL) is prepared in ice chilled 10 mM Tris acetate, 1mM EDTA buffer (*p*H 7.2) or any other suitable buffer may be used which dissolves RNA.
2. The isolated RNA is also dissolved in the above buffer to make an estimated concentration of 50 µg/mL.
3. Series of tubes containing 0.5mL, 1mL, 1.5 mL, and 3mL of isolated RNA; 0.5mL, 1.0mL, 1.5 mL and 3mL of 50µg standard RNA/mL are prepared.
4. Each tube is made up to 3mL with water. In addition a set of blank with 3mL of water is also prepared.
5. In each tube 6mL of Orcinol acid reagent is added.
6. Next in each tube 0.4mL of 6.0% alcoholic Orcinol is added. The tubes are shaken to mix the contents.
7. The tubes are heated in a boiling water bath for about 20 minutes.
8. The tubes are cooled. Absorbance is read at 660 nm against the blank.
9. A standard curve is prepared by plotting OD versus concentrations of the standard RNA solution.
10. The amount of isolated RNA is calculated through the standard curve.

Chapter 7.2.7(c)

Detection of Urea, Uric acid, Ammonia and Creatinine in the Test Sample Provided

A. Test for Urea :

(i) Effervescence Test :

Experiment	Observation	Inference	Explanation
3mL of unknown solution is taken in a test tube. A few drops of alkaline sodium hypobromide solution is added.	Effervescence results.	Urea is confirmed.	The effervescence is due to evolution of nitrogen in a very fast rate.

(ii) Urease Test :

Experiment	Observation	Inference	Explanation
(a) 5mL of the given solution to be tested is taken. Its pH is checked. If acidic this is neutralized with 2% Na_2CO_3.	—	—	This reaction is caused because of breakdown of urea into ammonium carbonate, by the action of urease. This increases the alkalinity distinctly, causing the reappearance of pink colour in the solution.
(b) Next in it added 4 drops of phenolphthalein indicator.	Light pink colours appears.	The solution is slightly alkaline.	
(c) To the above (b) solution drop by drop added just enough 1% acetic acid to decolourize the solution.	Pink colour disappears.	The solution is exactly neutral.	
(d) To above (c) added in a knife point (just a pinch) of urease powder. Mixed and allowed to stand for 5 minutes.	Pink colour reappears.	Urea is confirmed.	

B. Test for Uric acid :

(i) Oxidation test or murexide test :

Experiment	Observation	Inference	Explanation
(a) A small amount (about an mL) of given solution and added 2-3 drops of concentrated nitric acid in an evaporating disc.	—	—	Uric acid is oxidised to dialuric acid and alloxan which condense to form alloxanthin. This in turn reacts with ammonium or potassium hydroxide to form purple compound known as murexide.
(b) Carefully evaporated to dryness over a very low flame	Red or yellow coloured residue is left.	—	
(c) (b) is cooled and to a portion, a drop of very dilute ammonium hydroxide solution is added.	The colour changes to purplish red.	Uric acid is present in the solution.	
(d) (c) is warmed	Colour vanishes.	—	
(e) (b) is cooled and in it added dilute potassium hydroxide solution.	Purple violet colour appears.	Uric acid is confirmed.	
(f) (e) is warmed	Colour disappears.	—do—	

(ii) Benedict's test :

Experiment	Observation	Inference	Explanation
(a) 2-3mL of the given solution is taken, in it added 2 drops of Benedict's Uric acid reagent*	—	—	—
(b) In it added a pinch of anhy. sodium carbonate and mixed thoroughly.	The solution turns deep blue.	Uric acid is present in the solution.	

* **Benedict's Uric acid Reagent :** In a 1 litre flask, 100g of sodium tungstate is taken. In it added 600mL of distilled water. In it added 50g of pure arsenic acid, 25mL of syrupy phosphoric acid (85%) and 20mL of concentrated hydrochloric acid. Heated to boiling point for 20 minutes. Cooled. Added distilled water to make volume up to 1 litre.

C. Test for Ammonia :

Experiment	Observation	Inference	Explanation
(a) 2 test tubes are taken, in each 2-3mL of given solution is obtained.	—	—	The reaction is caused due to the catalytic activity of urease on urea, which is converted into ammonia.
(b) Added phenolphthalein indicator to each test tube.	The solution turns milky in each test tube.	Presence of Urea is indicated.	
(c) One of the test tube is taken from (b) and in it added a pinch of urease powder.	The solution turns red.	Ammonia is released.	

D. Test for Creatinine :

Experiment	Observation	Inference	Explanation
(a) 5mL of the given solution is taken in a test tube.	—	—	—
(b) Added 1mL of saturated solution of Picric acid and 1 mL of 10% Sodium hydroxide.	Solution turns red.	Creatinine is present in the solution.	The red colur is due to the formation of Creatinine picrate in cold alkaline solution.

Chapter 7.3
Haematology

Chapter 7.3.1

Preparation of Stained Blood Film and Identification of leucocytes (WBCs) in Blood Film

The blood cells bear granules that differ in their reaction. Some of them are basic and some are acidic in nature. So, to stain these sections, two types of stains are used. Acidic substances are basophilic and thus take basic stains in contrary basic substances which are acidophilic and thus stains with acidic stain. Various white blood cells (WBCs) are stained through a double stain (Leishmann) that contains both the acid and basic dyes. These WBCs vary in their size, shape and granules of special nature.

(a) Preparation of Leishmann stain :

Leishmann stain is prepared by dissolving 0.15 g of Leishmann powder in 100 mL of pure acetone free methyl alcohol and the mixture is warmed at 50° C for 15 minutes. It is then filtered and keeping the filtrate in 37°C refines the dye. The stain can be applied within an hour of two of preparation.

(b) Preparation of blood film :

Principle :

To count the different type of blood cells in the plasma, the smearing of blood is done on a grease free slide. The smearing of blood is known as drawing of blood and the whole thing after drawing is known as blood film.

Requirements :

1. Human candidate or a 10-12 weeks Sprague Dawley rat.
2. Clean grease free slides.

3. Disposable hypodermic syringe with 20-24G disposable needle.

4. Distilled water.

5. Absolute alcohol.

Procedure :

1. For preparation of blood film, the slides and coverslips must be made absolutely clean and free from grease by washing with distilled water and alcohol.

2. At first one of the fingertips after sterilization by rectified spirit is pricked by a 24G disposable needle. A drop of blood oozes out which is then touched at one of the edges of clean a slide (absolutely clean and grease free).

N.B. : While preparing blood film from a rat blood. Blood is collected from the animal by drawing blood through cardiac puncture with the help of a hypodermic syringe fitted with a 20-22G needle. From the syringe a drop of blood is put on a clean grease free slide.

3. The other slide which is known as spreader is kept at about 45° to the horizontal slide touching the horizontal slide about 2 cm from it's end. Then with the spreader at 30°- 45° angle the blood is spread through out the horizontal slide into a film in a fairly swift and steady motion.

4. The above technique will produce a thin uniform smear of blood occupying the middle portion of the horizontal slide (Figure 7.3.1.1).

5. The film is quickly air dried. An ideal film should be uniform in thickness and have tail of uniform and tongue shaped nature.

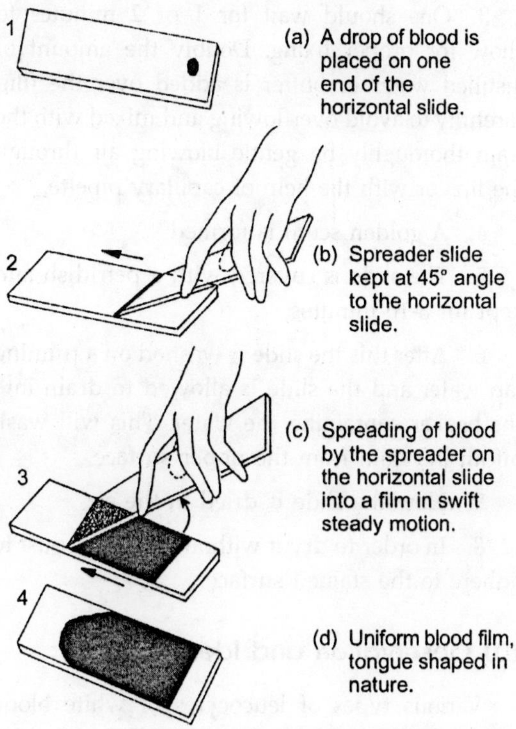

Fig. 7.3.1.1 : Preparation of blood film.

(c) Staining of blood film with Leishmann's stain :

Materials required :

1. Prepared Leishmann's stain in a drop bottle.

2. Distilled water (pH 7.0) or buffer solution (pH 6.4) (composition: 3.76g $Na_2HPO_4.2H_2O$ + 2.10g anhydrous KH_2PO_4 with maximum volume of 1L with double distilled water).

3. Fresh distilled water, which is slightly acid in nature.

4. Staining rack.

5. Glass beaker.

6. Capillary pipette.

7. Compound microscope with 10x, 45x and oil immersion lens.

Procedure :

1. The slide is put on the staining rack taking care that the slide with the blood film is faced upward.

2. Through a dropper sufficient amount of stain is poured on the prepared blood film so that it covers the whole of the film.

3. One should wait for 1 or 2 minutes to allow for proper fixing. Doubly the amount of distilled water or buffer is added over the film carefully to avoid overflowing and mixed with the stain thoroughly by gentle blowing air through the lips or with the help of capillary pipette.

4. A golden scum is formed.

5. The slide is covered with a petridish and kept for 8-10 minutes.

6. After this the slide is washed on a running tap water and the slide is allowed to drain into the beaker containing the water. This will wash off all the stain from the upper surface.

7. Now the slide is dried in the air.

8. In order to dry it without allowing dust to adhere to the stained surface.

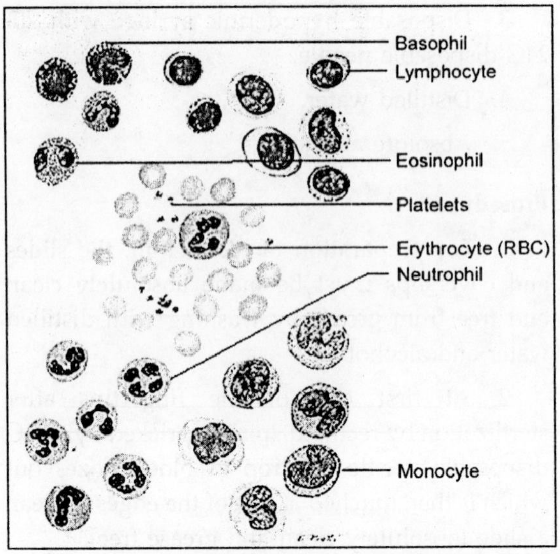

Fig. 7.3.1.2 : Various types of blood cells in a blood smear of human.

(D) Observation and Identification :

Various types of leucocytes or white blood cells are observed under 10x, 45x and 100x magnification (Table 7.3.1.1) under a compound microscope. Erythrocytes (RBCs) and Leukocytes (WBC) are identified using the pictorial key (Figure 7.3.1.2).

(E) Significance :

The blood film is made to identify the different types of W.B.C and R.B.C and platelets.

Table 7.3.1.1 : The different varieties of leucocytes in human are identified under oil-immersion lense of microscope are as follows :

	Type (% in blood)	Properties
GRANULOCYTES	1. Neutrophils or Polymorphonuclear leucocytes (40-70%)	Size 10-14μm. This type of WBCs posses' 2 to 7 lobed nucleus connected by chromatin thread. Nucleus stains blue-violet. Cytoplasm stains slate-blue. Cytoplasmic granules fine, neutrophil granules pinkish violet in colour.
	2. Eosinophils (1-6%)	Size 10-10μm. Nucleus stains blue violet. These cells have crescent-shaped 2-3 lobed nucleus connected by thick or thin chromatin band. Cytoplasm eosinophilic, light pink-red granules. Large coarse uniform sized granules brick red to orange in colour.
	3. Basophils or mast cells (0-1%)	Size 10-15μm. They have irregular shaped, may be S-shaped rarely bilobed blue-violet pale-staining nucleus. Cytoplasmic basophilic bluish and granular. Granules large and coarse, variable sized colour deep purple.
AGRANULOCYTES	4. Monocytes (5-10%)	Size 12-20μm. They possess big cell, comparatively large single horse-shoe or kidney or oval shaped nucleus in the centre that stains pale blue-violet. Cytoplasm abundant and frosty and slate blue in colour. No visible definite granules.
	5. Small Lymphocytes (20-40%)	Size 7-9μm. They are small cell slightly bigger than RBC, small rounded or oval deeply blue-violet stained nucleus occupying nearly the whole of the protoplasm of the cell. Condensed lumpy chromatin gives 'ink-spot' like appearance. Cytoplasm hardly visible and thin, crescent of clear and light blue staining leaving a very thin band of cytoplasm and without a definite visible granule.
	6. Large Lymphocytes (5-10.5%)	Size 10-15μm. They are medium sized cells. Nucleus large stains deep blue-violet. It is single rounded or oval and almost fills the cell. Centrally located and eccentric. Cytoplasm large and crescent shaped appears light blue. No visible granules.
See Figure 7.3.1.2		

Chapter 7.3.2

Differential Count (DC) of Leucocytes (WBCs)

Principle

There are various types of leucocytes or white blood cell or WBCs whose origin as well as number per 100 count is different. The function of each type of WBC is specific. Different physiological and clinical conditions cause variation in different WBC count. Hence, the deviation from the normal count of different WBCs per 100 WBC, counted in a blood film prepared from the blood of a particular person, helps to detect eosinophilia, tuberculosis, malaria and other diseases. This type of counting is known as **differential count**.

Procedure

2. First the tip of the finger is punctured and a drop of blood is taken on grease free, clean, regular slide. A film of blood is prepared. A blood film may also be prepared from rat blood *see* Chapter 7.3.1.

3. Care must be taken in drawing the film, as the film should be large, consistent, tongue shaped or flame shaped.

4. Next, it is stained with Leishmann's stain (refer Chapter 7.3.1).

5. Examined under the high power objective lens of the microscope. The film is observed end to end with the help of a movable stage.

6. 200 squares are drawn in the notebook for recording various WBCs as they are encountered and identified one after the other. These cells are entered by using letters 'N' for neutrophils, 'M' for monocytes ,'LL' for large lymphocytes, 'SL' for small lymphocytes, 'E' for eosinophils and 'B; for basophils.

One can also indicate these cells in a column (instead of 200 squares), and as one identify a cell, a short vertical stroke against that cell is made. In this way one can place different types of cells in groups of 5, a horizontal stroke representing the 5th cell (e.g. Neutrophils = 卌 or | | | etc.)

7. A drop of cedar wood oil is placed on the right upper corner of the film,a few mm. away from the head end. The oil immersion objective lens is brought in position till it enters the oil drop. The focus is adjusted. The slide is moved slowly from left to right as one encounter a leucocyte. The leucocyte is identified. As one is approaching the end of the slide. Two fields are moved down and the whole blood film is scanned in opposite direction. This process is contined and the film is traversed to and fro fashion till one have examined 200 cells (for better results one has to count 400 cells). This battlement procedure is shown in Figure 7.3.2.1.

Table 7.3.2.1 : Data sheet showing results of differential count of leucocyetes in Sprague Dawley rat.

Name of leucocytes	Normal Value (%)	Observed value
Neutrophil	11.3(male); 13.9 (female)	—
Lymphocyte	88.1 (male); 86(female)	—
Monocyte	0.44(male); 0.25(female)	—
Eosinophil	0.11(male); 0.13(female)	—
Basophil	0.0(male); 0.0(female)	—

N.B. : Normal differential count of human leucocytes can be supplemented from Table 7.3.1.1.

Fig. 7.3.2.1 : Movement of a well prepared stained blood film under oil immersion lens for differential counting of WBCs.

The counting is started from one end of the film to another and the different types of leucocytes in each field are noted and the total number is recorded. The counting is continued till the total number of leucocytes (WBCs) becomes 100.

Observation

1. The stained blood smear is examined in low and high power magnifications. One has to identify all the leucocytes (WBCs) with certainty.

2. A data sheet (Table 7.3.2.1) has to be constructed showing the type, number and percentage of WBCs in 200 or 400 WBCs identified and counted.

Chapter 7.3.3

Total Count (TC) of Blood Cells in Man and Rat

(a) Total count of WBC :

Principle :

The total numbers of leucocytes or WBCs per mm^3 of blood is determined by diluting known volume of blood 20 times with a fluid, which prevent haemolysis and coagulation. The blood cells are then counted in volume of diluted blood in counting chamber. The number of WBCs per mm^3 of the undiluted blood is then determined by this method.

Apparatus required :

1. **Neubauer's Haemocytometer :** The instrument consists of a special glass slide. The counting chamber is a heavy thick glass slide usually at the centre with three platforms, which is separated by wide grooves. The central platform is 0.1 mm lower than its neighbour. On the surface of the middle platform there are two sets of ruled area. The area has 9 large squares, each have 1 sq. mm area and are elaborately ruled. The four corner squares are divided into 16 equal sized squares. The central square is ruled into 25 groups of 16 small squares, each group is separated by triple lines. The ruled surface is 1/10 mm below the inner surface of the cover glass placed over the middle platform. Thus each smallest square of the corner has a volume of $1/4 \times 1/4 \times 1/10$ or $1/160$ mm^3.Whereas the smallest square in the centre has a volume of $1/20 \times 1/20 \times 1/10 = 1/4000$ mm^3. This central segment is used for RBC counting. In four big squares leucocytes or WBCs are counted which are placed at the four corners of the erythrocyte counting chambers. (Figure 7.3.3.1).

2. **WBC Pipette :** It comprises of a central bulb containing white bead and the stem having two marks one at the middle as 0.5 and other just little before the bulb as 1. The third mark is present beyond the bulb (other end of the bulb) as 11 mark. The bulb contains a white glass bead. The white bead acts as stirrer to mix. Blood is taken up to 0.5 or 1 marked and diluted to 11 marks by the diluting fluid. Then dilution becomes 20 or 10 times respectively (Figure 7.3.3.2a).

3. **Special cover slip :** It is very thick and specially designed rectangular shaped cover slip. It is perfectly plane in nature and just cover the groove portion of the counting chamber in such a way so that excess of the fluid in the chamber is removed.

4. **WBC Diluting fluids or Turk's fluid :** The fluid consists of glacial acetic acid, 1.5 mL

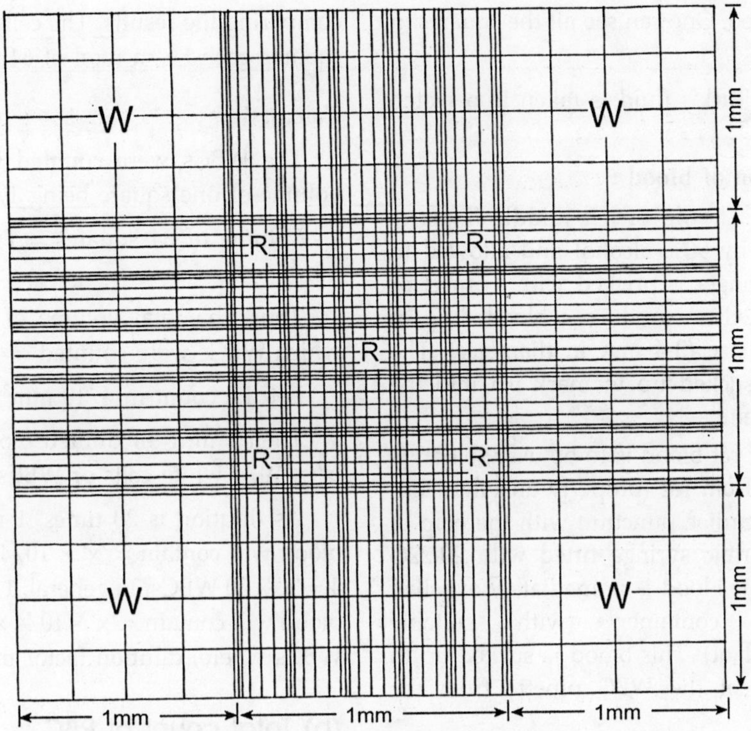

Fig. 7.3.3.1 : Neubauer's Haemocytometer.

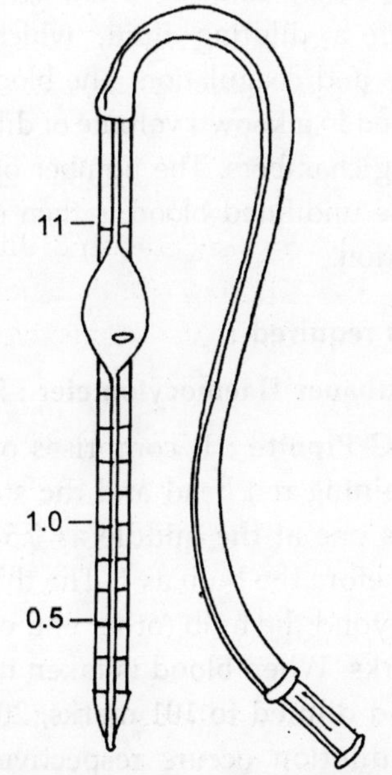

Fig. 7.3.3.2(a) : WBC pipette.

(for destruction and dissolution of RBCs and to fix WBCs); Gentian Violet solution, 1.5 mL (1% solution, stain WBCs). These components are mixed and diluted to 100 mL by distilled water.

5. Compound Microscope.
6. 90% ethyl alcohol.
7. Sterilized Lancet or needle (23G).
8. Hypodermic syringe with 20-22G needle (required for the blood to be drawn from rat).
9. Human candidate or Sprague Dawley rat.
10. Anaesthetic ether.
11. Anticoagulant – EDTA.

Procedure :

1. The counting chamber is placed on the microscope stage. The illumination is adjusted and the right upper group of the 16 squares for

WBC are focussed. One can see all the squares in alone field.

2. 1 mL of Turk's fluid is taken in a watch glass.

3. **Collection of blood :**

(a) *For man* : The tip of left middle finger is sterilized by 90% alcohol and allowed to dry. The finger is pricked with a lancet or needle (22G) to produce a blood drop of adequate size (3-4 mm in diameter). The blood is sucked up to mark 0.5 into the WBC pipette.

(b) *For rat* : If rat blood is to be used, blood is collected from rat (properly anaethesized) through cardiac puncture with the help of a hypodermic syringe fitted with 20-22G needle. The blood is immediately expelled into a container (with suitable anticoagulant). This blood is sucked up to mark 0.5 of the WBC pipette from the container.

4. The blood is diluted with Turk's fluid up to 11 marks by sucking the diluting fluid by the pipette kept in a watch glass. When the blood is taken up to the mark of 0.5, the blood is diluted only 20 times. The contents inside the bulb of the pipette is mixed for 3-4 minutes.

5. The first two drops of diluted blood from the pipette are discarded and then this mixture is run on the haemocytometer slide on both the chambers on both sides under a special coverslip.

6. The corpuscles are allowed to settle down for 3-4 minutes and then the number of WBCs are counted under low power of microscope in the four counting area.

7. The white cells are recognised by the refractile appearance and by the slight colour given to them by stain contained in the diluting fluid.

8. The cells are counted under high power lens. One may count the WBCs in 16 squares under low power and then under high power for comparing the results. The cells in 4 groups of 16 squares each i.e., a total of 64 squares.

Calculation :

The WBCs were counted in 64 squares the volume of one square being 1/64 mm^3.

Volume of 64 squares = $1/160 \times 64 = 4/10$ mm^3.

Thus, the total volume of diluted blood in which WBCs were counted = 4/10 mm^3.

Let the count in 4/10 mm^3 = 'x' WBCs.

Then 1 mm^3 of diluted blood will contain = 'x' × 10/4 white cells or WBCs.

As dilution is 20 times, 1 mm^3 of undiluted blood will contain = 'x' × 10/4 x 20 = 'x' × 200/4 = 'x' × 50 WBCs. In general, 1 mm^3 of undiluted blood will contain = 'x' × 10/4 × dilution factor of WBCs where, dilution factor may be 20 or 10.

(b) Total count of RBC :

Principle :

The total number of RBCs per mm^3 of blood is determined by diluting a known volume of blood with a diluting fluid, which prevents haemolysis and coagulation. The blood cells are then counted in a known volume of diluted blood in counting chambers. The number of RBCs per mm^3 of the undiluted blood is then determined by calculation.

Apparatus required :

1. **Neubauer Haemocytometer :** See 7.3.3.1.

2. **RBC Pipette :** It comprises of a central bulb containing red bead and the stem having two marks one at the middle as 0.5 and other just little before the bulb as 1. The third mark is present beyond the bulb (other end of the bulb) as 101 marks. When blood is taken up to 0.5 or 1 mark and diluted to 101 marks, 200 times or 100 time dilution occurs respectively (Figure 7.3.3.2b).

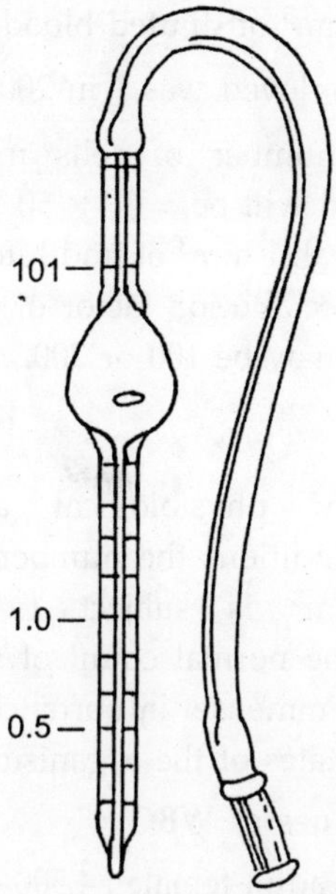

Fig. 7.3.3.2(b) : RBC pipette.

3. **Special cover slip :** It is very thick and specially designed rectangular shaped cover slip. It is perfectly plane in nature and just cover the groove portion of the counting chamber in such a way so that excess of the fluid in the chamber is removed.

4. **RBC Diluting fluids :** This is also known as the *Hayem's fluid.* It is made of mercuric chloride (acts as disinfectant and prevents fungal and bacterial growth in RBC when kept for prolonged use) – 0.25g, Sodium chloride (it maintains osmotic pressure and prevents haemolysis) - 0.5g, Sodium sulphate (it prevents **rouleaux** i.e., formation of rows of cells and coagulation of blood) - 2.5g and distilled water (used as solvent for above components)-100mL. After preparing the solution it is filtered several times through the same filter paper. The precipitate is discarded.

5. Compound Microscope.

6. 90% ethyl alcohol.

7. Sterilized lancet or needle (23G).

8. Hypodermic syringe with 20-22G needle (required for the blood to be drawn from rat).

9. Anaesthetic ether.

10. Human candidate or Sprague Dawley rat.

11. Anticoagulant : EDTA.

Procedure :

1. The counting chamber is placed on the microscope stage. The illumination is adjusted and the central 1mm square (RBC square on the counting grid) containing 25 groups of 16 smallest squares each. One can see all the squares in alone field.

2. 2mL of Hayem's fluid is taken in a watch glass.

3. **Collection of blood :**

(a) *For man :* The tip of left middle finger is sterilized by 90% ethyl alcohol and allowed to dry. The finger is pricked to produce a blood drop of adequate size (3-4mm in diameter) and the blood is sucked up to mark 0.5 into the RBC pipette.

(b) *For rat :* If rat blood is to be used, blood is collected from rat through cardiac puncture with the help of a hypodermic syringe fitted with 20-22G needle. The blood is immediately expelled into a container (with suitable anticoagulant). This blood is sucked up to mark 0.5 of the WBC pipette from the container.

4. The blood is diluted with Hayem's fluid up to 101 marks by sucking the diluting fluid by the pipette kept in a watch glass. When the blood is taken up to the mark of 0.5, the blood is diluted 200 times. The contents inside the bulb of the pipette are mixed for 3-4 minutes.

5. The first two drops of the pipette are discarded and then this mixture is run on the haemocytometer slide on both the chambers on both sides under coverslip.

6. The corpuscles are allowed to settle down for 3-4 minutes and then the number of RBC is counted under low power and then high power of microscope in the central 1 mm square (RBC counting area).

7. One has to switch over to high magnification lens and the distribution and count of the cells are checked.

8. The chamber is moved carefully and the left upper corner block of 16 squares in the field of the view. Cells in 5 square of the central area i.e., four corner squares and one central square are taken in to account . Thus the cells present in 80 smallest squares (5 × 16 = 80) are counted.

9. To avoid counting a RBC twice, those on a line are counted only when on the top and left lines or on the bottom and right lines.

Calculation :

It is known that 80 smallest squares have a volume of 1/50 mm^3. It is also known that cell count in 1 smallest square, which has volume of 1/4000 mm^3.

Let 'x' be the number of red blood cells in 1/50 mm^3 of diluted blood.

Cells in 1 mm^3 of diluted blood is 'x' × 50.

Dilution employed was 1 in 200.

Therefore, number of cells in 1 mm^3 of undiluted blood will be = 'x' × 50 × 200 = 'x' × 10,000. In general, 1 mm^3 of undiluted blood will have : 'x' × 50 × dilution factor of RBCs, where dilution factor may be 100 or 200.

Significance :

In different physiological as well as pathological conditions the number of RBC and WBC per mm^3 is subject to remarkable fluctuations. The normal count of the RBC and WBC are of immense importance in normal physiological states of the organisms.

Normal values of WBC :

Human : In adult female : 4,500 – 5,500/mm^3.

In adult male: 5,000 – 6,000/mm^3.

Normal value of WBC in rat : 3-17 × 10^3/μL.

Normal values of RBC :

Human : In adult female : 4,500,000-5,500,000 /mm^3.

In adult male : 5, 00,000 – 6, 00,000/mm^3.

Normal value of RBC in rat : 5-10 × 10^6/μL.

Chapter 7.3.4

Estimation of Haemoglobin (Hb) in Blood by Acid-haematin Method

Principle

The percentage of haemoglobin content of the given sample of blood or of one's own blood is estimated by Haldane's modification of Sahli haemometer.

N/10 HCl is used to dilute the blood so that the haemoglobin is converted into brownish yellow coloured acid haematin hydrochloride. The brown colour is then matched against the coloured prism of acid haematin of the comparator box of Sahil's haemometer, for the estimation of Hb. When the colour of Hb matches with prism the point is read on a scale and the Hb content gm/ 100 mL of blood is obtained directly.

Materials required

1. The following are the constituents of Sahlis haemometer (Haemoglobinometer) :
   (a) *Comparator box :* It is a rectangular plastic box with a slot in the middle which accommodates the Haemoglobin tube. The comparator also has a brown coloured prism on two sides for matching the colour of diluted mixture, Figure 7.3.4.1a.
   (b) *Double sided graduated diluting tube :* It is square or rounded tube which is calibrated in Hb (2-24g/dL) in yellow colour in one side, and in percentage Hb% (20-140%) in red colour, Figure 7.3.4.1b.

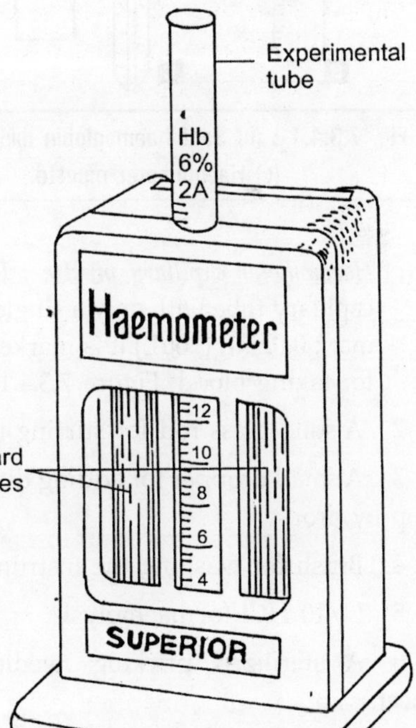

Fig. 7.3.4.1(a) : Haemometer.

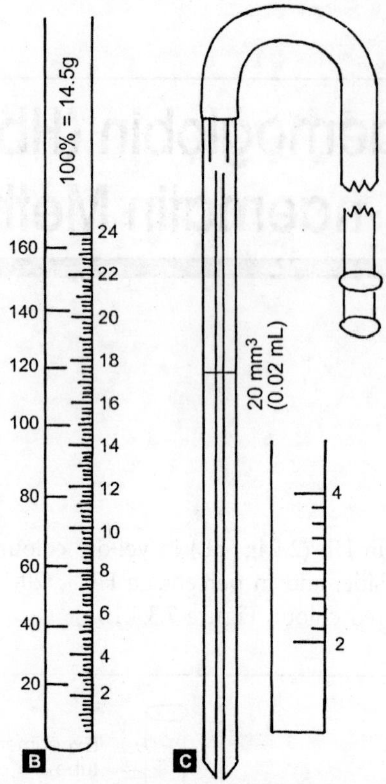

Fig. 7.3.4.1 : (b) Sahli-haemoglobin diluting tube, (c) Haemometer pipette.

(c) *Haemometer capillary pipette* : It is a glass capillary tube with only a single calibration mark 0.02mL (20µL) it is marked as 20mm^3 for taking blood, Figure 7.3.4.1c.

2. A solid glass rod for stirring the mixture.

3. A small dropper for adding distilled water drop by drop.

4. Brush for washing the instrument.

5. N/10 HCl for haemolysis.

6. A sterilized pricking needle (24G) or lancet.

7. A bottle with distilled water.

8. 90% ethyl alcohol.

9. Anaesthetic ether.

10. Human candidate/Laboratory white rate.

Procedure

1. N/10 HCl is taken in the graduated diluting tube with the help of a dropper and filled up to 20 mark.

2. The fingertip of the left middle finger is pricked after cleaning with 90% ethyl alcohol with the help of a sterilized pricking needle or lancet and the blood is sucked in the Haemometer pipette up to 20 mm^3. Care should be taken that no extra amount of blood is sucked in the pipette. In case of rat, blood can be drawn through intracardial route or through vein. Anaesthetisation of rat should be performed if required.

3. The blood from the pipette is transferred into the graduated diluting tube containing N/10 HCl and the pipette should be washed several times with a dilute acid solution.

4. With the help of the glass rod, the mixture in graduated tube is stirred several times. After a minute the haemolysis is completed and the blood turns brown.

5. The tube is now allowed to stand for 5 to 8 minutes or longer time till maximal colour develops.

6. The mixture is then gradually diluted with distilled water added drop by drop with continuous stirring after addition of each drop. The dilution is continued till the colour of the solution exactly matches with coloured prism inside the haemometer. The observation should be done under broad daylight.

7. After matching of colour, the glass rod is taken out of the graduated tube carefully without taking out any drop of the mixture with the rod.

8. The scale is read which gives Hb as percent and in g/100mL of blood.

Rules for observation during matching the colour

1. The instrument should be held between the eye of the observer and the light source(broad day light).

2. After addition of distilled water, stirring of the mixture should be done and then the colours are compared.

3. During matching of colours, stirrer should not be taken out but held above the level of the mixture in the graduated tube.

4. During matching the colours, the blank portion and not the graduated portion of the dilution tube should face the observer.

5. The mixture should appear clear and devoid of blood clots and brown in colour.

Observations and Result

Average of 3 readings in the Haemometer is taken and the result in average is expressed as below.

The amount of Hb in g/100mL of blood is ....

N.B. : In Human normal value of haemoglobin are as follows :

In adult male : 14.5g/dL or 100mL.

In adult female : 12.5g/dL or 100mL (with a range of 11.5-16g/dL or 100mL)

In rat normal range of haemoglobin are as follows : 11-19g/dL or 100mL.

Precautions

1. While taking blood from the finger, the finger should never be squeezed.

2. One should take care of the fact that blood is not passed out of the capillary tube.

3. Blood sucked inside the capillary pipette should not be more than 20 marks. If it happens then it should be carefully removed.

4. One should always wait for at least 5 minutes after taking the blood and N/10 HCl are thoroughly mixed because it takes 5 minutes time for acid haematin formation.

Chapter 7.3.5

Preparation of Haemin Crystals

Principle

Haemoglobin (Hb) readily splits into globin and haeme component on reaction with heat and glacial acetic acid. The small amount of NaCl present in the plasma is sufficient for the formation of haematin hydrochloride. In case of old blood, small amount of NaCl is added. The reaction takes place in the following way—

1. Glacial acetic acid + OxyHb ⟶ Haematin + globin.
2. Glacial acetic acid + NaCl ⟶ Na-acetate + Nascent HCl.
3. Nascent HCl + Acid haematin ⟶ Haematin hydrochloride or Haemin.

Materials required

1. 90% ethyl alcohol.
2. Sterilized pricking needle (24G) or lancet.
3. Clean and grease free slide and coverslips.
4. Glacial acetic acid.
5. Sodium chloride (NaCl).
6. Spirit lamp.
7. Anaesthetic ether.
8. Human candidate or laboratory white rat.

Procedure

1. One of the fingertips is washed and sterilized with 90% ethyl alcohol and allowed to dry.

2. The tip of the finger is the pricked by a sterilized needle and a small drop of blood is taken on the slide. In case of rat, blood may be collected through intra-cardiac route or through vein the rat must be suitably anaesthesized, if needed.

3. The blood is then spread over the slide to make a thick blood film.

4. It is then dried in the air or over flame of a spirit lamp.

5. The dried film is then scraped by means of a scalpel and scrapings are then collected in the centre of the slide.

6. They are turned into powder.

7. It is mounted on fresh glacial acetic acid and with one or two crystal of NaCl.

8. This is then warmed gently over the flame of a spirit lamp until bubbles of gas just comes out.

9. This is then allowed to cool.

Fig. 7.3.5.1 : Haemin crystals.

10. After cooling the slide with a cover slip it is observed under microscope. The diagram of the haemin crystal is drawn.

Observations

Minute dark brown rhombic crystals of hydrochlorate of haematin are observed It is important in medico-legal test for the detection of blood. It helps to give an opinion as to proof whether a stain is blood or some thing else. It is also useful in the differentiation of bloods of different species depending upon the shape of the crystal. Different shapes of haemin crystals are—Man: rhomboidal plates and prisms often arranged in star shaped clusters with round edges; Rat : narrow plates with varying width to needles, blunt at the ends; guineapig :triangular plates often arranged in the form of squares. (See Figure 7.3.5.1).

Chapter 7.3.6

Estimation of Erythrocyte Sedimentation Rate (E.S.R.) in Human

Principle

The Westergren Erythrocyte Sedimentation Rate instrument is used to calculate the ratio of volume of blood corpuscles and plasma of a given volume of blood. This provides information's not for diagnosis of any disease but for the prediction (prognosis) of the disease.

When a sample of blood treated with an anticoagulant is left standing for about one or two hour in a long tube placed vertically the red blood corpuscles (sp. gr. 1.095) being heavier than (denser) than the colloids plasma (sp. gr. 1.032) settles at the bottom of the tube leaving the blood plasma above. The height in millimetre of the clear plasma is the Erythrocyte Sedimentation Rate or E.S.R or suspension stability of the blood.

Materials required

1. Apparatus required (Erythrocyte sedimentation rate instrument)
   (a) *Westergren pipette or Graduated tube :* It is about 300mm long and has bore diameter of 2.5mm. It is calibrated in mm from 0 to 200, from above downwards.
   (b) *Vertical stand :* The Westergren stand can hold up to 6 tubes at a time. For each pipette there is a screw cap that slips over its top and at its lower end the pipette presses into rubber pad or cushion. When the pipette is fixed in position, there is enough pressure of the screw cap to stop leakage of the blood from its lower end.
2. Anticoagulant (0.5mL of 3.8% Sodium citrate solution).
3. 2 mL of human candidate's blood.
4. 90% alcohol.

Procedure

Westergren's Method :

1. With the help of venepuncture, 2mL of blood is taken from the vein in aseptic condition (site of puncture sterilised through 90 % alcohol).
2. Anticoagulant (0.5mL) is then taken in a dry empty vial.
3. The blood is stored as soon as possible in the anticoagulant vial.
4. Blood should be mixed thoroughly with the anticoagulant by rotating the vial gently between the palms.
5. The blood is the drawn into the Westergren's pipette up to the 0 mark.

6. Keeping one finger over the pipette it is transferred to the Westergren stand by firmly pressing its lower end against the rubber stopper on the base of the rack. It should be confirmed that there is no leakage of blood and the pipette stand vertically.

7. The elevation of the plasma column is the distance from '0' marks to the upper most level of the RBC column below. It is also the distance through which RBCs has settled down.

8. The pipette is left undisturbed for two hours.Corpuscles settle down because of their greater density, whereas the plasma keeps above.

9. Height in mm of the column is read at the end of 1 hour(hr.) and 2 hours.

Observation

If there is no haemolysis there is sharp line of demarcation between the RBCs and the clear cell free straw coloured plasma.

Calculation : E.S.R in mm/hour =

$$\frac{\text{1 hr. reading} + (\text{2nd hr. reading}/2)}{2}$$

Normal standards of ESR in human : The normal E.S.R values by Westergren's method are :

1. In men : 1st hr., 3 to 5mm and at the end of 2nd hr., 7 to 15mm.
2. In women : 1st hr., 4 to 7mm and at the end of 2nd hr., 12 to 17mm.
3. In new born children : 1st hr., 0.2 to 0.5mm and at the end of 2nd hr., 0.5 to 1.5mm.

Interpretations

E.S.R is a non specific reaction, gives information about the progress of a disease and is helpful in prognosis and differential diagnosis of a disease. It increases in all acute general infections in anaemia, pregnancy, nephrosis and malignancy. It is decreased in polycythaemia and congestive heart failure.

Significance of E.S.R.

It aids to prove out clinical findings and one can predict the progress of the disease. It is less important for diagnostic purpose since it is not useful for diagnosis of any particular disease. It helps to screen out functional groups of cases.

Chapter 7.3.7A

Determination of Bleeding Time in Human

Principle

Bleeding time is the time taken for complete stoppage of free flow of blood from a deep punctured vein.

Materials required

1. Stop watch.
2. Strips of blotting paper/filter paper.
3. Blood lancet or pricking needle sterilized.
4. 90% ethyl alcohol.
5. Human candidate.

Procedure (Duke's method)

1. A deep finger prick under aseptic condition (finger being sterilised with 90% ethyl alcohol and dried) is made with the help of a blood lancet or pricking needle at '0' time to get a free flowing blood.

2. Every thirty-second (sec) interval (measured through a stop watch) blood from the wound is removed through a blotting paper/ filter paper without touching or pressing the skin.

3. The blood spots on the paper should be numbered one onwards.

4. The process is continued until the stains on the blotting paper gradually decrease in size and becomes nonvisible and until finally the bleeding stops (i.e., the blotting paper is unstained).

5. The numbers of blood spots are counted and the time is noted with the help of a stopwatch.

6. The result is expressed in minutes or seconds.

7. Title of the experiment and student's name, roll and date of experiment should be written on the filter paper.

Observations

It is noted that the blood spot gets smaller with time after 1st and 2nd spots until there is no trace.

Result

See Table 7.3.7A.1.

Bleeding time of the candidate is .... seconds or .... minutes.

N.B. : The normal bleeding in human is 1-3 minutes. But it may vary from 2-5 minutes.

Table 7.3.7A.1 : Data sheet for calculation of coagulation time

Initial time (hr/min/secs)	Final time (hr/min/secs)	Difference seconds(secs)	Visibility of blood spot	Total No. of blood spots	Bleeding time = Total number of blood spots × 30 seconds
—	—	30 secs	Yes or No	—	
—	—	30 secs	Yes or No	—	
—	—	30 secs	Yes or No	—	
—	—	30 secs	Yes or No	—	
—	—	30 secs	Yes or No	—	

Note :

(a) Bleeding times depends not only on coaguability but power of injured capillary to contract. It is due to adhesion and clumping of the platelets sealing the capillary wounds. Bleeding time denotes integrity of capillary tubes and depends upon the platelets number and function.

(b) In *purpura* clotting time is normal but bleeding time is extended.

(c) In *Haemophilia* bleeding time is normal but clotting time is extended.

Bleeding time is abnormally long-standing in platelet deficiency and structural defect of the capillary caused due to ascorbic acid deficiency.

Chapter 7.3.7B

Determination of Coagulation Time in Human

Principle

Coagulation time is the time taken by the blood to coagulate subsequent to shedding.

Materials required

1. Capillary tube 10-11cm long and 0.8mm in diameter of uniform bore.
2. 90% ethyl alcohol.
3. Blood lancet or pricking needle (sterilized).
4. Stop watch.
5. Human candidate.

Procedure (Wright's capillary tube method)

1. A fine capillary tube is used for the purpose.
2. The tip of the finger is sterilised with 90% ethyl alcohol.
3. Next the sterilised finger is sharply punctured through a blood lancet or pricking needle.
4. The first drop of the blood is rejected and the capillary tube is touched with second drop of blood thus allowing the second drop of blood to readily fill the tube through capillary action.
5. The capillary tube is held between the palms of one's hand to keep the blood near body temperature.
6. At interval of every 30 seconds (measured through stop watch) small sections of the capillary tubes are broken off very gently and the time is noted until a fine strand of fibrin appears, as seen by fine strand trying to keep as a link between the broken ends of the capillary tube.

Result

See Table 7.3.7B.1.

Clotting time of the candidate is .... seconds or .... minutes.

N.B. : The normal coagulation time in man is about 3-5 minutes but can differ from 3-10 minutes.

Note :

(a) Coagulation time decreases when blood is pressed out from a skin puncture getting mixed up with tissue fluid.

(b) Due to deficiencies of factors V, VIII or IX coagulation time is abnormally prolonged. Clotting time is also prolonged in Haemophilia and other clotting disorders.

(c) Coagulation time varies with temperatures, so it is better to keep the capillary tubes in a water bath at 37°C mimicing body temperature.

Table 7.3.7B.1 : Data sheet for calculation of coagulation time

No. of breaks of capillary tube	Initial time (hour/min /secs)	Final time (hour/min /secs)	Difference seconds (secs)	Visibility of fine fibrin stands between the breaks	Total no. of breaks required to get the visible fibrin strand	Bleeding time = Total number of breaks × 30 secs
1	—	—	30 secs	Yes or No		
2	—	—	30 secs	Yes or No		
3	—	—	30 secs	Yes or No		....
4	—	—	30 secs	Yes or No		
5 etc.	—	—	30 secs	Yes or No		

Chapter 7.3.8

Blood Group and Rh Factor Determination

Principle

In 1900, Karl Landsteiner grouped human blood into four groups based on presence of two antigens on the surface of the RBCs. These groups are designated as A, B, AB and O. ABO system is an instance of multiple alleles. In this multiple allelic systems two of the alleles are dominant and is symbolized as I^A and I^B respectively. The only recessive allele is symbolized as I or I^O. ABO locus is located in chromosome 9 (9q34.1-9q34.2).An antigen called O or H antigen is produced by all individuals irrespective of their ABO blod group. In the ABO locus, an enzyme is coded which is known as galactosyl transferase that adds further sugar groups on O antigen.The I^A allele codes a specific enzyme that uses a cofactor UDP-N acetyl-galactose which converts O antigen into A type. Whereas I^B allele produce a specific enzyme that uses a cofactor UDP-galactose that converts O antigen into B antigen. The genotype; phenotypes, RBC antigens and compatible donors of various blood groups exhibited in the table (Table 7.3.8.1).

In addition to antigens of ABO system the red cells of 80-85% of the human also have an additional antigen called as Rh antigen or Rh factor.In 1940, Landsteiner and Weiner reported that rabbit sera contains antibodies against RBCs of the Rhesus monkey that agglutinates RBCs of some human being. The antigen was later named as Rh (Rhesus) factor. Later, it was found that D antigen is responsible for the Rh positive

Table 7.3.8.1 : Genotypes, phenotypes, RBC antigens and compatible donors of various blood groups

Phenotypes	Genotypes	RBC antigens (agglutinogens)	Plasma antibodies (agglutinins)	Compatible donor
A	$I^A I^A$ *or* $I^A I^O$	A	Anti-B	O, A
B	$I^B I^B$ *or* $I^B I^O$	B	Anti-A	O, B
AB	$I^A I^B$	A, B	Nil	A, B, AB, O
O	$I^O I^O$	Nil	Anti-A, Anti-B	O

condition. Person's RBCs whose red cells contain this additional antigen are called "Rh positive" whereas those persons who donot have this antigen is called as "Rh negative". There are several varieties of Rh antigen – C, D, E, c, d and e, but it is seen that D antigen is most common. Therefore Rh +ve persons are also referred as D +ve in contrary Rh –ve persons are referred as D –ve. There is no naturally occurring antibodies against Rh(D) antigen. Rh(D) antigen is absent in body fluids and tissues, but only present on RBCs (See Table 7.3.5).

Such an opposing property of Anti-A and Anti-B antibodies and antigens (A and B) are applied for the purpose of determination of ABO blood groups in man.

Again, during Rh determination blood Anti-D serum is mixed with blood and if there is agglutination then the blood has Rh(D) antigen and if it doesnot agglutinate then the blood does not have Rh(D) antigen.

Materials required

1. Human candidate.
2. Saline solution.
3. Absorbent cotton.
4. Disposable sterile needle.
5. Glass slide.
6. (i) Antiserum-A (contains monoclonal anti-A antibodies against human or α-agglutinins), (ii) Antiserum-B (contains monoclonal anti-B antibodies against human or β-agglutinins) (iii) Antiserum-D (contains monoclonal anti-Rh (D) antibodies against human called anti-D agglutinins.
7. Microscope.
8. 90% ethyl alcohol.
9. 3.8% sodium citrate solution.
10. Toothpick.

Procedure

1. Blood samples are collected by scrubbing the middle finger with a piece of cotton saturated with 90% ethyl alcohol and then pricking with a disposable sterile needle.

2. A drop of blood is placed on three points of a grease free slide (one on left side, one on right side and other on the centre of the slide) to each drop of blood added one drop of 3.8% sodium citrate solution, mixed thoroughly and to each added a drop of Antiserum-A (on left side), Antiserum-B (on right side) and Antiserum-D (the centre).

3. Each drop is the thoroughly mixed with a toothpick and then observed under a microscope.

4. One has to wait for 8-10 minutes. Then observations are made of the 3 antisera, first with naked eye to see whether there is any clumping and haemolysis has occurred or not, then it is confirmed under lower magnification (10x) of a microscope.

Observation

If any agglutination occurs it is visible to naked eyes. The haemolysed red blood cells are separated as dark cell masses of different sizes and shapes. There is a brick coloured serum formed through the haemoglobin released from haemolysed red blood cells. Under low power lens the clumps are visible as dark masses where the red cells have lost their definite outline.

(a) If agglutination occurs with Anti-A; then the group is blood group A.

(b) If agglutination occurs with Anti-B; then the group is blood group B.

(c) If agglutination does not occur with Anti-A or with Anti-B then the group is blood group O.

(d) If agglutination occur with Anti-A or with Anti- B then the group is blood group AB.

(e) If agglutination occurs with Anti-D, then the blood is Rh positive, if it does not the blood is Rh negative (refer Figure 7.3.8.1; Table 7.3.8.2).

Table 7.3.8.2 : Determination of Blood groups

(+) Denotes agglutination; (–) Denotes no agglutination

Agglutination		Candidates Blood group	Candidates RBC contain agglutinogens	Candidates Plasma contains agglutins	Candidates Plasma with agglutinate RBC of group
Anti-A serum	Anti-B serum				
+	–	A	A	Anti-B	B,AB
–	+	B	B	Anti-A	A,AB
+	+	AB	A,B	None	None
–	–	O	O	Both Anti-A, Anti-B	A,B,AB

In case of Rh(D) blood group :

- Agglutination + : When candidates RBCs contain Rh(D) antigen — One is Rh(D) +ve
- No agglutination – : When there is no Rh(D) antigen in the candidate's RBC — One is Rh(D) –ve

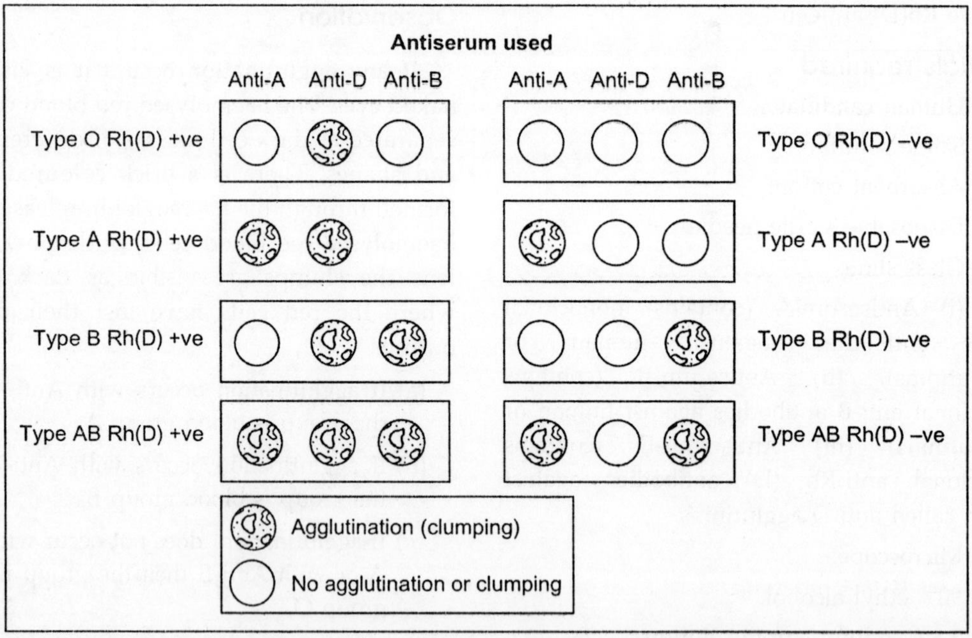

Figure 7.3.8.1 : Human blood group determination.

Note : The slides used should be dust free. The slides must be labelled A,B,D before putting three antisera i.e. Anti-A, Anti-B and Anti-D on the slides. Droppers supplied with the antisera should not be interchanged.

Chapter 7.3.9

Prothrombin Time (PT)

It is the time taken in stage two of coagulation i.e. conversion of Prothrombin to Thrombin in blood coagulation. It is connected to coagulation factors II, V,VII and X.

Materials required

1. Spirit, syringe, cotton, table lamp.
2. Water bath (37°C).
3. Stopwatch.
4. Test tubes.
5. Commercially available brain thromboplastin.
6. 3.8% Sodium citrate solution.
7. Calcium chloride solution.
8. Automated pipette (10mL-200mL) or 1mL pipette.
9. Blood plasma of rat.

Procedure

1. A test tube containing about 1 mL calcium chloride is placed in a water bath at 37°C.

2. 0.1mL of plasma* is pipetted in small tube (10mm × 75mm).

3. 0.1mL of Brain thromboplastin is added in (2) and mixed well.

4. One has to wait for 2 minutes at 37°C.

5. 0.1mL of prewarmed calcium chloride (see Step 1) is added to above solution through an automated pipette. Mixed and the stop watch is started.

6. The tube is held in front of a light source. At the first appearance of fibrin clots the watch is stopped immediately.

7. The time is recorded through a stop watch.

8. The test is repeated once again and the average reading is noted.

9. The test is repeated with normal plasma to find the control value.

Note :

Normal value is 14 to 25 seconds.

Prothrombin time is increased in case of :

1. Deficiency of factor II, V, VII and X.
2. Prolonged use of drugs like Coumarin.
3. Liver disease like viral hepatitis.

---

* Rat blood plasma is prepared by mixing 4 mL of rat blood with 3.8% sodium citrate solution (about 750µmL). The mixture is centrifuged and blood plasma is separated.

Chapter 7.4.1

Estimation of Arterial Blood Pressure and Heart Beat of Human During Normal and Stressful Exercise Condition

Principle

The cardiovascular system remains slightly overfilled with blood i.e., its contents are more than its capacity. As a result the blood exerts an outwards force against the blood vessels, this force is known as the **blood pressure (BP)**.

The apparatus employed for measuring blood pressure consisting of an inflatable rubber bag with rubber cuff which has to be put around the arm and connected to a manometer; air is pumped in till the pulse disappears at the wrist due to the pressure inside the artery producing a balancing air pressure of the blood inside the brachial artery and then estimating the former by means of a mercury manometer.

Stressful exercise affects blood pressure. BP depends on exercise in the following way :

(a) Type of exercise - aerobic or anaerobic, isotonic or isometric.

(b) Severity of exercise - whether mild or moderate or heavy.

(c) The duration of exercise

Box 7.4.1.1a. Types of Blood pressure
The **systolic pressure** is the pressure that is recorded in mm. of Hg by the sphygmomanometer, which signifies the maximum pressure exerted by blood during the maximum ejection phase of on the arterial wall at the heart. The **diastolic pressure** indicates the constant load against which heart has to work. It reflects on the degree of peripheral resistance and elasticity of aorta and large branches. The **pulse pressure** is the difference between systolic and diastolic pressures and signifies cardiac output.

Box 7.4.1.1b. Sphygmomanometer
Sphygmomanometer (Figure 7.4.1.1a) is an apparatus, which is used for measuring blood pressure (systolic, diastolic, pulse etc) in man of a peripheral artery the effect before and after exercise. It is derived by Riva Rucci in 1896 and subsequently improved.

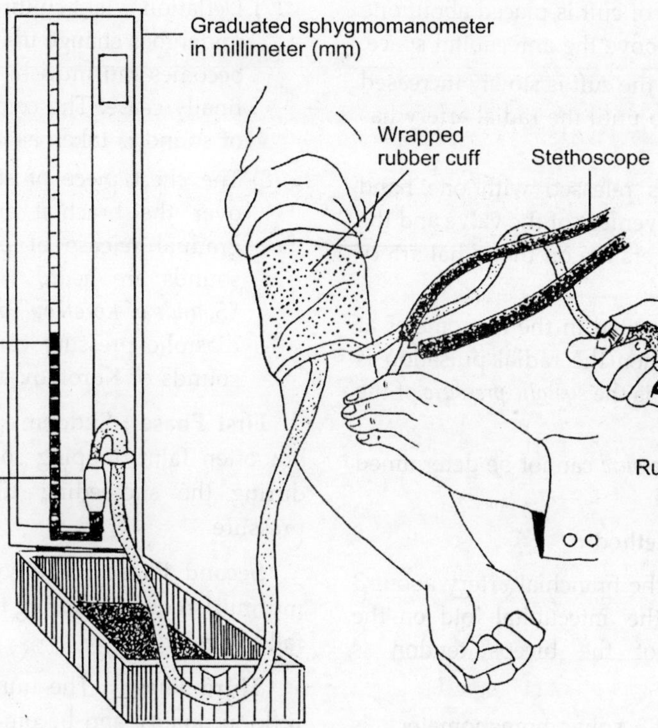

Fig. 7.4.1.1(a) : Sphygmomanometer.

Types of Blood pressure measurement procedure

(1) The subject is placed in recumbent position with the bend of elbow at the level of the heart.

(2) The arm cuff is wrapped firmly neither very loosen nor very tight and evenly round the arm about three or four inches above the elbow.

(3) The pressure is raised in the mercury manonmeter generally to 200mm of Hg. in normal subject or still there is no pulse at the wrist. The systolic and diastolic pressures may be measured by the following methods:

- **Oscillatory Method :**

(1) The pressure cuff is wrapped over the brachial artery and the instrument is kept at level of the heart.

(2) The air is pumped into the cuff to raise pressure above the systolic level until the oscillations in the manometer is absent.

(3) Next the air is released and the pressure is noted at which larger oscillations are seen in the manometer. This pressure coincides with the *systolic pressure* of the patient.

(4) The air is further released and the oscillation becomes smaller and disappears. This pressure is noted as the *diastolic pressure.*

- **Palpatory Method :**

(1) The forearm of the subject is placed over the table so that it is at the level of the heart.

(2) The cuff is deflated and wound round the arm beginning by placing on the medial aspect of upper arm.

(3) The lower edge of cuff is placed about one or two inches above the antecubital space.

(4) The pressure in the cuff is slowly increased by the air pump until the radial artery has no pulsations.

(5) The pressure is released with one hand through the movement of the valve and the other hand fingers are on the radial artery of the subject.

(6) The reading is noted in the manometer at the moment when the radial pulsation is perceived. This is the *systolic pressure* of the subject.

(7) The *diastolic pressure* cannot be determined by this method.

- **Ausculatory Method :**

(1) By palpation, the branchial artery about 3 inches above the antecubital fold on the medial side of the biceps tendon is identified.

(2) The cuff of sphygmomanometer is arranged and inflated to obliterate the pulse. The air pressure inside the inflated bag rise above the arterial pressure end obliterates the arterial lumen.

(3) The pressure is continuously reduced at a rate of 2 to 3mm per second by the release of valve until the arterial pressure just overcomes the air pressure and blood escapes and flows beyond the cuff into the peripheral section of the compressed artery.

(4) At this point, pressure is read from the manometer. As air pressure practically balances the systolic arterial pressure and the manometer reading must indicate the value of the latter.

(5) The chest piece of the stethoscope is then placed over the brachial artery just under the armlet. Next the pressure is released until a sound is heard with each heart beat.

(6) The quality of the sound is sharp, light and tapping. The pressure is noted as *Systolic pressure.*

(7) Deflation is continued until sound undergoes change in intensity and quality; becomes murmurish, dull, weakened and finally cease. The complete disappearance of sound is taken as *diastolic pressure.*

(8) The chest piece of stethoscope is placed over the brachial artery and with the gradual release of pressure, a series of sounds are heard which are known as *"Sounds of Korotkov"* by which systolic and diastolic pressure can be measured. The sounds of Korotkov are :

First Phase : Sudden appearance of a clear, but often faint, tapping sound growing louder during the succeeding 10mm of Hg fall in pressure.

Second Phase : The sound takes on a murmurish quality during the next 15mm of Hg. fall in pressure.

Third Phase : The murmur sound changes little in quality and becomes clearer and louder during the following 15mm of Hg. fall in pressure.

Fourth Phase : The loud sound suddenly becomes muffled in quality and rapidly starts to fade during the next 5 to 6mm of Hg. fall. After this, all sounds disappear.

The beginning of the first sound is taken as the *index of systolic pressure* and the fourth sound just before its complete disappearance is taken as the *index of diastolic pressure.*

According to British School, appearance of muffling sound is the index of diastolic pressure. U.S.A. School considers just disappearance of sound as diastolic pressure.

Methodology

Materials Required :

1. Subject (Human).
2. Sphygmomanometer.
3. Stethoscope (Fig. 7.4.1.1b).

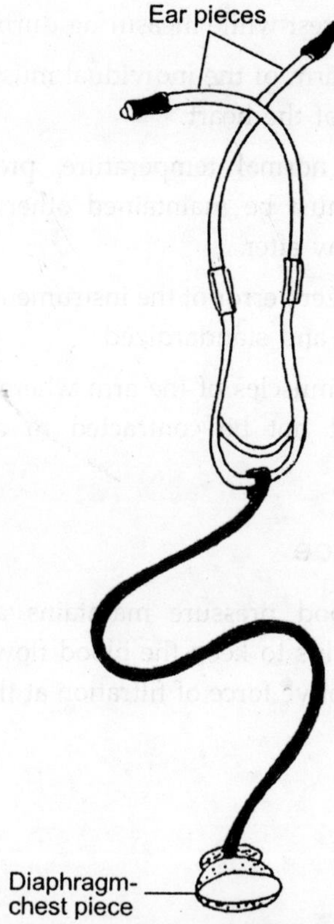

Fig. 7.4.1.1(b) : Stethoscope.

A. Basic Procedure for Blood pressure measurement :

1. The subject is made comfortable and the BP is recorded with arm at the level of the heart.

2. BP is measured through **Ausculatory Method** (refer to procedure mentioned before) during complete resting i.e. the subject should be relaxed and free from tension and anxiety. He/she should be assured and rested for 5 minutes.

3. The experimenter may employ any of the exercises: running in place (spot running) with thighs brought in place to horizontal alternatively for 3-5 minutes (min.); hopping on each foot for 3 minutes, raising the feet 12-15 inches off the ground, climbing up and down the stairs, jogging or **Harvard Step Test.**

4. BP is measured through **Ausculatory Method** is also recorded after exercise i.e., at 2 minutes, at 4 minutes and at 6 minutes.

B. Basic Procedure for heart beat measurement :

1. The subject is kept in a calm condition.The stethoscope is put on the left side of the chest of the subject and the 'lub-dub' heart sounds are

Table 7.4.1.2 : Measurement of blood pressure during rest and stressful exercise

Condition	No. of experiment	Systolic pressure in mm of Hg	Mean systolic pressure in mm of Hg	Diastolic pressure in mm of Hg	Mean diastolic pressure in mm of Hg	Pulse pressure (Systolic-diastolic) in mm of Hg	Mean heart beat/ minute
1. Rest (Before exercise at complete rest)	1. 2. 3.	— — —	—	— — —	—	—	—
2. After Exercise	1. At 2 min. 2. At 4 min. 3. At 6 min.	— — —	—	— — —	—	—	—
3. Blood pressure returned to resting level after ...... minutes.							
4. Heart beat returned to resting level after ...... minutes.							

heard. The heart beat is counted by counting the number of 'lub' in one minute.

2. This is repeated after any short of rigorous exercise or Harvard Step Test etc.

Result

Findings are to be entered in a tabulated form; Table 7.4.1.1 :

Note : Average systolic pressure of man of 120 mm of Hg. the diastolic pressure becomes 75 mm of Hg. and the pulse pressure becomes 45 mm of Hg.

Precautions necessary during measurement of blood pressure by sphygmomanometer

(1) The subject must be in complete bodily and mental rest while measuring during the rest.

(2) The arm of the individual must lie on the same level of the heart.

(3) The normal temperature, pressure and humidity must be maintained otherwise blood pressure may alter.

(4) The zero error of the instrument should be determined and standardized.

(5) The muscles of the arm where the cuff is tied should not be contracted to avoid false readings.

Significance

The blood pressure maintains a sufficient pressure helps to keep the blood flowing and to give the motive force of filtration at the capillary bed.

Chapter 7.4.2

Study of Oestrous Cycle by Vaginal Smear Technique in Rat, *Rattus rattus*

Introduction

The reproductive pattern and breeding behaviour amongst vertebrates differ outstandingly in relation to a particular season of the year. Such seasonal breeders go through a non-breeding or **anoestrus** (Gk. *an*, not; *oistrous*, gadfly) period, a stage where they remain sexually **dormant** (L. dormire, to sleep). Sexual cycles in such seasonal breeders are incredibly variable. All animals, apart from the higher primates, copulate only during a particular period in each periodic sexual cycle. These periods define an appropriate psychological and physiological condition, during which copulation is permitted, are designated as periods of heat or **oestrus** (Gk. *Oistros*, gadfly). The phase from the beginning of one heat to the beginning of the consequent one is called **oestrous cycle**. It is evident only during the breeding season.

Events in ovarian and vaginal histology of the rats (female) during oestrous cycle are tabulated below (Table 7.4.2.1b).

Oestrous cycle is under complex hormonal control, see Figure 7.4.2.1a.

Methodology

Principle :

Vaginal epithelium does not remain the same during the various phases of the oestrous cycle. Cyclical transformation in pituitary, ovarian and other hormones, and those of uterus, supports such changes in the histological characteristics of the vaginal epithelium. The vaginal epithelium is destroyed and reconstructed in a cyclic fashion, varying between the stratified epithelium and low cuboidal epithelium types. These cyclical changes can be observed by a serial study of

Table 7.4.2.1a : Classification of animals on the basis of oestrous cycle

Type	Description	Examples
Monoestrous	Animals have only one oestrous cycle during the year	Bears, foxes and wolves
Polyestrous	In many animals many oestrous cycles may be fired (with short intervals in between) during the same breeding season	Rat, mice, sheep, cattle and dogs (twice a year; dioestrous)

Table 7.4.2.1b : Stages of the oestrous cycle vis-a-vis ovarian and vaginal histology (after Nalbandov, 1958)

Name of the stage	Duration of the stage (in hours)	Ovarian events that occur	Types of cells that appear in the stage observed in the vaginal smears
Dioestrus	Constitutes half of the whole cycle (50-55 hours)	Corpus luteum	Presence of nucleated epithelial cells and leucocytes
Proestrus	About 12 hours span	Follicles grows fast	Presence of nucleated epithelial cells
Early oestrous	About 12 hours span	Follicles largest (mating occurs)	Presence of cornified squamous cells
Late oestrous	About 18 hours span	Ovulation occurs	Presence of cornified squamous cells
Metaestrous	About 06 hours span	Corpora luteum formed	Presence of leucocytes among cornified squamous
Initiation of dioestrus or anoestrus	—	Functional corpora luteum during early part	Gradual disappearance of cornified cells

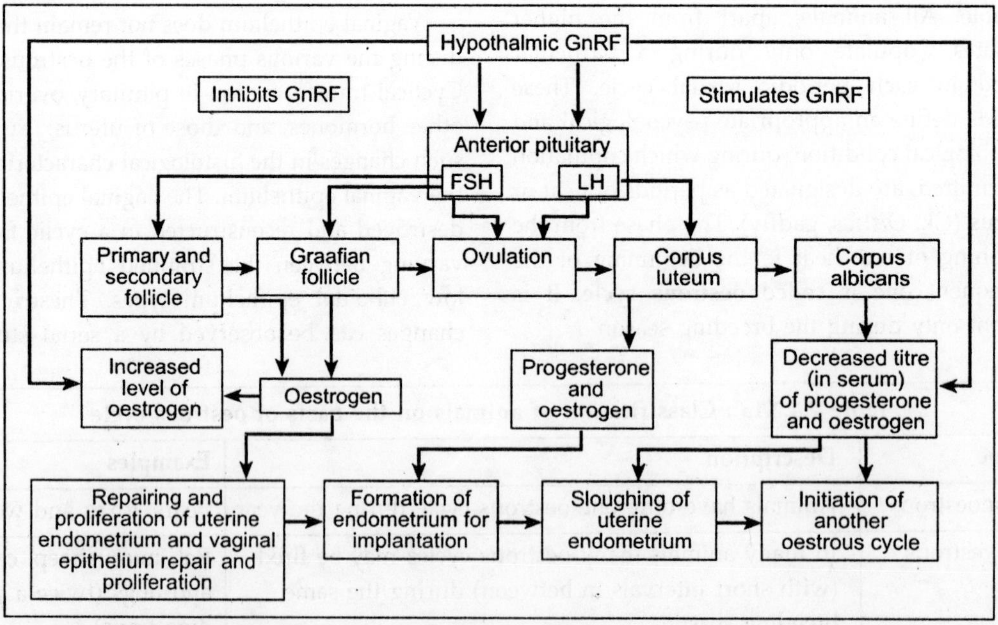

Figure. 7.4.2.1a : Diagrammatic representation of hormonal control of different stages of the oestrous cycle

stained vaginal smear preparations for consecutive days of oestrous cycle in rat *(Rattus rattus).*

Materials required :

1. Female Sprague-Dawley (albino) rats (12-14 weeks age group).
2. Normal saline (i.e., 0.9% NaCl solution).
3. Pasteur pipette (rubber teat fitted).
4. Microscopic glass slides (should be grease free).
5. Square cover slips (20 × 22mm).
6. Methanol.
7. staining set for haematoxylin-eosin double staining method.
8. DPX mountant.
9. Compound microscope (with 10X and 45X objective lens).
10. Glass marking pen.
11. Staining and fixing tray.

Procedure :

1. The female rats are acclimatized for 7 days in animal house by keeping them in a suitable mouse cage with recommended quality and quantity of food and water. The temperature and humidity of the animal house should also be regulated as per recommendation. Daily required light and dark phases should also be maintained.
2. One of the surface of glass slides are marked with serial numbers and date time with the help of glass marking pen.
3. Time schedule is fixed for taking the vaginal swab.
4. A little amount (about 0.3 to 0.5mL) of normal saline is introduced through the vagina of the female rat under study with the help of a Pasteur pipette. Care should also be taken to avert injury/damage in the reproductive tract.
5. The pipette takes out vaginal swab slowly.
6. A little amount of swab is put on the marked surface of the slides.A thin smear of that swab is made.
7. The smear is allowed to dry in air.
8. Few drops of methanol is put and left for 10 minutes in staining-fixing tray (properly covered) for fixation.
9. Fixed materials are then stained and, subsequently, dehydrated according to the routine haematoxylin-eosin double staining method (see section 4).
10. The stained smear is mounted with DPX and covered the same with cover slip.
11. The smear is observed under microscope under both 10X and 45X objective lenses.
12. Steps 4-12 are repeated at time intervals, according to predetermined time schedule for a complete study of the histological changes of vaginal smear at different stages of the oestrous cycle.

Observation :

1. Vaginal smear have nucleated (non-cornified) epithelial cells. This represents the proestrus stage of the oestrous cycle.
2. Vaginal smear contains cornified epithelial cells. This represents the oestrous stage of the oestrous cycle.
3. Vaginal smear contains leucocytes among cornified epithelial cells. This represents the metaoestrus stage of the oestrous cycle.
4. Vaginal smear contains (almost entirely) leucocytes. This represents the dioestrous stage of the oestrous cycle.

Refer Figure 7.4.2.1b; colour plates (Section 7).

Applied aspect :

Study of reproductive behaviour in animals is of immense importance in matters relating to animal breeding activities, designed for both academic as well as economic purposes.

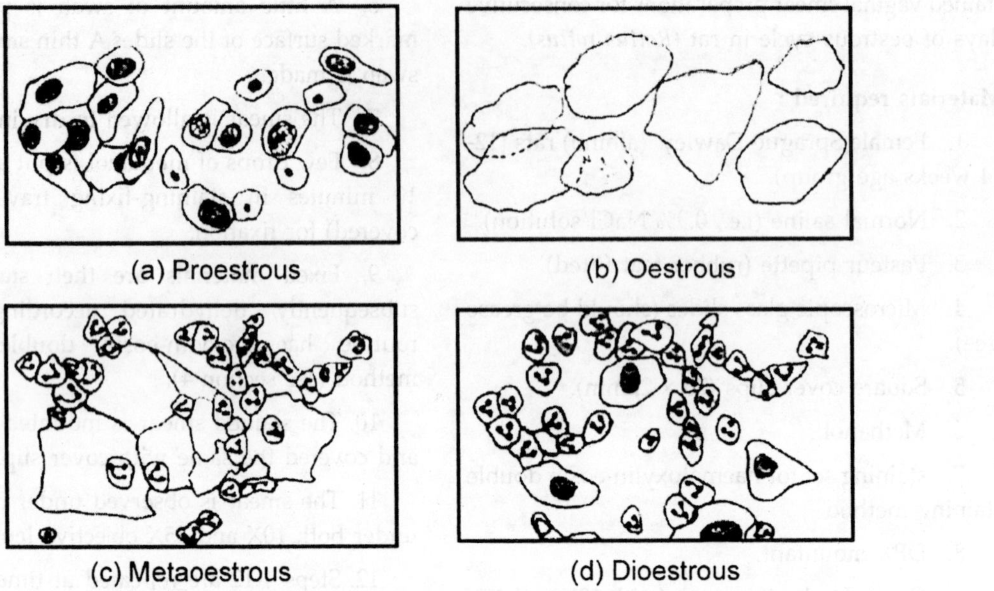

Figure 7.4.2.1b : Changes in the histological architectures as seen in the vaginal smear during different stages of the oestrous cycle (a) Proestrus, (b) Oestrus, (c) Metaoestrus, (d) Dioestrus. (Adapted from Grobman and Bern, 1962)

Chapter 7.4.3

Enzymatic Digestion of Carbohydrates and Proteins

A.1. Action of Ptyalin (Salivary amylase) :

Principle :

Saliva contains enzyme ptyalin, which hydrolyses boiled starch and certain other polysaccharides, the chief end product being maltose. The changes in hydrolysis are briefly as follows :

Starch → Soluble Starch → Higher Dextrins → Erythrodextrins → Achroodextrins → Maltose (hydrolysis product).

Ptyalin acts at a temperature of 37°C. When iodine solution is mixed with starch, blue colour is obtained. But after the starch is first hydrolysed with amylase and then mixed with iodine, blue colour solution is not formed. This occurs because starch is hydrolysed into hexose.

Materials required :

1. Starch powder.
2. 1% Iodine solution.
3. Benedict solution (for qualitative test).
4. Rubber band or paraffin wax.
5. Test tube.
6. Mortar and pestle.
7. Beaker and
8. Porcelain plate.

Procedure :

(a) **Collection of saliva :** The mouth is rinsed thoroughly with distilled water twice.

The flow of saliva is stimulated by chewing a rubber band (or paraffin wax) and a few mL of saliva is collected in a test tube.

(b) **Digestion of starch by ptyalin :**

(i) 1% starch paste is prepared by grinding 0.5g of starch powder in a mortar with a small quantity of water, adding the starch suspension to about 50mL of boiling water in a beaker, boiling again for 1 minute and allowing it to cool. An opalescent solution is formed.

(ii) 5 drops of saliva is added to 10mL of 1% starch paste in a test tube. Mixed well, the test tube is placed in a water bath at 37°C. A series of drops of 1% iodine solution is placed on a porcelain plate. At intervals of 1 minute a drop of the digesting mixture (saliva + starch) is removed and allowed it to fall on one of the iodine drops. The opalescence of the starch solution soon disappears due to the formation of soluble starch, which gives a blue colouration with iodine. With further drops of saliva on the starch iodine solution the blue colour gradually disappears.

(iii) Next, Benedict's test is performed with the digesting mixture. A yellowish precipitate will be formed, indicating the presence of reducing sugar.

Observations :

At first starch reacts with iodine, producing a blue colour. With the drops of digesting mixture added later the colour with iodine becomes red and finally shows no change due to formation of erythrodextrin and achroodextrin respectively. This point representing the moment when the last traces of eryrhrodextrin is converted to achroodextrin and maltose is known as the **achromic point**, the time taken being a measure of the activity of the enzyme.

Inference :

Salivary amylase in the saliva digests starch into maltose as reflected through the disappearance of blue colour of starch iodine complex to a colourless product(maltose).

A.2. Effect of Temperature on the Activity of Ptyalin (Salivary amylase) :

Principle :

Salivary amylase is protein in nature; the optimum temperature for its activity is 37°C. In extreme temperatures (cold and heat), the activity of the enzyme is inhibited (as high temperature causes denaturation of the enzyme being a protein).

Materials required :

1. Test tubes.
2. Previously prepared enzyme (saliva) solution (*see* A. 1).
3. 2% starch solution.
4. Benedict's solution (for qualitative test).
5. Ice bath
6. Water bath

Procedure :

1. 3 test tubes are taken and numbered as "1", "2" & "3".

2. 2 mL of the previously prepared enzyme solution(saliva) is put in each of the three test tubes.

3. To each test tube, added 2 mL of 2% starch solution and mixed well.

4. Immediately test tube "1" is put in an ice bath, test tube "2" in a 37°C water bath and test tube "3" in a 70°C water bath, each for 20 minutes.

5. Each tube is removed to room temperature and Benedict's test is carried out to check for the presence of the starch hydrolysis product "maltose" as follows: 2 mL of Benedict's reagent is added to the enzyme-starch mixture in each tube and put in a boiling water bath for 5 minutes.

Observations :

Benedict's reagent will maintain its clear blue color in test tubes "1" & "3" signifying the absence of maltose as a result of inhibited enzyme activity at cold and hot temperatures, whereas a yellow to red precipitate is formed in test tube "2" signifying the formation of maltose as a result of a retained action of salivary amylase at 37°C.

Inference :

Hot and cold temperature inhibited salivary amylase activity and Benedict reagent maintains its blue colour as there is no formation of maltose.

A.3. Effect of pH on the activity of Ptyalin (Salivary amylase) :

Principle :

The optimum pH for the activity of salivary amylase is the neutral pH i.e. about 7. In the acidic pH of the stomach, the action of salivary amylase gets inhibited.

Materials required :

1. Test tubes
2. Previously prepared enzyme (saliva) solution(see A.1)
3. 2 mL of acidic solution (pH 3 to 5)
4. 2 mL of neutral solution (pH 7)
5. 2 mL of alkaline solution (pH 9 to 11)
6. Benedict's solution (for qualitative test).
7. Water bath
8. 2% starch solution.

Procedure :

1. 3 test tubes are taken and numbered as "1", "2" & "3".

2. 2 mL of the previously prepared enzyme solution(saliva) is taken in each of the three test tubes.

3. To each test tube, added 2 mL of 2% starch solution and mixed well.

4. Added 2 mL of acidic solution to test tube "1", 2 mL neutral solution to test tube "2" and 2 mL alkaline solution to test tube "3" are put 3 test tubes in a 37°C water bath for 10 minutes.

5. Each tube is removed to room temperature and Benedict's test is carried out to check for the presence of the starch hydrolysis product "maltose" as follows: added 2 mL of Benedict's reagent to the content of each tube and then these tubes are put in a boiling water bath for 5 minutes.

Observations :

Benedict's reagent will preserve its clear blue color in test tubes "1" & "3" indicating the absence of maltose because of inhibited enzyme activity in acidic and alkaline pH, whereas a yellow to red precipitate is formed in test. tube "2" signifying the formation of maltose as a result of a preserved activity of salivary amylase in neutral pH.

Inference :

Acidic and alkaline pH inhibited salivary amylase action as a result no maltose is formed and thus Benedict's reagent preserve its blue colour.

B. Action of Pepsin on Proteins :

Principle :

Pepsin is the proteolytic enzyme of the gastric juice. It hydrolyses proteins to peptone in presence of HCl, the optimum pH and temperature being 2.0 and 40°C respectively. The changes in hydrolysis are ;

Protein → Acid metaprotein → Proteose → Peptone.

Materials required :

1. Test tubes.
2. Water bath.
3. Commercial pepsin.
4. Fibrin or coagulated egg white.
5. 0.4% HCl.
6. 0.5% Na_2CO_3.

Procedure :

1. Equal amounts of freshly washed fibrin or. pieces of coagulated egg white is placed in five test tubes labelled 'A', 'B', 'C', 'D' and 'E'.

   To 'A' added 3mL of 0.75% solution of commercial pepsin in water and 3mL of 0.4% HCl.

   To 'B' added 3mL of pepsin solution and 3 mL water.

   To 'C' added 3mL of water and 3mL of 0.4% HCl.

   To 'D' added 3mL of boiled pepsin solution and 3mL of 0.4% HCl.

   To 'E' added 3mL of pepsin solution and 3 mL of 0.5% Na_2CO_3

2. The test tubes are placed in a water bath at 40°C for half an hour.

Observations :

It is noted that in 'A' the fibrin swells up due to the action of acid and then dissolves due to the action of pepsin-hydrochloric acid. In 'B' and 'E' the fibrin is unaltered, showing that pepsin does not act in a neutral or alkaline medium. In 'C' the fibrin swells up and becomes transparent, but does not dissolve, as pepsin is absent. In 'D' the fibrin swells up but does not dissolve, because of the inactivation of pepsin by boiling.

Inference :

In tube 'A' pepsin digests fibrin or pieces of egg yolk due to action of pepsin in acidic environment. But in other tubes no digestion occurs due to absence(or destruction) of enzyme or acidic environment.

C. Action of Trypsin on Proteins (Pancreatic Digestion) :

Principle :

Trypsin is the major activated proteolytic enzyme of pancreatic juice, hydrolysing protein into polypeptides, at an optimum pH of 8.3, and temperature of 40°C.

Materials required :

1. Test tubes.
2. Water bath.
3. Neutral pancreatic extract* or solution of trypsin or pancreatin.
4. Fibrin or coagulated egg white.
5. 0.4 % HCl.
6. 0.5% Na_2CO_3.

Procedure :

1. Equal amounts of fresh washed fibrin or pieces of coagulated egg white are placed in 5 test tubes labelled 'A', 'B', 'C', 'D' and 'E'.

   'To 'A' added 3mL of neutral pancreatic extract or solution of trypsin or pancreatin and 3mL of water.

   To 'B' added 3mL of the extract and 3mL of 0.5% Na_2CO_3.

   To 'C' added 3mL of the extract and 3mL of 0.6% HCl.

   To 'D' added 3mL of water and 3mL of 5% Na_2CO_3.

   To 'E' added 3mL of boiled extract and 3 mL of 5% Na_2CO_3.

2. The tubes are placed in a water bath at 40°C for half an hour.

Observations :

Note that in 'B' the fibrin dissolves due to the activity of the trypsin. In 'C' the fibrin swells but does not dissolve. In 'A', 'D' and 'E' the fibrin remains unaltered.

Inference :

In tube 'B' pepsin digests fibrin or pieces of egg yolk due to action of trypsin in alkaline environment. But in other tubes no digestion occurs due to absence (or destruction) of enzyme or alkaline environment.

---

* **Preparation of pancreatic extract :** Fresh pancreas is taken from pig/sheep and all visible fats are removed. The same is ground in a mortar and pestle along with a little sand. A little water is added to it and filtered through a cheese cloth. Addition of few drops of chloroform prevents putrefaction and the extract can be stored for a considerable period of time.

Chapter 7.4.4(a)

Recording of Normal Cardiogram (heart tracing) of Toad

Principle

Heart is one of the most important organ of animal body. So it is important to know about the normal heart functioning i.e. how a heart beat, what is it's pattern? One can record the heart beat of toad in normal condition in form of a graph which is known a **normal cardiogram**. Toad heart is a convenient preparation as it continues to beat after the chest is open. Also it can function in a wide range of temperature.

Materials required

(1) A pair of large scissors and fine scissors.

(2) A pair of fine forceps.

(3) A pithing needle.

(4) Toad.

Apparatus Required

Kymograph assembly including—

(1) Crank's myograph board,

(2) Starling's heart lever,

(3) Sherrinton's recording drum Kymograph,

(4) Time marker,

(5) A pin hook,

(6) Thread

(7) Induction coil etc. (see Chapter 12 for Kymograph assembly details).

Circuit

No electrical stimulus is required hence there is no circuit.

Speed of Drum : Slow moving drum with slow gear and 1:4 pulley, i.e. 1.2mm/second.

Strength of Induced current : At first the secondary coil should be kept away and is brought gradually nearer the secondary till the inhibition of the heart is obtained. A very strong stimulation is generally avoided because strong stimulation develops electrolysis and destroys conduction in the nerve permanently.

Procedure

(1) A toad is pithed and positioned on a dissecting board. A midline incision is made in the skin of the abdomen running up to the mouth. The sternum is picked up and the muscles are cut through by scissors the transparent pericardial sac is exposed. The **xiphisternum** is removed along with a part of the anterior chest wall. A portion of the pericardium is snipped and the heart is exposed and seen to be beating.

(2) The toad is put on the **myograph board** and the bent hook is transfixed very carefully to the apex of the ventral aspect of the toad and it is attached to thread of the **startling heart lever.**

(3) Care should be taken that the hook gets a good bite of the ventricular musculature but at the same time does not cause a big puncture, giving rise to considerable bleeding.

(4) Heart lever is so adjusted that the maximum amplitude of contraction is obtained.

(5) The lever on the smoked drum (already being smoked) is so adjusted so that the writing point is horizontal when it is in the middle of its excursions. The writing point must be touching the drum lightly and uniformly as it moves up and down. The heart tracings are recorded on the slow moving drum.

(6) Time tracings are taken with a slowly vibrating **time marker** or **chronometer**. Time tracing of 5 seconds is recorded.

Observations

The heart shows regular rhythmic and uniform contraction (Figure 7.4.4a.1) of three parts viz., sinus venosus, auricle and ventricle under normal circumstances.

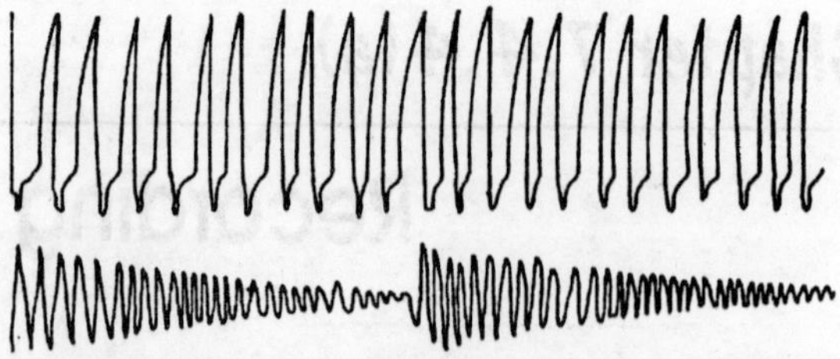

Fig. 7.4.4a.1 : Normal heart beat of toad.

Due to the contraction of the sinus venosus there is a short rise followed by a short descent, unless the adjustment is delicate and accurate, it is usually missed.

Due to contraction of the auricles there is a second rise followed by a slight descent for the relaxation of auricles.

Due to the ventricular contraction there is a third rise which is rapid and then slows, thus producing the systolic plateau. If the aorta are cut the systolic plateau disappears.

Chapter 7.4.4(b)

Perfusion of Isolated Toad Heart and Recording of Heart Beat (Demonstrate the Kymographic Record of the Perfused Heart of a Toad)

Principle

When a heart is taken out of the body it can carry on beating for a certain length of time but the condition slowly detoriate and lastly the heart stops due deficiency of ions, lack of appropriate temperature, deficiency of proper acid-base balance and lack of oxygen. In case of an amphibian heart all the conditions of the heart are not to be dynamically maintained and if the heart is perfused with a fluid (iso-osmotic with the blood), the oxygen supply and appropriate temperature are sustained. The, heart continues to beat for a much longer time period.

Materials required

(1) A pair of large scissors and fine scissors.

(2) A pair of fine forceps.

(3) A pithing needle.

(4) A perfusion set consisting of one Mariotta bottle or one reservoir with Syme's and venous perfusion cannulae filled with amphibian Ringer's-Locke fluid (0.60g NaCl, 0.010g $CaCl_2$ and 0.0075g KCl 0.01g $NaHCO_3$, 0.001g Na_2HPO_4 and 0.1g Glucose in 100 mL distilled water with pH adjusted to 7.4; please note that $NaHCO_3$ must be completely dissolved in water before $CaCl_2$; Glucose is to be added just before use).

(5) Toad.

Apparatus required

Kymograph assembly including—

(1) Crank's myograph board,

(2) Starling's heart lever,

(3) Kymograph,

(4) Time marker,

(5) A pinhook and

(6) Thread etc.

Circuit : Though the electrical stimulus is not necessary, therefore there is no circuit.

Speed of Drum : Slow moving drum with slow gear along with a 1:4 pulley, the sped 1.2 mm/sec at first then 2.5cm/sec.

Induced Current strength : Initially the secondary coil should be kept away and is brought gradually nearer the secondary till the inhibition of the heart is obtained. A very strong stimulation is generally avoided because strong stimulation develops electrolysis and destroys conduction in the nerve permanently.

Procedure

(1) A toad is pithed and positioned on a dissecting board. A midline incision is made in the skin of the abdomen running up to the mouth. The sternum is picked up and the muscles are cut through by scissors the transparent pericardial sac is exposed. The xiphisternum is removed along with a part of the anterior chest wall. A portion of the pericardium is snipped and the heart is exposed and seen to be beating.

(2) A thread is passed around the sinus and a loose knot is tied around it. A small slit is made in the sinus and a Syme's cannula is introduced into it.

(3) The aorta and other vessels are cut and the cannula is lifted along side of the heart. The startling heart lever is fitted, upside down, on a stand and the cannula is fixed in the clamp directly above the heart. The side arm of the cannula is connected to a reservoir or Mariotta bottle containing Ringer's-Locke fluid (pH of 7.4).The reservoir is raised about 30 cm above the heart to provide a pressure head.

(4) Just after the cannula is introduced, the heart including atrium and ventricle will be distended with perfusion fluid. The systemic aorta is immediately incised so as to permit the fluid to drain out. The Syme's cannula is kept 4 to 6 inches higher than the level of the myographic board. The Mariotta bottle or modified Syme's cannula has one outlet to facilitate drainage of excess fluid and thus maintains a constant pressure head. The heart will beat so long as the optimum pressure, pHs, ionic concentration of the perfusion fluid are maintained. An additional supply of oxygen is not required. If the perfusion fluid is bubbled with oxygen or the fluid is saturated with air by thoroughly shaking the bottle, improved result can be obtained.

(5) The heartbeat is carefully recorded as follows :

(a) The toad is transferred to the myograph board. The tip of ventricle is hooked to the lever.

(b) The writing point of the Starling's heart lever is brought in close and gentle contact with the smoked drum. Necessary adjustment is made and a few beat is recorded.

(c) The normal beat of the heart is recorded. The events of the cardiac cycle are marked (It is should be noted that upstroke is systole and down stroke is diastole). After the tracing has stabilized the reservoir is raised to increase the perfusion pressure. Writing point of the starling lever is lightly touched to the cylinder surface. The cardiac cycle is recorded for about 15 cm on the paper to get a **cardiogram.**

(d) A time tracing is taken with the slowly vibrating **time marker.**

Observations

1. To make comparison of the experimental cardiogram with the normal cardiogram shown in Figure 7.4.4.1a provided as a reference.

2. The number of peaks has to be recorded.

3. To identify the heart movement which is responsible for a particular peak.

4. The duration of cardiac cycle each of auricle and ventricle is independently studied.

Chapter 7.4.4(c)

Recording of Effect of Temperature, Acetylcholine, Nicotine and Adrenaline on Normal Heart Beat of Toad

Principle

At **high temperature** there is an increment in the metabolic activity of the pacemaker cells as a result there is generation of more cardiac impulses per unit time. Due to increased metabolic activity of the working cells of the atria and the ventricle there is an increase in the force of contraction. **Low temperature** has an opposite effect due to decrease in the metabolic activity.

Acetylcholine (Ach) is a **neurotransmitter**. It increases the permeability of K^+ ions, as a result more of these ions move out. This result into the **hyperpolarization** and decreases the phase-4 of AP. This results into the increase in the time interval between the APs thus decreasing the heart rate. Ach increases the permeability of K^+ ions, effux of K^+ shortens phase 2 of AP, which decreases the Ca^{++} influx thus the force of contraction i.e., negative **ionotropic effect**.

Nicotine increases heart rate at lower concentration, acts to be vagus positive. But at the higher concentration paralyses all autonomic ganglia and acts as vagus negative.

Adrenaline is also a neurotransmitter. It is **symtomimetic agent**. It is released by adrenal medulla. It acts directly on the cardiac cells. It increases their permeability mainly to the Ca^{2+} ions and to some extent to the Na^+ ions. Increase of Na^+ in pacemaker cells increases the slope of phase 4 of the AP. Thus the heart rate is increased. The large influx of Ca^{2+} ions increases the force of contraction.

Materials required

(1) A pair of large scissors and fine scissors.

(2) A pair of fine forceps.

(3) A pithing needle.

(4) A perfusion set consisting of one Mariotta bottle or one reservoir with Syme's and venous perfusion cannulae filled with amphibian Ringer's-Locke fluid (0.60g NaCl, 0.010g $CaCl_2$ and 0.0075g KCl 0.01g $NaHCO_3$, 0.001g $Na2HPO_4$ and 0.1g Glucose in 100 mL distilled water with pH adjusted to 7.4; please note that $NaHCO_3$ must be completely dissolved in water before $CaCl_2$; Glucose is to be added just before use.

(5) Toad Apparatus Required : (i) Crank's myograph board, (ii) Starling's heart lever, (iii) Kymograph,(iv) time marker,(v) A pin hook and (vi)Thread.

Circuit : Though the electrical stimulus is not necessary, therefore there is no circuit.

Speed of Drum : Slow moving drum with slow gear along with a 1:4pulley,the speed 1.2 mm/sec at first then 2.5 cm/sec.

Induced Current strength : Minimal stimulus.

Procedure

1. A toad is pithed and positioned on a dissecting board.

A midline incision is made in the skin of the abdomen running up to the mouth. The sternum is picked up and the muscles are cut through by scissors the transparent pericardial sac is exposed. The xiphisternum is removed along with a part of the anterior chest wall. A portion of the pericardium is snipped and the heart is exposed and seen to be beating.

2. The frog is transferred into the myograph board. The starling lever is fitted on the vertical rod of the strand directly above the heart. The sharp pin hook is passed through the apex of the ventricle, taking care that no puncture is made in its cavity.

3. The heart is lifted by raising the lever and its position is adjusted so that its movement is satisfactory and its mean position is horizontal.

4. The stand carrying the preparation and the lever so that the lever should be tangent to the drum and the writing point is lightly touching the drum surface. Some normal heart beats are recorded.

5.(a) **For effect of temperature :**

(i) Some amount Ringer solution at room temperature is poured on the heart to keep it moist.

(ii) With the drum running at slow speed some Ringer solution is warmed to about 40°c on the heart in a drop wise manner. Until there is obvious increase in its rate and force.

(iii) The Kymograph is stopped and Ringer solution is poured on the heart at room temperature until the heart resumes the previous rate and force.

(iv) The effect of Ringer solution at about 5°C is recorded in the same way (Figure 7.4.4.c.1a).

(b) **For effect of acetylcholine(Ach) :**

(i) After recording some normal heart beats. The drum is stopped and a few drops of 1 in 10,000 solution of acetylcholine are poured on the heart.

(ii) Nature of heart beat (heart rate and force of contraction) is observed.

(iii) The washed with Ringer solution and one has to wait till beats normalize which is to be recorded.

(c) **For effect of nicotine :**

(i) Few drops of 1% Nicotine solution is applied on the heart.

(ii) Nature of heart beat is observed.

(iii) The washed with Ringer solution and one has to wait till beats normalize which is to recorded. 2% Nicotine solution is also tried in the same way (Figure 7.4.4.c.1b).

(d) **For effect of Adrenaline :**

(ii) Few drops of 1 in 1000 adrenaline chloride is applied on the heart.

(iii) The effect is recorded.

(iv) Washed with Ringer solution and one has to wait till heart becomes normal again which is to be recorded.

6. 5 seconds time tracing is recorded below the graph obtained. The graphs are labeled accordingly for each type of treatment viz., temperature, acetylcholine, nicotine and adrenaline (Figure 7.4.4.c.1c).

Observations

	Treatment	Effect
1.	(a) Warm Ringer s solution (45ºC) (b) Cold Ringer solution (5ºC)	There will be a gradual increase in heart rate and its amplitude of contraction. The site of action is sinus venosus. 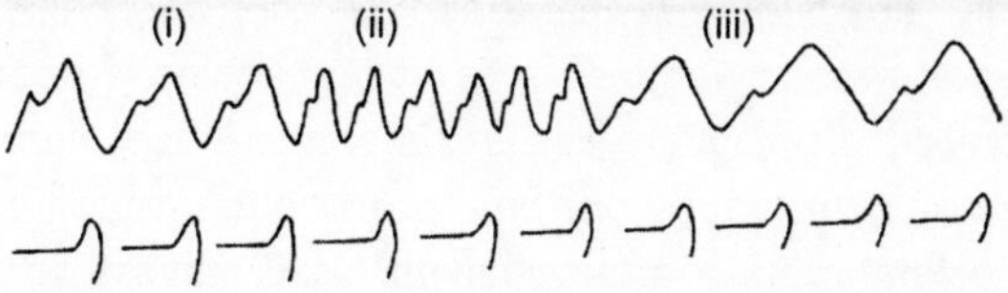 **Fig. 7.4.4c.1a : (i) Normal heart tracing, (ii) Effect of hot Ringer solution on heart beat, (iii) Effect of cold Ringer solution on heart beat; of toad.** The heart rate is decreased on cooling. The site of action is sinus venosus.
2.	Acetylcholine	Decrease the rate of the heart and also the force of contraction of the heart.
3.	(a) Nicotine (1%) (b) Nicotine (2%)	Heart rate decreases. Stimulates parasympathetic ganglia. Heart rat increases. Paralyses synapse in the ganglia. 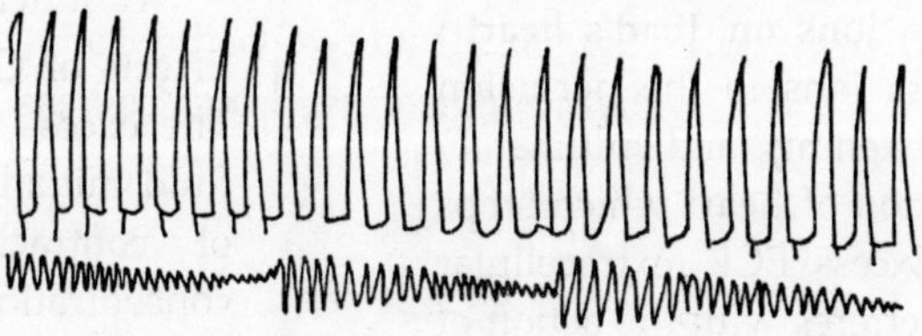 **Fig. 7.4.4c.1b : Effect of Nicotine on normal heart beat of toad.**
4.	Adrenaline	Increases rate of the heart and also the force of contraction of the heart. 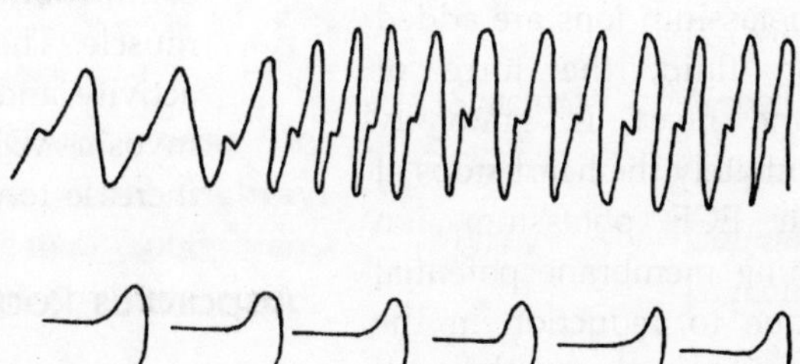 **Fig. 7.4.4c.1c : Effect of Adrenaline on heart beat of toad.**

Make comparison with normal heart beat recording (refer Figure 7.4.4a.1). The nature of heart beat is studied.

Chapter 7.4.4(d)

Effect of Different Ions on Toad's Heart

Principle

Sodium, potassium and calcium ions in specific concentrations are necessary for maintaining the normal activity of the heart.

(a) **Effects of Sodium ions on Toad's heart :** The excess sodium ions in the perfusion fluid causes weakening and ultimately failure of contraction of heart which stops at the diastole. Excess ECF (extracellular fluid); sodium interferes with the action of calcium which acts as link between the excitation and contraction.

(b) **Effects of excess Potassium ions on Toad's heart :** If excess potassium ions are added to the perfusion fluid, the force of contraction of the heart is gradually decreased and ultimately the heart stops at diastole. A high ECF potassium ion decreases the resting membrane potential of the muscle (due to reduction in the concentration gradient across the cell membrane).The excess potassium ion causes relaxation of the cardiac muscle. This effect is in relation to the degree of dissociation of actin and myosin by excess potassium ion. Besides, excess potassium ions causes—

(i) slowing of the diastolic depolarization phase in between heart beats,

(ii) depression of the SA and AV nodes and

(iii) a blocking of the conduction through conducting tissue in heart.

(c) **Effects of Calcium ions on Toad's heart :** The excess calcium ions in the perfusion fluid initially cause augmentation of force of contraction of heart. But if the concentration is increased further, then heart stops at systole and the condition is known as **calcium rigor**. The calcium ions bear some relation with the "Excitation-contraction coupling" of the cardiac muscle. The calcium ion triggers ATP-ase activity and facilitates formation of actin-myosin-ADP-complex, resulting into an increase tone of the cardiac muscle.

Apparatus Required

The apparatus which are required for this experiment are—

(1) Kymograph.

(2) Myograph board and Starling's heart lever.

(3) Time marker.

(4) Glass seeker.
(5) Pin hook.
(6) Thread.
(7) Perfusion set.
(8) Toad.

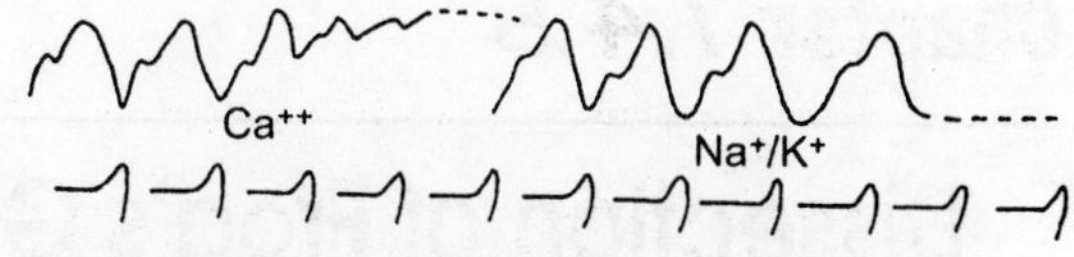

Fig. 7.4.4d.1 : Effect of Na^+, K^+ and Ca^+ ions on heart beat of toad.

Materials Required

The materials which are necessary for this experiment are 1.0% NaCl solution, 1.0% KCl and 1.0% $CaCl_2$ and Ringer's solution as perfusion fluid for optimal activity of the heart.

Circuit : Since the electrical stimulus is not be used, hence there no circuit.

Speed of Drum : Medium slow.

Strength of Induced Current : At first the secondary coil should be kept away and is brought gradually nearer the secondary till the inhibition of the heart is obtained.

Procedure

(1) The toad is dissected and heart is exposed as before (Chapter 7.4.4d.1).Added 1.0% NaCl solution via the side tube of the cannula to the perfusion fluid on toad's heart. The heart is washed with the perfusion fluid (perfusion of heart) and the effects of 1.0% KCl and 1.0% $CaCl_2$ in the same way.

(2) The toad is transferred to the myograph board. The tip of ventricle is hooked to the lever.

(3) The writing point of the Starling's heart lever is brought in close and gentle contact with the smoked drum.

(4) The normal beat of the heart is recorded. The events of the cardiac cycle are marked.

(5) The excess amount (higher concentrations) of sodium, potassium and calcium are added (if required) to the perfusion fluid on Toad's heart and their effects are also observed.

(6) A time tracing is taken with the slowly vibrating time marker.

(7) The events of the cardiac cycle are marked for each type of ions viz., Na, K and Ca and corresponding cardiograms are prepared.

Observations

After perfusion with NaCl, KCl and $CaCl_2$ the following observations are made :

1. To make comparison of the experimental cardiograms (Figure 7.4.4d.1) for effect of each type of ions with the normal cardiogram shown in Figure 7.4.4a.1 provided as a reference.
2. The number of peaks has to be recorded in the cardiograms. Any changes are to identified.
3. To identify the heart beat which is responsible for a particular peak.

Chapter 7.4.5

Dissection of Frog's Gastrocnemius Muscle with Sciatic Nerve for Recording Simple Muscle Twitch

Principle

Vertebrate skeletal muscle is a kind of muscle specified for balance and movement. The basic unit is the **motor unit**, which consists of a single motor neuron and the skeletal muscle fiber(s) it innervates. In vertebrates, motor neurons project from the central nervous system (CNS) to make synapses with skeletal muscle fibers at neuromuscular junctions. These synapses are always excitatory thus an action potential in a motor neuron should always results in the contraction of one or more skeletal muscle fibres.

A typical skeletal muscle is composed of many motor units. Each motor neuron tends to synapse onto only one type of muscle fibre. Stimulation of a single motor neuron tends to create contraction of only one muscle fibre type. Motor neurons that supply weak, slow, oxidative fibres have the lowest threshold; those innervating fast, intermediate-strength oxidative fibres have higher thresholds; and those that supply the fast, strong, glycolytic fibres have the highest thresholds. The CNS modulates the overall force of skeletal muscle contraction by altering the total number and type of motor units activated at any moment. As activity descending from the brain increases, progressively more motor neurons, and more of those innervating the stronger muscle fibre type, are activated into a response.

Force of skeletal muscle contraction is also augmented when the CNS increases the rate of electrical signalling to individual motor units. A brief skeletal muscle contraction pursued immediately by relaxation, otherwise known as a **twitch**, is usually the result of a single, supra-threshold stimulus of short duration. A single stimulus applied to the sciatic nerve results *muscle twitch* of **Gastrocnemius muscle**. The brief interval between the moment of stimulation and start of contraction is called latent period. It is followed by contraction and relaxation periods.

Apparatus required

1. Primary circuits consist of primary key, primary coils of the induction coils and current source. Kymograph (drum); drum with smoked paper or glazed paper for ink writing stylus.

2. The secondary circuit apparatus consists of secondary coils of the induction coil, electrodes, myograph board, myograph lever.

3. Instruments required for nerve muscle preparation.

4. Thread, tuning fork (See section 12).

Materials Required

1. Toads normal saline (0.65%, NaCl).
2. Pins.
3. Forceps.
4. Scissors.
5. Needle.
6. Anaesthetic ether.
7. Adult Frog.

Procedures

• **Gastronemius muscle and sciatic nerve preparation :**

(i) The frog is washed and caught by left hand from the vertebral column. It is pithed. The animal is completely anaesthetised i.e., when the limbs become flaccid, the corneal reflexes are abolished and there is a loss of cutaneous stimuli.

(ii) The pithed frog is fixed on the dissection tray with helps of pins. The skin of the hind limb is held on by forceps and a small cut is given by a vertebrate scissors which is extended up to the toe which results in the separation of the skin from the whole limb. The frog gastronemius muscle (Figure 7.4.5.1a) is inserted to **tendo-achillis.** The distal portion of the tendo-achillis is separated from its bonny attachment. The bone just above the knee is also cut to have a complete muscle. The muscle is detached at the back of the thigh by a needle and the sciatic nerve is will be seen accompanied by the femoral vessel. The sciatic nerve is traced from the knee to the exit from the vertebral column by separating the group of the muscles on the back of the thigh by help of two fingers and then with the help of needle and it is freed from the muscle and then the vertebral column. It is cut below and up the origin of the sciatic nerve. The nerve should be separated by a glass seeker. The nerve muscle preparation should be stored in toad's normal saline (0.65% NaCl) till it is transferred to the myographic board.

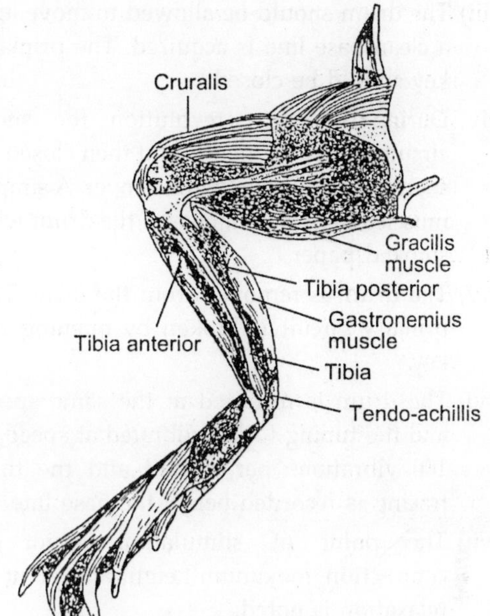

Fig. 7.4.5.1(a) : Frog gastronemius muscle.

• **Recording of muscle twitch :**

(i) A nerve muscle preparation is transferred to the myograph board. The tendon is hooked by a pin fastened by a thread and the other end of the thread remains attached to the fulcrum. The knee joint is fixed by a pin on the myographic board in such a way so that the thread becomes tightened. The electrode is put directly over the sciatic nerve.

(ii) The apparatus is arranged in a moderate speed. The lever should be horizontal and brought in gentle contact of the smoked paper wrapped round the drum first in tangent position to the drum. This results in the completion of the primary circuit.

(iii) The drum should be allowed to move and a clean base line is acquired. The primary key should be closed.

(iv) During the next revolution the short circuiting key is opened and then closed as soon as the muscle twitch is over. A simple muscle curve is obtained on the drum with smoked paper.

(v) The drum is removed from the lever. The primary circuit is broken by opening the key.

(vi) The drum is restarted at the same speed and the tuning fork is vibrated at speed of 100 vibrations per second and the time tracing is recorded below the base line.

(vii) The point of stimulation, point of contraction, maximum height and point of relaxation is noted.

(viii) The paper is removed from the drum. The nature of the experiment with date is to be noted.

(ix) The duration in seconds of the **latent period**, the **period of contraction** and **period of relaxation is noted.**

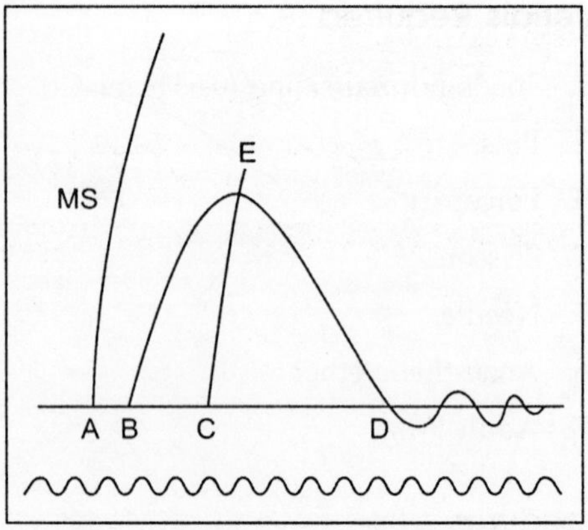

Fig. 7.4.5.1b : Simple muscle twitch. MS : movement of stimulation, AB : latent period , BC : contraction period, CD : relaxation period, BE : contraction phase, ED : relaxation phase. The curve below the twitch depicts the time tracing ,tuning fork = 100 Hz.

Observations and Results

A simple muscle twitch curve consists of phase of contraction and phase of relaxation. The phase of contraction begins with a latent period which is followed by period of contraction and then by the period of relaxation.

- The simple muscle twitch is the elementary unit of all muscular response.
- The tension of the muscle remains practically constant throughout and the contraction is isotonic.
- The lever shows a few oscillations at the end of relaxation before coming to rest due to elasticity of the muscle and instrumental inertia.
- The **latent period** of the curve is always constant, about 0.01 second.
- The **period of contraction** is always less than the period of relaxation
- The descent of the curve below the base line just after relaxation is due to instrumental inertia (Figure 7.4.5.1b).

Conclusion

When the nerve is stimulated it generates impulses which are transmitted to the muscles and the muscle contracts. The conduction of nerve impulses and the chemical events and electrical activity in the muscle head the mechanical events that cause the latent period.

Chapter 7.4.6

Demonstration of Reflex Action in Toad

Introduction

When the part of the muscle or skin is stimulated an afferent impulse initiates from there and reach to centre(s) in some part of the central nervous system. From there an efferent impulse originate down the efferent nerve fibres and contraction of a muscle or group of muscle occurs. The reflex response depends upon the integrity of this path and damage to any of its part causes disappearance of the reflex.

Materials required

1. Toad.
2. Chloroform.
3. 1 % HCl.
4. 1% KOH.
5. Stand and clamp.
6. Pithing needle.

Procedure

1. Frog is chloroformed and it is held by its hind legs. It is washed with water and held in the left hand with its head exposed. The head is put between the index and the middle finger and the thumb is placed in the back of the frog.

2. The head is bent forward and the pithing needle is placed on the head mid way between the eyes. Appling pressure the needle is drawn posteriorly towards the body until it enters the groves of occipito-vertebral junction. The pressure is applied and the needle is directed forward and a little downward through the skin and foramen magnum into the skull using a twisting side to side movement. This will destroy the brain.

3. The eyelid reflex is tested. If the eye shows no reflex, i.e. no raising of lower eyelid of the either eye the pithing is complete.

4. The pithed toad is suspended through a wire hook passed through the jars. Absence of normal shock is ascertained by pinching the toes.

5. Next the great toe is placed in 1% HCl solution. One should be careful that no part of the foot should come in contact with walls of the dish or HCl.

6. The refex time is noted.

7. This experiment is repeated thrice and the average is taken.

8. The same is repeated with 1% KOH solution.

Results

Sl. No.	Stimulus	Reflex time			Average
		1st	2nd	3rd	
1.	HCl				
2.	KOH				

Section 8

ECOLOGY, ETHOLOGY AND TOXICOLOGY

SECTION AGENDA

Chapter 8.1.1(a)

Determination of Minimum Size of Quadrats

Introduction

A quadrat is a sampling unit having four arms or a small enclosure for the community. It may be circular, rectangular or square in shape. These units are used to enclose a small area in the grass land and plants community. The quadrat used to sample the plants in a community must be of such size that all the species, which occur in that community, are fully represented. These units are used to enclose a small area in the grasslands and plants falling in it are studied.

Materials required

(1) Stakes, (2) Rope, (3) Graph Paper, (4) Hammer.

Procedure

1. A point is chosen at random in the study area and a stake is fixed at this point. A second stake is put at about 2 m (or less if the study area is a small one) from the first. A third stake is set at right angles to the second at the same distance. A rope is passed about the three stakes to make fig. 'L' (Figure 8.1.1a.1). Using a fourth stake and a length of rope 2 × 2 sq. m area is marked off at an angle of "L".

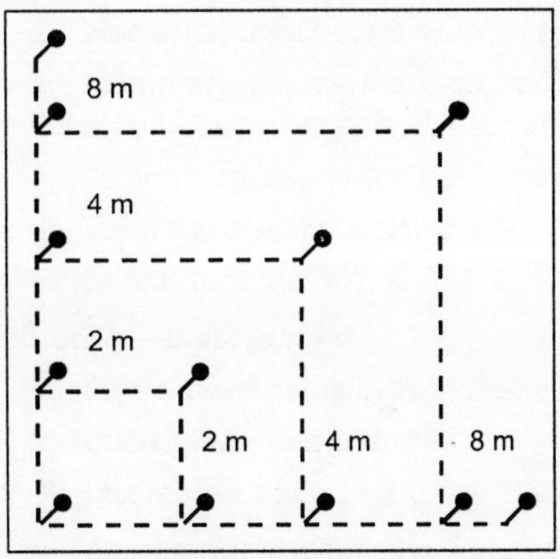

Figure 8.1.1a.1 : Preparation of quadrats.

2. Total number of species of plant is counted within it. The size of the square is doubled with using more stakes and more rope and the number of species is counted in it. The size of the square may be further enlarged in size. Recorded each of the number of species.

3. The number of species (in y-axis) is plotted against the size (area) of the square (quadrat) (in x-axis) in a graph paper. The curve is known as the **species-area curve** (Figure 8.1.1a.2).

The sizes of quadrat depend on the nature of vegetation present in the area. Larger quadrats are necessary to sample trees, while herbs and shrubs could be adequately sampled with smaller quadrats. If a separate tally for the species number of the trees, herbs and shrubs is made, it will be likely to arrive at suitable quadrat size for each of these structural types of vegetation.

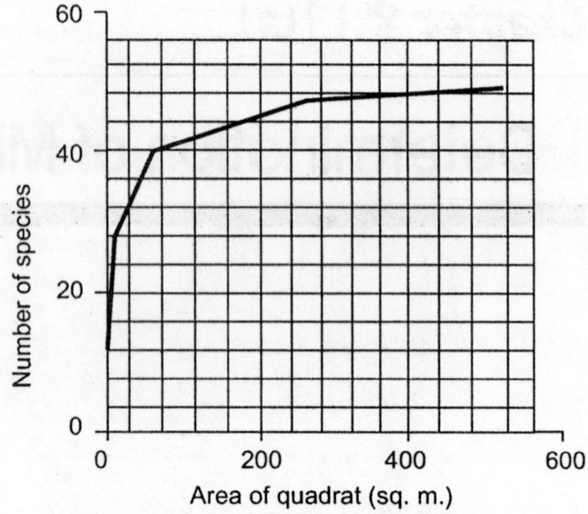

Fig. 8.1.1a.2 : Species area curve.

Observation and Result

Observing the species area curve, the point at which the curve start flattening up is noted. Through this point the area of the quadrat is found out from the X- axis of the graph.

Chapter 8.1.1(b)

Determination of Number of Quadrats Required

Introduction

The number of samples taken should be large enough to give a true measure of the density or abundance of the plants in any location. Inadequate sampling will result in either over or under estimation.

Requirements

1. Quadrat of definite size.
2. Graph paper.

Procedure

1. A number of quadrat is laid out of a size appropriate for the study in the chosen area. Number of species are recorded in the first quadrat.

2. Next quadrat is examined and occurrence of any additional species is recorded. Next quadrats are also examined in the same way.

Table 8.1.1 : Data sheet

Species Name	Quadrats									
	1	2	3	4	5	6	7	8	9	10
A										
B										
C										
D										
E										
F										
Total no. of species in each quadrat	X_1	X_2	X_3	X_4	X_5	X_6	X_7	X_8	X_9	X_{10}

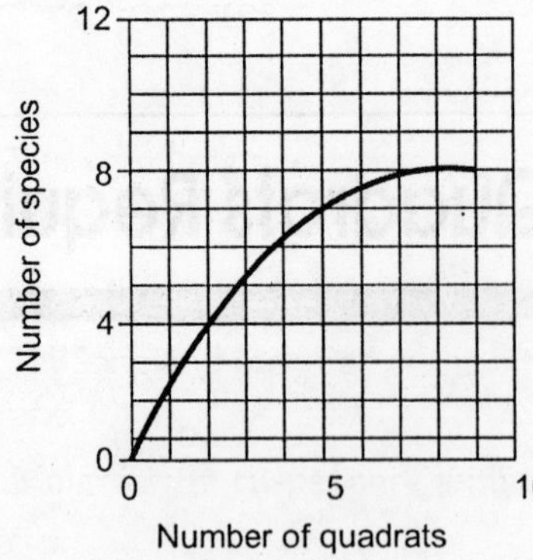

Fig. 8.1.1b.1 : Graph to assess the number of quadrats required for adequate sampling of an area.

3. The data is tabulated in a table (Table 8.1.1 as shown below). A graph (Figure 8.1.1b.1) is also drawn with number of quadrats on the horizontal axis (i.e. x-axis) and the number species on the vertical axis (i.e. y-axis). The point at which curve flattens depicts the number of quadrat required for adequate sampling of that area.

Observation and Result

The point, which the curve flattens out, points the number of quadrats required for adequate sampling of that area.

Chapter 8.1.2

Analysis of Structure of Community in Grassland Ecosystems

Introduction

The grassland communities consist of producers, herbivores, carnivores and decomposers. The chief producer component are the grasses that may be divided into tall (1.5-2.5 mtr.) viz: *Andropogon geredi, Sorghastrum sp.* etc.; medium (0.6-1.3 mtr.) viz. *Agropyron smithi, Orzyopsis* and short (< 0.5 mtr.) types. These grasses are important only in certain seasons. The consumer components of the grassland constitute very small animals except for large grazing animals native or domesticated by man. The carnivores are relatively small that includes foxes, coyotes, owls etc. The rodents include rabbits, rats and squirrels. The birds constitute the top carnivores. The insect comprises of the herbivores.

Materials required

(1) Quadrats of definite size.

(2) Collection vials or containers.

(3) Field note-book.

(4) Formalin.

(5) 70% alcohol.

Procedure

1. Required numbers of quadrats (1, 2, 3...,n) of appropriate size are laid down in grassland.

2. In each quadrat, different types of species are noted down and number of individuals in each species is also counted.

3. Smaller animal species which are to be identified are collected in vials or containers preserved in formalin or alcohol in required concentration and identified in authorized laboratory. Larger animal species are identified on the spot.

Calculation

1. Collected data are tabulated from each quadrat in the data sheet in observation and result section.

2. Frequency, Density and Abundance of different species in the community are calculated as below :

(a) Frequency (%) =

$$\frac{\text{Total number of quadrats in which species occurred}}{\text{Total number of quadrats used}} \times 100$$

(b) Density =

$$\frac{\text{Total number of individuals of species}}{\text{Total number of quadrats studied}}$$

(c) Abundance =

$$\frac{\text{Total number of individuals of the species}}{\text{Total number of quadrats in which species has occurred}}$$

Observation & Result

The data are revealed in form of Table 8.1.2 shown below :

Data collected are then put into different equations to find out frequency, density and abundance of different species.

Table 8.1.2 : Data sheet

Sl. No.	Species	Quadrats laid 1, 2, 3, 4,......n	Total number of individuals of species	Total number of quadrats in which species found	Total number of of quadrats used
1.					
2.					
3.					

Chapter 8.1.3

Estimation of Primary Production

The capture of solar energy by green plants and the transformation of a portion of this light energy through photosynthesis is called **primary production.** Photosynthesis entails intake of radiant energy and carbondioxide and release of oxygen. Respiration involves the uptake of oxygen and liberation of carbon dioxide and production of energy. It is thus a reverse process of photosynthesis. In the absence of light, photosynthesis is detained but respiration goes on. In the presence of light both processes take place simultaneously. These facts are used in devising ways of measuring primary production. Primary production is of three types :

1. **Gross Primary Production** : The total synthesis of organic matter results from exposure to light.
2. **Net Primary Production** : Amount of organic matter stored after expenditure due to respiration.
3. **Respiration** : Gaseous and heat exchange with the environment related with metabolic breakdown of organic molecules by living cells.

Primary production is the amount of organic carbon produced and primary productivity is the rate of such production occurs i.e. amount per unit of time.

The basic assumptions made in estimating primary productivity are—

1. The rate at which inorganic nutrients are used up have a steady relationship to the rate of energy fixation.
2. The energy content of organic molecules bears a constant relationship to its weight. The weight of the living materials in a community is called **biomass.** Biomass may be of three types: volume biomass, ash-free dry weight biomass or organo-biomass (total organic matter present both in the form of living biomass as well as the dead production present in the sediments).

Primary productivity may be calculated as gross productivity and net productivity. The relationship between these can be expressed as:

Net productivity (P_N) = Gross Productivity (P_G) – Respiration rate (R).

Estimation of primary productivity :

Different ways of measuring primary production are generally used. Each of these measures a different aspect of productivity. Two important ways of measuring primary productivity are :

(1) Harvest method :

Harvesting is the oldest techniques for determination of primary productivity. Plants are cut at ground level, on land or water. The harvested plant material is then dried out to constant weight and primary productivity is expressed in terms of dry weight biomass per unit area per unit time, for example gm/sq.m/yr. This method thus establishes production changes in dry weight of biomass over a time episode. For this method to be relevant, there should be no utilization of primary production by the next trophic level. The harvest method is usually in use within terrestrial ecosystem though it may be employed in aquatic systems also.

A method to determine phytoplankton production by the harvest method is explained below.

Materials required :

1. Crucible.
2. Diamond tipped pencil.
3. Hot air oven.
4. Muffle furnace.
5. Calcium chloride.
6. Ceatrifuge (laboratory) with centrifuge tubes.
7. Water from pond or aquarium.

Procedure :

1. An oven dried crucible is weighed in an analytical balance. The crucible is always handled using a pair of tongs.

2. An identification number is scratched using a diamond tipped pencil.

3. 25 ml of water sample from a natural pond or an aquarium (after stirring the water well) is placed in each of two centrifuge tubes (inclusions of algal mats are avoided). Spun for 10 minutes at 1500 rpm. The supernatant fluid is decanted.

4. Using a very small amount of water, the residue is transferred to be weighed in a dried crucible. The tube is rinsed with a small quantity of distilled water and this is added to the crucible. The crucible is placed in an oven set at 105-110°C and dried to constant weight. Weight of the crucible is subtracted from this and dry weight of sample is recorded.

5. The crucible is placed in a muffle furnace at 600-650°C for at least three hours, when the contents will burnt down to ash the crucible is removed and cooled in a desiccator over calcium chloride to avoid moisture absorption during the cooling process.

6. Reweighed. Difference between dry weight and ash weight gives the amount of total organic matter presents in the organisms.

7. The experiment is repeated after an interval of 24 hr. taking samples of same volume from the same place. The change in biomass is the production over that period of time.

Calculation :

1. 24 hr. production in g/mL =

$$\frac{\text{24 hr. production in sample}}{\text{Sample volume}}$$

The result multiplied by the volume of the aquarium or pond will give the production for the system for the period.

2. 24 hr. production g/m^2 =

$$\frac{\text{24 hr. production in aquarium or pond}}{\text{Surface area in } m^2 \text{ of aquarium or pond}}$$

3. 24 hr. production in g carbon/m^2

= Production in $g/m^2 \times 0.105$

Source of Error :

Presence of heterotrophs and non-inclusion of loss of grazing or sinking are the main sources of errors. An analysis of a fresh sample is necessary to determine the amount of heterotrophs present.

(2) Light and dark bottle method (Gaarder and Graan technique) :

This technique is used in measuring primary productivity of an aquatic community (phyto-plankton and macrophytes) and is based on the estimation of oxygen released by the producers over a time period. Oxygen generated is simultaneously being utilized up in respiration. It has to be borne in mind that photosynthesis occurs only in presence of light which will vary with different factors such as time of day, extent of cloud cover, conditions of water surface and clarity of water. These factors will fluctuate over the seasons.

Materials required :

1. Oxygen bottles (one transparent and other dark painted).
2. Bamboo poles or wooden planks.
3. Polythene floats.
4. Dissolved O_2 measuring chemicals and materials.
5. Secchi disk.

Procedure :

1. Two sets of "Oxygen bottles" are prepared. One set is painted black on the outside including the top of the lid. Other set is left transparent. One light (transparent) bottle and one dark (painted black) is used to sample water at each location.
2. Bamboo poles or wooden planks or polythene floats are used to suspend the bottles at desired depths.
3. Water taken in the sample bottles must be obtained from the depth at which productivity is measured. Precautions necessary for measuring dissolved oxygen (DO) by the **Winkler method** (see section 8.2.9) are to be followed.
4. A **secchi disc** is used to determine the light compensation level, where photosynthesis will equal respiration. The time at which light and dark bottles were suspended in the water is noted.
5. At this time the amount of dissolved oxygen contents of the water at the various locations where the light and dark bottles suspended are estimated. This initial value is recorded.
6. The bottles should be left suspended for one to several hours, up to 24 hr. A 6 hr. period is generally preferred.
7. The bottles are removed at the end of the test period and the dissolved oxygen content is estimated immediately.
8. If macro-vegetation is included in the bottles it should be removed and volume should be replaced with water from the pond. This will introduce an error which can however be overlooked. All values are recorded.

Calculation :

- *Respiration (oxygen consumed)* = Level of oxygen at the begining – Level of oxygen in the dark bottle at the end of experiment
- *Gross Primary Productivity (total oxygen produced)* = Level of oxygen in light bottles at the end of experiment – Level of oxygen in the dark bottles at the end of experiment
- *Net Primary Productivity* = Gross primary productivity – Respiration.

To convert the mg/L oxygen values to mg/carbon/m^3 the mg/L values are multiplied by 375.36. This will give mg carbon/m^3 for the duration of test period. This is converted to mg/carbon/m^3/unit time. Unit time could be per hour or per day. As there will be only 12 hour sunlight in a 24 hr. day the per hr. values should be multiplied by 12 not 24 to derive productivity values per day.

Chapter 8.1.4

Collection of Dust Particles in Ambient Air by Slide Trap Method

Introduction

In nature, due to imbalance of different components cause a negative effect on life which is a reflection of pollution. Pollution can be defined as an undesirable change in physical, chemical and biological characteristics of air, water and soil that may harmfully affect the life or create a potential health hazard of any living organism.

Pollution may be of different types such as water pollution, air pollution, noise pollution etc. Any substance which causes pollution is called as a pollutant.

Air pollution is caused by different pollutants, viz., gases, and chemicals, different types of dust particles, pollen grains, feathers and hairs of animals. Some of the pollutants remain as suspended particulate matter (SPM). These are causative factors behind chronic and acute diseases in humans.

Thus for combating environment pollution it is necessary to assess the amount of pollutants in the environment.

Paul and Nair (1958) described the 'slide trap' technique to collect and measure dust particles as pollutant in air.

Materials required

1. Slide.
2. Vaseline/Grease.
3. Thread.
4. Scalpel.
5. Microscope.

Procedure

1. A glass slide is cleaned thoroughly and on one side of slide is layered with vaseline.

2. These vaseline layered slides are hanged through a white thread horizontally at a optimum height in an open place. Such slides are to be hanged at different direction (north, south, east, and west).

3. After 24 hours these slides are observed under a microscope.

4. Under a microscope various types of particles such as fibre, pollen grains, feather parts, hair, micro arthropods, scales, dust particles, carbon particles etc.

These particles are then counted in a tabular form.

Results

Table 8.1.3 : Enumeration of different air pollutants (particles) collected through Slide Trap technique

Slide no.	Materials						Total no. of particles
	(a)	(b)	(c)	(d)	(e)	(f)	
1.							
2.							
3.							
4.							
...							
n							

Inference

From the result it becomes feasible to infer the types and number of each type of particles trapped in the 'slide trap'.

Chapter 8.1.5

Quantitative and Qualitative Analysis of Zooplanktons and Mounting of any Two Types of Specimens

Zooplanktons are floating animal forms inhabiting in fresh water ecosystem. The important groups of zooplankton include members of Protozoa, Rotifera, Copepoda, Ostracoda and Cladocera. Zooplankton may bear some locomotory structures but are unable of moving against the water currents. Planktons can be classified on the basis of its size.

Plankton individuals vary greatly in size. Generally animal plankton (**zooplankton**) are larger than the plant plankton (**phytoplankton**). Some phytoplankton, a few protozoa and bacteria are less than 1/100 mm. and bypasses even finest plankton net. Such planktonic forms are referred to as **nanoplankton**. The larger forms, which are retarded by standard plankton nets, are designated as **net plankton** or **filterable plankton**.

Dussart's classification of both fresh water and marine plankton is generally accepted.

1. Ultra nanoplankton below 2 μ
2. Nanoplankton 2-20 μ
3. Macroplankton 20-200 μ
4. Mesoplankton 200-2000 μ
5. Megaplankton above 2000 μ

Quantitative and qualitative analysis of zooplanktons helps to appreciate the ecological status of fresh water ecosystem from which the zooplanktons are collected. It also helps to identify different zooplanktons and prepare permanent slides of zooplanktons species.

Materials required

1. Plankton net.
2. 2-3% Formaldehyde solution or 70% ethanol.
3. Plastic bucket or jug.
4. Von dorn or Pump samplers.
5. Sample bottle.
6. Microscope.
7. Slide.
8. Cover slip.
9. Dropper.
10. Eosin or Borax carmine stain.
11. Alcohol grades.
12. Sedgwick-Rafter Cell: Sedgwick-Rafter Cell is a slide with a rectangular cavity (50 mm × 20mm × 1mm) of volume 1 cm^3 (i.e. 1 mL).
13. DPX.

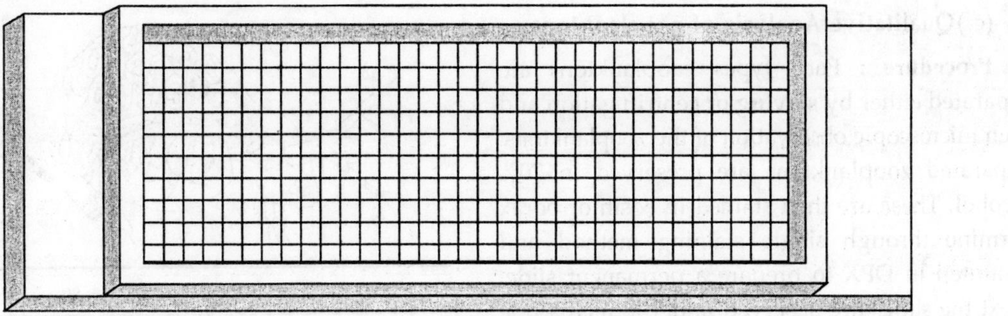

Fig. 8.1.5.1 : Diagram of Sedgwick Rafter cell

Procedure

(a) **Collection and Preservation of zooplankton :**

1. Zooplankton can be collected by filtering a known quantity (in volume) of water through zooplankton net made up of silk (No. 25, mesh size 55 μm). The surface water for this purpose may be collected with the aid of a plastic bucket or jug of specific volume but for sub-surface waters Von Dorn or Pump samplers are the most preferred ones.
2. Plankton net weighted at the base is lowered into the water to known depth with the help of a graduated rope. The net is then pulled gradually (say 10 cm/sec) out. The volume of water filtered through plankton net is computed using following formula :

$$V = \pi r^2 h$$

Where V= volume of water (L), π = a constant (3.142), r = radius of mouth of plankton net (m), and h= depth up to which the net was lowered (m).

When the net is pulled horizontally in water the 'h' in this formula represents the distance travelled by net (m).

Zooplankton collected is transferred to a small sample bottle (50-100ml) and preserved in 2-3% formaldehyde or with 70% ethanol.

(b) **Quantitative Analysis of Zooplankton Population (Sedgwick Rafter Method) :** Zooplankton is collected as described earlier and various species and total zooplankton is enumerated following the Sedgwick-Rafter cell method. The result is expressed in organisms/L.

Procedure

1. Concentrated sample is shaken well and without delay, 1 ml of it is transferred in the cavity of Sedgwick-Rafter cell (Figure 8.1.5.1) with the help of a dropper.

2. Cavity of the cell is covered with a cover slip avoiding any air bubble to get in. Planktons are allowed to settle and then counted under a microscope (10x) while moving the cell horizontally and vertically.

3. A good number of replicates (at least 10) are taken and the average count per ml is calculated.

Calculation

Zooplankton (Units/Litre) = N × C/Volume of sample in litre

Where N = number of zooplankton counted in 1mL of concentrate, C = total volume of concentrate (mL); volume of sample in Litre represents the total volume of water taken in bottle for sedimentation.

(c)**Qualitative Analysis of zooplanktons :**

Procedure : Each types zooplanktons are separated either by **sieving** or centrifugation and then microscopic observation of the zooplanktons. Separated zooplanktons are preserved in 70% alcohol. These are then stained in eosin or borax carmine through single staining method and mounted in DPX to prepare a permanent slide. Next the slides are observed under a microscope and different zooplankton's genus and species are identified (refer Figure 8.1.5.2 and Figures of Chapter 1.1.1 as referred).

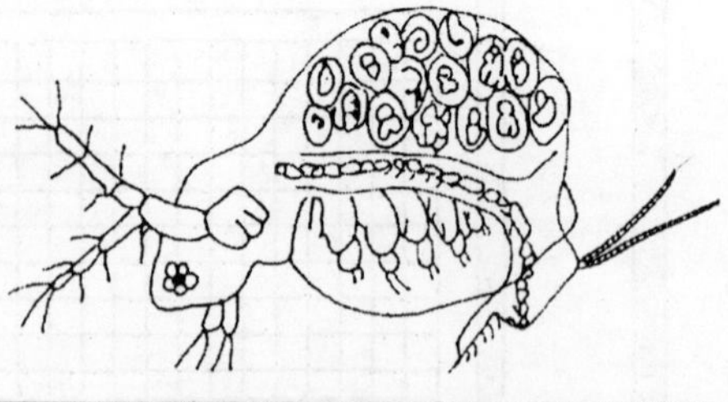

Fig. 8.1.5.2(a) : *Moina sp.*

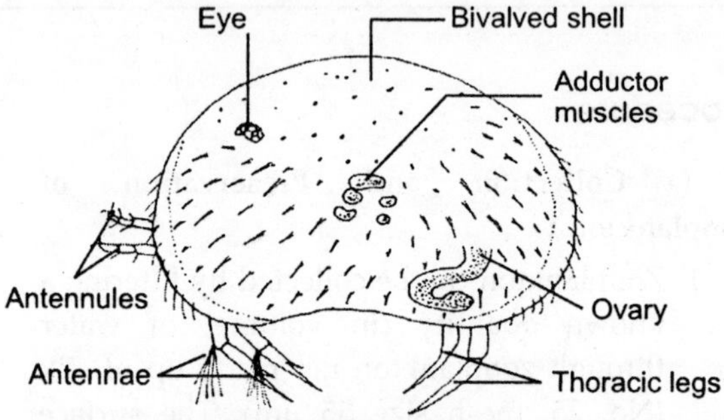

Fig. 8.1.5.2(b) : *Cypris sp.*

Identification

1. ***Daphnia sp.*** : See Chapter 1.1.1 Item No. 1
2. ***Moina sp.***

Systematic position :

Phylum	**Arthropoda**
Subphylum	**Mandibulata**
Class	**Crustacea**
Subclass	**Branchiopoda**
Type	***Moina sp.***

Distinctive features :

1. Body oval shaped.
2. Antennae are branched.
3. Rostrum is in horizontal arrangement.
4. Body terminates into a pair of caudal styles.

3. ***Cyclops sp.*** : See Chapter 1.1.1 Item No. 4
4. ***Cypris sp.***

Systematic position :

Phylum	**Arthropoda**
Subphylum	**Mandibulata**
Class	**Crustacea**
Subclass	**Ostracoda**
Type	***Cypris sp.***

Distinctive features :

1. The body is entirely enclosed within a bivalved carapace.
2. Well developed antennae used in swimming.
3. Gills enclosed within the valves.
4. Head bears four pairs of appendages.
5. Presence of three pairs of thoracic appendages.

5. ***Mysis sp.*** : See Chapter 1.1.1 Item No. 9
6. ***Nauplius larvae*** : See Chapter 1.2 Item No. 9.
7. ***Roifers*** :

Systematic position :

Phylum	**Rotifera**
Specimen	**Rotifer**

Distinctive features :

1. Elongated saccular body which is relatively cylindrical.
2. Body divisible into short anterior region, a large trunk composing the major part of the body and a terminal foot.

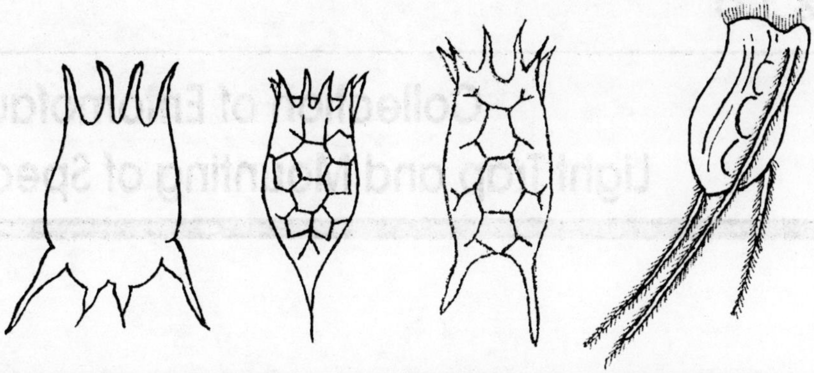

Fig. 8.1.5.2(c) : *Rotifers.*

3. Body covered by cuticle which may be ringed sculptured or ornamented in various ways.
4. The broad anterior region forms the head and bears a ciliated corona. Corona is of various types.

8. ***Nauplius larvae*** : See Chapter 1.2 Item No. 9.

9. ***Zoea larvae*** : See Chapter 1.2 Item No. 10.

10. ***Culex larvae*** : See Chapter 1.2 Item No. 13.

11. ***Anopheles larvae*** : See Chapter 1.2 Item No. 14.

Chapter 8.1.6

Collection of Entomofauna by Light Trap and Mounting of Specimens

Introduction

Collection of various insects by light trap is a conventional method. Many nocturnal insects are attracted towards a light source and show phototactic behaviour, such insect activity is utilized to trap them through a light source.

Light traps are used to trap and kill insect pests of crops. Apart from this light trap are also used to know about abundance and density of various insect species of a particular type.

Materials required

1. **A light trap :**

Structure of a simple light trap : Structure of the light trap (Figure 8.1.6.1) is of very simple in nature. A 100 watt electric bulb is fitted over a 45 cm diameter aluminium funnel. The bulb hangs 15 cm inside the funnel. A glass container is attached to the lower edge of the funnel. In this container tetrachloroethane (preservative) saturated chalk pieces are placed. Tetrachloroethane helps in killing the insect, which gets trapped in the light trap.

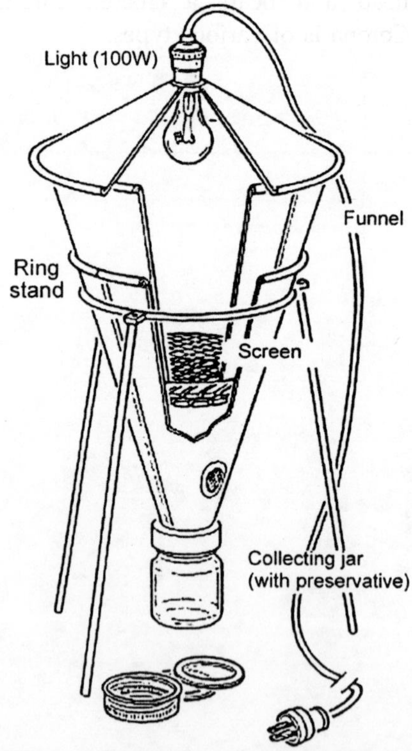

Fig. 8.1.6.1 : Light trap for collection of entomofauna.

Procedure

A light trap is placed in pest infested cropland or forest or grassland or any other terrestrial ecosystem. The lamp of the lighttrap is switched on in the evening; if electricity is not available hazack light is used. During rain the light trap is kept under a shade.

Observation and Result

Large number of insects gets attracted towards light trap such as gryllids, grasshoppers, moths, mantids, beetles etc.

Various insects being attracted to lights get trapped and killed and are then collected and mounted through an insect pin and labelled and fixed a cork for display. Each species and number of individuals in each species counted for each trapping schedule. This is done seasonally to find out insects variability and peak period of population etc. The data are exhibited in a table (Table 8.1.4) as shown below.

Table 8.1.4 : Seasonal incidence of entomofauna collected through light trap

Sl. no.	Species	No. of individuals/ Species in each seasons				Total no. of individuals per species in a year
		1	2	3	4	
1.						
2.						
3.						
4.						
5.						
6.						
7.						

N.B : *Season* : 1. Spring, 2. Summer, 3. Monsoon, 4. Autumn, 5. Winter.

Chapter 8.1.7

Study of Biotic Community of Water

Introduction

A **community** is a local association of several populations of different species. The plant community and the animal community which occur together in the same habitat constitute biotic community.

Biotic community in fresh water can be broadly divided into:

1. **Pelagic biotic community** : This community include the floating forms both the planktons and nektons.
2. **Benthic biotic community** : These are the bottom dwelling ones or remain attached to it. The term Benthos is used to describe those organisms living on or in the bottom of water mass or closely related to it. This includes a range of invertebrate species. The benthic system is characterized by its direct relationship to the type of substrate.

Materials required

1. Deeep water samplers (grab/dredge).
2. 4% to 5% neutral formalin or 70% alcohol.
3. Watch glass.
4. Scooping device.
5. Microscope.

Procedure

1. **Pelagic biotic community :**
   - These include the zooplanktons which are collected through plankton net (refer 8.1.6) and also nektons which can be collected through net.
   - The organisms are fixed in 4% neutral formalin or 70% alcohol.
   - All fixed samples are placed in a watch glass and the microarthropods are examined only under the microscope (*Daphnia*, *Cypris*, Mosquito larvae etc.)
2. **Benthic biotic community :**
   - Benthic biotic communities are collected by different method depending upon the depth of substratum. Small animals living on and in the bottom sediments may be collected by any container or by a scooping device. Deeper water samples can be collected by various types of bottom samples (dredge, grab, etc.).

- The materials are collected.
- The materials are washed in water and sieved.
- The organisms are taken in an enamel tray and sorted out.
- The organisms are fixed in 5% formalin.
- The various animals are examined and characterised.

Identification

- **Pelagic biotic community :**

1. ***Hirudinaria sp.*** : See Chapter 1.1.1 Nonchordates: Annelida, Item no. 9.

2. ***Glossiphonia sp.*** :

Systematic position :

Phylum	**Annelida**
Class	**Hirudinea**
Order	**Pontobdella**
Type	***Glossiphonia sp.***

Distinctive features :

1. Body divided by constrictions usually by large number of segments.
2. Prsence of suckers at both ends.
3. Body not divided externally into trachelosome and urosome.
4. Usually flattened and much wider than head.
5. Head sucker not freely expanded.
6. Complete somites mostly 3-annulate.

3. ***Daphnia sp.*** : See Chapter 1.1.1 Nonchordates : Arthropoda, Item no. 1.

4. ***Moina sp.*** : See Chapter 8.1.5, Item no. 2.

5. **Rotifer** : See Chapter 8.1.5 Item no. 7.

6. ***Triops sp.*** : See Chapter 1.1.1 Nonchordates; Arthropoda, Item no. 2.

7. ***Cyclops sp.*** : See Chapter 1.1.1 Nonchordates; Arthropoda, Item no. 4.

8. ***Mysis sp.*** : See Chapter 1.1.1 Nonchordates; Arthropoda, Item no. 9.

9. **Nauplius larvae** : See Chapter 1.2 Nonchordates: Arthropoda, Item no. 9.

10. ***Cypris sp.*** : See Chapter 8.1.5, Item no. 4.

11. **Zoea larvae** : See Chapter 1.2 Nonchordates; Arthropoda, Item no. 10.

12. **Culex larvae** : See Chapter 1.2 Nonchordates; Arthropoda, Item no. 13.

13. **Anopheles larvae** : See Chapter 1.2 Nonchordates; Arthropoda, Item no. 14.

14. **Amphipod :**

Systematic position :

Phylum	**Arthropoda**
Subphylum	**Mandibulata**
Class	**Crustacea**
Subclass	**Malacostraca**
Order	**Amphipoda**
Type	***Gammarus sp.***

Distinctive features :

1. Body is bilaterally symmetrical. Divided into head, thorax and abdomen.
2. First thoracic segment fused with head and is without carapace.
3. Head contains antennules, antennae, mandibles, maxillae and maxillulae.
4. Body slightly curved, second antenna not being bent when at rest. It swims sideways.

15. **Isopods (*Ascellus sp.*) :**

Systematic position :

Phylum	**Arthropoda**
Subphylum	**Mandibulata**
Class	**Crustacea**
Subclass	**Malacostraca**
Order	**Isopoda**
Type	***Ascellus sp.***

Distinctive features :

1. Body divisible into head thorax and abdomen.

2. Carapace absent. Body dorsoventrally flattened.
3. Antenna two pairs and jaws three pairs.
4. Presence of cephalothorax, absence of dorsal carapace.
5. Compound eyes sessile.
6. Second antenna at least three times long as the first one.
7. Seven pairs of walking legs.
8. First thoracic limbs modified as mouth parts.
9. Abdominal limbs branched and leaf like, constantly in motion serving as gills.
10. Eggs carried under bases of thoracic limbs.

16. *Nepa sp.* :

Systematic position :

Phylum	**Arthropoda**
Subphylum	**Mandibulata**
Class	**Insecta**
Subclass	**Pterygota**
Order	**Hemiptera**
Type	***Nepa sp.***

Distinctive features :

1. Body dorsoventrally compresed, divisible into head, thorax and abdomen.
2. Head with three jointed antenna.
3. Mouth parts form long rostrum adapted for piercing and sucking.
4. Presence of respiratory tube and consists of two spine like processes.
5. Three pairs of false spiracles.
6. Forewings not uniform, proximal portion hardened and distal portion membranous. At rest wing overlap each other.
7. Just below the thorax there is a scutellum.

17. *Gerris sp.* :

Systematic position :

Phylum	**Arthropoda**
Subphylum	**Mandibulata**
Class	**Insecta**
Subclass	**Pterygota**
Order	**Hemiptera**
Type	***Gerris sp.***

Distinctive features :

1. Antenna conspicuously longer than head. Animals dwelling on surface of water in open or among vegetations.
2. Legs particularly the posterior legs inserted towards or at the lateral margins of the thorax. Middle legs inserted nearer to the posterior one than the anterior ones.
3. Posterior femura extending well beyond the tip of the abdomen.

18. *Dytiscus sp.* :

Systematic position :

Phylum	**Arthropoda**
Subphylum	**Mandibulata**
Class	**Insecta**
Subclass	**Pterygota**
Order	**Coleoptera**
Type	***Dytiscus sp.***

Distinctive features :

1. Body covering hard and chitinous.
2. Head sunk in the thorax and eyes not protruding so that the front end has a smooth out line.
3. First visible abdominal sternite completely divides at the middle by hind coxal cavities.
4. Never squeak when caught.
5. Scutellum concealed or exposed.

19. ***Belostoma*** ***sp.*** : See chapter 1.1.1 Nonchordates; Arthropoda Item no. 34.

20. ***Lymnaea* sp.** : See chapter 1.1.1 Nonchordates; Mollusca Item no. 11.

21. ***Planorbis* sp.** : See chapter 1.1.1 Nonchordates; Mollusca Item no. 12.

- ***Benthic biotic community :***

1. ***Tubifex* sp.** : See chapter 1.1.1 Nonchordates; Annelida Item no. 8

2. ***Chironomid larvae* :**

Systematic position :

Phylum	**Arthropoda**
Subphylum	**Mandibulata**
Class	**Insecta**
Subclass	**Pterygota**
Order	**Diptera**
Type	***Chironomus* sp.**

Distinctive features :

1. Larvae dumb bell shaped.
2. Head capsule lacking labral fans.
3. Abdominal segments 5-8, not swollen. End of the abdominal segments not equipped with ring of hooks, extremely diverse in habitat.
4. Colour varies from bright red to pinkish yellow and green.

3. ***May fly larva* :**

(a) ***Coenis* sp. :**

Systematic position :

Phylum	**Arthropoda**
Subphylum	**Mandibulata**
Class	**Insecta**
Subclass	**Pterygota**
Order	**Ephemeroptera**
Type	***Coenis* sp.**

Distinctive features :

1. Sides of the abdomen with plate like, feather like or leaf like gills, usually with three terminal filaments.
2. Tarsal claws single.
3. Older larvae show developing wing pads.
4. First pair of gills reduced to tapering filaments, other gills stalked under large pair of gill filament.
5. Larvae found in mud and vegetation debris in still water and flowing water.

(b) ***Ephemera* sp. :**

Systematic position :

Phylum	**Arthropoda**
Subphylum	**Mandibulata**
Class	**Insecta**
Subclass	**Pterygota**
Order	**Ephemeroptera**
Type	***Ephemera* sp.**

Distinctive features :

1. Sides of the abdomen with plate like, feather like or leaf like gills, usually with three terminal filaments.
2. Tarsal claws single.
3. Older larvae show developing wing pads.
4. Gills feathery, consists of two branchiaes each thickly fringed with filaments.
5. Gills held over back during life.
6. Mandibles project beyond front of the head.
7. Larvae burrow in gravel, sand or mud in still and flowing water.

5. ***Nymphs of Odonates* :**

Systematic position :

Phylum	**Arthropoda**
Subphylum	**Mandibulata**
Class	**Insecta**
Subclass	**Pterygota**
Order	**Odonata**
Family	**Anisoptera**
Type	**Dragon fly nymph**
Family	**Zygoptera**
Type	***Damsel fly nymph***

(a) **Damsel fly nymph :**

1. Abdomen terminating in three caudal lamellae, longest more than one third length of abdomen.

(b) **Dragon fly nymph :**

1. Labium prominent, extensible forming a food capturing structure larger than head.

2. Abdomen terminating in three to five stiff pointed valves, longest less than one third of abdomen.

6. ***Lamellidens sp.*** : See chapter 1.1.1 Nonchordates; Mollusca Item no. 15.

See Figure 8.1.7.1 for pelagic and benthic animals of pond.

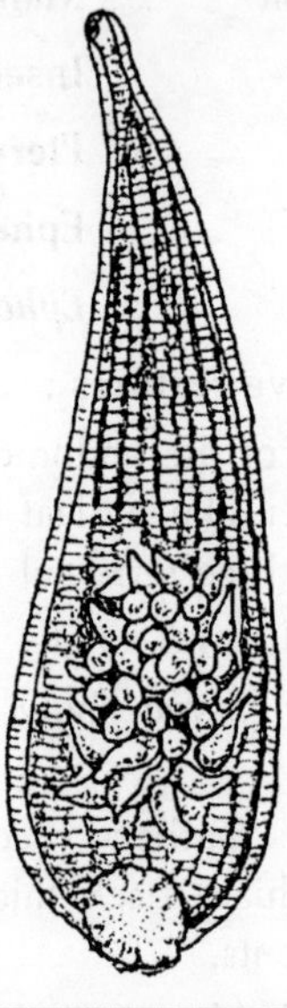

(a) : *Glossiphonia sp.*

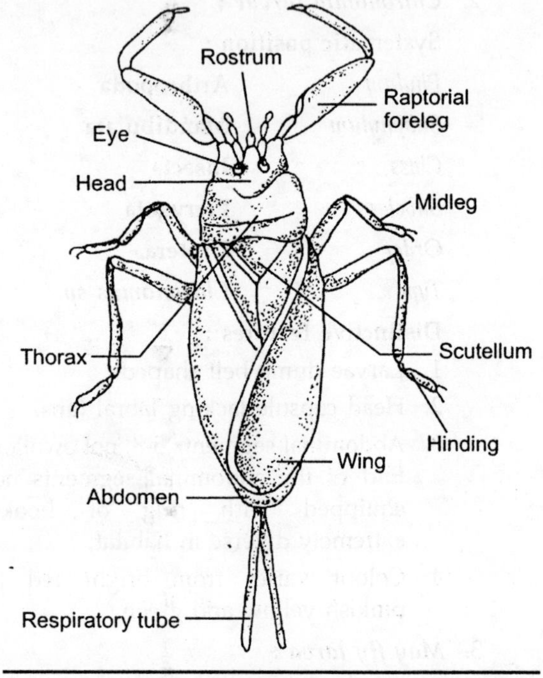

(c) : *Nepa sp.*

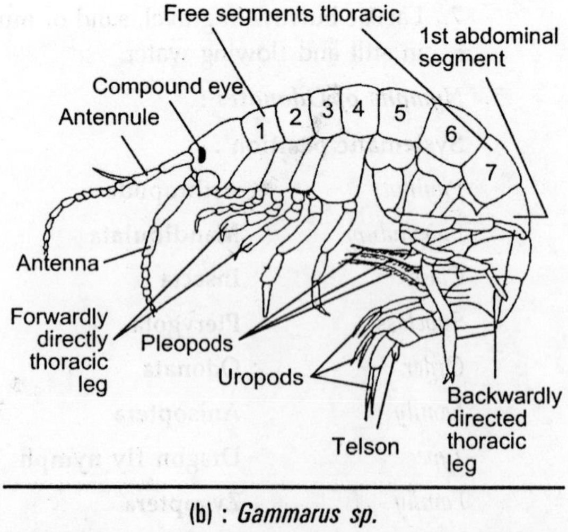

(b) : *Gammarus sp.*

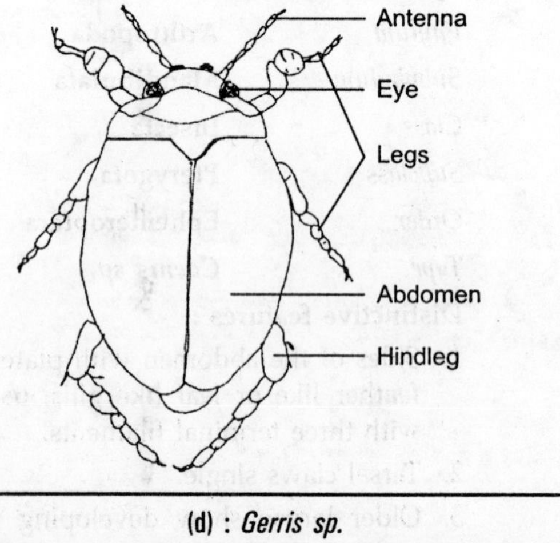

(d) : *Gerris sp.*

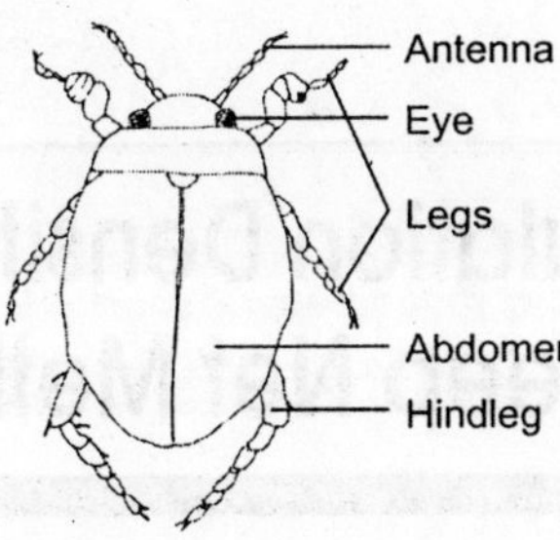

(e) : *Dystiscus sp.*

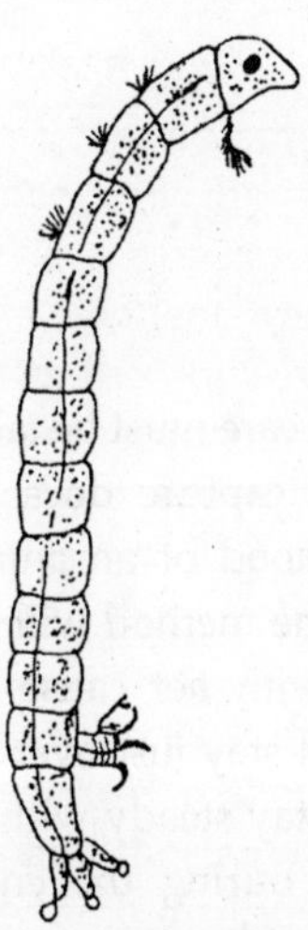

(f) : *Chironomid larvae.*

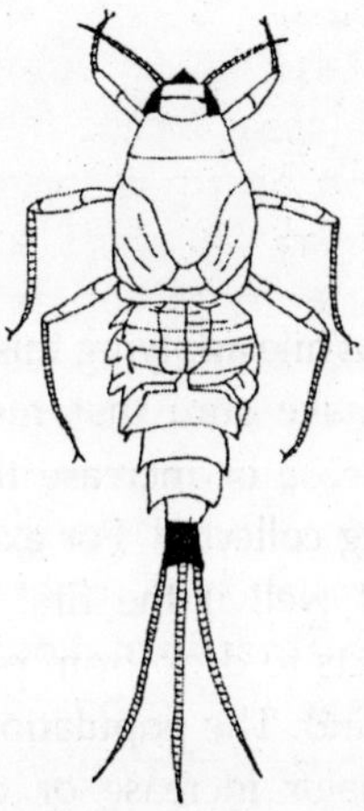

(g) : May fly *(Coenis sp.)*.

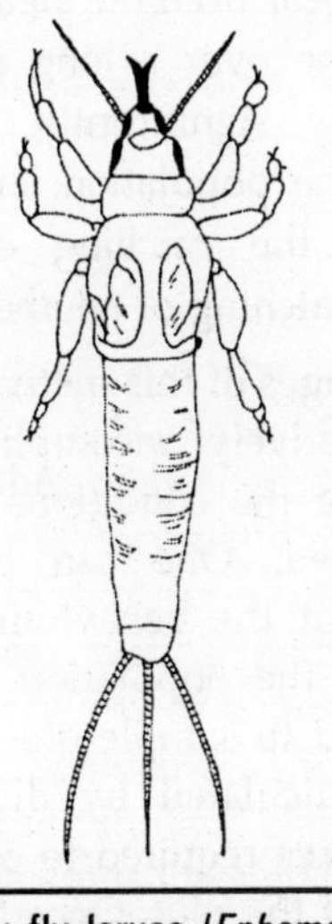

(h) : May fly larvae (*Ephemera sp.)*.

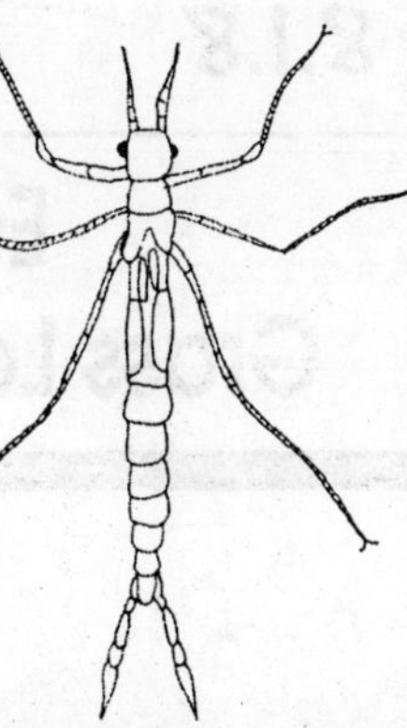

(i) : Damsel fly nymph.

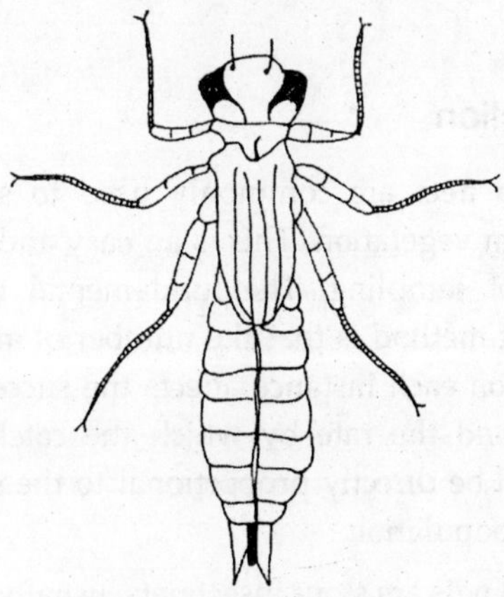

(j) : Dragon fly nymph.

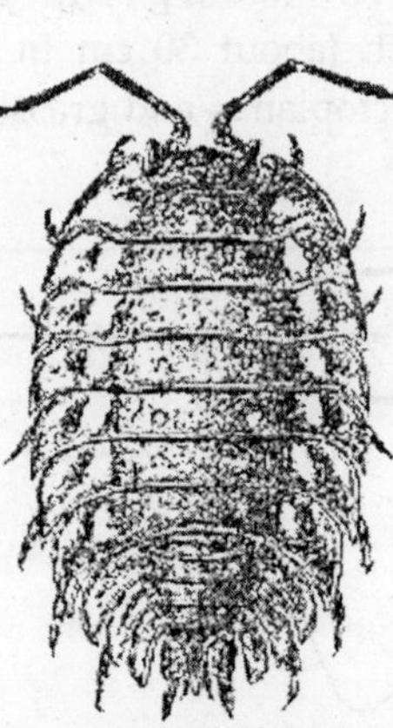

(k) : Isopod.

Fig. 8.1.7.1 : Biotic faunal community of fresh water lentic ecosystem.

Chapter 8.1.8

Estimation of Population Density of Grass Land Insect by Sweep Net Method

Introduction

Sweep nets are commonly used to sample insect from vegetation. This is an easy and rapid method of sampling. The fundamental to this sweep net method is that the number of animals removed on each instance affects the succeeding removal and the rate by which the catches go down will be directly proportional to the size of the total population.

Sweep nets are stout insect nets, usually made of linen, study cotton or synthetic material (Figure 8.1.8.1). The mouth generally round but a D-shaped mouth (about 30 cm in diameter) is ideal for use in croplands and grasslands or over short vegetation.

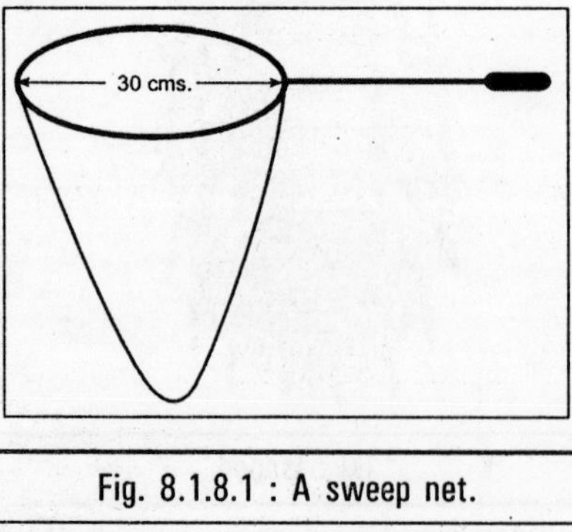

Fig. 8.1.8.1 : A sweep net.

While applying this method care must be taken to make sure that method of capture does not decrease or increase the likelihood of an animal being collected. For example, the method will not hold well if the first sweep with net cause the insects to drop from plants and stay just over the ground. The population must stay steady without a major increase or decrease during the entire phase of sampling. The sampling procedure should not disturb the animals to a degree such that they disappear from the area. Similarly, when trapping is done over a long period, the local population may significantly be changed by immigration. The population should not be too large so that the catching of one member influences the catching of another.

The advantages of this method are that, apart from being relatively uncomplicated and fairly precise, provide the conditions laid out earlier are met obliged. One can also understand something about the behaviour of the various components of the population. The number of sweeps required to sample one square metre or unit area is calculated by dividing the total number of strokes required to cover the area by the total area of the plot. Once this is found out (total number of stokes/total number of Quadrat) it is possible to sweep any area, without first measuring since the area sampled could be computed by dividing the total number of strokes used in the plot by the number of strokes needed to cover one square meter.

The drawbacks are that only insect which does not fall off or fly away when the collector approaches the vegetation will be caught. Alteration in the vertical distribution of the species, weather conditions, the diel cycle of vertical movement as well as changes made by the sweep net affects the procedure. Moreover, a sweep net cannot be successfully used on very short (grass) or very tall (tree, samplings) vegetation.

Materials required

(1) Sweep net.

(2) Quadrat sampling materials (see section 8.1.1).

(3) Killing jar.

(4) Graph paper.

Procedure

1. Fairly large Quadrat in one's study area is marked off using pegs and ropes.

2. The vegetation is swept by quickly moving the net through the plants while crisscrossing the quadrat at slow pace.

3. The animals collected from each sweep are removed and dropped into a killing jar. The number of strokes made with the net to cover the entire Quadrat is recorded for the first time.

4. All the captured and killed animals are counted and recorded and put in a bottle. Bottle is labelled as "first removal sample".

5. The process is repeated using the same number of strokes, the catch is bottled separately and their number is recorded. The total number of individual captured will reduce with every successive sampling.

6. The data is plotted on the graph with the number of captured insects on the Y-axis and the number of previously removed insects on the X-axis. The points will fall more or less in a straight line, if the probability of capture had remained constant. If the procedure is not working the points will not produce a straight line graph.

7. The number of sweeps necessary to sample 1 sq. m or unit area is obtained by dividing the total number of strokes needed to cover the area of the plot. Once this is determined i.e. **total number of strokes/total area of quadrats** it is feasible to sweep any area, without first measuring it, as because the area sampled could be computed by dividing the total number of strokes used in the plot by the number of strokes needed to cover 1 sq. m area.

Observation and Result

Observation and results are tabulated below :

Table 8.1.5 : Tabulated results of sweepnet method

Sl. No.	Species	Number of sweeps (X) required to cover one quadrat	Area of quadrat (Y m)	No. of quadrats 1,2,3,4...	Total no. of individuals/ species (n) in one quadrat	Total no. of individuals/m^2

Chapter 8.1.9

Extraction and Identification of Soil Microfauna (Soil Microarthropods only) through Tullgren Funnel

Introduction

Soil fauna is both plentiful and diverse. On the larger end of the scale, there are often 250,000 arthropods (insects and their relatives) in a single square yard of soil literally hundreds of species reside. Most of soil animals extracted by a Tullgren funnel are so-called microarthropods, arthropods with a body size of 0.1-5.0 mm. These animals are unable to dig their own way in soil; they inhabit soil crevices, pores, and hollows created by larger animals. As they consume plant and animal residues and dead organisms, graze on soil fungi and bacteria, and produce faecal pellets, these tiny arthropods promote the production of humus in the soil and aid in preserving soil structure. By scattering microbial propagules, they aid in the microbial colonisation of fresh litter material. After death they leave important nitrogenous waste. Through interactions with fungal and bacterial microflora and gives rise to a complex food web along with other soil fauna, these animals regulate decomposition rate, affect nutrient cycling and play an important part in soil fertility. Major soil microarthropods includes the mites, spring tails, termites, beetles, beetles larvae. For proper identification and other laboratory-based studies with these microarthropods it is the primary requirement to extract and identify these microarthropods. Extracting arthropods through their behavioural dynamics (positive geotaxis in response to dryness) easily does this. In this method, the animals are stimulated to leave the substratum(soil) through the application of heat.

Materials required

1. Tullgren Funnel (See Apparatus section).
2. Wooden stand.
3. Light arrangements.
4. Beaker.
5. 70% Methanol.
6. Soil corer.

Apparatus

Tullgren funnel (Fig. 8.1.9.1) is constituted of large funnel made from tin, inside which rests a fine wire sieve (mesh size; 6 square/sq.cm). Under the stem of the funnel a small beaker half filled with 70% ethanol is placed, used for extraction and collection of soil microarthropods.

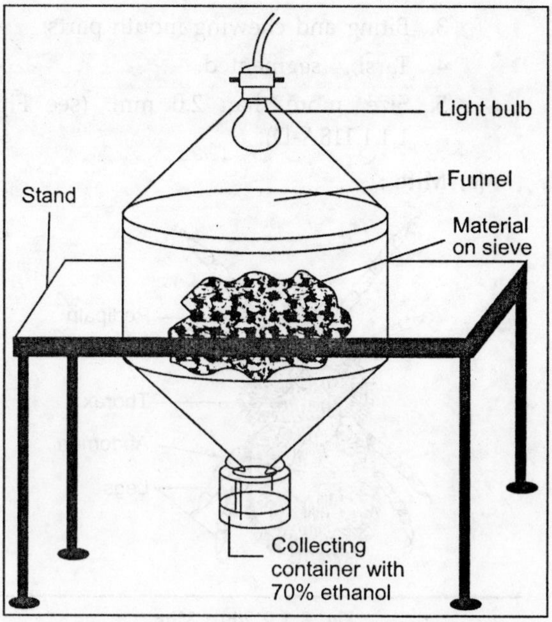

Fig. 8.1.9.1 : Tullgren funnel on a stand

A 60 watt bulb is placed about 15-20 cm above the funnel. All these are fitted to a wooden stand. More than one funnel along with associated structures can be used depending upon the volume of soil used for extraction.

Procedure

1. Soil samples are collected from a locality through a soil corer.

2. Collected sample is placed on the sieve that is fitted in funnel.

3. Electric bulb switched on and the temperature is noted. The system is left illuminated for 24 hours.

4. After few hours of illumination, the soil sample gets heated as result the soil inhabiting arthropods start moving away from direct heat downwards and fall in the collecting containers having 70% ethanol.

5. After complete extraction different types of microarthropods are taken in a watch glass and are segregated under a simple microscope.

6. Number of each type specimen is counted and is taxonomically identified.

7. This analysis helps to prepare a comparative analysis of soil microarthropods contents of various soil sample collected from various locations.

8. A set up of four (or may be more) are operated at a time.

Results

Table 8.1.6 : Extraction of soil microfauna through tullgren funnel

Species	Individual species extracted in funnel set up				Total number of individuals per species in all funnel (1+2+3+4)
	1	2	3	4	

Identification of soil microarthropods

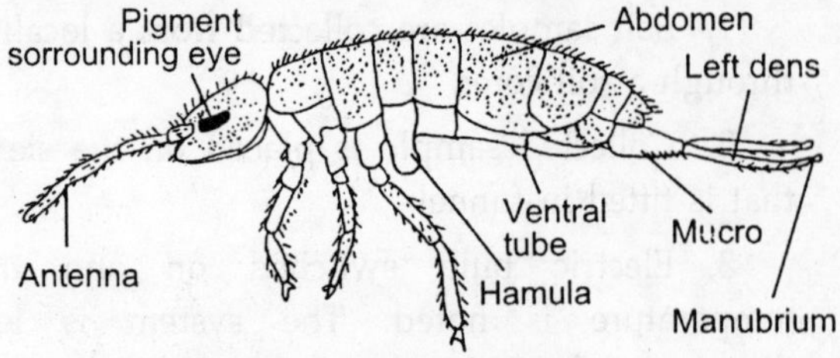

Fig. 8.1.9.2(a) : Collembola

(a) **Collembola or Spring tail :**

Systematic position :

Phylum	**Arthropoda**
Subphylum	**Mandibulata**
Class	**Insecta**
Subclass	**Collembola**
Specimen	***Collembola sp.***

Identifying characters :

1. Abdomen six segmented.
2. Usually three sets of appendages, maintaining a hook on 3rd segment that functios as powerful springing organ.
3. Size, ranges between 0.5-5.0 mm. (see Figure 8.1.9.2a).

(b) **Termite :**

Systematic position :

Phylum	**Arthropoda**
Subphylum	**Mandibulata**
Class	**Insecta**
Subclass	**Pterygota**
Order	**Hymenoptera**
Specimen	**Termite**

Identifying characters :

1. Contains a number of castes viz., winged fertile male and females and wingless sterile workers and soldiers.
2. Wings when present are long, similar, membranous and capable of being shed at basal fracture.
3. Biting and chewing mouth parts.
4. Tarsi, 4 segmented.
5. Size, more than 2.0 mm. (see Fig. 1.1.1.118A-D).

(c) **Mite :**

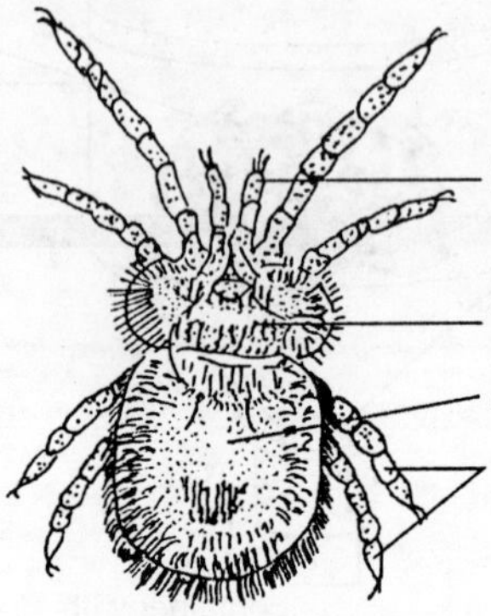

Fig. 8.1.9.2(b) : Mite

Systematic position :

Phylum	**Arthropoda**
Subphylum	**Mandibulata**
Class	**Arachnida**
Subclass	**Acarina**
Specimen	**Mite**

Identifying characters :

1. Unsegmented usually ovoid flattened body.
2. Six pairs of appendages.
3. Chelicerae are chelate.
4. Pedipalpi are leg like with 5 joints or less.
5. Usually four pairs of legs.
6. Size, about 1.0 mm. (see Fig. 8.1.9.2b).

(d) **Ants :**

Systematic position :

Phylum	**Arthropoda**
Subphylum	**Mandibulata**
Class	**Insecta**
Subclass	**Pterygota**
Order	**Hymenoptera**
Specimen	**Ant**

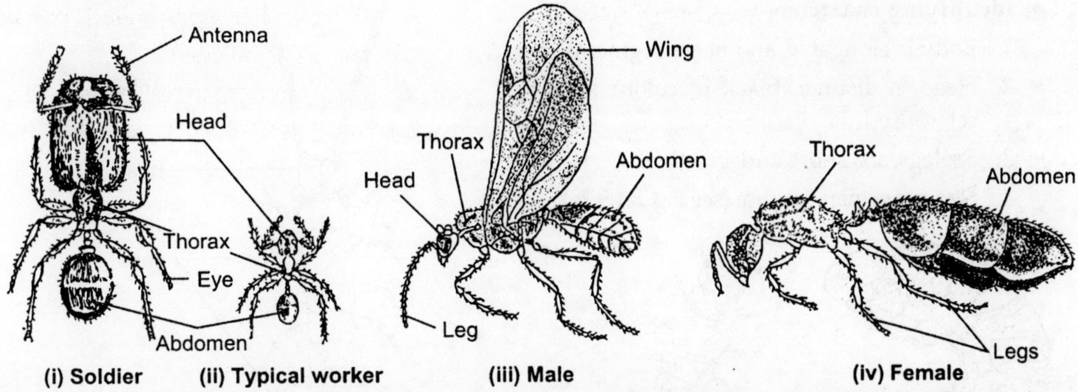

Fig. 8.1.9.2(c) : Ants

Identifying characters :

1. Contains members of castes that consist of winged fertile males and dealated fertile females and wingless sterile workers and soldiers.
2. Males have well developed genitalia and with smaller and rounded head.
3. The female bears a large gaster and well developed reproductive organs.
4. Workers and soldiers have well developed mandible and small gaster.
5. Junction between the abdomen and the thorax is very narrow and constricted.
6. Size, 2.5-8.0 mm. (see Figure. 8.1.9.2c).

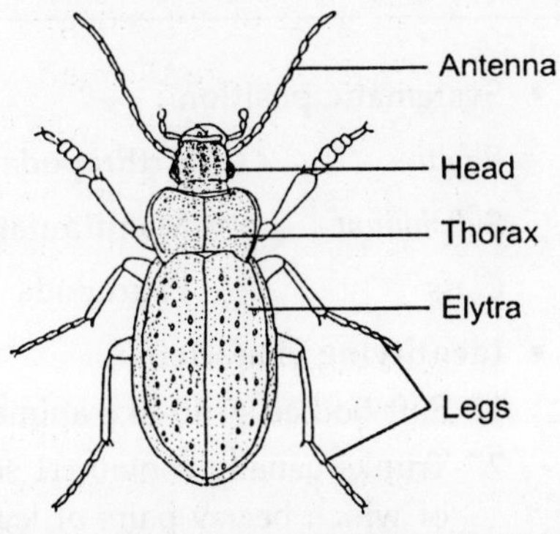

Fig. 8.1.9.2(d) : Beetle

(e) **Beetles :**

Systematic position :

Phylum	**Arthropoda**
Subphylum	**Mandibulata**
Class	**Insecta**
Subclass	**Pterygota**
Order	**Coleoptera**
Specimen	**Beetle**

Identifying characters :

1. Fore wings are modified into horny or leathery elytra.
2. Hind limbs membranous.
3. Bitting mouth parts.
4. Tarsi, 3 or 5 segmented.
5. Size, more than 2 mm. (see Fig. 8.1.9.2d).

(f) **Beetle larvae :**

- **Systematic position :**

Phylum	**Arthropoda**
Subphylum	**Mandibulata**
Class	**Insecta**
Subclass	**Pterygota**
Order	**Coleoptera**
Specimen	**Beetle larvae**

- **Identifying characters :**
  1. Body is elongated and bears segments.
  2. Head is distinct, black in colour and sclerotised.
  3. Prolegs are short and inactive.
  4. Size, more than 3.0 mm. (see Fig. 8.1.9.2e).

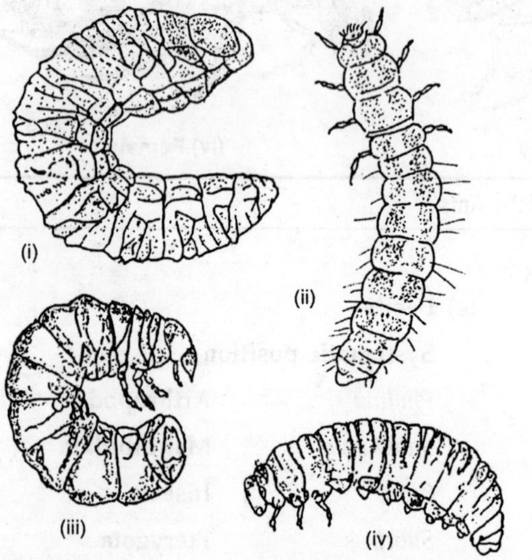

Fig. 8.1.9.2(e) : Beetle larvae. (i) Campodeiform (Cucujidae); (ii) Eruciform (Chrysomelidae); (iii) Scarabaciform (Scarabaeidae) and (iv) Apodous (Curculionidae)

(g) **Symphyllid :**

- **Systematic position :**

Phylum	**Arthropoda**
Subphylum	**Mandibulata**
Class	**Symphyla**

- **Identifying characters :**
  1. Body is divisible into head and trunk.
  2. Antenna laterally placed, unbranched.
  3. Trunk contains 12 leg bearing segments.
  4. Last segment contains a pair of spinnerets or cerci and a pair of sensory hairs.
  5. Presence of 12 pairs of legs.
  6. Size, 1.0-8.0 mm. (see Fig. 8.1.9.2f).

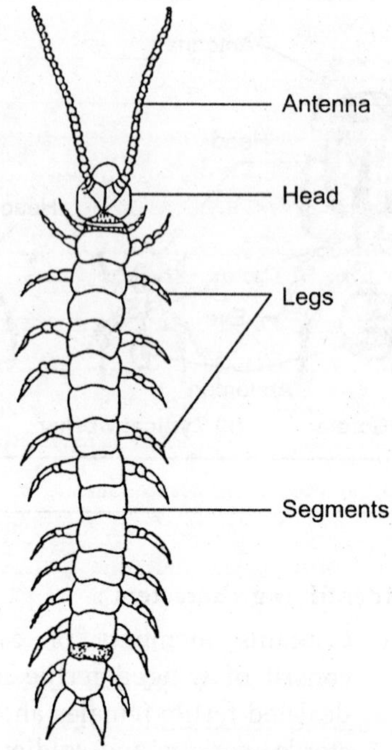

Fig. 8.1.9.2(f) : Symphyllid

(h) **Pauropoda :**

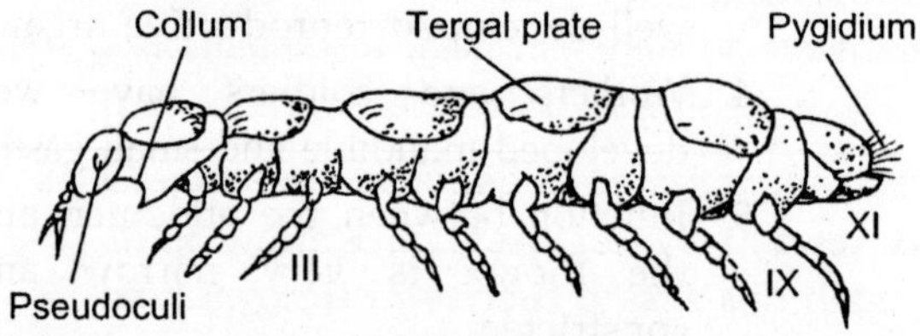

Fig. 8.1.9.2(g) : Pauropoda

- **Systematic position :**

Phylum	**Arthropoda**
Subphylum	**Mandibulata**
Class	**Pauropoda**

- **Identifying characters :**
  1. Soft bodied grub like animals.
  2. Trunks usually contain 11 segments, 9 of which bear 9 pairs of legs.
  3. Last segment has a telson.
  4. Antennae are two branched.
  5. Size, 0.5-2 mm. (see Fig. 8.1.9.2g).

Chapter 8.1.10

Hay Culture of *Paramoecium sp.*

Introduction

Paramoecium is commonly found in fresh water ponds, pools, ditches, lakes etc. It is specially found in abundance in stagnant ponds rich in decaying matter, in organic inclusions and in sewage. *Paramoecium* multiply rapidly in culture media.

Materials required

1. Hay Pieces.
2. Wheat Grains.
3. Beaker.
4. Tripod.
5. Bunsen burner.
6. Glass plates.
7. Glass jar.

Procedure

1. A beaker filled with 500 ml of distilled water together with some hay and 15-20 grains of wheat and then boiled for about 10 minutes. The water is allowed to cool and the beaker is covered with glass plate for about 4-5 days, until the water turns turbid due to bacterial growth.

2. In a glass jar little water along with some submerged weeds is collected from a fresh water pond. The jar is covered with a glass plate and allowed to decay for 1-2 days.

3. (1) is added to (2). The jar is partially covered with a glass plate so that there is enough aeration. Glass jar is then stored in the room temperature (25ºC) in a place having moderate light. *Paramoecium* feed on bacteria and grows very rapidly. After a week long incubation there is sufficient population growth of *Paramoecium* in the hay culture.

Observation and Result

Paramoecium (refer Chapter 1.1.11 for diagram and description) from the culture is sucked out through a pipette on a glass slide and preliminary examinations are made under low power of microscope.

Chapter 8.2.1

Estimation of Moisture Content of Soil

Introduction

Water plays a very important role in soil plant growth. It forms the most important part of the plant itself, necessary for physiological activities, acts as a solvent and nutrient carrier and maintains turgidity of the plants. Actually water is controller of physical, chemical and biological activities in the soil.

Classification of the Soil Moisture

It is known that there are number of different kinds of the soil. For practical purpose, soil water (moisture) is classified on the basis of tension and condition in which it exists in the soil (Table 8.2.1).

Measurement of soil moisture through Gravitation Method

Principle :

Soil sample is weighed and then placed in an oven at 105°C and it is dried to constant weight.

The weight difference is considered to be water present in the soil sample.

Percentage of moisture in soil =

$$\frac{\text{Loss in weight}}{\text{Oven dry weight of soil}} \times 100$$

Materials required :

(1) Sample auger.

(2) Moisture cans (numbered).

Table 8.2.1 : Classification of soil and physical properties

Soil water	Tension (pF)	Bars or Atmosphere	Potential (Mpa)
Capillary Water	4.5 to 2.5	-0.3 to –3.1	-0.03 to -3.1
Gravitational Water	2.5 to Zero	Zero to –0.3	Zero to -0.03
Ground Water	Tension Free	Zero	Zero
Hygroscopic Water	7.0 to 4.5	-3.1 to -10000	-3.1 to -1000
Water of Constitution or interlayer water	Above 7.0	Above -10000	Above -1000

(3) Drying oven.

(4) Desiccator.

Procedure

1. The empty moisture can is weighed.

2. About 100g of soil sample is taken from the required depth with the help of an auger.

3. Soil sample is placed immediately in the moisture can and the can is closed to prevent loss of moisture by evaporation.

4. Can containing the moist soil is brought to the laboratory and weighed immediately.

5. The lids are removed and the moisture cans are placed in oven to constant weight at 105°C. This takes approximately 46 hours.

6. The sample is allowed to cool for some time in oven. The cans are closed and are put into a desiccator for further cooling. The closed cans are weighed with the oven dry soil.

Observations

Wt. of empty moisture can = (x)g

Wt. of moisture can + moist soil = (y)g

Wt. of moisture can + oven dry soil = (z)g

Calculations

Moisture content in soil = $(y–z)$g

Weight of oven dry soil = $(z–x)$g

Percentage moisture in the soil = $\frac{y-z}{z-x} \times 100$

Note : An error can result from the oxidation of organic matter.

Results

Three such observations are taken into consideration and tabulated below.

Table 8.2.2 : Mean soil moisture

No. of Experiments	% of soil moisture	Mean soil moisture (%)
1.		
2.		
3.		

Chapter 8.2.2

Measurement of pH of Soil

Introduction

The pH value of soil is an estimate of the hydrogen ion activity of the soil water system and expresses the acidity and alkalinity of soil. The pH is a very essential property of soil as it establishes the availability of nutrients, microbial activity and physical condition of soil.

The pH of a solution has been defined as the negative logarithm of the hydrogen ion concentration, which in very dilute solution may be expressed as concentration, in gram mole per litre.

$$pH= -\log_{10}[H^+]$$

$$\text{or } [H^+] = 10^{-pH}$$

Note : In a concentrated solution, the ionic activity of ions has a propensity to suppress owing to high salt concentration. Thus in this solution the activity of ions is not in the accord of concentration in which they are present, that is why in definition of pH the word concentration is replaced by the word activity

$$pH= -\log_{10}[aH^+]$$

Where aH^+ is the hydrogen ion activity in g ion/L.

Measurement of soil pH by Electric pH Meter Method

The instrument generally used in this method is glass electrode pH meter with calomel reference electrode with a salt bridge. Modern digital pH meter at present have single (combined) electrode assembly. The instrument having a potentiometer circuit requires to be calibrated prior to use with buffer solutions of known pH values.

Principle

A glass surface in contact with hydrogen ions of the test solution (analyte) acquire an electric potential which depends upon the concentration of H^+ ions, thus is an reflection of electrical potential (emf). This in turn gives the value of H^+ ion concentrations i.e., pH of a solution.

Materials required

1. Glass electrode pH meter.
2. *Standard buffer solution:* These may be of pH 4.0, 7.0 or 9.2 in pure water. To prepare buffer solution, in case of buffer tablets (commercially available) a single tablet is dissolved in freshly

prepare double distilled (ddH_2O) water and volume is made up to 100 mL. It is necessary to prepare a fresh buffer more or less after every few days. In case a standard buffer is not obtainable, a saturated solution of potassium hydrogen tartarate (AR) may be used which gives a *p*H of 3.56 at 25°C.

3. Glass rod.
4. Beaker.
5. Double distilled water.

Procedure

• **Preparation of 1:2 soil water suspension :** To prepare 1:2 soil water suspensions, 40 g of soil is weighed into 250 mL of Erlenmeyer flask, added 80 mL of distilled water into it. The flask is stoppered and shaken on a reciprocating shaker for one hour.

• **pH Measurement :**

1. 1:2 soil water suspension is taken of which *p*H is to be determined.
2. On the *p*H meter, the temperature compensating knob is set and it is to be confirmed that the electrode is completely filled with saturated potassium chloride (KCl) solution. The *p*H meter is allowed to warm up for 15 minutes to eliminate asymmetric potential of the instrument.
3. Known standard buffer solution is placed in a beaker (say, *p*H 7), both the electrode and the single electrode (in case combination electrodes is provided) are immersed into the buffer solution. Electrode should not touch the wall of the beaker. With the help of knob the instrument reading is adjusted at the known *p*H of buffer (in this case at *p*H 7). The buffer is then removed and the electrodes are carefully cleansed with the distilled water. Another buffer solution of known *p*H say 9.2 is taken and reading is seen after immersing electrodes in it. The *p*H meter should read 9.2 (if, reading does not approach 9.2, the instrument should be readjusted by repeating above procedure). The second buffer is then removed and electrodes are again cleansed with distilled water.
4. The electrodes are then immersed in the beaker containing soil water suspension, and the *p*H is read on the dial (or digital panel) and recorded.
5. The electrodes from the soil suspension are removed and washed with ddH_2O and then dipped into a beaker of ddH_2O. The electrodes are maintained in working conditions by keeping them immersed in distilled water.

Results

Table 8.2.3 : Mean *p*H reading of soil sample measured through pH meter

No. of Experiments	pH value	Mean pH	Inference
1.			3-5 : Highly acidic 5-6.99 : Acidic 7.0 : Neutral 7.01-9.0 : Alkaline Above 9.0 : Highly alkaline
2.			
3.			

Chapter 8.2.3

Estimation of Soil Organic Carbon

Introduction

One of the most extensively used rapid soil analysis for the evaluation of available carbon is based upon the assessment of readily oxidisable organic carbon, which approximately presents roughly about 58% of the soil organic matter, the seat of carbon in soil. A variety of methods are available for the estimation of organic carbon based upon dry combustion and wet digestion. The dry combustion method provide absolute values and helpful for every precise estimation of organic and total carbon. For routine work and easily oxidizable carbon determination most commonly acceptable methods in India are modified Walkley-Black method (Walkley and Black, 1934) and colorimetric method (Datta et. al. 1962). These methods work on one basic principle i.e. wet oxidation (digestion) of organic carbon in an acid dichromate solution chased by back titration of remaining dichromate with ferrous ammonium sulphate or by photometric determination Cr^{3+}.

Walkley and Black rapid titration method

Principle :

Potassium dichromate and concentrated sulphuric acid oxidizes the organic matters within the soil utilizing the heat of dilution of H_2SO_4. The excess amount potassium dichromate that is not reduced by the organic matter of the soil is assessed by back titration with standard ferrous sulphate ($FeSO_4.7H_2O$) or ferrous ammonium sulphate [$FeSO_4.(NH_4)_2SO_4\ 6H_2O$].

Reactions :

$$K_2Cr_2O_7 + 4H_2SO_4 \longrightarrow K_2SO_4 + Cr(SO_4)_3 + 4H_2O + 3O$$

$$C \text{ (Organic Carbon)} + 2O \longrightarrow CO_2$$

$$2FeSO_4 + H_2SO_4 + O \longrightarrow Fe_2(SO_4)_3 + H_2O$$

Reagents required :

1. **Standard 1N potassium dichromate :** 49.04g of $K_2Cr_2O_7$ (AR grade) oven dried at 90°C is dissolved in distilled water and the volume is made up to one litre.

2. **0.5N ferrous ammonium sulphate :** 196g of the hydrated crystalline salt is dissolved in one litre of distilled water having 20mL of concentrated H_2SO_4. The solution is rather more stable and convenient to work than that of the ferrous sulphate, though it should be prepared fresh for each new set of sample.

3. **Diphenylamine indicator :** 0.5g diphenylamine is dissolved in a mixture of 20mL of ddH_2O (double distilled water) and 100mL concentrated H_2SO_4.

- *O-phenanthroline indicator :* Dissolved 3.0 g of O-phenanthroline monohydrate and 1.4g of ferrous sulphate heptahydrate ($FeSO_4.7H_2O$) in water. The solution is diluted to a volume of 200mL. This indicator is available under the commercial name Ferroin. At the end point change in colour is from greenish blue to reddish from brown.
- *N-phyenylanthranillic acid indicator :* Dissolved 0.1g of N-phenylanthranillic acid and 0.1g of Na_2CO_3 in 100mL of water. With this indicator, the colour changes from dark violet green to light green at the end point.

4. Concentrated sulphuric acid (sp.gr.1.84) containing 1.25% silver sulphate (if soil is free from chloride, application of Ag_2SO_4 can be avoided).

5. Ortho-phosphoric acid (85%) and or Sodium fluoride (AR grade).

Procedure :

1. 1.00g soil is taken in dry 500mL conical flask (Corning™/Pyrex™/Borosil make).

2. 10mL of 1N is $K_2Cr_2O_7$ is pipetted in and whirled a little.

3. The flask is kept over an asbestos sheet. Then 20mL of H_2SO_4 (containing 1.25% Ag_2SO_4) is added and whirled again two or three times.

4. The flask is allowed to stand for 30 minutes and then added 200mL of distilled water in it.

5. 10mL of phosphoric acid (85%) or and 0.5 g sodium fluoride and 1mL of diphenylamine indicator is added into it (it is better to use O-phenanthroline or N-phenylantranilic acid instead of diphenylamine indicator).

6. The contents are titrated with 0.5N ferrous ammonium sulphate solution till the colour changes from blue-violet to green.

7. At the same time, a blank is run without soil. If more than 7mL of the dichromate solution is consumed the determination must be repeated with a small quantity (0.25-0.50g) of soil.

Results :

See Table 8.2.4.

Observations :

Weight of soil taken = 'S'g

Volume of 0.5 N ferrous ammonium sulphate required to neutralize 10mL of 1N $K_2Cr_2O_7$ i.e. blank titration (blank reading) = BmL

Table 8.2.4 : Titration of soil with Ferrous ammonium sulphate

No. of experiment	Wt. of soil taken ('S'g)	Volume of ferrous ammonium sulphate (blank titration) (BmL)	Volume of ferrous ammonium sulphate (soil titration) (CmL)	(B-C)mL
1.				
2.				
3.				

Volume of 0.5N ferrous ammonium sulphates needed for titration of soil sample (reading with soil) = CmL

Difference in volumes of ferrous ammonium sulphate used in both titrations = (B – C)mL

Calculations :

Volume of 1N $K_2Cr_2O_7$ used for oxidation of soil carbon = {10(B – C)/B}mL soil carbon.

Here, 1L of 1N $K_2Cr_2O_7$ will be equal to: 12/4g carbon.

Or, 1mL of 1N $K_2Cr_2O_7$ will be equal to: 12/4 × 1/1000g carbon = 0.003g carbon.

(Where, 12 is the atomic wt. of carbon; 4 is the valency of carbon and 12/4=3 is the equivalent weight of carbon).

Therefore, amount of carbon present in 'S'g soil = {0.003 × 10 (B – C)}/'B'g

Amount of carbon present in 1g soil= 0.003 × 10(B – C)/B × S

Amount of carbon present in 100g soil= [{0.003 × 10(B – C)}/B × S] × 100

Percent organic carbon in the soil sample = [0.003 × 10 (Blank reading – Titration reading)/ Blank reading × weight of soil (g)] × 100

Or, alternatively the above formula can also be expressed as : % of carbon in soil= { N × (B – C)/weight of soil (g)} × 0.003 × 100

Where, N = normality of ferrous ammonium sulphate.

As there is an incomplete oxidation of the organic matter in this procedure as a result organic carbon obtained by above method is multiplied by a factor 1.3 based on assumptions that there is 77% recovery.

i.e. Organic carbon = Organic carbon estimated × 1.3.

Chapter 8.2.4(a)

Estimation of Total Alkalinity of Water

Introduction

Total alkalinity is the total concentration of bases in water expressed as parts per million (ppm) or milligrams per liter (mg/L) of calcium carbonate ($CaCO_3$). These bases are usually bicarbonates (HCO_3) and carbonates (CO_3), and they work as a buffer system that checks drastic changes in pH. For instance, in waters with low alkalinity, pH might fluctuate from 6 or lower to as high as 10 or above; while in high alkalinity waters, pH might fluctuate from about 7.5 to 8.5.

Greater production in bass-blue gill ponds is attained in high alkalinity waters because this pH buffering capacity makes phosphorus and other essential nutrients more available to the algae bloom.

Total alkalinity is not the same as hardness. Calcium (Ca^{++}) and magnesium (Mg^{++}) are primarily responsible for hardness. However, in most waters, alkalinity and hardness have comparable values because the carbonates and bicarbonates accountable for total alkalinity are normally brought into the water in the form of calcium carbonate or magnesium carbonate. Waters with high total alkalinity are not always hard, since the carbonates can be brought into the water in the form of sodium or potassium carbonate.

Principle

The measurement of total alkalinity is based upon simple acidimetric titration with the use of varied indicators, which function in pH range (above 8.2) or in acidic range (below 6.0). In a water sample analysis, Total alkalinity = Carbonate + Bicarbonate alkalinity.

Reagents required

1. Methyl orange indicator: 0.5% solution in 95 % ethanol.

2. Phenolphthalein indicator: 0.25 % solution in 60% ethanol.

3. Standard Sulphuric acid (0.02N): 2.8mL of H_2SO_4 (Concentrated) is diluted to 1 litre with distilled water. Further 200mL of this solution is diluted to 1 litre to get 0.02(N) H_2SO_4. The solution is standardized.

Procedure

1. 10mL of water sample is taken into a conical flask.

2. 2-3 drops of Phenolphthalein is added a, pink colour appears indicating the presence of carbonate. This is titrated with $0.02(N)H_2SO_4$ until the colour disappears this is Phenolphthalein indicator end point, where alkali carbonate is converted to bicarbonate. The titrant amount is designated as 'V_1'mL.

3. Next in the colourless solution, 1 to 2 drops of methyl orange indicator is added and titration is continued until colour changes from yellow to rose red. The titrant amount is designated as 'V_2'mL.

The volume of the acid required with Phenolphthalein is equal to 1/2 of $CaCO_3$ and the rest of acid required in the methyl orange is equal to rest of $CaCO_3$ and whole of $Ca(HCO_3)_2$.

Results

See Table 8.2.5(a).

Calculation

No. Total Alkalinity (TA) = $[HCO_3^-] + 2[CO_3^{--}]$ or TA = Phenolphthalein alkalinity + Methyl Orange alkalinity.

1. Volume of acid required to neutralize 1/2 $CaCO_3$ in 100mL = V_1mL.
2. Volume of acid required to neutralize whole $CaCO_3$ in 100mL = $2V_1$mL.
3. Volume of acid required to neutralize whole $Ca(HCO_3)_2$ and $1/2CaCO_3$ = V_2mL.
4. Volume of acid required to neutralize whole $Ca(HCO_3)_2$ = (V_2mL – V_1mL).

Concentration of CO_3^- equivalent to $CaCO_3$

i.e. 100mL of sample

$\equiv 2V_1$mL of 0.02(N) H_2SO_4

So, 1000mL of sample

$\equiv (2V_1 \times 0.02 \times 1000 \times f)/100$

$\equiv f \times 2V_1 \times 0.2$mL of H_2SO_4(N)

$\equiv f \times 2V_1 \times 0.2$mL of $CaCO_3$(N)

Now 1000mL of $CaCO_3$(N) contains 50g of $CaCO_3$ i.e. 1mL of $CaCO_3$(N) contains 50/1000 = 50mg of $CaCO_3$.

Therefore, 0.4 V_1f mL of $CaCO_3$(N) contains 50 × 0.4V_1f = 20V_1f mg.

Thus the **Carbonate concentration** is 20V_1f mg.

Concentration of bicarbonates as $CaCO_3$ equivalent :

100mL of sample is equivalent to ($V_2 - V_1$)mL of 0.02(N) H_2SO_4

i.e. 1000mL of sample

$\equiv (V_2 - V_1)f \times 0.02 \times 1000/100$

$\equiv 0.2\ (V_2 - V_1)f$ mL H_2SO_4(N)

$\equiv 0.2\ (V_2 - V_1)f$ mL $CaCO_3$(N)

1000mL of $CaCO_3$ (N) contains 50g of $CaCO_3$

1mL " " 50/1000 = 50mg of $CaCO_3$

i.e. 0.2($V_2 - V_1$)f mL $CaCO_3$(N) contains 50 × 0.2 × ($V_2 - V_1$)f = 10 × ($V_2 - V_1$)f mg

Thus **Bicarbonate** concentration

= 10 × ($V_2 - V_1$)f mg/L.

Table 8.2.5 : Titration of water for estimation of total alkalinity in presence of indicators

No. of experiment	Titrant volume with phenolphthalein indicator (V_1mL)	Mean (V_1mL)	Titrant volume with methyl organge indicator (V_2mL)	Mean (V_2mL)	Mean (V_2mL – V_1mL)
1.					
2.					
3.					

Chapter 8.2.4(b)

Estimation of Alkalinity of Soil

Principle

Total alkalinity of soil can be determined in the same way as that of water by direct titration of the soil solution with strong acid using methyl orange as an indicator.

Reagents required

1. **Methyl orange indicator; 0.5% :** 500mg of methyl orange is dissolved in distilled water to prepare 100mL of distilled water.
2. **Sodium carbonate; 0.1(N) :** 5.3g of Na_2CO_3 is dissolved in distilled water to prepare 1L of solution.
3. Hydrochloric acid 0.1(N); standardized with 0.1(N) Na_2CO_3.

Procedure

1:5 soil suspensions is prepared by shaking 20g of soil in 100mL of aerated distilled water for about one hour. The suspension is filtered through Whatman No. 50 filter paper using Buchner funnel and a vacuum pump. To 100mL of this soil solution 2-3 drops of methyl orange is added and titrated with 0.1(N)HCl. At the end point the colour will change to pink.

Results

Table 8.2.6 : Titration of soil for estimation of alkality in presence of methyl orange indicator

No. of experiment	Titrant volume with methyl orange indicator (V_1mL)	Mean (V_1mL)
1.		
2.		
3.		

Calculation

Total alkalinity in meq/100g = {(V_1mL × N) of HCl × 500}/mL of solution

Chapter 8.2.5

Estimation of Total Hardness of Water

Introduction

Calcium and Magnesium are present in aquatic ecosystem. The total concentration of calcium and magnesium expressed in equivalence with $CaCO_3$ is the gauge for total hardness.

Principle

Both calcium and magnesium ions in the water sample are titrated with ethylene diamine tetra acetic acid (EDTA) form Ca-EDTA and Mg-EDTA. A small volume of Erichrome-black-T indicator when added to a water sample and buffered at pH 10 forms a soluble wine red complex, by reacting with Ca^{++} and Mg^{++} and solution turns blue.

Ca^{2+} + Mg^{2+} + Mg Eriochrome (wine red) Black T + EDTA $\longrightarrow$ CaEDTA + MgEDTA + Eriochrome black-T(blue).

Reagents required

1. **Standard Versenate Solution** : 0.01(N) solution of ethylene diamine tetra acetic acid di-sodium salt (EDTA) or versenate is prepared by dissolving 2.0g of EDTA in distilled water in which 0.05g of magnesium chloride ($MgCl_2.6H_2O$) is further added, the volume is made up to one litre.

Then this versenate solution is standardized against 0.01(N) calcium carbonate solution (prepared by taking 0.5g $CaCO_3$ (AR grade) and dissolving in minimum excess of dilute hydrochloric acid and then diluted to 1 litre with distilled water.

2. **Ammonium chloride : Ammonium hydroxide buffer (pH 10)** : Pure ammonium chloride of amount 67.5g is dissolved in 570mL of ammonium hydroxide (conc.). The total volume is to adjusted up to 1 litre and the pH is adjusted to 10.

3. **Eriochrome black T (Solochrome black) indicator** : 4.5g of hydroxylamine hydrochloride (AR) and 0.5g solochrome black indicator is dissolved in 100mL of 95% ethanol.

Procedure

(1) 5mL of water sample is taken in a porcelain dish (about 8 cm diameter). It is further diluted to 25mL with distilled water.

(2) About 1mL of ammonium chloride-hydroxide buffer and 3-5 drops of solochrome black indicator is added.

(3) Above solution is titrated with standard versenate solution. Colour change occurs from wine red to blue or bluish green. At the end point of titration a tinge of red colour remains.

Results

See Table 8.2.7.

Calculation

The volume of titrant (EDTA/Versenate) be 'x'mL.

∴ Total Hardness or Ca^{++} plus Mg^{++} in g/L = [volume of EDTA ('x'mL) × N × 32.196]/mL of water sample taken.

N is the normality of standard Versenate solution determined by standardization against 0.01(N) calcium carbonate.

32.196 is a constant.

Table 8.2.7 : Titration of water sample with versenate for estimation of total hardness

No. of experiment	Burette reading		Volume of versenate required as titrant ('x'mL)	Mean volume of versenate required ('x'mL)
	Initial (AmL)	Final (BmL)		
1.				
2.				
3.				

Chapter 8.2.6

Estimation of Chlorinity of Water

Introduction

Water acts solvent of various salts. Salts are found in solution in natural waters. It is well known that the concentration of the nutrient salts is critical to plant growth. Amount of dissolved substances in natural water creates osmotic stress on the aquatic animals because of differences in the concentration of such substances between the internal and external media. The osmotic strength in the aquatic environment differs according to the salt content in the water body.

Natural waters may exist in two forms saline water (oceans, seas) and fresh water (lentic, lotic and underground system). Seawater differs from freshwater in having a higher concentration of dissolved salts. Sodium and chloride ions are most important in seawater whereas in hard fresh water, carbonate and calcium ions are significant. Chloride may occur in fresh waters as a result of salt deposits in the soil that is being dissolved. High chloride content in inland waters may be an indication of pollution.

Principle

The silver ion in the silver nitrate reacts with the chloride present in the water sample to produce a white precipitate. When the entire chloride ion combine with silver, the excess silver present combines with the chromate indicator to generate a pinkish orange coloured silver chromate. This change in colour of the precipitate is used to specify the end point in chloride determination.

Reagents required

(1) **Standard Silver Nitrate ($AgNO_3$) 0.014N :** Dissolved 2.395g $AgNO_3$ in 200mL of distilled water. The solution is made up to 1L in a volumetric flask. Stored in a brown bottle in the dark (as silver nitrate is affected by light). If the reagent is poured from the bottle, it should never be returned to it.

(2) **Potassium chromate indicator ($K_2Cr_2O_4$) :** Dissolved 50g potassium chromate in 500mL of distilled water. Added 0.0141N silver nitrate solution until a red precipitate is formed. It is allowed to stand overnight. Filtered and made up to 1L with distilled water.

(3) **Standard sodium chloride (NaCl) 0.014N :** Dissolved 824.1mg sodium chloride (dried at 60°C) in double distilled water (chloride free). Diluted to 1L using same double distilled water.

Procedure

1. 100mL of the water sample is taken in a conical flask. The sample is filtered first if it contains lot of suspended matter.

2. Added 1mL potassium chromate indicator solution in the water (test) sample.

3. Silver nitrate solution taken in a burette is run into test sample. The flask is swirled during the addition of silver nitrate.

4. The titration is continued until the precipitate is formed and solution turns the pink-orange colour.

Results

See Table 8.2.8.

Calculations

mg/mL of chloride ion = $(x - 0.2) \times N \times 354.5$.

Where, 'x' is mL of standard silver nitrate used. 0.2 is the blank value attributed to potassium chromate.

N is the normality of silver nitrate determined by standardization against sodium chloride; 354.5 is a constant.

Table 8.2.8 : Titration of water sample with Silver nitrate for estimation of chlorinity

No. of experiment	Burette reading		Volume of silver nitrate required	Mean volume of silver required
	Initial (AmL)	Final (BmL)	(A–BmL)	('x'mL)
1.				
2.				
3.				

Chapter 8.2.7

Determination of Salinity of Water

Method A

Principle :

Amount of inorganic compounds in water when expressed in grams/kilo or in ppt(‰) is known as **salinity**. It is determined by estimation of precipitable halide halogens in 10mL of water. Such estimation is done through the titration with silver nitrate using a chromate end point.

Reagents required :

1. 0.16N silver nitrate : 27.25g of silver nitrated dissolved in 1000mL of distilled water.

2. Potassium Chromate indicator : 5g of potassium chromate in dissolved in distilled water and the solution is diluted to 100mL.

Procedure :

10mL of the sample water is taken in a conical flask and titrated against 0.16N silver nitrate using potassium chromate indicator. The end point is indicator by persistent reddish brown colour.

Calculations :

The added volume of silver nitrate VmL is approximately equal to salinity (S) in parts per thousands.

Method B

Principle :

Salinity can also be measured through measuring Total Dissolved Salts and Electrical Conductivity. Total Dissolved Salts (TDS) is assessed by evaporating a known volume of water to dryness, then weighing the solid residue remaining. Electrical conductivity (EC) is measured by passing an electric current between two metal plates (electrodes) in the water sample and measuring how readily current flows (i.e. conducted) between the plates. The more dissolved salt in the water, the stronger the current flow and the higher the EC. Measurements of EC can be used to give an estimate of TDS. Measurement of TDS is tedious and cannot be carried out in the field. EC measurement is much quicker and simpler and is very useful for field measurement.

TDS is recorded in milligrams of dissolved solid in one litre of water (mg/L). Parts per million (ppm) is equivalent to mg/L but it is not a favoured unit. EC measures the charge carrying ability (ie conductance) of liquid in a measuring cell of specific dimensions. It is essential to clearly define the units of both conductance and length when talking ECs. To say a water sample is 2000 EC, is like saying a table is 2000 long, without specifying millimetres, centimetres or metres. The standard EC unit used by the Victorian Salinity Program and the Murray Darling Basin Commission is microSiemens per centimetre (μS/cm) at 25°C.

Relationship of total dissolved salts to EC where EC can be effectively converted to TDS for natural Victorian waters by the following relationship :

- TDS (mg/L) = EC (μS/cm at 25°C) × 0.6

Materials required :

1. EC meter.
2. Water collecting container (250mL).

Procedure :

- **Collecting of water :**

1. One has to be sure that the collecting container is very clean. Previous contents could affect the result. A container with an opening large enough to take the EC meter should be used. One should not use jars which smell (eg. vegemite, pickle jars) if samples are to be kept for a while.
2. A sample is chosen which is representative of the body of water being considered. It needs to be a sample which is like most of the water for which the experimenter wants to get information about. One should not try to take the sample too close to the surface, bottom or sides of the water body.
3. The container should be rinsed two or three times with some of the water to be sampled.
4. The sample is then collected.

- **Taking salinity reading :**

1. One has to ensure that the EC meter has been calibrated.
2. The protective cap is removed, the meter is switched on and the probe is inserted into the water sample up to the immersion level.
3. The probe is moved up and down to remove bubbles from around the electrodes.
4. This will ensure that good contact is achieved between water and electrodes (one should not swirl it around, as this may actually drive water out of the probe).
5. The probe is allowed to reach the temperature of the water before taking a reading.
6. Temperature has a significant impact on the salinity reading. EC units are standardised to a temperature of 25°C. Some meters automatically correct the reading taken at water temperature to a reading at 25°C.
7. If the meter has automatic temperature compensation, one has to wait about 30 seconds before taking the reading, if the water and probe are about the same temperature. If the water is much colder than the probe, a longer period is to be allowed, say two minutes before taking a reading.
8. If the meter has no temperature compensation the temperature of the sample is taken and a correction table is to be used to get the right value.
9. The display is read, and the result is recorded as mentioned below.
10. The probe is rinsed with tank water and any excess water is drained off, between each sample and at the end of sampling for the day. This will prevent false readings due to salt residues on the meter from the last sample.

Results :

The results of any sampling, should be recorded in a notebook for future reference. The information should include :

1. **Name of collector.**

2. **Date of sampling :** Salinity levels fluctuate throughout the year. The date of sampling becomes important when comparing readings.
3. **Sampling location :** A note should be made of where the sample was taken from. Further samples may then be taken from the same site in the future.
4. **Water source :** A note should be made of the water source. e.g. River, Creek, Lake, Dam, Swamp, Drain, Groundwater bore, etc.
5. **EC reading :** Readings should all be recorded as microSiemens per centimetre (μS/cm). Salinty units should be referred to convert readings (see formula above) in other units to μS/cm.
6. **Temperature reading :** If the meter has no automatic compensation, temperature recorded and resulting EC value is adjusted from a calibration table.

Method C

Salinity can be calculated from chloride determinations by applying formula :

Salinity ‰ = 0.03 + 1.805Cl‰.

Chapter 8.2.8

Determination of *p*H of Water

Introduction

Natural waters are generally alkaline in nature due to the presence of sufficient quantities of carbonates. *p*H of water gets drastically changed with time due to constant exposure to air, biological activity and temperature changes. In natural waters, *p*H changes diurnally and seasonally due to the variation in the photosynthetic activity, due to consumption of CO_2. *p*H has no direct effect upon health, still lower value i.e. if *p*H goes down below 5 due to excessive accumulation of acids the photosynthetic process is affected adversely. Higher *p*H causes scale formation in water heating apparatus and also reduces the germicidal capacity of chlorine. *p*H below 6.5 corodes pipes thereby releases toxic metals viz., Zn, Pb, Cd, Cu, etc.

Principle

Hydrogen ion concentration of water is very important chemical constituent of water. It both affects the biological as well as chemical system of water. For definition of pH refer chapter 8.2.2.

The measurement of *p*H is carried out electrometrically using a *p*H meter by means of either a general purpose combination electrode or a low ionic combination electrode, depending upon the ionic concentration of the sample. In low ionic strength solutions, a standard *p*H electrode will give a slow response and poor reproducibility. Still a low ionic electrode must not be used in solutions of high ionic strength (above 50μS) as this will result in damage to the electrode. Earlier the *p*H meter is regulated using a freshly made buffer solution (*p*H 7 etc.) and the slope of the electrode adjusted against a *p*H 4 buffer. Temperature compensation is adjusted manually according to the ambient sample temperature. The electrode is thoroughly rinsed with distilled water before each measurement.

Water samples should be collected in a clean glass beaker, well flushed with the sample, and the electrode is allowed to stand for several minutes without agitation before the *p*H value is determined.

Materials Required

(1) *p*H meter.

(2) Distilled water.

(3) Water Sample.

(4) Buffer tablets (*p*H 4.0,7.0,9.0) and correspondence buffer solutions.

Procedure

1. Buffer solution of *p*H 4.0, 7.0, and 9.0 are prepared using standard buffer tablets.

2. Electrodes are connected to *p*H meter.

3. Electrodes are washed with distilled water.

4. Electrodes are dipped in a buffer solution of *p*H 7.0

5. Temperature control of the *p*H meter is set as per the temperature of the sample water.

6. pH is recorded. The meter gives a reading near 7.0

7. The electrodes are then removed from the buffer solution and washed with distilled water and dried. The electrodes are dipped to buffer of *p*H 4.0 and 9.0 consecutively. The meter should correspondingly show a reading of 4.0 and 9.0 respectively.

8. Next the electrodes are dipped in the water sample (test sample). The sample is gently swirled twice and the *p*H is seen in the digital display or in the meter of the instrument. Three such replicates are taken.

9. Mean *p*H value is calculated.

Results

Table 8.2.9 : *p*H measurement of water

No. of experiment	*p*H value	Mean *p*H value
1.		
2.		
3.		

Inference

Depending upon the pH of the water one can state whether the water is acidic, neutral or alkaline.

Chapter 8.2.9

Estimation of Dissolved Oxygen in Water

Introduction

Fish, invertebrates, plants, and aerobic bacteria all need oxygen for respiration. Much of the dissolved oxygen in water comes from the atmosphere. After dissolving at the surface, oxygen is dispersed by current and turbulence. Algae and rooted aquatic plants also deliver oxygen to water through photosynthesis.

The major factor contributing to changes in dissolved oxygen levels is the build-up of organic wastes. Decay of organic wastes consumes oxygen and is often concentrated in summer, when aquatic animals require more oxygen to support higher metabolisms.

Depletions in dissolved oxygen can cause major shifts in the kinds of aquatic organisms found in water bodies.

Temperature, pressure, and salinity influence the dissolved oxygen capacity of water. The ratio of the dissolved oxygen content (ppm) to the potential capacity (ppm) gives the percent saturation, which is an indicator of water quality.

It is the amount of oxygen gas dissolved in a given volume of water at a particular temperature and pressure. It is often expressed as a concentration in parts of oxygen per million parts of water.

Principle

Dissolved oxygen is estimated as per Winkler's method.

Manganous sulphate reacts with sodium and potassium hydroxide to give a white precipitate of manganese hydroxide. Manganese hydroxide reacts with oxygen to form manganic basic oxide.

$$MnSO_4 + 2KOH \longrightarrow Mn(OH)_2 + K_2SO_4$$

$$2Mn^{+2}(OH)_2 + O_2 \longrightarrow 2Mn^{+4}O(OH)_2$$

When sulphuric acid is added to brown manganic oxide, it is dissolved and manganic sulphate is formed, which instantly react with iodide to form iodine.

$$Mn^{+4}O(OH)_2 + 2H_2SO_4 \longrightarrow Mn^{+4}(SO_4)_2 + 3H_2O$$

$$Mn^{+4}(SO_4)_2 + 2KI \longrightarrow Mn^{+2}SO_4 + K_2SO_4 + I_2$$

Iodine is determined titrimetrically with starch as indicator of end point.

Materials required

1. **Manganous Sulphate reagent** : 480g of $MnSO_4.4H_2O$ dissolved in 1 litre of distilled water.

2. **Alkaline iodine solution :** 500g of sodium hydroxide and 300g of potassium iodide (KI) in 1 litre of distilled water.

3. Concentrated sulphuric acid with Sp. gr 1.84.

4. **Sodium thiosulphate ($Na_2S_2O_3$) solution 0.025N :** 6.205g of sodium thiosulphate is dissolved in distilled water in boiling condition and the volume is made up to 1 litre. This solution is stored in black bottle with 2-3 pellets of sodium hydroxide.

5. **Starch Indicator :** 300mL of NaOH (20%) solution is slowly added to 3g of starch in 300mL of water (1%). The solution is stirred vigorously until it settles for 1-2 hours Next concentrated HCl is added until the solution turn just acid as revealed by litmus paper test. This is followed by addition of 2mL of glacial acetic acid.

Procedure

1. 250mL of water sample is taken in a conical flask (V_2).

2. To this sample 1mL of manganous sulphate followed by 1mL alkaline iodide is added (V_3).

3. Bottle is closed by a stopper and shaken well, so that the precipitated manganous-manganic hydroxide gets evenly dispersed.

4. The precipitate gets settled in 2-3 minutes with 1/3 clear supernatant.

5. 1mL of concentrated H_2SO_4 is added to the bottles covered by a stopper and again shaken well. The precipitate dissolves.

6. Iodine is liberated, due to oxidation of KI by oxygen dissolved in the water sample. The solution turns deep brown in colour.

7. The solution is allowed to stand for 15 minutes.

8. 100mL (V_4) of such treated sample is titrated with 0.025N sodium thiosulphate taken in a burette. The colour changes from brown to yellow.

9. At this point 5mL of starch indicator is added so that the solution turns blue black.

10. A blank is run parallel which does not have alkaline iodide added to it. The solution when titrated does not give blue colour. When the titrated by sodium thiosulphate solution (V_1) turns colourless.

This blank correction is subtracted from the titre obtained to the real titre 'V' mL and then the oxygen content of the sample is calculated.

Calculation

Dissolved oxygen (DO) in the sample (mg/L)

$$= \frac{(V_1 \times N \times 8 \times 1000 \times 0.698)}{V_4 \times \left(\frac{V_2 - V_3}{V_2}\right)}$$

where,

V_1 = Vol. of $Na_2S_2O_3$ required for titration.

V_2 = Vol. of sample water taken (eg. 250 mL).

V_3 = Vol. of alk. iodide & Manganous Sulphate aded.

V_4 = Vol. of sample water used for titration.

N = Normality of $Na_2S_2O_3$.

8 = Eqv. wt. of O_2

0.698 = Correction factor for mL to mg.

Chapter 8.2.10

Estimation of Dissolved and Free Carbon Dioxide of Water

Introduction

Carbon dioxide exists in the water in various forms which under a given set of equilibrium

$$CO_2 \rightleftharpoons H_2CO_3 \rightleftharpoons H^+ + HCO_3^- \rightleftharpoons 2H^+ + CO_3^{2-}$$

(Dissolved CO_2) *(Carbonic acid)*

Dissolved carbon dioxide (CO_2) is present naturally consequently of animal respiration, the decomposition of organic matter, and the breakdown of certain minerals. It is the major source of acidity in unpolluted water samples. Surface waters typically contain less than 10 ppm (mg/L) of dissolved CO_2, while ground waters, particularly if deep, may contain several hundred ppm (mg/L). Oil production brines also contain high CO_2 concentrations. High levels of CO_2 in surface waters can indicate abnormal organic or mineral decomposition. Measurement of CO_2 is a means for monitoring the quality of municipal water treatment systems. Dissolved CO_2 is corrosive to water handling equipment, particularly steam condensate systems. Some CO_2 is desirable as it helps maintain the carbonate equilibrium, thus avoiding the formation of calcite scale on exposed surfaces. Due to the delicate balance between corrosion and scale formation, CO_2 concentrations must be carefully monitored.

By free carbon dioxide we mean the amount of CO_2 in simple solution (dissolved) plus in form of bicarbonates. At acidic *p*H most of the forms are present as free carbon dioxide where as at neutral (7) *p*H most of the CO_2 is present in as CO_3^{2-} form.

(i) Estimation of dissolved carbon dioxide in water :

Materials required :

1. Phenolphthalein indicator: As described in following section.
2. N/10 Na_2CO_3 solution.
3. Beaker, Conical flask.
4. Burette.

Procedure :

1. 250mL of water sample is taken in a conical flask or a beaker.
2. 2-3 drops of phenolphthalein indicator are added into it.
3. The sample is titrated with N/10 Na_2CO_3 till a permanent pink colour appears.
4. Titre volume is recorded.

Calculation :

Volume of Na_2CO_3 (N/10) used for titration is VmL.

i.e. Dissolved CO_2 in the water sample = 8.8 × V mg/litre.

(ii) Estimation of Free Carbon dioxide of Water :

Free CO_2 can be assessed by titrating the sample using a strong alkali (such as free NaOH) to *p*H 8.3. At this *p*H all free CO_2 is converted into bicarbonates.

This should be done immediately after collection of the water sample as it diffuses in or out form the sampled water.

Materials required :

1. **(N/22) Na_2CO_3 solutions** : 0.482g of Na_2CO_3 (A.R. Grade) is dissolved in 200mL of distilled water.
2. **Phenolphthalein indicator** : 0.5g of Phenol-phthalein indicator in 100mL ethanol (50%).
3. Water sample.
4. Beaker, conical flask.
5. 20mL Burette.

Procedure :

1. 100mL of the sampled water is taken in a conical flask and a few drops of phenolphthalein indicator is added.
2. If there is no change of colour it is confirmed free CO_2 is present.
3. The water sample is titrated against Na_2CO_3 (N/22) taken in a burette till end point is indicated as slight pink.
4. The mean titre volume (V in mL) is recorded.

Equation :

$$Na_2CO_3 + H_2O + CO_2 \longrightarrow 2NaHCO_3$$

Results :

See Table 8.2.10

Calculations :

From the above equation it can be inferred that amount of Na_2CO_3 required in the reaction is equivalent to the amount of CO_2 in the water sample i.e.

$$CO_2 \equiv Na_2CO_3$$

i.e. 44g of $CO_2 \equiv$ 106g of Na_2CO_3

Since, 1000mL of Na_2CO_3 (N/22) contains 2.4091g

Therefore, VmL of Na_2CO_3 (N/22) = 2.4091 × f × V/1000 = 'n'g

(f = factor of sodium carbonate solution)

As 106g of $Na_2CO_3 \equiv$ 44g of CO_2

i.e. 'n'g of Na_2CO_3 = 44 × n/106 = 'y'g of CO_2

As 100mL of water sample is used in titration

i.e. amount of CO_2 present in a litre of water = y × 10 × 1000mg CO_2/litre.

Table 8.2.10 : Titration of water sample for estimation of free CO_2 of water

No. of experiment	Burette reading (mL)		Difference (A–BmL)	Mean amount of sodium carbonate required as titrant (VmL)
	Initial (AmL)	Final (BmL)		
1.				
2.				
3.				

Chapter 8.2.11

Estimation of Biochemical Oxygen Demand (BOD)

Introduction :

BOD is the amount of dissolved oxygen (DO) required by aerobic decomposers to break down the organic matter in a given volume of water at a certain temperature over a particular period of time.

Natural sources of organic matter take account of plant decay and leaf fall. However, plant growth and decay may be unusually accelerated when nutrients and sunlight are overly abundant due to human influence. Urban runoff carries pet wastes from streets and sidewalks; nutrients from lawn fertilizers; leaves, grass clippings, and paper from residential areas, which increase oxygen demand.

Oxygen consumed in the decomposition process deprives other aquatic organisms of the oxygen they need to live. Organisms that are more tolerant of lower dissolved oxygen levels may replace a diversity of more sensitive organisms.

Principle :

The method involves measuring the difference of the oxygen concentration between the sample and after incubating it for 5 days at 20°C temperature, keeping out any light from penetrating the sample to prevent photosynthesis. The Dissolved Oxygen (DO) in the sample is measured before and after the 5 days incubation period, and BOD is then calculated as the difference between initial and final DO measurements. BOD can be considered a more "natural" test in determining the oxygen required to oxidize organic matter, however it does not account for rapid changes in conditions.

Materials required :

1. **BOD bottles.**

2. **BOD incubator** : Having temperature control at 20°C temperature.

3. **Phosphate buffer** : 8.5g KH_2PO_4, 21.75g K_2HPO_4, 33.4g $Na_2PO_4.7H_2O$ and 1.7g NH_4Cl is dissolved in distilled water to prepare 1L of the solution. The *p*H has to be adjusted to 7.2.

4. **Magnesium sulphate** : 82.5g $MgSO_4.7H_2O$ is dissolved in distilled water to prepare 1 L of the solution.

5. **Calcium chloride** : 27.5g of anhydrous $CaCl_2$ is dissolved in distilled to prepare 1 L of solution.

6. Ferric chloride: 0.25g of $FeCl_3.6H_2O$ is dissolved in distilled water to prepare 1L of solution.

7. Sodium sulphite solution, 0.025(N) : 1.575g of Na_2SO_3 is dissolved and diluted to 1000 mL. Solution should be freshly prepared.

Procedure :

1. Dilution water is prepared in glass container by bubbling distilled water for about 30 minutes.

2. 1mL each of phosphate buffer, magnesium sulphate, calcium chloride and ferric chloride is added to each 1L of dilution water and mixed thoroughly.

3. The sample is neutralised to *p*H of 7.0 through the use of 1(N) NaOH or H_2SO_4.

4. As the Dissolved oxygen (DO) of the sample is likely to be exhausted; it is important to prepare a suitable dilution of the sample as per expected BOD range. Refer Table 8.2.11.

5. The dilutions are prepared in a bucket or a large glass trough. The contents are mixed thoroughly. Two sets of BOD bottles are filled.

6. One set of BOD bottle is kept in a BOD incubator at 20°C for 5 days, and the DO content in the other is determined immediately.

7. DO of the BOD bottle kept in a BOD incubator at 20°C is determined after 5 days.

Result :

See Table 8.2.10

Calculation :

BOD (mg/L) = (DO_0 - DO_5) × dilution factor

DO_0 = Initial (0 day) DO of the sample

DO_5 = DO after 5 days

Note :

If the sample has free residual chlorine it is important to remove chlorine. For the removal of chlorine, 50mL of sample is taken, acidified with 10mL acetic acid (1 + 1). 1g of KI is added. This is titrated with Na_2SO_3 (0.025N) using starch indicator. The whole volume of Na_2SO_3/mL is calculated and then added accordingly to the sample that is to be tested for BOD.

It is advisable to determine COD of those samples which may have toxic materials.

Table 8.2.11 : Dilutions of various ranges of BOD in the sample

Range of BOD mg/L O_2	Dilution(%)	Sample volume in 1L of mixture
0-6	No dilution	1,000
4-12	50	500
10-30	20	200
20-60	10	100
40-120	5	50
100-300	2	20
200-600	1	10
400-1200	0.5	5
1000-3000	0.2	2
2000-6000	0.1	1
Above 6000	0.05	0.5

Chapter 8.2.12

Estimation of Chemical Oxygen Demand (COD)

Introduction

COD is the measure of oxygen consumed during the oxidation of oxidisable organic matter by a strong oxidising agent.

COD is often measured as a swift indicator of organic pollutant in water. It is usually measured in both municipal and industrial wastewater treatment plants and gives an indication of the effectiveness of the treatment process. COD is measured on both influent and effluent water. The efficiency of the treatment process is generally expressed as COD removal, measured as a percentage of the organic matter purified during the cycle.

COD is often used as a measurement of pollutants in natural and waste waters and to determine the strength of waste such as sewage and industrial effluent waters. COD has further applications in power plant operations, chemical manufacturing, commercial laundries, pulp & paper mills, environmental studies and general education. In potable drinking water plants, COD values should be less than 10 mg/L at the end of the treatment cycle.

Principle

Potassium dichromate in presence of sulphuric acid generally used as oxidising agent in the determination of COD.The sample is refluxed with potassium dichromate and sulphuric acid in presence of mercuric sulphate to neutralize the effect of chlorides and silver sulphate (catalyst). The excess of potassium dichromate is titrated against ammonium sulphate using Ferroin indicator. The amount of potassium dichromate used is proportional to oxidisable organic matter present in the sample.

Materials required

1. **Potassium dichromate solution, 0.25N :** 12.259g of dried A.R. grade $K_2Cr_2O_7$ is dissolved in water to make 1L of solution.

2. **Potassium dichromate solution, 0.025N :** 0.25N $K_2Cr_2O_7$ is diluted 10 times.

3. **Ferrous ammonium sulphate, 0.1N :** 39.2g of $Fe(NH_4)_2(SO_4).6H_2O$ is dissolved in water, adding 20mL of concentrated H_2SO_4 to make 1L of solution. This solution is standardised with $K_2Cr_2O_7$. For standardisation 10mL of $K_2Cr_2O_7$ is

diluted to about 100mL and in it added 30mL of concentrated H_2SO_4 and titrated with ferrous ammonium sulphate using indicator.

4. **Ferrous ammonium sulphate, 0.01N :** 0.1 N Ferrous ammonium sulphate is diluted 10 times.

5. **Ferroin indicator :** 1.485g of 1, 10-Phenonthroline and 0.695g of Ferrous sulphate is dissolved in distilled water to make 100mL of solution.

6. Concentrated sulphuric acid.

7. $HgSO_4$.

8. Ag_2SO_4.

Procedure

1. 20mL of sample is taken in a 250-500mL COD flask (Round bottom or Erlermeyer flask with a ground joint for Leibig reflux condenser.

2. If the sample is expected to have COD more than 50 mg/L, 10mL of 0.025N Potassium dichromate solution is added to the sample. Extreme care should be taken in case of low COD samples. A small trace of organic matter in glassware may cause a significant error.

3. A pinch of $AgSO_4$ and $HgSO_4$ is added. If the sample contains chlorides in higher amount, $HgSO_4$ is added in the ratio of 10:1, to the chlorides. COD cannot be determined accurately if the sample contains more than 2000 mg/L of chlorides.

4. 30mL of sulphuric acid is added to it.

5. Refluxed at least for 2 hours on a water bath or hot plate. The refluxed water is made up to a final volume of 140mL.

6. Added 2-3 drops of Ferroin indicator mixed thoroughly and titrated with 0.1N Ferrous ammonium sulphate.

7. A blank is run with distilled water using same quantity of the chemicals.

Calculation

COD mg/mL = [(b–a) × strength of $K_2Cr_2O_7$ × 1000 × 8]/mL of sample

Where,

a = mL of titrant with sample

b = mL of titrant with blank

Chapter 8.3.1

Study of Learning in Rat through T-maze Technique

Introduction

The use of animal subjects in the study of learning began more than 80 years ago when Small (1899) tested rats in a reproduction of the famous maze at Hampton Court in England. This was the first use of a maze as a research device for studying animal learning. Although operant conditioning techniques are now generally preferred for such studies, many ethnologists and animal psychologists still use mazes for experiments involving a variety of animals.

Procedure

Different experiments proved that rats could master a very complicated maze; subsequent investigators simplified the maze pattern and standardized the experimental procedures.

T-maze technique can be used to study the learning behaviour in rats. T-maze is one of the simplest mazes to construct and use (Fig. 8.3.1.1).

1. The T-maze was kept on a flat surface and was cleaned with cotton dipped in alcohol,so that there remains no unnecessary smell or food materials, which may create error on rat.

2. Two rats are used for the experiment and was marked accordingly, these animals were

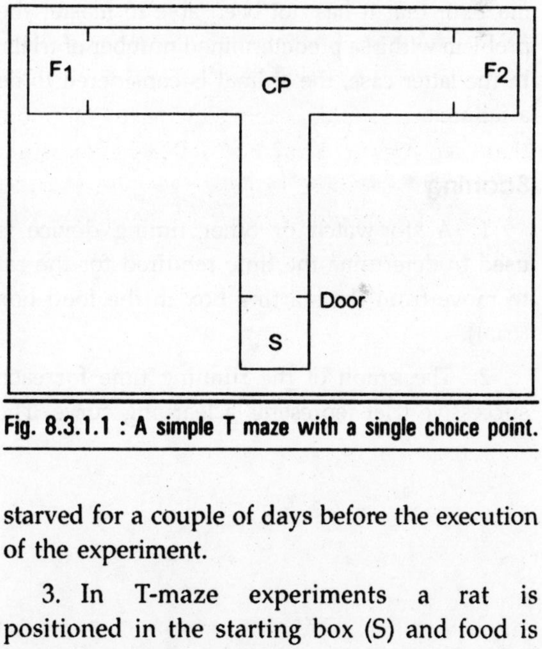

Fig. 8.3.1.1 : A simple T maze with a single choice point.

starved for a couple of days before the execution of the experiment.

3. In T-maze experiments a rat is positioned in the starting box (S) and food is put in one of the food boxes (F_1 or F_2). The first trial begins when the door leading from the starting box to the stem of the maze is opened. The choice point (CP) is noted where the animal be required to turn either left or right. If the food were present in F_1, the correct run order would then be SF_1. In other words, the animal must journey directly from the starting

box to the choice point, and then make a left turn and move directly to the food box. If the animal turns right at the choice point and goes to F_2, this is taken as an error (= fault: defined as any turn away from the correct path leading from the starting box to the food box). Since there is no food at F_2, the animal will ultimately leave that box and move toward F_1. Entry of F_1 completes the first trial. If the animal turns left at the choice point after departing F_2, another error would be scored.

4. After a short stay in the food box, the animal is again placed in the starting box, and after a rest period of one or two minutes, the second trial begins. The trials are recurred until the animal either reveals that it has learned the maze, or that it has not been able to master the problem within a predetermined number of trials. In the latter case, the animal is considered to be a failure.

Scorring

1. A stopwatch or other timing device is used to determine the time required for the rat to move from the starting box to the food box (trial).

2. The graph of the running time for each successive trial represents a learning curve. The learning curve is drawn by plotting trial time against each trial. Separate curves are compared for different rats.

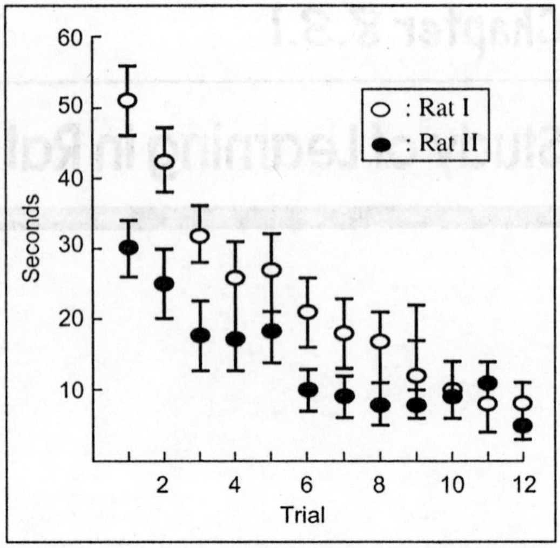

Fig. 8.3.1.2 : Typical learning curve of two rats.

3. A typical learning curves determined for two mice given 12 trials in a T maze is shown in Fig. 8.3.1.2.

Result

Result is interpreted from the learning curve drawn.

Chapter 8.3.2

Study of Aggressive Behaviour of Fishes

Introduction

Aggression may be described as the act of initiating hostilities or invasion or the accomplishment or habit of initiating attacks or antagonistic or destructive behaviour or activity. Aggression is one of the most essential and most controversial kinds of motivation.

Aggressive behaviour in animals is not rare. Such an activity is noticed in wide number of species of vertebrates. Aggression towards members of the same species is known as *social aggression.*

Such aggressive behaviours are manifested through *aggressive displays* and *aggressive activities.* These aggressive behaviour are inheritable i.e. individuals of same species exhibit a similar aggressive behaviour from one generation to other. Like other vertebrates, fishes also show different aggressive behaviour.

The biggest and healthiest fish normally wins these fights, but if size is not an obvious difference then the fish resort to fin flaring and rushing each other. The purpose of fin flaring is to make the fish look bigger to its opponent. Rushing at an opponent or waving one's body side-to-side in an S-shape is designed to produce a pressure wave in the water to be felt by the opponent.

Method

1. Fishes (*Channa punctatus*) were bought from local fish market. They were kept in glass aquarium provided with plenty of aquatic plants such as *Hydrilla*.The fishes were supplied with live and sliced earthworm; the fishes were aclimmatised in glass aqyuarium for atleast 7-15 days.

2. Details of the aggressive behaviour was noticed by separating the fish (socially and physically) in adjacent compartments of an aquarium with dimensions 8 × 12 × 9 cu. inch fitted with removable opaque partition.If the fishes are required to fight the partition is removed. The fish may be isolated for one day or three days.

Observations

The aggressive behaviour in fish viz., *Channa sp.* can be of various types :

- **Frontal display and approach :** The fishes adjust themselves towards each other.

Dorsal fin is lifted first then the ventral fin and pelvic fin, caudal fin is stretched. The colour turns darker and the black spot at base of the caudal fin fades out. The lower jaw is abducted as a result the size of the head looks bigger when viewed frontaly. The two fishes moves forward and stops coming ahead of each other. Some times face to face and collision may also occur. The fishes confront each other with their body arm approximately on straight line. From this stage mouth biting occurs. If they are still not ready for such biting behavior they turn away from each other slightly and moved forward and thus left to lateral spread. Sometimes it may also happen that one of the fish move ahead but the other stayed motionless or one of the fish rapidly orient itself towards the other and approached it. The opponent on the other hand remained on its original position for a moment, stretched its fins and oriented towards the rival approaching it.

- **Lateral spreading and tail biting :** In lateral spreading, a fish is oriented towards the side of another fish. All fins except the pectorals were fully expanded. Operculae are raised or closed. The mouth is kept occasionally open. The fish undulated its tail (tail beating), i.e. it moves forward and suppressed each other.

  The fishes turns around again, came side by side with their faces in conflicting direction (anti-parallel). Some times the fishes came alongside, facing the same direction (parallel).

- **Parallel running and flank attacks :** The inching in fishes are of two types such as horizontal and vertical. In the horizontal inching, one of the fish quickly turns restraining the rival from anti-parallel orientation in lateral spread and come alongside (parallel orientation) and swiftly turned around the head end of the rival.

  Sometimes vertical inching also takes place where the fish abruptly turns around the long axis of the body of it's rival maintained in a horizontal position.

  This occurred when two fishes tried to creep towards each other concurrently during parallel orientation. Initially, the fish moved alongside gradually then suddenly accelerated their movement and finally stopped abruptly together.

  This takes place after a tail beating; the body axis of the fish remained almost parallel. From this position one fish backed up and made an angle with the body of it's rival and then a peek attack was excluded towards the lateral surface of the opponent. Scales often became loose due to this kind of attack. Flank attack, may be done by both the opponents simultaneously.

- **Biting and jaw locking :** When the opponents bite on the body of another and if this is maintained, this event is called as biting. Biting attacks are mostly directed towards the pectoral fin lobe, dorsal, ventral and caudal fin.

  Sometimes mutual mouth to mouth biting along with the firm holding of each other by their jaws, this is called as the jaw locking. It persists for several seconds (10-50 seconds). During the time of jaw locking the pectoral fins fully raised. In a few seconds the fish jerked its head laterally and in most of the cases, this terminates the jaw locking.

- **Folding, fleeting and tilting :** There is no particular pattern: There are a variety of activities in the subordinate fish following competition of flight. In most case the dorsal, ventral and pelvic fin were tightly pressed against the body. The lobes of the median fins were compressed as far as possible and colour becomes pale. The defeated fish swims away for the dominant one. The subordinate fish maintained its body angle with head downward and tail upward. Sometimes the pelvic fin was raised mainly when dominant fish continued to follow it.

- **Ventral rolling, Dorsal rolling, stationary and sitting floor :** Generally, this occurred when attack from the dominant fish is confronting each other. The defeated fish assumed an angle with water surface with head upward presenting roughly it's ventral surface to the opponent. Occasionally, the fish laid it's body on one side in water following ventral rolling.Sometimes, the defeated fish laid over on one side, in such a way that it could present its dorsal surface in the opponent. The defeated fish came nearer to the water surface and remain motionless. The fins were kept folded. Sometimes pelvic fins were opened on seeing in the actions of the dominant fish. Sometimes defeated fish sat on the floor. Dorsal, ventral, pelvic and pectoral fins remained folded.
- **Chasing, Biting and Treating :** Chasing was the swift advance of a fish in the direction of its opponent, from a distance might be of one body length or more. The approaching fish was however able to touch the body of its opponent. During the straight forward rush, fins remained closed which are stretched on coming closer to the opponent, and touch the opponent with its snout. The intensity of biting is dependent on the spread and distance covered by the offensive fish. Occasionally, the dominant fish rose it's dorsal fin but remained stationary either on floor in water, seeing that the subordinate fish moved away. But when this is performed by the subordinate fish, immediately it was attacked by the dominant fish.
- **Course of Fight :** Aggressive behaviour was very complex. Particular posture and motor pattern and their order of occurrence are changeable. All motor patterns may not be seen in all encounters. In prolonged fight, most can be seen in specific order frontal display and approach – lateral spread-tail beating-flank attack biting- jaw locking-folding-fleeing-chasing or biting or threatening.

  Apart from that, some patterns can be called as manoeuvring, such as inching, parallel running and others, as tilting, ventral rolling and still others apparent appraisement postures such as stationary or sitting on the floor.

Results :

1. Each type of aggressive behaviour is noted in the record sheet with duration (in minutes or seconds) of different aggressive behaviours.

2. Total spending of time in each agreesive behaviour is converted into percentage. This data is used to draw a pie chart exhibiting the aggressive bahaviour.

Conclusion :

Conclusion is given on the following : One should conclude regarding the type of aggressive behaviour exhibited by the fishes and there time frame. Which fish is more aggressive? One has to find out why that fish turns to be more aggressive. Aggressive and normal behaviour of these fishes are to be differentiated.

Table 8.3.1a : Worksheet for cataloguing different category of aggressive behaviour in *Channa punctatus*

Position of the fish	Observed Behaviour	Starting time (1)	Ending time (2)	Total time spend (2–1)
Placed together after one day of isolation	1. Latency of orientation			
	2. Latency of bite			
	3. Tail parallel			
	4. Beating anti-parallel			
	5. Mouth biting			
	6. Circling			
	7. Air gulping			
	8. Jaw locking			
Placed together after three days of isolation	1. Latency of orientation			
	2. Latency of bite			
	3. Tail parallel			
	4. Beating anti-parallel			
	5. Mouth biting			
	6. Circling			
	7. Air gulping			
	8. Jaw locking			

Table 8.3.1b : Table for the total time spending by an individual in a particular aggressive behaviour of *Channa punctatus*

Position of the fish	Behaviour (Aggressive)	Total time spend in each type of aggressive behaviour in hours and minutes	% of time spend	Converted into angle = (% of time spend/100) x 360
Placed together after one day of isolation	1. Latency of orientation			
	2. Latency of bite			
	3. Tail parallel			
	4. Beating anti-parallel			
	5. Mouth biting			
	6. Circling			
	7. Air gulping			
	8. Jaw locking			

Table contd. next page

Position of the fish	Behaviour (Aggressive)	Total time spend in each type of aggressive behaviours in hours and minutes	% of time spend	Converted into angle = (% of time spend/100) x 360
Placed together after three days of isolation	1. Latency of orientation			
	2. Latency of bite			
	3. Tail parallel			
	4. Beating anti-parallel			
	5. Mouth biting			
	6. Circling			
	7. Air gulping			
	8. Jaw locking			

Note : Arbitrary numbers of behaviour are given in the worksheet; numbers can be more or less. Time may be in hours or minutes.

Chapter 8.3.3

Study of Sex Ratio

Introduction :

Sex is one of the most important characteristics of living organisms. It involves the transfer of genetic material between the individuals of the same species or between cell types produced by them. To test the ratio of observed number of individuals of two different sex (ratio of observed value) is statistically acceptable or not a χ^2 test (non-parametric test) is used. Ideally, sex ratio (ratio of expected value) of an organism is taken as 1:1; though there may be deviation from this ratio due to different reasons.

Procedure :

1. Specimens (may be insects, birds, lizards, fish etc.) whose sex ratio is to be analyzed are collected or trapped.

2. Males and females are segregated on the basis of sexually dimorphic characters in separate cage and their members are counted (observed value).

3. Null Hypothesis is constructed which states that: The specimen that is analyzed has equal number of individuals in each sex or equal sex ratio (1:1) against an Alternate Hypothesis: The specimen that is analyzed has unequal number of individual in each sex or unequal sex ratio deviating from 1:1 ratio.

4. χ^2 value has to be calculated, having a degrees of freedom (d.f) = no. of classes –1, in ratio study in sex there are only two type of sex or two classes i.e. male and female. Thus d.f. = 2-1 = 1.

Calculation :

Table 8.3.2 : Chi-square table for analyzing the sex ratio of an individual

Sex	Observed value (O)	Expected value (E)	(O-E)	$(O-E)^2$	$\frac{(O-E)^2}{E}$
Male					
Female					

$$\therefore \chi^2 = \Sigma \frac{(\text{Observed value} - \text{Expected value})^2}{\text{Expected value}}$$

Interpretation :

At 5% level of significance the calculated value of χ^2 value is found to be greater/lesser (any one) than the tabulated χ^2 value. Thus the Null hypothesis is accepted (if calculated value of χ^2 value is found to be lower than the tabulated χ^2 value) or rejected (calculated value of χ^2 value is found to be greater than the tabulated χ^2 value).

Chapter 8.3.4

Study of Morphological Modifications of Honey Bee

Systematic Position :

Phylum	**Arthropoda**
Class	**Insecta**
Order	**Hymenoptera**
Genus	**Apis**
Species	***Apis indica***

There are three types of leg modification in three pairs of legs of worker bees. For various types of activity different parts of the legs are modified in different ways.

Table 8.3.3 : Modifications of legs in Apis and related activities

Limb	Specialization	Position	Function
Fore limb	Antenna cleaner.	Near the joint of tibia and tarsus.	Fore limbs acts as an antennae cleaner. There is a curved groove within the operculum. Bees inserts its antenna in the grooves and close the operculum, drags the dirty antenna through it as a result it is cleaned.
Middle Limb	Tibial spur.	On the distal end of their tibia.	The spine like structure helps in the removal of wax from the abdominal glands.
Hind Limb	Pollen basket.	First tarsal segment.	It is noted that the first tarsal segment of the hind limb is enlarged and covered with hairs. It aids in the storage and transportation of pollen grains from the flower.

Chapter 8.3.5

Study of Behavioural Catalogue and Ethogram

Introduction

Behaviour may be regarded as external manifestation of the stimulus – response phenomenon which response occurs voluntarily and consciously.

We start ethological or behavioural study by compiling a catalogue of behaviours for the species. Thus **behavioural catalogue** is a list of all behaviours that we have observed, listened to, or has knowledge about it.

An **ethogram** is a catalogue of an animal's behavioral repetoire, detailing the different forms of behaviour that are displayed by an animal. In most cases, it is desirable to create an ethogram in which the categories of behaviour are objective, discrete, and do not overlap with each other. Definitions should be clear, detailed and distinguishable from each other.

Ethograms can be as specific or general as the study warrants. A research project dealing strictly with aggressive behaviour in chimpanzees can be composed simply of the various forms of aggression displayed by that species. Alternatively, some studies seek to describe all facets of a particular animal's behaviour, and the appropriate ethogram will comprehensively categorize any and all behaivours that may occur.

Requirements

1. Species to be observed.
2. Pen and paper.
3. Computer facility with MS excel software (if possible).

Procedure

1. During the first observation period, one has to meticulously observe the species concerned.

2. On a piece of paper, a list of all the different types of behaviour that one observe (observed behaviour) are made with time and duration.

3. Take the behaviours that are noticed, categorised in the particular worksheet (Table 8.3.4a) provided. For instance, one behaviour that one may observe is "scratching" and it could go in a "resting" or it could go in a 'grooming' category.

4. This helps in the construction of behavioural catalogue.

5. Next the total amount of time (seconds, minutes, hours) spend for a particular behaviour is recorded.

6. A table (Table 8.3.4b) is constructed for the total amount of time by an individual behaviour.

7. A pie diagram is constructed (through MS Excel or, manually) from the total amount of time use for an individual behaviour.

Observation

Observations are noted in the two different Tables (8.3.4a & 8.3.4b).

Conclusion

Maximum time of the day (observed behaviour) is ...... and the duration of it is "...... hours".

Minimum time of the day (observed behaviour) is ...... and the duration of it is "...... hours".

Table 8.3.4a : Worksheet for cataloguing different category of behaviour.

Sl. No.	Observed behaviour	Category of behaviour	Starting time (1)	Ending time (2)	Total time spend (2-1)
1.					
2.					
3.					
4.					
5.					

Table 8.3.4b : Table for the total spending time by an individual behaviour.

Category of behaviour	Total time spend in each category in Hours and minutes	% of time spend	Convert into angle = (% of time spend/ 100) × 360

Chapter 8.4.1

Assay of Acetylcholinesterase on Exposure to Pesticides

Introduction

Acetylchlorine esterase 1(AChE), an enzyme found in the free state mainly in brain, nerve cells, muscle, lung and erythrocytes. It plays an important role in the transmission of nerve impulse. The enzyme activity is strongly inhibited due to poisoning with organo-phosphorus compounds or pesticides. The determination of AChE in whole blood may be done when there is a question of probable poisoning with organophosphorous compounds. AChE is widely assayed in toxicological studies on insects when treated with organophosphorous chemicals or pesticides.

Principle

The hydrolyzed acetylcholine per unit time is measured by comparison of the initial concentration in a reference tube with the final concentration in the experimental tube. Acetylcholine is made to react with hydroxylamine to form the corresponding acylhydroxamic acid which forms a strongly coloured ferric hydroxamate with ferric salts and the colour of hydroxamate is measured at 490 nm.

Materials Required

1. 0.2mL blood is collected from the test animal and transferred to 5mL water. The haemolysate is used for estimating AChE enzyme.
2. *Veronal Buffer, 0.1M pH 8.6* : 4.92g sodium vernal and 3.24g sodium acetate is dissolved in about 300mL water, 3mL 1N HCl is added in it and diluted to 500mL with water. pH is noted.
3. *Acetylcholine stock solution (200mM)* : 18.2 g acetylcholine chloride (hygroscopic checked the purity) is placed in a 50mL volumetric flask dissolved in water and made up to marked volume.
4. *Substrate acetylcholine solution (1.33nM)* : 150mL Veronal buffer and 1mL of acetylcholine stock solution is mixed thoroughly.
5. *Sodium hydroxide (2.5N)* : 10g NaOH is dissolved in water and made up to 100mL.
6. *Hydroxylamine (1N)* : 7g hydroxylammonium chloride is dissolved in water and made up to a volume of 100mL. The solution is stored in a well-stoppered polyethylene flask in a refrigerator.
7. *Alkaline hydroxylamine solution* : Equal volumes of sodium hydroxide (2.5N) and hydroxylamine (1N) are mixed.

8. *Iron Solution (0.7M)* : 33.75g $Fe(NH_4)(SO_4)_2 \cdot 12H_2O$ is dissolved in about 70mL water with gentle warming. 2.5g potassium nitrate is added (dissolved separately in water) and then transferred to a 100mL volumetric flask and diluted to upto the mark.

9. *Citrate Buffer (1M) (pH 1.4)* : 2.10g citric acid and 0.8g NaOH is dissolved in minimum quantity of water in a 100mL volumetric flask, 89mL 1N HCl is added and diluted to the mark with water. 10mL of this solution is added in a volumetric flask and diluted to 100mL; *p*H of this solution must be between 1.4 and 1.2.

10. Volumetric flasks (100mL, 50mL).

11. Stoppered polyethylene flask.

12. Vis-spectrophotometer.

Procedure

1. Pipetted out the following in a 50mL volumetric flask

	Reference mL	Test mL	Blank mL
Sample (enzyme source)	—	2	—
Substrate solution	25	25	—
Mixed well and incubated at 37°C for 30 min.			
Alkaline hydroxylamine solution	5	5	5
Sample	2	—	—
Citrate buffer	5	5	5
Ferric solution	10	10	10

2. Ferric solution is allowed to run slowly down the wall of the flask.

3. Diluted with water to mark and shaken thoroughly.

4. Allowed to stand for 20 minutes at room temperature.

5. The solution is filtered through a double-folded filter paper and the first portion of filtrate is discarded.

6. The absorbance of the filtrate at 490nm against blank is measured in a visible spectrophotometer.

Calculation

The absorbance difference (ΔE) between reference (initial concentration of substrate) and test (final concentration of substrate) is used for calculation. For measurements at 490 nm the extinction of the dye is 0.961 per micromole. Hence the amount of dye formed from the non-hydrolyzed acetylcholine in 50mL :

$$C = E \times 100/\ 0.961 \times 1.0\ (\mu smole/50mL)$$

Chapter 8.4.2

Study of Toxicity of given Chemical thorough LC_{50} and LD_{50} Determination

Introduction

Toxicology is the study of the effect of poisonous toxic substances or **toxicant** on living organisms concluding the way in which these substances gain entry into the organism. Above certain conditions, the toxicant has detrimental effect on physiological and other biological function of the organism. The concentration at which detrimental effect occurs is determined by dose response. The critical and threshold doses varies from one species to other.

In particular dose or degree of exposure of an organism to a toxicant is influenced by – amount of toxicant\unit body mass of the organism; the rate of entry and duration of exposure.

Toxicity is commonly divided on the basis of duration of exposure as: acute (involves a single dose), chronic (exposure for a longer period of time) and sub chronic (doing over a small time period).

Some important terms related to dose and toxicity are

LD_{50} is **lethal dose 50**. This is the single dose of a toxic substance administered by any route (other than by inhalation) that causes the death of 50% of an animal population. LD_{Lo} is **lethal dose low**—the lowest dose of a substance reported to have caused death in humans or animals.

LC_{50} is **lethal concentration 50**. It is the concentration of a material in air that kills 50% of a group of test animals when administered as a single exposure in a specific time period (usually 1 hour).

Note that lethal concentration values are used when the route of administration is by inhalation (rather than oral, dermal, etc.) whereas lethal Dose; LD values are used for all routes except inhalation.

Among animals mice, rats, rabbits, hamsters, guineapigs are used as test species. This test can be well performed by using macroinvertebrates and fishes. The LD_{50} test is applied in assessing acute toxicity in a mammalian system.

TD_{Lo} is **toxic dose low**. It is the lowest dose of a substance reported to produce any toxic effect in humans or animals or to produce tumorigenic or reproductive effects in animals or humans. Like LD, it is used for all routes of administration except inhalation. The same subscripts, routes of administration, and species tested abbreviations are used as with LD and LC. TD_{50} is toxic dose 50, and means that 50% of the test subjects suffered from one or more toxic effects.

TC_{Lo} is **toxic concentration low**. It is the lowest concentration of a substance in air to which humans (or animals) have been exposed for any given period of time that has produced any toxic effect or produced a tumorigenic or reproductive effect to one or more members of the group of subjects. Toxic concentration values are for the inhalation toxicity of a compound—they are not used for any other route of exposure. TC values are of great importance to industrial workers and chemists exposed to fumes and dust.

Sometimes death is not easily assessable in case of the invertebrates, so in that case EC_{50} (Mean Effective Concentration) is estimated rather than LC_{50}. The effects in EC_{50} analysis are used by picking up cues such as immobilization (lack of movement). EC_{50} is used in shrimp, crayfish, crabs etc.

One of the most important methods for LD_{50} and LC_{50} determination is nomgram method of Litchfield and Wilcoxon (1949).

There is huge numbers of toxic chemicals having a particulars LD_{50}/LC_{50} value in vertebrates.

This is tabularized below :

Table 8.4.1 : Toxic elements or compound and their LD_{50}/LC_{50} value

Element/ Compound	LD_{50} in mammals (mg Kg^{-1})	LC_{50} in fish (mg L^{-1})
Cd	1.3	—
Co	50	—
Hg	1.5	—
Mn	18	—
Pb	70	—
Se	1.3	—
Chlorpyrifos	130-160	0.003
Glyphosate	60-80	0.17(blue gill)
Malathion	2800	0.1(blue gill)
Parthion	14-24	2.7(trout)

Adapted from Bower, 1979 and Alloway et.al. 1993.

(i) To determine LD_{50} of a hypothetical toxicant 'T' for rat

Principle :

In this type of test, the toxicant is administered in the test animal's body through oral route in a gradually increasing doses. Subsequent biological effect (response) is noted in a particular time frame.

Materials and conditions required :

1. Wistar rats in healthy condition of same of weight ±20% of mean body weight. 5 male and female rats must be taken for the procedure.
2. Test chemical or toxicant ("T") with proper brand name chemical name, solubility, appearance, odour, storage conditions should be recorded.
3. Optimum laboratory conditions with room temperature 22°C (±3°C), a relative humidity of 30-70% and proper artificial lighting (12 hours light and 12 hours dark cycle) should be maintained.
4. Conventional rodent diets with *ad libtum* supply of drinking water must be provided.

Procedure :

Step I : Toxicant administration :

(1) Animals are divided into 4 different groups which includes. One control (C) and three test groups (X,Y and Z). The test chemicals should be dissolved in water in different concentration and should be orally fed to animals in the 3 test groups where as animals in the control group should be fed with water (positive control). This is shown through Table 8.4.2.

Table 8.4.2 : Dose levels

Groups			
Control	Test		
C	X	Y	Z
Fed with water	Fed with toxicant 'T' as *a* mg/kg	Fed with toxicant 'T' as 2*a* mg/kg	Fed with toxicant 'T' as 3*a* mg/kg

Where '*a*' is a specified dose of toxicant 'T' per kg body weight of animal.

2. Animals are fasted prior to the feeding of the toxicant (no food and water in the previous night of treatment).

3. The animals are then weighed and the toxicant is administed in a single dose to the animals in test group (X, Y, and Z) in particular dose through gavage using a suitable stomach tube or cannula. The control group is fed only with water.

4. After the test substance is correctly administrated in the animal, observation should be made and data should be properly recovered and tabulated.

5. LD_{50} value is determined through Litchfield and Wilcoxan's (1949) logarathimic probit graph paper method. % mortality, adjusted % mortality and Probit value (Table 8.4.3).

6. Observations are made 15 minutes to 4 hours up to 24 hours. After such changes body weight, behavioural and gross **necropsy** may be observed up to a period of two weeks every 24 hours.

Percent (%) Mortality =

(No. of rats died/No. of rats administrated with toxicant) x 100

Adjusted % Mortality =

(a) For 0% death: 100 [(0.25/n)]

(b) For 100% death: 100[(n-0.25) n]

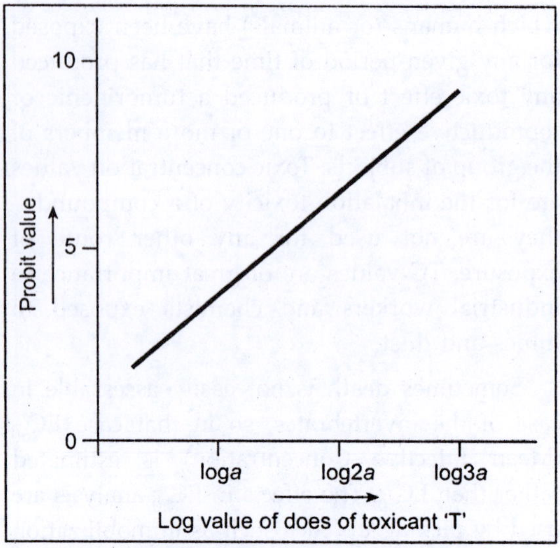

Fig. 8.4.2.1 : Probit graph showing LD_{50} toxicity study of toxicant 'T'.

n = no. of animals in the group.

The adjusted percent mortality is converted into Probit value from a % mortality-probit conversion table (Table 8.4.4).

7. The Probit value is plotted against log value of dose (Fig. 8.4.2.1) of the toxicant in a logarithmic probability graph paper and acute LD_{50} value of toxicant is determined.

Note that the Probit value 5 indicates the oral LD_{50} value (expressed in mg/kg of body weight)

Table 8.4.3 : Results : % mortality, % adjusted mortality and probit value

Group	No. of rats aministd. with	Dose (mg/1kg b.w.)	Log doses	No. of of rats died							Mortality %	Adjusted mortality %	Probit value
				15′	30′	1 hr.	2 hr.	4 hr.	24 hr.	total			
C	5	0.0		0	0	0	0	0	0	0	0	5.0	3.355
X	5	a	Log(a)										
Y	5	2a	Log(2a)										
Z	5	3a	Log(3a)										

Table 8.4.4 : Probit-conversion table

df (%)	0	1	2	3	4	5	6	7	8	9
0		2.67	2.95	3.12	3.25	3.36	3.45	3.52	3.59	3.66
10	3.72	3.77	3.82	3.87	3.92	3.96	4.01	4.05	4.08	4.12
20	4.16	4.19	4.23	4.26	4.29	4.33	4.36	4.39	4.42	4.45
30	4.48	4.50	4.53	4.56	4.59	4.61	4.64	4.67	4.69	4.72
40	4.75	4.77	4.80	4.82	4.85	4.87	4.90	4.92	4.95	4.97
50	5.00	5.03	5.05	5.08	5.10	5.13	5.15	5.18	5.20	5.23
60	5.25	5.28	5.31	5.33	5.36	5.39	5.41	5.44	5.47	5.50
70	5.52	555	5.58	5.61	5.64	5.67	5.71	5.74	5.77	5.81
80	5.84	5.88	5.92	5.95	5.99	6.04	6.08	6.13	6.18	6.23
90	6.28	6.34	6.41	6.48	6.55	6.64	6.75	6.88	7.05	7.33

of the test chemicals The calculated LD_{50} value of the toxicant is Y mg/kg.

Note : The lethal dose test does not accurately measure human health hazards and is very crude and imprecise. Lethal dose test results can be affected by the age and sex of the animals tested, their housing and nutritional conditions, temperature, time of day and year, and the exact method used to administer the substance. Different species react differently to substances, and reactions between individuals of the same species can also vary greatly. For example, nicotine is lethal to humans at 0.9 mg/kg, but lethal dose values of nicotine in dogs are a staggering 9.2 mg/kg; in pigeons, 75 mg/kg; and in rats, 53 mg/kg.

(ii) To determine LC_{50} of a hypothetical toxicant 'B' for *Channa sp.*

Materials required :

1. 6 glass aquaria of 40L volume
2. 60 live fish (*Channa punctatus*) of more or less equal weight.
3. Measuring cylinder (1L).
4. Balance.
5. Toxicant.
6. Semil-log coordinate paper.
7. Stop watch.

Procedure :

1. 6 glass aquaria are taken. These are arranged in a line on the experimental table. These are then marked as I, II, III, IV, V and VI.

2. Each aquaria are filled with 40 L of tap water.

3. Toxicant is added in the aquaria I, II, III, IV, V and VI in the concentration of 0 mg/L (no toxicant added, used as control),1 mg/L, 1.5 mg/L, 3 mg/L, 5 mg/L and 10 mg/L respectively. II, III, IV, V and VI aquaria are used as experimental ones.

4. 10 live and healthy fishes weighed and are released in each aquarium.

5. Mortality is observed for 96 hrs. The record is maintained as a table.

Table 8.4.5 : 48 hours and 96 hours mortality rate of *Channa punctatus* expressed as percentage exposed to a toxicant

Aquarium No.	No. of test animals (*Channa punctatus*)	Concentration of the toxicant (mg/L)	Mortality during 48 hrs.		Mortality during 96 hrs.	
			No. of fishes dead	% mortality	No. of fishes dead	% mortality
I	10	Control (no toxicant)	Nil	Nil	Nil	Nil
II	10	1.0	Nil	Nil	1	10
III	10	1.5	3	30	5	50
IV	10	3.0	5	50	7	70
V	10	5.0	7	70	9	90
VI	10	10.0	10	100	10	100

Note: *All value given in the table are hypothetical in nature. This only an illustration for calculating* LC_{50} *value of a toxicant in a test animal.*

Observations :

See Table 8.4.5

1. The LC_{50} value is determined through the interpolation which involves the plotting the data on a semi-log coordinate graph paper showing concentration of toxicant on log Y axis and percentage of mortality of the test animals (*Channa punctatus*) on the arithmetic scale on the X axis.

2. A straight line is drawn between two points indicative of percentage mortality at two successive points that represents percentage mortality at two successive concentrations.

3. The concentration at which line crosses the 50% lethality line will be the LC_{50} value. See Figure 8.4.2.2.

Results :

From the graph LC_{50} determined in this case for 48 hours is .... and for 96 hours is ....

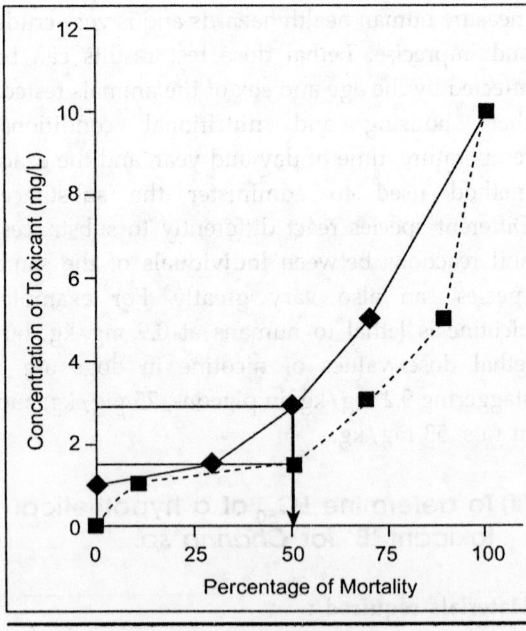

Fig. 8.4.2.2 : Determination of LC_{50} by straight line graphical interpolation

Chapter 8.4.3

Test of Food Adulterates

(a) Detection of argemone oil in mustard oil

Requirements :

1. Test tube.
2. Test tube holder.
3. Concentrated Nitric acid.
4. Concentrated Hydrochloric acid.
5. Water bath.
6. Dropper.
7. Mustard oil.
8. $FeCl_3$ solution.

Procedure :

Procedural part is tabulated below.

Table 8.4.6 : Test, observation and interference for detection of argemone oil adulteration in mustard oil

Test	Observation	Inference
(i) 5mL of Mustard oil is taken a test tube in which 3-5mL of HNO_3 is added and shaken vigorously. The whole mixture separates in two layers.	Lower layer is yellow or dark yellow or red in colour	Presence of argemone oil in mustard oil in good amount.If argemone oil is found in lesser amounts then the aqueous layer is colourless in that case the second test is carried on.
(ii) In a test tube in 5mL of mustard oil and 2mL of concentrated HCl is boiled for 5 minutes in a water bath. Next the mixture is cooled. The mixture separate in two layers. With the help of dropper the oily layer is removed. In the remaining aqueous layer 1mL of $FeCl_3$ solution is added. The mixture is shaken well and kept in boiling water bath for 10 minutes.	Reddish brown precipitate is formed.	Small amount of argemone oil is present.

(b) Detection of colouring substances in red chilli and turmeric powder

Materials Required :

1. Beaker.
2. Test tube.
3. Test tube holder.
4. Concentrated HCl acid.
5. $FeCl_3$ solution.
6. Distilled water.
7. Bunsen burner.

Procedure :

20mL of distilled water is taken in a beaker. In it one pinch of turmeric powder or chilli powder is added. Ingredients are mixed thoroughly for 2-3 minutes and boiled for 2-3 minutes over a burner. The mixture is cooled and filtered. Next following tests are carried out with the filtrate.

Table 8.4.7 : Test, observation and inference for detection of colouring substances in red chilli and turmeric powder

Test	Observation	Inference
(i) The colour of solution is observed.	If colour of the solution is dark yellow or reddish yellow.	Some colouring substances is present.
(ii) In 1mL of this solution, 2mL of distilled water is mixed in a test tube and intensity of the solution is reduced. In this solution 1mL of conc. HCl is mixed.	Solution becomes red or blue.	Colouring substance is present.
(iii)The solution is taken in a 5mL test tube. In it 2-3 drops of $FeCl_3$ solution are added.	Colour of the solution becomes red or blue or green or violet.	Colouring substance present.

(c) Detection of copper sulphate or malachite green solution on surface of green vegetables

Materials Required :

1. Distilled water.
2. Test tube.
3. Test tube holder.
4. Beaker.
5. Liquor ammonia solution.
6. $AgNO_3$ solution.
7. Concentrated HNO_3 acid.

Procedure :

Vegetable samples are washed thoroughly in distilled water in a beaker. This distilled water is used to carry on the following tests.

(d) Detection of chalk powder or sugar or metanil yellow colour

Materials Required :

1. Distilled water.
2. Test tube
3. Test tube holder
4. Conc. HCl acid

Table 8.4.8 : Test, observation and inference for detection of copper sulphate and malachite green on the surface of green vegetables		
Test	**Observation**	**Inference**
(i) Distilled water colour is observed.	If the colour is light blue.	Copper sulphate ($CuSO_4$) present.
(ii) Distilled water colour is observed.	If the colour is light green.	Malachite green present.
(iii)If the water solution is light blue, this solution is taken in a 5mL test tube, in which added excess amount of liquor ammonia solution.	The colours of the solution become dark blue.	$CuSO_4$ present.
(iv) If the water solution is light yellow, this solution is taken into 5mL test tube in which $AgNO_3$ solution is added.	Curdy white precipitate occurs. In green solution such precipitate is filtered out and washed in water. This white precipitate is taken into a test tube and liquor ammonia is added to it and precipitate become soluble. If in this solution conc. HNO_3 is added, white precipitate reappears.	Malachite green present.

5. Ether
6. Petri dish
7. Porcelain dish
8. Resorcinol (1 gm of resorcinol resublimed in 5 ml of conc. HCl)
9. Aniline chloride solution (7 ml of Aniline and 3 ml of 1 : 3 HCl)

Procedure :

Follow the table 8.4.9 given below :

Table 8.4.9 : Food articles, test, observation and inference for detection of adulteration in sweets			
Food articles	**Test**	**Observation**	**Inference**
1. Sugar	(i) The sample is taken in a test tube containing distilled water.	White precipitate is settle down in the form of a layer.	Chalk powder present.
	(ii) Few drops of conc. HCl is added in it.	Effervescence comes out.	Chalk powder present.
2. Gur (jaggery)	(i) The sample is taken in a test tube and 1 ml of HCl is added in it.	Colour become magenta red.	Matanil yellow colour present.

Table contd. next page

Food articles	Test	Observation	Inference
3. Honey	1. **Fieche's test :** (i) 5 ml of honey is taken in a test tube. 5 ml of solvent ether is added in it. The tube is shaken well and the ether layer is decanted in a petridish and the ether layer is completely evaporated by blowing. (ii) 2-3 ml of resorcinol is added in it.	The colour of the solution become cherry red.	Sugar or jaggery present.
	2. **Aniline chloride test :** (i) 5 ml of honey is taken in a porcelain dish. Aniline chloride solution is added in it and stirring well.	The colour of the solution become orange red.	Sugar present.
4. Sweet meet, ice creams and cold drinks	(i) Watery sample is taken in a test tube and 1 ml of HCl is added in it.	The colour of the solution become red.	Metanil yellow colour present.

Section 9

PARASITOLOGY, IMMUNOLOGY AND MICROBIOLOGY

SECTION AGENDA

Chapter 9.1

Parasitology

Parasitology is the study of parasites and their relationship to their hosts. The study of animal's parasites has a long and distinguished place in the history of biology. By definition **parasite** is the animal which thrives at the expense of another organism in which it lives and **host** is the animal which harbours the parasites. The parasites are classified as **Ectoparasites** and **Endoparasites**. Ectoparasites are those, which live outside on the surface of the body of the host. **Endoparasites** are those, which live inside the body of the host in the blood, tissues, body cavities, digestive tract and other organs. These parasites are again classified as pathogenic and non-pathogenic parasites. Non-pathogenic parasites, which live on superfluous fluids or solids in the host, are known as **commensals**. A pathogenic parasite lives inside its host and derive nourishment from the tissues and body fluids, inflicts injuries in the host tissues. Most of the endoparasites are pathogenic what ever may be, find suitable habitat inside its host and live in its food supply and feed upon it. It must have adaptations regarding infectiousness, establishment in a host and transmission to new hosts.

Both vertebrates and invertebrates serve as hosts. In case of vertebrates, study of parasitic infestation is important because it creates a threat that adversely affects their productivity and viability. Identification of these parasites make easier for the medical and veterinary personnel to treat the animals. Regarding invertebrate hosts, they are important in the context of their role as vectors or intermediate hosts to different types of parasites like protozoa, trematoda, cestoda and nematode. Identification of these parasites inside vertebrates gives a clear idea about them. The technical procedures that have been used in the present experiments also serve as first hand experience for the student who wishes to join in this field in near future.

9.1.1 STUDY ON SEMINAL VESICLE SMEAR OF EARTHWORM

Materials required :

1. Live Earthworm.
2. Dissecting tray.
3. Dissecting instruments like scissors, scalpels, needles, forceps etc.
4. Glass slides.

5. Fixatives, preservatives and stains (Mayer's albumen and Schauddinn's fluid used as fixatives; 50%, 70% and 90% alcohol used as preservatives; Haematoxylin and Eosin used as stains).

6. Petridish .

7. Dropper.

8. Coupling jars containing 50% 70%, 90% and 100% alcohol, Haematoxylin and Eosin.

9. Distilled water.

10. DPX as mountant.

11. Microscope.

Method :

1. The Earthworm is dissected out to locate the seminal vesicle (present in 11th and 12th segments).

2. The seminal vesicle is cut using a scalpel into two halves and the open portion is placed (length wise) on a Mayer's albumen coated slide and a thin smear is prepared by rubbing gently.

3. Immediately the slide is placed into a petridish and Schaudinn's fluid is dropped over it using a dropper and then it is transferred in 90% alcohol in a coupling jar and kept for 15 minutes.

4. The slide is then passed through 70% and 50% alcohol and then to distilled water.

5. The slide is then dipped into Haematoxylin and then into Eosin.

6. It is then dehydrated through 50%, 70%, 90% and 100% alcohol and mounted in DPX.

Observation :

Spindle shaped parasitic protozoa *Monocystis sp.* (See Chapter 1.1.1 Nonchordates : Protozoa Item no. 15) is observed under microscope.

9.1.2 STUDY ON GUT CONTENT SMEAR OF COCKROACH

Materials required :

1. Live Cockroach (male/female).

2. Dissecting tray.

3. Dissecting instruments like scissors, scalpels, needles, forceps etc.

4. Glass slides.

5. Fixatives, preservatives and stains (Mayer's albumen; Schauddin's fluid; 50%, 70% and 90% alcohol; Haematoxylin and Eosin stain).

6. Petridish.

7. Dropper.

8. Coupling jars containing 50%, 70%, 90%, and 100% alcohol; Haematoxylin and Eosin.

9. Distilled water.

10. DPX or canada balsam as mountant.

11. Microscope.

Method :

1. The cockroach is dissected out to open the hindgut.

2. The hindgut is incised longitudinally using a scalpel and the open portion is placed in a Mayer's albumen coated slide and a thin smear is prepared by rubbing gently.

3. To avoid dryness, immediately the slide is placed into a petridish and Schauddinn's fluid is spilled over it using a dropper and then it is transferred in 90% alcohol in a coupling jar and kept for 15 minutes.

4. The slide is than passed through 70% and 50% alcohol and then to distilled water.

5. The slide is then dipped into Haematoxylin and then into Eosin.

6. It is then dehydrated through 50%, 70%, 90% and 100% alcohol and mounted in DPX or canada balsam.

Observation :

Parasitic *Balantidium sp.* (See Chapter 1.1.1 Nonchordates: Protozoa Item no 10) *Nyctotherus sp.*(See Chapter 1.1.1 Nonchordates: Protozoa Item no 12)., *Gregarina sp. Leidynema sp.* are observed under microscope (Fig. 9.1.1).

9.1.3 STUDY ON RECTAL CONTENT SMEAR OF TOAD

Materials required :

1. Live Toad (male/female).
2. Dissecting tray.
3. Dissecting instruments like scissors, scalpels, forceps, needle etc.
4. Glass slides.
5. Fixatives, preservatives and stains (Mayer's albumen and Schauddinn's fluid; 0.7% sodium chloride solution; 50%, 70% and 90% alcohol; Haematoxylin and Eosin stain).
6. Petridish.
7. Dropper.
8. Coupling jars containing 50%, 70%, 90% and 100% alcohol; Haematoxylin and Eosin.
9. Distilled water.
10. DPX or canada balsam as mountant.
11. Microscope.

Method :

1. The Toad is dissected out to open the rectum.
2. The rectum is then incised using a scalpel and a few drops of 0.7% sodium chloride solution (normal saline) are dropped over it.
3. A little amount of rectal content is placed using a forceps and a thin smear is made on Mayer's albumen coated slide.
4. Immediately the slide is placed into a petridish and Schauddinn's fluid is dropped over it using a dropper and then it is transferred in 90% alcohol in a coupling jar and is kept for 15 minutes.
5. The slide is then passed through 70% and 50% alcohol and then to distilled water.
6. The slide is dipped into Haematoxylin and then into Eosin.
7. It is then dehydrated through 50%, 70%, 90% and 100% alcohol and mounted in DPX or canada balsam.

Observations :

Parasitic *Nyctotherus sp.* (See Chapter 1.1.1 Nonchordates: Protozoa Item no. 11), *Opalina sp.* (See Chapter 1.1.1 Nonchordates: Protozoa Item no. 6), and *Balantidium sp.*(See Chapter 1.1.1 Nonchordates: Protozoa Item no. 10), *Trichomonas sp.* are observed under microscope (Fig. 9.1.1a).

9.1.4 STUDY ON FAECAL SMEAR OF FOWL AND CATTLE

Direct Smearing technique :

Materials required :

1. Faecal matter of Fowl and Cattle (freshly passed out).
2. 0.7% sodium chloride solution (normal saline).
3. Forceps.
4. Glass slides.
5. Dropper.
6. Fixatives, preservatives and stains (Mayer's albumen, Schauddinn's fluid; 50%, 70%, and 90% alcohol; Haematoxylin and Eosin stain).
7. Petridish.
8. Coupling jars containing 50%, 70%, 90% and 100% alcohol; Haematoxylin and Eosin.
9. Distilled water.
10. DPX or canada balsam as mountant.

Method :

1. Freshly passed out faecals are collected and a small amount of faecal matter is placed on 3-4 glass slides coated with Mayer's albumen.
2. Few drops of normal saline is put on the faecal matter and mixed thoroughly to obtain a thin suspension.
3. The suspension is spread uniformly over a small area of the glass slide to make a thin smear.
4. Immediately the slide is placed into a petridish and schauddinn's fluid is put over it

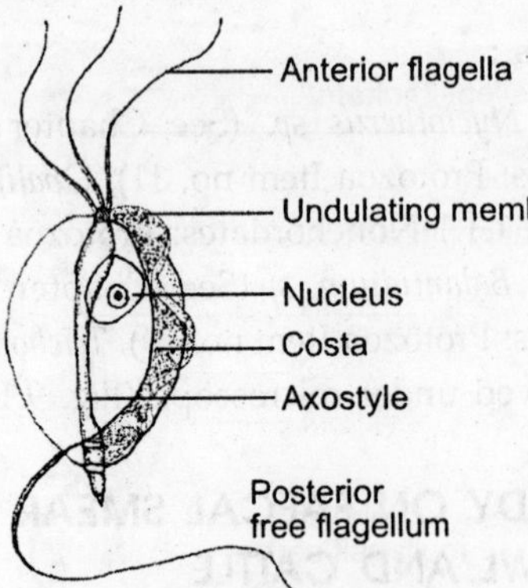

Fig. 9.1.1(a) : *Trichomonas sp.*

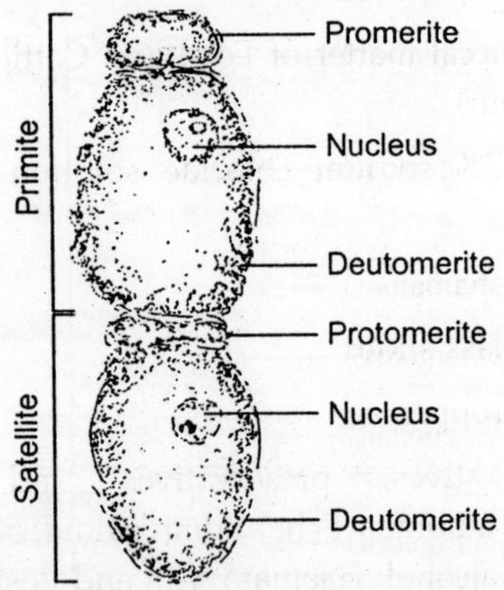

Fig. 9.1.1(b) : *Gregarina sp.*

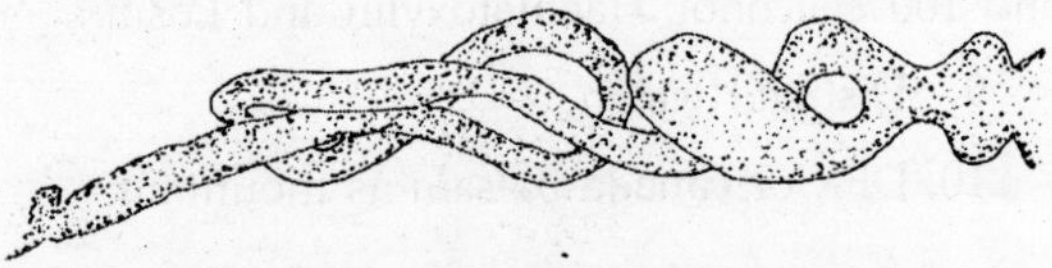

Fig. 9.1.1(c) : *Leidynema sp.*

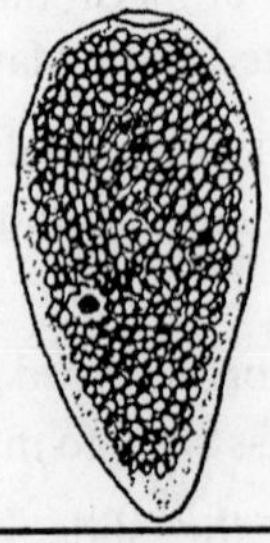

Fig. 9.1.1(d) : *Davainea sp.*

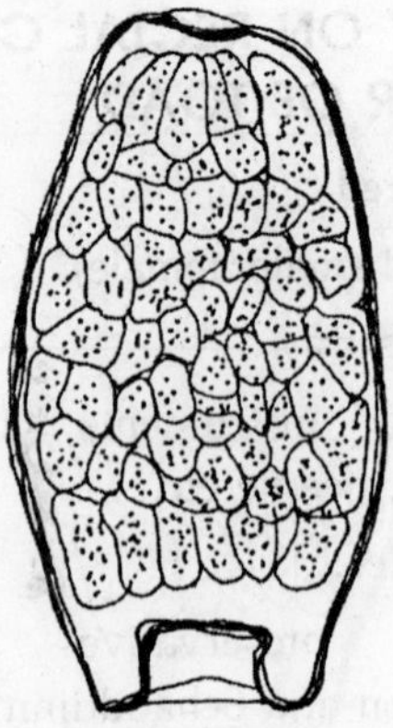

Fig. 9.1.1(e) : *Raillietina sp.*

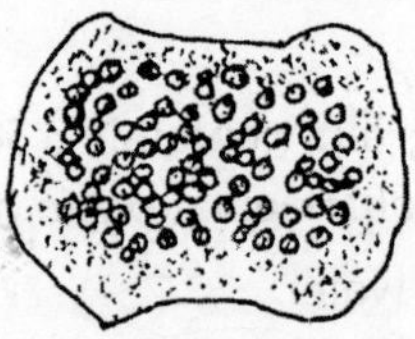

Fig. 9.1.1(f) : *Amoebotaenia sp.*

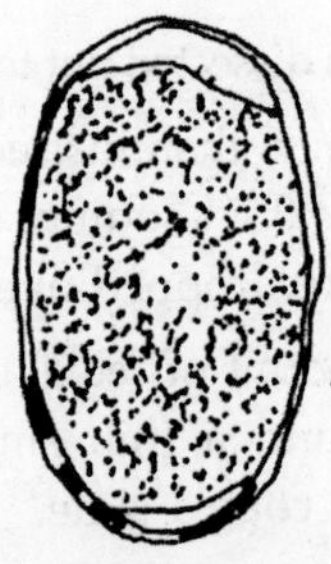

Fig. 9.1.1(g) : *Heterakis sp.*

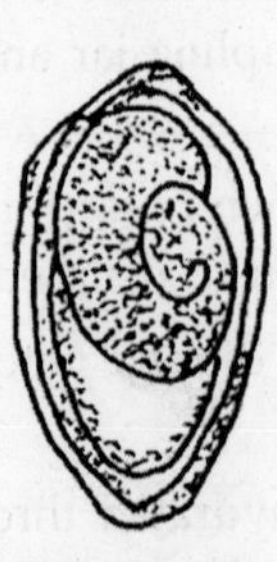

Fig. 9.1.1(h) : *Tetrameres sp.*

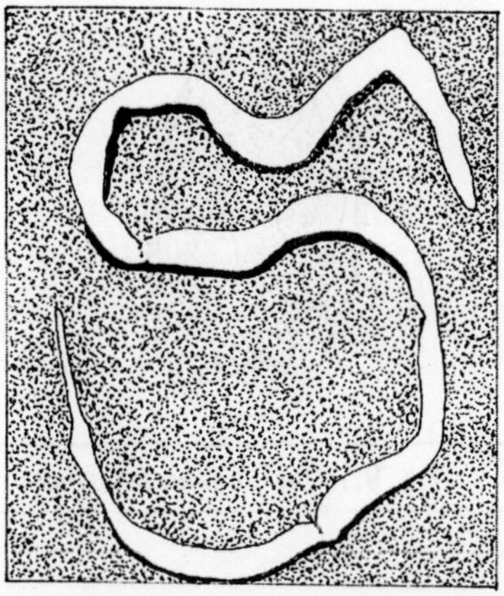

Fig. 9.1.1(i) : *Ascaridia sp.*

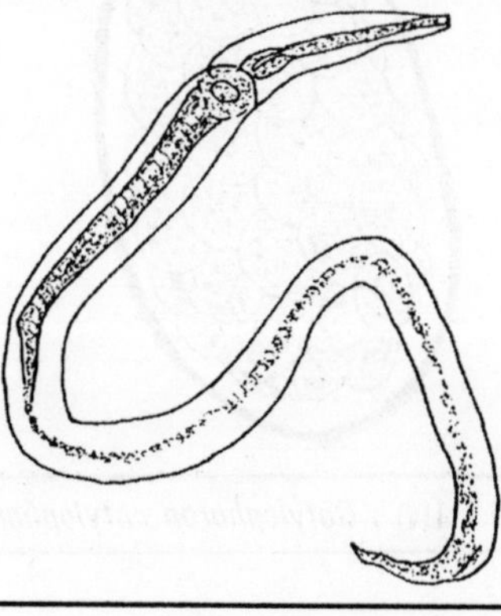

Fig. 9.1.1(j) : *Subulura sp.*

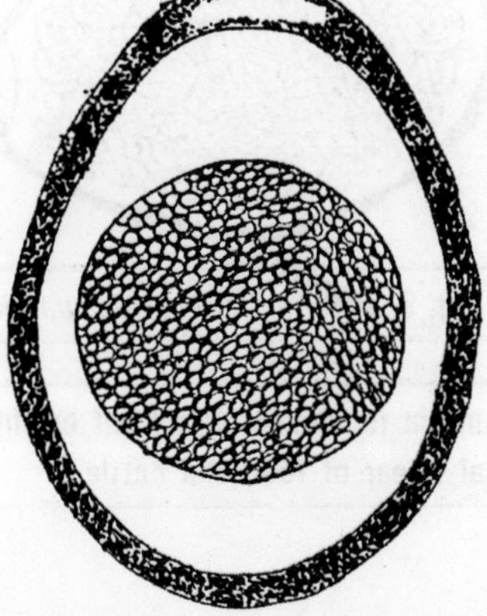

Fig. 9.1.1(k) : *Eimeria bovis*

Fig. 9.1.1(l) : *E. subspherica*

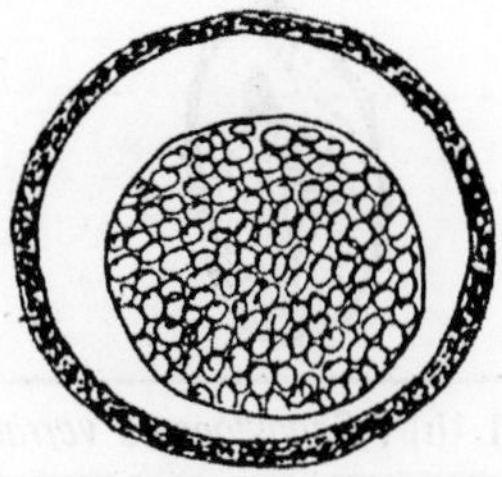

Fig. 9.1.1(m) : *E. Zurnii*

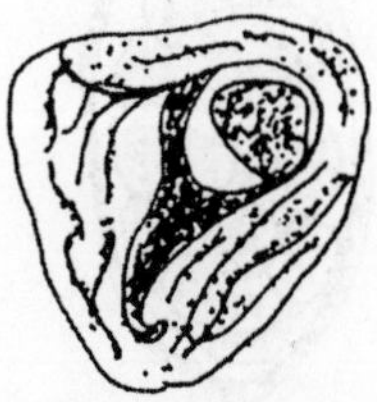

Fig. 9.1.1(n) : *Moniezia expansa*

Fig. 9.1.1(o) : *Avitellina centripunctata*

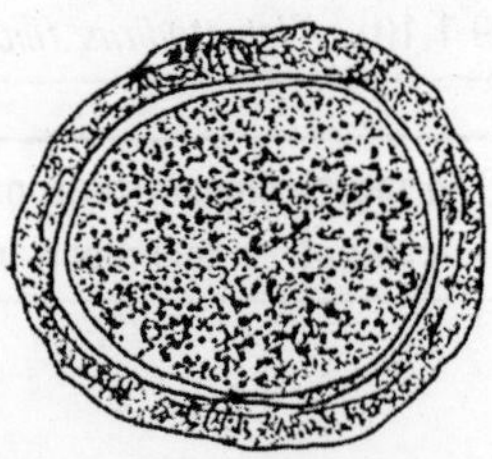

Fig. 9.1.1(p) : *Capillaria bilobata*

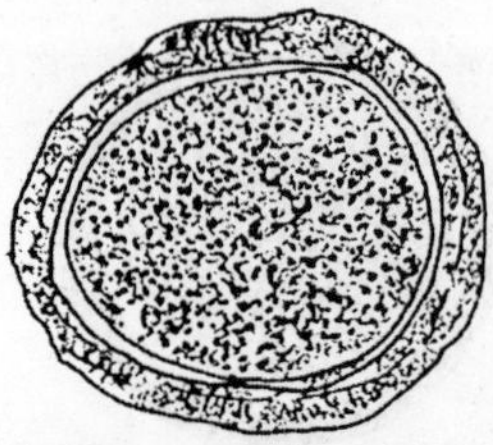

Fig. 9.1.1(q) : *Toxocara vitulorum*

Fig. 9.1.1(r) : *Gongylonema verrucosum*

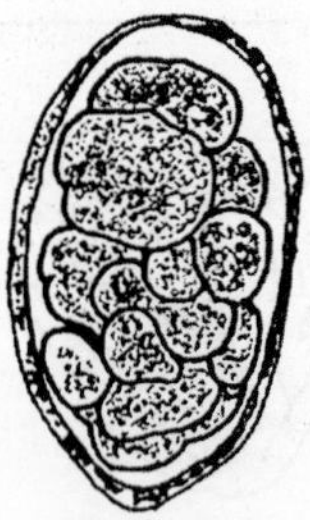

Fig. 9.1.1(s) : *Haemonchus pennata*

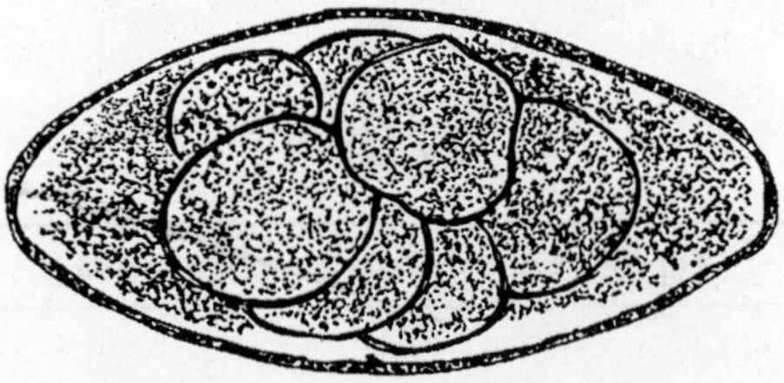

Fig. 9.1.1(t) : *Nematodirus filicollis*

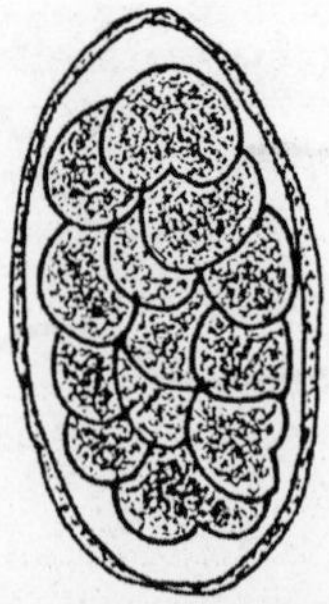

Fig. 9.1.1(u) : *Chabertia ovina*

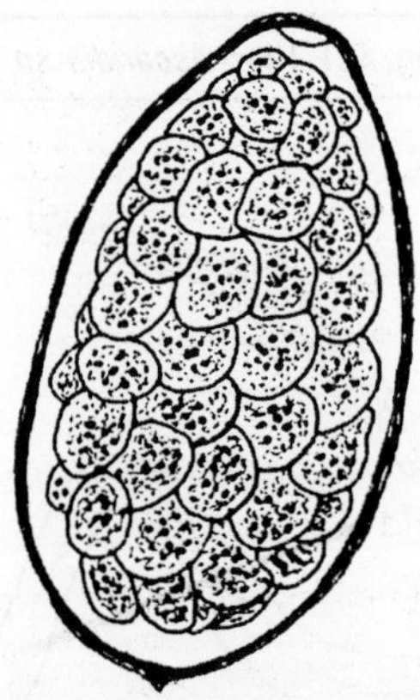

Fig. 9.1.1(v) : *Cotylophoron cotylophorum*

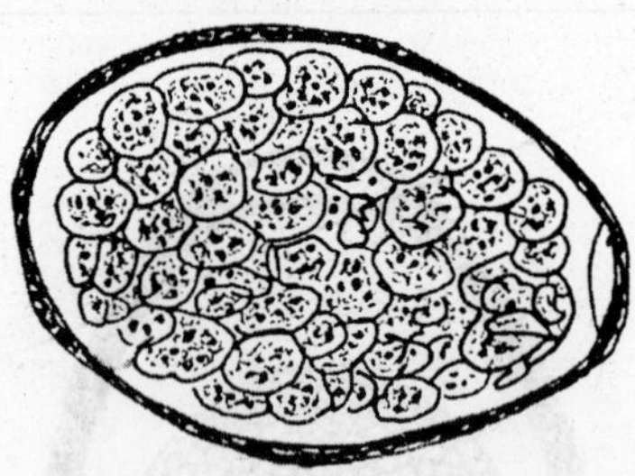

Fig. 9.1.1(w) : *Carmyerius spatiosus*

Fig. 9.1.1 : The most common form of parasite (egg/adult/cysts) found in te seminal vesicle of earthworm, gut content of cockroach, rectal smear of toad and faecal smear of fowl and cattle.

Table 9.1 : Various types of parasites found in fowl and cattle faecal sample

Animal	Protozoa	Castoda	Nematoda	Trematoda
Fowl faecal sample	*Eimeria sp.*((See Chapter 1.1.1 Nonchordates Item no 14), *Trichomonas sp.*	*Davainea sp., Raillientina sp., Amaebotaenia sp.*	*Heterakis sp., Ascaridia sp., Tetrameres sp., Subulura sp.*	—
Cattle faecal sample	*Eimeria bovis, E. zurnii, E. subspherica*	*Avitellina centripunctata, Moniezia expansa*	*Capillaria bilobata, Gongylonema verrucosum, Nematodirus filicollis, Toxocara vitulorum, Chabertia ovina, Haemonchus pennata*	*Carmyerius spatiosus, Cotylophoron cotylophorum*

using a dropper and it is then transferred to 90% alcohol in a coupling jar and is kept for 15 minutes.

5. The slide is then passed through 70% and 50% alcohol and then to distilled water.

6. The slide is dipped into haematoxylin and then into Eosin.

7. It is then dehydrated through 50%, 70%, 90% and 100% alcohol and mounted in DPX or canada balsam

Observation :

See Table 9.1 and Fig. 9.1.1.

9.1.5 COLLECTION OF HELMINTH PARASITES FROM VERTEBRATES, THEIR PRESERVATION AND STAINING

Introduction :

Phylum platyhelminthes includes animals usually known as flat worms. The flat worms, as parasites, inhabit all organ and systems of the host. Phylum platyheminthes is divided into four classes : Turbellaria, Monogenea, Trematoda and Cestoda.

Parasitic helminths belong to class Monogenea, Trematoda and cestoda. Monogenetic flatworms are usually parasitic on the skin or gills of fishes; that is, most are ectoparasites. Trematodans are endoparasitic, occurring in all classes of vertebrates. Cestodes are commonly known as tape worms are all internal parasites; that is endoparasitic in the various regions of the host digestive tract. They are primarily parasites of vertebrates.

Collection :

Living or freshly killed hosts should be examined for collection of parasitic helminthes. When examining the fish scales carefully the entire outer surface, specially the oral region, the gills, the operculum and the fins. On skin, fins, gills and in nasal chambers the Monogenea may be present. **Dactylogyrids, Gyrodactylids, Capsalids** and **Mazocraeids** are most commonly found as fish parasites (Fig. 9.1.2) one may find encysted metacercaria of digenetic trematodes on the body and fins of fishes.

After the external examination, the animal is kept in an enamel tray containing saline water. A long incision is made from anus to urinogenital opening on the mid-ventral surface and the body cavities are examined for parasites or cysts. Next the internal organs are removed, one at a time

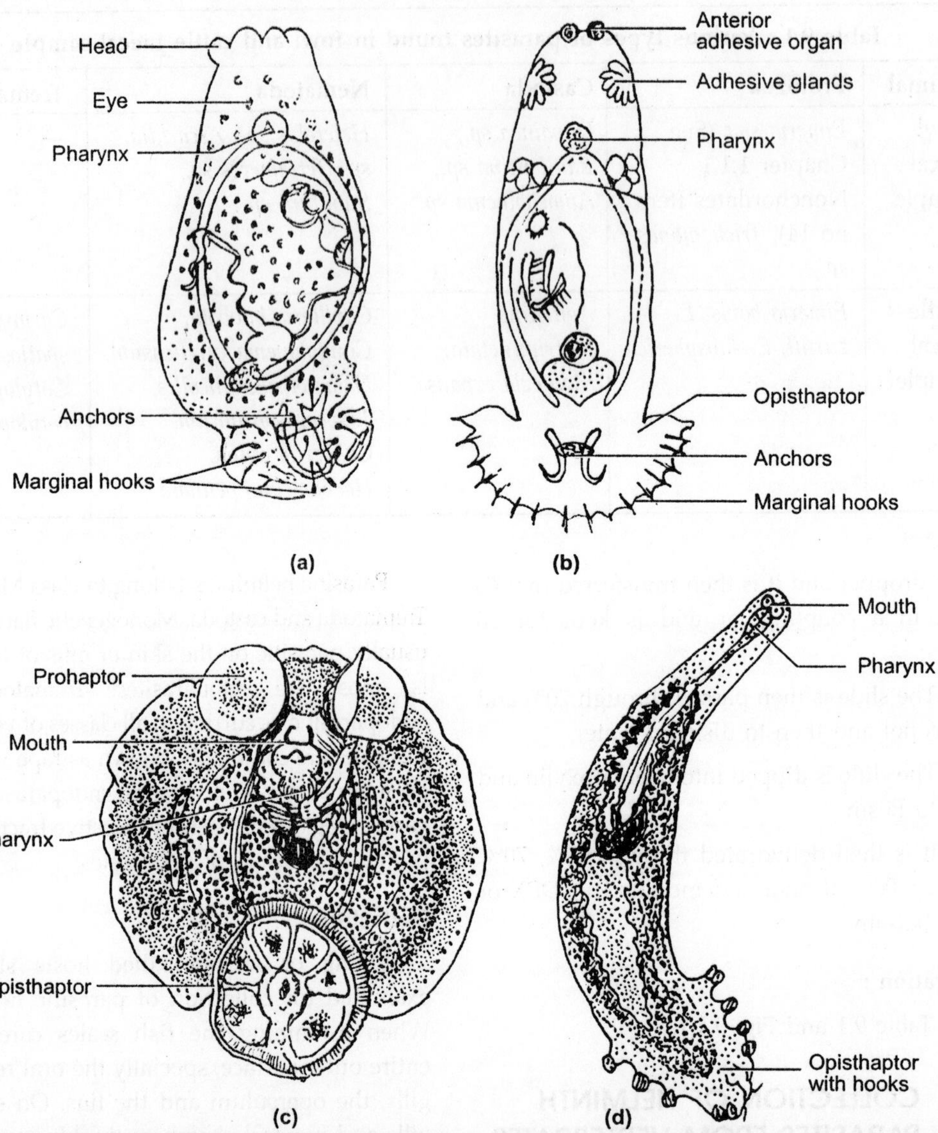

Fig. 9.1.2 : Fish monogenean parasites; (a) Dactylogyrids, (b) Gyrodactylids, (c) Capsalids and (d) Mazocraeids.

and placed separately in dishes containing saline water. Each organ is cut open and dipped in saline water so that the parasite is dislodged. The parasites are picked up with help of dropper or brush.

Fixation :

The worms are thoroughly washed in saline solution to free it from debris and mucus. After the worms are cleaned these are further washed in distilled water. Next the specimens are shaken vigorously in hot water or 4% hot formalin. Hot Bouin's fluid, hot 10% formalin can also be used as fixatives.

Presevation :

After fixation, the specimen are thoroughly washed and preserved in 70% ethyl alcohol. If the specimen is fixed in Bouin's fluid it should be washed in 70% ethyl alcohol till the yellow colour disappears. For long period of storage of the materials one should add 5% glycerine to the 70% ethyl alcohol. All the vials containing the specimens should be properly labelled.

Staining, dehydration and Mounting :

Many stains are used according to the need. The most commonly used stains for the whole mounts of helminths are carmine and haematoxylin. A carmine stain is used for alcohol fixed material, and haematoxylin stain is used for formalin fixed material (for procedure see chapter 4.5).

9.1.6 IDENTIFICATION OF SOME COMMON PARASITES OF MAN

I. *Entamoeba* : See chapter 1.1.1. Non chordate: Protozoa; Item no 2.

II. *Trypanosoma* : See chapter 1.1.1. Non chordate: Protozoa; Item no 7.

III. *Giardia* : See chapter 1.1.1. Non chordate: Protozoa; Item no 8.

IV. *Leishmania* : See chapter 1.1.1. Non chordate: Protozoa; Item no 9.

V. *Balantidium* : See chapter 1.1.1. Non chordate: Protozoa; Item no 10.

VI. *Plasmodium* : See chapter 1.1.1. Non chordate: Protozoa; Item no 13.

VII. *Schistosoma* : See chapter 1.1.1. Non chordate: Platyhelminthes; Item no 4.

VIII. *Taenia* : See chapter 1.1.1. Non chordate: Platyhelminthes; Item no 5.

IX. *Ascaris* : See chapter 1.1.1. Non chordate: Nematoda; Item no 1.

X. *Ancylostoma* : See chapter 1.1.1. Non chordate: Nematoda; Item no 2.

XI. *Wuchereria* : See chapter 1.1.1. Non chordate: Nematoda; Item no 3.

XII. *Trichinella* : See chapter 1.1.1. Non chordate: Nematoda; Item no 4.

XIII. *Pediculus* : See chapter 1.1.1. Non chordate: Arthropoda; Item no 32.

9.1.7 IDENTIFICATION OF PARASITE OF BIRD

1. Menopon

Habit and Habitat :

Wingless flattened ectoparasites infacting poultry, roam about freely in the hairy parts of the body; commonly feeding on feathers, hair and epidermal products; found world wide.

Systematic Position :

Phylum	**Arthropoda**
Subphylum	**Mandibulata**
Class	**Insecta**
Subclass	**Pterygota**
Order	**Mallophaga**
Type	**Menopon**

Distinctive Features :

1. Very broad head, always at least as broad as the thorax.
2. Heavily sclerotized mandibles which are brown or blackish found on the ventral surface of the head.
3. The legs are short and all much alike.
4. The tersi are longer and modified for cling to smooth surface.
5. Four segmented maxillary palp.
6. Antennae club-shaped and tucked into grooves.

Identification with reasons :

State classification with characters upto—

Sub class—**Pterygota** as in **Dragonfly.**

Order—**Mallophaga**

1. Head evenly expanded behind and broadly triangular.

2. Mouth parts modified biting type.
3. Prothorax well-developed, and free, meso and metathorax partially fused.
4. Four-segmented maxillary palpi.
5. Head evenly expanded behind and broadly triangular.
6. Abdominal segments with one row of bristles.
7. Eye loged in shallow sinus.

Type—**Menopon**

Economic Importance :

Menopon move from host to host only by direct body contact. They annoy their hosts principally by irretating the skin surface.

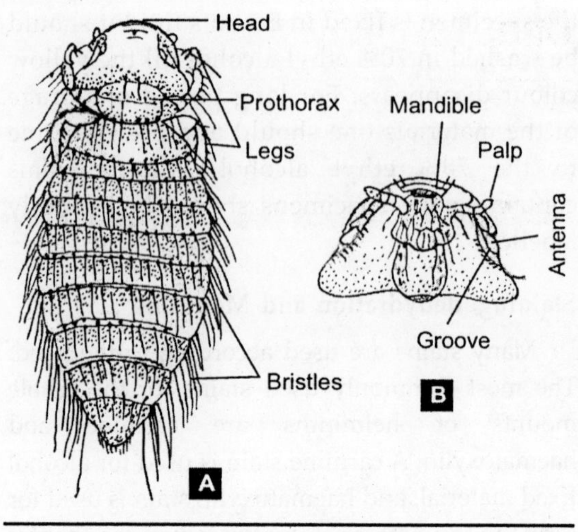

Fig. 9.1.3 : *Menopon sp.* A. Dorsal view, B. Ventral view.

Chapter 9.2

Immunology

9.2.1 PHAGOCYTOSIS OF R.B.C. BY MACROPHAGE AND *in vitro* ASSAY

A. Macrophage isolation and observation

Introduction :

The mononuclear phagocytic system consists of blood **monocytes** and **tissue macrophages.** In the hematopoietic process they arise from a common myeloid progenitor cells and differentiate into promonocyte, which leave bone marrow and the blood and further matures into monocytes.

These monocytes differentiate into **macrophages** when they migrate to different tissues. Some of these cells become tissue specific and convert into fixed macrophages; rest remain mobile and are designated as wandering or free macrophages.

Fixed macrophages are given different names depending upon their residence organs. These are viz., Histocytes in connective tissues; Alveolar macrophages in lungs; Kupffer cells in liver; Mesangial cells in kidney; Osteoclasts in bone and Microglial cells in the brain and is also easily available in the peritoneal cavity.

Macrophages are phagocytic cells and can adhere to the substratum artificially on glass surfaces such as the cover slips. Cytoplasm is basophilic and can be stained with basic stains such as Giemsa. Further these cells are able to produce humoral and cell mediated immune responses against foreign antigens. Thus considered as first line of defence against pathogens.

Materials required :

(1) Hypodermic syringe (5mL).

(2) Needle (22G).

(3) Laboratory centrifuge.

(4) Haemocytometer.

(5) Cover slips.

(6) Giemsa stain.

(7) Bacteriological incubator.

(8) Light microscope.

(9) White mouse.

(10) Phosphate buffered saline (PBS).

(11) Camera lucida.

Procedure :

Peritoneal mononuclear cells are isolated by peritoneal lavage of mouse in 5mL PBS, through a 22G needle fitted 5mL syringe. Cells are pelleted in 3500 rpm in a centrifuge, for 10 minutes at room temperature. Washed with PBS. Final pellet resuspended in 1-2mL PBS. Taken count in a haemocytometer. Cells are plated on cover slips @ 0.5×10^6 /mL in 800mL of PBS (made to 1mL final volume) on cover slips. Kept at 37°C for 1.5 to 2 hours in an incubator. Washed twice with 1mL PBS on the cover slips to remove lymphocytes. Stained with Giemsa and observed under a light microscope for peritoneal macrophage cells adhered on cover slips.

Observation :

The stained cells are observed under a light microscope at a magnification of 450X.

Result :

Cells are drawn through a Camera Lucida to know the morphology of the macrophages.

B. Phagocytosis assay

Introduction :

Macrophages show phagocytic activity. Phagocytosis involves recognition, binding, ingestion and digestion events. Macrophages are capable of ingesting and digesting exogenous antigens, which may be in form of whole microbe, insoluble particles injured or dead cellular remnants etc. In the initial phase of phagocytosis macrophages are drawn towards different substances produced in an immune reaction, this is called **chemotaxis**. The next phase is the **recognition** and **binding** of the antigen to macrophage membrane. After this occur the process of **ingestion**, where after the binding macrophage membrane produces protrusions called as **pseudopodia**, that elongates and surrounds the attached antigen and produces a membrane bound vesicle or **phagosome**. This phagosome is pinched of from the cell membrane and moves towards the lysosomes and fuses with it to form one structure called as the **phagolysosome**. The hydrolytic enzymes present in the lysosome bring about the digestion process. Then lysosome is designated as the **secondary lysosome**. Secondary lysosome now moves towards cell membrane, where the contents of the secondary lysosome are released out of the cell by the process of **exocytosis**.

Materials required :

(1) Isolated sheep RBC or chick RBC or human RBC.

(2) Cover slips.

(3) Giemsa stain.

(4) Light microscope.

(5) Heparinized tube.

(6) PBS.

(7) Bacteriological incubator.

(8) Laboratory centrifuge.

Procedure :

(a) **Preparation of RBC :** Taken 5mL blood in heparinised tube. Centrifuged as before. Washed with PBS. Suspended in 1mL PBS. Taken count, adjusted macrophage in RBC and ratio of 1:10, 1:5 as required in 800-1000mL PBS.

(b) **Phagocytosis assay :** Macrophage isolated as previous (see, 9.2.1). After washing of coverslips, added sheep RBC or chick RBC or human RBC 1:5 or 1:10 ratio on to macrophage adhered cover slips in 800-1000mL PBS. Kept in 37°C in a bacteriological incubator for 30 minutes, 60 minutes or 2 hours as required in PBS. Cover slips washed as required with 1mL PBS (2 times) stained with Giemsa.

Observation :

The phagocytosed RBCs are observed under a light microscope.

Results :

Macrophage phagocytose the RBCs and are seen in the phagosome.

9.2.2 PHAGOCYTOSIS OF YEAST BY MACROPHAGE AND *in vitro* ASSAY

Introduction :

Phagocytosis assay is simple. Briefly, phagocytic (antigenic; yeast in this case) cells are incubated with the macrophages. The cells were stained with stain, and attached and engulfed cells were counted with the help of microscope.

Reagents and materials required :

Potato dextrose broth and its preparation—

1. 200g of potatoes are to be peeled off and cut into small pieces and kept for boiling in 500mL water for half an hour.

   Potatoes are then minced and filtered through a muslin cloth.

   20g of dextrose are to be added, setting the *p*H to 5.6 and making volume to 1000 mL, and to be distributed in tubes or flasks.

   The medium is to be autoclaved at 121°C and 15 psi for 15 minutes.

   Agar is then added at the rate of 15g/L which will result in a solid culture medium potato dextrose agar.
2. DMEM Hams F-12 medium without phenol red 1% tannic acid solution.
3. Working solution of May Grunwald stain, freshly diluted with Giemsa buffer (1:2).
4. Yeast cells :
   (i) Yeast cells are to be used for determining phagocytosis. Yeast (*Saccharomyces cerevisiae*) NCDC40 strain was obtained from NCDC, NDRI.
   (ii) Yeast cells are to be maintained by sub-culturing fortnightly in potato dextrose broth and incubating at 30°C for 48 hours and stored under refrigeration between transfers.
5. Phosphate buffered saline (PBS).
6. 1% tannic acid.
7. Compound microscope.
8. Neubeur's chamber.
9. CO_2 incubator.
10. Petri-dish.

Procedure :

Preparation of yeast cells :

1. Yeast cells are to be cultured in potato dextrose broth for 48 hours at 30°C and autoclaved at 120°C for 45 min.
2. Then it is to be washed thrice with phosphate buffer saline (PBS) and stored at 4°C until use.
3. Just before use, yeast cells are to be sonicated gently to disrupt clumps and diluted to 10^8 cells/mL in DMEM Hams F-12 medium without phenol red.

Phagocytosis assay :

1. Peritoneal fluid of mice are to be collected by injecting 5mL of DMEM Hams F-12 medium (without phenol red) in peritoneal cavity by following a gentle massage of the abdomen.
2. Peritoneal exudate cell suspension (as a source of macrophages see, Chapter 9.2.1A) may be used for *in vitro* phagocytosis assays.
3. Macrophages are to be counted in peritoneal fluid using Neubeur's chamber.
4. About 1×10^6 cells/mL are to be distributed into 35 mm culture grade petri dishes.
5. It is then incubated in a humidified atmosphere of 5% CO_2 in air at 37°C for 2 hours to allow attachment of adherent cells.
6. Non-adherent cells are to be removed by washing 3 times with phosphate buffer saline.
7. After washing, cells are to be cultured in DMEM Hams F-12 medium and incubated in a humidified atmosphere of 5% CO_2 for 18 hours at 37°C.

8. After incubation, 100µl of yeast cells (1×10^8 cells/mL) are then added and incubated for 1 hour at 37°C.
9. Then washed with PBS.
10. 1mL of 1% tannic acid solution is to be added and kept for 1 minute.
11. After removal of tannic acid solution, again washed with PBS and dried in air.
12. Lastly, it is to be stained with May Grunwald stain freshly diluted with Giemsa buffer (1:2) for 2 minute.
13. Then it is to be washed with PBS and again dried.
14. After drying it is to be observed at 1000X magnification under oil immersion lens of a microscope.
15. Then percent phagocytosis are to be calculated by counting the number of yeast cells internalized per 100 macrophages.

Chapter 9.3

Microbiology

9.3.1 GRAM STAINING OF BACTERIA

Principle :

Bacteria are small, microscopic organisms and without any colour in any in natural condition and thus are not easily noticeable under a microscope. Thus bacteria require staining to become visible. Stains are artificial chemical colouring substances (dyes), which are used for biological purposes. There are several staining techniques used for bacterial examinations, differentiation and separation, helping in the study of there cytology and morphology.

Gram staining technique is one of the most popular staining procedure used for differentiation of two groups of bacteria viz., gram-positive and gram-negative. Christian Gram invented this method in 1884.

In this process of bacterial smear is drawn on a slide, and stained with crystal violet and then mordanted with iodine solution. The crystal violet and iodine complex gives a purple black colour to the bacterial cells. In Gram-positive bacteria, the chemical complex binds and is trapped in the cell wall followed by an ethanol treatment which may be due to decrement in the diameter of the pores in the cell wall peptidoglycan that stops release of crystal violet complex and thus get stained. Where as in the case of the Gram-negative bacteria, they have a much less amount of peptidoglycan and less extensively cross-linked. The pores of these bacteria remains open after ethanol treatment and thus allow crystal violet chemical complex to get extracted and thus these bacteria gets unstained.

Materials Required :

1. Bacterial slant culture of gram positive (*E.coli*) and gram negative rod (*Staphylococcus aureus*).
2. Crystal violet : 2g of crystal violet is mixed with 20mL ethanol (solution-A). 0.8g of ammonium oxalate mixed with 80mL of distilled water (solution-B). Solution A and B are mixed and then filtered.
3. Gram's iodine : 1g iodine, 2g potassium iodide is added to 3000mL distilled water slowly.
4. 95% ethyl alcohol.
5. Safranin 0 (counter stain) : 2.5g safranine in 100mL of 95% ethyl alcohol.
6. Inoculating loop

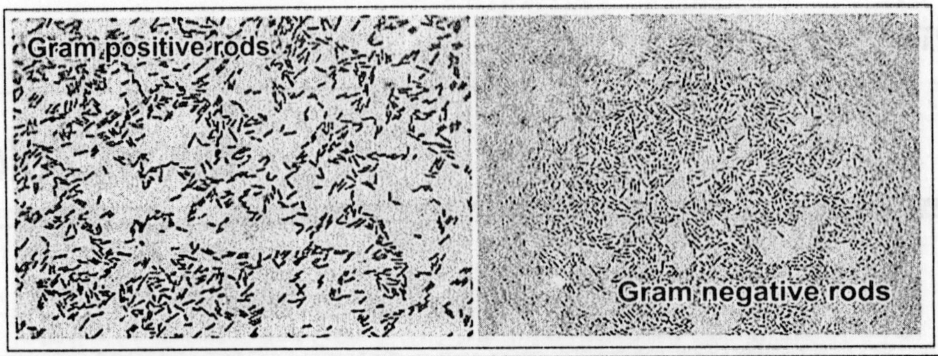

Figure 9.3.1.1 : Gram stained positive and negative rods (see coloured plates).

7. Bunsen burner or spirit lamp
8. Glass slides.
9. Staining tray.
10. Compound microscope with oil immersion lens.

Procedure :

1. Loopful of microbial cells are applied to the clean grease free slide with help of an inoculating loop from broth culture. To transfer microbial cells from culture growing on solid medium the culture is to be transferred first into sterile distilled water to dilute it.

2. It is then spread over large surface area to form a thin film.

3. The smear is dried completely. Generally heat fixation is carried out by quick passage of air dried smear through the flame of spirit lamp or Bunsen burner

4. On the smear 3-4 drops of Crystal violet is applied. Stained for 1 minute in a staining tray.

5. Washed under tap water. Then the smear is floated with Gram iodine for 1 minute and again washed under tap water.

6. The stain is decolourised with ethyl alcohol (95%) by dropping alcohol on the stained smear.

7. The slide is again washed with tap water.

8. Counter stained with Safranin (1:10 dilution) for about 45 seconds and again washed with water.

9. The slide is dried and studied under oil immersion lens of a compound microscope.

Observation :

Observing the stained bacterium under a microscope, they are classified on the basis of gram reaction. The bacteria, which stain dark violet or blue black in colour, are referred as gram-positive bacteria. Those staining pink or red in colour and do not take the stain is referred as Gram-negative bacteria (**Fig. 9.3.1.1**).

9.3.2 CARBOL FUCSHIN SIMPLE STAINING OF BACTERIA

Principle :

This is one of the simplest techniques recommended to study the morphology and arrangement of bacterial cells. Here Carbol fucshin, a single dye is used and hence referred as "simple staining" The basic stain with positively charged chromogen (dye) binds to the negatively charged cell wall components which is slightly acidic in nature.

Material required :

1. Bacterial culture or curd sample with bacteria.

2. Carbol fucshin : 3g Basic Fucshin is dissolved in 10mL of 95% ethanol and then in

5mL of Phenol in water. Mixed and kept for several hours. Filtered before use.

3. Acid alcohol : 5mL 3% v/v HCl in 95% ethyl alcohol + 95mL distilled water.

4. Distilled water.

5. Methylene blue.

6. Bunsen burner or spirit lamp.

7. Inoculation loop.

8. Glass slide.

9. Staining tray.

10. Spirit lamp or Bunsen burner.

11. Compound microscope.

Procedure :

1. Loopful of microbial cells are applied to the clean grease free glass slide with help of an inoculating loop from broth culture. To transfer microbial cells from culture growing on solid medium the culture is to be transferred first into sterile distilled water to dilute it.

2. It is then spread over large surface area to form a thin film.

3. The smear is dried completely. Generally heat fixation is carried out by quick passage of air dried smear through the flame of spirit lamp or Bunsen burner.

4. The smear is flooded with basic dye, Carbol fucshin for 15-20 seconds.

5. The stained slide is decolourized with acid-alcohol for 15-20 seconds.

6. Washed with distilled water to stop decolourization.

7. Counterstained with methylene blue for 30 seconds.

8. Then gently rinsed with tap water and dried with blotting paper. Slides are kept on a staining tray.

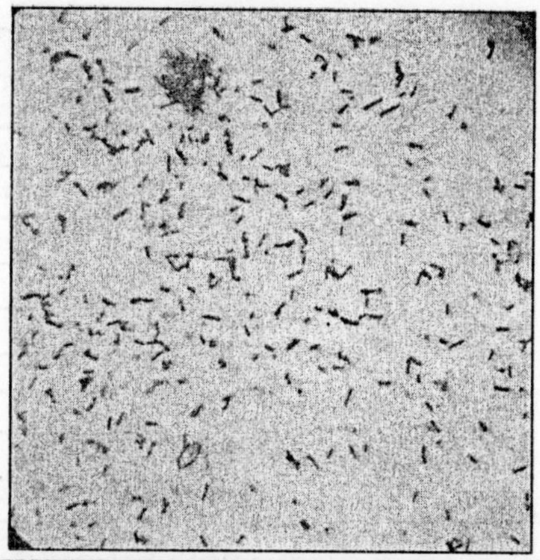

Fig. 9.3.2.1 : Carbol Fucshin simple staining of bacteria.
Courtesy : Ms. Anumita Das

9. Smear is examined under compound microscope.

N.B. : The smear may be treated with benzene to remove fatty substance before staining (in case of curd sample).

Observation :

Bacteria are observed as red against bluish white background under a microscope (Fig. 9.3.2.1, see coloured plate of Sec. 9).

9.3.3 DEMONSTRATION OF TECHNIQUE FOR PURE CULTURE OF MICROORGANISMS

Bacteria occur in nature as mixed populations. But for examination of particular bacterial strain it is necessary to isolate individual organism from a mixed culture. Such microbiological method of isolation of individual organism from mixed culture and consequent maintenance of these individuals and their successive progenies in an artificial media is known as **pure culture**. Such a culture requires a great deal of laboratory procedures and is a very complex affair. There are many methods for isolation of pure culture from a mixed population; one of the well known methods is the **streak plate method.** Such method

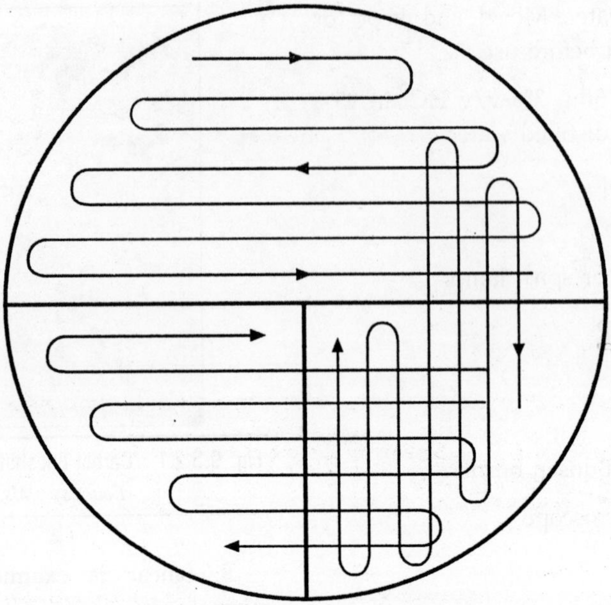

Fig. 9.3.3.1 : T- streak in agar plate.

is used to produce well isolated colonies of bacteria from concentrated cell suspension.

Materials required :

1. Mixed Culture.
2. Bunsen Burner.
3. Tripod and Wire gage.
4. Innoculating loop or needle.
5. Sterile petri dish.
6. Sterile nutrient agar medium (See Chapter 12.11).

Procedure :

1. Sterilized nutrient agar is liquefied in a boiling water bath and cooled at 45°C. Liquefied medium is poured into a sterile petri dish.

2. Agar is left undisturbed to solidify.

3. A bent innoculating loop or needle is flamed and then cooled by dipping it into the agar.

4. A drop of concentrated mixed bacterial suspension is taken at the loop and streaking is initiated.

5. The agar plates are streaked as T-streak as shown in Fig. 9.3.3.1. During streaking one half of the top cover of the Petri dish is raised to avoid contamination.

6. The plates are incubated at 30°C in an inverted position.

7. The growths of the colonies are observed after 2 days; well separated colonies are then transferred to agar stands as pure cultures.

Observation and Result :

Isolated colonies of bacteria strain to be examined on the plates will be visible.

Section 10

ECONOMIC ZOOLOGY

SECTION AGENDA

Chapter 10.1

Preparation of Pituitary Extract of Carp

Introduction :

A major problem of successful fish culture is concerned with the procurement of fish seeds of cultivable major carps. The major carps do not breed in stagnant water bodies. These fish come down to the low lying water bodies for the purpose of reproduction and spawning. Eggs and the fries are subsequently drained to the river tributaries and the rivulets by rain water,. The eggs are then collected through traditional methods through fine meshed nets. But this has certain set backs such as the catch is of seeds of mixed nature without getting egg of a single species and mortality of these delicate fish seeds are very high during the transport to the market .Thus in order to overcome the problem induced breeding of the fish by injecting pituitary hormone extract has been introduced.

The **pituitary gland** produces, accumulates, and stores the gonadotropic hormone(s) which plays a decisive role in ovulation. In so far as reproduction is concerned, the function of the pituitary gland is that of a liaison between the central nervous system and the gonads. The gonadotropic hormone(s) is produced by sexually mature fish and the cyclical changes in its concentration in the pituitary gland are correlated with the reproductive cycle of the fish. Its concentration is highest during the prespawning period, while it is very low or almost nil during and subsequent to spawning. The release of gonadotropin(s) by the pituitary gland is "under order" of the hypothalamus through the secretion of **gonadotropin-releasing hormone (GRH)**. The **gonadotropin**(s) is also responsible for inducing spawning migration, during which its concentration in the pituitary gland gradually decreases. The gonadal development during spawning migration is most probably directed by the continuously released gonadotropin(s).

The pituitary gland is located on the ventral side of the brain below the hypothalamus, which is linked to the pituitary gland by a funnel-like structure, the infundibulum. The part of the cranium where the pituitary gland is located is known as the **sella turcica**. The gland is usually embedded in fatty tissue. When the brain is taken out of the skull, the pituitary gland remains connected to the brain in some fishes, while in most fishes the infundibulum ruptures and the gland is left behind on the base of the skull. In the case of common carp, a small droplet of blood signifies the place where the pituitary gland was connected to the brain. The hypothalamus, below which one has to look for the pituitary, is a ventrally protruding part of the brain and is easy to distinguish.

In India, Khan (1938) successfully induced *Cirhinnus mrigala* to spawn by injecting mammalian pituitary hormone. The major Indian carp, the Chinese grass carp and silver carp were induced to breed successfully in confined stagnant water for the first time at Cuttack research station of CIFRI by Chaudhuri and Alikunhi in 1957.They were able to induce a number of major carp as well as minor carps to breed viz., *Labeo rohita, L. bata, C. mrigala, Puntius sarana*.

In the International arena the principle of hypophysation technique was first applied by pisciculturists of Brazil in 1934. The Russians were the next successful in 1937-38.

Materials Required :

(1) Absolute alcohol
(2) Forceps
(3) Knives
(4) Dark coloured phials
(5) Distilled water
(6) Refrigerator
(7) Glass Homogenizer
(8) Laboratory Centrifuge
(9) Electronics or electro-mechanical balance
(10) Filter paper
(11) Freshly killed carp

Procedure :

1. Pituitary gland is collected from the freshly killed fish. The pituitary gland is removed by exposing the brain of the fish with the help of a sharp knife. The pituitary gland can also be collected through the foramen magnum.

2. Just after the collection the pituitary gland is put in absolute alcohol through 3-4 changes. The gland is kept for at least 24 hours or it can also be stored in the dark coloured phials containing absolute alcohol.

3. The freshly collected or preserved gland (in alcohol) is put on a clean filter paper for removing most of the adsorbed alcohol. The gland is quickly weighed in a weighing balance.

4. The dealcoholized gland is put in a glass tissue homogenizer with a little amount of distilled water or 0.3% NaCl.

5. The tissue is gently macerated whilst keeping the lower part of the tissue homogenizer in ice (Figure 10.1.1).

6. The homogenate is diluted to about 20-40mg/ml of water.

7. The homogenate is centrifuged at 2000 rpm (in a fixed angle rotor).

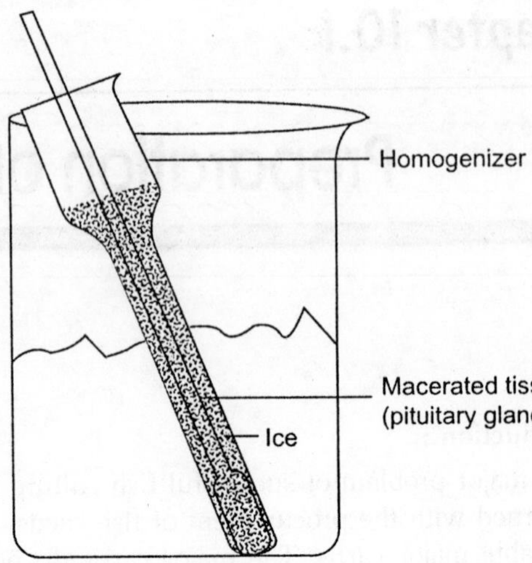

Fig. 10.1.1 : Homogenization of fish pituitary gland.

8. The clear supernatant is collected by drawing the fluid into hypodermic syringe.

9. This clear supernatant is transferred to sterile glass vials sealed immediately afterward by fusion.

Note :

The glands can also be preserved in absolute alcohol, in which case they are immersed in the absolute alcohol instantly after collection. After 24 hours, the glands are washed with absolute alcohol, and put in fresh absolute alcohol and stored in a cool, shady place at room temperature or under refrigeration until needed. To prevent moisture from getting into the phials, they should be sealed with wax or kept in desiccators. Another method of preserving glands is to subject them to instant freeze and storing them in a freezer. Fresh glands can also be used directly after collection to prepare a hypophysis solution. Pituitary glands are sometimes available in the market in some countries; e.g., the U.S.A., Hungary, and India. They are marketed in dried unbroken form. Glands are also marketed in powdered form and in the form of extract in glass ampoules ready for injection. But these are less reliable because of the possibility of adulteration.

Chapter 10.2

Identification of spawn, fry and fingerling of Rohu, Catla and Mrigel

Fish is found abundantly in all natural waters and is a valuable source of food. Culture of certain types of fishes in tanks or ponds is an old practice. The practice is to collect very young ones of carps – the fry about 4.25 cm. long and transfer them to tank, ponds and bunds. When these grow for a year or two, they are ready for the table. But now a day the idea of rearing fish is not only confined to ponds and tanks, instead various people are working co-operatively on a systematic plan for commercial purpose. **Pisiculture** has now become a chain of reaction which starts from eggs and finish with the production ready for table, for export and for stocks. The process in which eggs are hatched into spawns or fries is known as **hatching** and plant for hatching is known as **hatchery**.

For successful culture of major carps which are commercially important, one should have a first hand knowledge to segregate the eggs and spawns of *Rohu, Catla* and *Mrigel* fishes. The following table can give an idea about the same.

Identification of spawn, fry and fingerlings of *Rohu, Catla* and *Mrigel* fishes. (Table 10.1, Figure10.2.1 after Chakraborty and Murthy, 1972).

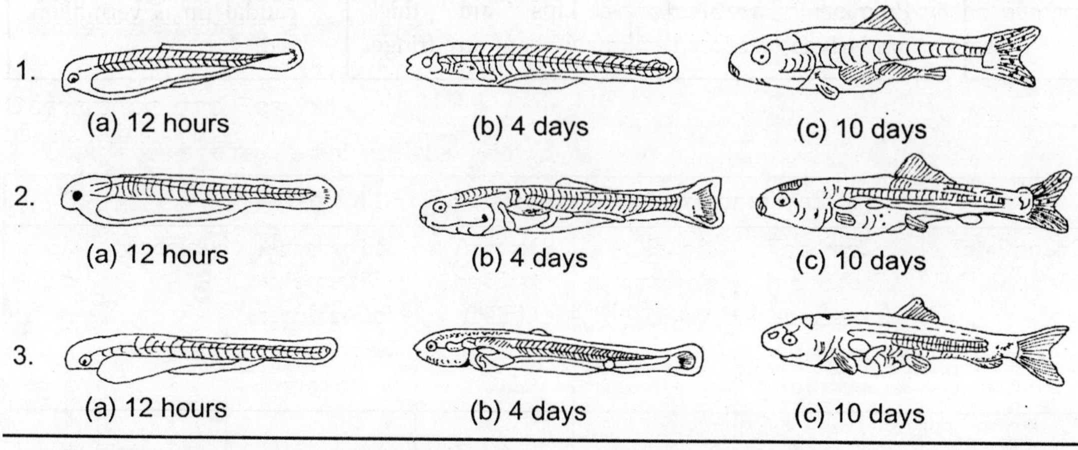

Fig. 10.2.1 : Spawn, fry and fingerling of Rohu, Catla and Mrigel fish—(a) Spawn, (b) Fry, (c) Fingerling. 1—Rohu fish, 2—Catla fish, 3—Mrigel fish.

Table 10.1

Stages	Rohu (*Labeo rohita*)	Catla (*Catla catla*)	Mrigel (*Cirrhinus mrigala*)
Spawn :	1. Average length is about 3.79 mm 2. Both swollen and narrow portion of yolk sac are equal. 3. Pre-anal myotome 26 and port-anal myotome 14.	1. Average length 4.8 mm 2. Other characters are like Rohu.	1. Average length is about 4.68 mm. 2. Narrow part of the yolk sac is not longer than the swollen portion. 3. Yolk sac club shaped. 4. 28 pre-anal and 14 post anal myotomes present.
Fry : (After 96 hrs. of hatching)	1. Average length 7.5 mm 2. There is a semi lunar black spot on the tail just at the terminal portion of the caudal end of the notochord. 3. At the origin of ventral fin there is a reddish groove. 4. Lips are fringed.	1. Average length 7.56 mm 2. There is a semi lunar black spot on the tail just at the terminal portion of the caudal end of the notochord. 3. Lips margin thick.	1. Average length is about 7.38 mm. 2. Black spots are seen at the base of the dorsal fin 3. Semi lunar black spot on the tail at the tip of the notochord. 4. Lips are very thin. 5. Dorsal fins separating out.
Fingerlings :	1. A black band at the base of the caudal fin. 2. Fins are red in colour. 3. Border of the caudal fin is black in colour. 4. Fringed lips. 5. Prominent bar at the upper lip.	1. Large head. 2. Body and tail lack spot. 3. Dorsal, caudal and anal fins are blackish. 4. Lips are thick, without any fringe.	1. A black diamond like large spot at the base of the caudal fin. 2. Body has longitudinal coloured lines. 3. The lower lobe of the caudal fin is vermillion red.

Chapter 10.3

Identification of Fish Diseases and Parasites

Different types of fish diseases and disease description is shown in table below.

Table 10.3.1 : Different types of fish diseases and causative agents

Disease/parasite	Causative agent	Description
"Ich" or "White Spot Disease"	*Ichthyophthirius multifiliis* (ciliate protozoan) 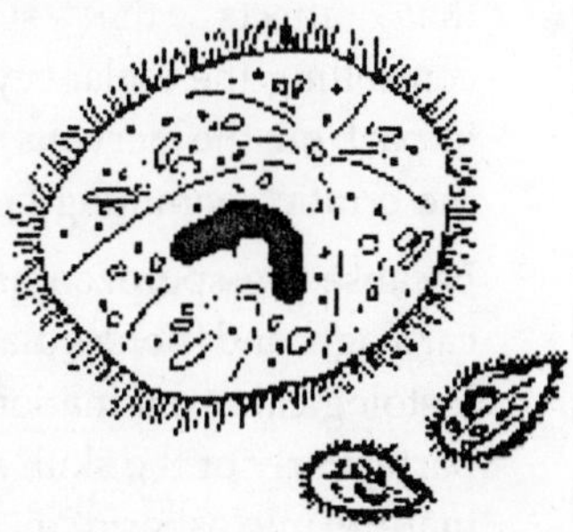	a) The white spots observed on fish are the adult **trophozoite** which is a feeding stage that are embedded in the skin or gill epithelium. b) The trophozoite breaks through the skin, leaving the host to form an encysted **tomont stage**, which may undergo numerous divisions. Numerous **theronts** break out of the tomont to search for a fish to infect. The theronts burrow into the epithelium and form the trophozoite. c) The pathology caused by the trophozoite include epithelial **hyperplasia** and **necrosis.** d) Diagnosis is made by observation of the numerous white spots on the surface of the fish. The presence of the trophozoite can be confirmed by microscopic examination of fresh skin scrapings, gill clippings, or fin clippings. The large ciliate protozoan with a characteristic C-shaped macronucleus will be seen.

Table contd. next page

Table 10.3.1 : Different types of fish diseases and causative agents

Disease/parasite	Causative agent	Description
Red Sore Disease	*Epistyis sp.*, ciliated protozoan 	a) Grossly, fish are observed to have white or hemorrhagic lesions on the body. The parasite causes ulceration at the site of attachment to the fish. This ulceration may serve as a sight of secondary bacterial or fungal infection. b) Diagnosis is made by the microscopic examination of fresh skin scrapings or gill clippings for the presence of the parasites.
Whirling disease	1. *Myxobolus cerebralis* 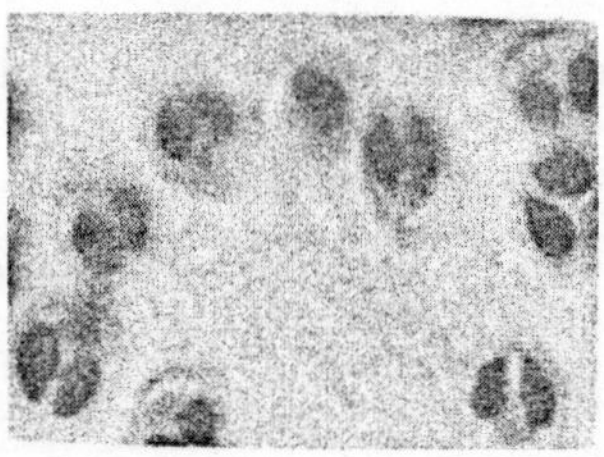 2. *Pleistophora sp.*– 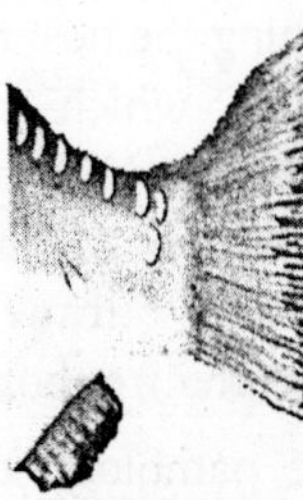	a) This myxosporidian has a tropism for cartilage and bony tissues. It is particularly problematic in young salmonid fishes, during the first three months of life before ossification is complete. b) The parasite feeds upon cartilagenous tissue. Disease signs include a "black tail" condition in young fish and a tail-chasing swimming behaviour. The "black tail" condition is caused by vertebral instability that affects the sympathetic nerves controlling the melanocytes of the tail. The impact on the nervous system also causes the erratic swimming behaviour. c) Because the sporozoa are embedded in the cartilage and bone, diagnosis is made by histological examination of those tissues, particularly of the skull and gill arches. The mature myxosporidian can be seen in the tissues. A giemsa stain will allow for an easier visualization of the parasites.
Gill rot disease (fungal disease)	*Saprolegnia sp.*– 	It is also a common myxsoporidian parasite that infects skeletal muscle. This disease is also known the **brachyomycosis** caused by fungus *Saprolegnia sp.* This fungus moves its hyphae to the gills and spread. The infected gill starts to rot. The colour becomes pale with red patches and in the advanced stages gills turns yellow brown and are partly destroyed.

Table contd. next page

Table 10.3.1 : Different types of fish diseases and causative agents

Disease/parasite	Causative agent	Description
Fin rot disease (bacterial disease)	Fin rot bacteria. 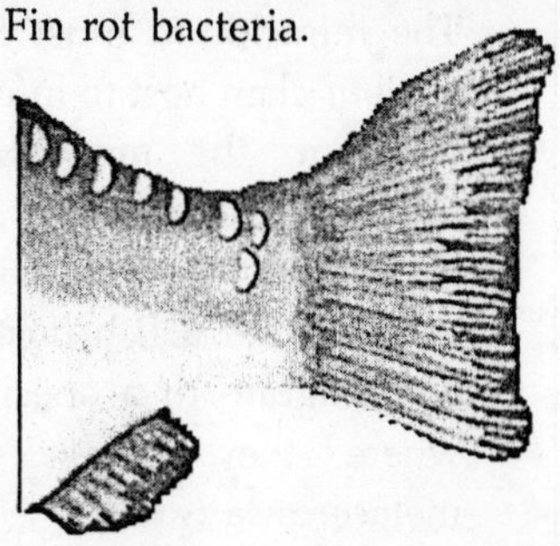	Fin rot can be the result of a bacterial infection (which causes a more ragged rotting), or as a fungal infection (which rots the tail more evenly and is more likely to produce a white 'edge'). Sometimes, both types of infection are observed together. Infection is commonly brought on by bad water conditions, injury, poor diet, or out wintering fancy goldfish (who should live in warmer temperatures). Fin rot starts at the edge of the fins, and destroys more and more tissue until it reaches the fin base. If it does reach the fin base, the fish will never be able to regenerate the lost tissue. At this point, the disease may attack the fish's body directly.
Diseases of monogenetic Trematode	*Dactylogyrus* *Gyrodactylus* 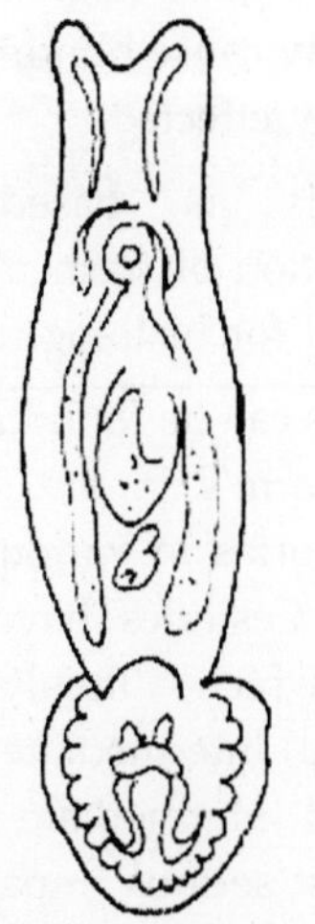 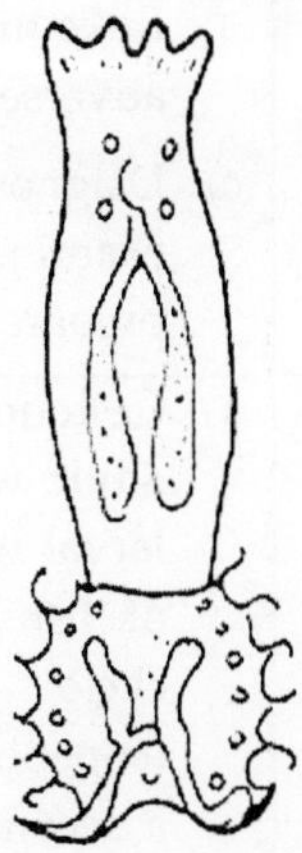	a) Important identifying features characteristic of these helminths is their organ of attachment, the **haptor**. This is at the posterior end and may have 0, 1 or 2 pair of anchors and a number of small hooklets around the periphery. Other identifying characteristics include the presence or absence of "eye spots." b) Lesions observed are primarily haemorrhage and necrosis of epithelial tissue of the gill and skin. c) Diagnosis is made by the microscopic examination of fresh skin scrapings or gill clippings for the presence of the parasites.
Diseases caused by digenetic Trematodes	*Diplostomum sp.* 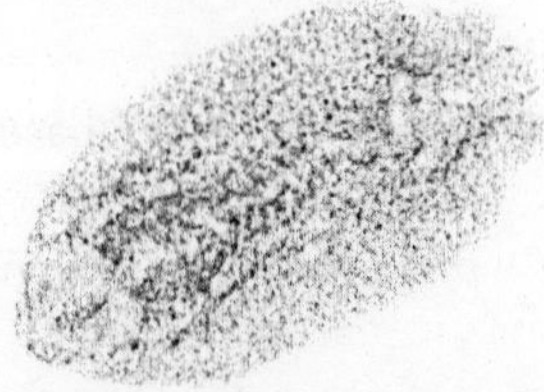	a) The majority of digenetic trematodes that infect fish do so as a metacercaria or "grub" stage. In general, the life cycle of these digenetic trematodes consists of the adult worm living in the gut of a fish-eating bird. Eggs from the worm are shed into the water with the bird faeces. These eggs hatch to form a free swimming larva (miracidium).

Table contd. next page

Table 10.3.1 : Different types of fish diseases and causative agents

Disease/parasite	Causative agent	Description
		The miracidium must find an appropriate snail or clam host to infect. Once in the snail or clam, the miracidium goes through several developmental stages and emerges as a free-swimming larva (cercaria). The cercaria find a fish host, penetrate the fish and migrate to a specific part of the fish where they encyst and become the metacercaria (grub). b) Some metacercaria are very specific with respect to the location (organ/tissue) they infect on the fish host. Others are not so specific. Usually the presence of digenetic trematodes does not have serious health consequences to the fish host. However, in very heavy infections or infections of specific organs (e.g. *Diplostomum*, the eye fluke may cause blindness), the fish may be adversely effected. c) Diagnosis is based on microscopic examination of fresh materials or materials prepared for histological examination.
Diseases caused by Cestodes	*Bothriocephalus acheilognathi* 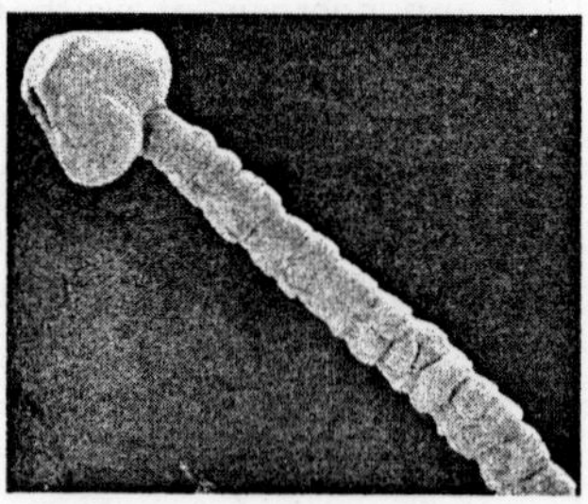	a) Cestodes can be found as parasites of fish as adult worm's in the digestive tract or as larval worms in various organs and tissues of fish. Cestodes have a complicated life cycle which involves an arthropod (copepod) intermediate host and sometimes a second intermediate host, such as a fish. The most serious impact on fish is usually due to a heavy infection of larval tissue dwelling forms. As these larvae migrate through the tissues, they leave a tract of scar tissue. b) Diagnosis is based on microscopic examination of fresh materials or materials prepared for histological examination.

Table contd. next page

Table 10.3.1 : Different types of fish diseases and causative agents

Disease/parasite	Causative agent	Description
Diseases caused by the Nematodes	*Camillanus sp., Capillaria sp.* 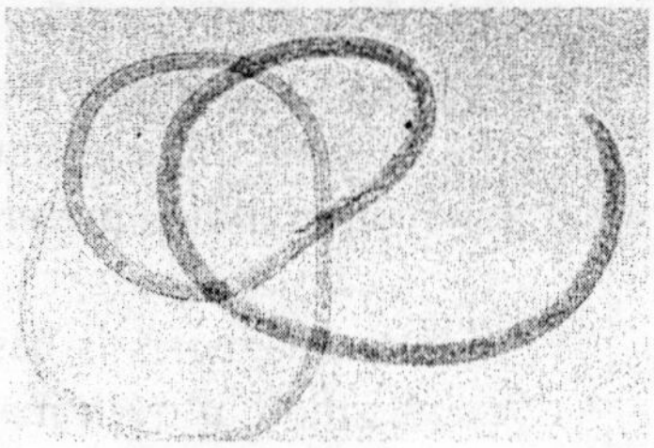	a) *Camillanus* is easily recognized as a small thread-like worm protruding from the anus of the fish. b) *Capillaria* is a large roundworm commonly found in the gut of angelfish often recognized by its double operculated eggs in the female worm. Heavy infestations are associated with debilitated fish, but a few worms per fish may be benign. c) Nematodes have a complicated life cycle which may require one or two intermediate hosts. Sometimes the fish is the final host, with the nematode being found in the intestine. Larval forms of nematodes are found encysted in various organs and tissues. d) Diagnosis is based on microscopic examination of fresh materials or materials prepared for histological examination.
Diseases caused by Crustaceans: Copepods and Branchiurans	1. *Lernea sp.* 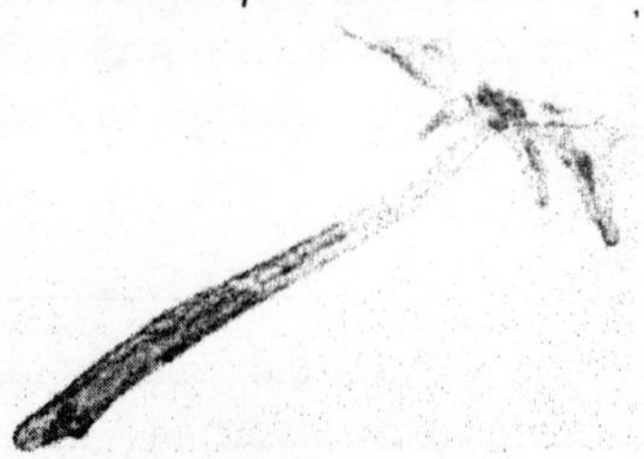 2. *Ergasilus sp.* 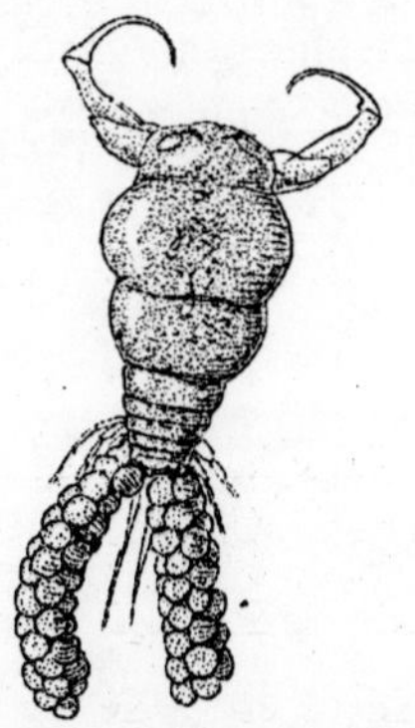	a) This is also known as the anchor worm, is the adult female of the genus *Lernea*. As immature forms, the male and female are free-swimming aquatic invertebrates. Once mating occurs, the male dies and the female searches for a fish host. The organism attaches to the fish by a "anchor" structure that develops below the skin, usually in the underlying muscle. The body of the parasite ("the worm") trails outside the fish. There may be two egg sacs on a gravid female. b) *Ergasilus sp.* usually infect the gills of fish. They attach by a pair of modified appendages that look like "ice tongs."

Table contd. next page

Table 10.3.1 : Different types of fish diseases and causative agents

Disease/parasite	Causative agent	Description
	3. *Argulus sp.* 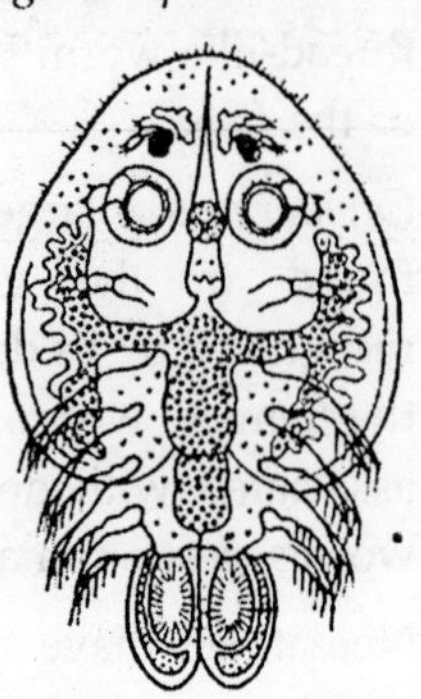	c) It is a branchiuran worm, also known as the fish louse. It has a flattened disk-like appearance. It infects the external surface of the fish, often being found underneath the operculum. When present in large numbers, it can cause extensive skin necrosis.Diagnosis is based on microscopic examination of fresh materials.

Chapter 10.4

Identification of Economically Important Crustacean and Molluscs

Crustaceans and molluscs are of great economic value. They are of advantage to man in several ways. They help for the well being of human beings.

(a) Crustacea :

Crustaceans are the most diverse of arthropods. They are extremely variable and abundant individually. They are important in fresh water and marine habitats, their disappearance would greatly disturb the natural balance of aquatic ecosystems.

Prawns, crabs, lobsters and shrimps are of great importance as human food. Several of these are freshly boiled or canned. Many small crustaceans, which are found in sea and fresh waters, are used as food by fishes, thus, are directly connected to us.

1. *Daphnia sp.* — See chapter 1.1.1. Non chordate: Arthropod, item no 1.
2. *Cyclops sp.* — See chapter 1.1.1. Non chordate: Arthropod, item no 4.
3. *Argulus sp.* — See chapter 1.1.1. Non chordate: Arthropod, item no 5.
4. *Balanus sp.* — See chapter 1.1.1. Non chordate: Arthropod, item no 6.
5. *Macrobrachium sp.* — See chapter 1.1.1. Non chordate: Arthropod, item no 10.
6. *Penaeus monodon* — See chapter 1.1.1. Non chordate: Arthropod, item no 11.
7. *Penaeus indicus* — See chapter 1.1.1. Non chordate: Arthropod, item no 12.
8. Crab — See chapter 1.1.1. Non chordate: Arthropod, item no 15.
9. ***Homarus sp.* (Lobster) :**

Habit and Habitat :

Marine; found in crevices in coasts of Europe from Norway to eastern Mediterranean; scavenger, tearing and crushing the shells of other mussels.

Systematic Position :

Phylum	**Arthropoda**
Sub Phylum	**Mandibulata**

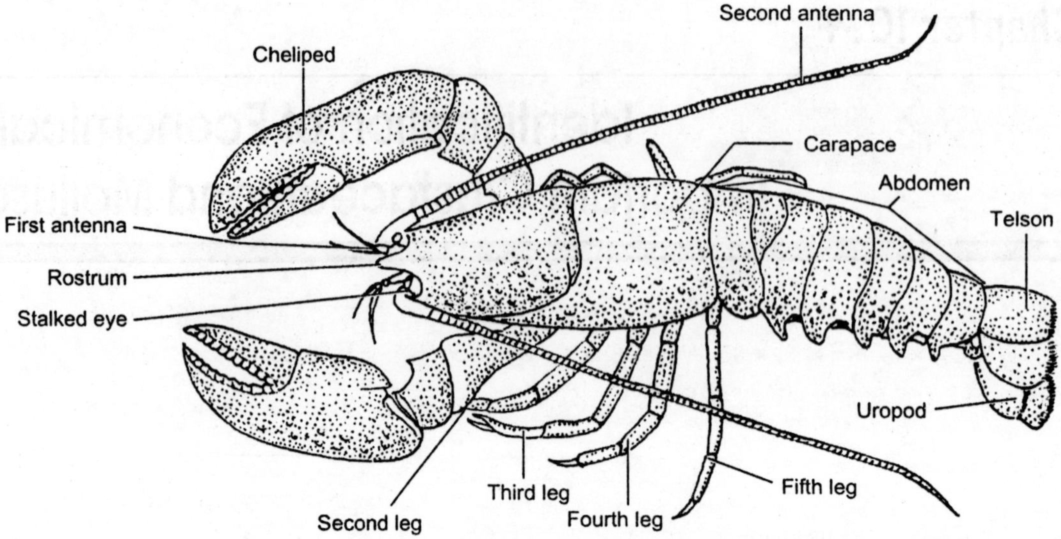

Fig. 10.4.1 : *Homarus sp.* (A lobster)

Class	**Crustacea**
Subclass	**Malacaostraeca**
Order	**Decapoda**
Type	***Homarus***

Distinctive features :

1. Body is elongated and divided into anterior cephalothorax, middle flexible abdomen and posterior telson.

2. Small spine like rostrum over cephalothorax and rostrum is serrate on lateral margins.

3. Much elongated second antennae.

4. First three pairs of pleopods are chelate and first pair is very large and pinching claws called chelipeds.

5. Eyes round, tailfin very broad.

Identification with reasons :

State classification and characters up to—

Order—**Decapoda** as those of *Macrobrachium* (Chapter 1.1.1).

Type—*Homarus* : State distinctive features as in above.

(b) Mollusca :

Molluscs are beneficial to man. Molluscs are found in abyssal tide line and are common in fresh water everywhere. Molluscan shells have always been fashioned into jewellery and buttons. Oysters, calms, squids and octopus are important food items and largely used in Europe, America and in Asia (Oriental countries). Some of the molluscs are harmful to man. *Achatina* are injurious to gardens and green houses. Ship worms cause great damage to the bottom of wooden vessels.

1. ***Haliotes sp.*** See chapter 1.1.1. Non chordate: Mollusca, item no 3.
2. ***Pila sp.*** See chapter 1.1.1. Non chordate: Mollusca, item no 4.
3. ***Cypraea sp.*** See chapter 1.1.1. Non chordate: Mollusca, item no 5.
4. ***Murex sp.*** See chapter 1.1.1. Non chordate: Mollusca, item no 6.

5. ***Achatina sp.*** See chapter 1.1.1. Non chordate: Mollusca, item no 9.
6. ***Lymnaea sp.*** See chapter 1.1.1. Non chordate: Mollusca, item no 11.
7. ***Planorbis sp.*** See chapter 1.1.1. Non chordate: Mollusca, item no 12.
8. ***Mytilus sp.*** See chapter 1.1.1. Non chordate: Mollusca, item no 13.
9. ***Pinctada sp.*** See chapter 1.1.1. Non chordate: Mollusca, item no 14.
10. ***Teredo sp.*** See chapter 1.1.1. Non chordate: Mollusca, item no 17.
11. ***Nautilus sp.*** See chapter 1.1.1. Non chordate: Mollusca, item no 19.
12. ***Sepia sp.*** See chapter 1.1.1. Non chordate: Mollusca, item no 20.
13. ***Loligo sp.*** See chapter 1.1.1. Non chordate: Mollusca, item no 21.
14. ***Octopus sp.*** See chapter 1.1.1. Non chordate: Mollusca, item no 22.

Chapter 10.5

Identification of Some Common Pests

A pest is an organism whose existence interferes in man's welfare, convenience and profits and whose increase in population above a certain level cause economic set back. It has been estimated that there are five million species in the animal kingdom and many of these species are pests of crops, domestic animals and house hold goods and articles. Practically all the pests are included in the different phyla, viz; Chordata, Mollusca, Nemathelminthes and Arthropoda. Phylum Arthropoda includes the largest number of pests of crops and domestic animals.

I. RICE PESTS

1. *Leptocorisa sp.*

See chapter 1.1.1. Non chordates: Arthropod, item no 33.

2. *Scirpophaga* sp. (= *Tryporyza*)

See chapter 1.1.1. Non chordates: Arthropod, item no 49.

3. *Hispa sp.*

.See chapter 1.1.1. Non chordates: Arthropod, item no 38.

4. *Nilaparvata sp.*

Systematic Position :

Phylum	**Arthropoda**
Subphylum	**Mandibulata**
Class	**Insecta (Hexapoda)**
Subclass	**Pterygota**
Order	**Hemiptera**
Type	***Nilaparvata lugens***

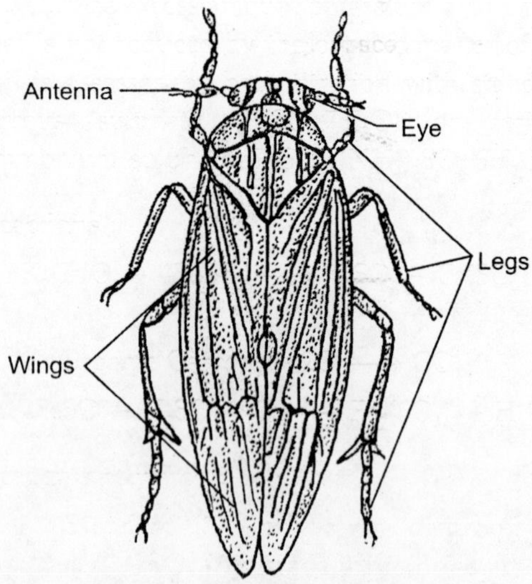

Fig. 10.5A : *Nilaparvata Lugens.*

Identification with reasons :

Characters same upto—

Subphylum—**Mandibulata** as in *Daphnia* (See chapter 1.1.1. Nonchordate; Arthropod, Item No. 1).

Class—**Insecta :**

1.Body divided into head, thorax and abdomen.

2.Three pairs of thoracic legs.

3.Antenna one pair.

Subclass—**Pterygota :**

1.With wings or secondarily wingless.

Order—**Hemiptera :**

1. Usually two pairs of wings, forewing modified as hemi elytra, with membranous apex and leathery base to cover membranous hind wings.
2. Legs absent or adapted for running, jumping, digging, grasping prey or swimming.
3. Piercing or sucking mouth parts.

Type—***Nilaparvata lugens*** :

1. Brownish or blackish about 3.4-4.5 mm long.
2. Short antennae with thick base and flagella like anterior end.
3. Large deep brown compound eyes.
4. Long legs, brown in colour with three jointed black tersi.
5. Wing veinations are deep black in colour.

Economic Importance :

Rice leaf hoper damaging paddy field.

II. COTTON PESTS

1. *Pectinophora sp.* (Pink boll worm)

Systematic Position :

Phylum	**Arthropoda**
Subphylum	**Mandibulata**
Class	**Insecta**
Subclass	**Pterygota**

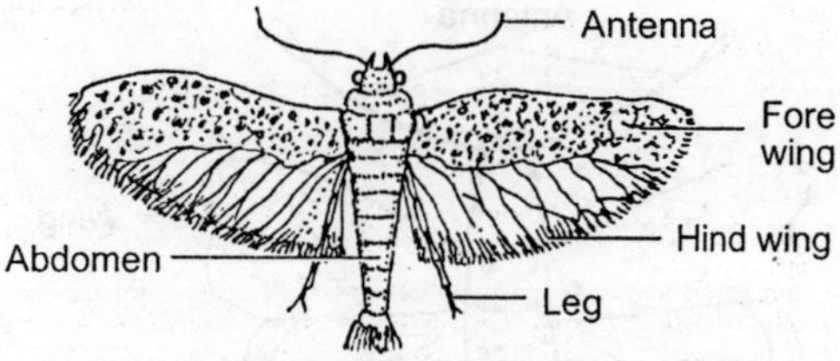

Fig. 10.5.1 : *Pectinophora gossypiella.*

Superorder	**Holometabola**
Order	**Lepidoptera**
Type	***Pectinophora gossypiella***

Identification with reasons :

Characters same up to—

Order—**Lepidoptera** as those of *Bombyx.* (See chapter 1.1.1. Non chordate; Arthropoda, item No. 44.)

Type—***Pectinophora gossypiella***

1. Adult insect is a deep brown moth, measuring 8-9 mm across the spread wings.
2. Blackish spot on the fore wings.
3. The margin of the hind wings are deeply fringed.

Economic Importance :

It is one of the most destructive **pests of cotton** in the world. It is highly destructive in the Punjab, Haryana and Pakistan. The damage is caused by the caterpillars only. Caterpillars are pink and found inside flower buds, Panicles and the bolls of cotton.

III. JUTE PESTS

1. *Anomis* (Semi looper)

Systematic Position :

Phylum	**Arthropoda**
Subphylum	**Mandibulata**
Class	**Insecta**
Subclass	**Pterygota**

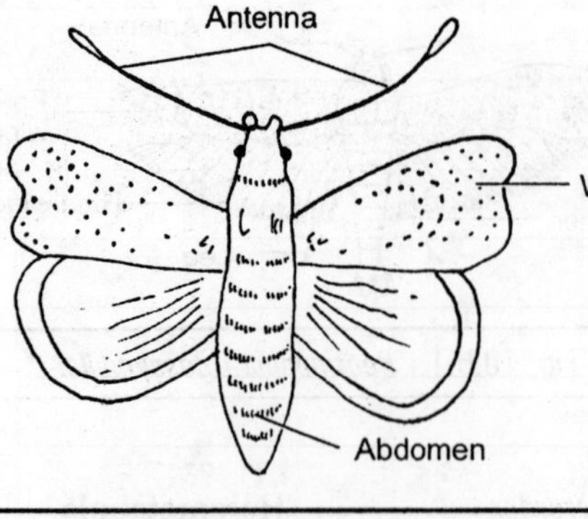

Fig. 10.5.2 : *Anomis sabulifera.*

Superorder	**Holometabola**
Order	**Lepidoptera**
Type	***Anomis sabulifera***

Identification with reasons :

Characters same up to—

Order—**Lepidoptera** as those of *Pectinophora.*

Type—***Anomis sabulifera***

1. Antennae with the shaft not dilated.
2. Hind wings edge circular.
3. Maxillary pulp minute.

Economic Importance :

This is specific **pest of jute** and is most destructive. Only the caterpillars cause damage by feeding on the foliage. Being green, the caterpillar camouflages.

2. *Spodoptera* (Indigo caterpillar)

Systematic Position :

Phylum	**Arthropoda**
Subphylum	**Mandibulata**
Class	**Insecta**
Subclass	**Pterygota**
Superorder	**Holometabola**
Order	**Lepidoptera**
Type	***Spodoptera exigua***

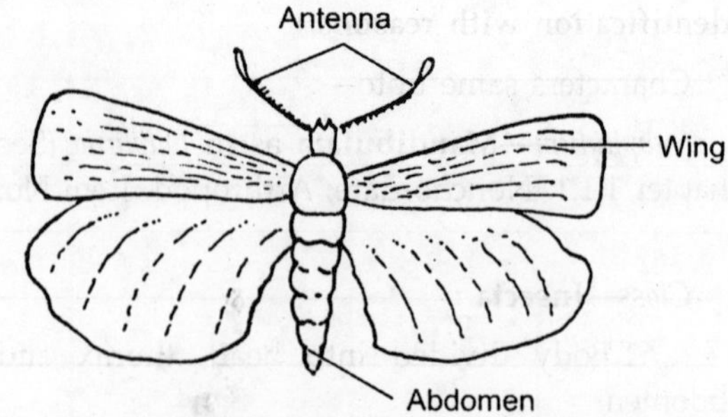

Fig. 10.5.3 : *Spodoptera exigua.*

Identification with reasons :

Characters same up to—

Order—**Lepidoptera** as those of *Pectinophora.*

Type—***Spodoptera exigua* :**

1. Moths have fore wings dark spotted and hind wings white.
2. Antennae with shaft not dilated.

Economic Importance :

It is a serious **pest of young jute plants**. The caterpillars feed on leaves and the damaged leaf gives a webbed appearance.

3. *Apion* (Stem weevil)

Systematic Position :

Phylum	**Arthropoda**
Subphylum	**Mandibulata**
Class	**Insecta**
Subclass	**Pterygota**
Superorder	**Holometabola**
Order	**Coleoptera**
Type	***Apion corchori***

Identification with reasons :

Characters same up to—

Order—**Coleoptera** as those of *Hispa.* (See chapter 1.1.1. Non chordate; Arthropoda, item No. 38.)

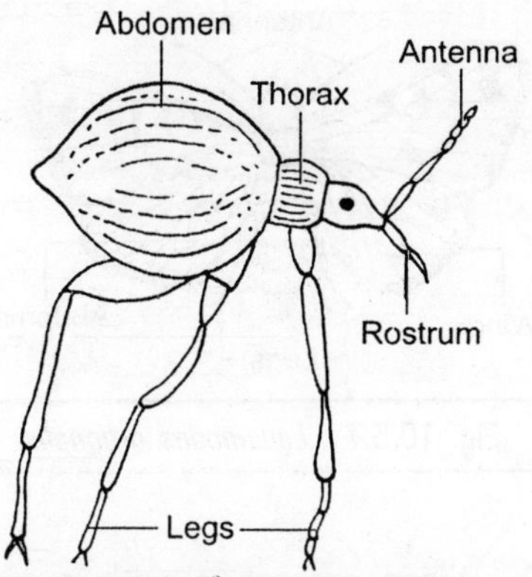

Fig. 10.5.4 : *Apion corchori.*

Type—Apion corchori :

1. Small weevil, body brown or dull black in colour, clothed with small whitish setae.
2. Conspicuous snout (rostrum) is present.

Economic Importance :

Adult causes appreciable damage to the **early-sown jute**. It bores holes with the rostrum in the jute stem and the larvae feeding inside the stem also injure the fibre.

IV. VEGETABLE PESTS

1. Pieris (Cabbage Caterpillar)

Systematic Position :

Phylum	**Arthropoda**
Subphylum	**Mandibulata**
Class	**Insecta**
Subclass	**Pterygota**
Superorder	**Holometabola**
Order	**Lepidoptera**
Type	***Pieris brassicae***

Identification with reasons :

Characters same up to—

Order—**Lepidoptera** as those of ***Pectinophora***.

*Type—**Pieris brassicae** :*

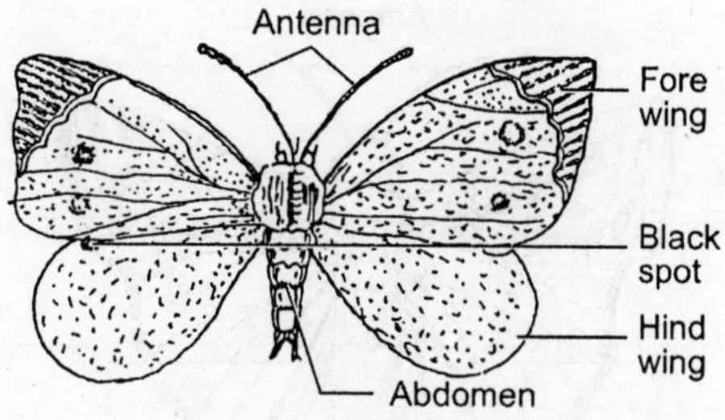

Fig. 10.5.5 : *Pieris brassicae.*

1. Butterflies are pale white with a smoky shade on the dorsal side of the body.
2. Wings are pale white, with a black patch on the apical angle of each fore wings.
3. Costal margin of each hind wing is with a black spot.
4. Male are smaller than female and with black spots on the under side of each fore wings.
5. Femalefly is with two conspicuous black circular dots on the dorsal side of each fore wing.

Economic Importance :

It is a serious **pest of Cabbage, Cauliflower, Knolkhol** and other **cruciferous** plants. Damage is caused by caterpillars only.

2. *Spodoptera* (Tobacco Caterpiller)

Systematic Position :

Phylum	**Arthropoda**
Subphylum	**Mandibulata**
Class	**Insecta**
Subclass	**Pterygota**
Superorder	**Holometabola**
Order	**Lepidoptera**
Type	***Spodoptera litura***

Identification with reasons :

Characters same up to—

Order—**Lepidoptera** as those of ***Pectinophora***.

*Type—**Spodoptera litura** :*

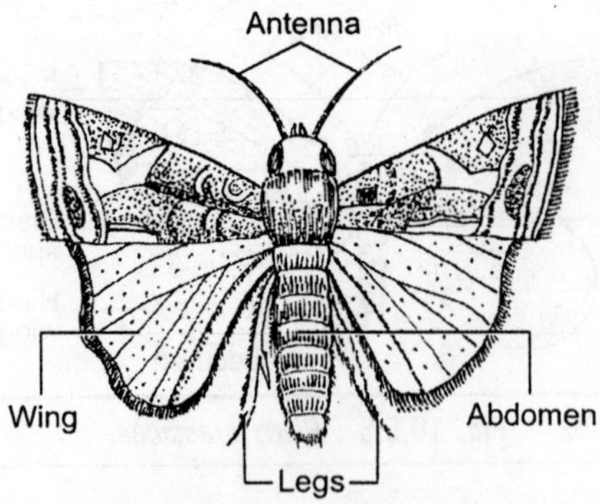

Fig. 10.5.6 : *Spodoptera litura.*

1. The fore wings of the moths have beautiful golden and greyish brown patterns.
2. Hind wings are pale white in colour.
3. Moth with hairy body.

Economic Importance :

It is a **pest of castor, ground nut, tomato, sunflower** and various **other cruciferous crops.** Only the catter pillars do damage. Larva feed on leaves and fresh growth.

3. *Leucinodes* (Brinjal fruit borer)

Systematic Position :

Phylum	**Arthropoda**
Subphylum	**Mandibulata**
Class	**Insecta**
Subclass	**Pterygota**
Superorder	**Holometabola**
Order	**Lepidoptera**
Type	***Leucinodes orbonalis***

Identification with reasons :

Characters same up to—

Order—**Lepidoptera** as those of ***Pectinophora.***

Type—***Leucinodes orbonalis :***

1. The moth is white but has pale brown or black spots on the dorsum of the thorax and abdomen.

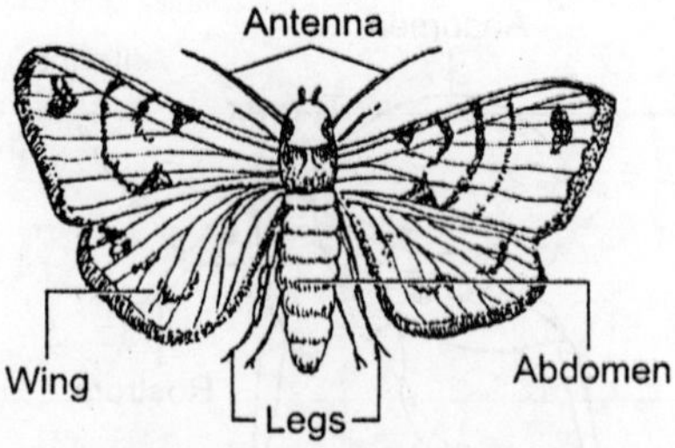

Fig. 10.5.7 : *Leucinodes orbonalis.*

2. Wings are white with a pinkish or bluish tinge and are ringed with small hairs along the apical and anal margin.
3. Fore wings are decorated with a number of black pale and light brown spots.

Economic Importance :

It is a **pest of brinjal** fruits and shoots. The caterpillars do the damage. Caterpillars bore into the tender shoots near the growing points, into flower buds or into fruits.

4. *Aulacophora* (Red pumpkin beetle)

Systematic Position :

Phylum	**Arthropoda**
Subphylum	**Mandibulata**
Class	**Insecta**
Subclass	**Pterygota**
Superorder	**Holometabola**
Order	**Coleoptera**
Type	***Aulacophora (Raphido palpa) foveicollis***

Identification with reasons :

Characters same up to—

Order—**Coleoptera** as those of *Apion.*

Type—***Aulacophora (Raphido palpa) foveicollis:***

1. Oblong beetle about 5-8 mm long.
2. Dorsal body surface is brilliant orange red and the ventral surface is black, being clothed in short white hairs.

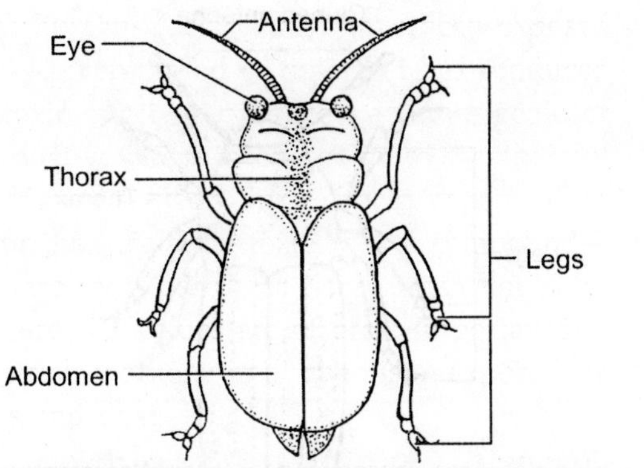

Fig. 10.5.8 : *Aulacophora (Raphidopalpa) foveicollis.*

Economic Importance :

It is a serious **pest of gourd, pumpkin, tinda, cucumber** and **melon.** Damage is caused by grubs as well as by beetles.

5. *Epilachna* (Hadda beetles)

Systematic Position :

Phylum	**Arthropoda**
Subphylum	**Mandibulata**
Class	**Insecta**
Subclass	**Pterygota**
Superorder	**Holometabola**
Order	**Coleoptera**
Type	***Epilachna sp.***

Identification with reasons :

Characters same up to—

Order—**Coleoptera** as those of ***Apion.***

Type—**(1) : *Epilachna vigintioctopunctata* :**

1. Deep red beetle with 7-14 black spots on each elytron.
2. Tip of the elytron is some what pointed.

Type—**(2) : *Epilachna dodecastigma* :**

1. Deep copper – coloured beetle with 6 black spots on each elytron.
2. The tip of the elytron is more rounded.

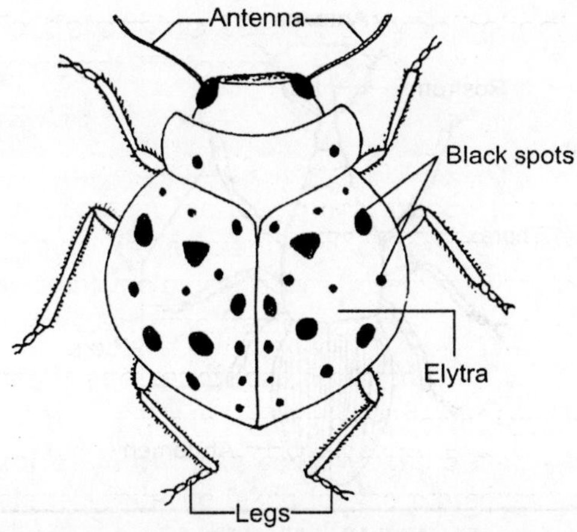

Fig. 10.5.9 : *Epilachna vigintioctopuntata.*

Type—**(3) : *Epilachna demurili* :**

1. Dull in appearance and light-coloured beetle.
2. Each elytron bears 6 black spot surrounded by yellowish rings.

Economic Importance :

It is a serious pest of brinjal, tomato, potato and cucurbitaceous vegetables. Damage is caused by the beetles as well as by the grubs.

V. STORED GRAIN PESTS

1. *Sitophilus* (Rice weevil)

Systematic Position :

Phylum	**Arthropoda**
Subphylum	**Mandibulata**
Class	**Insecta**
Subclass	**Pterygota**
Superorder	**Holometabola**
Order	**Coleoptera**
Type	***Sitophilus oryzae***

Identification with reasons :

Characters same up to—

Order—Coleoptera as those of ***Apion.***

Type—***Sitophilus oryzae* :**

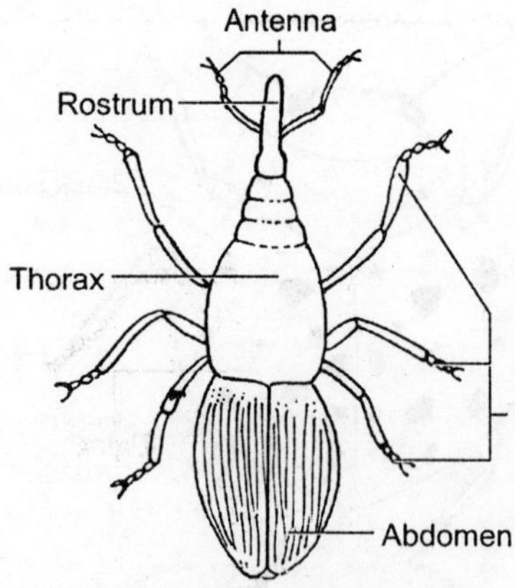

Fig. 10.5.10 : *Sitophilus oryzae.*

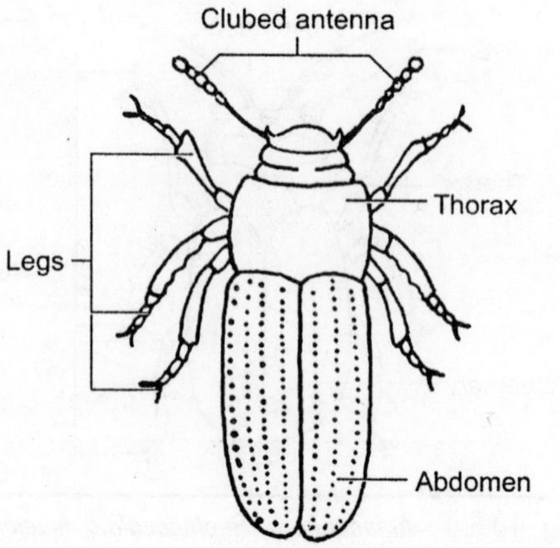

Fig. 10.5.11 : *Tribolium castaneum.*

1. Small, reddish brown beetle.
2. Cylindrical body with long, slender, curved rostrum.
3. The elytra bear four light reddish or yellowish spots.
4. Thorax is fitted with round depressions.

Economic Importance :

It is the most destructive **pest of stored grain.** It causes damage to wheat, rice, maize and sorghum grains particularly in the monsoon. Both the adults and grubs cause damage. They destroy more than what they eat.

2. *Tribolium* (Rust-red flour beetle)

Systematic Position :

Phylum	**Arthropoda**
Sub Phylum	**Mandibulata**
Class	**Insecta**
Subclass	**Pterygota**
Superorder	**Holometabola**
Order	**Coleoptera**
Type	***Tribolium castaneum***

Identification with reasons :

Characters same up to—

Order—**Coleoptera** as those of ***Apion.***

Type—***Tribolium castaneum*** **:**

1. Small reddish brown beetle about 3.5 mm in length and 1.2 mm in width.
2. The bent antennae bear a distinct club formed by three enlarged terminal joints.
3. The last antennal segment is transversely rounded.

Economic Importance :

It is the most common **pest of wheat-flour.** It also feeds upon dry fruits, pulses and cornflakes. Both larval and adults cause damage. In severe infestation, the flour turns greyish and mouldy, and has a pungent, unpleasant odour that makes it unfit for human consumption.

3. *Rhizopertha* (Lesser grain borer)

Systematic Position :

Phylum	**Arthropoda**
Subphylum	**Mandibulata**
Class	**Insecta**
Subclass	**Pterygota**
Superorder	**Holometabola**
Order	**Coleoptera**
Type	***Rhizopertha dominica***

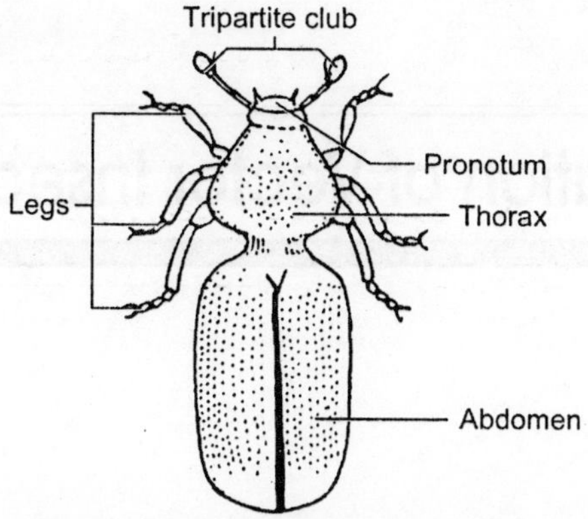

Fig. 10.5.12 : *Rhizopertha dominica.*

Identification with reasons :

Characters same up to—

Order—**Coleoptera** as those of ***Apion.***

Type—***Rhizopertha dominica :***

1. Small cylindrical beetle measuring about 3 mm in length and less than 1 mm in width.
2. Body is shining brown or black in colour with a deflexed head, covered by a crenulated hood shaped pronotum.
3. Ten jointed antennae terminate in a prominent tripartite club; the first two segments are triangular and the apical are oval.

Economic Importance :

It is a destructive **pest of wheat** and **other grains**. The adults and larvae cause serious damage to the grains by feeding inside them and reducing them to mere shells with many irregular holes.

VI. POLYPHAGUS PESTS

1. **Locusts :** See chapter 1.1.1. Non chordates: Arthropod, item no. 29.

2. **Grasshopper** : See chapter 1.1.1. Non chordates: Arthropoda, item no. 28.

3. **Termite :** See chapter 1.1.1. Non chordates: Arthropoda, item no. 27 and 27(a).

VII. HOUSEHOLD PESTS

1. ***Periplaneta sp.*** (Cockroach) : See chapter 1.1.1. Non chordates: Arthropoda, item no. 22.

2. ***Gryllus sp.*** (House cricket) : See chapter 1.1.1. Non chordates: Arthropoda, item no. 31.

3. ***Cimex sp.*** (Bed bug) : See chapter 1.1.1. Non chordates: Arthropoda, item no. 35.

4. ***Pediculus sp.*** (Human body louse) : See chapter 1.1.1. Non chordates: Arthropoda, item no. 32.

5. ***Culex sp.*** (Mosquito) : See chapter 1.1.1. Non chordates: Arthropoda, item no. 39.

6. ***Anopheles sp.*** (Mosquito) : See chapter 1.1.1. Non chordates: Arthropoda, item no. 40.

7. ***Aedes sp.*** (Mosquito) : See chapter 1.1.1. Non chordates: Arthropoda, item no. 41.

8. ***Musca sp.*** (House fly) : See chapter 1.1.1. Non chordates: Arthropoda, item no. 43.

9. ***Phlebotomus sp.*** (Sand flies) : See chapter 1.1.1. Non chordates: Arthropoda, item no. 42.

10. ***Xenopsylla sp.*** (Rat flea) : See chapter 1.1.1. Non chordates: Arthropoda, item no. 44.

11. ***Lepisma* sp.** (Silverfish) : See chapter 1.1.1. Non chordates: Arthropoda, item no. 20.

VIII. MAMMALIAN PESTS

Among the mammalian pest, rodents are responsible for causing enormous losses to crops and stored grains. Besides feeding on these products, they destroy a substantial quantity by spillage and contamination with their droppings, urine, hair etc.

1. ***Rattus sp.*** (Black rat) : See chapter 1.1.2. Chordates: Mammalia, item no. 9.

2. ***Bandicota sp.*** (Lesser bandicota) : See chapter 1.1.2. Chordates: Mammalia, item no. 10.

Chapter 10.6

Identification of Vector Insects

Insects transmit many human and animal diseases. The simplest method of disease transmission is indirect mechanical transmission, in which insects act as passive carriers of disease agents. Here they picking the disease agents upon the bodies or in the excretions of man or animals and depositing then on food. Important among these transmitters are houseflies and cockroaches.

Blood sucking insects pick up germs from the body of a diseased individual and directly transmit them into the blood or wounds of skin of other animal. Important among these transmitters are teste flies and tabanids (housefly).

Some time an insect acting as intermediate host allowing the organism to multiply or to cyclical changes or both with in its body and then transmit it to another host. Important transmitters are mosquitoes, sand flies, bedbugs and louse. Intermediate host, which transport developing individual from one definite host to another is called **vector**.

Identification :

1. ***Pediculus sp.*** : See chapter 1.1.1. Non chordate: Arthropod, item no 32.

2. ***Cimex sp.*** : See chapter 1.1.1. Non chordate: Arthropod, item no 35.

3. ***Culex sp.*** : See chapter 1.1.1. Non chordate: Arthropod, item no 39.

4. ***Anopheles sp.*** : See chapter 1.1.1. Non chordate: Arthropod, item no 40.

5. ***Phlebotomus sp.*** : See chapter 1.1.1. Non chordate: Arthropod, item no 42.

6. ***Xenopsylla sp.*** : See chapter 1.1.1. Non chordate: Arthropod, item no 44.

7. ***Aedes sp.*** : See chapter 1.1.1. Non chordate: Arthropod, item no 41.

Chapter 10.7

Identification of Same Beneficial Insects

Insects are the largest number of animal formed in this universe and are of great importance to man. Beneficial insects are very few in number. There are a few species, whose laborious products are used by the man for his own advantage. One of them is silkworm. The continuous silken thread which builds the cocoon serves as the source of the silk of our requirement. Second is honeybee. Honey and wax come from beehive; man for several purpose uses these two products. The third is lac–insects. It liberates a resinous product as exudate which forms a crust around the insect. From this exudate stick-lac of commerce is manufactured. Many insects are considered to be of great value of pollinators.

Identification :

1. ***Bombyx sp.*** **:** See chapter 1.1.1. Non chordate: Arthropod, item no 45.

2. ***Antheraea sp.*** **:** See chapter 1.1.1. Non chordate: Arthropod, item no 48.

3. **Butterfly :** See chapter 1.1.1. Non chordate: Arthropod, item no 50.

4. ***Apis sp.*** **(Honey bee) :**

Habit and Habitat :

Honeybee is a social insect. The nest of the honeybee is known as bee-hive. It is commonly seen on tall trees and ceiling of houses. They can be reared in artificial hives. Its food is nectar and pollen of flowers. Thousand of individuals, who live in the hive are of three different forms viz., **workers, drones** and **queens.**

Systematic Position :

Phylum	**Arthropoda**
Sub Phylum	**Mandibulata**
Class	**Insecta**
Subclass	**Pterygota**
Super order	**Holometabola**
Order	**Hymenoptera**
Type	***Apis***

Distinctive features :

1. Triangular head bears three ocelli in the middle.

2. Antennae two, short and many jointed.

3. Mouth parts both sucking and chewing type.

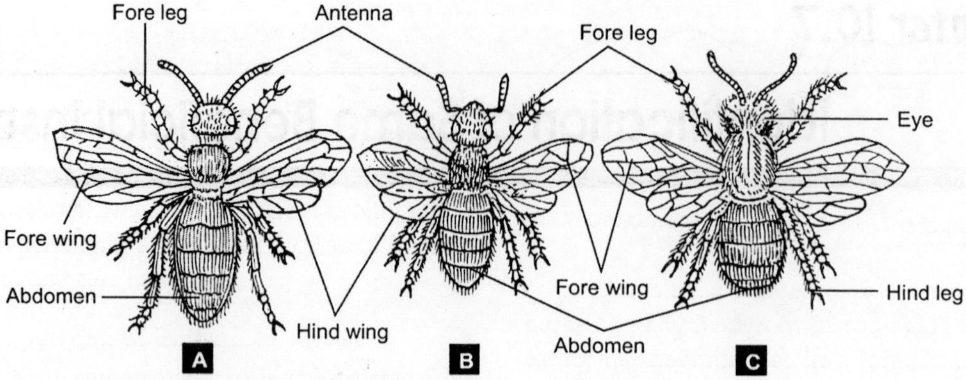

Fig. 10.7.1 : *Apis sp.* A–Queen, B–Worker, C–Drone.

4. Tibia of prothoracic leg is provided with soft hairs called **eye brushes** and on the first tarsal segment presents stiff bristles, **pollen brushes.**

5. In addition to pollen brushes, tibia of mesothoracic legs bears a **wax spine stick.**

6. The first tarsal segment of metathoracic legs bears at its inner part **pollen combs**. Near the junction of tibia and first tarsal segment **pollen packers** are present and the tibia bears **pollen basket**.

7. Abdominal segments seven in drones but six in workers and queen.

8. Queen and workers bear in the last abdominal segment a sting.

Worker :

1. Size small.

2. Body is densely covered with hairs.

3. Abdominal segments six and possess wax glands and stings.

Queen :

4. One and a half times as large as the worker bee.

5. Abdomen is greatly distended; wings are short.

6. Pollen baskets and wax glands are absent.

7. Abdominal segments seven and last segment bears sting.

Drone :

8. Medium size, larger than worker.

9. Body is comparatively broad and abdomen is blunt without any sting.

10. Head is quite round with long eyes.

11. Wax glands and pollen basket absent.

12. Abdominal segments seven.

Identification with reasons :

State classification and characters up to—

Superorder—**Holometabola** as those of ***Hispa*** (see chapter 1.1.1, item 38).

Order—**Hymenoptera :**

1. Mouth parts are of both sucking and chewing types.
2. Wings membranous, the hind wings smaller than the forewing and interlocked with the hind wings by means of hooklets.
3. The first abdominal segments fused with the metathorax.
4. Metamorphosis complete.
5. Ovipositor always present and modified for sawing, piercing or stinging.

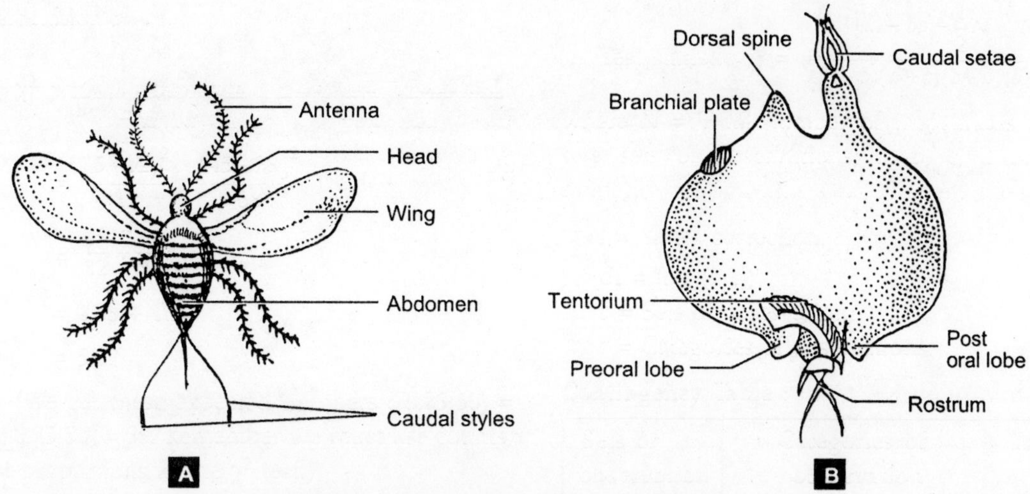

Fig. 10.7.2 : *Tachardia lacca.* A–Male, B–Female.

Type—***Apis*** : State distinctive features as in above.

1. *Tachardia* (Lac insect)

Habit and Habitat :

Live on certain specific trees like kusum (*Schleichera sp., Acacia sp.*), khair (*A. catechu)* and ber (*Zizyphirs sp.*), undergo two generation annually, males have dimorphic forms, **alate** and **apterous**; liberates a resinous product as exudates; distribution, tropical and sub tropical.

Systematic Position :

Phylum	**Arthropoda**
Sub Phylum	**Mandibulata**
Class	**Insecta**
Subclass	**Pterygota**
Order	**Homoptera**
Type	***Tachardia lacca***

Distinctive features :

Male :

1. Larger than female.
2. With or without wings.
3. Presence of caudal setae.
4. Presence of one pair of antennae and one pair of eyes.
5. Proboscis absent.
6. Presence of three pairs of legs.

Female :

1. Smaller then male.
2. Body globular and enclosed in dense resinous cells and with posterior bronchial plate.
3. Legs absent.
4. Antennae vestigial.
5. Proboscis present with pre and post oral lobes.
6. Prominent dorsal spine.

Identification with reason :

State classification and characters up to—

Order—**Homoptera** as those of ***Aphis*** (see chapter 1.1.1).

Type—***Tachardia*** : State distinctive features as in above.

Section 11

BIO-STATISTICS

SECTION AGENDA

Chapter 11.1

Data, Variables, Samples and Population

Introduction

In singular form, statistics is the scientific study, which deals with the theories and methods of collection, analysis and interpretation of such data in quantitative terms. In plural form, it refers to a set of numerical figures obtained by measurement or counting. The word *statistik* comes from the Italian word *statista* (meaning statesman). It was first used by Gottfreid Achenwall (1719-1772), a professor at Marlborough and Gottingen. In modern biology, statistics is a familiar word nowadays and is widely used in the laboratory or in the field wherever data is concerned.

Statistics is basically divided into two basic groups

(a) **Descriptive statistics** : Descriptive statistics provide devices for organising data and bringing into attention their essential characteristics, with the intention of reaching conclusions at a later stage.

(b) **Inferential statistics** : Inferential statistics provides the procedures to draw inferences about conditions that exist in a larger set of observations, from the study of a part of a set.

So the first and foremost important feature of statistics is statistical data or simply data. These data are of two types: 1. Primary data and 2. Secondary data.

1. Primary data :

It refers to those which are collected by a particular person or organization for a specific purpose directly from the field of enquiry and thus are original in nature.

2. Secondary data :

It refers to those which are collected by some other person or organization for one purpose and later used for another purpose.

Statistical Variables and Attributes

Different types of information are known as the **variables** of a dataset, e.g. age, weight, height, gender, marital status, annual income etc. for a set of individuals. The quantitative character of the statistical information collected from a group of units or objects are termed as *variables.*

Variables which are experimentally operated by a researcher are called **independent variables.** Variables which are measured from these independent variables are called **dependent**

variables. All other factors which may influence the dependent variable are called **confounding**, **extraneous** or **secondary variables**—unless these are the same for each group being tested, comparisons will be unreliable. Variables may be categorized as either quantitative or qualitative :

- **Quantitative variable** measures either **how much** or **how many** of something, i.e. a set of observations where any single observation is a number that represents an **amount** or a **count**.
- **Qualitative vairable** provides **labels**, or **names**, for categories of like items, i.e. a set of observations where any single observation is a word or code that represents a **class** or **category**.

Qualitative variable can be separated into :

Nominal variables : Variables with **no inherent order or ranking sequence**, e.g. numbers used as names (group 1, group 2...), gender, etc.

Ordinal variables : Variables with an **ordered series**, e.g. "greatly encouraged, moderately encouraged, indifferent, moderately discouraged, greatly discouraged". Numbers assigned to such variables indicate *rank order*, only the distance between the numbers has no meaning.

Interval variables : Equally spaced variables, e.g. length . The difference between a length of 66 feet and 67 feet is taken to be the same as the difference between 76 feet and 77 feet. Interval variables **do not have a true zero**, e.g. 88 feet is not necessarily double the length of 44 feet.

Ratio variables : Variables spaced at equal intervals with a **true zero point**, e.g. age.

Quantitative variable can be divided into :

Discrete variables : The set of all possible values which consist only of isolated (secluded) points, e.g. counting variables (1, 2, 3 ...).

Continuous variables : The set of all values within intervals, e.g. 0-9, 10-19, 20-29... etc.

The qualitative character is termed as an **attribute**. An attribute cannot be measured and can be classified only under different heads or categories; e.g. information about the habitat of a group of species is qualitative, viz., soil habitat, tree habitat, aquatic habitat etc.

Samples and Population

Researcher gathers data from a sample. They use this information to make inferences about the population that the sample represents. So, sample and population are relative terms. A population is a whole and the sample is a fraction or segment of that whole.

A **population** is a collection of all elements which are to be studied and about which a conclusion has to be drawn.

A **sample** is a collection of some, but not all, of the elements of a population. A representative sample contains the relevant characteristics of the population in the same proportion that they are included in that population. Collection of sample from a population is sometimes necessary for many analysis in biological sciences.

Sampling is a method by which one can select the part of the aggregate material to obtain information about the whole, i.e. select units from population. So a sample is that selected part of a population which is used to ascertain the characteristics of that population.

Types of Sampling : There are 5 different types of sampling, viz. (1) Simple random sampling, (2) Purposive sampling, (3) Stratified sampling, (4) Systematic sampling, (5) Multi-stage sampling. In biology, the first three types of sampling are widely used.

(1) ***Simple random sampling :*** It is a process by which a group of units can be selected in such a manner that every unit of the population has an equal chance of being included in the sample. This method of sampling can be drawn in two ways—

(i) *Simple random sampling with replacement* : It can be done when the sample organisms are drawn from the population one by one and after each drawing the selected one is noted and then returned to the population before the next one is drawn. In this method, at each stage, all the population units are considered for selection with equal probability and any of the population units may appear more than once in the sample.

(ii) *Simple random sampling without replacement* : In this process the sample organisms are drawn all at a time or drawn one by one in such a manner that after each drawing the selected one is not returned to the population when the next one is drawn. Here drawing is made with equal probability only from those units not selected in any of the earlier drawing.

(2) ***Purposive Sampling*** : In this method a sample is selected on the basis of individual judgment of the sampler. In this method there is a chance of bias. This type of sampling is useful when the sample size is small.

(3) ***Stratified sampling*** : In this method the population is subdivided into several parts or *strata* (each subgroup is homogeneous within themselves) in a purposive way, then a sub-sample is chosen from each of them. The combination of this sub-sample is known as *stratified random sampling*. This method is a mixture of both purposive and random sampling.

Chapter 11.2

Measures of Central Tendency

The '**average**' or '**measure of central tendency**' is the representative (or typical value) of a whole set of observations.

This value lies in a central position of the data so that some observations are larger and some others are smaller than it.

Generally, there are three measures of central tendency—

1. Mean or Arithmetic Mean.
2. Median.
3. Mode.

1. Mean or Arithmetic Mean :

Mean is the arithmetic average of the observed scores or values of a sample. It is the centre of gravity or centroid of the frequency distribution of the sample.

Computation of mean :

(a) **Arithmetic mean**

$$\overline{X} = \frac{1}{N}\sum_{i=1}^{n} x_i$$

where N = total number of observations,

x_i = i^{th} observations.

(b) **Arithmetic Mean of grouped data**

$$\overline{X} = \frac{1}{N}\sum_{i=1}^{n} f_i x_i$$

where N = total number of observations,

x_i = individual observation,

f_i = frequency of the ith observation.

$\sum_{i=1}^{n} f_i x_i$ = total of product of f_i of and x_i

(c) **Weighted Arithmetic Mean (average)**

$$\overline{X}_w = \frac{\sum (w_i \times x_i)}{\sum w_i}$$

where w = weight assigned to each observation;

$\Sigma(w_i x_i)$ = sum of all weights of each element times that element,

Σw_i = sum of all weights.

(d) **Population mean 'μ'** $= \frac{1}{N}\Sigma x$

where N = total number of observations,

x = individual observation.

(e) **Mean of ungrouped and grouped data (short cut method) :**

- For **mean of ungrouped data** some short cut methods can be employed. In such case a *provisional mean or assumed mean* is taken into consideration, which is denoted as *'A'*. Individual observations or variables are denoted by x_1, x_2, x_3 etc. The difference between the individual observations and *A* (which is taken as arbitrary, preferably the one corresponding to the maximum frequency, if it lies near the middle of the frequency distribution) is denoted by *'d'*, *i.e.*, $X - A = d$

$$\text{Therefore, } \overline{X} = A + \frac{\sum_{i=1}^{n} d_i}{N},$$

where N = total number of observations,

A = assumed or provisional mean, and

$\sum_{i=1}^{n} d_i$ = Sum of *'d'*.

- Now in case of **grouped data :**

$$\overline{X} = A + \frac{\sum_{i=1}^{n} f_i d_i}{N},$$

where f_i = frequency of each observation,

N = total number of observation,

A = assumed or provisional mean,

$\sum_{i=1}^{n} f_i d_i$ = Sum of product of frequency and corresponding difference.

In case of **mean of** grouped data with *equal class intervals* i.e., when the frequency table has an equal interval of width *'i'*, the formula that can be used is as follows :

$$\overline{X} = A + \frac{\sum_{i=1}^{n} f_i d_i}{N} \times i$$

where f_i = frequency of each observation,

N = total number of observation,

A = assumed or provisional mean,

$$d = \frac{X - A}{i},$$

i = class interval, and

f = frequency.

Problem 1 :

Simple arithmetic mean : The total length of 10 fishes, snout or rostrum to the tip of the tail is 15, 13, 17, 16, 14, 19, 21, 17, 21 and 15 cm.. Calculate the arithmetic mean of the length of fishes.

Solution :

Tabulation of data :

Serial Number	Length of Fishes in cm. ($x_1, x_2, \ldots x_{10}$)
1	15
2	13
3	17
4	16
5	14
6	19
7	21
8	17
9	21
10	15
$N = 10$	$\sum x_i = 168$

$$\text{Therefore, } \overline{X} = \frac{1}{N}\sum_{i=1}^{10} x_i = \frac{168}{10} = 16.8 \text{ cm}$$

Problem 2 :

Weighted Arithmetic Mean : The fresh weight of 31 crabs has been recorded as 1100g, 1150g, 1120g, 1210g, 1080g, 1100g, 1210g, 1120g, 1150g, 1100g, 1080g, 1150g, 1080g, 1100g, 1150g, 1100g, 1150g, 1120g, 1100g, 1150g, 1120g, 1120g, 1120g, 1150g, 1100g, 1120g, 1100g, 1100g, 1150g, 1120g, 1120g.

Calculate the Weighted Arithmetic Mean of weight of fresh crab from the above data :

Serial number	Weight of crab in g	Serial number	Weight of crab in g	Serial number	Weight of crab in g	Serial number	Weight of crab in g
1	1100	10	1100	19	1100	28	1100
2	1150	11	1080	20	1150	29	1150
3	1120	12	1150	21	1120	30	1120
4	1210	13	1080	22	1120	31	1120
5	1080	14	1100	23	1120		
6	1100	15	1150	24	1150		
7	1210	16	1100	25	1100		
8	1120	17	1150	26	1120		
9	1150	18	1120	27	1100		

Solution :

Tabulation of data :

From the above tabulated data it is to be noted that some figures have appeared more than once which means there is more than one frequency. For example, 1150g has been recorded eight times. It is to be said therefore that frequency of 1150g in the tabulated data is 8.

Tabulation of data (Problem 2) :

Weight in g. $x_1, x_2, x_3 ... x_n$	Frequency $f_1, f_2, ... f_n$	$f_i x_i$
1080	3	3240
1100	9	9900
1120	9	10080
1150	8	9200
1210	2	2420
	Total $N = 31$	**Total** $\Sigma fx = 34840$

Therefore, $f_1x_1 + f_2x_2 + f_3x_3 + f_4x_4 + f_5x_5 = 34840$

and Mean

$$= \frac{f_1x_1 + f_2x_2 + f_3x_3 + f_4x_4 + f_5x_5}{N} = \frac{\sum_{i=1}^{n} f_i x_i}{N}$$

(where N is the total number of observations)

$$= \frac{34840}{31} = 1123.87\text{g}.$$

The above example is a case of mean of grouped data.

Mean weight of crabs is 1123.87g.

Problem 3 :

Calculate the mean length of garden lizards (*Calotes sp.*) by both direct and short-cut methods.

Length in millimetres : 219, 216, 213, 210, 207, 204, 201, 198, and 195.

Number of house lizard : 2, 4, 6, 10, 11, 7, 5, 4, 1.

Solution :

Tabulation of data (Problem 3) :

Serial Number	Length in mm (x)	f_i	$f_i x_i$	$d_i = (x_i - A)$	$f_i d_i$
1	219	2	438	12	24
2	216	4	864	9	36
3	213	6	1278	6	36
4	210	10	2100	3	30
5	207	11	2277	0	0

Tabulation contd.

Serial Number	Length in mm (x)	f	fx	d = (x–A)	fd
6	204	7	1428	-3	-21
7	201	5	1005	-6	-30
8	198	4	792	-9	-36
9	195	1	195	-12	-12
	Total	**50**	**10377**	**0**	**Σfd = 27**

Here, assumed mean '*A*' has been taken as 207, which is the middle value.

By direct Method : $\bar{X} = \frac{\sum fx}{N} = \frac{10377}{50} = 207.54$

By short cut method :

$\bar{X} = A + \frac{\sum fd}{N} = 207 + \frac{27}{50} = \frac{19377}{50} = 207.54$

∴ Mean length of garden lizards is 207.54mm.

Problem 4 :

Calculate the Mean for the following distribution of species in a population of fish (*Channa punctatus*).

Length in mm	12.5 -17.5	17.5 -22.5	22.5 -27.7	27.5 -32.5	32.5 -37.5	37.5-42.5
Number of fish	2	22	10	14	3	4
Length in mm	42.5-47.5	47.5-52.5	52.5-57.5			
Number of fish	6	1	1			

Solution :

Tabulation of data (Problem 4) :

Class interval	Frequency (f_i)	Mid value(x_i)	$d_i = \frac{x_i - A}{i_i} = \frac{X - 30.0}{5}$	$f_i d_i$
12.5-17.5	2	15.0	-3	-6
17.5-22.5	22	20.0	-2	-44
22.5-27.5	10	25.0	-1	-10
27.5-32.5	14	30.0	0	0
32.5-37.5	3	35.0	1	3
37.5-42.5	4	40.0	2	8
42.5-47.5	6	45.0	3	18
47.5-52.5	1	50.0	4	4
52.5-57.5	1	55.0	5	5
Total	**63**	—	—	**Σfd = -22**

Please note, A = 30 and where i is class width.

Now, $\overline{X} = A + \frac{\sum fd}{N} \times i$

Therefore, $\overline{X} = 30 + \left(\frac{-22}{63}\right)5 = 30.00 - 1.75 = 28.25\text{mm}.$

So the mean of distribution of the population of fish (*Channa punctatus*) is 28.25mm.

Problem 5 :

Type of bacterial strain	Nutrient utilisation mg/hour	Progeny production number × 10^5/hour	
		Condition I	Condition II
Strain A	5.00	4	1
Strain B	7.00	3	2
Strain C	9.00	3	5

The investigator wants to know the weighted average nutrient requirement rate for each condition.

Solution :

The average nutrient requirement for *condition I* : One has to weight the hourly nutrient utilisation (x_i) for each strain by its proportion of total progeny number produced in a particular condition.

Strain A produces 4/10 of total production, strain B produces 3/10 of total production and strain C produces 3/10 of total production in condition I. These are the total weights assigned to each strain (w).

$$\therefore \quad \overline{X}_w = \frac{\sum_{i=1}^{n}(w_i x_i)}{\sum_{i=1}^{n} w_i}$$

$$= \left(\frac{4}{10}\times 5\right)+\left(\frac{3}{10}\times 7\right)+\left(\frac{3}{10}\times 9\right)$$

$$= 6.8 \text{ mg/hour}$$

∴ Weighted average or mean is 6.8mg/hour.

The average nutrient requirement for *condition II* : One has to weight the hourly nutrient utilisation (x_i) for each strain by its proportion of total progeny number produced in a particular condition.

Strain A produces 1/8 of total production, strain B produces 2/8 of total production and strain C produces 5/8 of total production in condition I. These are the total weight assigned to each strain (*w*).

$$\therefore \quad \overline{X}_w = \frac{\sum_{i=1}^{n}(w_i x_i)}{\sum_{i=1}^{n} w_i}$$

$$= \left(\frac{1}{8}\times 5\right)+\left(\frac{2}{8}\times 7\right)+\left(\frac{5}{8}\times 9\right) = 8 \text{ mg/hour}$$

∴ Weighted average or mean is 8mg/hour.

2. Median :

Median of a set of observations is the middle most value when they are arranged in order of magnitude. It can be calculated—

(i) when the observations are arranged either in the ascending order or descending order and when total number of observations is an odd number or an even number .

(ii) when the observations are in the form of a grouped series. Median is the real measure of central tendency as it gives the value of the most central observation.

Computation of Median :

(a) **Median of ungrouped data :**

1. When *N* is even,

Median = Average of $\left(\frac{N}{2}\right)^{\text{th}}$ and $\left(\frac{N}{2}+1\right)^{\text{th}}$ terms

2. When N is odd,

$$\text{Median} = \left(\frac{N+1}{2}\right)^{th} \text{ term}$$

(b) **Median of grouped data :**

1. When series is discrete,

$$\text{Median} = \left(\frac{N+1}{2}\right)^{th} \text{ term}$$

where $N = \Sigma f$, f = frequency

2. When series is continuous,

$$\text{Median} = l_1 + \frac{\frac{N}{2} - F}{f_m} \times C$$

where l_1= lower boundary of median class; N = total observations, F = cumulative frequency, f_m = frequency of median class; C = width of the median class (the class in which cumulative frequency of $N/2$ lies is called the median class.)

Problem 6 :

Calculate the median of the following observations obtained by measuring the length in inches of fishes. 12, 12, 10, 11, 9, 14, 13, 10, 11, 14, 12, 10, 13.

Solution :

Figures are first arranged in an ascending order 9,10,10,10,11,11,12, 12, 12, 13, 13, 14 and 14

As the 'N'or number of observation is odd,

$$\text{therefore Median} = \left(\frac{N+1}{2}\right)^{th} \text{ term.}$$

So the median of the observations is the 7th term i.e. 12 inches.

[**Note :** When the numbers of observations are odd the value of the middle most item is the median.]

Problem 7 :

Calculate the median of the following observations obtained by measuring the length in cm of insects. 3.1, 2.6, 5.0, 4.7, 2.4, 3.9, 5.1, 3.6

Solution :

Here the number observations is even, i.e 'n' is an even number. Data is first arranged in an ascending order 2.4, 2.6, 3.1, (3.6), (3.9), 4.7, 5.0 and 5.1. Here the two middle most items are 3.6, 3.9.

So the median is average of $(N/2)^{th}$ and $(N+1/2)^{th}$ terms.

$$\text{That is } \frac{3.6+3.9}{2} = 3.75.$$

So the median of the observations is 3.75cm.

[**Note :** When the number of observation is even, the arithmetic mean of the two middle most observation is taken as the median.]

Problem 8 :

Calculate median for the following data of occurrence of ciliates (Protozoa) in culture solution.

Number of days :	0	1	2	3	4	5	6
Number of ciliates :	7	44	35	16	9	4	1

Solution :

In this case it is required to calculate the cumulative frequencies as follows :

This is a discrete series.

Tabulation of data (Problem 8) :

Number of days (x_i)	Number of ciliates (f_i)	Cumulative frequency
0	7	7
1	44	(7+44) = 51
2	35	(51+35) = 86
3	16	(86+16) = 102
4	9	(102+9) = 111
5	4	(111+4) = 115
6	1	(115+1) = 116 = N

So cummulative frequency is N = 116

$$\text{Here, } \frac{N+1}{2} = \frac{116+1}{2} = 58.5.$$

From the above table it is found that (N+1)/2 = 58.5 is greater than cumulative frequency 51 but is smaller than the next cumulative frequency 86 corresponding to x = 2. So the median is 2.

∴ The median of the above data is 2.

[**Note :** In case of simple frequency distribution the cumulative frequency has to be calculated. If the total frequency be N, the value of the variable corresponding to cumulative frequency $\frac{N+1}{2}$ will be the median as in the above example.]

Problem 9 :

Find the median of the following distribution in a population of snakes :

Length of snakes in inches :	15-25	25-35	35-45	45-55	55-65	65-75
Number of snakes :	4	11	19	14	0	2

Solution :

In this case it is required to calculate the cumulative frequencies as follows :

This is a continuous series.

Tabulation of data (Problem 9) :

Length in inches (Class boundaries)	Number of snakes (frequency)	Cumulative frequency (F)
15-25	4	4
25-35	11	(11+4)=15
35-45	19=f_m (frequency of median class)	(15+19)=34
45-55	14	(34+14)=48
55-65	0	(48+0)=48
65-75	2	(48+2)=50=N

From the above $N/2$ = 25 which is more than cumulative frequency 15 but is less than the next cumulative frequency 34. Therefore the median class is 35-45. By applying the formula for grouped frequency distribution,

i.e. $$\text{Median} = l_1 + \frac{\frac{N}{2} - F}{f_m} \times c$$

where, l_1= lower boundary of median class;

N = total frequency,

F = cumulative frequency below l_1

f_m= frequency of median class;

c = width of the median class, (the class in which cumulative frequency $N/2$ lies is called the median class.)

Here l_1= 35, N =50, F=15, f_m= 19, c = 45-35=10.

So, Median = $35 + \frac{25-15}{19} \times 10 = 35 + 5.26 = 40.26$.

Hence the median of distribution of the population of snakes is 40.26 inches.

3. Mode :

Mode is the value that occurs with the maximum frequency in a given set of observations. From a simple series of distribution, mode can be determined by locating the value which occurs maximum number of times.

Computation of Mode :

(a) **In case of simple series :** The value which is repeated maximum number of times is the mode of the series.

(b) **In case of continuous frequency distribution (class of equal size) :**

$$\text{Mode} = l_1 + \left(\frac{d_1}{d_1 + d_2}\right) \times c$$

$$= l_1 + \left(\frac{f_0 - f_{-1}}{2f_0 - f_{-1} - f_1}\right) \times c$$

where, l_1 = lower boundary of the modal class i.e. the class containing the highest frequency;

d_1 =difference of the highest frequency and the frequency of class just preceding the modal class;

d_2 = difference of the highest frequency and the frequency of class just following the modal class;

c = common width of classes.

If f_0, f_{-1}, f_1 represent the frequencies of the modal class, the class just preceding and the class just following it respectively; then $d_1 = f_0 - f_{-1}$ and $d_2 = f_0 - f_1$.

(c) **If mean and median are known the mode can be calculated by using the relationship as follows :**

Mean – Mode = 3(Mean – Median)

Problem 10 :

The length (in cm) of *Lymnaea sp.* (snail) was noted as follows in a box of 100 species :

Length in cm	less than 2.5	2.8	3.0	3.2	above 3.3
Number of *Lymnaea sp.* :	20	30	20	20	10

Find the modal length of *Lymnaea sp.*

Solution :

Out of 100 species, 2.8 cm is the modal length, which occurs most frequently, and this value is the mode of these observations.

Problem 11 :

The monthly sampling in numbers of 100 sites is distributed as follows :

Sampling per sites	0-100	100-200	200-300	300-400	400-500	500-600
Number of sites	12	18	27	20	17	6

Calculate the modal value.

Solution :

Here the largest class frequency is 27, which lies in the class 200-300. Therefore this is the modal class.

By applying the formula, Mode $= l_1 + \left(\dfrac{f_0 - f_{-1}}{2f_0 - f_{-1} - f_1}\right) \times c$

$= 200 + [(27–18)/(27–18) + (27–20)] \times (300–200)$

$= 200 + [9/(9 + 7)] \times 100 = 256.25 \approx 256$

Therefore the modal length is 256 cm.

Chapter 11.3

Measures of Dispersion

The word '**deviation**' or '**dispersion**' denotes the degree of heterogeneity in the data. It shows the discrepancy from one observation to another. The deviation of a given set of observation is zero only when all of them are equal. These deviations can be studied with the help of the following measures of dispersion :

1. Range.
2. Standard deviation.
3. Variance.
4. Mean deviation.
5. Quartile deviation.
6. Coefficient of variation.

1. Range :

Range is defined as the difference between the largest and smallest observations.

Problem 1 :

A population of insects has following data.

Length (in cm) : 6 4 1 6 5 10 3

Calculate the range.

Solution :

Here, Maximum length=10 cm and minimum length=1 cm. Therefore, Range = 10–1 cm = 9 cm.

2. Standard deviation :

Standard deviation (S.D. or σ or *s*) is defined as the square root of the arithmetic mean of the squares of deviations of the observations from their arithmetic mean.

Computation of Standard deviation :

- For original and ungrouped observations

  Standard deviation (S.D.) or

$$s \text{ or } \sigma_x = \sqrt{\frac{\sum(x_i - \bar{x})^2}{N}} = \sqrt{\frac{\sum y_i^2}{N}}$$

- For small samples with $N < 30$,

  Standard deviation (S.D.) or

$$s \text{ or } \sigma_x = \sqrt{\frac{\sum(x_i - \bar{x})^2}{N-1}} = \sqrt{\frac{\sum y_i^2}{N-1}}$$

- A short-cut method for finding σ_x

1. $$\sigma_y = \sqrt{\frac{\sum y_i^2}{N} - \left(\frac{\sum y_i}{N}\right)^2}$$

...... for simple series

2. $$\sigma_y = \sqrt{\frac{\sum_{i=1}^{n} f_i y_i^2}{N} - \left(\frac{\sum_{i=1}^{n} f_i y_i}{N}\right)^2}$$

.... for frequency distribution

where x_i is the ith variable, $\bar{x}$ is the arithmetic mean and y is the altered value computed after changing of origin and scale of the original individual variable and N = total number of observations.

If d = class interval used for change of scale, then $\sigma_x = d\sigma_y$

- Standard deviation (σ) of combined group(s) :

$N\sigma^2=(n_1\sigma_1^2+n_2\sigma_2^2)+(n_1d_1^2+n_2d_2^2)$

$$\text{or, } \sigma^2 = \frac{(n_1\sigma_1^2+n_2\sigma_2^2)+(n_1d_1^2+n_2d_2^2)}{N}$$

$$\text{or, } \sigma = \left[\sqrt{\frac{(n_1\sigma_1^2+n_2\sigma_2^2)+(n_1d_1^2+n_2d_2^2)}{N}}\right]$$

where, $N\bar{x} = (n_1\bar{x}_1 + n_2\bar{x}_2)$

$$\text{or, } \bar{x} = \frac{(n_1\bar{x}_1+n_2\bar{x}_2)}{N}$$

where, $d_1 = \bar{x}_1 - \bar{x}$, $d_2 = \bar{x}_2 - \bar{x}$ and $\bar{x}$ is the mean of the composite group. $\bar{x}_1$, $\bar{x}_2$ are the means of group I and II respectively. N is the sum of n_1 (no. of observations of group I) and n_2 (no. of observations of group II). σ_1 and σ_2 are standard deviations of group I and II respectively.

Properties of Standard deviation :

1. For large sample that are drawn from bilaterally symmetrical, normally distributed population, standard deviation of the samples are randomly distributed around the population standard deviation to form a bilaterally symmetrical normal distribution obeying the laws of probability.

2. Addition or subtraction of a constant amount (change of origin) to each individual value or score leaves the standard deviation unaltered, whereas multiplication or division of each score by a constant (change of scale) causes the same change in standard deviation.

If $y_1, y_2,......, y_n$ represent the deviations of x_1, x_2,......and x_n from an arbitrary constant c, then

i.e., $y = x-c$, then S.D. of x = S.D. of y.

It means that S.D. does not depend on the origin.

If y_1, y_2,... and y_n represent the deviations of x_1, x_2,... and x_n from an arbitrary constant c, and divided by another constant d (change of scale), then

i.e. $y = \dfrac{x-c}{d}$

S.D. of $x = d$ (S.D. of y)

Problem 2 :

Calculate the standard deviation from the data in Problem 1.

Solution :

For the data in Problem 1, mean = 35/7 = 5 cm, and deviations from mean are 1, –1, –4, 1, 0, 5, –2. Therefore Standard deviation (S.D.).

$$= \sqrt{\frac{1^2+(-1)^2+(-4)^2+1^2+0^2+5^2+(-2)^2}{7}}$$

$$= \sqrt{\frac{48}{7}} = 2.62$$

Therefore, S.D. is 2.62.

Problem 3 :

In an experiment, fishes are kept in 9 jars for behavioural study. Calculate the standard deviation from the following frequency distribution :

Sampling jars :	1	2	3	4	5	6	7	8	9
No. of fishes :	92	49	52	82	102	60	35	24	4
Opercular movements :	5	6	7	9	4	3	10	9	5

Solution :

Tabulation of data (Problem 3) :

Sampling jar	Opercular movements x_i	No. of fishes f_i	$y_i = x_i - 4$ $c = 4$	$f_i y_i$	$f_i y_i^2$
1	5	92	1	92	92
2	6	49	2	98	196
3	7	52	3	156	468
4	9	82	5	410	2050
5	4	102	0	0	0
6	3	60	–1	–60	60
7	10	35	6	210	1260
8	9	24	5	120	600
9	5	4	1	4	4
		N = 500		$\Sigma fy = 1030$	$\Sigma fy^2 = 4730$

[**Note** : Since S.D. is unaffected by change of origin (c) we take $y = x - c = x - 4$ (where $c = 4$).

$$\therefore \sigma_y^2 = \frac{\sum_{i=1}^{n} f_i y_i^2}{N} - \left(\frac{\sum_{i=1}^{n} f_i y_i}{N}\right)^2$$

$$= \frac{4730}{500} - \left(\frac{1030}{500}\right)^2$$

$$= 9.46 - 4.24$$

$$= 5.22$$

$$\therefore \sigma_y = \sqrt{5.22} = 2.28$$

$$\therefore \sigma_x = \sigma_y = 2.28.$$

Problem 4 :

In a pond there are common carps (fishes) of different weights. Calculate the standard deviation from the following distribution :

Weights in Kg. :	20–29	30–39	40–49	50–59	60–69	70–79	80–89	90–99
No. of fishes :	5	12	15	20	18	10	6	4

Solution :

Tabulation of data (Problem 4) :

Class interval	f_i	Mid value (x_i)	$y_i = \frac{x_i - 54.5}{10}$	$f_i y_i$	$f_i y_i^2$
20-29	5	24.5	–3	–15	45
30-39	12	34.5	–2	–24	48
40-49	15	44.5	–1	–15	15

Table contd.

Class interval	f_i	Mid value (x_i)	$y_i = \frac{x_i - 54.5}{10}$	$f_i y_i$	$f_i y_i^2$
50-59	20	54.5	0	0	0
60-69	18	64.5	1	18	18
70-79	10	74.5	2	20	40
80-89	6	84.5	3	18	54
90-99	4	94.5	4	16	64
Total	N=90	—	—	**18**	**284**

Now, $y = \frac{x-c}{d}$, so $\sigma_x = d\sigma_y$; Thus $y = \frac{x-54.5}{10}$ (where c = 54.5 and d = 10)

where c = 54.5 and d = 10

$$\text{Now, } \sigma_y^2 = \frac{\sum_{i=1}^{n} f_i y_i^2}{N} - \left(\frac{\sum_{i=1}^{n} f_i y_i}{N}\right)^2$$

$$= \frac{284}{90} - \left(\frac{18}{90}\right)^2$$

$$= 3.156 - 0.04 = 3.116$$

Therefore $\sigma_y = \sqrt{3.116} = 1.77$

Therefore $\sigma_x = d\sigma_y = 10 \times 1.77 = 17.7$

♦ Problem 5 :

In a wetland for a flock of 50 pink-headed stork the mean number and the standard deviation of numbers on a secondary data are 59.5 and 8.38. For a flock of 40 spoonbills the mean and standard deviation are 54.0 and 8.23. Calculate the mean and S.D. of combined group of 90 migratory birds.

Solution :

Tabulation of data (Problem 5) :

Secondary data	Pink headed Stork	Spoonbill	Combined group
No of birds	n_1 = 50	n_2 = 40	N = 90
Mean	$\bar{x}_1$ = 59.5	$\bar{x}_2$ = 54.0	$\bar{x}$ = ?
Standard deviations	σ_1 = 8.38	σ_2 = 8.23	σ = ?

For combined group the formula are :

(1) $N\bar{x} = n_1\bar{x}_1 + n_2\bar{x}_2$

$$\text{or, } \bar{x} = \frac{n_1\bar{x}_1 + n_2\bar{x}_2}{N}$$

(2) $N\sigma^2 = (n_1\sigma_1{}^2 + n_2\sigma_2{}^2) + (n_1 d_1{}^2 + n_2 d_2{}^2)$

$$\text{or, } \sigma^2 = \frac{(n_1\sigma_1^2 + n_2\sigma_2^2) + (n_1 d_1^2 + n_2 d_2^2)}{N}$$

Where, $d_1 = \bar{x}_1 - \bar{x}$, $d_2 = \bar{x}_2 - \bar{x}$

By using formula; $N\bar{x} = n_1\bar{x}_1 + n_2\bar{x}_2$

$$\text{or, } 90\bar{x} = 50 \times 59.5 + 40 \times 54.0 = 5135.0$$

$$\text{Therefore, } \bar{x} = \frac{5135}{90} = 57.06$$

Therefore $d_1 = 59.5 - 57.06 = 2.44$ and $d_2 = 54 - 57.06 = -3.06$

By using formula; $N\sigma^2 = (n_1\sigma_1^2 + n_2\sigma_2^2) + (n_1d_1^2 + n_2d_2^2)$

$$\text{or, } 90\sigma^2 = \{50 \times (8.38)^2 + 40 \times (8.23)^2\} + \{50 \times (2.44)^2 + 40 \times (-3.06)^2\}$$

$$= (3511.22 + 2709.32) + (297.68 + 374.54) = 6892.76$$

$$\sigma^2 = \frac{6892.76}{90} = 76.59$$

$$\text{or, } \sigma = \sqrt{76.59} = 8.75$$

Therefore standard deviation of the combined group is 8.75.

3. Variance :

The square of standard deviation is variance. In other words, the average squared distance between the mean and each item of the population is known as variance, which is denoted by σ^2.

Computation of variance :

Thus variance is defined by the formula: (Standard deviation)^2

For simple series, variance $(\sigma^2) = \frac{1}{N}\sum_{i=1}^{n}(x_i - \bar{x})^2$

For frequency distribution, variance $(\sigma^2) = \frac{1}{N}\sum_{i=1}^{n}f_i(x_i - \bar{x})^2$

where x_i is the ith variable, $\bar{x}$ is the arithmetic mean,
N is the total number of observations and
f_i is the frequency of ith class in a frequency distribution.

4. Mean deviation :

Mean deviation or mean absolute deviation is defined as the arithmetic mean of absolute deviations from mean.

Computation of Mean deviation :

Mean deviation is defined by the formula : $\frac{1}{N}\sum|x_i - \bar{x}|$

where x_i is the ith variable, $\bar{x}$ is the arithmetic mean and N is the total number of observations.

Problem 6 :

Calculate the Mean deviation from Problem 1.

Solution :

For the data in Problem 1, mean = $\frac{35}{7}$ = 5 cm and deviations from mean are 1, –1, –4, 1, 0, 5, –2.

So absolute deviations (i.e. deviations without regard to signs) are 1, 1, and 4, 1, 0, 5, 2.

Therefore, Mean deviation= $\frac{1+1+4+1+0+5+2}{7} = \frac{14}{7}$ = 2 cm.

5. Quartile deviations :

Quartile deviation is defined as half the difference between the lower and the upper quartiles.

i.e., Quartile deviation = $\frac{Q_3 - Q_1}{2}$,

where Q_1 = value of variable corresponding to $\frac{N}{4}$ th observation and

Q_3 = value of variable corresponding to $\frac{3N}{4}$ th observation.

Problem 7 :

Calculate quartile deviation from the data in Problem 1.

Solution :

By arranging in order of magnitude the data in Problem 1 would be 1, 3, 4, 5, 6, 6, 10.

So, Q_1 = 3, Q_3 = 6. Therefore, Quartile deviation = $\frac{6-3}{2}$ = 1.5 cm.

Problem 8 :

For an experiment, benthic zone of a pond has been divided into many sampling sites. Among different benthic population, calculate the quartile deviations for snail (molluscs) from the following data :

Class interval of different sampling sites :	10-15	15-20	20-25	25-30	30-40	40-50	50-60	60-70	Total
Frequency of snail :	4	12	16	22	10	8	6	4	82

Solution :

For calculating Quartile deviation we have to find Q_1 and Q_3 i.e. values of variables corresponding to cumulative frequencies $\frac{N}{4}$ and $\frac{3N}{4}$. As the total frequency N = 82, Therefore, $\frac{N}{4}$ = 20.5 and $\frac{3N}{4}$ = 61.5.

Tabulation of data (Problem 8) :

Class boundary	Cumulative Frequency (less-than)
10	0
15	4 (i.e. 0+4=4)
20	16 (i.e. 4+12=16)
$\rightarrow Q_1$	$\leftarrow$ N/4=20.5
25	32 (i.e.16+16=32)

Contd.....

Class boundary	Cumulative Frequency (less-than)
30	54 (i.e.32+22=54)
$\rightarrow Q_3$	$\leftarrow 3N/4 = 61.5$
40	64 (i.e. 54+10=64)
50	72 (i.e. 64+8=72)
60	78 (i.e. 72+6=78)
70	82 (i.e. 78+4=82)=N

By applying simple interpolation, $\dfrac{Q_1-20}{25-20} = \dfrac{20.5-16}{32-16}$

or, $\dfrac{Q_1-20}{5} = \dfrac{4.5}{16}$

or, $Q_1 - 20 = 5 \times \dfrac{4.5}{16} = 1.4$

or, $Q_1 = 1.4 + 20 = 21.4$

Again, $\dfrac{Q_1-30}{40-30} = \dfrac{61.5-54}{64-54}$

or, $\dfrac{Q_1-30}{10} = \dfrac{7.5}{10}$

or, $Q_3 - 30 = \dfrac{7.5}{10} \times 10 = 7.5$

or, $Q_3 = 7.5 + 30 = 37.5$

Therefore, Quartile deviation = $\dfrac{Q_3 - Q_1}{2} = \dfrac{37.5-21.4}{2} = 8.0$

[**Note :**

Interpolation is the process of finding the intermediate value of a function from a set of given values of that function. For example, for the following problem, pairs of values of a and b are known :

a	12	17	22	27	32
b	2.8	5.1	8.8	12.0	15.2

If a =19.3 then what will be the value of b? In simple interpolation method, it is assumed that values of b change uniformly with those of a. The partial difference in the values of a and b are $d_1 = 19.3 - 17$ and $d_2 = b - 5.1$, since $a = 19.3$ between the tabulated values $a = 17$ and $a = 22$; hence the corresponding value of b, so b, lies between $b = 5.1$ and $b = 8.8$ as indicated by arrow marks in the following table :

a	b
12	2.8
17	5.1
$\leftarrow$ 19.3	$\leftarrow$ b
22	8.8
27	12.0
32	15.2

The total difference are $D_1 = 22 - 17$ and $D_2 = 8.8 - 5.1$

In interpolation method the formula would be

$$\frac{\text{Partial difference in } 'a'}{\text{Total difference in } 'a'} = \frac{\text{Partial difference in } 'b'}{\text{Total difference in } 'b'}$$

$$\text{i.e., } \frac{d_1}{D_1} = \frac{d_2}{D_2}$$

From the above data,

$$\frac{19.3-17}{22-17} = \frac{b-5.1}{8.8-5.1}$$

$$\text{or, } \frac{2.3}{5} = \frac{b-5.1}{3.7}$$

$$\text{or, } b - 5.1 = 2.3 \times \frac{3.7}{5} = 1.702$$

Therefore $b = 5.1 + 1.702 = 6.802$.]

6. Coefficient of Variation :

The *absolute measures of dispersion*, as they are called, viz., Range, Quartile deviation, Mean deviation and Standard deviation cannot be used for comparing the deviation of two or more distributions given in different units. **Coefficient of variation (C.V.)** is the relative measure of variability which is independent of units of measurement. C.V. is more suitable than variance or standard deviation in comparing variabilities of two groups of data expressed in two different units, as because both variance and standard deviation of the two groups being in different units cannot be compared, while comparison of two C.V.s are feasible as both of them are percentages only. Further C.V. is also more suitable than variance or standard deviation for comparison of two groups of data with widely divergent means even when the latter are in same units.

Computation of C.V. :

Coefficient of variation is defined by the formula : $(\text{C.V.}) = \frac{\text{Standard deviation}}{\text{Mean}} \times 100$

Problem 9 :

To find out soil protozoans from different sampling sites observations were made under microscope. Find the coefficient of variation from the following data :

Length of protozoans (μm) :	110-119	120-129	130-139	140-149
Frequency :	5	7	12	20
	150-159	160-169	170-179	180-189
	16	10	7	3

Solution :

Since, $\text{C.V.} = \text{S.D.} \times \frac{100}{\text{Mean}}$, therefore, we have to calculate both mean and standard deviation.

Tabulation of data (Problem 9) :

Length(mm)	Frequency(f_i)	Midvalue(x_i)	$y_i = \frac{x_i - 144.5}{10}$	$f_i y_i$	$f_i y_i^2$
110-119	5	114.5	−3	−15	45
120-129	7	124.5	−2	−14	28
130-139	12	134.5	−1	−12	12
140-149	20	144.5	0	0	0
150-159	16	154.5	1	16	16
160-169	10	164.5	2	20	40
170-179	7	174.5	3	21	63
180-189	3	184.5	4	12	48
Total	$N = 80$	—	—	$\sum_{i=1}^{n} f_i y_i = 28$	$\sum_{i=1}^{n} f_i y_i^2 = 252$

Here, Mean, $\bar{x} = 144.5 + 10 \times \frac{28}{80} = 148.0$ μm

$\sigma_y^2 = \frac{252}{80} - \left(\frac{28}{80}\right)^2 = 3.15 - (0.35)^2 = 3.027$

$\therefore$ σ_y (Standard deviation of y) = $\sqrt{3.027} = 1.74$

$\therefore$ σ_x (Standard deviation of x) = $10\sigma_y = 10 \times 1.74 = 10 \times 1.74 = 17.4$ μm

$\therefore$ C.V. = $\left(\frac{17.4}{148.0}\right) \times 100 = 11.76.$

Chapter 11.4

Simple Correlation, Regression and Kendall's Coefficient of Concordance

1. Simple correlation :

Correlation may be defined as the relationship between two variables such that a change in one causes a change in the other, and also a greater change in one variable results in corresponding greater or smaller change in the other variable.

Karl Pearson's coefficient of correlation 'r' (or simple correlation coefficient) is the most widely used method of measuring the degree of statistical relationship between two variables from bivariate data.

Karl Pearson's coefficient of correlation (r) assumes the following :

(1) that there is linear relationship between the two variables;

(2) that the two variables are indifferently related, which means that one of the variables is independent and the other one is dependent; and

(3) a large number of independent causes are in use in both variables so as to produce a normal distribution.

Karl Pearson's coefficient of correlation is also known as the *product moment correlation coefficient.*

It shows the magnitude of association between two variables in the same sample when association can be described by a straight line. Positive correlation indicates that the change in two variables is in the same direction, that is, as one increase the other increases; or if one variable decreases the other too decreases. Negative correlation indicates that variables change in opposite direction, that is, if one decreases the other increases and vice versa. A coefficient of 0 indicates no correlation.

Concept of Covariance :

Let the corresponding values of two variables be (x_1,y_1), (x_2,y_2), (x_3,y_3)......(x_n,y_n).

Then the covariance between x and y is denoted by cov(x,y). It is defined as

$$\text{Cov}(x_i,y_i) = \frac{1}{n}\sum_{i=1}^{n}(x_i-\bar{x})(y_i-\bar{y}),$$

where $\bar{x}$ and $\bar{y}$ are means of x_i and y_i respectively.

Another easier formula for covariance is

$$\text{Cov}(x_i, y_i) = \frac{1}{n}\sum_{i=1}^{n}x_iy_i - \left(\frac{1}{n}\sum_{i=1}^{n}x_i\right)\left(\frac{1}{n}\sum_{i=1}^{n}y_i\right)$$

Computation of Correlation coefficient :

Formula of Karl Pearson's coefficient of

correlation (r) = $\frac{\text{cov}(x, y)}{\sigma_x \sigma_y}$ i.e.

$$r = \frac{\frac{\sum_{i=1}^{n} x_i y_i}{n} - \left(\frac{\sum_{i=1}^{n} x_i}{n}\right)\left(\frac{\sum_{i=1}^{n} y_i}{n}\right)}{\sqrt{\frac{\sum_{i=1}^{n} x_i^2}{n} - \left(\frac{\sum_{i=1}^{n} x_i}{n}\right)^2} \cdot \sqrt{\frac{\sum_{i=1}^{n} y_i^2}{n} - \left(\frac{\sum_{i=1}^{n} y_i}{n}\right)^2}}$$

or simplifying :

$$r = \frac{n\sum_{i=1}^{n} x_i y_i - \sum_{i=1}^{n} x_i \sum_{i=1}^{n} y_i}{\sqrt{\left[\left\{n\sum_{i=1}^{n} x_i^2 - \left(\sum_{i=1}^{n} x_i\right)^2\right\}\left\{n\sum_{i=1}^{n} y_i^2 - \left(\sum_{i=1}^{n} y_i\right)^2\right\}\right]}}$$

where cov(x,y) is **covariance** of x and y;

σ_x is standard deviation of x_i

and σ_y is standard deviation of y_i .

Properties of '*r*' :

- *r* only measures the strength of a linear relationship. There are other kinds of relationships besides linear.
- *r* is always between –1 and 1 inclusive. –1 means perfect negative linear correlation and +1 means perfect positive linear correlation.
- *r* has the same sign as the slope of the regression (best fit) line.
- *r* does not change if the independent (x) and dependent (y) variables are interchanged.
- *r* does not change if the scale on either variable is changed. You may multiply, divide, add, or subtract a value to/from all the x-values or y-values without changing the value of *r*.
- *r* has a Student's *t* distribution.
- *r* is unaffected by change of origin and scale.

Problem 1 :

Calculate the correlation coefficient from the following data :

X(Length of testacids, μm) : 65 63 67 64 68 62 70 66

Y(Breadth of testacids, μm) : 68 66 68 65 69 66 68 65

Solution :

As the correlation coefficient is not affected by change of origin and scale, the origin of X and Y are changed to 65 and 67 respectively. Total no. of pairs of observation, '*n*' is 8.

Tabulation of Data (Problem 1) :

X_i	Y_i	$x_i = X_i – 65$	$y_i = Y_i – 67$	x_i^2	y_i^2	$x_i y_i$
65	68	0	1	0	1	0
63	66	-2	-1	4	1	2
67	68	2	1	4	1	2
64	65	-1	-2	1	4	2
68	69	3	2	9	4	6
62	66	-3	-1	9	1	3
70	68	5	1	25	1	5
66	65	1	-2	1	4	-2
$\sum_{n=1}^{n} X_i = 525$	$\sum_{n=1}^{n} Y_i = 535$	$\sum_{n=1}^{n} x_i = 5$	$\sum_{n=1}^{n} y_i = -1$	$\sum_{n=1}^{n} x_i^2 = 53$	$\sum_{n=1}^{n} y_i^2 = 17$	$\sum_{n=1}^{n} x_i y_i = 18$

$$\sigma_x = \sqrt{\frac{\sum x^2}{n} - \left(\frac{\sum x}{n}\right)^2}$$

$$= \sqrt{\frac{53}{8} - \left(\frac{5}{8}\right)^2}$$

$$= \sqrt{\frac{399}{64}}$$

$$\sigma_y = \sqrt{\frac{\sum y^2}{n} - \left(\frac{\sum y}{n}\right)^2}$$

$$= \sqrt{\frac{17}{8} - \left(\frac{-1}{8}\right)^2}$$

$$= \sqrt{\frac{135}{64}}$$

$$\text{cov}(x,y) = \frac{1}{n}\Sigma x_i y_i - \frac{\sum x_i}{n} \cdot \frac{\sum y_i}{n}$$

$$= \frac{18}{8} - \frac{5}{8} \cdot \frac{-1}{8} = \frac{149}{64}$$

$$\text{Therefore, } r_{xy} = \frac{\text{cov}(x, y)}{\sigma_x \sigma_y}$$

$$= \frac{\frac{149}{64}}{\sqrt{\frac{399}{64}} \cdot \sqrt{\frac{135}{64}}}$$

$$= 0.64$$

So, the required correlation coefficient $r = 0.64$.

2. Simple Regression Analysis :

Regression is the determination of a statistical relationship between two or more variables. In simple regression, we have only two variables, one variable (defined as independent) is the cause of the behavior of another one (defined as dependent variable).

Regression can only interpret what exists physically i.e. there must be a physical way in which independent variable X can affect dependent variable Y. The basic relationship between these two variables is given by equation :

$$\hat{Y} = a + bX$$

where the symbol $\hat{Y}$ denotes the estimated value of Y for a given value of X. This equation is known as the *regression equation of Y on X* (also represents the regression line of Y on X when drawn on a graph) which denotes that each unit change in X produces a change of b in Y, which is positive for direct and negative for inverse relationship.

In contrast, regression of X on Y can again only interpret what exists physically i.e. there must be a physical way in which independent variable Y can affect dependent variable X. The basic relationship between these two variables in this case is given by equation:

$$\hat{X} = a + bY$$

where the symbol $\hat{X}$ denotes the estimated value of X for a given value of Y. This equation is known as the *regression equation of X on Y* (also represents the regression line of X on Y when drawn on a graph) which means that each unit change in Y produces a change of b in X, which is positive for direct and negative for inverse relationship.

Regression Equation of X on Y is—

- $X - \overline{X} = b_{XY}(Y - \overline{Y}) = \frac{r\sigma_x}{\sigma_y (Y - \overline{Y})}$

Regression Equation of Y on X is—

- $Y - \overline{Y} = b_{YX}(X - \overline{X}) = \frac{r\sigma_y}{\sigma_x (X - \overline{X})}$

where X = Value of X, Y = Value of Y

$\overline{X}$ = Arithmetic mean of X series

$\overline{Y}$ = Arithmetic mean of Y series

σ_x = Standard deviation of X series

σ_y = Standard deviation of Y series

r = Correlation coefficient between X and Y

Regression Coefficients :

- The regression coefficients of Y on X is

$$b_{YX} = \frac{r\sigma_Y}{\sigma_X}$$

- The regression coefficients of X on Y is $b_{XY} = \frac{r\sigma_X}{\sigma_Y}$

Linear Regression Equation :

An expression in the form of a mathematical equation that expresses the relationship between the two variables X and Y, is relatively simple when the relationship is linear in character. Such equation is known as **Regression equation.**

The regression line is so located in relation to the points, that the sum of squared vertical distances of those points from the line is kept at a minimum (*method of least squares*).

Thus, the regression analysis is a statistical method to deal with the formulation of a mathematical model depicting relationship amongst variables, which can be used for the purpose of prediction of the values of dependent variable, given the values of the independent variable.

Computation of Regression Equation :

- For fitting a regression equation of the type $Y = a + bX$ to the given values of X and Y variables (where Y is the dependent variable and X is independent variable), we can find the values of the two constants viz., 'a' and 'b' by using the following two normal equations :

$$\Sigma Y_i = Na + b\Sigma X_i \qquad \text{... (1)}$$

$$\Sigma X_i Y_i = a\Sigma X_i + b\Sigma X_i^2 \qquad \text{... (2)}$$

Solving the two equations by ordinary methods of algebra gives the value of 'b' in general terms

$$b_{YX} = \frac{N\sum XY - \sum X \sum Y}{N\sum X^2 - (\sum X)^2}$$

$$= \frac{\sum XY}{\sum X^2} = \frac{r_{XY}\sigma_Y}{\sigma_X}$$

Dividing numerator and denominator of $\frac{N\sum XY - \sum X \sum Y}{N\sum X^2 - (\sum X)^2}$ by N^2

we have, $b_{YX} = \frac{\sum XY - N\overline{X}\overline{Y}}{\sum X^2 - N(\overline{X})^2}$

After finding the value of b_{YX}, values of ΣY, N, and ΣX can be substituted in equation (1) and the value of 'a' is calculated.

Thus, $\quad a_{YX} = \overline{Y} - b_{YX}\overline{X}$

where Y is dependent variable, X is independent variable, $\overline{Y}$ is mean of values of Y series, $\overline{X}$ is mean values of X series, b_{YX} is the slope of best fitting estimate line, a_{YX} is the Y-intercept and N is the number of total number of observations.

Substituting the value of a_{YX}, b_{YX} in the equation $\hat{Y} = a + bX$, the regression equation can be computed.

- The second alternative formula for calculation of b_{YX} which is useful while dealing with grouped data is :

$$b_{YX} = \frac{r_{YX}\sigma_Y}{\sigma_X} = \frac{\text{cov}(X,Y)}{\sigma_X^2}$$

The value of 'a' is calculated as before by substituting the appropriate figure in the equation

$$\Sigma Y_i = Na + b\Sigma X_i$$

The equation can also be written as

$$\frac{\sum Y}{N} = \frac{a + b\sum X}{N}$$

or, $\quad \overline{Y} = a + b\overline{X}$

Therefore knowing b_{YX}, $\overline{X}$ and $\overline{Y}$ it becomes easy to find the value 'a'.

or $\quad a_{YX} = \overline{Y} - b\overline{X}$

Simply knowing the value of b_{YX}, $\overline{X}$ and $\overline{Y}$ and substituting these values in the equation $Y - \overline{Y} = b_{YX}(X - \overline{X})$, one gets the required regression equation of Y on X.

Problem 2 :

Calculate the coefficient of correlation (r) from the following data :

X (Length of testacids) : 65 63 67 64 68 62 70 66

Y (Breadth of Testacids): 68 66 68 65 69 66 68 65

Find out the regression equation of Y on X.

Solution :

As the correlation coefficient is not affected by change of origin and scale, the origin of X and Y are changed to 65 and 67 respectively. Total no. of pairs of observation is 'N', where $N = 8$.

Tabulation of data (see below) :

For estimating regression equation Y on X, i.e., $Y - \overline{Y} = b_{yx}(X - \overline{X})$

$$\text{where } b_{YX} = \frac{\text{cov}(X,Y)}{\sigma_X^2}$$

$$\overline{X} = \frac{525}{8} = 65.63;$$

$$\overline{Y} = \frac{535}{8} = 66.88$$

As variance and covariance are not affected by change of origin,

$\text{cov}(X,Y) = \text{cov}(x,y)$

$$= \frac{\sum xy}{n} - \left(\frac{\sum x}{n}\right)\left(\frac{\sum y}{n}\right)$$

$$= \frac{18}{8} - \left(\frac{5}{8}\right)\cdot\left(\frac{-1}{8}\right)$$

$$= \frac{149}{64} = 2.33$$

$$\sigma_x^2 = \frac{\sum x^2}{N} - \left(\frac{\sum x}{N}\right)^2$$

$$= \frac{53}{8} - \left(\frac{5}{8}\right)^2 = \frac{399}{64}$$

$$\sigma_y^2 = \frac{\sum y^2}{N} - \left(\frac{\sum y}{N}\right)^2$$

$$= \frac{17}{8} - \left(\frac{-1}{8}\right) = \frac{135}{64}$$

$$= 2.10$$

$$b_{yx} = \frac{\text{cov}(x,y)}{\sigma_x^2}$$

$$= \frac{\frac{149}{64}}{\frac{399}{64}} = \frac{149}{399}$$

$$= 0.37$$

Tabulation of Data (Problem 2) :

X	Y	$x = X-65$	$y = Y-67$	x^2	y^2	xy
65	68	0	1	0	1	0
63	66	-2	-1	4	1	2
67	68	2	1	4	1	2
64	65	-1	-2	1	4	2
68	69	3	2	9	4	6
62	66	-3	-1	9	1	3
70	68	5	1	25	1	5
66	65	1	-2	1	4	-2
ΣX=525	ΣY=535	Σx = 5	Σy = –1	Σx^2=53	Σy^2=17	Σxy=18

Therefore the regression equation of Y on X :

$Y - \overline{Y} = b_{YX}(X - \overline{Y})$

$Y - 66.88 = 0.37(X - 65.63)$

i.e. $Y = 0.37X - 24.2831 + 66.88$

or $Y = 0.37X + 42.6$

Problem 3 :

Work out the regression equations with the help of the following data of two tests applied on same persons.

I.Q. tests scores :

mean = 18.2,

Standard deviation = 5.55

E.Q. tests scores :

mean = 42.8,

Standard deviation = 10.25

Correlation : = +0.68

Solution :

I.Q. tests scores :

$\overline{X} = 18.2,$

$\sigma_x = 5.55$

E.Q. tests scores :

$\overline{Y} = 42.8,$

$\sigma_y = 10.25$

$r = 0.68$

(a) **Regression of X on Y :**

$$b_{xy} = \frac{r\sigma_x}{\sigma_y} = 0.68 \times \frac{5.55}{10.25} = 0.37$$

$$a_{xy} = \overline{Y} - b_{xy}\overline{Y} = 18.2 - 0.37 \times 42.8 = 2.36$$

Therefore $\hat{X} = a_{xy} + b_{xy}Y$

or $\hat{X} = 2.36 + 0.37Y$

(b) **Regression of Y on X :**

$$b_{yx} = \frac{r\sigma_y}{\sigma_x} = 0.68 \times \frac{10.25}{5.55} = 1.26$$

$$a_{yx} = \overline{Y} - b_{yx}\overline{X} = 42.8 - 1.26 \times 18.2 = 19.87$$

$\overline{Y} = a_{yx} + b_{yx}X$

$\hat{Y} = 19.87 + 1.26X$

Problem 4 :

Some data regarding age of hamster (in months) and corresponding feeding amount/day of pellets (in grams) are shown below. Develop an estimating equation relating the two variable.

Age of hamster (in months) : 5 11 4 5 3 2

Feeding amount/per day (in grams) : 31 40 30 34 25 20

Solution :

Tabulation of data (see below) :

Therefore $\overline{X} = \frac{\sum X}{N} = \frac{30}{6} = 5$

$\overline{Y} = \frac{\sum Y}{N} = \frac{180}{6} = 30$

Tabulation of Data (Problem 4) :

No. of observations	Age of Hamster X	Feeding amount/day Y	XY	X^2
1	5	31	155	25
2	11	40	440	121
3	4	30	120	16
4	5	34	170	25
5	3	25	75	9
6	2	20	40	4
N = 6	**ΣX = 30**	**ΣY = 180**	**ΣXY = 1000**	**ΣX^2 = 200**

Therefore $b_{YX} = \dfrac{\sum XY - N\overline{X}\overline{Y}}{\sum X^2 - N(\overline{X})^2}$

$= \dfrac{1000-900}{200-150}$

$= 2$

Therefore $a_{YX} = \overline{Y} - b_{YX}\overline{X}$

$= 30 - 2 \times 5$

$= 20$

Now, $\hat{Y} = a + bX$

Substituting the value of a and b into the above equation :

We have, $\hat{Y} = 20 + 2X$

Therefore the estimating equation is :

$$Y = 20 + 2X$$

3. Kendall's Coefficient of Concordance :

Kendall's coefficient of concordance (W), symbolized by the symbol W, is an important *non-parametric measure of relationship*. It is in use for determining the degree of association among several (k) sets of ranking of N objects or individuals. Kendall's coefficient of concordance (W) is deemed as an appropriate measure of studying the degree of association among three or more sets of rankings.

The foundation of Kendall's coefficient of concordance is to envisage how the given data would look if there were no agreement between the several sets of rankings, and then to see how it would look if there were an ideal agreement among the several sets.

For instance, in case of, say , four inspector (judges) inspecting, say, six educational aspirants and assigning rank order on suitability for education, if there is observed perfect agreement (accordance) amongst the inspectors, then one aspirant would be assigned rank 1 by all the four judges and sum of his ranks would be 1+1+1+1= 4. Another aspirant would be allocated a rank 2 by all four and the sum of his ranks will be 2+2+2+2= 8. The sum of rank for the six aspirants would be 4, 8, 12, 16, 20, and 24 (not necessarily in this very order). In general, when perfect agreement exists amongst ranks assigned by k judges (inspectors) to N objects, the rank sums are k, 2k, and 3k, ...Nk. The entire sum of N ranks for k judges is kN (N +1)/2 and the mean rank sum is k (N +1)/2. The degree of agreement between judges reveals itself in the variation in the rank sums. When all judges concur, this sum is a maximum. Incongruity (disagreement) between judges reflects itself in a reduction in the variation of rank sums. For maximum incongruity, the rank sums will tend to be approximately equal. This provides the base for the definition of a Coefficient of Concordance. When perfect agreement exists between judges, W equals to 1. When maximum disagreement exists, W equals to 0. It may be noted that W does not take negative values because of the fact that with more than two judges, complete disagreement cannot take place. Thus, Coefficient of Concordance (W) is an index of divergence of the actual agreement shown in the data from the perfect agreement.

Computation of Kendall's Coefficient :

The method for computing and interpreting Kendall's coefficient of concordance (W) is as follows :

- All the objects, N, ought to be ranked by all K judges in the standard fashion and this information may be put in the form of a K by N matrix;
- For each object the sum of ranks (R_j) assigned by all the K judges is found out;
- $\overline{R}_j$ is found out and then the value of s is obtained as under;

$$s = \Sigma(R_j - \overline{R}_j)^2$$

- Value of W is worked out using the following formula;

$$W = \frac{s}{\frac{1}{12}K^2(N^3 - N)}$$

where, $s = \Sigma(R_j - \overline{R}_j)^2$

K = number of sets of rankings i.e., the number of judges;

N = number of objects ranked;

$\frac{1}{12}K^2(N^3 - N)$ = maximum possible sum of the squared deviations i.e., the sum s which would occur with perfect agreement among K rankings.

- **Case of Tied Ranks :** Where tied ranks arise, the average method of assigning ranks be accepted i.e., each member is assigned the average rank which the tied observations occupy. If the ties are not numerous, one may compute 'W' as stated above without making any adjustment in the formula; but if the ties are numerous, a correction factor (T) is calculated for each set of ranks. This correction factor is :

$$T = \frac{\Sigma(t^3 - t)}{12}$$

where t = number of observations in a group tied for a given rank.

For instance, if the ranks on X are 1, 2, 3.5, 5, 6, 3.5, 8, 10, 8, 8, we have two groups of ties, one of two ranks and one of there ranks. The correction factor (T) for this set of ranks for X would be

$$T = \frac{(2^3 - 2) + (3^3 - 3)}{12} = 2.5$$

A correction factor (T) is calculated for each of the K sets of ranks and these are added together over the K sets to obtain T. One can then use the formula for finding the value of 'W' as under:

$$W = \frac{s}{\frac{1}{12}K^2(N^3 - N)} - K\Sigma T$$

The application of the correction in this formula tends to increase the size of W, but the correction factor has a very limited effect unless the ties are quite numerous.

(i) The method for judging whether the calculated value of W is significantly different form zero depends on the size of N as stated below:

(ii) If N is 7 or smaller, the required Table is given in Appendix at the end of the book which gives critical values of s associated with 'W' at 5% and 1% significance levels. If an observed s is equal to or greater than that shown in the table for a particular level of significance, then H_0 (i.e., K sets of rankings are independent) may be rejected at that level of significance.

Problem 5 :

Learning performance of seven rats in a maze are ranked by four ethologists in an experiment as exhibited in the matrix.

	Rats						
	A	**B**	**C**	**D**	**E**	**F**	**G**
Ethologists 1	1	3	2	5	7	4	6
Ethologists 2	2	4	1	3	7	5	6
Ethologists 3	3	4	1	2	7	6	5
Ethologists 4	1	2	5	4	6	3	7

Is there significant agreement in ranked assignment by different ethologists? Test at 5% level of significance.

Solution :

There are four sets of ranking; one can work out Coefficient of Concordance (W) for judging significant agreement in ranking by different judges. For this purpose one has to construct the matrix as shown below.

Tabulation of data (Problem 5) :

	Rats							
K = 4	A	B	C	D	E	F	G	N = 7
Ethologists 1	1	3	2	5	7	4	6	
Ethologists 2	2	4	1	3	7	5	6	
Ethologists 3	3	4	1	2	7	6	5	
Ethologists 4	1	2	5	4	6	3	7	
Sum of Ranks (R_j)	**7**	**13**	**9**	**14**	**27**	**18**	**24**	**ΣR_j=112**
$(R_j - \bar{R}_j)^2$	**81**	**9**	**49**	**4**	**121**	**4**	**64**	**s=332**

$$W = \frac{s}{\frac{1}{12}K^2\left(N^3 - N\right)}$$

as, $s = 332$,

$$\text{Therefore, } W = \frac{332}{\frac{1}{12}(4^2)(7^3 - 7)}$$

$$= \frac{332}{448}$$

$$= 0.741$$

Conclusion :

To judge the significance one must look to the Table showing selected critical values of 's' in the Kendall's Coefficient of Concordance (see appendix H4) to find out the value of s at 5% level for $K = 4$ and $N = 7$. This value is 217.0 thus accepting the null hypothesis (H_0) that K sets of rankings are independent and the calculated value of 's' should be less than 217.0.

But calculated value of s is 332 which is greater than the table value which shows that $W = 0.741$ is significant. Thus we reject the Null hypothesis and infer that the Ethologists are applying the same standard in ranking the seven rats.

Chapter 11.5

Anova Analysis

ANOVA is an acronym of **Analysis of variance** which is a helpful technique used in biological fields when multiple sample cases are involved. This technique is used during examination of the significance of the difference amongst more than two sample means at the same time. As stated earlier the significance of difference between two means of two populations can be judged through either Z-test or t-test, but difficulty may arise when one happen to examine the significance of more than two populations at a time. ANOVA technique enables an experimenter to carry out simultaneous test. Variability in a sample can occur due to various reasons. Through ANOVA technique one can investigate any number of factors which are hypothesized or have an influence on the dependent variable. One can also investigate the differences amongst the various categories within each of these factors which may have large number of possible values. If one takes one factor and investigates the differences amongst its various categories having numerous possible values, one has to use the *One –way ANOVA* and when we use two factors at the same time, one has to use the *Two-way ANOVA*.

The basic principle of ANOVA is to test the difference among the means of the populations by examining the amount of variation within each of these samples, i.e. the relative amount of variation within each sample. Now to explain variation within the given population it is assumed that the values of (x_{ij}) differ from the mean of this population only due to random effects, whereas while examining the difference between the mean of jth population and the grand mean, is attributable to a 'specific factor' or technically can be stated as 'treatment effect'.

In short, one has to make two estimates of the population variance, one based on *between sample variance* and other based on *within sample variance.* Next these two estimates of the population variance are compared with F-test, where

$$F = \frac{\text{Estimate of population variance based on between sample variance (MSB)}}{\text{Estimate of population variance based on within sample variance (MSW)}}$$

where between samples variance is referred to as **Mean Squared Between (MSB)** and within the group variance is referred to as **Mean Squared Within (MSW)**.

Between sample variance represents the sampling error in the distributions and is also referred to as the *error variance* or *residual*. The between sample variance represents the influence of the variable of interest or group experimental value. If the between sample variance is not substantially greater than the within sample variance, the experimenter should conclude that the difference between the means is probably reflection of *sampling error*. Again, if F ratio is substantially greater than one, it would seem that the ratio between the between sample variance and within sample variance was probably attributable to sampling error.

This variability is the result of chance that occurs within each of the populations from which the samples have been taken. When one observes the means of more than two populations, there will be another kind of variability which is due to combined effect of the natural differences and possible real differences between the different groups. To test the differences among sample means one has to decide whether it has occurred only due to chance or whether the means of more than two populations are actually different. In both cases the hypothesis will be the means of the populations are equal. If this hypothesis is rejected one can conclude that at least one of the mean of the population is different from others.

In analysis of variance the difference is measured by "**Mean Square**" which is similar to variance and is defined by:

$$\text{Mean Square} = \frac{\text{Sum of squared deviations from mean}}{\text{Degrees of freedom}}$$

One mean square is used to measure difference *'within'* the groups. In such cases the deviations being taken from the respective group means and degree of freedom (df will be "total sample size minus number of samples"). Here mean square represents a measure of difference due to *chance* or *experimental error*.

Again, mean square is used to measure *group effect* or existing possible differences, if any, *'between'* the groups. It is based on sum of squared deviations of the individual sample or group means. Here deviations are taken from the grand mean of all observations, that is considered as one sample, and the degrees of freedom will be "number of samples minus one". If the means of all populations are equal, one can conclude that there is no group effect and the mean square between groups will represent difference due to chance alone. Again when the group means in the population are equal, the ratio of the mean square within groups and means square between groups are close to each other. If there is large ratio then it will indicate that the group means are not equal in the population.

For simplicity in the Graduate and Postgraduate level of elementary biostatistics syllabus in Zoology only One-way ANOVA is dealt with.

One-way (or single factor) ANOVA technique :

In a one way ANOVA, only one single factor is considered. This statistical technique is a way to test the equality of three or more means at one time by using variances.

Assumptions :

- The populations from which the samples were obtained must be normally or approximately normally distributed.
- The samples must be independent.
- The variances of the populations must be equal.

Hypotheses : The Null Hypothesis will be that all population means are the same; the alternative hypothesis is that at least one mean is different.

Computation of One Way ANOVA :

- **General Method :**

(i) Mean of each sample is obtained.

(ii) Mean of sample means is obtained.

(iii) Deviations of sample mean from mean of sample mean is calculated and their square value is multiplied by the number of items in each sample. Then their total is calculated. This is known as ***sum of squares for variance between samples (or SSB).***

$$SSB = n_1(\bar{x}_1 - \bar{\bar{x}})^2 + n_2(\bar{x}_2 - \bar{\bar{x}})^2 + \ldots\ldots + n_k(\bar{x}_k - \bar{\bar{x}})^2$$

SSB is divided by the degrees of freedom (df) between the samples (k-1) to find out ***mean square between samples (MSB).***

$$MSB = \frac{SSB}{k-1}.$$

(iv) Deviations of the values of sample items for all samples, from the corresponding means of the samples, are obtained, and results are squared and then summed. Each sum is totalled and this gives us the value of ***sum of squared variance within sample (SSW).***

$$SSW = \Sigma(x_{1i} - \bar{x}_1)^2 + \Sigma(x_{2i} - \bar{x}_2)^2 + \ldots + \Sigma(x_{ki} - \bar{x}_k)^2$$

(v) SSW is divided by df within samples $(N - k)$ to obtain the variance or ***mean square within samples (MSW)***

$$MSW = \frac{SSB}{N-k}.$$

where $(N - k)$ represents degrees of freedom within samples.

N = total no. of items in all the samples i.e. Sn_i

k = number of independent samples or group.

(vi) Now to check whether the total of the SSW and SSB is equal to sum of squares for total variance the following steps is worked out that can be written symbolically.

SS for total variance = $S(x_{ij} - \bar{\bar{x}})$

where $i = 1, 2, 3 \ldots$ and $j = 1, 2, 3\ldots$ and x_{ij} = j^{th} observation in the ith group

The d.f. for total variance would be equal to the number of items in all samples minus one which is equal to the total d.f. for between and within i.e. $(N-1) = (k-1) + (N-k)$. Thus, SS for total variance = SS (between) + SS (within).

(vii) Finally F ratio is worked out as follows:

F = MSB/MSW

This ratio is employed to judge whether the difference among many sample means is significant or is just a matter of sampling fluctuations. For this purpose one should look into the table for given degrees of freedom at different level of significance.

If calculated or observed F value is less than the tabulated F value, the difference is taken as significant i.e. due to the chance, and the null hypothesis (H_0) of no difference between sample mean is accepted.

Now, the information obtained through various steps stated above can be put as follows for setting up of *Analysis of Variance Summary Table* (see below).

- **Shortcut Method :**

(i) The total of the values of individual items in all the samples is taken i.e. Σx_{ij} (j^{th} observation in i^{th} group) where i =1, 2, 3... and j = 1, 2, 3... and it is called as T.

ANOVA Summary Table

Source of variation	Sum of squares (SS)	Degree of freedom (d.f.)	Mean square (MS)	F-ratio
Between samples	SSB	$(k-1)$	SSB/$(k-1)$	MSB/MSW
Within samples	SSW	$(N-k)$	SSW/$(N-k)$	
Total	TSS	$(N-1)$		

(ii) The correction factor is then worked out i.e., correction factor (c.f.) = $\frac{T^2}{N}$.

(iii) Total of the square of all the item values is then calculated and from this total C.F. is subtracted, and the result is the ***Sum of Squares for Total Variance or difference (TSS).***

$$\text{Total } SS(\text{TSS}) = \Sigma\Sigma x_{ij}^2 - \frac{T^2}{N}$$

where $i = 1, 2, 3...$ and $j = 1, 2, 3...$

(iv) The square of each sample total $(T_i)^2$ is then divided by the number of items in the concerning sample and total of the result is taken. From this total the c.f. is subtracted and the result is the Sum of Squares for variance between the samples (SSB).

$$\text{SSB} = \Sigma\left(\frac{T_i^2}{n_i}\right) - \frac{T^2}{N}; \text{ when } i = 1,2,3,...$$

(v) Now, SSW is found by subtracting the result of step (iv) from the result of step (iii).

(vi) SSW = TSS – SSB

$$= \left\{\Sigma\Sigma x_{ij}^2 - \frac{T^2}{N}\right\} - \left\{\Sigma\left(\frac{T_i^2}{n_i}\right) - \frac{T^2}{N}\right\}$$

$$= \Sigma\Sigma x_{ij}^2 - \Sigma\left(\frac{T_i^2}{n_i}\right)$$

(vii) Mean Square within groups (MSW) is calculated as = $\frac{\text{SSW}}{N-k}$.

(viii) Mean Square between groups (MSB) is calculated as = $\frac{\text{SSB}}{k-1}$.

(ix) The test statistic (F) is calculated as $\text{F} = \frac{\text{MSB}}{\text{MSW}}$.

Now, the table of ANOVA can be set up in similar way as described in the direct method.

Decision Rule :

The decision will be to reject the null hypothesis if the test statistic from the table (calculated F value) is larger than the F critical value (tabulated value) with $\frac{k-1}{N-k}$ of freedom.

Problem 1 :

A random sample of five bacterial samples is taken for each three strains maintained in a microbiological laboratory. The generation time (in minutes) of these samples is shown below. On the basis of the data, test whether the average generation time of three strains of bacteria are equal or not.

Generation time (in minutes) of bacterial strains (in minutes) :

Strain	A	B	C
	35	32	34
	34	32	33
	34	31	32
	33	28	32
	34	29	33

Solutions :

Null Hypothesis (H_0) is taken as the population means of generation time of the three strains of bacteria are the same.

Each observation is reduced by 30, as shown in the next page :

Tabulation of Data (Problem 1) :

Strain	A	B	C
	5	2	4
	4	2	3
	4	1	2
	3	–2	2
	4	–1	3

Total $T_1 = 20 \quad T_2 = 2 \; T_3 = 14; \; \Sigma T_i = T = 36 (= T_1 + T_2 + T_3)$

Σx_{ij}^2 82 14 42; $\Sigma\Sigma x_{ij}^2 = 138$

Sample size (n) $n_1 = 5 \quad n_2 = 5 \quad n_3 = 5; \; N = 15$ and $k = 3$

$$\text{C.F.} = \frac{T^2}{N} = \frac{(36)^2}{15} = 86.4$$

$$\text{TSS} = \Sigma\Sigma x_{ij}^2 - \text{C.F.} = 138 - 86.4 = 51.6$$

$$\text{SSB} = \Sigma\left(\frac{T_i^2}{n_i}\right) = \frac{T^2}{N} = \frac{(20)^2}{5} + \frac{(2)^2}{5} + \frac{(14)^2}{5} - 86.4 = 33.6$$

$$\text{SSE} = \text{TSS} - \text{SSB} = 51.6 - 33.6 = 18.0$$

Analysis of Variance Summary Table

				F values	
Source of Variation	**S.S**	**d.f.**	**M.S.**	**Observed**	**Tabulated**
Between groups	33.6	2	33.6/2 = 16.8	16.8/1.5 = 11.27	$F_{0.05} = 3.89$ $F_{0.01} = 6.93$
Within groups	18.0	12	18/12 = 1.5	—	—

Conclusion :

The calculated or observed F value is 11.27 > than 1% tabulated F value, so according to the decision rule, we reject the Null Hypothesis of equality of population means, and conclude that the generation time of the three strains of bacteria are not same.

Chapter 11.6

t-test

A t-test is any statistical hypothesis test in which the test statistic has a Student's-t distribution, if null hypothesis is true. The test was invented by W.S. Gossett (1908). The basic assumption used in the test is "The population from which sample(s) are drawn is **normally distributed**".

Assumptions for t-test :

- Samples are drawn from normal populations and are random.
- For testing the equality of two populations means, the population variances are regarded as equal.
- The population standard deviation may be unknown.
- In case of two samples an adjustment in the *degrees of freedom* are to be made.

Basic steps of t-value calculation :

- Assumption of Null Hypothesis H_0 and Alternative Hypothesis H_A.
- Calculation of test statistic (t).
- Deciding the *level of significance* (i.e. α = 0.05 or 0.01 or 0.1) as the case may be.
- Finding the *critical value*. This is found from the table value of $|t|$ at level of significance (α) for particular degrees of freedom (d.f.) i.e. $t_{\alpha,(d.f.)}$.
- Decision or conclusion : If the calculated $|t|$ value is < $t_{\alpha,(d.f.)}$ then the Null Hypothesis is accepted and if calculated $|t|$ value is > $t_{\alpha,(d.f.)}$ then the Null Hypothesis is rejected.

Box.11.4.1 : One tailed, Two-tailed tests and Critical Value

One tailed or Two tailed tests :

- *One-tailed test* : Used where there is some basis (e.g. previous experimental observation) to predict the direction of the difference, e.g. expectation of a significant difference between the groups.
- *Two-tailed test* : Used where there is no basis to assume that there may be a significant difference between the groups - this is the test most frequently used.

Critical value :

- **If the calculated value of t is greater than the critical value, H_0 is rejected,** i.e. there is evidence of a statistically significant difference between the groups.
- **If the calculated value of t is less than the critical value, H_0 is accepted,** i.e. there is no evidence of a statistically significant difference between the two groups.

1. t-Tests for independent (unpaired) samples :

The scores, values of variables that are obtained from two independent samples, constitute the unpaired observations. The means of these independent sets of observations are uncorrelated or independent. The sizes of the sample may be equal or unequal. The sample size may be large or small.

(a) **Small sample of unequal size ($N_1 \neq N_2$) :**

$$t = \frac{\bar{x}_1 - \bar{x}_2}{\text{S.E.}_{(\bar{x}_1 - \bar{x}_2)}}$$

$$= \frac{\bar{x}_1 - \bar{x}_2}{S_p\sqrt{\frac{1}{N_1} + \frac{1}{N_2}}}$$

$$= \frac{\bar{x}_1 - \bar{x}_2}{\sqrt{\frac{N_1 S_1^2 + N_2 S_2^2}{N_1 + N_2 - 2} \times \frac{N_1 + N_2}{N_1 N_2}}}$$

where $s_p^2 = \frac{N_1 S_1^2 + N_2 S_2^2}{N_1 + N_2 - 2}$

and d.f. = $N_1 + N_2 - 2$

(b) **Large sample of unequal size ($N_1 \neq N_2$) :**

$$S_1 = \sqrt{\frac{\sum(\bar{x}_1 - \bar{x}_2)^2}{N_1 - 1}}$$

$$S_2 = \sqrt{\frac{\sum(\bar{x}_1 - \bar{x}_2)^2}{N_2 - 1}}$$

$$t = \frac{\bar{x}_1 - \bar{x}_2}{\text{S.E.}_{(\bar{x}_1 - \bar{x}_2)}}$$

$$= \frac{\bar{x}_1 - \bar{x}_2}{\text{S.E.}\bar{x}_1^2 + \text{S.E.}\bar{x}_2^2}$$

$$t = \frac{\bar{x}_1 - \bar{x}_2}{\sqrt{\frac{S_1^2}{N_1} + \frac{S_2^2}{N_2}}}$$

and d.f. = $N_1 + N_2 - 2$

(c) **Samples of identical size ($N_1 = N_2$) :**

$$t = \frac{\bar{x}_1 - \bar{x}_2}{\text{S.E.}_{(\bar{x}_1 - \bar{x}_2)}}$$

$$= \frac{\bar{x}_1 - \bar{x}_2}{\sqrt{\text{S.E.}\bar{x}_1^2 + \text{S.E.}\bar{x}_2^2}}$$

$$= \frac{\bar{x}_1 - \bar{x}_2}{\sqrt{\frac{S_1^2 + S_2^2}{N}}}$$

and d.f. = $2(N - 1)$

where, $\text{S.E.}_{(\bar{x}_1 - \bar{x}_2)}$ is *standard error* of differences of mean, $\bar{x}_1$ and $\bar{x}_2$ are the mean of x_1 and x_2 variables respectively, S_p is the pooled S.D. of both samples; S_1^2 and S_2^2 ; S_1 and S_2; $\text{S.E.}\bar{x}_1^2$ and $\text{S.E.}\bar{x}_2^2$ are variances and standard deviations and standard error of two samples x_1 and x_2 respectively, N_1 and N_2 are size of the samples. N is sample size in case of identical size.

2. t-Test for paired observations of small samples by difference method :

In the paired observations of single group experiment that is composed of two sets of scores or values of the same variable, obtained by carrying on the same experiment on the same individuals, on two consecutive occasions, or may be under two different experimental conditions. These paired observations are correlated with each other.

Here the Null Hypothesis proposes that the mean difference ($\bar{D}$) between the paired observations does not deviate significantly from the difference between the population mean

$(\mu_1 - \mu_2)$. Null Hypothesis assumes that if the two sets of observation are taken from the same population, in that case, $\mu_1 - \mu_2 = 0$.

$$\bar{D} = \frac{\sum D}{N}$$

$$S_D = \sqrt{\frac{\sum(D-\bar{D})^2}{N-1.\text{S.E.}(\bar{D})}}$$

$$= \frac{S_D}{\sqrt{N}}$$

(a) $t = \bar{D} - \frac{\mu_1 - \mu_2}{\text{S.E.}(\bar{D})}$

$$= \frac{\bar{D}-0}{\text{S.E.}(\bar{D})}$$

or, $t = \frac{\bar{D}}{\text{S.E.}(\bar{D})}$

(b) $t = \frac{\sum\bar{D}}{\sqrt{\left\{\frac{N\sum D^2 - (\sum\bar{D})^2}{N-1}\right\}}}$

d.f. = N – 1

where, $\bar{D}$ is mean difference; S.E.(D) is S.E. of mean difference; S_D is standard deviation of differences, μ_1 and μ_2 are population means and N is number of pairs.

t can be computed by any of the two formulas stated above and can be used for both one-tail and two tail tests.

3. t-Tests for paired observations of large samples :

t-Distribution is close to normal distribution when samples are large i.e., $N \geq 30$. In such cases computation of t-tests involves the use of 'r' (correlation coefficient) between the sample means.

$$t = \frac{(\bar{x}_1 - \bar{x}_2) - (\mu_1 - \mu_2)}{\text{S.E.}(\bar{x}_1 - \bar{x}_2)}$$

where, $\text{S.E.}(\bar{x}_1 - \bar{x}_2)$

$$= \sqrt{\text{S.E.}\bar{x}_1^2 + \text{S.E.}\bar{x}_2^2 - 2r\text{S.E.}\bar{x}_1\text{S.E.}\bar{x}_2}$$

In this case, Null Hypothesis, Ho: $\mu_1 - \mu_2 = 0$, N is the number of pairs of observations and d.f. is N–1.

Problem 1 :

Following are the gain in length (in mm) of salamanders fed with two different diets A and B.

Diet A : 12 14 9 11 12 15 8 12 17 13 14

Diet B : 15 13 18 8 17 9 14 12

Test if the two diets differ significantly as regards to their effect on length.

Solution :

Null Hypothesis (H_0) : There is no significant difference between the mean increase in length due to diets A and B.

Alternate Hypothesis (H_A) : There is a significant difference between the mean increase in length due to diets A and B.

Tabulation of Data (see next page) :

$\bar{x}_1 = 12.45;\ \bar{x}_2 = 13.25$

Sample variance of $x_1(S_1^2)$

$$= \frac{\sum(x_1 - \bar{x}_1)^2}{N_1}$$

$$= \frac{66.7275}{11} = 6.07$$

Sample variance of $x_2(S_2^2)$

$$= \frac{\sum(x_2 - \bar{x}_2)^2}{N_2}$$

$$= \frac{87.50}{8} = 10.94$$

Unbiased estimate of common population standard deviation (S_p) :

$$S_p = \sqrt{\frac{\{(N_1-1)S_1^2 + (N_2-1)S_2^2\}}{N_1 + N_2 - 2}}$$

$$= \sqrt{\frac{\{(11-1)6.07 + (8-1)10.94\}}{11+8-2}}$$

$$= \sqrt{\frac{(10\times 6.07)+(7\times 10.94)}{17}}$$

Tabulation of Data (Problem 1) :

Diet A	x_1	$x_1 - \bar{x}_1$	$(x_1 - \bar{x}_1)^2$	Diet B	x_2	$x_2 - \bar{x}_2$	$(x_2 - \bar{x}_2^2)$
Mean	12	–0.45	0.2025	Mean	15	1.75	3.0625
$= \Sigma x_1/N_1$	14	1.55	2.4025	$= \Sigma x_2/N_2$	13	–0.25	0.0625
= 137/11	9	–3.45	11.9025	= 106/8	18	4.75	22.5625
= 12.45	11	–1.45	2.1025	= 13.25	8	–5.25	27.5625
	12	–0.45	0.2025		17	3.75	14.0625
	15	2.55	6.5025		9	–4.25	18.0625
	8	–4.45	19.8025		14	0.75	0.5625
	12	–0.45	0.2025		12	–1.25	1.5625
	17	4.45	20.7025		—	—	—
	13	0.55	0.3025		—	—	—
	14	1.55	2.4025		—	—	—
	Σx_1=137		$\Sigma(x_1 - \bar{x}_1)^2$ **= 66.7275**		Σx_2=106		$\Sigma(x_2 - \bar{x}_2)^2$ = 87.50

$$= \sqrt{\frac{60.7 + 76.58}{17}}$$

$$= \sqrt{8.0752} = 2.84$$

Standard error of difference, S.E. $(x_1 - x_2)$

$$= S_p\sqrt{\frac{1}{N_1} + \frac{1}{N_2}}$$

$$= 2.84\sqrt{\frac{1}{11} + \frac{1}{8}}$$

$$= 2.84 \times 0.22 = 0.62$$

Therefore the test statistic

$$t = \frac{\bar{x}_1 - \bar{x}_2}{\text{S.E.}_{(\bar{x}_1 - \bar{x}_2)}}$$

$$= \frac{12.45 - 13.25}{0.62}$$

$$= \frac{-0.8}{0.62}$$

$$= |-1.29| = 1.29$$

d.f. = 11 + 8 – 2 = 17

Conclusion : As Calculated 't' value, 1.29 < Tabulated $t_{0.05(18)}$ = 2.11, Null Hypothesis is accepted at 5% significance level i.e., there is no significant difference between the mean increase in length due to diets A and B.

Problem 2 :

A performance test was administered to 5 workers before and after they are trained. The results are shown below :

Workers :	I	II	III	IV	IV
Performance score before training :	110	120	123	132	125
Performance score after training :	120	118	125	136	121

Test whether there is any change in performance score after the training. It is given that $t_{0.01}$ = 4.6 for d.f =4.

Solution :

Null Hypothesis : There is no significant change in performance score after training.

Alternate Hypothesis : There is a significant change in performance score after training.

Tabulation of Data (Problem 2) :

Candidates	Performance score before training (y)	Performance score after training (x)	Difference $D = x - y$	D^2
I	110	120	10	100
II	120	118	–2	4
III	123	125	2	4
IV	132	136	4	16
V	125	121	–4	16
			$\Sigma D = 10$	$\Sigma D^2 = 140$

Tabulation of data (see below) :

$$t = \frac{SD}{N\sum D^2 - \dfrac{(\sum D)^2}{N-1}}$$

d.f. = $N - 1$

$$\overline{D} = \frac{\sum D}{N} = \frac{10}{5} = 2$$

Therefore, $$t = \frac{10}{\sqrt{\left(\dfrac{5\times140-100}{5-1}\right)}}$$

$$= \frac{10}{\sqrt{\dfrac{700-100}{4}}}$$

$$= \frac{10}{\sqrt{\dfrac{600}{4}}} = \frac{10}{12.25} = 0.816$$

Test statistic 't' = 0.816

Conclusion : The calculated value of 't' = 0.816 < Tabulated $t_{0.01(4)}$ = 4.16, therefore the Null Hypothesis is accepted, i.e., there is no significant difference in performance score after the training program.

Problem 3 :

	No. of samples	Mean life expectancy	Variance
Species I	9	600	121
Species II	8	640	144

Is there a significant difference between the two means? Value of 't' for 15 d.f. at 5% level is 2.131.

Null Hypothesis (H_0) : There is no significant difference between the two means.

Alternate Hypothesis (H_A) : There is significant difference between the two means.

Calculation : An estimate of common or pooled S.D. (S_p) is obtained as:

$$S_p = \sqrt{\frac{\{(N_1-1)S_1^2 + (N_2-1)S_2^2\}}{N_1+N_2-2}}$$

$$S_p = \sqrt{\frac{(8\times121)+(7\times144)}{9+8-2}}$$

$$= \sqrt{\frac{1976}{15}} = \sqrt{131.73} = 11.48$$

Therefore,

$$t = \frac{\bar{x}_1 - \bar{x}_2}{S_p\sqrt{\dfrac{1}{N_1}+\dfrac{1}{N_2}}}$$

$$t = \frac{600-640}{11.48\sqrt{\dfrac{1}{9}+\dfrac{1}{8}}}$$

$$= \frac{-40}{11.48\times\sqrt{0.24}}$$

$$= |-7.10| = 7.10$$

d.f. = 9 + 8 – 2 = 15

Conclusion : As the observed value falls in the critical region because $|t| = |-7.10| = 7.10$ is larger than the tabulated 't' value $t_{0.05(15)}$, 2.131, one can reject the Null Hypothesis at 5 % level of significance and conclude that there is significance difference between two the means.

Problem 4 :

The means of fecundity scores of 104 salamanders and 98 frogs were 5.60 (S.D. 1.7) and 5.20 (S.D. 1.5) respectively. Is the mean significantly higher in salamander than in frogs?

Salamanders : $x_1 = 5.60$, $S_1 = 1.7$, $N_1 = 104$

Frogs : $x_2 = 5.20$, $S_2 = 1.5$, $N_1 = 98$

$\mathbf{H_0}$: The mean score for salamanders is not significantly higher than that of frogs.

Now t-test is performed for this purpose

$$t = \frac{\bar{x}_1 - \bar{x}_2}{\sqrt{\frac{S_1^2}{N_2} + \frac{S_2^2}{N_1}}}$$

$$t = \frac{5.60 - 5.20}{\sqrt{\frac{(1.7)^2}{98} + \frac{(1.5)^2}{104}}}$$

$$= \frac{0.40}{\sqrt{0.029 + 0.022}}$$

$$= \frac{0.40}{0.23} = 1.74$$

d.f.. = $N_1 + N_2 - 2 = 104 + 98 - 2 = 200$

Conclusion : The computed t of 1.74 is higher than the critical value of t 0.05(α) = 1.645. Hence H_0 is rejected and the mean score is significantly higher in salamanders.

Problem 5 :

The mean systolic B.P. of 41 adult men was found to be 168.6 mm Hg (S.E. 8.7) and 160.5 mm BP. (S.E.7.58) respectively, before and after consumption of a particular drug. The correlation coefficient (r) between the initial and the final values was 0.68. Is the difference between the two means significant? (α = 0.05).

In this case a two tailed t-test for correlated means of large single group sample is applicable.

$\mathbf{H_0}$: The mean score of the paired observations for BP before and after consumption of a particular drug in men are not different.

$\bar{x}_1 = 168.6$, S.E. $\bar{x}_1 = 8.7$,

$\bar{x}_2 = 160.5$, S.E. $\bar{x}_2 = 7.58$

$N = 41$, $r = +0.68$

$$t = \frac{\bar{x}_1 - \bar{x}_2}{\sqrt{S.E.\bar{x}_1^2 + S.E.\bar{x}_2^2 - 2r.S.E.\bar{x}_1 S.E.\bar{x}_2}}$$

$$= \frac{168.6 - 160.5}{\sqrt{(8.7)^2 + (7.58)^2 - 2 \times 0.68 \times 8.7 \times 7.58}}$$

$$= \frac{7.5}{75.69 + 57.46 - 89.69}$$

$$= \frac{7.5}{43.46}$$

$= 0.17$

df = N − 1 = 40

For a two tailed test, critical value of $t_{0.05\,(40)}$ = 2.021.

Calculated t value of 0.17 is much lower than the critical value. So, $p > 0.05$. Therefore the difference between means is not significant at 0.05 level.

Chapter 11.7

Kruskal-Wallis One-Way Analysis of Variance

Kruskal-Wallis one-way analysis of variance is used as a test for determining whether three or more independent samples are significantly different. It is a *non-parametric test.*

This test is used to verify the Null Hypothesis that 'k' independent random samples come from identical universes against the Alternative Hypothesis that means of these universes are not equal.

Assumptions :

1. The p samples are random and independent.
2. There are five or more measurements in each sample.
3. p probability distributions from which the samples are drawn are continuous.

Computation of Kruskal-Wallis one-way analysis of variance :

- A Null Hypothesis (H_0) is constructed.
- All the measurement from the three or more populations is ranked as one group beginning with rank 1 for smallest measurement. If ties occur between two or more measurements, each measurement is assigned the mean of the ranks for which it is tied.
- Then the sum of the ranks are calculated for each column R_j.
- Square of each R_j s are divided by the number of the measurements in that sample i.e. $\frac{R_j^2}{N_j}$.
- Summation of $\frac{R_j^2}{N_j}$ made, i.e. $\frac{\Sigma R_j^2}{N_j}$.
- H statistics is calculated as

$$H = \left\{\frac{12}{N(N+1)} \times \frac{\Sigma R_j^2}{N_j}\right\} - 3(N+1)$$

- The calculated H value is compared with tabulated value in the Chi-square distribution table when all the sample size are at least 5, and if it is ≤ 5 the comparison is done with Kolmogorov-Smirnov two-sample, one tailed test.

Problem 1 :

Population density of deer in a sq. km, data of three habitats is shown in the table below :

Habitat I	Habitat II	Habitat III
17	25	41
10	27	31
23	22	35
11	15	30
14	33	39
19		37

Find out whether population density is significantly different in the three habitats through Kruskal-Wallis one-way analysis of variance.

Solution :

1. **Let the Null Hypothesis (H_0) :** There is no significant difference of population density in the three habitats.

2. **Alternative Hypothesis (H_A) :** There is significant difference of population density in the three habitats.

3. Ranking of the population density of the three habitats are made in the table below :

Tabulation of data (Problem 1) :

Habitat I	Habitat II	Habitat III
5	9	17
1	10	12
8	7	14
2	4	11
3	13	16
6		15
R_A = 25, where N_I = 6	**R_B = 43, where N_{II} = 5**	**R_C = 85, where N_{III} = 6**

4. $\frac{R_1^2}{N_1} = \frac{(25)^2}{6} = 104.17;$

$\frac{R_2^2}{N_2} = \frac{(43)^2}{5} = 369.80;$

$\frac{R_3^2}{N_3} = \frac{(85)^2}{6} = 1204.17$

5. $\Sigma \frac{R_j^2}{N_j} = 104.17 + 369.80 + 1204.17$

$= 1678.14$

6. $N = N_I + N_{II} + N_{III}$

$= 6+5+6$

$= 17$

7. $H = \left\{ \frac{12}{N(N+1)} \times \frac{\Sigma R_j^2}{N_j} \right\} - 3(N+1)$

$= \left\{ \frac{12}{17(17+1)} \times 1678.14 \right\} - 3(17+1)$

$= 11.80$

8. Calculated d.f. = 3 – 1 = 2 (i.e., no. of population–1)

Tabulated $\chi^2_{0.05(2)} = 5.99$

Since calculated H(11.80) is larger than the tabulated value, we reject our H_0 and conclude that there is significant difference of population density in the three habitats.

Chapter 11.8

Chi-square Test

Chi-square test, which is also described as the *'Goodness of fit'* is, used as a test statistic in which a hypothesis is tested that provides a set of theoretical frequencies with which observed frequencies are compared. Such a measure enables us to find out the degree of discrepancy between observed frequencies and theoretical frequencies. This test originated with Karl Pearson, one of the founders of the study of statistics.

With the Goodness of Fit test, we compare a list of observed frequencies to a list of the frequencies we would expect under some Null model. The list of the observed frequencies is written as O_i; the expected frequencies are E_i. *Important*: these frequencies are expressed as counts, not proportions.

The test statistic called χ^2 ("chi-square") can be calculated for the data as following :

$$\chi^2_{df} = \sum_{i=1}^{k} \frac{(O_i - E_i)^2}{E_i}.$$

where O_i = Observed frequency and E_i = Expected frequency; k = mutually exclusive classes.

One should notice that if the observed closely matches the expected, then the χ^2 value will be small, but if there is a great difference between the observed and expected, χ^2 will be large.

- **Assumption for the test statistic :**

The test statistic has a chi-square distribution when the following assumptions are met :

- The data are obtained from a random sample.
- The expected frequency of each category must be at least 5. This goes back to the requirement that the data be normally distributed.

- **Properties of the Goodness-of-fit test :**
  - The data are the observed frequencies. This means that there is only one data value for each category.
  - The degrees of freedom (df) are one less than the number of categories, not one less than the sample size.
  - It is always a right-tailed test.
  - It has a chi-square distribution.
  - The value of the test statistic does not change if the order of the categories is switched.

- The test statistic is $\chi^2_{df} = \sum_{i=1}^{k} \frac{(O_i - E_i)^2}{E_i}$

The "df" in the subscript of the χ^2 means "degrees of freedom", and the value of the χ^2 distribution that we expect under the Null model will depend on the number of degrees of freedom a particular test has.

- **Computation of Chi-square value :**

1. Null Hypothesis (H_0) is constructed : No association exists between the attributes. Alternative Hypothesis (H_A) is constructed : Association exists between the attributes.
2. The expected frequency is calculated.
3. The test statistics is calculated
$$\chi^2 = \sum_{i=1}^{k} \frac{(O_i - E_i)^2}{E_i}.$$
4. The table value of χ^2 for a given level of significance and particular d.f. is found out i.e., df = N–1, where N is no. of classes.
5. The calculated χ^2 value is compared with tabulated χ^2 value.
6. If calculated χ^2 value < tabulated χ^2 value then the Null hypothesis is accepted. If vice-versa, the Alternative Hypothesis is accepted.

Problem 1 :

A small aquatic crustacean called *Ceriodaphnia cornuta* has two distinct phenotypes, one with "horns" and one without i.e. "unhorned".

A Limnologist wanted to know whether the horns had any benefit in discouraging fish from eating the daphnia. So, in a large tank he put many *C. cornuta* in a 3 : 1 ratio i.e., 3 unhorned to every 1 with horns. He then introduced a natural predator of the daphnia, a small fish. After an hour, he removed the fish, "sacrificed" it, and looked at the daphnia in its belly. Of the 50 daphnia that the fish had devoured, 45 were unhorned and 5 were horned.

Is there any true pattern in which phenotype was eaten?

Solution :

The Hypothesis is constructed—

H_0 : Proportion unhorned eaten = 0.75.

H_A : Proportion unhorned eaten ≠ 0.75.

Tabulation of data (see below) :

This calculation can also be done in a non tabular form :

$$\chi^2 = \sum_{i=1}^{k} \frac{(O_i - E_i)^2}{E_i}$$

Tabulation of data (Problem 1) :

Phenotypes	Observed (O_i)	Expected (E_i)	O_i–E_i	$(O_i–E_i)^2/E_i$
Horned	5	12	7	49/12 = 4.08
Unhorned	45	38	7	49/38 = 1.29
Total				$\chi^2 \Sigma(O_i–E_i)^2/E_i$ = 5.37

Calculation :

$$\chi_i^2 = \frac{O_{\text{horned}} - E_{\text{horned}}}{E_{\text{horned}}} + \frac{O_{\text{unhorned}} - E_{\text{unhorned}}}{E_{\text{unhorned}}}$$

$$= \frac{(5-12)^2}{12} + \frac{(45-38)^2}{38}$$

$$= \frac{49}{12} + \frac{49}{38}$$

$$= 4.08 + 1.29$$

$$= 5.37$$

(We get these expected numbers : 0.25(50) = 12; 0.75(50) = 38. Remember we must use counts, not proportions in the χ^2 test.)

There are 2 classes here (unhorned and horned) and we have estimated no parameters from the data, so we have 2 – 1 – 0 = 1 d.f. We will therefore compare our results to a χ^2 distribution with 1 d.f.

Next, the test statistic is compared with the χ^2 distribution. Calculating the probability in the tail of the χ^2 distribution directly is difficult, but fortunately one can use a table (see appendix) to get an answer.

Critical value $\chi^2_{0.001}$ = 10.83. So the calculated χ^2 of 5.37 is lesser than the critical value. Thus the H_0 is significant; so the observed result does not diverge significantly ($p > 0.001$) from those expected in the hypothesis. Hence one can say that these data show that the phenotype of the Daphnia does affect their survivorship under predation.

Contingency Chi-square :

Sometimes it becomes necessary to compare one set of observation in a particular condition with that of another set taken under a different condition. Thus in these cases there are no expected values, what matters is that whether the results are dependent or contingent upon, or not dependent of the conditions under which they are observed. This sort of statistical test is known as **Contingency Chi-square test** .

$$\text{Test statistic } \chi^2 = \frac{\left[|ad - bc| - \frac{1}{2}N\right]^2 N}{\{(a+b)(a+c)(c+d)(b+d)\}}$$

where N = Total number of individuals of the two sets of observations; a, b, c, d are the numerical contributions of different classes, N;

$\frac{1}{2}N$ = Yate's correction.

df = $(c\text{-}1)(r\text{-}1)$;

c = Sets of observation,

r = Categories of observations.

Contingency Table :

Sets of observation	Categories of observation		Totals
	1	2	
A	a	b	$a + b$
B	c	d	$c + d$
Total	$a + c$	$b + d$	

Box 11.8.1 : Yates Correction

When there are only two categories (e.g. round/wrinkeled) or, more appropriately, when there is only one degree of freedom, the χ^2 test should not, strictly, be used. There have been different efforts to correct this insufficiency, but the simplest is to apply **Yates correction** to the data. To do this, one has to simply subtract 0.5 from each calculated value of "Observed - Expected", ignoring the sign (plus or minus). In other words, an "Observed - Expected" value of +6 becomes +5.5, and an "Observed - Expected" value of -5 becomes -4.5. To indicate that, one has to lessen the absolute value ignoring the sign, one uses vertical lines: | Observed - Expected |-0.5. Then one continues as usual but with the new (corrected) "Observed (O) - Expected (E)" values: one has to calculate (with the corrected values) $(O\text{-}E)^2$, $(O\text{-}E)^2/E$ and then sum the $(O\text{-}E)^2/E$ values to get χ^2. Yates correction only applies when we have two categories (one degree of freedom).

Problem 2 :

The following are the information obtained concerning number of surveys ($N = 50$) on genetic abnormalities conducted by Research Fellows and Scientists in cities and villages.

Tabulation of data (Problem 2) :

	Places		Total
	City (1)	Village (2)	
Research Fellows (A)	17	18	35
Scientists (B)	3	12	15
Total	20	30	50

Can it be inferred that the survey run by Scientists is more in Village than the city? Use χ^2 test.

Solutions :

Let us construct the null hypothesis H_0: There is no significant difference between Research Fellows and Scientists in the number of surveys conducted in cities and villages.

$$\chi^2 = \frac{\left[(ad - bc) - \frac{1}{2}N\right]^2 N}{\{(a+b)(a+c)(c+d)(b+d)\}}$$

$$= \frac{50\left[|17 \times 12 - 18 \times 3| - \frac{1}{2} \times 50\right]^2}{(35 \times 20 \times 15 \times 30)}$$

$$= \frac{50(125)^2}{315000}$$

$$= \frac{781250}{315000}$$

$$= 2.48$$

$$df = (c-1)(r-1)$$

$$= (2-1)(2-1) = 1.$$

The Tabulated χ^2 value for d.f. = 1 at 5% level of significance is 3.841. The calculated χ^2 value (2.48) < than the tabulated value. Hence the Null Hypothesis has to be accepted. So survey run by scientists is not more in villages than in cities in comparison to those run by the research fellows.

Section 12

BIOLOGICAL INSTRUMENT

SECTION AGENDA

Chapter 12.1

Light and Electron Microscopy

Microscopy may be defined as any technique that produces visible images of structures or details too small to otherwise be seen through a human eye. Microscopy has grown with the development of various types of microscopes. Hence there are three main branches of microscopy; optical, electron and scanning probe microscopy.

Microscopy may be classified into two basic types : **Optical (Light) microscopy** and **Electron microscopy**. Both the types involves the **diffraction, reflection**, or **refraction** of incident radiation falling upon the subject of study, and the consequent collection of this scattered radiation sequentially to build up an image.

I. LIGHT MICROSCOPY :

Light (or Optical) microscopy involves passing visible light transmitted in course of or reflected from the subject through a single lens or a sequence of lenses. The image formed can be sensed by the naked eye, represented on a photographic plate or digitally captured. The combination of the single lens with its attachments, or the system of lenses and imaging equipment, in conjuntion with the appropriate lighting equipment, sample stage and support, build up the light microscope.

Invention of compound microscope has a long history. In 1590, Hans Janssen and his son Zacharias Janssen, Dutch spectacle makers are often said to have invented the compound microscope. In 1609, Galileo Galilei developed an *occhiolino* or compound microscope with a convex and a concave lens thus he claims for the title of 'inventor of the microscope'. In late 1600s another Dutchman Christiaan Huygens, developed a simple two-lens ocular system that was achromatically corrected and therefore it was a huge step forward in microscope development.

Anton van Leeuwenhoek (1632-1723) developed a very simple microscope and is generally credited with bringing the microscope to the attention of biologists.

(a) Principles of Optical Microscopy :

The magnification of minute things is an essential feature of biological research, but apart from magnification the fine detail in cells and in subcellular components is also necessary that there must be imaging system that should be capable of providing spatial information across small distances.

Resolution is defined as the ability to distinguish two very small and closely-spaced objects as separate entities. If the distance separating the two tiny objects is small the resolution is the best. Resolution is measured by certain physical parameters that comprise the wavelength of light, and the light-gathering power of the objective and condenser lenses. A simple mathematical equation defines the smallest distance (h') separating the two very tiny objects:

The resolving power of a microscope is also given by the following equation:

$$h' = \frac{0.61\lambda}{n'\sin\theta}$$

where $n'\sin\theta$ = Numerical aperture (N.A.)

h' = smallest distance between the two points that is resolved

λ = Wavelength of light used

n' = Refractive index of the medium between object and the objective lens.

θ = Angle which the marginal rays form with the axis.

The numerical aperture of a lens is dependent upon two parameters, the **angle of the incidence** of light onto the lens, and the **refractive index** of the glass of which the lens is composed. The angle of incidence is also known as the **cone angle** and 1/2 of this value is designated by the symbol 'θ'. Half the cone angle is used to calculate the angle the light subtends relative to the light axis. The cone angle and thus 'θ' can be altered by inclusion of a substage condenser. If the condenser is moveable, the cone angle can be varied; the closer the substage condenser is to the object, the greater is the cone angle. This is a relatively inexpensive means of affecting the resolution of the microscope and thus nearly all microscopes are equipped with substage condensers. When 'θ' is 90°, N.A. cannot be more than one if the medium is air. But if the space between the object and objective lens is occupied by immersion oil then the value of N.A can increase over one.

In the above equation, N.A. is a mathematical calculation of the light-gathering capabilities of an objective lens. The N.A. of each objective lens is marked in the metal tube, and generally ranges from 0.25-1.4. Thus a lens with higher N.A. makes it better in the light gathering property thus makes a better resolution. Higher N.A. values also signify shorter working distances (one has to get the lens closer to the object). N.A. values greater than 1.0 also point towards that the lens is used with some immersion fluid, such as immersion oil.

From the equation above, one should be aware that the N.A. of the condenser is as important as the N.A. of the objective lens in determining resolution. Thus if there is a closure of the condenser diaphragm this may result into a loss of resolution. In fact, at full aperture and along with high-quality oil immersion lenses (N.A. 1.4 for both the condenser and the objective) it is possible to have a resolution slightly better than 0.2 μm.

From the equation above, it is also apparent that shorter wavelength light (i.e. bluer light) will provide the user with better resolution (i.e. lesser h' values). That is resolution is directly proportional to wavelength. The human eye is best accustomed for green light and our ability to visualize detail may be negotiated fairly with the use of blue or violet. Most manufacturers of microscopes correct their simplest lenses (**achromats**) for green light.

Box.12.1.1 Chromatic and spherical aberrations

Chromatic and Spherical aberrations ("distortions") are intrinsic in the design of a lens. Since the lens is a sphere, it projects an image that is spherical, while optical theory is based on images that are flat. Furthermore, because different wavelengths of light are refracted in a different way, the spherical image is still further distorted into multiple images, as each wavelength of light forms a separate image.

Thus a lens has to be corrected to yield flat fields rather than curved ones and thus is referred as a **plan lens**, while one which is being corrected for flat field and colour aberrations is termed a **plan achromat** lens. If the lens is corrected for chromatic aberrations for red and blue, whilst correcting spherically for green, the lens is an **achromat** lens. Increased resolution and increased price of the microscope are key factors in correction of these aberrations.

Alignment :

Proper use of a microscope demands that the optics and light source be aligned on the optical axis. All of the corrections for aberrations depend on proper alignment of the microscope components. There are two general techniques used for proper alignment of the microscope. The first, and perhaps best, is known as **critical illumination**. In this process an image of the light source (bulb filament) is projected into the plane of the object, thus superimposing the light source onto the object. But it has a distinct disadvantage, however, in that it calls for a flat constant light source, not in fact feasible with a tungsten filament bulb.

The second alignment procedure is known as **Koehler illumination**, after a pioneer in light optics, August Koehler. In this procedure, an image of the field diaphragm is projected onto the object plane. This procedure requires a field condenser lens equipped with a moveable (centerable) iris or diaphragm. Koehler illumination is the most commonly used alignment procedure.

Types of optical microscope :

(i) **Simple optical microscope** : A *simple microscope* is made up of only one lens for magnification. A simple microscope is made up of a single, small, convex lens mounted on a plate with a device to hold the material to be examined (that is the sample or specimen). The use of a lone, convex lens to enlarge objects for viewing is found today only in the magnifying glass or the hand-lens, and the loupe.

(ii) **Compound optical microscope** : The *compound microscope* uses a set of multiple lenses so as to maximize **magnification**. The diagram below iilustrate a **compound microscope** (Figure 12.1.1). In its simplest form for example, the compound microscope would have a single glass lens (with a short focal length) used as the objective, and another single glass lens used as the eyepiece or ocular and a mirror for reflecting light upon the object that is to examined. Modern compound microscopes whereas are usually more intricate , having multiple lenses in both objective and eyepiece assemblies. Such multi-component lenses are build in such a way so as to reduce **aberrations**, viz. **chromatic aberration** and **spherical aberration**. Todays modern microscopes the mirror is substituted by a lamp unit providing stable, controllable illumination.

Components of an Compound optical microscope :

- **The eyepiece or ocular** : a cylinder that contains two or more lenses to bring the image in to focus for the eye. The eyepiece is inserted into the top end of the body tube. Eyepieces are changeable and many different eyepieces having different magnifications (such as 5X , 10X and 20X) can be inserted. In some high performance microscopes, the optics of the objective (known as **apochromatic objective**) and eyepiece are matched to give the best possible optical performance.

- **The objective lens** : A cylinder that contains one or more lenses to accumulate light from the sample. At the lower end of the microscope tube one or more than one objective lenses are attached into a circular nose piece which may be rotated to opt for the required objective lens. Typical values of objectives are 4X, 5X, 10X, 20X, 40X, 80X and 100X. Usually, in a

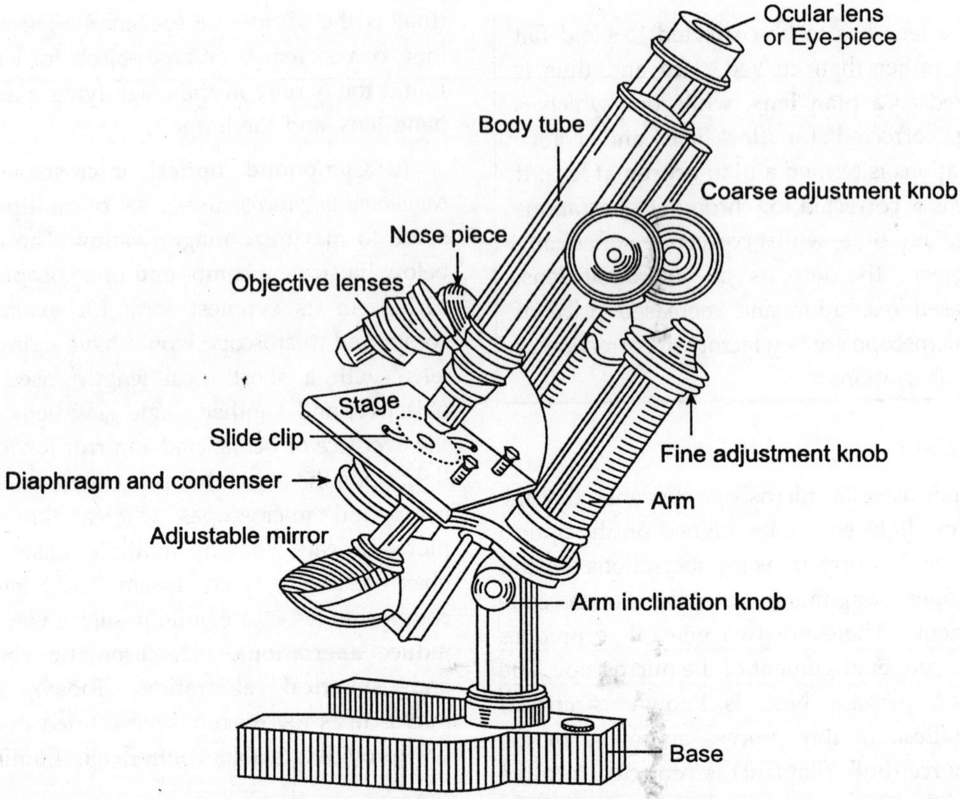

Figure 12.1.1 : Diagram of a compound microscope.

standard compound optical microscope, there are three objective lenses: a scanning lens (5X), low power lens (10X) or sometimes medium power lens (20X), and high power lens (40X). Advanced compound microscopes in most cases have a fourth objective lens, called an **oil immersion lens**. To make use of this lens, a drop of immersion oil is put on top of the cover slip, and the lens is very gradually and carefully lowered until the front objective element is dipped in the oil film. Such immersion lenses are designed in such a way so that the refractive index of the oil and of the cover slip are closely matched so that the light is transmitted from the specimen to the outer face of the objective lens with least refraction. An oil immersion lens usually has a power of 100X.

- **The stage :** It is a platform below the objective which bears the specimen being viewed. In the centre of the stage there is a circular hole through which light beams pass in to illuminate the specimen. The stage usually has arms to grip glass slides on which the specimen (sample) is mounted.

- **The illumination source :** Just below the stage, a light is provided that may be controlled in a variety of ways.Most simply, daylight is directed by means of a mirror. Mordern microscopes, usually, have their own controllable light source that is focused through an optical device known as the condenser (e.g. **Abbe condenser**) with diaphragms and filters which is important to control the quality and intensity of the light.

Operation of a Light compound microscope :

1. The microscope is placed in front, the power cord is connected and plugged it in.

2. The magnification power and the numerical aperture of the lenses which are on the microscope's nose-piece are noted. These values are stamped or painted onto the barrels of the objectives. The magnification power and numerical aperture of each lens are recorded in the space provided below (see box below) :

Magnification (x#) **Numerical Aperture (NA)**

The numerical aperture of the condenser ________

The magnification power of the oculars and whether they are normal or widefield ________

Maximum resolution will depend upon the highest effective numerical aperture of the system. The highest value is normally given by the 100X, or oil immersion lens.

To indicate the numerical aperture of the 100X lens =	________
To indicate the numerical aperture of the condenser =	________
The numerical aperture for an air interphase =	1.0
The numerical aperture for oil interphase =	1.3–1.5

Numerical aperture depends on the angle θ and thus on maximum positioning of the condenser. Using the lowest NA value as the working numerical aperture, the limit of resolution for a microscope is calculated, assuming violet light with a wavelength of 400 nm.

The limit of resolution = $0.61 \times \lambda/NA$, and therefore, the calculated value for your microscope is :

Limit of resolution = ______ μm.

3. A prepared microscope slide is obtained with the letter '**e**'. The slide is placed on the stage and one should ensure that it is locked in place with the slide holder. The condenser-focusing knob is rotated to move the condenser to its highest position of travel. Even though there is a best location for the condenser, the correct position of the condenser will vary slightly for each objective. Unless directed otherwise, it will not be necessary to move the condenser during any of the intended uses. If, on the other hand, one wishes to find the ideal location, the microscope is focused on any portion of a slide, and then simply the condenser aperture is closed down and the **condenser** is moved until one have a sharply focused view of the condenser aperture (usually with a slight blue hazy edge. The condenser should not be moved from this position.

4. Rotating the dimmer switch turns on the microscope and the light intensity is adjusted to a comfortable level. One has to be sure that the condenser aperture is open.

5. In case of binocular microscope the eyepieces are adjusted to one's interpupillary distance and dioptre. Many microscopes simply require pushing the eye tubes together or apart directly. The eye tubes are moved back or forth until you see one uniform field of view.

Focusing of the microscope on any object is done within the field of view.

- A suitably contrast location in the centre of the field is found and left eye is closed. Using the coarse and fine adjustments, the microscope is focused until one can obtain a sharp image with one's right eye only.
- Now right eye is closed and the focus of the left eyepiece is adjusted by rotating the dioptre adjusting ring located on the left eyepiece. All subsequent uses of the same microscope will involve use of the coarse and fine focus adjustments, without reference to the procedures in step 2. That is, step 2 need only be performed once at the beginning. It may, of course, be checked periodically if desired, and will need to be readjusted if someone else uses the microscope.

6. The microscope should always be focussed first with the 10X magnification. Even if one is going to use the 100X, it is more efficient to begin with the 10X and then move up to the power desired. The objective lenses are parfocal, which means that if one is focused, each of the others is approximately in focus when revolved into position.

With the slide from Step 4 in place, the coarse focus control are rotated until the slide is as close to the 10X objective as possible. The stage manipulators are moved until a portion of the slide is directly under the objective and carefully focused on the object in view. After adjusting the focus at 10X, the object to be viewed is centred, and the nosepiece is rotated to the next highest magnification. The fine focus is used to control only once the 40X or 100X objectives are in place.

The fine focus is manipulated to obtain the sharpest image. During use of the microscope, one hand should remain on the fine focus, as constant readjustment will be called for. The other hand is used to manipulate stage movements.

7. The 10X objective is returned and the slide is moved around until user locates the letter **'e'** in the view. The orientation of the letter **'e'** on one's slide is noted and in the field of view.

8. To use the 40X objective, the object user wish to view (the 40X will have a smaller field of view) is centred and the objective turret (referred to as the nosepiece) is rotated to bring the 40X objective into position.

Various types of Light microscopy :

(a) Bright field microscopy :

Bright field microscopy is the simplest of all the light microscopy techniques. Sample illumination is via transmitted white light, i.e. illuminated from below and observed from above.

Advantages and Limitations of Bright field microscopy :

Advantages :

- Very simple setup with only basic equipment required.
- Preparation of sample is not required, allowing viewing of live cells.

Limitations :

- Most biological samples have very low contrast.
- Apparent resolution very low due to the blur out of focus material.

(b) Dark Field Microscopy :

If the light image is reversed, then the microscope becomes a **dark field microscope**. All standard bright field microscopes can be readily converted to dark field by inserting a round opaque disk beneath the condenser. Dark field microscopy was first utilized to examine trans-filterable infectious agents, later to be termed viruses, and to determine that they were particulate in nature. Small objects, even those below the limits of resolution, can be detected easily with dark field microscopy, as the object appears to emit light on a dark field.

Advantages :

- Transparent objects are clearly shown in the sample.
- Simplicity of setup and just basic equipment required.
- Sample preparation not required, allowing viewing of live cells.

Limitations :

- Low light intensity in final image of many biological samples.
- Low apparent resolution due the blur out of focused objects.

(c) Phase Contrast Microscopy :

Dutch Physicist Frederick Zernicke in the 1930's as a means to generate contrast in biological specimens, changing these invisible phase differences into visible amplitude differences invented the Phase Contrast Microscope.

The phase contrast microscope is constructed in such a way that it carries on two unique tasks i.e. it splits the direct light (the back ground light from the field) from the light diffracted by the object, and it produces two types of waves to be approximately one – half of the wavelength out of phase of the other, so that they can destructively interact, and cause an alteration in intensity (**Figure 12.1.2**).The lights are separated in the following way: A disk is placed in the substage condenser that is solid apart from for the occurrence of a thin ring for allowing the light to pass through (Figure 12.1.3a). As a result the condenser produces a hollow cone of light to illuminate the field and penetrate the objective lens. As the illuminating light is in the form of a ring it is a ring of direct light that enters the objective transient through the slide. This can be seen evidently by removing the ocular and looking down the viewing tube at the back focal plane of the objective. The phase contrast objective lens are assembled in such a way that they have a ring designated as the **phase plate** at their back focal plane that match up geometrically with the position of the ring of direct light from the condenser. The entire direct light that is diffracted by the elements of the object is scattered from the points of diffraction across the entire surface of the lens as a result these two types of light have been separated to a large extent (Figure 12.1.3b).

When the light is diffracted by a transparent object it is retarded one-fourth wavelength relative to the undiffracted, direct background light. To acquire destructive interference it is essential to retard an additional one-quarter wavelength relative to the direct light. This is feasible through the employment of a phase plate present in the objective. This plate is placed in the lens where the glass is thinner and direct light passing through this decreased path of glass is advanced by another additional one – quarter

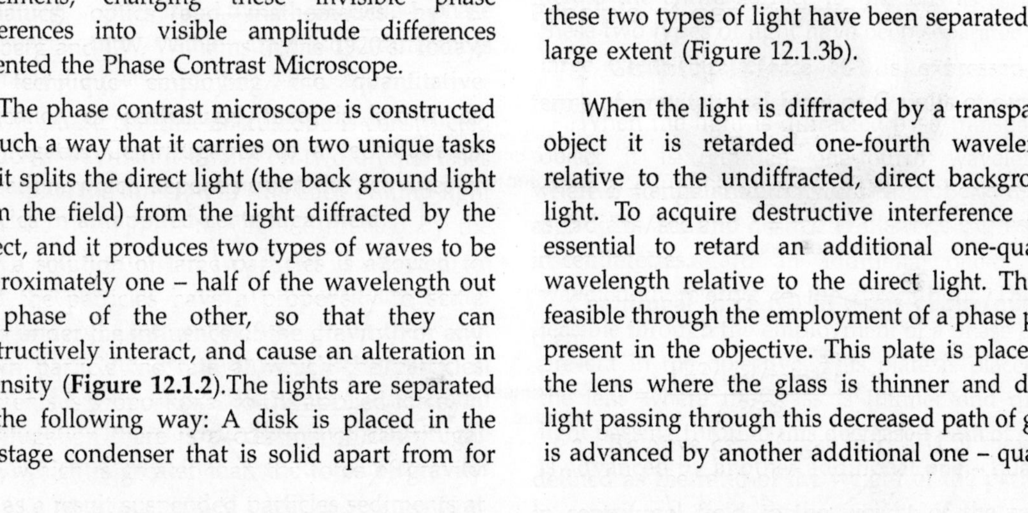

Fig. 12.1.2 : Destructive interaction of two waves which are out of phase of each other producing alteration in intensity.

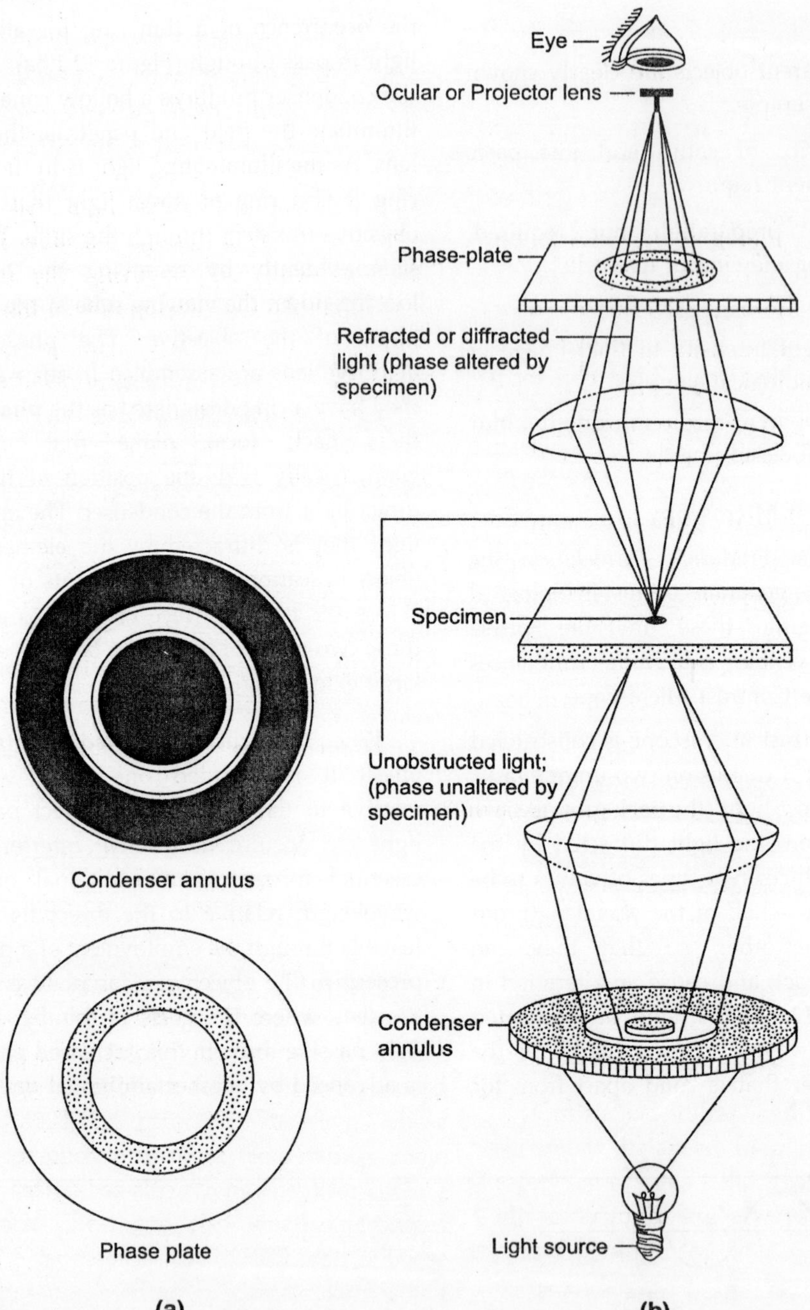

Fig. 12.1.3 : Phase contrast microscope assembly.
(a) Condenser annulus and phase plate;
(b) Optical diagram of phase contrast microscope.

of wavelength, which place it to one half of wavelength out of phase with the diffracted light. This is known as **Phase effect.**

Assuming for a very small part of the object being examined has precisely the same refractive index of the medium. All diffracted light and direct light would be one half wavelengths out of phase with one another and interference will effect in relative decrease in intensity. If the object has part with different refractive index all of which are in excess of the surrounding medium. When light ray crosses the surface between two transparent media of different refractive index an alteration in velocity occurs. A reduction in velocity results in elongation of each cycle with respect to time. When the ray emerges from the medium of higher refractive index it will be in a different phase of cycle than it would have been had it not entered more refractive substance, as a result it will leave behind a less refractive ray. Greater the refractive index more is the retardation. Again greater the thickness longer the ray will stay in the substance thus more will be the retardation. Difference in the refractive index will lead to shift in phase of light rays passing through different parts of the specimen and this will cause in the shift of phase of diffracted light relative to the direct light. Thus various parts of the specimen will appear with varying intensity.

Operation of any microscope in phase contrast mode :

1. Koehler illumination is established on the microscope. With the instructor (teacher's) approval, the phase annular ring and its corresponding phase plate is centred.

(a) A slide is placed on the stage of the microscope; the condenser is moved to its highest position and focused at 10X magnification.

(b) The condenser diaphragm is opened to its maximum setting, and the field diaphragm is closed completely.

(c) Using the condenser movement control, the condenser is moved until a sharp image of the field diaphragm is observed. To determine that the focus is indeed the field diaphragm, the field diaphragm is slightly opened and closed to see if its movement can be detected in the field of view. When focused, there will be a slight blue haze on the edge of the diaphragm.

(d) The field diaphragm is opened until it nearly fills the field, but can be still seen. The field diaphragm is centred in the field of view using the centering screws on the substage condenser. The field diaphragm is opened completely to fill the field of view.

(e) One of the oculars is removed from its tube and while peering down the tube, the condenser diaphragm is opened until it just fills the field of view at the bottom of the tube. The ocular in the tube is replaced.

Completing steps (a) to (d) establishes Koehler illumination, where the field diaphragm is superimposed onto the object and centres the major optical components of the microscope.

To check on the phase annulus and its corresponding phase plate, an ocular is removed and it is replaced with a telescopic ocular designed to focus on the rear lens of a phase objective. The phase objective is matched with its corresponding setting on the phase condenser and visually verified that the phase annulus (a clear ring) is perfectly matched to the phase plate (a darker ring). If it is not, the instructor is to be requested for assistance in centering the phase annulus. This is most often accomplished by adjusting a second set of centering screws attached to the phase condenser. The normal ocular is replaced before using the phase contrast optics. The phase condenser setting is returned to the normal bright field position.

2. A simple wet mount of any unicellular organism is made and observed under bright field microscopy at 10X.

3. An active organism is located and centered in the field of view. The condenser is rotated so that phase ring is matched to the 10X phase with the 10X objective.

4. The difference in the appearance of the organism is observed between normal bright field and phase contrast.

5. The organism viewed may be drawn under phase contrast.

6. The phase control is returned on the condenser to the normal bright field setting, switched to a higher magnification (20X or 40X) and the organism is observed at the higher magnifications with and without phase enhancement.

7. The view of the organism under phase contrast is compared to normal unstained bright field.

(d) Fluorescence Microscopy :

In some classes of atoms and molecules, electrons absorb light, get energized, and then quickly lose this energy in the form of heat and light emission. If the electron keeps its spin, the electron is said to go into a singlet state, and the type of light (larger wavelength) that is emitted as the electron come back to ground state is called **fluorescence**. This change in wavelength is known as **Stoke's shift**. If the electron alters its spin after being excited, enters the triplet state, and the kind of light that is emitted as the electron returns to ground state is known as **phosphorescence**. Phosphorescence has a much longer life than fluorescence. Both fluorescence and phosphorescence emissions have a particular wavelengths for specific excited electrons. Both types of emission depends on specific wavelengths of excitation light, and for both types of emission, the energy of excitation exceeds the energy of emission. Described in another way, wavelength of excitation light is shorter than of wavelength of emission light. In biological application, one can make use of fluorescence in localization reactions, so as to recognize particular molecules in complex mixtures or in cells. The basic advantage of fluorescence is of providing a very high signal-to-noise ratio, which facilitate the user to differentiate spatial distributions of rare molecules. To utilize fluorescence, one need to label the specimen (a cell, a tissue, or a gel) with a suitable molecule (a **fluorochrome**) whose distribution will become apparent after illumination. Many different fluorescent dyes or fluorochrome can be used to stain different structures or chemical compounds. One most important method is the combination of antibodies (can be tailored specifically for a chemical compound) coupled to a fluorochrome as in **immunostaining**. Most commonly used fluorochromes are **fluorescein** or **rhodamine**. For example, one strategy often in use is the artificial production of proteins, based on the genetic code (DNA). These proteins can then be used to immunize rabbits, which then form antibodies which bind to the protein. The antibodies are then tagged chemically to a fluorochrome and then employed to trace the proteins in the cells under study.

The fluorescence microscope is perfectly suited for the revealing of particular fluorochromes in cells and tissues. The basic technique of Fluorescence microscopy was described by Price and Schwartz (1956), Haitinger(1959) and Holborow (1970).

The fluorescence microscope that is in extensive use today follows the basic "incident-light" design of Ploem, who made use of a novel arrangement of filters with a **chromatic beam splitter**. Through the incident light, the object is illuminated with fluorescence excitation light through the objective lens. The object emits longer-wave length (λ) fluorescence in response to the shorter-λ excitation light. The objective

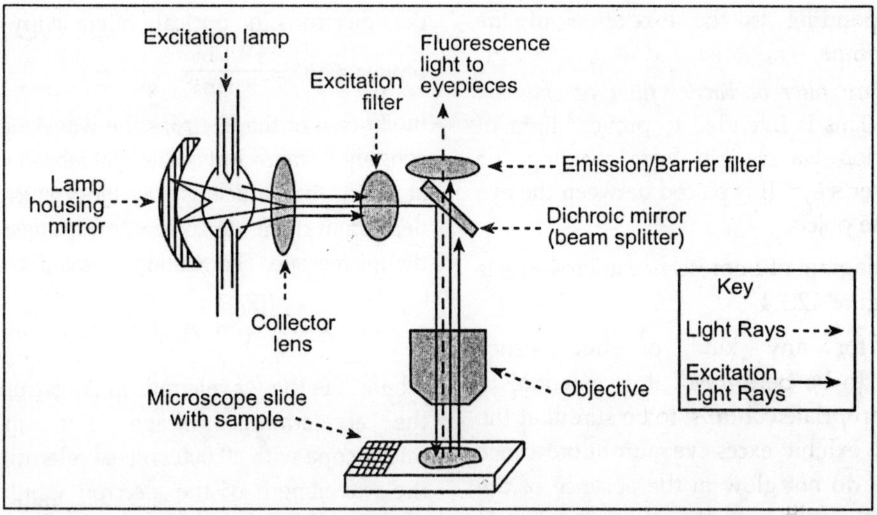

Fig. 12.1.4 : Optical diagram of a Fluorescence microscope.

lens then serves up both for illumination and imaging. The chromatic beam splitter transmits or reflects light, depending on its colour. In support of this application, shorter λ light is reflected and the longer λ light is transmitted by the splitter. The chromatic splitter is placed in the optical path between the objective lens and the ocular, at a 45° angle, so that it would reflect shorter λ light downward towards the objective. The longer λ fluorescence emission light would be transmitted through the chromatic beam splitter towards the ocular.

Light beams pass through a condenser lens system and provide illumination of an object at many points simultaneously. For incident light fluorescence microscopy, the objective lens also acts as a condenser for the incident radiation i.e. the excitation light beam. In its interaction with the object, some of this light is absorbed, some of this light is scattered, some of this light is reflected, and some of this light is slowed or retarded (relative to a beam of light that does not pass through the object). A part of the light that has interacted with the object subsequently passes through the imaging lens system of the microscope where it provides the user with visual or pictorial image information about the object. Like the process of illumination, the process of image generation functions in a parallel fashion, where large numbers of light beams add to the image at the same time. Resolution is limited by the closeness of overlapping points of brightness or darkness. In a practical sense, the limit of resolution is 0.18-0.2 μm with the best available objective lenses and a good specimen.

For the cells to be examined with the help of a fluorescence microscope, it is essential to recognize the spectral characteristics of the fluorochrome that has been employed. To excite the fluorochrome properly and then view its fluorescence emission, the suitable filter packages must be there in the microscope. The fluorochrome may not be able to fluoresce at all if the cells are illuminated with the improper filter pack present within the optical path.

Fluorescence microscope has two filters—

1. *Primary or exitation filter*: This is placed between the light source and the specimen. This filter allows light only over a comparatively narrow wavelength

corresponding to the excitation of the fluochrome.

2. *Secondary filter or barrier filter or emission filter* : This is intended to prevent light of excitation wavelength from reacting the observer's eye. It is placed between the eye and the object.

Optical diagram of fluorescence microscope is shown in Figure 12.1.4.

Finally, for any kind of fluorescence localizations to be performed, it is essential to have the appropriate controls; to be sure that the cells do not exhibit excessive autofluorescence (that is, they do not glow in the absence of the fluorochrome), and that the fluorochrome is responsible for the localization pattern observed.

II. ELECTRON MICROSCOPY :

The **electron microscope (EM)** is a type of microscope that uses electrons to create an image of the target. It has much higher magnification and resolving power than a normal light microscope, up to two million times, allowing it to see smaller objects in details.

Prototype of the the first electron microscope was built in 1933 by the German engineers Ernst Ruska and Max Knoll. It was based on the thoughts and discoveries of French physicist Louis de Broglie. Although it was primitive and not fit for practical use, the instrument was still capable of magnifying objects by four hundred times.

In 1931, Reinhold Rudenberg, the research director of Siemens, had patented the electron microscope. Siemens began developing the electron microscope in 1937, funding Ruska and Bodo von Borries to develop the instrument.

Resolution in electron microscopes :

The basic advantage of the electron microscope over light microscope as already stated is its greater resolution the instrument can deliver. This occurs due to the wave property of the electrons.In optical microscopy resolving power or $h' = \frac{0.61\lambda}{n'\sin\theta}$.

In case of the electrons the wavelength is not a constant distance, but is basically related to the speed at which the particle is travelling which in turn is dependent upon the accelerating voltage applied in the microscope. The relation is stated as :

$$\lambda = \sqrt{\frac{1509}{V}} \text{ or } \sqrt{\frac{1.5}{V}}$$

where λ is the wavelength in Angstrom and V is the accelerating voltage. For an electron microscope with 40,000 volts accelerating voltage, the wavelength of the electron would be 0.006 nm, i.e., $\sqrt{\frac{1.5}{40,000}}$. The precise practical resolution of today's electron microscopes is in range of 2-5 Angstrom.

Types of Electron microscopy :

(i) **Transmission Electron microscopy (TEM)** : TEM is made essentially of a tall empty cylindrical column inside which the electron beam is restricted and a console that have panels of dials which electronically control the operation within the column. On the top of the column there is a cathode specifically a tungsten wire filament which when heated acts as the source of electrons. The electrons are drawn from the hot filament and are speeded up as a fine beam through high voltage applied between it and the anode. The column is made vacuum through which electrons travel freely without any collision with gas molecules.

The lens of the EM is formed of electromagnets that are positioned in the walls of the column surrounding the evacuated core. The strength of the magnet can be manipulated by the current provided to them, which are carried on by the various dials of the console.

The condenser lenses are placed in between the electron source and the specimen and are necessary for focussing the electron beam on the specimen.

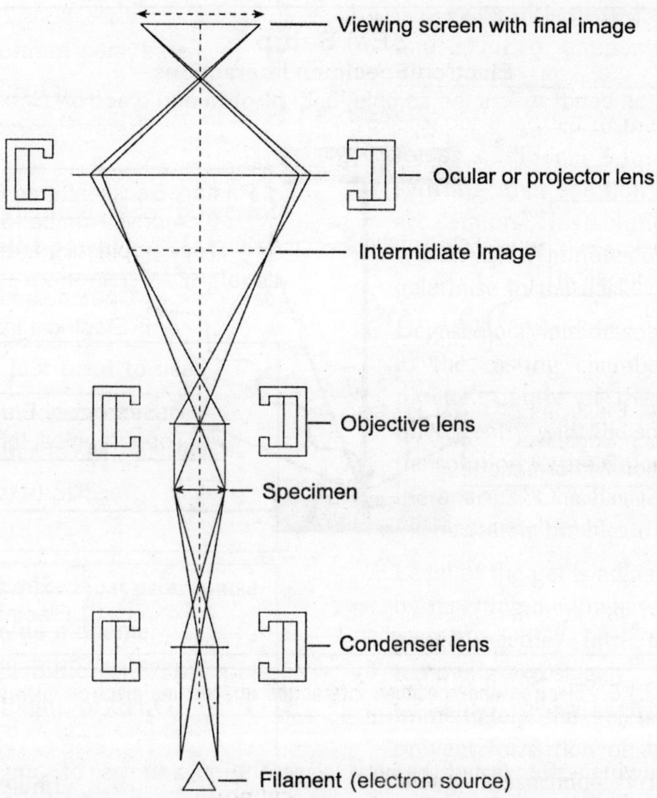

Fig. 12.1.5 : Lens system of Transmission electron microscope.

The specimen is held on a small, thin metal grid holder, which sequentially is placed into the middle region of the column of the microscope so that the grid is perpendicular to the electron beam.

The objective lens, which is situated beneath the specimen (object), focuses the electrons and is responsible for the formation of an image of the object. The objective lens is focussed through changing the current, which in turn cause a change in the strength of magnetic field and thus shifting its focal length.

The focal length of the lenses in the electron microscope varies subject to the current supplied to them. The image produced by the objective lens is only magnified about 100 times. Nevertheless in comparison to an optical microscope there are adequate and satisfactory details present in this image to magnify it an additional 10,000 times. In an electron microscope there are two extra lenses : the intermediate lens (enlarge the image another 20 times) and ultimately the projector lens which is capable of 100 times increase (Figure 12.1.5). Thus through the change of current applied to the various lenses (intermediate and projector lens) of the microscope, magnifications can vary from 1000 – 250,000 times. Thus much greater information of the image can be visualised by the eye.

Electrons that have passed through the specimen are brought to focus on a screen situated at bottom of the column. Electrons after hitting the screen excite a coating made of fluorescent crystals, which in sequence emit their own visible light that is perceived by the eye as an image of the specimen. Image photography is

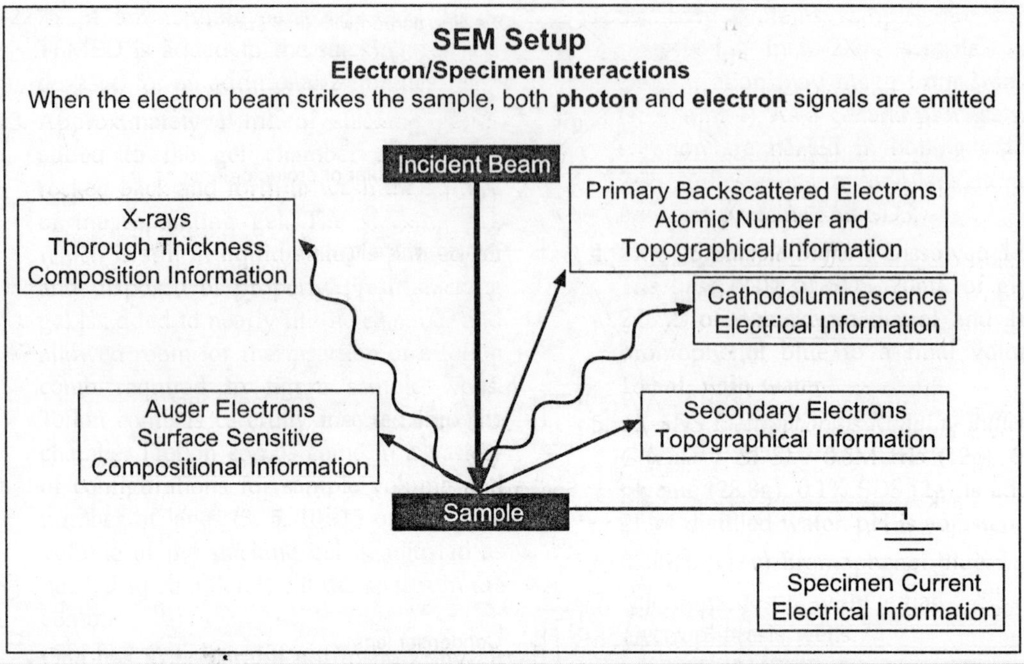

Fig. 12.1.6 : Electron and specimen interaction in scanning electron microscopy.

made by lifting the viewing screen out of the way and allowing the electrons to strike the photographic plate placed beneath the screen.

(ii) **Scanning Electron Microscopy (SEM) :** The Scanning Electron Microscope is a microscope that makes use of electrons rather than light to form an image similar to TEM. Scanning electron microscopy (SEM) visualizes details on the surfaces of cells and particles and gives a very nice 3D view. It produces results similar to the stereo light microscope and akin to that its most useful magnification is in the lower range than that of theTEM.

The SEM has a large depth of field, thus allowing a large amount of the sample to be in focus at one time. An image produced is of high resolution, which means that closely spaced features can be examined at a high magnification. Preparation of the sample is relatively easy since most SEMs only require the sample to be conductive. The combination of higher magnification, larger depth of focus, greater resolution, and ease of sample observation makes the SEM one of the most heavily used instruments in research areas today.

The electron beam comes from a filament, made of various types of materials. The most common is the tungsten hairpin gun. This filament is a loop of tungsten, which functions as the cathode. A voltage is applied to the loop, causing it to heat up. The anode, which is positive with respect to the filament, forms powerful attractive forces for electrons. This causes electrons to accelerate toward the anode. Some accelerate right by the anode and on down the column, to the sample. Other examples of filaments are lanthanum hexaboride filaments and field emission guns.

A beam of electrons is generated in the electron gun, located at the top of the column, which is pictured above. This beam is attracted through the anode, condensed by a condenser lens, and focused as a very fine point on the sample by the objective lens. The electron beam hits the sample, producing **secondary electrons or back-scattered electrons** (Figure 12.1.6) from the sample. These electrons are collected by a secondary detector or a backscatter detector, converted to a voltage, and amplified. The amplified voltage is applied to the grid of the Cathode-Ray Tube (CRT) and causes the intensity of the spot of light to change. The scan coils are energized (by varying the voltage produced by the scan generator) and create a magnetic field which deflects the beam back and forth in a controlled pattern. The varying voltage is applied to the coils around the neck of the Cathode-Ray Tube (CRT) which produces a pattern of light deflected back and forth on the surface of the CRT. The pattern of deflection of the electron beam is the same as the pattern of deflection of the spot of light on the CRT. The image consists of thousands of spots of varying intensity on the face of a CRT that correspond to the topography of the sample.

While all these signals are present in the SEM, not all of them are detected and used for information. The signals most commonly used are the **secondary electrons**, the **backscattered electrons** and **X-rays.**

Sample Preparation for Electron microscopy :

Materials to be viewed under an electron microscope may require processing to produce a suitable sample. The technique required varies depending on the specimen and the analysis required:

- *Cryofixation* : Specimen are frozen rapidly in liquid nitrogen or even liquid helium temperatures so that the water forms vitreous (non-crystalline) ice. This preserves the specimen in a snapshot of its solution state. An entire field known as cryo-electron microscopy has emerged from this technique. With the evolution of **cryo-electron microscopy of vitreous sections (CEMOVIS)**, it is now feasible to observe virtually any biological specimen similar to its native state.
- *Fixation* : Preservation of the sample to make it more realistic. Glutaraldehyde - for hardening and osmium tetroxide - which stains lipids black are used.
- *Dehydration* : It is the process of replacing water with organic solvents such as ethanol or acetone.
- *Embedding* : The process of infiltration of the tissue with a resin such as araldite or epoxy for sectioning.
- *Sectioning* : It is a technique that produces thin slices of specimen, semitransparent to electrons. These can be done through an ultramicrotome with a diamond knife to produce very thin slices. Glass knives are also used because they can be made in the laboratory and are less expensive.
- *Staining* : In staining, heavy metals such as lead, tungsten or uranium are used to scatter imaging electrons and thus give contrast between different structures, since most of the the biological materials are nearly "transparent" to electrons (weak phase objects). In biology, specimens are usually stained "en bloc" before embedding and also later on stained directly after sectioning through a brief exposure to aqueous (or alcoholic) solutions of the heavy metal stains.
- *Freeze-fracture or freeze-etch* : It is a preparation method specifically useful for studying lipid membranes and their inserted proteins in "face on" view. The fresh suspension of tissue or cell is frozen rapidly (**cryofixation**, in liquid freon,m.p. –150°C), then fractured by simply breaking or by using a knife while maintained at liquid nitrogen temperature. The cold fractured surface (sometimes "etched" by increasing the temperature to about –100°C for several minutes to let some ice sublime) is then shadowed with evaporated platinum or gold.The metal layer is deposited at an average angle of 45° in a high vacuum evaporator.This technique is termed as **Shadow casting**. A second coat

of carbon, evaporated perpendicular to the average surface plane is often performed to improve stability of the replica coating and also cements the patches of metal. The specimen is returned to room temperature and pressure, then the extremely fragile "pre-shadowed" metal replica of the fracture surface is released from the underlying biological material by careful chemical digestion with acids, hypochlorite solution or SDS detergent. The still-floating replica is thoroughly washed from residual chemicals, carefully fished up on EM grids, dried then viewed in the TEM.

- *Ion Beam Milling* : This is a technique that thins samples until they are transparent to electrons by firing ions (typically argon) at the surface from an angle and sputtering material from the surface. A particular type of this is **focused ion beam milling**, where gallium ions are employed to form an electron transparent membrane in a specific region of the sample, for example through a device within a microprocessor. Ion beam milling may also be applied for cross-section polishing before SEM analysis of materials that are difficult to prepare using mechanical polishing.
- *Conductive Coating* : This is an technique in which ultra thin coating of electrically-conductive material, deposited either through high vacuum evaporation or through low vacuum sputter coating of a particular sample. This procedure is done to check the building up of static electric fields at the specimen owing to the electron irradiation required during imaging. Such coatings are formed of gold, gold/palladium, platinum, tungsten, graphite etc. and are especially important for the study of specimens with the scanning electron microscope.
- *Specimen preparation in SEM* : There is two basic types of SEM's. The regular SEM, requires a conductive sample. An environmental SEM can be used to examine a non-conductive sample without coating it with a conductive material. Three requirements for preparing samples for a regular SEM are: (1) all water, solvents, or other materials that could vapourize while in the vacuum should be removed; (2) all the samples are firmly mounted and (3) non-metallic samples, such as bugs, plants, fingernails, and ceramics, should be coated so they are electrically conductive. Metallic samples can be placed directly into the SEM.

Chapter 12.2

Camera Lucida Handling and Drawing

Introduction :

The observations under a light microscope are sometimes required to be recorded. One of the method of such recording (i.e., through drawing) is done through the use of Camera Lucida. The British scientist William Hyde Wollaston discovered Camera Lucida in 1807. Camera lucida is Latin for "light chamber".

These drawings are made on paper with sharpened lead pencil as in the case of free hand drawings, but the difference in this case is that the sketch is much more precise and the relationship of a variety of components of the structure being studied is more accurate. Again, the sketch is a magnified version of the original structure on the slide, which depends upon the combination of eyepiece and the objective used.

Components of Camera Lucida :

Camera Lucida basically consists of the following parts :

(a) It has an adjustable ring that can be fitted to the tube of a microscope. This is possible by first removing the eyepiece, fitting the ring and then replacing the eyepiece into the tube of the microscope.

(b) The prism head which is made up of a prism fitted in such a way that it comes to lie exactly over the eye piece and the structure on the slide can be observed through this prism and the eye piece. The body on which the prism is fitted may be swivelled and the height is also changeable. The prism is so fitted that it also reflects the image of the object being observed by right angle on to a mirror. The body holding the prism is inserted and screwed on to a straight lever fixed to the adjustable ring. Though, in many models the adjustable ring and the body holding the prism are joined into one compact unit. In some model, prism is replaced by a mirror which is fitted at an angle of 45° to the body.

(c) The reflecting mirror is joined to an elongated lever that may be fitted and fastened to the body holding the prism. The image that is reflected by the prism falls on this mirror, which is arranged in such a manner that the back is uppermost and the reflecting surface face against the table. Adjustment of the angle of the mirror is essential and in several models an arrow mark is present on the clamp holding the mirror. When the mirror angle is adjusted to the arrow mark, the image is reflected down on the table, to the side of the microscope (see, Figure 12.2.1).

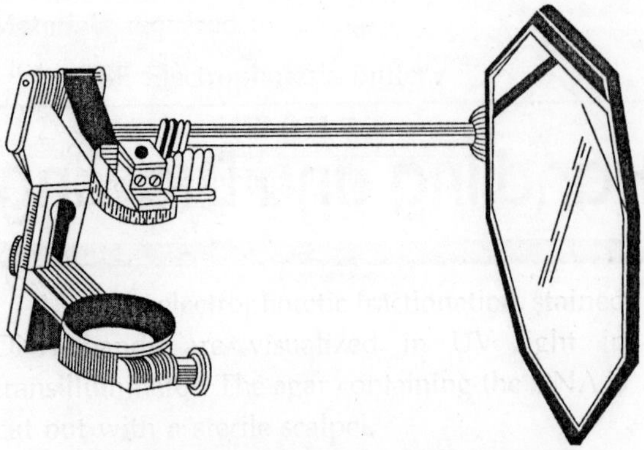

Figure. 12.2.1 : A mirror type Camera Lucida for reliable sketching of microscopic preparations.

Procedure :

1. The Camera Lucida is fitted to the microscope.
2. Slide bearing the object being studied is placed on the stage of microscope and the structure is focussed viewing through the prism of the Camera Lucida. (For drawing a zooplankton see Fig., 12.2.1A.)
3. Artificial light source has to be employed and other lights of the room should be dim and distant.
4. The distance of the light source has to be so adjusted that the light is neither bright nor too dim i.e., moderate.

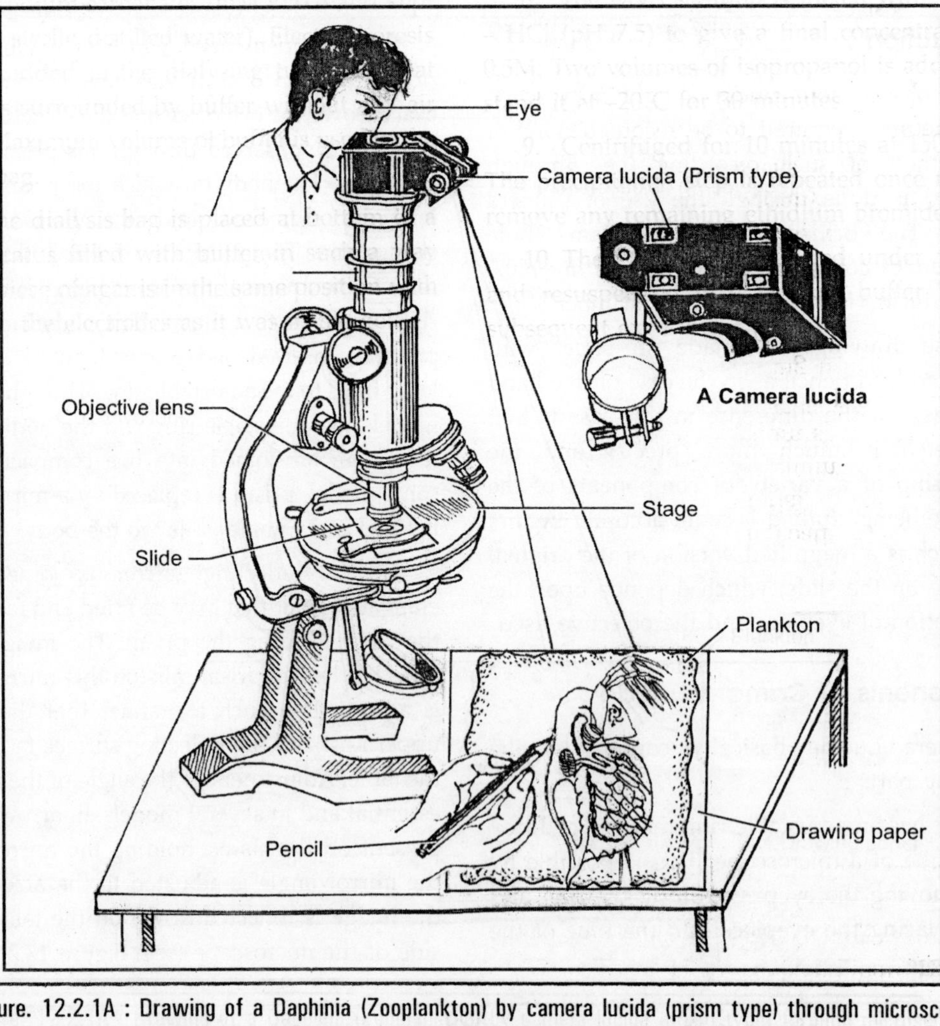

Figure. 12.2.1A : Drawing of a Daphinia (Zooplankton) by camera lucida (prism type) through microscope.

5. When the light is appropriately adjusted, a white paper is placed on the right hand side of the microscope on the table and a lead pencil is held in one's (observer) right hand.

6. The microscopic preparation as well as the white paper and the pencil (also the fingers holding it) should be visible to observer eye, although when one is gazing into the eye piece of the microscope and not to one side on the paper. It requires a little training. In new models filter slots are supplied so that filter can be used to reduce and regulate the light. For better focussing it may be some times required to raise the paper by a few inches from the surface of the table. This is done easily by placing any flat book or laboratory record book under the paper.

7. Once the object, as well as the paper and the pencil point are clearly focussed and visible, the outline and the details can be accurately drafted. Looking through the microscope, it seems that the pencil point is indeed moving over the object but actually the image of the object is being reflected down on the paper.

8. A permanent sketch can be made using India ink. The various parts of the diagram are then to be labelled and the legend giving particulars of the object stencilled on the bottom of the sketch. When final figures are being made, these should better be drawn directly on to white drawing paper. The magnifications such as 100X, 400X and 1000X, etc.) should invariably be mentioned after the legend.

Chapter 12.3

Homogenization

Each organelle has characteristics size, shape and density for example, which makes it different from other organelles within the same cell. To obtain pure organelles, the cells must be ruptured, so that the cell membrane is broken, but the organelle to be studied is not. The process of rupturing a cell is known as **homogenization** of the cell. Homogenization disrupts tissue and releases intracellular components to form a "brei" or homogenate. The major use of the process is a preliminary step for the separation of cellular components in investigations into metabolic compartmentalisation.

Homogenization techniques can be divided into various types, such as :

- **Osmotic alterations :** Many organelles are easier to separate if the cells are slightly swollen. The inbibition of water into a cell will cause osmotic swelling of the cell and/ or organelle, which can often assist in the rupture of the cell and subsequent organelle separation. The use of a hypo-osmotic buffer can be very beneficial, for example, in the isolation of mitochondria and in the isolation of mitotic chromosomes.
- **Mortars, Pestles :** Possibly the most frequent procedures employ Ten Broeck or Dounce homogenizers, both of which are glass mortar and pestle arrangements with manufactured, controlled bore sizes (Figure 12.3.1). The addition of a motor driven teflon pestle produce the Potter-Elvijem homogenizer. Ultrasonification is an useful accessory to this procedure, but is often adequate by itself.
- **French press :** Repeated high velocity compression and expansion is what is designated as a "French Press." The French press is very powerful and can disrupt bacteria and viral particles as well. This device forces slurry of the cells through an orifice (opening) at very high pressures (Figure 12.3.2). The rapid expansion of the pressure from within literally "blows" the cells apart. While this technique is not often required, it is the only way to break open some materials. The units have capacities from 1 to 40 mL and can reach pressures of 20,000-40,000 pounds per square inch (psi).
- **Blenders :** For molecular separations, mechanical blenders are often employed, varying in sophistication from household blenders to high speed blenders with purposely designed blades and chambers (e.g. a Virtis Tissue Homogenizer, Figure 12.3.3).

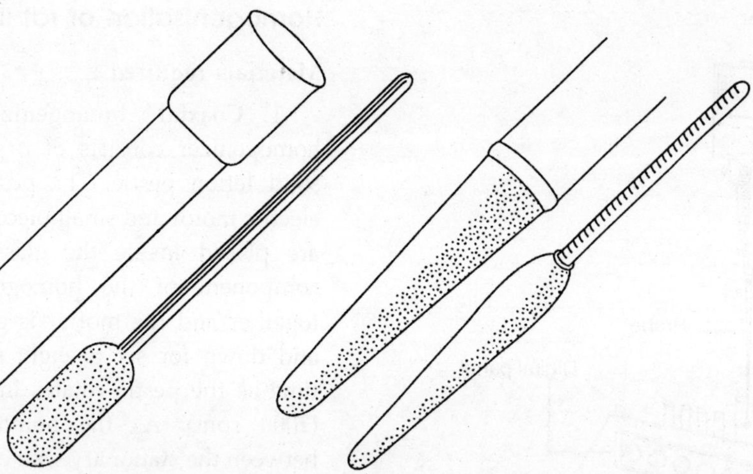

Fig. 12.3.1 : Homogenisers— (a) Glass pestle and mortar, (b) Teflon pestle and mortar.

Pressure Cells

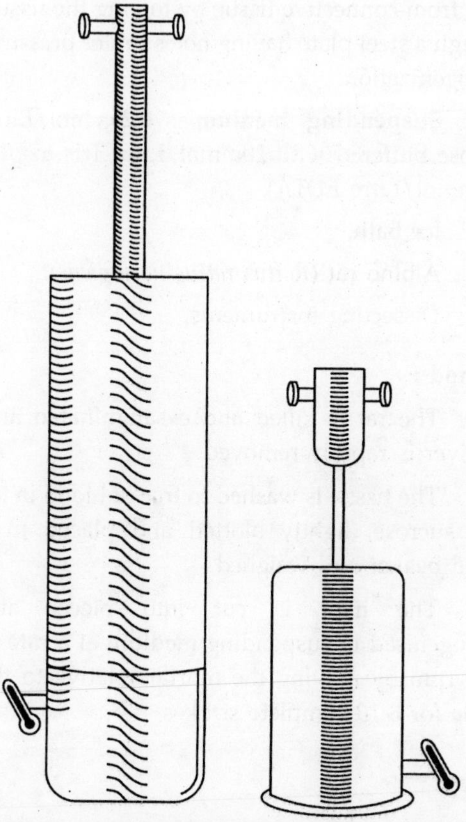

Fig. 12.3.2 : French press.

The mechanical procedures are improved by a variety of organic solvents (for phase separations) and/or detergents to assist the denaturation and separation of molecules (e.g. DNA from histones). When specific molecules are required, one must

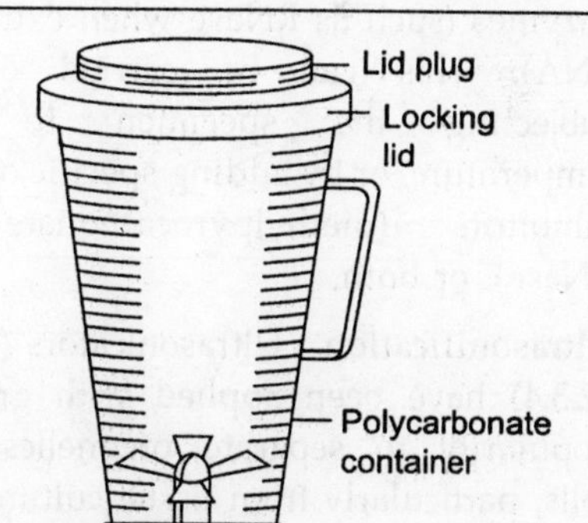

Fig. 12.3.3 : High speed blenders.

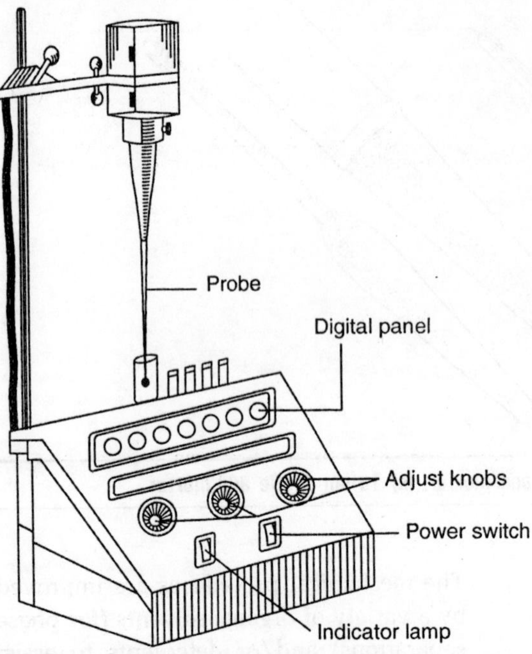

Fig. 12.3.4 : Ultrasonicator.

be careful to inhibit powerful degradation enzymes (such as RNase when extracting RNA). This can be carried on by subjecting the specimen to cold temperature, or by adding specific organic inhibitors (diethylpyrocarbonate for RNase), or both.

- **Ultrasonification :** Ultrasonicators (Figure 12.3.4) have been applied with growing popularity to separate organelles from cells, particularly from tissue culture cells. Light use of an ultrasonic wave can easily remove cells from a tissue culture substrate (such as the culture flask). It can also be attuned to merely separate cells, or to break open the plasma membrane and leave the internal organelles intact.

Homogenisation of rat liver

Materials required :

1. **Coaxial homogenizer :** A coaxial homogenizer consists of a glass mortar and a solid teflon pestle. The pestle is joined to an electric motor and small pieces of tissue materials are placed inside the glass mortar. The two component of the homogeniser are brought together and the mortar is gradually stirred up and down for six to eight strokes at the same time as the pestle rotates in a controlled speed (2000 rpm). As the homogenate is strained between the stationary wall of the mortar and the rotating pestle the tissue is subjected to a shearing force which is enough to rupture the cells excluding the organelles. These conditions are rather effective for liver but tougher tissues (kidney, skeletal muscle etc.) are to be primarily freed from connective tissue by forcing the tissue through a steel plate having holes under pressure homogenization.

2. **Suspending medium :** 0.25 mol/Litre Sucrose buffered with 20mmol/Litre Tris to pH, 0.1 mmol/Litre EDTA).

3. Ice bath.

4. Albino rat (*Rattus rattus norvegicus*).

5. Dissecting instruments.

Method :

1. The rat is killed and exsanguinated and the liver is rapidly removed.

2. The tissue is washed to free of blood in ice cold sucrose, lightly blotted and placed in a tarred beaker and weighed.

3. The liver is cut into pieces and homogenised in suspending medium at a rate of 2000 rpm by moving the mortar relative to the pestle for 8-10 complete strokes.

Chapter 12.4

Centrifugation

The most important advance in the use of centrifugal force to separate biologically important substances was the coupling of mechanics, optics and mathematics by T. Svedberg and J.W. Williams in the 1920's. Today, any technique employing the quantitative application of **centrifugal force** is known as centrifugation.Centrifugation is a technique of separation, which depends upon the activities of particles in an applied centrifugal field.

If a solution of large particles is allowed to stand the particles have a propensity to settle down under the influence of the gravity. For any known particle the rate at which the particles sediments is proportional to the applied force. In centrifugation there is a creation of centrifugal force, which is greater than the force of gravity thus as a result suspended particles sediments at a greater rate. Particles are generally suspended in a liquid medium and are kept within tubes or bottles which are positioned inside a rotor of the centrifuge. The rotor is placed centrally on the drive shaft of the centrifuge, particles that vary in density, shape or size can be separated since they sediment at different rates in the centrifugal field.

Svedverg (1920) discovered that the rate of sedimentation of the particle is directly proportional to the applied **centrifugal field, G** which act radially outwards.

Principle of centrifugation :

- **Centrifugal force (*G*)** is expressed in terms of gravitational force or G units of earth.

$$G = \omega^2 r,$$

where ω = angular velocity of the rotor expressed as radians/sec and r = radial distance expressed in centimetres.

Centrifugal force (G) in terms of rev/min is then:

$$G = \frac{4\pi^2 (\text{rev/min})^2 r}{3600}$$

- **Relative centrifugal force (RCF)** may be defined as the ratio of the weight of the particle in centrifugal field to the weight of the same particle when acted on any gravity alone.

$$\text{RCF} = \frac{4\pi^2 (\text{rev / min})^2 r}{3600 \times 981}$$

- **Net outward force (*F*)** experienced by a spherical particle (of known volume and density) at an angular velocity of ω radians/sec is given by:

$$F = \frac{4}{3}\pi r_p^3 \left(\rho_p - \rho_m\right)\omega^2 r$$

where $\frac{4}{3}\pi r_p^3$ = volume of sphere, with r_p = the radius of the sphere, ρ_p = density of the particle, ρ_m = the density of the suspending medium and r = distance of particle from the centre of rotation.

- **Frictional force (F_0)** opposing the motion is shown as :

$F_0 = vf$, where v = velocity or sedimentation rate of the particle, f = frictional coefficient of the particle.

'v' can also be given as :

$$v = \frac{dr}{dt} = \frac{2r_p^2(\rho_p - \rho_m)\omega^2 r}{9\eta}$$

where η = viscosity coefficient of the medium coefficient

- The sedimentation rate or velocity (v) of the particle can be expressed as sedimentation rate per unit of centrifugal field commonly designated as its **sedimentation coefficient(s)**. Sedimentation rate is proportional to $\omega^2 r$.

Therefore **sedimentation coefficient**,

$$s = \frac{v}{\omega^2 r} = \frac{dr/dt}{\omega^2 r}$$

Types of centrifuge:

Centrifuge can be classified into following groups :

(a) **Small bench centrifuge :** These are very simple and least costly. These are used to precipitate small amount of materials, which has a tendency to sediment rapidly. It has a speed range of 4000-6000 rpm (3000-7000 × g). These centrifuges operate in room temperature. Some modern design has also arrangement for cooling the rotors. These are also small microfuges having a speed of 8000-13000 rpm. These centrifuges are used to precipitate very small volumes (250mL – 1500mL).

(b) **Refrigerated large capacity centrifuge :** These have refrigerated rotor chambers and can separate particle in volume of 1050 cc and 100 cc tubes. Maximum speed is 6000 rpm (in a maximum relative centrifugal field of 6500 × g). The rotors of these centrifuges are generally mounted on a rigid suspension. These centrifuges are used to separate erythrocytes, coarse or bulky precipitate, nuclei, chloroplasts and other cells.

(c) **Refrigerated high speed centrifuge :** These centrifuges can rotate in a very high speed with a maximum speed of 25000 rpm producing a relative centrifugal force of about 60000 × g. They can hold a total volume of 1.5dL. These centrifuges can be used both to sediment small particles such as viruses, small organelle and ribosome's.

(d) **Continuous flow centrifuge :** These centrifuges are generally but high-speed forms. Rotors through which particles are suspended in medium flow continuously. In the process medium enters the rotating rotor particles get sedimented against its wall and excess, clarified medium overflows through an outlet port.

This type of centrifuge is used in precipitating yeast, bacteria in large volumes 10-100dL.

(e) **Ultracentrifuge :** Ultracentrifuges have a rotor which has a maximum speed of 80000 rpm (producing) a relative centrifugal field up to 600000 × g. The rotor chamber is refrigerated and evacuated so that there is no excess loss of rotor temperature caused by frictional resistance between air and spinning rotor.

Centrifugal tubes and their contents are accurately balanced to a range of 0.1g. As a safety measures the rotor chamber and ultracentrifuges are always enclosed in heavy amour plating.

These are used for separating major lipoprotein fractions from plasma, various cell organelles, nanoparticles proteins, cell organelles etc.

In this discussion we will restrict our self to the process of using a laboratory centrifuge.

(a) Separation of tissue constituents through centrifugation :

Materials required :

1. Fresh tissue.
2. Phosphate buffered saline (PBS).
3. 0.1N Perchloric acid.
4. 5% Trichlorocatic acid.
5. 80% alcohol.
6. Muslin cloth.
7. Refrigerator.
8. Refrigerated high speed centrifuge.
9. Beaker.
10. 5% Trichloroacetic acid.

Procedure :

1. About 5 g of fresh tissue is taken.
2. The tissue is chopped into small pieces.
3. The processed tissue is homogenized in 25 mL of PBS (pH 7.2) and then in 50mL of cold acid alcohol (80% alcohol in 0.1N Perchloric acid).
4. Homogenate is filtered through a 4-layer muslin cloth; filtrate is collected and kept under refrigeration.
5. Sample is taken in centrifuge tubes and balanced by putting equal amount of water in each tube so that each tube with its contents are of equal weight.
6. These tubes are placed in rotor of the centrifuge.
7. The lid of the centrifuge is closed.
8. The centrifuge is started. The speed is slowly raised gradually up to the required speed of 1000 rpm.
9. The centrifuge is run for required amount of time (10 minutes).
10. It is noted that the sample is precipitated as a pellet from the suspending medium. The supernatant is decanted in a clean labelled beaker.
11. The pellet is dissolved in another beaker having 20 mL of 5% trichloroacetic acid (TCA). This is the washing procedure.
12. The sample is again transferred to the centrifuge and the centrifuge is run at a high speed (say 5000 rpm) and for additional time (20 minutes).
13. The centrifuge is switched off after the required run. A clean well-formed pellet may be observed at the base of the centrifuge tube. The supernatant is decanted in a labelled beaker. This pellet can be used for further biochemical tests. Further the supernatant can also be used for biochemical assay.

(b) Low Speed Separation of blood Cells :

Materials required :

1. Clinical centrifuge tubes.
2. Laboratory centrifuge (see, Figure 12.4.1).
3. Phase contrast microscope.
4. Glass Pipette (attached to a dispenser) or automated pipette.
5. Hypodermic syringe with 23G needle.
6. Phosphate Buffered Saline (PBS) + 0.1% (w/v) Bovine Serum Albumin (BSA).
7. EDTA.
8. Albino rat (*Rattus rattus norvegicus*).

Procedure :

1. At least 5.0mL of blood is collected with a hypodermic syringe fitted with 23G needle containing EDTA as an anticoagulant through intracardial route of albino rat.
2. Added 5.0mL of EDTA anti-coagulated whole blood to a centrifuge tube. Mixed gently by inversion.
3. Centrifuged at 2,000 × g for 10 minutes at room temperature with a fixed angle rotor.
4. The centrifuge is decelerated slowly (one should not use a brake).
5. Keeping the centrifuge tube upright, an automated pipette or glass pipette is inserted into the tube and pushed until the MNC (mononuclear cells) band is displaced into the pipette.

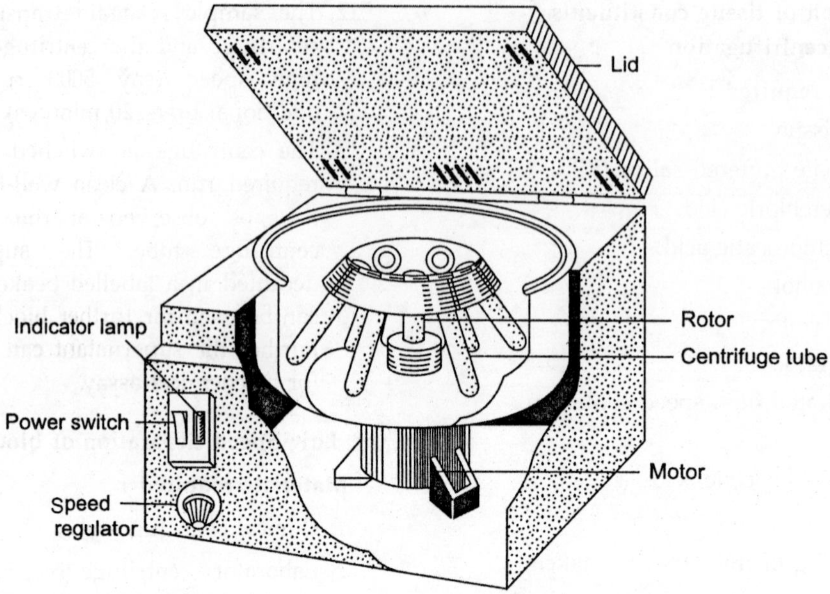

Fig. 12.4.1 : Laboratory centrifuge.

6. One should not collect more than 2.5mL into the pipette.
7. After the centrifugation of whole blood, there should be an opalescent compact band just below the meniscus of the tube. This band contains the mononuclear cells (lymphocytes and monocytes) and platelets. Below this band is the plasma, while the erythrocytes and polymorphonuclear cells will form a dark band at the bottom of the tube.
8. The contents are transferred through the pipette by inverting it into a 15 mL clinical centrifuge tube. Inside of the pipette is rinsed with 2.5mL of PBS and the washing is transferred to the clinical centrifuge tube.
9. The mononuclear cells are washed in the conical tube by adding 5mL of PBS containing 0.1% BSA for a total volume of 10 mL and mixed by inversion.
10. Centrifuged at 300 × g for 10 minutes at room temperature.
11. The mononuclear cells are resuspended in 5.0 mL of PBS. A simple wet mount of the suspended cells are made and the purity of the separation is examined using a phase contrast microscope.

Precaution :

1. Rotation of the centrifuge should not be stopped by hand.
2. Speed increments and decrement should be done gradually.
3. Pellet and supernatants should be properly labelled.

Chapter 12.5

Electrophoresis

Introduction :

Electrophoresis is one the most important technique for molecular separation of different macromolecules in a modern cell biology laboratory. It is extensively used as being easy and inexpensive. Generally in an electrophoretic apparatus molecules are allowed to migrate, all electrophoretic separations depend upon the charge distribution of the molecules being separated.

Electrophoresis can be :

1. *One dimensional* (i.e. single plane of separation), one dimensional electrophoresis is used for most routine protein and nucleic acid separations.
2. *Two dimensional* (i.e. double plane separation) used for finger printing, and when properly constructed can be extremely accurate in resolving all of the proteins present within a cell.

The support medium used in an electrophoresis can be shaped into a gel within a tube or it can be layered into flat sheets. The tubes are used for easy one dimensional separations (nearly anyone can make their own apparatus from inexpensive materials found in any lab), while the sheets have a larger surface area and are better for two-dimensional separations.

Electrophoretic mobility :

If a molecule bearing a charge '*q*' is present in an electric field of strength '*x*' then the force on the particle causing it to accelerate is *qx*. This is balanced by frictional resistance *f* to give a terminal velocity *v* : $qx = fv$

According to Stoke's law for spherical molecule of radius '*r*' moving through a medium of viscosity 'η', then, $f = 6\pi\eta r$

Combining these two equations, $qx = 6\pi\eta rv$

The electrophoretic mobility

$$\bar{v} = \frac{\bar{v}}{x} = \frac{q}{6\pi\eta r}$$

The mobility of the molecules as a result depends upon the viscosity of the medium (η) and the size, shape (*r*), and charge on the molecule (*q*).

Types of Electrophoresis :

- **Polyacrylamide gel electrophoresis (PAGE)** : Polyacrylamide gels are made through the polymerization of two compounds, acrylamide and N,N_1-methylene- bis-acrylamide

(Bis acts as a cross-linking agent for the gels). This polymerization is commenced by the addition of ammonium persulfate along with either β-dimethyl amino-propionitrile (DMAP) or N, N, N^1,N^1 tetramethylethylenediamine (TEMED). The gels are three-dimensional networks of long hydrocarbons cross linked by methylene groups which are of neutral, hydrophilic nature.

The separation of molecules within a gel is governed by the relative size of the pores formed inside the gel. The two factors, which determine the pore size, are, the total quantity of acrylamide present (known as %T) and the quantity of cross-linker (known as %C). Pore size is inversely proportional to the amount of polyacrylamide. Any increment or decrement in %C raises the pore size. Gels are measured as percent solutions. The total amount of acrylamide is stated as a % (w/v) of the acrylamide + the bis-acrylamide. Thus, an 7.5% T would indicate that there is a total of 7.5 g of acrylamide and bis per 100 mL of gel. A gel designated as 7.5T: 5%C is made up of a total of 7.5% (w/v) acrylamide + bis, and the bis would be 5% of the whole amount (with pure acrylamide composing the remaining 2.5%).

Proteins with molecular weight (MW) that range from 10,000 to 1,000,000 may be resolved and separated with 7.5% acrylamide gels, while proteins with higher MW need lower acrylamide gel concentrations. On the contrary, gels up to 30% have been used to separate small polypeptides. Smaller the pore size of the gel made from higher gel concentration is suitable for the separation of smaller molecules. 5% gels can be used for proteins ranging from 60kD to 200kD, 10% gels for a range of 16kD to 70kD and 15% gels for a range of 12kD to 45kD.

- **Sodium dodecyl sulphate PAGE (SDS – PAGE)** : Detergent SDS (sodium dodecyl sulphate) when used with proteins, proteins turn negatively charged by their tagging on to the SDS anions. The procedure of electrophoresis in presence of SDS tagged on PAGE is stated as SDS-PAGE (for Sodium Dodecyl Sulfate-Polyacrylamide Gel Electrophoresis). This technique has become a regular means for molecular weight determination of proteins.

- **Agarose acrylamide gel electrophoresis :** Acrylamide gels electrophoresis have become the standard procedure for protein separation but they are less appropriate for very high molecular weight nucleic acids (>200kD). So for proper separation of these large molecules, the acrylamide concentration needs to be lessened to a level where it remains liquid. On the other hand, by the addition of agarose (a naturally occurring linear polysaccharide), to the low concentration of acrylamide such electrophoretic separation of nucleic acid becomes feasible. In agarose when acrylamide concentrations of 0.5% is used molecular weights of up to 3.5×10^6 daltons can be separated. This is mainly useful for the separation of large sequences of DNA. As a result, agarose-acrylamide gels are used widely in today's molecular biology laboratory for the determination of gene maps.

For sake of simplicity only SDS-PAGE and Agarose gel electrophoresis are discussed in this chapter.

1. SDS PAGE electrophoresis of protein :

Step-I : Preparation of SDS-Polyacrylamide Gels :

Materials required :

1. Casting gel unit for electrophoresis (thoroughly clean and dry)
2. Siliconized pasteur pipettes
3. Syringes equipped with blunt, stub nosed, needles (easily manufactured from 22G needles with the points ground off. One should not use pointed needles when working with acrylamide)
4. Vacuum pump for degasing gels.
5. Micropipettes (10-300μL)
6. Stock 30% T : 0.8%C acrylamide monomer
7. 1.5M Tris-HCl buffer, pH 8.8.

8. 10% (w/v) SDS.
9. 10% (w/v) Ammonium persulfate.
10. Distilled water (d.H_2O).
11. TEMED, fresh from refrigeration.
12. Electrophoresis power pack and accessories.

Note : Temed acrylamide is a powerful neurotoxin. Do not breathe powder or otherwise come in contact with the monomer. Wear gloves at all times.

Stacking gel mixed just prior to use

- 2.66mL of Acrylamide monomer
- 5.0mL of Tris-buffer (0.6M), pH 6.9
- 0.2mL of 10% (w/v) SDS
- 12.4mL of d. H_2O

Separating gel (15%) mixed just prior to use

- 20mL of Acrylamide monomer
- 15mL of Tris-HCl Buffer (1.875M), pH 8.9 (22.7gm per 100mL of dH_2O)
- 0.6 mL of 10% (w/v) SDS
- Sterile d.H_2O : 4.40mL
- 24.1mL of d. H_2O

Procedure :

1. Slab gel unit is assembled with the glass sandwich set in the casting mode, and with 1.5mm (may vary) spacers in place. The assembly is held by bulldog clips. White petroleum jelly or 2% agar is applied around the edges of the spacers to hold them in place and to seal the chambers between the glass plates.
2. Separating gel is prepared from the ingredients listed above in sufficient volume (30mL for chamber of about 18 × 9 × 0.1cm).
3. Separating gel is added to a side arm flask; the flask is stoppered and attached to a vacuum pump equipped with a cold trap. The vacuum is turned on which degases the solution for approximately 10 minutes. During this period, the solution is gently swirled in the flask (in a magnetic stirrer).
4. The vacuum is turned off, flask is opened and 200μL of ammonium persulfate and 20μL of TEMED are added to the solution.
5. Stopper is added to the flask and degased for an additional 2 minutes while gently swirling the solution to mix the two accelerators This solution has to be used within a few minutes of mixing, or it will gelatinise in the flask.
6. Degased acrylamide solution is transferred to the casting chamber with a pasteur pipette. Centre of the glass chamber is filled gently with the solution by allowing the solution to run down the side of one of the spacers. One has to be careful not to introduce air bubbles during this step.
7. Level of the gel is adjusted in the chamber by inserting a syringe (equipped with a 22 gauge needle) into the chamber and removing excess gel.
8. Immediately the gel is water layered to prevent formation of a curved meniscus. Using a second syringe and needle, approximately 0.5mL of water is added to the chamber by placing the tip of the needle at an angle to a spacer and gently allowing the water to flow down the edge of the spacer and over the gel. Additional 0.5mL of water is further added to the chamber by layering it against the spacer on the opposite side of the chamber. Done correctly, the water will form a layer above the gel, and a clear line of segregation will be observed as the gel polymerises.
9. After 30 minutes, the gel should be polymerised. If degasing was inadequate, or the ammonium persulphate is not fresh in that case polymerization may need an hour or more. When the gel is polymerised, the gel is lifted in its casting chamber and tilted to decant the water layer.
10. **Stacking gel** is prepared from the listed ingredients (as stated above).
11. Stacking gel is degased as in step 3.

12. 75 μl ammonium persulfate and 10 μl TEMED is added to the stacking gel and degased for an additional 2 minutes.
13. Approximately 1 mL of stacking gel is added to the gel chamber and gently rocked back and forth to wash the surface of the separating gel. The stacking gel, which is still in liquid state, is poured off and disposed of properly. Fresh stacking gel is added to nearly fill the chamber and allowed room for the insertion of a teflon comb required to figure sample wells. Teflon comb is carefully inserted into the chamber [Teflon combs come in a variety of configurations for sample volume and number of lanes (3, 5, 10, 15 or 20)]. The volume of the stacking gel is adjusted as needed to completely fill the spaces in the comb.

    One has to be careful not to trap any air bubbles beneath the combs. Oxygen inhibits polymerization, and will subsequently result in poor protein separations.
14. The gels are allowed for polymerising for at least 30 minutes prior to use.

Note : Remember that liquid acrylamide is extremely hazardous!

Step-II : Separation of Protein Standards :

Materials required :

1. 10% SDS-polyacrylamide gel as prepared in step I.
2. *Standard Marker proteins includes :*
   (a) α-Lactalbumin,
   (b) Trypsin inhibitor soybean,
   (c) Trypsinogen,
   (d) Carbonic anhydrase,
   (e) G6PD rabbit,
   (f) Egg albumin, g.Bovine albumin.

   All the above proteins are added to 2X sample buffer at a concentration 1mg/mL.
3. Similarly proteins samples are similarly dissolved in 2X sample buffer concentration may range from 1-4mg/mL (see point 4). As a general practice sample solution are heated in boiling water for 2-3 minutes to ensure complete interactions between proteins and SDS.
4. *2X-SDS Sample Buffer :* Dissolved 1.52g of Tris base, 2.0g of SDS, 20mL of glycerol, 2.0mL of β-mercaptoethanol and 1mg of bromophenol blue to a final volume of 100mL with water.
5. *1X-SDS Electrophoresis Running Buffer (Tris-Glycine + SDS) :* 0.5M tris (12g), 0.192M glycine (28.8g), 0.1% SDS (2g) is added to 2L of distilled water. pH is adjusted to 8.3.
6. 0.001% (w/v) Bromophenol Blue.
7. Micropipettes with flat tips for electrophoresis wells.

Procedure :

1. Teflon combs from prepared gels are removed by gently lifting the combs from the chamber. The wells are rinsed (formed by the removal of the combs) with distilled water or 1X TGS buffer and the water is drained off.
2. Wells and the chamber are filled with running buffer.
3. Aliquots of a known protein standard are prepared by mixing equal parts of the protein standard with 2X sample buffer.
4. Using a micropipette, the standard marker protein sample and protein samples are added to the bottom of a well. (Note: The SDS/sample buffer contains glycerol and consequently the sample will have a higher density than the SDS/separating buffer. It is best to use a flat tipped micropipette designed specifically for this application, but any flat tipped syringe needle will also work. The volume of sample will depend on the size of the wells, but typically is 25-100μL. Some experimentation will be necessary to obtain the proper volume.

5. Same volume added as used in step 4 of SDS-sample buffer or Bromophenol blue to a separate well (Bromphenol blue is known as "tracking" dye and is used to monitor the progress of the electrophoresis separation). A drop of 0.1% phenol red may be substituted. Alternatively, prestained protein standards may be used without a tracking dye, although this is not recommended.
6. The gel is removed from its casting stand and assembled into the appropriate slab unit for running the electrophoresis. One has to ensure to follow the manufacturer's directions for assembly.
7. Sufficient quantity of running buffer is poured into both the lower and upper chambers of the electrophoresis apparatus until the bottom of the gel is immersed in buffer, and the top is covered, while the electrodes reach into the buffer of the upper chamber. One has to careful not to disturb the samples in the wells when adding buffer to the upper chamber. In many laboratory, samples are loaded after addition of buffers in the lane.
8. The top of the electrophoresis apparatus is assembled and connects the system to an appropriate power source. One has to be sure that the cathode (–) terminal is connected to the upper buffer chamber.
9. The power supply is turned and the gel is run at 20mA constant current per 1.5mm gel. For example, if two gels are run, each having 1.5mm spacers, the current should be adjusted to 40mA. One gel with 1.5mm spacers should be run at 20mA, while a gel with 0.75mm spacers should be run at 10mA.
10. When the tracking dye reaches the separating gel layer, the current is increased to 30mA per 1.5mm gel.
11. The current is continuously applied until the tracking dye reaches the bottom of the separating gel layer (approximately 4 hours).
12. The power supply is switched off and disconnected. The gel apparatus is disassembled and the glass sandwich containing the gel is removed. The sandwich flat is placed on paper towels and the clamps from the sandwich are carefully removed.
13. Working on one side of the sandwich, the spacers are carefully slid out between the two glass plates. Using the spacer or a plastic wedge as a lever, the glass plates are gently pried apart without damaging the gel enclosed within.
14. Bottom glass plate with the gel is lifted and the gel is transferred to an appropriate container (10″ × 10″ × 2″, glass baking dishes are useful for this purpose, or ordinary refrigerator containers of appropriate size. While the former will need a covering of plastic wrap, the latter come with lids) filled with buffer, stain or preservative. The gel may at this point be used for Coomasie Blue staining (described in the next part; step-III). The gel is placed into 50% methanol and the container is gently rocked for about 30 minutes prior to scanning. This can be accomplished by placing the gels into a flat dish and gently lifting the edge of the disk once every 30 seconds. There are commercially available rocker units for this purpose. If the gel is to be dried, one should use a commercial gel dryer. The manufacturer's directions are to be followed. The **relative mobility** of each protein is plotted against the log of its molecular weight.

Relative mobility is the term used for the ratio of the distance the protein has moved from its point of origin (the beginning of the separating gel) relative to the distance the tracking dye has moved (the gel front). The ratio is abbreviated as Rf. Molecular weight is expressed in daltons.

Step-III : Coomasie Blue Staining of Protein Gels :

Materials required :

1. Protein gel from above step II.
2. 25% (w/v) Coomasie Brilliant Blue R 250 in methanol-water-glacial acetic acid (5:5:1 v/v), filtered immediately before use. 7% (v/v) acetic acid: 2.5 g of Coomasie Brilliant Blue R250 is dissolved to a final volume of 1 liter with 20% (w/v) trichloroacetic acid (TCA). Some investigators use a 0.25% solution of Coomasie Blue in methanol-water-glacial acetic acid (5:5:1).
3. Commercial destaining unit (Optional).

Procedure :

1. Gel prepared is placed in at least 10 volumes of Coomasie Blue staining solution for 2-4 hours. Gently rocked to distribute the dye evenly over the gel.
2. At the conclusion of the staining, the gels re-washed with several changes of water.
3. The gels are placed into a destaining solution of methanol-water-glacial acetic acid (5:5:1 v/v) solution for at least 1 hour.
4. If the background is still deeply stained at the end of the hour, the gels are moved to fresh destaining solution as often as necessary. If a commercial destainer is available, this will decrease the time required for stain removal. One should follow the manufacturer's directions for use of the destainer. The gels are placed into containers filled with 7% acetic acid as a final fixative.
5. The gels are photographed or analysed spectrophotometrically.

Note : Coomasie Brilliant Blue R 250 is most commonly applied in staining procedure for the detection of proteins. It is the method of choice if SDS is employed in the electrophoresis of proteins, and is sensitive for a range of 0.5 to 20 mg of protein. Within this range, it also follows the Beer-Lambert law and thus can be qualitated and measured. The main drawback is the length of time for the method and the necessity for destaining (As overstaining results in a significant retention of stain within the gel, and thus a high background stain, which could wipe out the bands). If maximum sensitivity is required, one should try 2 hours for a 5% gel and 4 hours for a 10% gel. Destaining must be observed visually and adjusted accordingly.

2(a) DNA electrophoresis in agarose gel :

Agarose is a polysaccharide composed of galactose and a derivative of galactose, with the fibres crosslinked by H- bonds.

DNA molecules are separated by electrophoresis on the basis of size, shape and magnitude of net charge on the molecules. A base intercalating dye such as Ethidium bromide intercalates between the bases of the DNA and produces an orange fluorescence when exposed to UV radiation.

Different lengths of DNA fibre can be fractionated by using a variety of agarose concentrations.

One has to use 1.4% agarose for <0.5 kb (size of DNA)

One has use 1.0% agarose for 0.2-30 kb (- do -)

One has use 0.7% agarose for 10.0 kb (- do -)

One has use 0.3% agarose for 10.0 kb (- do -)

Materials required :

1. Tris-borate buffer : 0.9M Tris-HCl– 113.0 g, 0.025M EDTA Na_2– 9.3 g, 0.9M Boric acid– 55 g and distilled H_2O 1 Litre.

2. Agarose : 1.0 (w/v) in single strength, Tris borate running buffer. Autoclaved to dissolve agarose;then maintained at 50° C in a stoppered flask until used. Agarose used should be of very high quality(Sigma Type I grade)

3. Gel loading solution : Sucrose 30%, Bromophenol blue 0.25% and xylene cyanole FF 0.25% (all w/v) in a single strength Tris-borate buffer.

4. DNA preparation.

5. Standard DNA for size determination (λ DNA nicked with restriction endonuclease EcoRI and Hind III).

6. Gel casting plate, gel tank.

7. Power pack.

8. UV transilluminator with photography system, safety glasses (thick large plexiglass eye shields or special goggles when working with UV, as the latter damages the eyes and skin tissues).

9. Microsyringe.

10. Parafilm.

Procedure :

1. A wall is formed around a clean dry gel casting plates (10 × 6 × 0.5 cm) using zinc oxide tape. This helps in forming a leak proof wall about 1 cm high all round the plate. This setup is placed perfectly horizontal over a leveled plate.

2. 30 mL of 1% agarose solution (maintained at 50°C) is poured on to the casting plate. Immediately the suitable well-forming combs are placed about 1cm from one end of the plate. One should keep a note the teeth of the comb should never touch the glass plate. The gel is allowed to set of 1 hour.

3. The comb is removed from the gel carefully. The gel is transferred along with the glass plate to the electrophoresis tank such that the wells are near the cathode. Single strength Tris–borate buffer is poured into the tank until the gel is shrunk.

4. The electrodes are connected to the power supply with cathode terminal (electrical terminal) at the well end of the gel.

5. The DNA samples are loaded and the standards are taken into the gel loading solution using a microsyringe.

6. The power supply is turned on and run at 100V (10 -15 mA).The progress of the fast running tracking dye is to be monitored during the electrophoresis. The run is terminated when the tracking dye is about to leave the gel.

7. After the disconnection of the power supply the gel is transferred to a staining tray containing 250 μg of ethidium bromide (50μL of 5mg/mL solution) in 250 mL of used Tris buffer. The gel is stained for about 30 minutes. One must wear a gloves while handling ethidium bromide.

8. The gel is then transferred onto a thick plastic sheet and placed on a UV transilluminator and the gel is viewed in UV light. The UV light is shined on the gel in the dark. The DNA bands will fluoresce with a reddish orange colour.

The positions of the bands are noted relative to those in the marked lanes. A rough estimate of the size of the DNA in each band is made in this way. The distance traveled by each band from the wells is measured. The distances moved are plotted against the log molecular weight of standards to draw the calibration curve. The size of the DNA (and/or restricted fragments) is deduced using the calibration curve.

9. The gel is photographed using a UV light source and a red filter.

Note : One may preserve the gel for long periods submerged in dilute buffer solution in a covered large petridish. The edges of the cover are sealed with parafilm.

2(b) Recovery (elution) of DNA fragments from agarose gels :

Several methods have been developed in recent years for the recovery of DNA fragments from agarose and polyacrylamide gels following electrophoresis. However, no single method has been universally adopted. The following is one of the many methods for the recovery of DNA by electroelution that could work in the subsequent gene manipulation experiments.

Materials required :

1. TBE Electrophoresis Buffer.
2. 3M Tris – HCl (pH 7.5).
3. Dialysis Tubing.

Procedure :

1. After electrophoretic fractionation, stained DNA bands are visualized in UV light in transilluminator. The agar containing the DNA is cut out with a sterile scalpel.

2. The agar is transferred inside a piece of pre-treated dialysis tubing (dialysis tubing is sterilized for 30 minutes at 120°C in a solution of 5% w/v sodium carbonate, 1mM EDTA and kept at 4°C in sterile distilled water). Electrophoresis buffer is added in the dialysing tubing so that agarose is surrounded by buffer without any air bubble. Maximum volume of buffer is used in the dialysis bag.

3. The dialysis bag is placed at bottom of a gel apparatus filled with buffer in such a way that the piece of agar is in the same position with respect to the electrodes as it was in the gel.

4. 100-120 mA current is applied for nearly 2 hour to elute DNA out of the gel.

5. At the end, the current is reversed for 30 seconds in order to mobilize any DNA, which may get stuck to the dialysis membrane.

6. The buffer containing the electroeluted DNA is transferred through cotton-plugged 1mL tip into an Eppendorf tube by centrifuging for 2 minute. This process will remove contaminating agarose particles. .

7. The supernatant is extracted with phenol, then with chloroform: isoamyl alcohol (24:1 v/v) to remove any soluble contaminants.

8. The DNA solution is mixed with 3M Tris – HCl (pH 7.5) to give a final concentration of 0.5M. Two volumes of isopropanol is added and stand it at –20°C for 30 minutes.

9. Centrifuged for 10 minutes at 15000 × g. The precipitation step is repeated once more to remove any remaining ethidium bromide.

10. The DNA pellet is dried under vacuum and resuspended in a suitable buffer for any subsequent experiment.

Chapter 12.6

Autoradiography

Introduction :

Ionising radiations operates on photographic emulsion to create a latent image similar to visible light. For an autoradiograph a radiation source (radioactivity) emitting from within the body or any biological material or DNA/RNA/Protein gel (separated electrophoretically) to be imaged is required to strike through a sensitive emulsion. The emulsion have large amount of silver halide crystals embedded in as solid phase such as gelatin. As energy from the radioactive source is dissipated from the emulsion the silver halide becomes negatively charged and thus reduced to metallic silver thus forming particulate latent image. Photographic developers are designed to display these silver grains as blackening of the film and the fixers to remove any remaining silver halide. Thus a permanent image is of the location of the original radioactive event is formed. This process is known as **autoradiography**. Technique of autoradiography is very important in the field of molecular biology. In general weak β-emitting isotopes (e.g. ^{3}H, ^{14}C and ^{35}S) are most suitable for autoradiography. For location of DNA in tissue ^{3}H is used through metabolic labelling of organism.

Tritiated thymidine (^{3}H-thymidine) is frequently used, for instance, to study the synthesis and location of DNA. Thymidine is a soluble base which is specifically found in DNA. It is integrated into the macromolecular structure of DNA of cell nucleus of a tissue (viz., onion root tip) all through the synthesis and chromosomes replication. After fixing the cells for standard histological examination, the DNA molecules (with their incorporated, radioactive thymidine) are precipitated or cross-linked as permanent parts of the cell. Un-incorporated thymidine is removed from the cell, as it remains soluble and is thus removed in the tissue washing procedures.

After the tissues are sectioned and applied to a glass slide, they will have radioactive nuclei, that is those nuclei that were in the S-phase of division at the time of exposure of the cells to ^{3}H-thymidine. Radioactive sources can be detected directly, if a photographic emulsion is applied over the section, it will become exposed by the radioactive source.

After the photographic emulsion is subsequently developed, the exposed parts of the emulsion will contain reduced silver grains which are directly proportional to the amount of radiation being given off beneath it (in the nuclei as stated in the example).

If the exposed, developed slides are then examined under a microscope, there will be two aspects of interest. Firstly, focusing the microscope on the tissue itself will give an observation of the tissue and cell architecture. If the focus of the microscope is shifted upward, however, the cells will go out of focus, precisely as the photographic emulsion comes into focus. Inside the emulsion will be regions of reduced silver grains and clear areas containing no silver grains. By alternating the focus on the tissue and the emulsion, those nuclei that are radioactive can be readily observed and identified. If required, the number of silver grains can be calculated to provide a quantitative measure of ^{3}H- thymidine incorporation (and thus reflecting DNA synthesis). The quantity of silver grains produced depends on the type of photographic emulsion and the kind of ionizing particles emitted from the cell. **Alpha particles** produce linear, intense tracks a few micrometers in length. **Gamma rays** produce long haphazard tracks of grains and are ineffective for autoradiograms. **Beta particles** or electrons create single grains or tracks of grains. High energy Beta particles (such as those produced by ^{32}P) may move more than a millimetre before producing a grain. Low energy Beta particles (H and ^{13}C) generate silver grains within a few micrometers of the radioactive disintegration site and so offer very suitable resolution for autoradiography technique.

Electrophoretically separated radioactive macromolecules (DNA/RNA/Protein) in gel are placed on porous polyethylene or filter paper sheets and dried under vacuum. Autoradiograph of these dried gels is performed by sandwiching the dried gel and a piece of X-ray film between two pressure plates. After an appropriate exposure the X-ray film is separated from the gel and developed. For precise quantification the gel should be divided into small pieces and the amount of radioactivity per piece is determined with a scintillation spectrometer.

1. Autoradiography of Onion root tip ^{3}H-thymidine incorporated DNA :

Materials required :

1. ^{3}H-Thymidine, specific activity of 2.0 Curie/millimole.
2. Onion sets, jars and toothpicks.
3. Alcohol-acetic acid fixative.
4. Materials for Feulgen reaction.
5. Carnoy fixed, paraffin sections of tissue.
6. Oven at 60°C.
7. 1N HCl.
8. Schiff's Reagent.
9. Sulphurous acid.
10. Graded ethanol series, 50, 70, 95 and 100% (v/v) and distilled water.
11. Xylol, DPX and coverslips.
12. Microscope.
13. Paraffin embedding, sectioning equipment.
14. Kodak Nuclear Track Emulsion (or equivalent).
15. Darkroom.
16. Water bath at 42°C.
17. Kodak D19 Developer (to be kept in dark bottle).
18. Kodak Fixer (to be kept in dark bottle) and drierite.
19. Xylol, Permount.
20. Coverslips, beakers.
21. Microscope.
22. Alluminium foil.

Procedure :

1. Literature on precautions for the safe handling of radioactive materials has to be read first.

2. Toothpicks are placed into the slides of an onion and the onion is set into the top of a small

jar or beaker. The jar is filled with water and the onion is allowed to begin root development (3-7 days). Mild aeration of the water will assist in the growth of the roots.

3. Another beaker is set, containing a 10 mci /mL solution of ^{3}H-thymidine such that there are approximately 10 µci/mL, the onion and its growing roots are transferred to the jar containing the radioactive thymidine.

4. The roots are allowed to stay in the radioactive solution for 1 hour at room temperature.

5. Onion is removed and the bulb is held over a beaker containing water. The onion roots are rinsed by dipping several times in the beaker. The bulb is transferred to yet another beaker of aerated water and allowed to remain in this beaker for four hours.

6. Roots tips are cut off and fixed in alcohol-acetic acid (3:1) overnight. The fixative changed after the first four hours.

7. Roots are washed in water for several minutes and placed in 1N HCl at 60°C for 12 minutes.

8. The root tips are stained with the **Feulgen Reaction** as given in following steps:

- Slide is extracted in 1N HCl at 60°C (1 hour).
- Slide is extracted in distilled water at 60°C (1 hour).
- Rinsed gently in fresh distilled water (1 minute).
- Stained in Schiff's reagent (30 minutes).
- Bleached with sulphurous acid (10 minutes).

9. Root tips are dehydrated, embedded in paraffin and sectioned at 10-15 microns. Mounted on microscope slides.

10. Deparaffinized in xylol and sections are dehydrated by passing first through a series of alcohols and finally in two changes of water.

11. In a dark room some liquid autoradiographic emulsion is melted at a water bath at 42°C.

12. Two slides are placed back-to-back and dipped slowly into the melted emulsion. Removed, allowed to drain and placed in an appropriate light-proof container and to dry in a vertical position.

13. When dry, slides are placed into an opaque slide box containing drierite. Wrapped in aluminium foil and placed in a refrigerator.

14. The slides must stay in the refrigerator until a proper exposure has been made. This can vary from 5 days to over two weeks. After one week, a trial slide should be developed using Steps 15 and 16 below. The slides are examined at 10X with a bright field microscope and looked for the presence of black silver grains located over the cells. A correct exposure is determined by the appearance of silver grains over the cells, although few or no silver grains sited in areas without cells. If the appearance of the trial slide is acceptable, then the rest should be processed immediately.

15. Developed in the autoradiogram emulsion in the darkroom as follows:

- Developed in Kodak D-19 Developer at 20°C for 3 minutes.
- Washed for 10 seconds in distilled water.
- Fixed with Kodak Fixer for 3 minutes.
- Washed in running water for 15 minutes.
- Dehydrated by placing in 95% alcohol for 3 minutes.
- Placed in 100% alcohol for 3 minutes.
- Cleared through two changes of xylol for 3 minutes each.

16. Coverslips mounted with DPX.

17. Autoradiograms are drawn and labelled. The percent of cells actively undergoing DNA synthesis during the time of exposure to radioactive thymidine is calculated.

Note, that the tissue and the silver grains are in different planes of focus and you will need to constantly switch focus from one plane to the other.

2. Autoradiography of hybridized Southern blot

Materials required :

1. Dried hybridized Southern blot.
2. Autoradiogram cassettes.
3. X-ray film.
4. –70°C deep freezer.
5. Developer solution (to be kept in dark bottle).
6. Tray.
7. Fixer solution(to be kept in dark bottle).

Procedure :

1. Autoradiogram cassettes of appropriate size, the hybridized Southern blot, X-ray film etc. are all taken to the dark room.

2. The blot is first put on the lower lid of the cassettes followed by X-ray film. The X-ray film is then covered by intensifying screen provided in the cassette. The upper lid is then closed tightly. All this has to be done in the dark room under red light.

3. The cassette now can be brought in lighted room and should be kept in –70°C deep freezer for requisite time.

4. After the requisite exposure the cassette is brought out of –70°C box and kept at room temperature for some time and then developed in the dark room.

5. Developing is done as follows:

(a) The developer solution is poured in a plastic tray first. The X-ray film taken out from the cassette and sunk in developer solution.

(b) The tray is mildly shaken and after about 1 minute or so the X-ray film is taken out and washed in tap water.

(c) Afterwards the X-ray film is put in the fixer solution already poured in another tray.

(d) Shaken gently until the developer goes away.

(e) The X-ray film is then washed and air dried.

Rules for safe handling of radioactive isotopes :

1. All work with radioactive materials must be done in a tray lined with absorbent paper.

2. All glassware and equipment in contact with the radioactive material must be correctly labelled and kept within the tray. The only exception is that microscope slides of labelled cells may be removed from the tray after the drop of labelled cells has been applied to the slide and allowed to dry.

3. Plastic gloves should be worn when handling radioactive materials. All waste solutions containing radioisotopes, all contaminated gloves, paper, etc., must be placed in appropriate liquid or dry radioactive waste containers.

Note : H^3-thymidine can be used to trace sites of synthesis and utilization of DNA. Thymidine, the deoxyribose nucleoside of thymine, can be purchased with the tritium label conjugated to the methyl group of thymine. Thymidine is particularly incorporated into DNA. Some organisms can eliminate the methyl group from thymine, and incorporate the uracil product into RNA. Still in this case RNA would not be labelled because the tritium label would be removed with the methyl group. Methyl labelled thymidine, therefore, serves as a very specific label for DNA.

Chapter 12.7

Southern, Western and Northern Blotting Technique

A. SOUTHERN BLOTTING :

The Southern Blotting technique that encompasses the transferring of DNA fragments from a gel slab to nitrocellulose (cellulose nitrate) filter paper (0.45mm) either by capillary action (capillary blotting) or by electrophoresis (**electroblotting**). This is followed by the hybridization with a radioactive probe, and identifying the hybridized DNA by autoradiography. This "blotting" technique was first described by Southern in 1975 and is designated "southern blotting".

The technique relies on the strong affinity of DNA for nitrocellulose paper. The DNA to be investigated is fractionated by agarose or polyacrylamide gel electrophoresis. The gel slab is then placed in a Southern blotting assembly (**Figure 12.7.1**). The upward capillary movement of the buffer through the gel slab drives each DNA band in the gel into the overlying nitrocellulose paper. Thus the DNA get bound itself to the paper, while the buffer solution moves upwards and towards the dry filter and absorbent paper and evaporates. In this fashion, the DNA bands are blotted precisely on to the nitrocellulose paper.

The DNA bands in the gel slab are eluted into a closely apposed nitrocellulose paper through a traveling front of buffer solution. To fix the dissociated DNA fragments firmly, the blotted paper is baked at 65-80°C in a vacuum oven, for about 2-3 hours. The southern blot is now ready to be screened with a probe.

It can then be hybridized using a radioactively labeled (mRNA/nick-translated DNA) probe to identify the appropriate gene among the fragments.

Materials required :

1. Nitrocellulose filter papers.
2. Whatman 3 filter papers.
3. Denaturing solution (composed of)

NaOH (0.5M)	20g
NaCI (1.5M)	87.5g
Water to	1 Litre

4. Neutralizing solution (pH 7) (composed of)

Tris-HCl (0.5M)	60.5g
NaCl (3.0M0	175.3g
Water	1 Litre

5. Saline sodium citrate (SSC) pH 7 (coposed of)

Sodium citrate (0.3M)	88.2g
Sodium chloride (3M)	175.3g
Water	1 Litre

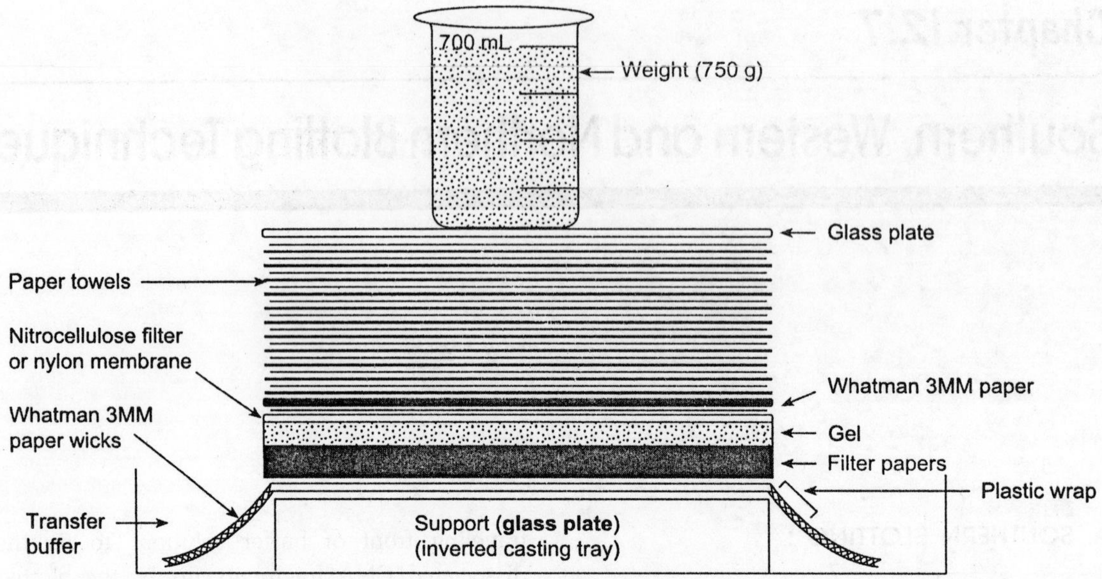

Fig. 12.7.1 : Southern blot assembly.

Prepared 20 fold concentrated SSC and store after autoclaving.

6. A boilable plastic lunch bag.

7. Hybridization solution (Formamide 50% SSC 4X).

8. Radioactive probe : ^{32}P-RNA or DNA denatured for 10 minutes at 100°C in SSC 0.1 X.

9. Water bath at 65°C.

10. Radiocative waste container.

11. SSC 0.5X, 2X.

12. RNaseA (10mg/mL) in SSC 2X.

This has to be boiled first at 80°C for 30 minutes and then cooled slowly.

13. Vacuum oven.

14. Glass tray.

15. Tris base (pH 9).

Procedure :

1. The agarose gel, following electrophoresis is transferred to an appropriate large dish. A thin sheet of plastic, a little larger than the gel is used for this transfer.

2. The gel is then immersed for 15 minutes in the denaturing solution.

3. Replaced with fresh solution and continued soaking for another 15 minutes.

4. The procedure is repeated twice with neutralizing solution.

5. In a suitable glass tray a glass plate is placed (somewhat larger than the gel) on a platform at a height of about 2 cm.

6. Three large pieces of Whatman 3MM paper is placed on the top to serve as a wick. Next added 10X SSC to the tray so that the wicks get wet completely.

7. Over the wick ten sheets of paper (that are cut exactly the same size of the gel to be blotted) are placed. These sheets are made wet and any air (between the sheets) are pressed out of it bubbles.

8. The pre- treated gel is placed on top of the wick (stack) between two plastic spacers across the wick.

9. The nitrocellulose filters are cut to the same width but slightly longer than the size of

gel and these are gently floated on 2X SSC, thoroughly made wet and immersed down. The surface of the gel is made wet with 2X SSC. The wet nitrocellulose is placed on the gel starting at the middle and extending along both the ends finally the ends lying over the spacers. Any air bubbles are removed between the gel and filtered carefully with the help of forceps.

10. Five sheets of Whatman 3MM paper (cut into gel size) are placed over the gel and then a pile of tissue papers (paper towels) are placed over Whatman paper. A glass beaker (with 700 mL of water or chemical catalogue) or about 700 g weights) is placed over the top.

11. The transfer is allowed to proceed overnight or even longer. The buffer moves upward by capillary action passing through the gel and filter. During this course the DNA fragments are transferred out of the gel and retained (blotted) on the nitrocellulose because of its higher affinity, thus making a replica of gel pattern.

12. The assembly is dismantled and the filter is recovered. Soaked in 2X SSC for 15 minutes, observed under UV illumination. The ethidium bromide will transfer along with the DNA so it would also be feasible to visualize some pattern.

13. The filter is dried by placing between two layers of 3MM paper and placing within a vacuum oven (2 hours at 80°C). Nitrocellulose filter should not be dried out in a hot air oven as the filter may explode.

14. The DNA will be immobilized by this drying.

15. The dried filter may be further analyzed by hybridization (southern hybridization) using a radioactivity labeled DNA as a probe to identify a particular gene.

16. **Hybridization** is done as follows :

(a) One end of the plastic bag is cut and the southern blotted paper is inserted.

(b) 15 mL of hybridization solution along with the probe solution is poured.

(c) All air bubbles are forced out of the filter paper.

(d) Three side of the bag is heat sealed.

(e) The bag is incubated in 65°C (temperature depends on homologous strand, T_m value of probe etc.) water bath for 18-48 hours.

(f) The bag is removed from the bath one end is cut open,the solution is poured into the radioactive waste container. The probe may be reused.

(g) The filter paper is washed at room temperature in 200 mL of SSC 0.5 X.

(h) If the probe that is used is RNA one has to remove all unhybridized RNA with RNaseA.

(i) The hybridized filter paper is washed in 200 mL of SSC 0.5 X for 15 minutes with gentle shaking.

(j) The step 16(i) is repeated.

(k) Washed in Tris base (pH 9) 2-3 times for 5 minutes each.

(l) The paper is dried.

(m) The hybridized Southern blot is autoradiographed to visualize the hybridized bands of DNA as exposed region.

Notes :

1. The gel and nitrocellulose filter should be grease free. One should wear disposable gloves and use forceps whenever handling the filter.

2. There various ways by which blotting is set. One has to be certain in any case that the buffer flows through the gel and filter. The upper layer of papers should not be in contact with the bottom layers.

3. Hybridization is feasible only in the single stranded DNA on the paper. Thus it is important for the gel to be treated before the transfer step, with a mild alkali (5M NaCl and 0.5M NaOH), and then neutralized using a NaCl solution in a tris HCl buffer (pH 7.4).

4. **Hybridization** with the probe is carried out by placing the blotted paper in a boilable plastic bag, containing a measured volume of the solution, with the probe (pre- hybridizing and hybridizing solutions). The prehybridizing solution have **Denhardt's reagent** and some denatured heterologous DNA. The hybridizing solution is composed of the the probe in SSC (standard sodium citrate) or SSPE (standard sodium phosphate EDTA) buffer. Denhardt's reagent which contains Ficoll polyvinyl pyrollidone and BSA (bovine serum albumin), stops ssDNA from binding itself to the filter paper without hampering hybridization of the bound ssDNA with the probe. The filter paper is washed and dried through appropriate methods and autoradiographed. As the patterns of bands on the nitrocellulose paper are identical to those which were on the agarose gel, the location of the DNA containing the wanted DNA sequences is detected and identified easily by finding for the band that has exposed the film.

B. WESTERN BLOTTING :

The technique for transfer of protein bands from an acrylamide gel onto a more stable and immobilizing support matrix is called as protein blotting or **Western blotting**. The method originated from the laboratory of George Stark at Stanford. The name **western blot** was given to the technique by W. Neal Burnette. Different types of supporting matrices such as nitrocellulose, diazobenyloxymethyl (DBM) cellulose sheets, nylon filters, PVDF membrane etc. are used for the purpose. In polyacrylamide gel proteins mixture can be separated on gel and therefore further analysis of proteins or their recovery is cumbersome. Proteins thus fractionated on polyacrylamide gels, are transferred to nitrocellulose or DBM paper that has been appropriately pretreated. One precaution has to be taken if DBM paper is used. As glycine reacts with the diazonium groups in DBM paper, the latter has to be washed free of glycine.

The analytical methods applied for such identification include: (i) immunodetection, (ii) binding of proteins to ligands immobilized on the filter paper, and (iii) binding of tagged molecules to proteins immobilized on the filter paper.

For the purpose of immunodetection, a protein on the DBM paper is treated with the antibody for the protein (antiserum), which can be labeled or unlabelled. If unlabelled, another labeled antibody that go with the immobilized one may be employed after the proteins are transferred to the paper. The labeled antibody probe will bind with the unlabelled ones. The complexes may be identified by a variety of ways.

The second method of detection engages the interaction of a ligand immobilized on the DBM paper with the protein to be identified. The ligand may be an antibody or an antigen or a lectin. The transferred protein has an affinity for its matching ligand and consequently forms a protein–ligand complex. The later may be detected through the process of autoradiography if the protein has a radioactive label. This immunodetection technique is a variant of the filter paper affinity one, where immunodiagnostic methods are employed to identify the wanted molecules.

In the third technique where the location of the protein and the binding molecule with respect to the blotting paper are reversed. In this case the protein is immobilized on the paper, and then treated with the labeled molecules (^{32}P – tagged nucleic acids or ^{125}I – tagged proteins). Protein-protein interaction and binding may be identified through labeled antisera, and protein-nucleic acid organization through autoradiography.

The effectiveness with which a particular antigen will be transferred to the blot depends upon the protein binding capacity of the membrane used, the transfer method and conditions employed as well as the nature of the antigen itself. To maximize transfer efficiency, some knowledge of the physical properties of the target antigen is beneficial. With respect to the efficiency of transfer, the most important properties are size i.e. molecular weight (MW) and hydrophobicity of the antigen. While SDS is required to facilitate migration of the protein out of the gel in response to an electric current, SDS can also interfere with binding of the protein to the membrane itself (this is particularly an issue with PVDF membranes).

Materials required :

1. Nitrocellulose paper (pore size 0.20-0.45mM).
2. Whatman No. 1 filter paper.
3. Blotting buffer (pH 8.3) :

0.02M Tris–HCl	2.42g
0.15M Glycine	10.25g
20% Methanol	200mL
Water make up volume to	1 litre

(This can be stored at 4°C for 2-3 weeks).

4. Protein stain: 0.01% Amido black 10B dissolved in Methanol: acetic acid: water (5:1:5 v/v).
5. Destaining solution: Methanol: acetic acid: water (5:1:5 v/v).
6. PBS.
7. PBS-Tween 20.
8. Antigenic proteins, antibodies, and horseradish peroxidase labeled antiglobulins.

Procedure :

1. A platform is a arranged by placing a 25 × 20 cm glass plate at a suitable height on the workbench.

2. Six layers of Whatman No.1 filter paper are assembled and placed over the platform. Short ends of the papers are dipped in two glass trays containing the blotting buffer placed on either side of the platform. The papers has to be wet thoroughly. The papers should be broad enough to accommodate the gel to be blotted.

3. After the separation of proteins in slab gels through SDS-PAGE, the stacking gel is discarded and carefully the separation gel to be blotted is laid on the wet filter paper. The required portion of the gel is cut with a scalpel blade if the whole gel is not required to be blotted, and then laid carefully.

4. A piece of nitrocellulose paper exactly the same size of the gel to be blotted is acquired. It is thoroughly made wet by floating over the blotting buffer. This wet nitrocellulose paper is carefully placed over the gel with out trapping any air bubbles (any air bubbles between the gel and nitrocellulose paper will result in a poor transfer of proteins). This is suitably done by lowering first the middle portion of the paper on the gel and then laying towards the ends. It is again desirable to wet the top of the gel with blotting buffer before layering the nitrocellulose paper.

5. The middle of a cling film is sliced out to the size of the gel, and layered over the nitrocellulose paper such that the film does not cover the nitrocellulose paper. In other words, the cling film will create a frame disclosing the nitrocellulose paper covering the surrounding portions. This shall stop by-pass (soaking) of buffer from the bottom layers to top layers of filter papers directly. This is to make certain that the buffer should pass only through the gel.

6. Over the top of the film, six layers of Whatman 3mm filter paper cut to the same size as the gel are laid followed by a pile of absorbent material (tissue paper may be used in this case) also of the same size of the gel.

7. Next a heavy weight (conical flask containing one litre water or chemical catalogues)

is placed over the sandwich and the set up is left for 1-2 days. It is important to note that there should be ample blotting buffer in the tanks during blotting.

Note that the duration of blotting basically depends upon a range of factors for instance the porosity of nitrocellulose, percentage of acrylamide and width of the gel, the ionic strength of the blotting buffer, the solubility of proteins etc. the low molecular weight (MW) proteins are easily transferred than the high MW proteins.

8. The buffer from the bottom layer papers slowly travels upward by absorption through the gel and nitrocellulose paper. During this progression of blotting, the proteins are transferred by capillary action from the gel to the nitrocellulose which has more affinity to the proteins.

9. After blotting for required time period the nitrocellulose paper is recovered by diassembling the set up. The nitrocellulose paper is preserved, pressed between a fold of filter papers until required for further analysis or may be hybridized or can be stained directly.

10. The protein blot is then immersed for 10-15 minutes with gentle shaking in the amido black dye solution. Afterward, the blot is put in the destaining solution with constant changes. The transferred proteins are then visualized as black bands. The blot is then dried flanked by several sheets of filter paper held flat with a heavy weight.

11. Hybridization of blot can be done as follows :

(a) The blot (the nitrocellulose sheet) is washed at least four times with 100 mL of PBS-Tween 20 for five minutes each on a rocking platform.

(b) The blot is cut into 0.5 cm strips.

(c) The serum is inactivated containing positive and negative antibody controls to the antigens under examination by treating at 56°C for 30 minutes. Dilutions of 1:100 and 1:1000 of the controls are made with PBS-Tween 20.

(d) 3 mL of the diluted sera or controls are placed onto a strip from Step 11(b) and incubated for 1 hour at room temperature while continuously rocking the sample.

(e) The strips are washed four times for 5 minutes each with 10 mL quantities of PBS-Tween 20. The first wash should be done at 50°C, but the last three may be done at room temperature.

(f) Added 3 mL of Horseradish peroxidase-labeled antiglobulin, optimally diluted in PBS-Tween and then incubated at room temperature for 1 hour with continuous agitation.

(g) The strips are washed four times for 5 minutes each with PBS- Tween 20 and one more time with PBS only.

(h) PBS is removed and 5 mL of substrate solution is added. Positive reaction bands usually become visible within 10 minutes. The reaction is stopped by washing with water.

Notes :

1. The nitrocellulose papers ought to be stored in an airtight condition at 4°C to prevent them being contaminated by volatile chemicals etc.

2. During assembling the sandwich for blotting, one should be careful so that there is no direct contact of nitrocellulose paper or the above layers of filter paper so as to circumvent the flow of buffer directly. In an ideal blot assembly it should only flow through the gel to the nitrocellulose paper.

3. The transfer of proteins by electroblotting is much quicker and efficacious than the capillary blotting method. Appropriate cassettes to assemble the sandwich for electroblotting are commercially available at the market.

4. Equilibration of the gel prior to blotting for renaturing of the separated proteins is recommended that depends upon the further analysis of the blot. The gel is equilibrated with constant shaking for 30-60 minutes prior to capillary blotting for renaturing the separated proteins. Following buffer is used:

1M Tris – HCI pH 7.0	5 mL
0.1M Dithiothreitol	0.5 mL
0.1M EDTANa_2	10 mL
5M NaCI	5 mL
Urea	120.12 g
Water to	350 mL

C. NORTHERN BLOTTING :

Like DNA and protein, RNA bands can also be transferred from electrophoresed gels to a solid substrate. Primarily nitrocellulose paper was used, and the technique obtained the customized name of **Northern blotting.** This technique was developed in 1977 by James Alwine and George Stark at Stanford University.RNA molecules of varying lengths (often naturally occurring mRNAs) are separated by size and blotted onto nitrocellulose. After transfer the membrane is either placed in UV-crosslinker or vacuum oven at 80°C to irreversibly bind the RNA to the nitrocellulose filter. The filter is then exposed to a solution containing a suitable probe (DNA or anti-sense RNA) and thus hybridization initiates. The probe is basically radioactively labeled as a result a hybridization signal can be visualized on a film.

The protocol is divided into three sections :

1. Electrophoresis of an RNA preparation under denaturing conditions in an agarose-formaldehyde gel.
2. Transfer of the RNA from the gel to a nylon or nitrocellulose membrane by upward capillary transfer.
3. Hybridization analysis of the RNA sequences of interest using a labeled DNA or RNA probe.

I. Agarose-Formaldehyde Gel Electrophoresis :

Materials Required :

1. Agarose.
2. Formaldehyde.
3. Running buffer.
4. 10xE : 0.2M MOPS, 0.05M NaAc and 0.005M EDTA, pH adjusted to 7.0 with NaOH.
5. Denaturation mix (100mL): 64.6 µL Formamide, 22.6 µL Formaldehyde and 13 µL 10xE.
6. Wash-buffer: 1% SDS + 0.1% 20xSSPE.
7. Loading dye mix: 50% Glycerol, 0.3% Xylene cyanol, 0.3% bromophenol blue and 1mM EDTA.
8. Gel apparatus with power pack.
9. Microfuge.
10. Fume hood.
11. Vortex.

Method :

Gel preparation : 0.75 g Agarose is dissolved in 36 mL water and cooled to 60°C in a water bath. After the flask is cooled to 60°C, it is placed in a fume hood and 5 mL of 10xE running buffer and 9 mL formaldehyde is added. The gel is poured and allowed to set. The comb is removed, the gel is placed in the gel tank , in it added sufficient 10xE running buffer to cover to a depth of ~ 1mm.

Sample preparation : The volume of each RNA sample is adjusted to 6 µL with water, and then added 2.5 x 6 µL freshly prepared sample denaturation mix. Mixed by vortexing, microcentrifuged briefly to collect liquid, and incubated for 15 minutes at 55°C. Cooled on ice for 2 minutes, then added in it 2 µL loading dye mix.

Gel run : The gel is run in 1x E running buffer at 100 volt for 10 minutes, then at 65 volt for 90 minutes

II. Transfer of RNA from Gel to Membrane :

Materials required :

1. 20X SSPE: 3M NaCl, 0.25M NaH_2PO_4 and 0.02M EDTA, adjusted to pH 7.4.
2. Diethylpyrocarbonate (DEPC).
3. Whatman 1 MM filter paper.
4. Nylon membrane.
5. Glass pipette.
6. Plastic wrap.
7. Deionized water.

Preparation of gel for transfer :

The gel is placed in an RNase-free dish (DEPC treated)and rinsed with changes sufficient deionized water to cover the gel for 4 × 20 minutes.

Transfer of RNA from gel to membrane :

1. The glass dish is filled with enough 20x SSPE.
2. 2 pieces of Whatman 1 MM paper is cut, it is placed on the glass plate and wetted with 20x SSPE.
3. The gel is placed on the filter paper and air bubble is squeezed by rolling a glass pipette.
4. Four strips of plastic wrap is cut and placed over the edges of the gel.
5. A piece of nylon membrane (MSI, Catalog #N00HY320F5) is cut just large enough to cover the gel and wetted in water. The wetted membrane is placed on the surface of the gel. One should try to avoid getting air bubbles under the membrane.
6. The surface of the membrane is flooded with 20x SSPE. 5 sheets of Whatman 3 MM paper is cut to the same size as membrane and placed on top of the membrane.
7. Paper towels are put on top of the Whatman 3MM paper to a height of ~6 cm, and a weight is added to hold everything in place.
8. Left overnight.

Preparation of membrane for hybridization :

Paper towels and filter papers are removed and the membrane and flattened gel is recovered. The position of the wells on the membrane is marked in pencil and one has to ensure that the up-down and back-front orientation are recognizable. The membrane is rinsed in 5x SSPE, it is placed on a sheet of Whatman 3MM paper and allowed to dry. RNA is placed on side-down on a UV transilluminator (254 nm wavelength), and irradiated for appropriate length of time.

III. Hybridization Analysis :

Materials required :

1. DNA sample.
2. ^{32}PdCTP.
3. 5X SSPE.
4. ProbeQuantTMG-50 microcolumns (Amersham-Pharmacia biotech).
5. Ridiprimer DNA labelling system (Amersham LIFE SCIENCE).
6. Wash-buffer: 1% SDS + 0.1% 20x SSPE.
7. 5XP : 0.25M Tris-HCl pH 7.5, 0.5% Sodium pyrophosphate, 1% Polyvinylpyrolidone, 1% Bovine serum albumin, 1% Ficoll and 5% SDS.
8. Pre-hybridization solution : 2.5 mL formamide, 1mL 5×P, 1.4mL H_2O, 0.292g NaCl and 0.1mL sperm DNA.
9. Wash-buffer: 1% SDS + 0.1% 20x SSPE.
10. Kodak XAR film.
11. X-ray intensifying screen.

Preparation of DNA or RNA probe (>108dpm/µg) :

The probe is labeled with Ridiprimer DNA labelling system (Amersham LIFE SCIENCE) :

1. The DNA to be labelled is diluted to a concentration of 2.5-25 ng in 45 µL of sterile water.

2. The DNA sample is denatured by heating to 95-100°C for 5 minutes in a boiling water bath.

3. Centrifuged briefly to bring the contents to the bottom of the tube, and put on ice for 10 minutes.

4. The denatured DNA is added to the labelling mix and the mix is reconstituted by gently flicking the tube until the blue colour is evenly distributed.

5. Added 5 µL of [^{32}P] dCTP and mixed by gently pipetting up and down. Centrifuged briefly to bring the contents to the bottom of the tube, and then incubated at 37°C for 30 minutes.

6. The probe is purified using ProbeQuantTMG-50 micro columns (Amersham-Pharmacia biotech).

7. G-50 micro column preparation. The resin in the column is resuspended by vortexing, the cap is loosened one-fourth turn and snapped off the bottom closure. The column is placed in a 1.5 mL screw-cap microcentrifuge tube for support, then the column is pre-spinned for 1 minute at 3000 rpm in a microfuge.

8. The column is placed in a new 1.5 mL tube and 50 µL of the sample is slowly applied to the top-centre of the resin, being careful not to disturb the resin bed. The column is spun at 3000 rpm for 2 minute. The purified sample is collected in the bottom of the support tube.

Hybridization :

Pre-hybridization : The membrane is wet in the 5 X SSPE and placed it RNA-side-up in a hybridization tube and 5 mL pre-hybridization solution is added, then the tube is placed in the hybridization oven and incubated with rotation for 6 hour at 42°C for DNA probe or 60°C for RNA probe.

Hybridization : Double-stranded probe was denatured by heating in a water bath for 10 minutes at 100°C, and then transferred to ice. The desired volume of probe is pipetted into the hybridization tube and continued to incubate with rotation overnight at 42°C for DNA probe or 60°C for RNA probe.

Autoradiography :

1. The membrane was washed twice for 5-10 minutes with wash-buffer at room temperature, and twice for 15 minutes at 65°C with prewarmed (65°C) wash-buffer.

2. Final wash solution is removed and the membrane is rinsed in 5x SSPE at room temperature. Excess liquid is blotted and covered in UV-transparent plastic wrap. One should not allow membrane to dry out if it is to be reprobed.

3. Blot was exposed at 80°C using Kodak XAR film and X-ray intensifying screens.

Chapter 12.8

Enzyme Linked Immunosorbant Assay (ELISA); Double Antibody Sandwich Technique

Introduction

The **ELISA** is a technique which permits the quantitative determination of antibody binding. Similar to the Western Blot, the ELISA involves a specific primary antibody against the substance to be tested. Variations of the ELISA exist. It is already used in medicine to detect the antigen or antibodies in serum samples. At present, it has application in the immunodiagnosis of several infectious diseases. Enzyme immunoassay techniques is basically the combination of specification of antibodies with the sensitivity of **spectrophotometry** in assaying enzymes by application of antigen-antibody complex coupled to an easily assayable enzyme having a high turn over number. This technique relies basically on the principles of enzyme substrate reaction, where a recognized enzyme is conjugated to an antibody, which when exposed to a coloured substrate complex generates a colour. Intensity of colour is then measured through a spectrophotometer, which is proportional to the enzyme activity. Engvall and Perlmann first introduced the ELISA technique in early 1970s.

The principle underlying the **double antibody sandwich technique** of ELISA is described below.

The unrecognized antigens are reacted with specific antibodies, which are adsorbed to a solid phase. Incubated and then washed with buffer to remove excess antigen. This antigen-antibody complex is then treated with enzyme labelled antibody (directed against another **epitope**, in case of monoclonal antibodies; Mab). The reactants are given a further wash and then enzyme substrate (**chromogenic** in nature) is added. This generates a colour. Intensity of colour which is proportional to the enzyme activity is measured through fibre optic multichannel spectrometer Amount of enzyme activity is directly proportional to the quantity of antigen present measured under standard conditions.

Materials required :

1. Flat-bottomed polystyrene Microtitre plates with 96 wells.

2. Micropipettes (Gilson/Finnpipettes) 0-250 µL (range).

3. Multi-channel pipette (0-250µL) for pipetting of all reagents (if available in the laboratory).

4. ELISA Reader (Multiscanphotometer), if available.

5. 0.1M Carbonate Buffer (pH 9.6) : 0.1M Na_2CO_3 solution is prepared and pH adjusted to 9.6 with NaOH. The chemicals should be of the best quality and water should be double distilled.

6. Wash solution : 90mL Tween 80 is mixed with 910 mL water.

7. BST : 0.2% (w/v) Bovine serum albumin, 0.01% Tween 80 and 0.9%(w/v) sodium chloride in distilled water.

8. Substrate solutions : It depends upon the enzymes in ELISA technique that is coupled with the conjugate. Two commonly used enzymes in ELISA technique are horseradish peroxidase (HRP) and alkaline phosphatase. For HRP, there are two substrate solutions and are prepared as shown below.

9. Solution 1 : 80 mg of 5-amino salicylic acid is dissolved (purple red brown colour) in 100mL 0.05M potassium phosphate buffer (pH 6.0) containing 0.001M EDTA. 20mL of H_2O_2 (30%) is added and mixed.

10. Solution 2 : A mixture is made up of :

(a) 24.3 mL 0.1M citric acid.

(b) 25.7 mL 0.2M Na_2HPO_4.

(c) 50 mL H_2O.

(d) 40 mg *ortho*-phenylenediamine (yellow in colour).

(e) 40 µl H_2O_2 (30%).

11. Stop Solution : 0.3M NaOH (in the case of solution 1) or 1M H_2SO_4 (in the case of solution 2) is used. If alkaline phosphate is the enzyme which is coupled, the substrate is then *p*-nitrophenyl phosphate and is released as yellow coloured compound *p*-nitrophenol.

12. A diluted solution of IgG against the antigen that is to be measured. The dilution is generally 1500-2500 folds that depend on the titre of IgG. Dilutions are carried out in 0.1M carbonate buffer.

13. Antigen solutions to be examined and standard antigen solutions.

14. Enzyme (HRP) labelled diluted IgG solution (see below). As a rule the conjugate solution should be diluted 500- 2000 times in BST.

Procedure :

A. The Double Antibody sandwich Technique :

1. 150µL of the diluted IgG solution is pipetted to each of the wells of a microtitre plate manually or using multichannel pipette.
2. The plate is covered and incubated overnight at room temperature.
3. The plates are washed with wash solution. The wells can be emptied by tapping the plate over a laboraty sink and then beating the plate upside down alongside a filter paper. The plates can be preserved for several months, in covered and cooled condition. A required volume of wash solution is pipetted into the wells and left for some minutes. The wells are then emptied as explained above. Washing is repeated more than ones to ensure good results.
4. After washing of 100µL of BST is added to each well.
5. 100µL of an antigen solution to be tested is added to the first well of each row.
6. Mixed carefully and thoroughly. Air bubbles or splashing of small drops should be avoided.
7. 100µL is taken from the first wells and transferred to the second wells in each row. The mixing procedure (step 5) is repeated. 100µL is taken from the second wells and added to the third wells and so on. By this way a two fold dilution series from wells 1-12 are made. Finally, 100µL excess is removed from wells.
8. The plates are incubated for 2 hour at 37°C, to allow the antigen to bind to the coated antiserum and then washed thoroughly.

9. Added 100μL of the diluted conjugate solution (*see* preparation of conjugate) is added to each well and incubated for 2 hour at 37°C, then the plate is washed thoroughly.
10. 100μL of substrate solution is added to each well and incubated for 1-2 hour at 37°C in the dark.
11. The reaction is stopped by adding 100μL of Stop solution.
12. The titre of the antigen solutions is read. This can be done either by using an ELISA Reader or visually by observing the last well that still gives some colour with the naked eyes.

B. **Preparation of conjugate :** A conjugate is the covalent complex made up of IgG and an enzyme. The coupling of HRP is described below.

1. 5mg of HRP is dissolved in 1mL 0.3M Na_2CO_3 (pH 8.1). This solution should be prepared fresh.
2. 0.1mL of 1% fluorodinitrobenzene is added in pure ethanol. If the HRP used is impure, a precipitate may occur that should be removed by centrifugation (18000 rpm for 10 minutes).
3. Mixed thoroughly and incubated for 1 hour at room temperature.
4. 1mL of 0.16M ethylene glycol is added, mixed and incubated for another hour at room temperature. The total volume is now 2.1mL.
5. The mixture is dialyzed against 0.01M sodium carbonate buffer (pH 9.5) for 25 hours. The buffer should be changed not less than three times.
6. IgG dissolved in 0.01M sodium carbonate buffer (pH 9.5) is added to the peroxidase aldehyde solution in the following ratio: one volume of IgG solution to one volume activated peroxidasealdehyde or 5mg purified IgG protein to 3mL peroxidase solution.
7. Mixed well and incubated for 2-3 hours but not longer at room temperature. If any precipitate is formed, it should be cleared by centrifugation (10 minutes, 10000 rpm).
8. Dialyzed extensively against 0.01M phosphate buffer (pH 7.2) containing 0.9% NaCl at 4°C. The conjugate should be stored in a refrigerator or freezer in small aliquots and should be used once only.

Chapter 12.9

Autoclave

The autoclave is a device used to sterilize (deep clean) medical and biological materials. In fact, the autoclave was originally invented and promoted as a method for preparing food by French physician Denis Papin. Papin called his invention a "steam digester." He described his "New Digester" in a 1681 pamphlet, emphasizing the benefits of using the device to process food for easier digestion (the process by which food is broken down for absorption into or elimination out of the body).It is a container with an airtight lid that traps steam from boiling water. The steam increases the pressure inside the container (due to the constant volume of the container), which raises the water's boiling point. The higher temperature kills bacteria and other pathogens much faster than at lower temperatures. The container includes a safety valve to prevent explosion if the steam pressure gets too high. In microbiology laboratories steam jacketed horizontal type autoclave (**Figure 12.9.1**) is used where the steam passes from below at the base. The side walls are heated by steam jacket.

The temperature and pressure relationship in autoclaving is shown in the following table.

Table 12.9.1 : Temperature and pressure relationship

Steam Pressure, Pressure per Square inch (psi)	Temperature (°C)				
	AIR DISCHARGE				
	100%	66%	50%	33%	0%
0	100.0				
1	101.9				
2	103.6				
3	105.3				
4	106.9				
5	108.4	100	94	90	72
6	109.8				
7	111.3				
8	112.6				
9	113.9				
10	**115.2**	109	105	100	90
11	116.4				
12	117.6				

	AIR DISCHARGE				
	100%	66%	50%	33%	0%
13	118.8				
14	119.9				
15	121.0	115	112	109	100
16	122.0				
17	123.0				
18	124.1				
19	125.0				
20	126.0	121	118	115	109
21	126.9				
22	127.8				
23	128.7				
24	129.6				
25	130.4	126	124	121	115
26	131.3				
27	132.1				
28	132.9				
29	133.7				
30	134.5	130	128	126	121

Around the working range of 15 psi, each psi raises the autoclave temperature by approximately 1°C.

Standard Operating Procedure of horizontal type autoclave :

• Preparation of Materials :

1. All materials autoclaved ought to be placed in an autoclavable tray with sides that are at least 5 inches in height. The utilization of trays should eliminate anything boiling over and spilling into the autoclave. Thus, all materials must fit within this tray. Only autoclavable materials may be put in the autoclave. If one is uncertain if something is autoclavable, one should check with the manufacturer's note. Liquids to be autoclaved must be placed inside an autoclavable vessel that is at least double the volume to be autoclaved (i.e. autoclaving 1 litre of media, one need to put it in a flask that holds at least 2 Litres). This allows for boiling during the cycle.

2. Caps must be slightly loose so that pressure created during the cycle does not cause the vessel to break. For screw cap containers user one can make the lid hand tight and then loosen the lid by one-half turn.

3. Dry goods must be put in an autoclavable tray with 5 inches sides. Garbage must be placed inside autoclavable bags and inside the autoclavable trays with 5 inches sides. Sharps containers must also go into the same type of autoclavable trays. To make sure about proper sterilization of both garbage and sharps containers, 1 litre of water should be added to the bags or sharps containers. This water will vapourize during the autoclave cycle to allow better penetration of steam and more efficient sterilization. Autoclave garbage bags must be tied off.

• Materials that can be autoclaved :

1. Culture and stocks of infectious agents.
2. Culture dishes and related devices.
3. Contaminated solids such as paper towels, cloth and plastic pipette tips, pipettes, vials, petridishes and gloves.
4. Discarded and attenuated vaccines.
5. Recombinant DNA, plant and animal specimens with Recombinant DNA.
6. Animal tissue specimens.
7. Pathological animal wastes.

• Autoclaving procedure :

1. All users must sign the autoclave log book and include all required information.
2. The lid of the autoclave is opened.
3. Water chamber is filled with water up to level indicated by level indicator.
4. Samples to be autoclaved are placed inside the sample chamber.

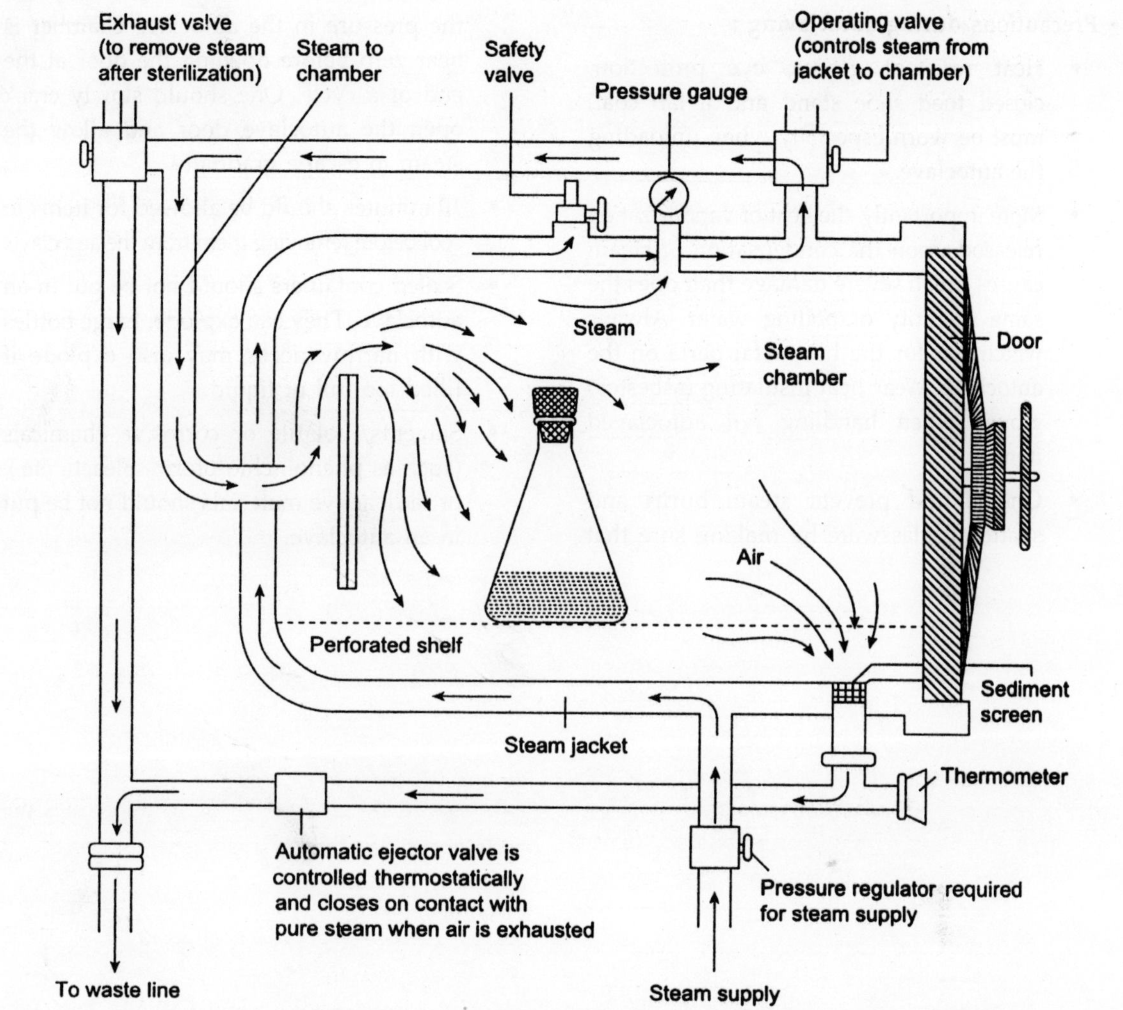

Fig. 12.9.1 : Steam jacketed horizontal type autoclave.

5. The lid of the autoclave is closed and the bolts are tightly fastened.

6. The autoclave is switched on, built in heater heats up water and converts into steam.

7. The water boils and steam accumulates in closed sample chamber.

8. The pressure is increased to 5 psi and then released, the pressure is then increased to 15 psi and maintained for 30 minutes. The pressure gauge attached to the autoclave indicates the pressure.

9. After the autoclaving is over the pressure is released through the steam outlet.

10. The pressure fall to zero as indicated by the pressure gauge.

11. The lid of the autoclave is opened by unfastening the bolts.

12. Autoclaved materials are removed from the autoclave for use.

- **Precautions during autoclaving :**
  - Heat resistant gloves, eye protection, closed toed shoe stand and a lab coat, must be worn especially when unloading the autoclave.
  - More importantly, the heat of vapourization released upon the condensation of steam causes much severe damage than does the same quantity of boiling water. Always watch out for the hot metal parts on the autoclave. Wear heat insulating (asbestos) gloves when handling hot autoclaved items.
  - One should prevent steam burns and shattered glassware by making sure that the pressure in the autoclave chamber is near zero before opening the door at the end of a cycle. One should slowly crack open the autoclave door and allow the steam to escape gradually.
  - 10 minutes should be allowed for items to cool before removing them from the autoclave.
  - Sealed containers should not be put in an autoclave. They can explode. Large bottles with narrow necks may also explode if filled too full of liquid.
  - Solvents, volatile or corrosive chemicals (such as phenol, chloroform, bleach, etc.), or radioactive materials should not be put in an autoclave.

Chapter 12.10

Narcotization

Animals chosen for preservation should be undamaged. **Narcotization** is performed by adding some anaesthetic substance very gradually in a very small doses to the water containing animals. A 5% solution of formalin (1 part commercial formaldehyde to 19 parts sea-water) is added to the water, drop by drop until the animals are anaesthetized and no longer respond to touch. During narcotization some form of aeration is essential. This can be done by slowly rotating the water by hand. Weak percentage of alcohol solution, magnesium sulphate, and chloral hydrate can also be used for **anaesthization** animals.

The organisms are first narcotized to prevent shrinking, distortion of shape etc. Different chemicals are used for narcotization of various organisms, some of these are:

1. **Menthol/Magnesium chloride** : The animals are kept in a clean basin or petridish with sea water and small quantity of menthol or magnesium chloride is sprinkled over water and covered with a lid. With in 4-6 hours animals are narcotized.

2. **Alcohol/Chloral hydrate** : 70% ethyl alcohol or 1% solution of chloral hydrate is added drop by drop at frequent intervals to sea water in which the animals are kept. The organisms are narcotised in a relaxed condition.

3. **5% formalin** can also be used for narcotization.

Organism specific narcotization :

• **Annelida :**

(i) **Polychaeta** : Narcotization is done by adding gradual amount of alcohol (5-10%) to a vessel of seawater containing the worms. Alcohol to water ratio should be 1:10. Cocaine-chloral hydrate and menthol may also be used for the above purpose. To avoid contraction of large sized species, they may be first kept well stretched in weak alcohol (30%-40%) and as soon the organism cease to react they should be transferred to 90% ethyl alcohol.

(ii) **Oligochaeta :** Earthworms are narcotised as the polychaetas but using fresh water. Aquatic species are washed and placed in a dish containing fresh water. After a while they tend to extend and killed by pouring concentrated formal into the water. It may also be carried on by adding 70% alcohol.

(iii)**Hirudinea :** Leeches may be relaxed prior to preservation. They are put inside a jar with a small amount of water and are narcotised very slowly by adding 5-20% ethyl alcohol or chloroform or chloral hydrate or weak nicotine or magnesium sulphate or carbon dioxide or even weak acids like citric acid. They are transferred in a shallow dish and straightened out before fixation.

- **Crustaceans :**

Decapods crustaceans viz., crabs, lobsters and shrimps require special care in killing as they have a tendency to shed their appendages when dropped in concentrated formalin or alcohol. This can be avoided if a weak solution of 1-2% of formaldehyde is used. After killing, large specimens should be fixed in 5% buffered formaldehyde solution for three to four days. After fixation the animals are washed thoroughly.

- **Molluscs :**

Aquatic molluscs after collection should be placed in an enamel tray filled with either fresh or seawater; the specimen should be entirely dipped in water. Then finely powdered menthol or magnesium sulphate or chloral hydrate should be sprinkled over the surface of the water and must be covered with water. Narcotization through magnesium sulphate and menthol may require 2-6 hours and 12-25 hours respectively. Totally narcotized specimens remain in fully extended condition.

Nudibranchs can be narcotized by 4% formaldehyde solution, 3-4 at a time repeating every half hour until the animal is narcotized.

Opisthobranchs and cephalopods should be injected with small amount of certain fixative before transferring them into fixative solution.

Fresh water snails and bivalves can be relaxed by dropping them into boiling water. Propylene phenoxetol (0.5%) is a good relaxing agent for marine bivalves.

Land snails can be killed by asphyxiation. A glass bottle or jar is fully filled with water, and the specimens are immersed into it, then the mouth of the bottle or jar is closed with a lid, making it air tight. Deoxygenated water or addition of a few drops of spirit at gap may aid in fast killing. After the animals are found fully dead or still, they should be removed and treated with mounting grades of ethyl alcohol (30%, 40%, 60%, 70%).

- **Echinoderms :**

Echinoderms like *Asterias, Echinus, Ophioderma and Antedon* are killed quickly in fresh water. Animals should be narcotized while in small volume of seawater, by adding crystals of magnesium chloride or crystals of menthol or by addition of 70% alcohol drop by drop. During narcotization the side bearing the tube feet should face up and the arms and the tube feet should not touch the bottom or sides of the jar.

In case of the Holothurians, the animals should first be narcotized as described in the above section. When the animal becomes motionless, 70% ethyl alcohol or 50% formalin (neutralised) is injected by a hypodermic syringe into the closed opening and the perivisceral coelom through the body wall.

- **Hemichordata and Cephalochordata :**

Narcotization is done in 5% magnesium chloride.

Chapter 12.11

Kymograph and Related Apparatus used in Experimental Physiology

In experimental physiology for the measurements of excitable tissues such as heart and muscle a set of instrument is used for this recording and is known as Sherrington Kymograph assembly (see **Figure 12.11.1**).

The assembly is grouped into two basic unit :

(a) **Recording unit** : It comprises Kymograph, Revolving recording drum, Crank myograph or the Muscle lever and the Time markers.

(b) **Stimulating unit** : It comprises of Obach Dry cell or Battery, Wires, Key, Pohl's commutator, Induction coil, Electrodes, Rheochord, Step-down arrangement etc.

(1) **Sherrington's Recording Drum or Kymograph** : This is made up of a strong vertical shaft of metal over which a metal drum is adjusted. The drum can be moved up or down as desired. The drum is enveloped over by a highly glazed paper, which is lightly smoked with a thin layer of smoke shoot to acquire good tracings. The smoke is done by revolving the drum over a kerosene flame containing benzene vapour. The

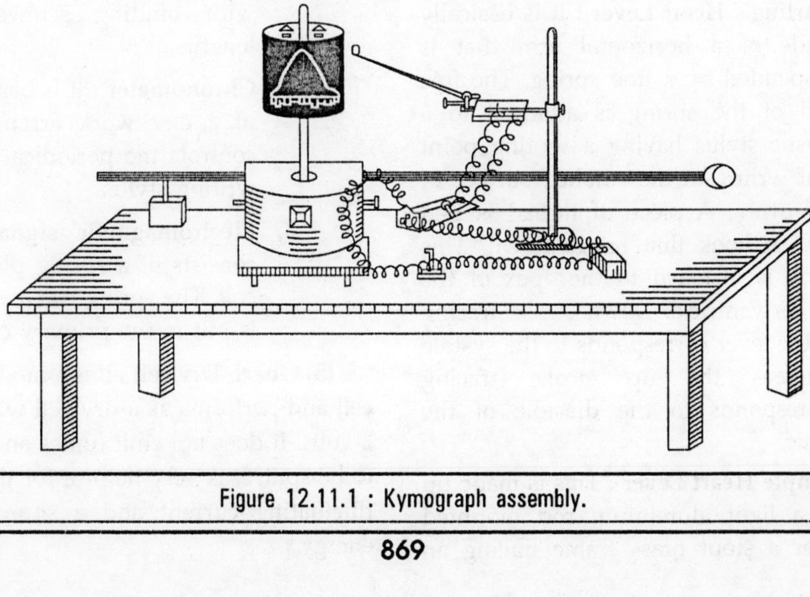

Figure 12.11.1 : Kymograph assembly.

recording drum is to revolve either directly by the electrical motor or by placing a pulley system in between motor and drum at regular rates. The body of the kymograph contains the following :

(i) *Gears:* It may be of fast, natural or slow.

(ii) *A switch* to initiate or stop the movement of the drum.

(iii) *Two terminals,* which wires in the inclusion of the kymograph into the primary circuit.

(2) **Crank Myograph or the Muscle Lever :** It comprises of a myograph board for griping muscle preparation with a crank lever fixed to it. The shorter arm of lever is designed for attachment to the muscle while the longer writing lever bears a writing point designed for recording the contraction on a smoked paper wrapped around the kymograph cylinder. The entire apparatus is held on a perpendicular stand, the arm of which is movable at the base, so that the writing point of the myograph can be sway towards or away from the drum without changing the position of the base of the stand. A screw is supplied with the vertical arm, which can be employed to make the load act or be free.

(3) **Heart Lever :**

(a) **Starling's Heart Lever :** It is basically made of a horizontal arm that is suspended by a fine spring. The free end of the spring is attached to a plastic stylus having a writing point that writes on the smoked surface of the drum. A piece of thread is tied with a hook that hangs down. This hook is fastened to the apex of the toad's ventricle. Down stroke tracing of the lever corresponds to the systole whereas the up stroke tracing corresponds to the diastole of the heart.

(b) **Simple Heart Lever :** This is made up of a light aluminium rod mounted over a stout brass frame having an adjustable screw. This screw helps in altering the length of the long and the short arm. The long arm is attached with a light celluloid writing point at the distal part. A piece of thread is tied with a hook that hangs down from the short arm of the lever. This hook is fastened to the apex of the toad's ventricle. Down stroke tracing of the lever corresponds to the diastole and up stroke tracing corresponds to the systole of the heart.

(4) **Time markers :** There are diverse categories of time markers, which are in frequent use in the laboratory. These are the following :

(a) **Tuning fork :** It is made up of two metallic prongs joined at one end with a projection. One metallic prong has a quill writer point. It is struck against a piece of wood or hand and the writing point is gently applied to the smoked paper of the moving drum. It produces hundred vibrations per second and every wave is equal to one hundredth of a second i.e. 1/100 or 0.01 sec.

(b) **Vibrating Reed :** These are calibrated reeds whose vibration frequency can be controlled by fastening the reed with binding screws at different lengths.

(c) **Chronometer :** It is basically made up of a clockwork arrangement, which controls the periodical flickering of a writing stylus.

(d) **Electromagnetic signal marker :** It consists of a spring placed above the coil. The spring has a writing point. It is put in the primary circuit.

(5) **Obach Dry cell :** It is a modified Leclanche cell and performs as a dry cell with an E.M.F. of 2 volts. It does not emit fumes and have no fluid to be split. It is very helpful for the flow of non-fluctuating current and a source of electrical energy.

(6) **Key or Switch :** It comprise of two thick metallic plates at two end each with two terminals. It is utilized for break up or diverting the course of a current. It is arranged to close and open a circuit being coupled together through a pool of mercury or by friction contact between two brass surfaces as in Du-Bois-Reymond's key.

(7) **Pohl's Commutator :** This consists of six mercury cups depressed out in a block of wood or vulcanite, each cup being connected to a binding screw. The four corner cups are connected diagonally by stout copper wire, which do not touch each other. Solid wires to a non-conducting crosspiece, which works as a handle, couples the two side cups. Each end of the handle carries a semi-circular copper wire, which is coupled to the wire going into the side cup and is of such length that will sink into the cup at either end by means of turning the handle over towards that end. If the handle is in such a position that a current inflowing the commutator by one of the side cup, emerges by an end cup of the same side, then, by turning the handle over, the cross-wires are brought to play and the current will now come out by the end cup of the side. The electrical current may be transmitted through this commutator to any of the two pairs of electrodes engaged in certain experiments, after the cross-wires are separated. The commutator is also used to reverse the course of the current in the external circuit, when the cross-wires are in arrangement. It is used in the experiment "velocity of never impulse".

(8) **Du-Bois-Reymond's Induction Coil :** It is used for flowing of induced current of high intensity and short interval. It facilitate in converting the galvanic current into induced or Farradic current. It is comprised of primary and secondary coil. The primary coil is wrapped above a solid iron rod. The secondary coil is wrapped on an empty wooden frame. During the normal flow of current in the primary coil no induced current is developed within the secondary coil since mainly no change takes place in the magnetic field. But there is an alteration in the magnetic field that results in the development of an induced current as soon as there is a "make" or "break". Keeping a proper distance between the primary and secondary coils can control the intensity of induced current. The greatest strength is of induced current when there is zero distance and least when the distance is maximum between primary and secondary coils.

(9) **Electrodes :** These are the piece of equipments to communicate the current to the nerve. They are applied to the open ends of the two wires which transmit the current to the tissue to be stimulated. They are made up of two insulated wires, the ends of which are clean and free from the insulating material and twisted round a piece of cork or rubber. These ends are usually designated as stimulating electrodes. These electrodes transmit current from the secondary coil to the nerve. The simple electrode is made from two thick wires with a non-conducting substance in between them.

(10) **Rheochord :** It is comprised of a single straight or zig-zag wire fixed on a wooden base, with a binding screw at either end or a movable contact linking them. The two ends are coupled to the battery poles. This apparatus is used for dividing a constant current by producing a circuit of comparatively small resistance, which, is capable of being varied, so that a variable part only of the current shall flow through the experimental circuit.

(11) **Step-down apparatus :** It is employed for converting the high voltage of current into low voltage specifically the voltage of the current used in domestic purpose is reduced to 1.5 to 2 volts which is essential for physiological experiment.

Electrical Circuit

The common map of electrical circuit is made up of primary circuit and secondary circuit. The primary circuit is concerned by means of the revolving drum, primary key and primary coils of the induction coil and the main source of current. The secondary circuit consists of secondary coils of induction coil, electrodes, nerve-muscle or heart preparation, muscle or heart lever and the smoked drum. The circuit can be used to obtain a single make or break shock by the help of primary key or short circuiting key.

Chapter 12.12

Biological Measurements

Introduction

While studying large organisms, it is often required to measure the length, breadth or height. In case of microorganisms, or in case of an organ or a cell, the diameter has to be recorded. In the first case, i.e., macro-measurements, no special apparatus is required while in the second case, *i.e.*, micro-measurements ocular and stage micrometers are used.

(a) Macro-measurements :

Procedure :

1. The animal is placed on a white paper spread on a plain surface.

2. With the help of a pointed pencil, points at the anterior and posterior tips of the animal are marked.

3. The animal is removed and the two points are connected by a straight line and the same is measured.

4. If the animal is small (say an insect) a divider may be used to mark the two tips.

5. The divider may then be placed on a standard graph and the distance between the two points of the divider are noted.

(b) Micro-measurements :

Science of measurement of microscopically visible organisms or cellular components is known as **micrometry**. A stage and an ocular micrometer (Figure 12.12.1a,b) are required for micrometry.

- **Stage micrometer :**

1. A stage micrometer is a microscopic slide, in the centre of which a line 1 mm long is drawn.
2. This line is divided into 10 parts with longer lines and marked 0, 1, 2, 3, 4 and so on. Each part is then subdivided into 10 smaller parts by small lines.
3. Thus 1 mm is divided into 100 parts each measuring 0.01 mm.
4. It is used to calibrate the ocular micrometer.

- **Occular micrometer :**

1. The ocular micrometer is a circular glass or plastic disc in the centre of which a line is drawn, which is subdivided into 100 parts, as in the case of the line on the stage micrometer.
2. The only difference in the case is that the value of each part is not given.

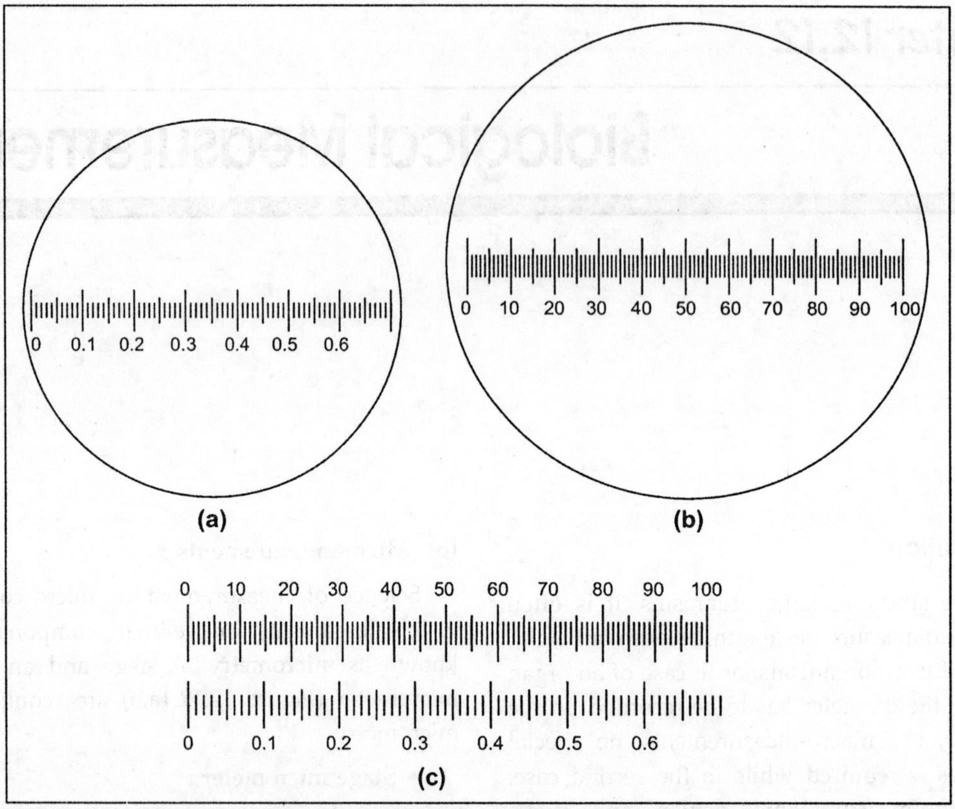

Fig. 12.12.1 : Diagram of (a) Stage micrometer; (b) Ocular micrometer and (c) Overlapping of stage micrometer and ocular scales.

3. It is placed inside the eyepiece of a microscope.
4. It is used to measure the accurate dimension of an microorganism, a cell etc.

Procedure :

1. An eye-piece of a microscope is removed and its magnification (say 10X) is noted.

2. After removing cap of the eye-piece, it is unscrewed from the ocular tube. A metal ring fits closely with the inside wall of the eye-piece. The ocular micrometer is inserted in the eye piece tube so that it rests flat on the metal ring. The cap of the eyepiece is screwed and the same is placed in its position on the microscope.

3. The stage micrometer is placed on the stage of the microscope and the microscope is focused. The magnification of the objective (say 10X) is noted. Both the lines (one on the stage micrometer and the other on the ocular micrometer) will be visible.

4. The stage micrometer is moved and the eye-piece is rotated so that the two lines coincide and exactly overlap (superimpose) each other at the starting point (**Figure 12.12.1c**). The line on the stage micrometer should coincide and overlap the starting-end point of the ocular micrometer.

5. Now the other end of the two lines are seen and where exact overlapping is visible is noted. For example, it may be noted that when '0'

of stage micrometer coincides with '0' of ocular micrometer, the other end of the latter coincides with 7 mark on the line of the former. It may sometimes be noted that instead of line of 10 of the ocular micrometer, line 9 coincides better with a corresponding line on the stage micrometer. This point where the two micrometers coincide at the other end should be noted.

6. One has to take 5-6 readings.

Calculations :

Let the number of divisions on the stage micrometer be *Y* and the number of divisions on the ocular micrometer be *X*. Since stage micrometer is graduated into small lines 0.01 mm apart, it can be said that the parts of lines on ocular and stage micrometers are equal at that magnification. In other words,

Number of divisions on ocular micrometer = Number of divisions on stage micrometer × 0.01 mm or *X* divisions = *Y* divisions × 0.01 mm.

Therefore, 1 ocular division = *Y* divisions/*X* divisions × 0.01 mm.

Supposing 100 ocular divisions (*X*) coincide with 60 stage divisions (*Y*), we can say that 100 ocular divisions (*X*) = 60 stage divisions (*Y*) × 0.0l mm.

Therefore, 1 ocular division = 60/100 × 0.01 mm = 6/1000 mm = 0.006 mm = 6 μm.

It must, however, be noted that this value of each ocular division will vary every time a new combination of eye-piece and objective is used. As a result it is, that the ocular micrometer should be calibrated at each feasible combination (like 10 eye piece (e.p.) × 10 objective (obj.), 10 e.p. × 40 obj and 10 e.p. × 100 obj.) and the values calculated and noted for future reference.

Once the value of each ocular division is known at a particular magnification, the microscopic preparation to be studied can be placed on the stage of the microscope and the particular organ or the cells can be focused. The length or the diameter of the cell can be noted as being equal to the number of ocular divisions.

As the ocular micrometer has already been calibrated at that combination of eye-piece and objective and the value of each ocular division is known, the exact diameter or the length (in μm) can be stated.

(c) Micrometry of blood cells :

Materials required :

1. Stained blood smear.
2. Ocular micrometer.
3. Stage micrometer.
4. Immersion oil.
5. Compound microscope.

Procedure :

1. Calibrated ocular micrometer is carefully put inside the eyepiece and placed inside the ocular tube.
2. A drop of immersion oil is placed in the centre of stained blood smear slide which is to measured. One of the blood cell which has to be measured is focused.
3. One has to measure 5-10 of such type cells and an average has to be taken.
4. The accurate size of the blood cell is assessed by multiplying with the calibration value of the ocular micrometer.

Results :

No. of blood cells measured	Ocular divisions occupied by each type of blood cells in the slide					
	Erythrocytes	Neutrophils	Basophils	Eosinophils	Lymphocytes	Monocytes
1						
2						
3						
4						
5						
Average						

Suppose one of the blood cells occupied 'A' ocular divisions and the calibration value of ocular micrometer at 100X magnification say 'B' μm. Therefore the size (diameter) of the blood cell will be A × B = C.

Section 13

FIELD TRAINING

SECTION AGENDA

Chapter 13.1

Introduction of Field Based Study (Research)

A. Field based study

Field based study can be an amazingly exciting and rewarding experience leading to significant discoveries and burst through ideas. Its aim is the same as research done in the library or on the Internet: to collect information that contributes to one's understanding of an topic or question and to systematize those findings in a cohesive and convincing document that proposes a new insight, answer or solution. The field study associates students' activities taking place in learning environments outside the usual (conventional) classroom, such as wetland environments, forests areas, sea shore and museums, national parks, zoos, pond and lakes , wild life areas, etc. It is based on the assumption that the most precious experiences of the students are achieved through images ingrained by the senses. It is associated with nearly all educational techniques and it frequently forms part of a project. It permits students to contribute in the design of the educational activity and to gain *in situ* know-how and knowledge through the research process.

B. Steps in Field based study (research)

(i) Identify or Define a Problem :

Research problem is the heart of the research project requirement, one has to state the problem with unwavering clarity, precision. Identification of the problem can be done from Journals, books, abstracts (library/trusted links) recommendation sections of theses and dissertations/journal articles, ideas from researcher's mentor or Professor ideas, from seminars, research colloquia and conferences personal/family experiences rare/interesting occurrences Characteristic of a research problem should be of great interest to the field of study and useful for the interested people in a particular field. It must possess novelty and lays basis for further study in the field, to be completed in the allotted time.

(ii) Formulation of Research Objective :

Research objectives reflect the questions whose answers the investigator wants to know. Such statement or a question serves as the steering wheel in the conduct of a research project serves as a lead in indicating variables till interpretation of results. For example, sample objectives help to develop an optimized procedure to detect presence of copepods in aquatic ecosystem and to define copepod diversity in zooplankton community.

(iii) Literature Review :

Following the research problem that has been identified and the objectives formulated, a review of related literature has to be done. This is because: to get familiar with the present studies related to the research to be conducted and to set up a rationale or a theoretical or conceptual framework based on earlier research studies done.

(iv) Building up or Formulation of testable Hypothesis :

A hypothesis is a tentative statement that proposes a possible explanation to some phenomenon or event. A useful hypothesis is a testable statement, which may include a prediction. A hypothesis should not be confused with a theory. Theories are general explanations based on a large amount of data.

A Good Hypothesis should be stated in declarative form and in definite terms. The relationship between variables should reflect the theory or literature. It should be brief and to the point and be testable. There are two types of Hypothesis:

(a) ***Null Hypothesis :*** A null hypothesis is a statistical hypothesis that is tested for possible rejection under the assumption that it is true (usually those observations are the result of chance).

(b) ***Alternative Hypothesis :*** The alternative hypothesis is the hypothesis applied in hypothesis testing that is contrary to the null hypothesis. It is generally taken to be that the observations are the result of a real effect (with some amount of chance variation superposed).

(v) Identification of Research variables :

Very simply, a variable is a measurable characteristic that varies. It may change from group to group, person to person, or even within one person overtime. Two important types of variables are:

(a) ***Independent Variable :*** A variable that is selected or controlled by the researcher, to determine its relationship to the observed outcome of the research—also called explanatory, predictor, or manipulated variable.

(b) ***Dependent Variable :*** The variable being measured as an outcome — also called outcome, response, criterion, or explained variable.

(vi) Construction of Research Design :

It represents the "plan of work" of the researcher in responding the research objectives. It is imporant acquiring all the pertinent data in relation to objectives and hypothesis. The precise areas of concern in the choice of a research design are the following: selection and number of subjects control and manipulation of relevant variables, establishment of criteria to evaluate outcomes, instrumentation maximization of internal and external validity.

Internal Validity refers to degree to which investigator is able to control the different biases influencing the study and in the end, measures what he really plan to measure.

External Validity refers to the degree to which the investigator is able to simplify the results of his study.

(vii) Experimentation and Data Collection :

In field based research, field based experiment has to be conducted. An experiment carried out outside the laboratory, in a 'natural' location. Field experiments are distinguished with laboratory experiments. Laboratory experiments are carried out in closed, controlled conditions. Field experiments are carried out in open, natural settings. Sometimes it is necessary to mimic a natural environment in laboratory setting. Field based experiment provides greater ecological validity and less sampling bias.

Data collection is the process of gathering and measuring information on variables of interest in an established systematic fashion that facilitates one to answer stated research questions, test hypotheses, and appraise outcomes. The data collection section of research is universal to all fields of study including physical and biological and social sciences, etc. The Data collection methodologies are:

- Indepth interview
- Observation methods
- Document review
- Questionnare

(viii) Data Processing :

Data processing is like the backstage of a theater, data processing critically affects an investigator's ability to carry out reliable, valid research. Data processing refers to the procedure of converting data from one format to another. It converts plain data into precious information and information into data. Researcher can gather data in a range of forms, be it spread sheets, audio devices, or plain printed material. Data processing services take the raw data and process it consequently to generate prudent information. The different applications of data processing can convert raw data into useful information that can be used further for required processes.

(ix) Data Analysis :

Data analysis is the procedure of assigning meaning to the collected information and settling on the conclusions, significance, and connotation of the findings. The steps involved in data analysis are generally function of the kind of information collected, though returning to the purpose of the assessment and the assessment questions will offer a structure for the organization of the data and a focus for the analysis. Analysis of numerical data is carried on through the application of statistical methods such as mean, frequency distribution, deviations, T-tests, F-tests, regression analysis, ANOVA etc.

(x) Drawing conclusion and recommendation :

Researcher sums up the discussion on the research findings and makes a comprehensible conclusion. Researcher categorizes major points that were not raised in the present study and could lay the outline for future undertakings. Data processing makes sure that the data is presented in a clean and systematic manner and is simple to understand and be used for further purposes.

(xi) Research Report :

Researcher write down a report of diverse activities one has carried out collectively with one's findings Report must be well-organized and presented in proper form and style and the basic principles of technical report writing are followed. A proper research report includes the following sections viz., Title page, Abstract, Introduction, Materials and Methods, Results, Discussion and Literature cited or References. Research reports usually, but not always, have "actionable" recommendations.

C. Sites of Field Training

Museums, Zoos, forests, pond and seashore are places in which students can learn about the variety of animal and plant life and can acquire knowledge about the distribution of biota over the face of the earth. Further visiting to a poultry, apiculture and sericulture farm etc. a student of zoology can know about important practices in economic zoology.

There are two major educational values acquired from museums, aquaria and forests. The first (apparent value) is to acquaint students with the subjects. The other value is the potential use of these institutions as information centre.

Species that have been identified as threatened or endangered can be preserved in suitable locations outside their natural habitat (***ex-situ* conservation**). Endangered species are being successfully conserved and propagated in botanical and zoological gardens, research centres, aquaria and other such locations. Besides conservations, such ex-situ areas provide ready study sites and opportunities for raising public awareness on conservations issues.

Chapter 13.1(a)

Visit to Zoological Garden or Park

Introduction

A **zoological garden, zoological park**, or **zoo** is an institution where mainly wild and exotic animals are restricted within enclosures, bred and displayed to the public. The term zoological garden refers to the biological discipline zoology, which derives from Greek words *zoos* ("animal"), and *logos* ("study"). The term was first used in 1828 for the *London Zoological Gardens*, soon shortened by the Londoners to the abbreviation "zoo".

The History of Zoos

Wild animals have, throughout history, intrigued the human race. We are amazed by their diversity of form and entertained by their actions. The first wild animal **menageries** are recorded as far back as Ancient Egypt where wild animals - regarded as objects of wealth - were presented as gifts to the pharaohs. The Romans also viewed animals as objects believing that they were created solely for the convenience of humans. Creating their own form of entertainment, the Romans flocked to see hundred of thousands of lions, tigers, elephants, leopards and bears forced to fight each other - as well as human gladiators - to the death. King Henry I created the first wild animal menagerie in Britain at Woodstock, England, which later was re-housed at the Tower of London by Henry III in the 13th Century. He introduced the first elephant to Britain, which was presented to him as a gift. Animal menageries remained private collections and it was not until the early part of the 19th Century that animal collections started to become accessible to the general public with the opening of zoological collections in Paris, Vienna, Dublin and later London. The 'zoo' concept had been established. At that time, these collections were nothing more than living museums exhibiting creatures as objects of curiosity from around the world, captured by the Victorian explorers or presented as gifts from parts of Africa and Asia in return for diplomatic favours.

The oldest still existing zoo, the Vienna Zoo in Austria, evolved from such an aristocratic menagerie founded in 1752 by the Habsburg monarchy and changed its face as well as its mission throughout the centuries. The first zoo founded primarily just for scientific and educational reasons was the *Ménagerie du Jardin des Plantes* in Paris (1794). The founders and members of the *Zoological Society of London* adopted the idea of the early Paris zoo when they initiated and established London Zoo in 1828. Until the 20th Century there had been little or no concern for the welfare of zoo animals. Animals were barely perceived to have physical needs, let alone mental and emotion needs, and many were kept in appalling conditions. The appearance of zoos however did begin to change in the 1970s and 1980s. Public opinion was changing and people started to question the role of zoos and their commitment to conservation, education and animal welfare. Zoos had no option but to begin to change. When ecology and conservation emerged as a matter of public interest through the 1970s, a few zoos began to consider making conservation their central role, with Gerald Durrell of the Jersey Zoo, George Rabb of Brookfield Zoo, and William Conway of the Bronx Zoo (Wildlife Conservation Society) leading the discussion. Since then, zoo professionals became increasingly aware of the need to engage themselves in conservation programmes and the American Zoo Association soon asserted that conservation had become its highest priority.

Aims of Zoological gardens or park

Most of today's non-profit and serious zoological gardens display wild animals not just for the amusement and the entertainment of their visitors but mainly for conservation of endangered species, for education and biological research. The concern of these institutions is to help save the diversity of life on Earth through applied conservation activities such as breeding endangered species.

The breeding of endangered species is coordinated by special cooperative breeding programmes containing international study books and coordinators, who evaluate the roles of individual animals and institutions from a global or regional perspective. There are various regional programmes for the conservation of endangered species:

Importance of Zoological gardens or park

1. Providing recreation and amusement for general people of the society.

2. Endemic and exotic varieties of animals and birds exhibited here.

3. Creating awareness among common people regarding the ecological importance of wild animals.

4. Provided easy study sites of wild animals in captivity.

5. Endanger species of animals and birds are successfully conserved and propagated here by *ex-situ* conservation.

6. It is a centre of captive breeding and opportunities for public education to enlightend awareness about endangered species.

Main objective of Zoological gardens or park (According to the National Zoo policy)

1. Coordinated conservation breeding of endangered species under ex-situ conditions (Box 13.1.1).

2. Conservation education.

3. Scientific studies for conservation.

4. Continue to function as rescue centre for orphaned wild animals.

Classification of Zoological garden or park (as per Central Zoo Authority, India)

For the purposes of deciding standards and norms for recognition of zoos and monitoring and evaluating their performance, the zoos, on the basis of number of animals, species, endangered species and number of animals of endangered species exhibited, shall be classified into four categories as specified below:

Table 13.1.1 : Classification of Zoological garden or park

Category of the Zoo	Large	Medium	Small	Mini
Number of animal exhibited	More than 750	500-750	200-499	Less than 200
Number of species exhibited	More than 75	50-75	20-49	Less than 20
Number of endangered species exhibited	More than 15	10-15	5-9	—
Number of animals of endangered Species exhibited	More than 150	100-149	50-99	—

Box 13.1.1 : Zoological gardens or park and Breeding of animals

In 1993 the World Association of Zoos and Aquariums (WAZA), formerly known as the International Union of the Directors of Zoological Gardens, produced its first conservation strategy. In November 2004, WAZA adopted a new strategy that sets out the aims and mission of zoological gardens of the twenty-first century.

In India every zoo shall participate in planned breeding programme of endangered species as permitted by the Central Zoo Authority in discussion with the Chief Wild Life Warden of the State. For this reason, the zoo operator shall exchange animals between zoos, by way of breeding, loans, gifts and the like as per the directions of the Central Zoo Authority.";. To safeguard against uncontrolled growth in the population of prolifically breeding animals, every zoo shall implement appropriate population control measures like separation of sexes, sterilization, vasectomy and implanting of pallets etc.

Box 13.1.2 : Zoological garden or park and Education and Research

1. Every enclosure in a zoo shall bear sign board exhibiting scientific information about the animals exhibited in it.
2. Every zoo shall bring out leaflets, brochures and guidebooks and make the same offered to the visitors, either free of cost or at a reasonable price.
3. Every large and medium zoo shall make arrangements for recording, in writing, the detailed observations about the biological behaviour, population dynamics and veterinary care of the animals exhibited as per directions of the Central Zoo Authority so that a detailed database could be developed. The database shall be exchanged with other zoos as well as Central Zoo Authority.
4. The zoos shall support study on the biology, behaviour, nutrition and veterinary aspects of animals in their collection. They shall also endeavour for creation of capability on zoo architecture and landscape designing, cooperation of recognised institutions already working in relevant fields in this regard shall be taken.

5. Zoos shall venture for transfer of technical skills available in the field for zoo personnel. The Central Government, Central Zoo Authority and State Government shall provide due support to zoos in these efforts. Support of Wildlife Institute of India (WII), Indian Veterinary Research Institute (IVRI) and other institutions within India and abroad, having proper expertise shall be taken in this regard.
6. Zoos shall also endeavour for distribution of information on scientific aspects of management through publication of periodicals, journals, newsletters and special bulletins. Help of non-governmental organization (NGOs) and government institutions shall also be availed in such efforts. The Central Zoo Authority shall provide technical and financial support to the Indian Zoos Directors Association (IZDA) and other institutions in this regard.

Box. 13.1.3 : Red Data List categories of animals:

International Union for the protection of nature and natural resources (IUCN) established in 1934 publishes its work in the Red data book, in which different threatened animal species are assigned the following categories (according to their conservation status in 1978):

Extinct (X) : No longer existing in the wild.

Endangered (E) : In immediate danger of extinction if factors causing decline continue operating.

Vulnerable (V) : Likely to become endangered soon if factors causing decline continue operating.

Critically rare (R) : Not endangered or vulnerable at present, but might be at risk by localized threats.

Indeterminate (I) : A temporary grouping when not enough information is available to put into endangered, vulnerable or critically rare categories, but it is reasonably certain that the species does indeed belong in one of these categories. Further data are needed to determine the status.

Uncertain (U) : A temporary grouping for unconfirmed reports suggesting inclusion in endangered, vulnerable or critically rare categories.

No longer threatened : No longer included in the above categories due to an increase in population sizes or numbers, or due to the discovery of more populations or more individuals.

New IUCN Red Data List Categories : The above categorizations are not precisely defined, and a new categorization has been made by the IUCN. The aim is to be uniform, objective, and transparent. The New Categories are :

Extinct (EX) : A taxon is Extinct when there is no reasonable doubt that the last individual has died.

Extinct in the wild (EW) : A taxon is Extinct in the wild when it is known only to survive in cultivation, in captivity or as a naturalized population (or populations) well outside the past range. A taxon is presumed extinct in the wild if exhaustive surveys in its known habitat in its historical range have failed to record an individual.

Critically endangered (CR) : A taxon is Critically Endangered when it is facing an extremely high risk of extinction in the wild in the immediate future.

Endangered (EN) : A taxon is Endangered when it is not Critically Endangered but is facing a very high risk of extinction in the wild in the near future.

Vulnerable (VU): A taxon is Vulnerable when it is not Critically Endangered or Endangered but is facing a high risk of extinction in the wild in the medium-term future.

Lower risk (LR) : A taxon is Lower Risk when it has been evaluated, does not fall into the categories Critically Endangered, Endangered or Vulnerable. There are three subcategories:

Conservation Dependent (CD) : Taxa which are the focus of a continuing conservation programme, and which would become threatened if such actions stopped.

Near Threatened (NT) : Taxa which do not qualify for Conservation Dependent, but which are close to qualifying for Vulnerable.

Least Concern (LC) : Taxa which do not qualify for Conservation Dependent or Near Threatened.

Data deficient (DD) : A taxon is Data Deficient when there is inadequate information to make an assessment based on its distribution and population status.

Not evaluated (NE) : A taxon is Not Evaluated when it is has not yet been assessed against the criteria.

Important Zoos of India

1. Nandan Kanan Zoological Park, Bhubaneshwar, Orissa.
2. Zoological Garden, Alipore, Kolkata, West Bengal.
3. Padmaja Naidu Himalyan Zoological Park, Darjeeling, West Bengal.
4. Nehru Zoological Garden, Hyderabad.
5. National Zoological Park, New Delhi.

Preparation of report of Excursion to a Zoological garden or Park :

Prepare the report on the following :

- Introduction
- Name of the Zoological Garden or Park.
- Location (City, District, State etc.).
- Period or time of visit.
- Weather conditions.
- Route followed (distance from the origin i.e. place from where visited; mode of travel; route followed.).
- Composition of the visitor, i.e., number, age, sex, status(teacher, non-teaching, students and others).
- History of the zoo visited.
- Size and zoo organization/Management.
- Animals conserved [Common name, scientific name, zoo-geographical distribution, sex composition, conservation status, description of the enclosure environment (IUCN status in red data list) etc.]. It may be reported in tabular form according to various taxonomic groups.
- In situ conservation and breeding program taken up by the respective zoo.
- Conclusion.

References of Zoo

1. Colin Tudge: *Last Animals in the Zoo: How Mass Extinction Can Be Stopped*, London 1991. ISBN 1559631570.

2. http://www.biaza.org.uk/resources/library/images/MANIFESTO.pdf John Regan Associates: *Manifesto for Zoos*, 2004.

3. http://www.waza.org/conservation/wzacs.php The World Zoo and Aquarium Conservation Strategy.

4. Vernon N. Kisling (ed.): *Zoo and Aquarium History*, Boca Raton 2001. ISBN 084932100X.

5. R. J. Hoage, William A. Deiss (ed.): *New Worlds, New Animals*, Washington 1996. ISBN 0801851106.

6. Elizabeth Hanson: *Animal Attractions*, Princeton 2002. ISBN 0691059926.

7. David Hancocks: *A Different Nature*, Berkeley 2001. ISBN 0520218795.

8. http://www.africultures.com/anglais/articles_anglais/43blanchard.htm].

9. Stephen St C. Bostock: *Zoos and Animal Rights*, London 1993. ISBN 041505057X.

10. Bryan G. Norton, Michael Hutchins, Elizabeth F. Stevens, Terry L. Maple (ed.): *Ethics on the Ark. Zoos, Animal Welfare, and Wildlife Conservation*, Washington, DC 1995. ISBN 1560985151.

11. Randy Malmud: *Reading Zoos. Representations of Animals and Captivity*, New York 1998. ISBN 0814756026.

Chapter 13.1(b)

Visit to a Museum

Introduction :

Museum is a place where all object of national importance such as antiquities, the arts, and science etc. are preserved. Here one can observed and study fossils, stuffed specimens, models and other things of past and present.

Indian museum in Kolkata is the oldest and largest museum in India. This multipurpose museum of national status was founded in 1841. Museum display various departments and galleries. Students of zoology should visit the palaeontological section/gallery, which is the largest and best collection of vertebrate, and invertebrate fossils in the world are exhibited. Visiting different galleries students of Zoology should keep a record for preparing there field report. While preparing report students should mention the date, time and guide name.

Importance of Museum :

1. Museum aims at acquisition, preservation, study in all subjects of national importance and dissemination of knowledge and reaction through them.

2. The collection highlights the culture, history and the natural sciences and knowledge (Source: Indian Museum 175th year; Publication).

3. Here one can observed and study fossils, stuffed specimens, and models etc. of past and present and deduced inference about their phylogeny, habit and habitat.

Preparation of report of Excursion to a Museum :

Prepare the report on the following :

- Introduction :
- Name of the Museum.
- Type of Museum (National or State museum, Museum associated with a college and University or an Institute.).
- Location (City, State etc.).
- Period or time of visit.
- Weather conditions.
- Route followed (distance from the origin i.e. place from where visited; mode of travel; route followed.).

- Composition of the visitor, i.e., number, age, sex, status(teacher, non-teaching, students and others).
- History of the Museum visited.
- Importance of the Museum.
- Types and number of Galleries (related to the interest of student of zoology, such as vertebrate, invertebrate, palaeontology etc.) in the museum.
- Description of each gallery (Animal displayed, way in which preserved, description about its distribution, ecology etc.).
- Museum administration.
- Conclusion.

Chapter 13.2 (a)

Visit to a Forest

Introduction :

Forest is any land managed for the diverse purposes of forestry whether covered with trees, shrubs, climbers etc. The forest biomes include complex assemblage of different kinds of biotic communities.

Before visiting a forest one should know about the forest communities.

Types of Forest :

Forests are composed of trees growing sufficiently close together to determinate the entire area of ground surface. Forest biomes of the world have been classified as: **Coniferous forest, tropical forest** and **temperate forest**. In cold climates most forests are **needle-leaved evergreen** (Deodar forest of Kashmir); in seasonally warm, moist climates, they are mostly **broad-leaved deciduous** (Oak forests of Himalayas), (Teak forest of central India); and in continuously warm and moist climates, they are **broad-leaved evergreen** (forest of central India).

Deciduous trees shed their leaves in the autumn, are bear over winter and grow new foliage in the spring. Evergreen coniferous trees, on the other hand, retain their foliage throughout the year, although dried leaves fall at all seasons. Difference in size, shape and structure of the leaves are important to many animals. Large areas of mixed grass, shrubs and trees may provide a good habitat for animals.

Forest in India :

Forests type in India can be broadly classified into coniferous forests and broad leaved forests. They can be further classified into nature of their tree species - **evergreen, deciduous, xerophytes, mangroves** etc. Various type of forests along with important inhabiting flora and fauna shown is in the Table 13.2.1.

Under the forest law of India the above mentioned forest are grouped into: **Reserved forest, Protected forests, village forests** and **unclassed forests.** In reserved forest every thing is prohibited unless specially permitted. Protected forest is not reserved, but every thing permitted unless specifically prohibited by the government. Village forest occurring in the vicinity of the villages served the need of concerned village with

Table 13.2.1 : Forests types in India and important inhabiting flora and fauna

Forests types	Flora	Fauna
Himalayan coniferous	Pine, deodar	Wild goats, sheep, snow leopard, Himalayan brown bear, Himalayan wolf, musk deer
Himalaya broad leaved	Maple, oak	Deer, bears, squirrels, gray foxes, wood-peckers
Evergreen north east, Western ghats and Andaman and Nicobar Islands	Jamun, *Ficus*, Dipterocarpus	Samber, Malabar whistling thrush, Malabar pied hornbill, Tree frogs, Rhino, pigmy hogs, Lion tailed macaque
Deciduous dry	Teak, Ain, *Terminalia*	Barking deer, Chital, Tiger, Fly catchers, Babblers, Hornbill
Deciduous moist	Sal	Arboreal amphibian chameleous, snakes, many species of birds
Thorn and scrub semi-arid forests	Babul, Ber, neem	Chinkara, four horned antelope, Partridge, Florican, Bustard, Monitor lizard
Mangrove delta forests	*Avicenia*	Tiger, Water monitor lizard, crocodiles, shore bird-sand pipers, plovers, fish, crustaceans

regards to the forest produce. All other forests are considered as unclassed.

Available animals :

(a) **Ground animals :** Ground animals are grouped on the basis of size into **megafauna**: large millipeds, centipedes and snails, amphibians, reptiles, small mammals; **microfauna**: large insects and spiders, small millipedes, centipedes and snail, earthworms; **mesofauna**: Pot worms, small arthropods such as spring tails, symphyla, pauropods. proturans, mites, insect larvae and **microfauna** : protozoans, rotifer a nematodes, tardigrades, turbellarians. In addition to invertebrates there are several vertebrates like mice, shrews, moles, amphibians and reptiles are considered as part of the groundfauna.

(b) **Foliage Arthropods :** Spider, ants, flies beetles, leaf hopes, bugs and larvae represented by large numbers.

(c) **Birds :** In a forest there are breeding and non-breeding bird populations.

(d) **Mammals :** Rodents (mice, chipmunks, squirrels) and insectivores (shrews, moles) are the most abundant small mammals of forest communities. Large mammals include small hopped cursorial ungulates, fast moving ungulates, insectivores, jungle cats, ant eater, giant flying squirrels, monkeys, sloths, tiger, elephant, samber deer, gaur, spotted deer, swamp deer etc.

Preparation of report of Excursion to a Forest :

Prepare the report on the following :

- Introduction :
- Name of the Forest.
- Location (Place, District, State etc.).
- Period or time of visit.

Box 13.2.1 : Glimpses of Indian Wild life heritage

Indian sub continent is unique in having immense natural beauty in its different biomes and also in possessing a rich and diverse wild life fauna. Various threatened and protected species along with other animals are kept in **Reserved forest**, **National Park** and **Sanctuaries**.

- Weather conditions.
- Route followed (distance from the origin i.e. place from where visited; mode of travel; route followed).
- Composition of the visitor, i.e., number, age, sex, status(teacher, non-teaching, students and others).
- Preparation of the map of the study area showing various land uses and where resources are collected from.
- To identify the forest type(plantation of natural forests; evergreen/semievergreen/deciduous/dry-deciduous/thorn forest).
- Whether natural forest or a plantation.
- Whether disturbed/partially disturbed/mildly disturbed/mildly degraded/severely degraded.
- Listing of natural resources, goods and services
  - o Goods: Food, fuelwood, fodder, water etc.
  - o Services: Water regime, climate control, removal of CO_2, oxygen and nitrogen cycle etc.
- Identifying the layers of the forest - ground, undergrowth, trunk, branches, and canopy.
- Listing of inhabiting flora and fauna in the forest (Diversity and abundance of life)
  - o If feasible appreciation of abundance of plant species(identify the most abundant, with count of individuals in different plant species)
  - o If feasible appreciation of abundance of animal species(identify the most abundant, with count of individuals in different animal species)
- Observation of food chains: To find out what all animals are feeding and reconstruct as many as food chains possible. To find out what a single species can play a role in different food chains. Thus food web may be constructed. Count the number (or biomass) of each component: autotrophic and heterotrophic (species wise) in each food chain. Thus a food pyramid can be constructed.
- Information on the forest administration.
- Conclusion.

Chapter 13.2(b)

Visit to a Grassland

Introduction :

There are some landscapes where the vegetation is predominantly the grasses and small annual plants. Grasslands are found where the annual rainfall is low or the soil depth and quality is poor. The low precipitation prevents growth of various trees and shrubs. Grasslands have a highly seasonal appearance. During the summer the above ground part of the grass dies. In the next monsoon the grasses grow back from the root and the seeds of the previous year. Grass land is inhabited by variety of grasses, herbs, and several species of insects, reptiles, birds and mammals. The grasses are the major producers of the biomass the grasslands. Each grass lands have a wide variety of flora and fauna. Grasslands are grazing areas in many rural communities. Cattle and goats are highly dependent on the grasslands. Grass is also used for thatching houses. The thorny branches and bushes are used as fuelwood. Some grass lands are sensitive to overgrazing and overused. Grasslands may also be destroyed by fire.

Type of grasslands of India :

- **Himalayan pasture belts :** This extends upto the snow line,there are grass lands at the lower form that patches along with the coniferous or broad leaved forests.

- **Terai :** This consists of tall grasslands interspersed with Sal forests. This constitutes the tall elephant grass (ranges above 5 metres.). The terai also includes marshes in the low lying depression.

- **Semi-arid plains of Western, Central India and Deccan :** This composed of grass lands tracts with patches of thorn forests. The scrublands of the Deccan plateau are covered with seasonal grasses and herbs.

- **Shola grasslands :** It consists of patches on hill slopes that occur along side of the shoal forests on the Western Ghats, the Nilgiri and Anamalai range.

Preparation of report of Excursion to Grassland :

Prepare the report on the following :

- Introduction :
- Type of Grassland and its location (Place, District, State etc.).
- Period or time of visit.
- Weather conditions.
- Route followed (distance from the origin i.e. place from where visited; mode of travel; route followed.).
- Composition of the visitor, i.e., number, age, sex, status (teacher, non-teaching, students and others).
- To get information on variety of plants and animals (if possible with seasonal variation, information on unique species of plants and animals and dominant species).
- Abundance of grassland species (Quadrat sampling method).
- Finding out the grassland carrying capacity i.e., number of animals the grassland can support.
- Observation of food chains: To find out what all animals are feeding and reconstruct as many as food chains possible. To find out what a single species can play a role in different food chains. Thus food web may be constructed. Count the number (or biomass) of each component: autotrophic and heterotrophic (species wise) in each food chain. Thus a food pyramid can be constructed.
- Usage of the grassland by local people, preparation of land use map during last few years. To observe differences in protected and degraded areas.
- Products that man use which come from the grassland viz., milk, meat etc.
- Threats to the grassland (degradation by overgrazing, burning, urbanization, conversion for other use etc.).
- Conclusion.

Chapter 13.3(a)

Visit to Sea Shore

Introduction :

Seashore is the ideal site for zoological field study as the sea contains multiudes of plants and animals. Such forms may vary from the minute algal plants (phytoplankton) larger animals such as the jelly fishes, plankton (Zoö plankton) and different phyla of invertebrata and Chordata.

The **Seashore**, with its numerous rock-pools exposed at low tide and usually covered with vast areas of seed weed, often arouses the desire to explore. Seashore is the ideal sites for the students of Zoology as the sea is the reservoir for multitude of animals, both micro and macro fauna.

Miles of coastline are washed by the sea. The shape of the coastline always undergoes changes, due to waves and wind action. The mass of water, called as the **swash** or **send** is powerful enough to carry sand and single, large stones – upon the beach. Some of the swash percolates downwards into the beach sand or shingle but the rest known as back wash, runs off the surface into the sea carrying beach material. This shingle further broken down into sand deposited to form a **Sandy beach**. The finest particles of material are deposited as mud, and this along with the mud carried down by river water, form a **muddy shore**. This type of shore is characteristics of estuaries.

A shore with rock pools is usually the most popular collecting ground for the zoologist. These provide a suitable substrate for hydroids and polyzoans, crustaceans, barnacles and a few molluscs such as *Patella* and *Mytilus*.

Sandy beaches afford easier passage to many animals that burrow into this medium and there the burrowing animals live, concealed and protected from desiccation at low tide. Typically inhabitants are the polychaete worms and many molluscs.

Estuaries muddy shores are termed **mudflats**. Mud contains much organic matter that forms the food for many burrowing forms, especially polychaete. A high and low tide occurs on coasts approximately every twelve hours. In September and March usually about 21st when the day is equal to the night, the tidal equinoxes occur. These are the periods of the highest and lowest spring tide of the year and it is only at these times that one can collect from the lowest part of the expressed shore. The average high and low tide levels occur between the spring and neap tides. The **lower shore** is only uncovered at low spring tides and the upper shore covered only at high spring tides. Above the upper shore is the splash zone.

Thus the nature of sea beach varies in different places. Student should note in their field report about the topography and nature of the beach they have visited along with date and time of visit, temperature chart, tide time, sunrise and set time and name of the guide.

A profile of a sea shore :

Digha seashore : Digha a well-known town is situated on the bank of Bay of Bengal in district of East Midnapur, 180 kilometres away from Kolkata. Digha is connected by both rail and bus service with Kolkata and other part of the country. The sea beach of Digha is provided with miles of hard golden sandy beach along with the blue deep of the sea-shady jhau (casuarinas) tree facing the cool breeze of the sea. The sea is clam and transparent in winter but rough in summer. A beauty spot of rear charm, Digha with its varieties of macro and micro fauna would continue to become attractive to the young explorer.

Work in the Field :

1. Observing, collecting, narcotizing and preserving sea shore animals :

Many sea-shore creatures are seen at their best when alive, either in their natural habitat surroundings or in some form of temporary receptacle, for prolonged observation an aquarium is desirable. For field study one can use a bucket or rectangular tray not affected by corrosive action of seawater. The observing receptacle should be filled to only half its capacity. Sea water collected from the sub littoral zone is suit able for the study.

2. **Collecting :** Collecting can only be done successfully at low tide. Always consult a tide table before going to the shore.

(i) **Collecting apparatus :**

1. **A wide mouth vacuum flask** — for transporting live animals.
2. **Plastic buckets** — for washing out animals from sea weds and later for anaesthetizing them.
3. **A stout knife** — for prising off chintons or limpets from rock faces and for cutting free hold fasts of seaweeds.
4. **A small hand net with a stout frame and muslin bag** — for sweeping under crevices of rock pools.
5. **A stout spade for muddy shore** — for digging up the burrowing forms.
6. **A selection of polythene bags** and **screw cap plastic jars** — for transporting the collection home.

(ii) **Narcotizing :** Animals chosen for preservation should be undamaged. Narcotization is performed by adding some anaesthetic substance very gradually in a very small does to the water containing animals. A 5% solution of formalin (1 part commercial formaldehyde to 19 parts sea- water) is added to the water, drop by drop until the animals are anaesthetized and no longer respond to touch. During narcotization some form of aeration is essential. This can be done by slowly rotating the water by hand. Weak percentage of alcohol solution, magnesium sulphate, and chloral hydrate can also be used for anaesthetizing animals.

(iii) **Preserving :** After anaesthetizing, the animals are properly preserved so that it may remains for a longer period. It must be kept in a well-stoppered bottle and clearly labelled. Labelling should be done on good quality paper and can be written with a soft pencil or in Indian ink.

Arthropods, Mollusca and Echinoderms are preserved in a mixture of alcohol, glycerine and water in about equal parts with addition of little aceticacid. (0.5%-0.75%); Coelenterate and Annelids are preserved in 5% formalin and little glycerine with addition of a little acetic acid (0.5%).

Animals collected from fishing boats/ trawlers or from shore dragnet can be preserved directly in 10% formalin and little glycerine with addition of a little acetic acid (0.5%). Only hard shell animals such as molluscs, small crabs and shore insects can be satisfactorily preserved in a dry state.

3. **List of Specimens collected** : The specimens collected from the sea beach should be properly identified. Students should write name, scientific name and systematic position of the identified animals in their field report.

4. **Important excursion sites in south India :**

(a) **Waltair** (Vizagapattanam) :

(i) **Nature of beach:** Mostly rocky around fishing port areas and sandy at RamKrishna Beach.

(ii) **Location:** On the East Coast of Bay of Bengal.

(b) **Marina Beach of Chennai :**

(i) **Nature of beach:** Completely sandy.

(ii) **Location:** South coast of Bay of Bengal.

(c) **Rameswaram :**

(i) **Nature of beach:** Mostly sandy, partly rocky.

(ii) **Location:** An island in the Palk-straits at extreme south eastern tip of the Indian Peninsula, 666 km from Egmore station of Chennai

(d) **Pamban :**

(i) **Nature of beach** : Partly rocky and sandy.

(ii) **Location** : 17 km from Rameswaram town and is connected by a bridge with Indian Peninsula.

(e) **Mandapam :**

(i) **Nature of beach** : Partly rocky and sandy.

(ii) **Location** : 9 km from Rameswaram town.

(f) **Krusadae Island :**

(i) **Nature of beach** : Sandy and muddy.

(ii) **Location** : 20 km from Rameswaram town and lies in the west of Pamban Bridge.

(g) **Kanyakumari :**

(i) **Nature of beach** : Western Part rocky and eastern part sandy.

(ii) **Location** : 714 km from Egmore station of Chennai and 294 km away from Mandapam by bus. Southern most tip of Indian Peninsula and in the confluence of Bay of Bengal, Indian Ocean and Arabian Sea.

(h) **Kovalum beach of Thiruvanthapuram :**

(i) **Nature of beach** : Mostly sandy and partly rocky.

(ii) **Location** : 13 km from Thiruvanthapuram town at the southern part of Arabian Sea.

Preparation of report of excursion to a Sea Shore :

Prepare the report on the following :

- Name of the Sea shore/beach.
- Type of sea beach (rocky/muddy/sandy).
- Location (City, State etc.).
- Period or time of visit.
- Weather conditions.
- Route followed (distance from the origin i.e. place from where visited; mode of travel; route followed).
- Composition of the visitor, i.e., number, age, sex, status (teacher, non-teaching, students and others).
- List of Specimens collected (should be enlisted as a table exhibiting their common name, scientific name, habit, and habitat).
- Conclusion.

Chapter 13.3(b)

Study of a Pond Ecosystem

Introduction :

A pond is a good example for a small ecosystem that presents a unity both in function and structure. It is highly dynamic. The four basic components: a biotic substances, producers, consumers or macro consumers and decomposers or micro consumers, can be sampled and studied in a pond. To study a pond one should cover all its phases i.e. monsoon, post-monsoon, winter and summer. Most ponds become dry after the rains are over. A pond begins to fill during the rains its life-forms such as the algae, the microscopic organisms, aquatic insects, snails and the worms emerges from the ponds floor. Gradually more complex animals such as the fish and the amphibians return to the pond. The aquatic vegetation consists of floating weeds and rooted vegetation on the periphery. After the monsoon is over the pond begins to dry up the peripheral aquatic plants which are colonized by terrestrial plants like grasses and herbs. As it shrinks more its aquatic flora and the fauna dies which is succeeded by terrestrial flora and fauna. The flora and fauna exists only in ditch and depression containing terrestrial forms and dormant aquatic invertebrates that wait for the next season.

Preparation of report of Excursion to a Pond :

Prepare the report on the following :

- Introduction :
- Location (City, State etc.)
- Period or time of visit(s)
- Weather conditions
- Route followed (distance from the origin i.e. place from where visited; mode of travel; route followed.)
- Composition of the visitor, i.e., number, age, sex, status(teacher, non-teaching, students and others)
- Preparation of the map of the study area
- To observe seasonally active pond if possible in several occasions before, during and after the monsoon
- Observation of vegetation zone of the pond
- Seasonal observation of pond

(i) Early stage (soon after the ponds fill with water in monsoon) : Observation of microscopic plants and animals

(ii) Fully active stage: Observation on submerged and emergent vegetation; fish, amphibians, molluscs, arthropods, annelids etc.

(iii) Shrinking phase: Observation on the drying up of aquatic plant life, with dead and decaying plant material and terrestrial plants growing on exposed mud of the pond.

(iv) Dry phase: Observation of grasses and shrubs and hidden dormant animals.

- Laboratory observations :

  (i) Counting and identification of phytoplanktons

  (ii) Counting and identification of zooplanktons

  (iii) Physicochemical analysis of water and sediments (temperature, colour, turbidity, pH, hardness, acidity, alkalinity, salinity etc. (see chapter 8.1.5)

- Making a checklist of all the visible aquatic flora and fauna. To identify the most abundant animal and plant species in different season. Estimation of abundance.
- Observation of food chains: To find out what all animals are feeding and reconstruct as many as food chains possible. To find out what a single species can play a role in different food chains. Thus food web may be constructed. Count the number (or biomass) of each component autotrophic and heterotrophic (species wise) in each food chain. Thus a food pyramid can be constructed.
- Conclusion.

Chapter 13.3(c)

Study of Terrestrial Ecosystem

Since man himself is terrestrial, the crop field or grass land represents the kind of ecosystem of primary concern to him. Complexity and variegation of terrestrial ecosystems, yet the study of a terrestrial ecosystem is convenient to work with autotrophic and heterotrophic strata separately and then to relate the two.

To study the terrestrial ecosystem it is better to select a simple terrestrial ecosystem, such as a one year abandoned crop field or grassland, provides important relations between structure and function at the ecosystem level. More over fewer instruments are required for such work.

To determine the numbers, biomass, and kinds of plants for autotrophic stratum in a field, one should remove all plants from one or more locations of known area, for example a square meter quadrates and places them in a large paper bag. The plants are sorted, identified and placed in a dry oven for determination of dry weight of the above and underground portions. From the plant data by applying statistics one can get the standard deviation (see chapter 11.2) to find out the difference between the lowest and highest value. The calculation give one point : the whole proves less variable than the part: that is, biomass of a whole community is less variable than the biomass of any one component species.

To estimate the standing crop of small animals (consumers) living in the above ground vegetation of a study field, three basic methods can be applied:

Total counts from area of known size, mark and recaptured procedure and removal sampling

1. **Quadrate Method :** Total counts are made from quadrate data which is available from limited sample areas usually known as quadrate. From total count one can get a good estimate (for some species) of the standing crop of animal in only a single square meter (see chapter 8.17).

2. **Marked and capture method :** Captured sample of animal from a known area are marked with dye, the individuals are released in the population. After few days the unmarked and marked individuals are recaptured. By counting the numbers of marked, unmarked and total population one can estimate the standing crops of a known area.

3. **Removal sampling :** A fairly large area can be sweeped off with a sweep net (see chaptér 8.18) to observed the populations of a motile

organisms. By recording the successive strokes and removable sample and putting them in x and y axis of graph, one can find the probability of captured invertebrates which remain constant (see chapter 8.18).

To estimate the standing crop of small soil animals one should used Tullgren funnel/Berlese funnel (see chapter 8.2.6). One can estimate the larger consumers (shrews, mice, rats) by removal trapping or by direct count(birds). One can study the decomposed groups by any kind of removal or dissection.

Preparation of report of excursion to a terrestrial area

Prepare the report on the following :

- Introduction
- Type of terrestrial field (crop field/grass land) and its location (place, district, state etc.).
- Period or time of visit(s).
- Weather conditions.
- Route followed (distance from the place from where visited; mode of travel; route followed.
- Composition of the visitor ie., number, age, sex, status (teacher, non-teaching, students and others).
- Sampling method for autotrophic stratum (count the numbers, biomass, kinds of plants to get the standard deviation to find out the difference between the lowest and highest value).
- Sampling methods for consumers (small animals) (Total count of known size, mark and recaptured method, removal sampling).
- Sampling methods for soil consumers (Tullgren funnel/Berlese funnel method).
- Sampling methods for large consumers (trapping/direct count method).
- Conclusion.

Chapter 13.4

Visit to Solid Waste Disposal Site

Introduction :

Increase in population, migration and rise in the standard of living there is a rise in the problem of solid waste. By 2011 the projected total quantity of urban solid waste would be 56.33 million tons in India. The urban solid wastes includes: domestic wastes (kitchen, faecal matter, paper wastes, metals, glass, leather and rubber, plastics, ashes and inert materials, hospital wastes and solid wastes from commercial buildings, street refuse (road sweepings, leaves, materials thrown away by people like match boxes, cigarettes buts, fruit peels, ground nut shells, paper, envelopes etc.). Further there are also agricultural solid wastes viz., rice husks, straw, bagasse. Industrial solid wastes include the materials from various industries and mines. These are produced during their manufacturing procedure. In India, Municipal solid waste has increased tremendously. This is due to: unregulated growth of urban areas, general apathy of cleanliness, wasteful consumerism, and lack of infrastructure for collection and transportation. The effect of indiscriminate disposal of solid wastes leads to serious health concern during the rainy season, ground water leachate percolation, spread of disease such as cholera, typhoid, encephalitis etc and damage to aesthetics. Thus it is extremely important to for a country to have a effective **solid waste management** strategy. For waste disposal it is important to remember the 'three Rs' i.e. **reduction, reuse** and **recycle**. The methods of wastes disposal are :

1. **Open dumps :** Open, unregulated dumps are still a predominant method though it can affect the aesthetics and cause pollution through leachate.
2. **Ocean dumping :** Every year 25,000 metric tons of solid wastes are dumped in the sea.
3. **Landfills :** In sanitary landfills solid waste disposal is regulated and controlled.
4. **Incineration :** Burning down of the garbage is an important method of waste disposal. Further such burning produce steam which is used as a useful resource.
5. **Composting :** Organic solid wastes can be can be degraded in oxygen rich environment into organic compost. This is nutrient rich and is used for soil

amendments, aids water retention and slows soil erosion and improves crop yield.

6. **Reuse** : In many cities glass and plastic bottles can be routinely returned to producers for washing and refilling. Bricks, wood works, window glass, brass fittings can be reused.
7. **Recycling** : It is the re-processing of discarded materials into new, useful materials, For example, glass bottle, aluminium cans etc. are melted and recast to new ones.

Preparation of report of Excursion to a solid waste disposal site :

Prepare the report on the following :

- Introduction :
- Name of the Solid waste disposal ground
- Location (Locality, City, District, State etc.)
- Period or time of visit
- Weather conditions
- Route followed (distance from the origin i.e. place from where visited; mode of travel; route followed.)
- Composition of the visitor, i.e., number, age, sex, status(teacher, non-teaching, students and others)
- Source(s) from where the solid wastes comes
- How it is collected, by whom and at what frequency
- Type and quantity of solid wastes disposed (degradable and non-degradable)
- Procedure of waste management and manpower requirement
- How much of the waste can be reused or recycled
- Conclusion

Chapter 13.5(a)

Visit to Poultry

Introduction :

In the past two decade poultry in India has gained surprising dimensions. Poultry farming as a profession and as hobby is fast developing. Andhra Pradesh, West Bengal, Tamil Nadu and Karnataka are the main suppliers of poultry. Small-scale producers include states like U.P, M.P, Punjab, Haryana, Maharastra, and Gujrat. Poultry includes several kinds of birds such as fouls, ducks, geese, turkey and guinea fouls. Poultry supplies meat, eggs, feathers and rich manure.

Work in the farm :

Proper selection of poultry breed, housing arrangements, location, food supply, market position, demand etc., is some of the main consideration for poultry farming. While visiting a farm one should minutely observed the different process of farming and note it down.

Important Indian breeds of poultry :

The most common breed is the small indigenous desi foul. The most common varieties of desi fouls are Aseel, Chittangong and Ghagus.

Aseel is ideal table bird and are found around Hyderabad in Andhra Pradesh and Rampur and Lucknow in U.P.

Chittagong is common in W. Bengal, Bihar and Assam.

Ghagus is found in Andhra Pradesh and Karnataka.

The exotic breeds of foul are the chief supplies of poultry meat and eggs. The eggs are large sized with thin shells.

Selection of breed :

Different breeds are used for raising eggs and meat. For eggs, hens having high laying qualities are to be selected. Similarly for meat, breeds which putting on more and more weight rather than laying eggs are to be chosen. Leghorn breeds are usually preferred for egg production. Hybrid chickens are considered for commercial boiler production.

Forms of Poultry farms :

There are two forms of poultry farm;

Table 13.5.1 : Standard Classes, breeds and varieties of Fowl

Class and its characteristics	Common breeds	Varieties
Asiatic Feathers shanks, skin yellow, earlobes red, massive size, loose plumage, eggs brown shelled.	1. **Brahma** Body circular shape, pea comb. 2. **Cochin**	Light, Dark and Buff. Buff
English Clean shanks, white skin, earlobes red, size medium, single comb, eggs brown shelled.	1. **Australorp** Deep oval body, shanks black 2. **Orpington** Low set, square body 3. **Sussex** Long rectangular body	Black Black, White, Buff and blue Deep rich red
American Clean shanks, skin yellow, medium size, brown-shelled eggs, for meat and eggs.	1. **Rhode Island Red** Long rectangular body 2. **New Hampshire** Board, deep and round body 3. **Plymouth Rock** Short, single comb	Single comb and Rose comb Single comb Barred and White Buff
Mediterranean Shanks clean, yellow or white skin, earlobes white, small size, feather light, white-shelled eggs, for eggs.	1. **Leghorn** Oval body, skin yellow 2. **Minorca** Long rectangular body, skin white	Single comb white, Single comb buff and Single comb light brown. Single comb. Rose comb. Black and white.

1. **Semi-intensive farm :** This is common in villages and small towns where plenty of open environment is available. Approximate area required 250 sq.mtrs, sufficient for 50-55 birds. Birds move freely in the open yard which is well protected from enemies and fire. Fencing with bamboo curtains or thatched walls is erected around the yard to protect the adult and chicken from strong cold or hot winds. Nets normally made of clay or wooden boxes are placed to house the birds at night.

2. **Intensive Poultry farms :** These farms are suitable to hold a large number of birds in short space. The birds are housed in (i) deep litter or in (ii) cages.

(i) **Deep litter :** Large pens, housed 250 birds each. The floor is covered with 15 cm thick bed of litter like rice husk, ground nut and cotton seed shells, bhusi, finely broken corn cobs, rice straw, sugar cane fibers and sawdust. The floor with the dropping of the birds and organic materials are exposed to the micro organism that causes decay and degeneration resulting into a highly porous and absorptive covering. Old litter is replaced at least 2-3 times in a year. Best period of renewal is May and November. This litter system is small enough to keep it clean and birds are safe. The litter serves as the food and is also a good manure. The litter materials are less costly.

(ii) **Battery system or Cage layer :** Hen is housed in cage. The cage is normally 45 x 35 x 45 cm and its floor is made of 2.5 x 2.5 cm galvanized wire net. The net is set in such a fashion that it slops from back down to the front, so that eggs roll out into a collecting tray. The cages are stacked in series 6-8 in a row, some time in two tiers. Below the cage a galvanized tray was placed for the droppings of the birds. The water and food bowls are kept inside the cages. Cages are housed under a tin shed with proper light and ventilation.

Food :

Food should be balanced and contain vitamins A and D., cod liver oil, yeast , dried milk powder, fish meal or other proteins and minerals. Feeds given to the poultry consists chiefly of cereals, cereal by products, animal proteins and green leafy vegetables. For raising boiler birds (table bird), the proportion of proteins is slightly more than for the pullets.

Causes of diseases in the poultry birds :

Farmers do not suffer much to market price fluctuations, but incur heavy losses due to diseases. Over crowding; unhygienic conditions, malnutrition, dampness, proper ventilation, sunlight and fresh air are some of the causes.

Preparation of report of Excursion to a Poultry farm :

Prepare the report on the following :

- Introduction
- Name of the Poultry farm.
- Location (Locality, City, District, State etc.).
- Period or time of visit.
- Weather conditions.
- Route followed (distance from the origin i.e. place from where visited; mode of travel; route followed.).
- Composition of the visitor, i.e., number, age, sex, status(teacher, non-teaching, students and others).
- Type of Poultry farm housing systems (semi intensive or intensive).
- Type of products (meat or egg).
- Type of poultry breed used and breeding system followed.
- Information on housing arrangements and equipments used.
- Food for the birds and poultry nutrition.
- Disease of the breeds and treatment.
- Manpower requirement and farm management and production.
- Marketing of products.
- Conclusion.

Chapter 13.5(b)

Visit to Sericulture Farm

Introduction :

The poor silk worm produces silk. Silk worm has been a creature of domestication, their eggs developed on mulberry leaves (*Morus alba*). Silk industry is based on a product released by the silk worm. The rearing of silk moth and production of silk is known as **sericulture**. Sericulture or the commercial production of silk is an important industry which has two definite phases, one is sericulture proper and the other is Raw silk industry. Common silk is produced by the silk worm *Bombyx mori*. Several other forms of *B. textor*, *B.fortunatus* and *B. meridionalis* are well known.

Work in the field :

Sericulture or commercial production of silk is an important industry. Silkworms are fed and tended from egg to cocoon very carefully. While visiting a farm one should minutely observe the different process of culture and note it down.

1. Selection of eggs or cocoon : Only pedigreed strain of silk worms is allowed to propagate from culture determined to be disease free. Creamy white moths are allowed to mate on a large hard board, where they are surrounded by a small glass bangle called **cellule.** Eggs are small poppy seed like, yellowish white color. Generally female laid nearly 500 eggs in 24 hours.

2. **Rearing of the silk worm :**

(a) **Rearing tray :** Made of bamboo mats or straw boards with eggs are placed on the tray with raised edges.

(b) **Rearing room :** Thatched hut or **Kachcha** house with proper ventilation. Temperature 20°c to 28°c, relative humidity 30% to 90%. Mats and trays regularly washed with copper sulphate solution and fumigated with sulphur.

(c) **Feeding tray :** Rearing tray arranged in 2-5 tiers supported by bamboo poles which are fixed into floor and looks like a machan. Hatched out larval fed upon soft sliced fresh mulberry leaves. Young larvae grow to 3-4 cm in length in 4-6 days.

(d) **Chandraki or Spinning tray :** After moulting 4 times larvae became fully grown. The dull and sluggish larvae are transferred to new trays or mats called

chandraki, made with bamboo split or straw. Here larva begin to pupate and forms the cocoon. Selected cocoon are allowed to mature and moths are bred for new generation.

3. **Processing Cocoon for raw silk :** After about 8 to 10 days of cocoon formation rearing cocoons are treated by steam or dry heat to kill the pupae. Dried cocoons are assorted according to colour and texture.

4. **Reeling and extraction of skill :** Cocoon is then soaked in warm water in large tubs. The outer loose terminals of thread is skill fully unreeled. Reeling is done by reeling machine. 4 to 5 threads are picked out and twisted together and passed it through the 'glass eye' on the reeling pool. 200-500 reeling machine form a unit called **filature reeling**. Reels of raw silk after complete drying are taken to another machine to form **'Skeins'** or **'lachchi'**, each having about 18000 meter of silk yarn. Left out threads is used as 'Silk waste' and formed into 'Spun silk'.

Non-mulberry sericulture :

It is known as **forest** or **wild sericulture**. The principal non-mulberry silk are **Tasar, Eri** and **Muga.**

(I) Tasar silk culture :

Reared in the tropical and temparate zone. The zones are Bihar, Madhya Pradesh, Orissa, West Bengal (Purulia district), Andhra Pradesh, Maharashtra, Karnataka, Manipur, Nagaland, Assam, Meghalaya, Jammu and Kashmir, Himachal Pradesh and Uttar Pradesh.

Two species are used for commercial production : *Antheraea mylitta* , (tropical tasar moth of India) and *A. proylei* (temperate tasar moth of India). Males are brown with a narrow abdomen, while females are grey and yellow with a distended abdomen.

(1) **Food plant :** *Terminalia tomentosa* (Asan), *T. arjuna* (Arjun), *Shorea robusta*(sal), *Eugenia jambolane*(jam), *Lagerstroenia parviglora* (sidha) *Zizyphus jujuba* (kul), and *Terminacia caiapa* (jungly badam).

(2) **Rearing :**

(a) **Egg collection :** Silk moth copulate in darkness at 24-26° in tropical zone. Female usually laid eggs for 6-7 days. In temperate zone moth starts laying after 24 hours of copulation and lay for 3-4 days. Eggs are deposited in batches of 5-10. Eggs are collected in leaf cup and tied to food plant.

(b) **Larva :** It is dull brownish yellow with a black head at first instar stage but turns green and head brown after 8 hours. Larva moult 4 times. Pupation period is 4-6 days.

(c) **Cocoon :** It is generally yellow or gray. Cocoon is taken for processing.

(3) **Reeling :** After sun dry one or two days, cocoon are subjected to degumming by boiling and then spun on spinning wheels.

(II) Muga culture :

Muga silk moth *Antheraea assama* is native of Assam; it is brown colour, polyphagous, semi-domesticated and multivolatine. Moths are confined to Brahmaputra valley of Assam and foot hill of East Garo hills to Meghalaya.

(1) **Host plant :** *Mychilus bombycina* (som), *Litsaea polyantha* (soalu), *L. citrita*(Mexankori) and *Michelia champaka*(Champa).

(2) **Cultivation :** Larval stages are wild. Larva undergoes 4 moults. Pupal stages 14 days in summer and 40 days in winter. Spinning of cocoon 3 days in summer and 7 days in winter.

(a) **Seed collection :** Farmers collects cocoons from villagers or from government farms. Cocoons are kept in the house in a bamboo basket where adult emerged and copulate.

(b) **Rearing :** Newly born active moths copulate with stationary female in darkness. With in 7-12 days female lays nearly 200 eggs. Emerging larva is tied in bunch to dry leaves of mango or jack fruits so that larvae can remain within the network of the leaves. Here a larva starts to spin cocoon. Spinning completed in 6 days.

(c) **Extraction :** Cocoons are boiled in solution of soda or potash, washed and rewashed in clean water, dried and spun and reel in vir or **Bauri**, which is a modified spinning wheel.

(III) Ericulture :

The eri silkworm, *Philosamia ricini* , is largely confined to Assam. The food plant is castar (*Ricinus commnis*). It is domesticated, multivoline, having 5-6 life cycles in a year. Male eri-moths are brown black and smaller than female. The life cycle of eri worm has the same pattern as the muga worm or tasar worm. Complete life cycle 44 days in summer and 85 days in winter. Spinning is completed in 3 days in summer and 5 days in winter. Larva life last for about 17 days.

Cocoons are open-mouthed and thus filaments are not continuous and reelable. Cocoons are exposed to sun dry for 1-2 days, them degumming by boiling in a solution of soda or potash. Filaments can be spun on **taklish** or on spinning wheels.

Sericulture in India : India is the only country producing all the four kinds of silk. Mysore is the biggest centre for production of raw silk in India.

Some important research organizations :

1. Central Sericulture Research station, Behrampur, West Bengal.

2. Sericulture Research Institute, Channapatnam, Karnataka.

3. Central Muga and Reasearch station, Titabar, Assam.

4. Central Tussar Reasearch station, Ranchi, Jharkahand.

5. Central silk Board, Ministry of textiles, Govt. of India, Bangalore and its various branches in different states of India.

Preparation of report of Excursion to a Sericulture farm :

Prepare the report on the following :

- Introduction
- Name of the Sericulture farm.
- Location (City, State etc.).
- Period or time of visit.
- Weather conditions.
- Route followed (distance from the origin i.e. place from where visited; mode of travel; route followed).
- Composition of the visitor, i.e., number, age, sex, status (teacher, non-teaching, students and others).
- Information of silk worm culture.
  - o Type of Sericulture (Mulberry or Non-mulberry).
  - o Species of silk worm cultured.
  - o Species of host plant used/cultured.
  - o Season of rearing (all the year round or particular season).
  - o Give description of seed collection, rearing.
  - o Grading of cocoon.
  - o Reeling and extraction etc.
- Farm production (Quantity of silk reeled, amount of silk waste).
- Information on any silk worm disease and predator.
- Manpower involved.
- Administration of the farm.
- Marketing of the product.
- Conclusion.

Chapter 13.5(c)

Visit to an Apiary

Introduction :

Honey bees (*Apis sp.*) are among mankinds most valuable domesticated insects. India has three varieties of honey bees: *Apis cerana indica, A. dorsatta, A. florae. A. cerana indica* is a greatly domesticated species even in artificial hives. Bee rearing or bee- keeping is called **apiculture.** Honey bees give us Honey and wax, apart from this they are important pollinators of fruit and seed crops. Details about the cost system, their anatomy, social life, reproduction, apiculture etc. are described in the text book of invertebrates and economic zoology book, here in this edition we have in a nut shell described about bee keeping or rearing. Students while visiting an apiary should minutely observed the various processes of bee- management in the hive.

A large number of apiaries are located in Maharastra, Tamil Nadu, Kerala, Karnataka, Himachal Pradesh and in the foot hills of Uttar Pradesh where honey is obtained on commercial basis. The central bee keeping research station was first established in the Punjab (1945).

The beehive :

Langstroth was the founder of Apiculture. He used the movable frame hive to provide honey bee colonies with suitable places in which they build their nests.

Modern Method or frame hive method :

A typical **Langstroth's hive** is formed with a large brood chamber adjusted on a wooden platform called bottom board with slit for the exit and entry of the bottom. In the chamber, there are a number of frames in each of which a wax sheet is arranged in a vertical position by a couple of wires. On the margin of these wires, bees rear their broad. A part of the broad chamber is excluded as the "queen – excluder", made from wood or sheet zinc perforated with slots large enough to allow worker bees to get through them but not the queen. Over the brood chamber is placed one or more extracting chambers called **super** containing combs in which the honey will be stored. Above the **super** is a **crown board** or **inner cover** with a feed hole in the centre. Above the inner cover is a well ventilated roof or inner cover with a feed hole in the centre. Above the inner cover is a well-ventilated roof or **top cover** flat or sloping.

Such artificial hives are kept in a garden with plenty of open space and sunshine.

Management of the bees in the hives :

The worker bees store the honey into the honey cells or combs and seal them with cappings. When the super is three quarters full of honey it is lifted and an empty super is replaced over the brood chamber. The first super is then placed on over the empty super. In this way a pile of supers is made. This arrangement facilitates removal of honey from the super. The bees can be excluded from reaching the first super by placing the inner cover beneath the extracting super. The extracting super is taken to the super clearing and extracting honey department of the apiary. Here the cappings are removed and the combs are centrifuged and thus the honey is collected in a container. The same comb can be used again and again. The empty combs are taken to the wax- collecting section of the apiary.

Equipment for handling of bees :

In the apiary the following equipments are used :

1. **Smoker :** To subdued or over powered the honey bees with smoke.
2. **Hive tool :** It is a flat piece of steel with one end broad and the other end sharpened it is used to scrap the bee resin and the extra growth of the hive.
3. **Overall :** Used to protect from the stinging of the bee.
4. **Bee-Veil :** It is used as face veil to protect the face and eyes from the sting of the bees.
5. **Bee-brush :** Used to brush aside the honey bees from the hive.
6. **High boots :** These are pair of gum boots that protects the farmers from attack of the bees of its ankles.
7. **Comb cutter :** With this old and worn out combs are cut away.
8. **Swarm bag :** The hives are stocked in the swarm bag.
9. **Conical queen cages :** Queens are transported in these cages.
10. **Division board feeder :** It is wooden trough of the regular Langstroth frame with its shoulders so made that it may hang in the hive boxes just like other frame. A wooded strip is pushed in an inclined fashion acts as float and a perch for the bees to sit and sip sugar solution.
11. **Wire embedded :** It is a device embedded in the comb foundation wax sheet on to the wires of the hive frame. Due to the presence of hexagonal impression on the foundation workers have ease to construct the cells to complete the comb.
12. **Cold uncapping knife :** It is a steel knife which is used to remove the seal of the honey cells.
13. **Steam uncapping knife :** Uncapping becomes easier with a double walled knife that is heated with steam.
14. **Uncapping basket :** It is a basket where caps of honey are collected.
15. **Rectangular Queen cage :** A simple rectangular queen introducing cage which is made up of large mesh wire gauge rolled into a rectangle or a cylinder.

Beekeeping Organisations in India :

1. All India Beekeepers' Association.
2. All India Khadi and Village Commission.

Training Centres :

1. **U.P.** : Goverenment Apiary, Jeolikote, Nainital.

2. **M.P.** : Maganwadi Apiary, Wardha.

3. **Maharastra** : Beekeeping centre, Mahabaleshwar.

4. **Tamil Nadu** : Agriculture College, Coimbatore.

5. **Uttarakhand** : Government Beefarm, Katrain, Kulu.

6. **West Bengal** : State Khadi and village industries Board; Baruipur bee keepers' Association, Baruipur, 24 Parganas; Midnapore Bee Keepers' Association, Midnapore.

Preparation of report of Excursion to a Apiculture farm

Prepare the report on the following :

- Introduction
- Name of the Apiculture farm.
- Location (City, State etc.).
- Period or time of visit.
- Weather conditions.
- Route followed (distance from the origin i.e. place from where visited; mode of travel; route followed).
- Composition of the visitor, i.e., number, age, sex, status (teacher, non-teaching, students and others).
- Information about silk worm culture.
  - o Type of Bee reared.
  - o Life history of bee reared.
  - o Method followed : Primitive or Modern method.
  - o Equipment used.
  - o Bee products produced : Honey, Beeswax.
  - o Extraction of bee products (write on technique followed).
  - o Season of rearing (all the year round or particular season).
  - o Give description of seed collection, rearing.
  - o Grading of cocoon.
  - o Reeling and extraction etc.
- Diseases and enemies of Honey bees such as Nosema diseases, Amoebic disease, Acarine disease, Brood disease; Moths, Beetles, Wasps, Ants, Lizards, Toads and Frogs, Birds and Mammals.
- Farm production (Quantity of bee product produced).
- Manpower involved.
- Administration of the farm.
- Marketing of the product.
- Conclusion.

APPENDIX

SECTION AGENDA

Appendix A

Preparation of Important Fixatives, Stains and Buffers

1. Fixatives :

The purpose of fixation is to preserve tissues permanently in as life-like a state as much as possible. Fixation should be carried out as soon as possible after removal of the tissues (in the case of surgical pathology) or soon after death (with autopsy) to prevent autolysis. A variety of fixatives is available for use, depending on the type of tissue present and features to be demonstrated.

A particular fixing agent may be better at preserving some tissue constituents than it is at preserving others. Fixative mixtures that contain this agent may be a better choice for fixation when this constituent is the target for examination. An example would be acetic acid, which is excellent for nucleoprotein. Fixatives that contain acetic acid often give sharp and clear preservation of nuclear structure as a consequence. For that reason an acetic acid fixative would be a good choice when nuclear structure is important. Formalin is one of the few fixers which are widely used as simple fixative. Alcohol is some times used as a simple fixative. Alcohol combined with glacial acetic acid is also used as fixatives. Picric acid in combination with acetic acid and formalin is a good fixative. Mercuric chloride in combination with glacial acetic acid and other fixers gives excellent result. Following are the compositions of some important fixatives :

(i) Formalin :

It is used for the preservation of whole mount specimen. 10 per cent formalin (4 per cent formaldehyde solution) is prepared as follows :

Formalin (40 percent formaldehyde)	10mL
Distilled water	90mL

Mixed the two thoroughly.

(ii) Carnoy's fluid (Acetic-alcohol) :

This fixative is used in some histochemical and cytological techniques.

Absolute alcohol	60mL
Glacial acetic acid	10mL
Chloroform	30mL

Mixed thoroughly.

(iii) Bouin's fluid (Aqueous) :

It is the most popular and gives excellent general fixation for both whole mount material and tissue preparation.

Picric acid (saturated L aqueous solution)	75mL
Formalin	25mL
Glacial acetic acid	5mL

Mixed thoroughly.

(iv) Bouin's fluid (Alcoholic) :

It is best used for animals having chitonous skeleton.

Picric acid (saturated solution in 70 percent alcohol)	75mL
Formalin	25mL
Glacial acetic acid	5mL

Mixed as usual.

(v) Schaudinn's fluid :

It is most popular fixative for protozoa.

Saturated aqueous mercuric chloride solution	66mL
95 percent alcohol	33mL
Glacial acetic acid	5mL

(vi) Sublimate-acetic (Corrosive sublimate) :

This fixative is good for animal tissues and small animals.

Saturated aqueous mercuric chloride (4.5g of mercuric chloride in 100mL distilled water)	100mL
Glacial acetic acid	0.5mL

Note : Mercuric chloride are extremely poisonous, it should be carefully manipulated. Hands should be washed thoroughly after using the fixative. Inhalation of the powder is to be avoided while making the solution. One should use wooden horn or glass forceps or spatula. Distilled water should always be used for preparation of the solution.

(vii) Zenker's fluid (Helly's fluid) :

Saturated aqueous mercuric chloride (10mL distilled water plus 5g mercuric chloride)	100mL
Potassium dichromate	2.5g
Glacial acetic acid	5.0mL

Three components mixed thoroughly and used. (The mercuric chloride and potassium dichromate may be kept as a stock solution to which acetic acid is added just before use).

(viii) Smith's fixative :

It is used to fix yolky amphibians or larvae.

Potassium bichromate	5g
Formalin (40%; v/v)	10mL
Glacial acetic acid	2.5mL
Distilled water	88mL

2. Stains :

Natural dyes are obtained from natural sources; they are not synthesized in the laboratory. Important nuclear dyes are carmine and haematoxylin. Combining alum with cochineal obtained from the dried cochineal bug produces carmine. Haematoxylin is obtained from the South American logwood tree.

Composition of some important stains are shown below :

(i) Borax carmine :

This is a good general stain for animal tissue.

Distilled water	100mL
Borax	4 g
Carmine	2 g
Alcohol 70 percent	100mL

Borax, carmine and distilled water are mixed, boiled for 30 minutes and allowed to stand for several days, stirring daily. Mixed 100mL of 70 percent alcohol and to stand for several days, stirring daily. Filtered before use.

(ii) Aceto-carmine (1%-2%) :

This is good cytological stain. This stains the chromosomes.

Carmine powder	1 or 2g
45% Glacial acetic acid	100mL

Mixed in a beaker and heated to boiling (about 80ºC). Allowed to cool. The solution is filtered through Whatman (No. 1) filter paper. The stain is stored in room temperature. Shelf life is 4 weeks.

(iii) Aceto-orcein (1%-2%) :

This is good cytological stain. This stains the chromosomes.

45% Glacial acetic acid	100mL
Orcein powder	1 or 2g

Mixed in a beaker and heated to boiling (about 80ºC). Allowed to cool, The solution is filtered through Whatman (No. 1) filter paper. The stain is stored room temperature. Shelf life is 4 weeks.

(iv) Ehrlich's acid haematoxylin :

This stain is used to stain animal tissue and *Monocystis.*

Distilled water	100mL
Absolute alcohol	100mL
Glycerine	100mL
Glacial acetic acid	10mL
Haematoxylin	2g
Aluminium ammonium sulphate (alum)	20g

Haematoxylin is dissolved in alcohol. Acetic acid, distilled water and glycerine are added to it. The solution is stirred and allowed to ripen in the sunlight for about 2 months opening the bottle occasionally. Alum is added before or after ripening the solution. When the solution ripens the solution is filtered and stored in a stoppered bottle away from the light.

(v) Delafield's haematoxylin :

A general stain for plant and animal nuclei.

Absolute alcohol	25mL
Glycerine	100mL
Methyl alcohol	100mL
Haematoxylin	4g
Alum (saturated solution)	400mL

Haematoxylin is dissolved in absolute alcohol, into this solution saturated solution of alum is added. The solution is left exposed to the light in an open bottle for 3-4 days. Filtered and to it added glycerine and methyl alcohol. The solution is allowed to stand until it assumes a dark colour. Filtered and left to ripen for at least 2 month before use.

(vi) Eosin (alcoholic) :

Stain for plant and animal tissue in general.

Eosin Y powder (s.s.)	1g
95 percent alcohol	100mL

The powder is dissolved in alcohol and is ready for use.

Eosin (aqueous) : Stain for plant and animal tissue in general

Eosin Y powder (w.s.)	1g
Distilled water	100mL

The powder is dissolved in alcohol and is ready for use.

(vii) Leishman's stain :

Used to stain blood and certain protozoa.

Leishman powder	0.15g
Absolute methyl alcohol	100mL

The powder is dissolved in alcohol with occasional stirring; mixture is warmed at 50°C for 15 minutes. It is then filtered and keeping the filtrate in 37°C refines the dye. The stain can be applied within an hour of two of preparation.

(viii) (a) Crystal violet :

Most accepted as spindle dye in Flemming's triple stain

Crystal violet	1g
Distilled water	100mL

Crystal violet is dissolved in distilled water.

(b) Crystal violet (for Gram stain) : 2g of crystal violet is mixed with 20mL 95% ethanol (solution-A). 0.8g of ammonium oxalate mixed with 80mL of distilled water (solution-B). Solution A and B is mixed and then filtered.

(ix) Toluidine blue :

Toluidine blue	1g
Distilled water	100mL

The powder is dissolved in water.

(x) Methylene blue :

Used to stain bacteria and blood.

Methylene blue	1g
Normal saline	100mL

The powder is dissolved in normal saline.

(xi) Trypan blue 0.2% (w/v) :

Used to study the living condition of a cell.

0.2g of trypan blue is dissolved to a final volume of 100mL with water.

(xii) Acid fucshin :

Stain is used for chitinous tissue.

Acid Fucshin	0.5g
Distilled water	100mL

Powder is dissolved in water.

(xiii) Carbol fucshin :

Used to stain bacteria.

A. Basic fucshin	1g-0.3g
Absolute alcohol	10mL
B. Phenol	5g
Distilled water	100mL

Solution A and B are mixed properly, allowed to stand overnight. Filtered and the final prepration is stored in dark brown bottle.

(xiv) Gentian violet :

Used to stain bacteria and epithelia.

Gentian violet	1.0g
Distilled water	100mL

3. Other reagents :

(i) Saline solutions :

(a) **Normal saline :**

Sodium chloride (A.R. grade)	0.78g
Distilled water	100mL.

Sodium chloride is dissolved in water.

(b) **Saline for amphibian tissue :**

Sodium chloride (A.R. grade)	0.75g
Distilled water	100mL

Sodium chloride is dissolved in distilled water.

(c) **Saline for mammalian tissue (except blood) :**

Sodium chloride (A.R. grade)	0.9g
Distilled water	100mL

Sodium chloride are dissolved in distilled water.

(d) **Ringer's solution :**

Sodium chloride (A.R. grade)	0.8g
Calcium chloride	0.02g
Potassium chloride	0.02g
Sodium bicarbonate	0.02g
Distilled water	100mL

All salts are dissolved in distilled water.

(e) **Lock's solution :**

Sodium chloride (A.R. grade)	0.9g
Calcium chloride	0.048g
Potassium chloride	0.042g
Sodium bicarbonate	0.02g
Glucose	0.2g
Distilled water	100mL

All salts are dissolved in distilled water.

(ii) Acid alcohol :

Alcohol (70%)	100mL
Concentrated HCl	2.5mL

These two are mixed.

(iii) Acid water :

Distilled water	100mL
Concentrated HCl	2.5mL

These two are mixed to prepare acid water.

(iv) Iodine solution :

To develop deep yellow colours added sufficient sublimed crystals to 20 % solution of KI in water.

(v) Grades of alcohol :

Different grades of alcohol are prepared from 90 percent by adding required distilled water.

- 30 percent alcohol: Added 206.22mL of distilled water to 100mL of 90 percent alcohol.
- 50 percent alcohol: Added 84.71mL of distilled water to 100mL of 90 percent alcohol.
- 70 percent alcohol: Added 31.05mL of distilled water to 100mL of 90 percent alcohol.

(vi) Mayer's albumen :

It is used to attach the paraffin sections into glass microscopic slides and also to stick small microscopic organism, protozoa etc. over glass slide.

Egg albumen	5mL
Glycerine	50mL
Sodium salicylate	1g

The white of egg is collected in a glass stoppered vial along with the reagents. Shaken together till they mix completely. Filtered in a clean glass-stoppered bottle, added a few crystals of thymol as preservative. Filtering is then performed very slowly.

(vii) Clearing agent :

After dehydration of tissue in histological preparation it is important to replace alcohol, as alcohol does not mix with embedding medium such replacement is made through clearing agent. Some important chemical used for this purpose are:

(a) **Berlese's medium :**

Chloral hydrate	16g
Gum Arabic	15g
Distilled water	20mL
Glucose syrup	10mL
Acetic acid (glacial)	5mL

(b) **Cedar wood oil :** Commercially available.

(c) **Xylene :** Commercially available.

(d) **Oil of cloves :** Commercially available.

(viii) Mountants :

(a) **Canada balsam :** Commercially available

(b) **DPX :** Commercially available

(c) **Glycerine jelly :** 10g of gelatine is soaked in 60mL distilled water. In it added 70mL of glycerine along with 1g of phenol. The mixture is kept in a water bath and heated gently. Mixed well and then cooled. Stored in a cool place. When required the jelly is melted in a hot water bath.

(d) **Dilute glycerine :** Glycerine is mixed with distilled water in ratio 1:1 (v/v). In it added few crystals of thymol.

4. Buffers :

A buffer, as defined by Van Slyke (1922), is "a substance which by its presence in solution increases the amount of acid or alkali that must be added to cause unit change in pH". Buffers are thus very important components in experiments planned to study biological reactions by retaining a constant concentration of hydrogen ions within the physiological range. The pH of mammalian blood is maintained close to 7.38 by buffer systems such as—

$H_2PO_4^- \Leftrightarrow HPO_4^{2-}$,

$CO_2 \Leftrightarrow H_2CO_3$,

$H_2CO_3 \Leftrightarrow HCO_3^-$,

In living plants, the normal range of pH in tissues is about 4.0-6.2. It is not as narrowly defined as in mammalian tissues.

Universally applicable buffers for biochemistry must exhibit :

- water solubility
- no interference with biological processes
- known complex-forming tendency with metal ions
- non-toxicity
- no intervention with biological membranes (penetration, solubilisation, adsorption on surface etc.)
- very low U.V. absorption at wavelength >260 nm.

General characteristics concerning Buffer Applications :

With few exceptions, studies of biochemical systems need the application of a buffer with the purpose of controlling the pH value. Therefore the action of the buffer is of prime importance. Factors controlling the action of buffer solutions and pH are :

- activity effects : concentration and electrical charge of the species involved.
- salt effects : added "indifferent" electrolytes.
- dilution effect : pH deviation on dilution of buffer solutions.
- buffer capacity : added base or acid.
- dependency on temperature.

The selection of the correct buffer for a particular biochemical system or technique depends on a number of additional factors, for eg. un-preferred interaction of the buffer with the biopolymer, redox stability, metal ion complexing properties and purity. One way to solve the complex problem of selecting the right buffer is to assess as many buffers as feasible. Reviews on the use of buffers in various areas are available. On the other hand they do not offer detailed information and a comprehensive treatise on the subject should be consulted.

Composition of different common buffers :

- **Krebs Phosphate Ringers (KPR) :** Each of the following components is prepared separately :

  0.90% (w/v) NaCl

  1.15% (w/v) KCl

  1.22% (w/v) $CaCl_2$

  3.82% (w/v) $MgSO_4 \cdot H_2O$

0.1M phosphate buffer, pH 7.4 (17.8g of $Na_2HPO_4 \cdot H_2O$ + 20mL of 1 N HCl, diluted to 1 litre)

To mix, 200mL of NaCl, 8mL of KCl, 6mL of $CaCl_2$ and 2mL of $MgSO_4$. Combined carefully, and with constant stirring, added to it 40mL of phosphate buffer.

- **Ca^{++}/Mg^{++} free Phosphate buffered saline-10X (10X PBSA) :** 80g of NaCl, 2.0g of KCl, 15.0g of dibasic sodium phosphate and 2.0g of monobasic potassium phosphate is dissolved in 1 litre of distilled water. This makes a 10X solution of Ca^{+}/Mg^{++} free phosphate buffered saline. Diluted 1:10 prior to use. Stored in a refrigerator.

- **Phosphate buffered saline (PBS) :** 100mL Ca^{++}/Mg^{++} free 10X PBSA is mixed with 800mL of distilled water. 0.1g of magnesium chloride and 0.1g of anhydrous calcium chloride is separately dissolved to make a final volume of 100mL with water. With constant stirring, magnesium/calcium chloride solution is slowly added to the diluted PBSA. If a precipitate forms, one has to start over, and addition should be made at slower pace with continuous stirring.

- **Phosphate buffered saline-Tween 20 (pH 4.8) :** In PBS added 0.1% (v/v) Tween 20.

- **Potassium phosphate buffer (0.01M - pH 7.4) :** 500mL of 0.01 M K_2HPO_4 and 500mL of 0.01 M KH_2PO_4 are prepared. K_2HPO_4 is placed onto a magnetic stirrer and a pH electrode is inserted. KH_2PO_4 is added slowly to adjust the pH to 7.4.

• **Protein Buffer** : 1.46g of KH_2PO_4 and 0.92g of K_2HPO_4 is dissolved in 80mL of distilled water. 2.5g of crystalline serum albumin is added to it and the final volume is adjusted to 100mL with distilled water.

• **Saline citrate buffer (SSC) 20X** : It is common to prepare this buffer as a stock 20X solution, to be diluted to 2X, 1X or 0.1X prior to use. To prepare a 20X stock solution, 174g of NaCl and 88.2g of sodium citrate are dissolved to a final volume of 1000mL of water. pH is adjusted to 7.0 with 1N HCl and brought to a final volume of 1 litre.

For use, as a 1X SSC, 1 part 20X stock is diluted with 19 parts distilled water. For a 2X SSC, 1 part 20X stock is diluted with 9 parts water and so on.

• **Sodium acetate buffer** (1M pH 5.7) : To 925mL of 1 M sodium acetate, added 75mL of 1 M acetic acid.

• **Sodium cacodylate buffer** (0.2M pH 7.4) : 0.2M solution of cacodylic acid is preapared, sodium salt (MW 159.91). 3.20g of cacodylic acid, sodium salt is dissolved to a final volume of 100mL with water. pH is adjusted to 7.4 with HCl.

Cacodylic acid contains arsenic. To be handled properly.

• **Sodium phosphate buffer** : These are the most common buffers used in biology. They are produced by adding equimolar solutions of KH_2PO_4 and Na_2HPO_4. Equal volumes of the two will yield a pH of 7.0, while sodium phosphate will increase the pH. Increased volumes of potassium phosphate will decrease the pH. The pH can be adjusted from 5.4 to 8.2. If a pH of 7.0 - 8.2 is desired, one should start with 500mL of sodium phosphate and potassium phosphate is added to it while stirring and monitoring the pH with a pH meter until the desired pH is reached. If a pH of 5.4-7.0 is desired, one should start with 500mL of potassium phosphate and sodium phosphate is added until the desired pH is reached.Typically, the molarity of the buffer will range from 0.01 to 0.1M. One should use the appropriate molarity of KH_2PO_4 and Na_2HPO_4. That is, if 0.05M buffer is desired, use 0.5M KH_2PO_4 and 0.5M Na_2HPO_4 as directed above.

• **Sodium potassium phosphate buffer** : Refer to Sodium phosphate buffer.

• **Tris buffer (1M)** : There are many variations on the basic Tris-HCl buffer combination, most of which are commercially available. Solutions with EDTA are known as TE buffers, while solutions with EDTA and acetic acid are known as TAE buffers. The terminology varies with the author, with Tris buffer being used to mean Tris-HCl solutions. Sigma Chemical Co., St. Louis, carries a full line of the buffers marketed under the trade name of Trizma (base and HCl).

The basic buffer is a combination of Tris (tris(hydroxymethyl)aminomethane) and HCl acid. These are sometimes referred to as Tris-base and Tris-HCl solutions. Tris buffers should not be used below a pH of 7.2 or above a pH of 9.0. Tris buffers are also extremely temperature sensitive. Directions are given for room temperature (25°C). The pH will decrease approximately 0.028 units for each degree decrease in temperature.

121g of Tris is dissolved in 800mL of distilled water. The pH is adjusted with concentrated HCl. Final volume is diluted of 1 litre. Lower required molarities can be diluted from this stock or mixed as combinations of lower molarities of Tris and HCl. It is important to measure the pH at the temperature and molarity that will be used in the final analysis.

• **Tris-Glycine buffer (5X)** : 15.1g of tris base and 72.0g of glycine are dissolved to a final volume of 1 litre. For use, 1 part 5X buffer is diluted with 4 parts water.

Appendix B

Normalities, Molality, Molarity, Percentage

1) Units of Concentration :

The quantitative association between chemical substances in a reaction is termed as **stoichiometry**. Avogadro (1776-1856) was an open up in this field of chemistry. Avogadro made an hypothesis that there was a definite number that would correspond to the number of atoms or molecules in a **mole** of that atom or molecule. The weight of that unit designated as a **mole** would be corresponding to the atomic or molecular weight of the atom or molecule in grams. In concurrence to this theory, one mole of carbon-12 would have a mass of 12g because carbon-12 has an atomic weight of 12 (6 neutrons and 6 protons). One mole of hydrogen would weigh one gram and would contain the equivalent number of atoms as one mole of carbon. The specific number was, in fact, discovered to be 6.023×10^{23}. It was termed the **Avogadro Number**. The modern definition of a mole is as follows: Exactly one mole represents the number of carbon atoms in exactly 12 grams of the carbon-12 isotope. To elucidate, for an atom a mole represents 6.023×10^{23} atoms. For a molecular compound, one mole represents 6.023×10^{23} molecules.

A mole of a substance is equivalent to the **gram-formula mass** or the **gram-molar mass**. This equals the total of all of the masses of all the elements in the formula of the substance. Fundamentally, if one were to enumerate all of the carbon atoms in one mole of carbon-12, there would be 6.02×10^{23} atoms and it would weigh 12g (note the atomic weight of carbon is 12).

There are five units of concentration that are particularly helpful to chemists. The first three: molality, molarity and normality are dependant upon the mole unit. The last two: percent by volume and percent by weight have nothing to do with mole, only weight or volume of the **solute** or substance to be diluted, versus the weight or volume of the **solvent** or substance in which the solute is diluted. Percentages can also be determined for solids within solids.

- **Molality :**

A molality is the number of moles of solute dissolved in one kilogram of solvent. Hence 1 molal solution = 1 mole per 1000g of solvent. It should be noted that the solvent must be weighed unless it is water (i.e. one litre of water has a specific gravity of 1.0 and weighs one kilogram) so one can quantify out one litre of water and add the solute to it. But most other solvents have a specific gravity more than or less than one. Accordingly, one litre of anything other than water is not likely to occupy a litre of volume. Thus to make a one molal aqueous (water) solution of sodium chloride (NaCl), one kilogram of water is measured and into it added one mole of the solute, i.e. NaCl to it. The atomic weight of sodium is 23 and the atomic weight of chlorine is 35. Therefore the formula weight for NaCl is 58, and 58g of NaCl dissolved in 1 kg water would result in a 1 molal solution of NaCl.

- **Molarity :**

The molar unit is most likely the most frequently used chemical unit of measurement. *Molarity is the number of moles of a solute dissolved in a litre (unit volume) of solution.* Thus, 1 molar solution or 1 mole/litre = mol.wt. in gram per litre of solution.A molar solution of sodium chloride is prepared by placing 1 mole of a solute into a 1 litre volumetric flask. (Taking data from the example above i.e. 58g of sodium chloride i.e. 58 is the mol. wt. of NaCl). Water is then added to the volumetric flask up to the one litre line to make up the volume of one litre. The result is a one molar solution of sodium chloride.

$$\text{Molarity} = \frac{\text{moles of solute}}{\text{L solution}} = \text{M (Molar)}$$

Thus A molar solution = 1mol/litre = 1mmol/mL = 1μmol/μL

- **Normality :**

Normality can only be calculated when we deal with reactions, because normality is a function of equivalents. *Normality is the number of equivalents of solute per litre.*

$$\text{Equivalents of solute} = \frac{\text{mass of solute}}{\text{equivalent weight}}$$

$$\text{Again, Equivalent Weight} = \frac{\text{Molecular weight}}{\text{K}}$$

where, K = Equivalents per mole, K is an integer constant ≥ 1.

Hence, N = K × M

K, for a particular species is defined by the reaction type and the balanced chemical reaction.

The example below uses potassium hydroxide (KOH) to neutralize arsenic acid. By studying the reaction it is possible to determine the proton exchange number to determine the normality of the arsenic acid.

In the equation $H_3AsO_4 + 2KOH \longrightarrow K_2HAsO_4 + 2H_2O$

$$\text{Equivalent weight} = \frac{\text{molar mass}}{(\text{H}^+ \text{ per mole})}$$

$$\text{Equivalent} = \frac{\text{mass of compound}}{\text{equivalent weight}}$$

$$\text{And Normality} = \frac{\text{(equivalents of solute)}}{\text{Litre}}$$

$$\text{Normality} = \frac{\text{equivalents of solute}}{\text{vol. of soln. in litres}} = \frac{\text{equivalents of solute}}{\text{solution L}} = \text{N}$$

The normality is related to molarity as: Normality = Molarity × n (where n = the number of protons exchanged in a reaction).

When a hydrogen atom is ionized and loses its electron, it is left with only a proton. So a hydrogen ion is basically a proton.

To find out the normality of 0.25M solution of H_3AsO_4 participating in reaction:

$$H_3AsO_4 + 2KOH \longrightarrow K_2HAsO_4 + 2H_2O$$

When H_3AsO_4 is neutralized by KOH, H_3AsO_4 supply two protons to form $2H_2O$.

It should ne noted that that H_3AsO_4 has three hydrogens, but K_2HAsO_4 only has one hydrogen. That means that 2 protons were exchanged,

Again Normality = Molarity × n

Normality of the solution is 0.25mol. H_3AsO_4 and there were two protons exchanged (2 equivalents/mole)

$$\frac{0.25 \text{ moles } H_3AsO_4}{1 \text{ L } H_3AsO_4} \times \frac{2 \text{ equivalents } H_3AsO_4}{1 \text{ L } H_3AsO_4}$$

$$\frac{0.50 \text{ equivalents } H_3AsO_4}{1 \text{ L } H_3AsO_4} = 0.5 \text{ normal } H_3AsO_4$$

So, in short, even as there is a relationship between the normality of a solution and the molarity of a solution (as already stated), the normality can only be determined by exploring reaction, determining the proton exchange and multiplying molarity by that number.

Normality is mainly helpful in titrations calculations.

where N = normality, V = volume, a = the substance on the left of the equation involved in proton exchange, and b = substance on the right of the equation involved in proton exchange:

$$NaVa = NbVb$$

● **Percentages** are simple to calculate since they do not require information about the chemical nature of the substance. Percentages can be determined as percent by weight or percent by volume.

● **Percent by weight :**

To make up a solution based on percentage by weight, one would simply determine what percentage was desired (for example, a 20% by weight aqueous solution of potassium chloride) and the total quantity to be prepared.

If the total quantity needed is 1 kg, then it would simply be a matter of calculating 20% of 1 kg which, certainly is: 0.20 × 1000 g/kg = 200 g i.e. 200 g of KCl/kg.

In order to bring the total quantity to 1 kg, it would be necessary to add 800 g water.

● **Percent by volume :**

Solutions based on percent by volume are calculated the same as for percent by weight, except that calculations are based on volume. Thus one would just determine what percentage was desired (for example, a 20% by volume aqueous solution of sodium chloride) and the total quantity to be prepared.

If the total quantity needed is 1 litre, then it would simply be a matter of calculating 20% of 1 litre which, of course is: 0.20 × 1000mL/litre = 200mL i.e. 200mL KCl/litre.

Percentages are used more in the technological fields of chemistry (such as environmental technologies) than they are in pure chemistry.

(i) To convert from % solution to molarity, the percent solution value is to be multiplied by 10 to get g/litre, and then divided by the formula weight (FW).

$$\text{Molarity} = \frac{(\text{\% of solution}) \times 10}{\text{FW}}$$

(ii) To convert from molarity to percent solution, the molarity is to be multiplied by the FW and divided by 10 :

$$\text{\% solution} = \frac{\text{molarity} \times \text{FW}}{10}$$

2. Other units of concentration :

There are a number of other different ways of expressing solute concentration that are frequently used. Some of these are listed below.

1. **Weight %, Wt %** = (mass of solute/mass of solution) × 100%
2. **Mass per volume, mg/L** = mass of solute/litre of solution
3. **mole fraction,** χ = moles of solute/total moles
4. **Parts per million :** The number of milligrams of solute per kg of solution = one ppm,

   Since, 1 mg = 10^{-3}g and 1kg = 10^{3}g.

   Believing the density of water is 1.00g/mL, 1 litre of solution = 1 kg and hence,

   1 mg/L = 1 ppm. This is generally true for freshwater and other dilute aqueous solutions.

Parts per million concentrations are essentially mass ratios (solute to solution) × a million (10^6). In this sense, they are similar to wt %, which could be thought of as parts per hundred (although nobody uses this term).

Other variations on this theme include :

- ppt – parts per thousand (used for common ions in sea water).
- ppb – parts per billion (used for heavy metals and organics).
- ppt – parts per trillion (used for trace metals and trace organics).

The following table summarizes common mass ratios for solutions and solids.

Table B.1 : Other units of concentration

Unit	Solutions		Solids	
ppm	mg/L	µg/mL	mg/kg	µg/g
ppb	µg/L	ng/mL	µg/kg	ng/g
ppt	ng/L	pg/mL	ng/kg	pg/g

Appendix C

Simple Solution and Dilutions

1. Simple Dilution (Dilution Factor Method) :

A *simple dilution* is one in which an *unit volume* of a liquid material of importance is combined with an suitable volume of a *solvent* liquid to get the preferred concentration. The *dilution factor* may be defined as the total number of unit volumes in which a material will be dissolved. The diluted material must then be cautiously mixed to reach the true dilution. For example, a 1:5 dilution involve combining 1 unit volume of diluent (the material to be diluted) + 4 unit volumes of the solvent medium (thus, 1 + 4 = 5 = dilution factor).

♦ **Illustration :** Frozen 100mL concentrated buffer is usually diluted in 4 × 100mL of cold water (the dilution solvent) giving a dilution factor of 5, i.e., the buffer concentrate represents an unit volume to which one have to add 4 more unit volume (same unit volumes) of water. So the buffer concentrate is now dispersed through 5 unit volumes. This would be called a 1:5 dilution, and the buffer is now 1/5 as concentrated as it was originally. Therefore, in a simple dilution, one should *add one less unit volume of solvent* than the desired dilution factor value.

2. Serial Dilution :

A *serial dilution* is basically a series of simple dilutions which augments the dilution factor quickly starting with a little original quantity of material (i.e., bacterial culture, a chemical etc.). The source of dilution material for each step comes from the diluted material of the previous. In a serial dilution the *total dilution factor* at any point is the *product* of the individual dilution factors in each step up to it.

So, Final dilution factor (DF) =

$$DF_1 \times DF_2 \times DF_3 \ldots\ldots$$

♦ **Illustrations :** In the microbiology laboratory the students perform a three step 1:100 serial dilution of a bacterial culture (see figure below).

1. The initial step combines 1 unit volume culture (10mΛ) with 99 unit volumes of broth (990mΛ) = 1:100 dilution.

2. In the subsequent step, one unit volume of the 1:100 dilution is combined with 99 unit volumes of broth now yielding a total dilution of 1:100 × 100 = 1:10,000 dilution.

3. Repeated again (the third step) the total dilution would be 1:100 × 10,000 = 1:1,000,000

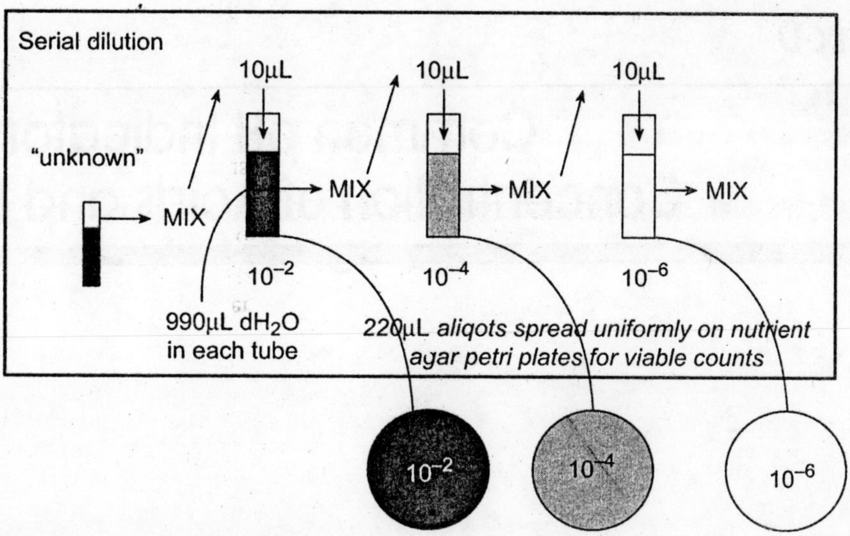

Figure C.1 : Illustration of serial dilution method.

total dilution. The concentration of bacteria is now one million times *less* than in the original sample. **Figure C.1.**

3. Making fixed volumes of specific concentrations from liquid reagents ($V_1C_1=V_2C_2$) method :

Very, frequently one needs to prepare a specific volume of known concentration due to scarcity of liquid materials (as some chemicals are very pricey and are only sold and used in minute quantities, e.g., micrograms) or to restrict the amount of waste. The formula below is a quick approach to calculating such dilutions where :

$\mathbf{V_1C_1}$ (source solution attributes) = $\mathbf{V_2C_2}$ (new solution attributes)

V = volume, C = concentration; in whatever units you are working.

♦ **Illustration 1 :** Suppose one have 3mL of a stock solution of 100mg/mL Ampicillin (= C_1) and one likes to make 200μL (= V_2) of solution having 25mg/mL (= C_2). You need to know what volume (V_1) of the stock to use as part of the 200mΛ total volume needed.

V_1 = the volume of stock you will start with. This is the unknown.

C_1 = 100mg/mL in the stock solution

V_2 = total volume needed at the new concentration = 200mL = 0.2mL

C_2 = the new concentration = 25mg/mL

By algebraic rearrangement :

$$V_1 = (V_2 \times C_2)/C_1$$

$$V_1 = (0.2mL \times 25mg/mL)/100mg/mL$$

Thus, V_1 = 0.05mL, or 50μL

So, one would take 0.05mL = 50μL of stock solution and dilute it with 150mL of solvent to get the 200mL of 25 mg/mL solution needed (one should remember that the amount of solvent used is based upon the final volume needed, so one has to subtract the starting volume form the final to calculate it).

Appendix D

Common pH Indicators and Concentration of Acids and Bases

Table D.1 : Common pH Indicators

Sl. No	Indicator	Preparation 0.1 g in 250mL of	Acid colour	Basic colour	pH range
1.	Cresol red	Water containing 2.62mL of 0.1N NaOH	Red	Yellow	0.2 – 1.8
2.	Thymol blue (acid range)	Water containing 2.15mL of 0.1N NaOH	Red	Yellow	0.5 - 2.8
3.	Bromophenol blue	Water containing 1.49mL of 0.1N NaOH	Yellow	Purple	2.8 – 4.6
4.	Bromocresol green (blue)	Water containing 1.43mL of 0.1N NaOH	Yellow	Blue	3.6 – 5.2
5.	Congo red	Water	Violet	Reddish range	3.0 – 5.0
6.	Bromocresol purple	Water containing 1.85mL of 0.1N NaOH	Yellow	Violet	5.2 – 6.8
7.	Litmus	Water containing Yellow (azolitmin)	Red	Blue	5.0 - 8.0
8.	Bromothymol blue	Water containing 1.6mL of 0.1N NaOH	Yellow	Blue	6.0 – 7.6
9.	Cresol red (Basic range)	Water containing 2.62mL of 0.1N NaOH	Yellow	Red	7.2 – 8.8
10.	Thymol blue	Water containing 2.15mL of 0.1N NaOH	Yellow	Blue	8.0 – 9.6
11.	Phenolpthalein	60% cellosolve	Colourless	Pink	8.3 – 10.0

Table D.2 : Concentrations of Acids and Bases

Common commercial acids and bases	Molecular weight	Moles/ litre	Grams/ litre	Percentage by weight	Specific gravity
Acetic acid glacial	60.05	17.4	1045	99.5	1.05
Acetic acid	60.05	6.27	376	36	1.045
Formic acid	46.02	23.4	1080	90	1.20
Hydrochloric acid	36.5	11.6 2.9	424 105	36 10	1.18 1.05
Nitric acid	63.02	15.99 14.9 13.3	1008 938 837	71 67 61	1.42 1.42 1.37
Perchloric acid	100.5	11.66 9.2	1172 923	70 60	1.67 1.64
Phosphoric acid	80.0	18.1	1445	85	1.70
Sulfuric acid	98.1	18.0	1766	95	1.84
Ammonium hydroxide	35.0	14.8	251	28	0.898
Potassium hydroxide	56.1	13.5 1.94	757 109	50 10	1.52 1.09
Sodium hydroxide	40.0	19.1 2.75	763 111	50 10	1.53 1.11

Appendix E

SI Units

The International System of Units, universally abbreviated SI (from the French *Le Systéme International d'Unités*), is the modern metric system of measurement. Long the dominant system used in science, the SI is rapidly becoming the dominant measurement system used in international commerce.

The definitive international reference on the SI is a booklet published by the International Bureau of Weights and Measures (BIPM, *Bureau International des Poids et Mesures*) and often referred to as the BIPM SI Brochure. Entitled *Le Systéme International d'Unités (SI)*, the booklet is in French followed by a text in English.

SI units are divided into two classes :

- *base* units.
- *derived* units.

The *Comité International des Poids et Mesures* (CGPM), considering the advantages of a single, practical, world-wide system of units for international relations, for teaching, and for scientific work, decided to base the International System on a choice of seven well-defined units which by convention are regarded as dimensionally independent: the meter, the kilogram, the second, the ampere, the kelvin, mole, and the candela. These SI units are called ***base units***. The second class of SI units is that of ***derived units***. These are units that are formed as products of powers of the base units according to the algebraic relations linking the quantities concerned. The names and symbols of some units thus formed in terms of base units may be replaced by special names and symbols which can themselves be used to form expressions and symbols for other derived units.

SI Base Units :

The SI is founded on seven *SI base units* for seven *base quantities* assumed to be mutually independent, as given in Table E.1 below.

Table E.1 : SI Base units

Base quantity	**SI base unit**	
	Name	**Symbol**
mass	kilogram	kg
time	second	s
electric current	ampere	A
Thermodynamic temperature	kelvin	K
amount of substance	mole	mol
luminous intensity	candela	cd
Length	meter	m

SI prefixes :

The 20 SI prefixes used to form decimal multiples and submultiples of SI units are given in Table E.2 below.

Table E.2 : SI prefixes

Factor	Name	Symbol	Factor	Name	Symbol
10^{24}	yotta	Y	10^{-1}	deci	d
10^{21}	zetta	Z	10^{-2}	centi	c
10^{18}	exa	E	10^{-3}	milli	m
10^{15}	peta	P	10^{-6}	micro	μ
10^{12}	tera	T	10^{-9}	nano	n
10^{9}	giga	G	10^{-12}	pico	p
10^{6}	mega	M	10^{-15}	femto	f
10^{3}	kilo	k	10^{-18}	atto	a
10^{2}	hecto	h	10^{-21}	zepto	z
10^{1}	deka	da	10^{-24}	yocto	y

It is important to note that the kilogram is the only SI unit with a prefix as part of its name and symbol. Because multiple prefixes may not be used, in the case of the kilogram the prefix names of Table E.1 are used with the unit name "gram" and the prefix symbols are used with the unit symbol "g" With this exception, any SI prefix may be used with any SI unit, including the degree Celsius and its symbol °C.

SI derived units :

Other quantities, called *derived quantities*, are defined in terms of the seven base quantities via a system of quantity equations. The *SI derived units* for these derived quantities are obtained from these equations and the seven SI base units. Examples of such SI derived units are given in Table E.3 and E.4, where it should be noted that the symbol 1 for quantities of dimension 1 such as mass fraction is generally omitted.

Table E.3 : Examples of SI derived units

Derived quantity	SI derived unit	
	Name	Symbol
area	square meter	m^2
volume	cubic meter	m^3
speed, velocity	meter per second	m/s
acceleration	meter per second squared	m/s^2
wave number	reciprocal meter	m^{-1}
mass density	kilogram per cubic meter	kg/m^3
specific volume	cubic meter per kilogram	m^3/kg
current density	ampere per square meter	A/m^2
magnetic field strength	ampere per meter	A/m
amount-of-substance concentration	mole per cubic meter	mol/m^3
luminance	candela per square meter	cd/m^2
mass fraction	kilogram per kilogram, which may be represented by the number 1	kg/kg = 1

Table E.4 : SI Derived Units

Name	Symbol	Unit of	Equivalent in Base Units	Other Equivalents
becquerel	Bq	activity (of a radionuclide)	1/s	—
coulomb	C	quantity of electricity, electric charge	A·s	F·V = J/V
degree Celsius	°C	Celsius temperature	K	K – 273.15
farad	F	capacitance	$A^2 \cdot s^4/kg \cdot m^2$	C/V=A·s/V
gray	Gy	absorbed dose, specific energy imparted, kerma	m^2/s^2	J/kg
hertz	Hz	frequency	1/s	—
joule	J	energy, work, quantity of heat	$kg \cdot m^2/s^2$	N·m = W·s = $Pa \cdot m^3$
katal	kat	catalytic activity	mol/s	—
lumen	lm	luminous flux	cd	cd·(4·π sr) = $lx \cdot m^2$
lux	lx	illuminance	cd/m^2	cd·(4·π sr)/m^2 = lm/m^2
newton	N	force	$kg \cdot m/s^2$	J/m = W·s/m = $Pa \cdot m^2$
ohm	Ω	electric resistance	$kg \cdot m^2/A^2 \cdot s^3$	V/A = 1/S
pascal	Pa	pressure, stress	$kg/m \cdot s^2$	$N/m^2 = J/m^3$
radian	rad	plane angle	1	1/(2·π) of a circle
siemens	S	electric conductance	$A^2 \cdot s^3/kg \cdot m^2$	A/V = 1/W
sievert	Sv	dose equivalent	m^2/s^2	J/kg
volt	V	electric potential difference, electromotive force	$kg \cdot m^2/A \cdot s^2$	W/A = J/C = Wb/s
watt	W	power, radiant flux	$kg \cdot m^2/s^3$	J/s = V·A = N·m/s

Units outside the SI :

Certain units are not component of the International System of Units, that is, they are outside the SI, but are significant and broadly used. Coherent with the recommendations of the International Committee for Weights and Measures (CIPM, *Comite International des Poids et Mesures*), the units in this category that are accepted for use with the SI are given in Table E.5.

Table E.5 : Units outside the SI that are accepted for use with the SI

Name	Symbol	Value in SI units
minute (time)	min	1 min = 60 s
hour	h	1 h = 60 min = 3600 s
day	d	1 d = 24 h = 86400 s
degree (angle)	°	$1° = (\pi/180)$ rad
minute (angle)	′	$1' = (1/60)° = (\pi/10\,800)$ rad
second (angle)	″	$1'' = (1/60)' = (\pi/648\,000)$ rad
litre	L	$1\ L = 1\ dm^3 = 10^{-3}\ m^3$
metric ton or tonne	t	$1\ t = 10^3$ kg
neper	Np	1 Np = 1
bel	B	1 B = (1/2) ln 10 Np[a], 1 dB = 0.1 B. (commonly used with the SI prefix deci)
electronvolt[b]	eV	$1\ eV = 1.60218 \times 10^{-19}$ J, approximately
unified atomic mass unit[c]	u	$1\ u = 1.66054 \times 10^{-27}$ kg, approximately
astronomical unit[d]	ua	$1\ ua = 1.49598 \times 10^{11}$ m, approximately

[a] Although the neper is coherent with SI units and is accepted by the CIPM, it has not been adopted by the General Conference on Weights and Measures (CGPM, *Conférence Générale des Poids et Mesures*) and is thus not an SI unit.

[b] The electronvolt is the kinetic energy acquired by an electron passing through a potential difference of 1 V in vacuum. The value must be obtained by experiment, and is therefore not known exactly.

[c] The unified atomic mass unit is equal to 1/12 of the mass of an unbound atom of the nuclide ^{12}C, at rest and in its ground state. The value must be obtained by experiment, and is therefore not known exactly.

[d] The astronomical unit is a unit of length. Its value is such that, when used to describe the motion of bodies in the solar system, the heliocentric gravitation constant is $(0.01720209895)^2\ ua^3.d^{-2}$. The value must be obtained by experiment, and is therefore not known exactly.

Other units outside the SI that are currently acknowledged for use with the SI by NIST are given in Table E.6. These units, which are subject to future evaluation, should be defined in relation to the SI in every document in which they are used; their constant use is not promoted. The CIPM currently accepts the use of all of the units given in Table E.6 with the SI except for the curie, roentgen, rad, and rem. Because of the continued extensive use of these units in the United States, NIST still accepts their use with the SI.

Table E.6 : Other units outside the SI that are currently accepted for use with the SI, subject to further review

Name	Symbol	Value in SI units
nautical mile		1 nautical mile = 1852 m
knot		1 nautical mile per hour = (1852/3600) m/s
are	a	1 a = 1 dam^2 = 10^2 m^2
hectare	ha	1 ha = 1 hm^2 = 10^4 m^2
bar	bar	1 bar = 0.1 MPa = 100 kPa = 1000 hPa = 10^5 Pa
ångström	Å	1 Å = 0.1 nm = 10^{-10} m
barn	b	1 b = 100 fm^2 = 10^{-28} m^2
curie	Ci	1 Ci = 3.7×10^{10} Bq
roentgen	R	1 R = 2.58×10^{-4} C/kg
rad	rad	1 rad = 1 cGy = 10^{-2} Gy
rem	rem	1 rem = 1 cSv = 10^{-2} Sv

Table E.7 : SI units and SI derivative units* and conversions :

Length	
1 metre	100 centimetres (cm)
1 millimetre (10^{-3} m)	1000 micrometres (mm)
1 micrometre (10^{-6})	1000 nanometres (nm)
*1 inch (in)	25.4 millimetres(mm)
*1 ångstrom	100 picometres (pm)
Mass	
1 kilogram (kg)	1000 grams (g)
1 gram (10^{-3} kg)	1000 milligrams (mg)
1 milligrams (10^{-6} kg)	1000 nanograms (ng)
*1 tonne (t)	1000 kilograms (kg) = 1 Megagram (Mg)
1 pound (lb)	0.454 kilograms (kg)
Volume	
1 cubic decimetre	1000 cubic centimetres (cm^3)
1 cubic centimetres (10^{-3} dm^3)	1000 cubic millimetre
*1 litre (L)	1 cubic decimetre = 10^{-3} m^3
*1 cubic inch (in^3)	16.38 cubic centimetres (cm^3)
Pressure	
1 kilopascal (kPa)	1000 Pascal (Pa)
*1 atmosphere (i.e. pressure 760 mm Hg at STP)	101.325 kilopascal (kPa)

Table contd. next page

Temperature	
Kelvin (K)	–273.15ºC
0º Celsius (ºC)	273.15 K = 32°F
*1º Farenheit (ºF)	5/9 × (t – 32)ºC
Time	
1 second (s)	1000 millisecond (ms)
Energy	
1 kilojoule (kJ)	1000 joules(J)
*1 calorie	4.187 joules (J)
Electrical potential difference	
1 volt (V)	1000 millivolts (mV)

Appendix F

Radiation and Radionuclides

Table F.1 : Units commonly used to describe radioactivity

Measurement	International (SI) Unit	U.S. Unit	Conversions
Radioactivity	Becquerel (Bq)	Curie (Ci)	3.7E+10 Bq = 1 Ci)
Absorbed Dose	Gray (Gy)	Rad	1 Gy = 100 rads
Equivalent Dose	Sievert (Sv)	Rem	1 Sv = 100 rems

1 Bq = 1 dps (disintegrations per second)

1 Ci = eqv. to/ dps i.e., 3.7×10^{10} d.p.s. 1 g radium

1 Gy = 1 J/kg (Joules absorbed per kg of target liquid or gas)

Radiation dose : Relationships

1 R (Rontgen) = ionizing energy emitted in air

1 rad (radiation absorbed dose in a liquid or gas; energy absorption of 0.01J/kg)

1 Sv = Q × Gy (amount of radiation that gives a dose in human eqv. to 1 Gy of X-rays)

1 rem = Q × rad (rem = Rontgen equivalents in man) $\left(= \frac{1}{100} \text{Sv}\right)$

(Q = "quality score" Q ~ 1 for § particles.)

To convert radioactivity to dose (e.g. 1 Ci ^{32}P in 1 L H_2O)

1 Ci = 4.12E-03 W/1 kg

= 4.12E – 3 Gy/sec

Table F.2 : Characteristics of common radionuclides in biology labs

Nuclide	Atomic number	Half life	Particle emitted	Energy of Particle (MeV)
^{3}H	1	1.233E+01 yr	β	0.018
^{14}C	6	5.730E+03 yr	β^-	0.155
^{32}P	15	14.282 days	β^-	1.71
^{35}S	16	87.51 days	β	0.167
^{125}I	53	60.14 days	EC*/γ	0.035
^{131}I	53	8.040 days	β^-	0.25

*EC : Electron capture.

Radiation risk :

Avg U.S. background radiation exposure = 0.36 rem

Maximum annual occupational dose = 5 rem

4 Sv causes a doubling of total cancer risk (baseline = 0.22)

1 Sv causes a doubling of leukemia risk (baseline = 0.0063)

References :

http://www.umich.edu/~radinfo/

http://www.radtexas.org/radunits.html

Appendix G

Laboratory Safety Guidelines

1. One should be patient while one is working. One should not start an experiment unless one ascertains all the steps.

2. Equipment should not be used unless one ascertains how to use. The laboratory equipment is to be used properly. If one does not know how to use, the equipment user's manual should be followed or others to be asked.

3. Any equipment should not be touched with contaminated gloves.

4. If there is spill or contamination by any toxic, antibiotics or carcinogens, the area or equipment should be decontaminated immediately.

5. Gloves and a laboratory coat should be worn while one handles any toxic or carcinogenic chemicals (e.g., EtBr, CsCl, Phenol, etc.).

6. Eye or face protection should be used while working under UV.

7. If any toxic chemicals, antibiotics or carcinogens are spilt on body or clothes, the contaminated clothing should be immediately removed and small spills are to be washed under water for at least 5 minutes Spills on legs can be rinsed in the sink. Do not hesitate to use the lab shower in the laboratory on large spill. Do not rinse contaminated skin with alcohol.

8. After any chemical, enzyme or solvent, etc. are used, it should be brought back to the place where it was.

9. Before one uses the last remains, a last bottle of a chemical, tubes, etc., indent for the purchase of the necessary items is to be made.

10. Most enzymes are expensive. The enzymes must be stored at –20°C. Quantity planned be small and convenient before taking them out of a freezer.

11. A new pipette tip should be used to remove the necessary amount of any enzyme.

Appendix H

Statistical Tables

Table H.1 : Student's t test

(If calculated t is greater than value shown, reject the null hypothesis)

SIGNIFICANCE LEVEL FOR ONE-DIRECTION TEST						
df	.10	.05	.025	.01	.005	.000
1	3.078	6.314	12.706	31.821	63.657	636.619
2	1.886	2.920	4.303	6.965	9.925	31.598
3	1.638	2.353	3.182	4.541	5.841	12.941
4	1.533	2.132	2.776	3.747	4.604	8.610
5	1.476	2.015	2.571	3.365	4.032	6.859
6	1.440	1.943	2.447	3.143	3.707	5.959
7	1.415	1.895	2.365	2.998	3.499	5.405
8	1.397	1.860	2.306	2.896	3.355	5.041
9	1.383	1.833	2.262	2.821	3.250	4.781
10	1.372	1.812	2.228	2.764	3.169	4.587
11	1.363	1.796	2.201	2.718	3.106	4.437
12	1.356	1.782	2.179	2.681	3.055	4.318
13	1.350	1.771	2.160	2.650	3.012	4.221
14	1.345	1.761	2.145	2.624	2.977	4.140
15	1.341	1.753	2.131	2.602	2.947	4.073
16	1.337	1.746	2.120	2.583	2.921	4.015
17	1.333	1.740	2.110	2.567	2.898	3.965
18	1.330	1.734	2.101	2.552	2.878	3.922
19	1.328	1.729	2.093	2.539	2.861	3.883
20	1.325	1.725	2.086	2.528	2.845	3.850
21	1.323	1.721	2.080	2.518	2.831	3.819
22	1.321	1.717	2.074	2.508	2.819	3.792
23	1.319	1.714	2.069	2.500	2.807	3.767
24	1.318	1.711	2.064	2.492	2.797	3.745
25	1.316	1.708	2.060	2.485	2.787	3.725
26	1.315	1.706	2.056	2.479	2.779	3.707
27	1.314	1.703	2.052	2.473	2.771	3.690
28	1.313	1.701	2.048	2.467	2.763	3.674
29	1.311	1.699	2.045	2.462	2.756	3.659
30	1.310	1.697	2.042	2.457	2.750	3.646
40	1.303	1.684	2.021	2.423	2.704	3.551
60	1.296	1.671	2.000	2.390	2.660	3.460
120	1.289	1.658	1.980	2.358	2.617	3.373
α	1.282	1.645	1.960	2.326	2.576	3.291

Contd. t test

	SIGNIFICANCE LEVEL FOR ONE-DIRECTION TEST					
df	.20	.10	.05	.02	.01	.001
1	3.078	6.314	12.706	31.821	63.657	636.619
2	1.886	2.920	4.303	6.965	9.925	31.598
3	1.638	2.353	3.182	4.541	5.841	12.941
4	1.533	2.132	2.776	3.747	4.604	8.610
5	1.476	2.015	2.571	3.365	4.032	6.859
6	1.440	1.943	2.447	3.143	3.707	5.959
7	1.415	1.895	2.365	2.998	3.499	5.405
8	1.397	1.860	2.306	2.896	3.355	5.041
9	1.383	1.833	2.262	2.821	3.250	4.781
10	1.372	1.812	2.228	2.764	3.169	4.587
11	1.363	1.796	2.201	2.718	3.106	4.437
12	1.356	1.782	2.179	2.681	3.055	4.318
13	1.350	1.771	2.160	2.650	3.012	4.221
14	1.345	1.761	2.145	2.624	2.977	4.140
15	1.341	1.753	2.131	2.602	2.947	4.073
16	1.337	1.746	2.120	2.583	2.921	4.015
17	1.333	1.740	2.110	2.567	2.898	3.965
18	1.330	1.734	2.101	2.552	2.878	3.922
19	1.328	1.729	2.093	2.539	2.861	3.883
20	1.325	1.725	2.086	2.528	2.845	3.850
21	1.323	1.721	2.080	2.518	2.831	3.819
22	1.321	1.717	2.074	2.508	2.819	3.792
23	1.319	1.714	2.069	2.500	2.807	3.767
24	1.318	1.711	2.064	2.492	2.797	3.745
25	1.316	1.708	2.060	2.485	2.787	3.725
26	1.315	1.706	2.056	2.479	2.779	3.707
27	1.314	1.703	2.052	2.473	2.771	3.690
28	1.313	1.701	2.048	2.467	2.763	3.674
29	1.311	1.699	2.045	2.462	2.756	3.659
30	1.310	1.697	2.042	2.457	2.750	3.646
40	1.303	1.684	2.021	2.423	2.704	3.551
60	1.296	1.671	2.000	2.390	2.660	3.460
120	1.289	1.658	1.980	2.358	2.617	3.373
α	1.282	1.645	1.960	2.326	2.576	3.291

Table H.2 : Chi-Square Table

(If x^2 calculated is greater than value shown at the predetermined confidence level, reject null hypothesis)

PROBABILITY UNDER H_o THAT X^2 CHI SQUARE														
df	.99	.98	.95	.90	.80	.70	.50	.30	.20	.10	.05	.02	.01	.001
1	.00016	.00063	.0039	.016	.064	.15	.46	1.07	1.64	2.71	3.84	5.41	6.64	10.83
2	.02	.04	.10	.21	.45	.71	1.39	2.41	3.22	4.60	5.99	7.82	9.21	13.82
3	.12	.18	.35	.58	1.00	1.42	2.37	3.66	4.64	6.25	7.82	9.84	11.34	16.27
4	.30	.43	.71	1.06	1.65	2.20	3.36	4.88	5.99	7.78	9.49	11.67	13.28	18.46
5	.55	.75	1.14	1.61	2.34	3.00	4.35	6.06	7.29	9.24	11.07	13.39	15.09	20.52
6	.87	1.13	1.64	2.20	3.07	3.83	5.35	7.23	8.56	10.64	12.59	15.03	16.81	22.46
7	1.24	1.56	2.17	2.83	3.82	4.67	6.35	8.38	9.80	12.02	14.07	16.62	18.48	24.32
8	1.65	2.03	2.73	3.49	4.59	5.53	7.34	9.52	11.03	13.36	15.51	18.17	20.09	26.12
9	2.09	2.53	3.32	4.17	5.38	6.39	8.34	10.66	12.24	14.68	16.92	19.68	21.67	27.88
10	2.56	3.06	3.94	4.86	6.18	7.27	9.34	11.78	13.44	15.99	18.31	21.16	23.21	29.59
11	3.05	3.61	4.58	5.58	6.99	8.15	10.34	12.90	14.63	17.28	19.68	22.62	24.72	31.26
12	3.57	4.18	5.23	6.30	7.81	9.03	11.34	14.01	15.81	18.55	21.03	24.05	26.22	32.91
13	4.11	4.76	5.89	7.04	8.63	9.93	12.34	15.12	16.98	19.81	22.36	25.47	27.69	34.53
14	4.66	5.37	6.57	7.79	9.47	10.82	13.34	16.22	18.15	21.06	23.68	26.87	29.14	36.12
15	5.23	5.98	7.26	8.55	10.31	11.72	14.34	17.32	19.31	22.31	25.00	28.26	30.58	37.70
16	5.81	6.61	7.96	9.31	11.15	12.62	15.34	18.42	20.46	23.54	26.30	29.63	32.00	39.29
17	6.41	7.26	8.67	10.08	12.00	13.53	16.34	19.51	21.62	24.77	27.59	31.00	33.41	40.75
18	7.02	7.91	9.39	10.86	12.86	14.44	17.34	20.60	22.76	25.99	28.87	32.35	34.80	42.31
19	7.63	8.57	10.12	11.65	13.72	15.35	18.34	21.69	23.90	27.20	30.14	33.69	36.19	43.82
20	8.26	9.24	10.85	12.44	14.58	16.27	19.34	22.78	25.04	28.41	31.41	35.02	37.57	45.32
21	8.90	9.92	11.59	13.24	15.44	17.18	20.34	23.86	26.17	29.62	32.67	36.34	38.93	46.80
22	9.54	10.60	12.34	14.04	16.31	18.10	21.24	24.94	27.30	30.81	33.92	37.66	40.29	48.27
23	10.20	11.29	13.09	14.85	17.19	19.02	22.34	26.02	28.43	32.01	35.17	38.97	41.64	49.73
24	10.86	11.99	13.85	15.66	18.06	19.94	23.34	27.10	29.55	33.20	36.42	40.27	42.98	51.18
25	11.52	12.70	14.61	16.47	18.94	20.87	24.34	28.17	30.68	34.38	37.65	41.57	44.31	52.62
26	12.20	13.41	15.38	17.29	19.82	21.79	25.34	29.25	31.80	35.56	38.88	42.86	45.64	54.05
27	12.88	14.12	16.15	18.11	20.70	22.72	26.34	30.32	32.91	36.74	40.11	44.14	46.96	55.48
28	13.56	14.85	16.93	18.94	21.59	23.65	27.34	31.39	34.03	37.92	41.34	45.42	48.28	56.89
29	14.26	15.57	17.71	19.77	22.48	24.58	28.34	32.46	35.14	39.09	42.56	46.69	49.59	58.30
30	14.95	16.31	18.49	20.60	23.36	25.51	29.34	33.53	36.25	40.26	43.77	47.96	50.89	59.70

Table H.3 : Critical values of F

(The values appearing on this page have been calculated via the built-in proprietary functions of Microsoft Excel, Within each cell, the first entry is the critical value of F for the 0.05 level of significance; the second entry (in boldface) is the critical value for the 0.01 level of significance.

df denominator	df numerator 1	2	3	4	5	6	7	8	9	10
2	18.51 **98.50**	19.00 **99.00**	19.16 **99.16**	19.25 **99.25**	19.30 **99.30**	19.33 **99.33**	19.3 **99.36**	19.37 **99.38**	19.38 **99.39**	19.40 **99.40**
3	10.13 **34.12**	9.55 **30.82**	9.28 **29.46**	9.12 **28.71**	9.01 **28.24**	8.94 **27.91**	8.89 **27.67**	8.85 **27.49**	8.81 **27.34**	8.79 **27.23**
4	7.71 **21.20**	6.94 **18.00**	6.59 **16.69**	6.39 **15.98**	6.26 **15.52**	6.16 **15.21**	6.09 **14.98**	6.04 **14.80**	6.00 **14.66**	5.96 **14.55**
5	6.61 **16.26**	5.79 **13.27**	5.41 **12.06**	5.19 **11.39**	5.05 **10.97**	4.95 **10.67**	4.88 **10.46**	4.82 **10.29**	4.77 **10.16**	4.74 **10.05**
6	5.99 **13.75**	5.14 **10.92**	4.76 **9.78**	4.53 **9.15**	4.39 **8.75**	4.28 **8.47**	4.21 **8.26**	4.15 **8.10**	4.10 **7.98**	4.06 **7.87**
7	5.59 **12.25**	4.74 **9.55**	4.35 **8.45**	4.12 **7.85**	3.97 **7.46**	3.87 **7.19**	3.79 **6.99**	3.73 **6.84**	3.68 **6.72**	3.64 **6.62**
8	5.32 **11.26**	4.46 **8.65**	4.07 **7.59**	3.84 **7.01**	3.69 **6.93**	3.58 **6.37**	3.50 **6.18**	3.44 **6.03**	3.39 **5.91**	3.35 **5.81**
9	5.12 **10.56**	4.26 **8.02**	3.86 **6.99**	3.63 **6.42**	3.48 **6.06**	3.37 **5.80**	3.29 **5.61**	3.23 **5.47**	3.18 **5.35**	3.14 **5.26**
10	4.96 **10.04**	4.10 **7.56**	3.71 **6.55**	3.48 **5.99**	3.33 **5.64**	3.22 **5.39**	3.14 **5.20**	3.07 **5.06**	3.02 **4.94**	2.98 **4.85**
11	4.84 **9.65**	3.98 **7.21**	3.59 **6.22**	3.36 **5.67**	3.20 **5.32**	3.09 **5.07**	3.01 **4.89**	2.95 **4.74**	2.90 **4.63**	2.85 **4.54**
12	4.75 **9.33**	3.89 **6.93**	3.49 **5.95**	3.26 **5.41**	3.11 **5.06**	3 **4.82**	2.91 **4.64**	2.85 **4.50**	2.80 **4.39**	2.75 **4.30**
13	4.67 **9.07**	3.81 **6.70**	3.41 **5.74**	3.18 **5.21**	3.03 **4.86**	2.92 **4.62**	2.83 **4.44**	2.77 **4.30**	2.71 **4.19**	2.67 **4.10**
14	4.60 **8.86**	3.74 **6.51**	3.34 **5.56**	3.11 **5.04**	2.96 **4.69**	2.85 **4.46**	2.76 **4.28**	2.70 **4.14**	2.65 **4.03**	2.60 **3.94**
15	4.54 **8.68**	3.68 **6.36**	3.29 **5.42**	3.06 **4.89**	2.90 **4.56**	2.79 **4.32**	2.71 **4.14**	2.64 **4.00**	2.59 **3.89**	2.54 **3.80**
16	4.49 **8.53**	3.63 **6.23**	3.24 **5.29**	3.01 **4.77**	2.85 **4.44**	2.74 **4.20**	2.66 **4.03**	2.59 **3.89**	2.54 **3.78**	2.49 **3.69**
17	4.45 **8.40**	3.59 **6.11**	3.20 **5.19**	2.96 **4.67**	2.81 **4.34**	2.70 **4.10**	2.61 **3.93**	2.55 **3.79**	2.49 **3.68**	2.45 **3.59**
18	4.41 **8.29**	3.55 **6.01**	3.16 **5.09**	2.93 **4.58**	2.77 **4.25**	2.66 **4.01**	2.58 **3.84**	2.51 **3.71**	2.46 **3.60**	2.41 **3.51**
19	4.38 **8.18**	3.52 **5.93**	3.13 **5.01**	2.90 **4.50**	2.74 **4.17**	2.63 **3.94**	2.54 **3.77**	2.48 **3.63**	2.42 **3.52**	2.38 **3.43**
20	4.35 **8.10**	3.49 **5.85**	3.10 **4.94**	2.87 **4.43**	2.71 **4.10**	2.60 **3.87**	2.51 **3.70**	2.45 **3.56**	2.39 **3.46**	2.35 **3.37**

df denominator	**df** numerator									
	1	2	3	4	5	6	7	8	9	10
21	4.32 **8.02**	3.47 **5.78**	3.07 **4.87**	2.84 **4.37**	2.68 **4.04**	2.57 **3.81**	2.49 **3.64**	2.42 **3.51**	2.37 **3.40**	2.32 **3.31**
22	4.30 **7.95**	3.44 **5.72**	3.05 **4.82**	2.82 **4.31**	2.66 **3.99**	2.55 **3.76**	2.46 **3.59**	2.40 **3.45**	2.34 **3.35**	2.30 **3.26**
23	4.28 **7.88**	3.42 **5.66**	3.03 **4.76**	2.80 **4.26**	2.64 **3.94**	2.53 **3.71**	2.44 **3.54**	2.37 **3.41**	2.32 **3.30**	2.27 **3.21**
24	4.26 **7.82**	3.40 **5.61**	3.01 **4.72**	2.78 **4.22**	2.62 **3.90**	2.51 **3.67**	2.42 **3.50**	2.36 **3.36**	2.30 **3.26**	2.25 **3.17**
25	4.24 **7.77**	3.39 **5.57**	2.99 **4.68**	2.76 **4.18**	2.60 **3.85**	2.49 **3.63**	2.40 **3.46**	2.34 **3.32**	2.28 **3.22**	2.24 **3.13**
26	4.23 **7.72**	3.37 **5.53**	2.98 **4.64**	2.74 **4.14**	2.59 **3.82**	2.47 **3.59**	2.39 **3.42**	2.32 **3.29**	2.27 **3.18**	2.22 **3.09**
27	4.21 **7.68**	3.35 **5.49**	2.96 **4.60**	2.73 **4.11**	2.57 **3.78**	2.46 **3.56**	2.37 **3.39**	2.31 **3.26**	2.25 **3.15**	2.20 **3.06**
28	4.20 **7.64**	3.34 **5.45**	2.95 **4.57**	2.71 **4.07**	2.56 **3.75**	2.45 **3.53**	2.36 **3.36**	2.29 **3.23**	2.24 **3.12**	2.19 **3.03**
29	4.18 **7.60**	3.33 **5.42**	2.93 **4.54**	2.70 **4.04**	2.55 **3.73**	2.43 **3.50**	2.35 **3.33**	2.28 **3.20**	2.22 **3.09**	2.18 **3.00**
30	4.17 **7.56**	3.32 **5.39**	2.92 **4.51**	2.69 **4.02**	2.53 **3.70**	2.42 **3.47**	2.33 **3.30**	2.27 **3.17**	2.21 **3.07**	2.16 **2.98**
31	4.16 **7.53**	3.30 **5.36**	2.91 **4.48**	2.68 **3.99**	2.52 **3.67**	2.41 **3.45**	2.32 **3.28**	2.25 **3.15**	2.20 **3.04**	2.15 **2.96**
32	4.15 **7.50**	3.29 **5.34**	2.90 **4.46**	2.67 **3.97**	2.51 **3.65**	2.40 **3.43**	2.31 **3.26**	2.24 **3.13**	2.19 **3.02**	2.14 **2.93**
33	4.14 **7.47**	3.28 **5.31**	2.89 **4.44**	2.66 **3.95**	2.50 **3.63**	2.39 **3.41**	2.30 **3.24**	2.23 **3.11**	2.18 **3.00**	2.13 **2.91**
34	4.13 **7.44**	3.28 **5.29**	2.88 **4.42**	2.65 **3.93**	2.49 **3.61**	2.38 **3.39**	2.29 **3.22**	2.23 **3.09**	2.17 **2.98**	2.12 **2.89**
35	4.12 **7.42**	3.27 **5.27**	2.87 **4.40**	2.64 **3.91**	2.49 **3.59**	2.37 **3.37**	2.29 **3.20**	2.22 **3.07**	2.16 **2.96**	2.11 **2.88**
36	4.11 **7.40**	3.26 **5.25**	2.87 **4.38**	2.63 **3.89**	2.48 **3.57**	2.36 **3.35**	2.28 **3.18**	2.21 **3.05**	2.15 **2.95**	2.11 **2.86**
37	4.11 **7.37**	3.25 **5.23**	2.86 **4.36**	2.63 **3.87**	2.47 **3.56**	2.36 **3.33**	2.27 **3.17**	2.20 **3.04**	2.14 **2.93**	2.10 **2.84**
38	4.10 **7.35**	3.24 **5.21**	2.85 **4.34**	2.62 **3.86**	2.46 **3.54**	2.35 **3.32**	2.26 **3.15**	2.19 **3.02**	2.14 **2.92**	2.09 **2.83**
39	4.09 **7.33**	3.24 **5.19**	2.85 **4.33**	2.61 **3.84**	2.46 **3.53**	2.34 **3.30**	2.26 **3.14**	2.19 **3.01**	2.13 **2.90**	2.08 **2.81**
40	4.08 **7.31**	3.23 **5.18**	2.84 **4.31**	2.61 **3.83**	2.45 **3.51**	2.34 **3.29**	2.25 **3.12**	2.18 **2.99**	2.12 **2.89**	2.08 **2.80**

df denominator	df numerator 1	2	3	4	5	6	7	8	9	10
41	4.08	3.23	2.83	2.60	2.44	2.33	2.24	2.17	2.12	2.07
	7.30	**5.16**	**4.30**	**3.81**	**3.50**	**3.28**	**3.11**	**2.98**	**2.87**	**2.79**
42	4.07	3.22	2.83	2.59	2.44	2.32	2.24	2.17	2.11	2.06
	7.28	**5.15**	**4.29**	**3.80**	**3.49**	**3.27**	**3.10**	**2.97**	**2.86**	**2.78**
43	4.07	3.21	2.82	2.59	2.43	2.32	2.23	2.16	2.11	2.06
	7.26	**5.14**	**4.27**	**3.79**	**3.48**	**3.25**	**3.09**	**2.96**	**2.85**	**2.76**
44	4.06	3.21	2.82	2.58	2.43	2.31	2.23	2.16	2.10	2.05
	7.25	**5.12**	**4.26**	**3.78**	**3.47**	**3.24**	**3.08**	**2.95**	**2.84**	**2.75**
45	4.06	3.20	2.81	2.58	2.42	2.31	2.22	2.15	2.10	2.05
	7.23	**5.11**	**4.25**	**3.77**	**3.45**	**3.23**	**3.07**	**2.94**	**2.83**	**2.74**
46	4.05	3.20	2.81	2.57	2.42	2.30	2.22	2.15	2.09	2.04
	7.22	**5.10**	**4.24**	**3.76**	**3.44**	**3.22**	**3.06**	**2.93**	**2.82**	**2.73**
47	4.05	3.20	2.80	2.57	2.41	2.30	2.21	2.14	2.09	2.04
	7.21	**5.09**	**4.23**	**3.75**	**3.43**	**3.21**	**3.05**	**2.92**	**2.81**	**2.72**
48	4.04	3.19	2.80	2.57	2.41	2.29	2.21	2.14	2.08	2.03
	7.19	**5.08**	**4.22**	**3.74**	**3.43**	**3.20**	**3.04**	**2.91**	**2.80**	**2.71**
49	4.04	3.19	2.79	2.56	2.40	2.29	2.20	2.13	2.08	2.03
	7.18	**5.07**	**4.21**	**3.73**	**3.42**	**3.19**	**3.03**	**2.90**	**2.79**	**2.71**
50	4.03	3.18	2.79	2.56	2.40	2.29	2.20	2.13	2.07	2.03
	7.17	**5.06**	**4.20**	**3.72**	**3.41**	**3.19**	**3.02**	**2.89**	**2.78**	**2.70**
51	4.03	3.18	2.79	2.55	2.40	2.28	2.20	2.13	2.07	2.02
	7.16	**5.05**	**4.19**	**3.71**	**3.40**	**3.18**	**3.01**	**2.88**	**2.78**	**2.69**
52	4.03	3.18	2.78	2.55	2.39	2.28	2.19	2.12	2.07	2.02
	7.15	**5.04**	**4.18**	**3.70**	**3.39**	**3.17**	**3.00**	**2.87**	**2.77**	**2.68**
53	4.02	3.17	2.78	2.55	2.39	2.28	2.19	2.12	2.06	2.01
	7.14	**5.03**	**4.17**	**3.70**	**3.38**	**3.16**	**3.00**	**2.87**	**2.76**	**2.68**
54	4.02	3.17	2.78	2.54	2.39	2.27	2.18	2.12	2.06	2.01
	7.13	**5.02**	**4.17**	**3.69**	**3.38**	**3.16**	**2.99**	**2.86**	**2.76**	**2.67**
55	4.02	3.16	2.77	2.54	2.38	2.27	2.18	2.11	2.06	2.01
	7.12	**5.01**	**4.16**	**3.68**	**3.37**	**3.15**	**2.98**	**2.85**	**2.75**	**2.66**
56	4.01	3.16	2.77	2.54	2.38	2.27	2.18	2.11	2.05	2.00
	7.11	**5.01**	**4.15**	**3.67**	**3.36**	**3.14**	**2.98**	**2.85**	**2.74**	**2.66**
57	4.01	3.16	2.77	2.53	2.38	2.26	2.18	2.11	2.05	2.00
	7.10	**5.00**	**4.15**	**3.67**	**3.36**	**3.14**	**2.97**	**2.84**	**2.74**	**2.65**
58	4.01	3.16	2.76	2.53	2.37	2.26	2.17	2.10	2.05	2.00
	7.09	**4.99**	**4.14**	**3.66**	**3.35**	**3.13**	**2.96**	**2.83**	**2.73**	**2.64**
59	4.00	3.15	2.76	2.53	2.37	2.26	2.17	2.10	2.04	2.00
	7.08	**4.98**	**4.13**	**3.65**	**3.34**	**3.12**	**2.96**	**2.83**	**2.72**	**2.64**
60	4.00	3.15	2.76	2.53	2.37	2.25	2.17	2.10	2.04	1.99
	7.08	**4.98**	**4.13**	**3.65**	**3.34**	**3.12**	**2.95**	**2.82**	**2.72**	**2.63**

df denominator	**df** numerator									
	1	2	3	4	5	6	7	8	9	10
61	4.00 **7.07**	3.15 **4.97**	2.76 **4.12**	2.52 **3.64**	2.37 **3.33**	2.25 **3.11**	2.16 **2.95**	2.09 **2.82**	2.04 **2.71**	1.99 **2.63**
62	4.00 **7.06**	3.15 **4.96**	2.75 **4.11**	2.52 **3.64**	2.36 **3.33**	2.25 **3.11**	2.16 **2.94**	2.09 **2.81**	2.03 **2.71**	1.99 **2.62**
63	3.99 **7.06**	3.14 **4.96**	2.75 **4.11**	2.52 **3.63**	2.36 **3.32**	2.25 **3.10**	2.16 **2.94**	2.09 **2.81**	2.03 **2.70**	1.98 **2.62**
64	3.99 **7.05**	3.14 **4.95**	2.75 **4.10**	2.52 **3.63**	2.36 **3.32**	2.24 **3.10**	2.16 **2.93**	2.09 **2.80**	2.03 **2.70**	1.98 **2.61**
65	3.99 **7.04**	3.14 **4.95**	2.75 **4.10**	2.51 **3.62**	2.36 **3.31**	2.24 **3.09**	2.15 **2.93**	2.08 **2.80**	2.03 **2.69**	1.98 **2.61**
66	3.99 **7.04**	3.14 **4.94**	2.74 **4.09**	2.51 **3.62**	2.35 **3.31**	2.24 **3.09**	2.15 **2.92**	2.08 **2.79**	2.03 **2.69**	1.98 **2.60**
67	3.98 **7.03**	3.13 **4.94**	2.74 **4.09**	2.51 **3.61**	2.35 **3.30**	2.24 **3.08**	2.15 **2.92**	2.08 **2.79**	2.02 **2.68**	1.98 **2.60**
68	3.98 **7.02**	3.13 **4.93**	2.74 **4.08**	2.51 **3.61**	2.35 **3.30**	2.24 **3.08**	2.15 **2.91**	2.08 **2.78**	2.02 **2.68**	1.97 **2.59**
69	3.98 **7.02**	3.13 **4.93**	2.74 **4.08**	2.50 **3.60**	2.35 **3.29**	2.23 **3.08**	2.15 **2.91**	2.08 **2.78**	2.02 **2.68**	1.97 **2.59**
70	3.98 **7.01**	3.13 **4.92**	2.74 **4.07**	2.50 **3.60**	2.35 **3.29**	2.23 **3.07**	2.14 **2.91**	2.07 **2.78**	2.02 **2.67**	1.97 **2.59**
71	3.98 **7.01**	3.13 **4.92**	2.73 **4.07**	2.50 **3.60**	2.34 **3.29**	2.23 **3.07**	2.14 **2.90**	2.07 **2.77**	2.01 **2.67**	1.97 **2.58**
72	3.97 **7.00**	3.12 **4.91**	2.73 **4.07**	2.50 **3.59**	2.34 **3.28**	2.23 **3.06**	2.14 **2.90**	2.07 **2.77**	2.01 **2.66**	1.96 **2.58**
73	3.97 **7.00**	3.12 **4.91**	2.73 **4.06**	2.50 **3.59**	2.34 **3.28**	2.23 **3.06**	2.14 **2.89**	2.07 **2.77**	2.01 **2.66**	1.96 **2.57**
74	3.97 **6.99**	3.12 **4.90**	2.73 **4.06**	2.50 **3.58**	2.34 **3.28**	2.22 **3.06**	2.14 **2.89**	2.07 **2.76**	2.01 **2.66**	1.96 **2.57**
75	3.97 **6.99**	3.12 **4.90**	2.73 **4.05**	2.49 **3.58**	2.34 **3.27**	2.22 **3.05**	2.13 **2.89**	2.06 **2.76**	2.01 **2.65**	1.96 **2.57**
76	3.97 **6.98**	3.12 **4.90**	2.72 **4.05**	2.49 **3.58**	2.33 **3.27**	2.22 **3.05**	2.13 **2.88**	2.06 **2.75**	2.01 **2.65**	1.96 **2.56**
77	3.97 **6.98**	3.12 **4.89**	2.72 **4.05**	2.49 **3.57**	2.33 **3.26**	2.22 **3.05**	2.13 **2.88**	2.06 **2.75**	2.00 **2.65**	1.96 **2.56**
78	3.96 **6.97**	3.11 **4.89**	2.72 **4.04**	2.49 **3.57**	2.33 **3.26**	2.22 **3.04**	2.13 **2.88**	2.06 **2.75**	2.00 **2.64**	1.95 **2.56**
79	3.96 **6.97**	3.11 **4.88**	2.72 **4.04**	2.49 **3.57**	2.33 **3.26**	2.22 **3.04**	2.13 **2.87**	2.06 **2.75**	2.00 **2.64**	1.95 **2.55**
80	3.96 **6.96**	3.11 **4.88**	2.72 **4.04**	2.49 **3.56**	2.33 **3.26**	2.21 **3.04**	2.13 **2.87**	2.06 **2.74**	2.00 **2.64**	1.95 **2.55**

Table H.4 : Selected critical values of S in the Kendall's coefficient of concordance

Values at 5% level of significance

K	N					Some additional values for N = 3	
	3	4	5	6	7	k	s
3			64.4	103.9	157.3	9	54.0
4		49.5	88.4	143.3	217.0	12	71.9
5		62.6	112.3	182.4	276.2	14	83.8
6		75.7	136.1	221.4	335.2	16	95.8
8	48.1	101.7	183.7	299.0	453.1	18	107.7
10	60.0	127.8	231.2	376.7	571.0		
15	89.8	192.9	349.8	570.5	864.9		
20	119.7	258.0	468.5	764.4	1158.7		

Values at 1% level of significance

K	N					Some additional values for N = 3	
	3	4	5	6	7	k	s
3			75.6	122.8	185.6	9	75.9
4		61.4	109.3	176.2	265.0	12	103.5
5		80.5	142.8	229.4	343.8	14	121.9
6		99.5	176.1	282.4	422.6	16	140.2
8	66.8	137.4	242.7	388.3	579.9	18	158.6
10	85.1	175.3	309.1	494.0	737.0		
15	131.0	269.8	475.2	758.2	1129.5		
20	177.0	364.2	641.2	1022.2	1521.9		

Appendix I

Synonyms of some Animals

Sypha(Sycon) sp.

Dugesia(Planaria) sp.

Neanthes(Nereis) sp.

Triops(Apus) sp.

Hippidea (Hippa) sp.

Cylindroiulus (Julus) sp.

Charonia (Triton) sp.

Doridium(Doris) sp.

Branchiostoma(Amphioxus) sp.

Torpedo(Narcine) sp.

Himantura(Dasyatis,Trygon) sp.

Puntius (Barbus) sp.

Colisa(Trichogaster) sp.

Monopterus (Amphipnous) sp.

Tenulosa(Hilsa) sp.

Hoplobatrachus (Rana) sp.

Duttaphyrnus (Bufo) melanostictus

Chelonia(Chelone) mydas

Note : Names in the brackets are the old generic names.

Glossary

This glossary provides an introduction to some basic and recurring terms in the text.

Abalone	The general name of *Haliotis* sp.
Abbe condenser	Condenser that causes the light to focussed on the object in a compound microscope.
Aboral surface	Surface not bearing the mouth.
Abortion	The removal or expulsion of an embryo or foetus from the uterus, resulting in or caused by its death. This can occur spontaneously or accidentally as with a miscarriage, or be artificially induced by medical, surgical or other means.
Abundance	Quotient of total number of individual species divided by total number of quadrats in which species has occurred.
Accessory respiratory organ	A system of air chambers formed by outgrowths from the mouth or gill region of those fish that occasionally leave the water.
Acellular	Non- multicellular organism.
Acellular	Not made up of cells or divided into cells.
Acetabulum	Cup-like suckers in the head of Taenia.
Acetylcholinesterase 1	Acetylcholinesterase-1(AChE-1), an enzyme found in the free state mainly in brain, nerve cells, muscle, lung and erythrocytes. It plays an important role in the transmission of nerve impulse. It breaks down acetylcholine into choline and acetic acid.
Acetylcholine (Ach)	A neurotransmitter found at the majority of synapses which occur where the nerve cells meet another
Achromic point	The point at which starch solution fails to give any color with iodine after acted upon by an amylolytic enzyme.

Acicula	The chitinous support rods in the parapodia.
Acid phosphatase	An enzyme that works under acid conditions and is made in the liver, spleen, bone marrow and the prostate gland. It is a phosphatase, a type of enzyme, used to free attached phosphate groups from other molecules during digestion.It is basically a phosphohomoesterase . It is stored in lysosomes and functions when these fuse with endosomes, which are acidified while they function; therefore, it has an acid pH optimum.
Acid-Haemin Method	It is the process for determination of haemoglobin(Hb) in blood. In the process N/10 HCl is used to dilute the blood so that the haemoglobin is converted into brownish yellow coloured acid haematin hydrochloride. The brown colour is then matched against the coloured prism of acid haematin of the comparator box, for the estimation of Hb. When the colour of Hb matches with prism the point is read on a scale and the Hb content gm /100 ml. of blood is obtained directly.
Acidosis	An abnormal physiological state characterised by plasma pH level below 7.5.
Acini	Large number of alveolar structure.
Acoelomate	Animal lack a body cavity between the gut and the outer body wall musculature.
Acrolein	Unsaturated aldehyde produced when glycerol is heated with dehydrating agents like acid potassium bisulphate
Adipose tissue	Loose connective tissue composed of adipocytes. Its main role is to store energy in the form of fat, although it also cushions and insulates the body. Two types of adipose tissue exist: white adipose tissue (WAT) and brown adipose tissue (BAT). Adipose tissue also serves as an important endocrine organ by producing recently-discovered hormones such as leptin, resistin and the cytokine TNF±.
Adrenal gland	A pair of glands situated above the kidney and produces the hormone adrenaline.
Adrenaline	A hormone when carried in the blood and a neurotransmitter when it is released across a neuronal synapse. It is a catecholamine, a sympathomimetic monoamine derived from the amino acids phenylalanine and tyrosine.

Affix	Gluing of the paraffin ribbon with the tissue sections on the slide with the help of an affixative (e.g. Meyer's albumin) for further staining.
Affixation	Process by which paraffin embedded tissues are trimmed for attachment into the microtome peg.
Agarose gel Electrophoresis	A electrophoresis carried out on gels where agarose (a naturally linear polysaccharide) is added to the low concentration of acrylamide. This enables electrophoretic separation of high molecular weight nucleic acids.
Aggressive behaviour	The act of initiating hostilities or invasion or the accomplishment or habit of initiating attacks or antagonistic or destructive behaviour or activity
Alate	Bearing wings.
Albumen	A proteinaceous substance that through the absorption of water forms the white of a egg
Aldehydes	An organic compound containing a terminal carbonyl group. This functional group, which consists of a carbon atom bonded to a hydrogen atom and double-bonded to an oxygen atom (chemical formula O=CH-), is called the aldehyde group. The aldehyde group is also called the formyl or methanoyl group.
Aldoses	Sugars that bear aldehyde groups or potential aldehyde group.
Alkalosis	An abnormal physiological state characterised by plasma pH level above 7.45.
Allantois	In tetrapods embryos a sac formed by outgrowth of the posterior ventral part of the gut as a precocious urinary bladder.
Alleles (allelomorphs)	Pair of genes that represent contrasting alternative of characters placed at the same locus in the homologous chromosomes.
Alpha particles	An alpha particle is a helium nucleus; it consists of two protons and two neutrons. Emission of an alpha particle causes decrease in atomic number by two and decrease in mass number by four. These particles have considerable energy (3-8) MeV. They cause ionisation of atom in their path. These particles produce linear intense tracks a few micrometers in length in autoradiograms.

Alveoli	Air sacs in the lungs of mammals. Alveoli occur in clusters at the end of each bronchiole. They have a thin moist walls and each is surrounded by network of capillaries that enables free exchange of gases between blood in the capillaries and the inspired air in the alveolus (singular of alveoli).
Ambulacral grooves	A deep linear depression found in the ambulacrum of star fish.
Amide linkage	The link between two amino acids. Formed by a reaction in which the carbonyl group of one amino acid reacts with the elimination of water.
Amino acids	Derivatives of carboxylic acids in which a hydrogen atom in an aliphatic acid has been replaced by an amino group. Thus from acetic acid glycine is derived.
Amnion	Epithelium surrounding the fluid filled embryo sac of the reptiles, birds and mammals.
Amoebiasis	Amoebiasis is amoebic dysentery, caused by *Entamoeba histolytica*.
Amphid	Paired lateral sensory organs on the head of Nematoda. It acts as a chemo receptor.
Ampullae	Small, mascular sacs attached to an echinoderm tube feet that bulges into the pevisceral coelom.
Anaesthization	The process of anesthetizing; also, the condition of the nervous system induced by anesthetics.
Anal cerci	Cylindrical or conical appendages at the posterior end of the abdomen.
Analysis of variance (ANOVA)	Powerful composite tests of the difference between variances. It divides the observed variance into independent components each caused by some specific independent factor and it estimates the relative magnitudes of the variance components due to uncontrolled random factors and fixed experimental treatments respectively and tests whether a particular component may be clarified away by null hypothesis.
Anaphase	The stage in mitosis or meiosis when chromatids are pulled towards opposite poles of the nuclear spindle. In mitosis the chromatids moving towards the poles represents a single chromosome. During Anaphase I of meiosis a pair of chromatids still connected at their centromere move to the spindle poles. During the Anaphase II the centromeres divide and single chromatids are drawn towards the poles.

Anaphase I	At this phase there is the partition of the homologous chromosomes. On the contrary to mitotic anaphase, the centromeres of a given homologue do not divide, and as a result each homologue progresses toward opposite poles. This results in a halving of the integer (number) of chromosomes, and is the root of the reduction division that characterizes meiosis.
Animalia	Multicellular with no cell walls and generally consume other organism or take in dissolved nutrients.
Anoestrus	In vertebrate seasonal breeders non-breeding period, a stage when they remain sexually dormant.
Antennae pilose	Antennae covered with fine hairs.
Apochromatic objective	A lens system whose secondary chromatic aberrations have been substantially reduced. It is designed to provide the same focal length for three wavelengths and freedom from spherical aberration for two wavelengths of light. The magnification can still vary with wavelength in which case compensating eyepieces used to cancel the coloured fringes.
Aponeurosis	A sheet like fibrous membrane, resembling a flattened tendon, that serves as a fascia to bind muscles together or as a means of connecting muscle to bone.
Appendicitis	Appendicitis is an inflammation of the appendix, which is the worm-shaped pouch attached to the cecum, the beginning of the large intestine. The appendix has no known function in the body, but it can become diseased. Appendicitis is a medical emergency, and if it is left untreated the appendix may rupture and cause a potentially fatal infection.
Apterous	Absence of wings.
Area opaca or Area vittelina	In the later part of the cleavage blastoderm differentiates into an area where the epithelium has not shed it yolky cells and renders that part of the blastodics as opaque.
Area vasculosa	This is formed from the non or extra-embryonic ectoderm which consists of the blood vessels.
Area pellucida	In the later part of the cleavage blastoderm differentiates into an area where the epithelium has shed it yolky cells and becomes thinner and renders that part of the blastodics as transparent.

Areolar tissue	Areolar tissue is formed throughout the body in loose sheets around blood vessels, nerves and tendons; as a soft pliable substance it helps to fill the spaces between larger organs. The tissue is made of a mixture of collagen, elastic and reticulin fibres.This can be seen in the liver substance and in the artery wall itself; here the circular muscle is held in place by a protective sheath of loose areolar tissue.
Aristotle's lantern	In regular and some irregular echinoids, the mouth is equipped with five teeth operated by a complex system of plates and muscles which is used to scrape food and scoop mud.
Arithmetic Mean	Arithmetic average of observed values or score of a sample.
Asconoid	The canal system is characterised by the presence of a complete continuous lining of choanocyte interrupted only by porocytes.
Atoke	The anterior or asexual part of the free swimming heteronereis.
Atriopore	Opening of a sac like chamber called atrium to the exterior of the urochordata and cephalochordata.
Attribute	The qualitative character of statistical information collected from the group. An attribute cannot be measured and classified only under different heads or categories; e.g. information about the habitat of a collection of species is qualitative, viz., soil habitat, tree habitat, aquatic habitat, moss habitat etc.
Ausculatory Method	A process of measuring blood pressure
Autoclave	The autoclave is a device used to sterilize (deep clean) medical and biological materials.
Autoradiography	A technique whereby thin slice of tissue or DNA/RNA/ Protein gel containing a radioactive isotope is placed in contact with a photographic plate. The image obtained on development shows the distribution of the isotope in the tissue or gel.
Autosomal inheritance	Inheritance through somatic chromosomes.
Autosomal recessive	An autosomal recessive means two copies of an abnormal gene must be present in order for the disease or trait to develop.
Autotomy	The voluntary severance by an animal of a part of its body.

Autotrophy	Obtaining carbon utilizing carbon dioxide.
Balbiani rings	In the polytene chromosomes of the dipteran salivary glands there is a distinct pattern of transverse banding. These bands occasionally forms reversible puffs or Balbiani rings or puffs which are associated with the differential gene expression.
Barbel	A fleshy protuberance near the month of fish it is provided with tactile and chemical receptor cells.
Barr body	Highly condensed ,darkly stained chromatin in the interphase somatic cells which is the inactivated X-chromosomes in the female mammals.
Basement membrane	A thin layer of jelly or cement that is found at the base of an epithelium.
Basophils	These white blood cells have irregular shaped, may be S-shaped rarely bilobed blue-violet pale-staining nucleus. Cytoplasmic basophilic (stain with basic dyes) bluish and granular. Granules large and coarse, variable sized colour deep purple.
Behavioral catalogue	List of all behaviors that can be observed, listened to, or has knowledge about it in an organism.
Benthic biotic community	Community that includes organisms that are the bottom dwelling ones or remain attached to it. The term Benthos is used to describe those organisms living on or in the bottom of water mass or closely related to it.
Beta particles	They are negatively charged particles and are very small and rapidly moving and carry a single negative charge. These particles have energy (near 0 – 2) MeV. They cause more penetration and less ionisation than the alpha particles. A beta particle emission the nucleus loses a neutron but gains a proton. These particles produce linear, intense tracks a few micrometers in length on photographic plate in autoradiograms.
Bilaterally symmetrical	Capable of division into symmetrical halves by only one longitudinal plane passing through the axis.
Bilateria	Structural body plan that is based on bilateral symmetry and that lacks any body cavities outside the gut and are slightly cephalized.
Binocellate mark	*Naja naja* has a distinct U-shaped marking behind its head (the *spectacle* mark).

Binomial	Paired naming of organism i.e., a generic name and a specific name.
Bipectinate	Ctenidia is provided with a central axis and lamellae on either side.
Biochemical Oxygen Demand (BOD)	Amount of dissolved oxygen (DO) required by aerobic decomposers to break down the organic matter in a given volume of water at a certain temperature over a particular period of time.
Bioindicator species	Species used to monitor the health of an environment or ecosystem. They are any biological species or group of species whose function, population, or status can be used to determine ecosystem or environmental integrity. An example of such a group of organisms are the copepods and other small water crustaceans present in many water bodies. Such organisms are monitored for changes (chemical, physiological, or behavioural) that may indicate a problem within their ecosystem.Eg.Lichen,chironomid larva eyc.
Bioluminescent	A chemical reaction in which much of the excess energy is given off as light rather than as heat.
Biotic community	It is a naturally occurring assemblage of plants and animals living in the same environment mutually sustaining and interdependent, constantly fixing, utilizing and dissipating energy.
Bipectinate	Double combed gill filaments extend from both sides of the ctenidial axis.
Biramous antennae	Two branched antennae.
Blastodisc	Disc-like area of active cytoplasm that contains zygote nucleus. Blastodiscs signifies the germinal area of a heavily yolked egg macrolecithal that produces the embryo.
Blastoderm	It is the primitive cellular plate of the blastula that consists of presumptive cells for ectoderm, mesoderm and endoderm and surrounds the blastocoele.
Blastogenesis	The development of a new individual from an asexual reproductive body.
Blastomeres	One of the cell produce during the cleavage of the egg.
Blastopore	This the mouth like opening of the gastrula which connect the archenteron of an embryo with the outside.
Blastozooid	Individuals that budded of from the oozoid.

Blood group	A classification of blood based on the presence or absence of inherited antigenic substances on the surface of red blood cells (RBCs).
Blood Pressure	A force exerted against the vascular walls by blood as the result of the push exerted by cardiac contraction and elasticity of the vessels walls.
Bone tissue	Bone tissue is composed of repeating, circular units called Haversian systems. Within each Haversian system is a canal, the Haversion canal. Arranged above the Haversion canals are concentric lamellae and cells.
Book gills	Book gills are flap-like appendages that are designed for gas exchange within water and seem to have their origin as modified legs. On the inside of each appendage there are attached over 100 thin leaf-like membranes called lamellae which appear as pages of a book, and are the areas of the gill where gas exchange takes place.It is found in King crab.
Botryoidal tissue	A peculiar type of connective tissue occurs between enteric canal and longitudinal layer of muscles of leech.
Bowman's capsule	The cup shaped end of a uriniferous tubules of vertebrate kidney. It surrounds a knot like blood capillaries (glomerulus) and together they form a Malpighian body.
Branchiae	Respiratory organ of aquatic animals that breathe oxygen dissolved in water.
Bright field microscope	In such a microscope, light from an incandescent source is aimed toward a lens beneath the stage called the condenser, through the specimen, through an objective lens, and to the eye through a second magnifying lens, the ocular or eyepiece.
Bright field microscopy	Simplest of all the light microscopy techniques. Sample illumination is through transmitted white light, i.e. illuminated from below and observed from above.
Bronchi	Air passages leading from the trachea and entering the lung in tetrapod. Each bronchus is a wide tube whose walls are stiffened by thick incomplete rings of cartilage and contains mucous secreting glands.
Brunner's glands	Submucosa of the initial portion of the duodenum consists of the clusters of the ramified coiled tubular glands known as the Brunner's gland. Their cells are of mucous type and produce in humans a neutral glycoprotein. The product of

	the gland is distinctly alkaline and probably acts to protect the duodenal mucosa against the intestinal contents to optimum pH for pancreatic enzyme action.
Byssal threads	A proteinaceous liquid secreted by byssal gland of mussels for attachment to hard substrates. The secretion solidifies to form many thin sturdy treads.
Byssus gland	A gland is situated at the base of the foot, within the mantle cavity of mussels that secrete a proteinaceous liquid.
Calcium rigor	A condition where the heart stops at systole in a perfusion experiment with excess calcium ions in the perfusion fluid that cause initially increase of force of contraction of heart but if the concentration is increased further, the heart stops.
Camera Lucida	An instrument that helps in the recording of the observation made under a light microscope.
Canal system	Numerous canals that traverse the body of sponges.
Capitulum	A sclerotised, compound part formed of the mouth parts and the first two pairs of prosomal appendages found in acarines. The broad part of the body of barnacles covered by carapace or mantle.
Captacula	Exsertile filamentous tactile organs with sucker-like ends near mouth of tusk shells (Scaphopoda).
Carapace	A carapace is a dorsal section of an exoskeleton or shell in a number of animal groups
Carbohydrates	A class of compounds occurring widely in nature and having the general formula $C_x(H_2O)_y$. Carbohydrates are widely found in the plant and the animal kingdom. Carbohydrates exists in the various forms viz., mono, di-, oligo, and polysaccharides.
Cardiac muscle	Type of *involuntary* striated muscle found within the heart. Its function is to "pump" blood through the circulatory system by contracting.
Carina	The dorsal calcareous plates of barnacles.
Carolus linnaeus	A Swedish naturalist who is real founder of science of taxonomy.
Cartilage	A tough, semi transparent, elastic, flexible connective tissue consisting of cartilage cells (chondrocytes and chondroblasts) scattered through a lipoprotein material strengthen by collagen fibers. The exterior part of cartilage

	is covered by dense fibrous membrane called the perichondrium. There are no nerves or blood vessels in cartilage, and when damaged, it does not heal readily. Cartilage has several functions. It covers the surface of joints, allowing bones to slide over one another, thus reducing friction and preventing damage; it also acts as a shock absorber. It forms part of the structure of the skeleton in the ribs, where it joins them to the breastbone (sternum). Cartilage is found in the tip of the nose, in the external ear, in the walls of the windpipe (trachea) and the voice box (larynx) where it provides support and shape. In an embryo, the skeleton is formed of cartilage which is gradually replaced by bone as the embryo grows. Cartilage is known as elastic cartilage, fibrocartilage or hyaline cartilage, depending on its different physical properties.
Caudal filament	A filament to wards the tail or back end of the body.
Caudal glands	They secrete adhesive material by which the nematodes can attach to a substrate. Usually there are three caudal gland cells but the number ranges from two to five.
Central disc	It is a disc like structure in star fish through which five arms extends.
Central tendency	An estimate of the "centre" of a distribution of values. There are three major types of estimates of central tendency: Mean, Median and Mode.
Centrifugal force	Term which may refer to two different forces which are related to rotation. Both of them are oriented away from the axis of rotation, but the object on which they are exerted differs.
Centrifugation	A technique in which suspensions may be rotated at a very high speeds in order to separate the components solids by centrifugal force.
Cephalothoracic carapace	A continuous shield –like exoskeletal covering , encloses the cephalothorax.
Cephalothorax	The cephalothorax (called prosoma in some groups) is an anatomical term used in arachnids and malacostracan crustaceans for the first (anterior) major body section.
Chalaza	The suspensory ligament of the yolk in a bird's egg.
Chelicerae	The Chelicerae are mouth parts of the Chelicerata, an arthropod subphylum that includes Arachnids, Merostomata (horseshoe crabs), and Pycnogonida (sea

	spiders). Chelicerae are pointed appendages which are used to grasp food, and are found in place of the chewing mandibles most other arthropods have.
Chemical Oxygen Demand (COD)	It is the measure of oxygen consumed during the oxidation of oxidisable organic matter by a strong oxidising agent.
Chemotaxis	During the initial phase of phagocytosis macrophages and other phagocytic cells are drawn towards different substances produced in an immune reaction.
Chi-square test	Test statistics in which a hypothesis is tested that provides a set of theoretical frequencies with which observed frequencies are compared. Such a measure enables us to find out the degree of discrepancy between observed frequencies and theoretical frequencies.
Chlorinity	Total content of chloride ion (mg/L) in water.
Choanocytes or collar cells	The flagellated cells surrounded by a collar like sheath of protoplasm.
Cholesterol	Esters of higher fatty acids with cyclic alcohol containing perhydrocyclopentanophenathrene ring.
Chondrocytes	Only cells found in cartilage. They produce and maintain the cartilaginous matrix, which consists mainly of collagen and proteoglycans.
Chorion	The outer cellular layer of the embryo sac of reptiles, birds and mammals.
Chromaffin cells	The adrenal medulla consists of paraganglia and granular cells termed as chromaffin cells.
Chromatic beam splitter	A beam splitter which removes aberrations such as coma and transverse chromatic aberration by subjecting all transmitted rays to equal and opposite amounts of induction of coma and transverse chromatic aberration, by curing residual astigmatism by providing a terminal astigmatism correcting lens in each transmitted ray path, and by avoiding ghost images by ensuring adequate separation between the main image and any ghost images.
Chromatography	Chromatography is one of the separation methods based on differential migration of components of a mixture.
Chromocenter	Heterochromatic regions of varying size which occur near the centromeres of proximal regions of chromosome arms. In dipteran salivary glands the chromocenters fuse to form a large chromocenter.
Chromogenic	Containing, or capable of forming, chromogen.

Cilia	Contractile filament surrounded by membrane that is continuous with the plasma membrane of the cell surface. Cilia take their origin in granules called kinetosomes
Ciliated epithelium	Any epithelium having motile cilia on the free surface.
Cirri	Small, tentacle like, cylindrical appendages found in the parapodium of a group of cilia that functions a single unit in ciliated protozoa; among barnacles, the thoracic appendages, which are modified for food collection; among crinoids, the prehensile appendages located aborally.
Cirriform limbs	Hairy thoracic appendages used to filter food particles from water.
Clasper	The pelvic fins of the male sharks, skates and rays, is connected with a copulatory organ called clasper. It has a rod like appearance having a dorsal groove leading to a siphon at its base.
Classification	Ordering of animals into groups on the basis of their relationship.
Clearing	During the preparation of the permanent microscopic slides the stage between dehydration and embedding. The purpose of which is to remove the dehydrating agents and replace it by substance that is miscible with embedding substances. Clearing renders tissue transparent.Clearing agents are benzene, xylene etc,
Clitellum	A swollen, glandular, saddle like region on the body of Oligochaeta and Hirudinea.
Cnidoblast cells	A rounded or oval, epidermal cell with a basal nucleus, that contains nematocysts.
Coefficient of Variation	A dimensionless number. For distributions of positive-valued random variables, it allows comparison of the variation of populations that have significantly different mean values. It is often reported as a percentage (%) by multiplying the above calculation by 100. The absolute value of the coefficient of variation expressed as a percentage is often referred to as the relative standard deviation (RSD).
Coelom	A liquid filled body cavity in the mesoderm, which develops in and is surrounded by the embryonic mesoderm.
Coelomates	The animals possessing coelom.
Colloid	Colloid are viscous liquid containing iodine in the form of thyroglobulin having active component called thyroxine that fills up the cavities of the vesicles.

Columnar epithelium	Epithelium formed of a single layer of prismatic cells taller than there width.
Commensals	Non-pathogenic endoparasites, which live on superfluous fluids or solids in the host,
Compound optical microscope	A particular type of microscope that uses a set of multiple lenses so as to maximize magnification. In its simplest form - for example, the compound microscope wouid have a single glass lens (with a short focal length) used as the objective, and another single glass lens used as the eyepiece or ocular and a mirror for reflecting light upon the object that is to examined. Modern compound microscopes whereas are usually more intricate, having multiple lenses in both objective and eyepiece assemblies.
Compound or conjugated Lipids	These are esters of fatty acids which on hydrolysis yield other substances in addition to fatty acids and an alcohol.
Conjugated proteins	They consists of simple proteins molecule attached with some non-protein prosthetic group (nucleic acid, phosphoric acid, lipids etc), eg. nucleoprotein, phosphoprotein, lecithoprotein lipoproteins etc.
Connective tissue proper	General term for mesodermally derived tissue that may be more or less specialised. Cartilage and bone are specialised connective tissue, as is blood, but the term is probably better reserved for the less specialised tissue that is rich in extracellular matrix (collagen, proteoglycan etc.) and that surrounds other more highly ordered tissues and organs.
Consanguineous marriage	It is the marriage or mating type that share a least one common ancestor in the preceding few generations.
Contingency Chi-square	A type chi-square test that compares of two groups with a dependent variable (roughly speaking; there is actually no mathematical distinction between dependent and independent variables in this test) that is divided into two sharply distinguished parts or classifications. It is used for frequency data (numerical counts).The contingency chi-square is based on the same principles as a simple chi-square analysis you observed in class in which we examine the expected vs. the observed frequencies. The computation is quite similar, except that the estimate of the expected frequency is a little harder to determine. The H_0 (null hypothesis) is that the variables are independent from one another.

Continuous variables	The set of all values which consists of intervals, e.g. 0-8, 9-18, 19-28...... etc.
Contractile vacuole	An organelle that expands as it fills and contracts as it emptied.
Copulatory spicules	Chitinoid spicules situated posteriorly in male nematode that are inserted into the female gonopore during sperm transfer.
Corallite	The skeleton of a solitary polyp.
Corallium	The skeleton of a colonial coral, made up from solitary corallites.
Corona radiata	Elongated part of the granulosa cell layer that remain in close contact of the ovum. This accompanies the ovum while it leaves the ovum.
Corpus albicans	Scar like appearance after the degeneration of the corpus luteum. It is made up of dense connective tissue.
Corpus luteum	Progesterone secreting temporary endocrine glandular mass of follicle cells that develops in the ovary after ovulation.
Correlation	It is the measure of the degree of statistical relationship between two variables from bivariate data.
Costae	Meridional swimming- plates of Hormiphora.
Crank Myograph or the Muscle Lever	It comprises of a myograph board for griping never muscle preparation with a crank lever fixed to it. The shorter arm of lever is designed for attachment to the muscle while the longer writing lever bears a writing point designed for recording the contraction on a smoked paper wrapped around.
Crossopterigian	Sub groups of sarcopterygians (second group of bony fish).
Critical illumination	This is a particular alignment of the microscope components i.e. lens and light source. In this process an image of the light source (bulb filament) is projected into the plane of the object, thus superimposing the light source onto the object.
Cryofixation	Specimen are frozen rapidly in liquid nitrogen or even liquid helium temperatures so that the water forms vitreous (non-crystalline) ice. This preserves the specimen in a snapshot of its solution state. This is sometimes required in the specimen preparation in the electron microscopy.

Crypts of Lieberkuhn	These are tubular glands in the epithelial cells of villi of intestine. Paneth cells with zymogen granules are seen at the base of these tubular glands.
Ctenidia	Comb like molluscan gills lie in the mantle cavity.
Ctenoid scale	A scale with many tiny, tooth like process (ctenii) in the outer segment.
Cuboidal epithelium	Simple epithelium with cells appearing as cubes in a vertical section but as polyhedral in surface view.
Cumlus oophorus	The mass of cells surrounding immediately the ovum are cumlus oophorus. The cumulus oophorus protrudes inside the interior of the antrum.
Cuttle bone	The inner chambered shell of cuttle fish that involves buoyancy regulation.
Cycloid scale	A type of scale whose edge being smooth all around.
Cystecercus stage	Bladder- like embryo of the tape worm having a small scolex.
Cysticercosis	An infection caused by the pork tapeworm, *Taenia solium*. Infection occurs when the tapeworm larvae enter the body and form cysticerci. When cysticerci are found in the brain, the condition is called neurocysticercosis.
Cytochemistry	Biochemistry of cells, especially that of the macromolecules responsible for cell structure and function. The branch of biochemistry that deals with the study of the chemical composition and activity of cells.
Cytopharynx	A short, passage like region through which ingested food passes.
Cytopyge	Anal spot situated posterior to the mouth.
Cytostome	The cell's mouth which food is ingested (Cyto = G : cell; stoma = G : mouth).
Dactylozooids	Simple hollow finger – like bodies specialized for defence, feeding and reproduction.
Dark Field Microscopy	An optical microscopy illumination technique used to enhance the contrast in unstained samples. It works on the principle of illuminating the sample with light that will not be collected by the objective lens, so not form part of the image. This produces the classic appearance of a dark, almost black, background with bright objects on it.
Defensive connective tissue	Various components of the connective tissue play roles in the defence or protection of the body including many of the components of the vascular and immune systems (plasma cells, lymphocytes, neutrophils, eosinophils, basophils, mast cells).

Dehydration	A process followed when tissue are prepared for permanent microscopic slides. Water is removed by immersing the tissue in increasing strength of ethyl alcohol. Dehydration is important as water does not mix with chemical that are involved in cleaning and mounting of sections.
Denatured proteins	These are the acid metaproteins and alkali metaproteins.
Density	Quotient of total number of individual species divided by total number of quadrats studied.
Denticles	A scale composed of dentine with a pulp cavity, which resembles teeth.
Dependent variables	Variable whose changes with the change of the value of another variable (independent variable) as seen in some association or existing relationship between two
Derived lipids	These are compounds formed in the hydrolysis of simple or compound lipids. This includes Fattyacids, Alcohols, Hydrocarbons, Fat soluble vitamins, Stererols and Steroids.
Derived protein	These are the intermediate product of protein hydrolysis through the action of heat, alkali, acid and enzymes. These may be of the following types:Primary protein derivative: These are the substances formed from some of the simple and conjugated proteins on denaturation. They are proteans, metaproteins.Secondary protein derivatives: They are substances formed during the hydrolysis of protein molecules. Eg. Proteoses, peptones, peptides and amino acids.
Dermis	Lower vascularised layer of the skin formed of connective tissue. The dermis contains the nerve fibres, tactile corpuscles, blood vessels and various glands such as sweat glands and sebaceous glands.
Descriptive embryology	A branch of embryology that is applied to the method of embryological study concerned with the direct observation and description of embryological development
Detorsion	Reversion to the changes that have occurred during torsion.
Deviations	As used in statistics, a variation from an expected number.
Diabetes	Clinical condition characterized by polyuria.
Diakinesis	This is a substage of Prophase I of meiosis. The chromosomes will carry on to coil tightly (reversing the

	slight uncoiling of the diplotene) and will attain their greatest state of contraction. As diakinesis carries on, chiasmata appear to move toward the ends of the chromosomes, a process known as "terminalization." As this stage is the conclusion of prophase, the nucleolus generally disappears, along with the nuclear envelope.
Diastema	A gap in teeth, most commonly between the incisors and the first premolar.
Diastolic pressure	Blood pressure that indicates the constant load against which heart has to work. It reflects on the degree of peripheral resistance.
Differential count	Counting of different white blood cells in stained blood smear to assess percentage of different types of white blood cells.
Dihybrid cross	A cross between two parents involving a single pair of contrasting pairs of two traits.
Dioestrus stage	In absence of fertilization the corpus luteum and endometrium diminish, ovarian hormone level falls and new follicle begins its growth.
Diphyodont	Milk teeth is shed and replaced by as second set of permanent teeth
Diploblastic	Organisms having two germinal layers i.e. ectoderm, and endoderm.
Diplotene	This is a substage of Prophase I of meiosis. The homologues have already paired during zygotene, recombined at pachytene stage and are at this stage beginning to repel (drive back) each other. As the chromosomes break up, they are seen to remain attached at points known as "**chiasmata.**" These are thought to be the locations where recombination of the genes has occurred.
Disaccharide	A carbohydrate that when hydrolysed produces two sugar units. Eg. Sucrose, maltose etc.
Discoblastula	It is a type of blastula that appears in the animal pole a small multilayered flat disc separated from the yolk by a narrow segmentation or subgerminal cavity.
Discrete variables	The set of all possible values which consist only of isolated (secluded) points, e.g. counting variables (1, 2, 3 ...10).
Dispersion	It is the difference between the largest and smallest observations.

Distal convulated tubule	Thick ascending limb of the Loop of Henle penetrates the cortex, it preserves its histologic structure but becomes torturous.
Dizygotic twin	Twins that are produced by simultaneous fertilization of two separate ova by two separate sperms. They contain separate sets of genes and are like any other brothers or sisters but just happen to grow in the same uterus. They may differ in their traits, development and often even sex.
Dominant trait	The trait or character that will appear in the offspring if one of the parents contributes it.
Double staining	Section of tissue that are usually stained with two dyes to bring contrast between different histological structures the process is known as double staining, e.g. Haematoxylin-Eosin stain.
Down Syndrome	A genetic disorder caused by the presence of all or part of an extra 21st chromosome. It is named after John Langdon Down, the British doctor who described it in 1866.
Ectoderm	The outermost layer of the germ layer.
Ectoparasites	Parasites which live outside on the surface of the body of the host
Ectoplasm	The outer gel like layer of the cell cytoplasm lies just beneath the cell membrane.
Ectothermic	An animal that maintains its body temperature within fairly narrow limits by basking or seeking shade.
Electrodes	Piece of equipment to communicate the current to the excitable tissues. They are applied to the open ends of the two wires which transmit the current to the tissue to be stimulated. They are made up of two insulated wires, the ends of which are clean and free from the insulating material and twisted round a piece of cork or rubber.
Electron microscope	Microscope that uses electrons to create an image of the target. It has much higher magnification and resolving power than a normal light microscope, up to two million times, allowing it to see smaller objects and details
Electrophoresis	The migration of the disperse phase particles of a sol towards one of the electrodes in a electric field.
Electrophoretic mobility	Mobility of charged particle through a electric field. It is the characteristic velocity of migration in a field with a fall of potential at 1 volt/cm.

Elephantiasis	A disease caused by *Wechereria* in which the limbs and other organs grow to enormous size.
Elisa	A technique which permits the quantitative determination of antibody binding.
Elytra	A series of overlapping protective plates on the body of Polychaeta.
Embedding	The technique of placing cells or tissue in a supporting medium so that thin sections can be cut using a microtome. The medium can be paraffin wax (paraffin embedding) or plastics (plastic embedding) such as epoxy resins.
Embryogenetic development	The development of a new individual from the fertilized egg or zygote.
Embryology	It is the division of life science that deals with the development of embryos.
Endoderm	The innermost layer of the germ layer.
Endomitosis	The duplication of the chromosome without the division of the nucleus.
Endomysium	A sheath of connective tissue that surrounds muscle fibres.
Endoparasites	Parasites which live inside the body of the host in the blood, tissues, body cavities, digestive tract and other organs.
Endoplasm	The inner layer of the cytoplasm.
Endosome or karyosome	A mass of chromatin near the center of a vesicular nucleus. The inner layer of certain sponges.
Eosinophilia	An absolute or relative increase in the normal number of eosinophils in the circulating blood. Various limits are given for example, absolute eosinophilia if the total number exceeds 500 mm^3 and relative if greater than 3% but total less than 500 mm^3. It may be associated with skin diseases, parasitic infestations, hay fever, asthma, angioneurotic edema, adrenocortical insufficiency, and Hodgkin's disease.
Eosinophils	These white blood cells have crescent-shaped 2-3 lobed nucleus connected by thick or thin chromatin band. Nucleus stains blue violet. Cytoplasm eosinophilic (stain with acidic dyes), light pink-red granules. Large coarse uniform sized granules brick red to orange in colour.
Epiblast	Outer or the upper layer of a double layered embryo such as in a chick

Epidermis	Upper layer of the skin composed of stratified squamous epithelium the cells of which are arranged in strata. It has four layers: (a) Stratum Corneum – Outermost keratinized layer of flat cells without nucleus comprising of amorphous matrix with prekeratin filaments. (b) Stratum lucidum – Next layer consists of dead cells without granules. (c) Stratum granulosum – Next to stratum lucidum and contains keratohyalin granules. (d) Stratum malpighii – The inner most layer, consisting of germinal layer it produces new cells
Epipodium	A ridge or fold on the lateral edges of each side of the foot of certain gastropod molluscs.
Epithilial tissue	A tissue consisting of a sheet or sheets of cells that covers a surface or lines a cavity. The cells are close to each other and have a little cement substances between them and they rest on a basement membrane.
Epitoke	A sexually mature male or female heteronereis that is highly specialized for swimming and sexual reproduction.
Epitope	The part of a macromolecule that is recognized by the immune system, specifically by antibodies, B cells, or T cells.
Epizoic	Non-parasitic animals that harbour themselves to the outer surface of other animals.
Equational division	Mitotic division in which chromosome number remains the same from mother cell to the daughter cell.
Equilibration	During the development of a chromatogram in thin layer chromatography the developing solvent is poured into the tank to a depth of 1.5cm and allowed to stand for no less than an hour with a cover plate over the top of the tank to make sure that the environment within the tank becomes saturated with solvent vapour. This is called equilibration.
Erythrocyte Sedimentation Rate (E.S.R)	When a sample of blood treated with an anticoagulant is left standing for about one or two hour in a long tube placed vertically the red blood corpuscles (sp. gr. 1.095) being heavier than (denser) the colloids plasma (sp. gr. 1.032) settles at the bottom of the tube leaving the blood plasma above. The height in millimetre of the clear plasma is the Erythrocyte Sedimentation Rate or E.S.R or suspension stability of the blood.

Ethogram	A catalogue of an animal's behavioural repertoire, detailing the different forms of behaviour that are displayed by an animal.
Ethology	The study of behaviour of animals.
Eusocial	Social insects colonies in which only one reproductive queen produces offspring, all other individuals attending to her needs cooperate in the care of the developing embryos and larva and performing other tasks for the welfare of the colony.
Exocytosis	The bulk transport of materials out of the cell across the plasma mambrane
Experimental embryology	A branch of embryology that uses experiment as method of investigation which seeks to find out why a process occurs at that specific time and in just that particular way.
Eye brushes	Soft hairs on the tibia of prothoracic leg of honey bee.
Eyepiece or ocular lens	Compound microscope bears lens that is inserted at the top of the cylinder of the compound microscope to bring the image to focus for the eye.
Fasciculi	The whole muscle, such as the biceps, is enclosed in a sheath of connective tissue, the epimysium. This sheath folds inwards into the substance of the muscle to surround a large number of smaller bundles, the fasciculi. These fasciculi consist of still smaller bundles of elongated, cylindrical muscle cells, the fibres. Each fibre is a syncytium, i.e. a cell that has many nuclei.
Fat soluble vitamins	They are mainly the vitamin A, D, E and K. They are hydrophobic and soluble in fat and fatty substances.
Fatty acid	A carboxylic acid often with a long unbranched aliphatic tail (chain), which is either saturated or unsaturated.
Fatty acids	They are straight chain acids of higher series and usually contain an even number of carbon atoms. These may be saturated (acetic acid, propionic acid,palmitic acid, stearic acid etc.) or unsaturated(oleic acid, arachidonic acid etc.).
Female gonopore	In female when the mascular vagina opens directly out side the body, called gonopore.
Fertilization	The fusion of male and female gamete to form zygote.
Feulguen's reaction	A specific staining procedure for DNA: mild acid hydrolysis makes the aldehyde group of deoxyribose available to react with schiffs reagent to give a purple colour.

Fibro-cartilage	Composed of bundles of thick clearly defined collagen fibers, along with intervening unicellular islands of cartilage arranged in small chains;- it has a low content of glycosaminoglycans (2% of dry weight).- cartilaginous matrix is small and is noticed only around each chondrocyte - collagenous fibers are usually arrayed parallel to tension on tissue;- it is found in symphyseal joints, intervertebral discs & in articular discs;
Fibroblasts	Connective tissue cells which secrete an extracellular matrix rich in collagen and other macromolecules.
Fibrous proteins	Long filamentous protein molecules that form one of the two main classes of tertiary structure protein (the other being globular proteins). Fibrous proteins are only found in animals. Fibrous proteins are usually used to construct connective tissues, tendons, bone matrix and muscle fiber. Examples of fibrous proteins include keratins, collagens and elastins
Filiform	Thread-like; long and slender.
Filopodia	Net-like system of branching and fusing pseudopodia.
Fin rays	A rod like support of the fin membrane.
Fingerling	When the fry becomes 5″-6″ in length is called fingering.
Fixation	The process by which tissue are rapidly killed by chemicals to preserve their morphology, as well as preventing deterioration, the process also render cell organelles and inclusion more visible and harden the tissue to prevent shrinkage and distortion during dehydration, embedding and sectioning.
Fixatives	The agents that are used in the fixing of cells are called as fixatives. The fixatives renders cell , organelles and inclusions more visible and the harden the tissue to prevent shrinkage and distortion during the dehydration, embedding and sectioning
Flagella	Contractile axial filament enclosed in a protoplasmic sheath which arises from granules called basal granules or blepharoplasts.
Fluid connective tissue	Some connective tissues of the body are fluids. Although it can be difficult to visualize fluids as "connective" components, an examination of their roles in body homeostasis provides insight into why they are considered tissues of this type. It might be better to think of fluid connective tissues as providing "physiological" rather than physical connections between tissues, Eg. Blood and lymph

Fluorescein	A fluorophore commonly used in microscopy, in a type of dye laser as the gain medium, in forensics and serology to detect latent blood stains, and in dye tracing. Fluorescein has an absorption maximum at 494 nm and emission maximum of 521 nm (in water).
Fluorescence	An emission phenomenon where a photon or quantum of light absorbed by a fluorescent molecule raises of its orbital electrons to a remoter orbit and consequently elevates the molecule to transient high energy excited state. As the electron returns to its original orbit the photon activated molecules regains its original low energy ground state by emitting part of absorbed energy as specific light wavelengths longer than the absorbed wavelengths.
Fluorescence Microscopy	Microscopy which is applied for the visualisation and detection of fluorescent molecules at a low concentration and pinpointing their intracellular position.
Fluorochrome	Molecules that exhibits fluorescence, eg. Flourescein and Rhodamine.
Food Adulterates	Food debased by mixing inferior or spurious ingredients.
Freeze-fracture or freeze-etch	It is a preparation method specifically useful for studying lipid membranes and their inserted proteins in "face on" view through an electron microscope.
French press	A device for homogenization of tissues and cells.
Frequency	Total no. of quadrats in which species occurred divided by Total no. of quadrats used multiplied by 100.
Fruit fly	Common name of *Drosophila sp.*
Fry	One inch length spawn of fish is called fry.
Fungi	Nucleated, usually filamentous, spore bearing organisms devoid of chlorophyll; typically reproducing both sexually and asexually; living as parasites in plants, animals, or other fungi, or as saprobes on plant or animal remains, in aquatic, marine, terrestrial, or subaerial habitats. Yeasts, mildews, rusts, mushrooms, and truffles are examples of fungi.
Galactosazone	Osazone produced by galactose that has a shape of strips and plates.
Gametogony	The development of merozoites into male and female gametes, which later fuse to form a zygote.

Gamma rays	These are electromagnetic radiation and therefore have no charge or mass. The energy amounts to about (0.5, 2.0 and 0.5) MeV. Gamma rays have low ionising power but high penetration. These particles produce long haphazard tracks of grains and are ineffective in autoradiograms.
Ganoine layer	Enamel like superficial layer found in the rhomboid shaped scale of some fossil as well as extinct fishes.
Gastronemius muscle	A muscle that originates from the femur and insert in the sole of the foot in the hind limb of the frog/toad.
Gene	Gene is a unit particle that control the appearance of character in an individual during development of the zygote and the gene is transmitted from one generation to the other.
Germinal disc	A flattened, disk like region of cells from which the embryo begins to develop in the fertilized ovum of many vertebrate species; also called blastodisc or embryonic disc.
Germinal epithelium	Layer of columnar cells which covers the surface of the ovary or testis. This give rise to follicles which become mature and form Graafian follicle in female. In male it give rise to spermatozoa.
Gill slits (stigmata)	Gills with individual openings rather than an outer cover. Cartilaginous fish such as sharks, rays, sawfish and guitarfish all have gill slits. Most have five pairs, but a few species have six or seven pairs. Bony fish have an outer bony gill covering called an operculum.
Glisson's capsules	Connective tissue surrounding the hepatic lobules.
Globular proteins	A folded protein that adopts an approximately globular shape.
Glomerulus	A small knot of capillaries within the vertebrate kidney which is surrounded by a cup shaped Bowman's capsule.The glomerulus is supplied with blood by afferent arteriole which branches from the renal artery.
Glucagons	A polypeptide hormone produced by the alpha cells of the Islets of Langerhans of the pancreas in response to the growth hormone. Its action causes an increase in blood glucose by promoting the breakdown of glycogen to glucose in the liver.
Glucosazone	Osazone produced by glucose and fructose and mannose which is fan, sleeves, and needle or cross shape.

Glycerol	Chemical compound with the formula $HOCH_2CH(OH)CH_2OH$. This colorless, odorless, viscous liquid is widely used in pharmaceutical formulations. Also commonly called **glycerin** or **glycerine**, it is a sugar alcohol, and is sweet-tasting and of low toxicity. Glycerol has three hydrophilic alcoholic hydroxyl groups that are responsible for its solubility in water and its hygroscopic nature.
Glycolipids	These are complex lipids containing fatty acid, carbohydrate in combination with long chain aliphatic alcohols or acids and nitrogenous base but devoid of phosphoric acid. Eg. Cerebrosides, Gangliosides, Cytolipins.
Glycosidic linkage	A bond between sugar and alcohol. A linkage that links to sugars to produce disaccharide, oligosaccharide or polysaccharide.
Graafian follicle	A fluid filled ball of cells inside the mature ovary in which an ovum develops. A Graafian follicle matures cyclically during active reproductive years from one of the huge number of follicles present in the ovary at birth. It enlarges and migrates to the surface of the ovary. Finally it burst and releases the ovum to the Fallopian tube.
Gram staining	Most popular staining procedure used for differentiation of two groups of bacteria viz., gram-positive and gram-negative. Christian Gram invented this method in 1884. In this process bacterial smear is drawn on a slide, and stained with crystal violet and then mordanted with iodine solution. The crystal violet and iodine complex gives a purple black colour to the bacterial cells. In Gram-positive bacteria, the chemical complex binding and is trapped in the cell wall followed by an ethanol treatment which may be due to decrement in the diameter of the pore in the cell wall peptidoglycan stops release of crystal violet complex and they get stained. Where as in the case of the Gram-negative bacteria, they have a much less amount of peptidoglycan and less extensively cross-linked. The pores of these bacteria remains open after ethanol treatment and thus allow crystal violet chemical complex to get extracted and thus these bacteria gets unstained.
Gram-negative	Bacteria which do not take Gram stain, eg. *E. coli.*
Gram-positive	Bacteria which take Gram stain, eg. *Staphylococcus aureus.*

Gregarious	Tending to move in or form a group with others of the same kind.
Gonadotropic hormone	These hormones are produced from anterior pituitary which stimulate gonad development.In fish it is also responsible for inducing spawning migration,
Gonadotropin-releasing hormone (GRH)	This hormone are secreted by the hypothalamus that controls the secretion of the Gonadotropic hormone from the pituitary.
Goodness-of-fit test	Means how well a statistical model fits a set of observations. Measures of goodness of fit typically summarize the discrepancy between observed values and the values expected under the model in question. Such measures can be used in statistical hypothesis testing, e.g. to test for normality of residuals, to test whether two samples are drawn from identical distributions.
Gross Primary Production	The total amount of organic matter production by producers in any community. It refers to the whole anabolic process.
Gular pouches	Pouch under the chin of the reptiles used for many physiological and sexual and defense display function.
Gynaecophoric canal	A tube-like ventral infolding of the thin, post-acetabular body margin of male Schistosoma.
Haematology	A branch of biology (physiology), pathology, clinical laboratory, internal medicine, and pediatrics that is concerned with the study of blood, the blood-forming organs, and blood diseases. Hematology includes the study of etiology, diagnosis, treatment, prognosis, and prevention of blood diseases.
Haemin Crystals	When glacial acetic acid reacts with oxyhaemoglobin it form haematin, Haematin reacts with nascent hydrochloric acid to form haematin hydrochloride or haemin. These appear as crystals and are important in medico-legal test for the detection of blood. The crystal shape is also species specific
Haemoglobin	The pigment of red blood cells that is responsible for the transport of oxygen from lungs to tissues. It consists of a basic protein, globin linked with haem. Its molecular weight is 64,500 and it contains four haem groups and hence four iron atoms per molecule.
Halteres	Second pair of wings of dipteran insects which is reduced to knobs.

Haptor	Organ of attachment of helminth parasites. This occur at the posterior end of the parasite and may have 0, 1 or 2 pair of anchors and a number of small hooklets around the periphery, eg. *Dactylogyrus sp., Gyrodactylus sp.*
Hardness	The total concentration of calcium and magnesium expressed in equivalence with $CaCO_3$ in water.
Harvest method	Harvesting is the oldest techniques for determination of primary productivity. Plants are cut at ground level, on land or water. The harvested plant material is then dried out to constant weight and primary productivity is expressed in terms of dry weight biomass per unit area per unit time for example gm/sq.mtr./yr. The harvest method is usually in use within terrestrial ecosystem though it may be employed in aquatic systems also.
Haversian system	Circular repeating units that composes the bone tissue. Within such a system there is a central canal where blood vessels and nerves can be found. This is surrounded by concentric layers of the matrix material called lamellae. The darkly-stained spots are spaces called lacunae which contain the osteocytes.
Heart Lever	It is basically made of a horizontal arm that is suspended by a fine spring. The free end of the spring is attached to a plastic stylus having a writing point that writes on the smoked surface of the drum. A piece of thread is tied with a hook that hangs down. This hook is fastened to the apex of the toad's ventricle. Down stroke tracing of the lever corresponds to the systole whereas the up stroke tracing corresponds to the diastole of the heart.
Hectocotylisation	Modified reproductive arm of male members of octopus, squid and sepia.
Heterocercal caudal fin	A tail or caudal fin in which the tip of the vertebral column turns upward, extending into the dorsal lobe of the tail fin; the dorsal lobe is often larger than the ventral lobe.
Heterodont	Possessing teeth that are differentiated into several forms e.g. incisors, canines, premolars and molars.
Heteromyrian	Unequal adductor masculative in bivalved mollusc.
Hibernation	A strategy for surviving winter cold ,the animal enters a deep sleep and reduced the metabolic rate to a minimum.
Hirudin	Anticoagulant found in the secretion of salivary gland of leech.

Histochemistry	It is a particular staining technique that is used to localize specific chemical compounds within the cells.
Holandric mode of transmission	The inheritance of Y linked genes localised in the non-homologous part of the Y-chromososme.
Holozoic	Nutrients obtained by consuming other organism.
Homocercal tail	Tail of fishes where the upper and lower lobes are of same size.
Homogenization	Process of rupturing of cell membrane to obtain the organelles or other intracellular materials
Host	Animal which harbours the parasites
Hybridization	Inter breeding of spices, races, varieties and so on, among plants or animals. A process of forming a hybrid by cross pollination of plants or by mating animals of different types.
Hydrocarbons	These are unsaturated compound without any carboxyl or alcohol groups containing isoprene rings which cannot be saponified. Eg. Carotenoids and squalene.
Hydrocaulus	The vertical portion bearing the zooids of obelia colony.
Hydrohiza	The horizontal portion of the *Obelia* sp. colony.
Hyperplasia	Hyperplasia (or "hypergenesis") is a general term referring to the proliferation of cells within an organ or tissue beyond that which is ordinarily seen. Hyperplasia may result in the gross enlargement of an organ, the formation of a benign tumour, or may be visible only under a microscope. Hyperplasia is considered to be a physiological response to a specific stimulus, and the cells of a hyperplastic growth remain subject to normal regulatory control mechanisms.
Hyperpolarization	An increase in the polarity of the potential difference across the membrane of a nerve or muscle cell, i.e. and increase in the resting potential.
Hypoblast	The inner or lower layer of the double layered embryo.
Hypognathus	The inclination of the long axis of the head of insect is vertical and the mouth parts ventral; the occipital foramen also lies in or near the vertical plane.
Hypostome	A conical elevation at the oral end of the body.
Immunostaining	Coupling of an antibody with a fluorochrome and consequent binding to an antigen and rendering the antigen to be visualised through fluorescence microscope.

Independent variables	Variable whose change in the value causes the change in the value of another variable (dependent variable) as seen in some association or existing relationship between two.
Infiltration	Replacing the dealcoholizing agent with paraffin during processing of tissue blocks.
Infundibulum	The stomach of Hormiphora from which a series of gastro vascular canals arise.
Instar	Each of the several stages between molts in arthropods.
Insulin	A hormone that controls the metabolism of glucose. It a polypeptide hormone produced by the beta-cells of the Islets of Langerhans of the pancreas.
Interphase	The stage in the cell cycle when the nucleus is not in a state of division. Interphase is divisible into various stages each characterized by a differing physiological activity.
Interstitial cells	Steroid secretory cells found in the interstitial tissue of the testis that produce male hormone testosterone.
Interval variables	Variable that are equally spaced e.g. height.
Introvert	Anterior narrow region of the body of *sipunculus.* This can be retracted in the anterior end of the trunk.
Inversion Test	An analytical test most commonly used to determine Reducing substances in the cane sugar i.e. sucrose. Basically it uses the reduction of copper (Cu) to determine the percent of Reducing Sugars in a sample.
Invert sugar	Mixture of glucose and fructose or levulose
Ionotropic effect	An effect that change the strength of contraction of the heart muscle.
Islets of Langarhans	Endocrine portion of the pancreas is known as the Islets of Langerhans.
Isoelectric point	Protein is positively charged at low pH and negatively charged at high pH. At the intermediate pH values a point is found where the sum of positive side chain charges exactly equals to the sum of negative charges so that the net charges on the protein is zero. This value is the isoelectric point.
Isomyarian	Similar adductor scar in givalred molluscs.
Isotonic contraction	In a muscle contraction where tension of the muscle remains practically constant throughout.
Iternative Hypothesis (H_A)	It is an opposite proposition to the Null hypothesis.
Karl Pearson's coefficient of correlation	It is the measure of the degree of statistical relationship between two variables from bivariate data.

Karyotyping	The process by which a karyotype is prepared.
Kendall's Coefficient of Concordance	An important non-parametric measure of relationship. It is in use for determining the degree of association among several (k) sets of ranking of N objects or individuals. Kendall's coefficient of concordance (W) is deemed as an appropriate measure of studying the degree of association among three or more sets of rankings.
Keratohyalin granules	Coarse basophilic non-membrane bound granules containing histidine rich protein found in stratum granulosum of the skin.
Ketoses	Sugars that bear ketone groups or potential ketone group.
Kidney	The kidneys are complex organs that have numerous biological roles. Their primary role is to maintain the homeostatic balance of bodily fluids by filtering and secreting metabolites (such as urea) and minerals from the blood and excreting them, along with water, as urine.
Kinetodesmata	A ciliary structure having a cord of fibres.
Kinetoplast	A region rich in DNA within the mitochondrion of the cell.
King crab	The big edible crab *lithodes* sp. is known as king carb. The horse-shoe crab, Limulus is also popularly known as king crab.
Koehler illumination	This is a particular alignment of the microscope components i.e. lens and light source. In this procedure, an image of the field diaphragm is projected onto the object plane. This procedure requires a field condenser lens equipped with a moveable (centerable) iris or diaphragm.
Kruskal-Wallis one-way analysis of variance	One-way analysis of variance is used as a test for determining whether three or more independent samples are significantly different. It is a non-parametric test. This test is used to verify the Null Hypothesis that 'k' independent random samples come from identical universes against the Alternative hypothesis that means of these universes are not equal.
Kupffer's cells	Mononuclear phagocytic cells in the liver sinusoids are called as the Kupffer's cells.
Kymograph	It is a instrument set used for the measurements of excitable tissues such as heart and muscle.
Lamina propia	A layer of connective tissue that forms the underlying covering of the epithelial tissue.

Large Lymphocytes	They are medium sized cells. Nucleus large stains deep blue-violet. It is single, rounded or oval and almost fills the cell. Centrally located and eccentric. Cytoplasm large and crescent shaped appears light blue.No visible granules.
Latent period	In the study of simple muscle twitch the short time gap between the application of stimulus and actual contraction of the muscle.
Law of priority	The generic or specific names first published are to be only recognized.
Layers of nacre	A thin innermost calcareous layer of shells of most molluscs.
LC_{50}	It is the concentration of a material in air that kills 50% of a group of test animals when administered as a single exposure in a specific time period.
LD_{50}	This is the single dose of a toxic substance administered by any route (other than by inhalation) that causes the death of 50% of an animal population.
Learning	The alteration of behaviour by an individual as a result of experience. Learning is the ability of the animal to discover which of its actions gives the best results in a particular set of circumstances and to modify its behaviour to its best advantage. Learnt behaviour is diverse and differs between individuals; it is best developed in animals with a long life span and a long period of parental care.
Learning curve	Refers to a relationship between the duration of learning or experience and the resulting progress. Initially introduced in the cognitive psychology, over time the term acquired a broader interpretation, and the expressions "experience curve", "improvement curve", "cost improvement curve", "progress curve"/"progress function", "start up curve", "efficiency curve" etc., may be used interchangeably, depending on the context. Some of these terms may also have other meanings.
Leishmaniasis or kala-azar	A disease of human whose causative agent is Leishmania sp.
Leptotene	This is a substage of Prophase I of meiosis. This stage is manifest by the first appearance of the chromosomes whilst the chromosomes are in their most extended form (except for during interphase). They appear to be a string with beads. The beads are designated as chromomeres.

Leuconoid	The canal system in which the choanocyte lining of the radial canals evaginates into many small chambers which repeat the same process to give rise to a cluster of small flagellated chambers.
Leukocyte	Blood cells that engulf and digest bacteria and fungi; an important part of the body's defense system.
Light microscope	A type of microscope which uses visible light and a system of lenses to magnify images of small samples. Optical microscopes are the oldest and simplest of the microscopes.
Lipids	A collective term used to describe a group of substances in the cells characterised by their solubility in organic solvents such as ether or benzene and insolubility in water.
Lipoproteins	These are lipid materials bound to proteins. The lipid component mainly consists of cholesterol esters and phospholipids containing stearic acid, palmitic acid and oleic acid.
Lithocysts	A small spherical sac- like body that encloses central calcareous mass and sensory cells.
Liver rot	The disease caused by the liverfluke.
Lobopodia	Blunt projected pseudopodia.
Loop of Henle	U-shaped convoluted tubule of the uriniferous tubule is termed as the Henle's loop. It has two parts the ascending loops and the descending loops.
Lophophore	A circumoral (i.e., around the mouth) body region characterized by circular or U – shaped.
Lung	The respiratory organs of the air breathing vertebrates.
Lymphatic nodules	Small, localized collection of lymphoid tissue, usually located in the loose connective tissue beneath wet epithelial (covering or lining) membranes, as in the digestive system, respiratory system, and urinary bladder. Lymph nodules form in regions of frequent exposure to microorganisms or foreign materials and contribute to the defence against them.
Lymphocyte	A type of white blood cell in the vertebrate immune system. By their appearance under the light microscope, there are two broad categories of lymphocytes, namely the large granular lymphocytes and the small lymphocytes. Most, but not all large granular lymphocytes are more commonly known as the natural killer cells (NK cells). The

	small lymphocytes are the T cells and B cells. Lymphocytes play an important and integral role in the body's defenses. An average human body contains about 10^{12} lymphoid cells, and the lymphoid tissue as a whole represents about 2% of the total body weight.
Macrolecithal Egg	Eggs with large amount of yolk, e.g. eggs of insects, reptiles, birds, prototherial mammals.
Macrophages	A large amoeboid white blood cell that can engulf, ingest and destroy bacteria and damaged cells and worn out red blood cells.
Magnification	The process of enlarging something only in appearance, not in physical size. Magnification is also a number describing by which factor of an object was magnified. Typically magnification is related to scaling up visuals or images to be able to see more detail.
Malatosazone	Osazone produced by maltose that has a shape of close tufts of short fine crystals.
Male gonopore	External opening of male reproductive system.
Mast cells	A cell found in the ground substance of the areolar connective tissue which it is thought to secrete. Mast cell is also found in the lining of the blood vessels and some occur in the blood as basophils. The lobed nuclei and cytoplasmic granules are stained with basic dyes. Mast cells secrete heparin and histamine.
Mean	The arithmetic average, the sum of all measurements or values of groups of objects divided by the number of objects.
Mean deviation	It is the arithmetic mean of absolute deviations from mean or any other specified value.
Median	Median is the middle of a distribution, half of the scores are above the median and half are below the median.
Megascleres	Large skeletal spicules of the sponge.
Meiocytes	The cell that undergoes meiosis.
Meiosis	The process of cell division leading to the production of daughter nuclei with half of the genetic complement of the parent cell. Cell formed through meiosis give rise to gametes and fertilization restores the correct chromosome complement.

Membrana granulosa	The cells lining the cavity of the follicle are called membrana granulosa.
Merogony	An asexual replication process used by some Protozoan parasites that increases the number of infective cells by multiple fission.
Mesenteries	Vertical partitions in the coelenterons: Folds in the dorsal portion of the membrane lining the abdominal cavity in which the abdominal organs are supported.
Mesoderm	The layer that lies between the outer ectoderm and inner endoderm.
Mesoglea	The non-cellular gelatinous layer, located between the epidermis and gastrodermis.
Mesoglea	A layer of jelly-like material between the external (ectoderm) and internal (endoderm) cellular layers of the body of a sponge or cnidarian.
Metamere	Any of the homologous segments, lying in a longitudinal series, that compose the body of certain animals, such as earthworms and lobsters.
Metamorphosis	It is an abrupt morphological, anatomical, physiological and hormonally regulated change in the larval form that transforms it into adult.
Metanephiridia	A **metanephridium** (pl. *metanephridia*) is a type of excretory gland or nephridium found in many types of invertebrates such as annelids, arthropods and molluscs. It typically consists of a ciliated funnel opening into the body cavity or coelom connected to a duct which may be variously glandularized, folded or expanded (vesiculate) and which typically opens to the organism's exterior.
Metaoestrus stage	A corpus luteum is formed from the follicle and progesterone is secreted.
Metaphase	The stage in mitosis and meiosis when the chromosomes become aligns along the equator of the nuclear membrane.
Metaphase I	The tetrads arrange in a line themselves in the middle of the cell; they fix to spindle fibres in a distinctive manner. The centromeres of a given homologue will attach to the spindles from only one pole.
Metapleural folds	These ventro-lateral ridges run from the oral hood to a position just posterior to the gonadal region of *Amphioxus sp.*

Metasoma	The metasoma is the posterior part of the body, or tagma, of arthropods
Metazoa	They are multicellular organisms (meta = after and zoon = animal). Metazoa arose by cellurisation of a syncytium.
Metazoa	Any member of the animal kingdom (Animalia or Metazoa) usually considered to include multicellular, heterotrophic eukaryotes in which (unlike plants) the cells lack cell walls.
Microfilariae	Minute larva of Wuchereria that in habit the large blood vessels during the day time in humans.
Micrometry	Science of measurement of microscopically visible organisms or cellular components
Microscleres	Small skeletal spicules of the sponge.
Microspectophotometry	It is a type of spectrophotometry that is applied in localization and quantitation of DNA and RNA in the cells, by means of ultraviolet light
Microtome	Microtome is rather a simple device consists of a stationary knife holder/blade and a specimen holder which advances by pre-set intervals sectioning paraffin embedded tissue specimen blocks.
Mitosis	The ordered process by which the cell nucleus divides into two. The chromosomes replicate prior to mitosis and are then separated during mitosis in such a way that each daughter cell inherits a genetic material identical to that of the parent cell.
Mode	Value that occurs most frequently in the data set or a probability distribution.
Monaxon	Spicules formed by the growth in one direction or both directions along a single axis.
Monera	Prokaryotic organisms that have no organized muscles.
Monocellate marks	*Naja kaouthia* has a single O shaped mark behind its head.
Monocytes	They possess big cell, comparatively large single horse-shoe or kidney or oval shaped nucleus in the centre that stains pale blue-violet. Cytoplasm abundant and frosty and slate blue in colour. No visible definite granules
Monoestrous	Animals having only one oestrous cycle during the year.
Monohybrid cross	A cross between two parents involving a single pair of contrasting characters.

Monomyarian	A shell having only one adductor scar; which is usually a centrally positioned posterior adductor scar.
Mononuclear phagocytic system	The cells performing these various functions such as phagocytosis, secretion of cytokines and antigen presentation have differing microscopic appearances but they are grouped together as the mononuclear phagocytic system.
Monopectenate	Ctenidia is provided with a single row of long series of thin triangular leaflets or lamellae.
Monopectinate	Gill filaments project from only one side of the supporting axis.
Monosaccharides	It is the simplest form of sugar that cannot be hydrolysed to simpler carbohydrates of smaller carbon content. Example Glucose, Fructose etc.
Monosex culture	Male and female pupas are separated on the basis of sex combs on tarsi of the first thoracic legs of the male pupa. Individuals from each sex are kept in separate food vials. Thus vials containing the virgin female pupae will contain all virgin females. It is one of the methods to obtain virgin individuals.
Monozygotic twin	Twins having no variability in their traits. They are produced from single fertilised ovum. These twins bear same genes and are of same sexes.
Motor unit	Vertebrate skeletal muscle is a kind of muscle specified for balance and movement. The basic unit which consists of a single motor neuron and the skeletal muscle fiber(s) it innervates.
Mucous gland	Glands, found in several different parts of the body, typically stain lighter than serous glands during standard histological preparation. Most are multicellular, but goblet cells are single-celled glands.
Muscle tissue	A very specialized tissue that has both the ability to contract and the ability to conduct electrical impulses. Muscles are are classified both functionally as either voluntary or involuntary and structurally as either striated or smooth. From this, there emerge three types of muscles: smooth involuntary (smooth) muscle, striated voluntary (skeletal) muscle and striated involuntary (cardiac) muscle. The names in the brackets are the common names given to the particular classification of muscle.

Myofibrillae	One of many contractile filaments that make up a striated muscle fibre.
Myograph board	It is flat horizontal board (a component of crank myograph) which is used for griping muscle preparation with a crank lever fixed to it.
Myotomes	The segment of a somite in a vertebrate embryo that differentiates into skeletal muscle.
Myxomatosis	A highly infectious, usually fatal disease of rabbits that is caused by a pox virus and is characterized by many skin tumors similar to myxomas.
Nannoplankton	Very minute plankton not secured by plankton net.
Narcotization	The process of placing under narcotics, i.e. to dull or deaden.
Nauplius eye	A single median eye comprising three- or four-pigment cup ocelli, sometimes with a lens, probably enabling the organism to determine the direction of a source of light for orientation. The median eye is typical of crustacean nauplius larvae. It may degenerate as the larva develops, or it may persist into the adult form.
Necrosis	Accidental death of cells and living tissue. Necrosis is less orderly than apoptosis, which is part of programmed cell death. In contrast with apoptosis, cleanup of cell debris by phagocytes of the immune system is generally more difficult, as the disorderly death generally does not send cell signals which tell nearby phagocytes to engulf the dying cell. This lack of signalling makes it harder for the immune system to locate and recycle dead cells which have died through necrosis than if the cell had undergone apoptosis. The release of intracellular content after cellular membrane damage is the cause of inflammation in necrosis.
Neoteny	Animal becomes sexually mature while retaining the larval morphology (see paedogenesis).
Nephridia	Simple or branched tubules, lined with cilia, which diffuse excretory products.
Nerve cord	The bundle of nerve fibres that runs along the longitudinal axis of the body.
Net plankton or filterable	Plankton that can be secured by plankton net.
Net Primary Production	It is the value of organic matter that remains after deducting Respiratory loss from Gross primary productivity

Neurenteric passage or canal	The canal that connects the blastopore and the archenteron with cavity of the neural tube.
Neurocoel	The system of cavities of the embryonic brain and spinal cord. The cavity of the chordate cerebrospinal system, consisting of the ventricles of the brain and the central canal of the spinal cord, regarded as an unit.
Neuropodium	Ventrally placed part of the biramous parapodium.
Neuropore	It is the opening or pore that communicates the front of the brain vesicle with the exterior.
Neurotransmitter	A chemical that is released from neurone endings to cause either excitation or inhibition of and adjacent neuron or muscle cell.
Neurula	Next stage after the gastrula where the embryo flattens laterally and extends in the anteroposterior direction. The dorsal side of the body arches, pushing the neural tube, notochord and somites more dorsally in the body.
Neutral lipids	These are esters of fatty acids and glycerol.
Neutrophils or Polymorphonuclear leucocytes	This type of white blood cells posses 2 to 7 lobed nucleus connected by chromatin thread. Nucleus stains blue-violet. Cytoplasm stains slate-blue. Cytoplasmic granules fine, pinkish violet in colour.
Nicotine	An alkaloid (*(S)-3-(1-Methyl-2-pyrroli-* dinyl)pyridine) found in the nightshade family of plants (*Solanaceae*), predominantly in tobacco, and in lower quantities in tomato, potato, eggplant (aubergine), and green pepper. Nicotine alkaloids are also found in the leaves of the cocoa plant. Nicotine constitutes 0.3 to 5% of the tobacco plant by dry weight, with biosynthesis taking place in the roots, and accumulating in the leaves. It functions as an antiherbivore chemical, being a potent neurotoxin with particular specificity to insects; therefore nicotine was widely used as an insecticide in the past, and currently nicotine derivatives such as imidacloprid continue to be widely used.
Nomenclature	Scientific naming of organisms.
Nominal variables	Is for mutual exclusive, but not ordered, categories. For example, any one's study might compare five different genotypes. One can code the five genotypes with numbers if one want, but the order is arbitrary and any calculations (for example, computing an average) would be meaningless.

Nonsaponifiable Lipid	Lipids that does not give saponification reaction.
Normal cardiogram	Record of the heart beat of animals in normal condition which is used as experimental model animal in form of a graph which is known a normal cardiogram.
Northern blotting	The technique transfer of RNA bands from an acrylamide gel onto a more stable and immobilizing support matrix.
Notochord	A flexible rod like structure composed of disc like turgid cells which extends the entire length of the body of adult and or larvae of chordate.
Notopodium	Dorsally placed part of the biramous parapodium.
Nuchal organ	A lobed pad, probably a chemoreceptor.
Nucleophilic attack	In chemistry, a nucleophile (literally *nucleus lover* as in nucleus and phile) is a reagent that forms a chemical bond to its reaction partner (the electrophile) by donating both bonding electrons. Because nucleophiles donate electrons, they are by definition Lewis bases (see *acid-base reaction theories*). All molecules or ions with a free pair of electrons can act as nucleophiles, although negative ions (anions) are more potent than neutral reagents. Neutral nucleophilic reactions with solvents such as alcohols and water are named solvolysis.
Null Hypothesis(H_O)	A hypothesis or proposition where an assumption is that the observed results have been produced solely by operation of chance inherent in random sampling.
Numerical Aperture (NA)	A mathematical calculation of the light gathering capabilities of an objective lens. The N.A. of each objective lens is marked in the metal tube, and generally ranges from 0.25-1.4. Thus a lens with higher N.A. makes it better in the light gathering property thus makes a better resolution. Higher N.A. values also signify shorter working distances (one has to get the lens closer to the object). N.A. values greater than 1.0 also point towards that the lens is used with some immersion fluid, such as immersion oil.
Objective lens	Compound microscope bears lens that is inserted at the the lower end of the microscope tube to accumulate light from the sample. One or more than one objective lenses are attached into a circular nose piece which may be rotated to opt for the required objective lens.

Occular micrometer	The ocular micrometer is a circular glass or plastic disc in the centre of which a line is drawn, which is sub-divided into 100 parts. This is used in micrometry.
Oestrous cycle	This comprises the recurring physiologic changes that are induced by reproductive hormones in most mammalian placental females. Humans undergo a menstrual cycle instead. Estrous cycles start after puberty in sexually mature females and are interrupted by anestrous phases. Typically oestrous cycles continue until death. Some animals may display bloody vaginal discharge, often mistaken for menstruation.
Oestruos stage	The period of heat. Ovulation occurs that is stimulated by rise of secretion of luteinizing hormone and copulation also takes place.
Oikoplastic epithelium	A monolayer of cells covering the trunk of the animal is responsible for secretion of the house.
Oils	They are liquid at room temperature. Eg. Fish oil, liver oil, rape seed oil, castor oil etc.
Oil immersion lens	A high power objective lens that is used after a drop of immersion oil is put on top of the cover slip(covering the object), and the lens is very gradually and carefully lowered until the front objective element is dipped in the oil film. Such immersion lenses are designed in such a way so that the refractive index of the oil and of the cover slip is closely matched so that the light is transmitted from the specimen to the outer face of the objective lens with least refraction. An oil immersion lens usually has a power of 100X.
Ommatophores	A movable stalk ending with an eye, as found is some snails.
Omphalomesenteric veins	In the early stage of the yolk sac development in the mesodermal component of the yolk sac these veins develop and traverse throughout it.
One-tailed test	t-tests that is used where there is no basis to assume that there may be a significant difference between the groups - this is the test most frequently used.
One-way (or single factor) ANOVA	ANOVA that investigates the effect of a single independent variable on dependent variable.
Ontogenic development	Origin and the development of an organism from the fertilized egg to its mature form.

Onychophora	Onychophora, the velvet worms (occasionally called walking worms or spitting worms), are segmented, caterpillar-like, terrestrial animals somewhat resembling both arthropods and annelid worms. They are generally regarded either as a separate phylum or a subphylum under phylum Arthropoda.
Ookinete	The motile, diploid zygote move to the mosquitoe's stomach.
Oozooid	The union of a sperm and egg that gives rise to an asexually reproducing morph that produces no gametes.
Opisthosoma	The opisthosoma is the posterior portion of the arachnids body behind the prosoma (cephalothorax).
Oral surface	The surface bearing the mouth.
Ordinal variables	Variables with an ordered series, e.g. "greatly motivated , moderately motivated, unrelated, motivated discouraged, greatly motivated "
Organ of Jacobsons	A special smell – taste organ in the hard palate of amphibians, lizards, snakes and many mammals.
Osazone	Aldoses and Ketoses when heated with excess amount of phenylhydrazine produces hydrazones and which exhibits osazones of definite crystalline shape
Osculum	Larger opening of the spongocoel of the summit of the sponge.
Osmotic alterations	Osmotic swelling of the cell and/or organelle, which can often assist in the rupture of the cell and subsequent organelle separation. This can be done by treating cell with hypo-osmotic buffer.
Osphradium	A chemoreceptor/tactile receptor generally located adjacent to ctenidium.
Osteoblasts	A mononucleate cell that is responsible for bone formation. Osteoblasts produce osteoid, which is composed mainly of Type I collagen. Osteoblasts are also responsible for mineralization of the osteoid matrix.
Osteoclasts	A type of bone cell that removes bone tissue by removing the bone's mineralized matrix. This process is known as bone resorption. An osteoclast is a large cell that is characterized by multiple nuclei and a cytoplasm with a homogeneous, "foamy" appearance. This appearance is due to a high concentration of vesicles and vacuoles.

Osteocytes	A star-shaped cell that is the most abundant cell found in bone. Cells contain a nucleus and a thin ring of cytoplasm.
Osteon	These are predominant structures found in some lamellar or compact bone. Osteons are found in many of the bones of many mammals, birds, reptiles, and amphibians running in a meandering way but generally parallel to the long axis of bones. It is this section of compact bone that is called matrix.
Ostia	Numerous minute opening on the body of the sponge.
Ovipositor	Specialised egg laying organ of insect formed from outgrowth of eight and ninth abdominal segments.
Pachytene	This is a substage of Prophase I of meiosis, the chromosomes become visible as "thick" strings, or pachynema. The chromosomes are about 1/4 the length they were in leptotene, and there are clearly two chromosomes, each with two chromatids in each bundle. The two chromosomes are referred to as a **"bivalent,"** while the same structure observed as four chromatids is known as a **"tetrad."**
Paedogenesis	Reproduction by larval or other immature forms (see neoteny).
Pancreas	A gland lying between the spleen and the duodenum and having a duct that enters the duodenum. It secretes pancreatic juice which contains enzymes such as trypsin, amylase and lipase. The gland also contains an endocrine part that produces insulin.
Paneth cells	These are exocrine serous cells that synthesize a complex of protein and polysaccharide found in the intestine.
Paper chromatography	Cellulose in the form of paper sheets makes for an ideal support medium where water is adsorbed between the cellulose fibres and acts stationary hydrophilic phase. Whatman filter paper or commercially available cellulose plates are used for the purpose. A small drop of the mixture is spotted on to the paper, and then dried. The chromatogram is by placing the filter paper in a tank containing a suitable solvent and allowed to flow along the sheet. The solvent front is marked and then after drying the paper the positions of the compounds thus separated are visualised and identified by a suitable staining reaction.

Paraffin embedding	The sealing of tissue which is prepared for permanent microscopic slides in a solid block of paraffin wax prior to sectioning.
Parafollicular cells	Clusters of pale staining cells found between the thyroid follicles. These cells are responsible for the synthesis and secretion of hormone calcitonin. This hormone promoter is the reduction of concentration of calcium in the blood.
Parapodia	Laterally extended body wall that forms a series of thin, flattened outgrowths.
Parasite	Animal which thrives at the expense of another organism in which it lives.
Parasitology	Parasitology is the study of parasites and their relationship to their hosts.
Parental care	Protection and maintenance of its own offspring by an animal.
Partition coefficient	If two solvent phases are in-contact with one another and if a solute is added, it will distribute itself among the two phases. The ratio of the concentrations of the solute in the two phases is called partition co-efficient.
Patagium	A fold of skin between the fore limb and the hind limbs stretched for gliding or flying.
Pearls	Pearl is the secretion of the mantle of pearl oysters, as a protection against foreign objects, usually the parasites.
Pedigree Analysis	Analysis of a pedigree chart to determine how the trait is inherited
Pedipalps	Pedipalps, the second pair of appendages of the cephalothorax in Arachnida.
Pelagic biotic community	Community that includes the floating forms both the planktons and nektons.
Peptones	They are produced by the action of dilute acids like HCl and H_2SO_4 or enzymes on natural proteins. They are soluble in water and does not heat coagulate. They cannot be salted out by ammonium sulphate or by means of other salt.
Perichondrium	A dense membrane composed of fibrous connective tissue that closely wraps (invests) all cartilage, except the cartilage in joints which is covered by a synovial membrane. Peri- is a prefix borrowed from the Greek meaning "around or about" and "-chondrium" also comes from a Greek word "chondros" meaning "cartilage or gristle."

Perimysium	The connective tissue sheath which surrounds a muscle, and sends partitions inwards between the bundles of muscular fibres.
Period of contraction	In a simple muscle twitch curve just after the latent period there is a period of fall in the curve that relates to the contraction of the muscle (shortening of muscle fibres).
Period of relaxation	In a simple muscle twitch curve just after the contraction period there is a period of rise in the curve that relates to the relaxation of the muscle.
Periodic acid – schiff (PAS) reaction	The Periodic acid–Schiff (PAS) reaction—the term was introduced by McManus—is one of the most widely used histochemical methods. A staining method used in histology and pathology. This method is primarily used to identify glycogen in tissues. The reaction of periodic acid selectively oxidizes the glucose residues, creates aldehydes that react with the Schiff reagent and creates a purple-magenta color. A suitable basic stain is often used as a counter stain. PAS staining is mainly used for staining structures containing high proportion of carbohydrate macromolecules (glycogen, glycoprotein, proteoglycans), typically found in connective tissues, mucus, and basal laminae.
Periosteum	The periosteum is a dense membrane composed of fibrous connective tissue that closely wraps (invests) all bone, except that of the articulating surfaces in joints which are covered by a synovial membrane.
Periostracum	Thin outer organic layer of the shells of most molluscs.
Perisarc	A transparent protective tube that encased the stolon and stalks.
Peristome	Oral disc, ventral to which to the mouth in invertebrates.
Peristomial palp	A sensory appendage situated near the mouth of many invertebrates.
Peristomium	The first body segment of earthworm which bears the mouth aperture on its ventral surface.
pH	Negative logarithm of hydrogen ion concentration in a solution or water.
Phagocytosis	The bulk transport of solid materials into cells across the plasma membrane.
Phagolysosome	A structure formed due to the fusion of the lysosome and the phagosome in a phagocytes.

Phagosome	Membrane bound vesicles formed from the protrusions or pseudopodia that is involved in surrounding the engulf antigen or pathogen.
Phase plate	A designated ring on the phase contrast objective lens at their back focal plane that match up geometrically with the position of the ring of direct light from the condenser. This plate present on the objective acquires destructive interference which is essential to retard an additional one-quarter wavelength relative to the direct light.
Phase Contrast Microscopy	Microscope that aids in the visualisation of highly transparent objects. The phase contrast microscope is constructed in such a way that it carries on two unique tasks i.e. it splits the direct light (the back ground light from the field) from the light diffracted by the object, and it produces two types of waves to be approximately one – half of the wavelength out of phase of the other, so that they can destructively interact, and cause a alteration in intensity and thus allows visualisation of transparent objects.
Phase effect	Effect of the phase plate that causes the direct light passing through this decreased path of glass (phase plate) is advanced/sometimes retarded by another additional one – quarter of wavelength, which place it to one half of wavelength out of phase with the diffracted light.
Phasmid	The chemosensory organ located at the posterior end of some nematodes.
Phosphorescence	An emission phenomenon where the electron alters its spin after being excited enters the triplet state, and the kind of light that is emitted as the electron returns to ground state.
Phospholipids or Phosphatids	Those lipids that on hydrolysis yields fatty acids, phosphoric acid and sometimes glycerol and nitrogenous base. Eg. Phosphatidic acid, Cephalins, Lecithins, Spingomyelins, Plasmalogens
Phosphorescent organs	Organs also known as *photophores* are characteristic structures in many of the deep-sea Teleostome fishes, and have been developed in widely different families (*Stomiatidae, Scopelidae, Halosauridae* and *Anomalopidae*), whilst numerous simple luminous organs have been detected in many species of Selachii. The number, distribution and complexity of the organs vary much in different fish. They

	are most frequent on the sides and ventral surface of the anterior part of the body and the head, and may extend to the tail. The simpler forms are generally arranged in rows, sometimes metamerically distributed; the more complex organs are larger and less numerous.
Phytoplankton	Small aquatic plants who has limited power of self locomotion.
Pilose	Unbushy branched antennae.
Pinacocytes	Large flat polygonal cell found in the epidermal lining.
Pinnae	An extension of the external ear of mammals to form a trumpet like structure supported by cartilage.
Pituitary gland	An endocrine gland in the vertebrate brain situated beneath the thalmencephalon behind the optic chiasma. It is regarded as the master endocrine gland. Anatomically pituitary consists of anterior lobe (pars distatlis and pars tuberalis) and posterior lobe (pars media and pars nervosa).
Pituitary extract	Extract produced after homogenising, centrifuging pituitary gland of fish that is applied in the induced breeding of fishes.
Placenta	A structure formed in part from the inner lining of the uterus and in part from the extra embryonic membranes. It serves as the connection between the mother and embryo during development and draw necessary substances from the maternal circulation.
Placoderms	A large and varied class of Paleozoic fishes characterized by complex bony armour covering the head and the front portion of the trunk.
Placoid scale	A scale consists of a hard base and a spiny process(cusp) which are covered by vitrodentine.
Plantae	Multicellular organisms with cell walls of chitin and they generally obtain nourishment by means of chemical break down of material (Osmotropic- autotropic).
Plantigrade	The entire foot makes contact with the ground, e.g. Human.
Plasteron	The lower portion of the shell of the chelonian, consisting of bony plates corresponding to part of the shoulder girdle and abdominal ribs of the primitive reptiles covered by horny epidermal plates.
Plumose antennae	Branched, bushy antenna.

Pneumatic sac	Swim bladder, or gas bladder; this is a synapomorphy of this group, called Actinopteri (not Actinopterygii). The sac of the gar, bowfin, and tetrapods has a folded respiratory surface, whereas in the others it is smooth.
Pneumatophore	An inverted medusoid bell devoid of mouth and tentacles.
Poikilothermous	Organisms whose body temperature varies according to temperature of its surroundings.
Poison fangs	The second cheliceral segment of arachnids, which contains the pore for poison gland.
Poison gland	Venom containing gland.
Poison spine	Some fishes have a spine on the dorsal or pectoral fin, which is poisonous. E.g. *Hydrologus, Callorhynchus, Heteorpneutus.*
Pollen collecting baskets	A basket which is placed on the outer side of the tibia of honey bee for depositing the pollens.
Pollen combs	Large flat brush like structures found on the inner part of the first tarsal segment of metathoracic legs of honey bee.
Polyacrylamide gel electrophoresis (PAGE)	An electrophoresis carried out on gels made up of Polyacrylamide which is a polymer of two compounds, acrylamide and N, N_1-methylene-bis-acrylamide. Mainly used in protein electrophoresis.
Polyestrous	Animals having many oestrous cycles (may be repeats with short intervals in between) during the same breeding season, eg. rat, mice, sheep and cattle.
Polylecithal egg	Eggs which contains huge amount of yolk material.
Polysaccharides	High molecular weight polymers of monosaccharides or sugars. Eg. Inulin, starch, glycogen, cellulose etc.
Polytene chromosomes	These are giant chromosomes that are thought to be composed of many chromatin strands fused together. For example in the larval salivary gland chromosomes.
Population	A population is a collection of all elements which are to studied and about which a conclusion has to be drawn.
Prehensile tail	Animal rapidly grasps with the tail.
Premeiotic interphase	Prior to meiosis, all chromosomes are duplicated in a process similar to chromosome duplication prior to mitosis.Outside the nucleus of animal cells are two centrosomes, each containing a pair of centrioles. The two centrosomes are produced by the duplication of a single

	centrosome during premeiotic interphase. The centrosomes serve as microtubule organizing centers (MTOCs). Microtubules extend radially from centrosomes, forming an aster.Plant cells do not have centrosomes. Different kinds of microtubule organizing centers serve as sites of spindle formation.
Primary data	It refers to those, which are collected by a particular person or organization for a specific purpose directly from the field of enquiry and thus are original in nature.
Primary oocytes	The product of oogonium formed through the first meiotic division.
Primary production	Amount of radiant energy converted into organic matter by photosynthetic and chemosynthetic activity of the producers.
Primary spermatocytes	It is the cell formed due to multiplication and growth of the spermatogonia.
Primitive streak	The thickened dorsal longitudinal strip in early avian, reptilian and mammalian embryos.
Principle of Independent Assortment	The segregation of one factor pair occurs independently of any other factor pair. But we know that this is only true for loci on non-homologous chromosomes. In case of Independent assortment the phenotypic ratios in the second filial generation is 9:3:3:1.
Principle of Segregation	Mendel observed that there was no blending or mixture of characters in the hybrids rather they remained separate. Thus each factor or pair of alleles is transmitted to the offsprings through the gamete,
Proband	The character which is expressed by a particular allele in presence of another allele at the same locus.
Proestrus stage	Graafian follicles grow in the ovary and secrete the hormone oestradiol which causes the uterine lining to proliferate and thicken.
Progesterone	A steroid hormone secreted by the corpus luteum. It initiates preparation of the uterus for implantation of the ovum, development of the placenta, and the development of the mammary gland.
Proglottids	A seemingly endless segments or sexual units of tape worm bud from the neck.
Pronotum	The dorsal, sclerotised cuticles of the first thoracic segmnent.

Prophase	The first stage of the cell division. During prophase the chromosomes becomes visible and the nuclear membrane dissolves.
Prophase I	The first phase of mitosis is marked by the early condensation of the chromosomes into visible structures. During the period of Prophase I, the chromosomes move in different ways. There is an increase in the volume of the nucleus. This stage of the meiosis is extremely extended when compared to the mitotic prophase.
Prophase I	The first stage in the second meiotic division of meiosis, following telophase I. In each of the two cells which were produced by the first meiotic division, the chromosomes migrate towards the cells centre. The stage ends when the chromosomes are aligned along a single plane in the centre of each cell.
Propositus	It is the clinically influenced male family member through whom concern has to be drawn to a pedigree of a particular interest.
Proposita	It is clinically influenced female family member through whom concern has to be drawn to a pedigree of a particular interest.
Prosoma	Fused head and thorax into a single functional unit of merostomatas.
Prostomium	A small fleshy lobe at the anterior most part of the body of earthworm.
Protein	Proteins are polypeptides i.e. they are made up of amino acids molecules joined together by peptide linkage. Their molecular weight can vary from few thousands to several million.
Proteoses	They are produced when hydrolysis continued even after formation of metaproteins. They are water soluble and does not heat coagulate. They can be precipitated through ammonium sulphate.
Prothrombin time	It is the time taken in stage two of coagulation i.e. conversion of Prothrombin to thrombin in blood coagulation. It is connected to coagulation factors II, V,VII and X..
Protista	Protists meaning the (most) first of all ones, are a diverse group of organisms, comprising those eukaryotes that cannot be classified in any of the other kingdoms as fungi, animals, or plants. They are usually treated as the kingdom

	Protista or Protoctista. Protoctists (or protists) are a paraphyletic grade, rather than a natural, (monophyletic) group, and so do not have much in common besides a relatively simple organization — either they are unicellular, or they are multicellular without highly specialized tissues.
Proloculum	It is the initial chamber of the test of foraminifera, e.g. *Elphidium* (*Polystomella*) sp.
Prototheria	A subclass under class mammalian that comprises the extinct orders Docodonta, tricodonta, multituberculata as well as the order monotremata, members of which still survive.
Proximal convoluted tubule	The tubule that leaves the neck of the glomerulus and lined with pyramidal cells with large spherical nuclei and brush border along the open surface.
Pseudocoele	The body cavity is not a cavity in the mesoderm. The body cavity is lined externally by fibrous processes of the longitudinal muscle cells and internally by cuticle covering the intestine.
Pseudocoelomate	The animal whose body cavity is not a coelom.
Pseudoctenoid	This is a type of scale in which spines do not project beyond the limit of the scale (see ctenoid).
Pseudopodia	Temporary finger – like protrusions of the body of amoeboid protozoan (Gr. Pseudo = false; podos = foot).
Ptyalin	An enzyme present in the saliva of man and some animals. It belongs to the group of carbohydrate hydrolysing enzymes known as the amylases. It catalyses the conversion of starch to maltose.
Puffs	*See,* Balbiani rings.
Pulmonary sac	A closed cavity which hangs from the dorsal wall of the mantle in the pulmonary chamber of pulmonates (Gastropoda, mollusca).
Pulse pressure	The difference between systolic and diastolic pressures and signifies cardiac output.
Pure culture	A population of cells or multicellular organisms growing in the absence of other species or types. A pure culture may originate from a single cell or single organism, in which case the cells are genetic clones of one another.
Purposive sampling	In this method a sample is selected on the basis of individual judgment of the sample. In this way there is a chance for bias.

Pygostyle	This refers to a number of the final few caudal vertebrae fused into a single ossification, to which the rectrices attach in birds. The pygostyle supports the tail feathers and musculature.
Quadarats	Sampling unit(s) having four main arms or a small enclosure for community. It may be circular, rectangular or square in shape. These units are used to enclose a small area in grassland and plants and animals falling in it are studied.
Quartile deviations	Measure of variability and in the half the difference between lower and upper quartiles.
Radial canal	Canals arising symmetrically around a central axis.
Radial canals	The out pushing of the wall into finger like projections at regular intervals in syconoid which are linned by choanocytes.
Radial symmetry	The arrangement of the body in which parts are arranged symmetrically around a central axis.
Radially symmetrical	Many lines of symmetry in one plane only i.e. body can be divided into a number of radial planes.
Radiata	Radially symmetric animals. Such as ctenophora and cnidaria.
Radula	A firm ribbon composed of chitin and protein along which are found two rows of sharp, chitinous teeth. It is a feeding structure.
Range	It is the difference between the largests and smallests observations.
Raptorial	Legs highly adapted for catching or tearing the prey.
Ratio variables	Variables spaced equal intervals with a true zero point e.g. age.
Recessive trait	The character controlled by and allele which remain unexpressed or latent in presence of another character controlled by another allele at the same locus.
Red pulp	Bulk of pulpy substance gives red colour to spleen is the red pulp. Red pulp is carried by network of reticular fibres and enormous number of blood cells found free in the interstices of the network.
Reducing sugars	Sugars having potentially free aldehyde or ketone group have reducing properties in alkaline solution.

Reflex time	The time that elapses between stimulation of a whole organism and production of detectable response.
Reflex action	Simplest expression of principle to which nervous system acts.
Regression	It is the method of the determination of a statistical relationship between two or more variables. Regression can only interpret what exists physically i.e. there must be a physical way in which independent variable can affect dependent variable. The basic relationship between these two variables is given by an equation
Renal corpuscles	Assemblage of Bowman's capsule and Glomerulus forms the renal corpuscles.
Reticulopodia	Pseudopodia in the form of a branching network.
R_f values	In a chromatographic separation of molecules, distance (cm) moved by the solute from the origin divided by the distance (cm) moved by the solvent from the origin.
Rh factor	Agglutinogen that may be present (Rh positive) or absent (Rh negative) from the surfaces of red blood cells. In 1940, Landsteiner and Weiner reported that rabbit sera contain antibodies against RBCs of the Rhesus monkey that agglutinates RBCs of some human being. The antigen was later named as Rh (Rhesus) factor. Later it was found that D antigen is responsible for the Rh positive condition.
Rhabdites	Many small, cylindrical hyaline structures found on the body surface of turbellarian.
Rheochord	An apparatus is used for dividing a constant current by producing a circuit of comparatively small resistance, which, is capable of being varied so that a variable part only of the current shall flow through the experimental circuit.
Rhinophores	A second pair of tentacle-like structures located dorsally in opisthobranchs head.
Rhodamine	A family of related chemical compounds, fluorone dyes. Examples are Rhodamine 6G and Rhodamine B. They are used as a dye and as a dye laser gain medium.
Ringer's solution	It is a saline solution made up of sodium chloride , calcium chloride, potassium chloride and sodium bicarbonate in dist. water.
Rostellum	The conical elevation at the tip of the head of Cestoda.

Rostral crest	A long projection of the carapace towards the anterior end.
Royal chamber	A chamber in bee hive where larva selected to become queen is developed.
Salinty	Amount of inorganic compounds in water when expressed in grams/kilo or in ppt (%). It is determined by estimation of precipitable halide halogens in 10 ml of water.
Salivary gland chromosomes	Giant polytene chromosomes found in the salivary glands of Dipteran genera such as *Drosophila sp., Chironomus sp.* and *Sciara sp.*
Samples	A sample is a collection of some but not all of the elements of a population. A representative sample contains the relevant characteristics of the population in the same proportion that they are included in that population.
Sanguivorous	Feeding the blood of vertebrates or invertebrates.
Saponifiable Lipids	Lipids that gives saponification reaction i.e. lipids on hydrolysis by NaOH or KOH leads to the formation of sodium or potassium salts of fatty acids.
Saponification	Hydrolysis of lipids by NaOH or KOH leads to the formation of sodium or potassium salts of fatty acids.
Sarcolemma	The cell membrane of a muscle cell. The membrane is designed to receive and conduct stimuli.
Sarcoplasm	Cytoplasm of a muscle fibre. It houses unusually large amounts of *glycosomes* (granules of stored glycogen) and significant amounts of myoglobin, an oxygen binding protein.
Scanning Electron Microscopy	A type of electron microscope that produces an image by scanning the surface of the specimen allowing the surface object to be seen with a great depth of field and giving a 3D appearance to the image.
Scapulets	In some medusae, fringed outgrowths on the outer surfaces of the arms near the bell.
Schistosomiasis	The disease caused by the blood schistosoma.
Sciatic nerve	A mixed nerve carrying motor or efferent fibres to muscle fibres and sensory nerve or afferent nerve fibres from sensory receptors such as muscle spindles ,joint and tendon receptors etc, to spinal cord.
Sclerous connective tissue	Cartilaginous, fibrous, and osseous tissue.

Scolex	Distinct anterior end of Cestoda which bears the hooks and suckers.
Scuta	Lower and lateral calcareous plates of barnacles forming lid or operculum.
Sebaceous glands	These are acinar glands lie embedded in the dermis. This is a holocrine gland that secretes lipoidal fluid known as the sebum.
Secondary data	It refers to those, which are collected by some other person or organization for one purpose and later used for another.
Secondary electrons	Reflected electrons from the sample that are produced after the sample is hit by a beam of electrons produced from the electron gun in a Scanning electron microscope.
Secondary spermatocytes	It is the cell formed due to the first meiotic division of the primary spermatocyte.
Sectioning	Cutting down of the paraffin embedded tissue block at a desired plane and thickness.
Sedimentation coefficient(s)	The sedimentation rate or velocity (v) of the particle can be expressed as sedimentation rate per unit of centrifugal field.
Sella turcica	Part of the cranium where pituitary gland is located.
Seminal receptacles	A small sac connected with the female reproductive organs of insects and many other invertebrates, serving to receive and retain the spermatozoa
Seminiferous tubules	A mass of minute coiled tubules found inside the vertebrate testis in which spermatozoa is produced.
Sepia ink	The dark pigmented fluid secreted by the ink sac of Sepia that is discharged through the anus.
Serous gland	Glands that contain serous acini, a grouping of serous cells that secrete a fluid, isotonic with blood plasma, that contains enzymes such as alpha amylase.Serous glands are most common in the parotid gland, but are also present in the submandibular gland and, to a far lesser extent, the sublingual gland.
Sertoli cells	Large pillar like cells in the germinal epithelium of the vertebrate testis which give support and nourish the spermatozoa.
Sex linked inheritance	The inheritance of X or Z or Y linked genes.

Sex linked recessive	A recessive trait that is inherited through X or Z or Y chromosome.
Sex Ratio	The ratio of observed number of individuals of two different sex of a bisexual animal.
Shadow casting	Layering of the freeze fractured surface with metal such as platinum or gold and subsequent carbon coating and release of the pre-shadowed metal replica from the underlying biological specimen. This is then viewed under an electron microscope.
Simple muscle twitch	A response that occurs through the application of a single stimulus applied to the sciatic nerve results in a sharp momentary contraction of the muscle followed by immediate relaxation
Simple lipids	Aliphatic monocarboxylic acids derived from or contained in esterified form in an animal or vegetable fat, oil or wax.
Simple optical microscope	A particular type of microscope that is made up of only one lens for magnification.. A simple microscope is made up of a single, small, convex lens mounted on a plate with a device to hold the material to be examined (that is the sample or specimen).
Simple random sampling	It is a process by which a group of units can select in such a manner that every unit of the population has an equal chance of being included in the sample.
Simple random sampling with replacement	It can be done when the sample organisms are drawn from the population one by one and after each drawing the selected one is noted and then returned to the population before the next one is drawn.
Simple random sampling without replacement	In this process the sample organisms are drawn all at a time or drawn one by one in such a manner that after each drawing the selected one is not returned to the population when the next one is drawn.
Single staining	Section of tissue that are usually stained with a single dye.
Siphonoglyph	Longitudinal grooves on the stomodaeum.
Sketetal (voluntary or striated) muscle	Skeletal muscle cells, also called fibres, are long multinucleated protoplasmic units arranged in parallel with their neighbours. These muscles are attached to and bring about the movement of the various bones of the skeleton, hence the name skeletal muscles. The whole muscle, such as the biceps, is enclosed in a sheath of connective tissue, the epimysium. This sheath folds

	inwards into the substance of the muscle to surround a large number of smaller bundles, the fasciculi. These fasciculi consist of still smaller bundles of elongated, cylindrical muscle cells, the fibres. Each fibre is a syncytium, i.e. a cell that has many nuclei. The nuclei are oval in shaped and are found at the periphery of the cell, just beneath the thin, elastic membrane (sarcolemma). The sarcoplasm also has many alternating light and dark bands, giving the fibre a striped or striated appearance (hence the name striated muscle). With the aid of an electron microscope it can be seen that each muscle fibre is made up of many smaller units, the myofibrils. Each myofibril consists of small protein filaments, known as actin and myosin filaments. The myosin filaments are slightly thicker and make up the dark band (or A-band). The actin filaments make up the light bands (I-bands) which are situated on either side of the dark band. The actin filaments are attached to the Z-line. This arrangement of actin and myosin filaments is known as a sacromere.
Small Lymphocytes	They are small cell slightly bigger than RBC, small rounded or oval deeply blue-violet stained nucleus occupying nearly the whole of the protoplasm of the cell. Condensed lumpy chromatin gives 'ink-spot' like appearance. Cytoplasm hardly visible and thin, crescent of clear and light blue staining leaving a very thin band of cytoplasm and without a definite visible granule.
Smooth (involuntary) muscle fibre	A type of muscle tissue that is made up of thin-elongated muscle cells, fibres. These fibres are pointed at their ends and each has a single, large, oval nucleus. Each cell is filled with a specialised cytoplasm, the sarcoplasm and is surrounded by a thin cell membrane, the sarcolemma. Each cell has many myofibrils which lie parallel to one another in the direction of the long axis of the cell. They are not arranged in a definite striped (striated) pattern, as in skeletal muscles - hence the name smooth muscle. Smooth muscle fibres interlace to form sheets or layers of muscle tissue rather than bundles. Smooth muscle is involuntary tissue, i.e. it is not controlled by the brain. Smooth muscle forms the muscle layers in the walls of hollow organs such as the digestive tract (lower part of the oesophagus, stomach and intestines), the walls of the bladder, the uterus, various ducts of glands and the walls of blood vessels.

Soaps	Basic salts produced when fats are reacted with basic hydroxides.
Sodium dodecyl sulphate-PAGE (SDS–PAGE)	It is a variant of polyacrylamide gel electrophoresis where detergent SDS (Sodium dodecyl sulphate) in buffer when used with proteins, proteins turn into negatively charged by their tagging on to the SDS anions masking the natural charge of the protein giving a constant charge mass ratio. Thus such negatively charged proteins are electrophoresed. This technique can be used as the regular means for molecular weight determination of such proteins.
Somatostatin or growth hormone inhibiting hormone (GHIH) or **somatotropin release-inhibiting hormone (SRIH)**	A peptide hormone that regulates the endocrine system and affects neurotransmission and cell proliferation via interaction with G-protein-coupled somatostatin receptors and inhibition of the release of numerous secondary hormones. Somatostatin is produced by neuroendocrine neurons of the periventricular nucleus of the hypothalamus. These neurons project to the median eminence, where somatostatin is released from neurosecretory nerve endings into the hypothalamo-hypophysial portal circulation. These blood vessels carry somatostatin to the anterior pituitary gland, where somatostatin inhibits the secretion of growth hormone from somatotrope cells. The somatostatin neurons in the periventricular nucleus mediate negative feedback effects of growth hormone on its own release; the somatostatin neurons respond to high circulating concentrations of growth hormone and somatomedins by increasing the release of somatostatin, so reducing the rate of secretion of growth hormone. Inhibit the release of growth hormone (GH) thus opposing the effects of Growth Hormone-Releasing Hormone (GHRH). Inhibit the release of thyroid-stimulating hormone (TSH). Suppress the release of gastrointestinal hormones Gastrin Cholecystokinin (CCK), Secretin Motilin Vasoactive intestinal peptide (VIP), Gastric inhibitory polypeptide (GIP) Enteroglucagon (GIP) lowers the rate of gastric emptying, and reduces smooth muscle contractions and blood flow within the intestine. Suppress the release of pancreatic hormones. Inhibit the release of insulin, inhibit the release of glucagon, suppress the exocrine secretory action of pancreas.
Somites	Block of dorsal mesodermal cells adjacent to the notochord during vertebrate organogenesis. These transient structures define the segmental pattern of the embryo, and

	subsequently give rise to vertebrae and ribs, dermis of the back, and skeletal muscles of the back, body wall and limbs.
Southern blotting	The technique transfer of DNA bands from an Agarose acrylamide gel onto a more stable and immobilizing support matrix.
Spawn	The hatching of egg of fish after fertilization.
Species	A group of inter-breeding natural population which are reproductively separated from other such groups.
Species area curve	A curve drawn by plotting the number of species (y axis) is plotted against the size (area) of the square (quadrat) (x axis) in a graph paper.
Spermatids	Reproductive cells resulting from the second meiotic division of the spermatocyte. It matures and undergoes a series of changes, as a result gets transformed into spermatozoa.
Spermatogenesis	The phenomenon of differentiation of male primitive germ cells to spermatozoa is called as spermatogenesis.
Spermatogonia	Reproductive cells in the testis. Situated in the germinal epithelium. That lines the seminiferous tubules. It undergoes period of multiplication and growth to gives rise to spermatocytes.
Sperms	Mature male reproductive cells (gamete) of motile nature formed in the testis. It basically has three parts, a head region containing a haploid nucleus; a middle region containing mitochondria and a long tail region containing a flagèllum.
Sphygmomanometer	Sphygmomanometer is an apparatus, which is used for measuring blood pressure (Systolic, diastolic, pulse etc) in man of a peripheral artery the effect before and after exercise. It is derived by Riva Rucci in 1896 and subsequently improved.
Spicules	Calcareous or siliceous skeletal framework supporting the soft, cellular mass.
Spingosine	An 18-carbon amino alcohol (2-amino-4-octadecene-1, 3-diol) with an unsaturated hydrocarbon chain, which forms a primary part of sphingolipids, a class of cell membrane lipids that include sphingomyelin, an important phospholipid.

Spiracle	A vestigial gill slit located between the mandibular and hyoid arch in adult sharks and rays.
Splanchnopleure	A fused embryonic layer formed from the union of visceral mesoderm and the endoderm.
Spleen	A lymphoid organ situated just beneath the stomach in vertebrates. It produces lymphocytes and destroys and stores red blood cells.
Splenic pulp	A soft pulpy substance in the interstices of the spleen.
Spongin fibres	The proteinaceous secretion of spongioblast, which forms a network.
Spongocoel	Empty space of the perforated bag, where inner surface is lined with flagellated cells.
Sporogony	Zygote nucleus divides into a number of nuclei to produce sporozoites.
Sporozoites	Spores produced by fusion of zygote.
Squamous epithelium	An epithelium characterized by its most superficial layer consisting of flat, scale-like cells called squamous cell.
Stage micrometers	A stage micrometer is a microscope slide used in micrometry. In the centre of which a line 1 mm long is drawn. This line is divided into 10 parts with longer lines and marked 0, 1, 2, 3, 4 and so on. Each part is then subdivided into 10 smaller parts by small lines.
Staining	A procedure that is designated to increase the contrast between different structures thus displays structural features of cell or tissue.
Standard deviation	A measure of variability in a population or sample of individuals.
Startling heart lever	It is used for recording the mechanical events of frogs/ toads heart. The frame of the lever carries a light, flat lever arm finely adjustable tension spring which supports the writing lever in horizontal position. A piece of thread tied to the writing lever carries a bent pin which can be hooked through the apex of the ventricle when its contraction is to be recorded. When the heart contract it pulls the lever down and when it relaxes the spring pulls the lever back to its horizontal position.
Statocyst	A small, white and spherical cuticular sac found inside the base of each antennules of prawn.

Sterols	These are esters of higher fatty acids with cyclic alcohol containing perhydrocyclopentanophenanthrene ring. Eg. Cholesterol, ergosterol, sitosterol etc.
Steroids	These are compounds having similar structure to sterols containing cyclopentanophenanthrene ring, eg. Adrenal corticosteroids, sex steroids, some alkaloids etc.
Stichosome	A column of stichocytes that surround and secrete into the oesophagus of Trichurata. A series of glands exterior to the oesophagus proper
Stoke's Law	If a small sphere of radius *r* and density *r′* falls vertically through a liquid of density *r* at a steady velocity μ in spite of the acceleration *g* due to gravity , then the coefficient of viscosity of the liquid is given by : $h = 2r^2g(r'-r)/9\mu$.
Stoke's shift	A change in wavelength of electrons that occurs in some classes of atoms and molecules where there electrons absorb light, get energized, and then quickly lose this energy in the form of heat and light emission. If the electron keeps its spin, the electron is said to go into a singlet state, and the type of light (larger wavelength) that is emitted as the electron come back to ground state.
Stolon	The stalk like structure of colonial invertebrates by which individuals anchor the substrate.
Stomochord	Extending into the proboscis is a rod of cells, which was initially assumed to be homologous to the notochord in Hemichordates.
Stomodaeum	The pharynx of hormiphora, which serves as a site of extra cellular digestion.
Stratified epithelium	An epithelium composed of multiple layers of cells, only the basal layer being in contact with the basal lamina. The basal layer is of stem cells that divide to produce the cells of the upper layers, in skin; these become heavily keratinised before dying and being shed as squames. Stratified epithelia usually have a mechanical or protective role.
Stratified sampling	In this method the population is subdivided into several parts or strata (each subgroup is homogeneous within themselves) in purposive way then a sub-sample is chosen from each of them.
Streak plate method	The streak plate method is a rapid and simple technique of mechanically diluting a relatively large concentration of

	microorganisms to a small, scattered population of cells. The goal is to obtain isolated colonies on a large part of the agar surface, so that desired species can then be brought into pure culture.
Stroma	Fibrous and vascular connective tissue of the ovary.
Sulpholipids	Lipids containing carbohydrate, cerebronic acid, sulphuric acid, potassium and nitrogenous base sphingosine.
Suprabranchial organ	The suprabranchial organ is a chamber above the gills, similar but more primitive than the labyrinth organ of the anabantoid. A membrane that absorbs oxygen from atmospheric air and supplies it to the blood stream covers this chamber. This adaptation enables the snakehead to survive in the most adverse conditions it encounters in its natural environment. The fish master low oxygen levels, poor water and living conditions, which most fish could not survive in, with ease. This adaptation also enables the snakehead to "walk" over land to another body of water, if the need should arise. Snakeheads need to breathe air at regular intervals to survive. Should access to the surface not be granted the fish would drown, just like their cousins the anabantoids.
Sweat glands	Simple coiled tubular glands found in the dermis of the skin that secretes a non-viscous fluid known as sweat.
Sweep net	Sweep nets are stout insect nets, usually made of linen, sturdy cotton or synthetic material commonly are used to sample insect from vegetation.
Syconoid	The canal system in which the choanocytes are only limited to radial canal.
Symtomimetic agent	Mimic actions of NE and EP, produce a variety of responses ,have many undesirable side effects, mimic noradrenergic transmission by activating adrenoceptors (eg. phenylephrine). Sympathomimetic drugs are substances that mimic the effects of the hormone epinephrine (adrenaline) and the hormone/neurotransmitter norepinephrine (noradrenaline). They all raise blood pressure and are all weak bases.
Syncytium	A multinucleated mass of cytoplasm that is not separated into individual cells.
Syngamy	The union of two gametes.
Synonyms	Synonyms are different scientific names that pertain to the same taxon, for example two names for the same species.

	The rule of zoological nomenclature is that the first name to be published is the **senior synonym**; any others are **junior synonyms** and should not be used.
Systolic pressure	Blood pressure that signifies the maximum pressure exerted by blood on the arterial wall at the heart. This is a variable pressure according to the physical condition of the body and is dependant on the height of the diastolic pressure.
Tagmata	Distinct sections of an arthropod, consisting of two or more adjoining segments, such as the cephalothorax of a spider.
Taxon	A group of real organism recognized as a formal unit at any level of hierarchic classification.
Taxonomy	Science which describe all known forms of animals to give names to them and establish their relationship or affinity (Gr. Taxis= arrangement)
TC_{Lo} (toxic concentration low)	Substances that affecting its quality nature, substance or quality of the food, the substance may be filthy, putrid, disgusting, rotten, decomposed or diseased animal or vegetable substance or being insect-infested, may be poisonous or deleterious, it may be prohibited colouring matter or preservative, or any permitted colouring matter or preservative in excess of the prescribed limits.
TD_{Lo} (toxic dose low)	Lowest concentration of a substance in air to which humans (or animals) have been exposed for any given period of time that has produced any toxic effect or produced a tumorigenic or reproductive effect one or more members of the group of subjects.
Tegmina	Tough and leathery fore wing of an insect.
Telolecithal egg	A type of egg where the yolk remains accumulated in one of the pole.
Telophase	The final stage in mitosis and meiosis before cells enter interphase. During this stage chromosomes uncoil and disperse, the nuclear spindle degenerates and a new nuclear membrane forms. The cytoplasm may also divide during this phase.
Telophase I	The stage in the first meiotic division of meiosis that follows anaphase I. The two sets of chromosomes have finished moving to opposite ends of the cell. cytokinesis occurs, forming two daughter cells. nuclear membranes do not form and the chromosomes do not decondense at this point, however, as they would in mitosis.

Telophase II	The last stage in the second meiotic division of meiosis which follows anaphase II. The two sets of chromosomes have finished moving to opposite ends of each of the two cells produced during the first meiotic division. Cytokinesis occurs, resulting in four haploid cells.
Telson	The last division of the body of a crustacean. It is not considered a true segment because it does not arise in the embryo from teloblast areas as do real segments. It never carries any appendages, but a forked "tail" called the caudal furca is often present.
Tendo-achillis	A tendon attached to the posterior end of the gastronemius muscle.
Tentaculozooids	Simple hollow tentacle – like bodies specialized for defence, feeding and reproduction.
Terga	Lateral and dorsal calcareous plates of barnacles forming lid or operculum.
Terminalization	Movement of chiasmata towards the end of the chromosomes.
Testis	The male reproductive organ of animals that produces sperms. In vertebrates there is a pair of testis that produces sex hormone called as the androgens. They develop in the abdominal cavity near the kidneys but in most mammals migrate downwards during foetal development and come to lie outside the body cavity within a pouch of skin (scrotum) situated behind the penis.
Testosterone	An androgen secreted by the testis under the influence of the Luteinizing hormone. Its function is to stimulate the development of the secondary sexual characters, male sex organs and spermatogenesis.
Tetrapoda	The vertebrate animal which have four limbs.
Tetraxon	Spicules composed of four rays radiating from the central point.
Tetrasaccharides	A carbohydrate that when hydrolysed produces four sugar units. Eg. Stachyose.
Theca folliculi	A layer of stroma around a follicle of the ovary is termed as theca folliculi. This layer subsequently differentiates into theca interna and theca externa.
Thecodont	Teeth are set in sockets in the bone.
Theria	A subclass under class mammalia that comprises of the three infraclass prototheria, metatheria and eutheria.

Theronts	These stages are formed from the tomont stage of some parasitic ciliate protozoans that search for a fish to infect. The theronts burrow into the epithelium and form the trophozoite, eg. *Ichthyophthirius multifiliis.*
Thin layer chromatography	A chromatographic method in which a glass plate covered with a thin layer of inert absorbent material (e.g. cellulose or silica gel) and the materials to be analysed are spotted near the lower edge of the plate. Then base of the plate is placed in solvent which rises above the solvent by capillary action separating the constituent of the mixtures.
Thyroid follicles	Thyroid follicle constitute of the thyroid lobule. The lobules are of various sizes lined with a single layer of cuboidal epithelium cells lying on a prominent basement membrane. The follicles are enclosed within a meshwork of connective tissue that supports a close network of anatomosing capillaries.
Thyroid gland	A gland situated in the neck consisting of two lateral lobes on either side of the trachea and larynx, giving the gland a butterfly appearance. Its main function is the regulation of metabolic rate by production of thyroid hormone.
Thyroxine	A hormone secreted by the thyroid gland. It is an iodine containing polypeptide and is essential for the normal cell metabolism.
Time marker	It is an accessory required for a time tracing on the recording surface in most cardiographic and myographic experiments.
Tomont stage	Encysted stage that is formed after the trophozoite of some parasitic ciliate protozoans breaks through the skin of the host, eg. *Ichthyophthirius multifiliis.*
Tongue	An organ responsible for taste situated in the buccal cavity of the vertebrates. In mammals it is made up of smooth and stripped muscle covered with stratified epithelium beneath which is loose connective tissue. The upper surface of the tongue is covered with projections called papillae containing chemoreceptors (taste buds) which are sensitive to the chemicals in food and fluid.
Total count	The total numbers of leucocytes or WBCs per mm^3 of blood that is determined by diluting known volume of blood 20 times with a fluid, which prevent haemolysis and coagulation.

Toxicant	A material or substance that is capable of producing harmful effects on a biological system.
Toxicology	The study of the effect of poisonous toxic substances or toxicant on living organisms concluding the way in which these substances gain entry into the organism.
Trabeculae	A elongated cell or line of cells across the cavity.
Tracking dye	A dye used to monitor the progress of the electrophoresis separation of molecules, eg. Bromophenol blue.
Tragus	A flap developed in mammalian from the lobe extending in front of the external opening of the ear.
Transmission Electron microscopy (TEM)	A type of electron microscope that produces an image by passing electrons through the specimen.
Triaxon	Spicules consists of three axes produce six rays from central point.
Trimming	Process by which paraffin embedded tissues are trimmed for attachment into the microtome peg.
Triploblastic	The body is composed of three cellular layers i.e. ectoderm, endoderm and mesoderm.
Triploblastic	Organisms having three germinal layers i.e. ectoderm, mesoderm and endoderm.
Trisaccharides	A carbohydrate that when hydrolysed produces three sugar units. Eg. Raffinose, Gentianose etc.
Trophozoite	A protozoan, especially of the class Sporozoa, in the active stage of its life cycle. A trophozoite is the activated, feeding stage in the life cycle of protozoan parasites such as the malaria-causing *Plasmodium falciparum* (the opposite of the trophozoite stage is the thick-walled cyst form).The trophozoite develops into a schizont which is a mother cell which asexually produces daughter cells known as merozoites. Another type of *Balantidium coli* is cysts.
True breeding	A plant, that when self-fertilized, only produces offspring with the same traits. The alleles for these types of plants are homozygous. For example, the gene for seed shape in pea plants exists in two forms, one form or allele for round seed shape (R) and the other for wrinkled seed shape (r). A true-breeding round seed plant would be (RR) for that trait and a true-breeding wrinkled seed plant would be (rr).

Trypsin	An enzyme that catalyses the partial hydrolysis of peptides. It catalyses the hydrolysis of peptide bonds formed from the carbonyl group of lysine and arginine residues.
t-tests	t-tests that is used where there is some basis (e.g. previous experimental observation) to predict the direction of the difference, e.g. expectation of a significant difference between the groups.
Tube feet	Many small tubular projections found most famously on the ventral face of a starfish's arms, but are characteristic of the water vascular system of the echinoderm phylum which also includes sea urchins, sand dollars and sea cucumbers and many other sea creatures.
Tularemia	An infectious disease caused by the bacterium *Francisella tularensis* that chiefly affects rodents but can also be transmitted to humans through the bite of various insects or contact with infected animals. In humans, the disease is characterized by intermittent fever and swelling of the lymph nodes. Also called rabbit fever.
Tullgren funnel	It is an apparatus constituted of large funnel made from tin, inside which rests a fine wire sieve applied to extract arthropods from a soil sample through their behavioural dynamics i.e. animals are stimulated to leave the substratum(soil) through the application of heat.
Tunic or test	Secreted protective external covering.
Tunica albuginea	A parallel arrangement of collagenous connective tissue that bounds the germinal epithelium
Tunica albuginea	Thick connective tissue capsule covering the testis is tunica albuginea.
Two-tailed test	It is the relationship between two variables such that a change in the other and also a greater change in one variable results corresponding greater or smaller change in the other variable.
Ultrasonification	Process in which tissues are exposed to ultrasonic waves that merely separate cells, or to break open the plasma membrane and leave the internal organelles intact.
Umbo	A conspicuous bulge in the shell of bivalves seen on the dorsal surface, adjacent to the hinge.
Unicellular	Body made up of a single cell only.
Uropod	The last pair of appendages of prawn.

Variables	The quantitative character of the statistical information collected from a group of units or objects.
Variance	It is the arithmetic mean of absolute deviations from mean.
Vector	Organisms that are able to carry a parasite in their body and pass them to a host. It must be susceptible to infection and a part of the parasitic life cycle can develop in it. *Anopheles* sp. mosquito is a vector of *Plasmodium sp.*
Vertebral column	The series of vertebrae forming the axis of the skeleton and protecting the spinal cord.
Vital or Supravital dyes	These are stain which is used to stain living materials.
Vitelline membrane	Non-cellular transparent mucoprotein layer forming the extracellular matrix which is both fibrous and amorphous. It has tensile and elastic property.
Water soluble vitamins	They are mainly the vitamin B and C. They are hydrophilic and soluble in water.
Waxes	Complex mixture of esters of long chain fatty acids and long chain monohydric alcohols (other than glycerol), free long chain alcohols and aldehydes and saturated high molecular weight hydrocarbons. The carbon atoms in the constitutional fatty acids and alcohols vary between 24 and 36.
Western blotting	The technique transfer of protein bands from an agarose gel onto a more stable and immobilizing support matrix.
Wet whole mounts	A name for a histological preparation where the whole specimen is mounted or spread on the glass slide; e.g., w. m. of retina; w.m. of mesentery.
White pulp	A rounded aggregation of lymphocytes in the splenic pulp.
White fibrous tissue	Typical connective tissue in which white elastic fibres predominate as distinguished from elastic tissue.
Wuchereriasis	Infection with worms of the genus *Wuchereria.* Also known as Bancroft's filariasis.
X-linked inheritance	The inheritance of X linked genes localised in the non-homologous part of the X- chromososme.
Yellow elastic fibres	Bundles of proteins (elastin) found in connective tissue and produced by fibroblasts and smooth muscle cells in arteries. These fibers can stretch up to 1.5 times their length, and snap back to their original length when relaxed. Elastic fibers include elastin, elaunin and oxytalan.

Yolk sac	Extra embryonic membrane which is filled with the yolk of the egg, function as first respiratory organ and as a digestive organ for the embryo.
Zona fasciculata	Middle layer of the adrenal cortex is made up of larger vacuolated cells that are arranged as one or two cell thick radially oriented columns having fenestrated capillaries in between.
Zona glomerulosa	Outer layer of the adrenal cortex that is made up of columnar cells.
Zona reticularis	Outer layer of the adrenal cortex is made up of consists of anastomosing network of pigmented reticular cells having pycnotic nucleus.
Zooplanktons	Small aquatic animals who has limited power of self locomotion.
Zygote	A fertilized ovum or a diploid cell formed due to the fusion of the male and female gamete.
Zygotene	This is a substage of Prophase I of meiosis. The homologous chromosomes are seen as paired units. The chromosomes are shorter and thicker than in leptotene, and in a number of cells they remain attached to the nuclear envelope at the points close to the aster. This gives rise to an image named the "**bouquet**". This attachment is unusual in invertebrates and absent in plants, where the chromosomes appear to be a tangled
Zymogen granules	The centro-acinar cells of the pancreas are serous cells, mature secretory granules the zymogen granules are membrane bound and accumulate in the apical portion of the cell.These vesicles contains inactive precursors of the enzyme, zymogen, e.g. trypsinogen of exocrine cells of pancreas.

References

Selected Readings :

1. Aberdeen , J.E.C. 1958. The effect of quadrat size, plant size, and plant distribution on frequency estimates in plant ecology. Australian Journal of Botany. 6(1): 47 – 58.
2. Ackefors, H. 1972.Plankton Methodology. FAO/SIDA Training course on marine pollution in relation to protection of living resources.FIR/TPLR/72/36.
3. Ahmed, E.1972. Coastal geomorphology of India, Orient Longman.pp.222.
4. Alexander, P. and Lundgren, H.P.1966-1969. A Laboratory Manual of Analytical methods of Protein Chemistry, Vols 1-5. Perganon Press, Oxford.
5. Alwine JC, Kemp DJ, Stark GR. 1977. "Method for detection of specific RNAs in agarose gels by transfer to diazobenzylo-xymethyl-paper and hybridization with DNA probes", *Proc Natl Acad Sci U S A.*, 74(12):5350-5354.
6. Anon. 1967. Forestry in India. Published by "Indian forester".
7. Ashwell, G.1957. In: Methods in Enzymol. 3.(Eds. Codowick, S.P. and Kaplan, N.O.) Academic Press New York p.99.
8. Ashwell; ibid, p.87.
9. Atwal, A.S. and Dhaliwal, G.S.2002. Agricultural Pests of SOUTH Asia and their Management.Kalyani Publishers, New Delhi.India.4th Edn. pp 498.
10. Balinsky, B.I. 1970. An Introduction to Embryology. W.B. Saunders Company, Philadelphia and London. UK. 3rd ed., xviii+725 pp
11. Bancroft, J.D. and Stevans, A. (Ed.).1990. Theory and Practice of histological techniques. Elsevier Science Health Science div.3rd. Edn.
12. Barrington, E.J.W. 1979. Invertebrate Structure and Function. (ELBS). Van Nostrand Reinhold (U.K.) Co. Ltd, England. UK. 2nd ed., pp. xiv+ 765.
13. Barrington, E.J.W.1965. The Biology of Hemichordata and Protochordata. Oliver and Boyd, Edinburgh. UK.pp.176.

14. Berman, Irwin. 2003. Color Atlas of Basic Histology. McGraw-Hill/Appleton & Lange. 3rd Edn. pp.377.

15. Berrie, A.D.1976. Detritus microorganisms, and animals in fresh water. In Anderson, J.M., Mcfadyen, A.(Eds.). The Role of Terrestrial and Aquatic organisms in Decomposition Processes. Blackwell Oxford.

16. Bhagyarekha, N. Pandav., Bhagyashri, A. Shanbhag., and Srinivas, K. Saidapur 2007. Ethogram of courtship and mating behaviour of garden lizard, *Calotes Versicolor,* Current Science, 93(8): 1164-1167.

17. Biggs, R. (Ed.). 1972. Human blood coagulation. Haemostasis and Thrombosis. J.B. Lippincott Co. Philadelphia,USA.

18. Black. C.A. (Ed.). 1965. Methods of soil analysis, Vol. 1 and 2, Amer. Soc. of Agronomy.

19. Blanchard, J. Intracellular esterase from **Lactococcus Lactis** subsp. Lactis NCDO 763 : Purification and Characterization. 1984. Methods Enzymol., 104:404.

20. Bloom, W. and Fawcett, D.W. 1975. A Text Book of Histology., Saunders Company, Philadelphia. USA. 10th Edn. pp.1033.

21. Bradfield, A.E. 1978. Life in sandy shores. Edward Arnold, London, pp.60.

22. Bradford, M.M.1976. A rapid and sensitive method for the quantification of microgram quantities of protein utilizing the principle of protein-dye binding.Anal. Biochem. 72:248-251.

23. BSCS Green Version, Students Manual Laboratory Field Investigations.1966. American Institute of Bio. Scieneces, Biological curriculum study, Rand Mc Nally and Co. Chicago. USA.

24. Burnette, W.N. 1981. Western blotting: electrophoretic transfer of proteins from sodium dodecyl sulfate — polyacrylamide gels to unmodified nitrocellulose and radiographic detection with antibody and radioiodinated protein A". *Anal Biochem*, 112 (2): 195-203.

25. Burton, K.1956. A study of the conditions and mechanism of the diphenylamine reaction for the colorimetric estimation of deoxyribonucleic acid.Biochem. J. 62:315-323.

26. Carpentar, C.B. and Miller, W.V. (Eds.) 1977. Clinical Histocompatibility Testing. Gruna & Stratton. New York, USA.

27. Carter, M.R., and Gregorich, E.G. (Eds.).2008. Soil Sampling and method of analysis. Boca Raton : CRCPress.USA.2nd. Edn. pp.1241.

28. Champion, H.G., and Seth, S.K.1968. A revised survey of forests types of India. Govt. of India Publication. New Delhi.

29. Chapman, H.D., and Pratt, P.F. 1961. Methods of analysis of soil, plants and water. Univ of California. pp.309.

30. Charmbach, A., and Rodbard, D.1971. Polyacrylmide Gel Electrophoresis, Science, 172: 440-451.

31. Chatterjee, A.K., and Chatterjee, A.K. 1999. Practical Zoology. Nirmala Library. Kolkata. India. 7th Edn. pp. 580.

32. Chaudhuri, H., and Alikunhi, K. H.1957. Observations on the spawning in Indian major carps by hormone injection.Curr. Sci., 26:381-382.

33. Chowdhury, Sujit K. 2004. Concise Medical Physiology. New Central Book Agency (P) Ltd. 5th Edn. pp.xxv+624.

34. Chhapgar, B.F. 1990. Nature Guides: Seashore Life of India.

35. Creighton, Thomas E. 1992. Proteins, Structures and Molecular Principles W.H. Freeman and Co., New York. 2nd Edn. pp.512.

36. Dabadghao, P.M., and Shankarnarayan, K.A.1973. The grass cover of India. ICAR, New Delhi. pp. 713.

37. Dawson, R.M.C., Elliot, D.C., Elliot, W.H., and Jones, K.M. 1986. Data for Biochemical Research, Oxford Science Publ. 3rd Edn.pp.1-131.

38. Dick. M. Heather.1980. Clinical Histocompatibility Testing. J Clin Pathol. 33(2): 203.

39. Duellman, W.E. and Trueb, L. 1986. Biology of Amphibians. McGraw-Hill Publishing Co. Ltd. New York. USA. pp.650.

40. Electron Microscopy Methods and Protocols. Hajibagheri, M.A.N.(Ed).1999. Methods in Molecular Biology, Vol. 117, Humana Press.

41. Ellis, K.J., and Morrison, J.F. 1982. Buffers of constant ionic strength for studying pH-dependent processes. Methods Enzymol. 87: 405-426

42. Everson Pearse, A.G. 1960. Histochenmistry:Theoritical and Applied. J & A Churchill, Ltd. Gloucester Place, England, UK. W.I. pp. x+ 998.

43. Fisher, R.A., and Yates F. 1973. Statistical tables for Biological, Agricutural and Medical Research.Longman & Co. 6th Edn. pp.155.

44. Furie, .and, Furie, B.C. 2005. "Thrombus formation in vivo". J. Clin. Invest. **115** (12): 3355–62.

45. Gaastra, W. 1984. In: Methods in Molecular Biology Vol 1 Proteins (Ed. Walker, J.M.) Humana Press, New Jersey. p. 349.

46. Gautheyrou, J. 2001. Soil Analysis. Taylor & Francis. London. UK. 1st Edn.

47. Ghai,C.L. 2005. A Textbook of Practical Physiology. Jaypee Medical Publishers Ltd. New Delhi, India, 6th Edn. pp.x+312.

48. Ghose, K.C., and Manna, B.1996. Practical Zoology. New Central Book Agency (P) Ltd. Kolkata. India. pp. ix+481.

49. Ghosh, A.B., Bajpai, J.C., Hasan, R., and Singh, D.1983.Soil and water testing methods, Laboratory manual. IARI, New Delhi, India..

50. Ginsburg, V., and Robbins, P. (Ed.). 1984. Biology of Carbohydrates. Wiley. New York. USA. Vol. 2.

51. Ginsburg, V. 1981. Biology of Carbohydrates, Vol 1. Wiley.New York. USA.

52. Good, N.E., and Izawa, S. 1972. Hydrogen ion buffers. In Methods Enzymol. 24: 62: 53-68.

53. Good, N.E., Winget, G.D., Winter, W., Conolly, T.N., Izawa, S., and Singh R.M.M. 1966. Hydrogen ion buffers for biological research. Biochemistry 5: 467-477.

54. Goswami, S.C.(Retd.)., Dhargalkar, V.K and X.N. Verlecar, X.N.(Ed). 2004. Zooplankton Methodology, Collection & Identification– a field Manual. National Institute of Oceanography. 1st Edn.

55. Griffiths, Anthony J.F. (Ed). 2005. An Introduction to Genetic Analysis.W. H Freeman & Co. 8th. Edn.pp.800.

56. Gurr, M.I., Harwood, J.L., and Fayn,K.N.2002. Lipid Biochemistry: An Introduction. Blackwell Science Ltd, London. 5th Edn. pp.320.

57. Guyton, A.C. 1981. Text book of Medical Physiology. W.B. Saunders Company. Philadelphia London Toronto, 26th Edn. pp.1485.

58. Hedge, J.E. and Hofreiter, B. T.1962. In: Carbohydrate Chemistry 17 (Eds. Whistler R. L. and Be Miller J. N.). Academic Press, New York.

59. Heezan, Bruce C., Trap M. and Ewing. E., 1959. The floor of the ocean.1. North Atlantic Geol.Soc. Amer. Spec. Paper. 65.p.122.

60. Hesse, P.R.1971.A Text Book of Soil Chemical Analysis. John Murry Ltd.London, UK. pp.520

61. Homberger, Dominique G., and Walker, Warren F. 2003.Vertebrate dissection. Brooks/Cole Pub Co. 9th Edn.pp.4 48.

62. Hyman, L.H.1951. The Invertebrates: Echinoderms. Vol IV. Mc. Graw-Hill. New York. USA.

63. Hyman, L.H.1951. The Invertebrates:Mollusca I . Vol VI. Mc.Graw-Hill. New York. USA.

64. Hyman, L.H.1951.The Invertebrates: Platyhelminthes and Rhynchocoela, the acoelomate Bilateria. Vol II.Mc. Graw-Hill. New York.USA.

65. Imms, A.D. 1957. A General Text Book of Entomology. 9th Edn. Richards, O.W., Davis, R.G. eds. Methuen & Co. Ltd. London. UK.pp.x+ 886

66. Issitt, P. D. 1979.Serology and Genetics of Rhesus Blood Group System. Cincinati, Montgomery Scientific Publications.

67. Iwine, J.C., Kemp D.J., and Stark G.R.1977. Method for detection of specific RNAs in agarose gels by transfer to diazobenzyloxymethyl-paper and hybridization with DNA probes", *Proc Natl Acad Sci, USA.*, 74(12): 5350-5354.

68. J.M. Graham., and D. Rickwood, D. 2001. Biological Centrifugation (The Basics) (Paperback).Graham, Bios Scientific Publishers Ltd. pp.210.

69. Jackson, M.L.1973. Soil Chemical Analysis. Prentice-Hall of India Ltd., New Delhi. India. pp.498.

70. Jain, S.K. 1990. "Botanical regions and flora of India". Everyman's Science, XXIV(6), 213-223.

71. James, J., and Tanke, H.J. 2008. Biomedical Light Microscopy (Hardcover) Springer, 2st Edn. pp.535.

72. Jhingram, V.G. 2002. Fish and Fisheries of India. Hindustan Publ. Crop. Delhi. India, pp.752.

73. John A. Smith, in *Current Protocols in Molecular Biology*, Section II, 10.2.2 "Electrophoretic Separation of Proteins". Greene Publishing Associates, Brooklyn 1987.

74. John S.A.1987. In *Current Protocols in Molecular Biology*, Section II, 10.2.2 "Electrophoretic Separation of Proteins". Greene Publishing Associates, Brooklyn.

75. Junqueira, L.C., and Carneiro, J. 1983. Basic Histology. Lange Medical Publications. Maruzen Asia, Singapore. 4th Edn., pp.xiii+ 510.

76. Junqueira, Luiz Carlos., and Carneiro, Jose. Basic Histology: Text & Atlas (Basic Histology).2005. McGraw-Hill Medical. 11th Edn. pp.544.

77. Kardong, K.V. 2002. Vertebrate-Comparative anatomy, Function, evolution.Tata McGraw-Hill Publishing Co.3rd. Edn.pp.xviii+782.

78. Kent, C. George, and Carr, K. Robert. 2001. Comparative Anatomy of the Vertebrates. McGraw-Hill. US. 9th Edn. p.544.

79. Kerese, I.1984. Methods of protein Analysis. E. Horwood, England. pp.371.

80. Khan, H. 1938. Curr. Sci. 7: 233

81. Kiernan, J.A. 1999. Histological and Histochemical Methods : Theory and Practice. Elsevier Science Health Science div. 3rd Edn. pp.502.

82. Kirchner, J.G. 1978. Thin layer Chromatography in Techniques of Chemistry. Vol.14 (Ed. Perry, E.S.) Wiley Interscience, New York.

83. Klug, W.S., and Cummings, M.R. 2005 Concepts of Genetics .Prentice Hall. 2005.8th Edn. pp.784.

84. Korf, Bruce. 2006. Human Genetics and Genomics. Blackwell Science.UK. 3rd Edn.p.288.

85. Kuchler, A. W.1983. World map of natural vegetation. In Goode's World Atlas. Rand Mcnally, New York.USA 16th Edn. p.16-17.

86. Lal,S.S. 1976. A Text Book of Practical Zoology - Vertebrates. Rastogi Publications.Merut.3rd Edn. pp.iii+318.

87. Lal,S.S. 1978. A Text Book of Practical Zoology - Invertebrates. Rastogi Publications. Merut. India.5th Edn. pp.iv+328.

88. Landsteiner K., and Wiener A.S. 1940.An agglutinable factor in human blood recognized by immune sera for rhesus blood. Proc Soc Exp Biol Med. 43:223-224.

89. Levine, N.D. *et.al.* 1980. A newly revised classification of Protozoa. Protozool., 27(1):17-58.

90. Litchfield Jr., J.T., and Wilcoxan, F. 1949. A simplified method of evaluating dose-effect experiments. J. Pharmacol. Exp. Ther. 96 2:99–113.

91. Lom. 1958. Practical Invertebrate Zoology. Compiled and Ed. Dales, R.P. 1981. Blackwell Scientific Publ. Oxford. London.

92. Long, J.A., and Evans, H.M. 1922. The oestrous cycle in rat and its associated phenomena. Memoirs of the Univ. Calif. 6 :1.

93. Lowry, O.H., Rosebrough, N.J., Farr, A.L., and Randall, R.J. 1951. Protein measurement with the Folin phenol reagent. *J. Biol. Chem.* 193, 265-275.

94. Lyon, M.F., and Russel, L.B. 1961. Gene Action in the X-chromosome of the Mouse (*Mus musculus* L.). Nature 190:372-373.

95. Macfayden, A. 1955. A comparison of methods for extracting soil arthropods. In Kevan D.K. Mc E (Ed.), Soil Zoology, University of Nottingham School of Agriculture. p.315-332.

96. Macfayden, A. 1962. Soil arthropod sampling. Adv. Ecol. Res. 1: 1-34.

97. Macfayden, A. 1968. Notes on methods for the extraction of small soil arthopods by the high gradient apparatus. Pedobiologia 8:401-406.

98. Mange, A.P., and Mange E.J.1980.Genetics: Human aspects. Saunders College. Philadelphia. pp. x+675.

99. Marshall, A.J., and Williams, W.D.(Eds.). 1972. Text Book of Zoology: Invertebrates. The Macmilan Press Ltd.7th Edn. pp.xix+874.

100. Meglitsch, P.A., and Schram, F.R.1991. Invertebrates Zoology. Oxford University Press, Inc. New York. USA.3rd Edn. pp.vi+623.

101. Michael, A., Ashburn, M.D., Golic, K.G., and Scott Hawley, R.2005. Drosophila: A Laboratory Handbook (Hardcover). Cold Spring Harbor Laboratory Press. New York. USA. 2nd Edn. pp.1409.

102. Michael, P. 1984. Ecological Methods for Field and Laboratory Investigations, Tata McGraw-Hill Co., New Delhi. India. pp.404.

103. Wim Gaastra 1984. In: Methods in Molecular Biology Vol 1 Proteins (Ed J M Walker) Humana Press New Jersey p 349.

104. Mohn, J.F. et.al.(Eds.). 1977. Human Blood Group, S.Karger, New York.

105. Moore, D.H.1968. Electrophoresis in Physical Techniques in Biological Research (Ed. D.H.Moore) Academic Press, New York.pp.325.

106. Morton, J.E. 1950 Collecting and Preserving Zoological Specimens. Tuatara, 3:3.

107. Murmer, J.1961. A procedure for the isolation of deoxyribonucleic acid from micro-organisms. J. Mol. Biol. 3: 208-213.

108. Nalbadev,A.V. 1958. Reproductive Physiol.,Freeman, San Francisco. USA.

109. Nelsen, O.E.1953. Comparative Embryology of the Vertebrates. McGraw-Hill Book Company, Inc., New York. USA. pp. xiii+982.

110. Neshyba, S. 1987. Oceanography: Perspectives on a fluid Earth. John Wiley & Sons, New York.

111. Newbould, P.J. 1967. Methods for estimating the Primary Production of Forests. International Biological Programme Handbook No. 2; Blackwell Scientific publications, Oxford and Edinburgh; (reprint 1970), pp.57.

112. Oakland, G.B. 1953. Determination of Sample size.Can.Ent.,85: 108-113.

113. Odum, E.P. 1971. Fundamentals of Ecology. W.B. Saunders and Co. Philadelphia,. 3rd Edn. pp.574.

114. Olson J.S. 1964. Gross and Net production of terrestrial vegetation. J. Ecol. 52 (Suppl.) : 99-108.

115. Parker, S.P.(Ed.). 1982. Classification and synopsis of living organisims, Vol.I. Mc.Graw-Hill. p.641-64 6(Sponges).

116. Parker, S.P.(Ed.). 1982. Classification and synopsis of living organisims, Vol.I. Mc.Graw-Hill. p.669-706 (Cnidarians).

117. Parker, S.P.(Ed.). 1982. Classification and synopsis of living organisms, Vol.I. Mc.Graw-Hill. p.707-715 (Ctenophores).

118. Parker, S.P.(Ed.).1982. Classification and synopsis of living organisms, Vol.I. Mc.Graw-Hill. p.880-929 (Nematodes).

119. Parker, S.P.(Ed.). 1982. Classification and synopsis of living organisms, Vol.II. Mc.Graw-Hill. p.946-1166 (Mollusca).

120. Parker, S.P.(Ed.). 1982. Classification and synopsis of living organisims, Vol.II. Mc.Graw-Hill.p.1-43 (Polychaetes and Clitellates); p.65-66 (Echiurians).

121. Parker, S.P.(Ed.). 1982. Classification and synopsis of living organisms, Vol.II. Mc.Graw-Hill. p.71-728 (Arthropods).

122. Parker ,S.P.(Ed.).1982.Classification and synopsis of living organisims,Vol.II. Mc.Graw-Hill. p.729-739(Tardigrada and Onychophora).

123. Parker, T.J. and Haswell,W.A.1962. Revised and Rewritten by Marshall,A.J.1962.a Text Book of Zoology. The Macmilan Co. Ltd. 7th Edn. London. UK. pp.xxiii+952.

124. Patten, B.M. 1958. Foundation of Embryology. McGraw-Hill Book Company. pp.xvi+: 578

125. Pearse, A.G.E. 1968. Histochemistry, Theoretical and Applied. (2 vols). Edinburgh. Churchill Livingstone. 3rd Edn.

126. Pearse, A.S., Humm, H.J., and Wharton, G.W.1942.Ecology of sand beaches at Beanfort, North Carolina. Ecological Monographs 12: 136-190.

127. Pechenik, J.A.2002. Biology of the Invertebrates. 4th Edn. India. Tata Mc Graw- Hill Publishing Company Ltd. India. pp.578.

128. Pelczar, Jr. M.J., Chan, E.C.S. and Kreig, N.R. 1993. Microbiology. Tata McGraw-Hill Publ. Co. Ltd. New Delhi, 2nd Edn.pp.870.

129. Perrin, D.D., and Dempsey, B. 1974. Buffers for pH and Metal lon Control. Chapman and Hall Laboratory Manuals. London, Chapman and Hall.

130. Pilz, Wolfgong. 1974. Acetylcholine esterase assay: In: Methods of Enzymatic Analysis (Ed. Bergmeyer) Academic Press Inc New York II 2nd. Edn.pp.840.

131. Plummer, David T.1983. An Introduction to Practical Biochemistry. Tata McGraw-Hill Edn. 3rd Edn. pp.xvii+ 332.

132. Prescott, L.M., Harley, J.P., and Klein, D.A. 2004. Microbiology (6th ed). McGraw-Hill Higher Education. pp.992.

133. Renart, J., Reiser J., and Stark G.R.1979.Transfer of proteins from gels to diazobenzyloxymethyl-paper and detection with antisera: a method for studying antibody specificity and antigen structure", *Proc Natl Acad Sci USA*, 76(7): 3116-20.

134. Robinson, H.S., and Robinson, P.J.M.1950. Some notes on observed behaviour of Lepidoptera in flight in the vicinity of light sources together with a description of a light trap designed to make entomological samples. Ent. Gaz. 1: 3-15.

135. Romer, Alfred Sherwood. 1971. The vertebrate body. W.B. Saunders. Philadelphia. USA. 4th Edn. pp.452.

136. Rosner, Bernard. 1995. Fundamentals of Biostatistics. Duxbury Pr; 4th Bk & Dsk Edn. Boston.

137. Ross Michael H., and Pawlina, W.2005. Histology: A Text and Atlas: With Correlated Cell and Molecular Biology. Lippincott Williams & Wilkins Philadelphia.. 5th Edn. pp.906.

138. Ross, M.H., and Reith, E.J.1985. Histology: A text and Atlas. Harper and Row, Publishers, J.B. Lippincott Company, New York. USA. pp.xvi+766.

139. Ruppert, E.E., and Barnes, R.D., 1994. Invertebrate zoology. Har Court Asia Pte Ltd. 6th Edn., pp. 1054.

140. Rowell, L.B. 1974. Human cardiovascular adjustments to exercise and thermal stress. Physiol. Rev., 54: 75.

141. Rugh, R. 1948. Experimental Embryology. Burgess, Mmneapolis. USA.

142. Russell, P.J. 2005.Genetics: A Mendelian Approach. Addison-Wesley. 2nd. Edn. pp.850.

143. Lewis, S.M., Bain, B.J., and Bates I (Eds.). 2001. Dacie and Lewis Practical Haematology. Churchill Livingstone. London. UK.9th Edn. pp.633.

144. Sadasivam, S. and Manickam, A.1996. Biochemical Methods.New Age International Publishers. 2nd Edn. pp.xv+256.

145. Saladin K.S.1998.Anatomy and Physiology, Mc.Graw-Hill Comp. Int.

146. Sangstad, E.S., Bram, R.A., and Nyquist, W.E. 1967. Factors influencing sweep-net sampling of alfalfa. J. Econ. Ent. 60: 421-426.

147. Sawhney, S.K., and Singh, R. 2005. Introduction to Practical Biochemistry. Narosa Publishing House Pvt. Ltd. New Delhi. 1st Edn. pp.262.

148. Schmidt, G.D., and Roberts, L.S. 1989. Foundations of Parasitology. Times Mirror Mosby College Publ. St. Louis. Missouri. USA. 4th Edn. pp.770.

149. Schneyer, L.H. 1974. Salivary secretion . In MTP International Review of Science : Physiology. Vol 4. Baltimore, University Park Press, pp.183.

150. Seegers, W.H. 1969. Blood clotting mechanisms: Three basic reactions. Annual Rev. Physiol., 31: 269.

151. Segel, I.H. 1968. Biochemical Calculations: How to Solve Mathematical Problems in General Biochemistry. John Wiley & Sons. New York. USA. 2nd Edn.

152. Sharma, I.J., and Singh, H.S. 2005. Students Laboratory manual of Veterinary Physiology. Kalyani Publishers, New Delhi, 2nd Edn.pp.262.

153. Smith, I. 1976. Chromatographic and Electrophoretic Techniques, Vol.I. Chromatography. Heinemann, London. 4th edn.

154. Snudsdtad, D.P. and Simmons, M.J. 2006.Principles of Genetics. John Wiley & Sons. Inc.4th Edn. pp.866.

155. Sokal, R.R., and Rohlf, F.J. 1995. Biometry: the principles and pratice of statistics in biological research. W.H. Freeman and Company. New York .3nd Edn. pp.887.

156. Soulsby, E.J.L. 1982. Helminths, Arthropods and Protozoa of Domestic Animals. ELBS. London. 7th Edn.

157. Southern, E.M.1975. Detection of specific sequences among DNA fragments separated by gel electrophoresis", *J Mol Biol.*, 98:503-517.

158. Southwood, T.R.E. 1971. Ecological Methods. Chapman and Hall, London. UK.pp.xxiv+524.

159. Spivak, J.L. (Ed.). 1980. Fundamentals of Clinical Hematology. Hagesstown, Md., Harper and Row. pp. x + 388.

160. Stephenson, T.A., and Stephenson, A. 1952. Life between tide-marks in North America: Northern Florida and Carolinas. Journal of Ecology 40: 1-49.

161. Stoll, V.S., and Blanchard, J.S. 1990. ibid. 182, 24.

162. Strachan, Tom., and Read Andrew. 2003. Human Molecular Genetics. Garland Science/Taylor & Francis Group Oxford. 3nd Edn. pp.696.

163. Subrahmanyam, N.S., and Sambamurty, A.V.S.S. 2006. Ecology. Narosa Publishing House Pvt. Ltd. New Delhi. 2nd Edn .pp.664.

164. Symth, J.D. and Wakelin, D. 1994. Introduction to Animal Parasitology. Cambridge University Press. 3rd Edn. pp.549.

165. Talwar and Srivastava (Ed.). (2003). Text book of Biochemistry and Human Biology, 3rd Edn. pp.xxxii+1294.

166. Tonapi, G.T. 1980. Freshwater animals of India(An Ecological Approach). Oxford & IBH Co. India. pp.341.

167. Van Slyke, J.1922. J.Biol. Chem. 52: 525.

168. Vander J.A., and Sherman J.H. and Luciano D.1975.Human Physiology-The mechanism of body function. Tata Mc Graw-Hill Publishing Company Ltd. New Delhi. India. pp.vii+614.

169. Wake Marvalee, H. (Editor).1992. Hyman's Comparative Vertebrate Anatomy. Chicago. University of Chicago Press. 3rd Edn. pp.795.

170. Weichert, C.K., and Presch, W. 1977. Elements of Chordate Anatomy. Tata McGraw-Hill Publishing Co. Ltd., New Delhi, India.4th Edn. pp.vii+526.

171. Welch, P.S. 1948. Limnological methods,. McGraw-Hill, New York. 2nd Edn.p.382.

172. Wessner, F.M. 1960. General Zoological Microtechniques. Indian ed, The Williams & Wilkins Company, Baltimore 2, Maryland, USA. xi+230pp

173. Wetzel, R.G. 1983. Limnology. Saunders. Philadelphia.USA.2nd Edn.pp.858.

174. Whyte, R.O. 1957. The grassland and fodder resources of India. ICAR monograph NO. 22. New Delhi.

175. Willims, B.L., and Wilson. K. 1974. In: Principles and Techniques of Practical Biochemistry, Edward Arnold, London. pp.xxix+586.

176. Young, J.Z. 1981. The Life of the Vertebrates. Oxford University Press. Oxford. UK.3rd Edn. pp.xv+645.

177. Zar, J.H. 1999. Biostatistical Analysis. Pearson Education India. Limited. India.4th Edn.pp. xxix+1266.

178. Devenport, L.D., Merriman, V.J., and J.A. Devenport. 1983. Effects of ethanol on forced spatial variability in the 8-arm radial maze. Pharm. Biochem. Behaviour. 18:55-59.

179. Golczewski J.A., Hiramoto R.N., and Ghanta V.K. 1981. Enhancement of maze learning in old C57BL/b mice by dietary lecithin. Neurobiology of Aging. 3:223-226.

180. Kessler, J., Markowitsch H.J., and Otto B., 1982. Subtle but distinct impairment of rats with chemical lesions in the thalamic mediodorsal nucleus, tested in a radial arm maze. J. Comp. Psych. 96:712-720.

181. Kinjoh, T. 1981. Effect of food deprivation on maze and discrimination learning in white rats. Ann. Animal Psychol. 31:11-24.

182. Small, W.S. 1899-1900. An experimental study of the metal processes of the rat. Amer. J. of Psych. 11:133-164.

183. Tryon, R.C. 1940. Genetic differences in maze learning in rats. National Soc. for the Study of Education, 39th Yearbook. Public School Publishing, Bloomington, Illinois.

184. Woodworth, R.S., and Scholsberg, H. 1954. Experimental Psychology. Holt, Rinehart and Winston, New York, 948 pp.

185. Southern, E.M. 1975 : "Detection of specific sequences among DNA fragments separated by gel electrophoresis", *J Mol Biol.*, 98:503-517.

186. W. Neal Burnette. 1981. "'Western blotting': electrophoretic transfer of proteins from sodium dodecyl sulfate — polyacrylamide gels to unmodified nitrocellulose and radiographic detection with antibody and radioiodinated protein A". *Anal Biochem* 112 (2): 195-203.

187. Renart J, Reiser J, Stark GR. 1979. "Transfer of proteins from gels to diazobenzyloxymethyl-paper and detection with antisera: a method for studying antibody specificity and antigen structure", *Proc. Natl. Acad. Scit. USA*, 76(7):3116-20.

Important Websites :

1. www.biologie.uni-hamburg.de/b-online/e03/03.htm.
2. www.cerlabs.com/experiments/10875404472.pdf. Frank, M.R. and Loffredo, W.M.M.1994. Qualitative testing for lipids. Modular Laboratory program in Chemistry. Chemical Education Resources, Inc.P.O. Box. 357220. Pennsylvania,.
3. www.coleparmer.com/techinfo/techinfo.asp?htmlfile=basic-centrifugation.htm.
4. www.des.umd.edu/biosafety/auto/autoclave.html.
5. www.estrellamountain.edu/faculty/farabee/BIOBK/BioBookhumgen.html
6. www.homepages.gac.edu/~cellab/appds/appd-f.html.
7. www.homepages.gac.edu/~cellab/chpts/chpt4/intro4.html
8. www.microscope-microscope.org/applications/pond-critters/culturing-protozoa.htm:Collecting and culturing protozoans. Microbus 2007.Retreived on 19.08.2008.
9. www.microscopy.com/articles/formulas/measurements.htm.
10. www.olympusmicro.com/
11. www.park.edu/bhoffman/courses/bi337l/laboratory/exercises/carboqual.htm.: Carbohydrate qualitative tests. Retreived on 19.08.2008
12. www.protocol-online.org/prot/General_Laboratory_Techniques/Centrifugation/
13. www.protocol-online.org/prot/Immunology/Macrophages/Index.html: In-vitro Phagocytosis Assay of Macrophages. Shalini Jain and Hariom Yadav. Animal Biochemistry Division, National Dairy Research Institute, Karnal-132001, Haryana, India.
14. www.radtexas.org/radunits.html
15. www.umich.edu/~radinfo/
16. www.visualhistology.com/Visual_Histology_Atlas/
17. www.vivo.colostate.edu/hbooks/genetics/biotech/gels/index.html.

Index

A

B

C

D

E

F

G

H

I

J

K

L

M

N

O

P

Q

R

S

T

U

V

W

X

Y

Z